Biotechnology

Second Edition

Fundamentals

Volume 1
Biological Fundamentals

Volume 2
Genetic Fundamentals and
Genetic Engineering

Volume 3
Bioprocessing

olume 4
easuring, Modelling, and Control

ducts

ne 5
ically Engineered Proteins and
lonal Antibodies

6
of Primary Metabolism

of Secondary Metabolism

mations

Special Topics

Volume 9
Enzymes, Biomass, Food and Feed

Volume 10
Special Processes

Volume 11
Environmental Processes

Volume 12
Modern Biotechnology:
Legal, Economic and
Social Dimensions

lschaft mbH, D-69451 Weinheim (Federal Republic of Germany), 1993

)1161, D-69451 Weinheim (Federal Republic of Germany)
P. O. Box, CH-4020 Basel (Switzerland)
d Ireland: VCH (UK) Ltd., 8 Wellington Court, Cambridge CB1 1HZ (England)
CH, 220 East 23rd Street, New York, NY 10010–4606 (USA)
Building, 10-9 Hongo 1-chome, Bunkyo-ku, Tokyo 113 (Japan)

Weinheim)
H, Weinheim)

ISBN 1-56081-153-6 (VCH, New York)
Set ISBN 1-56081-602-3 (VCH, New York)

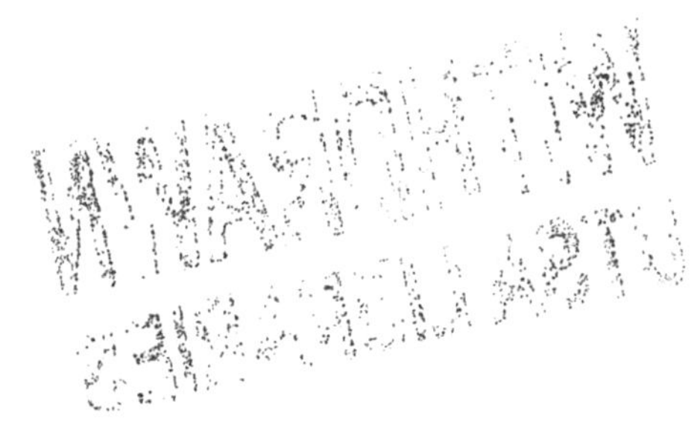

Biotechnology

Second Edition

Volume 3
Bioprocessing

WITHDRAW
UTSA LIBRAR

A Multi-Volume Comprehensive Treatise

Biotechnology

Second, Completely Revised Edition

Edited by
H.-J. Rehm and G. Reed
in cooperation with
A. Pühler and P. Stadler

Volume 3

Bioprocessing

Edited by
G. Stephanopoulos

Weinheim · New York
Basel · Cambridge · Tokyo

Series Editors:
Prof. Dr. H.-J. Rehm
Institut für Mikrobiologie
Universität Münster
Corrensstraße 3
D-48149 Münster

Dr. G. Reed
2131 N. Summit Ave.
Apartment #304
Milwaukee, WI 53202-1347
USA

Prof. Dr. A. Pühler
Biologie VI (Genetik)
Universität Bielefeld
P.O. Box 100131
D-33501 Bielefeld

Dr. P. J. W. Stadler
Bayer AG
Verfahrensentwicklung Biochemie
Leitung
Friedrich-Ebert-Straße 217
D-42096 Wuppertal

Volume Editor:
Prof. Dr. G. Stephanopoulos
Massachusetts
Institute of Technology
Cambridge, MA 02139
USA

This book was carefully produced. Nevertheless, authors, editors and publisher do not warrant the information contained therein to be free of errors. Readers are advised to keep in mind that statements, data, illustrations, procedural details or other items may inadvertently be inaccurate.

Published jointly by
VCH Verlagsgesellschaft mbH, Weinheim (Federal Republic of Germany)
VCH Publishers Inc., New York, NY (USA)

Editorial Director: Dr. Hans-Joachim Kraus
Editorial Manager: Christa Maria Schultz
Copy Editor: Karin Dembowsky
Production Director: Maximilian Montkowski
Production Manager: Dipl. Wirt.-Ing. (FH) Hans-Jochen Schmitt

Library of Congress Card No.: applied for

British Library Cataloguing-in-Publication Data:
A catalogue record for this book is available from the British Library

Die Deutsche Bibliothek – CIP-Einheitsaufnahme
Biotechnology : a multi volume comprehensive treatise / ed. by H.-J. Rehm and G. Reed. In cooperation with A. Pühler and P. Stadler. – 2., completely rev. ed. – Weinheim; New York; Basel; Cambridge; Tokyo: VCH.
NE: Rehm, Hans J. [Hrsg.]

2., completely rev. ed.
Vol. 3. Bioprocessing / ed. by G. Stephanopoulos
– 1993
ISBN 3-527-28313-7 (Weinheim)
ISBN 1-56081-153-6 (New York)
NE: Stephanopoulos, Gregory [Hrsg.]

©VCH Verlagsgesellschaft mbH, D-69451 Weinheim (Federal Republic of Germany), 1993

Printed on acid-free and low-chlorine paper.

All rights reserved (including those of translation into other languages). No part of this book may be reproduced in any form – by photoprinting, microfilm, or any other means – nor transmitted or translated into a machine language without written permission from the publishers. Registered names, trademarks, etc. used in this book, even when not specifically marked as such, are not to be considered unprotected by law.
Composition and Printing: Zechnersche Buchdruckerei, D-67330 Speyer.
Bookbinding: Fikentscher Großbuchbinderei, D-64205 Darmstadt.
Printed in the Federal Republic of Germany

Library
University of Texas
at San Antonio

Preface

In recognition of the enormous advances in biotechnology in recent years, we are pleased to present this Second Edition of "Biotechnology" relatively soon after the introduction of the First Edition of this multi-volume comprehensive treatise. Since this series was extremely well accepted by the scientific community, we have maintained the overall goal of creating a number of volumes, each devoted to a certain topic, which provide scientists in academia, industry, and public institutions with a well-balanced and comprehensive overview of this growing field. We have fully revised the Second Edition and expanded it from ten to twelve volumes in order to take all recent developments into account.

These twelve volumes are organized into three sections. The first four volumes consider the fundamentals of biotechnology from biological, biochemical, molecular biological, and chemical engineering perspectives. The next four volumes are devoted to products of industrial relevance. Special attention is given here to products derived from genetically engineered microorganisms and mammalian cells. The last four volumes are dedicated to the description of special topics.

The new "Biotechnology" is a reference work, a comprehensive description of the state-of-the-art, and a guide to the original literature. It is specifically directed to microbiologists, biochemists, molecular biologists, bioengineers, chemical engineers, and food and pharmaceutical chemists working in industry, at universities or at public institutions.

A carefully selected and distinguished Scientific Advisory Board stands behind the series. Its members come from key institutions representing scientific input from about twenty countries.

The volume editors and the authors of the individual chapters have been chosen for their recognized expertise and their contributions to the various fields of biotechnology. Their willingness to impart this knowledge to their colleagues forms the basis of "Biotechnology" and is gratefully acknowledged. Moreover, this work could not have been brought to fruition without the foresight and the constant and diligent support of the publisher. We are grateful to VCH for publishing "Biotechnology" with their customary excellence. Special thanks are due Dr. Hans-Joachim Kraus and Christa Schultz, without whose constant efforts the series could not be published. Finally, the editors wish to thank the members of the Scientific Advisory Board for their encouragement, their helpful suggestions, and their constructive criticism.

H.-J. Rehm
G. Reed
A. Pühler
P. Stadler

Scientific Advisory Board

Prof. Dr. M. J. Beker
August Kirchenstein Institute of Microbiology
Latvian Academy of Sciences
Riga, Latvia

Prof. Dr. J. D. Bu'Lock
Weizmann Microbial Chemistry Laboratory
Department of Chemistry
University of Manchester
Manchester, UK

Prof. Dr. C. L. Cooney
Department of Chemical Engineering
Massachusetts Institute of Technology
Cambridge, MA, USA

Prof. Dr. H. W. Doelle
Department of Microbiology
University of Queensland
St. Lucia, Australia

Prof. Dr. J. Drews
F. Hoffmann-La Roche AG
Basel, Switzerland

Prof. Dr. A. Fiechter
Institut für Biotechnologie
Eidgenössische Technische Hochschule
Zürich, Switzerland

Prof. Dr. T. K. Ghose
Biochemical Engineering Research Centre
Indian Institute of Technology
New Delhi, India

Prof. Dr. I. Goldberg
Department of Applied Microbiology
The Hebrew University
Jerusalem, Israel

Prof. Dr. G. Goma
Département de Génie Biochimique et Alimentaire
Institut National des Sciences Appliquées
Toulouse, France

Prof. Dr. D. A. Hopwood
Department of Genetics
John Innes Institute
Norwich, UK

Prof. Dr. E. H. Houwink
Organon International bv
Scientific Development Group
Oss, The Netherlands

Prof. Dr. A. E. Humphrey
Center for Molecular Bioscience and Biotechnology
Lehigh University
Bethlehem, PA, USA

Prof. Dr. I. Karube
Research Center for Advanced Science and Technology
University of Tokyo
Tokyo, Japan

Prof. Dr. M. A. Lachance
Department of Plant Sciences
University of Western Ontario
London, Ontario, Canada

Prof. Dr. Y. Liu
China National Center for Biotechnology Development
Beijing, China

Prof. Dr. J. F. Martín
Department of Microbiology
University of León
León, Spain

Prof. Dr. B. Mattiasson
Department of Biotechnology
Chemical Center
University of Lund
Lund, Sweden

Prof. Dr. M. Röhr
Institut für Biochemische Technologie und Mikrobiologie
Technische Universität Wien
Wien, Austria

Prof. Dr. H. Sahm
Institut für Biotechnologie
Forschungszentrum Jülich
Jülich, Germany

Prof. Dr. K. Schügerl
Institut für Technische Chemie
Universität Hannover
Hannover, Germany

Prof. Dr. P. Sensi
Chair of Fermentation Chemistry and Industrial Microbiology
Lepetit Research Center
Gerenzano, Italy

Prof. Dr. Y. H. Tan
Institute of Molecular and Cell Biology
National University of Singapore
Singapore

Prof. Dr. D. Thomas
Laboratoire de Technologie Enzymatique
Université de Compiègne
Compiègne, France

Prof. Dr. W. Verstraete
Laboratory of Microbial Ecology
Rijksuniversiteit Gent
Gent, Belgium

Prof. Dr. E.-L. Winnacker
Institut für Biochemie
Universität München
München, Germany

Contents

III. Product Recovery and Purification

IV. Process Validation, Regulatory Issues

Contributors

Dr. John G. Aunins
Biochemical Process Research and Development
Merck & Co., Inc.
P.O. Box 2000
Rahway, NJ 07065, USA
Chapter 11

Michael G. Beatrice
Associate Director for
Policy Coordination and Public Affairs
Center for Biologics Evaluation and Research
Food and Drug Administration
8800 Rockville Pike, HFM-10
Bethesda, MD 20892, USA
Chapter 28

Dr. Andreas S. Bommarius
Organic and Biological Chemistry Research and Development
Degussa AG
P.O. Box 1345
D-63403 Hanau
Federal Republic of Germany
Chapter 17

Dr. Barry C. Buckland
Biochemical Process Research and Development
Merck & Co., Inc.
P.O. Box 2000
Rahway, NJ 07065, USA
Chapter 1

Dr. Stuart E. Builder
Genentech, Inc.
460 Point San Bruno Blvd.
South San Francisco, CA 94080, USA
Chapter 4

Prof. Dr. Marvin Charles
Department of Chemical Engineering
Lehigh University
Iacocca Hall
111 Research Drive
Bethlehem, PA 18015-4791, USA
Chapter 16

Dr. Thomas Chattaway
Zeneca BioProducts
P.O. Box 1
Billingham, Cleveland, TS23 1LB
England
Chapter 14

Dr. Tzyy-Wen Chiou
Department of Chemical Engineering and Biotechnology Process Engineering Center
Massachusetts Institute of Technology
Cambridge, MA 02139, USA
Chapter 2

Dr. Jeffrey L. Cleland
Pharmaceutical Research and Development
Genentech, Inc.
460 Point San Bruno Blvd.
South San Francisco, CA 94080, USA
Chapter 20

Prof. Dr. Charles L. Cooney
Department of Chemical Engineering
Massachusetts Institute of Technology
Cambridge, MA 02139, USA
Chapter 9

Dr. Rajiv V. Datar
Pall Corporation
30 Sea Cliff Avenue
Glen Cove, NY 11542, USA
Chapter 18

Prof. Dr. Larry E. Erickson
Department of Chemical Engineering
College of Engineering
Durland Hall
Kansas State University
Manhattan, KS 66506-5102, USA
Chapter 13

Dr. Daniel Y. C. Fung
Department of Chemical Engineering
College of Engineering
Durland Hall
Kansas State University
Manhattan, KS 66506-5102, USA
Chapter 13

Dr. Robert L. Garnick
Genentech, Inc.
460 Point San Bruno Blvd.
South San Francisco, CA 94080, USA
Chapter 4

Dr. Christian F. Gölker
Verfahrensentwicklung Biochemie
Bayer AG
Friedrich-Ebert-Straße 217
D-42096 Wuppertal
Federal Republic of Germany
Chapter 26

Dr. Randolph L. Greasham
Biochemical Process Research and Development
Merck & Co., Inc.
P.O. Box 2000
Rahway, NJ 07065, USA
Chapter 7

Prof. Dr. Alan J. Grodzinsky
Department of Electrical Engineering and Computer Science
Massachusetts Institute of Technology
Cambridge, MA 02139, USA
Chapter 25

Prof. Dr. T. Alan Hatton
Department of Chemical Engineering
Massachusetts Institute of Technology
Cambridge, MA 02139, USA
Chapter 22

Dr. Hans-Jürgen Henzler
Verfahrensentwicklung Biochemie
Bayer AG
Friedrich-Ebert-Straße 217
D-42096 Wuppertal
Federal Republic of Germany
Chapter 11

Dr. James C. Hodgdon
Genentech, Inc.
460 Point San Bruno Blvd.
South San Francisco, CA 94080, USA
Chapter 4

Prof. Dr. Wei-Shou Hu
Department of Chemical Engineering and Materials Science
University of Minnesota
Minneapolis, MN 55455, USA
Chapter 6

Prof. Dr. Tadayuki Imanaka
Department of Biotechnology
Faculty of Engineering
Osaka University
Yamadaoka Suita
Osaka 565, Japan
Chapter 12

Prof. Dr. Jan-Christer Janson
Pharmacia BioProcess Technology AB
S-75182 Uppsala, Sweden
Chapter 23

Dr. Brian D. Kelley
Genetics Institute, Inc.
One Burtt Road
Andover, MA 01810, USA
Chapters 2 and 22

Dr. Konstantin Konstantinov
Department of Chemical Engineering
University of Delaware
Newark, DE 19716, USA
Chapter 15

Prof. Dr. Maria-Regina Kula
Institut für Enzymtechnologie
der Heinrich-Heine-Universität Düsseldorf
im Forschungszentrum Jülich
D-52425 Jülich
Federal Republic of Germany
Chapter 19

Dr. Daniel F. Liberman
Massachusetts Institute of Technology
Environmental Medical Service
Biohazard Assessment Office
18 Vassar Street
Cambridge, MA 02139-4307, USA
Chapter 29

Prof. Dr. Malcolm D. Lilly
Advanced Centre for Biochemical Engineering
Department of Chemical and Biochemical Engineering
University College London
Torringdon Place
London WCIE 7JE, England
Chapter 1

Dr. Gary A. Montague
Department of Chemical and Process Engineering
University of Newcastle upon Tyne
Newcastle upon Tyne NE1 7RU, England
Chapter 14

Prof. Dr. A. Julian Morris
Department of Chemical and Process Engineering
University of Newcastle upon Tyne
Newcastle upon Tyne NE1 7RU, England
Chapter 14

Dr. Jens Nielsen
Center for Process Biotechnology
Department of Biotechnology
Technical University
DK-2800 Lyngby, Denmark
Chapter 5

Dr. John R. Ogez
Genentech, Inc.
460 Point San Bruno Blvd.
South San Francisco, CA 94080, USA
Chapter 4

Gokaraju K. Raju
Department of Chemical Engineering
Massachusetts Institute of Technology
Cambridge, MA 02139, USA
Chapter 9

Prof. Dr. Matthias Reuss
Institut für Bioverfahrenstechnik
Universität Stuttgart
Böblinger Straße 72
D-70199 Stuttgart
Federal Republic of Germany
Chapter 10

Dr. Carl-Gustaf Rosén
ABITEC AB
Villavägen 36
S-64050 Björnlunda, Sweden
Chapter 18

Dr. Morris Rosenberg
Biogen, Inc.
14 Cambridge Center
Cambridge, MA 02142, USA
Chapter 2

Prof. Dr. Lars Rydén
Department of Biochemistry
Biomedical Centre
University of Uppsala
Box 576
S-75123 Uppsala, Sweden
Chapter 23

Prof. Dr. Athanassios Sambanis
School of Chemical Engineering
Georgia Institute of Technology
Atlanta, GA 30332-0100, USA
Chapter 6

Dr. Urs Saner
Department of Chemical Engineering
Massachusetts Institute of Technology
Cambridge, MA 02139, USA
Chapter 15

Prof. Dr. Dr. Karl Schügerl
Institut für Technische Chemie
Universität Hannover
Callinstraße 3
D-30167 Hannover
Federal Republic of Germany
Chapter 21

Prof. Dipl.-Ing. Horst Schütte
Technische Fachhochschule Berlin
FB 3, Seestraße 64
D-13347 Berlin
Federal Republic of Germany
Chapter 19

Dr. Ron Spears
Abbott Laboratories
Department 48F/R1/1040
1400 North Sheridan Road
North Chicago, IL 60064, USA
Chapter 3

Prof. Dr. Gregory Stephanopoulos
Department of Chemical Engineering
Massachusetts Institute of Technology
Cambridge, MA 02139, USA
Chapter 15

Srikanth Sundaram
Department of Chemical and
Biochemical Engineering
Rutgers University
Piscataway, NJ 08854-0909, USA
Chapters 24 and 27

Dr. Pravate Tuitemwong
Department of Chemical Engineering
College of Engineering
Durland Hall
Kansas State University
Manhattan, KS 66506-5102, USA
Chapter 13

Prof. Dr. John Villadsen
Center for Process Biotechnology
Department of Biotechnology
Technical University
DK-2800 Lyngby, Denmark
Chapter 5

Prof. Dr. Daniel I. C. Wang
Department of Chemical Engineering and
Biotechnology Process Engineering Center
Massachusetts Institute of Technology
Cambridge, MA 02139, USA
Chapters 2 and 20

John D. Wilson
ABEC
6390 Hedgewood Drive
Allentown, PA 18106, USA
Chapter 16

Dr. Richard A. Wolfe
Central Research Laboratories
Department of Cell Culture and Biochemistry
Monsanto Company
800 North Lindbergh Blvd.
St. Louis, MO 63167, USA
Chapter 8

Dr. David M. Yarmush
Department of Chemical and Biochemical Engineering and Center for Advanced Biotechnology
Rutgers University
Piscataway, NJ 08854, USA
Chapter 27

Prof. Dr. Martin L. Yarmush
Department of Chemical and Biochemical Engineering and Center for Advanced Biotechnology
Rutgers University
Piscataway, NJ 08854, USA
Chapters 24, 25 and 27

Prof. Dr. Toshiomi Yoshida
IC Biotechnology
Faculty of Engineering
Osaka University
Osaka 558, Japan
Chapter 15

Introduction

GREGORY STEPHANOPOULOS

Cambridge, MA 02319, USA

Two major changes have occurred in the field of bioprocessing with respect to the state-of-the-art reviewed in the first edition of *Biotechnology*. These changes are not due to generic advancements of equipment and processes. They were rather caused by the introduction of two new types of applications in the field, specifically the introduction of microbial recombinant fermentations and cell culture processes for the production of biochemicals and pharmaceutical proteins. In connection with these processes, new methods of production and new operating strategies have been introduced in the upstream as well as the downstream sections of the biotechnological plant. More traditional processes have benefited from these new manufacturing techniques in several ways, such as the use of improved types of equipment, better processing strategies and new methods for bioprocess monitoring and control.

The present volume reviews the state-of-the-art of bioprocessing as it applies in the case of traditional as well as the newly introduced processes. The structure of the volume follows to some extent the typical arrangement of bioprocessing equipment in a real process. Therefore, the usual distinction of upstream and downstream processing which one encounters in a biotechnological plant is also present in the arrangement of the chapters of this volume. In addition, we have included a group of chapters providing an overview of fermentation and cell culture processes and a closing group of chapters dedicated to the issues of process validation, measurement and regulation.

In keeping with the separation between microbial fermentations and cell culture processes, the first two chapters of *Part I* provide overviews of these two different types of cell cultivation. These are followed by a chapter providing a general overview of downstream processing. Since the sources and methods of synthesis of pharmaceutical proteins are of considerable importance, an additional chapter dedicated to this topic has also been included to illustrate the rationale for the development of the synthetic manufacturing methodologies that one encounters today.

Part II deals exclusively with fermentation and cell culture equipment as a means of biosynthesis of fermentation products and pharmaceutical proteins. It begins with an equipment description and presentation of design equations for microbial fermentors. This chapter reviews the predominant types of industrial fermentors as well as some experimental designs along with the design equations and the main characteristics of these units. Another chapter deals with cell culture bioreactors, their design equations and the predominant types which are presently in use in industry. The remaining chapters in Part II review the equipment and operation of other related steps in the way they are practiced in a plant. In this regard, chapters describing media formulation and the rationale of media design for microbial fermentations and cell culture are included. Issues of sterilization of air and fermentation media are considered next. Oxygen transfer and mixing in fermentors and the implications of these processes to the scale-up and control of microbial and cell culture bioreactors are examined in two separate chapters. Following these basic issues of fermentor design and operation, the specifics of particular fermentations are examined such as recombinant microbial fermentations and anaerobic fermentations. The topic of fermentor instrumentation, monitoring and control has been the subject of Volume 4 in this series. It was felt, however, that a review of this most important topic should be included in the present volume for the purpose of completeness. A summary of instrumentation techniques and conventional as well as more advanced control strategies is presented in the corresponding chapter. A new element in the present volume is a concise review of the various methods by which fermentation data can be analyzed and used for the diagnosis and control of fermentors. We hope this chapter will enhance the quality of information which can be obtained from the type of data typically collected in the course of a fermentation or cell culture process. Practical issues of fermentor construction and containment are addressed in a different chapter. The very important area of enzymatic reactions and biotransformations as well as the types of equipment which are used for this purpose are reviewed in the final chapter of Part II.

Part III is devoted to product recovery and purification. These chapters also follow the typical steps that one encounters in a plant. The first chapter addresses processes used for the removal of cells from the fermentation or cell culture medium. Specifically, centrifugation and cross-flow filtration are reviewed as a means of cell removal. Following cell removal, methods for the disruption of cells and isolation of non-secreted products are considered. In this regard, and in connection with microbial recombinant fermentations, the refolding of non-secreted proteins after they have been separated from the cell debris in the form of inclusion bodies is examined. Specifically, *in vitro* protein refolding and process parameters affecting this step are reviewed in the next chapter. Liquid-liquid extraction as it applies to the separation of small molecules follows. Other extraction processes geared towards the purification of proteins and enzymes are reviewed in another chapter, specifically techniques based on reversed micelles and two-phase systems are included. As mentioned earlier, the introduction of new biosynthetic technologies for the production of proteins necessitated the development of purification methods yielding products of very high purity. Chromatographic methods are used primarily for this purpose, and the fundamentals of chromatography as well as chromatographic applications for the purification of biopharmaceuticals are the subject of another chapter. In the same vein, affinity separations and electrokinetic separations have also been employed, and they are the subject of two additional chapters. As a final polishing step, lyophilization and spray-drying are examined in the last chapter of Part III.

Along with the development of the new exciting manufacturing technologies, new requirements for the containment of biologically active material, the satisfactory validation of biotechnological processes, and the means to monitor compliance with regulations have emerged. It is, therefore, fitting to devote *Part IV* to these critical issues. This part includes chapters on analytical protein chemistry devoted to regulatory issues as well as chapters on process validation criteria and methodologies for the treatment of biological waste.

There are some topics which have not been included in the present volume. They are in areas which have not yet been fully developed or areas which may be replaced in the future in favor of other competing processes. In the first category belong issues of optimization, in particular those dealing with the optimal structure of a train of equipment and the arrangement of process streams connecting such pieces of equipment. In the second category are processes for handling products from animal organs where the present trend is to replace such processes by the better definable and controllable systems based on microbial fermentations or cell culture processes. Depending on the way that these topics develop, they may become subjects of a future edition of *Bioprocessing,* or will be included in Volume 10 *Special Processes.*

Although the writing style of the various chapters is variable reflecting the individual authors, there are some common features that can be noted throughout the volume. Most chapters begin with a brief introduction of the fundamentals followed by a description of the equipment and processes as practiced in industry. After that, the discussion centers on the presentation of the most common procedures as well as problems and challenges furthering their industrial performance. A significant amount of process data is also included in the form of tables and figures. The reader is encouraged to search through the cited literature which provides information updated to the date of writing the chapter.

When reading the present volume, one should bear in mind the very diverse background of the readership that this volume tries to satisfy. The readership comprises the more mathematically oriented as well as those who favor a more qualitative approach in their field of biotechnological application. Although there was a conscientious effort to maintain a balance between mathematical complexity and process description, it is quite possible that the results may not please everybody. It is, however, the sincere hope of the editor and the authors that the product of their efforts will reach as broad an audience as possible and will satisfy the needs of such audience for a useful and at the same time rigorous treatment of the topic of bioprocessing. The extent to which this ambition is reached will determine the success of this work.

In closing, I would like to express my thanks to the authors of all chapters for a very professional and thorough job as well as the large number of reviewers who provided detailed reviews and critiques of the original manuscripts. The latter were instrumental in producing a volume of high quality. The staff of VCH has been extremely helpful in handling many technical and editorial matters and they have our deep appreciation as well.

Cambridge, MA, May 1993

G. Stephanopoulos

I. Nature and Issues of Bioprocessing

1 Fermentation: An Overview

Barry C. Buckland

Rahway, NJ 07065-0900, USA

Malcolm D. Lilly

London, UK

1 Introduction

Microbial fermentations are important sources of biological products used in the pharmaceutical, food, and chemical industries. During the last decade, there has been a large increase in the range of commercial products, especially secondary metabolites and recombinant proteins. There have also been significant changes in fermentor and facility design to improve performance and to ensure safe operation. Emphasis is now on the development of processes which are not only cost-effective but also meet the increasing demand by the public and regulatory authorities for greater reliability and reproducibility. This has increased the need for improved monitoring and control.

Whereas microbial transformations, like chemical reactions, often reach yields close to the theoretical maximum, many fermentation processes such as those for secondary metabolites are operating at much lower yields. Thus, there is great scope for advances in fermentation processes. Progress will result from an improved understanding of microbial physiology, of the interaction of microorganisms with the physical environment in fermentors and the ability to manipulate metabolic fluxes using molecular biology. In this chapter, we highlight many of these aspects of fermentation, although some of them will be covered more comprehensively in subsequent chapters of this volume.

The level of understanding of each type of fermentation processes is primarily a function of metabolic complexity. These categories of fermentation will be discussed in the next section. The scale of operation varies considerably depending on the type of product and the dose (Tabs. 1 and 2).

More recently the scope of fermentation technology has grown to include mammalian cell culture. If a mammalian cell culture line can be adapted to suspension culture, then similar approaches to process improvement can be utillized. Mammalian cells are bigger, more fastidious and more fragile than microbial cells, but the approach to process development can be remarkably similar.

Tab. 1. Some Typical Production Fermentor Sizes

Product	Volume (m^3)
Baker's yeast	100–250
Amino acids	100–250
Antibiotics	80–200
Industrial enzymes	80–250
Recombinant therapeutic proteins	0.5–50

Tab. 2. Scale of Operations for Therapeutic Drugs

Type	Dose	Annual Production 1000000 Doses
Immunomodulator or Vaccine	0.0001–0.1 mg	0.1–100 g
Hormone	1–10 mg	1–10 kg
Enzyme	100 mg	100 kg
Antibiotic	1 g	1000 kg

2 Fermentation Products

2.1 Cell as Product

Yeast fermentation for brewing and baking represents the most traditional form of fermentation technology. Techniques and cultures have been developed over the years for production of yeast at a large scale of operation as well as for simplified methods of cell harvesting (for example, by the deliberate enhancement of flocculation). This technology has become increasingly sophisticated, and some of the earliest examples of computer-enhanced fermentation technology were applied toward the more efficient production of yeast (WANG et al., 1977).

In the 1970s a number of companies became very interested in the production of single cell protein (SCP), and a whole technology was developed to grow microbial cells on very cheap substrates such as methanol and hydrocarbons. The driving force behind this interest was ultimately the price for other protein products such as soybeans and fish meal. Since the sudden price jump for protein in the early 1970s turned out to be a temporary phenomenon, the interest in SCP took a precipitous nosedive as soon as the protein price returned to normal. The economic question can be summarized by asking whether it is cheaper to generate protein by buying suitable land (for example, in Illinois) and planting soybeans, or is it better to construct and operate a fermentation facility. More recently, the emphasis has switched to generation of a product which can simulate meat (mycoprotein, marketed by Marlow Foods) and thereby compete with the more expensive end of the protein market on issues other than price (e.g., fat content and acceptability by vegetarians). Some of the technology developed proved to have applications in other areas, for example, large-scale DNA removal.

Other specific technologies (such as the 1.5 million liter ICI Pressure Cycle Fermentor) have not entered the mainstream. Interest in these innovative technologies will undoubtably resume at some future time, when the need is more urgent to develop chemicals from renewable resources.

2.2 Primary Metabolites

Process for products such as citric acid, ethanol, and glutamic acid are well developed in view of their historical importance and very large scale of operation. In addition, these are processes which are relatively simple metabolically and hence tend to be better characterized. Recombinant DNA technology is now used to enhance metabolic fluxes for products such as phenylalanine.

Stirred tank fermentors of around 200000 L in scale are generally used. If it is desired to go to greater volumes, then it is simpler to switch to an air-lift type design because of heat transfer limitations. An air-lift would generally have a height to diameter ratio of 10:1 and, therefore, has much more cooling surface area per unit volume than a stirred tank fermentor which normally has a height to diameter ratio of <3:1. Also, obviously mechanical mixing generates heat.

2.3 Secondary Metabolites

Since the widespread use of antibiotics began (around 1950), this area has represented the most commercially significant fermentation activity. Products such as penicillin, tetracycline, erythromycin, cephalosporins, cephamycins, and clavulanic acid are made at a very large scale in fermentors varying between 50000 to 200000 L in volume. In most cases these products serve a vital role in modern medicine, and their sales have continued to increase for decade after decade. Penicillin can now be considered a bulk chemical. To a large extent, the demand for more product has been matched by increases in productivity, and so companies have been able to obtain more and more product from an existing factory by advances made in process development (Fig. 1).

More recently, discovery groups have learned to use secondary metabolites as a way of generating leads for other therapeutic areas besides antibiotics. For example, avermectin was discovered by the Merck Research Labo-

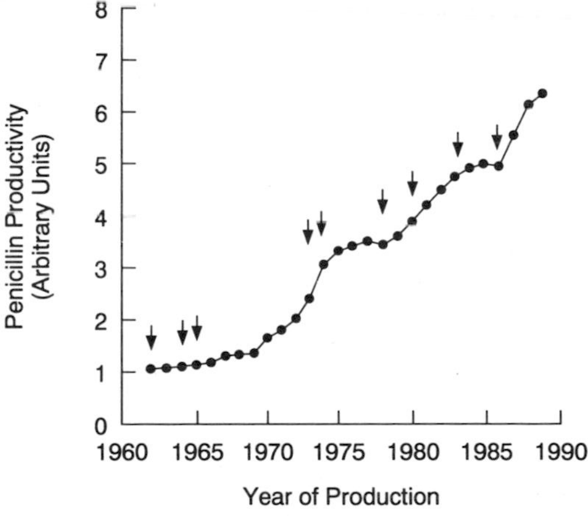

Fig. 1. Improvements in the productivity of penicillin fermentations. Arrows indicate the introduction of new strains (courtesy of Gist-Brocades) (redrawn from LILLY, 1992).

ratories (MRL) as an anti-parasitic drug. Lovastatin was discovered (also by MRL) as a potent inhibitor of HMGCoA reductase and has subsequently been used as an effective blood cholesterol lowering drug. Cyclosporin and FK506 were discovered as potent immunoregulants and are used as aids in tissue transplant therapies. Thus the 1980s have seen the commercial introduction of many important secondary metabolites (Tab. 3). This is the first decade in which a number of very important secondary metabolites were successfully introduced to the marketplace which are not antibiotics.

Tab. 3. New Pharmaceutical Products since 1980: Secondary Metabolites

Lovastatin (*Aspergillus terreus*)
Avermectin (*Streptomyces avermitilis*)
Cephamycin C (*Nocardia lactamdurans*)
Efrotomycin (*Nocardia lactamdurans*)
FK 506[a]
Cyclosporin
New Cephalosporins
Thienamycin (*Streptomyces catleya*)[b]
Clavulanic acid (*Streptomyces*)

[a] not yet approved in the USA
[b] now made synthetically

Secondary metabolite fermentations are very complex and, despite intensive investigation, have yet to yield many of their secrets even in the case of such well-known products as penicillin. Progress has been made in industry by a mixture of "art" and science (random mutation and selection, media development, development of nutrient feeding strategies, and improvements in oxygen transfer) (NALLIN et al., 1989). Applications of recombinant DNA technology to improve systematically a well established culture line have generally not been successful. Presumably, at some time in the future, most culture improvement will come from using recombinant DNA technology. However, this area is at a fascinating juncture. Tools via rDNA technology are now available which will allow a more complete understanding of these processes, and this in turn will provide further performance improvements or novel products. Despite the lack of fundamental knowledge, progress has often been remarkable with some product titers for secondary metabolites progressing more than a thousand-fold from <1 mg/L to tens of g/L.

2.4 Enzymes

Historically, industrial enzymes made by fermentation were mainly restricted to those produced extracellularly such as amylases and proteases. In recent years, other bulk enzymes such as lipases have been markteted for industrial and domestic applications.

With the advent of mechanical techniques for release of proteins from microorganisms on a large scale, intracellular enzymes have found wider application in the food, pharmaceutical, and chemical industries. Of particular note are glucose isomerase for production of high-fructose syrups and penicillin acylase for removal of the side chain of penicillins to allow subsequent manufacture of semi-synthetic penicillins. Other enzymes, such as glucose oxidase and cholesterol oxidase, are widely used for clinical analysis. Fermentor sizes for enzyme production generally range from 30 to 220 m^3.

In the future, it is likely that a large proportion of enzymes for commercial use will be synthesized using recombinant microorganisms. This approach has opened up the possibility of producing many different enzymes in substantial quantities, preferably using a small number of host/vector systems to minimize development costs. The increasing availability of enzymes will allow their exploitation as potent catalysts to introduce chirality and specificity into compounds in chemical processes to make, for instance, drugs and pesticides.

2.5 Therapeutic Proteins

In recent years, the ability to synthesize therapeutic proteins has improved rapidly. We can claim to have moved completely away from the "art" of fermentation to the science of fermentation, when we use a well characterized host such as *Escherichia coli* with a chemically defined medium with a known plasmid and promoter. This claim can be less easily made for a more complicated fungal host such as *Saccharomyces cerevisiae* and even less for animal cell culture.

For the fermentation technologist, the fact that there is a choice of hosts represents a new range of opportunities. For a microbial system one can choose primarily between *E. coli* or yeast. For a glycosylated protein it is generally necessary to use animal cells as host, but even here there is a choice between cell lines which predominantly attach to surfaces and those which can be grown in suspension culture. Also some cell lines can easily be adapted to serum-free conditions and others cannot. In these examples, decisions made at the initial stage of the project will have a huge impact on the eventual outcome of the manufacturing process because of the highly regulated environment surrounding the development of a *biological* as a product.

A decade ago, it was rare for recombinant microorganisms to be grown on a large scale. However, progress in the 1980s has been rapid, and by 1990 at least six recombinant products had annual sales in excess of one hundred million dollars (LILLY, 1992). These were human growth hormone, erythropoietin, alpha-interferon, human insulin, trypsin plasminogen activator, and hepatitis B vaccine. A list of rDNA therapeutic proteins commercialized since 1982 is given in Tab. 4.

Tab. 4. Recombinant DNA Therapeutic Proteins (since 1982)

Interferon (alpha and gamma)
tPA (tissue plasminogen activator)
EPO (erythropoietin)
Human growth hormone
Human insulin
Sargramostin (GM-CSF)
Filgrastim (rG-CSF)
OKT3 (muromonab-CD3)

2.6 Vaccines

Vaccine technology has evolved from using rather crude extracts from microbial cells (for example, *Bordetella*) as a vaccine to using more highly purified polysaccharides or membrane proteins such as HIB conjugate (Tab. 5). As a recent extension to the technology in this area, a vaccine for meningitis has been successfully introduced to the market by the Merck Research Laboratories (PedvaxHIB®) which is made by chemically conjugating polysaccharides from *Haemophilus influenzae* type b with an outer membrane protein complex from *Neisseria meningitidis*. This is one member of a whole new class of vaccines called "conjugate vaccines".

Tab. 5. Vaccines (since 1980)

Hepatitis B (plasma)
Hepatitis B (recombinant)
Hepatitis A[a]
Combination vaccines (various)
HIB conjugate
Varicella[a]
Pertussis (acellular)

[a] not yet approved in USA

Another recent example of a dramatic evolution in the vaccine area is illustrated by the development of a vaccine for hepatitis B. The original vaccine developed in 1980 was made by isolating virus particles from infected human blood and then by purifying the capsid protein followed by formulation with alum. The inactivation for viruses in plasma was performed at three different phases of purification. The recombinant version of the vaccine is made using *Saccharomyces cerevisiae* as the host. The yeast generates the subunits of the capsid protein which are then assembled to form the desired particle. This allowed vaccine production to move completely away from blood as the raw material toward the inherently safe recombinant yeast process and was the first example of a recombinant protein used as a human vaccine. The product is known as RECOMBIVAX® (Merck & Co., Inc.).

2.7 Gums

Xanthan gum has found an increasingly important role for use in the food industry as well as for applications in tertiary oil extraction. This product is made at a very large scale of operation using fermentors in the 50000 to

200000 L size range. This is a very competitive area and economics for production are critical. Technical issues relate to the fascinating problem of improving mass transfer to a very viscous fermentation broth. This is an even greater challenge for some of the newer gums with unusual rheological behavior (e.g., Gellan Gum).

3 Types of Fermentor

A wide range of fermentor types has been described in the literature. Nevertheless, it remains true that the standard of the fermentation industry is still the aerated agitated tank. There are several reasons for this. First, many companies installed fermentors of this kind in the 1960s and 1970s which still have many years of useful life and have been upgraded, where necessary, with improved agitation, instrumentation, and computer controlled systems. Second, stirred tanks, although not necessarily ideal for a particular fermentation, give good results for many different fermentations. This is important, especially for pharmaceutical companies, as several different products may be made in a fermentor during its lifetime. Another benefit arises from the fact that the capital investment in fermentors is normally recuperated from sales of the first product so that capital costs for subsequent products are greatly reduced.

3.1 Stirred Tank Fermentors

As new fermentation capacity has been installed over the last twenty years, there has been a gradual increase in the size of fermentors for particular products and this trend is likely to continue. It reflects the greater demand for such products, the need to reduce costs to remain competitive, and the improved understanding of fermentor design and operation by biochemical engineers.

At the smaller scale (10 L to 10000 L) there is very little incentive for a company to explore alternative designs to the stirred tank fermentors. Energy costs are usually inconsequential at this scale. Therefore, the large number of new therapeutic protein product candidates, which are typically run at the 100 L to 3000 L scale, have not generated a driving force for innovation in fermentor design in terms of performance except for the need to enhance oxygen transfer rates to meet the demands of *E. coli* cultures growing to 50 g dry weight per liter or more. Most recent improvements have been driven by good manufacturing practice (GMP) requirements and have resulted in tanks which are very easy to clean and which have good sterility performance. Cleaning is often done "in place" using CIP (clean in place) technology developed in the food industry. A modern fermentor design would include spray balls to allow for an automatic cleaning cycle.

In practice, agitation power/unit volume decreases with increasing scale of operation, and the lower agitator power input/volume results in longer mixing times and lower oxygen transfer rates. This is illustrated in Fig. 2 which shows that the mixing times (t_m) in aerated stirred fermentors increase with fermentor volume (V) according to

$$t_m = \text{constant} \cdot V^{0.3}$$

reaching about 100 s in a 100 m^3 vessel (Fig. 2). Thus, at a large scale there is likely to be

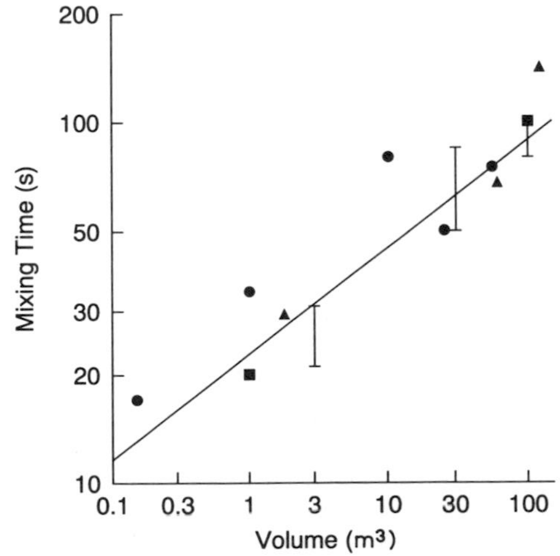

Fig. 2. Mixing times in agitated fermentors. Data taken from EINSELE (1978), (●); JANSEN et al. (1978), (▲); and CARLEYSMITH (personal communication), I—I.

considerable heterogeneity within the fermentation broth. Liquid additions to the broth will take time to disperse, and nutrient concentration gradients may occur. If the time constant for the rate of oxygen transfer is much shorter than the mixing time, then there will be a vertical dissolved oxygen concentration gradient, as observed by MANFREDINI et al. (1983) for chlorotetracycline and tetracycline production by strains of *Streptomyces aureofaciens* in a 112 m^3 fermentor. It is also possible in viscous fermentations to have radial dissolved oxygen (DOT) gradients in fermentors, especially in the stirrer region (OOSTERHUIS and KOSSEN, 1984). Thus the DOT readings for a large fermentation will depend on the location of the probe. Most companies determine empirically the measured DOT below which, for instance, product formation is reduced. The impact on cultures of the rapid fluctuations in nutrient concentrations, including DOT, which occur in large fermentors, is poorly understood.

There has been recent progress in agitator design. Multiple Rushton turbine radial flow impellers (Fig. 3) were the design of choice for fermentors from the early 1950s to the early 1980s. Now there is a number of attractive alternatives and these have been reviewed recently (NIENOW, 1992). The biggest design change has been the development of axial flow hydrofoil impellers (Fig. 4) which can give superior performance to the more traditional Rushton radial flow impellers for large-scale viscous mycelial fermentations (BUCKLAND et al., 1988). The specific advantages of the hydrofoil axial flow impeller design over the traditional Rushton radial flow impeller design for viscous mycelial fermentations are:

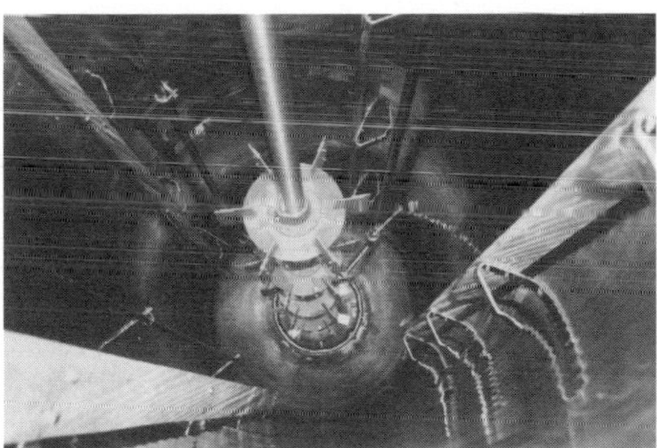

Fig. 3. Radial flow turbine impellers fitted to a 19000 L fermentor.

Fig. 4. Axial flow hydrofoil impellers fitted to a 19000 L fermentor.

- Improved oxygen transfer per unit power
- Lower maximum shear rates (thus improving fermentor versatility; i.e., a fermentor can be designed for use with cultures requiring a low shear environment, such as animal cell culture)
- Improved bulk mixing by elimination of compartmentalized flow resulting from use of multiple radial flow impellers
- Improved bulk mixing resulting in improved pH control and better control of nutrient feeding at the level of the individual cell. This results in a more homogeneous and stable environment for the microbial cells.

The comparison between radial flow and axial flow impellers is best summarized using the model of BAJPAI and REUSS (1982) (Fig. 5). For a viscous mycelial fermentation, the micromixer region is one of high shear, low apparent viscosity (due to the pseudoplastic rheology of mycelial broths), and high K_La whereas the macromixing region is one of low shear, high apparent viscosity, and low K_La. Presumably, the hydrofoil axial flow impellers radically increase the ratio between the micromixing region and macromixing region. In a

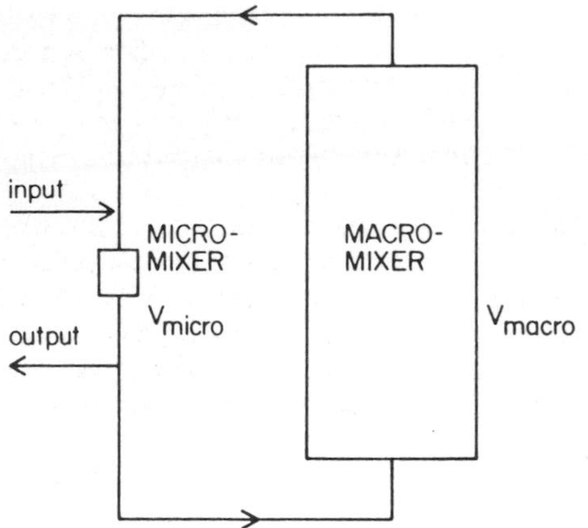

Fig. 5. Model of a fermentor as micromixer and macromixer regimes.

large fermentor, the macromixing region becomes much larger than the micromixing region.

3.2 Non-Mechanically Agitated Fermentors

Some companies have installed pilot and full-scale aerated tanks, with various internal configurations, to produce a range of microbial metabolites and biomass products. At least one large pharmaceutical company uses predominantly non-mechanically driven tanks, and other companies (including Merck & Co., Inc.) have built very large (250000 L) horizontal stirred tank fermentors. The huge air-lift design fermentor (1500000 L) pioneered by ICI for SCP has not been widely adopted. The air-lift and non-mechanically driven designs are best suited for very large-scale operation (>200000 L). As scale is increased beyond 200000 L, a mechanically driven tank becomes more and more difficult to design because of heat transfer (cooling) considerations. The knowledge of the performance of such aerated fermentors in these companies has allowed them to be used in the same versatile way as stirred fermentors. Thus, the choice between the two types depends very much on which are already installed in the company and, therefore, its familiarity with that type.

In contrast to stirred fermentors, oxygen transfer rates in aerated fermentors are normally low at a small scale and increase with scale. Thus, even where aerated fermentors are used on a large scale, it may be necessary to carry out small-scale development work in stirred fermentors.

Similar problems of heterogeneity to those described in Sect. 3.1 can occur in non-mechanically agitated fermentors. For instance, in air-lifts, where there is liquid circulation up one section and down the other the circulation times (t_c) may be long. The hydrodynamics of air-lift fermentors have been widely studied (CHISTI, 1989), and many values for circulation times are available. However, far fewer measurements have been made during fermentations where oxygen transfer must also be adequate. CARRINGTON et al. (1992) have reported circulation times of 9–12 s for a *Streptomyces* antibiotic fermentation in a 20 m^3 air-lift fermentor.

3.3 Facility Design

For many decades, antibiotics of high quality have been produced at very large multiproduct facilities. Because of the ease of final product definition, processes have been continuously improved over the years, and titers have often increased by two orders of magnitude. Process changes are allowed by regulatory authorities so long as there is proof that the drug made by the old and new process is the same. In most cases, the increase in productivity has been accompanied by improvements in final product quality (fewer by-products are being generated). Therefore, society has benefitted in two ways: pharmaceutical companies have been able to reduce prices and increase quality for this type of product. Increases in sales are often matched by rises in productivity without the need for additional fermentation capacity, and consequently many of the production facilities are rather old. The vibrant and fertile process development, which exists for secondary metabolite fermentations (regulated in the USA by CDER, the Center for Drug Evaluation and Research of FDA) long after these have been approved as new pharmaceuticals, results in a steady increase in productivity (see Fig. 1). From 1960 to 1990 penicillin productivity in the factory was increased six-fold. This is in contrast to the biologics area (regulated by CBER) in which the process itself and the equipment used is very tightly defined because of the historical difficulty of a rigorous analysis of the actual final product. As a consequence, it is difficult to make process changes, and the process development environment becomes stagnant. As analytical tools become more and more powerful and with the advent of biologic products which are small enough and pure enough to be well defined, it may be possible for the focus to switch away from regulation of the process to regulation of the final product characterization.

4 Fermentor Operations

A fermentor is usually surrounded by a web of auxiliary equipment: seed fermentors, nutrient feed tanks, pH control tanks, antifoam tanks, an area for media preparation, and one or more medium sterilization systems. Obviously, the types of operations involved are a function of the type of process being run.

Viewed simplistically, the fermentation step can be seen as a single unit operation. In practice, there are a multiple of operations which need to be tightly choreographed to result in a successful run: these include the seed train, medium preparation, and nutrient feeding. Everything has to be ready at exactly the right time. Maintaining asepsis is usually the single biggest and ongoing challenge, especially for a long-cycle secondary metabolite fermentation. Bacteriophage infections can also be a problem (PRIMOSE, 1990).

4.1 Fermentation Media

For recombinant cultures producing proteins, there is a growing trend toward tailoring a chemically defined (and soluble) medium for a particular host. This makes possible the use of filter sterilization and also allows for the tight control of the cell environment using nutrient feeding and pH control. The need for these makes the fermentation more complicated to operate and is probably only practical for a computer-controlled facility (BUCKLAND, 1984) with excellent alarm scanning capabilities to give instant warning of any process perturbation.

Efforts to generate satisfactory chemically defined media for secondary metabolites have been less successful. The desired biochemical environment has proven more difficult to define rigorously and empirically derived media using complex ingredients work remarkably well. The slow release of nutrients such as protein and phosphate from a material such as soybean meal often provides an excellent cell environment for high levels of product expression. For this type of medium, sterilization itself is an important variable, because the act of heating the medium changes its composition (CORBETT, 1980). For this reason, continuous sterilization is often preferred: it is usually cheaper to operate, control is much more precise than for batch sterilization, and scale-up is much easier than for batch sterilization (JAIN and BUCKLAND, 1988).

4.2 Mode of Operation

With the exception of sewage and industrial effluent treatment, very few industrial fermentations are operated as continuous-flow systems. Continuous operation in large fermentors has only been justified for low-cost products where the market is highly-competitive. Continuous culture is a useful developmental tool allowing investigation of the effects of limitation of growth and product formation by supply of individual nutrients. The information gained can be used to optimize feeding regimes for fed-batch fermentations.

It is equally true that few fermentations are operated in a truly batch mode. Intermittent or continuous feeding of nutrients may be performed for several reasons. First, the quantities of nutrients at the start of the fermentation may be restricted to avoid the oxygen demand during the growth phase exceeding the oxygen transfer capacity of the fermentor or because high concentrations are inhibitory or cause undesirable precipitates. Second, nutrient feeding regulates the metabolism of the organism and is one of the most important ways to enhance product formation. Finally, there are many fermentations where the nutrients required during the growth and product formation phases may be different. Because of the large volumes of nutrients which may be added during the fermentation, the fermentor will be only partly full at the start, and the impact of this on its operations needs to be understood.

In some cases, fed-batch fermentations may be extended by harvesting part of the broth on one or more occasions with the vessel being refilled through nutrient feeding.

4.3 Monitoring and Control

Computer process control has had a fundamental impact on the way a fermentation facility can be operated. Besides all the obvious benefits of tighter process control, alarm scanning capabilities are invaluable, and capabilities such as cascade control of dissolved oxygen (DO) (Fig. 6) allow for tighter control of the cell environment. In Japan, at Yamanouchi Pharmaceutical Company, this has been taken one step further (EIKI et al., 1992) toward "lights-out" production using computer control combined with very extensive automation. Many operations (including transfer steps) are carried out unattended at night and during the weekend. For example, the raw media stocked in the outdoor silos are automatically weighed and transferred into the interior material hoppers. Harvesting of the batch is initiated unattended very early in the morning. After broth transfer, the fermentor is automatically washed with alkaline hot water. By the time the operators come to the plant, these jobs are almost complete.

A production manager continually needs information to indicate whether the various processes are progressing normally. Any useful indication gained from on-line process monitoring is extremely valuable. In practice, we have found that the single most powerful tool available comes from on-line monitoring of vent gases using mass spectrometry (BUCKLAND et al., 1985). Subtle changes in oxygen uptake rate, carbon dioxide evolution rate, and respiratory quotient can be immediately detected and alarms activated.

4.4 New Measuring Techniques

During the last decade, a range of new measuring techniques have become available. Some of these allow better characterization of the fluid dynamical behavior in reactors of pilot or production scale (LÜBBERT, 1992).

Determination of biomass concentrations, particularly when undissolved nutrients are present, still poses problems. HARRIS et al. (1987) reported the possibility of using dielectric permittivity. There were some problems with early commercial devices, but the latest versions seem promising, particularly for the measurement of yeasts and mycelial organisms (FEHRENBACH et al., 1992) even in the presence of undissolved nutrients. The magnitude of the signal with non-mycelial bacteria is lower, but recently it has been used successfully to monitor cultures of *Pseudomonas putida*.

The development of image analysis techniques for measurement of the morphology of filamentous microorganisms over the last few years (PACKER and THOMAS, 1990) now allows the influence of the fermentation conditions on morphology to be determined more precisely. Furthermore, it is possible to measure biomass concentration and the extent of cell vacuolization during fermentations of fungi, such as *Penicillium chrysogenum* (PACKER et al., 1992).

Progress has been made with the automation of sampling and subsequent measurement

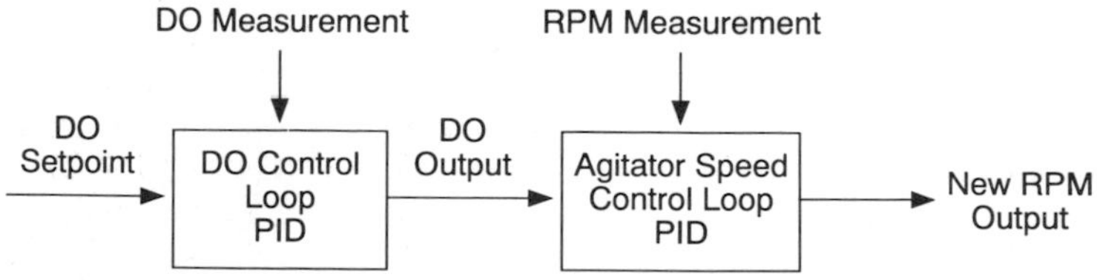

Fig. 6. Cascade control of dissolved oxygen (DO) by changes in agitator speed.

of metabolite concentration by analytical techniques such as HPLC and GLC. A novel approach has been the use of near infrared (NIR) spectroscopy, which may be adapted for online measurement using a fiber optic module (HAMMOND and BROOKES, 1992). The authors describe the use of NIR for determination of antibiotics (Fig. 7).

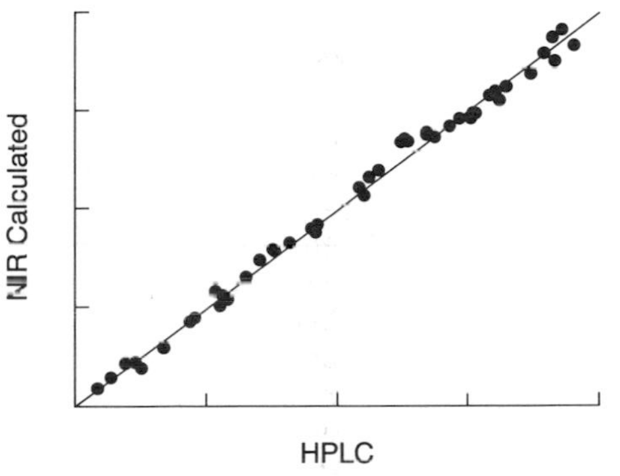

Fig. 7. Correlation plot for at-line antibiotic measurement using near infrared (NIR) spectroscopy and high performance liquid chromatography (HPLC) (HAMMOND and BROOKES, 1992).

A recent extension has been the enhancement of automation for sample preparation using robotics. This approach has been taken, because many of the most useful components to measure are intracellular or cell-associated. In one example, samples are automatically taken from the fermentor, accurately weighed, mixed with a calculated amount of methanol, centrifuged, and the clarified methanol transferred to HPLC for detailed analysis (REDA et al., 1991).

5 Influence on Product Recovery

There are many fermentation products, e.g., therapeutic proteins, where the cost of product recovery and purification exceed those for fermentation. Particularly in these cases, it is important to understand the impact of the fermentation conditions on the subsequent processing operations (FISH and LILY, 1984). Major improvements can be made by correct selection of the organism, culture medium, and growth conditions.

Many problems in purification can best be addressed by changing fermentation conditions. Some examples are summarized below:

- In a recombinant fermentation, if an exotic ingredient is added (for example, neomycin to help reduce contamination or to maintain plasmid stability), the burden will be on the producer to prove that this material is not in the final product. It is better to find a way to avoid its use in fermentation.
- In a recombinant fermentation, a very common problem is one of amino acid clipping (for example, by an amino peptidase) which may occur under certain conditions. This can present a severe problem to the manufacturer. It may then be necessary to separate a protein containing 189 amino acids from one containing 188 amino acids. This is a major challenge and best solved by a better understanding and control of the fermentation so that the amino acid clipping does not occur in the first place.
- In an enzyme fermentation, it proved to be much easier to harvest the host culture (*Pseudomonas aeruginosa*) by filtration using a chemically defined medium composition.
- Under certain culture conditions, one amino acid can be substituted for another (e.g., norleucine for methionine).

In these examples in our own process development work it has often proved easier to solve a purification problem at the fermentation stage.

6 Fermentation Process Development

6.1 Objectives

The objective of process development is the production of sufficient new or modified product to meet market demand in the quickest time possible, meeting all safety and quality requirements and by a cost-effective and reliable process. The role of biochemical engineers working on process development is to translate the process from the laboratory to full-scale production using knowledge provided by the biological scientists in the laboratory and their own expertise in process development and scale-up.

The various stages of process development are summarized in Fig. 8. Following identification of a new product candidate, a commitment to begin process development is made by the company. In parallel to the process development, there are other sequences which may need to take place. For instance, for therapeutic products it is necessary to provide materials for clinical trials. It may be possible to provide sufficient material for the pharmacological and toxicological studies which make up the pre-clinical trials by repeated use of the laboratory procedure. However, a pilot-scale or large-scale process must have evolved in time for the clinical trials which require large quantities of product. By the start of phase III trials it is essential that the process is well defined and understood as much as possible. Once phase III trials have commenced, it becomes difficult to make further process changes. In fact, the Center for Biological Evaluation and Research (CBER), FDA, prefers phase III material to be made in the final manufacturing facility using manufacturing staff.

There are many advantages gained by a company which does effective process development. First of all, especially for a biologic, it allows for the reproducible production of highly purified product which is a *sine qua non* for modern pharmaceutical manufacturing. Second, the impact on capital expenditure can be huge, for example, by quadrupling process performance both for new products and to meet increasing sales without having to build new factories. Third, there is an obvious benefit by reducing production cost thus releasing money for further investments by the company. Finally, the time between starting development of a process and introduction of the product to the market is minimized. This may result in reduction of development costs through, for example, lower interest charges and earlier sales income and resulting extension of sales under patent protection.

Successful scale-up of fermentations requires a good understanding of the interactions between microorganisms and the chemical and physical environments in the fermentor. The biochemical engineer endeavors to control the chemical environment by monitoring the fermentation and feeding nutrients intermittently or continuously. It is more difficult to control the physical environment during translation of scale as both agitation and aeration conditions change. Their impact on

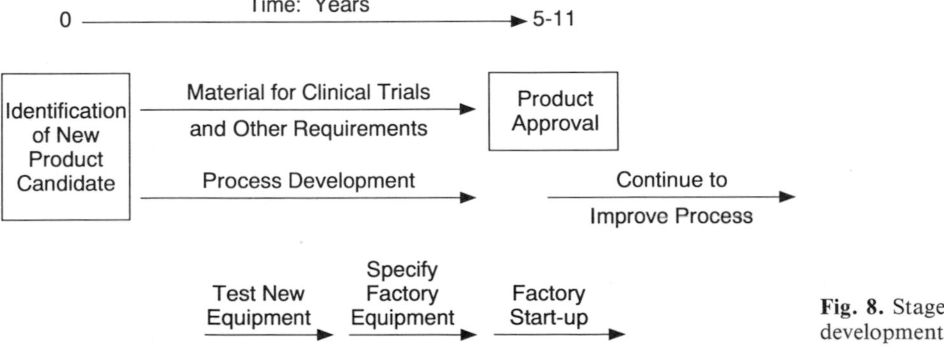

Fig. 8. Stages in process development.

various fermentation parameters is shown diagrammatically in Fig. 9. For instance, if agitation is low, mixing will be poor. If both agitation and aeration are low, then oxygen transfer will be poor and the culture will be oxygen-limited. Thus it is possible to define the operating boundaries within which the fermentation should be maintained. These boundaries are, of course, broad regions across which the operating problem increases in magnitude. The biochemical engineer needs to know how these boundaries shift during translation of scale so that fermentations can be operated without encountering these problems. This may not be possible for highly viscous fermentations such as those producing xanthan gums.

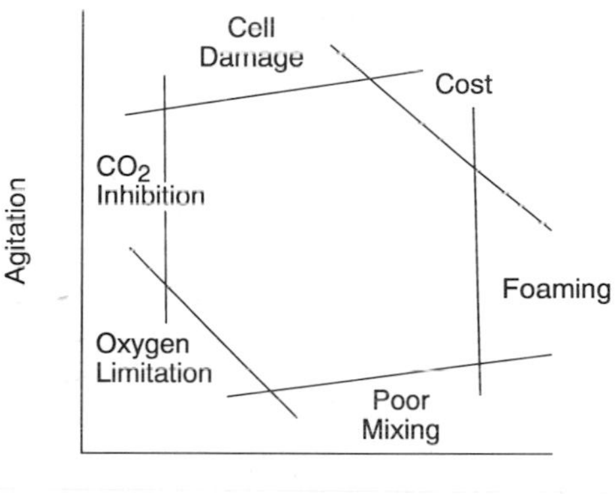

Fig. 9. Operating boundaries for fermentation scale-up (redrawn from LILLY, 1983).

Both agitation and aeration affect the distribution of nutrients, including oxygen, and relative fluid velocities in the fermentation broth, which in turn impact on culture morphology, broth rheology and product formation. It is important, therefore, to understand the interactions between these. For instance, increase in agitation rates can lead to greater fragmentation of mycelia and reduced penicillin production in *Penicillium chrysogenum* fermentations (SMITH et al., 1990).

During the initial scale-up and subsequent process development, there may be large increases in product yields or productivity. Each new strain which is introduced requires the biochemical engineer to modify appropriately the fermentation conditions. Substantial improvements are made as a result of changes in the seed train, mixing and aeration conditions, and feeding of nutrients. This is illustrated in Fig. 1 which shows the increases in productivity for penicillin fermentations at Gist-Brocades over the last three decades. It highlights the fact that improvements to fermentation processes are often made for many years after the product is first marketed (in this case for 50 years after the product was first marketed!).

6.2 Organism Selection

Many different microorganisms are already used for industrial fermentation processes. Nevertheless, this represents only a small proportion of the wide range of microorganisms around the world most of which, particularly in subtropical and tropical regions, have not been characterized. At the same time the number of vectors described for recombinant microorganisms is growing rapidly. The proliferation of choice is both exciting and challenging. In the case of recombinant organisms to produce proteins, it is beneficial to select a small number of hosts, vectors, and expression systems for new microbial processes to minimize the amount of work required and the time taken to develop each new process. This also requires evaluation of the advantages and disadvantages of hosts such as *Escherichia coli* and *Saccharomyces cerevisiae* and the location of the desired protein, i.e., intracellular, periplasmic, or extracellular.

6.3 Capital Investment

While process development is going on, it is also necessary to plan ahead for commercial operation and to estimate the capital investment required. The sequence shown in Fig. 8 implies that new equipment will be required. This will depend on whether or not there is existing fermentation capacity and recovery equipment available in the production plant. Although in recent years some companies have installed new fermentors, many new products are introduced without the need for new plants. Whenever possible, companies rely on improvements in titers of existing products to release sufficient fermentation capacity in time for introduction of the next new product. If this can be done, it minimizes the capital investment required for a new product. This approach does rely on the fact that by choosing in the past agitated aerated tanks or aerated towers, companies have very flexible if not ideal types of reactor.

6.4 Regulatory Issues

Secondary metabolites, such as antibiotics, can be well characterized analytically, and this has resulted in an environment in which process changes can be made, i.e., minor changes can be made on an ongoing basis as long as there is proof that product made by the new process is the same as that made by the old. The main challenge results from the fact that certain fungal and actinomycetes cultures can produce a remarkable array of very similar metabolites. Analytical techniques and purification techniques need to be very good, both to identify and remove minor levels of these similar metabolites.

By contrast, it is more difficult to characterize fully a protein molecule, and many of the same regulations developed historically for products such as vaccines (where the final product can be much harder to define) have been applied to recombinant proteins. In these examples, the process itself is much more rigorously defined which has resulted in a stagnant process development environment once a product is far advanced in clinical trials. In a recent talk, BADER (1992) made the interesting comment that, in order to improve product quality, changes have to be made to a process. These changes are inherently difficult to make for production of a "biologic" which is dominated by quality assurance and the emphasis is placed on installing safeguards to make sure that as few deviations as possible occur.

In addition, the process definition submitted to the regulatory authorities includes a detailed description of the equipment to be used as part of the Establishment License Amendment (Tab. 6). This has resulted in highly validated facilities, which are inherently less adaptable for other product candidates. Consequently, costs for recombinant proteins, both to operate a modern facility and also to construct a factory, are much higher than for secondary metabolites. In 1992 it costs as much to build a 3000 L scale biologics facility as it does to build a standard 200000 L scale antibiotics production facility. In addition to design for good manufacturing practice (GMP), there has also been an emphasis in recent years on containment of waste streams from a facility using recombinant cultures. These environmental restrictions have recently been considerably relaxed in the USA, as the evidence grows that the common microbial hosts used for recombinant DNA technology are indeed very safe and certainly present no credible threat to the environment. In fact, often these hosts represent the safest cultures being used in the pharmaceutical companies, because they have been deliberately adapted so that they will not survive in the general environment.

Tab. 6. Comparison – Drugs versus Biologicals

Area	Drugs	Biologicals
Release specs	Ycs	Yes
Process specs	General	Specific
Establishment license	No	Yes
Focus	Drug product	Drug substance
Change control	Flexible	Rigid

7 Conclusions

For some decades, the fermentation industry has made a major contribution to society through the provision of healthcare and food products, and the treatment of wastes. There is a growing demand by the regulatory authorities and the general public that products for healthcare and food products should be single enantiomers rather than the racemic mixtures often used in the past. Although some chemical and physical methods have been developed for separation of enantiomers, the biological route is increasingly attractive for this purpose.

The impact of rDNA technology is now being seen in the commercialization of therapeutic proteins. In the future, many enzymes will be made with recombinant organisms, and improvements in the synthesis of primary and secondary metabolites through metabolic engineering are likely. However, the excitement which these new developments have quite rightly caused should not lead to an underestimate of the major contribution which "traditional" secondary metabolite fermentations will make to the market into the next century.

For reasons explained earlier, in the past the emphasis during process development has been on yield improvement. While this will remain of prime importance, other criteria such as process reliability and reproducibility, safety and minimization of waste streams must be satisfied. All of these depend on the application of good biochemical engineering to the design, development and operations of fermentation processes in the future.

Acknowledgements
The authors wish to thank Drs. MAIGETTER, MASUREKAR and MONAGHAN for numerous helpful editorial comments.
The authors also wish to thank Professor N. W. KOSSEN (Gist-Brocades) and Dr. S. CARLEYSMITH (Smithkline Beecham) for data shown in Figs. 1 and 2, respectively.

8 References

BADER, F. (1992), Evolution in fermentation facility design from antibiotics to recombinant proteins, in: *Harnessing Biotechnology for the 21st Century* (LADISCH, M., BOSE, A., Eds.), pp. 228–231, Washington, DC: American Chemical Society.

BAJPAI, R. K., REUSS, M. (1982), Coupling of mixing and microbial kinetics for evaluation of the performance of bioreactors, *Can. J. Chem. Eng.* **60**, 384–392.

BUCKLAND, B. C. (1984), The translation of scale in fermentation processes: The impact of computer process control, *Bio/Technology* **2**, 875–884.

BUCKLAND, B. C., BRIX, T., FASTERT, H., GBEWONYO, K., HUNT, G., JAIN, D. (1985), *Bio/Technology* **3**, 982–988.

BUCKLAND, B. C., GBEWONYO, K., DIMASI, D., HUNT, G., WESTERFIELD, G., NIENOW, A. W. (1988), Improved performance in viscous mycelial fermentations of agitator retrofitting, *Biotechnol. Bioeng.* **31**, 737–742.

CARRINGTON, R., DIXON, K., HARROP, A. J. (1992), Oxygen in industrial air agitated fermentors, in: *Harnessing Biotechnology for the 21st Century* (LADISCH, M. R., BOSE, A., Eds.), pp. 325–333, Washington, DC: American Chemical Society.

CHISTI, M. Y. (1989), *Airlift Reactors*, Amsterdam: Elsevier.

CORBETT, K. (1980), Preparation, sterilization and design of media, in: *Fungal Biotechnology* (SMITH, J. E., BERRY, D. R., KRISTIANSEN, B., Eds.), pp. 25–41, New York: Academic Press.

EIKI, H., KISHI, I., GOMI, T., OGAWA, M. (1992), "Lights Out" production of cephamycins in automated fermentation facilities, in: *Harnessing Biotechnology for the 21st Century* (LADISCH, M. R., BOSE, A., Eds.), pp. 223–227, Washington, DC: American Chemical Society.

EINSELE, A. (1978), Scaling up bioreactors, *Process Biochem.*, July, 13–14.

FEHRENBACH, R., COMBERBACH, M., PETRE, J. O. (1992), On-line biomass monitoring by capacitance measurement, *J. Biotechnol.* **23**, 303–314.

FISH, N. M., LILLY, M. D. (1984), The interactions between fermentation and protein recovery, *Bio/Technology* **2**, 623–628.

HAMMOND, S. V., BROOKES, I. K. (1992), Near infrared spectroscopy – powerful technique for at-line and on-line analysis of fermentations, in: *Harnessing Biotechnology for the 21st Century* (LADISCH, M. R., BOSE, A., Eds.), pp. 325–333, Washington, DC: American Chemical Society.

HARRIS, C. M., TODD, R. W., BUNGARD, S. J.,

LOVITT, R. W., MORRIS, J. G., KELL, D. G. (1987), Dielectric permittivity of microbial suspensions at radio frequencies: a novel method for real-time estimation of microbial biomass, *Enzyme Microb. Technol.* **9**, 181–186.

JAIN, D., BUCKLAND, B. C. (1988), Scale-up of the Efrotomycin fermentation using a computer controlled pilot plant, *Bioprocess Eng.* **3**, 31–36.

JANSEN, P. H., SLOTT, S., GURTTER, H. (1978), Determination of mixing times in large-scale fermentors using radioactive isotopes, *Proc. 1st Eur. Congr. Biotechnol.*, Part II, 80–82.

LILLY, M. D. (1983), Problems in process scale-up, in: *Bioactive Microbial Products 2* (WINSTANLEY, D. J., NISBET, L. J., Eds.), pp. 79–89, London: Academic Press.

LILLY, M. D. (1992), Biochemical engineering research; its contribution to the biological industries, *Trans. Inst. Chem. Eng.* **70**, Part C, 3–7.

LÜBBERT, A. (1992), Advanced measuring techniques for mixing and mass transfer in bioreactors, in: *Harnessing Biotechnology for the 21st Century* (LADISCH, M. R., BOSE, A., Eds.), pp. 178–182, Washington DC: American Chemical Society.

MANFREDINI, R., CAVALLERA, V., MARINI, L., DONATI, G. (1983), Mixing and oxygen transfer in conventional stirred fermentors, *Biotechnol. Bioeng.* **25**, 3115–3132.

NALLIN OMSTEAD, M., KAPLAN, L., BUCKLAND, B. C. (1989), Fermentation development and process improvement, in: *Ivermectin and Abamectin* (CAMPBELL, W. C., Ed.), pp. 33–54.

NIENOW, A. W. (1992), New agitators vs. Rushton turbines: A critical comparison of transport phenomena, in: *Harnessing Biotechnology for the 21st Century* (LADISCH, M., BOSE, A., Eds.), pp. 193–196, Washington, DC: American Chemical Society.

OOSTERHUIS, N. M. G., KOSSEN, N. W. F. (1984), Dissolved oxygen concentration profiles in a production-scale bioreactor, *Biotechnol. Bioeng.* **26**, 546–550.

PACKER, H. L., THOMAS, C. R. (1990), Morphological measurements on filamentous microorganisms by fully automated image analysis, *Biotechnol. Bioeng.* **35**, 870–881.

PACKER, H. L., KESHAVARZ-MOORE, LILLY, M. D., THOMAS, C. R. (1992), Estimation of cell volume and biomass of *Penicillium chrysogenum* using image analysis, *Biotechnol. Bioeng.* **39**, 384–391.

PRIMROSE, S. B. (1990), Controlling bacteriophage infections in industrial bioprocesses, *Adv. Biochem. Eng.* **43**, 1–10.

REDA, K. D., THIEN, M. P., FEYGIN, I., MARCIN, C. S., CHARTRAIN, M. M., GREASHAM, R. L. (1991), Automatic whole broth multi-fermenter sampling, *J. Ind. Microbiol.* **7**, 215–220.

SMITH, J. J., LILLY, M. D., FOX, R. I. (1990), The effect of agitation on the morphology and penicillin production of *Penicillium chrysogenum, Biotechnol. Bioeng.* **35**, 1011–1023.

WANG, H., COONEY, C. L., WANG, D. I. C. (1977), Computer-aided baker's yeast fermentation, *Biotechnol. Bioeng.* **19**, 69.

2 Industrial Animal Cell Culture

Brian D. Kelley
Andover, MA 01810, USA

Tzyy-Wen Chiou
Cambridge, MA 02139, USA

Morris Rosenberg
Cambridge, MA 02142, USA

Daniel I. C. Wang
Cambridge, MA 02139, USA

1 Introduction

Recombinant proteins may be produced by either bacterial, yeast, or animal cell cultures, and the choice of host is determined by several factors. If the quantitiy of total protein required is quite large, hosts such as bacteria or yeast have an advantage because of their rapid growth and higher expression levels. Appropriate blood clearance rates and solubility of glycoproteins for parenteral therapeutics may require post-translational modifications similar to the native structures, which may not be provided by bacterial sources. The principles of refolding bacterially-derived proteins are becoming better understood, but not all proteins can be refolded at high concentrations with acceptable recoveries. These and other issues, including additional post-translational modifications, the presence of endotoxins, medium and bioreactor cost, and questions of viral contamination all figure into the choice of host for protein production. The current status of industrial research and development reflects the difficulty of this choice, as products are often produced by both prokaryotic and eukaryotic hosts during initial feasibility studies, before the final host selection is made.

Animal cells may now be cultured to very large volumes (up to 10000 L) to provide the necessary quantity of protein for new markets. This increase in scale brings the economics of animal cell production into focus, stressing previously less important questions such as the cost of medium, bioreactor choice, and maximum sustained cell density. These problems all motivate current research in animal cell culture, much of which will be summarized in the chapters in this book. This review will examine the historical perspective of cell culture, providing a basic understanding of animal cell culture within the development of biochemical engineering. Descriptions of the processes used to manufacture current products will provide a snapshot of the state of the art as it is practiced today, and will emphasize limitations of existing technology, equipment, and understanding. The basic questions which drive industrial and academic research will be addressed individually, with summaries of recent developments and solution methodologies. Finally, we will predict where animal cell culture is heading and what its position in the health care industry will be, by extrapolation of progress in both animal cell culture and competing technologies.

2 Historical Perspective

The history of industrial animal cell culture is a fascinating interplay of medicine, biology, and engineering. Cell culture products have progressed from vaccines for both human and animal diseases, to monoclonal antibodies and interferons, and now to recombinant proteins from genetically engineered cells. The first processes employed primary cells; later, normal cells maintained as a cell strain were approved, and now, transformed cell lines which have infinite life spans are used. Cell culture and propagation were initially conducted with roller bottles, then microcarriers, and now, many cell lines are cultured in stirred tank bioreactors. Cell lines are now being adapted to adverse culture conditions, including serum-free media, lack of anchorage support, and elevated ammonia and lactate levels. Table 1 gives a brief overview of the development of these products and various cell types used in industry.

Modern industrial cell culture began in the mid-1950s, with the use of animal cells for vaccine research and development (Spier, 1991). Industrial animal cell culture was first developed for the production of polio vaccines through propagation of normal diploid cells derived from tissue explants (Butler, 1987). These primary cells were trypsinized and grown to a confluent monolayer on roller bottles. The virus was inoculated into the confluent culture and harvested several days post infection. Vaccines could not be obtained by any other method, and so animal cell culture produced the inactivated or attenuated viruses necessary for vaccination. Because of the small amounts of material required, there was little pressure then to scale-up the process beyond roller bottle capabilities. Important principles of media preparation and formulation were developed during this stage, especially with re-

Tab. 1. Development of Mammalian Cell Culture for Industrial Use

1949	Virus growth demonstrated in cell culture (ENDERS)
1954	Salk polio vaccine produced in primary cells
1955	Chemically defined medium developed (EAGLE)
1962	Human diploid cell strain WI-38 established (HAYFLICK)
	BHK cells first grown in suspension (CAPSTICK et al.)
	Continuous VERO cell line established (YASUMURA)
1964	Foot-and-mouth disease vaccine produced from BHK cell line
1964–69	Vaccines for rabies, rubella, and mumps developed using WI-38
1967	Microcarrier culture developed (VAN WEZEL)
1970s	Large-scale suspension culture developed for FMD vaccine production
1975	Monoclonal antibody technology discovered (KOHLER and MILSTEIN)
1986	Interferon derived from continuous Namalwa cell line licensed
	Polio and rabies vaccines produced from VERO cell line
1987	First therapeutic monoclonal antibody (OKT3) licensed
1988–89	Recombinant products from continuous CHO cell lines licensed (tPA, EPO)

gard to understanding the basic metabolic requirements of the cell (EAGLE, 1959).

The susceptibility of human diploid cell strains to viral infection and propagation directed processes to be developed which avoided primary cell culture (HAYFLICK et al., 1963). These cell strains, while having finite life spans in culture unlike transformed, continuous cell lines, could be frozen and stored before use in production. Important controls were established to ensure reproducible and safe vaccine production, including the establishment of master and working cell banks, tests for genetic stability, and proof of non-tumorigenicity of both the products and the production host.

Several process improvements arose from the need to produce larger quantities of foot-and-mouth disease vaccine for cattle vaccinations. The large number of cattle treated combined with a regime entailing repeated administration resulted in world-wide demand for over a billion doses, which drove the scale of operation beyond the capacity of adherent cell lines. Early anchorage-independent cell lines derived from an immortal baby hamster kidney (BHK 21) cell line were shown to grow in submerged suspension culture (CAPSTICK et al., 1962); later, they were propagated in stirred tank reactors, mixed with turbine-type impellers and oxygenated by direct sparging. Typical process bioreactors had working volumes of 3000 L (RADLETT et al., 1985). With these principles established by the late 1960s, the large-scale cultivation of animal cells was possible, although the widespread application of such technology would wait for about a decade before the advent of both genetic engineering and hybridoma technology ushered in non-vaccine products requiring large scale cell culture.

A cell line which produces industrially feasible quantities of α-interferon (Namalwa, a human lymphoblastic cell line) became the next large-scale host, with 1000 L scale pilot plants built by 1978 (PULLEN et al., 1985). The non-recombinant cells were stimulated to produce the interferon in response to infection by Sendai virus. This was the first licensed use of a transformed cell line to produce a human biological. The largest manufacturing plant used 8000 L bioreactors, yielding grams of crude α-interferon per batch. The production medium employed a basal medium of glucose, amino acids, salts, and vitamins, and was supplemented with serum and serum-substitute materials.

Measurement and control in such bioreactors included pH (manipulated through CO_2 or sodium hydroxide addition), stirrer speed (via an indirect magnetic drive to reduce contamination), and dissolved oxygen (controlled through the addition of air to the culture). This process is in many ways a paradigm for current processes.

The ability to produce monoclonal antibodies by *in vitro* hybridoma culture opened up a wide range of uses for immunoglobulins. Applications in medical diagnosis, protein purification, and immunotherapy required an increase in the production scale for hybridoma cultures. This resulted in the development of large-scale airlift bioreactors currently operated at scales up to 2000 L (BLIEM et al., 1991). Antibody titers may reach 500 mg/L, and some hybridomas are cultivated with medium that is completely protein-free.

Bioreactor engineering then explored cell culture systems based on a continuous flow of medium to maintain high cell densitites without loss of viability. Cultivation in hollow fiber units, ceramic monoliths, or perfusion bioreactors demonstrated the ability to maintain high cell densities, increased product titers, and cell-free product streams. In some systems, cells reached densities nearly equal to those of body tissue, but began to show limitations in oxygen and nutrient supply or waste product removal.

Gene cloning and expression in mammalian cells now allow the production of heterologous proteins for the treatment of a wide variety of diseases, for use in diagnostic and research applications, or for production of subunit vaccines. The market size for these products varies widely, depending on the dosage, disease incidence rate, and number of administrations required. Stirred tank bioreactors have been built with 10000 L volumes, and are used to cultivate anchorage-independent cells grown in suspension (LUBINIECKI et al., 1989). The following section addresses current methodologies for large-scale cultivation of animal cells, emphasizing the diversity of process designs and operations.

3 Current Applications

Many therapeutic proteins derived from animal cells are currently in development or on the market. Most of these proteins are destined for use as *in vivo* human diagnostic or therapeutic reagents in markets that readily support high-value added products. Table 2 presents a summary of some of the major proteins either produced commercially today or in late stage clinical trials (FRADD, 1992). By the mid-1990s it is projected that human diagnostics and therapeutics manufactured by mammalian cell culture will result in revenues in excess of US $ 5 billion per year. This represents nearly half of the anticipated total revenues for biopharmaceuticals projected to be on the market by the mid-1990s.

Many proteins under development or on the market are made in mammalian cells, and this trend will increase during the coming decade. Proteins which are to be used as *in vivo* human reagents, thus, must be properly folded and

Tab. 2. Mammalian Cell Culture-Derived Products Produced Commercially or in Late Stage Clinical Trials

Product	Indication(s)	Current Status	Projected World Market (in million US $)
Tissue plasminogen activator	Heart disease	Commercialized	400
Erythropoietin	Kidney disease, cancer, AIDS	Commercialized	700
OKT3-MAb	Graft vs. host disease	Commercialized	20
Ceredase (CHO version)	Gaucher's disease	Phase II/III	200
Centoxin-MAb	Sepsis	PLA filed	500
Xomen E5-MAb	Sepsis	PLA filed	500
Oncorad-103-MAb	Ovarian cancer	PLA filed	50
Oncoscint-103-MAb	Ovarian cancer	Phase II	50
Hemolate factor VIII	Hemophilia	PLA filed	400
Bactolate factor VIII	Hemophilia	PLA filed	400

glycosylated in order to have maximal specific activity, be nonimmunogenic, and have the appropriate serum half-life. Improperly formed proteins with incorrect disulfide bridge formation, e.g., may have significantly reduced activity or present immunogenic determinants. Glycosylation provides proteins with a signature believed to dictate the serum half-life of the molecule as well as its specific activity in some cases. In other cases, most notably for antibodies, glycosylation provides key effector functions, such as complement fixation. Although there have been many advances in understanding refolding and glycosylation processes, it is unlikely that complex glycoproteins destined for use as human therapeutics will be manufactured in bacteria or yeast to a significant degree beyond the laboratory bench scale during the coming decade.

3.1 Cell Types

Many vaccines are still produced by cultivation of human diploid strains, although acceptance of continuous cell lines for vaccine production allows high potency vaccines to be economically manufactured for very large markets due to changes in the cultivation techniques. Both polio and rabies vaccines are produced with the VERO continuous cell line, which is grown on microcarriers in 1000 L bioreactors (MONTAGNON, 1988). Over 60 million doses per year are produced in the manufacturing of the polio vaccine.

During the past decade a relatively small number of host cell lines and expression systems have found their way into commercial development for recombinant protein production. The workhorse for the expression of recombinant proteins at a commercial scale has been the Chinese hamster ovary (CHO) cell line and has been used to produce such products as tPA, EPO, factor VIII, soluble receptors, and recombinant antibodies. The second predominant host cell line has been the murine myeloma cell line and related lines such as murine–murine and murine–human hybridomas. These cell lines have been used to produce a plethora of murine and recombinant monoclonal antibodies which are now in various stages of clinical trials. Less common host cells include the mouse C127 cell line, Syrian hamster ovary cells, hamster BHK, and human kidney 293 cells.

3.2 Types of Bioreactors

A wide variety of bioreactor configurations have been examined for mammalian cell culture (HU and PESHWA, 1991). The appropriate bioreactor reflects the characteristics of the production host, most notably the degree of anchorage dependence required for growth and viability. Some cells are able to grow in suspension, as single cells with no physical support; others require a surface to grow on and are called anchorage-dependent. Suspension cultures can be cultivated using bioreactors very similar to those for bacterial or yeast culture, and the change in size as the process is scaled up is usually accomplished by the fabrication of a single larger vessel, provided oxygen transfer limitations are not exceeded. Anchorage-dependent cultures, however, require bioreactors with a high surface-area-to-volume ratio, and have been difficult to culture at large scales. One common approach employs microcarriers to provide a surface for growth, while suspending the microcultures in stirred tank bioreactors. This allows anchorage-dependent cells to be grown in equipment similar to that used for suspension cells, which is generally simpler to design and operate.

Mammalian cell culture bioreactors can be divided into two categories; homogeneous systems in which the cells are in suspension culture or on microcarriers in a well-mixed stirred tank bioreactor; or heterogeneous systems where the cells are immobilized either within or on the surface of a biocompatible material. In homogeneous systems the cells are well mixed throughout the fermentor and it is relatively easy to obtain a sample of cells that is representative of the entire culture. This is an advantage, particularly from the viewpoint of validation and characterization of the cell morphology as required by the government regulatory authorities. There are no gradients within the reactor, preventing scattered suboptimal environments for cell growth. Examples of homogeneous systems are suspension stirred tank bioreactors, microcarrier bioreactors (where the microcarriers are suspended by gentle agitation in a stirred tank reactor), or airlift bioreactors. Heterogeneous systems are characterized by the fact that cells are generally entrapped or immobilized on a surface or within a microsphere (or macroporous microcarrier) and the bioreactor hydrodynamics imposes spatial gradients on the system. The most prevalent heterogeneous systems are fluidized bed systems with porous microcarriers, the ceramic monoliths, or hollow fiber bioreactors in which cells are immobilized in the extracapillary space. Heterogeneous bioreactors are often used for anchorage-dependent cells, but suspension cells can be cultured as well.

Although many different bioreactor systems have been developed at the bench scale, commercial scale cell culture products are almost exclusively produced in stirred tank bioreactors. The trend during the 1980s and into the early 1990s has been to develop and commercialize those production systems which are most easily understood and operated. As the time pressure to bring a product to market is obvious, any delay incurred from implementing new technologies often becomes critical in the final analysis.

3.3 Operating Strategies

Bioreactor operation strategies can be classified into one of three general modes; batch or fed-batch operations, the semi-continuous or cut-and-feed strategy, and lastly, perfusion culture. Batch culture is usually performed using suspension culture cells in a stirred tank bioreactor. Cells are inoculated into the vessel at a seeding density of roughly 0.5-2 $\times$ 10^5 viable cells/mL and the cells grow and secrete product over a period of 1–2 weeks. Product is harvested from the conditioned medium at the end of the batch cycle. The bioreactor is then cleaned and refurbished before the cycle is repeated. Fed-batch culture differs from batch culture in that nutrients are added either continuously or periodically during the batch cycle. Batch or fed-batch culture has proved to be quite popular in the biopharmaceutical industry. Many companies have pilot or commercial scale processes based on this operating strategy.

The semi-continuous or cut-and-feed strategy also employs stirred tank, homogeneously mixed bioreactors. In this operating strategy a bioreactor is inoculated at roughly 0.5-2 $\times$ 10^5 cells/mL and then allowed to grow until the culture is approaching early stationary phase. A large fraction of the cell culture broth is then harvested, usually on the order of 70–90%, and the bioreactor replenished with fresh medium. The cycle is then repeated. One can operate a bioreactor in this mode for several months. Usually the number of cut-and-feed cycles is limited by the stability of the expression rate and product quality during the production phase.

Perfusion operations retain cells within the reactor while allowing a cell-free sidestream to be removed; they can be subdivided into two categories, the homogeneous systems such as the perfusion chemostat or heterogeneous systems like hollow fiber or fluidized bed bioreactors. In a perfusion chemostat, cells are continuously recycled from the product stream either by an internal spin filter or via external tangential flow filtration. A second cell harvest stream continuously removes cells and conditioned medium at a rate sufficient to maintain a low cell debris load and high steady state cell viability. In the heterogeneous systems cells are immobilized or entrapped on a biocompatible surface thus preventing them from exiting in the product stream.

What distinguishes continuous perfusion systems from batch, fed-batch, or semi-continuous systems is that the former tend to use much smaller bioreactors and operate at roughly 10- to 50-fold higher cell density. Stirred tank bioreactors that operate in a perfusion chemostat mode have steady state viable cell densities of roughly 0.5–2.0 $\times$ 10^7 cells/mL. Hollow fiber and fluidized bed bioreactors typically report cell densities of 1–2 $\times$ 10^8 cells/mL fluid volume in the extracapillary space (hollow fiber units), or per mL of packed microcarrier volume (fluidized beds). Hollow fiber bioreactors and fluidized beds use a recirculating liquid stream to supply dissolved oxygen and other nutrients, adding volume to the total system. When cell density is calculated on this total volume basis, the densities are similar to those found in the homogeneous perfusion bioreactors.

An important point often overlooked is that while perfusion systems operate at much higher cell densities than batch bioreactors and thus require smaller fermentors, they still require as much if not more ancillary equipment to support the continuous operation. Perfusion systems will typically consume anywhere from 1–10 reactor volumes of media per day. These bioreactors therefore require extensive media preparation, filtration, and storage facilities as compared with batch or semi-continuous systems. A simple analysis of the medium requirements suggests that given equal cell yield (cells/L medium) and specific productivity (pg/cell/day), the same volume of medium will be required for high or low density cultivation, independent of the cell density in the bioreactor. On the other hand, if during perfused operations the specific productivity is reduced, this system would require a greater volume of medium per unit of product synthesized.

3.4 Purification of Animal Cell-Derived Products

Following the fermentation process, the protein product is purified through a combination of sequential unit processes. Cell conditioned media are usually pooled prior to or during the primary recovery operation. Primary recovery and process stream clarification are accomplished either through tangential flow filtration, depth filtration, or continuous centrifugation. Protein purification is achieved through a series of chromatographic and salt precipitation steps (SCHMIDT, 1989). Common forms of column chromatography applied at the pilot and commercial scale are size exclusion, ion-exchange, and hydrophobic interaction chromatography. High pressure liquid chromatography, while practiced extensively at the bench, is not often applied at the pilot or commercial scale. Occasionally phase extraction is used as a method of concentration and purification as in the case of α-interferon.

Product recovery validation must include demonstration of DNA and virus removal. While all protein products from recombinant processes must address the clearance of DNA, virus titer reduction is unique to animal cell processes where such infection is possible. Products from cell lines which could support adventitious viruses or have been shown to be tumorigenic must have downstream processes designed to prevent possible contamination from the parent cells (PRIOR, 1991). Typical virus removal should be 4–6 log reductions from the levels found in the conditioned media.

The bioreactor employed can have major impact on the initial steps of purification, but seldom on the later steps. Batch systems tend to have a higher ratio of product to total protein and a lower cell density as compared with continuous-flow perfusion systems. These two parameters can impact the primary recovery step and to a lesser extent the first purification column if it involves a capture of the product onto a resin. For example, during clarification using depth or tangential flow filtration, the greater the cell debris load the more membrane surface area will be required to process a given amount of cell conditioned medium. Therefore, one might expect that batch systems with high cell debris loads would have more expensive primary recovery operations on a per liter of process fluid basis as compared with a fluidized bed system.

3.5 Regulatory and Safety Issues

Animal cell culture for the production of pharmaceutical products is in most countries subject to a raft of regulations arising from the various federal, state, and local agencies which oversee both research and manufacturing (NELSON, 1991). In the U.S.A. the Food and Drug Administration's current Good Manufacturing Practices list the requirements for manufacturing biologics, the National Institute of Health regulates manipulation of recombinant organisms, and the Environmental Protection Agency is concerned with waste stream treatment and disposal. There are several unique concerns with the use of animal cell culture for protein or vaccine production (GARNICK et al., 1988). The slow growth of animal cells raises concern over contamination by microbes, as well as infection with mycoplasma. The genetic stability of the host cell must be validated by comparison of isoenzyme patterns and immunological markers on samples from the beginning to the end of the production run. Viral particles must be removed to prevent their carry over into the product; this is validated through "spiking" experiments measuring the clearance of each downstream processing step. Extensive cell lysis during the production phase should be avoided to reduce the DNA content of the culture fluid, as DNA removal must result in a final DNA concentration at or below the level detected in most available DNA assays. Lists of tests run on industrial processes using animal cells include these and many other analyses (MORANDI and VALERI, 1988).

4 Problems Motivating Current Research Interests

An examination of the process limitations, scale-up difficulties, and barriers to increased productivity of animal cell culture motivates the many areas of current research. These areas may be roughly divided into those addressing cell biology, where a basic understanding of the cell genetics, metabolism, and growth is needed, and bioreactor engineering, where manipulation of the cell environment and growth conditions affect protein production.

4.1 Cell Biology

4.1.1 Host Cell Development

The first step in developing a cell culture manufacturing process is in the appropriate selection of a host cell line for expression of the recombinant protein of interest. Desirable host cell characteristics include ease of transfection, ability to secrete active, recoverable product at high rates, good growth in cost effective cell culture media, and the lack of endogenous infectious agents such as retroviruses, mycoplasma, or prions.

There are several limitations in employing CHO and murine myeloma cell lines for production from a quality control standpoint. Neither cell line is of human origin and thus the products have a distinctly non-human glycosylation pattern. Studies are underway to engineer human glycosyltransferases into these cell lines in an attempt to produce a human glycoform. In the specific case of monoclonal antibodies, the murine pre-B cells from which most antibodies are currently derived, contain amino acid sequences which are distinctly murine and generate a well-characterized human-anti-mouse-antibody (HAMA) response in human individuals. The HAMA response occurs in most individuals given repeated doses of a murine antibody and it is suspected that this may reduce the efficacy of the drug (RIECHMANN et al., 1988). In order to circumvent the HAMA response, several companies are developing methods of "humanizing" murine sequences prior to transfection into rodent host lines (GORMAN, 1990). These humanization techniques are based on heuristic models derived from the growing databank of murine and human antibody gene sequences.

CHO cells usually have expression levels on the order of 1–20 pg/cell/day of active, recoverable product. Similarly, murine myeloma cell lines, either hybridoma derivatives or transfected myelomas, express antibodies or recombinant proteins at a rate of 1–50 pg/cell/day. These expression levels, though adequate by today's standards, are not sufficiently high to prevent mammalian cell culture from coming under increasing economic pressure from other non-protein modes of therapy or other non-cell culture modes of manufacture. Cell lines derived from tissues with well-developed secretory functions such as mammary cells could improve specific productivity. Extremely high productivities (>100 pg/cell/day) are demonstrated by hen oviducts, e.g., as they produce egg white proteins for the developing egg (SEAVER, 1991). Certain endocrine cells can be engineered to produce recombinant protein products, and secretion rates have been shown to be controlled by chemical factors in the environment (SAMBANIS et al., 1991). Such a strategy could allow cycling between synthesis and secretion, with media optimization for both distinct phases.

Finally, although CHO and murine myeloma cell lines are developing a track record of safety with the governing regulatory agencies, they are less than ideal. Murine myeloma cell lines have ubiquitous, endogenous murine retroviruses. CHO cell lines also contain retroviral-like particles over which regulatory agencies have expressed concern. The presence of these viral particles has forced manufacturers to implement extensive and costly viral clearance measures during purification of the drug. Ideally, future work would identify new hosts which would be devoid of potentially infectious viral agents (LUBINIECKI, 1987). Genetic stability of the host cell is also of primary importance as the manufacturing process needs to be reproducible, robust, and consistent. CHO cells have been shown to undergo a reduction in gene copy number, if selection pressure is released (KAUFMAN, 1990a). While batch and fed-batch operation is of short enough duration to avoid stability problems, long-term continuous cultures may have more severe problems. Better analytical methods for animal cell genetic make up and copy number will need to be developed, as well as stable, high copy number expression systems.

4.1.2 Media Formulation

The medium used to culture cells can have an enormous impact on maximum cell density, protein concentration, and ease of product recovery. The first media developed for cell culture were mixtures of a basal medium comprised of amino acids, carbohydrates, and vitamins, supplemented with serum. Serum stimulates growth by providing hormones, growth factors, and transport proteins (KITANO, 1991). The elimination or reduction of a medium's serum content reduces product variability, simplifies downstream processing and protein isolation, and can result in less costly media. There have been numerous serum-free media developed for mammalian cell lines in general, and CHO and murine myeloma cell lines in particular. Purified proteins may be added to a chemically defined basal medium, and act as a serum substitute. Typical protein supplements include transferrin and albumin as carriers for iron and fatty acids, respectively, and insulin or insulin-like growth factors. The use of serum-free media is not simply an issue of formulation and the correct balancing of metabolites and nutrients, but also involves the adaptation of cell lines to growth without serum. Indeed, the use of completely protein-free media has been shown for several applications, including the production of monoclonal antibodies by hybridomas (SCHNEIDER, 1989). Further work in medium optimization should provide protein-free formulations for many if not all cell lines.

The second area of development for media formulation is in the improvement of the basal nutrient mixture. Most cell culture media used today are variations on the pioneering work done by EAGLE and others during the 1950s and 1960s. Basal nutrient formulations need to be modified for specific bioreactor configurations. This issue is particularly germane to the optimization of media formulation and feeding strategy for continuous perfusion reactors. These reactors operate at a perfusion rate of 1–10 reactor volumes per day, resulting in a dilute product stream and high raw material costs at the commercial scale. By optimizing the nutrient spectrum and concentrations in the basal medium one should be able to significantly lower the cell specific perfusion rate resulting in higher product concentrations and significantly lower operating costs.

4.1.3 Genetic Engineering Approaches

While CHO and murine myeloma hosts have proved productive for the first generation of mammalian cell culture manufacturing processes they have presented several problems. Protein expression is related to gene copy number and transcription efficiency, and attempts to amplify gene expression through increases in both areas have met with success. Many amplifiable markers are based either on complementation of genetic deficiencies of the host strain or the introduction of drug resistance (KELLEMS, 1991). The most common system employs dihydrofolate reductase (DHFR) expression vectors combined with DHFR-deficient Chinese hamster ovary (CHO) cells. Host cells are transfected with expression vectors containing genes encoding for both DHFR as well as the product of interest. By passage in medium containing increasing levels of methotrexate, an inhibitor of DHFR, cells are selected for increased DHFR activity, which is often coamplified with the product gene. This may increase the product gene copy number up to 500 or 1000 copies per cell (KAUFMAN, 1990b). Single-cell cloning from these populations yields high-producing cell lines suitable for production in the absence of selective pressure. As an alternative, expression has been significantly increased by the use of the metallothionein promoter to drive the transcription of a desired gene, which is activated by high concentrations of zinc salts (FRIEDMAN et al., 1989). This strong promoter results in protein accumulation in excess of 200 mg/L, with purity above 90% when produced in serum-free medium.

What is called for are additional studies on expression systems to identify bottlenecks, and if these can be overcome, the identification of the absolute expression limits on a cell-specific basis. A small set of promoter/enhancer systems have been employed in industrial hosts and further work in this area might yield much stronger transcriptional activators. Relatively little work has examined translation efficiency in commercial cell lines. Lastly, there are several examples of protein expression limited by the host cell line's ability to efficiently secrete a recombinant protein. Manipulating the levels of BIP (an endoplasmic reticulum resident chaperone required for efficient protein secretion) through genetic engineering, one can increase or decrease the expression of an independently transfected recombinant gene several-fold (DORNER et al., 1988). This implies that mammalian cell hosts can be engineered for higher expression by manipulating not only the dosage of the recombinant gene but also the activity of genes critical for protein secretion. Further improvements in transcription rates, translation rates, and secretion efficiency via cellular engineering will, no doubt, occur over the next decade.

4.1.4 Glycoprotein Microheterogeneity

Glycoproteins produced in animal cells often have heterogeneous glycosylation patterns, with variation in the type and composition of the attached carbohydrate (PAREKH, 1991). The major class of oligosaccharides is attached to surface asparagines within a tripeptide sequence, with carbohydrate attached via an N-acetyl glucosamine residue (called an N-linked oligosaccharide). The oligosaccharide may contain two, three, or four antenna, each having a different monosaccharide sequence and composition. Heterogeneity occurs in both the types of carbohydrate residues and their linkage to each other, and arises from the spectrum and activity of the glycosyltransferases in the endoplasmic reticulum and Golgi apparatus of the host cell. The glycosylation pattern affects the clearance rates of glycoproteins from the bloodstream through the action of specific receptors which recognize non-human carbohydrate structures.

An understanding of the factors affecting glycosylation is necessary to control glycoform microheterogeneity. The cell lineage is of clear importance, as the carbohydrate retains patterns typical of the original cell species, although intra-species differences also exist. Culture conditions such as glucose levels and cytokine presence affect glycosylation also, and studies are being conducted to elucidate bioreactor effects on glycosylation (GOOCHEE et al., 1991). With knowledge gained from these studies, future bioreactor operation may be controlled to reduce heterogeneity, or selectively produce glycoforms which have beneficial properties for human therapy *in vivo*.

4.2 Bioreactor Engineering

4.2.1 Oxygenation of Bioreactors

Oxygenation plays an important role in animal cell culture because the dissolved oxygen level of the medium has a critical effect on cell metabolism. Due to its low solubility in the medium, oxygen becomes a limiting nutrient in high cell density cultures. Unlike most other nutrients which can be added in large amounts at a time, oxygen has to be supplied continuously to sustain culture viability and productivity. Membrane oxygenation, medium saturation in an external device, and direct sparging may all be employed for oxygenation.

Membranes with high oxygen permeability made from silicone and other polymers provide efficient oxygen transfer without gas bubble encounters with the cells. For scale-up at constant power per unit volume in membrane oxygenation systems, the loss of interfacial area per unit volume greatly affects the mass transfer rate achievable in the bioreactor, limiting them to moderately-sized bioreactors. The most likely solutions to these limitations are increases in the interfacial area and driving force for transfer (AUNINS et al., 1986).

Saturating the medium with oxygen through external contactors has the advantage that the aeration system and other coarse environmental adjustments such as sparging and agitation can be carried out in a cell-free chamber and thus do not interfere with the cell growth. However, on a large scale, a very high medium flow rate or a high oxygen tension in the oxygenated stream will be required to maintain adequate bioreactor oxygen levels and may be detrimental to cells.

Sparging has been proven to be a simple, efficient, and scaleable aeration method in microbial fermentation. Since both liquid mixing and gas transfer can be provided by sparging, successful airlift bioreactor operation has been reported for a variety of animal, insect, and plant suspension cell lines. However, physical damage to cells exposed to bubbling has been reported. An understanding of the physical forces responsible for cell damage is necessary to design processes which reduce cell death during sparging. It appears that cell damage is associated with gas–liquid interfacial effects at the region of bubble disengagement and not with shear stresses occurring during bubble rise in the column (HANDA et al., 1987). A combination of appropriate media formulation and altered sparger and bioreactor design should be considered to minimize bubble break up at the free surface. Examples to prevent such damage include decreasing the surface-area-to-volume ratio, sparging with very small (micron-sized) bubbles, or adding surface active agents (such as Pluronic F-68 nonionic polymer series) at concentrations which result in more stable foams at the culture surface (HANDA-CORRIGAN, 1990). It is not clear, however, how stable foams will affect product stability, as excess foaming may lead to protein denaturation at the gas–liquid interface. In addition, the antifoam added to suppress foaming may be problematic in downstream processing.

Alternatively, a bioreactor configuration which separates the cells from the bubbles can avoid exposing the cells to damaging sparging. This has been done previously in the form of "caged aeration" of microcarriers, and has recently been incorporated into commercial stirred tank bioreactors by New Brunswick Scientific (Celligen), and by Chemap (Chemcell). In an alternate design, sparged aeration has been successfully conducted in a fixed bed system by the design of a modified airlift fiber-bed bioreactor (CHIOU et al., 1991). Oxygenation takes place in the cell-free draft tube, preventing bubble disengagement in the presence of cells.

4.2.2 Agitation and an Understanding of Shear Damage

Cell suspension and microcarrier cultures are commonly agitated to increase oxygen transfer and provide a homogeneous environment for cell growth. Excess agitation damages mammalian cells due to their fragility and relatively large size; cell death or detachment from microcarriers is often observed in conjunction with vigorous agitation. Therefore, understanding the bioreactor hydrodynamics and the mechanisms of cell injury in agitated bioreactors is essential for bioreactor engineering and design. Significant progress has been made toward elucidating the hydrodynamic forces acting on animal cells present on microcarriers (CROUGHAN and WANG, 1991).

Cell death in microcarrier culture is dependent on eddy size, with eddies which are smaller than the microcarriers causing the greatest damage. The size of the smallest eddies depends on the Kolmogorov length scale, decreasing with kinematic viscosity or with power input. Larger eddies do not generally cause damage as they entrain the entire microcarrier so that it is not subject to a large velocity gradient. Additionally, if the time-averaged velocity components vary considerably over a small length scale, strong hydrodynamic forces could result and kill cells. Cell damage is rarely seen as a result of time-averaged flow fields, but may be considerable when there is a jet region off the impeller or a close clearance between the impeller and the vessel wall. In microcarrier cultures, two neighboring microcarriers might collide with each other due to the action of turbulent eddies. By understanding the cell damage mechanisms, bioreactor design can be altered to reduce shear stress and maximize mixing. Controlling factors include liquid-height-to-vessel-diameter ratio, bottom shape, clearance between the impeller and vessel, and impeller shape and size.

For suspension cells, damage in agitated bioreactors is the result of two distinct phenomena; at low agitation rates, with sparged aeration, damage results from air entrainment and subsequent bubble break up. Damage in gas-free systems occurs at much higher agitation, as the Kolmogorov eddy length decreases to that of the single cell. Therefore, modified bioreactor designs which eliminate bubble entrainment should minimize cell damage even at moderate agitation rates (PAPOUTSAKIS, 1991).

4.2.3 High Cell Density Bioreactors

The cell densities achieved by either airlift bioreactors or stirred tank bioreactors are typically below 10^7 cells/mL. In addition, batch or semi-batch operation results in a significant fraction of a production run required for cell growth, during which time only moderate product formation has occurred. Perfusion bioreactors operate at higher cell densities, resulting in higher product titers; they also allow continuous product formation, which is not possible in batch systems. Continuous operation thus reduces unproductive time spent growing cells in the start-up period. These advantages of perfusion culture continue to drive development of new cell retention mechanisms, bioreactor design, and oxygenation methods capable of meeting the increased oxygen demand (HU and PESHWA, 1991).

While high density systems have made inroads into small-scale and non-therapeutic applications, problems still remain before such continuous systems are widely used for production of human therapeutics. Processes utilizing continuous culture are nearing approval in the U.S.A., however (FOX, 1992). Homogeneous systems such as stirred tank bioreactors ensure that each cell experiences the same environment and culture conditions. Heterogeneous bioreactors, however, may have culture conditions which vary within the bioreactor. For example, axial gradients of nutrients and growth factors have been found as the result of Starling flow within hollow fiber bioreactors. Heterogeneity of cell metabolism could result in decreasing product quality. Sampling of cell populations grown in ceramic monolithic and hollow fiber bioreactors is impossible during the production phase, and, in combination with questions of genetic stability over extended production, may limit the length of culture time. Future research must address these aspects of gradients, and reduce them in order to eliminate questions of product consistency.

4.2.4 Application of Biosensors

Because cells are very sensitive to metabolites and substrates in the medium, proper monitoring and control of the pertinent biochemical parameters is necessary to maintain proper cell metabolism and high rates of product synthesis. On-line process sensors include pH, temperature, and dissolved oxygen, but samples are usually monitored through off-line analyses for specific medium components, enzyme levels which indicate cell lysis or product concentrations. Biosensors may provide a more direct means to monitor bioprocesses, without the need for off-line analysis. Biosensors are being developed to determine the biomass level (by ATP and DNA measurements), nutrient concentration (including carbohydrates and amino acids), waste levels (lactic acid or ammonia), and the presence of secreted products (monoclonal antibodies and enzymes). Detailed descriptions of the construction and application of many biosensors are in the literature (MERTEN, 1988; WISE, 1989).

Biosensors incorporating biological reagents such as enzymes may have high selectivity, but have several problems associated with their construction. Usually, the biosensors are not steam sterilizable or autoclavable due to their heat sensitivity. In addition, their working lifetime is limited, depending on the biocatalyst used, the immobilization method, and the quantity and the purity of the immobilized biocatalyst. Membranes used for immobilizing the biocatalyst or for separating the biosensor from the bioreactor can be fouled, clogged, and poisoned, causing drift and requiring recalibration. Because of these drawbacks, biosensors are difficult to apply directly to cell culture systems. Therefore, sampling and filtration systems are usually employed (SCHEIRER and MERTEN, 1991).

In contrast to biosensor measurement of the extracellular components, NMR spectroscopy can provide fundamental intracellular metabolic information (DALE and GILLIES, 1991). The current limitations on whole cell NMR are its lack of sensitivity and the requirement for compatible bioreactors. The low sensitivity can be overcome by increasing the cell density in the sample or by growing cells to sufficiently high density within a vessel compatible with NMR geometry and conditions. However, these bioreactors represent heterogeneous catalytic systems, and they may be diffusion-limited. To generate reliable information for metabolic studies, diffusion must be considered along with reaction when monitoring by NMR.

5 Future Applications

Proteins for human therapeutic use are the principal products of animal cell culture processes today. The advantages offered by the use of animal cells include secretion of a properly folded, glycosylated product and the ability to culture cells at reasonable densities in large bioreactors. Inroads are being made by bacterial and yeast cultures as hosts for recombinant proteins, because of their reduced production costs and increased product concentration and volumetric productivities. Attempts to understand the refolding of aggregated proteins produced as inclusion bodies are yielding promising results, and may one day dominate the therapeutic protein market, especially when products eventually lose patent protection and competition will drive rapidly toward lowered production costs. Even the production of glycoproteins by animal cell culture may be superseded by the use of yeasts or insect cells capable of producing glycoforms close to the natural pattern, or by the PEGylation of bacterially derived proteins (GOODSON and KATRE, 1990). While some proteins currently produced by animal cell culture may eventually switch to bacterial or yeast hosts (as some vaccines already have; STEPHENNE, 1990), it is difficult to envision the more complex, multisubunit proteins being efficiently produced without higher eukaryotic hosts.

Recent advances in the production of recombinant proteins in transgenic animals suggest that this technology could eventually replace traditional fermentation technology. Complex glycoproteins such as tPA and α-1-antitrypsin have been produced in transgenic mice and more recently in transgenic goats, sheep, and dairy cattle (DENMAN et al., 1991; HENNINGHAUSEN et al., 1990). Transgenic animals are now able to produce several grams per liter of protein in their milk and purification yields of up to 25% have been reported for such processes. Given the fact that a dairy cow can produce 10000 L milk a year one can readily see the economic benefits of such a production method. Several obstacles still remain, however. Successful implantation of genetically engineered embryos and the subsequent expression of active product on a consistent basis remain problematic. In many cases, it is difficult to purify a recombinant product from the complex milieu found in milk. Lastly, there will be many regulatory hurdles that will have to be surmounted and this will, no doubt, take time.

One area where cell culture will be impossible to replace is for the case when the product is much more complex than proteins, and unobtainable from other sources – the cells themselves! The culture of cells for the repopulation of damaged human tissues will require a new focus in biotechnology, as it shifts from protein products to cells. Should individual cells retain their immunological identity, this will drive a scale-down from large bioreactors to smaller, more flexible operations tailored for the culture of cells derived from an individual patient. If immunologically neutral cells can be obtained, large bioreactors could provide a single cell type for all patients. Current efforts to culture hematopoietic stem cells illuminate some of the great potentials and problems associated with this technology (EMERSON et al., 1991). The *in vitro* culture of pluripotent stem cells would be of great value in treating patients who have lost the regenerative capabilities of their white blood cell population due to disease or disease treatments. By culturing the patient's own stem cells from bone marrow explants, a supply of the most primitive cells could be reintroduced into the patient following the loss of the native population, with subsequent colonization and reestablishment of a functional immune system. Culture of these stem cells is a difficult task, as they must be maintained in an undifferentiated state while multiplying to generate a substantial quantity for transplantation. The biological control of this process *in vivo* has only recently been elucidated, but several aspects are still to be explained. It will be through an understanding of these effectors (the family of cytokines, including colony-stimulating factors and interleukins) that the true potential of the cultivation of human or animal cells, i.e., for the production of the cells themselves as therapeutic agents, will be realized.

Further use of cell cultures will arise for the cultivation of tissues or organs capable of serving as implants or transplants. Hepatocytes are being cultivated *in vitro* for examination of cell transplantation and as an extracorporeal device (JAUREGUI and GANN, 1991). Skin cells are being grown in sufficient quantity to allow grafting, with retention of the proper structure and organization of the native tissue (DELUCA et al., 1988). Such technologies combine aspects of cellular biology, for the maintenance of the differentiated state and reduction of immunogenicity, materials engineering, to provide a proper structure for tissue growth, and biochemical engineering, to design bioreactors capable of sustaining growth to sufficient densities without nutrient limitations and death.

6 Conclusions

Industrial use of animal cell culture has progressed from the early use in vaccine manufacture to the production of proteins of therapeutic interest. Genetic engineering now allows gene expression in a variety of animal cell lines, with adequate expression levels for commercial use. Problems remain to be solved in the optimization of the protein yield and quality, prompting investigations into aspects of the cell biology and genetics, as well as bioreactor design and operation. Proteins derived from animal cell culture have advantages over those derived from microorganisms, including less immunogenic glycosylation patterns, proper conformation, and proteolytic processing. Progress in protein production from other sources has led to higher refolding efficiencies, immunological alteration due to polyethylene glycol addition, and transgenic production by animals, all of which may eventually successfully compete with animal cell processes. The lasting use of animal cell culture may be for the manufacture of the cells themselves, not for the proteins they secrete, which would be used for tissue regeneration, blood cell repopulation, and implantation therapy.

Acknowledgements

The authors wish to acknowledge the financial support of the National Science Foundation to the M.I.T. Biotechnology Process Engineering Center under the cooperative agreement CDR-8803014 which made this work possible.

7 References

AUNINS, J. G., CROUGHAN, M. S., WANG, D. I. C., GOLDSTEIN, J. M. (1986), Engineering developments in homogeneous culture of animal cells: oxygenation of reactors and scale-up, *Biotech. Bioeng.* **17**, 699–723.

BLIEM, R., KONOPITZKY, K., KATINGER, H. (1991), Industrial animal cell reactor systems: aspects of selection and evaluation, *Adv. Biochem. Eng.* **44**, 1–26.

BUTLER, M. (1987), *Animal Cell Culture: Principles and Products,* Hyattsville, MD: Open University Press of America.

CAPSTICK, P. B., TELLING, R. C., CHAPMAN, W. G., STEWART, D. L. (1962), Growth of a cloned strain of hamster kidney cells in suspended cultures and their susceptibility to the virus of foot-and-mouth disease, *Nature,* **195**, 1163–1164.

CHIOU, T.-W., MURAKAMI, S., WANG, D. I. C., WU, W.-T. (1991), A fiber-bed bioreactor for anchorage-dependent animal cell cultures: Part 1. Bioreactor design and operations, *Biotechnol. Bioeng.* **37**, 755–761.

CROUGHAN, M. S., WANG, D. I. C. (1991), Hydrodynamic effects on animal cells in microcarrier bioreactors, in: *Animal Cell Bioreactors* (HO, C. S., WANG, D. I. C., Eds.), pp. 213–249, Stoneham, MA: Butterworth-Heinemann.

DALE, B. E., GILLIES, R. J. (1991), Nuclear magnetic resonance spectroscopy of dense cell populations for metabolic studies and bioreactor engineering: a synergistic partnership, in: *Animal Cell Bioreactors* (HO, C. S., WANG, D. I. C., Eds.), pp. 107–118, Butterworth-Heinemann.

DELUCA, M., D'ANNA, F., BONDANZA, S., FRANZI A. T., CANCEDDA, R. (1988), Human epithelial cells induce human melanocyte growth *in vitro* but only skin keratinocytes regulate its proper differentiation in the absence of dermis, *J. Cell Biol.* **107**, 1919–1926.

DENMAN, J., HAYES, M., O'DAY, C., EDMUNDS, T., BARTLETT, C., HIRANI, S., EBERT, K. M., GORDON, K., MCPHERSON, J. M. (1991), Transgenic expression of a variant of human tissue-type plasminogen activator in goat milk: purification and characterization of the recombinant enzyme, *Bio/Technology* **9**, 839–843.

DORNER, A. J., DRANE, M. G., KAUFMAN, R. J. (1988), Reduction of endogenous GRP78 levels improves secretion of a heterologous protein in CHO cells, *Mol. Cell Biol.* **8**, 4063–4070.

EAGLE, H. (1959), Amino acid metabolism in mammalian cell cultures, *Science* **130**, 432–43.

EMERSON, S. G., PALSSON, B. O., CLARKE, M. F. (1991), The construction of high efficiency human bone marrow tissue *ex vivo, J. Cell. Biochem.* **45**, 268–272.

FOX, J. L. (1992), FDA panel okays two Factor VIIIs, *Bio/Technology* **10**, 15.

FRADD, B. M. (1992), Biopharmaceutical firms put emphasis on biology to cure human diseases, *Gen. Eng. News* **12**(1), 26–2.

FRIEDMAN, J. S., COFER, C. L., ANDERSON, C. L., KUSHNER, J. A., GRAY, P. P., CHAPMAN, G. E., STUART, M. C., LAZARUS, L., SHINE, J., KUSHNER, P. J. (1989), High expression in mammalian cells without amplification, *Bio/Technology* **7**, 359–362.

GARNICK, R. L., SOLLI, N. J., PAPA, P. A. (1988), The role of quality control in biotechnology: an analytical perspective, *Anal. Chem.* **6**, 2546–2557.

GOOCHEE, C. F., GRAMER, M. J., ANDERSEN, D. C., BAHR, J. B., RASMUSSEN, J. R. (1991), The oligosaccharides of glycoproteins: bioprocess factors affecting oligosaccharide structure and their effect on glycoprotein properties, *Bio/Technology* **9**, 1347–1355.

GOODSON, R. J., KATRE, N. V. (1990), Site-directed PEGylation of recombinant interleukin-2 at its glycosylation site, *Bio/Technology* **8**, 343–346.

GORMAN, C. H. (1990), Mammalian cell expression, *Curr. Opin. Biotechnol.* **1**, 36–43.

HANDA, A., EMERY, A. N., SPIER, R. E. (1987), On the evaluation of gas-liquid interfacial effects on hybridoma viability in bubble column bioreactors, *Dev. Biol. Stand.* **66**, 241–253.

HANDA-CORRIGAN, A. (1990), Oxygenating animal cell cultures: the remaining problems, in: *Animal Cell Biotechnology* (SPIER, R. E., GRIFFITHS, J. B., Eds.), Vol. 4, pp. 122–132, London: Academic Press.

HAYFLICK, L., MOOREHEAD, P. S., POMERAT, C. M., HSU, T. C. (1963), Choice of a cell system for vaccine production, *Science* **140**, 766–768.

HENNINGHAUSEN, L., RUIZ, L., WALL, R. (1990), Transgenic animals–production of foreign proteins in milk, *Curr. Opin. Biotechnol.* **1**, 74–78.

HU, W.-S., PESHWA, M. V. (1991), Animal cell bioreactors–recent advances and challenges to scale-up, *Can. J. Chem. Eng.* **69**, 409–420.

JAUREGUI, H. O., GANN, K. L. (1991), Mammalian hepatocytes as a foundation for treatment in human liver failure, *J. Cell. Biochem.* **45**, 359–365.

KAUFMAN, R. J. (1990a), Vectors used for expression in mammalian cells, *Methods Enzymol.* **185**, 487–511.

KAUFMAN, R. J. (1990b), Selection and co-amplification of heterologous genes in mammalian cells, *Methods Enzymol.* **185**, 537–566.

KELLEMS, R. E. (1991), Gene amplification in mammalian cells: strategies for protein production, *Curr. Opin. Biotechnol.* **2**, 723–729.

KITANO, K. (1991), Serum-free media, in: *Animal Cell Bioreactors* (HO, C. S., WANG, D. I. C., Eds.), pp. 73–106, Stoneham, MA: Butterworth-Heineman.

LUBINIECKI, A. S. (1987), Safety considerations for cell culture derived biologicals, in: *Large-Scale Cell Culture Technology,* (LYDERSON, B. B., Ed.), pp. 232–247, München: Hanser.

LUBINIECKI, A., ARATHOON, R., POLASTRI, G.,

THOMAS, J., WIEBE, M., GARNICK, R., JONES, A., VAN REIS, R., BUILDER, S. (1989), Selected strategies for manufacture and control of recombinant tissue plasminogen activator prepared from cell cultures, in: *Advances in Animal Cell Biology and Technology for Bioprocesses* (SPIER, R. E., GRIFFITHS, J. B., STEPHENNE, J., CROOY, P. J., Eds.), pp. 442–451, London: Butterworth.

MERTEN, O.-W. (1988), Sensors for the control of mammalian cell processes, in: *Animal Cell Biotechnology* (SPIER, R. E., GRIFFITHS, J. B., Eds.), Vol. 3, pp. 75–140. London: Academic Press.

MONTAGNON, B. J. (1988), Polio and rabies vaccines produced in continuous cell lines: a reality for VERO cell line. *Dev. Biol. Stand.* **70**, 27–47.

MORANDI, M., VALERI, A. (1988), Industrial scale production of β-interferon, *Adv. Biochem. Eng. Biotechnol.* **37**, 57–72.

NELSON, K. (1991), Biopharmaceutical plant design, in: *Recombinant DNA Technology and Applications* (PROKOP, A., BAJPAI, R. K., HO, C., Eds.), pp. 509–565, New York: McGraw-Hill.

PAPOUTSAKIS, E. T. (1991), Fluid-mechanical damage of animal cells in bioreactors, *Trends Biotechnol.* **9**, 427–437.

PAREKH, R. B. (1991), Mammalian cell gene expression: protein glycosylation, *Curr. Opin. Biotechnol.* **2**, 730–734.

PRIOR, C. P. (1991), Large-scale process purification of clinical product from animal cell cultures, in: *Animal Cell Bioreactors* (HO, C. S., WANG, D. I. C., Eds.), pp. 445–478, Stoneham, MA: Butterworth-Heinemann.

PULLEN, K. F., JOHNSON, M. D., PHILLIPS, A. W., BALL, G. D., FINTER, N. B. (1985), Very large-scale suspension cultures of mammalian cells, *Dev. Biol. Stand.* **60**, 175–177.

RADLETT, P. J., PAY, T. W. F., GARLAND, A. J. M. (1985), The use of BHK suspension cells for the commercial production of foot-and-mouth disease vaccines over a twenty year period, *Dev. Biol. Stand.* **60**, 163–170.

RIECHMANN, L., CLARK, M., WALDMAN, H., WINTER, G. (1988), Reshaping human antibodies for therapy, *Nature* **332**, 323–327.

SAMBANIS, A., LODISH, H. F., STEPHANOPOULOS, G. (1991), A model of secretory protein trafficking in recombinant AtT-20 cells, *Biotechnol. Bioeng.* **38**, 280–295.

SCHEIRER, W., MERTEN, O.-W. (1991), Instrumentation of animal cell culture reactors, in: *Animal Cell Bioreactors* (HO, C. S., WANG, D. I. C., Eds.), pp. 405–443, Stoneham, MA: Butterworth-Heinemann.

SCHMIDT, C. (1989), The purification of large amounts of monoclonal antibodies, *J. Biotech.* **11**, 235–252.

SCHNEIDER, Y.-J. (1989), Optimization of hybridoma cell growth and monoclonal antibody secretion in a chemically defined, serum- and protein-free culture medium, *J. Immunol. Methods* **116**, 65–77.

SEAVER, S. S. (1991), An industrial cell biologists's wish list, *Can. J. Chem. Eng.* **69**, 403–408.

SPIER, R. E. (1991), An overview of animal cell biotechnology: the conjoint application of science, art, and engineering, in: *Animal Cell Bioreactors* (HO, C. S., WANG, D. I. C., Eds.), pp. 3–18, Stoneham, MA: Butterworth-Heinemann.

STEPHENNE, J. (1990), Production in yeast vs. mammalian cells of the first recombinant DNA human vaccine and its proved safety, efficacy, and economy: hepatitis B vaccine, *Adv. Biotechnol. Processes* **14**, 279–299.

WISE, D. (1989), *Appied Biosensors,* Stoneham, MA: Butterworth-Heinemann.

3 Overview of Downstream Processing

RON SPEARS

North Chicago, IL 60064, USA

1 Introduction

A key segment of the production and marketing of any pharmaceutical product, whether it be an antibiotic, a peptide, or a complex protein, is the processing of the material from its initial milieu (tissue, fermentation broth, etc.) to a pure form suitable for its intended use. This key segment, termed downstream processing can be, and often is, a complicated series of isolation and purification steps which is usually quite costly. The advent of genetic and protein engineering technologies has provided new and powerful routes for the production of therapeutic and industrially significant products utilizing a variety of production vectors that include cellular, bacterial, and yeast expression vectors. These cellular manufacturing plants are efficient and relatively inexpensive sources of product when compared to natural sources.

Product recovery methods which are effective, efficient, and well designed are essential in developing a downstream process which can deliver a marketable product to the public, and a financial return to the company. Biotechnology manufacturing concerns must be able to design and implement purification schemes that are repeatable, reliable and meet current good manufacturing practice (GMP) guidelines. FDA regulations for documentation and validation of the manufacturing process must also be met before the product can be brought to market.

Hence, it is essential that a manufacturing strategy that encompasses production, recovery, isolation, and purification technologies be considered at the outset of process design. This will effect proper selection and design of the expression vector and subsequent purification techniques that are simple, complementary, and logical.

The new generations of products from the biotechnology industry are more complex and are often quite labile. Thus, a strategic overall approach of the entire process from fermentation, through isolation and purification, with due consideration to product stability and yield, becomes imperative. As many process development departments have experienced, designing a large-scale purification scheme from one developed piecemeal by other departments (at much smaller scales) is a series of adaptations and compromises that could have been done more efficiently, and less expensively, if only process development had the chance to provide input at the beginning stages.

This chapter is intended to be an overview of downstream processing techniques and design schemes, and will explore some of the salient issues to be considered when designing a production and manufacturing process for a therapeutic product. Novel and innovative isolation and purification procedures will also be presented to provide some insight to the reader into future directions and developments in this most exciting and integral aspect of the new biotechnology.

2 Upstream Considerations: The Source Vector

Typically, the less complex the starting material, the simpler the recovery and purification process. The more steps required to generate the desired purity of the product, the lower the product yields (Fig. 1). Therefore, the use of an expression vector which has the potential to ease and simplify the downstream processing of the product merits serious consideration.

Prior to the use of molecular biological techniques to insert or modify genes for specific proteins within a host cell (vector), the only viable means for obtaining large quantities of desired therapeutic materials were from natural sources such as serum, blood, tissue, urine, or from native strains of bacteria or yeast. These sources are by far the most complex starting materials by virtue of the diversity of the components present. Moreover, the product of interest is often present in only small or even trace quantities, making extraction and purification difficult and expensive, and perhaps unattractive commercially. Minimally, this would require freezing the material during

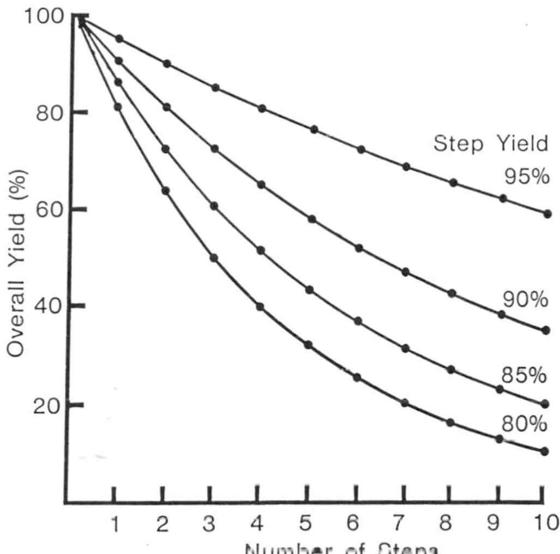

Fig. 1. Reduction in yield of product as a function of the number of purification steps.

shipment and large-scale storage on site for the large amounts of raw material needed, unless a consistent source of fresh material is available.

Animal sources have generally been consistent in supply but can display some variance among subspecies. Human sources are less readily available, far more expensive, and have the potential of viral contamination (e.g., HIV or hepatitis). The cost and contamination issues (and the resulting containment requirements) involved in deriving therapeutic products from natural sources have steered many companies towards natural or transformed cell lines, or bacterial, insect, or yeast sources. The ability to enhance the product generation levels in these eukaryotic and prokaryotic vectors has often made them the only viable choice.

Selection of a production vector involves consideration of a multitude of issues such as complexity of the fermentation or culture media and growth requirements, the cost of both the equipment and media (and containment) for the fermentation, the product expression level in the vector, and the ease and yield of recovering the desired product from the fermentation broth. In the light of downstream processing, it is essential that both the fermentation development and process development teams interact in order to strike a balance among all the aforementioned factors. For example, an *Escherichia coli*/plasmid combination could be selected as the production vector for its ease of growth and high expression level though the product may be generated as an inclusion body. The relatively high density of inclusion bodies can allow rapid initial isolation of the product provided a method for refolding of the protein, which must be subject to denaturing conditions for extraction and recovery, can be found.

Downstream processing requirements of an intracellular product are quite different from those of an extracellular or secreted product. Systems where the product is secreted from the cell (GRAY et al., 1984; OKA et al., 1985) or into the periplasmic space (GRAY et al., 1985) are gaining favor. Cell disruption methods are milder for products secreted into the periplasmic space, and eliminated entirely in the case of extracellular secretion. Additionally, protein transport across cellular membranes is usually accompanied by cleavage of precursor forms to yield mature N-termini (OGEZ et al., 1989).

The selection of a low expression level vector which secretes its product into the growth medium, and the associated substantial capital outlay required for the cellular fermentation equipment on the front end of the process, could easily be justified by the savings in time and equipment required for downstream purification, and the more desirable form of the product obtained from this type of fermentation.

It has recently been demonstrated that vectors in which a desired product is synthesized intracellularly, in its native glycosylated form and packaged in membrane vesicles via the Golgi apparatus, have the ability to secrete product only after the infusion of media containing a substance which can induce exocytotic release of the membrane vesicle contents (STEPHANOPOULOS, 1993). An analogous system has been described for *E. coli* (YU and SAN, 1992). The induced release of product, together with the ability to substantially simplify the media into which the product was released, makes this technology quite attractive from a downstream processing point of view.

More frequently, eukaryotic expression systems are selected by virtue of their ability to produce proteins more similar to the native state. Engineered proteins have been successfully produced in yeast (SINGH et al., 1984), insect cells (SMITH et al., 1985), as well as in mammalian cells, and recent work (WILSON, 1984; KAUFMAN et al., 1985; WALLS and GAINER, 1991) indicates progress towards greater expression levels in these cellular vectors.

Vector modification where the gene for the desired protein is fused with genetic material coding for a polypeptide with properties facilitating downstream purification has been described (MOKS et al., 1992; BEITLE and ATAAI, 1993; ENFORS et al., 1992; WINSTON et al., 1993). These fused protein products allow modification of the product to fit existing purification techniques thereby greatly shortening the purification process. To be effective, however, cleavage of the fusion product to generate native protein, in high yield, is essential for this technology to be commercially viable. Nonetheless, this procedure exemplifies the kind of interaction between fermentation and process development groups that is so crucial in generating an overall efficient and cost-effective manufacturing process.

3 Isolation and Recovery

Recovery of the product is the next step in the manufacturing process following fermentation. The techniques used for initial recovery and isolation depend entirely on the nature of the product, its own unique properties and characteristics, and the state of the product, soluble or insoluble, intracellular or extracellular, as it leaves the fermenter. Regardless of the physical state of the product, all early isolation and recovery steps seek to remove cellular debris, concentrate the product and perhaps attain a significant degree of purification, all while maintaining high yield. The effects of shear and temperature sensitivity of the product, as well as potential degradation of the product by indigenous proteases, must also be considered at this stage.

3.1 Clarification

3.1.1 Centrifugation

Initial clarification has been effected using a variety of technologies. Centrifugation has been widely employed at this point since it is a rather straightforward application, it is gentle, and has little effect on the overall yield. Centrifugation can, however, be quite costly at large scale and is often inefficient at achieving the desired clarification of the feed stream. This has led many companies to seek less expensive alternatives or ways in which to enhance centrifugal method efficiency.

Soluble and insoluble products demand different applications of centrifugal separation technology. Insoluble products can be intracellular and require isolation of the entire cellular suspension. Alternatively, the product may be soluble but require that debris load be reduced before it can be purified further.

Insoluble products are often in the form of inclusion bodies generated within *E. coli* (insoluble products could also be deliberately precipitated material which will be discussed later in this chapter). Tough and relatively insensitive to the shear forces required for cell rupture, they can easily be separated from cellular debris due to their relatively high density (BUILDER and OGEZ, 1985). The centrifugal needs for this application are fairly straightforward. Following centrifugal recovery, the inclusion bodies should be washed (MARSTON, 1986; FISHER et al., 1993), the product extracted using chaotropic agents such as urea or guanidine HCl, and the resulting denatured product should be refolded to an active state which can be taken to later purification steps. In one case, addition of anion exchange resin actually effected solubilization of protein from inclusion bodies in an *E. coli* lysate (HOESS et al., 1988).

Concentration of cellular suspensions (i.e., dewatering) or reduction of particulate load, has been effected using a continuous flow form of centrifuge such as the disk stack (ERIKSON, 1984) or the scroll decanter (WARD and HOARE, 1990). These offer continuous operation at the price of complex equipment. Tubular bowl centrifuge designs such as a

Sharples provide a high centrifugal force and can be cooled. The solid capacity of the tubular bowl is limited, however, requiring periodic dismantling and cleaning of the system which may be a significant disadvantage (BELTER et al., 1988). Satisfactory operation of any of these systems, however, requires that the debris has a sedimentation coefficient high enough for efficient recovery and dewatering of the suspension, or clarification of the feed stream in the case of soluble product.

Enhancement of the sedimentation properties of cellular debris and undesirable proteinaceous material increases the efficiency of centrifugal separations and often allows the use of lower centrifugal forces (WARD and HOARE, 1990) which significantly reduces operational and equipment costs. Reduced pH, elevated temperature, or both, can boost the level of contaminant precipitation. Use of these relatively harsh conditions, in some cases, has had a small effect on product yield (SPEARS, 1982) as the contaminants, suspended particulate matter, and the overall contents of the complex solution provided a protective effect on the desired product.

Flocculating agents have been shown to increase contaminant sedimentation rates by as much as 2000 times over those observed in untreated broths (ROSÉN, 1984). Flocculation agents such as synthetic polyelectrolytes (VINCENT, 1974; JENDRISAK, 1987), polyvalent cations (ROSÉN, 1984), those derived from bacterial sources (KURANE, 1990), or inorganic salts have been used and in many cases, scaled up successfully (BENTHAM et al., 1990).

Commercially available bioprocessing aids have effectively been used for feed stream pretreatment to enhance centrifugal efficiency (FLETCHER et al., 1990). Moreover, these cationic bioprocessing aids, typically cellulosic (Whatman) or polymeric (TosoHaas) in composition, have been shown to reduce pyrogen, nucleic acid, and acidic protein loads which can foul chromatography columns. Cellulosic products, such as CDR from Whatman, have the advantage of being quite inexpensive and designed for disposal after use (e.g., incineration).

3.1.2 Extraction

In the case of small molecules or peptides, the recovery process is greatly simplified as these materials have no secondary or tertiary structure, so characteristic of proteins, that would be affected adversely by the harsh conditions created by many of the available extraction methods. Solvent extraction techniques have been successfully applied to antibiotics (HARRIS et al., 1990), organic acids, steroids, peptides, and simple organic acids and bases using batch, staged, differential and fractional extraction methodologies (BELTER et al., 1988, and references therein).

Liquid–liquid extraction has been applied to recovery of proteins via aqueous two-phase systems using polyethyleneglycol (PEG) (HUDDLESTON et al., 1990; RIVEROS-MORENO and BEESLEY, 1990; ENFORS et al., 1992; DIAMOND et al., 1990, and references therein). PEG has also been used effectively for product precipitation and removal of feed stream contaminants (FERNANDES and LUNDBLAD, 1980). Protein engineering and gene fusion technologies (DIAMOND et al., 1990; ENFORS et al., 1992; WINSTON et al., 1993; LUTHER and GLATZ, 1993) have recently been used to modify the amino acid sequence, resulting in an increase in the solubility and thus the recovery of the protein of interest in aqueous two-phase systems composed of PEG/salts, PEG/dextran/salt or charged polymers.

3.1.3 Filtration

Crossflow filtration (CFF), also known as tangential flow filtration (TFF), has been employed for product recovery and often concomitant feed stream clarification (PRITCHARD et al., 1990; ZAHKA and LEAHY, 1985; COONEY et al., 1990; GABLER and RYAN, 1985; VAN REIS et al., 1992; TAYLOR et al., 1992). Available in a wide array of filter medium compositions and system configurations, the common theme of all designs involves rapid passage of the feed stream across the filter (crossflow), under sufficient pressure (transmembrane pressure; TMP) to cause permeable solutes to pass through (passage) with the flow of solvent (flux). Since filter media range from micron and sub-micron size porosities down to membrane filters with very small pores which can retain small peptides, the potential number of applications is tremendous. The reader is referred to CHERYAN (1986) for a more substantive treatment of the theory and principles of CFF.

Although applicable to crude suspensions, CFF rapidly loses efficiency under high debris and particulate loads. This is due, in part, to the build-up of the gel layer or cake of material that forms at the membrane surface during CFF system operation (Fig. 2). Particulate matter can also foul (essentially plug) the membrane pores, further reducing system efficiency. Since the formation of a gel layer, to some degree, is inevitable, the key to successful CFF is manipulation and/or minimization of this gel layer by varying a combination of conditions including filter selection, system configuration, operating pressures and temperatures, and the flow properties of the feed stream.

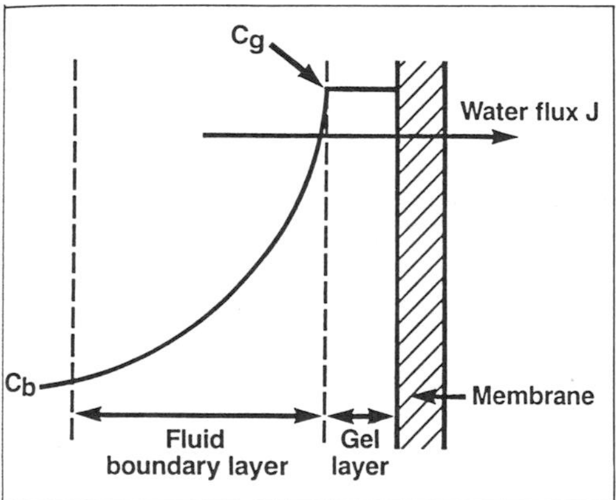

Fig. 2. Scheme representing dynamic formation of a gel layer during crossflow filtration where C_g is the solute concentration in the gel layer and C_b is the solute concentration in the bulk solution.

CFF filter medium and membranes are available in a wide variety of porosities and compositions which must be matched to the application. This is often done empirically by testing a number of membrane materials and system configurations on the feed stream, since the parameters involved in a dynamic filtration of a particular feed stream, especially a crude one, are generally too complex to deal with practically on a theoretical basis. Large microporous membranes (microfilters) have more commonly been applied to clarification and cell suspension concentration and are available in materials such as ceramic and steel which can be aggressively cleaned and sterilized in place. Polymeric materials such as polyvinyldifluoride (PVDF) and polyethersulfone (PES) are also used for microfilters and have been used for clarification and suspension concentration successfully (SPEARS, 1991). These polymeric materials, while effective, can be more difficult to clean and may require chemical rather than steam sterilization (which can be a validation problem).

Lower porosity filtration membranes, rated by the nominal molecular weight range of product retention (termed ultrafiltration) are generally applied to less crude feed streams. Recent developments in membrane technology have yielded filtration membranes in the porosity range between those of ultra- and microfilters that are less often subject to the mechanical fouling of their larger porosity cousins, and are fairly stable mechanically and chemically. These filter membranes provide additional alternatives in recovery stages of the purification process.

As was alluded to earlier, mechanical fouling of the filter surface is a major concern with application of CFF to early feed stream clarification. Fouling can only be reversed after process completion (which, due to the gross reduction in the flux associated with fouling, takes many times longer than expected), by an aggressive cleaning regimen. It is therefore wise to select a filter membrane that provides lower initial flux, but is substantially less subject to fouling.

The second concern with CFF is the inevitable formation of the gel layer that forms at the membrane surface under positive TMP. Under pressure, solute molecules are drawn to the membrane surface and those that are small enough have the chance to pass through. Their passage is often hindered by those solute molecules to which the membrane is impermeable. They, too, are driven to the membrane surface forming a gel-like layer (see Fig. 2) which can, and does, alter the membrane's designed permeability range. This gel layer, as one might expect, builds up as the feed stream concentrates with a concomitant decrease in flux, and passage of permeable solute. Flux reduction, as a function of feed stream concentration, is a known and well documented phenomenon. It is not uncommon, however, for passage of a desired soluble product to drop off quickly upon the formation of the boundary layer, often much faster that the flux. Recent investigations indicate that optimization of operational parameters such as pH, crossflow rate and TMP can minimize denaturation and aggregation of proteins (YEN and YANG, 1993) resulting in maintenance of higher flux and permeable solute passage.

Membrane system manufacturers offer a wide array of configurations all designed to improve efficiency (Fig. 3). In all cases, the goal is to increase flux and passage by disruption of the gel layer which is generally accomplished by generating turbulent flow. Turbulent flow enhances the shear effects generated at the membrane surface by laminar flow. Shear at the membrane surface is required to maintain flux and passage and to minimize fouling (MURKES and CARLSSON, 1988). Some of the configurations shown in Fig. 3 use screens or other means of generating turbulent flow. Rotation of the filter to generate Taylor vortices can also generate turbulent flow and has recently been applied to clarification and/or concentration of product feed streams, with promising results (COONEY, 1992; PARNHAM and DAVIS, 1993).

Hollow-fiber systems allow fast crossflow rates, are easily cleaned, but must rely on laminar flow to minimize the gel layer. Spiral-wound systems utilize screens to provide turbulence but are often difficult to clean due to the dead spaces in the winding (crossflow is perpendicular to the winding). This makes spiral designs less applicable to crude suspensions.

Plate and frame systems have no dead spaces and can offer the same high crossflows as hollow fibers. More recent designs employ short crossflow path lengths (i.e., from feed inlet to outlet) and offer configurations capable of generating turbulent flow (with membranes packaged in a cassette-type format). These systems also offer the greatest membrane surface area in the smallest amount of floor space, when compared to either spiral-wound or hollow-fiber membrane systems. These properties suggest that plate and frame systems may be the best choice at the recovery stage, and perhaps later stages as well.

Overall, CFF offers lower equipment and utility costs when compared to centrifugation and may be more efficient, since CFF systems are contained and can yield sterile filtrates (ZAHKA and LEAHY, 1985). CFF systems do, however, require regular membrane maintenance and cleaning (and validation), and periodic membrane replacement. Recent advances in membrane and system technology by a number of manufacturers such as Millipore and Filtron have further increased the efficiency of CFF and allow a more simple scale-up of the system, coupled with greater chemical resistance and easier cleaning and validation to comply with GMP guidelines.

3.2 Purification

Following treatment of the crude product feed stream to remove debris and particulate matter, the next task awaiting the process development team is to reduce the working volume of the feed and achieve at least a threefold purification. It is good separation strategy to employ a high resolution purification step as early in the process as possible. Combining other tasks such as product concentration reduces product handling thus preserving product yield, reducing process steps and, therefore, process costs.

Most typical downstream purification processes utilize combinations of chromatography, product concentration, and buffer exchange steps. The types of separation modes selected, and the order in which they are applied will have a profound effect on the product yield and the cost and efficiency of the separation.

Small molecules such as peptides or antibiotics are usually present in a clarified and substantially purified state at this point (typically by extractive processes) and can be applied directly to chromatographic media (e.g., ion exchange media). This provides product concentration and, with optimization, a substantial degree of further purification.

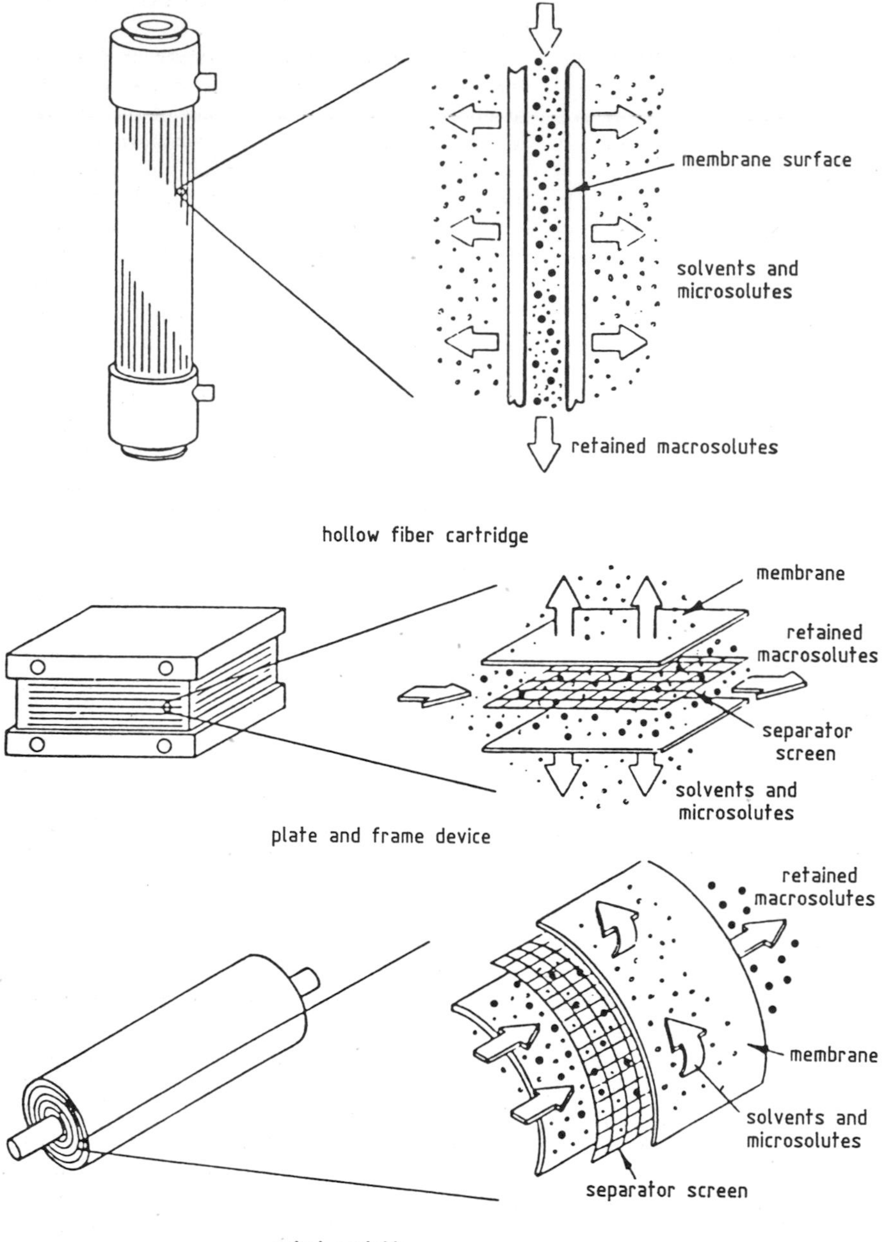

Fig. 3. Diagrams of the three major configurations for crossflow filtration systems.

The case for proteins, however, is not often as straightforward. Precipitative and chromatographic purification steps can be applied at this point which can offer both concentration and purification of the product.

3.2.1 Initial Fractionation

Use of agents to selectively fractionate proteins of interest has a long history. One of the first widely used large-scale processes to employ precipitation was fractionation of plasma (COHN et al., 1946). Although not a high resolution technique (usually limited to 2–3-fold per step purification), precipitation remains a useful technique for inexpensive and large-scale processes of products such as plasma or industrial enzymes or food proteins (HOARE and DUNNILL, 1984).

The obvious benefit of a precipitation step is that it is an easy and inexpensive way to quickly concentrate and partially purify the product. The precipitated product is typically collected by centrifugation at fairly low centrifugal force. To maximize yield, the pelleted precipitate should be washed and re-centrifuged which, unfortunately, increases process time. Ultrafiltration has been explored as an alternate means for recovery and washing of protein precipitates with some success (DEVEREUX et al., 1984).

For many protein products, precipitation is unacceptable due to the irreversible denaturation often observed when the product is exposed to the harsh ionic and chaotropic conditions required for precipitation (BELTER et al., 1988). Nonetheless, in cases where precipitation is applicable (e.g., gamma globulin precipitation with PEG; FERNANDES and LUNDBLAD, 1980), the benefit to the downstream purification process is significant. To bring the precipitate to the next step, it must be redissolved in the appropriate mobile phase.

Whether or not precipitation is used, the separation technique most often employed after recovery and/or clarification is an adsorptive one. Ion exchange (IEX) has proved to be the most versatile, since it offers high resolution and product concentration concurrently using media that are moderate in cost and quite rugged allowing many purification cycles reducing overall cost of the process in the long run.

Due to the level of contaminants and microparticulates in a clarified feed stream, most adsorptive IEX processes have been carried out in batch mode. The IEX medium (e.g., Amberlite®, DEAE-Sephadex® or cellulosic media) is stirred with the product for the adsorption, after which the slurry is transferred to a column where a buffer is used to elute selectively the product from the column. Though an inefficient application of IEX, this method has proved very successful for initial purification steps of a clarified feed stream.

Recent developments in media technology have made available chromatographic adsorbents that are better suited for clarified feed streams. Rigid agarose media, from companies such as Pharmacia (Sepharose® Big Beads) and Sterogene, offer high flow rates and lower pressure drops than conventional media (HEDMAN and BARNFIELD FREJ, 1992). This is due in part to the physical rigidity of the beads, and to the superior flow properties of large (100 to 300 microns and greater) spherical rigid beads.

These superior flow characteristics allow clarified feed streams to be applied directly to packed chromatographic beds at high flow rates while maintaining column efficiency (HEDMAN and BARNFIELD FREJ, 1992). The obvious advantage is the superior resolution that can be achieved with an optimized elution scheme.

Further enhancement of this application has been achieved by employing large bead chromatographic media in a limited form of fluidized beds, called expanded beds (upward flow) and first described by DRAEGER and CHASE (1990). Expanded beds are more stable than fluidized beds and allow operation similar to a packed chromatographic column. The media's binding capacity is also better utilized. After capture of the product following application of the crude (and often unclarified) sample, the particulates and non-binding proteins are washed out of the column in the expanded mode. The product is subsequently eluted in the downward direction as in a packed column, affording the efficiency of any packed column (JANSON and RYDÉN, Chapter 23, this volume).

3.2.2 Purification

Once initial fractionation has been accomplished, higher resolution and perhaps more specific separation techniques can be applied. In the case of biotechnology, these techniques are most often chromatographic, often supported by CFF. Most chromatographic separations of proteins and other biologically active polypeptides are carried out in aqueous medium and often at reduced temperatures due to the labile nature of these materials. Smaller peptides can often withstand exposure to organic phases during purification. Notably, large-scale reversed-phase chromatography (RPC) is used in the purification of human insulin and generates product with a potency equal to that of insulin purified by more conventional aqueous methods (KROEFF et al., 1989). Since it is unusual to successfully apply RPC to proteins, the remainder of discussion of chromatography will deal with aqueous based separation systems.

The first chromatographic separation technique widely applied to proteins was gel filtration chromatography (GFC; also known as size-exclusion chromatography or SEC; PORATH and LINDNER, 1961). Elution chromatography (EC), which encompasses IEX, displacement (DC), hydrophobic (HIC) and affinity (AFF) methodologies, has long since eclipsed GFC due to its greater versatility, capacity, and resolving power (BONNERJEA et al., 1986).

To merit inclusion in a separation scheme, a chromatographic method must offer significant product enrichment and should have a relatively high capacity for the product (or the contaminants, allowing a more pure product to pass through the column unimpeded) keeping the size and cost of the purification step within reasonable limits. Regeneration and sanitization protocols should be simple and inexpensive and consistent with good manufacturing practice (GMP) guidelines (FRY, 1985; CENTER FOR DRUGS AND BIOLOGICS, 1987). The separation media, to offer good process economy, should remain chemically and mechanically stable through many cycles of operation.

The sequence of purification steps should exploit the unique nature of the product at each point in the process to minimize process stream manipulation and maintain product yield. Chromatographic media properties such as capacity, recovery, resolution, and selectivity, with respect to the product, are essential elements in assembling an efficient, economic and sensible purification scheme. For example, HIC would be an excellent candidate to follow a salt precipitation step where the product remains in the supernatant. Since HIC requires high salt content in the loaded sample, and is relatively insensitive to sample volume as it is a concentrating technique, the product stream could be applied directly to the HIC column with minimal handling.

As was mentioned in Sect. 3.2.1, IEX is often employed at the front end of a purification sequence. Both the resolving and concentrating properties warrant this application which together provide both purification of the product and volume reduction of the feed stream. GFC is often used at the end of a purification sequence as a polishing step, often following an IEX or AFF step. The product would already be in a very concentrated state, and could be applied to a GFC column with little or no adjustment. Moreover, a buffer exchange could be carried out concurrently with minimal effort and cost. Examples of purification sequences are shown in Fig. 4.

In addition to IEX, HIC and AFF are also good choices for early application in the purification sequence. Caution is urged when considering AFF at early points in the sequence as these media are generally very expensive and could be rendered useless, if fouled by particulate or contaminants whose removal could destroy the AFF medium.

HIC is just beginning to gain favor in downstream processing, as the industry develops a better understanding in applying this technique. Better and more stable HIC media are now commercially available, the better known products being manufactured by Tosohaas and Pharmacia. HIC is analogous to RPC, but relies on comparatively weak hydrophobic interactions with the approximately 45% of surface amino acid residues that are hydrophobic. This results in little or no loss in biological potency upon elution. High ionic strength condi-

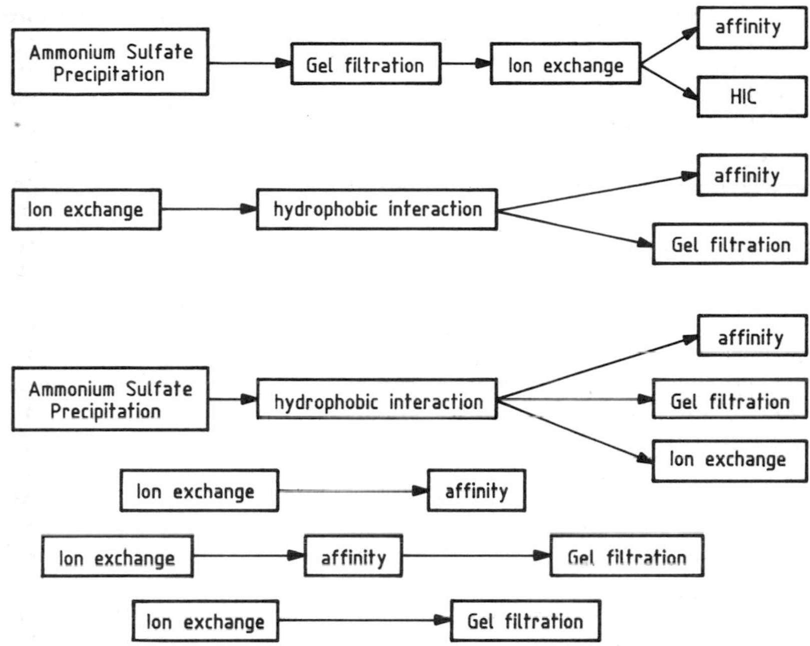

Fig. 4. Examples of logical and efficient purification sequences.

tions are required for protein binding making HIC best applied when the product stream is high in salt (e.g., following IEX or precipitation). Product is often eluted by simply reducing the ionic strength of the elution buffer resulting in a concentrated product fraction at low or moderate ionic strength.

Potentially the most powerful technique available to downstream processing is affinity chromatography (AFF). The uniquely specific nature of the affinity adsorbent for the product affords a very high product enrichment in the process stream. In some cases, affinity adsorbents applicable to the product are commercially available. Attachment of a unique ligand to a pre-activated resin bead is, however, often the only option (for a more detailed discussion of specific ligands used in AFF, the reader is referred to Chapter 23 by Janson and Rydén). In either case, an AFF step in a purification sequence can minimize the number of actual chromatography steps between product recovery and polishing.

AFF media are unfortunately quite expensive and may have a limited number of effective production cycles which would increase the cost of the operation. Leakage of ligand from the column, due to instability of the ligand–support matrix coupling, can reduce column capacity and contaminate the purified product (Johansson, 1992). In addition, ligand lability can limit the ability to clean the column, further reducing effective column life. Superior ligand binding chemistries on preactivated and derivatized gels offered by companies such as Sterogene, Pierce, Bio-Rad, Bioprocessing Ltd., and Pharmacia have effectively minimized ligand leakage enhancing the versatility of affinity media.

With all these potential drawbacks, AFF may still be an excellent choice for the purification scheme. Taken in conjunction with the entire process, the costs associated with implementing an AFF step may be well worth considering (Janson, 1984). Also, it may be fairly easy to remove any ligand leached from the AFF column in a later purification step and, with a well designed and sensitive assay, adequately documented.

GFC is best applied at later and final stages of a purification scheme. GFC can effect buffer exchange together with a size-based separation of the product from contaminants or aggregates. New developments in GFC media composition and construction have made this chromatographic separation technique more accessible to large-scale processes with improvements in both stability and selectivity, allowing good resolution at attractive flow rates. More aggressive cleaning and sanitization protocols can be applied to these more stable separation matrices.

Basic principles, however, cannot be ignored and GFC remains limited by small sample volumes that can be applied to the column bed to achieve maximum resolution (HAGEL, 1989). This is, of course, a "rule of thumb" and can be adjusted according to the resolution required. GFC remains an excellent step for final product polishing, removing the last of the trace contaminant and product aggregates formed during the purification, and for transfer of the product into a buffer that is suitable for product formulation and packaging.

Once the only choice for buffer exchange, GFC is rapidly being supplanted by crossflow filtration (CFF) which has been increasingly utilized in support of downstream purification processes. The ability to concentrate product, exchange buffer, and sometimes crudely fractionate based on size, all at the same time, makes CFF an indispensable tool in process development. Moreover, the speed at which these operations can be carried out make CFF one of the best choices for dewatering a dilute product stream, specifically when initial purification steps must be performed quickly to minimize proteolysis or other degrading reactions, and loading large volumes of dilute product onto a concentrating adsorbent simply requires too much time.

CFF has been applied in bacterial or viral load reduction, where reusable membranes offer long-term cost savings over traditional dead-end depth filters. CFF has also been used extensively in diafiltration mode (i.e., buffer exchange) for removal of non-aqueous solvent and PEG from mainly aqueous feed streams (SPEARS, unpublished results).

CFF diafiltration for buffer exchange is faster and much more reliable than either dialysis (avoiding catastrophic product loss due to dialysis bag rupture) or GFC. It is substantially less expensive to set up and operate than GFC, and presents the option of product concentration in the same apparatus. Furthermore, CFF cleaning and regeneration cycles are easier and more rapid than GFC.

The wide variety of polymeric and cellulosic membranes from manufacturers such as Amicon, Millipore, Ag, Sepracor, Gelman, and Filtron has expanded the application range of CFF. The enhanced chemical and mechanical stability of many recently introduced ultrafiltration and microfiltration membranes increases the suitability of CFF in processes that must meet validated cleaning protocols according to GPM guidelines.

4 A Look Ahead

What does the future hold for downstream processing? Perhaps we can better visualize it by examining some of the difficult hurdles that must be overcome in process development and some of the current research and pilot studies in progress to address these concerns.

The earlier a high resolution step can be introduced into a process, the more efficient a process can be. Advances in chromatographic media technology may allow more processes to utilize higher resolution separation methods to fractionate crude process streams, often with little more than gross clarification. Products and technologies from companies such as Pharmacia and Sterogene seem to offer the greatest promise at this time.

Applied electric field potentials on conducting polymers (PRZYBYCIEN et al., 1993) and microfiltration membranes (BRORS and KRONER, 1992) have enhanced performance of these systems and may offer improvements in the ability to more efficiently work with crude cellular suspensions and lysates. RUDGE and TODD (1990) have provided a detailed examination of applied electric fields on a number of downstream process separation methods, from both a theoretical and applied perspective.

The application of more selective chromatographic adsorbents, coupled with protein engineering to provide purification handles, is a new and expanding technology and may be the best route for future process development, provided all the necessary tools and an effective means for removing the "handle" are available (CARTER, 1990). Immunoaffinity chromatography, although not new to the world of affinity separations, remains a viable separation tool for development in industrial separations (BAILON and ROY, 1990).

Metal-affinity adsorbent matrices offer new and novel means for chromatographic separations and may hold the promise of more selective adsorbents that are stable and inexpensive (PLUNKETT et al., 1992; SMITH et al., 1990; SUH et al., 1990; ANDERSSON and PORATH, 1986; ARNOLD, 1991). Furthermore, engineering of proteins for greater specificity for these types of separations has recently been reported (BEITLE and ATAAI, 1993).

More rapid means of separations, without sacrificing efficiency and selectivity have been explored in a variety of venues. Use of microfiltration membranes for affinity separations (CHAMPLUVIER and KULA, 1992) and for IEX (BRIEFS and KULA, 1992) offer novel supports for well-known separation techniques and could have powerful if limited application in downstream processing. LIGHTFOOT (1990) describes a variety of novel and useful separation modes that will play varying roles in downstream processing in the years ahead and has suggested that chromatography may not play a major role in downstream processing in the future.

Displacement chromatography (DC) is essentially a modification of IEX in which a large mixed-solute load is fed to the column, and then displaced by a more strongly adsorbed material. Upon elution, the individual solutes emerge from the column, in the order of decreasing adsorptive strength, separated by comparatively short mixed zones where only the adjacent solute pairs are present (FRENZ, 1992). DC offers fairly high resolution with the operational simplicity of step elution. It does not offer the complete separation of IEX, but may be an acceptable cost-effective alternative when considered along with the entire process.

Perfusion chromatography (media and technology marketed by PerSeptive Biosystems, Inc.), which utilizes highly substituted and very porous rigid beads, offers high resolution at flow rates often as much as 3000 cm/h (AFEYAN et al., 1991). The potential for process development with this technology is tremendous by virtue of the speed in which analytical separations can be accomplished. This effectively eases the development chemist's burden presented by the often empirical investigations into the appropriate chromatographic application for a product. Large-scale operations with perfusion chromatography media are limited, however, both by the cost of the media and the equipment required to contain the media, and to deliver the flow rates necessary to achieve the rapid separations that perfusion chromatography offers.

5 Conclusion

It is hoped that this chapter has provided a broad overview of the issues facing the process development in the design of a purification scheme. More detail has been provided in some sections than in others to stimulate the reader to consider applications previously unexamined. Although not a comprehensive overview, the work cited identifies many of the groups active in new and exciting separation chemistries and applications.

The author wishes to strongly encourage a greater and more substantial communication between process development and R & D groups both in the area of vector selection and in the creation of purification schemes. The use of an encompassing strategy will make better use of existing resources and generate a more cost-effective approach to downstream process development.

6 References

AFEYAN, N. B., FULTON, S. P., REGNIER, F. (1991), Perfusion chromatography packing materials for proteins and peptides, *J. Chromatogr.* **544**, 267–279.

ANDERSSON, L., PORATH, J. (1986), Isolation of phosphoproteins by immobilized metal (Fe^{3+}) affinity chromatography, *Anal. Biochem.* **154,** 250–254.

ARNOLD, F. H. (1991), Metal-affinity separations: a new dimension in protein processing, *Bio/Technology* **9,** 151–156.

BAILON, P., ROY, S. K. (1990), Recovery of recombinant proteins by immunoaffinity chromatography, in: *Protein Purification, From Molecular Mechanisms to Large Scale Processes* (LADISCH, M. R., WILLSON, R. C., PAINTON, C. C., BUILDER, S. E., Eds.), pp. 150–167, *ACS Symp. Ser.* **427,** Washington, DC: American Chemical Society.

BEITLE, R. R., ATAAI, M. M. (1993), One-step purification of a model periplasmic protein from inclusion bodies by its fusion to an effective metal-binding peptide, *Biotechnol. Prog.* **9,** 64–69.

BELTER, P. A., CUSSLER, E. L., HU, W.-S. (1988), *Bioseparations: Downstream Processing in Biotechnology,* New York: John Wiley & Sons.

BENTHAM, A. C., BONNERJEA, J., OSBORN, C. B., WARD, P. N., HOARE, M. (1990), The separation of affinity flocculated yeast cell debris using a pilot plant scroll decanter centrifuge, *Biotechnol. Bioeng.* **36,** 397–401.

BONNERJEA, J., OH, S., HOARE, M., DUNNILL, P. (1986), Protein purification: The right step at the right time, *Bio/Technology* **4,** 954–958.

BRIEFS, K.-G., KULA, M.-R. (1992), Fast protein chromatography on analytical and preparative scale using modified microporous membranes, *Chem. Eng. Sci.* **47,** 141–149.

BRORS, A., KRONER, K. H. (1992), Electrically enhanced cross-flow filtration of biosuspensions, in: *Harnessing Biotechnology for the 21st Century* (LADISCH, M., BOSE, A., Eds.), pp. 254–257, Washington, DC: American Chemical Society.

BUILDER, S. E., OGEZ, J. R. (1985), Purification and activity assurance of precipitated heterologous proteins, *US Patent* No. 4,511,502, April 16, 1985.

CARTER, P. (1990), Site-specific proteolysis of fusion proteins, in: *Protein Purification, From Molecular Mechanisms to Large Scale Processes* (LADISCH, M. R., WILLSON, R. C., PAINTON, C. C., BUILDER, S. E., Eds.), pp. 181–193, *ACS Symp. Ser.* **427,** Washington, DC: American Chemical Society.

CENTER FOR DRUGS AND BIOLOGICS (1987), *Guidelines on General Principles of Process Validation,* Rockville, MD: FDA.

CHAMPLUVIER, B., KULA, M.-R. (1992), Dye-ligand membranes as selective adsorbents for rapid purification of enzymes: a case study, *Biotechnol. Bioeng.* **40,** 33–40.

CHERYAN, M. (1986), *Ultrafiltration Handbook,* Lancaster, PA: Technomic Publishing Co.

COHN, E. J., STRONG, L. E., HUGHES, W. L., MULFORD, D. J., ASHWORTH, J. N., MELIN, M., TAYLOR, H. L. (1946), Preparation of components of human plasma, *J. Am. Chem. Soc.* **68,** 459–475.

COONEY, C. L. (1992), Presentation at *Downstream Processing Symposium VI:* Session B, at the 1992 ACS Biochemical Technology Division Meeting in Washington, DC.

COONEY, C. L., HOLESCHOVSKY, U., AGARWAL, G. (1990), Vortex flow filtration for ultrafiltration of protein solutions, in: *Separations for Biotechnology 2* (PYLE, D. L., Ed.), pp. 122–131, New York: Elsevier Science Publishing.

DEVEREUX, N., HOARE, M., DUNNILL, P. (1984), The development of improved methods for the industrial recovery of protein precipitates, in: *Solid-Liquid Separation* (GREGORY, J., Ed.), pp. 143–160, Chichester, UK: Ellis Horwood Ltd.

DIAMOND, A. D., YU, K., HSU, J. T. (1990), Peptide and protein partitioning in aqueous two-phase systems, in: *Protein Purification, From Molecular Mechanisms to Large Scale Processes* (LADISCH, M. R., WILLSON, R. C., PAINTON, C. C., BUILDER, S. E., Eds.), pp. 52–65, *ACS Symp. Ser.* **427,** Washington, DC: American Chemical Society.

DRAEGER, N. M., CHASE, H. A. (1990), Modeling of protein adsorption in liquid fluidized beds, in: *Separations for Biotechnology 2* (PYLE, D. L., Ed.), pp. 325–334, New York: Elsevier Science Publishing.

ENFORS, S.-O., KOHLER, K., VEIDE, A. (1992), Recovery of fused proteins by liquid-liquid extraction, in: *Harnessing Biotechnology for the 21st Century* (LADISCH, M., BOSE, A., Eds.), pp. 280–283, Washington, DC: American Chemical Society.

ERIKSON, R. A. (1984), Disk stack centrifuges in biotechnology, *Prog. Chem. Eng.*, 51–54.

FERNANDES, P. M., LUNDBLAD, J. L. (1980), Preparation of a stable intravenous gamma-globulin: process design and scale-up, *Vox Sang.* **39,** 101–112.

FISHER, B., SUMNER, I., GOODENOUGH, P. (1993), Isolation, renaturation and formation of disulfide bonds of eukaryotic proteins expressed in *Escherichia coli* as inclusion bodies, *Biotechnol. Bioeng.* **41,** 3–13.

FLETCHER, K., DELEY, S., FLEISCHAKER, R. J., JR., FORRESTER, I. T., GRABSKI, A. C., STRICKLAND, W. N. (1990), Clarification of tissue culture fluid and cell lysates using Biocryl® bioprocessing aids, in: *Separations for Biotechnology 2*

(PYLE, D. L., Ed.), pp. 142–151, New York: Elsevier Science Publishing.

FRENZ, J. (1992), Frontiers of biopolymer purification: *Displacement Chromatogr. LCGC* **10**, 668–674.

FRY, E. M. (1985), Process validation: The FDA viewpoint, *Drug Chem. Ind.,* July, 46–51.

GABLER, R., RYAN, M. (1985), Processing cell lysate with tangential flow filtration, in: *Purification of Fermentation Products: Applications to Large Scale Processes* (LEROITH, D., Ed.), pp. 1–20, *ACS Symp. Ser.* **271**, Washington, DC: American Chemical Society.

GRAY, G., MCKEOWN, A., JONES, A., SEEBURG, P., HEYNEKER, H. (1984), *Pseudomonas aeruginosa* secretes and correctly processes human growth hormone, *Bio/Technology* **2**, 161–164.

GRAY, G., BALDRIDGE, J., MCKEOWN, K. S., HEYNEKER, H., CHANG, C. N. (1985), Periplasmic production of correctly processed human growth hormone in *E. coli*: Natural and bacterial signal sequences are interchangeable, *Gene* **39**, 247–254.

HAGEL, L. (1989), Gel filtration, in: *Protein Purification, Principles, High Resolution Methods and Applications* (JANSON, J.-C., RYDÉN, L., Eds.), pp. 63–106, New York: VCH Publishers Inc.

HARRIS, T. A. J., KHAN, S., REUBEN, B. G., SHOKOYA, T. (1990), Reactive solvent extraction of *beta*-lactam antibiotics, in: *Separations for Biotechnology 2* (PYLE, D. L., Ed.), pp. 172–180, New York: Elsevier Science Publishing.

HEDMAN, P., BARNFIELD FREJ, A.-K. (1992), Adapting Chromatography for initial large-scale protein recovery, in: *Harnessing Biotechnology for the 21st Century* (LADISCH, M., BOSE, A., Eds.), pp. 271–274, Washington, DC: American Chemical Society.

HOARE, M., DUNNILL, P. (1984), Precipitation of food proteins and their recovery by centrifugation and ultrafiltration, *J. Chem. Tech. Biotechnol.* **34B**, 199–205.

HOESS, A., ARTHUR, A. K., WANNER, G., FANNING, E. (1988), Recovery of soluble, biologically active recombinant proteins from total bacterial lysates using ion exchange resin, *Bio/Technology* **6**, 1214–1217.

HUDDLESTON, J. G., OTTOMAR, K. W., NGONYANI, D., FLANAGAN, J. A., LYDDIATT, A. (1990), Aqueous two-phase fractionation: practical evaluation for productive biorecovery, in: *Separations for Biotechnology 2* (PYLE, D. L., Ed.), pp. 181–190, New York: Elsevier Science Publishing.

JANSON, J.-C. (1984), Large scale affinity purification – state of the art and future prospects, *Trends Biotechnol.* **2**, 31–38.

JENDRISAK, J. (1987), The use of polyethyleneimine in protein purification, in: *Protein Purification: Micro to Macro* (BURGESS, R., Ed.), pp. 76–97, New York: Alan R. Liss, Inc.

JOHANSSON, B. L. (1992), Determination of leakage products from chromatographic media aimed for protein purification, *Biopharm* **5**, 34–37.

KAUFMAN, R., WASLEY, L., SPILIOTES, A., GOSSELS, S., LATT, S., LARSON, G., KAY, R. (1985) Coamplification and coexpression of human tissue-type plasminogen activator and murine dihydro-folate reductase sequences in Chinese hamster ovary cells, *Mol. Cell. Biol.,* July, 1750–1759.

KROEFF, E. P., OWENS, R. A., CAMPBELL, E. L., JOHNSON, R. D., MARKS, H. I. (1989), Production scale purification of biosynthetic human insulin by reversed-phase high-performance liquid chromatography, *J. Chromatogr.* **461**, 45–61.

KURANE, R. (1990), Separation by biopolymers: Separation of suspended solids by microbial flocculants, in: *Separations for Biotechnology 2* (PYLE, D. L., Ed.), pp. 48–54, New York: Elsevier Science Publishing.

LIGHTFOOT, E. N. (1990), Separations in biotechnology: The key role of adsorptive separations, in: *Protein Purification, From Molecular Mechanisms to Large Scale Processes* (LADISCH, M. R., WILLSON, R. C., PAINTON, C. C., BUILDER, S. E., Eds.), pp. 35–51, *ACS Symp. Ser.* **427**, Washington, DC: American Chemical Society.

LUTHER, J. R., GLATZ, C. E. (1993), Using genetic engineering to enhance protein recovery in aqueous two-phase systems, *Abstract* **95**, *Book of Abstracts,* ACS Division of the Biochemical Technical Section of the 205th National Meeting.

MARSTON, F. A. O. (1986), The purification of eukaryotic polypeptides synthesized in *Escherichia coli, Biochem. J.* **240**, 1–12.

MOKS, T., SAMUELSSON, E., UHLEN, M. (1992), Protein engineering to facilitate purification and folding of recombinant proteins, in: *Harnessing Biotechnology for the 21st Century* (LADISCH, M., BOSE, A., Eds.), pp. 238–240, Washington, DC: American Chemical Society.

MURKES, J., CARLSON, C. G. (1988), *Crossflow Filtration,* New York: John Wiley & Sons, Ltd.

OGEZ, J. R., HODGDON, J. C., BEAL, M. P., BUILDER, S. E. (1989), Downstream processing of proteins: recent advances, *Biotechnol. Adv.* **7**, 467–488.

OKA, T., SAKAMOTO, S., MIYOSHI, K.-I., FUWA, T., YODA, K., YAMASAKI, M., TAMURA, G., MIYAKE, T. (1985), Synthesis and secretion of epidermal growth factor by *Escherichia coli, Proc. Natl. Acad. Sci. USA* **82,** 7212–7216.

PARNHAM, C. C. III, DAVIS, R. H. (1993), Rotary and tangential flow microfiltration for protein recovery from cell debris, *Abstract* **43,** *Book of Abstracts,* ACS Division of the Biochemical Technical Section of the 205th National Meeting.

PLUNKETT, S., DHAL, P. K., ARNOLD, F. H. (1992), Novel metal-affinity adsorbents prepared by template polymerization, in: *Harnessing Biotechnology for the 21st Century* (LADISCH, M., BOSE, A., Eds.), pp. 238–240, Washington, DC: American Chemical Society.

PORATH, J., LINDNER, E. B. (1961), Separation methods based on molecular sieving and ion exclusion, *Nature* **191,** 69–70.

PRITCHARD, M., SCOTT, J. A., HOWELL, A. J. (1990), The concentration of yeast suspensions by crossflow filtration, in: *Separations for Biotechnology 2* (PYLE, D. L., Ed.), pp. 65–73, New York: Elsevier Science Publishing.

PRZYBYCIEN, T. M., LAM, P., WNEK, G. E., ELLIKER, P., PREZYNA, L. A. (1993), Conducting polymer chromatography for protein separations, *Abstract* **118,** *Book of Abstracts,* ACS Division of the Biochemical Technical Section of the 205th National Meeting.

RIVEROS-MORENO, V., BEESLEY, J. E. (1990), Purification by two-phase partitioning of a hepatitis core protein-pertussis epitope fusion, in: *Separations for Biotechnology 2* (PYLE, D. L., Ed.), pp. 181–190, New York: Elsevier Science Publishing.

ROSÉN, C.-G. (1984), Cell harvest – advances in centrifugal separations, in: *Biotech '84,* pp. 317–330, Pinner, UK: Online Publications.

RUDGE, S. R., TODD, P. (1990), Applied electric fields for downstream processing, in: *Protein Purification, From Molecular Mechanisms to Large Scale Processes* (LADISCH, M. R., WILLSON, R. C., PAINTON, C. C., BUILDER, S. E., Eds.), pp. 244–270, *ACS Symp. Ser.* **427,** Washington, DC: American Chemical Society.

SINGH, A., LUGOVOY, J., KOHR, W., PERRY, J. (1984), Synthesis, secretion and processing of alpha factor-interferon fusion proteins in yeast, *Nucleic Acids Res.* **12,** 8927–8938.

SMITH, G. E., GU, G., ERICSON, B. L., MOSCHERA, J., LAHM, H. W., CHIZZONITE, R., SUMMERS, M. D. (1985), Modification and secretion of human interleukin-2 produced in insect cells by a Baculovirus expression vector, *Proc. Natl. Acad. Sci. USA* **82,** 8404–8409.

SMITH, M. C., COOK, J. A., FURMAN, T. C., GESELLCHEN, P. D., SMITH, D. P., HSIUNG, G. (1990), Chelating peptide-immobilized metal-ion affinity chromatography, in: *Protein Purification, From Molecular Mechanisms to Large Scale Processes* (LADISCH, M. R., WILLSON, R. C., PAINTON, C. C., BUILDER, S. E., Eds.), pp. 168–180, *ACS Symp. Ser.* **427,** Washington, DC: American Chemical Society.

SPEARS, R. M. (1982), *Ph. D. Thesis,* University of Maryland, College Park, MD.

SPEARS, R. M. (1991), unpublished results.

STEPHANOPOULOS, G. (1993), Dept. Chem. Eng., MIT, Cambridge, MA, *personal communication.*

SUH, S.-S., VAN DAM, M. E., WUENSCHELL, G. E., PLUNKETT, S., ARNOLD, F. H. (1990), Novel metal-affinity protein separations, in: *Protein Purification, From Molecular Mechanisms to Large Scale Processes* (LADISCH, M. R., WILLSON, R. C., PAINTON, C. C., BUILDER, S. E. Eds.), pp. 139–149, *ACS Symp. Ser.* **427,** Washington, DC: American Chemical Society.

TAYLOR, G., GRANT, I., ISON, A., HOARE, M. (1992), Membrane processing of concentrated biological suspensions, in: *Harnessing Biotechnology for the 21st Century* (LADISCH, M., BOSE, A., Eds.), pp. 267–270, Washington, DC: American Chemical Society.

VAN REIS, R., CLAPP, E. M., YU, A. W., WOLK, B. M. (1992), Industrial scale recovery of proteins by tangential flow filtration, in: *Harnessing Biotechnology for the 21st Century* (LADISCH, M., BOSE, A., Eds.), pp. 262–266, Washington, DC: American Chemical Society.

VINCENT, B. (1974), The effect of adsorbed polymers on dispersion stability, *Adv. Colloid Interface Sci.* **4,** 193–277.

WALLS, E. L., GAINER, J. L. (1991), Increased protein productivity from immobilized recombinant yeast, *Biotechnol. Bioeng.* **37,** 1029–1036.

WARD, P. N., HOARE, M. (1990), The dewatering of biological suspensions in industrial centrifuges, in: *Separations for Biotechnology 2* (PYLE, D. L., Ed.), pp. 93–101, New York: Elsevier Science Publishing.

WILSON, T. (1984), More protein from mammalian cells, *Bio/Technology* **2,** 753–755.

WINSTON, S., BENTLEY, W. E., PULLIAM, T. R., GROSFELD, H., FLASHNER, Y., WHITE, M., SHALITA, Z., REUVENY, S., MARCUS, D., PAPIR, Y., ROSENBERG, H., BINO, T., COHEN, S., SHAFFERMAN, A. (1993), Optimization of production of four HIV-peptide *beta*-galactosidase fusion proteins in *E. coli, Abstract* **7,** *Book of Abstracts,* ACS Division of the Biochemical Technical Section of the 205th National Meeting.

YEN, J. W., YANG, S. T. (1993), Denaturation and aggregation of *beta*-galactosidase during tangential flow ultrafiltration, *Abstract* **98,** *Book of Abstracts,* ACS Division of the Biochemical Technical Section of the 205th National Meeting.

YU, P., SAN, K.-Y. (1992), Protein release in recombinant *Escherichia coli* using bacteriocin release protein, *Biotechnol. Prog.* **8,** 25–29.

ZAHKA, J., LEAHY, T. (1985), Practical aspects of tangential flow filtration in cell separations, in: *Purification of Fermentation Products: Applications to Large Scale Processes* (LEROITH, D., Ed.), pp. 51–69, *ACS Symp. Ser.* **271,** Washington, DC: American Chemical Society.

4 Proteins and Peptides as Drugs: Sources and Methods of Purification

STUART E. BUILDER
ROBERT L. GARNICK
JAMES C. HODGDON
JOHN R. OGEZ

South San Francisco, CA 94080, USA

1 Introduction

Peptides and proteins have been used as pharmacologic agents for many years. This chapter focuses on their sources and methods of preparation. A historical perspective is used to show how the types of polypeptides and their respective sources and methods of purification have evolved as a result of major technical achievements.

The skin, digestive system, and mucous membranes acted as barriers to the use of polypeptides and proteins until the development of the hypodermic needle at the end of the last century. Prior to this invention, only polypeptides that could survive the acidic and proteolytic conditions of the gut and those agents that were permeable to the skin and mucous membranes were useful. Although not considered a therapeutic molecule, perhaps the first recorded use of a polypeptide as a pharmacologic agent was alpha-amanitin, the principal agent in the poisonous death cap mushroom *(Amanita phalloides)*. This molecule surives the gut at least partly because it is a cyclic polypeptide. We will see later how this structural feature reappears in recent times with the synthesis of cyclic polypeptides specifically designed to produce orally active analogs of otherwise easily digestible peptides.

Jumping forward 2000 years to JENNER's time, in the 18th century, we find a different way of packaging proteins to circumvent natural biological barriers and achieve a pharmacologic effect, namely live viruses. Cow pox virus was scratched into the skin which caused its protein coat to elicit crossreacting antibodies that were protective against smallpox. Today, live viruses continue to be an effective way to get proteins into the body. Perhaps the best known current example is the Sabin live oral polio vaccine. However, given the problems associated with live viruses, there is an increasing trend toward the use of subunit vaccines that do not have the potential to cause the disease they were designed to prevent. Success in this approach was first achieved for the A_{24} strain of foot-and-mouth disease virus (KLEID et al., 1981). More recently, recombinant subunit vaccines for hepatitis virus have been licensed to two companies. These can replace the previous vaccine which was obtained from infectious plasma of hepatitis carriers. It is unlikely that even a killed virus vaccine will be acceptable for AIDS and most investigators are pursuing recombinant subunit vaccines.

In this chapter, we discuss the sources of proteins and peptides in five groups ordered in an approximate chronology with respect to their first use. The sources are grouped as follows: (1) animal and plant tissue; (2) conventional fermentation; (3) chemical synthesis; (4) recombinant fermentation; (5) semi-synthetics and conjugates. It will be seen that there is a significant correlation between the sources, the final quality desired and the purification methodology. Table 1 shows examples of pharmacologic polypeptides derived from each of these sources.

2 Proteins and Polypeptides Derived from Animal and Plant Tissue Sources

The advent of the hypodermic injection, around the turn of this century, opened the modern era in the use of proteins and peptides as drugs. All of the early pharmacologic polypeptides were derived from tissue sources. The protein of interest already existed in the starting material and thus no knowledge of its sequence or structure was needed. These sources still supply the bulk of medicinal proteins on a mass basis. The group includes human blood, which supplies albumin (the largest volume protein pharmaceutical), factor VIII (also called anti-hemophilic factor or AHF), gamma-globulin/immune sera and plasminogen. It also includes bovine and porcine pancreas, from which much insulin is still obtained, and human placenta, which is the source for human chorionic gonadotropin (hCG). Bovine lung continues to be the main source of trypsin inhibitor and of complex surfactants, while papain and bromelains are still derived from the juices of papaya and pineapple, respective-

Tab. 1. Pharmacologic Polypeptides

Source	Name	Use	Year Introduced
I.	**ANIMAL TISSUE AND FLUID**		
	A. BLOOD		
	1. Albumin	Hypoproteinemia; plasma extender for shock, burns	1942
	2. Human IgG	Augmentation of immune system capacity, especially during hepatitis or tetanus infection	
	3. Rh IgG	Prevention of Rh immunization	
	4. Equine antivenins (snake, house ant, black widow spider, etc.)	Antitoxin, antianaphylactic	
	5. Human factor VIII	Clotting factor for hemophilia	
	6. Human factor IX	Clotting factor for hemophilia	
	7. Bovine thrombin	Topical for control of oozing after injury or surgery	1944
	8. Bovine plasmin	Wound debridement	
	9. Human Cohn fraction fibrinolysin	Wound debridement	1959
	10. Hepatitis vaccine	Specific immunity	1981
	B. PANCREAS		
	1. Porcine lipase and amylase	Pancreatitis; to increase absorption of dietary fats	1960
	2. Pepsin	Wound debridement	1939
	3. Trypsin (and chymotrypsin)	Cataract surgery; to reduce inflammation and edema after trauma	1951 1960
	4. Insulin	Control of glucose metabolism, diabetes	1939
	5. Bovine DNAse	Wound debridement	~1968
	6. Glucagon	To terminate severe hypoglycemia as a result of insulin overdose	1960
	C. PITUITARY		
	1. Human growth hormone	Induction of bone growth in hypopituitary dwarfism	1976
	D. LUNG		
	1. Trypsin inhibitor (Trasylol)	Pancreatitis, antidote after streptokinase treatment	
	E. SKIN		
	1. Bovine collagen	Control of bleeding, cosmetic surgery	~1974
	F. URINE		
	1. Hormones (FSH and LH) from postmenopausal women	Induction of ovulation; stimulation of spermatogenesis	1971
	2. Chorionic gonadotropin from pregnant women	Cryptorchidism in males; stimulation of spermatogenesis	1971
	G. MISCELLANEOUS		
	1. Hyaluronidase (bovine testicular)	Facilitation of subcutaneous administration of drugs	1958
	2. Glucocerebrosidase (placental)	Gaucher's disease	1991
	H. PLANT		
	1. Papain	Reduction of soft tissue inflammation	1962
	2. Bromelains (proteases) from pineapple	Reduction of soft tissue inflammation	1963

Tab. 1. Pharmacologic Polypeptides (Continued)

Source	Name	Use	Year Introduced
II.	**CONVENTIONAL FERMENTATION**		
	A. BACTERIA		
	1. Streptokinase	Dissolution of blood clots	1980
	2. Asparaginase	Antineoplastic: cancer therapy	1978
	3. Bacitracin	Antibiotic	1967
	4. Collagenase	Wound debridement	1980
	5. *Bacillus* proteases	Wound debridement	1969
	6. *Bacillus* penicillinase	To terminate anaphylaxis after penicillin administration	1958 1957
	7. *Bacillus* amylase	Digestive aid	1963
	B. MOLDS AND FUNGI		
	1. *Aspergillus* lactase	Digestive aid	
	2. Amylase	Digestive aid	
	C. MAMMALIAN CELLS		
	1. Urokinase	Dissolution of blood clots	1979
	2. Whole-viral vaccines DPT, MMR, polio, rabies, yellow fever, etc.	Confer specific immunity	1957–63
III.	**SYNTHETIC**		
	1. Oxytocin	Induction of labor, reduction of post-partum bleeding	1962
	2. Vasopressin	Antidiuretic	~1967
	3. Gastrin (pentagastrin)	Diagnostic for evaluation of gastric acid secretion	1974
	4. ACTH	IM as diagnostic to distinguish between adrenal cortical insufficiency and hypopituitarism	1958
	5. Salmon calcitonin	Paget's disease, hypercalcemia, osteoporosis	1975
	6. Human calcitonin	Paget's disease	1986
	7. Octreotide	Treatment of carcinoid tumors and certain vasoactive intestinal tumors	1988
IV.	**RECOMBINANT**		
	A. BACTERIA		
	1. Insulin	Diabetes	1982
	2. Human growth hormone	Hypopituitary dwarfism	1985
	3. Alpha-interferon	Hairy cell leukemia; genital warts; Kaposi's sarcoma; hepatitis C	1986
	4. Gamma-interferon	Chronic granulatomous disease	1991
	B. Yeast		
	1. Hepatitis vaccine		1987
	C. MAMMALIAN CELLS		
	1. tPA	Thrombolytic; heart attack and pulmonary embolism	1987
	2. Erythropoietin	Anemia in end-stage renal failure	1989
	3. G-CSF	Neutropenia	1991
	4. GM-CSF	Bone marrow transplant	1991
	D. HYBRIDOMAS		
	1. OKT-3	Suppression of graft rejection	1986

Tab. 1. Pharmacologic Polypeptides (Continued)

Source	Name	Use	Year Introduced
V.	**SEMISYNTHETICS AND CONJUGATES**		
	1. Insulin	Diabetes	1983
	2. APSAC	Thrombolytic	1991
	3. PEG-adenosine deaminase	ADA-severe combined immune deficiency	1990

General references: AMA Drug Evaluations, first through sixth editions
American Hospital Formulary Service Drug Information
Physicians' Desk Reference
PMA *Biotechnology Medicines in Development,* 1991 Survey

ly. Until the development of recombinant DNA technology, human growth hormone was derived exclusively from human cadaver pituitaries and human urine was the primary source of urokinase. Urine of postmenopausal women is still a source of FSH (follicle-stimulating hormone) and LH (luteinizing hormone). Fibrinolysin and DNAse are derived from bovine plasma and pancrease, respectively, and are used in combination for debriding wounds.

There are a number of properties common to most tissue sources which affect the purification technology used to extract peptides and proteins from them. The concentration of the protein of interest is usually fixed within some normal range, as are the concentration and identity of most contaminants. In contrast to recombinant sources, tissue sources often lack the opportunity for titer improvement or contaminant alteration. Tissue sources are subject to individual variations of the animal, including its age, health, and nutritional status. Infections may alter the concentration of a desired product but often, more importantly, may add undesirable contaminants. It is most serious when the contaminant is not detected.

Two recent and important examples are worth noting. Until 1984, all marketed human growth hormone (hGH) was derived from human cadaver pituitaries. The observation of the very rare Creutzfeld-Jacob (C-J) disease in several patients who had undergone treatment with pituitary-derived hGH led to an investigation that implicated the product as the cause. C-J virus is carried in the brain and pituitary tissue of infected individuals, and growth hormone derived from these tissues was contaminated with it. Because it is a slow virus, it was not detected in the product by normal testing. The product was removed from the market because of the continued risk to patients. Because an assay for the presence of the virus in the product was not in place, it was not clear how many more patients were being exposed to it.

Virtually all tissue sources are pooled from many individuals. Therefore, the contamination of even a single donor contaminates the entire pool. Various pooling and testing schemes have been developed to minimize the chances of an undetected contamination, while also attempting to minimize costs due to testing and the losses resulting from discarded tissue. These preventative measures have been developed to a great degree by the plasma fractionation industry because blood is known to be a source of hepatitis virus and HIV (human immunodeficiency virus). Almost all hemophiliacs who have received AHF from pooled human serum test positive for antibodies to hepatitis. In addition, many have died of AIDS. This is not the case with recipients of albumin because it is pasteurized. Thus, safe final product can be derived from such starting materials with proper purification technique.

There is a significant desire today to not rely solely on testing to produce virus-free products. Increasingly, inactivation and removal procedures are being incorporated into purification processes. This approach has become a central part of production of recombinant products, especially those derived from mammalian cells. In general, bacterial contamination is not a direct infective risk to patients because it can easily be eliminated by terminal sterile filtration of products. However, toxins, pyrogens and various proteases or other degradative enzymes may be introduced into a tissue source by microbial contamination. The best ways to deal with this problem are to more closely link recovery to harvest and to use preservation techniques such as reduced temperature and antimicrobials.

The fixed impurity profile of tissue sources can either be an asset or a liability, depending on the circumstances. The impurities found with tissue-derived product sources can have related pharmacologic activities that are difficult to separate from the product of interest. On the other hand, tissue sources can supply very complex mixtures of materials that may be difficult to make by other means. A good example of this is the complex of lung surfactants derived from bovine lung that is used to aid the breathing of neonates (FUJIWARA et al., 1980). Tissue sources are also excellent sources of very complex individual species, such as AHF, which is a large and highly glycosylated protein. In time, it may be replaced by simpler species made by genetic engineering.

In general, while tissue costs can be low, they are typically at least partly out of the control of the manufacturer compared to the other sources. This is especially true in the case of bovine and porcine tissue which is collected as a by-product of the meat packing industry and, as such, its availability and cost are dictated by forces unrelated to its use as a pharmaceutical protein.

Blood, including plasma and serum, is one of the oldest and still most common sources of human pharmaceutical proteins. There are many useful substances in blood which are often fairly easy to extract and purify. In addition, it is a plentiful source since it may be taken from normal humans in large amounts. Blood collection programs established around World War II are still the backbone of this industry. A number of proteins derived from blood have traditionally been only partially purified for human use. AHF, as a cryoprecipitate, may be only 1% pure and albumin may be only 97–99% pure. Since the expected impurities are also naturally found in blood, they have not been of too much concern. However, with the increasing incidence and awareness of viral transmission by blood products, the availability of more powerful separation methods, and the advent of competing technologies, purities of blood products have generally risen. A good example is affinity-purified AHF, which can be as much as 100 times more pure than cryoprecipitate.

The use of salt, pH, temperature, and solvents to differentially precipitate biopolymers goes back nearly a century, to the time of EMIL FISCHER. For nearly the first half of this century, purification schemes relied upon such fractional precipitation methods, including crystallization. Insulin, oxytocin, and vasopressin were initially purified using these techniques. The purification of albumin from human serum by the Cohn fractionation method (COHN et al., 1946) is a classic example of such a scheme, which is still in use today. Its success was due to its simplicity, reliability, accuracy and detail of the original description, and the economical supplies and equipment required.

About the same time, MARTIN and SYNGE (1941) discovered partition chromatography, for which they received the Nobel Prize. This discovery has led, over the succeeding years, to a vast number of applications. Virtually every protein purification scheme developed today uses at least one form of chromatography (BECKER et al., 1983; OGEZ et al., 1989). A good comparison of conventional and modern methods of plasma fractionation can be found in "Separation of Plasma Proteins" and "Methods of Plasma Protein Fractionation", both edited by JOHN CURLING (1980, 1983).

The second major modern method used for protein purification is ultrafiltration. It is included in most large-scale protein fractionation schemes, at least in part because it is economical, reliable, and readily scaled. An excellent review of this technology is the "Ultrafiltration Handbook" by MUNIR CHERYAN (1986), while many examples of its use can be found in "Membrane Separations in Biotechnology", edited by W. COURTNEY MCGREGOR (1986).

3 Polypeptides and Proteins from Conventional Fermentation and Cell Culture Sources

Fermentation has been used as a method of bioconversion and biosynthesis since before recorded history. Up until the last 100 years, it was predominantly used to produce small molecules such as alcohol and organic acids, as well as to aid in cheese aging and the production of fermented foods such as yogurt. During World War II, the drive to produce large quantities of penicillin gave rise to the modern pharmaceutical fermentation industry, which is based on deep tank culture. Since that time, many pharmaceuticals, including a number of proteins, have been derived from cultured broths. This group includes asparaginase from *Escherichia coli,* streptokinase from *Streptomyces* and urokinase from cell culture. The switch to fermentation broths as a feedstock freed the producer from the problems of tissue collection and distribution, quality problems, stability after collection, and seasonal variability. It permitted centralized production of the feedstock and the final product, and more control over the impurities and contaminants. It also allowed closer coupling of the initial purification to the harvest. Strain improvement and media development soon led to ever-improving titers. This improved the economy of production and often permitted a simpler purification. Fermentation and recovery engineering led to further enhancements in economy, capacity, and reliability.

For non-mammalian sources, there is always concern about the possibility of antibody formation to not only the impurities but also to the foreign protein product itself. For example, asparaginase is used to catalyze the breakdown of asparagine in the blood of patients having lymphocytic leukemia. This strategy is based on the dependence of the malignant cells for asparagine, whereas normal cells are capable of synthesizing it. Asparaginase is purified from *E. coli* by a combination of extraction, precipitation and chromatographic steps (WHELAN and WRISTON, 1969). While its purity is at least 98–99%, significant allergic reactions have been reported, including death by anaphylaxis. Further purification is not likely to reduce this problem because the product itself is the immunogen.

Streptokinase is an enzyme that activates plasminogen to plasmin and can be used to accelerate the lysis of blood clots. It is purified from beta-hemolytic *Streptococcus* (TORREY, 1983). Because it, like asparaginase, is a non-mammalian protein, it is also associated with immune complications. This problem, as well as its tendency to lead to hypotension and fibrinogenolysis, has been the basis for the significant shift to the use of the mammalian cell-derived tissue-type plasminogen activator, which is free of these problems.

While not traditionally included in the group of pharmaceutical proteins and peptides, antibiotics are a cornerstone of the modern pharmaceutical industry. Many of them, such as gramicidin and bacitracin, have peptide bonds. Because they are much smaller than proteins and lack the secondary and tertiary structural features of proteins, they can be purified by simpler and harsher methods, such as solvent extraction, as well as the methods already mentioned. The successful application of two-phase liquid extraction for the purification of large polypeptides has only recently been demonstrated (KULA et al., 1982). It was made possible primarily by the development of systems having low interfacial surface tension and high dielectric constant. Two-phase liquid–liquid extraction is particularly useful as an economic method for product enrichment from whole broth or cell homogenates.

Normal human fetal kidney cells have largely replaced urine as a source of urokinase. Normal mammalian cell lines are also used as a substrate for viral production. In contrast, today the majority of pharmaceutical proteins produced in mammalian cells are recombinant molecules from immortal lines. Transformed lines are required because the cloning, selection, amplification, and banking require a significant number of passages relative to the longevity of a normal cell. The cell line would not have an economically useful remaining life unless it was immortal.

4 Peptide Drugs Derived from Chemical Synthesis

The pioneering work of SANGER, in 1953, on the elucidation of the sequence of insulin opened the door for the sequence analysis of a number of important peptides such as adrenocorticotropin (ACTH) and ribonuclease. With this new knowledge of protein structure, organic chemists turned their attention to the formidable task of developing strategies for the synthesis of naturally occurring peptides. While the work of EMIL FISCHER early in this century led to the synthesis of small homopolypeptides, it was not until 1953 that V. DU VIGNEAUD succeeded in the synthesis of the hormones oxytocin and vasopressin, each containing nine amino acid residues. MERRIFIELD's initial report on the solid-phase synthesis of peptides, first described in 1959, gave organic chemists the ability to prepare naturally occurring peptides and their analogs in high yield and in sufficient quantity for commercial use.

Over the past 50 years, studies of biological phenomena of commercial interest have led to some of the most significant and important developments in peptide chemistry. For example, the significance of human and porcine insulin in the treatment of diabetes in man has contributed enormously to the advancement of the science as well as to the benefit of treatment of this disease. Peptide drugs such as ACTH, DDAVP (1-desaminocysteine-(8-D-arginine)-vasopressin) and salmon calcitonin have pioneered the way for the regulatory approval of synthetic peptides as drugs throughout the world.

While in theory the synthesis of peptides of any size can be performed by solid-phase or solution chemistry, the practical limit appears to be 3000 to 4000 daltons because of the errors which accumulate when the number of sequential reactions is large. For larger peptides, recombinant DNA technology is the method of choice. Recombinant DNA methods for the production of fusion proteins (from which small peptides may be obtained by suitable cleavage techniques such as selective proteolysis or chemical hydrolysis) now offer a significant alternative to the chemical synthesis of smaller peptides as well. An example of this approach is currently being developed for the peptide IGF-1, which can be produced in genetically modified *E. coli* as a fusion protein that can later be chemically cleaved to yield IGF-1 in an acceptable yield.

The development of peptide synthesis methodology has been paralleled by the development of methodology for peptide analysis. After SANGER elucidated the structure of insulin, it became possible to begin to develop strategies for the chemical synthesis of such peptides. Advances in amino acid analysis, Edman sequence analysis, thin layer chromatography, ion exchange chromatography, high performance liquid chromatography and most recently, capillary electrophoresis, have made it possible to isolate reasonable quantities of peptides for sequence analysis. It is also possible to prepare these products in high yield with reasonable purity for clinical evaluation. These advances, along with the isolation and structural elucidation of receptors for specific biological functions of many peptides, have made the search for more active peptide analogs more efficient. Because of this, the future of peptide synthesis may reside in the area of the production of synthetic analogs whose function is to mimic the action of large proteins (peptidomimetics).

While a number of strategies to synthesize peptides in solution have been developed, none has been commercially successful. This is due to the difficulties involved in the preparation and selective removal of both suitable protecting groups and the removal of impurities. The first practical method for peptide synthesis was developed by MERRIFIELD in 1959 (STEWART and JOUNG, 1984; MERRIFIELD, 1963). The main advantage of this method is the ease of removing soluble reagents and impurities. The basis of the Merrifield synthesis is the stepwise building of a peptide from the C-terminus. First, the C-terminus is attached to a solid polymeric support. Next, the elongation of the immobilized polypeptide chain is accomplished by stepwise coupling and deprotection of the appropriate amino acids (BARANY et al., 1987). When the desired peptide has been completed, it is cleaved from the polymer support and all the side chain protecting groups are quantitatively removed.

The choice of the side chain protecting group is critical to a successful synthesis. There are many available protecting groups for side chains, and the strategies for their removal at the appropriate time must be carefully thought out in order to avoid the complications and loss of yield from branching and other side chain reactions. The production of required disulfide bridges is usually accomplished by metal-catalyzed oxidation of the free cysteine residues or by nucleophilic substitution of a suitably modified cysteine residue. The free peptide is then purified using a combination of chromatographic techniques. Often, this includes one method usually not used for proteins, namely reversed-phase HPLC.

Following the basic solid-phase methodology described above, a number of commercially valuable peptides have been produced. These products include oxytocin, vasopressin, somatostatin, and both human and salmon calcitonin. Calcitonin, which contains 32 amino acids, is among the largest commercially produced peptides made to date. Calcitonin also represents the basic practical limit for peptide synthesis in a commercially feasible yield. The use of automated peptide synthesizers has made it possible to produce large peptides in highly reproducible yield.

The advances in solid-phase peptide synthesis made it possible to produce large quantities of rare peptides or their analogs in order to study their effects in man. One example is DDAVP, which was designed specifically for intranasal treatment of diabetes insipidus. The structural modifications developed for this product resulted in a marked increase in antidiuretic activity. Other results include a reduction in side-effects and the ability to obviate the parenteral mode of administration by removing the recognition site of nasal proteases through the introduction of a D-arginine residue (VAVRA et al., 1968). Another example of a synthetic analog is a derivative of eel calcitonin in which the disulfide bridge of the molecule has been replaced with a methylene bridge, resulting in improved activity and simplicity of production. More recently, researchers have created a peptidomimetic analog of somatostatin for treatment of certain intestinal tumors. This cyclic peptide analog also incorporates D-amino acids and an amino acid alcohol in order to reduce its degradation by proteolytic enzymes (LONGNECKER, 1988).

The approval of peptides of the same primary sequence as the natural product has been much easier than that of synthetic analogs. Synthetic analogs may require extensive clinical trials and toxicity data in man and animals before their safety can be established. As a result, many companies have not pursued the development of peptide analogs. With the recent approvals of drugs such as eel calcitonin and DDAVP, however, more interest in analogs has occurred despite the regulatory hurdles described above.

The issue of the purity of synthetic peptides, and the difficulties involved in determining what an acceptable level of purity is for regulatory approval has plagued synthetic peptides since their introduction. The difficulty lies in the broad spectrum of possible impurities that can occur in solid-phase synthesis and, most importantly, their potential effect on activity, antigenicity, and safety in man. Some examples of these impurities are listed below:

- amino acid deletion peptides
- open or scrambled disulfide bridges
- residual protecting groups
- oxidized methionine residues
- racemization
- beta-aspartyl peptides
- acylated glutamic acid residues
- chain termination impurities
- alkylated or halogenated tyrosine residues
- dehydration of asn and gln to nitriles

In the past, separation of the impurities described above has been difficult or impossible to accomplish. The absolute purity of many commercial peptide products was only in 70–90% (w/w) range. Typically, they contained significant quantities of the impurities described above. In most cases, single amino acid deletion products represented the largest percentage of the impurities. With the recent development of analytical methodology such as reversed-phase HPLC and high performance ion exchange resins for use in large-scale purification, it is now possible to prepare peptides having purity levels greater than 95%. These highly purified peptides should markedly speed the regulatory approval process by obviating many of the questions about purity.

Historically, the analytical methods used for the evaluation of the homogeneity of peptides have included amino acid analysis, Edman sequence analysis, thin layer chromatography, optical rotation, electrophoresis (SDS-PAGE), and isoelectric focusing. In addition, a whole animal or cell line-based bioassay was used to evaluate the activity or potency of the peptide. With the intensive application in the early 1980s of more sophisticated analytical methods such as HPLC, mass spectroscopy, and other techniques for conformational analysis, there has been a shift away from reliance on lower resolution methods. It is now possible to characterize synthetic peptides to a degree not thought possible ten years ago, and to do so on a routine basis. This advance in analytical methodology has also benefited the process scale separations scientists. The newer and more powerful separation methods, such as reversed-phase HPLC, can be successfully incorporated into the purification of synthetic peptides. The resultant increases in purity and specific activity have been instrumental in convincing regulatory agencies of the safety of new peptide drugs and analogs.

5 Proteins and Peptides from Recombinant Fermentation

The discovery of techniques for creating hybridomas (KOHLER and MILSTEIN, 1976) and recombinant organisms (MORROW et al., 1974) made possible the commercial production of otherwise unattainable molecules. It removed the limitations characteristic to tissue sources, such as low titer, limited supply, and variable quality. It also avoided the difficulties associated with chemical synthesis of large polypeptides and glycoproteins. Finally, it made available proteins that could not be obtained economically through conventional fermentations.

Initially, little was known about the safety of products derived from recombinant sources. Special precautions were therefore taken to minimize any real or perceived risks to the environment, workers, or recipients of the products. For example, stringent killing procedures were devised to avoid releasing live recombinant cells into the environment. DNA and protein impurities coming from the host cell were reduced to unprecedented low levels. In the case of mammalian substrates, the cell banks were exhaustively characterized with respect to genetic identity and stability and possible contamination by such adventitious agents as viruses, retroviruses, or mycoplasma (WIEBE and MAY, 1990). With the accumulation of experimental data and clinical experience, concern over many of these hypothetical risks has lessened considerably. For example, naked DNA, and even oncogenes, have been shown to be unable to induce tumors after injection into normal animals (PALLADINO et al., 1987). Process validation studies have been carried out to demonstrate that a variety of purification methods are capable of significant reduction of nucleic acids. Anion exchange chromatography resins bind polynucleotides very tightly, and usually provide several logs of nucleic acid removal. Using a sensitive hybridization assay to measure process fluids directly as well as spike-removal studies, LUBINIECKI et al. (1990) demonstrated reduction of DNA to approximately 0.1 pg per dose of a plasminogen activator preparation. This is at least two orders of magnitude lower than that recommended by the 1984 FDA "Points to Consider" (see BUILDER et al., 1989). Similar validation was also done to demonstrate removal and inactivation of model viruses and retroviruses. Multiple steps utilizing independent mechanisms are essential. By employing inactivation methods such as treatment with acid, chaotropes, or detergents along with removal methods such as microfiltration, cumulative viral reduction of >15 logs can be achieved.

Because of concern about the safety of proteins from non-human sources with respect to the generation of immune responses, these proteins are generally being brought to unprecedented levels of purity. Today it can be as difficult to quantify and prove these levels of purity as it is to achieve them. For example, whereas the purity of albumin preparations is commonly about 95–99%, the purity of recombinant human growth hormone (Protropin®) and recombinant human insulin (Humulin®) is greater than 99.99% with respect to host proteins. In order to measure impurities at this level, two major analytical strategies have been developed. The first method, which is uniquely applicable to recombinant products, is the use of a "blank run". This is accomplished through fermentation and recovery using a host cell containing the selectable marker but lacking the gene for the product. In this way, one may be able to specifically prepare and quantitate the host cell-derived impurities (BUILDER et al., 1989). The second approach is the direct measurement of impurities. The most general method uses an immunoassay based on antibodies to the host cell proteins (JONES, 1988). Although this type of assay is complex in both its development and composition, it provides an extremely sensitive way to quantitate protein impurities in each batch of product. When properly validated, such assays can usually quantify host cell protein impurities at the level of 1–100 ppm.

Initially, most recombinant polypeptides were cloned and expressed in *E. coli* because its genetics were well understood, restriction enzymes were available, and it could be grown cheaply. The first generation of recombinant therapeutics, which includes human insulin, growth hormone and alpha-interferon, was produced in this bacterial host (BOLLON, 1983) New molecules that are produced in *E. coli* are currently moving toward the marketplace in various stages of clinical trials. These include gamma-interferon, interleukins, colony stimulating factors, tumor necrosis factor, insulin-like growth factor, and various animal growth hormones.

Recombinant proteins which are expressed intracellularly in bacteria such as *E. coli,* possess methionine as the N-terminal amino acid. For proteins which have this amino acid as the natural N-terminal residue, this is obviously not an issue. For others, the structure, activity, and safety of the resulting polypeptide must be evaluated on a case by case basis. Several approaches have been used to avoid an undesired N-terminal methionine. Cyanogen bromide cleavage is one method for removing methionine after synthesis, especially when the product must be separated from a peptide leader sequence (WETZEL et al., 1981). This approach, however, is only useful when the desired product has no internal methionine residues. Also, with certain N-terminal sequences, an endogenous aminopeptidase may be used to process away the terminal methionyl residue (BEN-BASSAT et al., 1987).

In a different approach, an expression sequence that allows secretion into either the periplasmic space or the extracellular medium can be used. The leader sequence, or "signal peptide", may be cleaved off by the organism as the product passes through the cell membrane or wall, yielding the mature protein with the correct N-terminus (HSIUNG et al., 1986). As an alternative, the fusion protein can be engineered to remain intact, with the leader serving as an aid in the purification of the product. For example, in the EcoSec® expression system developed by KabiGen, the leader has homology with Staphylococcal protein A, which binds tightly to the Fc region of antibodies. This allows rapid affinity purification of the fusion protein on an immobilized IgG column. The desired product is then separated from the leader sequence by specific cleavage using hydroxylamine (MOKS et al., 1987).

Another problem which has been overcome relates to the stability of the newly synthesized protein. For small peptides, intracellular degradation can be so fast that even with high synthesis rates, little product accumulates. This problem was first observed with molecules such as somatostatin and insulin, and was solved by expressing the peptide in a "fusion" protein which contained a larger *E. coli*-derived leader sequence (RIGGS et al., 1985). Subsequently, the leader could be cleaved away, as in the case of insulin chains, or the fusion protein could be left intact for use as an immunogen, as with foot-and-mouth disease viral capsid proteins (KLEID et al., 1984). "Substrate-assisted catalysis" has recently been developed as a novel and extremely specific way to cleave fusion proteins (CARTER and WELLS, 1987). This method uses a mutant protease, subtilisin, whose active site histidine residue has been genetically replaced by an alanine. The enzyme is thus active only against peptide substrates that can supply the missing histidine residue. This strategy was used to cleave a "protein A" – alkaline phosphatase fusion protein from *E. coli,* and should be applicable to the cleavage of almost any fusion of this leader with a desired peptide (CARTER et al., in press).

A problem that was not originally expected in the production of certain mammalian proteins in prokaryotes is the formation of improperly folded and insoluble aggregated product. Under conditions which promote rapid expression, the non-native protein aggregates to form dense, insoluble particles in the bacterial cytoplasm. These particles are known as "inclusion" or "refractile" bodies because they are visible by transmission electron microscopy or phase-contrast light microscopy, respectively. In traditional biochemistry, product in such a denatured state would normally have been discarded. However, at the time it was the only starting material available, so recovery methods were developed to deal with the problem. Initially, the methods relied on the ability to selectively extract and purify the small percentage of properly folded material (OLSON et al., 1981). Later, methods were developed which allowed the recovery of molecules which were improperly folded or had mispaired disulfide bonds, which dramatically increased the process yields (BUILDER and OGEZ, 1986; RAUSCH, 1986). These processes first carried out a refractile body isolation utilizing extraction and homogenization in a non-denaturing buffer followed by low-speed centrifugation. The product was then extracted in a strongly denaturing solvent, purified, and refolded by reducing the strength of the denaturing conditions thus allowing the protein to reacquire its normal three-dimensional conformation.

For some polypeptides, especially large proteins or glycoproteins, mammalian expression systems may be an attractive alternative to bacteria since they can secrete cloned mammalian proteins which are properly folded and similar or identical to those derived from natural sources. It was initially feared that mammalian cells would not be economically viable production systems. In the last few years, however, improved expression levels, progress in developing large-scale cell culture production technologies (ARATHOON et al., Eur. Patent Application), and formulation of more highly defined growth media have combined to dramatically improve the economic feasibility of mammalian cell substrates.

In addition, the development of recovery operations which overcome the safety concerns associated with the use of continuous cell lines has made these systems viable from a regulatory standpoint. Perhaps the best known example is tissue plasminogen activator (tPA), which is used to stop ongoing myocardial infarction. It was the first recombinant human therapeutic produced in mammalian cells to receive marketing approval. Other cell culture-derived products that are approved or are nearing approval include erythropoietin, colony stimulating factors, monoclonal antibodies for cardiac imaging and treatment of sepsis, AIDS therapeutics, factor VIII, and human growth hormone.

The number of passages required for cloning, selection, amplification, and banking prior to production clearly necessitates the use of immortal cell lines, since primary cells cannot be propagated long enough to give an economically useful time in the production stage. Initially, there was concern about the safety of these immortal cell lines since, by definition, they were transformed and therefore probably contained oncogenes. Furthermore, mammalian cells are susceptible to viral contamination and many established cell lines (especially rodent cell lines) are known to exhibit the presence of retrovirus-like particles in electron micrographs. Such concerns have been ameliorated by exhaustive characterization of the cell banks (LUBINIECKI and MAY, 1985) in which no functional retrovirus has been found by electron microscopy, reverse transcriptase assay, or by plaque-formation assays. Furthermore, recovery steps to remove or inactivate putative retroviruses provide an additional level of safety.

With recent advances in hybridoma and cell culture technology, it is now possible to produce large amounts of monoclonal antibodies in highly purified form, which can be used as probes, diagnostic agents, imaging agents, aids in purifying protein therapeutics, and as therapeutics themselves. The first therapeutic monoclonal antibody approved for use in humans was OKT3, a T-cell blocker. It is used to suppress rejection of donor tissue following bone marrow or kidney transplantation

(GOLDSTEIN, 1987; COSIMI, 1987). Even though it is a murine monoclonal which is produced in ascites fluid, purification and testing procedures have established that it is a safe therapeutic for the approved indications.

As we saw in Sect. 3, the development of more sophisticated analytical techniques played an important part in improving the chemical synthesis of small peptides. Subsequently, the ability to create novel peptides has made it possible to correlate the primary sequence with function. This, in turn, has opened the door to the rational design of second-generation peptide drugs. The same kind of evolution is now occurring in the development of new recombinant protein therapeutics. The development of analytical tools and a theoretical framework in which to identify, understand, and manipulate the structure and function of proteins has extended the level of understanding beyond the primary sequence. X-ray crystallography, computer modeling, and other techniques are now used to understand the function of the protein's domains and to design second generation molecules which possess enhanced characteristics such as higher solubility, greater activity or specificity, or altered pharmacokinetics. For example, when the gene structure for tPA was compared to the tPA protein sequence and the sequences of related proteases, it was discovered that separate exons encoded distinct structural domains. Subsequently, these domains were shown to have distinct and essentially autonomous functions (van ZONNENFELD et al., 1986). As a result, it has been possible to rationally direct the incorporation of mutations into the regions of the protein that affect stability, fibrin binding, catalytic activity, and clearance (HARRIS, 1987).

Recombinant DNA techniques have been used to produce a new "single chain" type of antibody molecule in bacteria (BIRD et al., 1988). The gene construction fuses the variable segments of a light chain and a heavy chain, giving in essence a recombinant monoclonal Fab fragment. Such "antibodies" might be superior therapeutic or imaging agents because they cannot be contaminated with residual free Fc fragments, and they might be produced more economically in bacteria than in hybridoma cells.

Another example of the power of genetic engineering technology is the creation of a third-generation CD4 molecule for use in AIDS therapy (CAPON et al., 1989). CD4 is an integral membrane protein present on a subset of mature T cells which has been implicated as the cell surface receptor for HIV. It was initially shown that a soluble version of CD4, possessing only the extracellular domain, could block HIV infectivity *in vitro*. This form of CD4 is currently in clinical trials in ARC and AIDS patients. In the third generation hybrid molecule, the HIV binding portion of CD4 was fused to the Fc domain of an immunoglobulin heavy chain, creating a chimeric called an "immunoadhesin". The hybrid has a much longer serum half-life (about 200-fold) than truncated CD4, which permits the molecule to remain at high levels using much lower and less frequent doses. In addition, the presence of the Fc region causes it to bind to the Fc receptors on phagocytes. Therefore, it should participate in antibody-dependent cell-mediated cytotoxicity and lead to active killing of the HIV.

Genetic engineering has also been used to produce a unique toxin that is directed toward cells that are infected with HIV. The hybrid protein (CHAUDHARY et al., 1988) possesses a portion of CD4, the HIV receptor protein, as well as two domains of the *Pseudomonas* exotoxin A, a protein which is similar to ricin. Directed toxins such as this, which would be difficult or impossible to produce without the aid of recombinant DNA technology, may prove to be highly specific antiviral or antitumor agents.

6 Semisynthetics and Conjugated Molecules

As we have seen in Sect. 3, one way to obtain non-natural polypeptide drugs is by chemically synthesizing the entire molecule from amino acid building blocks. For small peptides this is an easy route to both the natural sequence as well as to a large number of non-natural analogs. For large proteins and glycoproteins, the gene for the desired protein can be synthesized and placed into a bacterial or mammalian cell expression system for production (Sect. 4). There is a set of molecules that can be made efficiently by a different approach, in which an existing molecule is chemically altered to produce a new pharmaceutical agent. These molecules are the semisynthetic and conjugated polypeptides. In one example, semisynthetics, single amino acids or peptide fragments are replaced while the remainder of the protein is left intact. This technique usually involves (1) the limited proteolysis of a naturally occurring polypeptide to yield a set of workable fragments; (2) the chemical synthesis of at least one new fragment; and (3) the ligation of the synthetic and native partners to create a new molecule. Semisynthetic strategies, reviewed by CHAIKEN (1981), have been applied to a variety of polypeptides, including the generation of native and altered sequences of growth hormone and ACTH. Probably the most prominent commercial application of semisynthesis involves the conversion of porcine insulin to human insulin. Human and porcine insulins differ only in the C-terminal amino acid (position 30) of the B chain. Under the proper conditions, the C-terminal alanine of pork insulin can be selectively replaced by threonine using a transpeptidation catalyzed by trypsin in the presence of carboxyl-protected threonine (MORIHAWA et al., 1979).

In another strategy, domains within a protein can be selectively modified by conjugating certain amino acid side chains with non-protein groups. If the modification is properly controlled, it may be possible to preserve the protein's desirable activities while altering some of the less-desirable effects. This may include enhancing its effectiveness by prolonging its lifetime in the circulation. The potency of many protein drugs is limited either by their susceptibility to proteolysis or by their rapid clearance from the circulation. Clearance is sometimes mediated through the action of liver cell receptors against specific structural features of the polypeptide chain or toward certain types of attached carbohydrates. When the terminal sialic acid residues are removed from complex-type carbohydrate side chains to expose the penultimate galactose moieties, the protein is rapidly removed by hepatocyte receptors. For many proteins, it has been possible to extend the circulating half-life, increase stability, or reduce immunogenicity by covalently attaching long-chain dextrans or polyethylene glycols (APLIN and WRISTON, 1981). Polyethylene glycol-modified adenosine deaminase (PEG-ADA) was recently approved for treatment of children suffering from a form of severe combined immune deficiency resulting from ADA insufficiency. A PEGylated form of L-asparaginase is currently being tested for treatment of certain leukemias. The modified form has a serum half-life of two weeks (versus 18 hours for the unmodified form) and appears to be much less immunogenic than the natural unmodified *E. coli*-derived protein. PEG-modified hemoglobin is also being developed as a blood substitute in emergency situations. The modified protein may remain stable during storage for much longer than whole blood (POOL, 1990). The serum half-life of superoxide dismutase, an enzyme which may be useful for treating reperfusion injury and certain types of inflammation, was increased from a few minutes to over nine hours by PEG derivatization, while retaining virtually all of its enzymatic activity (BEAUCHAMP et al., 1983).

Another strategy for prolonging the serum half-life of a therapeutic protein has been exploited in the synthesis of APSAC ("anisoylated plasminogen-streptokinase activator complex"; BEEN et al., 1986). For several years, streptokinase has been used as a thrombolytic agent to treat ongoing myocardial infarction. One of the drawbacks of this therapy is that large amounts of circulating plasmin are generated, which rapidly react with and consume the natural plasmin inhibitor alpha-2-antiplasmin, as well as fibrinogen. This limits the efficacy of the drug as well as adversely affecting the balance of the body's hemostatic system. By acylating the active site serine of the plasminogen, the zymogen is protected from inactivation by antiplasmin. Inactivation occurs only after spontaneous hydrolysis of the anisoyl moiety. By choosing acyl groups having different hydrolysis rates, one can control the rate of inactivation, and thus the serum half-life of the drug. The product was recently approved for treatment of acute myocardial infarction.

By fusing together two different proteins, or portions of them, one may be able to create a new drug that possesses the best properties of the parent components. One approach in this area has focused on the coupling of monoclonal antibodies to intact natural toxins such as diptheria toxin or ricin (VITETTA et al., 1983). Because of their great specificity, monoclonal antibodies can be used to target otherwise toxic chemotherapeutic agents in a tissue-specific manner. Targeting decreases the nonspecific cytotoxicity, permitting much higher local doses and thereby increasing the efficacy of the chemotherapeutic. Plant toxins such as ricin are composed of two disulfide-bonded subunits. The "B" subunit is usually a galactose-specific lectin which confers binding of the toxin to the cell membrane. Once the toxin has been translocated into the cytoplasm, the disulfide bond connecting the two chains is reduced and the "A" chain exerts its toxic effect by enzymatically (ADP-ribosylating) inactivating the large ribosomal subunit. Because of its catalytic nature, a single molecule of the A chain may be sufficient to kill a cell.

In initial attempts using antibodies coupled to intact ricin, the compounds had unacceptable general cytotoxicity because of the presence of the ricin B chain. Research then turned to adducts between antibodies and the pure A chain. These molecules were found to be more specific in their action, but were limited by their short serum half-life due to the presence of liver cell receptors that bind the high-mannose carbohydrate moiety attached to the A chain of the toxin. Thus, third-generation immunotoxins that employ either intact antibody or Fab' fragments which are coupled to deglycosylated A chain or recombinant A chain synthesized in *E. coli* have recently been produced and are now being tested (VITETTA et al., 1987).

In another version of this approach, TILL et al. (1988) have created a protein - toxin by attaching deglycosylated ricin A chain to a soluble, truncated form of CD4. The hybrid molecule selectively binds to and kills HIV-infected cells *in vitro,* and may be used therapeutically to either prevent or delay the onset of AIDS in humans.

Alternatively, tumor-directed monoclonal antibodies have been conjugated to small chemotactic peptides, which induce the migration of macrophages into the tumor. Such adducts represent novel ways of stimulating the body's own defenses against cancerous cells (OBRIST et al., 1986).

7 References

APLIN, J.D., WRISTON, J. C. (1981), Preparation, properties and applications of carbohydrate conjugates of proteins and lipids, *CRC Crit. Rev. Biochem.* **10**, 259–306.

ARATHOON, W. R., BUILDER, S. E., LUBINIECKI, A. S., VAN REIS, R. D. Process for Producing Biologically Active Plasminogen Activator, *Eur. Patent Application* 0 248 675.

BARANY, G., KNEIB-CORDONIER, N., MULLEN, D. G. (1987), Solid-phase peptide synthesis: a silver anniversary report, *Int. J. Pept. Protein Res.,* **30** 705–735.

BEAUCHAMP, C. O., GONIAS, S. L., MENAPACE, D. P., PIZZO, S. V. (1983), A new procedure for the synthesis of polyethylene glycol - protein adducts; effects on function, receptor recognition and clearance of superoxide dismutase, lactoferrin and α-2-macroglobulin, *Anal. Biochem.* **131,** 25–33.

BECKER, T., OGEZ, J. R., BUILDER, S. E. (1983), Downstream processing of proteins, *Biotechnol. Adv.* **1**, 247–261.

BEEN, M., DE BONO, D. P., MUIR, A. L., BOULTON, F. E., FEARS, R., STANDRING, R., FERRES, H. (1986), Clinical effects and kinetic properties of intravenous APSAC - anisoylated plasminogen-streptokinase activator complex (BRL 26921) in acute myocardial infarction, *Int. J. Cardiol.* **11,** 53–61.

BEN-BASSAT, A., BAUER, K., CHANG, S.-Y., MYAMBO, K., BOOSMAN, A., CHANG, S. (1987), Processing of the initiation methionine from proteins: properties of the *Escherichia coli* methionine aminopeptidase and its gene structure, *J. Bacteriol.* **169**, 751–757.

BIRD, R. E., HARDMAN, K. D., JACOBSON, J. W., JOHNSON, S., KAUFMAN, B. M., LEE, S. M., LEE, T., POPE, S. H., RIORDAN, G. S., WHITLOW, M. (1988), Single-chain antigen binding proteins, *Science* **242**, 423–426.

BOLLON, A. P. (Ed.) (1983), *Recombinant DNA Products: Insulin, Interferon and Growth Hormone,* Boca Raton, FL: CRC Press.

BUILDER, S. E., OGEZ, J. R. (1986), Purification and Activity Assurance of Precipitated Heterologous Proteins, *U. S. Patent* 4 620 948.

BUILDER, S. E., VAN REIS, R., PAONI, N. F., FIELD, M., OGEZ, J. R. (1989), Process Development in the Regulatory Approval of Tissue-Type Plasminogen Activator, in: *Proc. 8th Int. Biotechnol. Symp.* (DURAND, G., BOBICHON, L., FLORENT, J., Eds.) Vol. 1, pp. 644–658.

CAPON, D. J., CHAMOW, S. M., MORDENTI, J., MARSTERS, S. A., GREGORY, T., MITSUYA, H., BYRN, R. A., LUCAS, C., WURM, F. M., GROOPMAN, J. E., BRODER S., SMITH, D. H. (1989), Designing CD4 immunoadhesins for AIDS therapy, *Nature* **337**, 525-531.

CARTER, P., WELLS, J. A. (1987), Engineering enzyme specificity by substrate assisted catalysis, *Science* **237**, 394–399.

CARTER, P., NILSSON, B., BURNIER, J., BURDICK, D., WELLS, J. A. Designing a site-specific protease, in press.

CHAIKEN, I. M. (1981), Semisynthetic peptides and proteins, *CRC Crit. Rev. Biochem.* **11**, 255–301.

CHAUDHARY, V. K., MIZUKAMI, T., FUERST, T. R., FITZGERALD, D. J., MOSS, B., PASTAN, I., BERGER, E. A. (1988), Selective killing of HIV-infected cells by recombinant CD4-*Pseudomonas* exotoxin hybrid protein, *Nature* **335**, 369–372.

CHERYAN, M. (Ed.) (1986), *Ultrafiltration Handbook,* Lancaster, PA: Technomic Publishing Co.

COHN, E. J., STRONG, L. E., HUGHES, W. L., MULFORD, D. J., ASHWORTH, J. N., MELIN, M., TAYLOR, H. L. (1946), Preparation and properties of serum and plasma proteins. IV. A system for the separation into fractions of the protein and lipoprotein components of biological tissues and fluids, *J. Am. Chem. Soc.* **68**, 459–475.

COSIMI, A. B. (1987), OKT3: first-dose safety and success, *Nephron* **46**, Suppl. 1, 12–18.

CURLING, J. M. (Ed.) (1980), *Methods of Plasma Protein Fractionation,* London: Academic Press.

CURLING, J. M. (Ed.) (1983), Separation of plasma proteins, in: *The Joint Meeting of the 19th Congress of the International Society of Hematology and the 17th Congress of the International Society of Blood Transfusion*. Pharmacia Press.

FUJIWARA, T., MAETA, H., CHIDA, S., WATABE, Y., ABE, T. (1980), Artificial surfactant therapy in hyaline-membrane disease, *Lancet*, 55–59.

GOLDSTEIN, G. (1987), Monoclonal antibody specificity: orthoclone OKT3 T-cell blocker, *Nephron* **46**, Suppl. 1, 5–11.

HARRIS, T. J. R. (1987), Second-generation plasminogen activators, *Protein Eng.* **1**, 449–458.

HSIUNG, H. M., MAYNE, N. G., BECKER, G. W. (1986), High-level expression, efficient secretion and folding of human growth hormone in *Escherichia coli, Bio/Technology* **4**, 991-995.

JONES, A. J. S. (1988), Sensitive detection and quantitation of protein contaminants in rDNA products, in: *The Impact of Chemistry on Biotechnology. Multidisciplinary Discussion. ACS Symp. Ser.* **362**, (PHILLIPS, M., SHOEMAKER, S. P., MIDDLEKAUFF, R. D., OTTENBRITE, R. M., Eds.), American Chemical Society, Washington, D.C., pp. 193-201.

KLEID, D. G., YANSURA, D., SMALL, B., DOWBENKO, D., MOORE, D. M., GRUBMAN, M. J. MCKERCHER, P. D., MORGAN, D. O., ROBERTSON, D. H., BACHRACH, H. L. (1981), Cloned viral protein for foot-and-mouth disease: responses in cattle and swine, *Science* **214**, 1125–1129.

KLEID, D. G., YANSURA, D. G., DOWBENKO, D., WEDDELL, G. N., HOATLIN, M. E., CLAYTON, N., SHIRE, S. J., BOCK, L. A., OGEZ, J., BUILDER, S., PATZER, E. J., MOORE, D. M., ROBERTSON, B. H., GRUBMAN, M. J., MORGAN, D. O. (1984), Cloned viral protein for control of foot-and-mouth disease, *Dev. Ind. Microbiol.* **25**, 317–325.

KOHLER, G., MILSTEIN, C. (1976), Derivation of specific antibody-producing tissue culture and tumor lines by cell fusion, *Eur. J. Immunol.* **6**, 511-519.

KULA, M.-R., KRONER, K. H., HUSTEDT, H. (1982), Purification of enzymes by liquid - liquid extraction, *Adv. Biochem. Eng.* **24**, 73–118.

LONGNECKER, S. M. (1988), Somatostatin and octreotide: literature review and description of therapeutic activity in pancreatic neoplasia, *Drug Intell. Clin. Pharm.* **22**, 99–106.

LUBINIECKI, A. S., MAY, L. H. (1985), Cell bank characterization for recombinant DNA mammalian cell lines, *Dev. Biol. Stand.* **60**, 141–146.

LUBINIECKI, A. S., WIEBE, M. E., BUILDER, S. E. (1990), Process validation for cell culture - derived pharmaceutical proteins, in: *Large-scale Mammalian cell Culture Technology,* (LUBINIECKI, A. S., Ed.), pp. 515-542, New York: Marcel Dekker, Inc.

MARTIN, A. J. P., SYNGE, R. L. M. (1941), Separation of the higher monoaminoacids by countercurrent liquid - liquid extraction: the amino acid composition of wool, *Biochem. J.* **35**, 91–121.

MCGREGOR, W. C. (Ed.) (1986), *Membrane Separations in Biotechnology,* New York: Marcel Dekker, Inc.

MERRIFIELD, R. B. (1963), *J. Am. Chem. Soc.* **85**, 2149.

MOKS, T., ABRAHAMSEN, L., OSTERLOF, B., JOSEPHSON, S., OSTLING, M., ENFORS, S.-O., PERSSON, I., NIELSSON, B., UHLEN, M. (1987), Large-scale affinity purification of human insulin-like growth factor from culture medium of *Escherichia coli, Bio/Technology* **5**, 379-382.

MORIHAWA, K., OKA, T., TSUZUKI, H. (1979), Semi-synthesis of human insulin by trypsin-catalysed replacement of ala-B30 by Thr in porcine insulin, *Nature* **280**, 412.

MORROW, J. F., COHEN, S. N., CHANG, A. C., BOYER, H. W., GOODMAN, H. M. (1974), Replication and transcription of eukaryotic DNA in *Escherichia coli, Proc. Natl. Acad. Sci. USA*, 1743–1747.

OBRIST, R., SCHMIDLI, J., OBRECHT, J. P. (1986), Conjugation behavior of different monoclonal antibodies to *m*-methionyl-leucyl-phenylalanine, *Int. J. Immunopharmacol.* **8**, 629–932.

OGEZ, J. R., HODGDON, J. C., BEAL, M. P., BUILDER, S. E. (1989), Downstream processing of proteins: recent advances, *Biotechnol. Adv.* **7** (4), 467–488.

OLSON, K. C., FENNO, J., LIN, N., HARKINS, R. N., SNIDER, C., KOHR, W. H., ROSS, M. J., FODGE, D., PENDER,G., STEBBING, N. (1981), Purified human growth hormone from *E. coli* is biologically active, *Nature* **293**, 408–411.

PALLADINO, M. A., LEVINSON, A. D., SVEDERSKY, L. P., OBIJESKI, J. F. (1987), Safety issues related to the use of recombinant DNA-derived cell culture products; I. cellular components, *Dev. Biol. Stand.* **66**, 13–22.

POOL, R. (1990), Research News, *Science* **248**, 305.

RAUSCH, S. K. (1986), Purification and Activation of Proteins from Insoluble Inclusion Bodies, *Eur. Patent Application,* Publ. No. 0 212 960.

RIGGS, A. D., ITAKURA, K., BOYER, H. (1985), From somatostatin to human insulin, in: *Recombinant DNA Products: Insulin, Interferon and Growth Hormone*, (BOLLON, A. P., Ed.) Boca Raton, FL: CRC Press.

STEWART, J. M., YOUNG, J. D. (1984), *Solid Phase Peptide Synthesis.*

TILL, M. A., GHETIE, V., GREGORY, T., PATZER, E. J., PORTER, J. P., UHR, J. W., CAPON, D. J., VITETTA, E. S. (1988), HIV-infected cells are killed by rCD4-ricin A chain, *Science* **242**, 1166–1168.

TORREY, S. (Ed.) (1983), *Enzyme Technology: Recent Advances*, Park Ridge, NJ: Noyes Data Corp.

VAVRA, I., MACHOVA, A., HOLECEK, V., CORT, J. H., ZAORAL, M., SORM, F. (1968), Effect of a synthetic analogue of vasopressin in animals and in patients with *diabetes insipidus, Lancet* **1**, 948–952.

VITETTA, E. S., KROLICK, K. A., MIYAMA-INABA, M., CUSHLEY, W., UHR, J. W. (1983), Immunotoxins: a new approach to cancer therapy, *Science* **219**, 644–649.

VITETTA, E. S., FULTON, R. J., MAY, R. D., TILL, M., UHR, J. W. (1987), Redesigning nature's poisons to create anti-tumor reagents, *Science* **238**, 1098–1104.

WETZEL, R., KLEID, D. G., CREA, R., HEYNEKER, H. L., YANSURA, D. G., HIROSE, T., KRASZEWSKI, A., RIGGS, A. D., ITAKURA, K., GOEDDEL, D. V. (1981), Expression in *Escherichia coli* of a chemically synthesized gene for a "mini-c" analog of human proinsulin, *Gene* **16**, 63–71.

WHELAN, H. A., WRISTON, J. C., Jr. (1969), Purification and properties of asparaginase from *Escherichia coli* B., *Biochemistry* **8**, 2386–2393.

WIEBE, M. E., MAY, L. H. (1990), Cell Banking, in: *Large-Scale Mammalian Cell Culture Technology* (LUBINIECKI, A. S., Ed.), pp. 147–160, New York: Marcel Dekker, Inc.

ZONNENFELD, A. J., VAN VEERMANN, H., PANNEKOEK, H. (1986), Autonomous functions of structural domains on human tissue-type plasminogen activator, *Proc. Natl. Acad. Sci. USA* **83**, 4670–4674.

II. Product Formation (Upstream Processing)

5 Bioreactors: Description and Modelling

JENS NIELSEN
JOHN VILLADSEN

Lyngby, Denmark

List of Symbols

a	specific interfacial area (m^{-1})
$b(z, t)$	breakage function
c_i	concentration of i'th species (kg/m^3)
$c_{f,i}$	concentration of i'th species in the feed (kg/m^3)
$c_{e,i}$	concentration of i'th species in the effluent (kg/m^3)
c_i^*	saturation concentration (kg/m^3)
d_s	stirrer diameter (m)
D	dilution rate (h^{-1})
$D_{i,eff}$	effective diffusion coefficient for species i (m^2/s)
$f(z, t)$	cellular distribution function
$h(z, t)$	rate of production of cells with property z
k_1a	volumetric mass transfer coefficient (s^{-1})
K	consistency index
K_s	saturation constant in the Monod model (kg/m^3)
m	degree of mixing
m_s, m_p	maintenance coefficients (kg/kg DW/h)
n	power index in the rheological model, Eq. (14)
N	agitation rate (s^{-1})
N_f	flow number
N_p	power number
N_A	aeration number
$p(z', z, t)$	partitioning function
$\boldsymbol{p}$	metabolic product concentration vector (kg/m^3)
P	power input (W)
P_g	power input at gassed conditions (W)
q_i	volumetric formation rate of the i'th species ($kg/m^3/h$)
q_i^{mit}	volumetric mass transfer rate ($kg/m^3/h$)
$\boldsymbol{r}$	vector of intracellular reaction rates (kg/kg DW/h)
r_p	specific metabolic product formation rate (kg/kg DW/h)
r_s	specific metabolic substrate utilization rate (kg/kg DW/h)
r_x	specific growth rate of the biomass (kg DW/kg DW/h)
$\boldsymbol{R}$	vector of net formation rates of biomass components (kg/kg DW/h)
Re	Reynolds number
$\boldsymbol{s}$	substrate concentration vector (kg/m^3)
t_m	mixing time (s)
$\boldsymbol{u}$	rate vector for metamorphosis reaction (h^{-1})
u_s	superficial gas velocity (m/s)
v_e	liquid effluent flow rate (m^3/s)
v_f	liquid feed flow rate (m^3/s)
v_g	gas flow rate (m^3/s)
v_{pump}	pumping capacity (m^3/s)
V	liquid volume (m^3)
V_z	cell property space
x	biomass concentration (kg/m^3)
$\boldsymbol{X}$	vector of intracellular components (kg/kg DW)
Y_{ij}	yield coefficient (kg j/kg i)
$\boldsymbol{z}$	cellular property vector
$\boldsymbol{Z}$	vector of concentration of morphological forms (kg/kg DW)

$\boldsymbol{A}$	stoichiometric matrix for substrates
$\boldsymbol{B}$	stoichiometric matrix for metabolic products
$\dot{\gamma}$	shear rate (s^{-1})
$\boldsymbol{\Gamma}$	stoichiometric matrix for biomass components
$\boldsymbol{\Delta}$	stoichiometric matrix for morphological forms
η	viscosity (kg/m s)
η_{eff}	effectiveness factor
Θ	residence time (h)
μ	specific growth rate of the biomass (kg DW/kg DW/h) ($=r_x$)
μ_{max}	maximum specific growth rate of the biomass (kg DW/kg DW/h)
μ_q	specific growth rate of the q'th morphological form (kg q/kg DW/h)
ρ_l	liquid density (kg/m^3)
τ_s	shear stress (N/m^2)
Φ_{gen}	generalized Thiele modulus

1 Introduction

The bioreactor is the central piece of equipment in any industrial bioprocess. It is preceded by units in which the medium is prepared and by smaller reactors in which inoculum is grown, and it is followed by a chain of units – the downstream processes in which the desired product is separated from numerous impurities and thereafter purified, often to the extent that no impurities can be detected even with the best analytical methods. Undoubtedly, the cost of downstream processes is a substantial fraction of the total production costs, especially since the desired product is often present in only minute amounts in the effluent from the reactor. Therefore, an *integration* of the bioreaction with the upstream processes (usually by selecting a particularly suitable strain) and with the downstream processes (e.g., membrane reactors where a separation of desired products from other reaction compounds occurs during the bioreaction) is becoming increasingly important.

Still the dominant position of the bioreaction and, therefore, of the bioreactor remains unchallenged. The promised results of molecular biology have repeatedly been postponed, because it was difficult to transfer the results obtained in the test tubes to industrial-scale or even laboratory-scale production in bioreactors. What is sorely missing today is a great deal of quantitative description of bioreactions – all the way from a proper understanding of metabolic fluxes in the main pathways and how these fluxes are influenced by industrial enzyme kinetics to an understanding of the effect of non-instantaneous admixture of a feed stream into the bulk of the reactor medium.

Modelling of bioreactors has to be done on many levels: The cell kinetics, mass transfer and liquid mixing are very different but equally important items. What can be done in the present context is to give a mainly qualitative account interspersed with a few quantitative results of the many bits of models which have to be joined together in order to understand both the importance and the difficulty of a mathematical or numerical treatment of bioreactor design.

2 Bioreactors

Bioreactors are used both for fermentations (microbial conversions) and for bioconversions (based on enzymatic processes), and often the reactor design is virtually the same which makes a further distinction between the two applications unnecessary. Traditionally the term *fermentor* was used for any reactor designated for a microbial conversion, but today this term is becoming synonymous with a not very specified vessel equipped with basic stirring and air dispersal devices. *Bioreactors* are equipped with a large number of hardware and software controllers which strive to keep the environment of the microorganism between tightly set limits. In our description of bioreactors we will distinguish between *high-performance bioreactors*, which are used in the laboratory for physiological studies of the microorganism of interest, and *industrial bioreactors*, which are used in the production process.

2.1 High-Performance Bioreactors

In studies of living organisms at conditions closely approaching those encountered in an industrial reactor it is very important to distinguish between the dynamics of the cell reactions and the dynamics of the environment, specifically of the reaction vessel. Thus in so-called Physiological Engineering Studies it is desired to minimize the influence of the bioreactor dynamics on the overall performance of the process by a careful control of the environmental conditions. This is achieved in high-performance bioreactors, which are practically ideal bioreactors with a very low mixing time and a very high gas–liquid mass transfer. In these bioreactors the microorganisms are subjected to an unchanging environment when they are circulated throughout the liquid medium, and the response to imposed variations in the environment is therefore a consequence of the microbial behavior only.

SONNLEITNER and FIECHTER (1988, 1992) described compact loop bioreactors which satisfy the demands for a very low mixing time and a high gas–liquid mass transfer. These bioreactors are equipped with an internal loop positioned around the impeller shaft on which marine impellers are mounted. This gives a very high axial liquid circulation resulting in a very low mixing time (below one second). The bioreactors are equipped with a large number of *in situ* sensors, and to control operating variables a flexible direct digital control (DDC) system is used rather than classical single-purpose controllers. Hereby a precise control of many operating variables is ensured, and these variables can therefore be assumed to be so-called *culture parameters*. The major advantage of these high-performance bioreactors in physiological studies is that reproducible experiments can be carried out (see, e.g., LOCHER, 1991). This is of paramount importance for the future exploitation of microbial processes, since reproducibility is the key to the model-based process development which is characteristic of Biochemical Engineering Science. The value of high-performance bioreactors in physiological studies is well illustrated by the studies at ETH Zurich (by the group of SONNLEITNER and FIECHTER) of spontaneous oscillations of the yeast *Saccharomyces cerevisiae* in continuous cultures. These spontaneous oscillations occur as the result of a partial synchronization of the culture, and it is only with the well controlled environmental conditions in the high-performance bioreactors that synchronization can be maintained for many generations – something that permits detailed studies of the cell cycle of *S. cerevisiae* (see, e.g., MÜNCH et al., 1992).

Modern commercially available laboratory bioreactors (volume less than 10 L) normally fulfill the requirements for being high-performance bioreactors. Several bioreactor companies offer systems with flexible DDC controllers which allow a precise control of many operating variables (either single-phase or cascade control). Furthermore, with proper design of the stirrer and sparger in these laboratory bioreactors, it is possible to ensure a very low mixing time and a high gas–liquid mass transfer. Tab. 1 summarizes the characteristics of a 9 L commercially available bioreactor (from Chemap AG) which is used in Physiological Engineering Studies of filamentous fungi in our laboratory. The bioreactor has an aspect ratio (the ratio between tank height and tank diameter) of 2 and it is equipped with two standard Rushton turbines. The mixing time is less than 1 second, even with a high viscosity medium (containing up to 45 g/L biomass of *Penicillium chrysogenum*), and the $k_l a$ for oxygen is around 1000 h^{-1} for aqueous media (and about 200 h^{-1} for a high-viscosity medium). Thus, it is possible to maintain a practically homogeneous medium and a constant and sufficiently high level of dissolved oxygen for most microbial systems.

In addition to sensors for the culture parameters, modern high-performance bioreactors are normally equipped with a number of sensors for monitoring *culture variables*. These sensors can be grouped into three types: (1) Head space gas analysis, which measures the composition of the head space gas (or the exhaust gas), (2) *in situ* sensors, which are positioned directly in the bioreactor and give a continuous reading of the measured variable, and (3) on-line sensors, which measure medium component concentrations on an automatically withdrawn sample. Presently there are no reliable, autoclavable *in situ* sensors available for measuring medium components, and one must rely on on-line sensors if continuous monitoring of the important culture variables is to be carried out. Tab. 2 lists sensors for culture variables applied in our laboratory. Head space gas analysis, i.e., measurement of exhaust O_2 and CO_2, is fairly standard, and we use sensors based on paramagnetic detection of O_2 and infrared detection of CO_2, which have the advantage of having a well proven reliability and are relatively inexpensive compared to mass spectrometers. Our *in situ* sensors are optical sensors for measuring the biomass concentration (as the optical density) and the physiological state of the biomass (as the culture fluorescence). These optical sensors (in slightly different versions) are the only commercially available *in situ* sensors for monitor-

Tab. 1. Characteristics of a 9 L High-Performance Bioreactor (Commercially Available from Chemap AG) used in Physiological Engineering Studies of Filamentous Fungi

Culture Parameter	Sensor	Range	Accuracy/Precision
Temperature	Pt-100	0–150 °C	0.1 °C
Stirring speed	Tacho generator	0–1000 min^{-1}	1 min^{-1}
Pressure	Piezo resistor	0–2 bar	20 mbar
Medium mass	Electronic balance	0–30 kg	20 g
Liquid fluxes	Electronic balance	0–20 kg	0.1 g
Dilution rate	Electronic balance	0–1 h^{-1}	0.005 h^{-1}
Gas flux	Thermal mass flow meter	0–20 L min^{-1}	20 mL min^{-1}
pH	pH electrode	2–12	0.02
pO_2	Polarographic Clark electrode	0–400 mbar	2 mbar
pCO_2	Membrane-covered pH electrode	0–100 mbar	2 mbar

Tab. 2. Sensors for Measuring Culture Variables in High-Performance Bioreactors

Variable	Sensor	Frequency	Range	Accuracy/Precision (in % of Range)
Head Space Gas				
O_2	Hartmann Braun	1 min^{-1}	18–21%, 19–21%	0.5
CO_2	Hartmann Braun	1 min^{-1}	0–3%, 0–5%	0.5
In situ				
OD	Mettler	>1 min^{-1}	0–10 V	0.1
Fluorescence	Ingold	>1 min^{-1}	0-20 mA	0.1
On-line				
Glucose	FIA	30 h^{-1}	1–50000 mg/L	<2
Ethanol	FIA	12 h^{-1}	2–25000 mg/l	<2
Lactic acid	FIA	12 h^{-1}	2–25000 mg/L	<2
Glycerol	FIA	12 h^{-1}	1–50000 mg/L	<2
Ammonia	FIA	12 h^{-1}	2–20000 mg/L	<2
Phosphate	FIA	30 h^{-1}	5–50000 mg/L	<2
Nitrate/Nitrite	FIA	30 h^{-1}	5–10000 mg/L	<2
Penicillin	FIA	30 h^{-1}	1–10000 mg/L	<2
α-Amylase	FIA	12 h^{-1}	5 10000 U/L	<2
Biomass	FIA	12 h^{-1}	1–10000 mg/L	<2

ing culture variables, whereas a number of *in situ* sensors are available for measuring culture parameters as listed in Tab. 1. In on-line analysis a liquid sample is automatically withdrawn, e.g., using an *in situ* membrane module (CHRISTENSEN et al., 1991), and analyzed in an automated analytical system. For this purpose we apply flow injection analysis (FIA), which is well suited for on-line monitoring due to its high speed, good precision, and good reliability (NIELSEN, 1992). Application of FIA for on-line monitoring of medium components is illustrated in Fig. 1, where a high-performance bioreactor system with an on-line glucose analyzer is shown together with results from a batch process with *Aspergillus oryzae*. Using the almost continuous stream of precise glucose measurements it is possible to obtain a reliable estimate of the rate of glucose uptake throughout the various stages of the batch experiment. Other analytical systems, e.g., MS, HPLC or GC, may be applied in on-line analysis (see, e.g., SCHÜGERL, 1988); but in general it is difficult for other techniques to match the high measurement frequency and the reliability of FIA.

An important aspect of high-performance bioreactor systems with on-line analyzers is automation. Most bioreactor manufacturers sell software packages for PCs which may be useful for monitoring and control of bioreactors. However, integration of control algorithms for on-line analyzers, e.g., FIA systems, is often impossible in these commercial systems. In a

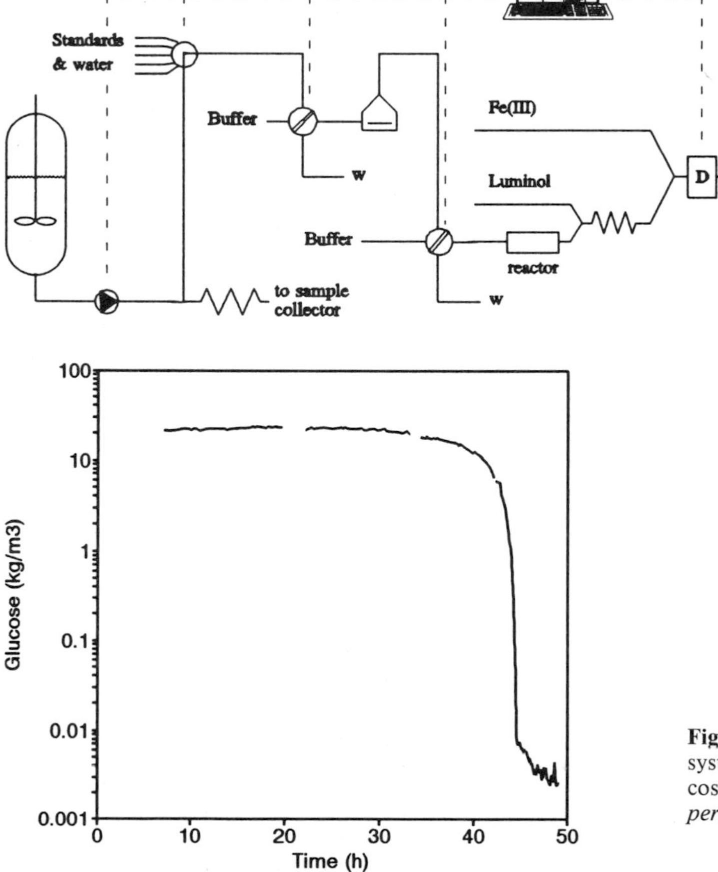

Fig. 1. High-performance bioreactor system with on-line monitoring of glucose during a batch process with *Aspergillus oryzae*.

research environment where the bioreactors are constantly put on different tasks, it will probably be best to design an automation system which is more or less specific for the individual laboratory (LOCHER et al., 1992; JOHANSEN et al., 1992).

2.2 Industrial Bioreactors

With the recent advances in molecular biology the scale of an industrial bioprocess has become extremely dependent on the nature of the process. A sizable share of the global demand for a high-value product like human growth hormone can be produced in a 200 L bioreactor, whereas traditional end products of microbial conversions in the pharmaceutical industry, e.g., penicillin, are produced in 200 m^3 bioreactors. Furthermore, in wastewater treatment plants one finds bioreactors in volumes up to 15000 m^3 (SCHÜGERL, 1991). A complete discussion of all issues pertinent to industrial-scale bioreactors is therefore not possible within this chapter, and we shall restrict ourselves to some general comments on industrial bioreactors.

The first large-scale microbial process in the pharmaceutical industry, the penicillin production, was developed in a stirred tank bioreactor, and the stirred tank is still the preferred bioreactor design. With the high capital costs of bioreactors, the industry prefers to modify existing bioreactors, e.g., replace the stirrers, rather than introduce completely new bioreactor designs. Furthermore, application of the same general bioreactor design to different microbial processes gives a large flexibility for the company. This is the reason why the stirred tank bioreactor is often the first choice when a new process is developed, and it is only when the requirements for, e.g., gas–liquid mass transfer cannot be met in this type of bioreactor that other designs will be considered.

Stirred tank bioreactors are typically cylindrical with a slightly curved or almost flat bottom. The aspect ratio may vary somewhat, but it is typically 1:1, 2:1, or 3:1. In small tanks temperature control is achieved by a heating/cooling jacket, whereas for larger tanks internal or external heat exchanger loops are necessary to keep the temperature constant. It should be emphasized that especially in aerobic processes large quantities of heat are to be removed – perhaps 50–80 kJ for a production of 1 kg biomass per h – and that the requirements for temperature control are very strict. Typically a microorganism functions only in a temperature range of about 5 °C.

In aerated processes the gas is normally supplied through a sparger placed at the bottom of the tank below the lowest stirrer. Typically the sparger is formed as part of a ring or a plate, or it may simply be an open tube end with only one orifice hole. Normally stirred tanks are equipped with baffles to prevent the formation of a large central vortex due to the centrifugal forces introduced by the stirrer action. Stirring is by one or more impellers mounted on a centrally lined impeller shaft. There are many different stirrer designs, but these can roughly be divided into two groups: (1) turbine impellers and (2) axial flow impellers. The turbine impellers are modifications of the simple paddle design where a vertically placed plate rotates around its axis. Often the blades are placed equidistantly on a horizontal circular plate as illustrated by the much used Rushton turbine. With axial flow impellers the main flow pattern is upwards at the tank wall and downwards at the center around the stirrer axis, and a much better axial mixing is generally obtained with these impellers compared with the turbine impellers which tend to give significant dead zones in the reactor. For low-viscosity liquids, the marine impeller may give sufficient axial flow, but for highly viscous media other designs are preferred, e.g., the Intermig (NIENOW, 1990).

Where very large bioractors are used for single cell protein production or for wastewater treatment, the high energy demands and the requirement for heat removal cannot be met in a stirred tank bioreactor, and other bioreactor designs are, therefore, used. In these designs a major economic design criterion is to obtain a high gas–liquid mass transfer at a more moderate power input than that required in a stirred tank bioreactor. This can be achieved if the energy for dispersal of the gas into the liquid is supplied by means of an external pump combined with effective gas nozzles rather than by mere mechanical agitation and standard gas dispersers (SCHÜGERL, 1991). Natural flow bubble columns are simple examples of such reactors. They have a large aspect ratio, and the energy input is entirely by means of sparging with compressed gas. Bubble columns are characterized by relatively low capital costs (mainly due to their simple mechanical configuration). The operating costs are normally lower than for stirred tank bioreactors due to lower energy requirements, but in their standard form they are not very efficient for mass transfer. Bubble columns are well known in chemical processes, and a typical application in the bioindustry is for beer production. The main liquid circulation is obtained by the upflow of gas bubbles; and to ensure good mixing of the liquid phase the column may be designed as a loop reactor (or an air-lift), where the liquid phase flows up through the so-called riser together with the gas bubbles and flows down in the so-called down-comer. The loop may be either internal, i.e., present within the column, or external, i.e., the down-comer is a separate column. To give small bubbles, plates of different designs, e.g., perforated plates, may be positioned at various locations through the riser. Bubble columns have been treated in great detail by DECKWER (1992) and by SCHÜGERL (1991).

For processes where a high gas–liquid mass transfer is required, one may use an injector nozzle (or ejector nozzle) for aeration. Here, the liquid is circulated in a loop by means of an external pump, and it re-enters the bioreactor through a nozzle at which it is mixed with compressed gas. As a result of the high rates of momentum transfer from the liquid to the gas, a dispersion of very small bubbles is formed in the continuous liquid phase. Hereby a high volumetric mass transfer coefficient can be obtained but only in the close proximity of the nozzle, because the small bubbles tend to coalesce. A repeated break-up of gas bubbles in a slender loop reactor can also be achieved by static mixers installed at appropriate intervals. This design and a loop reactor with multiple nozzles for injection of gas are both possible solutions for the high mass transfer demand required in a single cell protein process based on natural gas and oxygen as substrates.

In Tab. 3 different bioreactor designs are compared with respect to specific interfacial area obtained at a given value of the so-called power input measured in kW per m^3 medium. It is observed that at the same energy input higher specific interfacial areas can be obtained in bubble columns than in stirred tank bioreactors. Also by using nozzle injectors/ejectors or static mixers it is possible to obtain very high specific interfacial areas with a moderate power input. The power input does not increase alarmingly, but no design can circumvent the problem that a high mass transfer necessarily requires a high power input.

3 Modelling of Bioprocesses

Microbial conversions involve multiphase systems with many different interactions, i.e., cellular reactions, gas–liquid mass transfer, li-

Tab. 3. Comparison of Specific Interfacial Areas for Dispersed Bubbles in Different Bioreactor Designs (Data from SCHÜGERL et al., 1977)

Bioreactor	Aerator	a (m^{-1})	P/V (kW/m^3)
Stirred tank		120–2200	1–10
Bubble column	Perforated plate	650	0.5
	Porous plate	2000	0.4–0.9
Nozzle reactor	Injector nozzle	6000	1.5
	Ejector nozzle	8000	2.2
Internal loop nozzle reactor	Ejector nozzle	1300–2500	0.9–7.2

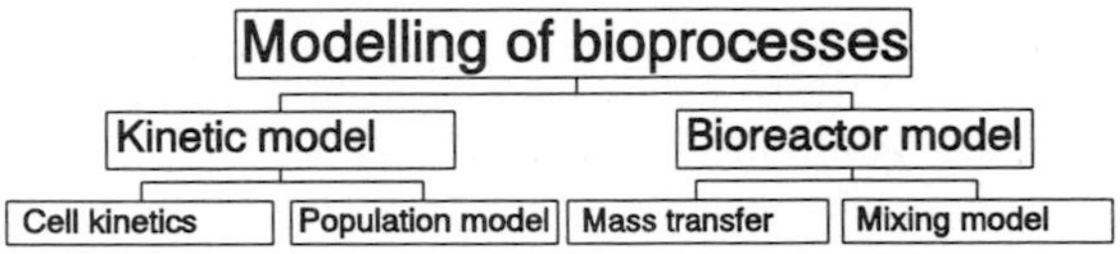

Modelling of bioprocesses
Kinetic model | Bioreactor model
Cell kinetics | Population model | Mass transfer | Mixing model

Fig. 2. Different aspects of modelling bioprocesses.

quid mixing, and substrate diffusion into cell aggregates. A complete quantitative description of a bioprocess, therefore, involves many different bits of modelling, and a final model for the whole system may be very complex. In an attempt to reduce the complexity of the problem, one normally studies the effect of different mechanisms separately. Thus, the microbial kinetics is studied in a high-performance bioreactor where effects of mass transfer and liquid mixing are eliminated (or at least minimized) and the mass transfer processes are studied using model media. An overview of the different aspects of modelling microbial reactions is given in Fig. 2. To describe the bioreaction kinetics we have to consider both the single cell kinetics and possible distributions in the cellular population. To describe the bioreactor performance we have to consider both mass transfer and the flow pattern in the bioreactor. In the following the individual aspects of bioprocess modelling are considered in greater detail.

3.1 Bioreaction Kinetics

The overall bioreaction kinetics of a cell culture is a representation of the influence of many intracellular reactions in a large population of individual cells, and the kinetic model should never be interpreted as a complete mechanistic description of the system. However, for practical purposes many reactions can be lumped together, and often the cell population can be assumed to be homogeneous. Hence, a reasonably simple kinetic model is often adequate for interpretation of the influence of key variables. Always the complexity of the model should be chosen with a view to its application. Thus if the aim is to obtain a model which can simulate transient growth experiments, it is often possible to apply a simple structured cell model (see, e.g., NIELSEN et al., 1991a), whereas if the aim is to study, e.g., the influence of foreign protein production on the cell behavior a more detailed cell model is required (see, e.g., LEE and BAILEY, 1984). Furthermore, if it is desired to study the influence of cell property distribution on the overall population kinetics, a structured (or segregated) population model has to be applied (see, e.g., SEO and BAILEY, 1985).

3.1.1 Cell Kinetics

Cellular growth involves the conversion of substrates to metabolic products and biomass components in intracellular reactions. With J cellular reactions, where substrates $\boldsymbol{s}$ are converted to metabolic products $\boldsymbol{p}$ and biomass components $\boldsymbol{X}$, the stoichiometry can be summarized (NIELSEN and VILLADSEN, 1993) as

$$\boldsymbol{A}\,\boldsymbol{s}+\boldsymbol{\Gamma}\boldsymbol{X}+\boldsymbol{B}\,\boldsymbol{p}=0 \qquad (1)$$

where $\boldsymbol{A}$, $\boldsymbol{B}$ and $\boldsymbol{\Gamma}$ are stoichiometric matrices. Normally the same symbols are used for the species and their concentration, e.g., s_i is the concentration of the i'th substrate. The concentrations of intracellular components are given in g/g DW (dry weight) whereby the sum of all intracellular concentrations equals 1. In detailed models the stoichiometric coefficients can be deduced by consultation of metabolic pathway schemes, whereas in lumped models they often take the form of model parameters to be estimated by fitting the model to experimental data. With the forward reaction rates for the J reactions collected in the rate vector $\boldsymbol{r}$, the specific rates, i.e., the rates per cell mass, of substrate uptake ($\boldsymbol{r}_s$) and product formation ($\boldsymbol{r}_p$) can be calculated:

$$\boldsymbol{r}_s=-\boldsymbol{A}^T\boldsymbol{r} \quad \boldsymbol{r}_p=\boldsymbol{B}^T\boldsymbol{r} \qquad (2)$$

Similarly the specific growth rate for the biomass can be calculated from

$$r_x=\mu=\sum_{i=1}^{L}\boldsymbol{\Gamma}_i^T\boldsymbol{r} \qquad (3)$$

where the summation is over the L intracellular components. Finally, for the intracellular biomass components the net rate of formation, $\boldsymbol{R}$, (NIELSEN and VILLADSEN, 1993) is

$$\boldsymbol{R}=\boldsymbol{\Gamma}^{\mathrm{T}}\boldsymbol{r}-\mu\boldsymbol{X} \tag{4}$$

For a homogeneous bioreactor with sterile feed the net rate of formation of intracellular components equals the accumulation (NIELSEN and VILLADSEN, 1993), i.e.,

$$\frac{\mathrm{d}X_{\mathrm{i}}}{\mathrm{d}t}=R_{\mathrm{i}} \tag{5}$$

From the specific rates given by Eqs. (2)–(4) the volumetric net rates of formation for the substrates and for the metabolic products can be calculated:

$$q_{\mathrm{s}}=-r_{\mathrm{s}}x=\boldsymbol{A}^{\mathrm{T}}\boldsymbol{r}x \quad q_{\mathrm{p}}=r_{\mathrm{p}}x=\boldsymbol{B}^{\mathrm{T}}\boldsymbol{r}x \tag{6}$$

where x is the biomass concentration. Similarly for the biomass:

$$q_{\mathrm{x}}=r_{\mathrm{x}}x=\mu x \tag{7}$$

The volumetric net rates given by Eqs. (6)–(7) appear in the mass balances for the considered bioreactor as illustrated in Sect. 4 (see, e.g., Eq. (30)), and they hereby couple the microbial kinetics to the bioreactor performance.

Each of the components in the vector $\boldsymbol{r}$ of the specific rates is taken to be a function of substrate and product concentrations and of the biomass composition:

$$\boldsymbol{r}=f(\boldsymbol{s},\boldsymbol{p},\boldsymbol{X}) \tag{8}$$

In the so-called *balanced growth* situation the cell composition does not change during the experiment. This means that the rates of the intracellular reactions have to be proportional with proportionality constants given by the initial composition of the cell mass. When the cell composition is constant the growth of biomass, formation of metabolites, and consumption of substrates can all be described in terms of only one cellular variable, and the natural choice is the biomass concentration x. This is the basis for the so-called *unstructured models*, of which the Monod model is a well known example. With a simple proportionality between rates assumed in the Monod model one neglects the substrate utilization and concomitant metabolite formation associated with maintenance, and maintenance terms are therefore normally added to the rate expressions for substrate uptake and metabolic product formation (as shown, e.g., in Eq. (41)).

Whenever an unstructured model is used appropriately, i.e., to interpret steady-state chemostat experiments or the exponential growth phase of a batch process, its success has been proven. It fails when the underlying assumption of balanced growth is not satisfied, i.e., when the state of the cell must be defined via the internal composition vector $\boldsymbol{X}$. This often happens at the start of batch growth where an "active" part of the cell must be preferentially synthesized to get the cells out of the so-called lag phase. It also happens when an essential growth substrate (e.g., NH_3) is almost used up at the end of a batch process and the cell composition changes. In the case of a depleted N-source the cell may start to accumulate carbohydrates. Finally in the interpretation of transients in chemostats (changing from one dilution rate to another or pulsing the chemostat with a limiting substrate) one encounters dramatic changes in the biomass composition, and an unstructured model may give results which are qualitatively different from the experimental results.

All *structured models* are in one sense or the other improvements of the unstructured models, since some basic mechanisms of the cellular behavior are at least qualitatively incorporated. Thus, the structured model may have some predictive strength, i.e., it may describe the growth process at different operating conditions with the same set of parameters, and it can be applied to, e.g., optimization of the process. But one should bear in mind that "true" mechanisms of the metabolic processes are of course never revealed by a structured model, even if the number of parameters is quite large. In structured models all the biomass components are lumped into a few key variables, i.e., the vector $\boldsymbol{X}$, which is hopefully representative of the state of the cell. Hereby, the microbial activity becomes not only a function of the abiotic variables, which may change with very small time constants, but also of the cellular composition, and consequently of the "history" of the cells, i.e., the environmental conditions which the cells have experienced in the past. The biomass can be structured in many different ways: in simple structured models only a few cellular components are considered, whereas in highly structured models up to 20 intracellular components are considered. The choice of structuring depends on the aim of the modelling exercise as discussed previously, but often one starts with a simply structured model onto which more and more structure is added as new experiments are added to the data base. Even in highly structured models many of the cellular components included in the model represent "pools" of different enzymes, metabolites, or other cellular components. The cellular reactions considered in structured models are, therefore, empirical in nature, since they do not represent the conversion between "true" components. Consequently, it is permissible to write the kinetics for the individual reactions in terms of reasonable empirical expressions, with a form which is judged to fit the experimental data with a small number of parameters. Thus, Monod type expressions are often used since they summarize some fundamental features of most cellular reactions, i.e., being first order at low substrate concentration and zero order at high substrate concentration. Despite their empirical nature, structured models are normally based on some well known cell mechanisms, and they have the ability to simulate certain features of experiments quite well.

Since the first structured models appeared in the late 1960s (RAMKRISHNA et al., 1967; WILLIAMS, 1967), many different structured models have been presented (for reviews see HARDER and ROELS, 1982; NIELSEN and VILLADSEN, 1992). Many of these models have aimed at describing the growth of the yeast *Saccharomyces cerevisiae*, which is probably the most studied microorganisms in the field of Physiological Engineering. Thus, NIELSEN and VILLADSEN (1992) discussed 12 different yeast models, and this probably only represents a fraction of the total number of structured/mechanistic yeast models which appear in the open literature. Despite differences there are, however, many similarities between the structured models, and in a discussion of different models it is convenient to classify these into four groups (NIELSEN and VILLADSEN, 1992): (1) Simple structured models, (2) models describing growth on multiple substrates, (3) models describing non-growth associated product formation (e.g., recombinant protein formation), and (4) single cell models. An overview of the application of the different types of models and representatives from each group of models is given in Tab. 4.

3.1.2 Population Models

Often it is assumed that the population of cells is homogeneous, i.e., all cells behave identically. Although this assumption is cer-

Tab. 4. Classification of Different Structured Models

Type	Typical Application	References
Simple models	Simulation of dynamic conditions	AGRAWAL et al. (1983) SWEERE et al. (1988) NIELSEN et al. (1991a)
Multiple substrate models	Simulation of growth on complex (industrial) media	VAN DEDEM and MOO-YOUNG (1975) KOMPALA et al. (1984)
Product formation models	Simulation of recombinant protein formation	BENTLEY and KOMPALA (1989) NIELSEN et al. (1991c) COPPELLA and DHURJATI (1990)
Single cell models	Theoretical studies of cellular activity, e.g., host–plasmid interactions	SHULER and DOMACH (1982) STEINMEYER and SHULER (1989) PERETTI and BAILEY (1987)

tainly crude if a small number of cells is considered, it gives a very good picture of certain properties of the cell population because there are billions of cells per mL medium. Furthermore, the kinetics is often linear in the cellular properties, e.g., in the concentration of a certain enzyme, and the overall population kinetics can therefore be described as a function of the average property of the cells (ROELS, 1983). There are, however, situations where cell property distributions influence the overall culture performance, and here it is necessary to consider the cellular property distribution.

In terms of the state or property vector z, the distribution function is $f(z, t)$, i.e., $f(z, t)\mathrm{d}z$, is the number of cells with properties in the space between z and $z+\mathrm{d}z$. The balance for $f(z, t)$ is given by the so-called *population balance equation* (PBE) as described by RAMKRISHNA (1979):

$$\frac{\partial f(z,t)}{\partial t}+\sum_{\mathrm{i}}\frac{\partial}{\partial z_{\mathrm{i}}}(R_{\mathrm{i}}(z,t)f(z,t)) = h(z,t)-Df(z,t) \tag{9}$$

$R_{\mathrm{i}}(z, t)$ is the net rate of increase in z_{i} for a cell in the state z and $h(z, t)$ is the net rate of production of cells with property z formed upon cell division. With binary fission $h(z, t)$ is given by (RAMKRISHNA, 1979):

$$h(z,t)=2\int_{V_z} b(z',t)p(z',z)f(z',t)\mathrm{d}z' - b(z,t)f(z,t) \tag{10}$$

where $b(z, t)$ is the breakage function describing the rate of division of cells with property z, and $p(z, z)$ is the partitioning function describing the probability for formation of a cell with property z upon division of a cell with property z'. Normally only one cellular property is considered, e.g., cell age or single cell mass, and the distribution function becomes one-dimensional (see, e.g., HJORTSO and BAILEY, 1982, 1984). Through comparison of a calculated distribution function and experimental measurements of single cell properties, e.g., using flow cytometry or image analysis, important information of the single cell behavior can be extracted.

Another approach to model cellular distributions in a cell culture are the so-called *morphologically structured models* (NIELSEN and VILLADSEN, 1993). Here, the cells are divided into a finite number of cell states $\boldsymbol{Z}$ (or morphological forms), and conversion between the different cell states is described by a set of empirical metamorphosis reactions with the stoichiometry:

$$\boldsymbol{\Delta Z}=0 \tag{11}$$

where $\boldsymbol{\Delta}$ is a stoichiometric matrix. Z_{q} represents both the q'th morphological form and the fractional concentration (g q'th morphological form per g DW). The mass balance for Z_{q} in a homogeneous bioreactor with sterile feed is:

$$\frac{\mathrm{d}Z_{\mathrm{q}}}{\mathrm{d}t}=\boldsymbol{\Delta}_{\mathrm{q}}^{\mathrm{T}}\boldsymbol{u}+(\mu_{\mathrm{q}}-\mu)Z_{\mathrm{q}} \tag{12}$$

where $\boldsymbol{u}$ is the forward reaction rate for the Q metamorphosis reactions. μ_{q} is the specific growth rate of the q'th morphological form, and μ is the overall specific growth rate of the biomass which is given by:

$$\mu=\sum_{\mathrm{q}=1}^{\mathrm{Q}}\mu_{\mathrm{q}}Z_{\mathrm{q}} \tag{13}$$

In a morphologically structured model growth of the individual cell forms is normally described with a simple unstructured model, e.g., μ_{q} is given by the Monod model, since the complexity becomes substantial when a biochemically structured model is introduced for each morphological form (NIELSEN and VILLADSEN, 1993). Furthermore, it is experimentally difficult to verify the model by comparison of simulations and measurements of intracellular composition in the different morphological forms. The concept of morphologically structured models is especially suited for modelling the growth of filamentous fungi, as illustrated by MEGEE et al. (1970) and NIELSEN (1993).

3.2 Mass and Energy Transfer in Bioreactors

Two factors of dominating influence on the performance of a bioreactor are highlighted in Fig. 2: (1) mass transfer and (2) liquid mixing. Both of these factors are intimately related to the rheology of the microbial medium, and our discussion of bioreactor performance will therefore be preceded by a short review of major rheological properties of microbial media.

3.2.1 Rheology

With unicellular microorganisms the medium is normally Newtonian, and except for very high biomass concentrations the viscosity is close to that of pure water. Microorganisms which produce extracellular polysaccharides exert a significant effect on the rheology of the medium. The effect originating from the cells themselves is, however, negligible, and the change in viscosity can be ascribed to the formation of the polymer. The rheology of these media can therefore normally be described with a rheological model analogous to those used for a pure polymer solution, e.g., the power law model:

$$\tau_s = -K\,\dot{\gamma}^n \tag{14}$$

K is the consistency index and n is the power law index which is smaller than 1 for pseudoplastic fluids (the most common non-Newtonian property). Fluids with this property exhibit shear thinning, i.e., the viscosity decreases with increasing shear rate, $\dot{\gamma}$.

With filamentous microorganisms the medium gradually becomes very viscous, and the non-Newtonian behavior is caused by the mycelial network. The rheological properties depend on the macroscopic morphology, i.e., whether mycelia or pellets (which are aggregates of mycelium) are present, but the power law model gives an adequate description of the rheology of both morphological forms (ALLAN and ROBINSON, 1990; PEDERSEN et al., 1993). Normally, both the degree of shear thinning and the viscosity are higher in a medium containing a mycelium than in a medium where pellets are formed (at the same biomass concentration), i.e., the power law index is smaller and the consistency index is higher for media with mycelia than for media containing pellets. Other models, e.g., the Casson model, have also been used to describe the rheology of microbial media containing filamentous microorganisms, but for reasonable shear rates in a bioreactor ($\dot{\gamma} > 20\ s^{-1}$) it is not possible to distinguish between the power law model and the Casson model (ROELS et al., 1974). In Fig. 3 the power law parameters are shown as functions of the biomass concentration in processes with filamentous fungi (*Penicillium chrysogenum* and *Aspergillus niger*) for both types of macroscopic morphology: (1) pellets and (2) mycelium. It has been observed that the power law index is approximately constant, $n \approx 0.45$, for pellet morphology and slightly lower for mycelium. Furthermore, the consistency index increases with the biomass concentration for both morphologies, but the increase is much more rapid with mycelium than with pellets.

In a stirred bioreactor there are large variations in the shear rate throughout the tank. It is, therefore, not possible to specify a "viscosity" for the medium with non-Newtonian properties. An approximate value for the average shear rate can, however, be calculated from Eq. (15) where k is a characteristic constant for the considered system and N is the stirring speed.

$$\dot{\gamma}_{aver} = k\,N \tag{15}$$

The value of the empirical constant k has been reported to be in the range of 4–13 depending on the system, and with a standard Rushton turbine impeller $k = 10$ may be used (NIENOW and ELSON, 1988).

3.2.2 Mass Transfer

Mass transfer phenomena of importance in cellular processes can be divided into two categories: (1) gas–liquid mass transfer and (2) molecular diffusion of medium components into pellets or cell aggregates.

Gas–liquid mass transfer is normally synonymous with transfer of oxygen from gas bub-

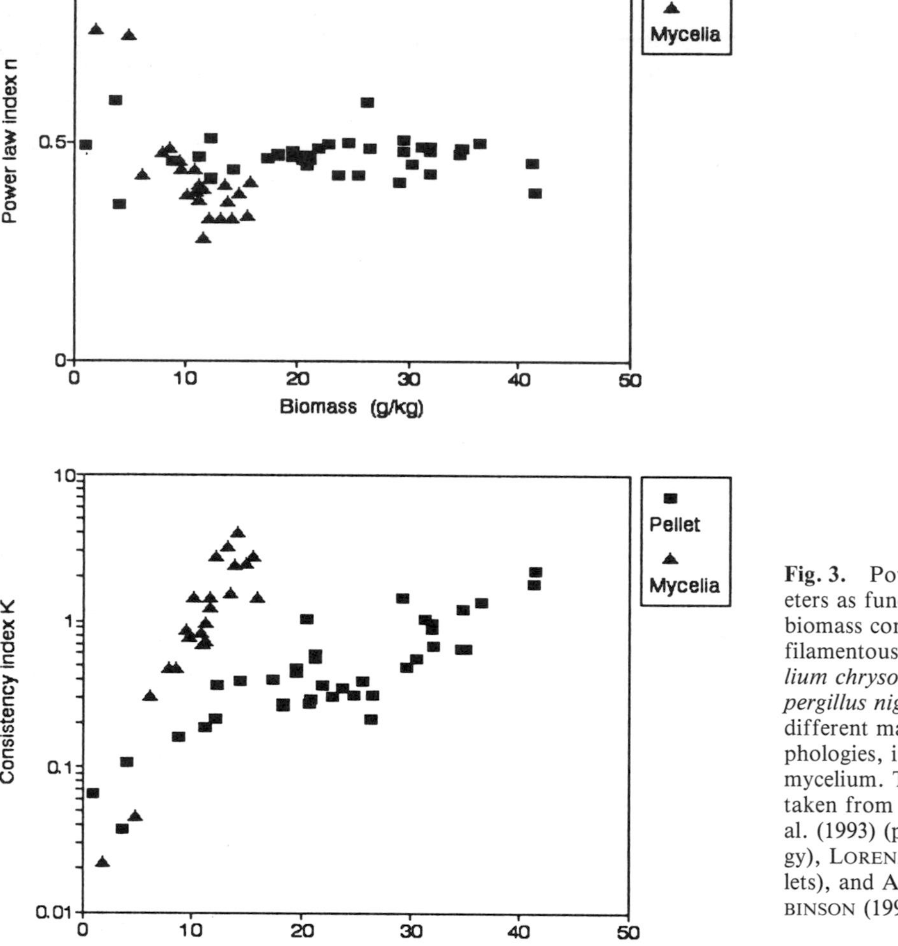

Fig. 3. Power law parameters as function of the biomass concentration of filamentous fungi (*Penicillium chrysogenum* and *Aspergillus niger*) with two different macroscopic morphologies, i.e., pellets and mycelium. The data are taken from PEDERSEN et al. (1993) (pellet morphology), LORENZ (1985) (pellets), and ALLEN and ROBINSON (1990) (mycelium).

bles to the bulk liquid, since many industrially important bioreactions are aerobic processes, and with the low oxygen solubility in aqueous solutions a continuous transfer of this substrate from the gas phase to the liquid phase is decisive for maintaining the oxidative metabolism of the cells. Other sparingly soluble gases are also of interest in connection with microbial processes. Thus, methane and other light hydrocarbons are used for the production of single cell protein. Finally, gas–liquid mass transfer is not only of importance for aerobic processes. A constant removal of methane and carbon dioxide is necessary for successful operation of some anaerobic wastewater treatment processes.

Gas–liquid mass transfer is quantified as the product of the volumetric mass transfer coefficient $k_l a$ and the driving force, i.e.,

$$q_{\mathrm{A}}^{\mathrm{mt}} = k_l a (c_{\mathrm{A}}^* - c_{\mathrm{A}}) \tag{16}$$

where c_{A}^* is the saturation concentration in the bulk liquid. The volumetric mass transfer coefficient is the product of the mass transfer coefficient k_l and the specific interfacial area

a, i.e., the total gas–liquid interfacial area per liquid volume. To a certain extent the two factors k_l and a can be varied independently – an increase of $k_l a$ can be achieved by an increase of the gas dispersion (higher a) but also by an increase of k_l (e.g., by surface active compounds). In the literature one may find many different theoretical and empirical correlations for both k_l and a (see, e.g., MOO-YOUNG and BLANCH, 1981). Generally, a given correlation holds for a specific reactor system and in certain regimes of operating conditions. Similarly a large number of different empirical correlations for the volumetric mass transfer coefficient $k_l a$ have been published. Most of these correlations can be written in the form

$$k_l a = k\, u_s^{\alpha} \left(\frac{P_g}{V}\right)^{\beta} \tag{17}$$

where u_s is the superficial gas velocity (m/s), P_g is the power input (W) at gassed conditions, and V is the liquid volume (m^3). The parameters in Eq. (17) are specific for the considered system, and some of the parameter values reported in the literature for stirred tanks are listed in Tab. 5. A closer study of the parameter values reveals that for any particular agitator type the mass transfer coefficient for a non-coalescing medium is predicted to be about twice as high as for a coalescing medium at the same operating conditions. These overall correlations are, however, very rough simplifications since they are constructed as average data fitters for many different bioreactor designs.

The liquid viscosity has a significant influence on the flow properties and, therefore, also on the gas–liquid mass transfer. The mass transfer coefficient k_l decreases with increasing liquid viscosity, and since the degree of turbulence is less in a high viscosity liquid, the maximum stable bubble diameter is higher resulting in a lower specific interfacial area a. To account for the influence of liquid viscosity on the volumetric mass transfer SCHÜGERL (1981) specified the correlation (18), which is similar in form to Eq. (17) except for the viscosity term η (units: kg m^{-1} s^{-1}).

$$k_l a = k\, u_s^{0.4} \left(\frac{P_g}{V}\right)^{0.6} \eta^{-0.7} \tag{18}$$

Often the influence of viscosity on the mass transfer is, however, accounted for in terms of dimensionless parameter groups (SCHÜGERL, 1981).

From the correlation (17) it is observed that the power input is a determining factor for the gas–liquid mass transfer. The power input is also one of the major operating costs for aerobic processes, and it is therefore an important design factor for the agitation system. Thus, the ideal stirrer system is one where the power input is minimized for a given mass transfer and liquid mixing effciency (see also Sect. 3.2.3). The power input at ungassed conditions P(W) from an agitator is correlated to the stirrer diameter d_s (m) and the stirring speed N (s^{-1}) through:

$$P = N_p \rho_l N^3 d_s^5 \tag{19}$$

where N_p is the power number (dimensionless) – a characteristic parameter for the applied stirrer. As illustrated in Fig. 4, the power num-

Tab. 5. Parameter Values for the Empirical Correlation Eq. (17). The parameter values are specified with all variables being in SI units, i.e., the power input in W/m^3 and the superficial gas flow rate in m/s.

Medium	k	α	β	Agitator	Reference
Coalescing	0.025	0.5	0.4	6 Bladed Rushton turbines	MOO-YOUNG and BLANCH (1981)
	0.00495	0.4	0.593	6 Bladed Rushton turbines	LINEK et al. (1987)
	0.01	0.4	0.475	Different agitators	MOO-YOUNG and BLANCH (1981)
	0.026	0.5	0.4	Not specified	VAN'T RIET (1979)
Non-coalescing	0.0018	0.3	0.7	6 Bladed Rushton turbines	MOO-YOUNG and BLANCH (1981)
	0.02	0.4	0.475	Different agitators	MOO-YOUNG and BLANCH (1981)
	0.002	0.2	0.7	Not specified	VAN'T RIET (1979)

ber is a function of the Reynolds number for the stirrer:

$$Re_s = \frac{\rho_l N d_s^2}{\eta} \tag{20}$$

For low Re_s numbers ($Re_s < 10$), i.e., in the laminar flow regime, the power number is approximately proportional with Re_s^{-1} whereas for high Re_s numbers ($Re_s > 10^4$), i.e., in the turbulent flow regime, the power number is approximately constant. In a flow regime between these limits there is no general correlation between N_p and Re_s. The correlation shown in Fig. 4 is only valid for a particular system – a Rushton turbine in a bioreactor equipped with baffles. Without baffles much smaller values of N_p are obtained at high Re_s numbers, and for other stirrer designs the profile of N_p versus Re_s looks quite different (see, e.g., SCHÜGERL, 1991 for N_p versus Re_s for different stirrers). The power number at high Re_s numbers and with different stirrer designs is listed in Tab. 6. In Sect. 3.2.3 different stirrer designs are compared with respect to power input and liquid mixing capabilities.

Eq. (19) only holds for single impeller systems. For systems with multiple impellers the power input is approximately:

$$P = n_i P_{\text{single impeller}} \tag{21}$$

Tab. 6. Power Numbers N_p at High Re_s Numbers for Different Stirrer Designs in a Bioreactor System Equipped with Baffles

Impeller Type	N_p	Reference
Rushton	5.20	NIENOW (1990)
Intermig	0.35	NIENOW (1990)
Prochem	1.00	NIENOW (1990)
Marine impeller	0.35	SCHÜGERL (1991)

where n_i is the number of impellers and $P_{\text{single impeller}}$ is given by Eq. (19).

When gas is sparged to a tank, gas bubbles are drawn to regions of low pressure. This results in the formation of gas-filled areas (called cavities) behind the stirrer blades. The formation of these cavities depends on the ratio between the gas flow rate and the stirring speed, often expressed through a dimensionless group, the so-called aeration number N_A:

$$N_A = \frac{v_g}{N d_s^3} \tag{22}$$

Here v_g is the volumetric gas flow rate (m^3/s). For low aeration numbers, i.e., a low aeration rate or high stirring speed, a homogeneous gas–liquid dispersion is obtained throughout the tank. For increasing values of N_A the gas–

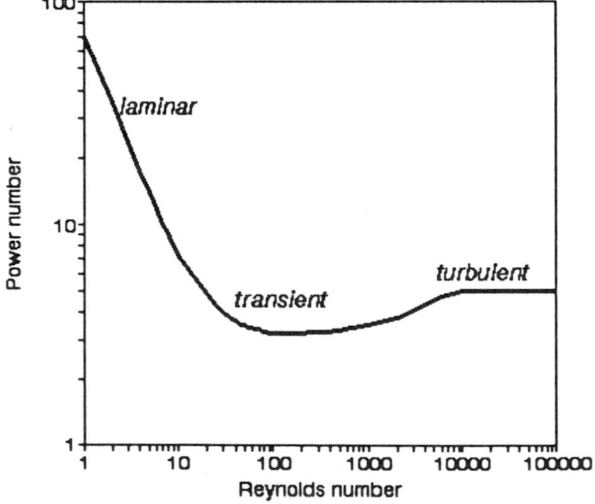

Fig. 4. Correlation between the power number and the Reynolds number for a bioreactor system with baffles and a Rushton turbine impeller.

liquid dispersion becomes less efficient and the best gas–liquid dispersion is just above the bottom stirrer. Very high aeration numbers lead to flooding conditions at the stirrer, i.e., the liquid flow is driven by the gas flow resulting in a gas–liquid flow upwards in the middle of the tank and a downwards liquid flow at the walls.

Due to the formation of cavities behind the stirrer blades in a gas sparged bioreactor, the power input at gassed conditions P_g is lower than the power input P at ungassed conditions (all other conditions are the same). This is illustrated in Fig. 5 where the ratio P_g/P is depicted as a function of the aeration number N_A (in some references one finds the power number plotted as a function of the aeration number). It has been observed that with the Prochem impeller a much better power dissipation at high aeration numbers is obtained than with the Rushton turbine. In viscous media the gas-filled cavities are very stable once they are formed, and they can often be maintained even after the gas flow rate is reduced significantly. The influence of aeration rate, therefore, counteracts the effect of increasing viscosity on the power input leading to an almost constant gassed power consumption even in highly viscous media (SCHÜGERL, 1981).

Transport of medium components in and out of pellets or cellular aggregates is normally by molecular diffusion, and it involves two steps: (1) external mass transfer, i.e., transport of the species from the bulk liquid to the pellet surface, and (2) intraparticle diffusion, i.e., the molecular diffusion of species in the pores of a pellet. External mass transport is normally described with an expression similar to Eq. (16), i.e., as the product of a volumetric mass transfer coefficient (often called $k_s a$ in the case of transport to a solid particle) and the driving force (the difference between the concentration in the bulk liquid and the concentration at the pellet surface). This mass transfer term couples the mass balances for species in the bulk liquid with the reaction at the pellet surface.

With unidimensional molecular diffusion into a spherical pellet the transient mass balance for the i'th species becomes:

$$\frac{\partial c_i}{\partial t} = D_{i,\text{eff}} \frac{1}{z^2} \frac{\partial}{\partial z} \left(z^2 \frac{\partial c_i}{\partial z} \right) + q_i \tag{23}$$

where q_i is the volumetric rate of formation of species i, $D_{i,\text{eff}}$ is the effective diffusion coefficient of the species, and t may either be real time or the ratio between a volume and a volumetric flow (a residence time) for a distributed steady-state system. Instead of solving the partial differential equation to obtain the concentration profile in the pellet as a function of time, the effect of mass transfer on the overall conversion at any given time is normally quan-

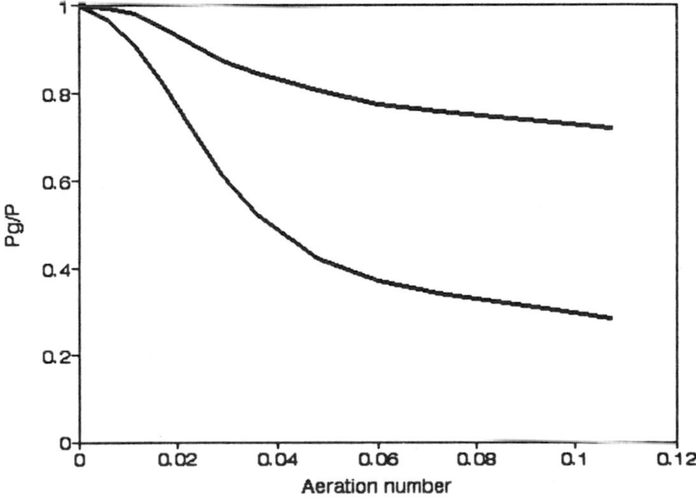

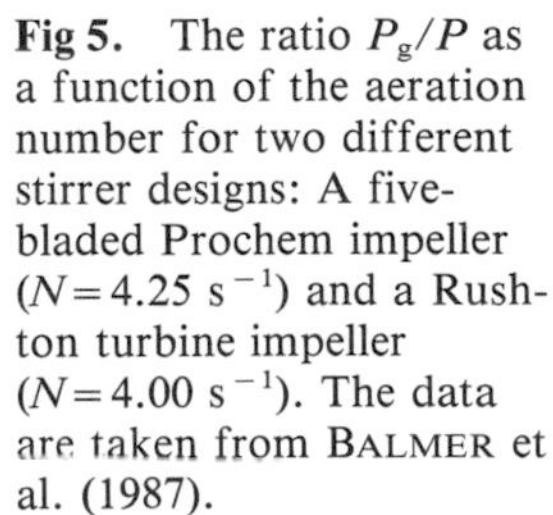

Fig 5. The ratio P_g/P as a function of the aeration number for two different stirrer designs: A five-bladed Prochem impeller ($N = 4.25\ \text{s}^{-1}$) and a Rushton turbine impeller ($N = 4.00\ \text{s}^{-1}$). The data are taken from BALMER et al. (1987).

tified by means of the so-called effectiveness factor $\eta_{\text{eff}}(t)$, which is defined as the ratio of the observed reaction rate and the reaction rate without mass transfer resistance, i.e.,

$$\eta_{\text{eff}} = \frac{q_{\text{i,obs}}}{q_{\text{i}}(c_{\text{i,bulk}})} \tag{24}$$

ARIS (1975) demonstrated that the effectiveness factor is satisfactorily approximated by Eq. (25) for any reasonable kinetics and any particle shape. Φ_{gen} is the so-called generalized Thiele modulus which is given by Eq. (26) where L is defined as the ratio between the pellet volume and the pellet exterior surface area.

$$\eta_{\text{eff}} = \frac{\tanh \Phi_{\text{gen}}}{\Phi_{\text{gen}}} \tag{25}$$

$$\Phi_{\text{gen}} = \frac{-q_{\text{i}}(c_{\text{i,bulk}})L}{\sqrt{2D_{\text{i,eff}} \int_0^{c_{\text{i,bulk}}} -q_{\text{i}}(c_{\text{i}})\text{d}c_{\text{i}}}} \tag{26}$$

For most design purposes it is sufficient to apply Eq. (25) to calculate the effectiveness factor and this reduces Eq. (23) to an ordinary differential equation which may subsequently be solved as a function of time or distance travelled through the apparatus. Intraparticle transport in cellular systems is fraught with unanswered questions – what is the effective diffusivity, and what is the effect of mass transport on the growth within a pellet? Thus, it does not appear to be reasonable to use a detailed model to calculate, e.g., the concentration profile in a pellet of *Penicillium chrysogenum* – at least not for design purposes.

3.2.3 Liquid Mixing

Microbial media are three-phase systems, i.e., they consist of a gas phase, a liquid phase, and a solid (biotic) phase; and "true" homogeneity can therefore not be obtained for the whole system. By rigorous mixing one may, however, obtain a "pseudohomogeneity" without any (or only very small) concentration gradients in either the gas or the liquid phase. This "ideal" situation is referred to as *maximal mixedness*, and it is obtained in the high-performance bioreactors described in Sect. 2.1. The other extreme of mixing is *total segregation* where no interaction occurs between different volume elements in the bioreactor. This can be illustrated by the ideal plug flow reactor. A system that does not exhibit any of these two extremes in behavior is called *partially segregated*.

The process of mixing can be divided into two parts. *Macromixing* is the distribution of volume elements into each other on a scale which is very small compared with the total reactor volume, but still large enough to contain thousands of cells and billions of substrate molecules. *Micromixing* is the exchange of material on the molecular level (or cell level) between these small volume elements. Macromixing is mainly caused by convection and turbulence, while micromixing occurs by molecular diffusion. Turbulence does, however, indirectly effect the micromixing due to its influence on the contact area between the volume elements.

Micromixing can be modelled in terms of the time t_{m} needed for mixing of a liquid phase to a certain degree of mixing m:

$$m = \frac{c - c_0}{c_\infty - c_0} \tag{27}$$

where c is the concentration at time t_{m}, c_0 the initial concentration, and c_∞ is the concentration for $t \to \infty$. In the literature one may find mixing times for different values of m, e.g., $m = 0.99$, but it is generally suggested to use the mixing time for $m = 1 - 1/e = 0.632$, since t_{m} hereby represents the reciprocal of the rate constant when mixing is approximated as a first-order process. Hereby, the mixing time becomes the characteristic time for the mixing process. Several correlations between mixing time in stirred tank reactors and the stirring speed N (s^{-1}) are given in the literature. For unaerated systems these correlations are of the type

$$t_{\text{m}} \propto \frac{1}{\text{N}} \tag{28}$$

and the proportionality constant is in the range of 8 to 24 (PEDERSEN, 1992). For aerated systems there is no well defined correlation between the mixing time and the stirring speed, but the mixing time is generally found to decrease with increasing gas flow rate (PEDERSEN, 1992).

For continuous flow systems the macromixing is characterized by the *residence time distribution* (*RTD*), where the residence time represents the time spent within the boundaries of the system, i.e., the time between inlet and outlet of a defined liquid volume. Except for an ideal plug flow reactor all volume elements leave the system with different residence times, and the resulting distribution gives information on the macromixing. Key characteristic parameters for the residence time distribution function are the mean residence time (or the space time) and the variance, which can be found from the first and the second moments of the distribution function, respectively. The residence time distribution is often modelled using the dispersion model or the tank-in-series model, both of which are one-parameter models. It is not to be expected that the residence time distribution in an arbitrary bioreactor design can be fitted with a one-parameter model, and various multiparameter models have to be used. The more complicated mixing models belong to either of the following four groups:

- *Loop models* are based on the assumption of a well mixed center and a number of loops through which the medium is circulated. The circulation of the medium through the loops is normally described with the tank-in-series model or with the dispersion model. An example of a multi-loop recirculation model has been given by VAN DE VUSSE (1962).
- *Compartment models* are based on a division of the total tank volume into a network of separately described sub-volumes which are assumed to be ideal reactors (CSTRs and plug flow reactors). The compartment number is normally low, i.e., below five. An example of a simple compartment model is that of OOSTERHUIS and KOSSEN (1983).
- *Network of zone-models* are extensions of the compartment model concept, where a very large number of compartments is used, e.g., 200 (MANN et al., 1981).
- *Turbulence models* are based on a complete set of balances for the system, i.e., there is a complete description of the hydrodynamics in the reactor. With these models it is, in principle, possible to calculate the flow velocities of any point in the tank, but a large hurdle in the application of these complex models is the problem of correct description of the bubble behavior in aerated systems. For a review of these models see, e.g., RANADE and JOSHI (1989).

For batch and fed-batch processes the residence time is the same for all volume elements in the reactor, and the residence time distribution can therefore not be used for quantification of macromixing. For these systems macromixing is characterized by the *circulation time distribution* (*CTD*), i.e., volume elements follow different circulation patterns in the tank. The *CTD* can be experimentally determined by means of a small, neutrally buoyant radio transmitter and a monitoring antenna placed in the tank. The *CTD* can be fitted to a normal distribution of $\ln(t)$ with mean $\mu = \overline{\ln(t)}$ and variance σ^2 (BRYANT, 1977). The mean circulation time is related to the mixing time, and the relationship $t_m = 5\, t_c$ can be used if no further information is available (MOSER, 1988).

The degree of mixing obtained with a given stirrer design is mainly determined by its pumping capacity v_{pump} (m^3/s). This quantifies the ability of the impellers to circulate the liquid around the bioreactor. The pumping capacity of an impeller is characterized by a flow number N_f for the medium, and v_{pump} is given by

$$v_{pump} = N_f N\, d_s^3 \tag{29}$$

For a Rushton turbine and a low-viscosity liquid $N_f = 0.72$, and this value can also be used as an approximation for other stirrer designs (NIENOW, 1990). The flow number is a function of the medium, and for a viscous medium (*CMC* solution with $\eta = 0.059$ kg/m s) CROZIER (1990) found $N_f = 0.43$. NIENOW (1990) reviewed the application of different stirrer designs for tank bioreactors. For high Re_s numbers he found the power numbers listed in Tab. 6 for three different designs. Going from one agitator design to another with a smaller power number, it is possible to obtain the same power input at a given stirring speed for a larger impeller diameter. Since the pumping capacity is a strong function of the impeller diameter, it is possible with a better stirrer design to obtain a higher pumping capacity for the same power input and stirring speed (NIENOW, 1990). Thus, with an Intermig impeller it is possible to have a diameter which is 72% greater than for the Rushton turbine impeller with the same power input and stirring speed. This gives a five-fold increase in the pumping capacity (if the flow number is assumed to be the same for the Rushton turbine and for the Intermig impeller). With the Intermig impeller a much better mixing may be obtained with the same power input compared to the Rushton turbine, and this is especially important for highly viscous media.

4 Performance of Standard Bioreactors

When the models for bioreaction kinetics and for the interfacial mass transport discussed in Sect. 3 are combined in mass balances, it becomes possible to study the performance of a given bioreactor design. Issues of general importance are: Which type of reactor should be preferred for a given bioreaction process? Is there a particular advantage attached to a certain operational mode? How robust is a certain mode of operation to process disturbances? These issues will be addressed in the following, and the reactor designs to be discussed are: (1) the stirred tank reactor and (2) the tubular reactor.

By far the most popular bioreactor design is the stirred tank reactor. It is very difficult to move the cell mass through the reactor in any fashion resembling plug flow – except perhaps in biological treatment of solid waste – and, therefore, the biomass or a solid substrate is usually present in a more or less homogeneous phase, either stirred internally in a tank or rapidly recirculated by means of external pumps in a loop reactor to give a better control of the environment of the microorganisms. Sparingly soluble substrates such as oxygen will, however, often move through the reactor in a complex pattern which can neither be classified as ideally mixed or plug flow. When in an attempt to reduce the substrate concentration to a very low level the medium is made to flow through a series of tanks, the mode of operation becomes a hybrid between well mixed and plug flow.

The complexities arising from non-ideal mixing in the stirred tank reactor cannot be included in a description of the reactor without completely clouding the general issues of comparing the different modes of operation. Likewise any attempt to include back-mixing and other non-idealities in the treatment of tubular reactors will introduce a host of parameter combinations which will anyhow be difficult to determine, unless the bioreaction kinetics is known with an unrealistic degree of accuracy. Consequently, the following treatment will assume that the tank reactor is ideal, i.e., the concentration of biomass, substrates, and products is the same in every point of the reactor and in the outlet. Also the tubular reactor model will be presented for a system which moves in plug flow with absolutely no mixing between neighboring elements – the total segregation of Sect. 3.2.3.

4.1 The Stirred Tank Reactor

The general mass balance for the ideal stirred tank is

$$\frac{\mathrm{d}(V\boldsymbol{c})}{\mathrm{d}t} = V(\boldsymbol{q}^{\mathrm{mt}} + \boldsymbol{q}) + v_{\mathrm{f}}\boldsymbol{c}_{\mathrm{f}} - v_{\mathrm{e}}\boldsymbol{c}_{\mathrm{e}} \tag{30}$$

where $\boldsymbol{c}$ is the vector of reactant concentrations, $\boldsymbol{q}^{\mathrm{mt}}$ is given by Eq. (16), and $\boldsymbol{q}$ is given by Eqs. (6)–(7). Reactants flow into the reactor ($v_{\mathrm{f}}\boldsymbol{c}_{\mathrm{f}}$) and leave the reactor ($v_{\mathrm{e}}\boldsymbol{c}_{\mathrm{e}}$). In the ideal stirred tank the reactant concentration $\boldsymbol{c}$ is the same in the reactor (in $\boldsymbol{q}$), in the accumulation term, and in the effluent $\boldsymbol{c} = \boldsymbol{c}_{\mathrm{e}}$; but in certain modes of operation the biomass concentration in the tank is higher than in the effluent (sedi-

mentation in the tank or recirculation of cells after separation from spent medium). Three different standard modes of operation for the tank reactor will be reviewed:
(1) Batch operation, i.e., no flow into or out of the reactor: $v_f = v_e = 0$.
(2) Continuous operation – the chemostat: $v_f = v_e \neq 0$.
(3) Fed-batch operation (or semi-batch in standard reaction engineering terms): $v_f \neq 0$, $v_e = 0$.

4.1.1 Batch Operation

For the case of a simple, unstructured kinetic model where the components of q are proportional, one obtains

$$q_x = \mu x \quad q_s = Y_{xs}\,\mu x \quad q_p = Y_{xp}\mu x \tag{31}$$

where μ is given by Eq. (3) and Y_{xs} and Y_{xp} are yield coefficients which express the mass of substrate (s) or metabolic product (p) associated with the formation of one unit biomass. When these expressions are inserted in Eq. (30) and the flow terms as well as the mass transfer term q^{mt} are set to zero, the following set of mass balances appears:

$$\frac{dx}{dt} = \mu x \qquad \frac{ds}{dt} = -Y_{xs}\,\mu x \qquad \frac{dp}{dt} = Y_{xp}\,\mu x \tag{32}$$

where $(x, s, p) = (x_0, s_0, p_0)$ at $t = 0$. When the rates are proportional, it is possible to reduce the three coupled ordinary differential equations (32) to a single differential equation in x and two simple algebraic equations:

$$\begin{aligned} \frac{dx}{dt} &= \mu x \\ s &= s_0 - Y_{xs}(x - x_0) \\ p &= p_0 + Y_{xp}(x - x_0) \end{aligned} \tag{33}$$

or with dimensionless variables

$$\begin{aligned} \frac{dX}{dt} &= \mu X \\ S &= 1 - Y_{xs}(X - X_0) \\ P &= P_0 + Y_{xp}(X - X_0) \end{aligned} \tag{34}$$

where

$$S = \frac{s}{s_0} \quad X = \frac{x}{Y_{sx}s_0} \quad P = \frac{p}{Y_{sp}s_0} \tag{35}$$

An analytical (or numerical) solution of the mass balances (33) is possible once the function μ has been specified. For the classical Monod kinetics

$$\mu = \frac{\mu_{max}s}{s + K_s} = \frac{\mu_{max}S}{S + a} \quad a = \frac{K_s}{s_0} \tag{36}$$

the solution to the differential equation in Eq. (34) is

$$\begin{aligned} \mu_{max}t = {} & \left(1 + \frac{a}{1 + X_0}\right) \ln\left(\frac{X}{X_0}\right) \\ & - \frac{a}{1 + X_0} \ln(1 + X_0 - X) \end{aligned} \tag{37}$$

The last term of Eq. (37) dominates the solution for large values of t, but more important for practice is that the major part of the batch reaction is described quite accurately by the first term. Typically, the saturation constant K_s is much smaller than the initial substrate concentration s_0, i.e., $a \ll 1$, and in most cases it is found that for about 90–95% of the total batch operation the solution (37) is indistinguishable from (NIELSEN and VILLADSEN, 1993)

$$\mu_{max}t = \ln\left(\frac{X}{X_0}\right) \tag{38}$$

which describes the well known *exponential growth phase*.

Other forms of $\mu(s)$ can be used in place of the Monod kinetics Eq. (35). Thus substrate inhibition (i.e., a decrease of the substrate uptake rate for substrate concentrations higher than a certain value – typically observed for the oxygen uptake and also for the uptake of many organic substrates) can be modelled by Eq. (39), and an inhibition due to a metabolic product (the negative influence of ethanol on the growth of yeast is the best known example) is typically modelled by Eq. (40).

Substrate inhibition:

$$\mu = \frac{\mu_{max} S}{bS^2 + S + a} \tag{39}$$

Product inhibition:

$$\mu = \frac{\mu_{max} S}{S + a}\left(1 - \frac{P}{P_{max}}\right) \tag{40}$$

In Eq. (39) $b = s_0/K_i$ is a parameter which describes the severity of the substrate inhibition, and in Eq. (40) P_{max} is a maximum product level at which growth stops.

As long as the rates are proportional, both S and P can be found from Eq. (34) as functions of X, and the cell mass balance is easily integrated. Analytical integration of Eq. (30) is, however, not possible when q_s and q_p are linear functions of μ rather than proportional to μ as in Eq. (31). In Eq. (41) typical rate expressions which include a *maintenance* term besides the growth-related term are shown:

$$\begin{aligned} q_s &= -(Y_{xs}\mu + m_s)x \\ q_p &= (Y_{xp}\mu + m_p)x \end{aligned} \tag{41}$$

Quite often the constants m_s and m_p are related by simple stoichiometric relations, e.g., for homofermentative lactic acid fermentation the extra glucose used for maintenance of the cells is quantitatively converted to the product lactic acid and $m_s = m_p$.

Rate expressions derived from structured kinetic models can also be used in Eq. (30). An extra mass balance, Eq. (5), for the cell composition has to be integrated together with mass balances for $\boldsymbol{x}$, $\boldsymbol{s}$, and $\boldsymbol{p}$, but apart from the fact that a numerical solution has to be used (and this is a routine job on a PC equipped with standard software), the use of structured models does not introduce any new difficulties compared with Eq. (32).

Both for scientific studies and for production the main advantage of the batch reactor is that the process is easy to set up and easy to run. The initial conditions – sterilization of the medium, inoculation, etc. – can be standardized, and the operating conditions are easily controlled without danger of infection when proper sampling procedures are used. In the pre-production phase a number of strains and/or operational conditions (e.g., different medium compositions) can easily be screened in parallel batch reactors. For production purposes the batch reactor may be less suitable. There is usually a lag phase (which is not explained by any unstructured kinetic model, but easily simulated even with a crude structured model, NIELSEN and VILLADSEN, 1993), and a high biomass concentration is obtained only at the end of the exponential growth phase where the batch reaction abruptly stops due to substrate depletion. If a high cell density is desired, the continuous reactor is much more effective, and if a secondary (non-growth associated) metabolite is the desired product, then a fed-batch process is mandatory. Finally it should be noted that the batch reactor is unsuited for scientific studies of the metabolism, since the cell growth is at balanced conditions almost throughout the whole process. No fine details in the substrate uptake can be discerned, and usually it is not even possible to find the saturation constant K_s in a crude unstructured model, because there are too few measurements available in the short time interval (5–10 minutes) between exponential growth and no growth at all.

4.1.2 Continuous Operation – the Chemostat

With no mass transfer terms and $v = v_f = v_e$, one obtains the following mass balances from Eq. (30):

$$\begin{aligned} \frac{dx}{dt} &= q_x + D(x_f - x) \\ \frac{ds}{dt} &= q_s + D(s_f - s) \\ \frac{dp}{dt} &= q_p + D(p_f - p) \end{aligned} \tag{42}$$

where $D = v/V$ is the *dilution rate* in the bioreactor. Usually both x_f and p_f are equal to zero. The dimensionless variables, Eq. (35), (with s_f rather than s_0 as scale factor) are introduced. For Monod kinetics, Eq. (36), and steady-state conditions the following algebraic equations result when $x_f = p_f = 0$:

$$D = \frac{\mu_{max} S}{S + a} \quad X = 1 - S \quad P = 1 - S$$
$$S = \frac{s}{s_f} \quad X = \frac{x}{Y_{sx} s_f} \quad P = \frac{p}{Y_{sp} s_f} \tag{43}$$

The steady-state chemostat will cease to function when $D > \max(\mu)$, since the rate of production of biomass is smaller than the rate at which the biomass is washed out of the reactor. For Monod kinetics, $D_{max} = \mu_{max}/(1+a)$, a value only slightly smaller than μ_{max} since a is usually quite small. For any value of $D < D_{max}$, the steady-state design of a chemostat based on Monod kinetics is obtained from Eq. (43). There is only one design variable which may be D, S, X, or P. Maximum productivity of biomass or product or some other design criterion can be satisfied by suitable manipulations with the model. Thus, the maximum of DX or DP, i.e., the maximum of $F(S) = S(1-S)/(S+a)$ when Eq. (36) is used for the kinetics, yields the value of S (the outlet substrate concentration) for which maximum productivity is obtained.

The possibility of having a biomass concentration x different from the effluent biomass concentration x_e opens the possibility for a higher productivity in a given reactor volume. With rates given by Eq. (31) one obtains

$$\mu x = D x_e \quad Y_{xs} \mu x = D(s_f - s_e) \quad Y_{xp} \mu x = D p_e \tag{44}$$

A filter installed internally or outside the reactor, or a cell centrifuge is a means of operating with $x > x_e$ while $s_e = s$ and $p_e = p$. Let $f = x_e/x < 1$ due to a hydrocyclone or a centrifuge with separation factor $\beta = x_R/x > 1$. Then from mass balances taken for the reactor and separation unit (NIELSEN and VILLADSEN, 1993):

$$Df = \mu \quad x = \frac{x_e}{1 - R(\beta - 1)} \quad \alpha = \frac{\beta R}{1 + R} \tag{45}$$

Eq. (45) shows that for given values of x_e and s_e, which are related by the total mass balance, Eq. (33), for the combined system, one is able to work with a high volumetric throughput $D = v/V$, if a recycle stream $R = v_r/v$ with a cell concentration x_R higher than x is returned to the reactor. The fraction of produced cells which is returned is given by α. In theory recycling works well, but operation with a high cell density x in the reactor and an even higher cell density x_R in the recycle stream may create a hostile environment for the cells, and if there is a noticeable maintenance demand, then a relatively high substrate consumption for this purpose is unavoidable.

The solution to the unsteady-state balances, Eq. (42), is of considerable interest both for modelling the control of a steady state and because steady-state processing is always preceded by a start-up period. A batch is inoculated and is run for a certain time in which a sufficiently high cell concentration is built up to approach the conditions of high cell density and low substrate concentration which is characteristic of continuous chemostat operation. Unless the process dynamics is modelled with a reasonable accuracy, it may be very tricky to preprogram the switch-over from batch to continuous operation, and the control parameters cannot be properly set unless the (often very high) time constants of the cell processes are estimated. In both cases suboptimal continuous operation results (large fluctuations in outlet conditions or even complete wash-out of the culture).

A step in dilution rate brings the culture from one steady state to another with different outlet concentrations of biomass, substrate and product. If the step is made from just below a dilution rate where there is a shift in metabolism to a higher value (e.g., from a dilution rate where no appreciable ethanol is produced by *Saccharomyces cerevisiae* to a value where significant amounts of ethanol are produced by an overflow metabolism), then important information on the synthesis of key enzymes can be obtained.

A step in feed concentration s_f at constant D will not lead to any permanent change in the effluent substrate concentration, but the biomass concentration will change. This type of experiment is, therefore, well suited to test whether the cellular activity is independent of the biomass concentration – and at the same time the efficiency of the sample retrieval system is also being tested. If a step-up in feed concentration is made at a very low D value, the transient overshoot in effluent substrate concentration may be used in the same way as pulse experiments in a study of regulatory bottlenecks in the pathways.

Finally, a substrate pulse addition is an excellent means of testing the stability of a given steady state, and also the rate of consumption of the pulse may be used together with measurements of key metabolites to study the steady-state flux distribution at branches in the metabolic pathways. Thus, a pulse of glucose added to a slow growing culture of *S. cerevisiae* will lead to a redistribution of energy flux between the Embden–Meyerhof–Parnas pathway and the pentose phosphate pathway, and the ATP level of the cells will stay low (due to consumption to build biomass precursors) long after the pulse has been metabolized.

The chemostat is the ideal system for Physiological Engineering Studies. The steady state is used to obtain reaction kinetics at external conditions which can be maintained for many generations of the cells. Any long-time change in the outlet composition can be ascribed to real physiological phenomena – a genetic adaptation of the microorganism to the specific medium, or a loss of a property, e.g., the ability to over produce a certain enzyme due to reversion of a genetically engineered microorganism to the wild type. Pulse experiments of short duration are used to study the fast response of key metabolic pathways by measurement of transient metabolite concentrations, and the gradual change of cell composition with a time constant of 2–3 h is also measurable from step-up experiments in dilution rate or substrate pulse experiments (NIELSEN et al., 1991b). Some of the characteristic properties of chemostats – especially their response to disturbances – are, however, troublesome in an industrial environment. They are able to produce high concentrations of biomass and product at a very high conversion of substrate, but the low specific growth rate prevailing at these conditions also makes the reactor very sensitive to disturbances – wash-out is never far away from the optimal operation conditions. An expectation of a long period of continuous production may be abruptly curtailed, if even for a short period the feed contains a trace of a contaminant which at the conditions prevailing in the reactor grows faster than the desired strain. Then wash-out of the desired strain is the unavoidable final result (NIELSEN and VILLADSEN, 1993).

4.1.3 Fed-Batch Operation

Some of the most important bioproducts are almost exclusively manufactured in stirred tanks where after a certain batch growth period the production of the desired species (baker's yeast, antibiotics, recombinant proteins, hormones, or biopolymers) is initiated. At the time of switch-over from batch to fed-batch operation, the biomass density has reached a reasonably high level and the concentration of the key substrate has decreased to almost zero. This low level of the limiting substrate is retained by slow addition of a feed stream with a constant, high substrate concentration. Again the general model, Eq. (30), is used to obtain the mass balance ($v_e = 0$):

$$\frac{d(V\boldsymbol{c})}{dt} = V\frac{d\boldsymbol{c}}{dt} + \boldsymbol{c}\frac{dV}{dt} = V\frac{d\boldsymbol{c}}{dt} + v_f(t)\boldsymbol{c} = V\boldsymbol{q} + v_f(t)\boldsymbol{c}_f \tag{46}$$

or

$$\frac{d\boldsymbol{c}}{dt} = \boldsymbol{q} + \frac{v_f(t)}{V(t)}(\boldsymbol{c}_f - \boldsymbol{c}) \tag{47}$$

Eq. (47) is formally the same as Eq. (42), the transient mass balance for the chemostat, but now $v(t)/V(t)$ changes with time according to a given production strategy:

- keep the substrate concentration at a constant, low level
- keep a certain volumetric production rate, e.g., q_x below a certain level.

The first criterion is the most frequently applied: If biomass is the desired product, then a parallel production of an overflow metabolite (e.g., ethanol or acetic acid) should be minimized. The second criterion is applicable if the capacity for supply of a key substrate is limited (e.g., oxygen in aerobic processes) or if removal of the heat of reaction is a problem.

Both production criteria can be studied using the analytical solution of Eq. (47) for simple kinetic expressions such as those in Eq. (31). Thus, for $s(t)=s_0$ the feed strategy $v(t)$ and the increase of biomass concentration and of medium volume are given by (NIELSEN and VILLADSEN, 1993):

$$v(t)=b\mu(s_0)x_0V_0\exp(\mu(s_0)t) \tag{48}$$

$$x(t)=\frac{x_0\exp(\mu(s_0)t)}{1-bx_0+bx_0\exp(\mu(s_0)t)} \tag{49}$$

$$V(t)=V_0(1+bx_0(\exp(\mu(s_0)t)-1)) \tag{50}$$

Here x_0 and V_0 are the biomass concentration and the volume at the start of the fed-batch operation, i.e., at $t=0$. The parameter $b=Y_{xs}/(s_f-s_0)$ and it is seen that with the exponential feed addition, Eq. (48), the biomass concentration increases from x_0 at $t=0$ to $1/b=Y_{sx}(s_f-s_0)$ at $t\rightarrow\infty$. In principle, x_0 can be very small whereby virtually no substrate is diverted to (an undesirable) metabolic product in the batch phase which precedes the fed-batch, and a high overall yield of biomass may therefore be obtained.

The so-called *repeated fed-batch* operation consists of series of fed-batch periods with removal of a portion of the reactor content at certain time intervals. The biomass concentration can now be constantly kept at a high level, and the productivity of the reactor is high. This mode of operation is, however, victimized by an undesirable accumulation of potentially toxic metabolites in the medium, and the culture is almost certain to become senescent. An alternative is to insert a batch growth period after each emptying of the tank, i.e., to dilute the reactor contents with substrate before restart of the process. Now the culture is more resistant, and the extended batch/fed-batch operation can continue much longer than a single fed-batch process (e.g., 600 hours of sustained penicillin production versus only 200 hours for a single fed-batch operation).

4.2 The Tubular Reactor

The continuous operation of a microbial process in a tubular reactor is often – by analogy with "normal chemical processes" with positive-order kinetics – mistakenly supposed to be the optimal solution. Considering the autocatalytic nature of microbial processes this is generally far from the truth. First, the tubular reactor requires that cells are introduced with the feed. This is certainly no problem, if agricultural waste is to be processed to cattle feed in a primitive conveyer type apparatus, but for submerged processes in an industrial environment a sterile feed is an absolute requirement. Second, the continuous tank reactor has a much higher productivity than the plug flow reactor (and also the fed-batch reactor) unless the substrate level is very low. A combined reactor system with a plug flow reactor downstream of a tank reactor can be an attractive solution, if extremely low substrate effluent concentrations are desired (e.g., removal of pesticides from the wastewater of a malathion factory).

The design formula for the steady-state plug flow tubular reactor is derived from Eq. (30):

$$\frac{\mathrm{d}c}{\mathrm{d}\theta}=q \tag{51}$$

where

$$\mathrm{d}\theta=\frac{\mathrm{d}V}{v} \tag{52}$$

is the time it takes a fluid element to travel through the fraction $\mathrm{d}V$ of the reactor volume. In Eq. (51) it is supposed that at the inlet to the reactor there is some biomass present. This is provided by a recycle loop from the reactor exit – either directly or through a separation process in a hydrocyclone or a centrifuge. Integration of Eq. (51) is from this inlet condition (x_f, s_f, p_f) at $\theta=0$ to $\theta=V/v$, where V is the total reactor volume. This gives formally the same result as that obtained by integration of the non-steady-state batch reactor model, but there is certainly no guarantee that the cells follow the same metabolism in the plug flow reactor design as they do in the a very different reactor construction with intimate mixing of the reaction medium and possibly with a much better dispersion of gaseous substrates.

4.3 Conclusion

Of the reactor designs which are shortly reviewed in this chapter the chemostat is probably endowed with the best qualities for microbial reactions. It is far the best design for fundamental studies, and in an idealized world it would also be the most economical apparatus for industrial processes. Lack of deep insight into the process model and worries about potential unstable operation or possible sources of infection will, however, tend to count against the use of the chemostat in an industrial environment – but various types of loop reactors with very high retention times for liquid substrates and biomass are seen to win acceptance for new processes. The fed-batch process is the typical work-horse in the bio-industry, whereas the plug flow reactor has little to speak for itself, and the batch reactor with its high labor costs and low production capacity is recommended mostly for screening studies of for once-only productions.

5 References

AGRAWAL, P., LIM, H. C., RAMKRISHNA, D. (1983), An extended bottleneck model accounting for the metabolic turnover effect in microorganisms, *J. Chem. Technol. Biotechnol.* **33B**, 155–163.

ALLAN, D. G., ROBINSON, C. W. (1990), Measurement of rheological properties of filamentous fermentation broths, *Chem. Eng. Sci.* **45**, 37–48.

ARIS, R. (1975), *The Mathematical Theory of Diffusion and Reaction in Permeable Catalysts*, Vol. 1: *The Theory of the Steady State*, Oxford: Clarendon Press.

BALMER, G. J., MOORE, I. P. T., NIENOW, A. W. (1987), Aerated and unaerated power and mass transfer characteristics of Prochem agitators, in: *Biotechnology Processes: Scale-up and Mixing* (HO, C. S., OLDSHUE, J. Y., Eds.) pp. 116–127, New York: American Institute of Chemical Engineers.

BENTLEY, W. E., KOMPALA, D. S. (1989), A novel structured kinetic modeling approach for the analysis of plasmid instability in recombinant bacterial cultures, *Biotechnol. Bioeng.* **33**, 49–61.

BRYANT, J. (1977), The characterization of mixing in fermenters, *Adv. Biochem. Eng.* **5**, 101–123.

CHRISTENSEN, L. H., NIELSEN, J., VILLADSEN, J. (1991), Delay and dispersion in an *in-situ* membrane probe for bioreactors, *Chem. Eng. Sci.* **46**, 3304–3307.

COPPELLA, S. J., DHURJATI, P. (1990), A mathematical description of recombinant yeast, *Biotechnol. Bioeng.* **35**, 356–374.

CROZIER, D. B. A. (1990), A detailed study of the flow characteristics in a tall stirred tank, in: *Laser Anemometry – Proc. 2nd Int. Conf.*, pp. 359–368, Heidelberg–New York: Springer-Verlag.

DECKWER, W.-D. (1992), *Bubble Column Reactors*, Chichester: John Wiley & Sons.

HARDER, A., ROELS, J. A. (1982), Application of simple structured models in bioengineering, *Adv. Biochem. Eng.* **21**, 55–107.

HJORTSO, M. A., BAILEY, J. E.. (1982), Steady-state growth of budding yeast populations in well-mixed continuous-flow microbial reactors, *Math. Biosci.* **60**, 235–263.

HJORTSO, M. A., BAILEY, J. E. (1984), Plasmid stability in budding yeast populations: Steady state growth with selection pressure, *Biotechnol. Bioeng.* **26**, 528–536.

JOHANSEN, C. L., CHRISTENSEN, L. H., NIELSEN, J., VILLADSEN, J. (1992), Monitoring and control of fed-batch penicillin fermentation, *Comp. Chem. Eng.* **16**, S297–S304.

KOMPALA, D. S., RAMKRISHNA, D., TSAO, G. T. (1984), Cybernetic modelling of microbial growth on multiple substrates, *Biotechnol. Bioeng.* **26**, 1272–1281.

KOSSEN, N. W. F., OOSTERHUIS, N. M. G. (1985), Modelling and scaling-up of bioreactors, in: *Biotechnology* (REHM, H.-J., REED, G., Eds.), Vol. 2, pp. 571–605, Weinheim: VCH.

LEE, S. B., BAILEY, J. E. (1984), Analysis of growth rate effects on productivity of recombinant *Escherichia coli* populations using molecular mechanism models, *Biotechnol. Bioeng.* **26**, 1372–1382.

LINEK, V., VACEK, V., BENES, P. (1987), A critical review and experimental verification of the correct use of the dynamic method for the determination of oxygen transfer in aerated agitated vessels to water, electrolyte solutions and viscous liquids, *Chem. Eng. J.* **34**, 11–34.

LOCHNER, G. (1991), Bioprocess automation: Equipment, methodology and benefits, *Ph. D. Thesis*, ETH Zurich, Switzerland.

LOCHNER, G., SONNLEITNER, B., FIECHTER, A. (1992), Software and implementation for automated decision making in bioprocess control, *Proc. Control Qual.* **2,** 257–274.

LORENZ, T. (1985), Grundlagen zur Penicillinproduktion in Schlaufenreaktoren, *Ph. D. Thesis,* Universität Hannover, Germany.

MANN, R., MAVROS, P. P., MIDDLETON, J. C. (1981), A structured stochastic flow model interpreting flow follower data from a stirred vessel, *Trans. Inst. Chem. Eng.* **59,** 271–278.

MEGEE, R. D., KINISHITA, S., FREDRICKSON, A. G., TSUCHIYA, H. M. (1970), Differentiation and product formation in molds, *Biotechnol. Bioeng.* **12,** 771–801.

MOO-YOUNG, M., BLANCH, H. W. (1981), Design of biochemical reactors: Mass transfer criteria for simple complex systems, *Adv. Biochem. Eng.* **19,** 1–69.

MOSER, A. (1988), *Bioprocess Technology,* New York: Springer-Verlag.

MÜNCH, T., SONNLEITNER, B., FIECHTER, A. (1992), The decisive role of the *Saccharomyces cerevisiae* cell cycle behaviour for dynamic growth characterization, *J. Biotechnol.* **22,** 329–352.

NIELSEN, J. (1992), On-line monitoring of microbial processes by flow injection analysis, *Proc. Control Qual.* **2,** 371–384.

NIELSEN, J. (1993), A simple morphologically structured model describing the growth of filamentous microorganisms, *Biotechnol. Bioeng.* **41,** 715–727.

NIELSEN, J., VILLADSEN, J. (1992), Modelling of microbial kinetics, *Chem. Eng. Sci.* **47,** 4225–4270.

NIELSEN, J., VILLADSEN, J. (1993), *Bioreaction Engineering Principles,* (textbook, preliminary version), Lyngby: Technical University of Denmark.

NIELSEN, J., NIKOLAJSEN, K., VILLADSEN, J. (1991a), Structured modelling of a microbial system 1. A theoretical study of the lactic acid fermentation, *Biotechnol. Bioeng.* **38,** 1–10.

NIELSEN, J., NIKOLAJSEN, K., VILLADSEN, J. (1991b), Structured modelling of a microbial system 2. Experimental verification of a structured lactic acid fermentation model, *Biotechnol. Bioeng.* **38,** 11–23.

NIELSEN, J., PEDERSEN, A. G., STRUDSHOLM, K., VILLADSEN, J. (1991c), Modelling fermentations with recombinant microorganisms: Formulation of a structured model, *Biotechnol. Bioeng.* **37,** 802–808.

NIENOW, A. W. (1990), Agitators for mycelial fermentations, *TIBTECH* **8,** 224–233.

NIENOW, A. W., ELSON, T. P. (1988), Aspects of mixing in rheologically complex fluids, *Chem. Eng. Res. Des.* **66,** 5–15.

OOSTERHUIS, N. M. G., KOSSEN, N. W. F. (1983), Oxygen transfer in production scale bioreactor, *Chem. Eng. Res. Dev.* **61,** 308–312.

PEDERSEN, A. G. (1992), Characterization and modelling of bioreactors, *Ph. D. Thesis,* Lungby: Technical University of Denmark.

PEDERSEN, A. G., BUNDGÅRD, M., NIELSEN, J., VILLADSEN, J., HASSAGER, O. (1993), Rheological characterization of media containing *Penicillium chrysogenum, Biotechnol. Bioeng.* **41,** 162–164.

PERETTI, S. W., BAILEY, J. E. (1987), Simulations of host-plasmid interactions in *Escherichia coli:* Copy number, promoter strength, and ribosome binding site strength effects on metabolic activity and plasmid gene expression, *Biotechnol. Bioeng.* **29,** 316–328.

RAMKRISHNA, D. (1979), Statistical models of cell populations, *Adv. Biochem. Eng.* **11,** 1–47.

RAMKRISHNA, D., FREDERICKSON, A. G., TSUCHIYA, H. M. (1967), Dynamics of microbial propagation: Models considering inhibitors and variable cell composition, *Biotechnol. Bioeng.* **9,** 129–170.

RANADE, V. V., JOSHI, J. B. (1989), Flow generated by pitched blade turbines: I. Measurements using laser doppler anemometer, *Chem. Eng. Commun.* **81,** 197–224.

ROELS, J. A. (1983), *Energetics and Kinetics in Biotechnology,* Amsterdam: Elsevier Biomedical Press.

ROELS, J. A., VAN DEN BERG, J., VONCKEN, R. M. (1974), The rheology of mycelial broths, *Biotechnol. Bioeng.* **16,** 181–208.

SCHÜGERL, K. (1981), Oxygen transfer into highly viscous media, *Adv. Biochem. Eng.* **19,** 71–174.

SCHÜGERL, K. (1988), On-line analysis and control of production of antibiotics, *Anal. Chim. Acta* **213,** 1–9.

SCHÜGERL, K. (1991), *Bioreaction Engineering,* Vol. 2, New York: John Wiley & Sons.

SCHÜGERL, K., LÜCKE, J., OELS, U. (1977), Bubble column bioreactors, *Adv. Biochem. Eng.* **7,** 1–84.

SEO, J.-H., BAILEY, J. E. (1985), A segregated model for plasmid content on growth properties and cloned gene products formation in *Escherichia coli, Biotechnol. Bioeng.* **27,** 1668–1674.

SHULER, M. L., DOMACH, M. M. (1982), Mathematical models of the growth of individual cells, in: *Foundations of Biochemical Engineering: Kinetics and Thermodynamics in Biological Systems,* pp. 93–133, Washington, DC: American Chemical Society.

SONNLEITNER, B., FIECHTER, A. (1988), High-performance bioreactors: A new generation, *Anal. Chim. Acta* **213**, 199–205.

SONNLEITNER, B., FIECHTER, A. (1992), Impacts of automated bioprocess systems on modern biological research, *Adv. Biochem. Eng.* **46**, 143–159.

STEINMEYER, D. E., SHULER, M. L. (1989), Structured model for *Saccharomyces cerevisiae, Chem. Eng. Sci.* **44**, 2017–2030.

SWEERE, A. P. J., GIESSELBACH, J., BARENDSE, R., DE KRIEGER, R., HONDERD, G., LUYBEN, K. CH. A. M. (1988), Modelling the dynamic behaviour of *Saccharomyces cerevisiae* and its application in control experiments, *Appl. Microbiol. Biotechnol.* **28**, 116–127.

VAN DEDEM, G., MOO-YOUNG, M. (1975), A model for diauxic growth, *Biotechnol. Bioeng.* **17**, 1301–1312.

VAN DE VUSSE, J. G. (1962), A new model for stirred tank reactor, *Chem. Eng. Sci.* **17**, 507–521.

VAN'T RIET, K. (1979), Review of measuring methods and results in non-viscous gas–liquid mass transfer in stirred vessels, *Ind. Chem. Process Dev.* **18**, 357–364.

WILLIAMS, F. M. (1967), A model of cell growth dynamics, *J. Theor. Biol.* **15**, 190–207.

6 Cell Culture Bioreactors

ATHANASSIOS SAMBANIS

Atlanta, GA 30332-0100, USA

WEI-SHOU HU

Minneapolis, MN 55455, USA

1 Introduction

Only a decade ago, the use of cultured animal cells was limited to production of human and animal vaccines and few other biologicals. Currently, however, animal cells are used for the production of several protein therapeutics because of their unique abilities to perform post-translational modifications on endogenous and recombinant proteins. Complex molecules, such as tissue plasminogen activator (tPA), erythropoietin (EPO), Factor VIII, and Protein C, require post-translational modifications for biological activity and adequate stability in the blood stream of patients. Such modifications involve, but are not limited to, disulfide bond formation, glycosylation, proteolysis, phoshorylation, sulfation, amidation, and acetylation (MAINS et al., 1987). Animal cell culture products that have reached the market include, besides human and animal vaccines which have existed for decades, various monoclonal antibodies for diagnostic use, tPA and EPO. Products currently under clinical trials include therapeutic anti-AIDS virus agents, Factor VIII, and monoclonal antibodies for imaging and therapeutic use.

Microbial cells are useful as host organisms when either the protein does not require post-translational modifications for activity or when the inactive protein can be chemically or enzymatically processed to active product. An example of the former is the production of unglycosylated γ-interferon in *Escherichia coli* for treatment of humans. The latter is true primarily for low-molecular weight polypeptide hormones. For example, in a process developed by Eli Lilly for the manufacturing of insulin, each of the two insulin chains is expressed in a different strain of recombinant *E. coli*. Following solubilization of inclusion bodies and protein unfolding, the two chains are chemically combined to form biologically active insulin (JOHNSON, 1983). In an alternative scheme developed by Novo, proinsulin expressed in yeast and secreted as such is cleaved proteolytically to mature insulin in a single enzymatic conversion step (BARFOED, 1987). With regard to complex molecules, progress has been made in assembling immunoglobulin fragments in bacterial (SKERRA and PLÜCKTHUN, 1988; CONDRA et al., 1990) or plant systems (HIATT et al., 1989); however, the processes are still inefficient (VAN BRUNT, 1990), and animal cells remain indispensable for production of antibodies.

Mammalian cell culture poses significant challenges for a number of reasons. Animal cells lack a rigid cell wall and are thus more susceptible than microorganisms to shear and sparged aeration. Cells may require an appropriate surface on which they can attach and spread in order to grow and function properly. Depletion of nutrients and accumulation of metabolites generally inhibits cell growth and productivity. Cell culture medium is expensive, so efficient medium utilization is essential for reasonable product cost. Animal cells grow at slow rates and in many cases secrete products in a growth-independent fashion; retainment of high densities of slow-growing but live, functional cells could thus constitute a suitable approach for several production applications.

Mammalian cell bioreactors are used for both research and production purposes. The intense competition among biotechnology companies to produce sufficient amounts of new proteins for clinical trials is one reason why basic and well-tested bioreactor designs are most often used at production scale. At present, novel bioreactors are evaluated primarily at the lab scale (HU and PESHWA, 1991). Innovative designs aim at increasing cell density (for increased product titers), cell viability, and productivity, by providing an improved chemical and hydrodynamic environment to cells. It is expected that, when competition shifts from synthesis of new products to optimized manufacture of existing products, novel, improved bioreactor designs will be implemented at the large scale as well.

Besides mammalian cells, cells derived from insects have been employed to produce recombinant proteins (LUCKOW and SUMMERS, 1988). These cells lack some abilities for post-translational modifications that exist in mammalian cells. The patterns of protein glycosylation in insect cells also differ from those synthesized in mammalian cells (GOOSEN, 1991). Nevertheless, in cases where these considerations are not important, insect cell culture is an attractive means of production.

In this chapter, we present a comprehensive

description and critical analysis of basic bioreactor configurations used for animal cell culture. Cell culture is first discussed at the microscale, i.e., with regard to the requirements for cultivating single cells or cell aggregates in suspension, cells on solid or porous microcarriers, and cells entrapped in biocompatible polymers. Bioreactor design utilizes the various microscale techniques for large-scale culture. For instance, cells grown on porous microcarriers can be propagated in stirredtank bioreactors with or without retention devices, or in packed bed perfusion units. Bioreactor configurations are thus described after microscale culture techniques are discussed.

2 Culture at the Microscale

Depending on their requirements for growth, animal cells are divided into two major categories. *Suspension cells* can be grown as single units in a medium that provides all necessary nutrients. *Adherent cells* require, in addition to nutrient medium, an appropriate surface on which they can attach in order to maintain metabolic activity and grow. Cell adhesion to a surface can be thought of as involving the following steps: contact of cells and surface; attachment of cells to surface mediated via adhesion molecules provided by the medium (e.g., serum fibronectin) and/or produced by the cells; and spreading of attached cells. Adherent cells may express a greater or lesser degree of surface dependence. Normal diploid primary cells derived from tissues usually have a strict requirement for adhesion and die when no proper attachment surface is provided. Tumor cells, on the other hand, are less adhesion-dependent compared to their normal, primary counterparts. In the absence of an attachment surface, certain tumor cell lines may form multicellular aggregates (*spheroids*) of various sizes and grow as such. Some normal diploid cells may also form aggregates in culture.

Suspension cells and cell aggregates can be cultured suspended in medium. Adherent cells can be cultured on surfaces that are flat or curved and of various chemical compositions. Both suspension and adherent cells can be immobilized in macroporous polymeric or ceramic matrices, or entrapped in gels or capsules. Each of these different configurations may offer distinct advantages in terms of cell growth, cell metabolism, expression of specific cell properties, especially differentiated ones, and achievable cell densities in a certain bioreactor.

Insect cells are typically cultured in suspension. Manufacturing is commonly carried out in batch mode, since the production of protein is usually initiated by virus infection and terminated after a short product formation period. Therefore, the reactors described below for suspensions of mammalian cells are also applicable to insect cells. Currently, the majority of insect cells are cultured in simple stirred tank and air-lift bioreactors.

2.1 Solid Microcarriers

Culturing of adherent animal cells on solid microcarriers was first introduced by VAN WEZEL (1967). The technology matured and found extensive applications in the late 70s and early 80s (MEIGNER, 1978, VAN WEZEL et al., 1979). The original microcarriers were based on cross-linked dextran. Subsequent developments have concentrated on the preparation of surface-derivatized, dextran-based carriers, and on employing other materials, such as polystyrene (KUO et al., 1981; JOHANSSON and NIELSEN, 1980), cellulose (REUVENY et al., 1982), gelatin (VARANI et al., 1985, NILSSON et al., 1988), and glass (VARANI et al., 1983).

Microcarrier particles range from 100–230 µm in diameter. Smaller particles have curvatures too big for effective cell spreading, and larger particles are difficult to keep in suspension. It is desirable that the microcarrier optical properties allow microscopic observation of cells on them, so cell attachment and spreading can be easily monitored. For suspension cultures, the microcarrier density should be slightly higher than that of the surrounding medium, i.e., around 1.04 g/mL. Surface areas per unit culture volume recommended for optimal cell growth vary with the type of cells and microcarriers considered. Research in microcarrier technology aims at improving the surface properties for cell attachment and growth, at simplifying preparation procedures for culture, and at reducing cost. Examples include the development of polystyrene, glass and ceramic microcarriers which cost significantly less than dextran-based carriers and do not require hydration, and the development of surface-derivatized (e.g., sulfonated or amino-modified polystyrene) and surface-coated (e.g., collagen-coated) microcarriers for improved cell adhesion.

A microcarrier culture is typically inoculated with cells detached by trypsinization from another surface culture. For the culture to succeed, cells should exhibit a sufficiently high attachment rate to the surface of particles. If attachment does not occur or occurs very slowly, cells may die in suspension or form aggregates. Depending on the cell type and culture conditions, cells may form monolayers (e.g., FS-4 cells; HU et al., 1985) or multilayers on the microcarriers (e.g., Chinese Hamster Ovary (CHO) cells; CROUGHAN and WANG, 1990), or even bridges between particles resulting in cell/microcarrier aggregates of various sizes (e.g., bovine embryonic kidney cells; CHERRY and PAPOUTSAKIS, 1988). Serial propagation of cultures requires that cells be dissociated from confluent microcarriers without significant loss of viability, and then be seeded into a new culture. Detachment is accomplished with proteolytic enzymes, treatment with a calcium chelator such as EDTA, or a combination of the two procedures. It is recommended that new microcarriers be used with each passage, although it has been suggested that used (or "conditioned") particles could be employed with certain cell lines (TAO et al., 1988).

The vast majority of microcarrier cultures are carried out in stirred tank bioreactors (STBs), which provide a relatively homogeneous culture environment that can be assessed and controlled (CROUGHAN and WANG, 1991). Fluid-lift reactors have found limited use with microcarriers. In these systems, microcarriers are retained in a small culture volume and kept suspended by incoming medium pumped from a separate reservoir (CLARK and HIRTENSTEIN, 1981; CROUGHAN and WANG, 1991). There are no published reports on the use of air-lift reactors in microcarrier culture. Although hydrodynamic damage to cells could occur at the point of bubble disengagement, cells on microcarriers may also be damaged from the power dissipated near rising bubbles (CROUGHAN and WANG, 1991).

STBs with microcarriers can be operated in either of three modes: in a batchwise fashion; with semi-continuous medium replenishment, where part of spent medium is periodically replaced with fresh without removal of microcarriers; or with continuous medium replenishment without removal of microcarriers, also referred to as perfusion operation. The effect of the hydrodynamic environment on cell growth and death in such systems has been studied rather extensively. Hydrodynamic damage, which will be discussed further in Sect. 3.2, may impose a burden on the scaling of microcarrier systems to commercial production.

2.2 Macroporous Microcarriers

Macroporous microcarriers constitute a more recent development in microcarrier technology. These microcarriers may support growth of adherent cells both on outside and inside surfaces. Cells inside the carriers are protected from potential hydrodynamic damage. Furthermore, since internal surfaces are also utilized, porous microcarriers offer a larger surface area for cell adhesion than solid ones per unit microcarrier mass, and thus can support higher cell densities at comparable carrier loadings (NILSSON et al., 1986; LOOBY and GRIFFITHS, 1990). Materials that have been used for porous microcarriers include gelatin (NILSSON et al., 1986; REITER et al., 1990), collagen (DEAN et al., 1987a, b), collagen-glycosaminoglycan copolymers (CAHN, 1990), glass (LOOBY and GRIFFITHS, 1990), and, more recently, reticulated polyvinyl formal resin (YAMAJI and FUKUDA, 1991), polyethylene (REITER et al., 1991), polystyrene (LEE et al., 1992), cellulose (SHIRAGAMI et al., 1991), and ceramics (PARK et al., 1990; GRAMPP and STEPHANOPOULOS, 1992).

Cultures are initiated by the addition of trypsinized cells. Cells attach first to the external surface of microcarriers and subsequently migrate inward to colonize internal surfaces. For internal surfaces to be effectively utilized, cells on the external surface should migrate inward before any extensive hydrodynamic damage occurs. The rate of migration and the final cell density depend on the particular cell strain or line used and on the amount and properties of surface accessible to cells. Colonization of the interior of porous gelatin microcarriers has been confirmed by thin sectioning followed by microscopic observations (NIKOLAI and HU, 1992).

Cells in the interior of porous carriers are not directly in contact with the flow field and are thus protected from shear damage and collisions. Transport of nutrients occurs by diffusion and possibly by convection (STEPHANOPOULOS and TSIVERIOTIS, 1989; PARK et al., 1990). If convective flows are present, cells can remain viable and functioning in the interior of larger microcarriers than if transport occurred by diffusion alone. However, convective transport requires the presence of voids. This is not always possible, since many cell lines form multilayers and completely fill the interconnecting pores within the microcarriers.

Porous microcarriers can be used in various bioreactors, including fixed and fluidized beds and stirred tanks. These configurations will be discussed further in Sect. 3.

2.3 Aggregate Culture

Some adherent cells exhibit a tendency for spontaneous aggregation into three-dimensional structures termed spheroids. Spheroids may form when trypsinized cells are added to culture medium in a vessel offering no surface proper for cell attachment (TOLBERT et al., 1980; PAPAS and SAMBANIS, 1992). Aggregation can be promoted by manipulating the calcium concentration in the medium, agitating the culture at a moderate to high rate, or, as shown by GOETGHEBEUR and HU (1991), by using 10–60 µm microspheres as "nucleating sites". Such particles, considerably smaller than conventional microcarriers, can support cell attachment but not cell spreading due to their high curvature. In experiments with CHO, 293, Vero, and ST cells, aggregates of cells and microspheres were formed shortly after cell attachment. Each structure contained from two to ten microspheres (GOETGHEBEUR and HU, 1991).

Spheroids may range from approximately 100 µm in diameter to macroscopic aggregates up to 4000 µm in size (FREYER, 1988). Because of diffusional limitations, significant gradients of nutrients and metabolites exist in large spheroids resulting in heterogeneous populations of cells. An outer layer consists of viable, proliferating cells, a second sublayer contains nutrient-limited (primarily hypoxic) quiescent cells which become reproductive when removed from such environment, and a central necrotic core consists of dead cells. Generally, proliferating cells are located in the outer 3–5 cells layers, i.e., within approximately 75 µm thickness. The distance from the periphery of the spheroid where necrosis begins may vary from 50 to 300 µm depending on the cell type and packing density, the nutrient consumption rates, and the nutrient concentrations in the medium (FREYER and SUTHERLAND, 1986; SUTHERLAND, 1988). In the three-dimensional spheroid structure, cells are in close association with each other both directly, by means of desmosomes and tight and gap junctions, and indirectly, by means of the extracellular biomatrix secreted by cells. This association may cause cells in spheroids to exhibit different properties than the same cells in monolayers. There exists morphological, biochemical, and immunological evidence that the tissue-like environment in spheroids may contribute to the expression of differentiated functions. For instance, spheroids of human colon adenocarcinoma cells differentiate in culture to develop pseudoglandular structures that possess features of tumors *in vivo* (SUTHERLAND et al., 1986; SUTHERLAND, 1988).

Another advantage of aggregate versus conventional microcarrier culture is the high fraction of culture volume occupied by biomass. In microcarrier cultures, considerable volume is taken by the microcarriers themselves. Potential disadvantages of aggregate culture are the long time that may be needed for effective adaptation of cells to culturing as spheroids, and the significant fraction of hypoxic or necrotic cells that may be present in cultures of large spheroids. Although spheroids have been cultured primarily in stirred tank vessels, fluidized-bed and air-lift bioreactors could also be used.

2.4 Cell Entrapment

Cell entrapment in biocompatible polymers is used primarily with suspension cells, such as hybridomas (NILSSON et al., 1983; RUPP et al., 1987) and aggregates of adherent cells (RONEN and DEGANI, 1989; PAPAS and SAMBANIS, 1992). Cells or aggregates are entrapped by being surrounded with a thin membrane (*encapsulation*) or by being imbedded in a polymeric gel. Entrapped cells are protected from potential hydrodynamic damage, and this may result in increased cell density and volumetric productivity (μg protein per mL culture medium per day) in bioreactors (SINACORE et al., 1989). Small-molecular weight nutrients and metabolites, such as glucose and lactate, generally diffuse freely through the polymers. Proteins, on the other hand, may not pass through if their molecular weight is larger than the molecular weight cutoff imposed by the polymer (KING et al., 1987). In such cases, the product may reach high concentrations in capsules (DUFF, 1985).

Polymers commonly used for animal cell entrapment are agarose (NILSSON et al., 1987), chitosan (OVERGAARD et al., 1991; YOSHIOKA et al., 1990), polyacrylate (GHARAPETIAN et al., 1986), and calcium alginate (LIM and SUN, 1980; GOOSEN et al., 1989). Cell entrapment in calcium alginate is accomplished by preparing a cell suspension in sodium alginate and adding it, in a dropwise fashion, into a solution of calcium chloride. Calcium cross-links alginate instantaneously forming beads containing entrapped cells. Alginate may be coated with polylysine for increased mechanical and chemical stability, but such treatment decreases the molecular weight cutoff imposed by the preparation, and thus prohibits large-molecular weight proteins in the medium from reaching the cells. To prepare hollow spheres (capsules), the alginate gel inside a bead coated with polylysine is liquified through treatment with a calcium chelator, such as EDTA or citrate. In a different approach implemented by WANG et al. (1991), alginate capsules were formed by suspending cells in a solution of calcium chloride and dextran and by adding droplets of this suspension into a solution of sodium alginate. Alginate entrapment is particularly attractive. It can be carried out at physiological temperature and pH, does not involve potentially toxic organic solvents, and the molecular weight cutoff imposed by the polymer can be adjusted by varying certain preparation parameters, in particular the polylysine molecular weight and the polylysine/alginate reaction time (GOOSEN et al., 1989). Depending on the alginate type and concentration, calcium alginate gels may have sufficient cavities to allow for growth of entrapped hybridomas (SINACORE et al., 1989). Cell densities may be increased by coentrapping cells with particles of gelatin, which dissolves at 37 °C creating cavities into which cells can grow (FAMILLETTI and FREDERICKS, 1988).

Significant concentration gradients of oxygen or other nutrients and metabolites may exist in beads of entrapped cells. This is especially true for high cell density preparations, since effective diffusivities decrease as the density of entrapped cells or aggregates increases. If a preparation is reasonably homogeneous, the system of entrapped cells may be modeled using the Thiele modulus and effectiveness factor approach (KAREL et al., 1985; MURDIN et al., 1988). However, if cells are not homogeneously distributed in beads, if spheroids rather than single cells are entrapped, or if aggregates form during growth of entrapped cells, the systems are difficult to model.

Entrapped cells may be cultured in STBs and fixed- and fluidized-bed reactors. Air-lift reactors could be used as well, provided that the beads are small and not significantly denser than the medium.

3 Mammalian Cell Bioreactors

Key objectives in the design of mammalian cell bioreactors are the attainment of high cell productivities and product titers, and the abili-

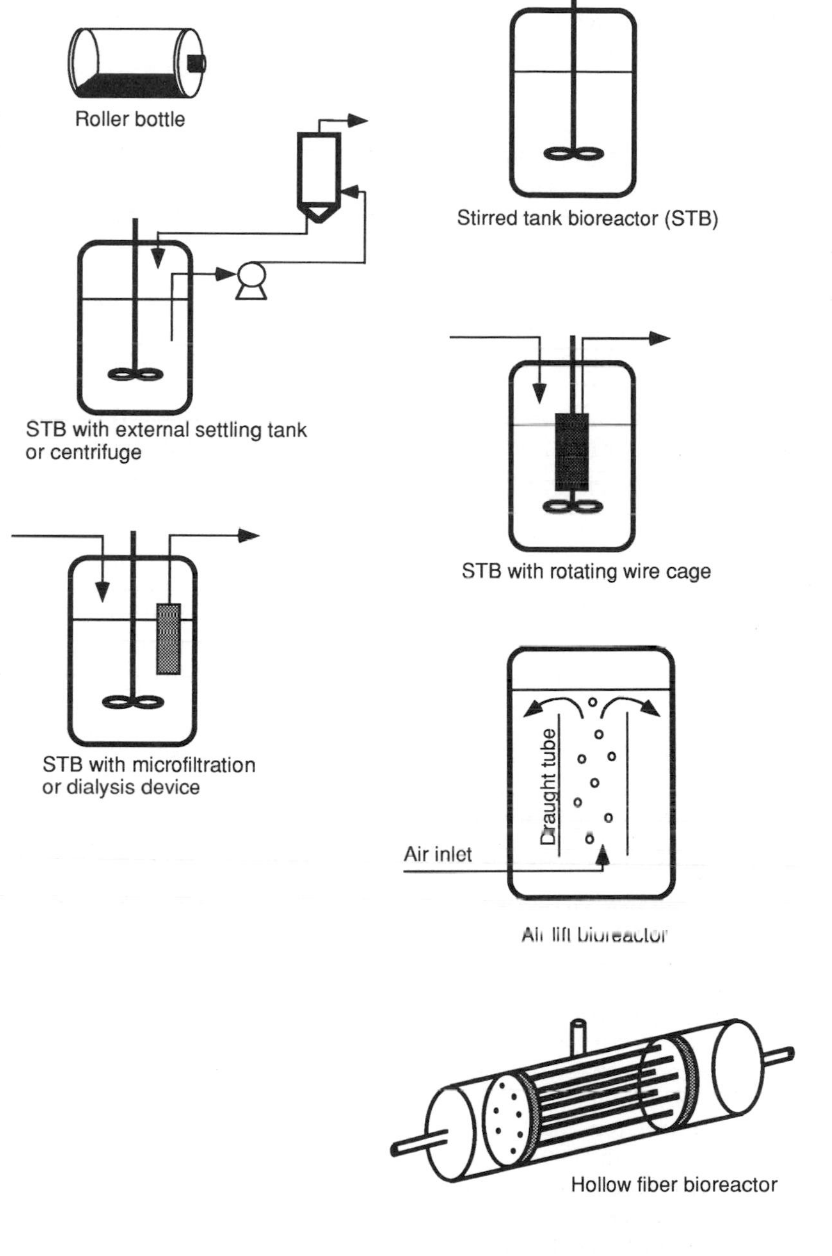

Fig. 1. Basic bioreactor configurations.

ty to scale up. Bioreactors utilize the microscale culture possibilities in various macroscopic configurations in order to achieve the foregoing objectives.

Fig. 1 shows the basic types of bioreactors used in mammalian cell culture. Depending on the design and operation, the bioreactor environment may be homogeneous or heterogeneous at the micro- or macroscale. Microscopic and macroscopic homogeneity prevail in well-mixed suspension cultures of single cells, of cells on solid microcarriers if only monolayers are formed, and of single entrapped cells if the polymer imposes a minimal diffusional resistance. Suspension cultures of solid microcarriers with cell multilayers, of cells on and in macroporous microcarriers, of large spheroids, and of cells entrapped in a polymer imposing significant diffusional limitations are all macroscopically homogeneous but microscopically heterogeneous. Ceramic-matrix and hollow-fiber bioreactors are macroscopically heterogeneous. The various configurations and their advantages and limitations are discussed in more detail below.

3.1 Roller Bottles

Roller bottles are cylindrical vessels usually made of disposable plastic. Adherent cells grow on the inside of the cylindrical surface. Bottles are positioned on horizontal rollers, where they are rotated at speeds of the order of 1 rpm. At a given time, the medium pool covers only a fraction of cells, so cells are exposed alternatively to medium and air.

Traditional roller bottle facilities constitute a rudimentary approach for culturing adherent animal cells. They require a relatively small capital investment for incubators, laminar flow hoods, and equipment for medium harvesting and culture feeding. This system is labor intensive and cannot be effectively monitored and controlled. On the other hand, cell culturing in roller bottles is very flexible, and total capacity can be easily adjusted to meet market needs. Because of that, roller bottles are used extensively in vaccine manufacture (HU and PESHWA, 1991). Recent improvements include the use of robotics in automating roller bottle facilities. Implementation of this technology should make production-scale roller bottle processes more competitive by reducing labor cost and contamination frequency (HU and PIRET, 1992).

3.2 Stirred Tank Bioreactors (STBs)

STBs are widely used for culturing suspension cells, such as hybridomas, usually free in suspension, possibly also entrapped in polymers. Adherent cells can be cultured in STBs when they are grown on solid or porous microcarriers, or as free or entrapped spheroids in suspension.

Conventional STBs are stainless steel vessels with height-to-diameter ratios ranging from 1:1 to about 3:1. Mixing is accomplished with agitators having one or more pitched-blade paddles or marine-type impellers (HU and DODGE, 1985). In a different design developed by KLEIS et al. (1990), mixing was provided under low shear stress conditions by a rotating disc agitator contoured at the bottom of the vessel. A flow pattern of toroidal shape was obtained, and adequate agitation occurred at shear stresses below those reported to cause damage to cells.

Hydrodynamic damage of cells on solid microcarriers in STBs is relatively well understood. Possible causes of cell damage include (a) interactions between microcarriers and turbulent eddies, (b) collisions between microcarriers, and (c) collisions of microcarriers with moving and stationary solid parts of the vessel, such as the impeller and the vessel walls (CHERRY and PAPOUTSAKIS, 1986, 1988). Cell damage from mechanism (a) above may be severe when the Kolmogorov length scale for the smallest turbulent eddies becomes comparable to the diameter of microcarriers (CROUGHAN et al., 1987). Microcarrier–microcarrier collisions are expected to increase in frequency and severity in the presence of eddies of size equal to the average interparticle distance. Since particle size and interparticle distance are virtually equal in typical microcarrier cultures, cell damage from both mechanisms (a) and (b) could be maximized at the same Kolmogorov length scale or agitation rate (CHERRY and PAPOUTSAKIS, 1986). If cell/microcarrier aggregates are formed, cellular bridges may be broken due to collisions or interactions with eddies, resulting in additional damage to cells (CHERRY and PAPOUTSAKIS, 1988). On the other hand, cells grown on the inside of porous microcarriers are protected from potential hydrodynamic damage (ADEMA et al., 1990; NIKOLAI and HU, 1992).

Suspended cells are generally less prone to hydrodynamic damage than cells on microcarriers. In experiments with freely suspended hybridomas, KUNAS and PAPOUTSAKIS (1990) found that growth retardation and cell damage were caused primarily by large air bubbles and vortices entrained from the culture surface. In the absence of vortex and bubble entrainment, cells were damaged only at high agitation rates, above 700 rpm, where the Kolmogorov length scale was comparable to the size of suspended particles, in this case single cells. In general, at the agitation rates needed to ensure homogeneous suspensions, single cells and cells in porous microcarriers suffer no significant hydrodynamic damage, whereas damage to cells on solid microcarriers may be considerable.

Surface aeration is often sufficient for oxygenation of small-scale cultures in STBs. If necessary, the overall oxygen transport rate can be increased by pumping air through semipermeable silicone tubing coiled inside the reactor. Oxygenation by direct sparging is avoided at a small scale, due to potential cell damage at the surface of cultures where bubbles rupture (HANDA et al., 1987; CHALMERS and BAVARIAN, 1991). Cells may be partially protected by addition of polyols in the culture medium, such as Pluronic F-68 (MURHAMMER and GOOCHEE, 1990). Industrial-scale STBs, however, are commonly aerated by direct sparging without serious compromise of cellular viability during the period of a production cycle. Due to the longer residence times of air bubbles in large bioreactors, effective oxygenation can be achieved at smaller volumetric air flow rates per unit culture volume; this results in a lower frequency of bubble rupture at the culture surface and thus reduced cell damage. Foaming at the culture surface can be controlled by addition of an antifoam.

Dissolved oxygen (DO) concentration may be monitored on-line with DO probes. Electrodes are also available for monitoring the culture redox potential. In some cases, redox probes have been used to control the DO concentration after determining the redox potential corresponding to the optimal DO value for the particular cell type and culture conditions employed (ARATHOON and BIRCH, 1986). The pH of the culture medium is usually controlled by the addition of carbon dioxide, for acidification, and by the addition of sodium hydroxide or sodium carbonate, or by the removal of carbon dioxide, for alkalinization (HU and OBERG, 1990).

STBs can be operated in continuous, batch, semi-batch, or perfusion mode. Continuous chemostat operation is used primarily for lab-scale bioreactors and has found limited application in production-scale systems. In batch operation, all nutrients are added at the beginning of a production cycle, except for oxygen which may be provided continuously during the culture. In semi-batch operation, certain nutrients, such as amino acids, may be added during cultivation, or spent medium with or without cells may be harvested periodically and replaced with fresh medium. During perfusion operation, spent medium is continuously replaced with fresh without any significant removal of cells, which are retained in the reactor using one of the devices described in the following section. Perfusion operation offers the advantages that repeated growth cycles are avoided and cells can be maintained productive and at a high density for prolonged periods of time. Potential problems include loss of product consistency and increased susceptibility of culture to contamination during long-term operation. Typical densities achieved in STBs without cell retention are $1\text{–}3 \cdot 10^6$ cells/mL and $1\text{–}5 \cdot 10^6$ cells/mL for suspension and solid microcarrier cultures, respectively. These densities can be increased by approximately an order of magnitude in systems with retention devices. Cells on porous microcarriers typically reach densities of $5\text{–}20 \cdot 10^6$ cells/mL.

3.2.1 Retention Devices

Simple gravitational settling of cells is too slow for continuous perfusion operation of STBs. Although this is true primarily for single cells in suspension, it may also be the case for cells grown on solid or porous microcarriers, as aggreates in suspension, or entrapped in polymers. Reversible flocculation of hybridomas and other single suspended cells reduces settling time significantly but may pose problems pertaining to the consistency of secreted glycoproteins (AUNINS and WANG, 1989). Cell retention devices, on the other hand, should permit easy and rapid separation of cells from spent medium and allow for long-term operation, thus compensating for the increased complexity in the system.

Retention devices may be physically positioned inside or outside the bioreactor vessel. In the first case, the bioreactor is segregated into two compartments, one containing medium and cells and the other essentially cell-free medium. An outside device may be directly attached to the bioreactor or connected to it via medium circulation lines. In the latter configuration, cells and spent medium are pumped to the retention device, where part of the medium is replaced with fresh medium; cells and medium are then returned to the bioreactor.

External devices connected to the reactor with circulation lines have been used exclusively with suspension cells, since microcarriers or beads with entrapped cells are easily damaged by pumping. Single cells in suspension can be damaged, too, and this is a potential problem that should be considered when designing the bioreactor and retention device configuration. External devices that have been used include settling tanks (KITANO et al., 1986), centrifuges (TOKASHIKI et al., 1990), and membrane filtration units (TAKAZAWA et al., 1990). Microfiltration appears to be the best choice, but membrane fouling due to protein deposition and clogging by cells remain potential problems. Internal and directly attached retention devices are more promising and are described below.

3.2.1.1 Rotating Wire Cage

Rotating cages are cylindrical in shape with a solid bottom and a cylindrical surface made of a wire screen. The device may be directly connected to the impeller so that a single shaft is used, or the cage and the impeller may be driven by different shafts. Fresh medium is added to the culture outside the cage, and spent medium is withdrawn from within the cage. Rotating cages can be used with single suspended cells, cells on microcarriers or entrapped in polymers, and cell aggregates. The openings of the screen are in the range of 25–60 µm for cultures of suspension cells; somewhat larger screen openings are used for microcarrier cultures (VARECKA and SCHEIRER, 1987; HIMMELFARB et al., 1969).

Since typical cell diameters are 10–15 µm, single cells in suspension can pass through the screen. However, the cell density inside the cage is much lower than outside the device, so retention is achieved. The hydrodynamic mechanisms responsible for cell retention are, as yet, poorly understood. In experiments with polystyrene particles of size and density similar to cells, retention increased as the rotation speed increased from 50 to 100 rpm but decreased as the rotation speed was increased further (FAVRE et al., 1990). Thus, centrifugation may not be the dominant retention mechanism. Fouling can become a problem in long-term operation. Screens made of polyamides appear less prone to fouling than screens made of stainless steel (ESCLADE et al., 1991). Recent results indicate that fouling follows classical cake filtration laws (FAVRE and THALER, 1992). To avoid or at least delay fouling, ESCLADE et al. (1991) recommended that the screen be made of synthetic instead of metallic material and be as flat as possible without hills and valleys. Also, use of serum should be avoided and serum-free media or media with serum substitutes should be used instead.

Microcarriers, capsules or beads with entrapped cells, and cell aggregates, are larger than the screen openings, so in these applications size exclusion filtration is at least one mechanism of retention. However, centrifugation may still play a role (HU and PESHWA, 1991).

Rotating cages can accomplish effective cell retention on a large scale. Nevertheless, for the rational design of large-scale systems, a fundamental understanding of the cell retention mechanism is essential.

3.2.1.2 Dialysis Unit

Dialysis membrane units have been placed inside bioreactors for rapid exchange of low-molecular weight nutrients and metabolites while excluding high-molecular weight compounds and cells. This device is analogous to hollow-fiber bioreactors (see Sect. 3.5), except that the extracapillary space is well mixed and the cell density lower. High-molecular weight nutrients that cannot permeate the dialysis membrane, such as certain growth factors, need to be added directly to the cell side of the culture vessel. Thus, dialysis systems may be very economical with regard to serum proteins necessary to achieve a certain amount of growth, but they also require the development and implementation of unique feeding strategies. Reported hybridoma densities and IgG concentrations in STBs with dialysis units are of the order of $8-10 \cdot 10^6$ cells/mL and 400–500 mg/L, respectively. These values are significantly higher than the $1-3 \cdot 10^6$ cells/mL and 50–100 mg IgG/L typically observed in conventional batch cultures (COMER et al., 1990).

Besides single suspended cells, dialysis units can be used with cells on microcarriers or entrapped in biopolymers, and with cell aggregates. The device is scaleable, but the high membrane surface area needed for large bioreactors complicates the system and may cause problems of fouling and mechanical stability.

3.2.1.3 Inclined Sedimenter

Inclined sedimenters are based on the enhanced sedimentation phenomenon first observed by BOYCOTT (1920) and analyzed rigorously by HERBOLZHEIMER and ACRIVOS (1981). In a vertical sedimenter containing a dilute particulate suspension between two parallel plates, the settling velocity v_0 of a particle at low Reynolds numbers is given by Stokes' law:

$$v_0 = \frac{2}{9} \frac{R^2 g (\rho_p - \rho_f)}{\mu_f}$$

where R is the Stokes radius of the particle, g the acceleration of gravity, ρ_p the particle density, and ρ_f and μ_f are the density and viscosity of the fluid. If the two parallel plates are tilted an angle θ from the vertical, the settling velocity becomes

$$v_s = v_0 \left(1 + \frac{H}{b} \sin\theta\right)$$

where H is the vertical overall height of the suspension in the sedimenter and b the distance between the two parallel plates. Sedimentation is thus enhanced by a factor $[1 + (H \cdot \sin\theta)/b]$ (STEPHANOPOULOS et al., 1985).

A commonly used geometry for sedimenters is that of an inclined tube with rectangular cross section. Inverted conical sedimenters, possibly equipped with lamellar devices, have also been used with success in perfusion hybridoma cultures (TYO and THILLY, 1989). Besides suspension cells, inclined sedimenters can be used with cells on microcarriers, entrapped in polymers, or grown as aggregates. It has been argued that in cultures of hybridomas and recombinant CHO cells it may be possible to withdraw preferentially through the sedimenter smaller, dead cells, and leave viable cells in the bioreactor (BATT et al., 1990; SEARLES and KOMPALA, 1992).

The scaleability of inclined sedimenters is doubtful when they have the geometry of a tube. Inverted conical lamellar separators, however, may be more amenable to large-scale applications. Lamellar settlers have indeed been used to recycle yeast cells in commercial fermentations (TABERA and IZNAOLA, 1989). The residence of cells in the sedimenter should be of concern; non-optimal conditions resulting from the absence of mixing and oxygenation, as well as the formation of a thick cell layer on the lower inclined surface of the sedimenter, may contribute to increased cell death.

3.3 Air-Lift Bioreactors

Air-lift bioreactors are of simple design, consisting of a central draught tube (riser) and an outer cylindrical annulus (downcomer). Air is introduced into the central tube, where it decreases the bulk liquid density causing the liquid to rise. Liquid flows downward in the outer annulus, thus creating circulation. Air-lift reactors have been used successfully with suspension cultures of BHK 21, human lymphoblastoid, CHO, hybridomas, and insect cells. Attained cell densities are of the same order as in STBs without retention devices. Hybridomas propagated in air-lift and STBs under otherwise similar conditions exhibit the same antibody productivity (KATINGER et al., 1979; ARATHOON and BIRCH, 1986).

The simple design and construction of airlift reactors, which avoid drive mechanisms and shaft bearing seals, result in reduced capital and maintenance costs and reduced contamination risks. The successful use of these reactors in cell cultivation suggests that direct air sparging may not cause excessive cell damage, at least under certain culture conditions. Since the hydrodynamic and mass transfer characteristics are relatively well understood, these bioreactors can be rationally scaled to volumes of at least 2000 liters. Models indicate that in the central draught tube DO concentration reaches a maximum at some point above the sparger and then declines. In the downcomer, DO decreases in a linear fashion. Dissolved oxygen is typically measured at the end of the downcomer, where it usually attains its minimum value. The ratio of the downcomer to the riser cross-sectional area (A_D/A_R) is an important design parameter, since it strongly affects the mixing and oxygen transfer characteristics in the bioreactor. As A_D/A_R increases, the volumetric oxygen transfer coefficient and the mixing time decrease to varying extents. Scale-up calculations may be simplified by keeping A_D/A_R constant (RHODES et al., 1991). Dissolved oxygen and pH are monitored and controlled using the devices and methods described in Sect. 3.2 for STBs.

3.4 Fluidized-Bed Bioreactors

Fluidized-bed systems provide enhanced convective transport of dissolved compounds to particles in suspension. However, the size and density of single cells, cell aggregates and cells on conventional microcarriers are such that fluidization velocities are too low for effective operation of fluidized bed systems. To circumvent this problem, Verax developed macroporous collagen microcarriers of roughly 500 μm in diameter, weighted to a specific gravity of approximately 1.6 with non-cytotoxic heavy metals. Fluidization of these carriers enhanced transport of dissolved oxygen, CO_2, and other nutrients relative to conventional stirred reactors (DEAN et al., 1987b).

In the Verax system, the bioreactor is part of a recycle loop also containing a gas exchanger, a heater, and pH, dissolved oxygen, and temperature probes. Dissolved oxygen can be controlled through the gas exchanger, and pH using one of the methods described for STBs. The microcarriers have been used to culture a variety of cell lines, including hybridomas and recombinant CHO cells. A small fraction (1–5%) of the total cell population ends up suspended in the medium, whereas the rest of the cells are attached to or entrapped in the porous microcarriers. The system is scaleable to various sizes (DEAN et al., 1987a). Potential problems include excessive growth of cells, which may result in agglomeration of microcarriers in the reactor, and a large number of suspended cells in the recirculating medium.

In a different fluidized-bed system developed by REITER et al. (1991), medium recirculation takes place internally with a low shear stress impeller, instead of externally in a recirculation loop with a pump. CHO cells were successfully grown in this bioreactor on polyethylene macroporous carriers of 1.2–1.5 mm in size. Maximum cell densities achieved were of the order of $1.2 \cdot 10^8$ cells/mL of carrier bed.

Besides increased mass transfer rates, fluidized-bed systems also accomplish higher loadings of cultures with microcarriers. It has been postulated that the maximum loading of STBs with microcarriers is 15 g/L, or 25% of total culture volume. In fluidized-bed reactors, on the other hand, carrier concentrations of 50% or more can be effectively utilized (GRIFFITHS, 1988; REITER et al., 1991).

3.5 Hollow-Fiber Bioreactors

A hollow-fiber unit consists of a bundle of hundreds of thousands of capillary fibers sealed at both ends in a cylindrical housing. In most applications, medium flows in the fiber lumen and cells grow in the extracapillary space or shell side of the bioreactor. Hollow-fiber units thus simulate vascularized tissues and can accommodate high cell densities of the order of 10^8 cells/mL (PIRET and COONEY, 1990a). Fibers are made of several materials, such as polysulfone and cellulose esters (recently reviewed by HEATH and BELFORT, 1990). Fiber membranes impose molecular weight cutoffs ranging from 10000 to 100000 Dalton. Low-molecular weight nutrients, including dissolved oxygen, diffuse through the fiber membrane, but high-molecular weight growth factors that cannot permeate the membranes need to be added directly to the shell side. Secreted high-molecular weight proteins, such as immunoglobulins, are retained in the extracapillary space, so high product concentrations can be achieved. The bulk of medium is in an external reservoir and is continuously circulated through the unit. The medium is oxygenated in the reservoir or in separate oxygenation devices positioned in the perfusion loop. Dissolved oxygen and pH may be monitored and controlled as in fluidized bed systems. Spent medium in the reservoir is replaced with fresh in a continuous, semi-batch, or batch mode.

Hollow-fiber reactors have been used with both suspension and adherent cells. Depending on the cell line and medium composition, adherent cells may attach on the shell wall, grow as aggregates, or be entrapped in the fiber pores. There have also been attempts, mainly with primary cells from tissues, such as hapatocytes, to load conventional microcarriers in the shell side to promote cell attachment (ARNAOUT et al., 1990). In this case, however, most of the extracapillary space is occupied by microcarriers rather than cells.

At the entrance of hollow-fiber reactors, there exists a pressure drop from the lumen into the extracapillary space causing medium to permeate through the membrane in this direction. This pressure difference is reversed at the exit of the reactor, where medium flows back into the lumen. The longitudinal pressure drop in the shell side produces axial convective flow, denoted as Starling flow. Nuclear magnetic resonance imaging experiments have indicated the existence of this flow in the extracapillary space of cell-free reactors containing a microporous polysulfonate fiber or several hollow fibers with a molecular weight cutoff of 100000 (HAMMER et al., 1990; HEATH et al., 1990). Starling flow causes axial polarization of cells and product in hollow-fiber units operated under unidirectional medium flow. PIRET and COONEY (1990b) found significantly higher monoclonal antibody concentrations at the downstream part of hollow-fiber reactors containing hybridomas. Cells also concentrated downstream, because of either convective transport of the inoculum, polarization of growth factors, or both. Alternating the direction of medium flow in the fiber lumen reduced polarization and increased antibody productivity (PIRET and COONEY, 1990b).

The maximum length of hollow-fiber units is determined by the requirement of adequate nutrient concentrations at the bioreactor exit. The limiting factor is the concentration of dissolved oxygen which, in order for cells at the downstream position to remain viable, should not become exhausted in a single pass. It appears that the axial gradients of other nutrients and metabolites can be easily minimized. Radial gradients depend on the distances between hollow fibers. PIRET et al. (1991) measured radial accumulations of ammonia and lactate in commercially available hollow-fiber reactors containing hybridomas. The maximum concentrations of these metabolites were significantly lower than those inhibitory to cell growth and product formation. On the other hand, when cells in the extracapillary space reach tissue-like densities, anoxic domains are likely to develop. Cells under anoxic conditions stop producing and eventually die. Scale-up of hollow-fiber reactors is thus accomplished by increasing the diameter of units while keeping the cross-sectional density of capillary tubes approximately constant. Scale-up to production levels can also be accomplished by increasing the number of units operating in parallel.

As with other macroscopically heterogeneous culture systems (i.e., ceramic-matrix bioreactors, Sect. 3.6), cell density and viability in hollow-fiber reactors cannot be measured by direct microscopic observations. Instead, indirect methods based on metabolic indicators need to be used. Such methods include measuring the consumption rate of glucose, glutamine or dissolved oxygen, the production rate of lactate, ammonium or alanine, and calculating the respiratory quotient (MERTEN, 1987). The extent of cell death and lysis can also be estimated by measuring the supernatant concentration of lactate dehydrogenase, an enzyme released only by lysed cells or dead cells with permeable membranes. It should be kept in mind, however, that all these methods offer estimates at best, and that the stability of lactate dehydrogenase in spent culture medium varies greatly from one cell line to another.

3.6 Ceramic-Matrix Bioreactors

Ceramic bioreactors consist of a cylindrical or rectangular monolith with square channels passing longitudinally through the matrix. Cells inoculated into the channels adhere to the material or become entrapped in the pores of the ceramic. For biocompatibility, the ceramic material usually needs to be treated with nitric acid to leach out heavy metals or other toxic substances introduced by the manufacturing process. Like hollow-fiber reactors, ceramic matrices are also plug-flow units. There is an important difference, however. In ceramic reactors the medium flowing through the matrix comes in direct contact with cells, so cells are bathed in all medium components, including high-molecular weight compounds (LYDERSEN et al., 1985; LYDERSEN, 1987).

As with hollow-fiber units, ceramic reactors are usually parts of perfusion loops. The bulk of medium is in an external reservoir and is oxygenated either there or in separate oxygenation units positioned in the loop. A fraction of cells is generally sloughed off from the ceramic matrix by the medium flow. This is especially true in the case of suspension cells entrapped in the pores. Cells circulating in the loop are prone to death and lysis due to mechanical stress caused by the pumping action. These cells can be removed by periodic settling in the medium reservoir or a sedimentation device. The requirement for adequate nutrient concentrations, primarily dissolved oxygen, at the exit of the bioreactor is the scale-up criterion for length. The diameter of the cylinder is limited by the requirement for even distribution of medium flow among the matrix channels.

Recent developments with lab-scale systems may promote the use of ceramic bioreactors in research and production. In a modified system described by APPLEGATE and STEPHANOPOULOS (1992), silicone tubing was threaded through each square channel of a ceramic matrix. Hybridomas entrapped in the ceramic matrix were oxygenated by pumping air through the silicone tubing. The remaining nutrients were provided by flowing medium through the channels, as in conventional ceramic units. Due to the continuous supply of oxygen through the silicone tubing, this configuration could be operated in a single-pass (rather than a recycle) mode. Depletion of nutrients other than oxygen and metabolite accumulation affected only cells proximal to the exit, instead of the entire culture as in conventional recycle operation. In addition, cells close to the exit possibly exhibited some adaptation to the stressful conditions prevailing at that point. Final antibody titers were 80% higher in the single-pass unit compared to the recycle bioreactor (APPLEGATE and STEPHANOPOULOS, 1992).

3.7 Fixed-Bed Bioreactors

This section examines fixed-bed bioreactors with adherent or suspension cells immobilized on solid or porous microcarriers or entrapped in polymers. These systems are to be distinguished from the ceramic-matrix units discussed in the previous section, which also can be considered fixed-bed reactors, but in which the solid support is in the form of a monolith. Fixed-bed reactors are parts of perfusion loops similar to those employed with hollow-fiber, ceramic, and fluidized-bed units. Compared to fluidized beds, fixed-bed configurations have the advantage of stable operation at a much wider range of flow rates, and the disadvantages of reduced tranport rates and of potential flow channeling problems.

A study comparing the performance of fixed- and fluidized-bed reactors was reported by LOOBY et al. (1989). CHO cells cultured on porous Siran glass spheres (Schott Glaswerke, Mainz, Germany) grew both on the surface and in the internal structure of beads in the fixed bed, but only in the internal bead structure in the fluidized-bed system. A density of $2 \cdot 10^7$ cells/mL bed volume was attained in the fixed-bed configuration, which was roughly twice that achieved per unit settled bed volume in a fluidized-bed reactor operated under similar conditions. Hybridomas entrapped in the same glass spheres and cultured in a fixed bed reached significantly higher cell densities compared to air-lift reactors and STBs (LOOBY and GRIFFITHS, 1988; LOOBY et al., 1989).

3.8 Other Bioreactor Developments

Certain other types of bioreactors reported in the literature do not clearly fall into one of the above categories, so they are described in this section. SCHOLZ and HU (1990) developed a two-compartment, three-zone bioreactor in both flat-bed and hollow-fiber configurations, and used it to culture Protein C-producing recombinant 293 cells. The cell and medium compartments were separated by an ultrafiltration membrane. The cell chamber was loaded with a cell suspension in collagen or collagen/chitosan which, upon gelation and contraction, formed two zones: one consisting of the gel with entrapped cells, and a second consisting of a liquid phase containing the product. Low-molecular weight nutrients and metabolites were exchanged between cells and medium through the membrane and product zone. High-molecular weight products accumulated in the product zone in the cell chamber. Different holding times could be achieved for the cells, the low-molecular weight, and the high-molecular weight compounds.

YAMAJI and FUKUDA (1991) cultured mouse myeloma cells immobilized in reticulated formal resin particles in a circulating-bed reactor. Agitation was provided by gas sparged at the bottom of the vessel. The circulation pattern attained was similar to that prevailing in air-lift units, except that no central draught tube was used. Following some initial adaptation, leakage of cells from the particles diminished and long-term cultivation was effectively accomplished.

A modified bioreactor of the packed-bed type was developed and used for culturing adherent cells by CHIOU et al. (1991). The bioreactor consisted of a central draught tube and an outer annulus containing a fixed bed of glass fibers where cells became attached. Oxygen-containing gas was sparged through the inner draught tube, and bubble-free medium flowed downward through the packed bed providing oxygenation and convective transport of dissolved nutrients to cells. In this respect, the bioreactor also resembled an air-lift unit. Theoretical considerations based on a simple hydrodynamic model indicated that the bioreactor could be scaled from 10 to 67000 liters with the cell density maintained in the range of $1\text{–}5 \cdot 10^7$ cells/mL (MURAKAMI et al., 1991).

4 Concluding Remarks

Cultivation of mammalian cells poses unique engineering problems due to the fragility of cells, their slow growth rates, low titers of secreted products and, in some cases, adherence requirement. Tab. 1 summarizes the basic bioreactor configurations described in Sect. 3, along with the typical cell densities attained in them, methods of medium oxygenation, operating sizes and modes, and general advantages and disadvantages of each system. Innovative designs aimed at alleviating some of the production difficulties in mammalian cell cultures have been implemented and evaluated mainly at a small scale. However, as competition shifts from the manufacture of new products to the economical production of clinically evaluated proteins, it is expected that novel designs will start being used at production scale as well. In any case, before characterizing and implementing an innovative bioreactor for large-scale protein production, the additional burden of validation for product consistency should be considered.

The general trend appears to be towards systems of high cell density accomplished by cell retention. Product retention with the cells is worth considering when (a) product breakdown or feedback inhibition of biosynthesis or secretion is not a problem; (b) the product is shear-sensitive; or (c) purification is economi-

Tab. 1. Basic Bioreactor Configurations and Their Characteristics

Bioreactor Type	Typical Cell Densities	Oxygenation Methods	Operating Sizes	Operating Modes	Advantages; Disadvantages
Stirred tank (STB) without retention devices	Suspension cells: $1–3 \cdot 10^6$ cells/mL; solid microcarriers: $1–5 \cdot 10^6$ cells/mL; porous microcarriers: $5–20 \cdot 10^6$ cells/mL	Surface aeration, direct sparging, silicone tubing	Suspension cells: up to few thousand liters; microcarriers: up to 1000 L	Batch, semi-batch, continuous	Simple design, scaleability, straightforward instrumentation and control, significant existing information from microbial fermentation and chemical industries; relatively low cell densities attained, microcarriers required for adherent cells.
Stirred tank (STB) with retention devices	$1–3 \cdot 10^7$ cells/mL	Surface aeration, direct sparging, silicone tubing	Up to 1000 L	Perfusion	High cell densities attained, possible extended production phase, decreased equipment turn-over time; increased mechanical complexity, potential problems with long-term operation of retention device and consistency in product quality.
Air-lift	$1–3 \cdot 10^6$ cells/mL	Direct sparging	Up to 2000 L[a]	Primarily batch; semi-batch and continuous modes also feasible[a]	Simple design, scaleability; relatively low cell densities attained, potential problems in use with adherent cells.
Fluidized-bed	$1–3 \cdot 10^8$ cells/mL beads; $3–5 \cdot 10^7$ cells/mL fluidized bed[b,c]	External oxygenator	Up to 84 L fluidized-bed volume[c]	Batch, semi-batch, continuous[f]	High cell densities, high liquid-to-biomass mass transfer coefficients, use with both suspension and adherent cells; require microcarriers with high sedimentation velocity, which are not widely available.
Packed-bed	Solid microcarriers: 10^6 cells/mL bed volume; porous microcarriers: $1–4 \cdot 10^7$ cells/mL bed volume[d]	External oxygenator	Up to 100 L; potentially scaleable to more than 200 liters[d]	Batch, semi-batch, continuous[f]	Simple design and operation, use with both adherent and suspension cells; low mass transfer coefficients, possible flow channeling, gradients in bioreactor.

Tab. 1. Basic Bioreactor Configurations and Their Characteristics (Continued)

Bioreactor Type	Typical Cell Densities	Oxygenation Methods	Operating Sizes	Operating Modes	Advantages; Disadvantages
Hollow-fiber	10^8 cells/mL	External oxygenator	30 mL to a few liters	Batch, semi-batch, continuous[f]	High cell densities, easy to inoculate and operate, product may be segregated from bulk of medium and thus obtained in relatively pure and concentrated form, use with both suspension and adherent cells; gradients within reactor, poor scaleability.
Ceramic matrix	10^7–$5\cdot10^8$ cells/mL ceramic matrix (Opticell system)[e]	External oxygenator	A few mL to 39 L (Opticell system monolithic core volumes)[e]	Batch, semi-batch, continuous[f]	Use with both suspension and adherent cells, reusable matrix; expensive matrix materials, inhomogeneities in reactor.

[a] RHODES et al. (1991) [b] DEAN et al. (1987b) [c] REITER et al. (1991) [d] GRIFFITHS and LOOBY (1991) [e] NICHOLSON et al. (1991)
[f] In all these systems, the cells in the bioreactor are continuously perfused by medium; during a production cycle, the medium in the entire perfusion loop may be replenished continuously, or periodically (semi-batch operation), or not be replenished at all (batch operation).

cal only if the product is at high concentration (MERTEN, 1987). Strict validation requirements have made batch operation the method of choice, since it is relatively straightforward to demonstrate reproducibility of batch production runs. As the issue of product consistency in long-term operation is addressed, however, bioreactors with slow- or non-growing but secreting cells may be implemented in production, as they offer the advantage of avoiding repeated cell growth cycles.

Bioreactor monitoring is easier for macroscopically homogeneous systems (e.g., STBs) for which representative samples of both cells and the abiotic phase can be removed. With macroscopically heterogeneous systems (e.g., hollow-fiber reactors), sampling of the abiotic phase may be straightforward, but representative cell samples cannot be withdrawn. Cellular density and viability can be assessed only indirectly by following metabolic indicators such as glucose consumption and lactate production. Currently available information on cellular viability and the distribution of nutrients and metabolites in heterogeneous systems has been derived by sectioning small-scale bioreactors at the end of experimental runs. On-line monitoring of either type of system is generally limited to temperature, pH, and dissolved oxygen concentration. As reliable sensors for glucose, lactate, other nutrients and metabolites, possibly also immunosensors, are developed for long-term operation, on-line monitoring is expected to become more sophisticated for both laboratory- and production-scale systems.

Another difficulty associated with the operation and control of animal cell bioreactors is the lack of dependable models relating cellular metabolism and productivity to the chemical and physical characteristics of the milieu in which cells are cultured. This is mainly the result of the remarkable complexity of cell culture systems: the cellular environment is determined by the composition of the medium provided to cells, the compounds secreted by cells and not rapidly diffusing away (e.g., extracellular matrix proteins), the physical and chemical nature of the solid substrate for adherent cells, and the hydrodynamic conditions to which cells are exposed. Considering cell cycle processes and internal cell structure introduces additional complexity. Thus, even for the relatively simple case of single cells in suspension, accounting for cell cycle phenomena and for some limited internal cell structure results in equations that cannot be handled even by supercomputers (FREDRICKSON, 1992). As a result, operation and control of mammalian cell bioreactors are currently guided by educated guesses, rather than models based on a fundamental, mechanistic understanding of the system.

Scaling up of bioreactors should be carried out in a cautious fashion as well. Since cellular metabolism and productivity depend strongly on the cell culture method and conditions, results obtained in the laboratory with, e.g., T flasks, may not be suitable for calculating the volume of a microcarrier bioreactor required for obtaining a certain productivity. Instead, scale-up should be conducted by studying systems which are qualitatively as similar as possible to the bioreactors that will be used eventually for production.

5 References

ADEMA, E., SHNEK, D., CAHN, F., SINSKEY, A. J. (1990), Use of porous microcarriers in agitated cultures, *BioPharm,* July/August, 20–23.

APPLEGATE, M. A., STEPHANOPOULOS, G. (1992), Development of a single-pass ceramic matrix bioreactor for large-scale mammalian cell culture, *Biotechnol. Bioeng.* **40,** 1056–1068.

ARATHOON, W. R., BIRCH, J. R. (1986), Large-scale cell culture in biotechnology, *Science* **232,** 1390–1395.

ARNAOUT, W. S., MOSCIONI, A. D., BARBOUR, R. L., DEMETRIOU, A. A. (1990), Development of bioartificial liver: bilirubin conjugation in gunn rats, *J. Surg. Res.* **48,** 379–382.

AUNINS, J. G., WANG, D. I. C. (1989), Induced flocculation of animal cells in suspension culture, *Biotechnol. Bioeng.* **34,** 629–638.

BARFOED, H. C. (1987), Insulin production technology, *Chem. Eng. Prog.* **83** (10), 49–54.

BATT, B. C., DAVIS, R. H., KOMPALA, D. S. (1990), Inclined sedimentation for selective retention of viable hybridomas in a continuous suspension bioreactor, *Biotechnol. Prog.* **6,** 458–464.

BOYCOTT, A. E. (1920), Sedimentation of blood corpuscles, *Nature* **104,** 532.

CAHN, F. (1990), Biomaterials aspects of porous microcarriers for animal cell culture, *Trends Biotechnol.* **8,** 131–136.

CHALMERS, J. J., BAVARIAN, F. (1991), Microscopic visualization of insect cell–bubble interactions. II: The bubble film and bubble rupture, *Biotechnol. Prog.* **7,** 151–158.

CHERRY, R. S., PAPOUTSAKIS, E. T. (1986), Hydrodynamic effects on cells in agitated tissue culture reactors, *Bioprocess Eng.* **1,** 29–41.

CHERRY, R. S., PAPOUTSAKIS, E. T. (1988), Physical mechanisms of cell damage in microcarrier cell culture bioreactors, *Biotechnol. Bioeng.* **32,** 1001–1014.

CHIOU, T.-W., MURAKAMI, S., WANG, D. I. C., WU, W.-T. (1991), A fiber-bed bioreactor for anchorage-dependent animal cell cultures: Part I. Bioreactor design and operations, *Biotechnol. Bioeng.* **37,** 755–761.

CLARK, J. M., HIRTENSTEIN, M. D. (1981), Optimizing culture conditions for the production of animal cells in microcarrier culture, *Ann. NY Acad. Sci.* **369,** 33–46.

COMER, M. J., KEARNS, M. J., WAHL, J., MUNSTER, M., LORENZ, T., SZPERALSKI, B., KOCH, S., BEHRENDT, U., BRUNNER, H. (1990), Industrial production of monoclonal antibodies and therapeutic proteins by dialysis fermentation, *Cytotechnology* **3,** 295–299.

CONDRA, J. H., SARDANA, V. V., TOMASSINI, J. E., SCHLABACH, A. J., DAVIS, M.-E., LINEBERGER, D. W., GRAHAM, D. J., GOTLIB, L., COLONNO, R. J. (1990), Bacterial expression of antibody fragments that block human rhinovirus infection of cultured cells, *J. Biol. Chem.* **265,** 2292–2295.

CROUGHAN, M. S., WANG, D. I. C. (1990), Reversible removal and hydrodynamic phenomena in CHO microcarrier cultures, *Biotechnol. Bioeng.* **36,** 316–319.

CROUGHAN, M. S., WANG, D. I. C. (1991), Hydrodynamic effects on animal cells in microcarrier bioreactors, in: *Animal Cell Bioreactors,* (HO, C. S., WANG, D. I. C., Eds.), pp. 213–249, Boston: Butterworth–Heinemann.

CROUGHAN, M. S., HAMEL, J.-F., WANG, D. I. C. (1987), Hydrodynamic effects on animal cells grown in microcarrier cultures, *Biotechnol. Bioeng.* **29,** 130–141.

DEAN, R. C., JR., KARKARE, S. B., PHILLIPS, P. G., RAY, N. G., RUNSTADLER, JR., P. W. (1987a), Continuous cell culture with fluidized sponge beads, in: *Large Scale Mammalian Cell Culture Technology* (LYDERSEN, B. K., Ed.), pp. 145–167, New York: Hanser Publishers.

DEAN, R. C., JR., KARKARE, S. B., RAY, N. G., RUNSTADLER, JR., P. W., VENKATASUBRAMANIAN, K. (1987b), Large-scale culture of hybridoma and mammalian cells in fluidized bed bioreactors, *Ann. NY Acad. Sci.* **506**, 129–146.

DUFF, R. G. (1985), Microencapsulation technology: a novel method for monoclonal antibody production, *Trends Biotechnol.* **3**, 167–170.

ESCLADE, L. R. J., CARREL, S., PÉRINGER, P. (1991), Influence of the screen material on the fouling of spin filters, *Biotechnol. Bioeng.* **38**, 159–168.

FAMILLETTI, P. C., FREDERICKS, J. E. (1988), Techniques for mammalian cell immobilization, *Bio/Technology* **6**, 41–44.

FAVRE, E., THALER, T. (1992), Investigations on the basic physical behavior of rotating sieves for cell retention, presented at *Engineering Foundation Conference on Cell Culture Engineering,* Palm Coast, Florida, February 1992.

FAVRE, E., PÉRINGER, P., WOLDEN, C., KYUNG, Y. S., HU, W.-S. (1990), Retention of animal cells in a rotating wire-cage bioreactor: a mechanistic study, in: *Trends in Animal Cell Culture Technology* (MURAKAMI, H., Ed.), pp. 127–132, Tokyo: Kodansha Publishing.

FREDRICKSON, A. G. (1992), Incorporation of cell cycle phenomena into distributed models of cell population growth, presented at *Engineering Foundation Conference on Cell Culture Engineering,* Palm Coast, Florida, February 1992.

FREYER, J. P. (1988), Role of necrosis in regulating the growth saturation of multicellular spheroids, *Cancer Res.* **48**, 2432–2439.

FREYER, J. P., SUTHERLAND, R. M. (1986), Regulation of growth saturation and development of necrosis in EMT6/Ro multicellular spheroids by the glucose and oxygen supply, *Cancer Res.* **46**, 3504–3512.

GHARAPETIAN, H., DAVIES, N. A., SUN, A. M. (1986), Encapsulation of viable cells within polyacrylate membranes, *Biotechnol. Bioeng.* **28**, 1595–1600.

GOETGHEBEUR, S., HU, W.-S. (1991), Cultivation of anchorage-dependent animal cells in microsphere-induced aggregate culture, *Appl. Microbiol. Biotechnol.* **34**, 735–741.

GOOSEN, M. F. A. (1991), Large-scale insect cell culture: methods, applications and products, *Curr. Opin. Biotechnol.* **2**, 365–369.

GOOSEN, M. F. A., KING, G. A., MCKNIGHT, C. A., MARCOTTE, N. (1989), Animal cell culture engineering using alginate polycation microcapsules of controlled membrane molecular weight cutoff, *J. Membr. Sci.* **41**, 323–343.

GRAMPP, G. E., STEPHANOPOULOS, G. (1992), Development and scale up of controlled secretion processes for improved product recovery in animal cell cultures, *Ann. NY Acad. Sci.,* in press.

GRIFFITHS, J. B. (1988), Overview of cell culture systems and their scale up, in: *Animal Cell Biotechnoloy,* Vol. III, (SPIER, R. E., GRIFFITHS, J. B., Eds.), pp. 179–220, London: Academic Press.

GRIFFITHS, B., LOOBY, D. (1991), Fixed immobilized beds for the cultivation of animal cells, in: *Animal Cell Bioreactors* (HO, C. S., WANG, D. I. C., Eds.), pp. 165–189, Boston: Butterworth–Heinemann.

HAMMER, B. F., HEATH, C. A., MIRER, S. D., BELFORT, G. (1990), Quantitative flow measurements in bioreactors by nuclear magnetic resonance imaging, *Bio/Technology* **8**, 327–330.

HANDA, A., EMERY, A. N., SPIER, R. E. (1987), On the evaluation of gas-liquid interfacial effects on hybridoma viability in bubble column bioreactors, *Dev. Biol. Stand.* **66**, 241–253.

HEATH, C. A., BELFORT, G. (1990), Membranes and bioreactors, *Int. J. Biochem.* **22**, 823–836.

HEATH, C. A., BELFORT, G., HAMMER, B. E., MIRER, S. D., PIMBLEY, J. M. (1990), Magnetic resonance imaging and modeling of flow in hollow-fiber bioreactors, *AIChE J.* **36**, 547–558.

HERBOLZHEIMER, E., ACRIVOS, A. (1981), Enhanced sedimentation in narrow tilted channels, *J. Fluid Mech.* **108**, 485–499.

HIATT, A., CAFFERKEY, R., BOWDISK, K. (1989), Production of antibodies in transgenic plants, *Nature* **342**, 76–78.

HIMMELFARB, P., THAYER, P. S., MARTIN, H. E. (1969), Spin filter culture: the propagation of mammalian cells in suspension, *Science* **164**, 555–557.

HU, W.-S., DODGE, T. C. (1985), Cultivation of mammalian cells in bioreactors, *Biotechnol. Prog.* **1**, 209–215.

HU, W.-S., OBERG, M. G. (1990), Monitoring and control of animal cell bioreactors: biochemical engineering considerations, in: *Large Scale Mammalian Cell Culture Technology* (LUBINEICKI, A. S., Ed.), pp. 451–481, New York: Marcel Dekker Inc.

HU, W.-S., PESHWA, M. V. (1991), Animal cell bioreactors – recent advances and challenges to scale up, *Can. J. Chem. Eng.* **69**, 409–420.

HU, W.-S., PIRET, J. M. (1992), Mammalian cell culture processes, *Curr. Opin. Biotechnol.* **3**, 110–114.

HU, W.-S., GIARD, D. J., WANG, D. I. C. (1985), Serial propagation of mammalian cells on microcarriers, *Biotechnol. Bioeng.* **27**, 1466–1476.

JOHANSSON, A., NIELSEN, V. (1980), Biosilon: a new microcarrier, *Dev. Biol. Stand.* **46**, 125–129.

JOHNSON, I. S. (1983), Human insulin from recombinant DNA technology, *Science* **219**, 632–637.

KAREL, S. F., LIBICKI, S. B., ROBERTSON, C. (1985), The immobilization of whole cells: engineering principles, *Chem. Eng. Sci.* **40**, 1321–1354.

KATINGER, J. W. D., SCHEIRER, W., KRÖMER, E. (1979), Bubble column reactor for mass propagation of animal cells in suspension culture, *Ger. Chem. Eng.* **2**, 31–38.

KING, G. A., DAUGULIS, A. J., FAULKNER, P., GOOSEN, M. F. A. (1987), Alginate-polylysine microcapsules of controlled membrane molecular weight cutoff for mammalian cell culture engineering, *Biotechnol. Prog.* **3**, 231–340.

KITANO, K. Y., SHINTANI, Y. I., TSUKAMOTO, K., SASAI, S., KIDA, M. (1986), Production of human monoclonal antibodies by heterohybridomas, *Appl. Microbiol. Biotechnol.* **24**, 282–286.

KLEIS, S. J., SCHRECK, S., NEREM, R. M. (1990), A viscous pump bioreactor, *Biotechnol. Bioeng.* **36**, 771–777.

KUNAS, K. T., PAPOUTSAKIS, E. T. (1990), Damage mechanisms of suspended animal cells in agitated bioreactors with and without bubble entrainment, *Biotechnol. Bioeng.* **36**, 476–483.

KUO, M. J., LEWIS, JR., C., MARTIN, R. A., MILLER, R. E., SCHOENFELD, R. A., SCHECK, J. M., WILDI, B. S. (1981), Growth of anchorage-dependent mammalian cells on glycine-derivatized polystyrene in suspension culture, *In Vitro* **17**, 901–906.

LEE, D. W., PIRET, J. M., GREGORY, D., HADDOW, D. J., KILBURN, D. G. (1992), Polystyrene macroporous bead support for mammalian cell culture, in: *Biochemical Engineering VII* (DIBIASIO, D., MUTHARASAN, R., Eds.), New York: New York Academy of Sciences, in press.

LIM, F., SUN, A. M. (1980), Microencapsulated islets as bioartificial endocrine pancreas, *Science* **210**, 908–910.

LOOBY, D., GRIFFITHS, J. B. (1988), Fixed bed porous glass sphere (porosphere) bioreactors for animal cells, *Cytotechnology* **1**, 339–346.

LOOBY, D., GRIFFITHS, J. B. (1990), Immobilization of animal cells in porous carrier culture, *Trends Biotechnol.* **8**, 204–209.

LOOBY, D., RACHER, A. J., GRIFFITHS, J. B., DOWSETT, A. B. (1989), The immobilization of animal cells in fixed and fluidized porous glass sphere reactors, in: *Physiology of Immobilized Cells, Proc. Int. Symp.*, Wageningen, Netherlands, 10–13 December, 1989 (DE BONT, J. A. M., VISSER, J., MATTIASSON, B., TRAMPER, J., Eds.), pp. 255–264, Amsterdam: Elsevier Science Publisher.

LUCKOW, V. A., SUMMERS, M. D. (1988), Trends in the development of *Baculovirus* expression vectors, *Bio/Technology* **6**, 47–55.

LYDERSEN, B. K. (1987), Perfusion cell culture system based on ceramic matrices, in: *Large Scale Cell Culture Technology* (LYDERSEN, B. K., Ed.), pp. 169–192, New York: Hanser Publications.

LYDERSEN, B. K., PUGH, G. G., PARIS, M. S., SHARMA, B. P., NOLL, L. A. (1985), Ceramic matrix for large scale animal cell culture, *Bio/Technology* **3**, 63–67.

MAINS, R. E., CULLEN, E. I., MAY, V., EIPPER, B. A. (1987), The role of secretory granules in peptide biosynthesis, *Ann. NY Acad. Sci.* **493**, 278–291.

MEIGNIER, B. (1978), Cell culture on beads used for the industrial production of foot and mouth disease virus, *Dev. Biol. Stand.* **42**, 141–145.

MERTEN, O.-W. (1987), Concentrating mammalian cells. I. Large-scale animal cell culture, *Trends Biotechnol.* **5**, 230–237.

MURAKAMI, S., CHIOU, T.-W., WANG, D. I. C. (1991), A fiber-bed bioreactor for anchorage-dependent animal cell cultures: Part II. Scaleup potential, *Biotechnol. Bioeng.* **37**, 762–769.

MURDIN, A. D., KIRKBY, N. F., WILSON, R., SPIER, R. E. (1988), Immobilized hybridomas: oxygen diffusion, in: *Animal Cell Biotechnology,* Vol. 3. (SPIER, R. E., GRIFFITHS, J. B., Eds.), pp. 55–73, London: Academic Press.

MURHAMMER, D. W., GOOCHEE, C. F. (1990), Sparged animal cell bioreactors: mechanism of cell damage and Pluronic F-68 protection, *Biotechnol. Prog.* **6**, 391–397.

NICHOLSON, M. L., HAMPSON, B. S., PUGH, G. G., HO, C. S. (1991), Continuous cell culture, in: *Animal Cell Bioreactors* (HO, C. S., WANG D. I. C., Eds.), pp. 269–303, Boston: Butterworth–Heinemann.

NIKOLAI, T. J., HU, W.-S. (1992), Cultivation of mammalian cells on macroporous microcarriers, *Enzyme Microb. Technol.* **14**, 203–208.

NILSSON, K., SCHEIRER, W., MORTON, O. M., OSTBERG, L., LIEHL, E., KATINGER, H. W. D., MOSBACH, K. (1983), Entrapment of animal cells for production of monoclonal antibodies and other biomolecules, *Nature* **302**, 629–630.

NILSSON, K., BUZSAKY, F., MOSBACH, K. (1986), Growth of anchorage-dependent cells on macroporous microcarriers, *Bio/Technology* **4**, 989–990.

NILSSON, K., SCHEIRER, W., KATINGER, H. W. D., MOSBACH, K. (1987), Entrapment of animal cells, *Methods Enzymol.* **135**, 399–410.

NILSSON, K., BIRNBAUM, S., MOSBACH, K. (1988), Microcarrier culture of recombinant chinese hamster ovary cells for production of human interferon and human tissue type plasminogen activator, *Appl. Microbiol. Biotechnol.* **27**, 366–371.

OVERGAARD, S., SCHARER, J. M., MOO-YOUNG, M., BOLS, N. C. (1991), Immobilization of hybridoma cells in chitosan alginate beads, *Can. J. Chem. Eng.* **69**, 439–443.

PAPAS, K. K., SAMBANIS, A. (1992), The characterization of the metabolic and secretory behavior of suspended free and entrapped insulin-producing AtT-20 spheroids, presented at *1992 Annual Meeting of the American Institute of Chemical Engineers,* Miami Beach, Florida.

PARK, S., SINGHVI, R., WANG, D. I. C., STEPHANOPOULOS, G. N. (1990), Porous ceramic beads for animal cell culture, presented at *1990 Annual Meeting of the American Institute of Chemical Engineers,* Chicago, Illinois.

PIRET, J. M., COONEY, C. L. (1990a), Immobilized mammalian cell cultivation in hollow fiber bioreactors, *Biotechnol. Adv.* **8**, 763–783.

PIRET, J. M., COONEY, C. L. (1990b), Mammalian cell and protein distributions in ultrafiltration hollow fiber bioreactors, *Biotechnol. Bioeng.* **36**, 902–910.

PIRET, J. M., DEVENS, D. A., COONEY, C. L. (1991), Nutrient and metabolite gradients in mammalian cell hollow fiber bioreactors, *Can. J. Chem. Eng.* **69**, 421–428.

REITER, M., HOHENWATER, O., GAIDA, T., ZACH, N., SCHMATZ, C., BLÜML, G., WEIGAN, F., NILSSON, K., KATINGER, H. (1990), The use of macroporous gelatin carriers for the cultivation of mammalian cells in fluidized bed reactors, *Cytotechnology* **3**, 271–277.

REITER, M., BLÜML, G., GAIDA, T., ZACH, N., UNTERLUGGAUER, F., DOBLHOFF-DIER, O., NOE, M., PLAIL, R., HUSS, S., KATINGER, H. (1991), Modular integrated fluidized bed bioreactor technology, *Bio/Technology* **9**, 1100–1102.

REUVENY, S., SILBERSTEIN, L., SHAHAR, A., FREEMAN, E., MIZRAHI, A. (1982), Cell and virus propagation on cylindrical cellulose based microcarriers, *Dev. Biol. Stand.* **50**, 115–123.

RHODES, M., GARDINER, S., BROAD, D. (1991), Scaleup of animal cell suspension culture, in: *Animal Cell Bioreactors* (HO, C. S., WANG, D. I. C., Eds.), pp. 253–268. Boston: Butterworth–Heinemann.

RONEN, S., DEGANI, H. (1989), Studies of the metabolism of human breast cancer spheroids by NMR, *Magn. Res. Med.* **12**, 274–281.

RUPP, R., GILBRIDE, K., OKA, M. (1987), Cellular microencapsulation for large-scale production of monoclonal antibodies, in: *Large-scale Culture of Microencapsulated Cells* (LYDERSEN, B. K., Ed.), pp. 81–94, New York: Hanser Publications.

SCHOLZ, M., HU, W. S. (1990), A two-compartment cell entrapment bioreactor with three different holding times for cells, high and low molecular weight compounds, *Cytotechnology* **4**, 127–137.

SEARLES, J., KOMPALA, D. S. (1992), A high viable cell density suspension bioreactor for recombinant mammalian cultures, presented at *Engineering Foundation Conference on Cell Culture Engineering,* Palm Coast, Florida, February 1992.

SHIRAGAMI, N., OHIRA, Y., UNNO, H. (1991), Anchorage-dependent animal cell growth in porous microcarrier culture, in: *Animal Cell Culture and Production of Biologicals, Proc. Third Annual Meeting of the Japanese Association for Animal Cell Technology,* Kyoto, Japan, 11–13 December 1990 (SASAKI, R., IKURA, K., Eds.), pp. 121–126, Dordrecht, Netherlands: Kluwer Acad. Publisher.

SINACORE, M. S., CRESWICK, B. C., BUEHLER, R. (1989), Entrapment and growth of murine hybridoma cells in calcium alginate gel microbeads, *Bio/Technology* **7**, 1275–1279.

SKERRA, A., PLÜCKTHUN, A. (1988), Assembly of a functional immunoglobulin F_v fragment in *Escherichia coli, Science* **240**, 1038–1041.

STEPHANOPOULOS, G., SAN, K.-Y., DAVISON, B. H. (1985), A novel bioreactor–cell precipitator combination for high-cell density, high-flow fermentations, *Biotechnol. Prog.* **1**, 250–259.

STEPHANOPOULOS, G., TSIVERIOTIS, K. (1989), Effect of intraparticle convection on nutrient transport in porous biological pellets, *Chem. Eng. Sci.* **44**, 2031–2039.

SUTHERLAND, R. M. (1988), Cell and environment interactions in tumor microregions: the multicell spheroid model, *Science* **240**, 177–240.

SUTHERLAND, R. M., SORDAT, B., BAMAT, J., GABBERT, H., BOURRAT, B., MUELLER-KLIESER, W. (1986), Oxygenation and differentiation in multicellular spheroids of human colon carcinoma, *Cancer Res.* **46**, 5320–5329.

TABERA, J., IZNAOLA, M. A. (1989), Design of a lamella settler for biomass recycling in continuous ethanol fermentation process, *Biotechnol. Bioeng.* **33**, 1296–1305.

TAKAZAWA, Y., TOKASHIKI, M., HAMAMOTO, K., MURAKAMI, H. (1990), High cell density perfusion culture of hybridoma cells recycling high molecular weight components, *Cytotechnology* **1**, 171–178.

TAO, T.-Y., JI, G.-Y., HU, W.-S. (1988), Serial propagation of mammalian cells on gelatin-coated microcarriers, *Biotechnol. Bioeng.* **32**, 1037–1052.

TOKASHIKI, M., ARAI, T., HAMAMOTO, K., ISHIMARU, K. (1990), High density culture of hybridoma cells using a perfusion culture vessel with an external centrifuge, *Cytotechnology* **3**, 239–244.

TOLBERT, W. R., HITT, M. M., FEDER, J. (1980), Cell aggregate suspension culture for large-scale production of biomolecules, *In Vitro* **16**, 486–490.

TYO, M. A., THILLY, W. G. (1989), Novel high density perfusion system for suspension culture metabolic studies, presented at *Annual Meeting of the American Institute of Chemical Engineers,* San Francisco, California, November 1989.

VAN BRUNT, J. (1990), Monoclonal fine-tuning continues in *E. coli, Bio/Technology* **8,** 276.

VAN WEZEL, A. L. (1967), Growth of cell strains and primary cells on microcarriers in homogeneous culture, *Nature* **216,** 64–65.

VAN WEZEL, A. L., VAN HERWAARDEN, J. A. M., VAN DE HEUVEL-DE RIJK, E. W. (1979), Large-scale concentration and purification of virus suspension from microcarrier culture for the preparation of inactivated virus vaccines, *Dev. Biol. Stand.* **42,** 65–69.

VARANI, J., DAME, M., BEALS, T. F., WASS, J. A. (1983), Growth of three established cell lines on glass microcarriers, *Biotechnol. Bioeng.* **25,** 1359–1372.

VARANI, J., DAME, M., FEDISKE, J., BEALS, T. F., HILLEGAS, W. (1985), Substrate-dependent differences in growth and biological properties of fibroblasts and epithelial cells grown in microcarrier culture, *J. Biol. Stand.* **13,** 67–76.

VARECKA, R., SCHEIRER, W. (1987), Use of a rotating wire-cage for the retention of animal cells in a perfusion fermentor, *Dev. Biol. Stand.* **66,** 269–272.

WANG, H. Y., EISFELD, T., SAKODA, A. (1991), A new approach for cell encapsulation, in: *Biologicals from Recombinant Microorganisms and Animal Cells: Production and Recovery, Proc. 34th Oholo Conference,* Eilat, Israel, 1990 (WHITE, M. D., REUVENY, S., SHAFFERMAN, A., Eds.), pp. 173–182, New York: VCH Publishers, Inc.

YAMAJI, H., FUKUDA, H. (1991), Long-term cultivation of anchorage-independent animal cells immobilized within reticulated biomass support particles in a circulating bed fermentor, *Appl. Microbiol. Biotechnol.* **34,** 730–734.

YOSHIOKA, T., HIRANO, R., SHIOYA, T., KAKO, M. (1990), Encapsulation of mammalian cell with chitosan-CMC capsule, *Biotechnol. Bioeng.* **35,** 66–72.

7 Media for Microbial Fermentations

RANDOLPH L. GREASHAM

Rahway, NJ 07065, USA

1 Introduction

Many medium formulations have been developed for desired microbial growth and production of high levels of commercially important metabolites. These formulations are usually not published but guarded as company trade secrets. This practice clearly illustrates the importance fermentation media have on the commercial success of microbial-derived products. Advanced analytical capabilities for measuring medium components and cell metabolites coupled with statistical optimization strategies and biochemical studies have proven to be key factors in developing media that fully support highly productive, consistent and economical fermentation processes.

2 Nutritional Groups

There is an enormous diversity of microorganisms that exhibit a spectrum of nutritional patterns. At one end of this spectrum are microorganisms that synthesize all their complex chemical structures (proteins, fats, carbohydrates, vitamins, cell walls, nucleic acids, etc.) from atmospheric carbon dioxide or carbonates and a few simple inorganic compounds such as ammonium sulfate, magnesium sulfate, ferric chloride, potassium phosphate, and sodium chloride. These microbes are referred to as autotrophs. The autotrophs may be subdivided on the basis of their ability to utilize the energy of light (phototrophs) or the energy of chemical reactions (chemotrophs) for cell growth. At the other end of the nutritional spectrum are the microorganisms that utilize organic compounds both as a source of energy and organic material for synthesis of cellular components. These microbes are called heterotrophs. This autotrophic and heterotrophic grouping of microorganisms is being replaced by another nutritional grouping that better categorizes the photosynthetic microbes. The lithotrophs use carbon dioxide as the sole carbon source and gain their energy from light (photolithotrophs) or from the oxidation of inorganic substrates (chemolithotrophs). The organotrophs prefer oxidizable organic substrates and utilize the energy of light to assimilate carbon dioxide and organic compounds (photoorganotrophs) or oxidize/ferment organic compounds (chemoorganotrophs) for growth. Most microbes belong to this latter nutritional group, including the industrially important ones such as yeasts, molds, and actinomycetes. Also, the popular host for the expression of "foreign" proteins, *Escherichia coli*, is a member of this group. The fermentation media that support these important microorganisms will be the focus of this chapter.

3 Nutritional Requirements

The fermentation medium must be formulated to supply appropriate energy as well as meet the elemental and specific nutrient requirements of the microbe, represented as

$$\text{C-source} + \text{N-source} + \text{minerals} + \text{oxygen} \rightarrow \text{cell mass} + \text{product} + CO_2 + H_2O + \Delta H$$

3.1 Carbon Source

As indicated above, reduced carbon compounds are utilized as carbon sources for building cell mass and forming product, as well as acting as an energy source. Although glucose is frequently added to both growth and production media to serve these roles, there are a number of other naturally occurring organic compounds (glycerol, lactose, etc.) that may be employed. The level of the carbon source in media is quite important. If the glucose concentration in the medium is increased to 50 g/L or more, bacterial growth begins to decrease due to dehydration of the cell (plasmolysis). Much higher levels of glucose (greater than 200 g/L) are tolerated by yeast and molds due to their lower dependency on water. Also, at certain concentrations, the carbon source may repress one or more of the enzymes responsible for product formation; this is referred to as carbon catabolite repression or carbon source repression (DEMAIN, 1982, 1989). HEIM et al. (1984) found that in *Cephalosporium acremonium*, a key enzyme (deacetoxycephalosporin C synthase) in the biosynthetic pathway for cephalosporin C formation, was sensitive to carbon catabolite repression. This same nutrient control mechanism was also demonstrated in the biosynthesis of cephamycin C in *Nocardia lactamdurans* where glucose was shown to repress the biosynthetic enzymes δ(α-aminoadipyl)-cysteinyl-valine synthetase and deacetoxycephalosporin C synthase (CORES et al., 1986). One approach to avoid carbon catabolite repression is to feed the carbon source at a rate equal to its consumption, thus keeping the substrate below the repression level. Another approach is to use carbon sources that do not exhibit catabolite repression. These are often monosaccharides other than glucose, polysaccharides, oligosaccharides, or oils.

The energy generated during carbon metabolism is captured in high-energy phosphate compounds such as adenosine triphosphate (ATP) either by substrate level phosphorylation (anaerobic respiration in which organic compounds serve as hydrogen or electron acceptors in a series of oxidation–reduction reactions) or by oxidative phosphorylation (aerobic respiration in which molecular oxygen is the ultimate acceptor). This captured energy is utilized in subsequent energy-requiring biosynthetic reactions and in cell maintenance. Some of the energy generated during carbon metabolism is also given off as heat, an important issue that frequently requires attention during fermentation scale-up.

3.2 Nitrogen Source

Most of the bacteria, molds, and yeasts utilize ammonia or nitrate for the synthesis of nitrogenous compounds such as amino acids, purines, and pyrimidines. Many also form utilizable ammonia from the degradation of organic nitrogenous compounds such as proteins, peptides, or amino acids. Under circumstances where a carbon source becomes limited in the medium, amino acids may be utilized as a carbon source, resulting in an accumulation of ammonia. As previously described for carbon sources, certain nitrogen sources, frequently ammonium ions, can also repress the synthesis of secondary metabolites (SHAPIRO, 1989). Cephalosporin production in *Streptomyces clavuligerus* was repressed by several inorganic ammonium salts, including 40 mM ammonium chloride (AHARONOWITZ and DEMAIN, 1979). Similarly, ammonia (200 mM NH_4Cl) repressed lincomycin production in *Streptomyces lincolnensis* without affecting cell growth (YOUNG et al., 1985). The addition of ammonium-trapping agents such as natural zeolites or magnesium phosphate to control the level of nitrogen was reported to increase tylosin production by *Streptomyces fradiae* (MASUMA et al., 1983) and leucomycin production by *Streptomyces kitasatoensis* (TANAKA et al., 1981), respectively. The nitrogen level may also be controlled by appropriate feeding strategies, such as in the production of penicillin (SWARTZ, 1979).

3.3 Minerals

The mineral requirements of most microorganisms are met by the addition of phosphate, sulfate, magnesium, manganese, iron, potassium, and chloride salts. Frequently trace amounts (0.1–100 μM) of copper, zinc, cobalt, molybdenum, and calcium are required and are usually supplied by the water and as contaminants of major medium constituents. Many of these metal ions influence the biosynthesis of secondary metabolites (WEINBERG, 1989). For example, patulin biosynthesis in *Penicillium urticae* has an essential requirement for manganese (SCOTT and GAUCHER, 1986; SCOTT et al., 1986). Inorganic phosphate is of special interest, since it has been found to repress the synthesis of several secondary metabolites (MARTIN, 1989). Phosphate was shown to repress at least one enzyme (*p*-aminobenzoic acid synthetase) in the biosynthesis of candicidin in *Streptomyces griseus* (MARTIN et al., 1981).

3.4 Oxygen

Oxygen is a key nutrient for aerobic microbes and is commonly found as a constituent of cellular water and organic compounds. Carbohydrates are excellent sources of oxygen for many microorganisms. As discussed above, microorganisms that are dependent on respiration for generating energy require molecular oxygen as the final hydrogen or electron acceptor. Because of the low solubility of gaseous oxygen in water (6.99 ppm at 35 °C), continual transfer of oxygen from air–liquid surfaces to the growth medium is required to meet the oxygen demand of most microbial cultures. Oxygen transfer in flask culturing is achieved by shaking the flask and is enhanced by employing flasks with baffles. In fermentors, aeration is achieved by sparging and agitation. As medium optimization proceeds at shake-flask scale, the liquid volume is frequently reduced to achieve adequate aeration. The potential problem of medium evaporation under reduced volume cultivation is minimized by maintaining a high relative humidity level in the incubation chamber.

3.5 Specific Nutrients

Many microorganisms exhibit specific requirements for nutrients that they are not able to synthesize, such as amino acids, purines, pyrimidines, and vitamins. Compounds required by a specific microorganism are referred to as growth factors for that microbe. Biotin is required for the growth of many yeasts and molds. The heme portion of hemoglobin, and nicotinamide adenine dinucleotide are required by the fastidious microbe, *Hemophilus influenzae* type B, used to produce polysaccharide for the *Hemophilus* conjugate vaccine PedvaxHIB®. Initial medium formulations usually have one or more complex nutrients such as yeast extract to supply a specific unknown nutrient(s) that the microbe may require.

4 Environmental Requirements

Both the hydrogen ion concentration of the medium and the temperature of cultivation are important environmental parameters to consider in developing and evaluating new medium formulations. The pH of the nutrient medium often affects growth and product formation by affecting uptake of nutrients and other physiological activities. Bacteria exhibit an optimal pH range of 6.5–7.5 for growth, whereas a lower range of 4.5–5.5 is optimum for yeasts and molds (MOAT, 1988). The optimal pH for growth may not be the optimum for product formation. Maximum penicillin production in industrial strains of *Penicillium chrysogenum* (a mold) was achieved at a pH of > 6.0 and < 7.3 (PAN et al., 1972). Control of pH in shake-flask fermentations can be quite difficult, as acidic and alkaline metabolites are produced during cultivation. Frequently, buffers are used at shake-flask scale to control pH. Organic buffers such as TRIS or MOPS are often used when phosphate is repressive to the process.

Most of the industrially important microorganisms are mesophilic, with optimum growth and product formation occurring at temperatures of 20–40 °C. Early in medium development, the temperatures for optimal cell growth and product formation are determined and again confirmed as medium development proceeds.

5 Medium Formulation

Nutrient media may be formulated with chemically defined or chemically undefined ingredients (such as yeast extract, peptones, corn steep liquor, soybean meal, cottonseed flour, fish meal, etc.). The former formulations are referred to as synthetic or more properly, chemically defined media and the latter as complex, rich, or crude media.

5.1 Chemically Defined Media

Chemically defined media are commonly used at laboratory scale to study the effects nutrients may have on cell growth and biosynthesis of a particular metabolite of interest (MASUMA et al., 1983; YOUNG et al., 1985). For example, the chemically defined medium employed to study cephamycin C formation in *Nocardia lactamdurans* is shown in Tab. 1 (GREASHAM and INAMINE, 1986). The economics of using chemically defined media for production of fermentation products are usually unattractive – too expensive, sub-maximal productivity, etc. However, in fermentations where product yield and process consistency are of prime importance (such as the production of biologics), chemically defined media become quite attractive. For these relatively high-cost products, production media may be formulated with higher-priced ingredients.

Currently, the bacterium *Escherichia coli* and the yeast *Saccharomyces cerevisiae* are commercially important hosts for production of recombinant proteins, notably insulin, human growth hormone, and hepatitis B surface antigen. Chemically defined media to support the growth of these microbes as well as their expression of foreign genes may be formulated initially based on the elemental composition of these cells (Tab. 2). For example, the minimum level of nitrogen (supplied as ammonium sulfate) to support growth of *Escherichia coli* to a concentration of 50 g dry cell weight (DCW) per liter may be calculated as follows:

$$\frac{50\text{ g DCW}}{\text{L}}\cdot\frac{0.125\text{ g N}}{\text{g DCW}}\cdot\frac{132\text{ g }(NH_4)_2SO_4/\text{mol }(NH_4)_2SO_4}{28\text{ g N/mol }(NH_4)_2SO_4/\text{L}}=\frac{29.5\text{ g }(NH_4)_2SO_4}{\text{L}} \quad (1)$$

This approach may be used to determine the minimum required level of the other elements in Tab. 2 except carbon. As discussed previously, the carbon source serves cellular functions in addition to production of cell mass. Thus, the minimum level of carbon can be calculated using the cell yield coefficient (g DCW per g of nutrient). If unknown, the coefficient may be determined experimentally by plotting cell mass achieved against increasing levels of carbon. The slope of the line would be the cell yield coefficient for carbon. Additionally, recombinant cultures for commercial use are usually constructed with two to three nutritional markers, requiring the medium to have special growth factors such as a specific amino acid or purine. Once the initial medium is formulated and evaluated, it is subjected to optimization studies.

Tab. 1. Chemically Defined Medium for Cephamycin C Biosynthesis

Ingredient	Concentration (g/L)
Glucose	10.0
Sodium L-glutamate	4.25
NH_4Cl	1.0
Inositol	0.2
K_2HPO_4	2.0
NaCl	0.5
$MgSO_4\cdot 7H_2O$	0.5
$CaCO_3$	0.25
$FeSO_4\cdot 7H_2O$	0.025
$ZnSO_4\cdot 7H_2O$	0.01
$MnSO_4\cdot H_2O$	0.005
p-Aminobenzoic acid	0.001

Tab. 2. Typical Elemental Composition of Bacteria and Yeast

Element	Bacteria	Yeast
	(average % dry cell weight)	
Carbon	48	48
Nitrogen	12.5	7.5
	(g/100 g dry cell weight)	
Phosphorus	2.5	0.4
Sulfur	0.6	0.3
Potassium	2.3	1.4
Magnesium	0.3	0.2
Sodium	0.8	0.3
Calcium	0.6	0.8
Iron	0.01	0.15
Copper	0.02	–
Manganese	0.01	–

5.2 Complex Media

Complex media are by far the most frequently used in the fermentation industry. These media usually contain inexpensive raw materials that are by-products of meat, cheese, grain, and fiber processing (MILLER and CHURCHILL, 1986; CEJKA, 1985). Some of these raw materials are listed in Tab. 3, grouped by their major nutritional value for cellular activities. Also, as shown in Tab. 4 (ZABRISKIE et al., 1988), many of these materials provide more than one nutrient.

Usually, complex media for production of new secondary metabolites are empirically formulated, since (in most cases) little is known about the microbe or biosynthesis of the desired compounds. From previous experience or a search of the literature, media may be formulated and adjusted by a few laboratory-

Tab. 3. Nutrients Frequently Used in Commercial Fermentation Processes

Carbon Sources:
Cerelose (commercial glucose)
Glycerol
Molasses (Black strap or cane)
Oils (soybean, corn, and cottonseed), methyl oleate
Corn starch, dextrins, and hydrolysates
Whey (65% lactose)
Alcohols (such as methanol)

Nitrogen Sources:
Cottonseed flour
Soybean meal, flour, or grits
Peanut meal
Dried distillers' solubles
Whole yeast, yeast extract, or yeast hydrolysates
Corn steep liquor or its powder
Corn gluten meal
Linseed meal
Fish meal
Urea
Ammonium sulfate
Ammonia gas

Tab. 4. Fermentation Media Ingredient Analysis

Ingredient	Major Components						Mineral Content					
	Dry matter (%)	Protein (%)	Carbohydrates (%)	Fat (%)	Fiber (%)	Ash (%)	Calcium (%)	Magnesium (%)	Phosphorus (%)	Available phosphorus (%)	Potassium (%)	Sulfur (%)
Cerelose (commercial glucose)	91.5	–	91.5	–	–	–	–	–	–	–	–	–
Corn	82.0	9.9	69.2	4.4	2.2	1.3	0.02	0.11	0.28	0.1	0.31	0.08
Corn germ meal	93.0	22.6	53.2	1.9	9.5	3.3	0.3	0.16	0.5	0.16	0.34	0.32
Corn gluten feed, 21%	90.0	22.3	57.1	3.8	8.0	2.5	0.44	0.29	0.57	0.19	0.57	0.16
Corn gluten meal, 41%	91.0	42.0	40.2	2.5	4.3	2.0	0.14	0.05	0.46	0.16	0.03	0.50
Corn gluten meal, 60%	90.0	62.0	20.0	2.5	1.6	1.8	0.0	0.06	0.54	0.19	0.05	0.83
Corn steep liquor	50.0	24.0	5.8	1.0	1.0	8.8	–	–	–	–	–	–
Corn steep powder	95.0	48.0	–	0.4	–	17	0.06	1.5	3.3	1.1	4.5	0.58
Dried distillers' solubles	92.0	26.0	45.0	9.0	4.0	8.0	0.30	0.65	1.3	1.2	1.75	0.37
Edamine (protein hydrolysates)	–	7.5	–	–	–	6.0	–	–	–	–	–	–
Fish meal (anchovy), 65%	92.0	65.0	–	3.8	1.0	21.3	4.0	0.25	2.6	2.60	0.74	0.54
Fish meal (herring), 70%	93.0	72.0	–	7.5	1.0	–	2.0	0.14	1.5	1.50	1.12	0.62
Fish meal (menhaden), 60%	92.0	61.0	–	7.5	1.0	–	5.2	0.14	2.9	2.9	0.73	0.45
Fish solubles	50.0	32.0	7.0	3.0	1.0	10.0	0.05	0.04	0.49	0.49	1.48	0.13
Linseed meal	92.0	36.0	38.0	0.5	9.5	6.5	0.4	0.56	0.9	0.3	1.22	0.39
Molasses, beet	77.0	6.7	65.1	0.0	0.0	5.2	0.16	0.23	0.02	0.01	4.71	0.47
Peanut meal and hulls	90.5	45.0	23.0	5.0	12.0	5.5	0.15	0.32	0.55	0.2	1.12	0.28
Pharmamedia (cottonseed flour)	99.0	59.2	24.13	4.02	2.55	6.71	0.25	0.74	1.31	0.31	1.72	0.6
Proflo (cottonseed flour)	98.8	61.1	23.2	4.1	3.19	6.73	0.25	0.75	1.31	0.35	1.73	1.56
Soybean meal, expeller	90.0	42.0	29.9	4.0	6.0	6.5	0.25	0.25	0.63	0.16	1.75	0.32
Soybean meal, solvent	90.0	45.0	32.2	0.8	6.5	5.5	0.25	0.27	0.6	0.15	1.92	0.32
Whey, dried	95.0	12.0	68.0	1.0	0.0	9.6	0.9	0.13	0.75	0.75	1.20	1.04
Yeast, brewers	95.0	43.0	39.5	1.5	1.5	7.0	0.1	0.25	1.4	1.4	1.48	0.49
Yeast, hydrolysate	94.5	52.5		0.0	1.5	10.0	–	–	–	–	–	–

Modified from ZABRISKIE et al. (1988)

scale experiments. Although many of these initial formulations may support low product formation, they are usually sufficient to make material for initial isolation and product evaluation studies. If the product candidate continues to look attractive after preliminary evaluation, medium optimization is initiated in concert with culture improvement and biochemical studies.

6 Medium Optimization

Fermentation processes usually employ two media, one for inoculum development and the other for growth and production of the product. The function of the former is that of supporting good cell growth with little, if any, production of the product. Although optimization of inoculum media usually focuses on cell growth, there are instances where other parameters are used. For example, spore inocula are desirable for some fungal fermentation processes (CALAM, 1976), requiring medium optimization to focus on maximal spore formation in submerged culture. The function of the production medium is self-explanatory. In addition to optimizing productivity, one may want to minimize the synthesis of compounds structurally similar to the product (to enhance downstream processing). Through careful selection and optimization of the appropriate parameters, highly productive and economical media can be developed.

Vitamin Content							Amino Acids															
Biotin (mg/kg)	Choline (mg/kg)	Niacin (mg/kg)	Pantothenic acid (mg/kg)	Pyridoxine (mg/kg)	Riboflavin (mg/kg)	Thiamine (mg/kg)	Arginine (%)	Cystine (%)	Glycine (%)	Histidine (%)	Isoleucine (%)	Leucine (%)	Lysine (%)	Methionine (%)	Phenylalanine (%)	Threonine (%)	Tryptophan (%)	Tyrosine (%)	Valine (%)
–	–	–	–	–	–	–	–	–	–	–	–	–	–	–	–	–	–	–	–
	528	22.0	5.72	7.6	1.1	–	0.50	0.09	0.43	0.20	0.40	1.10	0.20	0.17	0.50	0.40	0.10	–	0.40
–	1760	41.8	4.4	–	3.74	–	1.3	0.4	–	–	–	–	0.9	0.57	–	1.1	0.18	–	–
–	2420	74.8	17.16	–	2.42	–	1.0	0.5	–	–	–	–	0.6	0.2	–	0.4	0.1	–	–
–	1320	77.0	10.34	–	2.2	–	1.4	0.8	–	–	–	–	0.8	1.2	–	1.4	0.2	–	–
–	220	81.4	2.86	–	2.2	0.1	1.9	1.1	–	–	–	–	1.0	1.9	–	2.0	0.25	–	–
0.88	–	–	–	19.36		0.88	0.4	0.5	1.1	0.3	0.9	0.1	0.2	0.5	0.3	–	–	0.1	0.5
–	5.6	0.16	0.03	0.02	0.01	0.01	3.3	1.9	5.1	2.8	3.6	11.3	2.5	1.9	4.4	4.0	–	3.4	5.8
2.86	4400	110.0	19.8	–	15.4	5.5	1.0	0.6	1.1	0.7	1.6	2.1	0.9	0.6	1.5	1.0	0.2	0.7	1.5
–	–	–	–	–	–	–	3.1	2.2	1.7	1.9	5.4	10.0	10.0	1.8	3.4	3.8	2.1	3.3	4.1
–	3740	93.5	9.68	–	9.46	–	3.6	0.6	6.31	1.25	2.43	4.27	4.7	1.9	2.37	2.80	0.7	1.91	2.83
–	3960	88.0	8.8	–	9.02	–	4.2	0.7	3.53	1.34	2.86	4.70	5.7	2.0	2.52	2.96	0.8	1.76	3.61
–	2860	55.0	8.8	–	4.84	–	3.6	0.5	–	–	–	–	4.6	1.7	–	2.96	0.6	–	–
	2860	165.0	35.2	12.54	1.32	5.5	2.2	1.4	5.3	2.4	1.5	2.2	7.6	2.4	1.3	1.1	0.7	–	1.4
–	1848	35.2	17.6	–	3.08	8.8	2.5	0.6	0.23	0.5	1.3	2.1	1.0	0.8	1.8	1.4	0.7	1.7	1.8
–	880	39.6	4.62	–	2.2	–	–	–	–	–	–	–	–	–	–	–	–	–	–
–	1672	167.2	48.4	–	5.28	7.26	4.6	0.7	3.0	1.0	2.0	3.1	1.3	0.6	2.3	1.4	0.5	–	2.2
1.52	3270	83.30	12.40	16.40	4.82	3.99	12.28	1.52	3.78	2.96	3.29	6.11	4.49	1.52	5.92	3.31	0.95	3.42	4.57
0.792	33.44	83.82	12.47	0.88	5.1	4.36	7.15	1.45	2.75	1.98	2.42	6.46	3.3	1.4	4.17	2.46	0.86	2.53	3.25
–	2420	30.36	14.08	–	3.08	–	2.9	0.62	–	–	–	–	2.8	0.59	–	1.72	0.59	–	–
–	2673	26.4	14.52	–	3.3	–	3.2	0.67	2.92	1.06	2,25	3.42	3.0	0.65	2.14	1.9	0.63	1.71	2.44
–	2420	11.0	48.4	2.86	19.8	3.96	0.4	0.4	0.7	0.2	0.7	1.2	1.0	0.4	0.5	0.6	0.2	0.5	0.6
–	4840	498.3	121.44	49.72	35.2	74.8	2.2	0.6	3.4	1.3	2.7	3.3	3.4	1.0	1.8	2.5	0.8	1.9	2.4
–	–	–	–	–	–	–	3.3	1.4	–	1.6	5.5	6.2	6.5	2.1	3.7	3.5	1.2	4.6	4.4

6.1 Statistical-Mathematical Approaches

The classical approach to medium optimization in which only one medium component is changed at a time (single-dimensional search) is rapidly being replaced by powerful statistical approaches. These approaches have proven to be quite efficient not only in optimizing media for the original production culture but also for subsequently improved cultures (MONAGHAN and KOUPAL, 1989; OMSTEAD et al., 1989). Two of several statistical-mathematical methods of optimization that have proven valuable in expediting medium development are presented.

6.1.1 Plackett–Burman Design

The initial stages of medium optimization frequently involve a large number of independent variables such as potential nutrients. Use of a full factorial search to examine these variables would require an impractical number of experimental trials, making such an approach unattractive. For example, if only eight potential nutrients were screened at two concentrations in a full factorial search (i. e., every possible combination of nutrients at two levels being examined), a total of 2^8 or 256 experimental trials would be required. An alternative and attractive approach to screening only a few of the important variables from a large number of possible variables is the use of a design that is a fraction of the two-factorial design, such as the Plackett–Burman designs (PLACKETT and BURMAN, 1946; STOWE and MOYER, 1966). These designs are available for 8 to 100 experiments (at increments of four); designs of 12, 20, and 28 experiments have proven to be most useful. Although the maximum number of variables which these designs can handle are 11, 19, and 27, respectively (i. e., $N-1$ variables in N experiments), the actual number of "assigned" variables, in practice, is usually less since at least three variables in each design should be "unassigned" (used to estimate the experimental error). The Plackett–Burman design for 12 experiments (trials) is shown in Tab. 5. The factors (x_1–x_{11}) are identified across the top of the table. For convenience, the first eight are "assigned" independent variables, and the last three are "unassigned". The trials are numbered in the left-hand column. Each variable is tried at a high (H) and low (L) level. Selection of the levels should be bold; however, too large of a level differential

Tab. 5. Plackett–Burman Design for Twelve Experimental Trials

Trial	Random Order[a]	Assigned Variables								Unassigned Variables		
		x_1	x_2	x_3	x_4	x_5	x_6	x_7	x_8	x_9	x_{10}	x_{11}
1		H	H	L	H	H	H	L	L	L	H	L
2		L	H	H	L	H	H	H	L	L	L	H
3		H	L	H	H	L	H	H	H	L	L	L
4		L	H	L	H	H	L	H	H	H	L	L
5		L	L	H	L	H	H	L	H	H	H	L
6		L	L	L	H	L	H	H	L	H	H	H
7		H	L	L	L	H	L	H	H	L	H	H
8		H	H	L	L	L	H	L	H	H	L	H
9		H	H	H	L	L	L	H	L	H	H	L
10		L	H	H	H	L	L	L	H	L	H	H
11		H	L	H	H	H	L	L	L	H	L	H
12		L	L	L	L	L	L	L	L	L	L	L

[a] Use random number table
H, high level; L, low level

for sensitive variables may mask the results of the others. Inspection of the matrix shows that each variable is tested six times at its high and low levels. Also, each time that independent variable x_1 is tested at its high level, x_2 is tested three times at its low and high levels. Likewise, when x_1 is tested at its low level, x_2 is again tested three times at its low and high levels. This matrix design cancels the effect of changing x_2 in calculating the effect of x_1. Since this same design holds true for the other variables, each variable is independently evaluated, preventing any estimate of the interactions between variables. This characteristic of the design is acceptable in situations where the interactions between variables are not apparent.

After the experimental trials are completed and the responses (such as final product concentration) determined, the effect of each variable on the response is easily determined by calculating the difference between the average responses at its high level and those at its low level. For this 12-trial example, the effect (E) variable x_1 had on the response is calculated as follows:

$$E_{X1} = \frac{\begin{array}{c}\text{total of}\\ \text{6 responses}\\ \text{at high level}\end{array}}{6} - \frac{\begin{array}{c}\text{total of}\\ \text{6 responses}\\ \text{at low level}\end{array}}{6} \qquad (2)$$

The variance of an effect (V_{eff}) is estimated by averaging the square of the "unassigned" variable effects (in this example variables x_9, x_{10}, and x_{11}).

$$V_{eff} = \frac{(E_{X9}^2 + E_{X10}^2 + E_{X11}^2)}{3} \qquad (3)$$

The number of "unassigned" variables represents the degrees of freedom for error.

The standard error (SE) of an effect is determined by taking the square root of the variance.

$$SE = \sqrt{V_{eff}} \qquad (4)$$

The significance level (P value) of each effect is determined by the familiar Student t test:

$$t_x = E_x / SE \text{ (at 3 degrees of freedom)} \qquad (5)$$

where t_x is the effect of variable x. The t test provides for an evaluation of the probability of finding the observed effect by chance. If the probability is found to be small, the idea that the observed effect is a result of varying the level of the variable is accepted. The important variables are identified based on the significance level of their effects (i. e., the ones with the higher values).

Other Plackett–Burman designs may be generated from the first row of the matrix opposite the number of experiments (N) shown in Tab. 6. The matrix is completed in a cyclic manner by shifting each subsequent row one place for N minus two times. The last row of the matrix is added as all "low" elements.

Once the key variables have been identified, determination of their optimal levels is required. This objective may be achieved by using response surface optimization techniques.

6.1.2 Response Surface Design

The response surface type of experimental program involves generating a contour plot (or map) from the linear, interaction and quadratic effects of two or more variables. Usually, a practical limit is five variables. These effects

Tab. 6. First Row of Three Plackett–Burman Designs

Number of Elements (N)	Element
16	H H H H L H L H H L L H L L L
20	H H L L H H H H L H L H L L L L H H L
24	H H H H H L H L H H L L H H L L H L H L L L L

H, high level; L, low level

are described for three independent variables in the following full quadratic polynomial model:

$$Y = b_0 + b_1 x_1 + b_2 x_2 + b_3 x_3 + \\ + b_{12} x_1 x_2 + b_{13} x_1 x_3 + b_{23} x_2 x_3 + \\ + b_{11} x_1^2 + b_{22} x_2^2 + b_{33} x_3^2 \quad (6)$$

where Y is the dependent variable (such as predicted product yield); b_0 is the regression coefficient at center point; b_1, b_2, and b_3 are linear coefficients; x_1, x_2, and x_3 are independent variables; b_{12}, b_{13}, and b_{23} are second-order interaction coefficients; and b_{11}, b_{22}, and b_{33} are quadratic coefficients.

Since curvature effects are to be estimated in identifying the optimum, the experimental design must have at least three levels for each independent variable. Although full three-level factorials are feasible and can identify the optimum for each independent variable, the large number of experimental trials required usually make these factorials unattractive for rapid medium optimization. Therefore, the experimental designs that are commonly used employ fewer points than those in the corresponding full three-level factorials. Such designs include the Box–Wilson design (BOX and WILSON, 1951) and the Box–Behnken design (BOX and BEHNKEN, 1960). Since the Box–Wilson design employs five levels of each independent variable and the Box–Behnken design only three levels, the latter is frequently selected. An example of the Box–Behnken design for three independent variables is presented in Tab. 7. The independent variables (x_1–x_3) are listed at the top of matrix and the experimental trials in the left-hand column. The letters H, M, and L represent the high, middle, and low levels, respectively, of each independent variable. Usually, the values assigned to these levels are equally spaced.

Once the experimental trials are completed and the parameter of interest (dependent variable) determined, the coefficients of the quadratic polynomial model are calculated using regression analysis. Since there are several computer programs that can perform these analyses (such as the PROC RSREG routine in the SAS package; SAS Institute Inc., Cary, N.C.), consultation with a statistican proves helpful in selecting one. The calculated equations may be put into a program that generates contours of the responses against two independent variables. For more than two independent variables, additional contour plots are required. Although the objective is to identify the optimum, these contour plots may only suggest the region where the optimum may be located. A second set of experimental trials may be required to identify the optimum, indicating the sequential nature of this experimentation. The response surface technique was successfully employed by MCDANIEL et al. (1976) to expeditiously optimize the medium for candidin production by *Streptomyces viridoflavus*. As shown in Fig. 1, the optimum yield point was 3.2% soybean meal and 6.2% cerelose.

Tab. 7. Box–Behnken Design for Three Independent Variables

Trial	Random Order[a]	Variables		
		x_1	x_2	x_3
1		H	H	M
2		H	L	M
3		L	H	M
4		L	L	M
5		H	M	H
6		H	M	L
7		L	M	H
8		L	M	L
9		M	H	H
10		M	H	L
11		M	L	H
12		M	L	L
13		M	M	M

[a] Use random number table
H, high; L, low; M, medium

6.2 Biochemical Methods

Certain biochemical data are essential for rationally and systematically improving fermentation media (GREASHAM and INAMINE, 1986). Such data include information on precursors, biosynthetic pathways, specific nutrient requirements, and on the regulation of primary and secondary metabolite production,

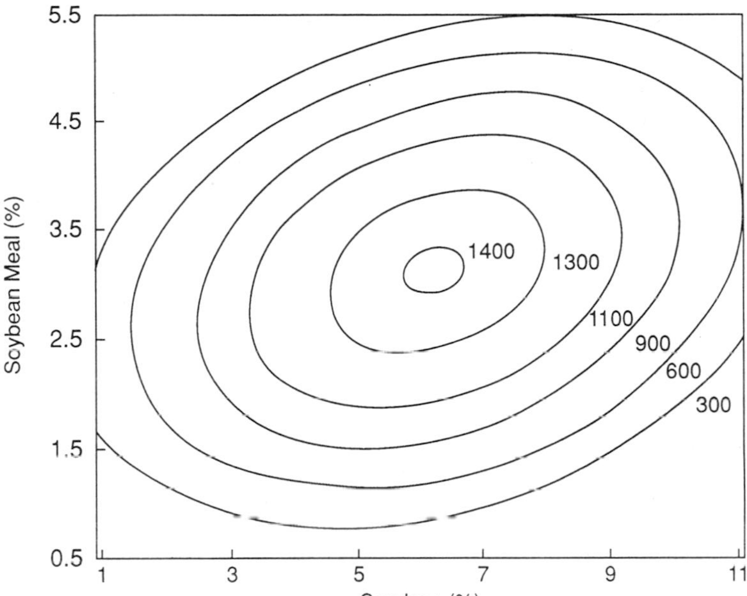

Fig. 1. Contour plot for independent variables, soybean meal and cerelose (redrawn from McDaniel et al., 1976).

especially by carbon, nitrogen, phosphate, and trace elements, as discussed earlier. These data are usually generated from precise biochemical studies that employ chemically defined media (Williams and Katz, 1977), avoiding the complicated interactions of complex media. Typical biochemical studies employed to enhance production of secondary metabolites include isotope incorporation studies, enzymatic studies, and chemical stimulation/inhibition studies. Use of these studies greatly improved (over 100%) the production of cephamycin C by *Nocardia lactamdurans* (Greasham and Inamine, 1986).

7 Economics

The major objectives in developing a fermentation process are to maximize product formation and minimize manufacturing costs, including raw materials. Since the most expensive ingredient of production media is usually glucose, costs may be lowered by using less-expensive complex sources (vegetable oils, corn starch, etc.). However, these complex sources which are usually poorly defined and variable, may increase fermentation variability, product isolation costs, and waste disposal costs, easily offsetting initial savings. In addition to the price of the medium ingredients, consideration should also be given to other associated costs, such as the cost of transporting, handling, storing, pretreating (e. g., enzymatic hydrolysis of starch), and special sterilization conditions. Also, nutrient availability and stability of the nutrient price are important.

Acknowledgement

I thank Professor Arnold Demain for his critical review of the manuscript and suggestions.

8 References

Aharonowitz, Y., Demain, A. L. (1979), Nitrogen nutrition and regulation of cephalosporin production in *Streptomyces clavuligerus, Can. J. Microbiol.* **25**, 61–67.

Box, G. E. P., Behnken, D. W. (1960), Some new three level designs for the study of quantitative variables, *Technometrics* **2**, 455–475.

BOX, G. E. P., WILSON, K. B. (1951), On the experimental attainment of optimum conditions, *J. R. Stat. Soc.* **B 13**, 1–45.

CALAM, C. T. (1976), Starting investigational and production cultures, *Process Biochem.* **11**, 7–12.

CEJKA, A. (1985), Preparation of media, in: *Biotechnology* (REHM, H.-J., REED, G., Eds.), 1st Ed., Vol. 2, pp. 629–698, Weinheim–Deerfield Beach, FL–Basel: VCH.

CORTES, J., LIRAS, P., CASTRO, J. M., MARTIN, J. F. (1986), Glucose regulation of cephamycin biosynthesis in *Streptomyces lactamdurans* is exerted on the formation of α-aminoadipyl-cysteinyl-valine and deacetoxycephalosporin C synthase, *J. Gen. Microbiol.* **132**, 1805–1814.

DEMAIN, A. L. (1982), Catabolite regulation in industrial microbiology, in: *Overproduction of Microbial Products* (KRUMPHANZI, B., SIKYTA, B., VANEK, Z., Eds.), pp. 3-20, New York: Academic Press.

DEMAIN, A. L. (1989), Carbon source regulation of idiolite biosynthesis in actinomycetes, in: *Regulation of Secondary Metabolism in Actinomycetes* (SHAPIRO, S., Ed.), Chap. 4, Boca Raton: CRC Press.

GREASHAM, R., INAMINE, E. (1986), Nutritional improvement of processes, in: *Manual of Industrial Microbiology and Biotechnology* (DEMAIN, A. L., SOLOMON, N. A., Eds.), pp. 41–48, Washington, DC: American Society for Microbiology.

HEIM, J., SHEN, Y.-Q., WOLFE, S., DEMAIN, A. L. (1984), Regulation of isopenicillin N synthetase and deacetoxycephalosporin C synthetase by carbon source during the fermentation of *Cephalosporium acremonium, Appl. Microbiol. Biotechnol.* **19**, 232–236.

MARTIN, J. F. (1989), Molecular mechanisms for the control of phosphate of the biosynthesis of antibiotics and other secondary metabolites, in: *Regulation of Secondary Metabolism in Actinomycetes* (Shapiro, S., Ed.), Chap. 6, Boca Raton: CRC Press.

MARTIN, J. F., ALEGRE, M. T., GIL, J. A., NAHARRO, G. (1981), Characterization of mutants of *Streptomyces griseus* deregulated or blocked in candicidin biosynthesis: molecular mechanism of control of antibiotic biosynthesis, *Adv. Biotechnol.* **3**, 129–134.

MASUMA, R., TANAKA, Y., OMURA, S. (1983), Ammonium ion-depressed fermentation of tylosin by the use of a natural zeolite and its significance in the study of biosynthetic regulation of the antibiotic, *J. Ferment. Technol.* **61**, 607–614.

McDANIEL, L. E., BAILEY, E. C., ETHIRAJ, S., ANDREWS, H. P. (1976), Application of response surface optimization techniques to polyene macrolide fermentation studies in shake flasks, *Dev. Ind. Microbiol.* **17**, 91–98.

MILLER, T. L., CHURCHILL, B. W. (1986), Substrates for large-scale fermentations, in: *Manual of Industrial Microbiology and Biotechnology* (DEMAIN, A. L., SOLOMON, N. A., Eds.), pp. 122–136, Washington, DC: American Society for Microbiology.

MOAT, A. G., FOSTER, J. W. (1988), *Microbial Physiology*, 2nd Ed., Chap. 1, New York: John Wiley & Sons.

MONAGHAN, R. L., KOUPAL, L. R. (1989), Use of Plackett & Burman technique in a discovery program for new natural products, in: *Novel Microbial Products for Medicine and Agriculture* (DEMAIN, A. L., SOMKUTI, G. A., HUNTER-CEVERA, J. C., ROSSMOORE, H. W., Eds.), pp. 25–32, New York: Elsevier.

OMSTEAD, M. N., KAPLAN, L., BUCKLAND, B. C. (1989), Fermentation development and process improvement, in: *Ivermectin and Abamectin* (CAMPBELL, W. C., Ed.), pp. 33–54, New York: Springer-Verlag.

PAN, C., HEPLER, L., ELANDER, R. (1972), Control of pH and carbohydrate addition in the penicillin fermentation, *Dev. Ind. Microbiol.* **13**, 103–112.

PLACKETT, R. L., BURMAN, J. P. (1946), The design of optimum multifactorial experiments, *Biometrika* **33**, 305–325.

SCOTT, R. E., GAUCHER, G. M. (1986), Manganese and antibiotic biosynthesis. II. Cellular levels of manganese during the transition of patulin production in *Penicillium urticae, Can. J. Microbiol.* **32**, 268–272.

SCOTT, R. E., JONES, A., GAUCHER, G. M. (1986), Manganese and antibiotic biosynthesis. III. The site of manganese control of patulin production in *Penicillium urticae, Can. J. Microbiol.* **32**, 273–279.

SHAPIRO, S. (1989), Nitrogen assimilation in actinomycetes and the influence of nitrogen nutrition on actinomycete secondary metabolism, in: *Regulation of Secondary Metabolism in Actinomycetes* (SHAPIRO, S., Ed.), Chap. 5, Boca Raton: CRC Press.

STOWE, R. A., MOYER, R. P. (1966), Efficient screening of process variables, *Ind. Eng. Chem.* **58**, 36–40.

SWARTZ, R. W. (1979), The use of economic analysis of penicillin G manufacturing costs in establishing priorities for fermentation process improvement, in: *Annual Reports of Fermentation Processes* (PERLMAN, D., Ed.), Chap. 4, New York: Academic Press.

TANAKA, Y., TAKAHASHI, Y., MASUMA, R., IWAI, Y., TANAKA, H., OMURA, S. (1981), Enhancement and cultural characteristics of leucomycin production by *Streptomyces kitasatoensis* in the presence of magnesium phosphate, *Agric. Biol. Chem.* **45**, 2475–2481.

WEINBERG, E. D. (1989), Roles of micronutrients in secondary metabolism of actinomycetes, in: *Regulation of Secondary Metabolism in Actinomycetes* (SHAPIRO, S., Ed.), Chap. 7, Boca Raton: CRC Press.

WILLIAMS, W. K., KATZ, E. (1977), Development of a chemically defined medium for the synthesis of actinomycin D by *Streptomyces parvulus*. *Antimicrob. Agents Chemother.* **11**, 281–290.

YOUNG, M. D., KEMPE, L. L., BADER, F. G. (1985), Effects of phosphate, glucose, and ammonium on cell growth and lincomycin production by *Streptomyces lincolnensis* in chemically defined media, *Biotechnol. Bioeng.* **27**, 327–333.

ZABRISKIE, D. W., ARMIGER, W. B., PHILLIPS, D. H., ALBANO, P. A. (1988), *Trader's Guide to Fermentation Media Formulation*, Trader's Protein, Memphis, Tenn.

8 Media for Cell Culture

RICHARD A. WOLFE

St. Louis, MO 63167, USA

1 Introduction

In developing any process utilizing mammalian cell culture all of the aspects of upstream and downstream processing must be considered. The goal is to produce conditioned medium containing a high concentration of an intact, uniformly processed, crude product which can then be readily isolated from other cellular products and medium components. One must emphasize the interrelationship between the cultured cells (and their associated metabolic activities), the aqueous environment, the substratum, the bioreactor design (3-dimensional geometry, materials, mass transfer limitations, "shear stresses", etc.), and product of interest (including the associated downstream processing considerations). This concept dictates the approach for producing any product through mammalian cell culture technology.

In Tab. 1 many of the subjects discussed in detail elsewhere in this volume are grouped within these five categories to illustrate their interrelationships. This chapter focuses on the aqueous environment, which includes low molecular weight nutrients, hormones, transport proteins, buffers, and reagents intended to either reduce stresses imposed on the cultured cells in the bioreactor environment or stabilize the product. The medium, in many cases, also includes components such as attachment proteins and/or hormones that absorb to the surface(s) of the bioreactor and mediate interaction with the cultured cells.

The cells also produce proteins and other metabolites which modify the environment (both aqueous and substratal), and in many cases these can be detrimental to both the cells and the product of interest. Thus it is common practice to include in media antioxidants, protease inhibitors, and buffers. Furthermore, the cell-specific rates of medium utilization and metabolite production impact many of the values for process control parameters, which are in turn influenced by the medium design. Media components that destabilize or copurify with the product must, of course, be avoided.

While it is true that many factors of the production process influence medium design, it is, in practice, usually the case that the medium design dictates many aspects of the overall process strategy. In addition, the medium composition defines the requirements for most of the raw materials and their storage and quality control (QC). Accordingly, much of the overall cost of manufacturing the final product can depend on the medium composition and its degree of optimization.

Therefore, a common first step toward process design is the definition of the hormonal, nutritional, and substratum requirements of the cell line. In most cases, these are independent of the bioreactor design and process parameters. Evaluation of these data in light

Tab. 1. Mammalian Cell Bioreactor Systems – Interrelationship of Components

Cells	Aqueous Environment	Substratum Composition	3-D Geometry/ Structure	Product
Growth rate	Basal nutrients	Surface charge	Adherent/suspension	Isolation procedure
Shear sensitivity	Dissolved gases	Smooth/rough	Curvature of surface	Copurified supplements
Hormonal regulation	Hormones	Hydrophobicity	Cell immobilization	Interference by media
Nutritional needs	Growth inhibitors	Adsorbed proteins	Uniform perfusion	Copurified cell products
Toxic wastes	Inducers	Peptide analogs	Shear	pH stability
Attachment dependence	Redox state	Stability	Waste removal	Redox stability
Lifetime	pH/buffers	Sterility	Dissolved gasses	Temperature stability
Protease	Viscosity	Product adsorption	Temperature	Proteases
Induction of product	Cell products	Cell modification	pH control	Structural assay
Stability of construct	Carrier proteins	Functional group	Product removal	Functional assay
Secretion rate	Trace elements	Density of groups	Nutrient feed	Contaminant assays
Post-translational changes	Stability	Uniformity	Sterility	Sterility
Respiration rate	Safety	Safety	Safety	Safety
Safety	Regulatory issues	Regulatory issues	Regulatory issues	Regulatory issues
Regulatory issues	QA/QC[a]	QA/QC	QA/QC	QA/QC

[a] QA: quality assurance, QC: quality control

of the restrictions defined by the desired product suggest appropriate conditions, bioreactors, processes, and the course for medium optimization. It should be noted that even small modifications in the medium can influence cell line stability, product quality and yield, operational parameters of the bioreactor, and downstream processing.

2 Medium Composition

The aqueous environment to which the cells are exposed results from a combination of many of the factors discussed above. The "conditioned medium" in the bioreactor differs significantly from the feed stock. In addition to the control of dissolved gasses and pH, metabolites and other cell products may be either stimulatory or inhibitory to the cultured cells. Almost all of the medium components have both lower and upper limits in which the cells perform optimally, and these ranges are interdependent. The effective concentration range of a given amino acid, e.g., is dependent on the concentrations of other amino acids that are taken up by the cells through the same pumping mechanism. The medium and process parameters must therefore be designed such that no component either exceeds or falls below its effective concentration range (both net utilization and biosynthesis of specific amino acids are observed, dependent on the metabolite and the conditions). This concept of "metabolic balance" has been clearly expounded in relation to static culture systems (T-flasks) by R. HAM and his colleagues in several forums (HAM and MCKEEHAN, 1979; HAM, 1980).

In practice, an analytical comparison of fresh and conditioned medium provides considerable insight into medium optimization. Most of the components routinely included in media are discussed in detail below, and can be grouped into four categories: water, low molecular weight nutrients, proteins, and supplements of a non-hormonal/non-nutritional nature. Tab. 2 lists the components of four common medium formulations.

2.1 Water Quality

Water from any natural source is full of contaminants, and mammalian cells are exceedingly sensitive to many compounds that often contaminate the water used for media preparation (WAYMOUTH, 1981). This is particularly noticeable with "reduced protein" formulations, and contaminants not removed during the purification process or introduced during storage are common causes of aberrant cultures. These contaminants can be either

Tab. 2. Typical Basal Nutrient Media for Mammalian Cell Culture

Compound (g/mol)	DME (mg/L)	F-12 (mg/L)	Coon's F-12 (mg/L)	1:1 DME/Coon's F-12 (mg/L)	DME (mol/L)	F-12 (mol/L)	Coon's F-12 (mol/L)	1:1 DME/Coon's F-12 (mol/L)
Bulk ions and trace elements								
$CaCl_2$ (anh) (111.00)	200.0	33.220	124.300	151.035	1.80×10^{-3}	2.99×10^{-4}	1.12×10^{-3}	1.36×10^{-3}
$CuSO_4\cdot6\ H_2O$ (249.68)	0	0.002	0.002	0.001	0	9.97×10^{-9}	9.97×10^{-9}	4.01×10^{-9}
$FeSO_4\cdot7\ H_2O$ (278.02)	0	0.834	0.834	0.400	0	3.00×10^{-6}	3.00×10^{-6}	1.44×10^{-6}
$Fe(NO_3)_3\cdot9\ H_2O$ (404.00)	0.10	0	0	0.050	2.48×10^{-7}	0	0	1.24×10^{-7}
KCl (74.5)	400.00	223.600	305.000	347.000	5.37×10^{-3}	3.00×10^{-3}	4.09×10^{-3}	4.65×10^{-3}
KH_2PO_4 (anh) (136.09)	0	0	61.240	29.700	0	0	0	2.18×10^{-4}
$MgCl_2$ (anh) (95.22)	0	57.200	46.650	24.710	0	6.01×10^{-4}	4.90×10^{-4}	2.60×10^{-4}
$MgSO_4$ (anh) (120.40)	97.87	0	25.300	58.430	8.11×10^{-4}	0	0	4.85×10^{-4}
NaCl (58.44)	6400.00	7599.000	7517.000	6975.000	1.10×10^{-1}	1.30×10^{-1}	1.29×10^{-1}	1.19×10^{-1}
Na_2HPO_4 (142.1)	0	142.100	132.400	62.500	0	1.00×10^{-3}	0.93×10^{-3}	0.44×10^{-4}
$NaH_2PO_4\cdot2\ H_2O$ (156.00)	125.00	0	0	62.500	8.01×10^{-4}	0	0	2.86×10^{-4}
$ZnSO_4\cdot7\ H_2O$ (287.54)	0	0.863	0.863	0.072	0	3.00×10^{-6}	3.00×10^{-6}	2.50×10^{-7}
Buffers and other components								
Carbon dioxide (%)	10	5		5				
$NaHCO_3$ (84.01)	3700.00	1176.000	2676.000	1176.000	4.40×10^{-2}	1.40×10^{-2}	11.40×10^{-2}	1.40×10^{-2}

Tab. 2. Typical Basal Nutrient Media for Mammalian Cell Culture (Continued)

Compound (g/mol)	DME (mg/L)	F-12 (mg/L)	Coon's F-12 (mg/L)	1:1 DME/ Coon's F-12 (mg/L)	DME (mol/L)	F-12 (mol/L)	Coon's F-12 (mol/L)	1:1 DME/ Coon's F-12 (mol/L)
Essential amino acids								
L-ARG (HCl) (210.70)	84.00	211.000	422.000	115.750	3.99×10^{-4}	1.00×10^{-3}	2.00×10^{-3}	5.49×10^{-4}
L-CYS HCl·H_2O (175.60)	62.57	36.120	70.260	32.000	3.56×10^{-4}	2.00×10^{-4}	4.00×10^{-4}	1.82×10^{-4}
L-GLN (146.10)	584.00	146.100	292.100	438.000	4.00×10^{-3}	1.00×10^{-3}	2.00×10^{-3}	3.00×10^{-3}
L-HIS HCl·H_2O (209.70)	42.00	20.960	42.000	40.000	2.00×10^{-4}	1.00×10^{-4}	2.00×10^{-4}	1.91×10^{-4}
L-ILE (131.20)	105.00	3.940	7 800	56.500	8.00×10^{-4}	3.00×10^{-5}	6.00×10^{-5}	4.31×10^{-4}
L-LEU (131.20)	105.00	13.100	26.200	65.500	8.00×10^{-4}	9.98×10^{-5}	2.00×10^{-4}	4.99×10^{-4}
L-LYS HCl (182.70)	146.00	36.540	73.000	110.000	7.99×10^{-4}	2.00×10^{-4}	4.00×10^{-4}	6.02×10^{-4}
L-MET (149.20)	30.00	4.480	9.000	19.000	2.01×10^{-4}	3.00×10^{-5}	6.03×10^{-5}	1.27×10^{-4}
L-PHE (165.20)	66.00	4 960	10.000	38.000	4.00×10^{-4}	3.00×10^{-6}	6.50×10^{-5}	2.30×10^{-4}
L-THR (119.10)	95.00	11.910	23.810	59.500	7.98×10^{-4}	1.00×10^{-4}	2.00×10^{-4}	5.00×10^{-4}
L-TRP (204.20)	16.00	2.040	4.000	10.000	7.84×10^{-5}	9.99×10^{-6}	2.00×10^{-5}	4.90×10^{-5}
L-TYR (diNa salt)·2 H_2O (237.20)	103.79	7.780	15.860	59.105	4.38×10^{-4}	3.28×10^{-5}	6.68×10^{-5}	2.49×10^{-4}
L-VAL (117.20)	93.80	11.700	23.400	59.000	8.00×10^{-4}	9.98×10^{-5}	2.00×10^{-4}	5.03×10^{-4}
L-Cystine 2 HCl (310.80)	64.57	0	0	31.290	2.02×10^{-4}	0	1.01×10^{-4}	1.01×10^{-4}
Non essential amino acids								
L-ALA (89.09)	0	8.900	18.000	9.000	0	2.00×10^{-4}	9.99×10^{-5}	1.01×10^{-4}
L-ASN·H_2O (150.10)	0	15.010	30.000	13.000	0	2.00×10^{-4}	1.00×10^{-4}	8.66×10^{-5}
L-ASP (133.10)	0	13.300	26.000	13.000	0	2.00×10^{-4}	9.99×10^{-5}	9.77×10^{-5}
L-GLU (147.10)	0	14.710	30.000	15.000	0	2.00×10^{-4}	1.00×10^{-4}	1.02×10^{-4}
GLY (75.07)	30.00	7.500	16.000	23.000	4.00×10^{-4}	2.00×10^{-4}	9.99×10^{-5}	3.06×10^{-4}
L-PRO (115.10)	0	34.500	70.000	35.000	0	6.00×10^{-4}	3.00×10^{-4}	3.04×10^{-4}
L-SER (105.10)	42.00	10.510	21.000	32.000	4.00×10^{-4}	2.00×10^{-4}	1.00×10^{-4}	3.04×10^{-4}
Amino acid derivatives								
Putrescine 2 HCl (161.10)	0	0.161	0.161	0.150	0	9.99×10^{-7}	9.99×10^{-7}	9.31×10^{-7}
Water soluble vitamins and coenzymes								
Ascorbic acid (176.12)	0	0	15.000	7.500	0	8.52×10^{-5}	8.52×10^{-5}	4.26×10^{-5}
Biotin (244.30)	0	0.007	0.007	0.035	0	2.99×10^{-8}	2.99×10^{-8}	1.43×10^{-7}
D-Ca pantothenate (238.30)	4.00	0.238	0.238	2.250	1.68×10^{-5}	1.00×10^{-6}	1.00×10^{-6}	9.44×10^{-6}
Folic acid (441.40)	4.00	1.300	1.300	2.500	9.06×10^{-6}	2.95×10^{-6}	2.95×10^{-6}	5.66×10^{-6}
Niacinamide (122.10)	4.00	0.037	0.037	2.020	3.28×10^{-5}	3.03×10^{-7}	3.03×10^{-7}	1.65×10^{-5}
Pyridoxal HCl (203.60)	4.00	0	0	2.000	1.96×10^{-5}	0	0	9.82×10^{-6}
Pyridoxine HCl (205.60)	0	0.062	0.062	0.030	0	3.02×10^{-7}	3.02×10^{-7}	1.46×10^{-7}
Riboflavin (376.40)	0.40	0.038	0.038	0.220	1.06×10^{-6}	1.01×10^{-7}	1.01×10^{-7}	5.84×10^{-7}
Thiamine HCl (337.30)	4.00	0.340	0.340	2.150	1.19×10^{-5}	1.01×10^{-6}	1.01×10^{-6}	6.37×10^{-6}
Vitamin B_{12} (1355)	0	1.357	1.357	0.680	0	1.00×10^{-6}	1.00×10^{-6}	5.00×10^{-7}
Carbohydrates and derivatives								
D-Glucose (180.16)	4500	1802	1082	3151	2.50×10^{-2}	1.00×10^{-2}	1.00×10^{-2}	1.75×10^{-2}
Na pyruvate (110.00)	110	110	220	165	1.00×10^{-3}	1.00×10^{-3}	2.00×10^{-3}	1.50×10^{-4}
Purine derivatives								
Hypoxanthine (Na salt) (146.10)	0	4.2	4.2	2.34	0	2.87×10^{-5}	2.87×10^{-5}	1.60×10^{-5}
Pyrimidines								
Thymidine HCl (337.30)	0	0.730	0.730	0.350	0	2.16×10^{-6}	2.16×10^{-6}	1.04×10^{-6}
Lipids and derivatives								
Choline chloride (139.60)	4.00	13.960	13.960	9.000	2.87×10^{-5}	1.00×10^{-4}	1.00×10^{-4}	6.45×10^{-5}
i-Inositol (180.20)	7.20	18.000	118.000	2.600	4.00×10^{-5}	9.99×10^{-5}	9.99×10^{-5}	6.99×10^{-5}
Linoleic acid (280.40)	0	0.080	0.080	0.040	0	2.85×10^{-7}	2.85×10^{-7}	1.43×10^{-7}
Lipoic acid (Thioctic) (206.30)	0	0.210	0.210	0.100	0	1.02×10^{-6}	1.02×10^{-6}	4.85×10^{-7}
Calculated total osmolarity (mosm)	—	—	—	—	367	316	360	367
Calculated total sodium (mosm)	—	—	—	—	156	147	164	157
Calculated total potassium ion (mM)	—	—	—	—	5	3	5	5

growth stimulatory or growth inhibitory (with the latter being more common). Unpurified water can contain organics (from detergents or degraded biomass), inorganics (such as chlorine and heavy metals), microorganisms (which often result in endotoxin and pyrogen contamination), or particles. It is recommended that water of at least the pharmaceutical "WFI" (water for injection) quality be utilized for mammalian cell culture, and it should be noted that such preparations are not necessarily pure: they simply meet standards for injection. Agents present in water are diluted throughout the organism with injection, but remain in a relatively concentrated form in cell culture.

It is recommended that multiple techniques be utilized to purify water for cell culture. In the author's facility water is filtered, deionized, adsorbed with carbon, purified by RO (reverse osmosis), passed through a Milli-Q™ polishing unit (Millipore Corporation, deionization and carbon adsorption), depyrogenated via ultrafiltration, and then stored at 80 °C in stainless steel tanks until use. Routine maintenance and monitoring of the various stages of the purification stream are performed. Other techniques such as distillation and UV oxidation are commonly utilized.

2.2 Low Molecular Weight Nutrients

Bulk ions and trace elements should be considered nutrients because their adsorption, secretion, and concentration in a given organism are finely controlled (ZOMBOLA et al., 1979). The concentrations of these ions maintain the osmotic pressure of the microenvironment, and their distribution within the tissue (primarily Na^+ and K^+) maintains the membrane potential, and therefore both the absolute and relative amounts of sodium and potassium are important. The transient alterations in the availability of other ions such as Ca^{2+} and Mg^{2+} mediate transmembrane signaling events, and these ions are also necessary for integrin mediated cell adhesion. Amongst other biochemical roles, phosphate and bicarbonate ions buffer the acidity of the microenvironment. Other ions such as selenium, vanadium, copper, zinc, iron, molybdenum, and manganese are necessary as cofactors for certain enzymes. Depending on their concentration, trace elements can be either stimulatory or inhibitory in cell culture.

Trace elements are contained as contaminants in the reagents used to prepare media, and care must be taken to assure that such unknown materials remain consistent throughout a production operation. In order to insure this consistency, it is common practice to "ball mill" large batches of powdered nutrients, and perform QC on random aliquots. The batch is then packaged in appropriately sized containers since the uniformity of the powder decays with time. Preparation of media from these large lots then requires minimal QC (provided a good source of water is utilized), and verification of osmolarity, pH, O_2, CO_2 (Blood Gas Analyzer), and sterility is usually sufficient.

There are two major energy sources utilized by cultured mammalian cells; six carbon sugars and glutamine (REITZER et al., 1979). Although glucose is most commonly provided (2–20 mM) as a source of energy, cells can frequently be adapted to growth in galactose or other six carbon surgars (LEIBOVITZ, 1963). In this case, a sometimes desirable reduction in the rate of lactate production and six carbon sugar utilization is observed, presumably via a decrease in the steady state concentration of glycolytic intermediates. However, a concomitant reduction growth rate is often seen. Pyruvate can also function as an energy and carbon source. Although glutamine (2–10 mM) can serve as a source of energy and carbon, the byproducts of its metabolism are frequently toxic. Ammonia produced via the conversion of GLN to GLU can be detrimental at concentrations exceeding 2 mM, and cultures of cells like murine hybridomas tend to build up significant concentrations of this metabolite. Furthermore, GLN is unstable in aqueous solution even at 4 °C. Thus, long-term storage of media containing this amino acid should be avoided (GLN half-life is approximately 100 days). If necessary, GLN-free formulations can be supplemented just prior to use.

Amino acids provide the major nitrogen source in these cultures, and have been grouped into two categories as a result of tissue culture experiments; the essential and non-essential amino acids (see Tab. 2). However, most cultured cells perform better in the presence of the non-essential amino acids, and increases in the quantitative requirements for the essential amino acids are observed in the deprived cultures. A notable exception to this classification are the CHO cell-derived lines which are auxotrophic for proline (0.1–1 g/L generally required). Other amino acids such as SER and GLY have been shown to be beneficial supplements for low density cultures, but relatively unimportant under conditions generally employed in large-scale mammalian cell cultures.

Vitamins are enzymatic cofactors that are necessary for cellular metabolism, and they are reutilized (not metabolized and/or degraded) by the cells. They have classically been grouped according to their solubility; water soluble (B_{12}, biotin, folic acid, niacinamide, pantothenic acid, pyridoxine, riboflavin, and thiamine) and fat soluble (vitamins A, E, K) compounds. The latter vitamins are usually added to the medium from a concentrated ethanolic stock, and are not particularly stable in solution. Vitamin A (a retinoid) can have no effect, stimulate growth, or induce differentiation (dependent on the cell type and culture conditions). Vitamin E serves as an antioxidant, and vitamin K is necessary for the gamma-carboxylation of proteins. The water soluble vitamins function in multiple metabolic roles, and are generally necessary components of a medium formulation. Vitamin C, a water soluble antioxidant has been demonstrated to be beneficial in culture, particularly with cells that tend to synthesize collagen. It is readily oxidized and continuous supplementation is advisable if beneficial effects are observed (SENOO et al., 1989).

Many cultured cells require supplementation with fatty acids, phospholipids (and/or their precursors) and other hydrophobic compounds such as cholesterol. Much has still to be discerned about the interconversion and biosynthesis of lipids by cultured cells, but a few generalizations can be made. *Cis*-unsaturated fatty acids (e.g., linoleic, oleic, arachidonic) have been shown to be beneficial for many cell lines, but are relatively insoluble and inherently unstable in aqueous solutions. Thus, they are usually supplied as conjugates with delipidated albumin. The quantitative and qualitative requirements vary with cell type, culture conditions, and duration of exposure to the supplement (WOLFE and SATO, 1982). Certain lots of albumin or improperly balanced albumin conjugates can be quite toxic. Liposomes have been utilized to provide cells with both fatty acids and phospholipids such as sphingomyelin, phosphotidyl choline, phosphotidyl ethanolamine and phosphotidyl inositol. This approach is difficult to utilize in large-scale cell culture, and in many cases the need for supplementation with the more complex lipids can be circumvented by providing the cells with precursors such as choline, ethanolamine, and inositol. A few cells such as NS-1 derived hybridomas have a deficiency in cholesterol biosynthesis and supplementation is necessary, usually as an albumin conjugate (SATO et al., 1984).

Nucleic acid precursors such as the nucleosides and nucleotides are usually not required in a medium formulation, provided at least one source of pyrimidine and purine is supplied. Therefore, hypoxanthine and thymidine are included in most media. Particular care must be taken with mutant lines lacking enzymes of the biosynthetic pathways or with cultures grown without glucose. In the latter case a source of five carbon sugars or small concentrations of glucose is advisable.

2.3 Proteins

Historically, the first cultures were established in mixtures of plasma and tissue extracts (HARRISON, 1908; LEWIS and LEWIS, 1912; CARREL, 1913). Nutrient media designed in the late 1950s to the early 1960s were developed using cultures supplemented with dialyzed serum. Serum provides a host of factors not contained in many basal nutrient formulations, including low molecular weight nutrients (sugars, lipids, vitamins, amino acids, trace elements, etc.), hormones, blood proteins, metabolites, and protective compounds. This rather complex mixture has many functions in cell culture, but also is the root of many problems associated with the production of a product via this technology. Supplementation with 10% serum results in the contamination of the product-rich conditioned medium with mixtures of hundreds of proteins totaling about 4 g/L. This significantly complicates and increases the expense of downstream processing. Furthermore, the composition of serum is undefined and significant inter-lot variation exists.

Serum is expensive, and the availability of high quality serum has been (and is likely to be) a problem for large-scale efforts. On top of this, extensive labor-intensive prescreening of a particular serum lot must be performed. Long-term growth, cellular morphology, cellular productivity, and the possibility of adventitious contaminants must be examined. Once identified, a large lot of serum is usually purchased. Long-term storage of costly, unstable biologicals like serum in a monitored freezer facility is very expensive. Taken together, these factors have driven research toward developing the technology to reduce the need for serum supplementation in mammalian cell culture.

It is now well accepted that the major role played by serum in cell culture is to provide hormones, growth factors, transport proteins, and attachment factors. Although more expensive, fetal bovine serum is most frequently utilized because it contains relatively low concentrations of growth inhibitory substances and high concentrations of growth stimulatory compounds (as compared to bovine serum). Many of these individual "factors" have been identified in the last decade, and several are discussed below (see Tab. 3). A large fraction of these factors has been discovered and characterized by investigators attempting to reduce the level of serum supplementation in their particular culture system by providing their cultured cells with purified molecules (McKEEHAN et al., 1978).

Serum has many other beneficial effects in culture. It supplies "generic" protein that coats the system, preventing non-specific adsorption of essential factors to the bioreactor walls and filters. Enzymes contained in serum modify media components to molecules the cells can more readily utilize and also detoxify other materials. Serum also provides essential nutrients that are not included in the older, less complex basal media. Recently, the importance of the numerous protease inhibitors present in serum has also become apparent, and several reports of increased product integrity in cultures supplemented with protease inhibitors are present in the literature (tPA/aprotinin). In addition, the physical properties of medium (colloid osmotic pressure, diffusion rates, viscosity) are altered by serum supplementation.

Tab. 3. "Factors" Used in Mammalian Cell Culture

Supplement	Structure	Source	Activities
Growth factors			
Nerve growth factor-β	2 chain protein, mw = 26000	Submandibulary gland	Cell growth, neuron development and survival
Epidermal growth factor	Protein, mw = 6100	Submandibulary gland	Stimulates broad range of responses including growth
Platelet derived growth factor	Dimeric glycoprotein, mw = 28–31000	Platelets	Stimulates broad range of responses including growth
Insulin-like growth factor I	Protein, mw = 7649	Liver	Stimulates growth and insulin-like activity
Insulin-like growth factor II	Protein, mw = 7471	Liver	Stimulates growth and insulin-like activity
Basic fibroblast growth factor	Protein, mw = 16415	Brain, hypothalamus, pituitary, kidney	Stimulates growth of mesodermal cells

Tab. 3. "Factors" Used in Mammalian Cell Culture (Continued)

Supplement	Structure	Source	Activities
Acidic fibroblast growth factor	Protein, mw = 15883	Brain, hypothalamus	Stimulates growth of mesodermal cells
Transforming growth factor-β	2 chain protein, mw = 25000	Kidney, platelets	Stimulates or inhibits growth or differentiation
Transforming growth factor-α	2 chain protein, mw = 6000	Transformed cells	Similar to EGF
Liver cell growth factor	GLY-HIS-LYS	Plasma	Synergistic with Cu^{2+} to stimulate growth and survival
Interleukins (IL)			
IL-1	Protein, mw = 17000	Macrophages	Stimulates T-cell growth
IL-2	Protein, mw = 20000	T-cells	Stimulates T-cells, natural killer cells and TIL-cells
IL-3	Protein, mw = 28000	T-lymphoma	Stimulates stem cell growth and differentiation
IL-4	Glycoprotein, mw = 15500	Thymoma	Stimulates B-, T-, and mast cells
IL-5	Glycoprotein	T-cells	Eosinophil differentiation, T-cell replacing factor
IL-6	Protein, mw = 26000	Monocytes, fibroblastoid cells	Stimulates β-cell growth, hepatic acute phase, hybridoma survival
IL-7	Protein, mw = 17000		Early B- and T-cell growth
IL-8	Protein, mw = 10000	Neutrophils	Neutrophil stimulation, chemotaxis
Cytokines			
Erythropoietin	Glycoprotein, mw = 45000	Kidney	Stimulates growth and differentiation of red cell precursors
Interferon-α	Protein, mw = 19–21000	Monocytes, macrophages	Inhibits growth of B- and C-cells
Interferon-γ	Protein, mw = 35000	T-cells	Stimulates NK- and T-cell growth, differentiates myeloid cells
Granulocyte colony stimulating factor	Protein, mw = 23000	Macrophages	Growth and differentiation
Ganulocyte macrophage colony stimulating factor	Protein, mw = 23000	Fibroblasts, T-cells	Growth and differentiation
Macrophage colony stimulating factor	Protein, mw = 70000	Fibroblasts	Stimulates macrophage differentiation and growth
Tumor necrosis factor	Dimer, mw = 17000	Macrophages	Stimulates fibroblast growth and activates a variety of cells

Tab. 3. "Factors" Used in Mammalian Cell Culture (Continued)

Supplement	Structure	Source	Activities
Attachment/substratal proteins and compounds			
Cell Tak™	Polyphenolic protein	Marine mussel	Effectively "glues" most cells to most surfaces
Collagen types I–V	Multisubunit protein	Most tissues, distribution specific for type	Binds and presents adhesion molecules and stimulates attachment and growth
Extracellular matrix (solubilized)	Mixture of proteoglycans, collagens, and attachment factors	Extracts of EHS mouse sarcoma and human placenta available	Promotes attachment, spreading, mitosis, and differentiation of many epithelioid cells
Fetuin	α-1-Globulin	Fetal serum	Enhances attachment and growth
Fibronectin	Dimeric protein, mw = 440000	Plasma	Promotes attachment and spreading via RGD and other motifs of most cells from mesenchyme
Laminin	Dimeric protein, mw = 900000	Extracellular matrix	Stimulates attachment and growth of cells derived from endoderm and ectoderm; stimulates neutrophil oxidative burst
Brain, pituitary, or retinal extracts	Undefined mixture	Specified tissue	Contains a variety of activities sometimes necessary for certain cells
Poly-D-lysine	Polypeptide, polymers, mw = 50000 common	Synthesis	Reduces net negative charge of surface enhancing attachment
Vitronectin	Protein, mw = 70000	Plasma	Stimulates attachment and growth of many cell types
Transport proteins			
Albumin	Protein, mw = 68000	Plasma	Carries fatty acids and trace elements and can detoxify
Ceruloplasmin	Protein, mw = 135000	Plasma	Transports copper
Hemoglobin	Quaternary protein, mw = 65000	Red cells	Binds oxygen carbon dioxide and can detoxify
High density lipoprotein	Particle containing lipid, cholesterol and multiple protein subunits	Plasma	Transports cholesterol, cholesterol esters, and other lipids
Low density lipoprotein	Particle containing lipid, cholesterol and protein Apoprotein B subunits	Plasma	Transports cholesterol, cholesterol esters, and other lipids
Transferrin	Protein, mw = 78000	Plasma	Carries iron and can detoxify

Hormones and growth factors bind to cellular receptors and, in the process, signal the cells to perform specific tasks. Mammalian cells evolved in the presence of a mixture of these factors, only some of which are well characterized today. The many components of basal cellular physiology are regulated by the concentrations of hormones to which the cells are exposed (BETTGER and MCKEEHAN, 1976). In addition, a given cell (depending upon the receptors that the cell expresses) can be signaled to grow, die (apoptosis), or differentiate by exposure to different regulatory factors. Hormones such as the steroids, insulin, and thyroid hormone regulate basal metabolic activities. In many cases such signals are necessary for continued viability of cells in culture. In fact, nearly every cultured cell requires stimulation by insulin (or an insulin-like factor) for survival (0.5–10 μg/mL).

Transport proteins such as albumin, transferrin, low and high density lipoproteins (LDL, HDL), transcobalamin, hemoglobin, and thyroid hormone binding protein bind necessary nutrients and hormones which would (in their unbound form) be toxic to the cultured cells at high concentrations. These compounds are either slowly released in a controlled manner (fatty acids, thyroid hormone) or are transported into cells via binding protein and cell receptor mediated processes (Fe^{2+} and transferrin). In "reduced serum" cultures, most cells require supplementation with additional transferrin (0.5–10 μg/mL) unless frequent supplementation of the medium with ferrous citrate/sulfate and iron chelators is performed (some success utilizing hemoglobin as a transport moiety has also been reported). Transport proteins also serve to help detoxify medium by binding contaminants (heavy metals, detergents) and solubilizing essential nutrients that do not readily dissolve in aqueous solutions (cholesterol/lipids and lipoproteins, vitamin B_{12} and transcobalamin). Lipids, e.g., when added to cultures above the critical micellular concentration rapidly form micelles which can lyse cultured cells. Albumin and serum lipoproteins are frequently used to supply lipids and cholesterol to cultured cells (POLET and SPIEKER-POLET, 1975). It should be noted that the lipid composition of these supplements can vary considerably between preparations, and care must be taken to select one appropriate for the cell of interest.

Attachment proteins (fibronectin, laminin, vitronectin) adsorb to culture surfaces and then interact with the cultured cells via specific receptor molecules. The attachment, spreading, and growth of many cells is abnormal (and/or limited) in the absence of these proteins. These molecules (along with proteoglycans) are secreted by cells *in vivo* to form the extracellular matrix (ECM) with which the cell and its neighbors interact. The composition of the ECM varies with the tissue, and has been shown to regulate certain aspects of cellular physiology (YAMADA, 1991). Thus, it is sometimes necessary to provide the cultured cells with the appropriate molecules in order for the cells to perform specific desired tasks *in vitro*. These molecules can be either preadsorbed to the surfaces, or added to the medium at concentrations of a few μg/mL during inoculation of adherent cell cultures. Precoating microcarriers with serum proteins is also a common practice, and some success utilizing the attachment motifs in synthetic peptides has been achieved.

2.4 Non-Hormonal/ Non-Nutritional Supplements

These compounds are frequently included in medium formulations to reduce specific detrimental conditions that are generated in the artificial microenvironment of the bioreactor, or to facilitate control and monitoring of the system. Also included in this group are antibiotics, which although frequently utilized are not necessary or recommended. These compounds interfere with cellular metabolism, and have been developed by selecting molecules that are more active with microbial enzymes than their analogous mammalian counterparts. Nevertheless, most antibiotics compromise some aspects of mammalian metabolism when present in their effective concentration range, and all are toxic at relatively high dosages. Serum and other supplements can, in some cases, reduce the detrimental effects of antibiotics on cultured cells, and the inhibitory effects of a specific antibiotic is greatly dependent on the cell type. Therefore, if antibiotics must be utilized in a specific culture system, an extensive dose-response analysis should be performed under conditions resembling (as close as possible) those of the bioreactor. Potential copurification with the desired product or interference with downstream processing must also be examined.

In the bioreactor pH is stabilized with buffers (phosphate and bicarbonate), and usually maintained through control of dissolved CO_2 and base addition. Phosphate has a pK of 7.21 and is present in media (2–3 mM) both as a nutrient and as a buffer. The concentration of phosphate is limited by the solubility of calcium phosphate, and higher concentrations of phosphate are only practical in reduced calcium formulations. The principal pH buffer in mammalian organisms is the bicarbonate/carbon dioxide system, and a similar control strategy is utilized with most bioreactor systems:

$$H_2O + CO_2 \leftrightarrow H_2CO_3 \leftrightarrow H^+ + HCO_3^-$$
$$pK' = 6.1 \text{ @ } 37\,^\circ\text{C},\ 760 \text{ mm Hg}$$

where $[H_2CO_3]$ (mM) $= 0.0308\ p_{CO_2}$ (mm Hg).

Applying the Henderson–Hasselbalch equation

$$pH = pK' + \log \frac{[HCO_3^-]}{[H_2CO_3]}$$
$$= 6.1 + \log \frac{(\text{total } CO_2) - 0.0308\ p_{CO_2}}{0.0308\ p_{CO_2}}$$

it is apparent that increases in the partial pressure of CO_2 result in decreases of pH. Equilibrium in a bioreactor is essentially dependent on the rate of CO_2 dissolution, diffusion, and bulk mixing since the rate of formation of H_2CO_3 from water and dissolved carbon dioxide is rapid ($T_{1/2} = 11.5$ s; MOUNTCASTLE, 1980). At pH 7.4 the relative ratios are approximately

$$20000:1000:1\ (HCO_3^- : CO_2 : H_2CO_3).$$

The amount of bicarbonate present in the medium is chosen to be close to equilibrium conditions for the specified operating conditions. This is usually about 23 mM (similar to circulating blood), and will be at equilibrium with 5–10% CO_2. Bicarbonate in excess of this equilibrium point is used to titrate acid produced by the cultured cells. Carbon dioxide is also a nutrient, and aqueous solutions equilibrated with air at 37 °C contain only 0.21 mm CO_2 (7 µM). Therefore, if air equilibrated cultures are desired, it is recommended to use an additional organic buffer such as HEPES (N-2-hydroxyethylpiperazine-N′-2-ethanesulfonic acid) and 1–2 mM bicarbonate to maintain adequate concentrations of dissolved CO_2.

Although relatively expensive, HEPES is frequently used as an additional buffer since, unlike bicarbonate, its pK falls in the physiological range (7.31 at 37 °C). The pK is however extremely temperature sensitive (−0.014 per change of 1 °C), and thus a room temperature solution at pH 7.55 would be pH 7.31 when warmed to operating temperature. HEPES is utilized at concentrations ranging from 10 to 50 mM. The concentration of sodium chloride needs to be reduced to compensate for changes in osmolarity at the higher levels, and a reduction in the level of potassium ions should be made to maintain similar Na^+/K^+ ratios. At [HEPES] >15 mM the buffer is sometimes growth inhibitory (dependent on cell type via stimulation of the production of detrimental oxygen metabolites).

Phenol red has classically been included in culture media as a pH indicator. There is no compelling reason for its use with instrumented systems, and in fact it has been shown to have estrogenic activity with cultured cells (C127, MCF-7) and to copurify with proteins (tPA, IFN-γ).

Many protective agents are often added to the medium to protect the cells from damage caused by osmotic gradients, shear, and interaction with air/liquid interfaces (bubble damage). The agents include dextran (30–50 g/L, 80000 mw), polyethylene glycol (20000 mw), methyl cellulose (0.1–0.2% w/v), and pluronic polyol F-68 (0.025–0.3%). The latter, a block copolymer of polyoxypropylene and polyoxyethylene of 8300 molecular weight, has been demonstrated to reduce bubble induced cell damage and is currently used more frequently (MURHAMMER and GOOCHEE, 1990).

Antioxidants are sometimes beneficial. Superoxide radicals and hydrogen peroxide are generated by normal cellular metabolism and enzymes such as xanthine oxidase present in serum supplements. These compounds can also be generated by photooxidation of compounds like TRP and riboflavin, and their production can be stimulated by several factors such as hyperoxic conditions, general cell stress, and compounds like HEPES. These highly reactive metabolites can cause damage to both the cultured cells and the cell products. Enzymes (superoxide dismutase, catalase), vitamins (C, E, and *β*-carotene), carrier proteins (transferrin, hemoglobin), amino acids, trace elements (Se^{2+} as a cofactor for glutathione peroxidase), bilirubin, and reduced glutathione are all antioxidants useful in mammalian cell cultures. All except the enzymes and glutathione are components of serum.

Reducing agents (*β*-mercaptoethanol) can help to prevent oxidative damage, primarily by restoring the reduced form of glutathione. When added to cultured cells increases in cystine uptake (presumably by the mixed disulfide) and antibody secretion rates (hybridomas) have been observed.

Many cells actively synthesize and secrete proteases, and even the limited cell death (inherently a part of the microenvironment of any bioreactor) causes the release of intracellular proteases into the medium. Therefore, sup-

Tab. 4. Biocompatible Protease Inhibitors

Inhibitor	g/mol	mg/L	Purchase Unit (L of Media)	US $/L	Properties
Pepstatin	685	0.7	142.8	1.96	Inhibits acid proteases, especially pepsin, renin, cathepsin D, chymosin (2 mM stock solution in alcohol)
Aprotinin	6511	6.8	146.3	4.00	Inhibits serine proteases (trypsin, kallikrein; 40 U/mL = 6.8 μg/mL) At pH 7–8 aqueous solutions are stable for only 1 week at 4 °C
ε-Amino-caproic acid (EACA)	131.2	918.4	1088.9	0.04	Inhibits plasmin
Leupeptin	460	2.0	25.0	4.62	Reversible competitive inhibitor of serine and thiol proteases; plasmin, trypsin, kallikrein, cathepsin B, thrombokinase. 2 μg/mL non-toxic, plasmin inhibition 50% at 8 μg/mL
Ovoinhibitor		2.5	40.0	4.41	Inhibits trypsin and chymotrypsin equally on a weight basis; 5 μg/mL usually non-toxic
Egg white trypsin inhibitor	28000	50.0	20000.0	<0.01	At 10–100 μg/mL inhibits serine proteases (trypsin, plasmin, coagulation factor Xa, kallikrein; no anti-chymotrypsin activity)
Bovine pancreas trypsin inhibitor		2.0	12.5	16.94	Inhibits trypsin and chymotrypsin; usually 2–5 μg/mL non-toxic
Soybean trypsin inhibitor	20100	50.0	100000.0	<0.01	At 10–100 μg/mL inhibits serine proteases (trypsin, plasmin, coagulation factor Xa, kallikrein; no anti-chymotrypsin activity)
Albumin	68000	1000.0	1000.0	2.98	Serves as a non-specific substrate

plementation with protease inhibitors that interact with the particular set of enzymes generated in a specific culture system is recommended. Cultures that have been established or maintained in serum-supplemented medium are particularly sensitive due to large quantities of adsorbed plasminogen (which is readily activated to plasmin by a variety of proteases). Ten to forty KIU of aprotinin will effectively inhibit the plasmin (and kallikrein) activity. ε-Amino caproic acid also inhibits plasmin at 2–7 mM concentrations (higher levels are somewhat toxic). Several other biocompatible inhibitors are now commercially available at non-prohibitive costs, see Tab. 4. Pepstatin inhibits acid proteases (at 0.7–1 μg/mL). Leupeptin is a reversible, competitive, polypeptide inhibitor of serine and thiol proteases that shows little toxicity at 5 μg/mL. In addition, both soybean and egg white trypsin inhibitors inhibit serine proteases. They are effective between 10 and 100 μg/mL, and cost only a few cents per liter.

Heparin or heparin sulfates (10–100 μg/mL) can help to reduce cell clumping when a normally adherent cell line is grown in suspension. Heparins also stabilize certain growth factors and enhance their bioactivity (ECGF, BFGF, VPF). Heparins are crude mixtures available from many vendors and are not likely to be identical. Preparations are available with different nominal molecular weights, and those below 5000 lack some of the biological activities present in other preparations.

3 Media Design and Selection

It should be emphasized that the goal is to produce conditioned medium containing a high concentration of an intact, uniformly processed, crude product which can then be readily isolated from other cellular products and medium components. The choice of approach is therefore highly dependent on the particular product and cellular system. Since much of the expense associated with the overall process is dependent on the composition of the medium utilized, considerable care should be taken in first defining the requirements of the system. Processes can be developed based on a "reduced-serum" formulation, a serum-free medium, a chemically defined formulation, or a combination that involves a medium exchange. In the latter case the cell mass is usually generated in medium supplemented with serum or a reduced-serum formulation, and then the production of conditioned medium proceeds in a more defined, generally less expensive formulation. This is possible because many mammalian cell lines survive and produce product at similar rates in a non-proliferative "maintenance state" when cultured in the absence of the hormonal signals that normally tell the cell to grow. Although impractical in batch mode processes, perfusion systems can be utilized for this approach. The drawback is that survival is generally not 100%, the cell mass decreases with time, and considerable damage to the cells can occur during the washing process (releasing the intracellular contents, including proteases, into the conditioned medium).

3.1 Reduced-Serum and Serum-Free Formulations

A basal synthetic medium must first be selected that contains all of the nutrients that the cells will need. Energy sources, amino acids, vitamins, nucleotide precursors, buffers, ions, and trace elements must all be considered. Many media formulations have been developed for the culture of specific cell types in the presence of a variety of supplements. Most of the older formulations were developed for the mass culture of cells in the presence of 10% serum supplementation (see DME, Tab. 2; DULBECCO and FREEMAN, 1959). Some of the most complete formulations have been developed by R. HAM and his colleagues by performing sequential dose-response analyses for all the components of the medium (HAM, 1965). In most cases, the endpoint in these experiments was clonal growth after 14 days. Thus, these formulations are likely to contain all the materials that low density cultures need to survive and proliferate, but are rapidly depleted of essential nutrients by higher density cultures.

Tab. 2 compares the composition of DME and a medium developed by HAM for fibroblasts, F-12. Also shown is COON's modification of this medium in which the concentrations of the amino acids are doubled (and Ca^{2+} is increased) to accommodate for denser cultures. Neither of these formulations are appropriately balanced for most cells. An adequate starting formulation can usually be obtained by simply mixing appropriate commercially available media (BARNES and SATO, 1980a, b) and in many cases the cells grow better in these mixtures than in either medium by itself (when supplemented with either serum or defined factors; WOLFE et al., 1980). It is common practice to empirically determine "good mixtures" for a particular cell type. Formulations based on mixtures of DME and modified F-12 are probably the most frequently utilized for the culture of a variety of cell types (see Tab. 2; ISCOVE and MELCHERS, 1978; BUTLER, 1986; HIGUCHI and ROBINSON, 1973). The subsequent analysis of the amino acid and carbohydrate consumption by cells cultured in the specified bioreactor system utilizing a "good mixture" often suggests appropriate modifications of the resulting basal nutrient formula.

The approach toward "reduced-serum" and ultimately "chemically defined" media was first suggested by a series of publications by G. H. SATO and his colleagues in the mid-1970s that demonstrated that serum could be replaced by mixtures of defined and/or partially defined factors, and that the necessary replacements were qualitatively and quantitatively different dependent on the cell type (BOTTENSTEIN et al., 1979).

Therefore, the amount of serum supplementation can frequently be reduced by adding critical components to the medium. Undefined mixtures of serum fractions and purified factors (that do not contain components that copurify or modify the product) frequently support growth and survival in the presence of considerably reduced concentrations of whole serum. Many cell lines will grow in 0.5–2% serum if the medium is supplemented with additional insulin and transferrin (considerable species specificity exists in the interaction of transferrin with its receptor, and therefore a preparation isolated from a similar species is recommended). Further supplementation with hormones, growth factors, crude preparations of other transport proteins such as LDL and HDL, undefined serum fractions, or tissue extracts (appropriately adsorbed to prevent interference with downstream processing) can produce additional reductions in the serum requirement. Significant reductions can also occur through establishing and maintaining higher density cultures (thus increasing the presence of beneficial autocrine factors). A formulation is considered "serum-free" if sufficient factors have been added to the media to eliminate the need for whole serum.

3.2 Chemically Defined Media

The feasibility for chemically defined media increases as the critical components necessary for many varied culture systems become defined. Only a few such systems are utilized in large-scale cultures today. The need, from a regulatory viewpoint, to demonstrate control of all aspects of the synthesis and processing of proteins destined for licensing as pharmaceutical agents is a major force promoting the more common application of these formulations. The practicality of this approach will increase as more purified factors become available (at significantly reduced prices) through recombinant technology. Even today, many hormone-supplemented, chemically defined formulations can be prepared at less expense than medium supplemented with 10% serum.

Tab. 5 shows typical costs of several supplements and media components. As indicated, 100 L of medium supplemented with 10% fetal bovine serum cost approximately $ 2500 for raw materials. Many murine hybridomas grow well in mixtures of DME/F-12/RPMI supplemented with bovine insulin, bovine transferrin, ethanolamine, selenium, and F-68 (MURAKAMI et al., 1982; MURAKAMI, 1989; GLASSY et al., 1988). The indicated formulation can be prepared for about $ 150/100 L (excluding labor and water costs), and the only problematic contaminant is transferrin (2 mg/L which can copurify with IgG. 10% fetal bovine serum contains about 4 g/L total protein, and the

Tab. 5. Expense of Common Medium Components

Components	US $/100 L
"Custom ball milled" basal nutrient powder	25–50
Fetal bovine serum (10%)	2500
Newborn calf serum (10%)	500
HEPES buffer (25 mM)	50
Pluronic Polyol F-68	<1
Insulin, bovine (5 µg/mL)	25
Insulin, human (5 µg/mL)	75
Transferrin, bovine (2–3 µg/mL)	125
Transferrin, human (2–3 µg/mL)	12
Albumin, bovine serum (250 µg/mL)	12
Albumin, human serum (250 µg/mL)	75
Lipoproteins (Excyte™), bovine (0.1%)	25
Lipoproteins (Excyte™), human (0.1%)	150
Epidermal growth factor, human recombinant (8 ng/mL)	200
Fibroblast growth factor, basic, human recombinant (8 ng/mL)	650
Insulin-like growth factor 1, human recombinant (10 ng/mL)	540
Fibronectin, human (1 µg/mL)	$ 33/mg
Aprotinin (40000 KIU/L)	300
Trypsin inhibitor, soybean IS, purified (50 µg/mL)	117
Heparin (10 µg/mL)	3

concentration of transferrin is 180±50 mg/L. Product (IgG) concentrations frequently range from 5–100 mg/L depending on the particular hybridoma. Thus, considerable savings in both expense and downstream processing can be achieved with chemically defined media.

4 Post-Translational Modifications and Medium Composition

There are numerous tissue specific post-translational modifications that occur; e.g., glycosylation, palmitylation, acetylation, glutamylation, hydroxylation, sulfation, and myristylation. These modifications can influence further processing of the protein, its pharmacokinetics, and its efficacy in transmembrane signaling. For the case of proteins produced as therapeutic agents, the exact structure of the protein does not matter. The material has only to be consistent, and fit a well defined pharmacological profile. On the other hand, if the protein is being produced as a reagent for rational drug design, its structure should be as close as possible to the natural species. The enzymatic cofactors and substrates required for these post-translational processes need to be supplied to the cultured cells.

Recent data demonstrate that the medium composition can influence the type and degree of post-translational modifications that a protein undergoes. N-linked glycosylation is particularly sensitive to changes in both the hormonal and nutrient milieu. It is clear that both the type and degree of a recombinant protein's glycosylation is dependent on the cell in which it is expressed. Furthermore, the phenotype (with respect to glycosylation) of a recombinant cell is not absolute and invariant. Hormone-dependent regulation of both glycosidases and glycosyltransferases have been documented with several types of cultured cells, and glucose concentration-dependent changes of glycosylation occur in some glucose starved cells (80 mg/L or less; GOOCHEE and MONICA, 1990). Therefore, both the medium composition and process conditions can influence the chemical structure of the cellular product. This should be considered during process development and medium design.

5 References

BARNES, D., SATO, G. H. (1980a), Methods for growth of cultured cells in serum-free medium, *Anal. Biochem.* **102**, 255–270.

BARNES, D., SATO, G. (1980b), Serum-free cell culture: a unifying approach, *Cell* **22**, 649–655.

BETTGER, W. J., MCKEEHAN, W. L. (1976), Mechanisms of cellular nutrition, *Physiol. Rev.* **66**(1), 1–35.

BOTTENSTEIN, J., HAYASHI, I., HUTCHINGS, S., MASUI, H., MATHER, J., MCCLURE, D. B., OHASA, S., RIZZINO, A., SATO, G., SERRERO, G., WOLFE, R., WU, R. (1979), The growth of cells in serum-free hormone-supplemented media, *Methods Enzymol.* **45**, 94–109.

BUTLER, M. (1986), Serum-free media, in: *Mammalian Cell Technology* (THILLY, W. G., Ed.).

CARREL, A. (1913), Artificial activation of the growth *in vitro* of connective tissue. *J. Exp. Med.* **17**, 14–19.

DULBECCO, R., FREEMAN, G. (1959), *Virology* **8**, 396.

GLASSY, M. C., THARAKAN, J. P., CHAU, P. C. (1988), Serum-free media in hybridoma culture and monoclonal antibody production. *Biotechnol. Bioeng.* **32**, 1015–1028.

GOOCHEE, C. F., MONICA, T. (1990), Environmental effects on protein glycosylation. *Bio/Technology* **8**, 421–427.

HAM, R. G. (1965), Clonal growth of mammalian cells in a chemically defined, synthetic medium, *Proc. Natl. Acad. Sci. USA* **53**, 288–293.

HAM, R. G. (1980), Survival and growth requirements of non transformed cells, in: *Handbook of Experimental Pharmacology,* pp. 13–88, New York: Springer-Verlag.

HAM, R. G., MCKEEHAN, W. L. (1979), Media and growth requirements, *Methods Enzymol.* **58**, 44–93.

HARRISON, R. G. (1908), Embryonic transplantation and development of the nervous system, *Anat. Rec.* **2**(9), 385–410.

HIGUCHI, K., ROBINSON, R. C. (1973), Studies on the cultivation of mammalian cell lines in a serum-free, chemically defined medium. *In Vitro* **9**(2), 114–121.

ISCOVE, N. N., MELCHERS, F. (1978), Complete replacement of serum by albumin, transferrin, and soybean lipid in cultures of lipopolysaccharide-reactive B lymphocytes, *J. Exp. Med.* **147**, 923–933.

LEIBOVITZ, A. (1963), The growth and maintenance of tissue-cell cultures in free gas exchange with the atmosphere, *Am. J. Hyg.* **78**, 173–80.

LEWIS, M. R., LEWIS, W. H. (1912), The cultivation of tissues from chick embryos in solutions of NaCl, $CaCl_2$, KCl, and $NaHCO_3$, *Anat. Rec.* **5**(6), 277–293.

MCKEEHAN, W. L., GENEREUX, D. P., HAM, R. G. (1978), Assay and partial purification of factors from serum that control multiplication of human diploid fibroblasts, *Biochem. Biophys. Res. Commun.* **80**(4), 1013–1021.

MOUNTCASTLE, V. B. (Ed.) (1980), *Medical Physiology,* 14th Ed., St. Louis: C. V. Mosby Co.

MURAKAMI, H. (1989), *Monoclonal Antibodies: Production and Application,* pp. 107–141, New York: Alan R. Liss, Inc.

MURAKAMI, H., MASUI, H., SATO, G. H., SUEOKA, N., CHOW, T. P., KANO-SUEOKA, T. (1982), Growth of hybridoma cells in serum-free medium: ethanolamine is an essential component, *Proc. Natl. Acad. Sci. USA* **69**, 1158–1162.

MURHAMMER, D. W., GOOCHEE, C. F. (1990), Structural features of nonionic polyglycol polymer molecules responsible for the protective effect in sparged animal cell bioreactors, *Biotechnol. Prog.* **6**, 142–148.

POLET, H., SPIEKER-POLET, H. (1975), Serum albumin is essential for *in vitro* growth of activated human lymphocytes, *J. Exp. Med.* **142**, 949–55.

REITZER, L. J., WICE, B. M., KENNELL, D. (1979), Evidence that glutamine, not sugar, is the major energy source for cultured HeLa Cells, *J. Biol. Chem.* **264**(8), 2669–2676.

SATO, D. J., KAWAMOTO, T., MCCLURE, D. B., SATO, G. H. (1984), Cholesterol requirement of NS-1 mouse myeloma cells for growth in serum-free medium, *Mol. Biol. Med.* **2**, 121–134.

SENOO, H., TSUKADA, Y., SATO, T., HATA, R. (1989), Co-culture of fibroblasts and hepatic parenchymal cells induces metabolic changes and formation of a three-dimensional structure, *Cell Biol. Int. Rep.* **13**, 197–206.

WAYMOUTH, C. (1981), Major ions, buffer systems, pH, osmolality, and water quality, in: *The Growth Requirements of Vertebrate Cells in vitro* (WAYMOUTH, C., HAM, R. G., CHAPPLE, P. J., Eds.), Cambridge: University Press.

WOLFE, R. A., SATO, G. H. (1982), Continuous serum-free culture of the N1BTG-2 neuroblastoma, the C6BU-1 glioma, and the NG108-15 neuroblastoma X glioma hybrid cell lines, in: *Growth of Cells in Hormonally Defined Media* (SIRBASKU, D., Ed.), pp. 1075–1088, New York: Cold Spring Harbor Press.

WOLFE, R. A., SATO, G. H., MCCLURE, D. B. (1980), Continuous culture of rat C6 glioma in serum-free medium, *J. Cell Physiol.* **87**, 434–441.

YAMADA, K. M. (1991), Adhesive recognition sequences, *J. Biol. Chem.* **266**(20), 12809–12812.

ZOMBOLA, R. R., BEARSE, R. C., KITOS, P. A. (1979), Trace element uptake by L-cells as a function of trace elements in a synthetic growth medium, *J. Cell Physiol.* **101**, 57–66.

9 Media and Air Sterilization

Gokaraju K. Raju
Charles L. Cooney
Cambridge, MA 02139, USA

1 Introduction

1.1 Definitions

Generally, culturing of microorganisms needs to be carried out in an aseptic manner. Asepsis refers to the continued exclusion of undesirable microorganisms. This need for asepsis motivates the need for disinfection. Disinfection is the removal or destruction of living organisms that may decrease the yield or quality of the product.

However, although disinfection requires the destruction or removal of only those microorganisms that can produce an unwanted outcome, it is, in general, difficult to selectively remove only such organisms. Hence, the need for media and air sterilization. *Media sterilization* is the destruction or removal of all life forms from the medium. Thus, sterilization of growth medium is a pretreatment prior to inoculation with the desired culture. After inoculation the system is no longer sterile, but is maintained aseptic by excluding unwanted organisms. *Air sterilization* involves the continuous removal of life forms from the air stream being fed into the fermentor during the course of the fermentation.

1.2 Modes of Media Sterilization

Media sterilization may be accomplished either by physical removal of all living organisms or by killing or inactivation of the organisms. Whatever method is employed, the purpose of sterilization is to make sure that no living organisms remain. Both the physical removal and the inactivation of microorganisms can be reached by many different methods. In choosing between them, one needs to evaluate the methods based on the following four main criteria:

1. Effectiveness in achieving an acceptable level of sterility
2. Reliability and ease of validation
3. Positive or negative effects on medium components
4. Capital expense and operating cost

When evaluated against these four criteria, there are two main methods of sterilization most suitable for large-scale industrial sterilization: *filtration* for removal and *thermal treatment* for inactivation. Thermal treatment is suitable because of its relatively low cost, its effectiveness in achieving an acceptable level of sterility, and the ease of validation. Moreover, it is often the only method applicable in sterilizing media containing suspended solids. Filtration, on the other hand, being comparatively expensive, has the advantage of minimizing adverse effects on the medium quality. As these are the two methods used most widely in industry, we will focus mainly on them in this chapter. Our goal is to use these two methods to demonstrate the basic principles and techniques involved in large-scale media sterilization. For additional discussions on this subject, we recommend reading RICHARDS (1968), SOLOMONS (1969), AIBA et al. (1973), BAILEY and OLLIS (1977), WANG et al. (1979), or COONEY (1985).

1.3 Modes of Air Sterilization

Similar to media sterilization, air sterilization can be accomplished through either removal or destruction of the microorganisms. Because of the large quantities of air involved, contaminant removal by filtration is preferred to thermal destruction by heat. Conventional fibrous filters were the most common means to sterilize air in the past. They have now been replaced by cartridge filters (MELTZER, 1987).

2 Media Sterilization by Removal

2.1 Methods of Removal

A large number of different methods have been used in the bioprocess industry for the removal of particulate matter from fluids. Tab. 1 summarizes those most commonly used. These methods can be viewed as be-

Tab. 1. Methods of Removal

Method	Type
Filtration	Absolute Depth
Adsorption	Ion exchange Electrostatic Charcoal Affinity Molecular sieve
Sedimentation	Centrifugation Flocculation Flotation

longing to three classes of techniques: filtration, adsorption, and sedimentation. However, while adsorption and sedimentation techniques are quite valuable for the removal of particulate matter from fluids in general, they are of little or no value for sterilization. None of the methods is sufficiently effective in achieving a level of removal that would be acceptable for the sterilization of most fermentation media. These methods can at best reduce the number of microorganisms by a factor of about 10^3. This reduction is not sufficient for media sterilization where the number of microorganisms often has to be reduced in a larger volume of media by a factor of 10^{15}. Hence, filtration is the only "effective" way of physically removing microorganisms from fluids. Filtration is particularly useful for sterilizing heat-labile materials. Thus, sterilization by filtration is quite common when it is essential to process nutrients or products in liquid form. Examples include sterilization of complex organic media such as those used in animal cell culture and final product sterilization. Another important example is the use of microporous filtration and ultrafiltration as means for media sterilization in the preparation of pyrogen-free process water (MULLER, 1990).

2.2 Physical Characteristics of Microorganisms

In order to design an adequate procedure for removal of microorganisms from liquid media by filtration, it is important to understand the physical properties of these microorganisms. Filamentous mycelial organisms, such as molds, typically have a minor dimension of greater than 3 µm and often a major dimension that can range from 20 µm up to several millimeters. As a consequence, these organisms are relatively easy to remove by filtration. Of more concern are the single-cell organisms. Tab. 2 summarizes the size of these microorganisms. Fortunately, many fermentation processes are not susceptible to phage contamination and, as a consequence, one usually designs a filtration process to remove bacteria. Although it is possible to assign a size to most organisms, they should not be considered as rigid spheres or rods. Cell size will change with the specific growth rate; slowly growing cells are much smaller. In addition, under pressure, many cells exhibit flexibility and are extruded through small spaces which are typically less than the minor dimension of the organism. An important example are mycoplasms which are often found in animal cell culture media.

Tab. 2. Sizes of Microorganisms to be Removed

Microorganism	Size
Yeast	3.00–5.0 µm
Bacteria	0.30–2.0 µm
Bacteriophage/Virus	0.02–0.1 µm

2.3 Types of Filtration

2.3.1 Depth Filtration

Depth filters are usually constructed from fairly porous of fibrous materials such that the characteristic pore size or dimension between fibers in these filters is greater than the minimum size of the materials to be removed. Particle removal is based on the probability that a particle will be retained in the filter. Tab. 3 shows different types of depth filters and different mechanisms of retention. The first three mechanisms are most important and are shown graphically in Fig. 1.

Tab. 3. Depth Filters

Types	Mechanisms
Porous (ceramic)	Inertial impaction
Celite (diatoms)	Interception
Glass wool	Diffusion
Cellulose fiber	Sedimentation
Sintered metal	Electrostatic interaction

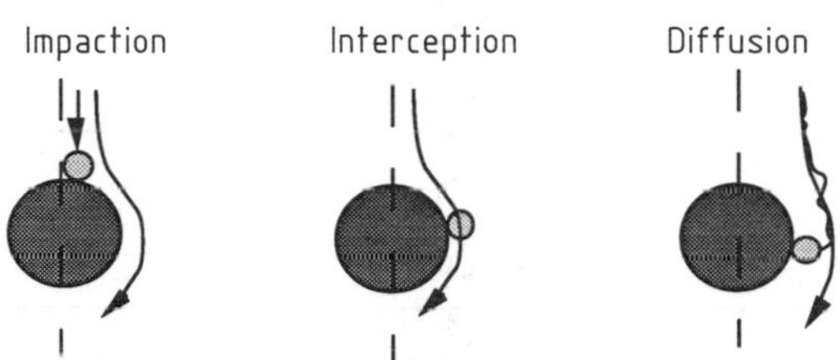

Fig. 1. Schematic representation of the three main modes of particle removal in a fibrous depth filter.

Inertial *impaction* is based on the fact that a particle in motion will tend to stay in motion in a straight line. As a consequence, as the fluid passes around the fiber, a particle will leave the flow stream and impact upon the fiber. Not surprisingly, the larger the organism and the greater its velocity, the greater its momentum and the higher the probability that it will impact on a fiber.

Interception occurs when a particle flowing around a fiber comes too close to a surface and enters the stagnant liquid boundary layer where the liquid velocity is zero. If it comes in contact with the particle or the fiber itself, it will usually 'stick'. This removal method is important in all flow regimes; however, if the velocity of the fluid becomes too high, it is possible to re-entrain intercepted or impacted particles into the fluid stream.

A third removal mechanism is *diffusion*. As a particle enters the stagnant boundary layer, it exhibits diffusion or Brownian motion. If the fluid velocity is not too high, the retention time of the organism is sufficiently long for the particle to diffuse and there is some finite probability that it will 'bump' into and adhere to the fiber.

Two other relevant mechanisms of removal (not shown in Fig. 1) are *sedimentation* or gravitation (settling) and *electrostatic interaction*. Sedimentation is caused by the density difference between the microorganism and the fluid and results in the microorganism's falling or rising relative to the fluid. Electrostatic interaction accounts for a relatively minor fraction of the total effect, unless the fibers have a strong net charge.

The relative importance of these mechanisms depends strongly on particle size and velocity. For example, diffusion is the predominant mechanism in the removal of small particles. Also, there usually exists some optimal velocity for minimum particle retention. Thus, in the evaluation of alternative media for depth filters, it is important to identify the critical velocity at which the filtration efficiency is minimal. By using this value for design, any increase or decrease from the critical velocity will improve filtration efficiency (YAO et al., 1971). Further details of depth filtration as it is applied to air sterilization are discussed in Sect. 4.

2.3.2 Absolute Filtration

One filtration mechanism is the absolute removal of organisms based on *size exclusion*. Tab. 4 depicts the types of absolute filters and the different mechanisms of filtration. As shown, it is possible to use membranes such as ultrafiltration, microporous or even macroporous membranes whose maximum pore size is less than the minimum size of the particles to be removed. The mechanism of filtration is primarily absolute size exclusion. In addition, particle accumulation around an open pore can cause *bridging*. These particles can then provide a filtration medium that is more efficient than the original filter itself. Lastly, there can be non-specific *adsorption* of particles to the solid portion of the filter. These three dif-

Tab. 4. Absolute Filters

Types	Mechanisms
Ultrafiltration	Size exclusion
Microporous	Bridging
Macroporous	Adsorption

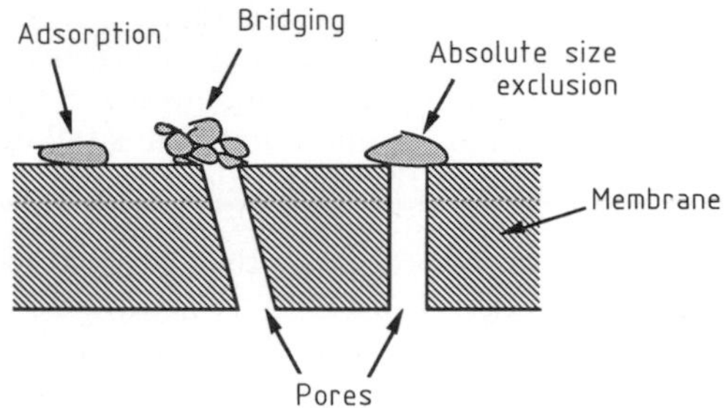

Fig. 2. Absolute filters: mechanisms of filtration.

ferent mechanisms are schematically shown in Fig. 2.

It is to be noted that while absolute filters are quite effective in physically separating microorganisms from media, they are not very efficient with media containing suspended solids. If the suspended particle concentration is above 10^6 mL^{-1}, it is not feasible to use absolute filters. On the other hand, depth filters can be used quite effectively with media containing suspended solids. As a consequence, for media sterilization one often employs prefiltration with a depth filter followed by an absolute sterilizing filter.

3 Media Sterilization by Inactivation

3.1 Introduction

3.1.1 Methods of Inactivation

An alternative to physical removal is inactivation of cells by thermal, chemical, or physical means. The goal is to simply kill or inactivate without necessarily disintegrating the cells completely. In fact, for most practical purposes the inability of a bacterial population to maintain or increase its numbers is sufficient to prevent contamination. Tab. 5 summarizes different modes by which viable cells and spores can be inactivated. Of these only heat is used for large-scale sterilization. The application of dry heat, for example by heating air in heat exchangers, is generally much less effective than moist heat and used only for instruments and stable materials. Many spores are highly resistant to dry heat. Hence, moist heat in the form of steam under pressure is much more effective. The use of other forms of energy like ultraviolet light is common in laboratories, sterile rooms, and spaces where work must be carried out under aseptic conditions. However, its intensity is greatly diminished beyond surface sterilization. Radiation is used mainly for sterilizing packed pharmaceutical and surgical materials (Richards, 1968). Chemical agents are useful mostly as disinfectants rather than sterilizing agents. Since *thermal treatment* is of primary importance in processing most raw materials and media used in biochemical processes, we will focus only on that mode of inactivation in the following section.

Tab. 5. Modes of Inactivation of Viable Cells and Spores

Methods	Forms
Heat	Wet Dry
Incineration	
Chemical agents	
Electromagnetic radiation	Ionizing gamma X-ray Ultraviolet
Sonic radiation	

3.1.2 Thermal Death Kinetics

We define microbial death as the inability of an organism to replicate itself in a given environment. For this reason, it is important to measure cell viability in an environment that is representative of the medium to be sterilized. Furthermore, death does not mean that all enzymatic activity in the cell is destroyed; although a cell may not replicate, it may still have the potential for catalyzing one or more reactions. Usually, this is not of importance in media sterilization; however, in food processing the presence of active enzymes in a moist food for long periods of time can promote the lowering of quality. Except in cases where obvious physical destruction of microorganisms is apparent, the mechanism by which microbial death occurs is not always clear. Both microbial growth and death are the results of a complex series of reactions each of which is influenced by temperature. It is common to describe the change in viable cell numbers (N) as a consequence of growth and death by:

$$\frac{dN}{dt} = \mu N - k N \tag{1}$$

where t is time, μ is the specific growth rate, and k is the specific death rate.

Here the series of reactions that result in growth are lumped into a specific growth rate, μ, and the series of reaction that result in death are lumped into a specific death rate, k. That is:

$$\frac{1}{N}\frac{dN}{dt} = \mu - k \tag{2}$$

In other words, the observed growth rate, $\mu - k$, is a balance between growth and death. Fig. 3 shows the surviving fraction of microorganisms, N/N_0, as a function of time at temperatures exceeding those used for growth. As shown in the figure, some organisms will exhibit a brief non-logarithmic phase. There are a number of explanations and models for such deviations from logarithmic death kinetics:

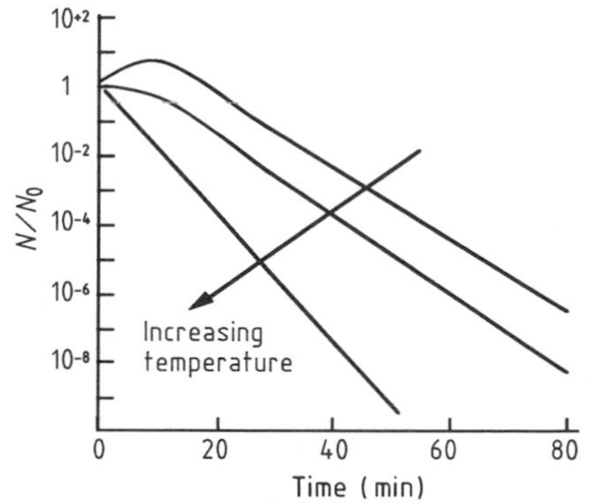

Fig. 3. Observed death rate kinetics at different temperatures.

spore germination, poor experimental technique, requirement of multiple "hits" for inactivation, and induction or sequential events of death. At very high temperatures there is usually only a logarithmic phase. While both μ and k are functions of temperature, k is usually a much stronger function of temperature. At lower temperatures the growth rate is greater than the death rate. However, because of its larger temperature sensitivity, at high temperatures, the specific death rate k is much higher than the specific growth rate μ. In other words, for $k \gg \mu$, Eq. (1) can be simplified to:

$$\frac{dN}{dt} = -k N \tag{3}$$

This equation may be integrated with respect to time over a period during which the number of viable organisms is reduced from an initial value N_0 to N. The resulting equation is:

$$\ln\left(\frac{N}{N_0}\right) = -k t \quad \text{or} \tag{4}$$

$$N = N_0 \, e^{-kt} \tag{5}$$

In other words, logarithmic death kinetics can be described by a simple first-order rate equation. Logarithmic death kinetics are most representative for vegetative cells. Non-logarithmic death kinetics are often observed for bacterial spores. While bacterial spores are much more resistant to thermal degradation than vegetative cells, in both cases the non-logarithmic profiles become more logarithmic at higher temperatures.

The concentration and the type of microbial contamination in the medium to be sterilized are usually not known. However, the relative thermal resistance of different microorganism types offers a conservative means for designing sterilization cycles. In fact, as depicted in Tab. 6, the relative resistance differs by approximately six orders of magnitude. As a consequence, sterilization processes need to be designed around the most resistant organism. For the case of media preparation for fermentation or bioconversion or for the sterilization of foods, the criterion is based on thermal destruction of microbial spores such as from the genera *Bacillus* or *Clostridium*. If one designs

Tab. 6. Relative Resistance of Various Microorganisms to Moist Heat

Type of Microorganisms	Relative Resistance
Vegetative bacteria and yeast	1
Viruses and bacteriophages	1– 5
Mold spores	2–10
Bacterial spores	$3 \cdot 10^6$

a sterilization cycle based on the destruction of bacterial spores, one can be reasonably assured that other contaminants will be destroyed simultaneously. As indicated in Fig. 3, it is possible to achieve a particular level of sterility by maintaining an appropriate temperature for a specified time. A sterility level of 10^{-3} means that one can expect that less than one batch in a thousand will become contaminated as a consequence of inadequate sterilization.

3.1.3 Effect of Temperature on Death Kinetics

If one considers the thermal death of microorganisms in a manner analogous to that of chemical reactions, the effect of temperature on thermal destruction may be described by the Arrhenius relationship as shown by Fig. 4 and Eq. (6):

$$k = A\,e^{\frac{-E_a}{RT}} \tag{6}$$

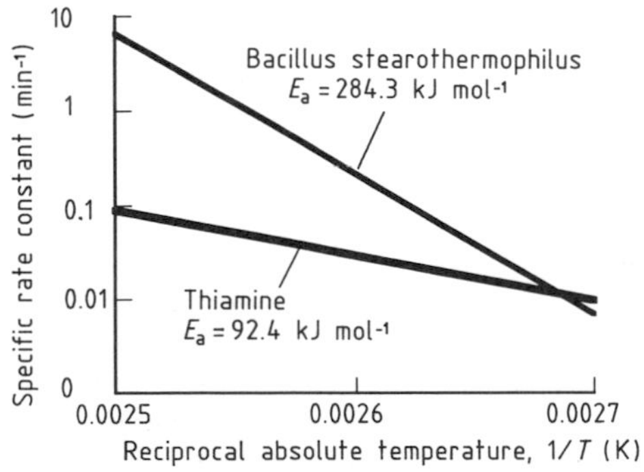

Fig. 4. Arrhenius plot for *Bacillus stearothermophilus* and thiamine.

where k is the thermal death constant, A is the Arrhenius constant, E_a is the activation energy, R is the universal gas constant, and T is the absolute temperature (K). Typical E_a values for microorganisms are 250 to 290 kJ mol^{-1} (60 to 70 kcal mol^{-1}).

During thermal treatment of the medium, there is not only the possibility of killing microbial spores and cells, but also of destroying ingredients in the medium. The sensitivity of nutrients to thermal destruction is illustrated in Tab. 7. The characteristic activation energy for vitamins and amino acids is typically 84 to 92 kJ mol^{-1}. By comparison, protein inactivation is about 165 kJ mol^{-1}. An observation from Tab. 7 and Fig. 4 is that the high activation energy for thermal destruction of microorganisms, e.g., 250–290 kJ mol^{-1}, means that a small increase in temperature has a greater effect on cell death than on nutrient destruction. This fact becomes the basis for the use of high temperature short time (HTST) sterilization discussed later in this chapter.

Tab. 7. Some Values of Activation Energy for Vitamins and Other Nutrients (COONEY, 1985)

Compound	Activation Energy (kJ mol^{-1})
Folic acid	70.2
D-Panthothenyl alcohol	87.8
Cyanocobalamin (B12)	96.6
Thiamine HCl (B6)	92.0
Riboflavin (B2)	98.7
Hemoglobin	321.4
Trypsin	170.5
Peroxidase	98.7
Pancreatic lipase	192.3

3.2 Batch Sterilization

With an understanding of thermal death kinetics and their dependence on temperature, it is possible to proceed with the design of a batch sterilization cycle for prevention of contamination while minimizing overheating and destruction of the medium components. We define a sterilization criterion, ∇, which represents the extent of death that occurs during some time interval. The required sterilization criterion may be calculated by a modification of Eq. (4) by expressing the initial number of spores in terms of a contamination concentration, N_i (spores/mL), and a vessel volume, V_F.

$$\nabla_{total} = \ln\left(\frac{N_0}{N}\right) = \ln\left(\frac{N_i V_F}{N}\right) \tag{7}$$

It is important to remember that the total sterilization criterion is a function of the fermentor volume. The initial number of viable organisms, N_i, is assumed to be the initial spore contamination. Because of the difficulty in measuring spore levels rapidly, one usually assumes a conservative value, typically 10^6 bacterial spores/mL. This is a high level of spore contamination, but it adds a safety margin to the design criteria. The final level of contamination, N, represents some acceptable risk in sterilization. Typically, a value of 10^{-3} spores per fermentor is used. This suggests that one will tolerate one batch in a thousand to become contaminated.

3.2.1 Temperature–Time Profile During Sterilization

The design objective of batch sterilization is to calculate a temperature–time profile that will ensure adequate sterilization. A typical profile is shown in Fig. 5. The heat-up and cool-down periods are dependent on equipment design, steam temperature, cooling water temperature, and flow rate (AIBA et al., 1973). In a new bioreactor it is possible to manipulate the various design parameters. However, often one is interested in designing a more effective sterilization cycle for an existing piece of equipment or is simply trying to understand why a bioreactor is not adequately sterilized or allows substantial destruction of essential nutrients. Thus, the heat-up and cool-down periods are usually considered to be fixed by the equipment and operating conditions in a given plant, and most of the design evaluation is based on calculating the required holding time at the maximum temperature in order to assure adequate sterilization. The equations that may be used to calculate the temperature–time profiles during heat up and cool down have been described by AIBA et al. (1973). For the purpose of the discussion here, it will be assumed that the temperature profile is measured experimentally for the equipment of interest.

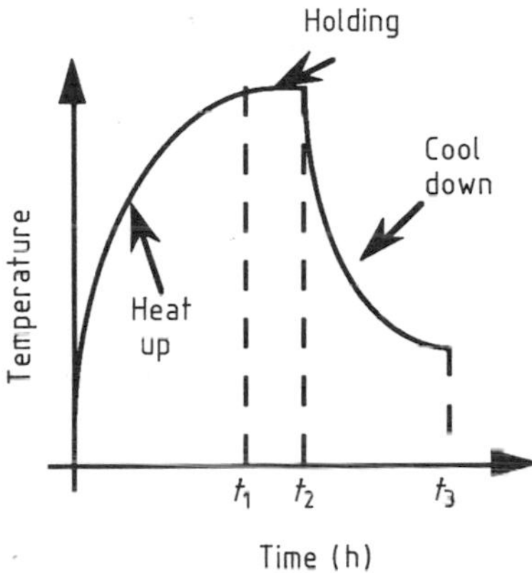

Fig. 5. Temperature–time profile for a batch sterilization of an industrial fermentor.

The overall sterilization criterion, ∇_{total}, may be broken down into three component parts: heating, holding, and cooling periods in the cycle

$$\nabla_{total} = \nabla_{heating} + \nabla_{hold} + \nabla_{cooling} \tag{8}$$

Because the heating and cooling temperature–time profile is fixed by the equipment, it is possible to calculate from this profile the extent of cell death using the Arrhenius relationship. From temperature–time data, such as in Fig. 5, it is possible to construct a plot of the thermal death constant k as a function of time; the area under the curve is:

$$\ln\left(\frac{N_0}{N}\right) = \nabla_{total} = \int_0^{t_3} k\,dt \tag{9}$$

Thus, by graphical integration, one can readily calculate the extent of thermal death during the heat-up and cool-down period. Rearrangement of Eq. (8) gives the following equation for sterilization during the holding period:

$$\nabla_{hold} = \nabla_{total} - \nabla_{heating} - \nabla_{cooling} \tag{10}$$

Once a temperature is chosen, k is fixed, and ∇_{hold} is calculated by

$$\nabla_{\text{hold}} = k(t_2 - t_1) \tag{11}$$

This holding time $(t_2 - t_1)$ is usually much smaller than the time required for heat up or cool down (see (Fig. 5). While the degree of sterilization during each of these periods depends on the temperature profile, most commonly sterilization is achieved during the holding period. Tab. 8 illustrates the impact of holding temperature on sterilization by showing the time required to achieve a 15-fold reduction in viable spores.

Tab. 8. Typical Values of k for *Bacillus stearothermophilus* Spores

Temp. (°C)	k (min^{-1})	Holding Time (min)
100	0.02	1730
110	0.21	164
120	2.00	17
130	17.50	2
140	136.00	0.25
150	956.00	0.04

3.2.2 Nutrient Degradation During Sterilization

One of the goals of the sterilization process is to minimize overall nutrient degradation while ensuring sterility. LA VERNE et al. (1989) described a kinetic procedure designed to evaluate and control the effects of temperature and heating on chemical reactions occurring in the media. This is particularly important during scale-up. During the scale-up of a batch sterilization cycle, an increase in the liquid volume affects both the overall sterilization criterion as well as the equipment design, leading to alterations in the temperature–time profile for heat up and cool down. This is illustrated in Fig. 6. Of particular concern is the effect of scale-up on nutrient destruction (MULLEY et al., 1975), for example, Maillard reactions.

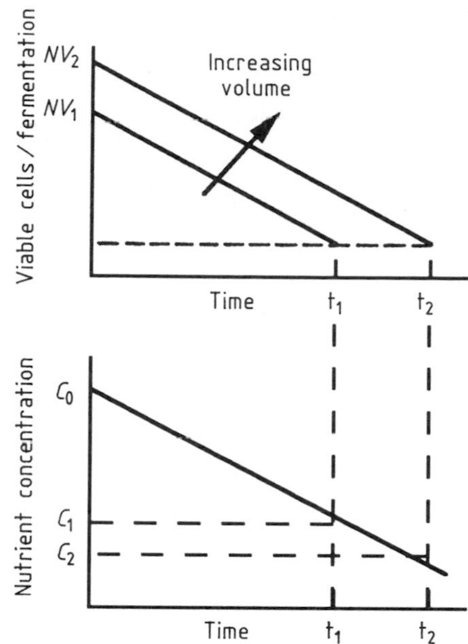

Fig. 6. Effect of scale-up on nutrient sterilization.

When the fermentor volume is increased from V_1 to V_2, it is necessary to hold the broth for a longer period of time; this results in greater destruction of nutrients in the broth. If the availability of a nutrient is critical to product formation, over-sterilization may lead to reduced product titers upon scale-up. This phenomenon is often seen in fermentations utilizing complex nutrients. As a consequence, medium optimization for large-scale fermentation processes must use sterilizing conditions that reflect those in larger vessels. SINGH et al. (1989) described a calculation procedure for optimization and scale-up of batch sterilization cycles in large-scale fermentors. Their technique determines the sterilization temperature and holding time necessary to minimize nutrient damage.

The example problem (RAJU and COONEY, in press) in Sect. 3.2.3 demonstrates the design of a typical sterilization cycle and shows the effect of sterilization on nutrient degradation.

3.2.3 An Example Problem

A batch sterilization cycle has been designed for a 1500 L pilot plant fermentor containing 1000 L of medium. It is assumed that the initial medium is contaminated with 10^5 spores/mL and 10^7 vegetative cells/mL and the sterilization cycle is designed for a level of 10^{-3} spores/fermentor. The medium also contains 500 mg/L of thiamine which is essential for the fermentation. The temperature during the holding time of the cycle is 121 °C, and the fermentation medium is finally cooled down to 37 °C. In addition, the following information is provided:

The fermentor has a height to diameter ratio of 2 and the vertical portion is covered by a jacket. Saturated steam maintains steam in the jacket at 25 psig.

The overall heat transfer coefficient when the fermentor is agitated is 153.7 J (36.6 cal)/h/cm²/°C. Cooling water is available at 17 °C.

For spores, at 125 °C, $k = 0.07\ \text{s}^{-1}$, and $\Delta E = 284.3$ kJ (67.7 kcal) mol^{-1}. For thiamine, at 102 °C, $k = 0.014\ \text{min}^{-1}$, and $\Delta E = 92.4$ kJ (22 kcal) mol^{-1}.

a) Calculate the degree of sterilization associated with each segment of a sterilization cycle designed to meet the specified criteria.

We know,

$$D_T = \frac{H_T}{2}, \quad \text{and } H_L = \left(\frac{V_L}{V_T}\right) H_T$$

where D_T = diameter of fermentor (cm)
H_T = height of fermentor (cm)
V_T = volume of fermentor (cm³)
V_L = volume of liquid (cm³)
H_L = height of liquid (cm)

Area for heat transfer,

$$A = \pi D_T H_L = 4.06 \cdot 10^4 \text{ cm}^2$$

Energy balance

$$\rho V_L C_p \left(\frac{dT}{dt}\right) = UA(T_s - T)$$

where U = overall heat transfer coefficient
A = area available for heat transfer
ρ = density of liquid
C_p = heat capacity of liquid
T_s = temperature of steam (K)
T = temperature of liquid (K)

that is

$$\frac{dT}{dt} = \frac{(36.6)(4.06 \cdot 10^4)}{(1000 \cdot 10^3)(1)(1)} (T_s - T)$$

Integrating,

$$\int_T \frac{dT}{(T_s - T)} = \int_t (1.49\ \text{h}^{-1})\, dt$$

Fig. 5 (Sect. 3.2) shows a temperature profile during sterilization. During heat up,

$$\ln\left(\frac{403 - T}{403 - 310}\right) = -1.49\, t$$

Hence, $T = 403 - 93\, e^{-1.49t}$ is the temperature profile during heat up.

Knowing that the holding temperature is 121 °C, we can calculate the total time for heat up.

$$t_1 = \frac{-1}{1.49} \ln\left(\frac{403\text{–}394}{403\text{–}310}\right)$$

i.e., total time of heating $t_1 = 1.57$ h

During cool down,

$$\ln\left(\frac{T - 290}{394\text{–}290}\right) = -1.49(t - t_2)$$

Hence, $T = 290 + 104\, e^{-1.49(t - t_2)}$ is the temperature profile during cool down.

Knowing that the final cool-down temperature is 37 °C, we can calculate the total time for cool down.

$$t_3 - t_2 = \frac{-1}{1.49} \ln\left(\frac{310 - 290}{394\text{–}290}\right)$$

i.e., total time of cooling $t_3 - t_2 = 1.11$ h

The process will be designed around the killing of the spores which are the most resistant to heat.

We know,

$$\nabla_{total} = \ln\left(\frac{N_0}{N}\right)$$

where N_0 = initial number of viable organisms per fermentor (number of organisms)
N = number of viable organisms per fermentor (number of organisms)

that is

$$\nabla_{total} = \ln\left(\frac{N_0}{N}\right) = \ln\left(\frac{(10^5)(10^3 \cdot 10^3)}{10^{-3}}\right) = 32.24$$

We know, rate constant

$$k = A \exp\left(\frac{-\Delta E}{R\,T}\right)$$

where R = universal gas constant (J/mol/K)
ΔE = activation energy for death (J/mol)
A = frequency factor (min^{-1})
k = specific death rate constant (min^{-1})

We are given that at 125 °C, $k = 0.07\ \text{s}^{-1}$; and $\Delta E = 284.3$ kJ (67.7 kcal)/mol^{-1}. Hence,

$$A = \frac{k}{\exp\left(\frac{-\Delta E}{R\,T}\right)} =$$

$$= \frac{4.2\ \text{min}^{-1}}{\exp\left(\frac{-67\,700}{1.987 \cdot 398}\right)} = 3.80 \cdot 10^{39}\ \text{h}^{-1}$$

We are now ready to calculate the required degrees of sterilization. We know that

$$\nabla_{total} = \nabla_{heating} + \nabla_{hold} + \nabla_{cooling}$$

where ∇_{total} = total degree of sterilization during sterilization cycle
$\nabla_{heating}$ = degree of sterilization during heat up
∇_{hold} = degree of sterilization during hold up
$\nabla_{cooling}$ = degree of sterilization during cool down

$$\nabla_{total} = \int_0^{t_1} A \exp\left(\frac{-\Delta E}{R\,T}\right) dt + \int_{t_1}^{t_2} A \exp\left(\frac{-\Delta E}{R\,T}\right) dt + \int_{t_2}^{t_3} A \exp\left(\frac{-\Delta E}{R\,T}\right) dt$$

The ratios

$$\left(\frac{\nabla_{heating}}{\nabla_{total}}\right), \left(\frac{\nabla_{hold}}{\nabla_{total}}\right), \text{ and } \left(\frac{\nabla_{cooling}}{\nabla_{total}}\right)$$

give us the degrees of sterilization in each segment. Need to calculate $\nabla_{heating}$, ∇_{hold}, and $\nabla_{cooling}$.

$$\nabla_{heating} = 3.80 \cdot 10^{39} \int_0^{1.57} \cdot \exp\left[\frac{-67\,700}{1.987(403 - 93\exp(-1.49t))}\right] dt =$$

$= 25.78$ (**80%** of sterilization)

$$\nabla_{cooling} = 3.80 \cdot 10^{39} \int_0^{1.11} \cdot \exp\left[\frac{-67\,700}{1.987(290 + 104\exp(-1.49t))}\right] dt =$$

$= 3.17$ (**9.8%** of sterilization)

$$\nabla_{hold} = \nabla_{total} - \nabla_{heating} - \nabla_{cooling} =$$
$$= 32.24 - 25.78 - 3.17$$

$= 3.29$ (**10.2%** of sterilization)

At 121 °C, k_{hold} = specific death rate constant at holding temperature = $1.75\ \text{min}^{-1}$

Therefore, holding time

$$= \frac{\nabla_{hold}}{k_{hold}} = \frac{3.29}{1.75} = \mathbf{1.9\ min}$$

b) How much thiamine remains after sterilization?

$$\frac{dc_{Th}}{dt} = -k_{Th}\, c_{Th}$$

where c_{Th} = thiamine concentration (g/L)
k_{Th} = specific decomposition rate of thiamine (min^{-1})

We are given that, for thiamine $k = 0.014$ min^{-1} at 102 °C, and $\Delta E = 92.4$ kJ (22 kcal) mol^{-1}.
Hence,

$$A = \frac{k_{Th}}{\exp\left(\frac{-\Delta E}{R\,T}\right)} =$$

$$= \frac{0.84\ h^{-1}}{\exp\left(\frac{-22000}{1.987 \cdot 375}\right)} = 5.58 \cdot 10^{12}\ h^{-1}$$

In general, at any temperature,

$$k_{Th} = 5.58 \cdot 10^{12} \exp\left(\frac{-22000}{1.987\,T}\right)$$

$$\nabla_{Th_{heating}} = 5.58 \cdot 10^{12} \int_0^{1.57} \cdot$$

$$\exp\left(\frac{-22000}{1.987(403 - 93\exp(-1.49\,t))}\right) dt = 1.81$$

$$\nabla_{Th_{cooling}} = 5.58 \cdot 10^{12} \int_0^{1.11} \cdot$$

$$\exp\left(\frac{-22000}{1.987(290 + 104\exp(-1.49\,t))}\right) dt = 0.34$$

At 121 °C,

$$k_{Th_{hold}} = 5.58 \cdot 10^{12} \exp\left(\frac{-22000}{1.987 \cdot 394}\right) =$$

$$= 3.49\ h^{-1} = 0.058\ min^{-1}$$

For a 2.9 min hold, $\nabla_{Th_{hold}} = 1.9 \cdot 0.058 = 0.11$

Thus,

$$\nabla_{Th_{total}} = \nabla_{Th_{heating}} + \nabla_{Th_{hold}} + \nabla_{Th_{cooling}} = 2.26$$

Thus, after, sterilization, residual thiamine concentration $= \frac{c_{Th_0}}{\exp(\nabla_{Th})} = \frac{500\ mg/L}{\exp(2.26)}$

where c_{Th_0} = initial thiamine concentration (g/L)
i.e., residual thiamine concentration = **52.2 mg/L**

c) If a 15 min holding time for sterilization is used instead of the described design, how much thiamine will be present?

$$\nabla_{Th_{hold}} = k_{Th_{hold}}(\text{holding time})$$
$$= (0.058\ min^{-1})\ (15\ min) = 0.87$$

$$\nabla_{Th_{total}} = \nabla_{Th_{heating}} + \nabla_{Th_{hold}} + \nabla_{Th_{cooling}}$$

Thus, after sterilization, residual thiamine concentration

$$= \frac{c_{Th_0}}{\exp(\nabla_{Th})} = \frac{500\ mg/L}{\exp(3.02)} = \mathbf{24.4\ mg/L}$$

d) If the sterilization is carried out by direct steam injection instead of steam in the jacket, (1) how long would heat up and hold take?

The time required for heat up depends on how quickly the steam is injected into the medium.
Given no other information, we assume this time is negligible. i.e., $\nabla_{heating} = 0$.
Designing for the same sterility condition, we have $\nabla_{total} = 32.24$.
Assuming the cooling profile is unaffected, we have $\nabla_{cool} = 3.17$ (same as before).
We know,

$$\nabla_{hold} = \nabla_{total} - \nabla_{heating} - \nabla_{cooling}$$

Hence,

$$\nabla_{hold} = 32.24 - 3.17 - 0.0 = 29.07$$

Therefore, holding time

$$= \frac{\nabla_{hold}}{k_{hold}} = \frac{29.07}{1.75} = \mathbf{16.6\ min}$$

(2) what would be the remaining thiamine?

For thiamine,

$\nabla_{\mathrm{Th}_{\mathrm{hold}}} = (16.6)(0.058) = 0.96$

i.e., for thiamine,

$\nabla_{\mathrm{Th}_{\mathrm{total}}} = \nabla_{\mathrm{Th}_{\mathrm{heating}}} + \nabla_{\mathrm{Th}_{\mathrm{hold}}} + \nabla_{\mathrm{Th}_{\mathrm{cooling}}}$

i.e., $\nabla_{\mathrm{Th}_{\mathrm{total}}} = 0.0 + 0.96 + 0.34 = 1.3$

Thus, remaining thiamine

$$= \frac{c_{\mathrm{Th}_0}}{\exp(\nabla_{\mathrm{Th}})} = \frac{500 \text{ mg/L}}{\exp(1.3)} = \mathbf{136.3\ mg/L}$$

(3) what would be the volume change in the fermentor?

For steam at 130 °C, ΔH_{vap} = heat of vaporization = 2.18 kJ (518 cal)/g
Ignoring evaporation losses and other heat losses through the equipment,
Q = heat required to heat up the media = $M\,C_p\Delta T$, where M is the mass of the medium (in g), and ΔT is the change in temperature (°C)
Assume: $C_p = 4.2$ kJ (1.0 cal) per gram per °C and density of fermentation medium $\rho_m = 1$ g/cm^3, i.e., $Q = (1 \cdot 10^6 \text{ g})(1 \text{ cal/g/°C})(121 - 37 \text{ °C}) = 8.4 \cdot 10^7$ cal.
Applying an energy balance for direct steam injection (ignoring evaporation and other losses), the mass (g) of steam added, F_s, is given by

$$F_s = \frac{Q}{\Delta H_{\mathrm{vap}} + C_p(T_s - T_f)}$$

where T_s = temperature of steam (°C)
T_f = final temperature of fermentation medium (°C)

$$F_s = \frac{8.4 \cdot 10^7}{518 + 1.0(130 - 121)} =$$

$$= 15.94 \cdot 10^4 \text{ g steam/batch}$$

Hence, total steam added is approximately **159 liters.**

3.3 Continuous Sterilization

Continuous sterilization involves the thermal destruction of microorganisms as the medium flows through the holding section of a continuous sterilizer. It is based on the concept of high temperature short time (HTST) treatment. JACOBS et al. (1973) described HTST destruction of spores. HTST treatment takes advantage of the fact that an increase in temperature has a greater effect on thermal destruction of cells than on nutrient. The rationale for using continuous versus batch sterilization is based not only on the beneficial effect on heat-sensitive media which can be sequentially sterilized, but also on improved steam economy, easier and more reliable scale-up, ease of interfacing with continuous processes and more efficient fermentor use. The steam usage in continuous sterilization is usually one-fifth to one-fourth of that for a batch cycle. In addition, the total time required for continuous sterilization is usually only two to three hours compared with a typical batch sterilization cycle of five to six hours for a large fermentor. The basic ideas of continuous sterilization are described here together with some practical implications. For details on the underlying theory, the reader is referred to LIN (1975) and SVENSSON (1988).

3.3.1 Fluid Flow

It is important to understand the flow pattern through a continuous sterilizer. The Reynolds number, Re, gives an indication of the type of flow. The Reynolds number in the sterilizer tube is given by:

$$Re = \frac{d\,\bar{U}\rho}{\mu} \tag{12}$$

where d is the tube diameter, $\bar{U}$ is the average velocity, ρ is the density, and μ is the medium viscosity. Fig. 7 shows the different types of flow. Ideally, one would like perfect plug or piston flow as this would allow for uniform residence times in the holding section of the continuous sterilizer.

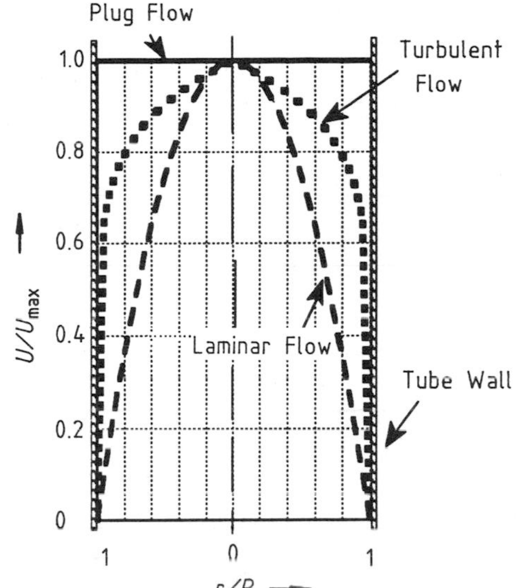

Fig. 7. Distribution of velocities in fluids exhibiting different types of flow inside round pipes.

In the case of laminar flow, the flow profile and the average velocity are given by (BIRD et al., 1960)

$$\frac{U}{U_{\max}} = \left[1 - \left(\frac{r}{R}\right)^2\right] \quad \text{and} \quad \bar{U} = 0.5\, U_{\max} \tag{13}$$

For turbulent flow, the flow profile and average velocity are given by

$$\frac{U}{U_{\max}} = \left(1 - \frac{r}{R}\right)^{1/7} \quad \text{and} \quad \bar{U} = 0.82\, U_{\max} \tag{14}$$

Typically, the flow will be somewhere between laminar and fully turbulent flow, such that the average velocity is between 0.5 and 0.82 times the maximum velocity. In order to minimize overheating of the medium while still meeting the design criteria for sterilization, it is desirable to approach fully turbulent flow. This occurs when the Reynolds number is at least $2.5 \cdot 10^3$ and preferably above $2 \cdot 10^4$. In practice, however, pipes are short and have bends. This induces turbulence at lower Reynolds numbers. Hence, the criterion described above is conservative.

Axial dispersion is one method to characterize the flow pattern in the tube. The Peclet number, *Pe*, signifies the importance of convective mass transport relative to dispersive mass transport, i.e.,

$$Pe = \frac{\bar{U} L}{D_z} = \frac{\bar{U} N}{\left(\frac{D_z N}{L}\right)} = \frac{\text{convective mass transport}}{\text{dispersive mass transport}} \tag{15}$$

where D_z is the axial dispersion coefficient, a modeling parameter which provides a means to represent the mixing effects of more than one physical phenomenon. It is not usually equal to molecular diffusivity.

The degree of dispersion (as indicated by the Peclet number) affects the residence time of organisms in the continuous sterilizer. A *Pe* close to 0 indicates a large amount of dispersion or mixed flow. A *Pe* close to ∞, on the other hand, indicates very little dispersion and hence depicts plug flow. Fig. 8 shows the effect of dispersion on the residence time of a tracer in a continuous sterilizer. The experiment involves a step change in the tracer concentration at the inlet from an initial tracer concentration C_0 to an inlet tracer concentration of zero. C represents the outlet tracer concentration. Fig. 8 shows the normalized outlet tracer concentration as a function of time for different *Pe*. A step change of tracer concentration made on fluid entering the reactor that is being operated in perfect plug flow ($Pe \rightarrow \infty$), causes no change in outlet concentration until after precisely one residence time when the concentration falls precipitously to zero. In such a situation all the organisms would be exposed to sterilization for the same amount of time. However, this ideal situation is usually not achieved. Rather, Peclet numbers ranging from 3 to 600 are typical. As can be seen from Fig. 8, the tracer concentration, C, begins to decline from the initial C_0 value for residence times significantly less than 1. Hence, part of the medium is exposed to the high temperatures for a considerably shorter time. Therefore, deviations from plug flow must be taken into account in the design of the sterilizer.

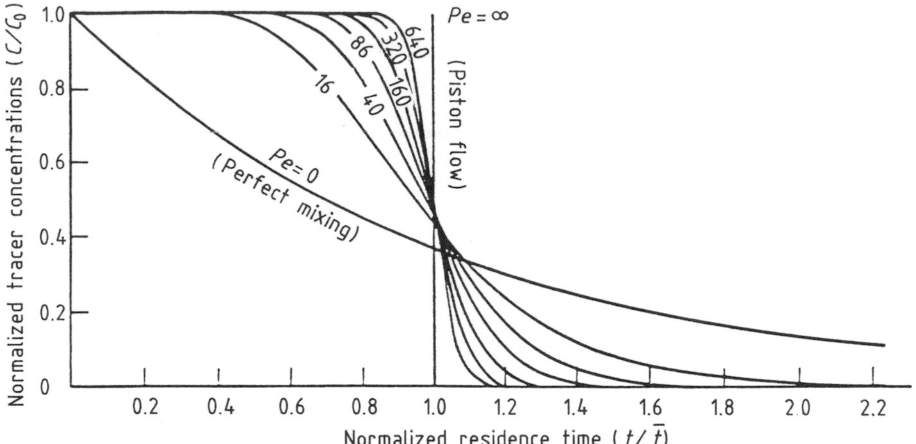

Fig. 8. Effect of different types of flow as shown by different values of the Peclet number (*Pe*) and residence times ($t/\bar{t}$) in continuous medium sterilization (modified from AIBA et al., 1973).

3.3.2 The Dispersion Model

Many models have been proposed to describe the sterility level of the medium as it flows through the sterilizer. Of these, the dispersion model is most commonly used to describe non-ideal tubular sterilizers. In this model, the transport of microorganisms is governed by an analogy to Fick's law of diffusion superimposed on the bulk flow. As shown in Fig. 9, in addition to the bulk flow, microorganisms in the medium are transported through any cross-section of the sterilizer by axial dispersion. We use the term "axial" to distinguish mixing in the direction of flow from mixing in the radial direction which is not our primary concern here.

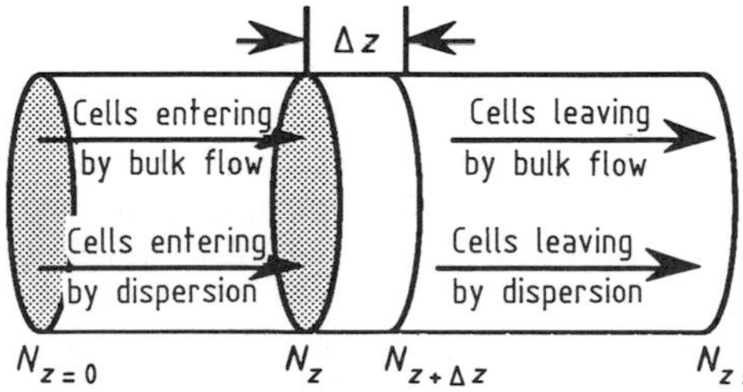

Fig. 9. Flow through a continuous sterilizer depicting a balance on cells.

At steady state, a material balance based on the cell number over the elementary section shown in Fig. 9 is given by:

$$\begin{pmatrix}\text{cells}\\ \text{leaving by}\\ \text{bulk flow}\end{pmatrix} - \begin{pmatrix}\text{cells}\\ \text{entering by}\\ \text{bulk flow}\end{pmatrix} + \begin{pmatrix}\text{cells}\\ \text{leaving by}\\ \text{axial dispersion}\end{pmatrix} - \begin{pmatrix}\text{cells}\\ \text{entering by}\\ \text{axial dispersion}\end{pmatrix} + \begin{pmatrix}\text{rate of}\\ \text{thermal destruction of}\\ \text{cells within the section}\end{pmatrix} = 0$$

That is, for constant area,

$$N_{z+\Delta z}\bar{U}A - N_z\bar{U}A - \left(D_zA\frac{dN}{dz}\right)_{z+\Delta z} + \left(D_zA\frac{dN}{dz}\right)_z + k\,N(A\Delta z) = 0 \tag{16}$$

or

$$D_z\frac{\left[\left(\frac{dN}{dz}\right)_{z+\Delta z} - \left(\frac{dN}{dz}\right)_z\right]}{\Delta z} - \bar{U}\frac{(N_{z+\Delta z} - N_z)}{\Delta z} - k\,N = 0$$

Taking limits as $\Delta z \to 0$, we have

$$D_z \frac{d^2N}{dz^2} - \overline{U}\frac{dN}{dz} - k\,N = 0 \tag{17}$$

where D_z = axial dispersion coefficient (cm^2/s)
N = microorganism concentration (number/cm^3)
z = distance along the axial direction (cm)
$\overline{U}$ = average medium velocity (cm/s)

Before proceeding to solve this equation, we will first put the equation in dimensionless form. Let

$$\overline{N} = \frac{N}{N_0} =$$

dimensionless microorganism concentration

$$\bar{z} = \frac{z}{L} =$$

dimensionless distance along the axial direction

Putting this into Eq. (17), we get

$$\left(\frac{D_z}{\overline{U}\,L}\right)\frac{d^2\overline{N}}{d\bar{z}^2} - \frac{d\overline{N}}{d\bar{z}} - \left(\frac{k\,L}{\overline{U}}\right)\overline{N} = 0 \tag{18}$$

The coefficients within parentheses in Eq. (18) represent well known dimensionless numbers, i.e.,

$$\frac{\overline{U}\,L}{D_z} = \text{axial Peclet number } Pe$$

and

$$\frac{k\,L}{\overline{U}} = \text{Damköhler number } Da$$

The Damköhler number is regarded as a measure of the rate of destruction of microorganisms by thermal inactivation relative to the rate of convective mass transport, i.e.,

$$Da = \frac{k\,L}{\overline{U}} = \frac{(k\,N\,L)}{(\overline{U}\,N)} = \frac{\text{rate of destruction of cells by thermal inactivation}}{\text{rate of mass transport by convection}} \tag{19}$$

Substituting for the dimensionless quantities, Eq. (18) can be written as

$$\frac{1}{Pe}\frac{d^2\overline{N}}{d\bar{z}^2} - \frac{d\overline{N}}{d\bar{z}} - Da\,\overline{N} = 0 \tag{20}$$

The second-order differential equation, Eq. (20), describes the concentration of the microorganisms as a function of distance along the holding section in the continuous sterilizer. It can be solved analytically when first-order kinetics are assumed. Using the Danckwerts boundary conditions (DANCKWERTS, 1953):

$$\frac{-1}{Pe}\frac{d\overline{N}}{dz} + \overline{N} = 1 \quad \text{at } z = 0$$

$$\frac{d\overline{N}}{d\bar{z}} = 0 \quad \text{at } z = 1$$

At the end of the continuous sterilizer where $\bar{z} = 1$, the solution to Eq. (20) given the above boundary conditions is

$$\overline{N}_{\bar{z}=1} = \left(\frac{N}{N_0}\right)_{\bar{z}=1} = \frac{4\,a\exp\left(\frac{Pe}{2}\right)}{(1+a)^2\exp\left(\frac{a\,Pe}{2}\right) - (1-a)^2\exp\left(\frac{-a\,Pe}{2}\right)} \tag{21}$$

where $a = \sqrt{1 + \frac{4Da}{Pe}}$

This solution can be conveniently expressed in graphical form as shown in Fig. 10 which shows the sterility level (N/N_0) for various Damköhler numbers (Da) and various Peclet numbers (Pe). From this plot, it is clear that as ideal plug flow is reached ($Pe \to \infty$), the desired degree of medium sterility can be achieved with the shortest possible sterilizer.

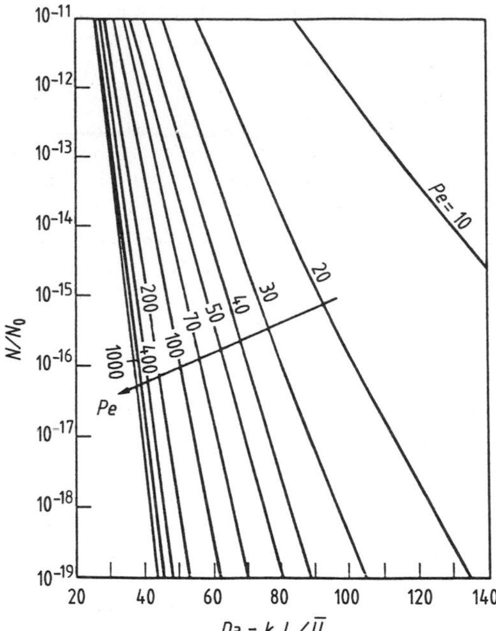

Fig. 10. Effect of different types of flow as shown by different *Pe* values on the destruction of organisms (N/N_0) at different rates of destruction indicated by the Damköhler number $Da = k\,L/\overline{U}$ (modified from AIBA et al., 1973).

3.3.3 The Design Approach

The objective in sterilizer design is to adequately kill viable spores while minimizing the overheating of the medium. Given below are the steps involved in designing a continuous sterilizer.

Step 1. Determine the type of flow

a) Choose a pipe diameter, *d*, and liquid velocity, $\overline{U}$, to achieve fully turbulent flow ($Re > 2 \cdot 10^4$) and a liquid flow rate that permits to fill the fermentor in 2–4 hours.

b) By calculation (LEVENSPIEL, 1958) or measurement from a tracer experiment (on a sterilizer of known length) calculate the Peclet number and hence the degree of axial dispersion at this Reynolds number.

Step 2. Specify a sterilization criterion N/N_0

a) *Batch fermentation:*

Here, it is typical to assume an initial spore concentration (N_i) of 10^6 spores per mL and design for a desired tolerance (N) of 10^{-3} spores per fermentor (i.e., a contamination of one batch in a thousand). Thus, for a batch fermentation

$$\frac{N_0}{N} = \frac{N_i V_F}{N} = V_F \cdot 10^{12}$$

where V_F is the volume of the fermentor (in liters).

b) *Continuous fermentation:*

The equivalent expression for a continuous fermentation is

$$\frac{N_0}{N} = \frac{F\,t}{n} \cdot 10^9$$

where F is the medium flow rate (L/h), and t is the operation time for the fermentation (in hours). In the case of continuous fermentation, n reflects the number of fermentations that one would tolerate to become contaminated (typically 10^{-3}).

Step 3. Calculate the Damköhler number

Having chosen the sterilization criterion (from step 2), one can now use Fig. 10 along with knowledge of the Peclet number (from step 1) to estimate the Damköhler number *Da*.

Step 4. Calculate the length *L*

Choose a holding temperature T to calculate the value of k. Then, from the Damköhler number, $Da = k\,L/\overline{U}$, one can calculate L, the length of the sterilizer.

Step 5. Repeat steps 1 to 4

Reiterate the above calculations using different values of d and T to obtain reasonable values for L and pressure drop. A trade-off can be made between the length of the sterilizer and the operating temperature. This represents a trade-off between capital investment and operating costs.

It is to be noted that for sterilizing viscous media like starch-containing media, it is important to compensate for viscosity by increasing the fluid velocity or changing the characteristic dimension to maintain an acceptable Reynolds number. With non-Newtonian fluids it may be possible to take advantage of any shear thinning effect.

3.3.4 Effect of Solids

Because the time to reach sterilization temperature in a continuous sterilizer is much shorter than in batch sterilization, there is a potential problem associated with the presence of particles. Shown in Tab. 9 are the typical times to reach 99% of the final temperature for particles ranging in diameter from 1 μm to 1 cm. As a consequence, it is important to exclude particles with a dimension greater than 1 or 2 mm from continuous sterilizers (see DE RUYTER and BRUNET, 1973). ARMENANTE and LESKOWICZ (1990) described a mathematical model that was developed to analyze the performance of a continuous sterilizer fed with a liquid fermentation medium containing suspended solids.

Tab. 9. Effect of Particle Size on Heat-Up Time of Solids

Diameter (cm)	Time to Reach 99% of Final Temperature
$1 \cdot 10^{-4}$	1 μs
$1 \cdot 10^{-3}$	0.1 ms
$1 \cdot 10^{-2}$	10.0 ms
$1 \cdot 10^{-1}$	1 s
1	100 s

3.3.5 Equipment for Continuous Sterilization

The essential components of the continuous sterilizer system include a heat exchanger for recovery of heat from the sterilized fluid and preheating fresh medium, a heat exchanger or steam injector for heating the medium to sterilizing temperatures, a holding section for sterilization, a heat exchanger for cooling the medium to the fermentation temperature and a number of peripheral pieces of equipment including a medium mixing tank, sterile medium receiver, which is often the fermentor, and a set of pumps, valves and controllers. The system is usually started up by passing water through the system at sterilizing temperatures. This serves to avoid having any regions in the equipment below the critical temperature for sterilization. The time required to sterilize and fill a batch fermentor is typically two to four hours.

Figs. 11 and 12 show two different modes of heating up the medium. Continuous steam injection, as illustrated in Fig. 11, is used to rapidly heat up the medium without the use of a heat exchanger. Continuous plate or spiral heat exchangers are used for media sterilization to avoid direct steam injection and are depicted in Fig. 12. Fig. 13 shows the typical temperature–time profiles of each of these two modes of heat transfer.

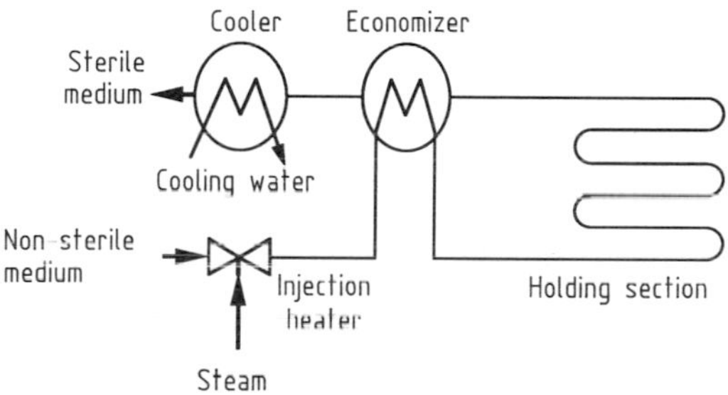

Fig. 11. Schematic view of continuous steam injection for media sterilization.

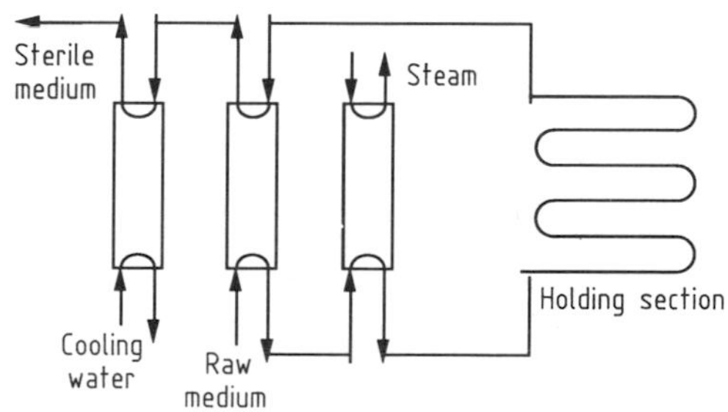

Fig. 12. Continuous plate or spiral heat exchangers used for media sterilization, avoiding direct steam injection.

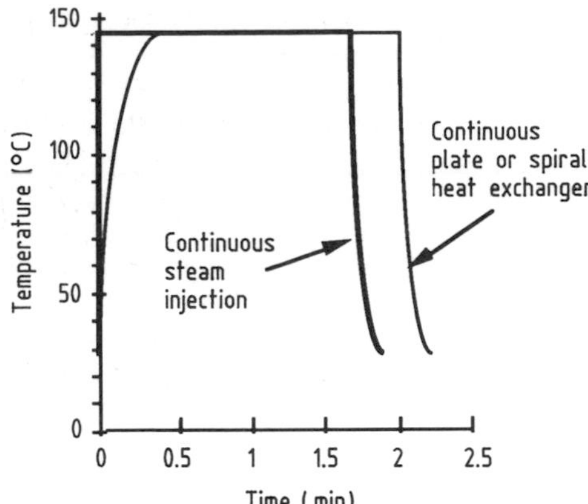

Fig. 13. Typical temperature profiles for direct steam injection and plate or spiral heat exchanger system.

The steam injection approach is particularly effective with media that tend to foul heat exchange surfaces. In using a particular steam injection approach (see PICK, 1982), it is important to avoid two-phase flow. Therefore, one operates a continuous sterilizer at 5 to 10 psi above the bubble point. The disadvantage of the steam injection approach is dilution of the medium with condensed steam and the difficulty in controlling pressure and temperature due to variation in medium viscosity. It also is important to use clean steam. This serves to avoid the possibility of transferring materials from the steam into the medium. Heat exchangers are used for preheating and cooling of the medium in order to achieve efficient energy usage.

On the other hand, disadvantages of the heat exchanger approach are higher capital costs and the potential for fouling of hot heat exchange surfaces and for leaks around the exchanger gaskets and seals. The spiral heat exchanger is particularly effective with media containing suspended solids and heat-sensitive organic material. The wide gap, curved walls, and high liquid velocity create secondary flow patterns which minimize surface fouling. In addition, such heat exchangers efficiently exploit the available space and minimize the gasket area that could potentially leak. Plate and frame heat exchangers are often used for media with low suspended solid concentrations. They are very flexible in their design since the heat transfer area can be easily altered by adjusting the number of plates. It is also possible to achieve high heat transfer coefficients in these heat exchangers.

4 Air Sterilization

4.1 Introduction

Large quantities of air are required to pass through a fermentor during the course of a fermentation. This is due to the low solubility of oxygen in water. Typically 0.5 to 1.5 vvm (volume of air per volume of liquid per minute) are used in aerobic fermentations.

The incoming air contains a mixture of microorganisms, a typical distribution of which is shown in Tab. 10. Air is a source of potential contamination and must be sterilized before it enters the fermentor. There may also be a need to sterilize exhaust gas from the fermentor in order to maintain containment.

Tab. 10. A Typical Distribution of Microorganisms in Air (approximate)

Microorganism	Distribution of Population
Molds	50%
Yeast	<1%
Gram-positive bacteria	10%
Gram-negative bacteria	40%
Phages	very low

To design an air sterilization system one needs an estimate of the total microbial content of the air. Tab. 11 highlights some typical numbers obtained by the Medical Research Council (BOURDILLON et al., 1948). As seen from this table, the total bacterial content of the air depends on location. Samples of outside air around various pharmaceutical and food-canning plants have been bound to have between 750 and 3000 microorganisms per cubic meter. It is important to note that con-

Tab. 11. Colonies Obtained/m^3 of Air Sampled in Different Locations (adapted from BOURDILLON et al., 1948)

Location	Colonies/m^3 (approximate range)
City street	60 to 180
School classroom	1200 to 3000
"Sterile" laboratory	30 to 60
Hospital ward	180 to 18000

centration varies greatly, not only from place to place, but also at any one location on different days. Microorganisms are more numerous in the air during dry weather than just after a rain. Their concentration also depends on the height at which the sample is taken. The number of microorganisms at a particular location decreases as the height above ground is increased. Because of this variability, it is often advisable to determine contaminant loadings by actually measuring them for any particular plant. When data are not available, a loading of 3000 contaminants/m^3 is typically assumed.

Although heat is sometimes used as a means to sterilize the air, filtration is by far the most common method. In contrast to heat sterilization of media, where sterilization is based on thermal destruction of the most heat-sensitive organism (the bacterial spore), air sterilization by filtration is usually based on the removal of the entire population rather than a resistant subset of the population.

4.2 Depth Filtration Using Fibrous Materials

When submerged culture of aerobic fermentation became the method of choice for manufacturing in the 1940s, the problem of sterilizing large quantities of air became a challenge. The technology that evolved employed depth filters with fibrous materials. Later, the use of packed carbon, ceramic and sintered metal filters was also considered.

4.2.1 Single-Fiber Efficiency

As described in Sect. 2, four primary mechanisms for depth filtration are inertial impaction, interception, diffusion, and sedimentation. In designing filters for sterilization, attempts have been made to base the particle removal efficiency of a filter on models for the efficiency of a single-fiber collector. The single-fiber collector efficiency (η) is defined to be the rate at which particles strike the collector divided by the rate at which particles flow toward the collector. This single-fiber collector efficiency is the combination of the single-fiber efficiencies due to inertial impaction η_{imp}, interception η_{int}, diffusion η_{diff}, and gravitational sedimentation η_{sed}. Tab. 12 summarizes the functional forms of the single-fiber collector efficiency for pure impaction (HUMPHREY and GADEN, 1955), and pure interception, diffusion and gravitational sedimentation (YAO et al., 1971).

The efficiency of a single fiber is obtained by summing the efficiencies due to each of these four mechanisms of filtration, i.e.,

$$\eta = \eta_{imp} + \eta_{int} + \eta_{diff} + \eta_{sed} \tag{22}$$

The overall efficiency of the filter η_{filter} is then expressed in terms of the single-fiber efficiency, η.

Tab. 12. Functional Forms of Single-Fiber Collector Efficiency

Mechanisms of Filtration	Single-Fiber Collector Efficiency
Inertial impaction	$\eta_{imp} = 0.15 \left(\frac{\rho_p v_0 d_p^2}{18\mu d}\right)^{1/2}$
Interception	$\eta_{int} = \frac{3}{2}\left(\frac{d_p}{d}\right)^2$
Diffusion	$\eta_{diff} = 0.9 \left(\frac{k T}{\mu d_p d v_0}\right)^{2/3}$
Gravitational sedimentation	$\eta_{sed} = \frac{(\rho_p - \rho) g d_p^2}{18 \mu v_0}$

ρ, density of particle; v_0, velocity; d_p, diameter of particle; μ, viscosity of fluid; d, diameter of fiber; k, Boltzmann's constant; T, absolute temperature; ρ, density of fluid; g, acceleration due to gravity

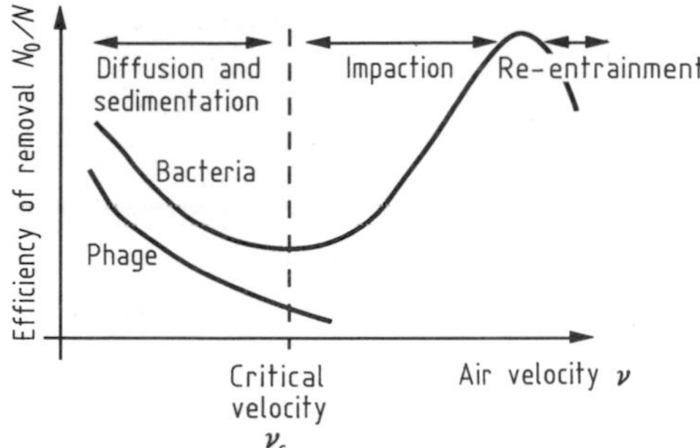

Fig. 14. Effect of air velocity on filter efficiency.

The effect of these mechanisms strongly depends on particle size and velocity. The effect of air velocity on filter efficiency is shown qualitatively in Fig. 14. As can be seen, the relationship is non-linear and goes through a minimum. From Tab. 12 we see that the efficiencies due to the diffusion and sedimentation vary as $v_0^{-2/3}$ and v_0^{-1}, respectively, while the efficiency due to impaction varies as $v_0^{1/2}$. Also, collection efficiency due to interception is independent of velocity. These opposing velocity dependencies result in a minimum efficiency level corresponding to a critical velocity. Below this critical velocity, diffusion and sedimentation dominate and hence, efficiency decreases with increasing air velocity. Above the critical velocity, impaction dominates and results in increasing efficiency with increased air velocity. In any event, whether by impaction, interception, or diffusion, when the particle is captured by the fiber and sits on the surface, the stagnant boundary layer around the fiber minimizes the chance for re-entrainment. If the fluid velocity is increased enough, then re-entrainment occurs and the efficiency decreases. When viruses and phages are in the air stream, their removal efficiency declines continuously with increasing air velocity. This is because they are too small for inertial impaction to be significant.

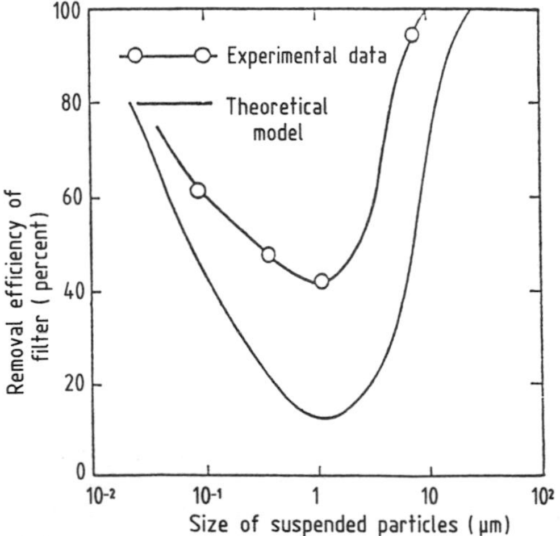

Fig. 15. Removal efficiency as a function of particle size: theoretical vs. observed (YAO et al., 1971).

Fig. 15 (YAO et al., 1971) shows the removal efficiency of a filter as a function of the particle size. Once again, the functionalities given in Tab. 12 are successful in describing the qualitative behavior of the relationship. As can be seen from Tab. 12, efficiency due to diffusion varies as $d_p^{-2/3}$, while the efficiencies due to impaction, interception, and sedimentation vary as d_p, d_p^2, and d_p^2, respectively. Due to these opposing particle size dependencies, the efficiency goes through a minimum. Diffusion dominates at low particle sizes, while interception and sedimentation dominate at larger particle size.

Also shown in Fig. 15 is a comparison between the theoretical and actual removal efficiencies of the filter. The considerable discrepancy is typical of single-fiber efficiency-based models, and this makes them unsuitable for use in most design situations. This discrepancy is due to many possible reasons. There are other removal mechanisms, and in most situations the presence of other fibers alters the flow field neighboring a given fiber and affects its collection efficiency. While theory provides a means to understand the qualitative behavior of fibrous filters, it is usually inadequate to predict quantitative performance.

4.2.2 Log Penetration Theory

The design of fibrous air filters was standardized by HUMPHREY et al. (1955) with the use of the log penetration theory. HUMPHREY recognized that particle removal could be described by

$$\frac{dN}{dL} = -K N \qquad (23)$$

where K is a constant and N is the concentration of contaminant at a depth L of the filter. In other words, this equation assumes that each layer of the filter reduces the concentration entering by the same fraction. The value of K depends on the superficial velocity, filter void fraction, fiber diameter, particle diameter, and particle density. Hence, given a particular system, K depends only on the superficial velocity. This equation can now be integrated over the length of the filter.

$$\int_{N_0}^{N} \frac{dN}{N} = -\int_{0}^{L} K\, dL$$

or,

$$\ln\left(\frac{N}{N_0}\right) = -K L \qquad (24)$$

Eq. (24) can also be written as:

$$\ln\left(\frac{N}{N_0}\right) = -\left(\frac{2.3}{L_{90}}\right) L \qquad (25)$$

where L_{90} is the length of a fibrous bed that will remove 90% of incident particles. This relationship is shown graphically in Fig. 16

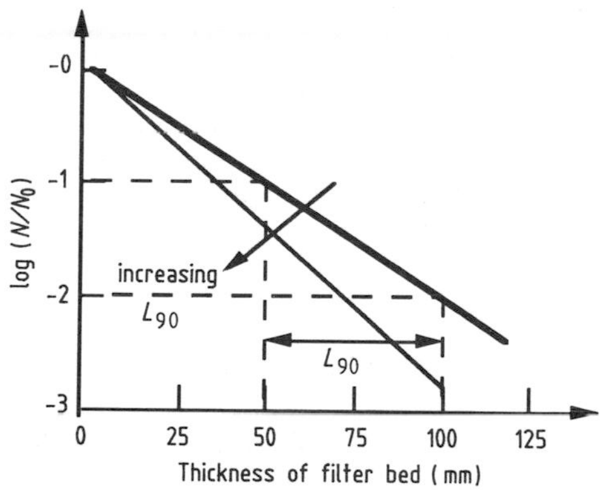

Fig. 16. Reduction of contaminant concentration over filter length.

where L_{90} is the length of filter required to reduce the contaminant concentration by a factor of ten. Smaller L_{90} calls for a shorter filter, and L_{90} values vary with filter type, particle and velocity. Some typical values are shown in Tab. 13.

For a filtration system an L_{90} value allows one to compute the desired filter length. Since the efficiency of the filtration system is lowest at the critical velocity, it is common to design a filter based on the critical velocity.

4.2.3 Design Approach

This background allows us to formulate a design for fibrous filters.

Step 1. Determine the sterilization criteria

In order to calculate the total loading or challenge to the filter, it is necessary to have an estimate of the microbial concentration in the entering air. If experimental data are not available, a value of $3000/m^3$ may be used for the contaminant concentration N_c. Then the total number of organisms entering a filter (N_0) during the operating period may be computed using the equation

$$N_0 = Q_{air}\, N_c\, t_f$$

where Q_{air} is the volumetric air flow rate, N_c is the contaminant concentration in entering air, and t_f is the operating period. A penetration probability needs to be chosen. Typically an allowable penetration probability of 10^{-3} is assumed. This corresponds to the design value N.

Step 2. Determine the critical velocity

A particular microorganism (spores, bacteria, or phages) is chosen to be the basis for design. This determines the particle diameter d_p (typically 1 µm for bacteria). An appropriate fiber material is then selected for the filtration. This determines the fiber diameter d_f. We are now ready to determine the approximate critical velocity for this system. The critical velocity that corresponds to the minimum efficiency of a filter can be determined by using the

Tab. 13. Typical L_{90} Data (adapted from HUMPHREY, 1960)

Filter Material	d_f (mm)	Organism	V (cm/s)	L_{90} (cm)
Owens-Corning IMF glass fiber mat	16	*Bacillus subtilis* spores	3.1	4.06
			15.3	9.14
			30.5	11.68
			152.5	1.52
			305.0	0.38
Corning glass No. 800 fiber	8.5	*E. coli* T2-phage	3.1	0.43
			15.3	0.61
			30.5	0.71
			152.5	0.86
			305.0	1.12
Norik Pk 0.5–1 carbon 15–30 mesh	–	*Bacillus cereus* spores	1.43	1.75
			6.5	1.6
Howe and French 4×8 mesh activated carbon	–	*Bacillus subtilis* spores	18.3	155.25
			28.98	88.9
Glass wool filter	18.5	*Serratia marcescens*	39.65	3.2
			79.3	3.15

equation (AIBA et al., 1973)

$$v_c = \left(\frac{1.125\,\mu}{C\,\rho_p}\right)\frac{d_f}{d_p^2}$$

where μ is the viscosity of air, ρ_p is the density of the particles and C is a correction factor for deviations from Stokes' law. As shown in Fig. 14, any other velocity above or below this critical velocity will result in a higher filter efficiency.

Step 3. Calculate the filter depth

For a given filter system, L_{90} values can be obtained from Tab. 13. The filter depth, L, can now be calculated using the log penetration relationship:

$$\log\left(\frac{N}{N_0}\right) = -\frac{L}{L_{90}}$$

This calculation is done for various values of velocity above and below the critical velocity.

Step 4. Calculate the filter diameter

Once the depth of the filter is determined, the next step is to determine the filter diameter. For a chosen volumetric air flow rate, the diameter of the filter can be computed directly by dividing the air flow rate Q_{air} (at filter inlet pressure) by the superficial velocity corresponding to the critical velocity of the filter.

Step 5. Other considerations

Other important considerations include the pressure drop across the filter. Many empirical pressure drop equations can be found in the literature. A typical one is the equation proposed by WONG and JOHNSTONE (1955):

$$\frac{\Delta P}{L} = \frac{2\rho\,\alpha\,C_D\,v_0^2}{\pi\,d_{fiber}}$$

where C_D is a drag coefficient depending on the bed characteristics and α is the fiber volume fraction or the volume of fibers per volume of bed.

Choosing between possible filters often involves economic considerations, e.g., a trade-off between capital expense related to the filter size and operating cost related to the operating pressure.

4.2.4 Difficulties Associated with Fibrous Depth Filters

While packed beds of fibrous filters were the standard method of air sterilization from the 1950s through about the 1970s, they are used much less frequently today. Some of the problems are due to the use of glass fibers as the primary medium of column packing. The glass fibers would retain moisture after sterilization and took considerable time to dry; for this reason they often were not sterilized after every fermentation batch. If wet, the pressure drop across the filter was high. Hence, not only did the cost of air compression increase, but also the air flow to the fermentor decreased. In addition, if the fibrous filters were wet, contaminants would accumulate in the filter. This would allow passage of bacteria into the fermentor, and thus contamination would follow. Moisture retention can be prevented by keeping the filter hot, but this adds additional cost. Two of the important considerations in setting up a fibrous filtration system are the issues of channeling and internal movement of fibers. Both must be minimized. The integrity of the bed is usually inspected visually during plant shut-downs. Some or all of the beds would need to be replaced to remove captured dirt and reduce pressure drop. Glass fiber material is hazardous; hence, considerable care is necessary in its handling.

4.3 Cartridge Air Filtration

After the early 1970s experience with manufacturing and application of cartridge air filters improved. These have become the standard method of filtration today, and essentially all new facilities install pleated cartridge filters. These filters consist of preformed filter media in an easy-to-handle form and provide highly predictable performance. They feature high particle-collection efficiency and good capacity to hold solids.

Cartridge filters can be either fibrous media-based depth filters or membrane-based absolute filters. Fibrous media-based cartridge filters can be either wound or pleated (SHUCOSKY, 1988). Wound filters create the same effect as a stack of woven cloth, but the yarn is wound around a mandrel rather being interwoven as a cloth. These fibers can be made of staple fibers (wool or cotton) or monofilaments. Pleated filters, on the other hand, are fabricated from sheets of porous non-woven fabrics that have been pleated to provide additional surface area and strength. The efficiency of pleated filters is more sensitive to particle size. Wound cartridges are frequently used for removing larger particles or as a prefilter ahead of cartridge filters designed to filter smaller particles. Membrane-based cartridge filters, on the other hand, are typically produced by the controlled precipitation of a continuous polymeric sheet from a solvent phase (CONACHER, 1976).

Cartridge filters have a number of advantages over conventional fibrous filters. They typically have a longer life and are often used for more than 50 sterilization cycles. In contrast to most fibrous materials, membranes are typically hydrophobic in nature and reject moisture. They are safe to use and are not fiber-releasing. Their operation cost can be lower, because they do not require heat-tracing to keep the filter medium dry. The pressure drop across the filter is low, typically 2–3 psi. An important additional advantage of cartridge filters is their ease of validation.

The design approach for cartridge is quite straightforward. The first step is to decide on a particle size rating. This is typically 0.2 μm for a sterilizing filter. The next decision is to determine the maximum flow rate and initial clean pressure drop. From the flow rate per cartridge one calculates the number of cartridges required to sterilize the air. For most membrane cartridge filters the pressure drop varies linearly with flow rate. It is usual to have multiple cartridge filters together in one filter housing. This way the flow rate through each filter is considerably smaller than the total flow rate. Pressure drops of 2–3 psi are typical for cartridge filters. The criterion for changing a cartridge filter is usually an increase in the observed pressure drop beyond this range.

4.4 Typical Air Sterilization Set-Up

A typical flow scheme for the air train is shown in Fig. 17. Inlet air into the plant passes through coarse filters to remove particulates prior to compression. The air leaving the compressor typically has a temperature of 130–160 °C because of adiabatic compression. A holding time of even a few seconds at this temperature is sufficient to achieve log reduction in the content of viable cells. The air is cooled to about 80 °C. A coalescing pre-filter is used to remove particles and oil and moisture. This pre-filter is often a depth filter. The air then passes through the sterilizing filter which is either a depth filter or an absolute filter.

After leaving the fermentor, the gas has a larger volume because of the lower pressure at the exit. For example, if a fermentor is operated with an overpressure of one atmosphere and the liquid depth is 10 m, then the air is fed into the sparger of the reactor at 2 atm pressure. There is additional pressure drop associated with the sparger and associated piping. The air temperature on filtration is typically 80 °C, while the fermentation may be carried out at 35 °C. Assuming that the respiratory quotient is about one, then the volume of the exhaust air is about 60% greater than the air to the inlet sterilizer. In addition, the air is saturated with moisture and contains mist with cells; thus, the challenge to the exhaust filter is much greater than to the inlet filter. Sterile filtration of exhaust gas is generally not required unless there is a need for containment.

5 Validation Issues

5.1 Motivation

Biologicals are different from pure chemical drugs, and these differences affect both the overall design of the plant and its operating procedures. Some of the main differences are that biologicals are typically

- derived from living organisms,
- heat-labile,
- good substrates for supporting microbial growth,
- complex mixtures of proteins and components that cannot be easily identified/quantified.

These differences have motivated strict U.S. Food and Drug Administration (FDA) guidelines for the design, operation, and validation of bioprocessing plants. Hence, no discussion of media sterilization is complete without a basic understanding of the implications of these guidelines (HILL and BEATRICE, 1989).

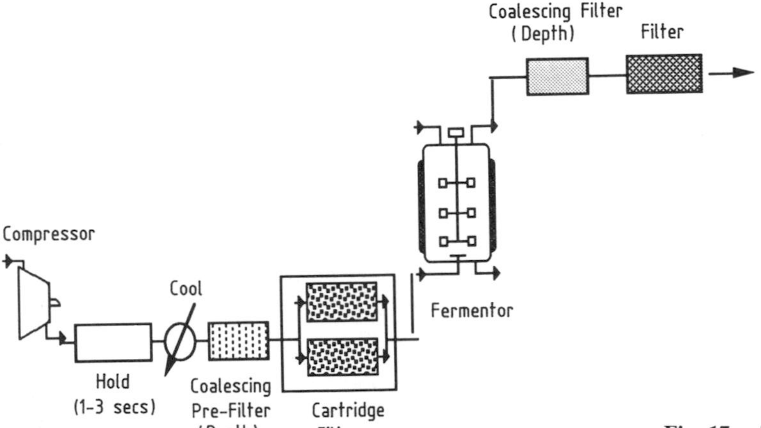

Fig. 17. Typical air sterilization set-up.

5.2 Validating Media Sterilization

While media sterilization is an early step in bioprocessing, its effectiveness can have profound effects on the final product quality. If under-sterilized media are used, contamination can occur in the fermentation that not only reduces yield and productivity, but also may introduce contaminants that are not normal and may not be removed in the purification. Alternatively, if the medium is over-sterilized, the excessive thermal destruction of nutrients may create impurities that also are not normal and difficult to remove. Thus, it is essential in process development to ensure that the sterilization protocol whether by thermal or filtration methods is adequate to consistently sterilize the medium while not creating variable chemical composition. Standard operating procedure needs to be developed and complied with; temperature-time profiles need to be monitored for operation within acceptable limits. Early in process development, the effectiveness of sterilization needs to be validated to ensure adequate sterilization and fixed so as not to produce variable quality medium. On scale-up, the performance of sterilization needs to be questioned again to assure consistency with the protocols employed in research; that is the product quality should be unaffected.

5.3 Validating Air Sterilization

Efficacy testing of filters for air sterilization is made by challenging the filter with a known amount of very small bacteria, e.g., *Pseudomonas diminuta* (0.3×0.8 μm) and then measuring their concentration at the outlet from the filter. With concern about the efficiency of filters for virus and bacteriophage removal, a small phage T_1-bacteriophage has been used in a challenge test for air sterilization (CONWAY, 1985). Such challenge tests are not suitable for use *in situ*, and other validation procedures such as the "bubble point" measurements have to be used (KESTING et al., 1981).

6 References

AIBA, S., HUMPHREY, A. E., MILLIS, N. F. (1973), *Biochemical Engineering*, Tokyo: University of Tokyo Press.

ARMENANTE, P. M., LESKOWICZ, A. A. (1990). Design of continuous sterilization system for fermentation media containing suspended solids, *Biotechnol. Prog.* **6**, 292–306.

BAILEY, J. E., OLLIS, D. F. (1977), *Biochemical Engineering Fundamentals*, New York: McGraw-Hill.

BOECK, L. D., ALFORD, J. S., PIEPER, R. L., HUBER, F. M. (1989), Interaction of media components during bioreactor sterilization: definition and importance of R_0, *J. Ind. Microbiol.* **4**, 247–252.

BOURDILLON, R. B., LIDWELL, O. M., LOVELOCK, J. E., RAYMOND, W. F. (1948), Airborne bacteria found in factories and other places, *Studies in Air Hygiene, Medical Research Council Special Report* **N. 262**, London: H.M.S.O.

CONACHER, J. C. (1976), Membrane filter cartridges for fine particle control in the electronics and pharmaceutical industries, *Filtration Society's Conference on Liquid-Solid Separation*, September 1975.

CONWAY, R. S. (1985), Selection criteria for fermentation air filters, in: *Comprehensive Biotechnology* (MOO-YOUNG, M., Ed.), New York: Pergamon Press.

COONEY, C. L. (1985), Media sterilization, in: *Comprehensive Biotechnology* (MOO-YOUNG, M., Ed.), New York: Pergamon Press.

DANCKWERTS, P. V. (1953). *Chem. Eng. Sci.* **2**, 1.

DE RUYTER, P. W., BRUNET, R. (1973). Estimation of process conditions for continuous sterilization of foods containing particulates, *Food Technol.* **7**, 46–51.

GADEN, E. L., HUMPHREY, A. E. (1955), Fibrous filters for air sterilization, *Ind. Eng. Chem.* **8**, No. 12.

HILL, D., BEATRICE, M. (1989), Biotechnology facility requirements, part II – Operating procedures and validation, *BioPharm*, November/December, 28–32.

HUMPHREY, A. E. (1960), Air sterilization, *Adv. Appl. Microbiol.* **2**, 300.

HUMPHREY, A. E., GADEN, E. L. (1960), Air sterilization by fibrous media, *Ind. Eng. Chem.* **47**, 924.

JACOBS, R. A., KEMPE, L. L., MILONE, N. A. (1973), High temperature short time (HTST) processing of suspensions containing bacterial spores, *J. Food Sci.* **38**, 168–172.

KESTING, R., MURRAY, A., JACKSON, K., NEWMAN, J. (1981), Highly anisotropic microfiltration membranes, *Pharm. Technol.* **5** (5), 52–60.

LA VERNE, D. B., ALFORD, J. S., PIEPER, R. L., HUBER, F. M. (1989), Interaction of media components during bioreactor sterilization: Definition and importance of R_0, *J. Ind. Microbiol.* **4**, 247–252.

LEVENSPIEL, O. (1958), Longitudinal mixing of fluids flowing in circular pipes, *Ind. Eng. Chem.* **50**, 343.

LIN, S. H. (1975), A theoretical analysis of thermal sterilization in a continuous sterilizer, *J. Ferment. Technol.* **53** (2), 92–98.

MELTZER, T. H. (1987), *Filtration in the Pharmaceutical Industry*, New York–Basel: Marcel Dekker.

MULLER, K. (1990), Steam-sterilizable ultrafiltration for pyrogen-free water in pharmaceutical production, *Swiss Contamination Control* **3**, No. 4a, 146–148.

MULLEY, E. A., STUMBO, C. R., HUNTING, W. M. (1975), Kinetics of thiamine degradation by heat, *J. Food Sci.* **40**, 985–996.

PICK, A. E. (1982), Consider direct steam injection for heating liquids, *Chem. Eng.* **6**, 87–89.

PROKOP, A., HUMPHREY, A. E. (1970), in: *Disinfection*, pp. 61–84, New York: Marcel Dekker.

RAJU, G. K., COONEY, C. L. (1993), in: *Process Computations in Biotechnology*, India: Tata-McGraw Hill.

RICHARDS, J. W. (1968), *Introduction to Industrial Sterilization*, New York: Academic Press.

SHUCOSKY, A. C. (1988). Select the right cartridge filter, *Chem. Eng.*, January.

SINGH, V., HENSLER, W., FUCHS, R. (1989), Optimization of batch fermentor sterilization, *Biotechnol. Bioeng.* **33**, 584–591.

SOLOMONS, G. L. (1969), *Materials and Methods in Fermentation*, New York: Academic Press.

SVENSSON, R. (1988), Continuous media sterilization in biotechnical fermentation, *Dechema-Monogr Vol.* **113**, pp. 225–237, Weinheim: VCH Verlagsgesellschaft.

WANG, D. I. C., COONEY, C. L., DEMAIN, A. L., DUNNILL, P., HUMPHREY, A. E., LILLY, M. D. (1979), *Fermentation and Enzyme Technology*, New York: Wiley-Interscience.

WONG, J. B., JOHNSTONE, H. F. (1955), *Collection of Aerosol by Fiber Materials*, University of Illinois, Technical Report II, AEC Contract AT (30-3)-28.

WONG, J. B., RANZ, W. E., JOHNSTONE, H. F. (1956), Collection of aerosol particles, *Appl. Phys.* **27**, 161.

YAO, K. M., HABIBIAN, M. T., O'MELIA, C. R. (1971), Water and waste filtration: concepts and applications, *Curr. Res.* **5**, 1105–1112.

10 Oxygen Transfer and Mixing: Scale-Up Implications

MATTHIAS REUSS

Stuttgart, Federal Republic of Germany

List of Symbols

a	specific surface area gas–liquid
A	surface area gas–liquid
c_i	concentration of component i in the liquid phase
c_L	concentration of oxygen in the liquid phase
c_L^*	equilibrium concentration of oxygen at the liquid interface
c_{crit}	critical concentration of oxygen
c_X	biomass concentration
$c_{G,m}$	dimensionless concentration of oxygen in the *m*-fraction of the gas phase
$c_{G,p}$	dimensionless oxygen concentration in the *p*-fraction of the gas phase
d	stagnant fraction of gas phase
d_B	bubble diameter
D_{O_2}	diffusivity coefficient of oxygen in the liquid phase
Da	Damköhler number
De	Deborah number
D_I	impeller diameter
D_T	tank diameter
Fl	gas flow number
g	acceleration of gravity
H	Henry coefficient
Ha	Hatta number
k	reaction rate constant
K^*	dimensionless $k_L a$ value
k_L	mass transfer coefficient characterizing the intensity of transport from the gas–liquid interface into the bulk liquid
$k_L a$	volumetric mass transfer coefficient gas–liquid
K_C	consistency index
m	*m*-fraction of well mixed gas phase
n	flow index
N	agitation speed or number of cells in the plug flow
Ne	power (Newton) number
N_{O_2}	total flux of oxygen from the gas phase into the biosuspension
p	fraction of plug flow in the gas phase
P	system pressure
p_{O_2}	partial pressure of oxygen in the gas phase
Q	gas flow rate
Q_{O_2}	volumetric oxygen uptake rate
r	radius of the cavern
R	gas constant
Re	Reynolds number
Sc	Schmidt number
Sh	Sherwood number
t	time
T	temperature
u_r	relative velocity
V_B	terminal bubble-rise velocity
$\dot{V}_G$	gas flow rate
V_L	liquid volume
V_S	superficial gas velocity
V_R	reaction volume = gas volume + liquid volume

X_{O_2}	mole fraction of oxygen in the liquid phase
Y_{O_2}	mole fraction of oxygen in the gas phase
$Y_{O_2}^{\alpha}$	mole fraction of oxygen in the gas phase at the inlet
$Y_{O_2}^{\omega}$	mole fraction of oxygen in the gas phase at the outlet
z	axial coordinate for plug flow fraction in the gas phase
$\dot{\gamma}$	shear rate
ε	energy dissipation rate
ε_G	gas hold-up
η	dynamic viscosity
θ	dimensionless residence time
θ_C	circulation time
θ_m	terminal mixing time
λ	natural timc
μ	specific growth rate
ν	kinematic viscosity
ρ	density
σ	surface tension
τ	shear stress
τ_y	yield stress

1 Introduction

The only uncontroversial fact about scale-up is that it is one of the most complicated tasks and challenging endeavors in the field of biochemical engineering. Although an enormous number of book chapters, reviews, and original papers have been devoted to the subject this activity does not necessarily mirror a corresponding increase of knowledge for a reliable prediction of process translation from laboratory to production scale. In contrast to an optimistic point of view, the state of the art is still at a frustrating low level. The necessary breakthroughs for developing new and innovative concepts related to this topic are not available and do not come to view in the foreseeable future. Some of the important reasons for this unsatisfactory state of the art have been elaborated by CHARLES (1985b), who critically concluded that the existing scale-up results "cannot meet adequately the technical and economic demands of modern biotechnology industry".

This chapter is not aimed at trying to present new and meaningful solutions to the problems. Rather, attention will be drawn to some new developments and small progresses which have occurred mainly in the last ten years. These progresses are pieces of a puzzle from which more rational and reliable scale-up strategies should develop. Obviously, significant advances in this area will only come in time.

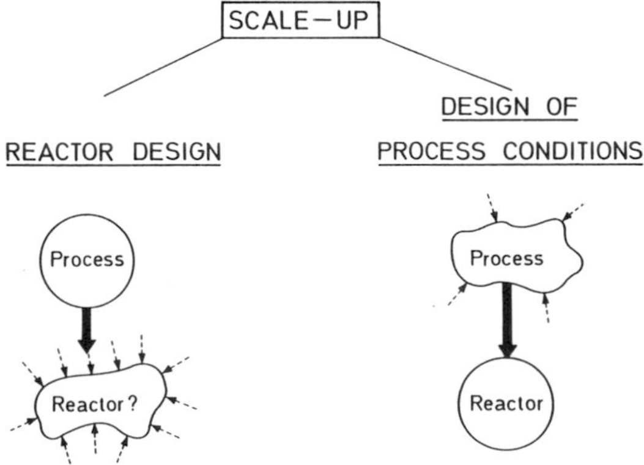

Fig. 1. Strategies for scale-up of bioprocess systems (bioreactor + bioprocess).

Faced with the enormous complexity of scale-up, it appears worthwhile to arrange the manifold facets into some order. Thus, instead of displaying the whole background of scale-up strategies and assessing their possible impacts on particular processes one after the other, the intent here is to break down the collective and amorphous term "scale-up" into two basic problems. This is schematically illustrated in Fig. 1.

1. The first problem is associated with the design of a fermentor or a production plant, respectively. It is assumed that the designer has the flexibility to select and/or develop a system to meet the process requirements.
2. The second problem is concerned with design of process conditions which are compatible with an existing full-scale fermentor. This task seems to be very common in industrial practice, particularly when many different and sometimes unrelated fermentation products are produced in a limited facility.

It is evident that because of the different flexibilities involved we cannot look on these two different topics in the same way. That is to say, we cannot try to solve the problems with the same set of tools.

Thus, while it may be true that one or the other of the strategies illustrated in Tab. 1 has proven useful for the scale-up of a particular process, it does not necessarily mean that such a strategy can always be applied in practice. In fact, it might happen that a seemingly optimal solution from this list will be far from that in actuality, because of the constraints of the existing plants.

Keeping these basic distinctions in mind, the generally accepted goals for scale-up must be considered with caution and the solution to the task may not always result in application of a straight-forward strategy. Thus, if successful process scale-up depends upon the extent to which the system characteristics at different scales of operations resemble each other, it is not necessarily wise to try to duplicate a seemingly optimal solution from the pilot plant. Suitable operation conditions which are established at a small scale in the laboratory most

Tab. 1. Common Criteria for Scale-Up of Stirred Bioreactors

Volumetric oxygen transfer coefficient	$k_L a$
Volumetric power input	P/V_L
Volumetric gas flow	$\dot{V}_G/V$
Impeller tip speed	Nd_i
Agitation speed	N
Terminal mixing time	θ_∞

often correspond to the state under which perfect mixing is a valid presumption. However, because of practical limitations during process scale-up, it is almost impossible to hold exactly the same fluid dynamic conditions as in smaller reactors. As a result of inevitable compromises, mixing and/or mass and heat transfer may be inferior to that obtained in smaller laboratory reactors. This translates to a change in the intensity of transport processes. The impact of this difference upon the outcome of the scaled-up process determines, if the expectations of laboratory optima can be duplicated under new conditions. Because we very seldom can reproduce in a large commercial plant the optimal conditions found earlier in a pilot plant, it is crucial to find strategies which will cover the entire range of key variables. And it is imperative that experiments in pilot plants should not concentrate on finding an "optimum" which might be incongruous of that in the production plant.

To overcome these deficiencies, it is useful to think of the pilot plant as a permanent satellite of the production plant and to iteratively coordinate the processes of scale-up and scale-down. Such strategies have been suggested and described in more detail by AIBA et al. (1973) and CHARLES (1985b). While this approach may support the scale-up procedure for a particular process, the concept remains empirical and cannot compensate for the need to develop rational strategies based upon a better understanding of the fundamentals.

If one thinks about a fermentation process in terms of microorganisms interacting with their environment, it becomes clear that strategies for scale-up should rest upon transport phenomena as well as structured metabolic reaction kinetics. Such approaches will inevitably result in complex mathematical models. If we knew more about how to structure these models, the design of process conditions at large scale might be much improved. However, even if sufficient knowledge were available to substantiate such an approach, time constraints and lack of skilled specialists limit the necessary research in the industrial environment.

Because of these restrictions, scale-up procedures are still based on empirical attempts to maintain certain selected parameters. The logic behind selecting the different criteria is to preserve each of these and thus maintain the constancy of the corresponding extracellular environment at different scales. If this environmental property sustains the one that most critically influences the desired microbial productivity, a successful scale-up might result.

The success of any of these strategies is dependent on (a) selecting the appropriate property or combination of parameters and (b) the predictability of the scale dependencies and their effect on the selected parameters. For a variety of reasons, these two requirements are most often not satisfied. In particular, with respect to the two most important transport phenomena in bioreactors – mixing and mass transfer – most of the published correlations cannot be recommended. There are three reasons which are mainly responsible for this unsatisfactory situation: (1) incorrect measurements, (2) physical properties of the systems of interest are not considered, and (3) the majority of correlations presented in the literature are based on measurements at only very small scale (ZLOKARNIK, 1990).

This chapter is aimed at reviewing some of these problems and, in particular, to discuss some of the more important developments which have occurred in this field.

2 Mass Transfer

2.1 Fundamentals

The design of a bioreactor for aerobic processes, including reliable predictions of its operation variables, should rest on a thorough understanding of the transport of oxygen from the dispersed bubbles to the utilization sites inside the biotic phase (Fig. 2). Because of the low solubility of oxygen in fermentation media, the mechanism of oxygen transfer for the bulk of the gas phase into the bulk liquid suspension of single cell organisms is controlled by the liquid phase mass transfer resistance. This process is schematically illustrated in Fig. 3. The oxygen flux is given by

$$N_{O_2} = k_L A (c_L^* - c_L) \tag{1}$$

where k_L is the liquid mass transfer coefficient in m/s, A is the interfacial area for mass transfer in m^2, and c_L^* and c_L are the equilibrium and bulk concentrations of oxygen, respectively. The oxygen transfer rate per unit of reactor volume Q_{O_2} is then given by

$$Q_{O_2} = \frac{N_{O_2}}{V_L} = k_L a (c_L^* - c_L) \tag{2}$$

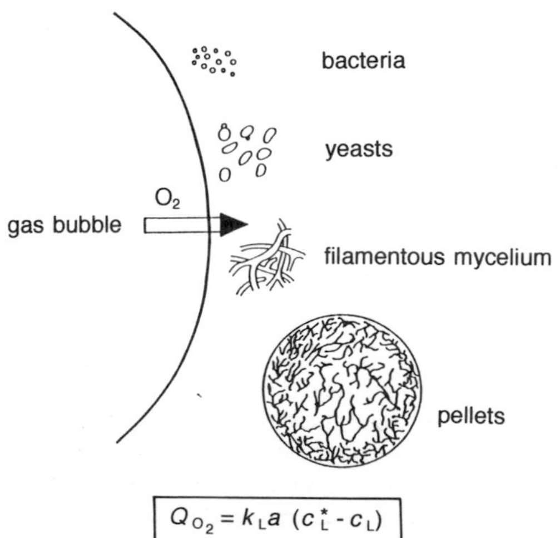

Fig. 2. Oxygen supply to suspended microorganisms.

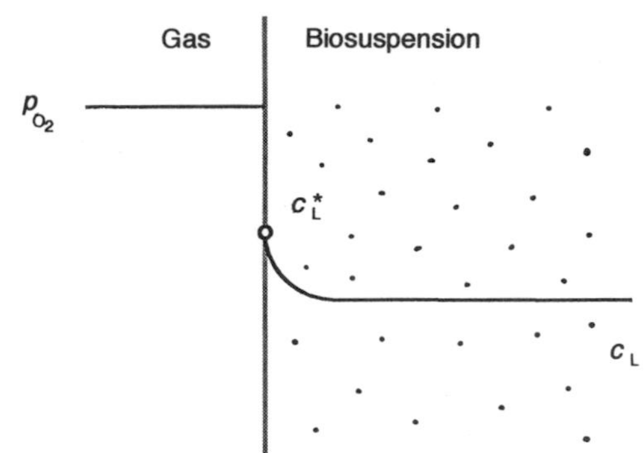

Fig. 3. Oxygen profiles for transport from the inside of the gas phase into the bulk of the biosuspension.

Before discussing application of Eq. (2), it should be mentioned that additional mass transfer resistances such as liquid–solid mass transfer and intraparticle diffusion, may be important for reactions involving larger microbial agglomerates (pellets, biofilms, flocs, immobilized biomass and/or enzymes). Tackling of this problem has been extensively described in several reviews and textbooks (ATKINSON, 1974; MIURA, 1976; BAILEY and OLLIS, 1986; KARGI and MOO-YOUNG, 1985). A complete numerical solution of the molecular diffusion and microbial reaction in a sphere, considering the sum of mass transfer resistances gas/liquid and liquid/solid via boundary conditions at the surface of the sphere, has been presented and discussed in Volume 4 of this series (REUSS and BAJPAI, 1991).

For most of the fermentation processes the problem of oxygen transfer into the biosuspension is, however, adequately represented by Eq. (2). The use of this relationship will be illustrated with the aid of a simple example. Consider the case in which a microbial culture is characterized by its specific growth rate μ, biomass concentration c_X, and the yield coefficient $Y_{O/X}$ in mol O_2/kg biomass. The oxygen demand of the culture is given by:

$$(Q_{O_2})_{\text{demand}} = \frac{\mu\, c_X}{Y_{O/X}} \tag{3}$$

At steady state the oxygen supply absorption rate, Eq. (2), and demand must balance:

$$k_L a(c_L^* - c_L) = \frac{\mu\, c_X}{Y_{O/X}} \tag{4}$$

Let us next assume that the concentration of oxygen in the bulk of the liquid should always be greater than the critical value c_{crit}:

$$(k_L a)_{SP} = \frac{\mu\, c_X}{Y_{O/X}(c_L^* - c_{crit})} \tag{5}$$

Eq. (5) sets the minimum value of $k_L a$ required for process design. To apply this estimation procedure, an equation of the form: $k_L a = f$ (type of reactor, operation conditions, material properties) is required. Some equations of this kind will be considered later in this chapter.

Some reliable guidelines regarding the design of bioreactors for aerobic processes can be extracted from Eq. (2). In order to increase the oxygen transfer rate, we may either increase the mass transfer coefficient k_L or the specific service area a. Both actions require an increase in the turbulence intensity which is in some way related to the reactor power input. Other possibilities include an increase in the systems pressure or oxygen content of the gas phase. These two operation parameters are proportional to the saturation concentration c^*. This can be shown with the aid of Henry's law:

$$H = \frac{p_{O_2}}{(X_{O_2})^*} = \frac{y_{O_2} P}{\dfrac{c_L^*}{\sum c_i}} \tag{6}$$

or

$$c_L^* = \frac{y_{O_2} P}{\sum c_i H} \tag{7}$$

where H is the Henry coefficient (in bar), y_{O_2} is the volume fraction of oxygen in the gas phase, P is the system pressure, and $\sum c_i$ is approximately equal to the molar concentration of water (55.55 mol/L). The influence of various medium components on the solubility has been investigated by POPOVIC et al. (1979) and SCHUMPE et al. (1982).

For practical application, usually the mass transfer coefficient k_L and the specific service area a are not separated and the product $k_L a$ is used as a key parameter for characterizing the oxygen transfer. However, as a useful guide for practical design and operation it is quite advisable to know how the individual terms are influenced by the operating parameter. The correlation for the mass transfer coefficient k_L is usually presented in the dimensionless form:

$$Sh = f(Re, Sc) \tag{8}$$

with Sherwood number $Sh = k_L d_B / D_{O_2}$, Reynolds number $Re = u_r d_B / \nu$ and Schmidt number $Sc = \nu / D_{O_2}$. For extensive discussion of these equations see the reviews of BLANCH and BHAVARAJU (1976), BRAUER (1971, 1985), KARGI and MOO-YOUNG (1985), and MOO-YOUNG and BLANCH (1981). For estimating k_L values in turbulent flow fields, modifications of Higbie's penetration theory have been combined with Kolmogoroff's theory of isotropic turbulence, resulting in equations of the form:

$$k_L \sim \left(\frac{\nu}{D}\right)^{-\frac{2}{3}} (\varepsilon\, \nu)^{1/4} \tag{9}$$

KAWASE and MOO-YOUNG (1988) have suggested a modification of this approach for non-Newtonian fluids. The specific service area a is related to gas hold-up ε_G and bubble diameter d_B:

$$a = \frac{6\varepsilon_G}{d_B(1 - \varepsilon_G)} \tag{10}$$

CALDERBANK (1958) proposed a correlation for the specific surface area which is also based on Kolmogoroff's theory of turbulence and is written as:

$$a = C_1 \left(\frac{\varepsilon^{2/5} \rho^{3/5}}{\sigma^{3/5}}\right) \left(\frac{V_S}{V_B}\right)^{\frac{1}{2}} \tag{11}$$

with V_S, V_B superficial gas velocity and terminal velocity of bubbles in free rise, respectively.

Since the separate calculation of k_L and a is not widely accepted for real fermentation fluids, correlations for $k_L a$ are most often based on experimental observations.

The overwhelming majority of measurements of $k_L a$ are still based on dynamic methods. If not carefully performed and critically analyzed, these measurements may be corrupted by large errors and the resulting equations for predicting k_La may be questionable. Therefore, the following section is devoted to a comprehensive description of the measurement methods and discussion of their possible errors.

2.2 Measurement Methods for Determination of Volumetric Mass Transfer Coefficients

2.2.1 Quasi-Steady State Method

The most accurate method for predicting k_La values is by definition, to measure the variables on the right side of Eq. (2).

$$k_La = \frac{Q_{O_2}}{\Delta c_L} \tag{12}$$

The essential prerequisite of this method is an oxygen sink, an assumption which is fulfilled during aerobic growth.

The oxygen uptake is predicted with the aid of gas phase balance measurements in the exhaust gas:

$$Q_{O_2} = \frac{\dot{V}_G}{V_L}\left[y_{O_2}^{\alpha} - \frac{1 - y_{O_2}^{\alpha} - y_{CO_2}^{\alpha}}{1 - y_{O_2}^{\omega} - y_{CO_2}^{\omega}}\, y_{O_2}^{\omega}\right] \tag{13}$$

The concentration of oxygen in the bulk of the liquid is calculated from partial pressure as measured with a polarographic electrode according to Henry's law. An additional assumption must be made for the driving force Δc. If the gas phase is assumed to be perfectly mixed, the driving force is calculated from

$$\Delta c_L = c_L^* - c_L \tag{14}$$

with c_L^* predicted from the partial pressure as measured by exhaust air analysis. Alternatively, we may assume plug flow behavior and apply the logarithmic driving force:

$$\Delta c_m = \frac{c_{L,in}^* - c_{L,out}^*}{\ln \dfrac{c_{L,in}^* - c_L}{c_{L,out}^* - c_L}} \tag{15}$$

2.2.2 Dynamic Methods

The most common methods for determination of k_La in aerated vessels involve dynamic experiments. Because the only apparatus required for these experiments is an oxygen electrode, this method is relatively low cost and, thus, often the popular method of choice. The simplicity of the experiments should not, however, obscure the fact that these methods include some serious limitations which have lasting influence on the applicability and/or reliability of the results (LINEK and VACEK, 1982; LINEK et al., 1982; TATTERSON, 1991; VAN'T RIET and TRAMPER, 1991). Therefore, k_La correlations which rest upon these methods should be closely investigated.

The method is usually based on either interchanging sparged gases (nitrogen/air) or stopping aeration in the case of an oxygen-consuming biosuspension. The mass balance for oxygen in the liquid phase can be written as:

$$V_L \frac{dc_L}{dt} = k_LA\,(c_L^* - c_L(t)) - Q_{O_2}V_L \tag{16}$$

or

$$\frac{dc_L}{dt} = k_La(c_L^* - c_L(t)) - Q_{O_2} \tag{17}$$

At the end of the dynamic experiment, we may assume quasi-steady state conditions,

$$\frac{dc_L}{dt} = 0 \tag{18}$$

and

$$Q_{O_2} = k_La(c_L^* - c_L(t \to \infty)) \tag{19}$$

Substituting Eq. (19) into Eq. (17) and integrating from t_1 to t_2, gives

$$k_La = \frac{\ln\left(\dfrac{c_L(t\to\infty) - c_L(t\to t_1)}{c_L(t\to\infty) - c_L(t\to t_2)}\right)}{t_2 - t_1} \tag{20}$$

For application of the dynamic gassing out method without oxygen consumption, $c(t\to\infty)$ is replaced by c_L^*.

One of the aforementioned problems of the dynamic oxygen measurements is related to the response characteristics of the electrode. The dynamics of the sensor are influenced by the molecular diffusion of oxygen through the membrane and, sometimes, also by the transport of oxygen from the fluid bulk to the surface of the membrane. The available models for correction of the probe signal have been reviewed by VAN'T RIET (1979).

More serious problems include the improper assumptions regarding gas phase concentrations and depletion. Various models have been suggested to correct these effects (DUNN and EINSELE, 1975; DANG et al., 1977; LINEK and VACEK, 1982; LINEK et al., 1982). LINEK et al. (1982) studied different dynamic measurement methods (gas interchange: $N_2 \rightarrow O_2$, $O_2 \rightarrow N_2$, $N_2 \rightarrow$ air, air $\rightarrow N_2$, 0 (degassed liquid) $\rightarrow$ air and $0 \rightarrow O_2$) and compared the k_La values obtained from different models with the results from steady state methods. The k_La values obtained differed both significantly and systematically. Only the interchange $0 \rightarrow O_2$ yielded k_La values which agree with the results from steady state measurements. LINEK concluded that an ideally mixed gas phase, as well as the concept of piston flow for the gas, will lead to underestimations of k_La.

This conclusion can be supported by more carefully analyzing the dynamic method with the aid of measured gas residence time distributions and mixing models for the gas phase derived from these measurements (POPOVIC et al., 1983). The model for mixing of the gas phase in liquids of low and high viscosity (biosuspensions of *Penicillium chrysogenum*) are schematically illustrated in Figs. 4 and 5. For analysis of the dynamics of oxygen transport, the liquid phase is assumed to be ideally mixed.

The dimensionless balance equations for oxygen in the gas phase are given by

$$\frac{d\bar{c}_{G,m}}{d\theta} = \frac{1}{m}(1-\bar{c}_{G,m}) - K^* \frac{V_L}{\dot{V}_G} \frac{RT}{H} (\bar{c}_{G,m} - \bar{c}_L) \tag{21}$$

for the m-fraction (ideally mixed) and

$$\frac{\partial \bar{c}_{G,p}}{\partial \theta} = -\frac{1}{p} \frac{\partial \bar{c}_{G,p}}{\partial \bar{z}} - K^* \frac{V_L}{\dot{V}_G} \frac{RT}{H} (\bar{c}_{G,p} - \bar{c}_L) \tag{22}$$

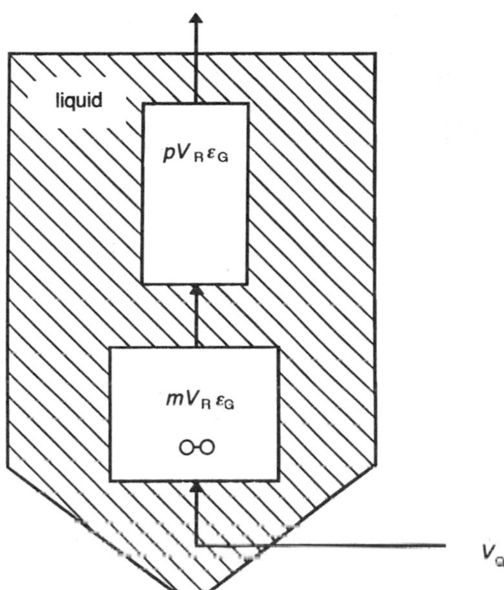

Fig. 4. Schematic representation of the proposed model for gas residence time distribution in low viscous Newtonian fluids (POPOVIC et al., 1983).

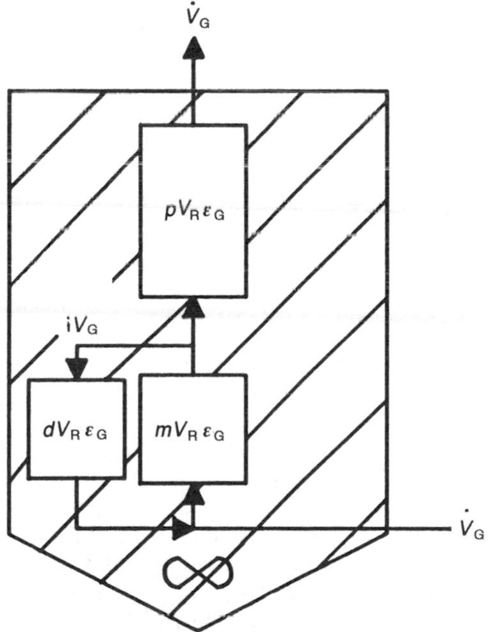

Fig. 5. Schematic representation of the proposed model for gas residence time distribution in highly viscous non-Newtonian fluids (POPOVIC et al., 1983).

for the p-fraction (plug flow). The mass balance for oxygen in the liquid phase consists of the transfer from the well-mixed and plug flow fraction and can be written as

$$\frac{d\bar{c}_L}{d\theta} = K^* m\, \bar{c}_{G,m} + p\, K^* \int_0^1 \bar{c}_{G,p} d\bar{z} - K^* \bar{c}_L - Da_0 \tag{23}$$

where $\bar{c}_G = c_G/c_G^0$ with c_G^0, the inlet concentration; $\bar{c}_L = c_L H/(c_G^0\, R\, T)$ with R, the gas constant, T, the temperature, H, the Henry coefficient; $\theta = t/\tau_G$ with τ_G, the mean residence time of the gas phase; $\tau_G = (V_R \varepsilon_G)/\dot{V}_G$ with V_R, the gassed volume, $\dot{V}_G$, the gas flow rate, ε_G the gas hold-up, V_L, the liquid volume; $K^* = k_L a\ \tau_G$; $Da_0 = r_{O_2}\ \tau_G\ H\ /(\ c_G^0\ R\ T)$; $\bar{z} = z\ V_L\ \varepsilon_G / H_0 V_G\ P$ with z, the axial coordinate, H_0, the liquid height, and P, the system pressure.

For numerical integration of these equations, it is convenient to divide the plug flow volume into N cells. Thus, Eqs. (22) and (23) can be approached by

$$\frac{d\bar{c}_{G,p^j}}{d\theta} = \frac{N}{p} (\bar{c}_{G,p^{j-1}} - \bar{c}_{G,p^j}) - \\ - K^* \frac{V_L}{\dot{V}_G} \frac{R\, T}{H} (\bar{c}_{G,p} j - \bar{c}_L) \tag{24}$$

$$\frac{d\bar{c}_L}{d\theta} = K^* (m\, \bar{c}_{G,m} + \frac{p}{N} \sum_{j=1}^{N} \bar{c}_{G,p} j) - K^* \bar{c}_L - Da_0 \tag{25}$$

Initial conditions for integration are

$$\theta = 0:\ \bar{c}_{G,m} = 0\ \bar{c}_L = \bar{c}_0 \tag{26}$$

$$\bar{c}_{G,p}(\bar{z} = 0),\quad \text{or}\quad \bar{c}_{G,p^{j=0}} = \bar{c}_{G,m} \tag{27}$$

Regarding the mixing model for highly viscous fluids (Fig. 5), the additional balance equation for oxygen in the d-fraction is given by

$$\frac{d\bar{c}_{G,d}}{d\theta} = \frac{i}{d} (\bar{c}_{G,m} - \bar{c}_{G,d}) - K^* \frac{V_L}{\dot{V}_G} \frac{R\, T}{H} (\bar{c}_{G,d} - \bar{c}_L) \tag{28}$$

For comparison of the different fermentation models dynamic measurements have been performed in biosuspensions of *Saccharomyces cerevisiae* as a model system for low viscous fluids and *Aspergillus niger* as a model system for highly viscous non-Newtonian fluids.

The measured signals of the oxygen electrode have been corrected with a first-order transfer function. By minimizing the error square between model predictions and measurements, $k_L a$ values were estimated. Results for the low viscous system are summarized in Tab. 2a by showing $k_L a$ values predicted with:

- oxygen concentration in the gas phase = const. (DYN 1)
- a well-mixed gas phase (DYN 2) (DUNN and EINSELE, 1975)
- a model from measured gas residence time distribution (DYN 3).

Additionally, $k_L a$ values determined with the steady state method are presented in Tab. 2a. It can be concluded from this comparison that only with the aid of intrinsic gas mixing are $k_L a$ values obtained from dynamic measurements which can be compared with the steady state method. Similar results were obtained for the highly viscous non-Newtonian fluids (Tab. 2b).

An interesting method for analysis of experimental data from dynamic measurements has been suggested by CHAPMAN et al. (1982). The following expression for the volumetric mass transfer coefficient has been derived.

$$k_L a = \frac{dc_L^*/dt}{(Q/V_G) \int_0^t (1 - c_0^*) dt - c_L^* [(HV_L/V_G) + 1]} \tag{29}$$

where c_0^* and c_L^* are recorded concentrations divided by their new steady state values. The disadvantage of this equation is that both liquid and gas phase concentrations must be measured. Applying the L'Hospitale rule at $t = 0$ to Eq. (29) leads to

$$k_L a = \\ = \frac{d^2 c_L^*/dt^2}{[(Q/V_G)(1 - c_0^*)] - (dc_L^*/dt)[(HV_L/V_G) + 1]} \tag{30}$$

and, since at $t = 0$: $c_0^* = 0$ and $dc_L^*/dt = 0$, Eq. (30) reduces to

$$k_L a = \frac{V_G d^2 c_L^*}{Q dt^2} \tag{31}$$

Tab. 2a. Comparison between Dynamic and Steady-State $k_L a$ Measurements (*Saccharomyces cerevisiae*, $V_L = 55$ L)

Operation Parameter					
Agitation (rpm)	Aeration $\frac{\dot{V}_G}{V_L}$ (min^{-1})	$(k_L a)_{stat}^{log}$ (h^{-1})	$\frac{\text{DYN 1}}{(k_L a)_{stat}^{log}}$	$\frac{\text{DYN 2}}{(k_L a)_{stat}^{log}}$	$\frac{\text{DYN 3}}{(k_L a)_{stat}^{log}}$
300	0.33	71	1.10	1.20	1.10
300	0.44	103	0.93	1.20	1.10
300	0.60	109	0.62	1.07	0.90
400	0.44	165	0.76	1.20	1.01
400	0.60	176	0.66	1.20	1.08
500	0.60	275	0.52	1.30	1.02
600	0.33	294	0.45	1.30	0.95
600	0.44	280	0.53	1.30	1.00
600	0.60	291	0.60	1.50	1.10

$(k_L a)_{stat}^{log}$ = $k_L a$ predicted from steady-state measurements with logarithmic mean driving force
DYN i (i = 1, 2, 3)/$(k_L a)_{stat}^{log}$ = ratio between $k_L a$ value predicted from dynamic measurements with different models and steady-state value
DYN 1: dynamic model with $c_L^* \neq f(\text{time})$
DYN 2: dynamic model of DUNN and EINSELE (1975)
DYN 3: dynamic model based on measured residence time distributions of the gas phase

Tab. 2b. Comparison between Dynamic and Steady-State $k_L a$ Measurements (*Aspergillus niger*, $C_X = 10$ kg dry weight/m^3, $K_C = 5.6$ Pa s^n, $n = 0.2$, DYN 4: dynamic model based on measured residence time of the gas phase in Fig. 5)

Operation Parameter					
Agitation (rpm)	Aeration $\frac{\dot{V}_G}{V_L}$ (min^{-1})	$(k_L a)_{stat}^{log}$ (h^{-1})	$\frac{\text{DYN 1}}{(k_L a)_{stat}^{log}}$	$\frac{\text{DYN 2}}{(k_L a)_{stat}^{log}}$	$\frac{\text{DYN 4}}{(k_L a)_{stat}^{log}}$
400	0.57	86	1.04	1.00	0.98
500	0.57	189	0.58	0.83	1.10
600	0.57	205	0.54	0.80	1.08
400	0.96	110	0.86	0.93	0.98
500	0.96	250	0.50	0.72	0.90
600	0.96	323	0.50	0.72	1.10

Thus, $k_L a$ is obtained from the initial second derivative of the liquid concentration. The use of this method for initial response analysis has been successfully demonstrated for measurements in gas sparged stirred vessels by GIBILARO et al. (1985). NIENOW et al. (1988) reported pronounced deviations between the $k_L a$ values predicted from Eq. (29) and those in which the raw data from the electrode were treated according to the simple "no depletion" model,

$c_L^* \neq f$(time). The difference between the $k_L a$ values increases at higher agitation speed where a large amount of air is recirculated (NIENOW et al., 1977).

2.2.3 Chemical Reactions

These methods, in which absorbed gas (O_2 or CO_2) is reacting with chemicals added to the liquid phase, have been widely used in the characterization of multiphase reactors. Because many equations for $k_L a$ reported in the literature rest on such measurements, it is also important to know the advantages and limitations of these methods. The following discussion is restricted to sulfite oxidation.

$$\frac{1}{2} O_2 + SO_3^{2-} \xrightarrow{Co^{2+}} SO_4^{2-} \tag{32}$$

The measurement can be performed either by taking liquid samples and performing iodometric titrations or alternatively, by predicting the oxygen consumption with exhaust analysis. If the catalyst concentration is high, the reaction rate is fast enough to reduce the dissolved oxygen concentration to zero. Under these conditions no measurement of oxygen content is required and,

$$Q_{O_2} = k_L a \, c_L^* \tag{33}$$

The important constraint for the performance of these experiments is that the reaction should be fast enough to reduce c_L in the bulk to zero, but not so fast as to cause an enhancement of $k_L a$. Assuming sulfite oxidation kinetics zero order for sulfite concentration and second order for oxygen concentration, the necessary criteria for $k_L a$ measurements (REITH and BEEK, 1973) are that

$$k_L a \ll k \, c_L^* \tag{34}$$

with k, the reaction rate constant (m^3 mol^{-1} s^{-1}) as a function of catalyst concentration, pH, and temperature, and that

$$\text{Hatta number } Ha = \left(\frac{2 k c_L^* D}{3 k_L^2}\right)^{1/2} < 0.3 \tag{35}$$

Several critical reviews (VAN'T RIET, 1979; LINEK and VACEK, 1981; VAN'T RIET and TRAMPER, 1991) have shown that the variation of the reaction rate constant can be enormous. Even the quality of the water may result in drastic effects (LINEK and VACEK, 1981). For these reasons quantitative comparison of such $k_L a$ values measured by different authors is usually not valid.

RUCHTI et al. (1985) discussed some practical guidelines and gave an empirical procedure by which the appropriate reaction conditions can be easily determined. The procedure employed is to vary the Co^{2+} concentration and to identify the region in which $k_L a$ is independent of Co^{2+} concentration. The same authors studied the influence of catalyst induction time and found that during the first 20 to 40 minutes, variations of oxygen uptake rates are observed. This effect has been often overlooked and may lead to erroneous results.

2.3 Correlations for the Volumetric Mass Transfer Coefficient in Stirred Tank Reactors

2.3.1 Low Viscous Systems

Attempts to correlate $k_L a$ with the operating variables of gas-liquid reactors go back to COOPER et al. (1944) who suggested a correlation for the absorption coefficient K_V (lb moles oxygen absorbed)/(cu.ft. sulfite solution) (atm oxygen partial pressure) (hr) based on sulfite oxidation measurements in gas sparged stirred vessels equipped with vaned disks:

$$K_V = C \left(\frac{P}{V}\right)^{0.95} V_S^{0.67} \tag{36}$$

The years and decades thereafter have seen numerous correlations based on the same or similar structures as Eq. (36). For an extensive compilation and description of these correlations, the reader is referred to one of the various reviews of the subject (WANG et al., 1979; MOO-YOUNG and BLANCH, 1981, TATTERSON, 1991). A very impressive comparison of some earlier correlations has been presented by CHARLES (1985b). The results of these computations are presented in Tabs. 3 and 4. The large discrepancies between oxygen uptake rates predicted from the various correlations are remarkable and do not require any further comment.

The research of LESSARD and ZIEMINSKI (1971), ROBINSON and WILKE (1972, 1973, 1974), ZLOKARNIK (1978) and VAN'T RIET (1979) was foremost in establishing beyond any doubt the pronounced influence of gas bubble coalescence upon the surface area. This was reflected in the exponents of the $k_{L}a$ correlations. VAN'T RIET (1979) detailed a general form of $k_{L}a$ correlations as

Tab. 3. Oxygen Transfer Correlations for Newtonian Fluids[a] (CHARLES, 1985a)

Correlation	Reference
(A) $K_V - K_1 \left(\frac{P_G}{V_L}\right)^{0.95} V_S^{0.67}$	COOPER et al. (1944)
(B) $K_V = K_1 \left(\frac{P_G}{V_L}\right)^{\alpha}$	OLDSHUE (1966)
(C) $K_V = K_1 \left(\frac{P_G}{V_L}\right)^{0.4} V_S^{0.5} N^{0.5}$	RICHARDS (1961)
(D) $K_V = K_1 (2.0 + 2.8 N_i) \left(\frac{P_G}{V_L}\right)^{0.77} V_S^{0.67}$	FUKUDA et al. (1968)

[a] All correlations used as presented in WANG et al. (1979).

Tab. 4. Oxygen Transfer Rates Predicted by Various Correlations for Newtonian Fluids (CHARLES, 1985a)

P_G/V_L (kW m^{-3})	V_S (cm min^{-1})	N (min^{-1})	N_i	Oxygen Transfer Rate (mmol L^{-1} h^{-1}) A	B	C	D
4.00	180	60	1	1200	572	26	206
		180	1	1200	572	46	206
		60	2	1200	572	26	327
		180	2	1200	572	46	327
		60	3	1200	572	26	448
		180	3	1200	572	46	448
2.65	180	60	1	816	432	21	151
		180	1	816	432	32	151
		60	2	816	432	21	239
		180	2	816	432	32	239
		60	3	816	432	21	327
		180	3	816	432	32	327
1.32	180	60	1	422	266	16	88
		180	1	422	266	29	88
		60	2	422	266	16	140
		180	2	422	266	29	140
		60	3	422	266	16	192
		180	3	422	266	29	192

$$K_L a = K \left(\frac{P}{V}\right)^{\alpha} V_S^{\beta} \tag{37}$$

The exponents show a great variation ($0.4 < \alpha < 1$ and $0 < \beta < 0.7$). The same author (VAN'T RIET, 1979; VAN'T RIET and TRAMPER, 1991) provided correlations for $k_L a$ data depending upon the nature of the solution:

$$k_L a = 2.6\ 10^{-2} \left(\frac{P_G}{V_L}\right)^{0.4} V_S^{0.5}\ [s^{-1}] \tag{38}$$

for coalescing non-viscous fluids, such as water, and

$$k_L a = 2.0\ 10^{-3} \left(\frac{P_G}{V_L}\right)^{0.7} V_S^{0.2}\ [s^{-1}] \tag{39}$$

for non-coalescing, non viscous fluids, such as water–salt solutions.

ZLOKARNIK (1978) suggested the use of dimensionless groups for correlating $k_L a$ values which were originally based on a relevance list:

$$k_L a = f(P/V_L, V_G/L_L, \ldots) \tag{40}$$

By carefully analyzing $k_L a$ data for tank volumes ($2.5\ L < V < 900\ m^3$), JUDAT (1982) showed that, in agreement with the equations of VAN'T RIET (1979), the superficial gas velocity V_S should be used in the relevance list for $k_L a$ instead of volumetric gas throughput $\dot{V}_G/V$ (or vvm). Following the same line, HENZLER (1982) was able to evaluate a large number of experimental results from water and water-salt solutions with various tank sizes by employing the relationship

$$\left(\frac{k_L a}{V_S}\right)\left(\frac{\nu^2}{g}\right)^{1/3} = K_2 \left(\frac{P_G}{V_L V_S \rho g}\right)^{a} \tag{41}$$

The same dimensionless groups have been applied by REUSS et al. (1986) to correlate steady state $k_L a$ measurements during fermentations of *Saccharomyces cerevisiae* (fed-batch on substrates glucose or molasses at different scales of operation and with different geometrical properties). Figs. 6 and 7 show the experimental results along with the correlation

$$\left(\frac{k_L a}{V_S}\right)\left(\frac{\nu^2}{g}\right)^{1/3} Sc^{0.3} = 5.5\ 10^{-4} \left(\frac{P_G}{V_L V_S \rho g}\right)^{0.7} \tag{42}$$

The Schmidt number *Sc* has been introduced to apply the same correlation to data from gluconic acid fermentations with *Aspergillus ni-*

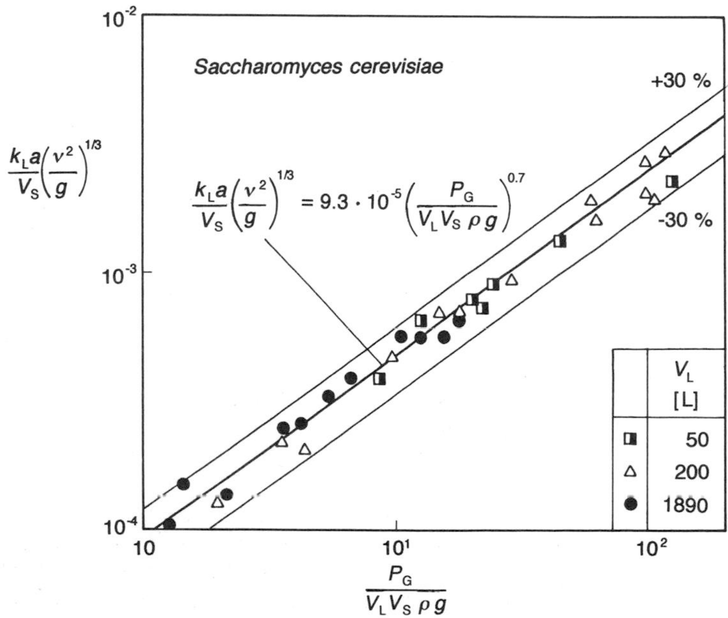

Fig. 6. Dimensionless correlation for volumetric mass transfer coefficients during fermentation of *Saccharomyces cerevisiae*.

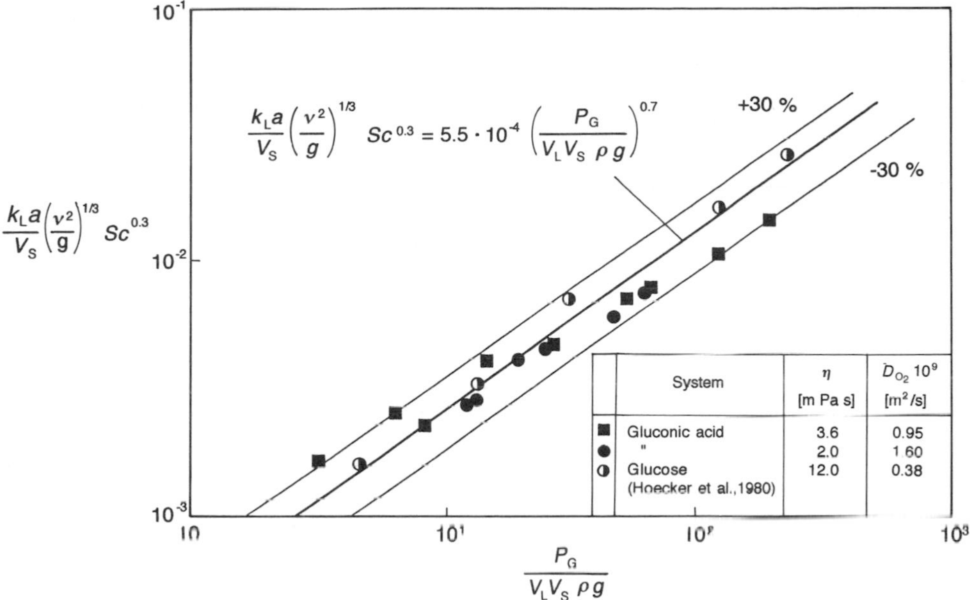

Fig. 7. Comparison of the $k_L a$ correlation, Eq. (42), with measurements from gluconic acid fermentation and data for glucose solutions (from HOECKER et al., 1980).

ger. High sugar and gluconic acid concentrations (up to 300 kg/m³) resulted in an increase of viscosity and a decrease of molecular diffusivity, which required the addition of this dimensionless group in order to account for the effects of these molecular transport properties on the mass transfer coefficient. It is worthwhile to stress that Eq. (42) reduces to Eq. (41) for $Sc = (Sc)_{\text{water}}$ and that introducing the physical properties of water into Eq. (41) results in

$$k_L a = 3.0\ 10^{-3}\left(\frac{P_G}{V_L}\right)^{0.7} V_S^{0.3} \tag{43}$$

This is very close to the correlation suggested by VAN'T RIET (1979) for non-coalescing systems.

LINEK et al. (1988) reviewed the literature dealing with various aspects of coalescence. The account was aimed at developing a correlation which covers the whole range of coalescing and non-coalescing properties. The equation suggested by LINEK et al. (1988) is given by

$$k_L a = \exp(-4.18\, n_1 - 4.36)\left(\frac{P_T}{V_L}\right)^{n_1} V_S^{0.4} \tag{44}$$

with P_T as the total power input

$$n_1 = \frac{1.5 X_2}{1/10^6 + 3.05 X_2} + 0.56$$
$$X_2 = C\left(\frac{d\sigma}{dC}\right)^2\left(\frac{1 + d\ln\sigma}{d\ln C}\right)^{-1} \tag{45}$$

In agreement with the observations of KEITEL and ONKEN (1981), it was reported as doubtful that ionic strength is sufficiently suitable for a coalescence parameter, even for aqueous electrolyte solutions. This approach seems to be attractive and promising. However, further work is necessary to incorporate data from real fermentation processes. It would also be desirable to include the influence of η, ρ, σ and D in the correlation. It is indeed a very difficult problem of how coalescence of the bubbles should be taken into account. For correlating such a complex property as $k_L a$, it is worthwhile to critically examine the oversimplification of using a product of two variables (P/V and V_S) or corresponding dimensionless groups raised to various powers. The fundamental work of SMITH and NIENOW (most of their work has been critically reviewed in

Chapter 6 of TATTERSON, 1991), regarding the dispersion of gas in a stirred tank, has considerably increased the fundamental understanding of this process. It would be rather surprising to see that such complex phenomena could be manifested in a simple catch-all relationship for $k_{L}a$ which covers the entire range of agitation and aeration.

Work in the direction of interrelating impeller flow regimes and $k_{L}a$ has been initiated. SMITH and WARMOESKERKEN (1985) provided two $k_{L}a$ correlations:

$$\frac{k_{L}a}{N} = 1.1\ 10^{-7}\ Fl^{0.6}\ Re^{1.1} \quad (46)$$

one for the region before the 3-3 structure (this structure consists of three large cavities and three clinging cavities in a symmetrical pattern around a six-blade turbine impeller), and the other

$$\frac{k_{L}a}{N} = 1.6\ 10^{-7}\ Fl^{0.42}\ Re^{1.02} \quad (47)$$

after formation of the large cavity (3-3 structure) with gas flow number $Fl = \dot{V}_{G}/N\ D_{I}^{3}$ and impeller Reynolds number $Re = N\ D_{I}^{2}\ \rho/\eta$. For a more detailed description, the different flow regimes are shown in Fig. 8 (WARMOESKERKEN and SMITH, 1988). These fundamen-

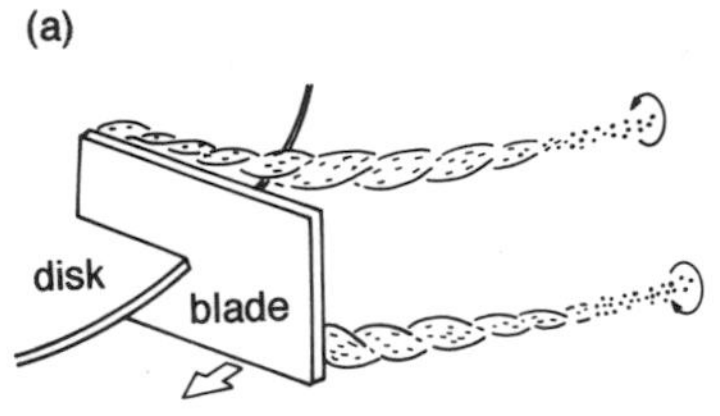

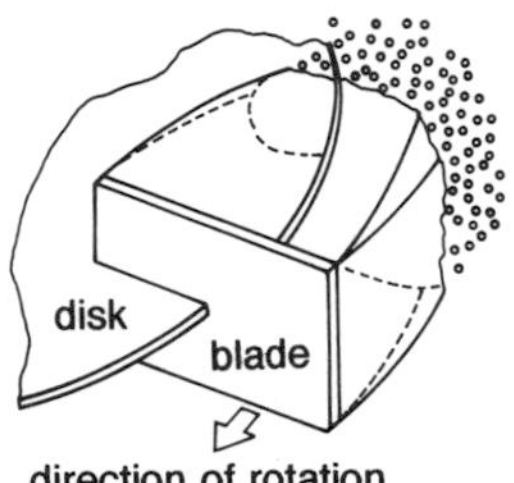

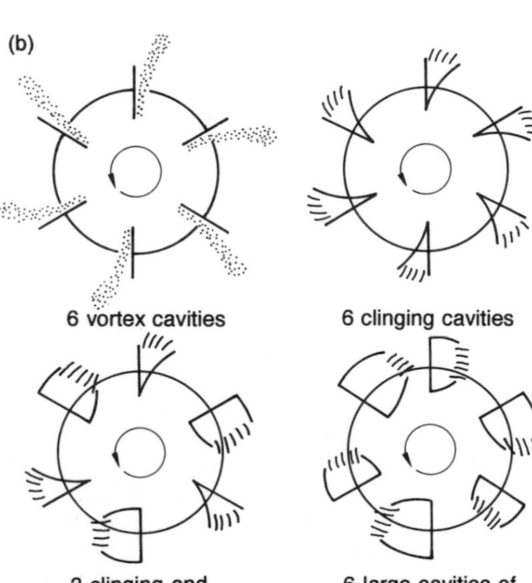

Fig. 8. Vortex and large ventilated cavities behind blades (a) and successive cavity configuration with increasing gas flow number (b) (reprinted from WARMOESKERKEN and SMITH, 1988, with permission).

tal observations have largely contributed to the understanding of gas dispersion with standard Rushton turbine impellers. Generalized flow regime charts have also been developed (SMITH and WARMOESKERKEN, 1985) in order to predict the different flow regimes. The idea to use different correlations, because the flows associated with the different regimes are so different, is tempting for those cases in which the volumetric mass transfer coefficient is influenced to a greater extent by the flow condition next to the impeller. NIENOW et al. (1988), however, reported that these observations and representation in the two correlations may be largely corrupted by the experimental determination of $k_L a$. When comparing $k_L a$ values which are predicted from simultaneous measurements of dissolved oxygen and oxygen content of exhaust air, Eq. (29), during unsteady state experiments these authors observed large deviations from the results according to the simple "no depletion" model. These deviations are manifested in a change of the exponent on power/volume and are particularly magnified at higher $k_L a$ values where the 3-3 structures are formed. Such diverging of $k_L a$ correlations due to different treatment of the experimental data from dynamic measurements, has also been described by MIDDLETON (1985).

Another aspect of flow regions has been brought into discussion by NISHIKAWA et al. (1981a, b). These authors suggested an approach for correlating $k_L a$ measurements in aerated mixing vessels which is based on the different relative intensity of aeration/agitation conditions. The correlations include two different terms for consumption per unit mass of liquid.

For aeration

$$P_{sm} = V_S \, g \tag{48}$$

and for agitation

$$P_m = \frac{Ne N^3 D_i^5}{V_L} \tag{49}$$

with Ne, the power number of the impeller. The dimensions of P_{sm} and P_m are cm^2/s^3. Under conditions in which flow is primarily controlled by bubbling (high aeration rates and lower stirrer speeds), $k_L a$ was correlated as

Regime I:

$$k_L a = 1.25 \; 10^{-5} \, P_{sm} \tag{50}$$

with $k_L a$ in units s^{-1}. For the agitation controlled regime the following equations were obtained

Regime II:

$$k_L a = 3.92 \; 10^{-6} \, P_{sm}^{0.33} \, P_m^{0.8} \tag{51}$$

for a disk turbine. The $k_L a$ values for the intermediate region, in which flow is influenced by both agitation and aeration, were correlated as:

$$k_L a = [3.92 \; 10^{-6} \, P_m^{0.8} + \\ + 1.25 \; 10^{-5} \, W_P \, P_{sm}^{0.66}] \, P_{sm}^{0.33} \tag{52}$$

where W_P is a weighting factor given by:

$$W_P = \frac{P_{sm}}{(P_m/Ne + P_{sm})} \tag{53}$$

The boundaries for the two regions have been described in two correlations. If P_{sm} is defined, the P_m at which agitation regime controls are calculated is given as

$$P_m^{0.8} \left(1 + \frac{P_m}{Ne \, P_{sm}}\right) = 12.7 \, P_{sm}^{0.66} \tag{54}$$

And the minimum value at which gas sparging dominates, is predicted from

$$P_m = 0.514 \, P_{sm}^{0.84} \tag{55}$$

There are also results showing that the scale of operation exerts a strong influence. CHANDRASEKHARAN and CALDERBANK (1981) measured absorption rates in water and suggested the following correlation:

$$k_L a = \left(\frac{C}{D_T^4}\right) \left(\frac{P_G}{V_L}\right)^A (\dot{V}_G)^{A/\sqrt{D_T}} \tag{56}$$

where $C = 0.0248$ and $A = 0.551$ in SI units.

2.3.2 Highly Viscous Systems with Non-Newtonian Flow Behavior

Faced with the previously mentioned problems regarding reliable k_La correlations for scale-up application over the entire range of operation parameters, it is not surprising to see that the problems are all the more pronounced in the case of highly viscous, non-Newtonian fluids. The well known difficulties involved in obtaining a reliable and completely quantitative characterization of the flow behavior of fermentation broths undoubtedly contribute to these uncertainties. This is particularly true for the characterization of the dynamic effects of elasticity at the gas–liquid interface. However, even if only viscous flow behavior which has been correlated with the power law according to

$$\tau = K_c \gamma^n \tag{57}$$

is taken into account most of the empirical results are questionable. This analysis has been critically reviewed by BLANCH and BHAVARAJU (1976), CHARLES (1978), METZ et al. (1979), PACE (1980), REUSS et al. (1982), and REUSS (1982). CHARLES (1985b) pointed out that in many cases results which have been published are not only confusing and misleading, but are "just plain wrong". Therefore, it is advisable to look at the published equations for k_La cautiously particularly when real fermentation broths are concerned.

First attempts toward consideration of the viscosity factor are expressed in equations of the form:

$$k_La = C \left(\frac{P_G}{V_L}\right)^a (V_S)^b (\eta_{app})^c \tag{58}$$

Examples of this type of correlation have been presented by RYU and HUMPHREY (1972) and HICKMAN and NIENOW (1986), with exponents $c = -0.86$ and -0.5, respectively.

PEREZ and SANDALL (1974), using k_La measurements from CO_2 absorption in water and carbopol solutions, suggested a correlation of the following form

$$\frac{k_La\, D_i^2}{D} = 21.2 \left(\frac{\rho N D_i^2}{\eta_{app}}\right)^{1.11} \left(\frac{\eta_{app}}{\rho D}\right)^{0.5} \left(\frac{V_S D_i}{\sigma}\right)^{0.447} \left(\frac{\eta_{gas}}{\eta_{app}}\right)^{0.694} \tag{59}$$

It must be emphasized that the group $V_S D_i/\sigma$ is not dimensionless. MOO-YOUNG and BLANCH (1981) have substituted this group with $\eta_{app} D_i/\sigma$. The apparent viscosity η_{app} is estimated from the correlation of METZNER and OTTO (1957) for fluids with viscous flow behavior as correlated with the power law

$$\eta_{app} = K(11N)^{n-1} \left(\frac{3n+1}{4n}\right)^n \tag{60}$$

YAGI and YOSHIDA (1975) correlated k_La data from measurements in glycerol and millet jelly (Newtonian flow behavior) as well as sodium polyacrylate, PAA, and sodium carboxymethyl cellulose, CMC (non-Newtonian flow behavior) as

$$\frac{k_La\, D_i^2}{D} = 0.060 \left(\frac{\rho N D_i^2}{\eta_{app}}\right)^{1.5} \left(\frac{N^2 D_i}{g}\right)^{0.19} \cdot \left(\frac{\eta_{app}}{\rho D}\right)^{0.5} \left(\frac{\eta_{app} V_S}{\sigma}\right)^{0.6} \left(\frac{N D_i}{V_S}\right)^{0.32} \tag{61}$$

For viscoelastic fluids the right-hand side of this equation was multiplied by the factor $(1 + 2.0\,(\lambda N)^{0.5})^{-0.67}$, where λ is a characteristic time (PREST et al., 1970).

Based on measurements in PAA and CMC solutions, RANADE and ULBRECHT (1978) suggested the equation:

$$\frac{k_La\, D_T^2}{D} = 2.5\; 10^{-4} \left(\frac{\rho N D_i^2}{\eta_{app}}\right)^{1.8} \cdot \left(\frac{\eta_{app}}{\eta_W}\right)^{1.39} (1 + 100 De)^{-0.67} \tag{62}$$

with η_W, the viscosity of water, and De, a modified Deborah number.

In another attempt to correlate k_La in viscous solutions, NISHIKAWA et al. (1981b) combined the previously mentioned approach which considered different regions of agitation and aeration with the equations of YAGI and YOSHIDA (1975) which considered the agitation controlled region, and the formula of NA-

KANO (1975) which considered the sparging controlled region.

NIEBELSCHÜTZ (1982) performed steady state $k_L a$ measurements in different non-Newtonian fluids including fermentation broths. Fig. 9 summarizes attempts to correlate the results of these measurements with a dimensionless correlation suggested by ZLOKARNIK (1978). Similar plots for non-Newtonian fluids have been presented by HOECKER et al. (1980) and JURECIC et al. (1984) who correlated $k_L a$ data from actual fermentation processes on a 67 m^3 scale. The data presented in Fig. 9 include measurements for CMC solutions (HOECKER et al., 1980), CMC II, as well as Newtonian fluids, glucose and glycerol solutions. Observations from xanthan fermentation with *Xanthomonas campestris* and processes with mycelial organisms (*Aspergillus niger* and *Penicillium chrysogenum*) have also been included. Some of the measurements (CMC I and some of the xanthan data) have been obtained from $k_L a$ predictions made during baker's yeast fermentation, in which the polymer was added to the batch during a computer-controlled feeding. This turned out to be an interesting system because of the presence of an oxygen sink for steady state $k_L a$ measurements. Compared to fermentation processes with non-Newtonian broth characteristics, these experiments are easier and faster to perform.

It is evident from Fig. 9 that each of the fluids studied resulted in a different correlation. A careful analysis of the data in which a prediction is made according to the different properties of the materials, leads to the conclusion that the problem is caused by an inappropriate representation of the apparent or effective viscosity. The calculation of the apparent viscosities for the data in Fig. 9 is based on a combination of power law and the correlation of METZNER and OTTO (1957). Thus,

$$\eta_{app} = K_c (11.5N)^{n-1} \quad (63)$$

The problem with this estimation procedure is that for two fluids with quite different values for the consistency index K_c and the flow index n eventually a similar apparent viscosity at a given impeller speed N can be predicted. This situation is illustrated in Fig. 10 for the two different CMC solutions of Fig. 9 with different molecular weight distribution and flow be-

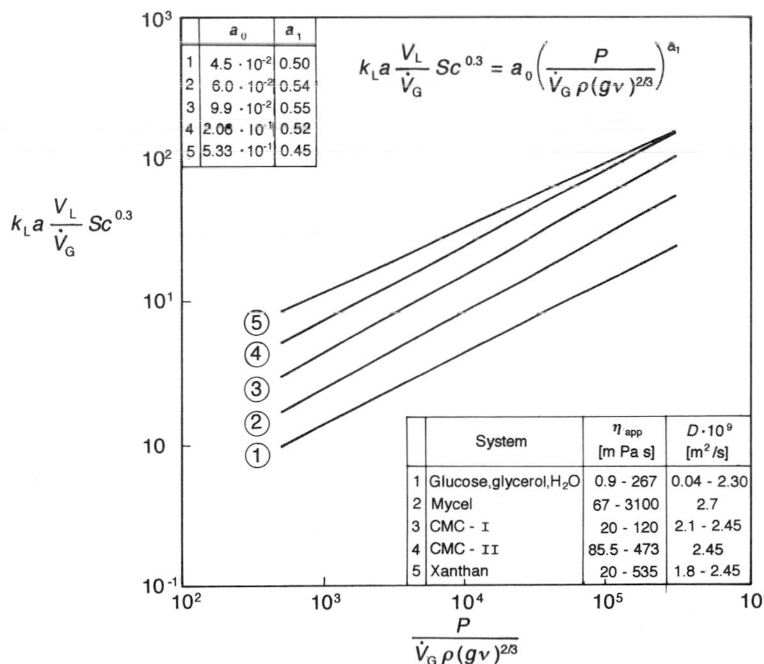

	a_0	a_1
1	$4.5 \cdot 10^{-2}$	0.50
2	$6.0 \cdot 10^{-2}$	0.54
3	$9.9 \cdot 10^{-2}$	0.55
4	$2.06 \cdot 10^{-1}$	0.52
5	$5.33 \cdot 10^{-1}$	0.45

	System	η_{app} [m Pa s]	$D \cdot 10^9$ [m^2/s]
1	Glucose, glycerol, H_2O	0.9 - 267	0.04 - 2.30
2	Mycel	67 - 3100	2.7
3	CMC - I	20 - 120	2.1 - 2.45
4	CMC - II	85.5 - 473	2.45
5	Xanthan	20 - 535	1.8 - 2.45

Fig. 9. Dimensionless correlation for $k_L a$ in highly viscous Newtonian and non-Newtonian fluids (NIEBELSCHÜTZ, 1982).

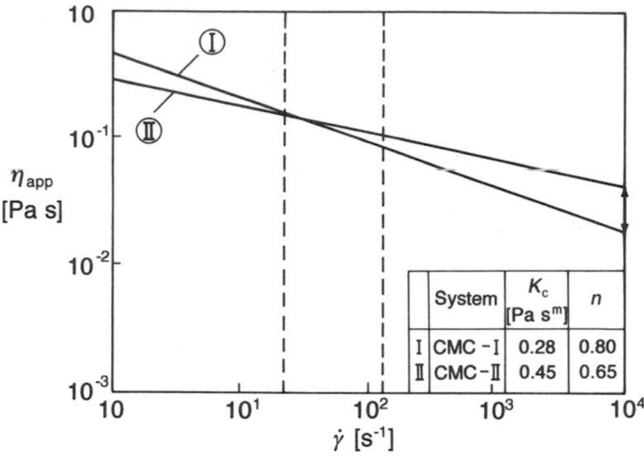

Fig. 10. Apparent viscosities as a function of shear rate for two CMC solutions differing in molecular weight distribution (NIEBELSCHÜTZ, 1982).

havior. RANADE and ULBRECHT (1978) also pointed to this problem, suggesting the use of two different characteristic shear rates. For those processes controlled by microscale turbulence, the characteristic shear rate should be of the order of the frequency of dissipative eddies. Other phenomena, like bubble break-up and coalescence, take place on a much larger scale and, thus, specific surface area a may be influenced by a shear rate which is determined by the impeller speed. The frequency of the dissipative eddies has been estimated in the order of 10^4 s^{-1} (MASHELKAR, 1973; RANADE and ULBRECHT, 1978).

Fig. 11 shows the result of the attempt to fit the k_La measurements from three different CMC solutions with one correlation, in which the viscosity in the Schmidt number is calculated with

$$Sc_D = \nu_{app}(\dot{\gamma} = 10^4 \text{ s}^{-1})/D_{O_2} \tag{64}$$

The apparent viscosity in the group $P/\dot{V}_G\rho(g\,\nu)^{2/3}$ is still predicted with METZNER's and OTTO's relationship. This approach is supported by the findings (Fig. 12) that k_La values in xanthan solutions and biosuspensions correlate with the same equation as the one used for CMC solutions. Although the concept of two different representative viscosities, as originally suggested by RANADE and ULBRECHT (1978), considerably improves the quality of the correlation of data, it is still empirical and, therefore, it is uncertain if this approach should be recommended for scale-up. The data presented in Figs. 11 and 12 have

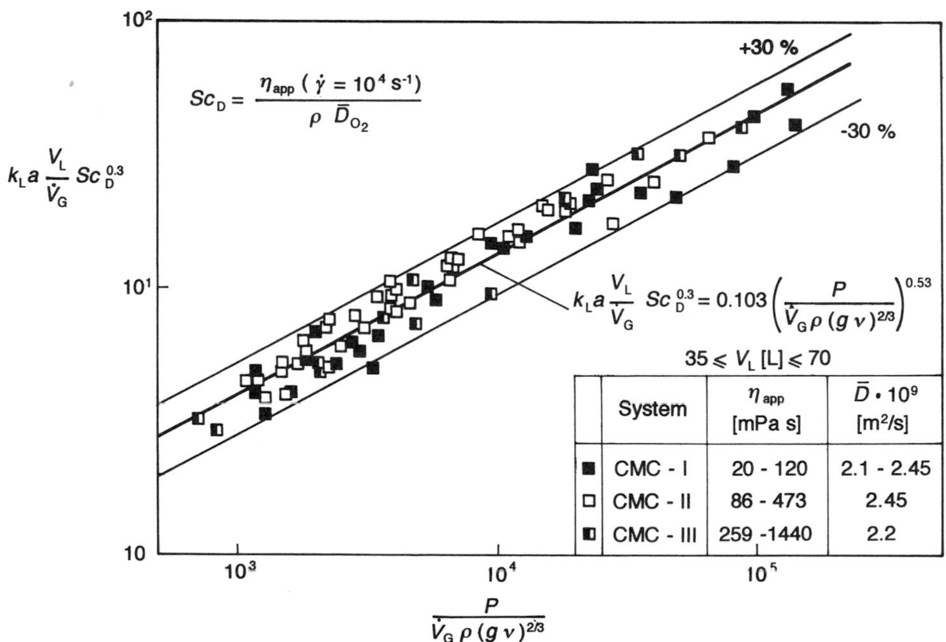

Fig. 11. Dimensionless correlation for k_La in three different CMC solutions (NIEBELSCHÜTZ, 1982).

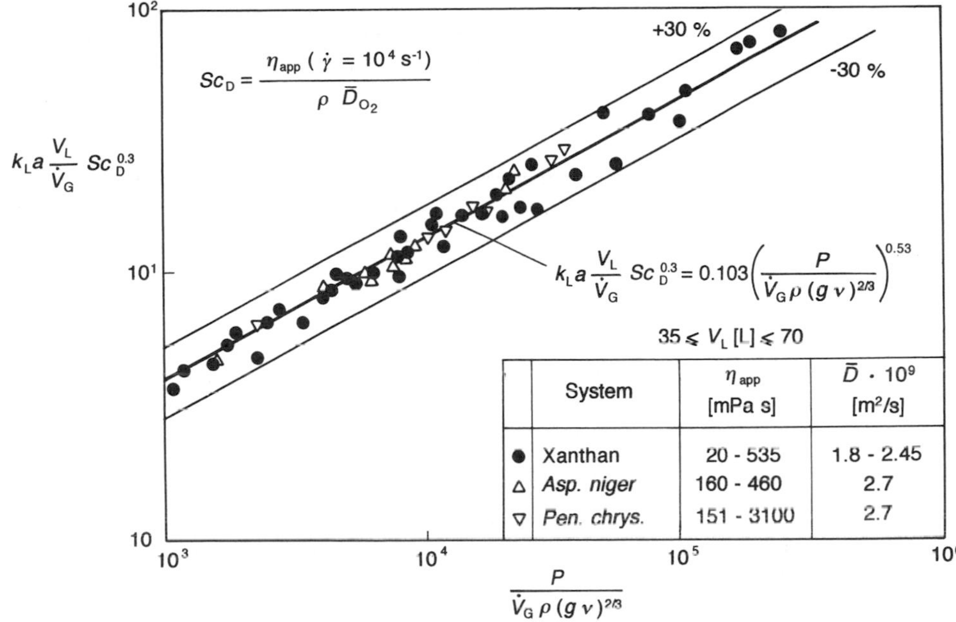

Fig. 12. Dimensionless correlation for $k_L a$ in different fermentation broths with non-Newtonian flow behavior (NIEBELSCHÜTZ, 1982).

been determined at only two scales: 50 and 300 liters.

Alternative relationships for estimation of the characteristic shear rate $\dot{\gamma}$ include, CALDERBANK and MOO-YOUNG (1961):

$$\dot{\gamma} = 11.5 N \left(\frac{4n}{3n+1} \right)^{\frac{n}{1-n}} \tag{65}$$

STEIN (1984):

$$\dot{\gamma} \sim \left(\frac{P_G}{V_L} \right)^{0.335} V_S^{0.245}\ \eta_{app}^{-1.115} \tag{66}$$

HENZLER and KAULING (1985):

$$\dot{\gamma} = \left(\frac{P_G/V_L}{\eta_{app}} \right)^{0.5} \tag{67}$$

and WICHTERLE et al. (1984):

$$\dot{\gamma} = N(1+5.3\,n)^{1/n} (N^{2-n} D_i^2 \rho_L / K_c)^{\frac{1}{1+n}} \tag{68}$$

These different approaches are considered in Fig. 13 (HERBST et al., 1992), which shows a comparison of the calculated apparent viscosites for a xanthan solution. Obviously, the differences between the various equations are

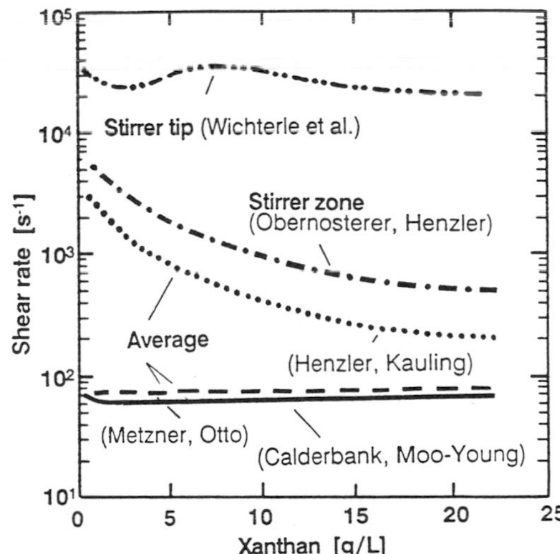

Fig. 13. Comparison of shear rate relations using the flow behavior in xanthan fermentations (HERBST et al., 1992).

remarkable. A universal relationship remains undefined, and the entire field requires more fundamental research to improve our understanding and knowledge.

Based upon separated correlations for k_L and a, KAWASE and MOO-YOUNG (1988) developed a new correlation for evaluating volumetric mass transfer coefficients in aerated, stirred tank reactors with Newtonian or non-Newtonian media. This equation is given by

$$k_L a = C_2 \sqrt{D} \frac{\rho^{3/5} \varepsilon^{9+4n/(10(1+n))}}{(K_c/\rho)^{1/2(1+n)} \sigma^{3/5}} \left(\frac{V_S}{V_B}\right)^{0.5} \left(\frac{\eta_{app}}{\eta_W}\right)^{-0.25} \quad C_2 = 0.675 \qquad (69)$$

The correlation was compared with earlier equations for Newtonian and non-Newtonian fluids. In the case of Newtonian media, the equation fits the data of 8 different sources with an error of approximately ±50%. The comparison with non-Newtonian fluids was restricted to data from CMC and carbopol solutions and resulted in a reasonable fit. HERBST et al. (1992) applied this equation to data from xanthan fermentations and found an accuracy in the order of 30%.

The problem of mass transfer in non-Newtonian fluids will be further commented in the next section, in the context with fluid mixing.

3 Mixing

3.1 General Considerations and Mixing in Low Viscous Systems

The intensity of mixing will determine the spatial variations of concentrations and temperature in bioreactors. Obviously, mixing will play a role in the outcome of a process only if the metabolism of the organisms is sensitive to these variations. The challenge for reliable scale-up is, therefore, twofold. First, the mixing process in terms of spatial variations for the large tank should be replicated either in small scale or via mathematical computer model and, secondly, the response of the microbial culture to these variations should be studied or simulated, if enough knowledge about the dynamic response is available. Although in the last decade work has been guided in this direction, knowledge is still rudimental and there is a great need for further research in this field.

As far as dissolved oxygen is concerned, spatial variations have long been known to exist in large-scale bioreactors (STEEL and MAXON, 1962, 1966). However, systematic studies regarding experimental observations of defined fluctuations have only been initiated in the last decade. Fundamental studies addressing this problem include investigations of the influence of periodic pressure variations in *Penicillium chrysogenum* fermentation (VARDAR and LILLY, 1982), periodic changes of air and nitrogen during gluconic acid fermentation with *Gluconobacter oxydans* (OOSTERHUIS et al., 1985) and *Saccharomyces cerevisiae* (SWEERE et al., 1988a, b), periodic changes from air to nitrogen in a stirred-tank loop reactor with *Penicillium chrysogenum* and *Escherichia coli* (LARSSON and ENFORS, 1988), as well as oscillating oxygen concentrations during gluconic acid production with *Gluconobacter oxydans* and *Aspergillus niger* in airlift reactors (TRÄGER, 1990) and production of pullulan with *Aureobasidium pullulans* in a loop reactor (KRISTIANSEN and MCNEIL, 1987). The state of the art concerning the mathematical modeling has been summarized and discussed in Volume 4 of this series (REUSS and BAJPAI, 1991).
In the following discussion, the physical rather than the biochemical or reaction engineering point of view of mixing will be considered. This contribution, however, will not be aimed at presenting the state of the art for operation in chemical engineering forms. Only those phenomena which are closely linked to application of the aerated stirred tank as a bioreactor will be emphasized. The problems involved with incomplete mixing of dissolved oxygen will be accentuated. For more fundamental information the interested reader is referred to the standard textbooks (UHL and GRAY, 1966; NAGATA, 1975; VAN'T RIET and TRAMPER, 1991; TATTERSON, 1991) and published reviews (MIDDLETON, 1985; NIENOV and UL-

BRECHT, 1985; SMITH, 1985). In Chapter 10 of Volume 4 of this series various measurement methods and the characterization of mixing in stirred tank reactors are presented as an interplay to mathematical modeling and coupling with microbial reactions.

Mixing, which is defined as the process that decreases inhomogeneities to a given degree (HIBY, 1981), is typically characterized in single- and multi-phase batch reactors with the help of a terminal mixing time. There are three problems with terminal mixing times: (1) They do not give any evidence about the mechanism of mixing, (2) their experimental prediction is only apparently simple, and (3) due to an undefined degree of homogeneity, they are often "subjective". The first problem particularly hampers any progress in studying the interaction of spatial variations in concentration with microbial response, because terminal mixing times do not provide the information necessary to replicate circulation distribution, which has the largest impact on the cells. In aerobic bioprocesses, where mass transfer gas–liquid is known to be most intensive in the impeller regions, the frequency of broth circulations may be much more important than a terminal mixing time which is a superposition of many different phenomena.

In the case of single-phase mixing, it is common to assume that the terminal mixing time θ_m and the circulation time θ are related by (VONCKEN et al., 1964):

$$\theta_m = 4\,\theta_{circ} \tag{70}$$

Thus, it may be justified to estimate the circulation time from one of the published correlations for terminal mixing times. For two-phase systems these equations are rather scarce. PANDIT and JOSHI (1982) suggested the following correlation:

$$N\,\theta_m = 20.41 \left(\frac{a\,H_L + D_T}{D_T}\right) \left(\frac{D_T}{D_i}\right)^{\frac{13}{6}} \left(\frac{W}{D_i}\right) \left(\frac{\dot{V}_G}{N\,V_L}\right)^{\frac{1}{12}} \left(\frac{N^2\,D_i^4}{g\,W\,V_L^{2/3}}\right)^{\frac{1}{15}} \tag{71}$$

with W as the blade width and a, a factor depending on the size of the recirculation loop. It must be stressed that estimation of the recirculation time according to Eq. (70) is risky because of the following reasons: in the case of aerated systems, although the circulation time increases due to a decreased pumping capacity of the impeller, at the same time mixing on the circulation paths can be more intensive, indicating that bubbles contribute to the mixing process. These phenomena have been observed by BRYANT and SADEGHAZEDEH (1979) and also by REUSS and BAJPAI (1991) when applying radio- and magneto-flow follower measurement techniques, respectively. The superposition of circulation and dispersion manifests itself with only a minor influence on aeration of the terminal mixing time (exponent of $1/12$ in Eq. (71)). The effect on circulation time alone is much more pronounced and was correlated (RIESMEIER, 1984; REUSS and BAJPAI, 1991) by:

$$\theta_{circ} \sim \left(\frac{1}{Ne_G}\right)^{0.35} \tag{72}$$

for the non-flooded turbine impeller. The exponent in the power number in very close to $-(1/3)$, which is theoretically and experimentally (VAN'T RIET and TRAMPER, 1991; MERSMANN et al., 1975; KIPKE, 1978) verified for the influence of power number on terminal mixing times in unaerated fluids. Eq. (72) has also been confirmed in the form:

$$\theta_{90}\,Ne_G^{1/3}\,N = \text{constant} \tag{73}$$

by COOK and MIDDLETON (1988), who studied mixing times in sparged agitated vessels (30 L $< V_L <$ 4.3 m^3).

There are uncertainties related to the experimental predictions. An undefined degree of homogeneity obviously often contributes to the inconsistency of observations for terminal mixing times. JANSEN et al. (1978), for example, reported a mixing time predicted in a 60 m^3 sparged and aerated vessel which is approximately the same as the one estimated from an equation for single-phase flow in stirred vessels. At the same scale of operation, the measured mixing time reported by EINSELE (1978) is twice the estimated value. This example has been elaborated by VAN'T RIET and TRAMPER (1991) and clearly demonstrates the limitation of these experiments. It is difficult to conclude if estimation based on reasonable assumptions inspires more confidence than experiments performed under dubious conditions.

3.2 Highly Viscous Non-Newtonian Fluids

The problems of mixing in highly viscous non-Newtonian fluids in stirred tank bioreactors have been reviewed in a number of papers (TAGUCHI, 1971; BLANCH and BHAVARAJU, 1976; CHARLES, 1978; MARGARITIS and ZAJIC, 1978; PACE and RIGHELATO, 1980; MARGARITIS and PACE, 1985; NIENOV, 1984). For more fundamental aspects of mixing in high-viscosity material, the interested reader is again referred to the textbooks (UHL and GRAY, 1966; NAGATA, 1975; TATTERSON, 1991). The only data available for circulation time distributions are those measured by MUKATAKA et al. (1980) and FUNAHASHI et al. (1987) with the help of magneto-flow follower techniques. MUKATAKA et al. (1980) reported the following empirical equation for the mean circulation time:

$$\theta_{\text{circ}} = 7.8 \left[n^3 \left(\frac{0.35 \left(\frac{P}{V} \right)}{D_T^2} \right)^{\frac{2-n}{3}} (10^2 D_i)^{1-n} \right]^{-1} \tag{74}$$

with n being the flow index in the power law. These measurements were performed in broths of *Penicillium chrysogenum* ($4.5 < c_X < 9.1$ kg/m^3) and suspensions of paper pulp ($7 < c < 15$ kg/m^3). The observed flow indices were in the range of $0.37 < n < 0.64$, and the consistency indices varied between $0.3 < K_c < 3$ Pa s^n. These experimental observations were performed in only small vessels, up to a 10 liter volume, and the coils for detecting the magneto-flow follower were located outside the vessel wall.

Experiments performed in the author's laboratory (RIESMEIER, 1984), with xanthan solutions in an 80 liter reactor with 40 liter working volume, point to limitations in the applicability of flow follower techniques to these systems. At concentrations of 3 and 5 kg/m^3 xanthan, the observed mean circulation times were even smaller than those measured in water at the same speed of agitation. The comparison between the results of the stimulus response experiments using acid as a tracer and measuring the response with the aid of a pH electrode in water and xanthan solutions, respectively, shows the expected differences in the mixing behavior (Fig. 14). The high viscosity of the xanthan solution along with its non-Newtonian flow behavior considerably prolongs the mixing process. This result is in accordance to what has been reported in the literature for mixing in xanthan solutions (CHARLES, 1978). The observation that the flow follower is unable to sufficiently detect this mixing behavior, is probably caused by a superposition of two effects. First, due to its size the sphere is unable to follow any diffusive transport path at the microscale. This transport mechanism, however, may have a pronounced influence on the exchange of mass between the convective bulk circulation flow, characterized by lower apparent viscosities, and possible dead regions in the vessel. Secondly, well-known secondary flow effects of viscoelastic fluids (CHARLES, 1978) may, in addition, prevent any centrifugal acceleration of the sphere in the impeller region. Under certain conditions the normal stresses may dominate the inertial effects and lead to a flow reversal (ULBRECHT, 1974). In the light of such complex phenomena, the flow follower technique can no longer be recommended for studying the circulation time distribution in these fluids.

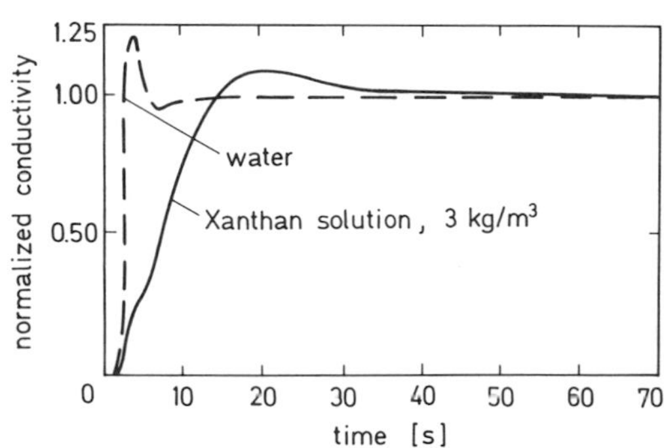

Fig. 14. Comparison between stimulus response experiments in water and xanthan solutions.

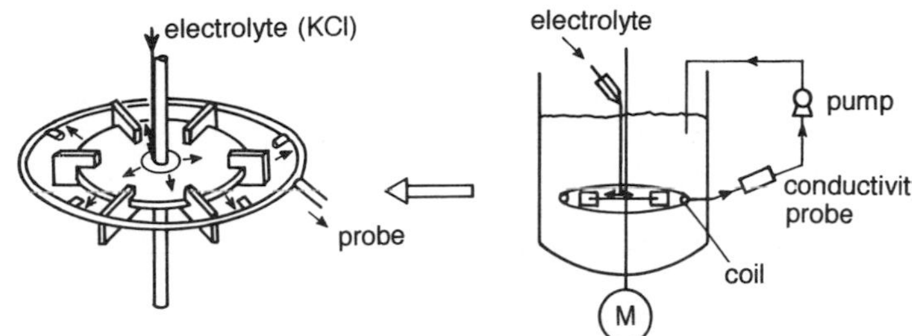

Fig. 15. Experimental set-up for tracer experiments in highly viscous systems.

The results of the primary tracer experiment presented in Fig. 14 suggest that at least two different flow regions need be considered in order to explain the observed mixing behavior. One is a faster circulation flow, manifested in the overshooting of tracer concentration, and the other is probably a dead region which is responsible for the long tail in the signal. To gain a better insight into the fluid flow, the experimental conditions for introducing the tracer and measuring the signal have to be modified. When designing an appropriate flow model, tracer input and measurement of the signal response should be as close as possible to the impeller region. A configuration as suggested by KHANG and LEVENSPIEL (1976), using several porous glass injection diffusers and conductance probes positioned closely around the impeller, would probably be appropriate for measurements in the laboratory and for pilot-scale operation. If, however, such experiments are to be made on a larger scale, this method may create some problems. In any case, a special design of the conductance measurement system is required, as described by KHANG and FITZGERALD (1975). A primary attempt was made to follow an alternative concept, in which a representative sample is continuously withdrawn out of the impeller region and where measurement of the tracer response is performed with a conventional electrical conductivity sensor located in a bypass (Fig. 15). The tracer is still injected into the impeller region with the aid of a syringe pipette connected to a circular distributor. On a larger scale, one would probably prefer an injection of the tracer from the outside. This could be achieved with an additional bypass having plug flow characteristics.

Fig. 16 shows an example of the response signal using concentrated KCl solution as a tracer in an 80 liter stirred tank reactor containing 40 liters xanthan solution at a concentration of 5 kg/m^3. The pronounced tail in the

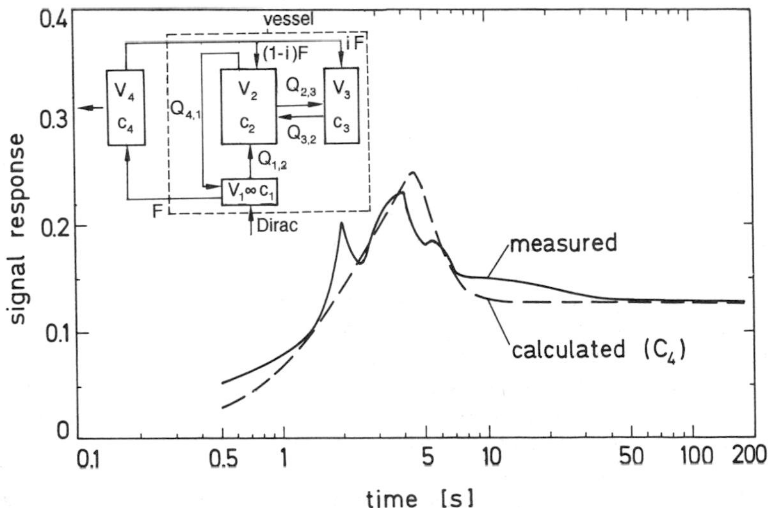

Fig. 16. Comparison between measured and calculated signal response in a xanthan solution (3 kg/m^3) with measurement probe in a bypass ($V_L = 47$ L; identified parameters: $V_1 = 7.6$ L, $V_2 = 37.2$ L, $V_3 = 2.0$ L, $V_4 =$ volume of the bypass for measurement).

response curve provides us with some reasonable information about the dead regions in the tank. Also, a first attempt has been made to consider various flow models for a quantitative description of the observed phenomena. An example is schematically illustrated in Fig. 16. This model assumes an impeller region and a circulation flow which exchanges mass with an additional dead zone. The second circulation flow stands for the external loop in which the measurement is performed.

Similar observations have been reported by SOLOMON et al. (1981), ELSON et al. (1986), GALINDO et al. (1988), FUNAHASHI et al. (1987) and HENZLER and OBERNOSTERER (1991). SOLOMON et al. (1981) showed that a well-mixed region of high velocity liquids, called a "cavern" surrounded the impeller. The equation for estimating the radius of this cavern was given as

$$r^3 = \frac{Ne\,\rho\,N^2\,D_i^5}{2\,\pi^3\,\tau_y} \tag{75}$$

with τ_y as the yield stress of the fluid. ELSON et al. (1986) using X-rays to measure the cavern size suggested a geometric model similar to a cylinder. GALINDO et al. (1988) reported about such caverns in stirred vessels containing xanthan solutions. Incorporating the first observation of fluid movement at the wall, an experimentally determined cavern size at $r = D_T/2$ was compared with the predicted value of Eq. (75). Reasonable agreement was found for non-aerated and aerated (Ne_G instead of Ne in Eq. (75)) fluids.

In a systematic study on mixing and mass transfer in xanthan solutions, HENZLER and OBERNOSTERER (1991) identified three different zones in the stirred vessel: a mixing volume V_M, a poorly mixed volume V_{end}, and a dead volume V_{dead}. To quantitatively determine the size of these zones, they applied an interesting temperature equalization method. From systematic studies with two fluids (PAA and xanthan), and different tank to impeller diameter ratios, they were able to illustrate that mixing can be improved by increasing aeration rates at lower impeller speeds. At higher speeds, the mixing intensity is impaired. The stronger elastic properties of PAA relative to xanthan, resulted in comparatively smaller mixing volumes. An improved correlation for $k_L a$ was achieved by replacing the volume in P/V_L and the cross-section for superficial gas velocity by the geometric properties of the well-mixed region (Fig. 17). Finally, some interesting conclusions regarding the scale of operation have been made. By using the Reynolds number for a correlation of the mixing and dead volume, HENZLER and OBERNOSTERER (1991) showed that because the Reynolds number increases with increasing scale of operation, the problem of the dead regions should be less pronounced at the larger scale.

4 Scale-Up Tools

The enormous number of papers, reviews and chapters in books dealing with the scale-up problems of bioreactors suggest both the importance of the problem as well as the difficultities in finding simple and approved solutions. A comprehensive overview of the available tools, and critical considerations of the state of the art, can be found in the reviews of WANG et al. (1979), KOSSEN and OOSTERHUIS (1985), and CHARLES (1985a, b). The complexity and diversity of bioprocesses limit the flexibility of strategies which may be applied to practical scale-up.

Much has been written about the principle of "similarity" for performing scale-up. This concept, however, can seldom be directly applied, because the different and important similarity states (geometric, kinematic, and dynamic) are virtually impossible to maintain when going from laboratory- to large-scale operation.

The most common scale-up rules are still based on attempts to maintain one or the other performance indices (see Tab. 1) constant at different scales. Most of them are more or less closely related to the oxygen supply of the biosuspension. A scale-up founded on one single operating condition will have diverse effects on the other operating parameters. The drastic effects are usually documented with the help of Tab. 5 from the famous paper of OLDSHUE (1966) (see also UHL and VON ESSEN, 1987).

The process of scale-up, therefore, involves the following requirements: (1) selection of the

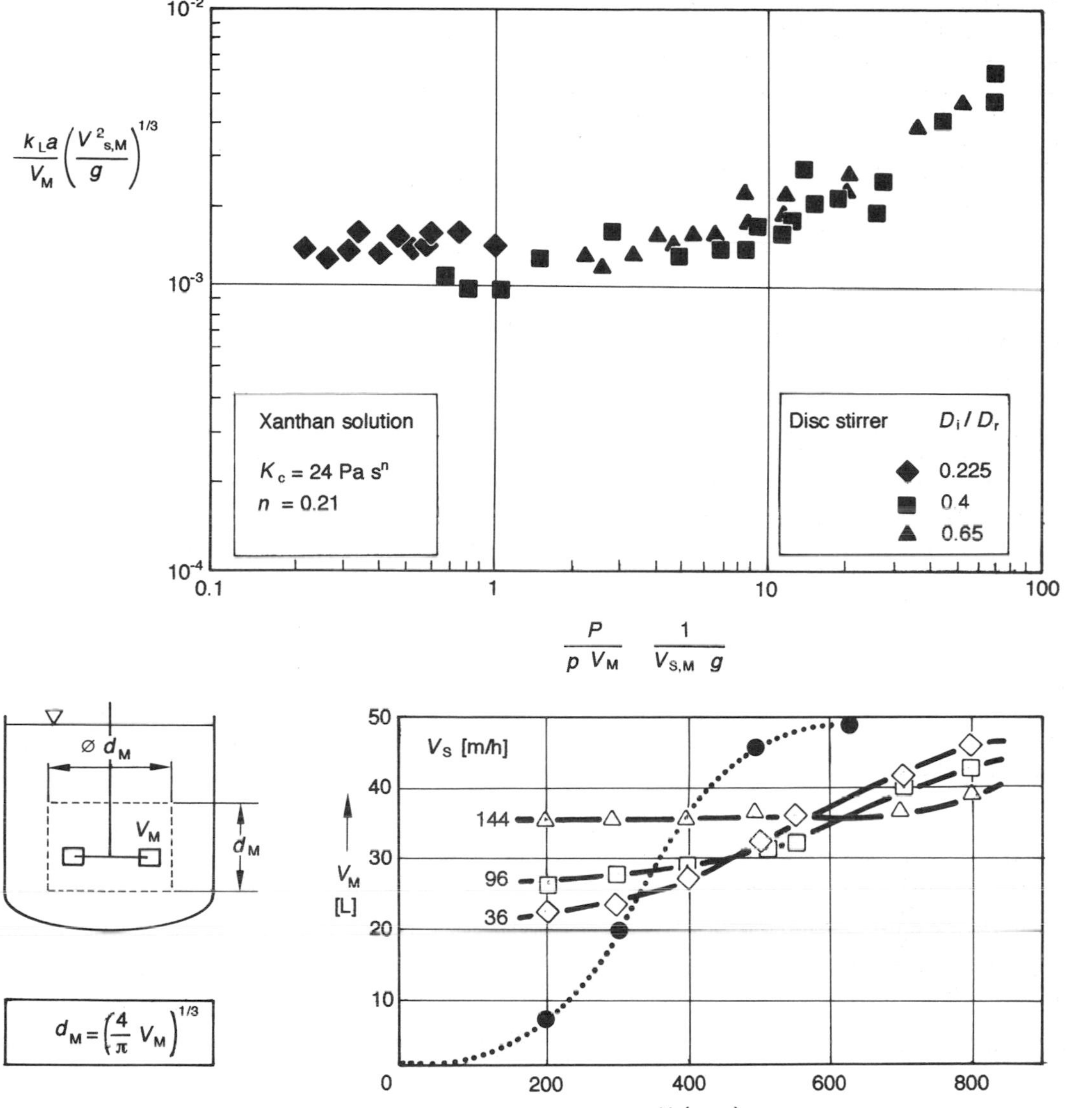

Fig. 17. Sorption characteristics formed with mixing volume V_M and mixing area = $\pi/4\ d_M^2$ (reprinted from HENZLER and OBERNOSTERER, 1991, with permission).

performance index, (2) choice of an appropriate equation for estimating the operating conditions and guaranteeing maintenance of the selected property at different scales, and (3) estimation of the consequences on other key variables with the help of suitable translation equations. Obviously, this is not a straightforward approach but rather an iterative process for which a prerequisite is an investigator with a good feeling for the physical nature of the processes involved and laboratory experience. Of particular importance is the advice to perform scale-down experiments at intervals during the scale-up process (KOSSEN

Tab. 5. Effects of Different Criteria in the Linear Scaling-Up by a Factor of 5 (volume scaling factor of 125), (OLDSHUE, 1966)

	Laboratory Scale	Scale-Up Criterion			*Re*
		P_G/V_L	N	N_D	
Diameter D_T	1	5	5	5	5
Specific power input P_G/V_L	1	1	25	0.2	$1.6 \cdot 10^{-3}$
Power input P_G	1	125	3125	25	0.2
Rotational speed of stirrer N	1	0.34	1	0.2	0.04
Tip speed of stirrer $N\,D_i$	1	1.71	5	1	0.2
Reynolds number *Re*	1	8.55	25	5	1

and OOSTERHUIS, 1985; KOSSEN, 1991) to check the influence of the various interdependent effects upon the outcome of the process.

As far as step (2) in this sequence is concerned, it should be clear from the previous sections that the required and reliable correlations are unavailable. Thus, the engineer must exercise great caution.

One of the most popular methods for scaling-up is based on maintaining k_La constant. Application of one of the k_La correlations introduced in Sect. 2, could lead to the recommendation to keep P_G/V_L and V_S at constant values. There are two apparent consequences when applying this concept. First, mixing times at the large scale would significantly increase (about five-fold when scaling-up from 10 L to 10 m^3). Secondly, constancy of superficial gas velocity implies an increase in the mean gas residence time with increasing scale because

$$\tau_G = \frac{V_L}{\dot{V}_G} \sim \left(\frac{D_T^3}{D_T^2}\right)_{V_S=\text{const.}} = \frac{1}{D_T} \tag{76}$$

This obviously produces a reduction of the driving force for oxygen mass transfer from the gas into the liquid phase. Alternatively, the more common approach (vvm = constant):

$$\tau_G = \frac{V_L}{\dot{V}_G} = \text{idem} \tag{77}$$

leads to an increase of the superficial gas velocity, implying a more difficult dispersion problem for the impeller (NIENOW et al., 1977).

Irrespective of these interdependent and unalterable effects, it is worthwhile to guide the selection process of the key operation parameters with some basic rules of thumb. A very valuable aid for weighing one mechanism against the other, and estimating the outcome of scale-up effects, is the concept of characteristic times. This topic is only shortly touched upon in this chapter. For a more complete description and impressive examples of application, the interested reader is referred to KOSSEN and OOSTERHUIS (1985) and SWEERE et al. (1988a, b).

Time constants, or more generally speaking, characteristic times of a particular process, may be derived from differential equations, rules of thumb, and the ratio of capacity/rate.

The characteristic times necessary to compare the importance of the two processes, oxygen mass transfer and mixing, are

- the characteristic time for oxygen consumption ($c_L{}^\circ/r_{O_2}$)
- the characteristic time for mass transfer ($1/k_La$)
- the circulation time (θ).

For example, if the characteristic times for oxygen consumption and transfer are of the same order of magnitude, oxygen limitation may occur. Oxygen gradients are likely to occur if circulation time is of the same order of magnitude.

Application of this analysis regime is extremely useful to estimate the order of magnitude of the respective condition at full scale.

Because estimation of the characteristic times in most practical cases rests upon empirical correlations (k_La, circulation time), the success of the approach also depends on the quality of these correlations. These are, however, oftentimes questionable.

5 Outlook

After almost 50 years of research on gas dispersion and mass transfer gas–liquid in sparged agitated vessels, we are beginning to understand the real complexity of these processes and to learn something about the reasons why dreams for small compact catch-all correlations for satisfying scale-up of these processes will probably never be fulfilled. Even if encouraging progress is made, there is still a long way toward significant success regarding reliable solution to scale-up tasks. This is particularly the case, if we step beyond the generic aspects of mixing and mass transfer and try to attack the more specific problems related to bioprocessing.

Although there are several directions from which important contributions can be expected, it is the firm belief of the author that this complex area is particularly animated by

1. ongoing basic engineering research on the fundamentals of gas dispersion and mass transfer, when real fermentation fluids are taken into account,
2. improvements in modeling by applying multiphase turbulence or compartment models, as described by REUSS and BAJPAI (1991).

Another important contribution can be expected from the reduction of the various sources of error in measuring the important parameters, such as k_La and mixing times. This would significantly contribute to reduce uncertainties and confusion, and increase the reliability of the correlations.

As a final note, it should be stressed that engineering methods for tackling the problems of mass transfer and mixing should in no way be restricted to analysis of data in terms of correlating and modeling for the already existing design. An ingenious solution will also include suggestions for change of the design (type of impeller, geometric properties) to overcome such problems as spatial variations in the concentrations or limitations in oxygen supply. Improved performance of fermentation processes, following replacement of traditional Rushton turbines in stirred tank fermenters, has recently been reported (BALMER et al., 1987; GBEWONYO, et al., 1987; NIENOW, 1990). More systematic research should assure further improvements in the design of basic innovations. At the same time, it should be guaranteed that characterization of these new, hopefully better impellers does not require another 50 years of research.

6 References

AIBA, S., HUMPHREY, A. E., MILLIS, N. (1973), *Biochemical Engineering*, 2nd Ed., Tokyo: University of Tokyo Press.

ATKINSON, B. (1974), *Biochemical Reactors*, London: Pion, Ltd.

BAILEY, J. E., OLLIS, D. F. (1986), *Biochemical Engineering Fundamentals*, 2nd Ed., New York: McGraw-Hill.

BALMER, G. J., MOORE, I. P. T., NIENOW, A. W. (1987), Aerated and unaerated power and mass transfer characteristits, in: *Biotechnology Processes, Scale-up and mixing* (HO, C. S., OLDSHUE, J. Y., Eds.), pp. 116–127, New York: American Institute of Chemical Engineers.

BLANCH, H. W., BHAVARAJU, S. M. (1976), Bioengineering report: Non-Newtonian fermentation broths: Rheology and mass transfer, *Biotechnol. Bioeng.* **18**, 745–790.

BRAUER, H. (1971), *Stoffaustausch*, Aarau–Frankfurt am Main: Verlag Sauerländer.

BRAUER, H. (1985), Transport processes through the interface of particles, in: *Biotechnology* (REHM, H.-J., REED, G., Eds.), Vol. 2, pp. 77–111, Weinheim–Deerfield Beach/Florida–Basel: VCH Verlagsgesellschaft.

BRYANT, J., SADEGHZADEH (1979), Circulation rates in stirred and aerated tanks, *Proc. 3rd Eur. Conf. on Mixing*, 4–6 April, York, Vol. I, pp. 325–336.

CALDERBANK, P. H. (1958), Physical rate processes in industrial fermentation, Part I: The interfacial area in gas-liquid contacting with mechanical agitation, *Trans. Inst. Chem. Eng.* **36**, 443–463.

CALDERBANK, P. H., MOO-YOUNG, M. (1961), The power characteristics of agitators for the mixing of Newtonian and non-Newtonian fluids, *Trans. Inst. Chem. Eng.* **39**, 337.

CHANDRASEKHARAN, K., CALDERBANK, P. H. (1981), Further observations on the scale-up of aerated mixing vessels, *Chem. Eng. Sci.* **36**, 819–823.

CHAPMAN, C. M., GIBILARO, L. G., NIENOW, A. W. (1982), A dynamic response technique for the estimation of gas–liquid mass transfer coefficients in a stirred vessel, *Chem. Eng. Sci.* **37**, 891–896.

CHARLES, M. (1978), Technical aspects of the rheological properties of microbial cultures, *Adv. Biochem. Eng.* **8**, 1–61.

CHARLES, M. (1985a), Fermenter design and scale-up, in: *Comprehensive Biotechnology* (MOO-YOUNG, M., Ed.), Vol. 2, pp. 57–75, Oxford–New York–Toronto–Sydney–Frankfurt: Pergamon Press.

CHARLES, M. (1985b), Fermentation scale-up: problems and possibilities, *Trends Biotechnol.* **3**, 134–139.

COOKE, M., MIDDLETON, J. C. (1988), Mixing and mass transfer in filamentous fermentations, *2nd Int. Conf. on Bioreactor Fluid Dynamics* (KING, R., Ed.), BHRA Fluid Engineering Centre, London, New York: Elsevier Appl. Sci. Publ.

COOPER, C. M., FERNSTROM, G. A., MILLER, S. A. (1944), Gas-liquid contactors, *Ind. Eng. Chem.* **36**, 504–509.

DANG, N. D. P., KARRER, D. A., DUNN, I. J. (1977), Oxygen transfer coefficients by dynamic model moment analysis, *Biotechnol. Bioeng.* **19**, 853–865.

DUNN, I. J., EINSELE, A. (1975), Oxygen transfer coefficients by the dynamic method, *J. Appl. Chem. Biotechnol.* **25**, 707–720.

EINSELE, A. (1978), Scaling up bioreactors, *Process Biochem.*, July, 13.

ELSON, T. P., CHEESMAN, D. J., NIENOW, A. W. (1986), X-ray studies of cavern sizes and mixing performance with fluids processing a yield stress, *Chem. Eng. Sci.* **41**, 2555–2562.

FUKUDA, H., SUMINO, Y., KAUSAKI, T. (1968), Scale-up of fermentors. I. Modified equations for volumetric oxygen transfer coefficient, *J. Ferment. Technol.* **46**, 829–837.

FUNAHASHI, H., HARADA, H., TAGUCHI, H., YOSHIDA, T. (1987), Circulation time distribution and volume of mixing regions in highly viscous xanthan gum solution in a stirred vessel, *J. Chem. Eng. Jpn.* **20**, 277–282.

GALINDO, E., NIENOW, A. W., BADHAM, R. S. (1988), Mixing of simulated xanthan gum broths, *2nd Int. Conf. on Bioreactor Fluid Dynamics* (KING, R., Ed.), pp. 65–78, BHRA Fluid Engineering Centre, London, New York: Elsevier Appl. Sci. Publ.

GBEWONYO, K., DIMASI, D., BUCKLAND, B. C. (1987), Characterization of oxygen transfer and power absorption of hydrofoil impellers in viscous mycelial fermentations, in: *Biotechnology Processes, Scale-up and Mixing* (HO, C. S., OLDSHUE, J. Y., Eds.), pp. 128–134, New York: American Institute of Chemical Engineers.

GIBILARO, L. G., DAVIES, S. N., COOKE, M., LYNCH, P. M., MIDDLETON, J. C. (1985), Initial response analysis of mass transfer in a gas sparged stirred vessel, *Chem. Eng. Sci.* **40**, 1811–1816.

HENZLER, H.-J. (1982), Verfahrenstechnische Auslegungsunterlagen für Rührbehälter als Fermenter, *Chem. Ing. Tech.* **54**, 461–476.

HENZLER, H.-J., KAULING, J. (1985), Scale-up of mass transfer in highly viscous liquids, *Proc. 5th Eur. Conf. on Mixing*, Würzburg, Germany.

HENZLER, H.-J., OBERNOSTERER, G. (1991), Effect of mixing behavior on gas-liquid mass transfer in highly viscous, stirred non-Newtonian liquids, *Chem. Eng. Technol.* **14**, 1–10.

HERBST, H., SCHUMPE, A., DECKWER, W.-D. (1992), Xanthan production in stirred tank fermentors: oxygen transfer and scale-up, *Chem. Eng. Technol.* **15**, 425–434.

HIBY, J. W. (1981), Definition and measurement of the degree of mixing in liquid mixtures, *Int. Chem. Eng.* **21**, 197.

HICKMAN, A. D., NIENOW, A. W. (1986), Mass transfer and hold up in an agitated simulated fermentation broth as a function of viscosity, *Proc. Int. Conf. on Bioreactor Fluid Dynamics*, Cambridge, April 1986, pp. 301–316, BHRA Fluid Engineering Centre, Cranfield.

HOECKER, H., LANGER, G., WERNER, U. (1980), Gas-Flüssigkeit-Stoffaustausch in gerührten newtonschen und nicht-newtonschen Flüssigkeiten, *Chem. Ing. Tech.* **52**, 752.

JANSEN, H., SLOT, S., GÜRTLER, H. (1978), Determination of mixing times in large scale fermenters using radioactive isotopes, *Preprints First Eur. Congr. on Biotechnology*, Part II, pp. 80–83, poster papers, Frankfurt: DECHEMA.

JUDAT, H. (1982), Stoffaustausch Gas/Flüssigkeit im Rührkessel – eine kritische Bestandsaufnahme, *Chem. Ing. Tech.* **54**, 520–521.

JURECIC, R., BEROVIC, M., STEINER, W., KOLOINI, T. (1984), Mass transfer in aerated fermentation broths in a stirred tank reactor, *Can. J. Chem. Eng.* **62**, 334–339.

KARGI, F., MOO-YOUNG, M. (1985), Transport phenomena in bioprocesses, in: *Comprehensive Biotechnology* (MOO-YOUNG, M., Ed.), Vol. 2, pp. 5–56, Oxford–New York–Toronto–Sydney–Frankfurt: Pergamon Press.

KAWASE, Y., MOO-YOUNG, M. (1988), Volumetric mass transfer coefficients in aerated stirred tank reactors with Newtonian and non-Newtonian media, *Chem. Eng. Res. Des., Trans. Inst. Chem. Eng.* **66**, 284–288.

KEITEL, G., ONKEN, U. (1981), Inhibition of bubble coalescence by solutes in air/water dispersions, *Chem. Eng. Sci.* **37**, 1635–1638.

KHANG, S. J., FITZGERALD, T. J. (1975), A new probe and circuit for measuring electrolyte conductivity, *Ind. Eng. Chem. Fund.* **14**, 208–213.

KHANG, S. J., LEVENSPIEL, O. (1976), New scale-up and design method for stirred agitated batch mixing vessels, *Chem. Eng. Sci.* **31**, 569–577.

KIPKE, K. (1978), Verfahrenstechnische Fortschritte beim Mischen, Dispergieren und bei der Wärmeübertragung in Flüssigkeiten, *Preprints*, pp. 21–36, VDI (GVC), Düsseldorf.

KOSSEN, N. W. F. (1992), Scale-up in biotechnology, in: *Recent Advances in Industrial Application of Biotechnology* (VARDAR-SUKA, F., SUKA SUKAN, S., Eds.), pp. 147–182, Dordrecht–Boston–London: Kluwer Academic Publishers.

KOSSEN, N. W. F., OOSTERHUIS, N. M. G. (1985), Modelling and scaling-up of bioreactors, in: *Biotechnology* (REHM, H.-J., REED, G., Eds.), Vol. 2, pp. 572–605, Weinheim–Deerfield, Beach/Florida–Basel: VCH Verlagsgesellschaft.

KRISTIANSEN, B., MCNEIL, B. (1987), The design of a tubular loop reactor for scale-up and scale-down of fermentation processes, in: *Bioreactors and Biotransformations* (MOODY, G. W., MOO-YOUNG, M., Eds.), pp. 321–334, London–New York: Elsevier Appl. Sci. Publ.

LARSSON, G., ENFORS, S.-O. (1988), Kinetics of microbial response to insufficient mixing in bioreactors, *6th Eur. Congr. on Mixing*, Pavia, Italy, pp. 443–450.

LESSARD, R. R., ZIEMINSKI, S. A. (1971), Bubble coalescence and gas transfer in aqueous electrolytic solutions, *Ind. Eng. Chem. Fund.* **10**, 260–269.

LINEK, V., VACEK, V. (1981), Chemical engineering use of catalyzed sulfite oxidation kinetics for the determination of mass transfer characteristics of gas-liquid contactors, *Chem. Eng. Sci.* **36**, 1747–1768.

LINEK, V., VACEK, V. (1982), Consistency of steady state and dynamic methods of k_La determination in gas-liquid dispersions, *Chem. Eng. Sci.* **37**, 1425–1425.

LINEK, V., BENES, P., VACEK, V., HOVORKA, F. (1982), Analysis of differences in k_La-values determined by steady-state and dynamic methods in stirred tanks, *Chem. Eng. J.* **25**, 77–88.

LINEK, V., BENES, P., HOLECEK, O. (1988), Correlation for volumetric mass transfer coefficient in mechanically agitated aerated vessel for oxygen absorption in aqueous electrolyte solutions, *Biotechnol. Bioeng.* **32**, 482–490.

MARGARITIS, A., PACE, G. W. (1985), Microbial polysaccharides, in: *Comprehensive Biotechnology* (MOO-YOUNG, M., Ed.), Vol. 3, pp. 1006–1044, New York: Pergamon Press.

MARGARITIS, A., ZAJIC, J. E. (1978), Mixing, mass transfer and scale-up of polysaccharide fermentations, *Biotechnol. Bioeng.* **20**, 939–1001.

MASHELKAR, R. A. (1973), Drag reduction in rotational flows, *AIChE J.* **19**, 382–384.

MERSMANN, A., EINENKEL, W.-D., KAEPEL, M. (1975), Auslegung und Maßstabsvergrößerung von Rührapparaten, *Chem. Ing. Tech.* **47**, 953–964.

METZ, B., KOSSEN, N. W. F., SUIJDAM, J. C. (1979), The rheology of mould suspensions, *Adv. Biochem. Eng.* **11**, 104–156.

METZNER, A. B., OTTO, R. E. (1957), Agitation of non-Newtonian fluids, *AIChE J.* **3**, 3–11.

MIDDLETON, J. C. (1985), Gas-liquid dispersion and mixing, in: *Mixing in the Process Industries* (HARNBY, N., EDWARDS, M. F., NIENOW, A. W., Eds.), pp. 322–355, London: Butterworth.

MIURA, Y. (1976), Transfer of oxygen and scale-up in submerged aerobic fermentation, *Adv. Biochem. Eng.* **4**, 1–40.

MOO-YOUNG, M., BLANCH, H. W. (1981), Design of biochemical reactors, mass transfer criteria for simple and complex systems, *Adv. Biochem. Eng.* **19**, 1–69.

MUKATAKA, S., KATAOKA, H., TAKAHASHI, J. (1980), Effects of vessel size and rheological properties of suspensions on the distribution of circulation times in stirred vessels, *J. Ferment. Technol.* **58**, 155–161.

NAGATA, S. (1975), *Mixing Principles and Applications*, New York: John Wiley & Sons.

NAKANO, S. (1976), *M. Eng. Thesis*, Kyoto University, Kyoto, Japan.

NIEBELSCHÜTZ, H. (1982), Sauerstoffübergang Gas/Flüssigkeit in gerührten Bioreaktoren, *Dr.-Ing. Thesis*, Technische Universität Berlin, Germany.

NIENOW, A. W. (1984), Mixing studies on high viscosity fermentation processes – Xanthan gums, *The World Biotechnology Report*, Vol. I, *Europe*, pp. 293–304, London: Online.

NIENOW, A. W. (1990), Agitation for mycelial fermentations, *Trends Biotechnol.* **8**, 224–233.

NIENOW, A. W., ULBRECHT, J. J. (1985), Gas-liquid mixing and mass transfer in high viscosity liquids, in: *Mixing of Liquids by Mechanical Agitation* (ULBRECHT, J. J., PATTERSON, G. K., Eds.), Chap. 6, New York: Gordon & Breach.

NIENOW, A. W., WISDOM, D. J., MIDDLETON, J. C. (1977), The effect of scale and geometry on flooding, recirculation, and power in gassed stirred vessels, *2nd Eur. Conf. on Mixing*, Cambridge, BHRA Fluid Engineering Centre, London, F1-1-F1-16, New York: Elsevier.

NIENOW, A. W., HUOXING, L., HAOZHUNG, W., ALLSFORD, K. V., CRONIN, D., HUDCOVA, V. (1988), The use of large ring spargers to improve the performance of fermentors agitated by single and multiple standard Rushton turbines, *2nd Int. Conf. on Bioreactor Fluid Dynamics* (KING, R., Ed.), pp. 159–177, BHRA Fluid Engineering Centre, London, New York: Elsevier Appl. Sci. Publ.

NISHIKAWA, M., NAKAMURA, M., YAGI, H., HASHIMOTO, K. (1981a), Gas absorption in aerated mixing vessels, *J. Chem. Eng. Jpn.* **14**, 219–226.

NISHIKAWA, M., NAKAMURA, M., HASHIMOTO, K. (1981b), Gas absorption in aerated mixing vessels with non-Newtonian liquid, *J. Chem. Eng. Jpn.* **14**, 227–232.

OLDSHUE, J. Y. (1966), Fermentation mixing scale-up techniques, *Biotechnol. Bioeng.* **8**, 3–24.

OOSTERHUIS, N. M. G., KOSSEN, N. W. F., OLIVER, A. P. C., SCHENK, E. S. (1985), Scale-down and optimization studies of the gluconic acid fermentation by *Gluconobacter oxydans, Biotechnol. Bioeng.* **27**, 711–720.

PACE, G. W. (1980), Rheology of mycelial fermentation broths, in: *Fungal Biotechnology*, The British Mycological Soc. Symp. Ser. No. 3 (SMITH, J. E., BERRY, D. R., KRISTIANSEN, B., Eds.), pp. 95–110, London–New York: Academic Press.

PACE, G. W., RIGHELATO, R. C. (1980), Production of extracellular microbial polysaccharides, *Adv. Biochem. Eng.* **15**, 41–70.

PANDIT, A. B., JOSHI, J. B. (1983), Mixing in mechanically agitated gas-liquid contactors, bubble columns and modified bubble columns, *Chem. Eng. Sci.* **38**, 1189–1215.

PEREZ, J. F., SANDALL, O. C. (1974), Gas absorption by non-Newtonian fluids in agitated vessels, *AIChE J.* **20**, 770–775.

POPOVIC, M., NIEBELSCHÜTZ, H., REUSS, M. (1979), Oxygen solubilities in fermentation fluids, *Eur. J. Appl. Microbiol. Biotechnol.* **8**, 1–15.

POPOVIC, M., PAPALEXIOU, A., REUSS, M. (1983), Gas residence time distribution in stirred tank bioreactors, *Chem. Eng. Sci.* **38**, 2015–2025.

PREST, W. M., PORTER, R. S., O'REILLY, J. W. (1970), Non-Newtonian flow and the steady state shear compliance, *J. Appl. Polym. Sci.* **14**, 2697–2706.

RANADE, V. R., ULBRECHT, J. J. (1978), Influence of polymer additives on the gas-liquid mass transfer in stirred tanks, *AIChE J.* **24**, 796–809.

REITH, T., BEEK, W. J. (1973), The oxidation of aqueous sodium sulphite solutions, *Chem. Eng. Sci.* **28**, 1331–1339.

REUSS, M. (1982), Biochemical engineering: Physical properties of biotechnological process materials, *VDI Fortschritte der Verfahrenstechnik – Progress in Chemical Engineering*, 20/F, pp. 519–536, Düsseldorf: VDI Verlag.

REUSS, M., BAJPAI, R. K. (1991), Stirred tank models, in: *Biotechnology 2nd Ed.* (REHM, H.-J., REED, G., PÜHLER, A., STADLER, P.), Vol. 4, pp. 300–348, Weinheim–Deerfield Beach/Florida–Basel: VCH Verlagsgesellschaft.

REUSS, M., DEBUS, D., ZOLL, G. (1982), Rheological properties of fermentation fluids, *Chem. Eng.* **381**, 233–236.

REUSS, M., FRÖHLICH, S., KRAMER, B., MESSERSCHMIDT, K., POMMERENING, G. (1986), Coupling of microbial kinetics and oxygen transfer for analysis and optimization of gluconic acid production with *Aspergillus niger, Bioprocess Eng.* **1**, 79–91.

RICHARDS, J. W. (1961), Studies in aeration and agitation, *Prog. Ind. Microbiol.* **3**, 143–172.

RIESMEIER, B. (1984), Einfluß der Begasung auf das Rezirkulationsverhalten der Flüssigphase in gerührten Bioreaktoren, *Diploma Thesis*, Technische Universität Berlin, Germany.

ROBINSON, C. W., WILKE, C. R. (1972), Simulating fermentation media, in: *Fermentation Technology Today, Proc. IVth Int. Ferm. Symp.* (TERUI, G., Ed.), pp. 73–82, Society of Fermentation Technology Japan.

ROBINSON, C. W., WILKE, C. R. (1973), Simultaneous measurement of interfacial area and oxygen absorption in stirred tanks: A correlation for ionic strength effects, *Biotechnol. Bioeng.* **15**, 755–782.

ROBINSON, C. W., WILKE, C. R. (1974), Simultaneous measurement of interfacial area and mass transfer coefficients for a well-mixed gas dispersion in aqueous electrolyte solutions, *AIChE J.* **20**, 285–294.

RUCHTI, G., DUNN, I. J., BOURNE, J. R. (1985), Practical guidelines for the determination of oxygen transfer coefficients (k_La) with the sulfite oxidation method, *Chem. Eng. J.* **30**, 29–38.

RYU, D. Y., HUMPHREY, A. E. (1972), A reassessment of oxygen transfer rates in antibiotics fermentation, *J. Ferment. Technol.* **50**, 424–431.

SCHUMPE, A., QUICKE, G., DECKWER, W.-D. (1982), Gas solubilities in microbial media, *Adv. Biochem. Eng.* **24**, 1–38.

SMITH, J. M. (1985), Dispersion of gases in liquids, in: *Mixing of Liquids by Mechanical Agitation*

(ULBRECHT, J. J., PATTERSON, G. K., Eds.), pp. 139–202, New York: Gordon & Breach.

SMITH, J. M., WARMOESKERKEN, M. M. C. G. (1985), The dispersion of gases in liquids with turbines, *Proc. 5th Eur. Mixing Conf.*, Würzburg, Germany, BHRA Fluid Engineering Centre London, Paper 13, pp. 115–126.

SOLOMON, J., ELSON, T. R., NIENOW, A. W., PACE, G. W. (1981), Cavern sizes in agitated fluids with a yield stress, *Chem. Eng. Commun.* **11**, 143–164.

STEEL, R., MAXON, W. D. (1962), Some effects of turbine size on novobiocin fermentation, *Biotechnol. Bioeng.* **4**, 231–240.

STEEL, R., MAXON, W. D. (1966), Dissolved oxygen measurements in pilot- and production scale novobiocin fermentations, *Biotechnol. Bioeng.* **8**, 97–108.

STEIN, W. A. (1984), Berechnung des charakteristischen Schergefälles für begaste nicht-Newtonsche Flüssigkeiten, *Chem. Ing. Tech.* **56**, 422–423.

SWEERE, A. P. J., MESTERS, L., JANSE, L., LUYBEN, K. CH. M., KOSSEN, N. W. F. (1988a), Experimental simulation of oxygen profiles and their influence on baker's yeast production: I. One-fermentor system, *Biotechnol. Bioeng.* **31**, 567–578.

SWEERE, A. P. J., JANSE, L., LUYBEN, K. CH. M., KOSSEN, N. W. F. (1988b), Experimental simulation of oxygen profiles and their influence on baker's yeast production: II. Two-fermentor system, *Biotechnol. Bioeng.* **31**, 579–586.

TAGUCHI, H. (1971), The nature of fermentation fluids, *Adv. Biochem. Eng.* **1**, 1–30.

TATTERSON, G. B. (1991), *Fluid Mixing and Gas Dispersion in Agitated Tanks*, New York: McGraw Hill, Inc.

TRÄGER, M. (1990), Einfluß oszillierender Sauerstoffkonzentrationen auf die Gluconsäurefermentation in Hinblick auf die Maßstabsvergrößerung, *Fortschrittsberichte VDI*, Reihe 17: Biotechnik, Nr. 60, pp. 1–161, Düsseldorf: VDI-Verlag.

UHL, V. W., GRAY, J. B. (1966), *Mixing*, Vols. 1 and 2, New York: Academic Press.

UHL, V. W., VON ESSEN, J. A. (1987), Scale-up of fluid mixing equipment, in: *Biotechnology Processes, Scale-up and Mixing* (HO, C. S., OLDSHUE, J. Y., Eds.), pp. 155–167, New York: American Institute of Chemical Engineers.

ULBRECHT, J. (1974), Mixing of visco-elastic fluids by mechanical agitation, *Chem. Eng.* **286**, 347–353.

VAN'T RIET, K. (1979), Review of measuring methods and results in nonviscous gas-liquid mass transfer in stirred vessels, *Ind. Eng. Chem. Proc. Des. Dev.* **18**, 357–363.

VAN'T RIET, K., TRAMPER, J. (1991), *Basic Bioreactor Design*, New York–Basel–Hongkong: Marcel Dekker, Inc.

VARDAR, F., LILLY, M. D. (1982), Effect of cycling dissolved oxygen concentrations on product formation in penicillin fermentation, *Eur. J. Appl. Microbiol. Biotechnol.* **14**, 203–211.

VONCKEN, R. M., HOLMES, D. B., DEN HARTOG, H. W. (1964), Fluid flow in turbine-stirred, baffled tanks-II. Dispersion during circulation, *Chem. Eng. Sci.* **19**, 209–213.

WANG, D. I. C., COONEY, C. L., DEMAIN, A. L., DUNNILL, P., HUMPHREY, A. E., LILLY, M. D. (1979), Aeration and agitation, Chapter 9, in: *Fermentation and Enzyme Technology*, pp. 157–193, New York–Chichester–Brisbane–Toronto: John Wiley & Sons.

WARMOESKERKEN, M. M. C. G., SMITH, J. M. (1988), Impeller loading in multi-turbine vessels, *2nd Int. Conf. Bioreactor Fluid Dynamics*, pp. 179–197, BHRA Fluid Engineering Centre London, New York: Elsevier Appl. Sci. Publ.

WICHTERLE, K., KADLEC, M., ZAK, L., MITSCHKA, P. (1984), *Chem. Eng. Commun.* **26**, 25–32.

YAGI, H., YOSHIDA, F. (1975), Gas absorption by Newtonian and non-Newtonian fluids in sparged agitated vessels, *Ind. Eng. Chem. Proc. Des. Dev.* **14**, 488–493.

ZLOKARNIK, M. (1978), Sorption characteristics for gas-liquid contacting in mixing vessels, *Adv. Biochem. Eng.* **8**, 133–151.

ZLOKARNIK, M. (1990), Trends and needs in bioprocess engineering, *Chem. Eng. Prog.*, April, 62–67.

11 Aeration in Cell Culture Bioreactors

JOHN G. AUNINS

Rahway, NJ, 07065, USA

HANS-JÜRGEN HENZLER

Wuppertal, Federal Republic of Germany

List of Symbols

A	absolute mass transfer surface area
A_s	cross-sectional area of sparged compartment
A_u	upriser cross-sectional area
A_v	surface area of screen cage
a	specific interfacial surface area
a_m	specific membrane interfacial area
a_s	specific interfacial area in the sparged compartment
C^*	liquid phase dissolved gas concentration in equilibrium with the gas phase
C_m^*	membrane phase dissolved gas concentration in equilibrium with the gas phase
C_R^*	saturation concentration in the reactor in Eq. (34)
C_L	dissolved gas concentration in the liquid
C_m	microcarrier concentration
D	reactor diameter
D_{max}	maximum shake flask diameter
D_L	draught tube diameter
d	impeller diameter
d_1	inside diameter of tubular membrane
d_2	outside diameter of tubular membrane
d_b	bubble diameter
$d_{b,max}$	maximum stable bubble diameter
d_h	tubing cage diameter
d_L	diameter of sparger orifices
d_m	microcarrier diameter
d_p	particle diameter
d_s	screen diameter for indirect aeration
d^*	transition diameter for orifice spargers, Eq. (47)
d'	stretched diameter of tubular membrane
E	amplitude of Vibromixer® oscillations
$E(d_b)$	liquid aerosolization rate
E_p	activation energy for permeation
E_s	surface elasticity
E_Y	Young's modulus for elastic tubing
e	eccentricity (throw)
F	convective exchange flow
Fr	Froude number for orifice sparger ($=v_L^2/g\,d_L$)
Fr_i	impeller Froude number ($=n^2\,d/g$)
f	frequency of Vibromixer® oscillations
Δf	coalescence frequency per bubble rise distance
G	mass flow gas transfer rate
g	acceleration due to gravity
H	liquid depth
Hy	Henry's law constant for the liquid–gas system ($=P_{gas}/C^*$)
Hy_m	Henry's law constant for the membrane–gas system ($=P_{gas}/C_m^*$)
h	impeller depth below liquid surface
h_h	tubing cage height
k	overall mass transfer coefficient
k_d	cell death rate constant
k_d^*	modified cell death rate constant

k_L	liquid phase mass transfer coefficient
k_m	membrane phase mass transfer coefficient
k_v	screen mass transfer coefficient
$k_L a_c$	airlift loop reactor calculated mass transfer coefficient
L	membrane tubing length, unstressed/unstretched
L'	membrane tubing length, stressed/stretched
M	solvent molecular weight
M_i	solute molecular weight
M_w	average molecular weight
Ne	Newton number ($=P/\rho\, n^3\, d^5$)
n	rotation speed
P	impeller power input
P_{adm}	admissible tubing transmembrane pressure
P_{gas}	gas partial pressure
P_o	impeller power input in the absence of gassing
$(P)_R$	pressure in the reactor in Eq. (34)
P_{O_2}	O_2 partial pressure in the gas or membrane phase
Δp	transmembrane pressure
q	gas flow rate
q_L	liquid flow rate
q_{O_2}	cell specific oxygen consumption rate
R	ideal gas constant
Re_i	impeller Reynolds number ($=n\, d^2/\upsilon$)
Re_t	tubing Reynolds number ($=u\, d_t/\upsilon$)
r_p	pore radius
Sc	Schmidt number ($\upsilon/\mathscr{D}$)
Sh	Sherwood number ($=k_L\, d_{char}/\mathscr{D}$)
s	spacing between impeller and membrane ($=(d_h-d)/2$)
T	temperature Kelvin
t	time
u	characteristic or maximum liquid velocity
u_i	agitator tip speed ($=\pi\, n\, d$)
u_{screen}	screen velocity ($=\pi\, n_{screen}\, d_{screen}$)
$\sqrt{\overline{u'^2}}$	root mean square velocity of turbulent fluctuations
V	reactor volume
V_A	solute volume at the normal boiling point
V_f	liquid volume in shaker flask
V_s	sparged compartment volume
v	superficial gas velocity for the reactor cross-section
v_b	bubble rise velocity
v_L	superficial gas velocity through the sparger orifice
v_s	superficial gas velocity in the sparged compartment
We	Weber number for orifice sparger ($=v_L^2\, d_L\, \rho/\sigma$)
We'_{crit}	critical modified Weber number for sparger orifice ($=v_L^2\, d_L\, \rho_g/\sigma$)
w	tubing pitch or spacing per turn
X	cell concentration
X_s	surface cell concentration (cells/area)
Δx	diffusion layer thickness
Y	sorption coefficient ($=k_L\, a(\upsilon^2/g)^{1/3}$)
z	number of tubes or holes

ε	specific energy dissipation rate ($=P/\rho\, V$)
ϕ	association parameter
ϕ_p	volume fraction of solids
η	liquid dynamic viscosity ($=\rho\, \upsilon$)
η_g	gas dynamic viscosity ($=\rho_g\, \upsilon_g$)
ϑ	temperature Celsius
λ	Kolmogorov eddy microscale
μ	cell growth rate
ρ	liquid density
ρ_g	gas density
σ	gas–liquid surface tension
σ^*	dimensionless surface tension ($=\sigma/\rho(\upsilon^4\, g)^{1/3}$)
σ_{adm}	admissible tubing pressure stress
τ	shear stress
τ_{crit}	critical shear stress for bubble breakage
τ_m	membrane pore tortuosity
θ_m	membrane void or pore fraction
υ	liquid kinematic viscosity
$\mathscr{D}$	diffusion coefficient in the liquid phase
$\mathscr{D}_m$	diffusion coefficient in the membrane phase
$\mathscr{D}_p$	diffusion coefficient in the membrane pore
f	geometric correction factor in Eq. (28)

1 Introduction

Since the advent of recombinant DNA technology and cell fusion techniques, the use of animal and insect cell culture for production of post-translationally modified proteins has made cell culture a much more important part of the industrial realization of biotechnology. Cell culture is no longer limited to small- and medium-scale vaccine production processes of a few hundred liters vessel volume. Refinements in molecular and cell biology have resulted in cell lines productive enough to consider making kilogram quantities of proteins in reactors greater than the 10000 L scale. Unlike many recombinant microbial fermentations, a large-scale culture step can be the dominant cost in production; hence the importance of culture optimization. Given the relative fragility of many cells in culture, reactor design becomes an important issue in optimizing process economics. It is the purpose of this chapter to describe the design and scale-up of aeration for cell culture reactors, for which no comprehensive reviews exist.

Industrial cell culture in large vessels has been practiced since the 1950s, with the introduction of veterinary and human vaccines in deep tank culture systems using the transformed cell lines VERO and BHK. The use of diploid animal cell cultures in other reactor systems has also been widely seen; However, many of these reactors, e.g., roller bottles, are scaled by increasing the number of unit vessels rather than the size of the unit. Although many specialty reactors have been devised for animal cell culture, industrial large-scale reactors to date have mainly employed traditional vessel designs – the stirred tank or airlift reactor. Despite their popularity, and despite the experience level with microbial stirred tanks, relatively little published data exist on these vessels under cell culture conditions of operation and configuration. Of the data that exist, few correlations have been offered. For these reasons, and to bring forth a coherent picture of the operational characteristics of these reactors, this chapter will focus on aeration design in those reactors which have seen industrial acceptance, and are scaleable to the anticipated volumes necessary for eventual production of therapeutic and diagnostic proteins. We purposely avoid in-depth description of specialty reactors, but give a review of the general considerations from which these reactors might be designed. To a large extent, these specialty reactors are of a loop design, and are aerated by either gas-permeable membranes or stirred tank conditioning vessels. The same design correlations and criteria discussed here apply to these aeration systems as well.

2 Aeration Objectives and Constraints

2.1 Oxygen Supply

The simplest objective of cell culture aeration is to supply oxygen. Liquid film resistance dominates in cell cultures, so for a batch reactor, the mass balance reduces to:

$$\frac{G}{V} = k\,a(\Delta C) = q_{O_2}X \tag{1}$$

Since liquid mixing is usually ideal for typical scales and viscosities, and gas phase concentration change is often negligible, the concentration difference ΔC driving the process is usually given by:

$$\Delta C = C^* - C_L \quad \text{where } C^* = \frac{P_{O_2}}{Hy} \tag{2}$$

If either assumption is violated, the logarithmic concentration difference applies

$$\Delta C = \frac{(C_1^* - C_{L1}) - (C_2^* - C_{L2})}{\ln\left(\dfrac{C_1^* - C_{L1}}{C_2^* - C_{L2}}\right)} \tag{3}$$

In Eqs. (1–3), G/V is the gas mass flux per reactor volume, k the overall mass transfer coefficient, $a = A/V$ the specific interfacial area, C^* the saturation concentration, C_L the liquid phase concentration, P_{O_2} the gas phase O_2 partial pressure, and Hy the Henry's law constant. The indices 1 and 2 refer to inlet and outlet locations in the transfer apparatus.

Beyond O_2 transport lie other goals and constraints complicating aerator design. The most widely recognized of these is the need to avoid damage of the cells. Damage can be physical or chemical, as discussed below. Recently, CO_2 ventilation has been recognized to affect recombinant culture performance, and this may practically limit C^*. Culture physical properties also constrain the aeration problem to ranges of parameters which are sometimes inalterable within the physiological context of growing cells.

2.1.1 Oxygen Demand of Cultures

The O_2 requirement of animal cells is low compared to microorganisms, with $q_{O_2}X$ typically 20–200 mg/L/h for X less than 10^7 cells/mL. Oxygen demands (*OUR*) for common cell types are shown in Tab. 1; other types are cataloged extensively in several review articles (McLimans et al., 1968; Spier and Griffiths, 1984). *OUR* varies with culture conditions, and for scale-up and design, *OUR* determination under representative culture conditions is imperative.

2.1.2 Oxygen Concentration

Satisfying *OUR* is not sufficient to specify aeration conditions, as cultures are sensitive to extremes of oxygen concentration C_L. Low C_L affects protein secretion and growth, at least partially via effects on primary energy metabolism. Growth effects from hypoxia are typically below $C_L = 5\%–20\%$ of C^* for air at 1 bar. Miller et al. (1988) recorded energy pathway changes near 0.1% of C^*. These authors, and Ozturk and Palsson (1990) found that hybridoma cells can adapt to zero or near-zero C_L. O_2 tensions above 100% of C^* are detrimental to cell growth (Kilburn and Webb, 1968; Mielhoc et al., 1990), presumably by creating superoxide radicals which directly affect cells, or by oxidizing medium components (Glacken et al., 1989). Protein formation effects often occur at the same C_L as growth rate changes; Mielhoc et al. (1990) reported no effect above 5%, and Ozturk and Palsson (1990) found no effect above 1%. The C_L for maximum growth rate and protein expression rate can, however, differ widely (Miller et al., 1987; Phillips et al., 1987; Reuveny et al., 1986). Like *OUR* values, the limiting C_L should be measured under typical operating conditions.

Tab. 1. Reported Oxygen Demand for Animal Cell Types of Industrial Interest

Cell Line	Cell Type	Reference	*OUR* ($mg \cdot 10^{-8}$/cell/h)
WI-38	Human embryonic lung	CRISTOFOLO and KRITCHEVSKY, 1966	0.2
WI-38	Human embryonic lung	BALIN et al., 1976	0.5–1.6
MRC-5	Human embryonic lung	AUNINS, unpublished data	0.7
CEF	2° Chick embryo fibroblast	SINSKEY et al., 1981	0.1–0.3
BHK 21	Baby hamster kidney	RADLETT et al., 1972	0.03–0.5
VERO	Green monkey kidney	HU and OBERG, 1990	0.1–0.5
HO-323-8	Human myeloma	YAMADA et al., 1990	0.24
HB4C5	Human hybridoma	YAMADA et al., 1990	0.32
TIB 18	Murine myeloma	MIELHOC et al., 1990	0.18
r-NS/0	Murine myeloma	AUNINS, 1992	0.32–0.54
HB44	Murine hybridoma	WOHLPART et al., 1991	0.6–1.0
NB 1	Murine hybridoma	BIRCH et al., 1987	0.6–1.1
CHO	Chinese hamster ovary	GREGG et al., 1968	0.39
γIFN-CHO	Chinese hamster ovary	AUNINS et al., 1986	0.45
tPA-CHO	Chinese hamster ovary	AUNINS et al., 1986	0.48
r-CHO	Chinese hamster ovary	CROUGHAN, 1992	0.4–0.8
Sf-21	*Spodoptera frugiperda*	TRAMPER and VLAK, 1986	2.4
Sf-9	*Spodoptera frugiperda*	KAMEN et al., 1991	0.64

2.2 Carbon Dioxide Removal

Cells both produce and consume CO_2 in various reactions. CO_2 and bicarbonate ion, HCO_3^-, have optimal ranges for cells. It can be important to match oxygenation and ventilation by aeration control strategy to avoid CO_2 build-up or excessive stripping. Most media employ bicarbonate buffers, however, so CO_2 and HCO_3^- are present independently of cellular metabolism. Treatment of mass transfer for carbon dioxide is complicated by the reaction network leading to gas phase CO_2.

$$[CO_2]_{gas} \leftrightarrow [CO_2]_{liquid} \xleftrightarrow{H_2O} [H_2CO_3] \leftrightarrow [H^+] + [HCO_3^-] \tag{4}$$

Reaction enhancement of CO_2 absorption, and inhibition of desorption, can occur via reaction kinetics as well as changes in local film conditions (ISHIZAKI et al., 1971). In most culture situations, medium pH does not change rapidly, and the CO_2 evolution rate (*CER*) is slow relative to the desorption and acid–base reactions. Here, ventilation may be treated as a first-order process, utilizing Eqs. (1–3) (ROYCE and THORNHILL, 1991). Then Eq. (4) may be assumed to be at pseudo-steady state, and the liquid phase equilibrium is described by

$$pH = pK_1 + \log_{10}\left(\frac{(HCO_3^-)}{(CO_2) + (H_2CO_3)}\right) \tag{5}$$

where $pK_1 = 6.1$ at 0.15 mol/L ionic strength and $\vartheta = 37\ °C$ for human plasma (ALBERTY and DANIELS, 1980). For cell concentrations less than 10^7 cells/mL, the above assumptions hold fairly well. For this situation, CO_2 desorption speed is proportional to O_2 absorption speed by the ratio of the diffusion coefficients to the 1/2–2/3 power (see Sect. 6; CALDERBANK, 1959); this factor is approximately 0.9.

Ventilation rates and CO_2 concentrations change with scale and operating procedures due to pH control strategy, gas phase composition and flow rate, and reactor total (and hydrostatic) pressure. For caustic pH control, $NaHCO_3$ can build up in the reactor. Acid neutralization by $NaHCO_3$ or Na_2CO_3 exacerbates this problem. Upward pH control performed by letting CO_2 outgas (independent of oxygen demand) strips $NaHCO_3$. For bubble column or airlift reactors, a constant total gas flow in excess of oxygenation (ventilation) demand is often used to drive reactor mixing (BORASTON et al., 1984); this can also strip

CO_2 if the cells are producing lactic acid. For a large-scale airlift operation at constant air sparge rate, liquid phase P_{CO_2} was observed to decrease from a starting value of 0.08 bar (60 mm Hg) to 0.008 bar (6 mm Hg) at the end of the culture (BIRCH et al., 1987).

The gas O_2 content and flow rate, and the dissolved O_2 control strategy used also affect ventilation. Conceptually, sparging tiny bubbles of pure O_2 on demand to maintain C_L can maximize gas utilization, since much oxygen dissolves before the bubble reaches the top. This minimizes cell damage from bubble bursting and coalescence (see Sect. 2.4), but also minimizes ventilation of CO_2. If high or low CO_2 tensions affect a culture greatly, it may be necessary to employ sophisticated reactor control strategy to maintain dissolved oxygen, CO_2/bicarbonate and pH set points simultaneously (SMITH et al., 1990).

Industrial reactors accumulate CO_2 in excess of small cultures due to pressure. Increasing operating pressure to improve O_2 solubility or to discourage contamination increases CO_2 solubility. Hydrostatic pressure also increases solubility of CO_2 for large reactors. Although pressure effects can be remedied to some extent by decreasing the initial medium bicarbonate concentration, it is possible that the ultimate scale-limiting factor for some culture processes may be buildup of CO_2 due to these effects, rather than O_2 supply limitation.

2.2.1 Carbon Dioxide Production Rate

CO_2 evolution rates (*CER*) are scarce compared to *OUR* data. For uninfected Sf-9 insect cells, a rate of $0.7 \cdot 10^{-8}$ mg/cell/h was found by KAMEN et al. (1991) using exhaust gas analysis. Both *OUR* and *CER* increased by 30–40% for these cells after baculovirus infection. LOVRECZ et al. (1992) have recently determined a rate of $1.3 \cdot 10^{-8}$ mg/cell/h for recombinant Chinese hamster ovary (CHO) microcarrier cultures, also using exhaust gas analysis. Using a clinical blood gas analyzer, AUNINS (unpublished data) measured *CER* from 0.17 to $0.42 \cdot 10^{-8}$ mg/cell/h, for a recombinant NS/0 myeloma cell line over the course of a fed-batch culture. This measurement entailed simple respirometry using a tightly capped syringe incubated without a gas headspace; periodic injection of the sample into the gas analyzer allowed off-line measurement of both *OUR* and *CER*.

Coupling the *OUR* and *CER* measurements yields the respiratory quotient of the cells, useful in determining the extent of ventilation at a given aeration rate. Fig. 1 shows the correla-

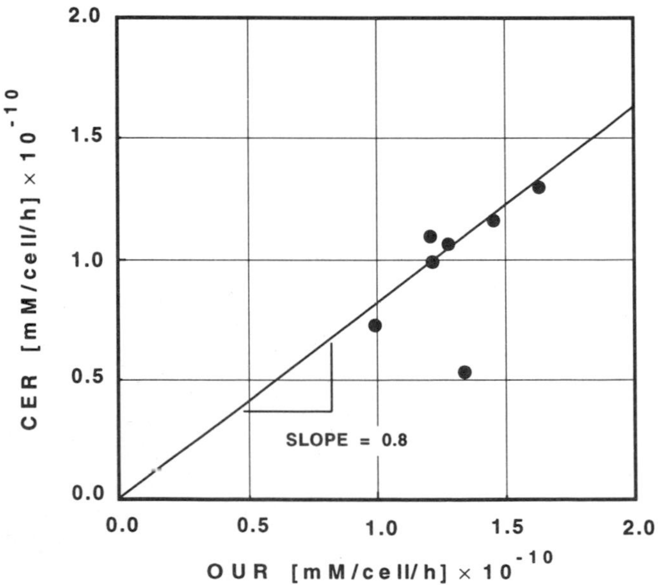

Fig. 1. Respiratory quotient determination for a rNS/0 myeloma cell line (AUNINS, unpublished data).

tion of *CER* with *OUR* according to AUNINS (unpublished data). This measurement and the data of KAMEN et al. (1991) yield respiratory quotients of 0.8. In the study of LOVRECZ et al. (1992) measured respiratory quotients of 0.6 via mass spectrometry. ZUPKE (personal communication to J. AUNINS, 1992) found respiratory quotients of 1.2 for murine hybridoma cells. In general, it is clear that respiratory quotients are not uniform, and may be variable such that it is useful to track CO_2 evolution over the course of a cultivation.

2.2.2 Carbon Dioxide Concentration

Despite considerable basic research on the roles of CO_2 and HCO_3^- in metabolism, incomplete information is available on the range of CO_2 known to affect growth and protein/virus productivity. This is partly hampered by the difficulty of distinguishing between the effects of CO_2, bicarbonate, and pH. Carbon dioxide participates in the *de novo* synthesis of purines and pyrimidines, and also appears to be necessary to prime energy metabolism reactions. Conversely, excess CO_2 can act to inhibit respiration reactions, and may act as an acid poison by crossing membranes and changing intracellular compartment pH. In human serum, bicarbonate is typically 10 mM (mmol/L), and the partial pressure of carbon dioxide is around 0.053 bar. Equilibrium intracellular concentration with human serum is 28 mM bicarbonate, and 0.059 bar P_{CO_2} pressure (GUYTON, 1966). Culture media contain typically 35 mM of bicarbonate ion, giving a P_{CO_2} of 0.079 bar at pH 7.2. Minimal levels of CO_2 for growth are in the range 0.005–0.02 bar (MCLIMANS, 1972). Transformed cells appear to be less stringent than primary cells; KILBURN and WEBB (1968) found mouse LS cell growth to be insensitive to P_{CO_2} from 0.001 to 0.09 bar. MITAKA et al. (1991) have recently reported the optimal levels of HCO_3^- as 25–30 mM at pH 7.6–8.0 for a maximal DNA synthesis rate in cultured rat hepatocytes, corresponding to ca. 0.01–0.04 bar P_{CO_2}.

Negative effects of excess CO_2 can be reflected in industrial culture performance. Fig. 2, from DRAPEAU et al. (1990), shows the effect of increasing P_{CO_2} on productivity and growth of CHO cells producing recombinant

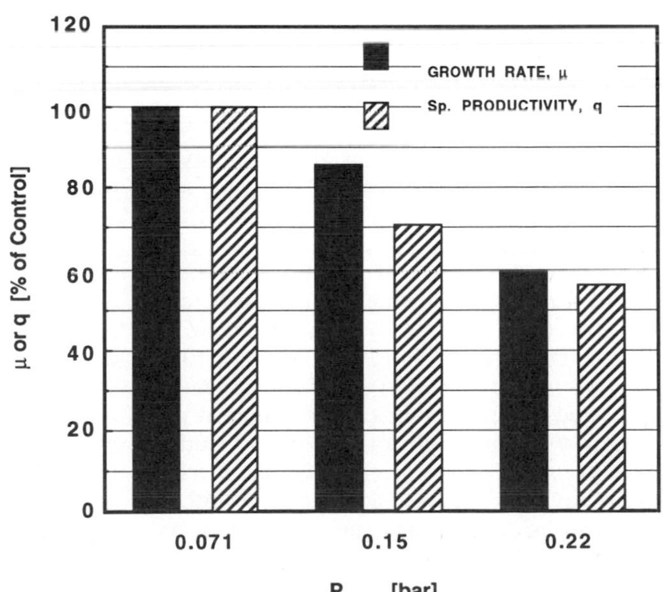

Fig. 2. Effect of CO_2 partial pressure on rM-CSF CHO cultures at constant pH and P_{O_2} (from DRAPEAU et al., 1990).

human macrophage colony stimulating factor. In this experiment, dissolved oxygen and pH as well as other parameters were held constant, and the reactors were inoculated from a common stock culture. The graph shows three-day endpoints from batch cultures, demonstrating deleterious effects of increasing P_{CO_2}. Growth rate and cell-specific productivity are depressed to as much as 59% and 56% of control, respectively. These experiments were conducted to simulate the levels of CO_2 experienced in 2500 L bioreactors, where scale effects and operating strategy combined to increase P_{CO_2} to levels not seen in development scale cultures. In 200 L scale recombinant NS/0 myeloma cell cultures, yield of IgG decreased by 20% with maximum P_{CO_2} increase from 60–120 mm Hg (AUNINS et al., 1991a). In this case, sparging small amounts of pure oxygen through a stainless steel frit sparger to achieve small bubbles resulted in very efficient oxygen transfer, but poor ventilation compared to a simple ring sparger ($d_L = 2$ mm) producing large bubbles. This increased P_{CO_2} in the frit sparged culture to double the levels seen in the ring sparged tank. While specific antibody productivity was not affected, the additional CO_2 suppressed net cell growth, reducing the MAB titer from 320 to 270 mg/L.

2.3 Agitation-Induced Cell Damage

In conjunction with the relatively low oxygen demand of animal cell cultures, it is their lack of a cell wall and the ensuing sensitivity to shear forces, that practically distinguish animal cells from microbes. No discussion of cell culture aeration is complete without reference to the operating space of physical conditions within which it is possible to aerate cells and still grow them successfully. Hydrodynamic effects on cultured cells have been explored since AUGENSTEIN et al. (1971) examined suspended cells, and SINSKEY et al. (1981) examined microcarrier cultures. Since then, myriad studies have been performed to elucidate the mechanisms governing damage and their prevention. For suspension cells, damage is usually incurred when gas is introduced into the reactor, either by sparging or surface vortexing. For microcarrier cultures of attached cells, additional mechanisms of damage come into play due strictly to agitation. This impacts directly on culture aeration, as it sets limits on the mixing intensities possible for these cultures.

2.3.1 Microcarrier Culture

Although microcarriers are used at the 1000 L scale for poliovirus production (MONTAGNON et al., 1984), the lack of design criteria for mixing and aerating microcarrier reactors has thwarted the promise of process intensification via high microcarrier concentrations. For practical purposes of aeration design, hydrodynamically caused cell damage in microcarrier cultures may be treated as an all-or-nothing death phenomenon for diploid cell lines (CROUGHAN and WANG, 1989). The same cannot generally be said for transformed cells (CROUGHAN and WANG, 1990), but there are instances where these cells behave like their diploid counterparts (AUNINS et al., 1991b).

Adherent animal cells are removed from surfaces at steady laminar shear stresses of the order of 0.5–10 N/m^2 (STATHOPOULOS and HELLUMS, 1985; CROUCH et al., 1985). Modifications in monolayer morphology can be demonstrated at steady or pulsatile laminar shear forces of the order of 0.1–1.0 N/m^2 (LUDWIG et al., 1992), and altered gene expression and cell permeability can be demonstrated under similar conditions (DIAMOND et al., 1989). Virus productivity decreases can also be demonstrated at increasing, albeit absolutely low shear stresses in small-scale microcarrier cultures (SINSKEY et al., 1981). Although many sublethal events are detectable, insufficient biological information is currently available to present rigorous guidelines for aeration system design. An inadequacy of the literature is the lack of data for adherent cell behavior in well-defined turbulent flow apparatus. Along with the lack of biological information, the interaction of turbulent fluid forces with particles has not been measured, although models have been proposed (see Sects. 2.3.2 and 7.2.1). System behavioral changes such as microcarrier aggregation (JUNKER et al., 1992) are also uncharacterized.

A major cause of agitation damage appears to be turbulent eddy dissipation, as proposed by CROUGHAN et al. (1985), and later shown in a series of papers (CHERRY and PAPOUTSAKIS, 1986; CROUGHAN et al., 1987; MCQUEEN and BAILEY, 1989). When the microcarrier (or suspended cell, or cell aggregate) size is on the same scale as the Kolmogorov microscale of turbulence, defined as

$$\lambda = \left(\frac{\upsilon^3}{\varepsilon}\right)^{1/4} \tag{6}$$

the particle no longer sees a steady shear environment, but experiences chaotic eddy currents which dissipate energy against the surface in high-energy bursts. In Eq. (6), λ is the Kolmogorov microscale, υ the medium viscosity, and ε the global energy dissipation rate. Eddy dissipation causes local transient shear rates sufficiently high that cell stripping from the microcarrier can occur, although time-averaged shear rates (and the associated mean stress applied to the microcarrier surface) in the culture can be calculated to be harmless per the studies above. Although it is difficult to estimate the energy transmitted to a particle (see Sect. 7), several design correlations have appeared. All are concerned with microcarrier–eddy encounter frequency, and are more or less complicated by the energy dissipated in the interaction.

CROUGHAN et al. (1985, 1987) proposed that for energy inputs which produce reactor-averaged Kolmogorov microscales below a certain critical length, eddy encounters kill cells with a constant probability. By postulating that a constant volume fraction contains eddies of the critical length, and that homogeneous, isotropic turbulence occurs in the vessel, they developed a model for cell death rate:

$$\frac{\mathrm{d}X}{\mathrm{d}t} = \mu X \quad \text{for } \lambda_{\text{global}} > d_{\text{microcarrier}} \tag{7}$$

$$\frac{\mathrm{d}X}{\mathrm{d}t} = (\mu - k_\mathrm{d})X \quad \text{for } \lambda_{\text{global}} < d_{\text{microcarrier}} \tag{8}$$

where k_d is given by

$$k_\mathrm{d} = K\left(\frac{\varepsilon}{\upsilon^3}\right)^{3/4} \tag{9}$$

In Eqs. (7–9), μ is the specific growth rate, k_d the specific death rate, and K a cell and system dependent constant. These authors were able to show death rate dependence on both power input (CROUGHAN et al., 1987) and viscosity (CROUGHAN et al., 1988) according to the prediction of Eq. (9). Although differences in K are expected for different cell lines, for geometrically similar reactors K should be fairly constant with scale if the impeller discharge region encompasses a similar volume fraction. CROUGHAN et al. (1987) obtained values of $K = 3.6 \cdot 10^{-9}$ and $K = 1.6 \cdot 10^{-8}$ (cm^3/h) for FS-4 and VERO cells on Cytodex 1 microcarriers, respectively. CHERRY and PAPOUTSAKIS (1989) also showed that significant death rates only occurred when λ was smaller than the microcarrier diameter. They used an impeller region volume, $V_{\text{impeller}} = d^3$, rather than the reactor volume, $V = \pi D^2 H/4$, to calculate ε (see Sect. 7.2.1 and Eq. (60) for further discussion). This gives a more conservative and perhaps more accurate estimate for the allowable operation of a reactor. Inspection of the various literature sources shows that in actuality, there is not an absolute cutoff value of λ beyond which Eq. (7) applies. However, for an individual system, the death rate increase has a fairly sharp drop with decreasing λ. Further refinements await more powerful modeling and measurement of reactor hydrodynamics.

CROUGHAN et al. (1988) also showed a death rate component from interparticle collision. Assuming turbulence involvement, they proposed a model which used a modified death rate constant with the same form as Eq. (9), where K is substituted for by $K = K' \, C_\mathrm{m}$. By this theory, bead collision induced by energy dissipation affects death rates in the same way as for turbulence-induced death, and hence Eqs. (7–9) still apply. Using HINZE's (1971) theory of turbulent particle collision, CHERRY and PAPOUTSAKIS (1989) defined "turbulent collision severity" factors based on either small ($\lambda < d_\mathrm{m}$) or large ($\lambda > d_\mathrm{m}$) eddy involvement in microcarrier collisions. The former predicts k_d proportional to $(\varepsilon\,\upsilon)^{3/4}$, while the latter predicts k_d proportional to $(\varepsilon/\upsilon)^{3/2}$. Both factors correlated death rates within the error of the authors' data, rendering model choice difficult.

Damage of plant cell aggregates by impact with stirred tank agitators was considered by PROKOP and ROSENBERG (1989) by an "impeller collision severity" factor. This theory proposed a dependence of k_d on (ε/d), which would imply that for scale-up at constant ε and tank geometry, the death rate would decline by $V^{1/3}$. CROUGHAN et al. (1987) were unable to correlate impeller tip speed with cell death in spinners, although they did not rule out steady shear damage to cells; the data examined were limited to below 30 cm/s tip speed.

Unfortunately the regions of high microcarrier concentration and large (and varying) tank scale are precisely those where necessary data are needed most, and mostly lacking. In fairness to researchers in the field, it is extremely difficult to perform the elegant experiments, and even more difficult to acquire precise data that are required to give reliable, predictable scale-up. As a design guide, all of the models revolve around medium viscosity and power input. Since the former does not change for most cultures (except high C_m; see Sect. 2.5.2), design and scale-up for aeration can be reasonably based on constant power input at levels where the critical turbulence microscale is larger than the microcarrier size. The practitioner should be aware of situations where the relative size of the impeller discharge region or the distribution in energy dissipation change (see e.g. Sect. 5.2.3, energy dissipation within an aeration support).

2.3.2 Suspension Culture

Suspension cell damage under normal culture conditions is almost completely due to the introduction, dispersion, and disengagement of gas bubbles. Although cells can be damaged strictly by fluid agitation (AUGENSTEIN et al., 1971; TRAMPER et al., 1986; C. G. SMITH et al., 1987; MCQUEEN and BAILEY, 1989), the turbulent shear stresses necessary to give significant death rates are quite high, from 0.1 to 1 N/m^2 (100–1000 s^{-1} shear rate) for insect cells (TRAMPER et al., 1986; GOLDBLUM et al., 1990), and greater than 15 N/m^2 for hybridomas (MCQUEEN and BAILEY, 1989). These controlled studies have been corroborated by reactor experiments showing that extreme agitation is necessary to induce damage in the absence of bubbles (OH et al., 1989; KUNAS and PAPOUTSAKIS, 1990). As an example of shear damage considerations for suspended cells, Cherry and KWON (1990) have estimated that the maximum transient turbulent shear stress ranges from 0.05 to 5 N/m^2 where $d_{cell} \ll \lambda$, and $0.1 < P/V < 100$ W/m^3. This maximum turbulent stress is lower than steady laminar stresses known to affect cell monolayers (see Sect. 2.3.1), and could be reasonably presumed not to affect cells. A long-range goal of cell damage research would be to define the relationship between reactor cell death and shear tolerance in defined apparatus. With such knowledge, correlating equations (such as Eqs. (59 and 61) might be used to accurately design reactor agitators, internals, and upper limits of energy dissipation.

There are commonly encountered situations where shear tolerance is not easily defined or is changing. The first, and most likely to be encountered in industrial culture, is aggregating cells, such as CHO, BHK, VERO or Sf-9. Aggregates break or shed cells erosively when absorbing energy, potentially causing cell death in the process. Although aggregate cell culture death mechanisms, rates, and methods for prevention are unexplored, it appears that turbulent eddies control maximum aggregate size in these situations (MOREIRA et al., 1992).

A second exception is the adaptation of formerly stagnant cultures to agitated suspension (MATHER, 1990), or from low to higher levels of agitation (VOGELMANN, 1981; KRAMER et al., 1991; SCHMID et al., 1992). Whereas the ultimate suspension culture is not shear-sensitive, in the initial stages of adaptation, low viability is experienced and patience is required before proceeding to larger scales of operation and more intense agitation. For stagnant cultures, a progression from unbaffled shaker flasks to conventional paddle-agitated spinners under increasing agitation rates can be used to adapt the cells. For transformed cell lines which must be weaned from surface attachment, it may be necessary to grow the cells as aggregates before a transition can be made to small aggregates or single cells by increasing agitation (TOLBERT et al., 1980). For transition from low to high agitation (generally also with gas bubble presence), cells can adapt to shear in a commercial reactor, resulting in improved culture performance (VOGELMANN, 1981; KRAMER et al., 1991). Adaptation may be a practical way to remove agitation limits on many cell cultures. The requirement for shear adaptation will of course be related to the energy intensity of the reactor and aeration design. This might be minimized by the use of airlift or membrane aerated reactors.

The third exception occurs when the cell is under infective attack by virus, is unable to maintain itself, and may be undergoing cytoplasmic membrane changes. Unlike most suspended mammalian cell cultures, insect cell reactors for baculovirus expression systems have definite aeration and agitation design limitations imposed upon them. Under infection, insect cells become very shear-sensitive (SILBERKLANG, 1987). As mentioned above, virus-infected animal cells on microcarriers are especially susceptible to agitation damage (TYO and WANG, 1981). Compounding the problem for insect cells is the increase in oxygen demand during baculovirus infection (see Sect. 2.2.1), and the fact that insect cells aggregate under certain conditions.

Independent of global power input, infected insect cell culture performance can be affected by the impeller type, as shown in Fig. 3 from JUNKER et al. (1990). This graph shows the cell growth of two otherwise identical 50 L cultures with either a six-blade Rushton turbine ($d/D=0.31$) or hydrofoil impellers ($d/D=0.42$, Lightnin A315, Mixing Equipment Co.), and the impellers are sized to draw the same power. The growth curves are similar up to infection, after which the hydrofoil impeller is superior, despite its similar energy input and higher tip speed. Although both cultures produce the same total of cells, the viability in the A315-equipped culture is higher during infection.

2.4 Bubble-Induced Cell Damage

In addition to agitation, gas–liquid interfaces created by sparged aeration can be detrimental to cell health. Bubble-induced damage is a major concern for suspension cultures, since this is the easiest aeration technique for cell cultures, and the reactor operating parameters (primarily gas flow rate) are not as affected by scale as other methods. Quantitative information on bubble damage to microcarrier cultures is scarce (AUNINS et al., 1986), and no thorough studies have emerged to describe the effects of sparging on microcarrier cultures. As such, the discussion below is necessarily limited to suspension cultures.

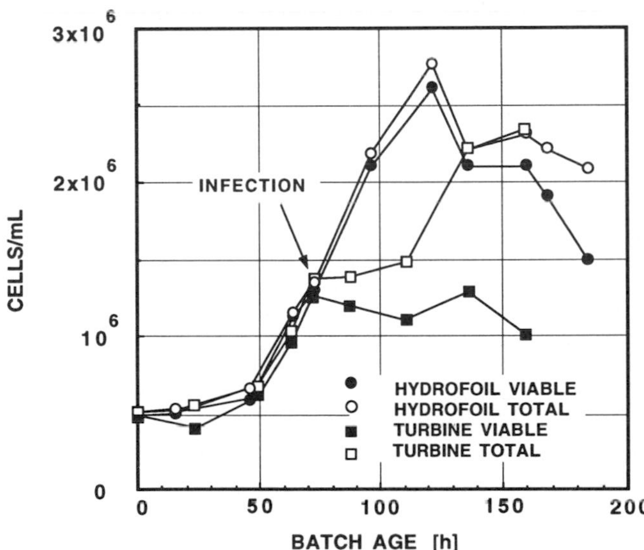

Fig. 3. Effect of impeller type on Sf-21 culture viability under baculovirus infection (from JUNKER et al., 1990).

2.4.1 Mechanisms of Bubble-Induced Damage

Although damage is a direct result of gas introduction, it appears to be mechanically transmitted via fluid shear, as opposed to being a surface phenomenon such as cell lysis due to interfacial adsorption and deformation. This has been determined by a series of experiments visualizing cell–bubble interactions during bubble rise and disengagement from the liquid (HANDA et al., 1987; BAVARIAN et al., 1991; CHALMERS and BAVARIAN, 1991; ORTON and WANG, 1991). Current observations show that for certain medium compositions, cells are loosely adsorbed to gas–liquid interfaces. A fraction of the adsorbed cells is killed when the bubble forms, bursts or coalesces, due to the formation of transient high shear boundary layers and jets.

From existing data, bubble bursting causes most of the cell death from sparging. TRAMPER et al. (1986) first correlated cell death in bubble columns with a "killing volume" associated with a bubble. CHALMERS and BAVARIAN (1991) and BAVARIAN et al. (1991) showed that this volume involves the lamella above a bubble, and the jet produced upon bursting and lamella contraction. ORTON and WANG (1990) estimated the killing volume via the correlations of NEWITT et al. (1954) and GARNER et al. (1954). These authors determined the water volume aerosolized by bursting bubbles of different diameters. Using the aerosolization measurements, ORTON and WANG (1990) unified their own observations and literature data (TRAMPER et al., 1986, 1988; GARDNER et al., 1989; MURHAMMER and GOOCHEE, 1988) on cell death by bubbles of varying sizes. Fig. 4 shows the bubble diameter effect on E, the liquid entrainment rate and k_d^*, a modified cell death rate defined by

$$k_d^* = k_d \left(\frac{\rho V}{\rho_g q}\right) \qquad (10)$$

Here q is the gas flow rate and ρ_g is the gas density. The modified death rate transforms the cell death rate per unit time to a death rate per mass of gas sparged per mass of liquid, permitting the plot of Fig. 4. Both $E(d_b)$ and k_d^* decrease dramatically for bubble diameters greater than 5.5 mm. Their correlation, $k_d^* = E(d_b)$, gives rise to the design equation

$$k_d = \frac{\rho_g v}{\rho H} E(d_b) \qquad (11)$$

Here, v is the superficial gas velocity, and H is the liquid height, Eq. (11), along with Fig. 4, can be used to estimate the maximum death rate of a culture for a given sparge rate and bubble size. This rate can be reduced dramatically by the use of surfactants, discussed in Sect. 2.4.2. Interestingly, the data in Fig. 4 are derived from both insect and animal cell cultures, which differ widely in their shear suscep-

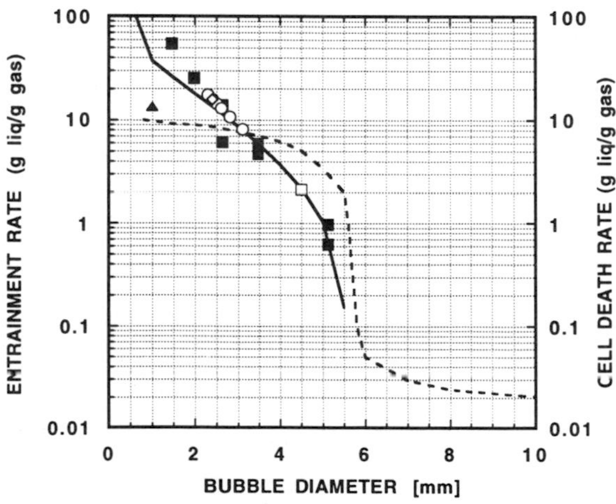

Fig. 4. Liquid entrainment rate and modified cell death rate versus bubble diameter (ORTON and WANG, 1990). Culture data: ■ ORTON and WANG (1990), ▲ TRAMPER et al. (1988), ○ TRAMPER et al. (1986), □ GARDNER et al. (1990). Entrainment data: --- GARNER et al. (1954), — NEWITT et al. (1954).

tibility in defined shear experiments. The ability to correlate both to the simple and universal parameter of the liquid entrainment rate suggests that the shear forces involved in bubble bursting are so large as to give an all-or-nothing death phenomenon, which has been suggested for turbulent microcarrier cultures.

Although k_d decreases with increasing d_b, the bubble surface area decreases as well. This necessitates increasing gas sparge rates to satisfy oxygen demand. There will be an optimum bubble diameter which satisfies oxygen demand with minimum cell death, one large and one small; the large diameter will lie close to 6 mm, as shown by the data and curves in Fig. 4. A very small bubble would minimize death rates by giving very high $k_L a$, minimizing gas throughput and hence aerosolization rates. ORTON (pers. comm. to J. AUNINS, 1992) has calculated this size to be less than 1 mm. However, there are practical problems in designing spargers to achieve very small bubbles at a large scale. In addition, CO_2 ventilation may be a problem, see Sect. 2.2.2.

The above discussion holds where bubble coalescence is negligible – low gas holdup and low power input to the reactor. Here, bubble size can be controlled via sparger design and orifice flow rate. When surface vortexing, agitator energy, fractional gas holdup, or bubble instability (via large bubble diameter, d_b) cause bubble coalescence or breakup, secondary effects may occur. KUNAS and PAPOUTSAKIS (1990) showed that in a surface aerated bioreactor, significant cell damage occurred only when agitation caused a vortex to occur at the gas–liquid interface. The experiments of OH et al. (1989) showed that the extent of cell damage under sparged aeration is dependent on the sparger location relative to the impeller. For a tube sparger placed outside and above the impeller jet, little cell death was observed compared to spargers located in the jet or under the impeller. Although it has been suggested that in agitated tanks bubble coalescence and breakup are major causes of cell death (YANG and WANG, 1992), the evidence is indirect. In many of these cases, the agitation intensity and low gas holdup may cause bubble size reduction into the lethal size regime of Fig. 4. Like the microcarrier collision problem, elegant experiments to confirm effects of bubble coalescence/breakup are difficult to perform. At this time it is only possible to suggest that the tank design and operation be arranged so that vortexing, and bubble dispersion induced by agitator or gas holdup are avoided; these regimes of operation are described in Sects. 4.3 and 6. To avoid the latter while achieving the necessary mass transfer coefficients, it may be desirable to use multiple spargers which provide gas distribution without intense mixing. For airlift reactors, it may be useful to employ low-level mechanical agitation to achieve desired liquid circulation rates without excessive sparging and gas holdup.

2.4.2 Effect of Surfactants

Sparging damage can be reduced significantly by the presence of surfactant additives to the medium. Serum and a number of synthetic polymers have been shown to reduce interfacially-induced cell damage. As most sparging problems are encountered in serum-free and low-protein media, the synthetic, non-nutritional polymers will be briefly reviewed. These additives and their reported effective concentrations are shown in Tab. 2.

The mechanism(s) of polymer protection are not completely determined at this time; reviews of most available hypotheses can be found in PAPOUTSAKIS (1991), and CHALMERS (1993). Existing evidence suggests that polymers reduce damage in several ways. Pluronics, the most widely used polymers, affect cells directly by adsorption to or insertion into the cell membrane (C. M. SMITH et al., 1987; BENTLEY et al., 1989; MURHAMMER and GOOCHEE, 1990; RAMÍREZ and MUTHARASAN, 1990). Many other surfactants lower the static surface tension of the medium and alter viscosity, as discussed in Sect. 2.5. These static property changes are in general not large enough to account for the protective effects of the polymers, when combined with theory on their effects on the various parameters, such as bubble size (e.g., Eq. (49)), rise velocity, and so on. It is rather more likely that the dynamic elastic properties of the polymers, on the timescale of bubble film thinning and rupture, contribute significantly to their protective effects

Tab. 2. Commercial Surfactants Successfully Employed for Cell Damage Reduction

Polymer	Cell System	Reference	Concentration (g/L)
Pluronic F38	Sf-9	MURHAMMER and GOOCHEE, 1990	2
Pluronic F68	BHK21	KILBURN and WEBB, 1968	1
Pluronic F68	Murine hybridoma	GARDNER et al., 1990	1–4
Pluronic F68	Sf-9, TN-368	GOLDBLUM et al., 1990	2–5
Pluronic F88	Human lymphoblast	MIZRAHI, 1984	0.5–2.0
Pluronic 10R-5	Sf-9	MURHAMMER and GOOCHEE, 1990	2
PEG 1400	Murine hybridoma	MICHAELS and PAPOUTSAKIS, 1991	0.5
PEG 8000	Sf-9	MURHAMMER and GOOCHEE, 1990	2
PVA 10000	Murine hybridoma	MICHAELS and PAPOUTSAKIS, 1991	2
Edifas B50 (Na-CMC)	BHK21	TELLING and ELLSWORTH, 1965	0.8
Hydroxyethyl starch	RPMI7430	MIZRAHI, 1984	2.0
Methocel E4M	Sf-9, TN-368	GOLDBLUM et al., 1990	5

(MICHAELS and PAPOUTSAKIS, 1991; ORTON, personal cummunication to J. AUNINS, 1992).

A consistent observation of polymer-treated suspensions is that cells no longer cling to the bubbles, leaving few victims present during bursting (HANDA et al., 1987; CHALMERS and BAVARIAN, 1991; ORTON and WANG, 1991). This has led to the proposal (CHALMERS, 1993; GARCIA-BRIONES et al., 1992) that polymer adsorption to the cell and/or interface changes the thermodynamic equilibrium adsorption of cells to bubbles. This may also explain how small changes in gas–liquid surface tension determine whether damage occurs or not.

Hopefully, thermodynamic modeling will provide rational protectant selection in the future. At present, there are no design guidelines available for selection of polymers based on cell protection, foaming properties, and interference with product purification. Pluronic F68 in the concentration range of 0.5 to 3 g/L possesses a wide range of desirable properties, and is commercially available in pure preparations that have removed toxic low molecular weight components (KING et al., 1988). Polyvinyl alcohol reportedly gives better protection than F68 at high agitation, although it is more inhibitory to cell growth at low agitation (MICHAELS and PAPOUTSAKIS, 1991). From the data in the following sections, it appears that Pluronic F68 has only minor effects on aerator performance. The effects of traditional antifoams, e.g., simethicone and polypropylene glycol antifoams, on mass transfer, and cell death due to bubble aeration are not well known, although several types have no detrimental effects on cell growth in spinner culture (AUNINS et al., 1986). Of the antifoams investigated in that study, a 5:1 v/v mixture of Dow–Corning medical-grade antifoam emulsions A and C suppressed fetal calf serum foam best (AUNINS, unpublished data).

2.5 Culture Physical Properties

The properties of live cultures are not precisely known, since they change quickly under non-sterile conditions, and are therefore difficult to measure directly. Rough values can be obtained from medium for low density cultures, and where cell lysis or secreted proteins are not significant. Many situations are emerging where the secreted product is the major medium protein component (BROWN, 1992; SILBERKLANG et al., 1992; SLIWKOWSKI et al., 1992). This may necessitate property measurements of product-containing broth for accurate knowledge of medium characteristics. A large amount of the aeration information available has been obtained for water or salt solutions as the medium, and at temperatures other than culture temperature. For this reason, it is relevant to review medium properties and the corrections that may pertain to the data presented below.

2.5.1 Solubility and Diffusivity

Oxygen solubility and diffusivity measurements in culture are virtually non-existent. However, media do not usually have large amounts of organic components which alter the equilibrium, nor do they usually contain a large cell mass fraction that would alter partitioning. Calculated O_2 solubility in DMEM (Dulbecco's modified Eagle's medium) is 6.38 mg/L at $\vartheta=37\,°C$ and $P_{O_2}=0.21$ bar, after SCHUMPE et al. (1982). O_2 solubility of 1:1 DMEM:F12 with 0.3 g/L of albumin at $\vartheta=37\,°C$ and $P_{O_2}=0.21$ bar was measured to be 6.5 mg/L (HENZLER and KAULING, 1993), corresponding to 91% of pure water solubility, and giving a Henry's law constant of $Hy_{O_2}=3.2\cdot10^6$ N m/kg ($1/Hy=31$ mg/L/bar).

Data on O_2 diffusivity in cell culture media are lacking. Calculated diffusivity at $\vartheta=37\,°C$ is $2.88\cdot10^{-5}$ cm²/s using the Wilke–Chang correlation (REID et al., 1977):

$$\mathscr{D}=7.4\cdot10^{-8}\,\frac{(\phi M)^{1/2}\vartheta}{\eta V_A^{0.6}} \tag{12}$$

Here, ϕ is an association parameter, V_A is the solute volume at the normal boiling point in cm³/g/mol, and M is the solvent molecular weight. For the oxygen–water system, $\phi=2.26$ and $V_A=25.6$ cm³/g/mol.

For carbon dioxide, SCHUMPE et al. (1982) predicted a Henry's law coefficient of $Hy_{CO_2}=9.4\cdot10^7$ N m/kg ($1/Hy=1.06$ g/L/bar) in DMEM medium. Diffusivity is reported to be $2.28\cdot10^{-5}$ cm²/s in water at $\vartheta=37\,°C$ (HO et al., 1987). Eq. (12) gives a value of $2.53\cdot10^{-5}$ cm²/s using $V_A=34.0$ cm³/g/mol (REID et al., 1977).

2.5.2 Viscosity

Although basal medium viscosity is close to water, complete medium viscosity varies with protein content. Fig. 5, bottom panel, shows

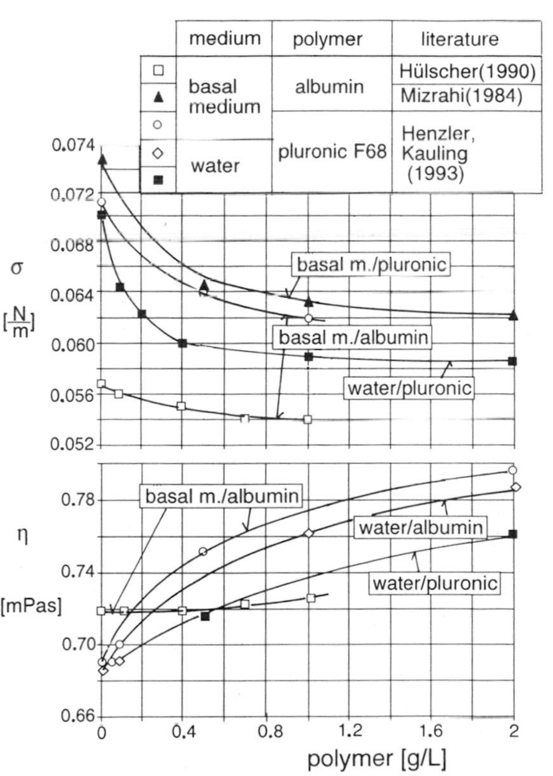

Fig. 5. Influence of polymers on surface tension (σ) and viscosity (η), $\vartheta=37\,°C$.

measured viscosities of water and DMEM/Ham's F12 solutions containing albumin and Pluronic F68 at $\vartheta = 37\,°C$. GOLDSTEIN (1986) measured viscosity of DMEM/fetal bovine serum mixtures at $\vartheta = 37\,°C$, and MICHAELS and PAPOUTSAKIS (1991) measured a variety of media with additives; these values are shown in Tab. 3. To correct for temperatures other than 37 °C, the correlation of MAKHIJA and STAIRS can be employed (REID et al., 1977)

$$\log^{10} \eta = A' + \frac{B'}{T - T'} \quad (13)$$

In this equation, η is viscosity in mPa s, and T is absolute temperature in Kelvin. Values of A', B' and T' for water are −1.567 K, 230.3 K, and 146.8 K, respectively, for this system of units.

Culture viscosity changes for several reasons. For microcarriers or high density cultures, a modified Stokes–Einstein relation may be used to relate suspension viscosity to the particle volume fraction

$$\upsilon = \frac{\eta_L (1 + 2.5\phi_P + 10\phi_P^2)}{\rho_L (1 - \phi_P) + \rho_P \phi_P} \quad (14)$$

In this equation, ϕ_P is the particle volume fraction, and the subscripts L and P denote liquid and solid properties. Cell lysis may release DNA to produce a slimy, viscoelastic solution under some conditions, such as pumping the cells during external loop aeration or filtration. Small amounts of DNA can viscosify solutions; intrinsic viscosity of $M_w = 6 \cdot 10^6$ DNA at $\vartheta = 25\,°C$ is 5000 cm^3/g (CANTOR and SCHIMMEL, 1979). Using this number, 200 mg/L of DNA doubles medium viscosity, corresponding to lysis of 10^7 cells/mL (DARNELL et al., 1986). As human genomic DNA is on average $2 \cdot 10^{11}$ M_w, corresponding intrinsic viscosity is higher and micrograms per liter may give high viscosities and lead to non-Newtonian flow behavior if not bound inside nuclei or nucleosomes, condensed by proteins, or degraded. High cell density processes which entail accumulation of DNA, RNA, and cell debris may encounter significant viscosity increases, and will certainly encounter interfacial effects due to cell component release.

2.5.3 Surface Tension

Surface tension affects coalescence and size of bubbles, and gas–liquid mass transfer in

Tab. 3. Medium Viscosities and Surface Tensions at $\vartheta = 37\,°C$ (from MIZRAHI, 1984; GOLDSTEIN, 1986; MICHAELS and PAPOUTSAKIS, 1991)

Medium	Additive	Concentration (g/L)	Viscosity (mPa s)	Surface Tension (dyn/cm)
Deionized water	—	—	—	72.8
DMEM	—	—	0.717 ± .007	—
DMEM	FBS	5 (% v/v)	0.720 ± .006	—
DMEM	FBS	10 (% v/v)	0.727 ± .002	—
RPMI 1640	—	—	—	74.7
RPMI 1640	Pluronic F68	0.5	—	64.5
RPMI 1640	Pluronic F68	1.0	—	63.1
RPMI 1640	Pluronic F68	2.0	—	62.0
DMEM:RPMI 1640	—	—	0.71	70
DMEM:RPMI 1640	Albumin	1.0	0.73	53
DMEM:RPMI:albumin (1 g/L)	Pluronic F68	1.0	0.73	47
DMEM:RPMI:albumin (1 g/L)	Pluronic F68	2.0	0.75	47
DMEM:RPMI:albumin (1 g/L)	PEG 8000	0.5	0.73	54
DMEM:RPMI:albumin (1 g/L)	PEG 8000	1.0	0.73	52
DMEM:RPMI:albumin (1 g/L)	PEG 8000	2.0	0.76	53
DMEM:RPMI:albumin (1 g/L)	PEG 8000	4.0	0.77	53
DMEM:RPMI:albumin (1 g/L)	PVA 10000	1.0	0.75	47
DMEM:RPMI:albumin (1 g/L)	PVA 10000	1.0	0.77	45

general. An unavoidable detriment to mass transfer in cell cultures is the presence of surfactant and protein films. The resistance to gas–liquid mass transfer offered by such films has not been well quantified under cell culture conditions, although the films can reduce mass transfer coefficients substantially, as shown in Fig. 22. As discussed in Sect. 2.4.2, static surface tension may not be relevant for calculating cell death criteria; however, it suffices for mass transfer calculations. Measured surface tensions of culture media and solutions (MIZRAHI, 1984; MICHAELS and PAPOUTSAKIS, 1991) are shown in Fig. 5, top panel, and in Tab. 3.

3 Survey of Aeration Methods

The major aeration methods for the submerged cultivation of animal cells are shown in Fig. 6, and industrially relevant examples are given below. These aerator designs will be discussed in the following sections in detail.

Surface aeration is transfer of gas through the free gas–liquid interface at the reactor top. In most reactors, surface mass transfer will always be present, unless special arrangements are made. This aeration mechanism is the sole method for small vessels such as T-flasks, roller bottles, shaker flasks, and spinner cultures. Surface aeration is not usually sufficient to support cell growth except on the small scale, and at low cell densities. An upper reactor volume limit for surface aeration effectiveness is usually seen in the range from 10 to 100 L. Below this it can suffice for most batch cultures of cells. Examples of industrial tank processes which rely on surface aeration are not known to the authors. However, the vast majority of viral vaccine processes, which are typically performed in disposable T-flasks, roller bottles (MATHEWS et al., 1992), Nunc Cell Factories (SIEGL et al., 1984), or other low cell density reactors, are aerated in this manner. β-Interferon is also produced in surface-aerated Nunc Cell Factories (JOHANSSON, 1988) at Bioferon, and erythropoietin is currently produced in roller bottles by Amgen and Ortho. Seed trains for large-scale processes also often go through a spinner or shaker flask stage which is surface-aerated, and hence it is useful to characterize these vessels. A SmithKline-RIT stirred tank microcarrier process for Aujeszky vaccine production at the 150 L scale was aerated by a combination of surface and sparged aeration (BAIJOT et al., 1987).

In membrane aeration, gas diffuses through a permeable membrane which can be either microporous or made of a material with high gas solubility, such as silicone. Most often, the membrane is arranged as a coil or bank of tubes inside or outside the reactor, analogous to heat exchangers. Membrane aeration use is not yet extensive in industry, especially at the large scale. Although the technology possesses process advantages, it can be complex to execute in the factory, and design data have previously been scarce. Some of the complexity can be rectified by the use of fiber-reinforced tubing. This technology is most appropriate for small- and intermediate-scale processes, 10–500 L, which have high oxygen demand, but also possess sensitivity to fluid shear stress or sparged aeration. This situation is likely to be encountered in anchorage-dependent microcarrier processes, virally-infected systems, and situations where antifoams or shear protectants cannot be used. Although *in situ* membrane aerated reactors have been investigated by academicians at least up to the 150 L scale (BÜNTEMEYER et al., 1987) and are available for purchase (e.g., Braun Biotech), only scattered reports suggest their use in industry (VARECKA and BLIEM, 1990; TOLBERT and SRIGLEY, 1987). The only industrial *in situ* membrane-aerated process known to the authors is the Bayer/Miles/Cutter Factor VIII process.

In contrast to *in situ* aeration, *ex situ* technology is used widely in loop configuration specialty reactors. One example of silicone tubing aeration is the Verax CF-IMMO fluidized bed reactor, which employs a gas-filled chamber with a silicone tubing coil inside (DEAN et al., 1987). Medium runs in the tubing lumen, and the shell can be pressurized to enhance gas transfer. Virtually all of the commercially available hollow fiber reactors are

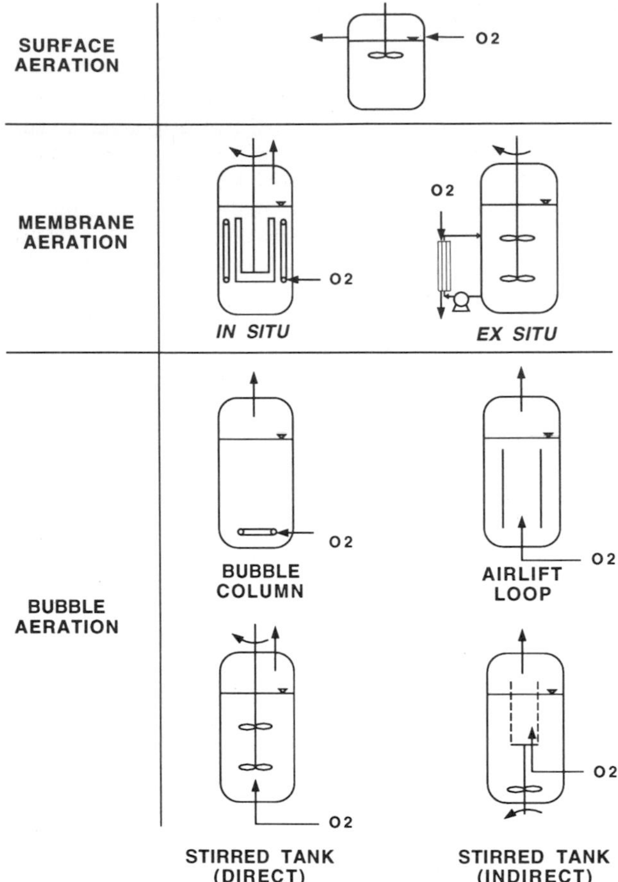

Fig. 6. Methods of oxygen supply for submerged culture of animal cells.

equipped with membrane aerators, which can be either silicone or microporous tubing. Processes using hollow fiber reactors are not yet as popular for production as once predicted, although examples exist, such as Xoma Corporation's antibody processes.

Sparged aeration uses the introduction of gas bubbles at the base of the reactor to create a mass transfer surface area. As in microbial culture, sparged aeration is a very popular and convenient method for scales of 1 L or more. The simplest type of sparged reactor is a bubble column, where gas is introduced into an otherwise featureless reactor; bubble motion provides liquid circulation. To the authors' knowledge, bubble column reactors have not been reported in an industrial application. This is apparently due to the rather high sparge rates at which a bubble column reactor typically achieves acceptable liquid circulation, and the turbulent, unstructured flow which results.

More recently, airlift reactors have been employed for insect and animal cell cultivation. An airlift reactor possesses a draught tube which structures the bubble-induced liquid flow, and allows operation at lower superficial gas velocities. These reactors require less maintenance by virtue of possessing no internal moving parts, and intrinsically avoid agitator power input. However, sparging above that strictly needed for aeration is required in the early (and late?) culture stages to achieve desired liquid circulation, and this may be more or less harmful than agitator power input depending on the culture. It is important to note that liquid circulation is necessary not only for bulk mixing, but also for heat transfer, so that a certain level of sparging is necessary at all times. This can be avoided with modest mechanical power input. Firms employing airlift reactors at the 1000–2000 L scale are Celltech (BIRCH et al., 1987) and Chiron (MAIORELLA et al., 1988). Celltech's major use is for antibody production from hybridomas, recombinant myelomas and CHO cells. This also appears to be the case at Chiron.

Stirred tanks can achieve higher mass transfer rates than other sparged reactors due to the extra power input from the agitator. This serves to increase liquid velocities, and in some cases bubble surface area for transfer. In the earlier days of deep tank culture, simple stirred tanks were employed by Burroughs Wellcome at an impressively large 1000–8000 L scale for veterinary vaccines against foot-and-mouth disease and rabies (PHILLIPS et al., 1985). Improvements on these tank designs, such as axial flow impellers and improved process control, have subsequently been utilized for numerous recombinant culture processes at Genentech, Boehringer Ingelheim (Karl Thomae), and Genetics Institute. Although myriad processes are in various stages of clinical development, few products have been licensed. These include tPA at Genentech and Karl Thomae (LUBINIECKI et al., 1989), and β-interferon at WellGen.

In indirect sparging, gas is introduced into a compartment which is cell-free, and gas exchange to the cell compartment occurs due to a convective liquid exchange flow. The cells are thus protected from bubble bursting damage. To the authors' knowledge, indirect sparging techniques have also not been reported for an industrial process, although they have been shown to be useful on the small scale for microcarrier vaccine processes (GRIFFITHS et al., 1987). Like membrane aeration, this method has most utility for bubble-sensitive, high oxygen demand cultures at the small to intermediate scale. An added potential advantage of indirect aeration is that the devices can also be used to achieve medium exchange, since the separating screen creates a cell-free supernate.

4 Surface Aeration

In surface aeration, mass transfer occurs through the surface of the liquid only. This method is used for culturing in T-flasks, shaker flasks and roller bottles, and in bioreactors at low cell concentrations. For a stationary phase interface, the area A is known, mass transfer occurs via diffusion, and k in Eq. (1) is given by the quotient of the diffusion coefficient $\mathscr{D}$ and the diffusion layer thickness of the liquid phase, Δx. Thus, $k = k_L = \mathscr{D}/\Delta x$.

If the liquid is in motion, gas exchange increases because Δx reduces to the thickness of the boundary layer. If the phase interface is not deformed, the specific area a is known, and k can be estimated from an eddy model derived by LAMONT and SCOTT (1970) for the case of fully developed isotropic turbulence

$$k = k_L = C\,Sc^b(\varepsilon\,\upsilon)^{1/4} \tag{15}$$

In this equation, ε is the local energy dissipation at the interface. This can be substituted for by the global energy dissipation rate in the reactor in many cases, and by the surface impeller power input for reactors with surface aerators. For mobile interfaces (no surface-active compounds present), the exponent of the Schmidt number $Sc = \upsilon/\mathscr{D}$ is $b = -0.5$. According to a series of experiments (THEOFANOUS et al., 1976; RISSE et al., 1985), the constant C in Eq. (15) is about 0.2 for turbulent channel flow. KAWASE and MOO-YOUNG (1989) suggested values of $b = -0.667$ and $C = 0.138$ for submerged-impeller bioreactors using the impeller power input for ε; these authors compared values obtained with Eq. (15) with the literature, and found good agreement for data using cell culture (LAVERY and NIENOW, 1987) and fermentation (YAGI and YOSHIDA, 1974) medium. Their estimation of b reflects a rigid interface. In vessels with rapidly turning impellers, a higher exponent than 0.25 is often found for the term $(\varepsilon\,\upsilon)$ in Eq. (15) (BIN, 1984) because the liquid surface becomes deformed as a result of the higher stirring and/ or the location of the impeller in the vicinity of the surface. For wave motion of the surface the effective phase interface is greater than the projected surface area of the liquid and it becomes appropriate to use the empirically-derived product $ka = k_L a$, rather than independent estimations of k_L and a. Alternatives to the LAMONT and SCOTT development are presented below based on experiments under cell culture conditions.

4.1 Surface Aeration in Laboratory Vessels

Although myriad publications have appeared characterizing mass transfer in laboratory-scale cell culture vessels of a few liters volume or less, most report data in terms of the impeller RPM (often without citing system dimensions), or for unobtainable reactors with non-standard geometries. A few correlations for surface aeration are available for laboratory reactors which may be applied generally, however.

4.1.1 T-Flasks, Roller Bottles

The simplest vessel available for culture is the stagnant T-flask, for which aeration occurs by diffusion only. For these vessels, consideration of steady-state diffusion (see Eq. (1) with $dC_L/dt = 0$) produces the simple relation for oxygen concentration at the cell layer

$$C_{\text{layer}} = C^* - \frac{H q_{O_2} X_s}{\mathscr{D}} \tag{16}$$

In this equation, H is the liquid depth to the surface, and X_s is the surface cell concentration (cells/area). More extensive models of diffusion into other than flat-plate geometries have also been considered in the literature (MURDIN et al., 1987). No observations are available for roller bottle cultures, where the cells spend a significant portion of time submerged in the liquid phase.

4.1.2 Shaker Flasks

For mass transfer in unbaffled shaker flasks, a model of general validity is available (HENZLER et al., 1986; HENZLER and SCHEDEL, 1991):

$$k_L a \left(\frac{\upsilon}{g^2}\right)^{1/3} = B \left(\frac{\upsilon}{\mathscr{D}}\right)^{-1/2} \left(\frac{D_{\max}}{V_F}\right)^{8/9} \cdot \left(\frac{\upsilon^2}{D_{\max}^3 g}\right)^{8/27} \left(\frac{n^2 \sqrt{e D_{\max}}}{g}\right)^{c} \tag{17}$$

In this equation, $D_{\max}$ is the maximum diameter of the Erlenmeyer-style flask, V_F is the liquid volume in the flask, n is the rotation speed, and e is the eccentricity of the shaker table. Constants c and B are system-dependent, and both should equal 0.5 for culture media. This correlation was developed in DIN 12380 conical flasks of 50–200 mL working volume, and should be applicable to common flasks.

4.2 Surface Aeration in Agitated Vessels

4.2.1 Subsurface Impellers

For subsurface impellers without gas entrainment, AUNINS et al. (1986) measured $k_L a$ versus stirrer speed for baffled 0.5 and 10 L laboratory culture vessels equipped with two four-bladed paddle impellers. AUNINS et al. (1989) measured $k_L a$ versus stirrer speed and ε for commonly used 0.5 L spinner vessels (Corning Glass, Corning, NY), unbaffled and agitated by two-blade paddle impellers. The results of both of these experimental sets are shown combined in Fig. 7. Despite some differences in reactor geometry, the different studies can be correlated to yield the Sherwood number Sh as a function of impeller Reynolds number Re:

$$Sh = \frac{k_L D}{\mathscr{D}} = 1.4\, Re_i^{0.76} \tag{18}$$

where $Re_i = n d^2/\upsilon$. For the Corning spinner experiments, power input correlations were obtained as well; even for impeller conditions that were not fully turbulent ($2000 < Re < 20000$), the Sherwood number correlated with power input to the 0.3 power, similar to Eq. (15). To use Eq. (15) for these un-

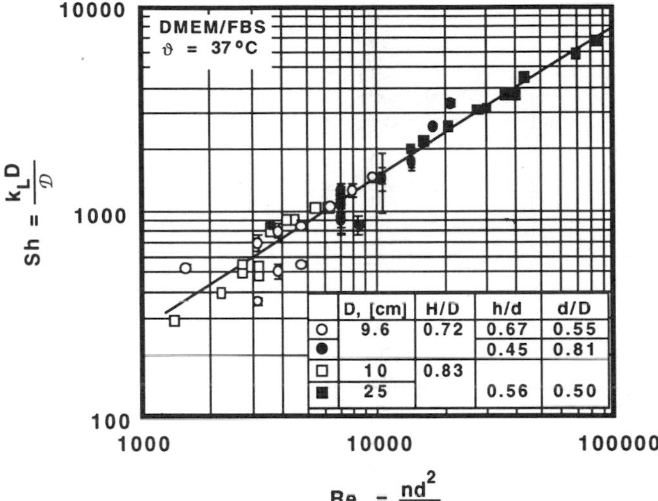

Fig. 7. Sherwood number versus power input for Corning 0.5 L spinners, 0.5 L and 10 L bioreactors (from AUNINS et al., 1986, 1989).

baffled $D = 96$ mm vessels, the power inputs as a function of the Reynolds number for the two impeller sizes are:

$Ne = 52\,Re_i^{-0.45}$ $d/D = 0.55/5.25$ cm impeller), $2000 < Re_i < 15\,000$ (19a)

$Ne = 22\,Re_i^{-0.42}$ $d/D = 0.81$ (7.8 cm impeller), $4000 < Re_i < 20000$ (19b)

$Ne = 0.33$ $d/D = 0.81$ (7.8 cm impeller), $20\,000 < Re_i < 35\,000$ (19c)

In these equations, the power or Newton number is defined as in Eq. (53).

Since dependence on the Schmidt number was not investigated, its effect on these experiments is unknown, although it must be accounted for when considering other gas–liquid systems. It is an unsolved matter whether surface rigidity occurs in cell culture surface aeration with protein surfactants present, and contradictory results have been obtained by many different authors. Note that the reduction of experimental $k_L a$ values to k_L is only valid for a relatively undeformed interface.

Increasing the mass transfer rate by increasing impeller speed results in vastly disproportionate power input change for the added aeration rate obtained. This is patently undesirable for microcarrier cultures and suspensions undergoing shear adaptation. The mass transfer coefficient may be improved without a power increase by placing the impeller at the liquid surface (HU et al., 1986; AUNINS et al., 1989). Impeller height effects are shown in Fig. 8 for the Corning spinner with the 7.8 cm impeller. Transfer is relatively unaffected until the impeller is within $h/d = 0.13$ of the liquid surface. Transfer can be increased by a factor of 50% by using a surface-located impeller in these vessels. HU et al. (1986) reported a four-fold increase using a surface impeller in addition to the normal impeller.

4.2.2 Surface Impellers

ZLOKARNIK's (1979) investigations of surface aerators in sewage treatment suggest that the product for the mass transfer coefficient and the aspect ratio $k_L a \cdot H/D$ should be the correlating parameter for surface aerator performance. If the material properties remain constant, this parameter will generally depend on d, d/D, n, and g. HENZLER and KAULING (1990, 1993) found in the low power input range that the sorption characteristic is the simple relation

$$k_L a \left(\frac{H}{D}\right) = 400 \left(\frac{d}{D} - 0.13\right) \frac{n^2 d}{g} \tag{20}$$

The dependence of the impeller Froude number, $Fr_i = n^2 d/g$, and the diameter ratio dependence for $d/D < 0.3$ is very similar to the

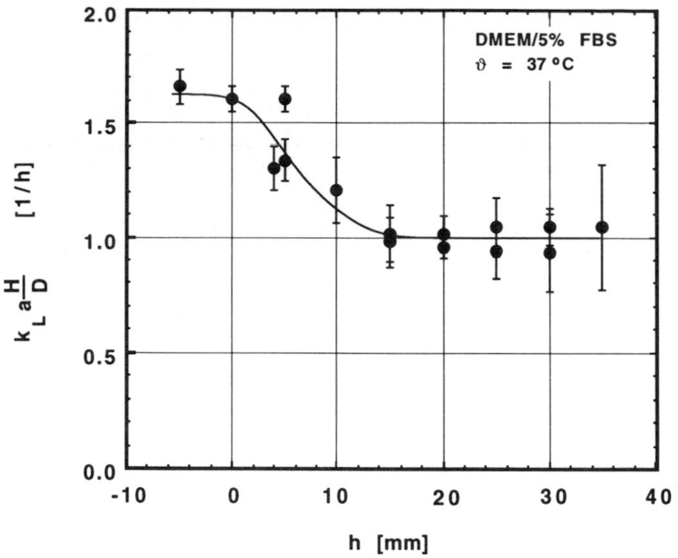

Fig. 8. Surface mass transfer versus impeller depth, Corning 0.5 L spinner (from AUNINS et al., 1989).

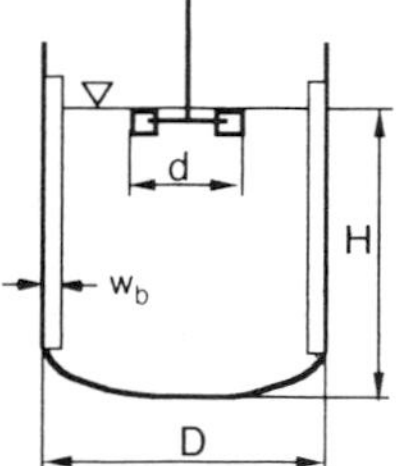

	medium	D [m]	H/D	d/D	lit.
×	water	0.4	1	0.15	Henzler, Kauling (1993)
□	water	0.4	1	0.3	Henzler, Kauling (1993)
■	N/1g/L albumin	0.4	1	0.3	Henzler, Kauling (1993)
◇	water	0.4	1	0.4	Henzler, Kauling (1993)
✳	water	0.4	1	0.65	Henzler, Kauling (1993)
⊙	water	1	0.6	0.26	Henzler, Kauling (1993)
△	water	0.9	0.2	0.2	Zlokarnik

4 baffles, $w_b/D = 0.1$; N: nutrient solution

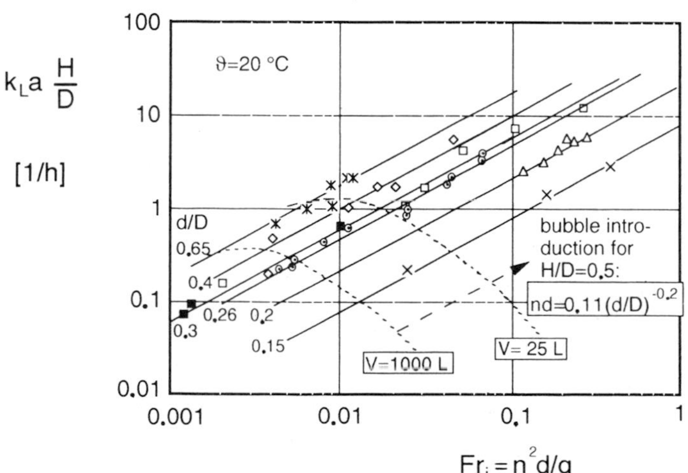

Fig. 9. Sorption characteristic for surface aeration in stirred vessels and limiting curves for the start of bubble incorporation (from HENZLER and KAULING, 1990, 1993).

findings of ZLOKARNIK (1979). The data supporting this relation are shown in Fig. 9. For changing medium properties, i.e., viscosity, diffusivity, and surface tension, the complete sorption characteristic is required. This is not yet known, since in the existing investigations known to the authors the material parameters were not changed.

Investigations with various impellers show that the standard disc turbine impeller with six blades produces intense surface agitation. Increasing the number of blades, or setting the blades at an angle, leads to poorer mass transfer at a given speed. This is due to the lower power input to the liquid. An alternate representation of the sorption characteristic with specific reference to the surface impeller power input is

$$k_L a \approx \left(0.84 - \frac{d}{D}\right) (\varepsilon \rho)^{3/4}$$
$$\text{for } 0.03 < \varepsilon\rho < 20 \qquad (21)$$

This equation gives $k_L a$ in 1/h, if the specific impeller power $\varepsilon\rho$ (W/m^3) is used. This equation is valid for $0.5 < H/D < 1.0$, and $0.15 < d/D < 0.65$. It should be noted that the mass transfer rate is much more sensitive to power input for surface impellers due to wave formation, which increases the phase interface area.

In the ideal case, surface aeration mass transfer remains unaffected by additives and proteins of the type used in nutrient solutions (see albumin-medium data in Fig. 9) because changes in viscosity and surface tension should have no appreciable influence on mass transfer in accordance with the small-eddy model of surface aeration, Eq. (15). Literature data are confused on the issue, although surfactants generally reduce the mass transfer rate (KAWASE and MOO-YOUNG, 1989). Surface-active agents have definite effects when bubble entrainment is encountered, as discussed in Sect. 4.3.3.

4.3 Bubble Entrainment

Uncontrolled entrainment of bubbles into the culture can be detrimental to cell health, as suggested in several studies (see Sect. 2.4.1). Bubble incorporation by a powered impeller leads to bubbles in the dynamic equilibrium size range (see Sect. 6.1.3), and these bubbles can have the most lethal sizes, given cell culture medium properties and power inputs. In addition, since the bubbles originate at the liquid surface rather than deep in the tank, they are inefficient at aeration relative to sparged bubbles. This is because the agitation power typically used does not disperse bubbles homogeneously through the reactor. Since entrainment should be avoided, oxygen input cannot be increased indefinitely by surface aeration.

4.3.1 Subsurface Impellers

Complex relations for entrainment have been proposed for sparged bioreactors with submerged impellers by TOPIWALA (1972) and MATSUMURA et al. (1977). The latter authors proposed a correlation for the rate of surface gas entrainment for a baffled, turbine-agitated tank, finding strong dependences on impeller speed ($\propto n^{3.4-5.2}$), impeller diameter ($\propto d^{4.3-6.6}$), and impeller geometry ($\propto (d/D)^{4.2-6.4}$). The value of their correlation is unknown for cell culture conditions, as it was derived for sparged, high-power input situations. It differs significantly from the correlation proposed in Sect. 4.3.2 for surface impellers under known relevant conditions. More work is necessary in this area to delineate cell culture operating regimes.

4.3.2 Surface Impellers

According to the experimental findings of HENZLER and KAULING (1990, 1993), the limiting condition for bubble incorporation by surface aerators is given by the simple relation $(Fr_i\, We)^{1/4} \approx nd = \text{constant}$. With disc turbine impellers located at the surface of the liquid, the condition for the beginning of bubble incorporation in water is given by

$$nd = \frac{0.11}{(d/D)^{0.2}} \text{ (m/s)} \qquad (22)$$

Because of the increasing influence of d/D on the sorption characteristic equation (Eq. 20 or 21), there is an optimum diameter ratio, $d/$

$D \approx 0.5$, at which maximum mass transfer can be achieved without bubble incorporation (see limiting curves in Fig. 9).

4.3.3 Bubble Entrainment Effects on Mass Transfer

Fig. 9 shows that bubble entrainment has no effect on the mass transfer correlations presented by Eqs. (15, 20 and 21), when water is the aerated fluid. However, when proteins are present and agitation is higher than the limit of bubble incorporation, the results can be disastrous not only from the standpoint of bubble incorporation and cell death, but also for aeration. Fig. 10 shows aeration past the vortex entrainment limit in a pilot-scale vessel containing medium with 5% fetal calf serum, equipped with subsurface turbine (6-blade) or hydrofoil (Prochem Maxflo T, 6-blade) impellers. At high agitation rates, entrained bubbles form a surface froth which decreases $k_L a$. The turbine impeller is located closer to the surface, and hence aerates better at lower Fr_i than the hydrofoil impeller. However, the limit of agitation before bubbles are entrained is lower than with the deeper hydrofoil.

4.4 Surface Aeration Design

Although the sorption equations (Eqs. 18, 20 and 21) would suggest that oxygen input could be increased indefinitely by surface aeration, it is limited in practice by the power dissipation level at which bubble entrainment occurs. Due to the decreasing influence of the diameter ratio d/D on the sorption characteristic, there is an optimum diameter ratio $d/D \approx 0.5$, at which maximum mass transfer can be achieved without bubble incorporation.

For surface impeller operation at the optimized geometry and maximum tip speed without bubble incorporation, Eqs. (21) and (22) lead to the following scale rule:

$$\frac{(k_L a)_l}{(k_L a)_s} = \frac{H_s}{H_l} = \left(\frac{(H/D)_s}{(H/D)_l}\right)^{2/3} \left(\frac{V_s}{V_l}\right)^{1/3} \tag{23}$$

Using the mass transfer relation $Sh \propto Re_i^{3/4}$ (per Eq. 18) for subsurface impellers and optimal geometry conditions, a scale rule similar to Eq. (23) results:

$$\frac{(k_L a)_l}{(k_L a)_s} = \left(\frac{(H/D)_s}{(H/D)_l}\right)^{23/36} \left(\frac{V_s}{V_l}\right)^{13/36} \tag{24}$$

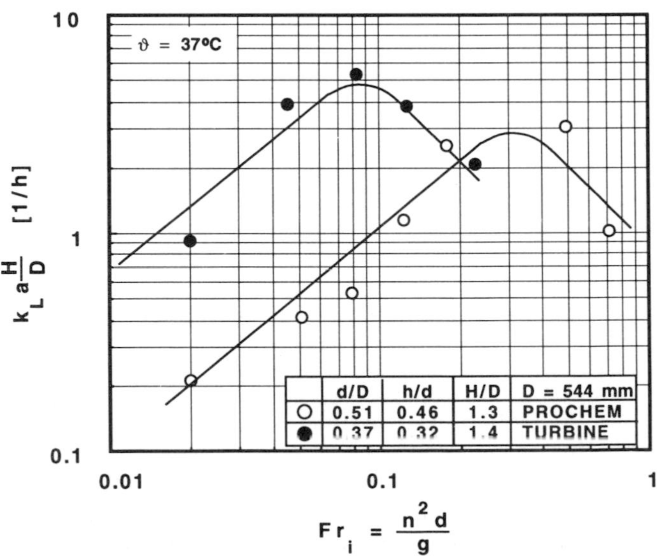

Fig. 10. Surface mass transfer with vortex formation in pilot-scale tanks containing DMEM/5% FBS (AUNINS et al., 1991a).

In both Eqs. (23) and (24), the subscripts refer to small and large scales of operation. This unfavorable scale-up relationship is clear from the limiting curves for the start of bubble incorporation for volumes $V=25$ and 1000 L shown in Fig. 9. As shown by this figure, it is difficult even on the pilot scale to provide sufficient oxygenation for cell cultures via the surface of the liquid. However, studies to delineate entrained gas flows and bubble size distributions and their effect on cells are needed to determine whether these limits are unnecessarily stringent.

Eq. (22) and the work of MATSUMURA et al. (1977) suggest that impeller tip speed be held constant on scale-up. For a single-shaft reactor with multiple impellers this will entail sizing the top impeller according to a different criterion than the lower impellers (see Sects. 2.3.2, 6.2.4 and 7.2 for alternative criteria). Entrainment can also be discouraged by positioning the top impeller deeper beneath the liquid, and by using axial flow impellers pumping upward towards the liquid surface. Although the above studies suggest that the local surface power dissipation rate governs mass transfer rather than impeller design, currently not enough is known about impeller design and placement to maximize surface aeration while minimizing entrainment. Baffles with smaller width and not extending to the liquid surface are also useful to discourage vortex formation and cavitation behind the baffles.

5 Membrane Aeration

Membrane aeration is used to aerate cultures without damaging bubbles or power inputs. The membranes used are open-pore polypropylene (PP) or polytetrafluoroethylene (PTFE), and polydimethylsiloxane (silicone) diffusion types. The two types are shown in Fig. 11. In the microporous membranes, a gas–liquid interface is held stationary inside, or just outside the pores of a hydrophobic membrane. For silicone diffusion membranes, the gas dissolves in the hydrophobic membrane, which has no pores. Mass transfer for either type is through series resistances from

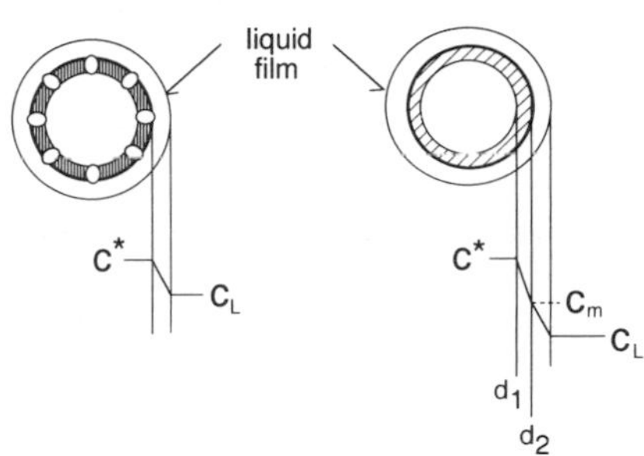

Fig. 11. Aeration membrane structures and concentration gradients.

the liquid film and the membrane itself, and is described by

$$\frac{1}{k}=\frac{1}{k_{\mathrm{L}}}+\frac{1}{k_{\mathrm{m}}} \tag{25}$$

Here, k_{m} is the membrane mass transfer coefficient. In general, both *in situ* and *ex situ* aeration devices employ hollow fibers or tubing, so the following discussion revolves around the cylindrical geometry.

5.1 Membrane Transport Properties

5.1.1 Microporous Membranes

For microporous membranes, the pore does not usually pose a significant resistance to mass transfer; indeed, it is only when the pores are gas-filled that these membranes are superior to diffusion membranes. In the general case, however, the effective diffusivity and the void fraction appear in the mass transfer coefficient

$$k_{\mathrm{m}}=\frac{2\,\mathscr{D}_{\mathrm{p}}\,\theta_{\mathrm{m}}}{\tau_{\mathrm{m}}\,d_2\ln(d_2/d_1)} \tag{26}$$

Here $\mathscr{D}_{\mathrm{p}}$ is the pore diffusivity, θ_{m} the membrane void fraction, τ_{m} the pore tortuosity, and d_1 and d_2 are the membrane tubing inner and outer diameters. As PTFE and PP gas permeabilities are low, transfer occurs only through the membrane pores. Pore size and gas–medium–membrane surface tensions determine the bubble point of the membrane before capillary pressure is overcome and the membrane effectively becomes a sparger (MATSUOKA et al., 1992). This pressure is usually quite low, in the mbar range. For the Accurel® tubing used below, the bubble point is approximately 13 mbar. Properties for some commercially available materials are shown in Tab. 4. For micron-sized pores, $\mathscr{D}_p$ is the free gas diffusivity, and $\tau_m = 1$. For the sub-micron pore Celgard® fibers, however, gas diffusion is in the Knudsen regime where diffusivity is given by

$$\mathscr{D}_p = \frac{8}{3}\left(\frac{RT}{2\pi M_i}\right)^{1/2} r_p \tag{27}$$

Here, $R = 8814.3$ J/kmol K is the ideal gas constant, T is absolute temperature, M_i the gas molecular weight, and r_p the mean pore radius. In this regime, diffusivity differences between O_2 and CO_2 are slightly greater than in liquids.

Properties of microporous membranes can change over the course of usage in culture. Due to protein and cell debris deposition, the pore walls can become hydrophilic and fill with liquid. This difficulty is exacerbated with small pore diameters and long culture times. QI and CUSSLER (1985) determined membrane resistances for NH_3 mass transfer through Celgard® X-20 fibers. In this study, as in any study with these $r_p = 0.015$ µm fibers, visual inspection does not reveal whether the pores were wetted; if the pores are assumed to be gas-filled, their data suggest a tortuosity of $\tau_m = 13$ using Eq. (27) for $\mathscr{D}_p$. If the pores are assumed to be wetted, they calculated $\tau_m = 3$ using $\mathscr{D}_p = \mathscr{D}_{liquid}$. Thus, tortuosity is not negligible for these particular fibers, whether wetted or not. With large-pore, thin membranes the membrane mass transfer resistance is usually negligible and need not be extensively considered unless catastrophic wetting occurs. However, the drawbacks of potential pore wetting and low bubble point make microporous membranes difficult to use *in situ*. They are best applied to short-term cultures, or in an external loop configuration which allows cartridge exchange during the culture if wetting occurs.

5.1.2 Solution Diffusion Membranes

Non-reinforced, and pressure-resistant reinforced silicone tubes and silicone-coated tubular membranes are used as diffusion membranes, as described by a host of authors (MILTENBURGER and DAVID, 1980; FLEISCHAKER and SINSKEY, 1981; BRÄUTIGAM, 1985; AUNINS et al., 1986; BRÄUTIGAM and SEKULOV, 1986). Here, the membrane mass transfer coefficient is modified by the solubility of the gas in the membrane phase. Differences in solubilities and diffusivities between O_2, CO_2, N_2, and H_2O (and NH_3) give rise to selective permeation through the membrane, so that transfer of each gas must be considered independently. Information is relatively unavailable on gas counterdiffusion effects in

Tab. 4. Commercial Microporous Membrane Properties

Manufacturer	Material	Pore Size (µm)	Fractional Pore Vol.	Membrane Thickness (mm)	Membrane O.D. (mm)
Hoechst Celanese	Celgard® X10 (PP)	0.05	0.30	0.3	0.30
Hoechst Celanese	Celgard® X20 (PP)	0.03	0.35	0.0265	0.45
ENKA	Accurel® (PP)	0.3	0.75	0.40	2.60
Gore & Assoc.	Goretex® (PTFE)	2.0	0.50	0.50	4.00
Gore & Assoc.	Goretec® (PTFE)	3.5	0.70	0.50	1.00
Sumitomo Denko	Poreflon® (PTFE)	0.6	0.68	0.55	2.80
Sumitomo Denko	Poreflon® (PTFE)	0.8	0.68	0.55	2.80
Simutomo Denko	Poreflon® (PTFE)	1.0	0.70	0.50	3.00

these membranes, hence they are not considered here, although they definitely occur (BLAISDELL and KAMMERMEYER, 1972; STERN et al., 1977; THORMAN et al., 1975). The mass transfer coefficient is generally described by

$$k_m = \frac{2\,\mathscr{D}_m Hy\,\theta_m f}{Hy_m d_2 \ln(d_2/d_1)} \tag{28}$$

Here $\mathscr{D}_m$ is the diffusion coefficient *in* the membrane, and $Hy_m = P_{gas}/C_m^*$ is the Henry's law constant for the membrane. The group $\mathscr{D}_m/Hy_m$ is known as the membrane permeability. The term θ_m is the tubing void fraction available for gas transport. Resistance increases for tubing which is fiber-reinforced to allow pressurization; here, the void fraction becomes less than unity. For the reinforced tubing of Membrantechnik Hamburg used below, $\theta_m = 0.9$. f is a correction factor which accounts for tubing dimension changes from elastic deformation of the silicone rubber.

Unlike the microporous membranes, silicone tubing, especially reinforced tubing, can be internally (or externally) pressurized to increase O_2 transfer driving force. When non-reinforced tubing is pressurized, modifications in the apparent dimensions are necessary to account for tubing elasticity even at moderate transmembrane pressures (VARGA, 1966; STERN et al., 1977). Radial expansion under pressure affects calculation of k_m, and can be accounted for by f, defined by

$$f = \frac{d_2 \ln(d_2/d_1)}{d_2' \ln(d_2'/d_1')} \tag{29}$$

Here the prime denotes the pressure stressed state. f is unity with unstretched tubing and no transmembrane pressure; in the stressed state it is a non-dimensional function of the ratio of the unstressed inner and outer tube diameters, $n = d_1/d_2$, and the transmembrane pressure divided by Young's modulus, $\Delta p/E_Y$. Fig. 12 shows a graphical representation of f for tubing undergoing internal pressurization. For tubing being pressurized from the outside (as in Verax-style aerators), the analogous plot is available from STERN et al. (1977). The Young's modulus, which can be anisotropic in the radial and axial dimensions, was measured as an average 30.6 bar for Dow-Corning Silastic® Medical Grade by the above authors. For reinforced tubing, the membrane modulus is large and anisotropic; elasticity is smaller, and Fig. 12 does not apply.

For either pressurizing tubing or stretching it over supports, surface area expansion must

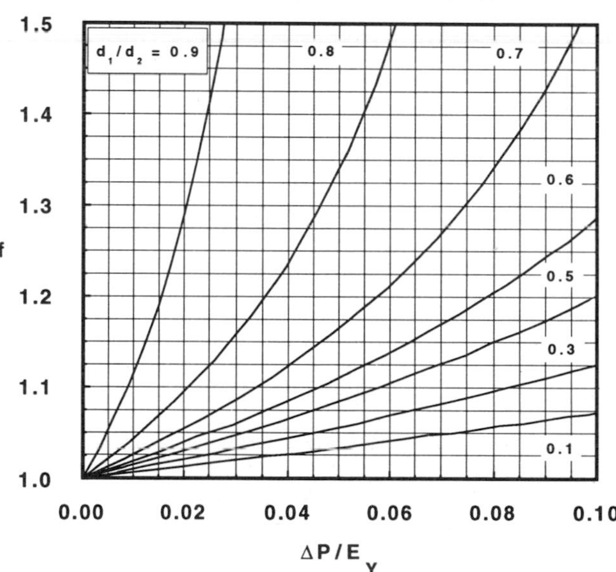

Fig. 12. Tubing coefficient correction for expansion due to transmembrane pressure.

be accounted for independent of the effect on k_m, as it also affects liquid side transfer:

$$a' = \frac{L' d_2'}{L d_2} a \tag{30}$$

Here L is the tubing length, and the prime L' denotes the pressured and/or stretched state.

Silicone rubber is particularly favorable for oxygen transfer, since the O_2 diffusion coefficient is slightly higher than in water, and solubility is about 4 to 5 times that in water. According to BRÄUTIGAM (1985), the temperature dependence of the system constant $\mathscr{D}_m Hy/Hy_m$ for oxygen–silicone–water in the temperature range $10 \leq \vartheta(°C) \leq 40$ is:

$$\frac{\mathscr{D}_m Hy}{Hy_m} = 7.3 \cdot 10^{-9} \exp(0.024\,\vartheta) \;\; (m^2/s) \tag{31}$$

For the other culture gases, Tab. 5 shows permeabilities in dimethylsiloxane rubber from ROBB (1968). Although silicone rubber is very permeable to O_2 relative to other polymers, it is even more permeable to CO_2 and NH_3, and is also permeable to water when presented with a liquid water driving force. Reactors which employ silicone tubing aeration will clearly experience differential oxygenation and ventilation, and reactor operation will require modification accordingly to maintain both O_2 and CO_2 at optimal levels. A further ramification is that water vapor pressure inside the tubing reaches equilibrium for low gas flow rates, so the transfer driving force should be modified according to Eqs. (3) and (38) where necessary (this is also true of microporous tubing; see gas flow rate effects in BÜNTEMEYER et al., 1987). Dependence on temperature at constant pressures generally follows the law

$$\frac{\mathscr{D}_m}{Hy_m} = c \exp\left(\frac{E_p}{RT}\right) \tag{32}$$

where c is a solute-dependent constant, E_p an activation energy of permeation, R the ideal gas constant, and T absolute temperature. STERN et al. (1977) showed E_p values of 265 and 337 J/kmol for O_2 and N_2, while ROBB (1968) showed that CO_2 permeability is insensitive to temperature. The silicone tubing type also influences k_m via permeability changes. Modification of the common polydimethylsiloxane by other aliphatic or aromatic pendant groups decreases permeability by orders of magnitude for all culture gases. Silica filler, added to enhance material strength properties, also decreases permeability. Silastic® Medical Grade tubing, with a silica content of 32.3 vol %, possesses an oxygen permeability 74.5% of the value shown in Tab. 5, which was measured with a polymer containing 18.2 vol % filler (STERN et al., 1977). It is a practical problem to accurately determine membrane properties, as dimensions can vary along the tubing, and can change by deformation due to installation. As an aside, anchorage-dependent cells are capable of adhering to silicone rubber (HARRIS et al., 1980), and this can pose an additional mass transfer resistance. Although modifications in solution membrane resistance and permeability are a science in themselves, only in cases where k_m is the dominant resistance to mass transfer it needs to be known exquisitely. This does occur in many cases, where liquid velocities across or within the tube face are large, making the liquid film resistance small.

Tab. 5. Silicone Membrane Permeabilities at $\vartheta = 25\,°C$ for Poly(dimethylsiloxane), 18 vol.% Silica Filler (from ROBB, 1968)

Gas	Permeability [kg m/s m² (N/m²)]
Oxygen	$6.4 \cdot 10^{-15}$
Nitrogen	$2.6 \cdot 10^{-15}$
Carbon dioxide	$4.8 \cdot 10^{-14}$
Ammonia	$3.4 \cdot 10^{-14}$
Water vapor [a]	$2.2 \cdot 10^{-13}$

[a] The value for water vapor was determined against a driving force of liquid water at a 1 bar and 25 °C. Although this results in an apparent permeability which is much too large given the hydrophobicity of the membrane, it illustrates the relative flux of water which occurs in the reactor environment.

5.2 *In situ* Membrane Aeration

The Sherwood number describing liquid phase transport from the membrane face should strictly depend only on the Schmidt number and the membrane Reynolds number, $Sh_t = f(Sc, Re_t)$. Here the Sherwood number $Sh_t = k_L d_2 / \mathscr{D}$ and the Reynolds number $Re_t = u d_2 / \upsilon$ are functions of the outer diameter d_2 of the membrane and a characteristic velocity u at the tube face. This velocity ideally will be known from the tubing velocity itself, or will be proportional to the impeller tip speed. For non-ideal *in situ* aeration situations, the characteristic velocity can be ill-defined due to spacial constraints on the tube support necessary to accomodate tank internals, or agitator design compromises necessary to satisfy other design criteria. In these cases, a mean circulation flow will determine the characteristic velocity at the tube as discussed in Sect. 5.2.3.

5.2.1 Microporous Membrane Aeration

Some experimental $k_L a$ measurements in reactors with open-pore Accurel® tubes are shown in Fig. 13 in the form $Sh_t = f(Re_t)$ (HENZLER and KAULING, 1993). The pore area, equal to 75% of the total area and roughly corresponding to the gas–liquid interfacial area, is used to determine k_L values. Membrane resistance is negligible in these experiments as the pores are large and unwetted. Various definitions are appropriate for the Reynolds number, so that direct comparison of the efficiencies of the different reactor types is impossible in this form. In the case of the Diessel reactor, an eccentrically moving tube is the mixing mechanism, and the Reynolds number is derived directly from the velocity of the surface of the tube, $u = 2\pi e n$ (LEHMANN et al., 1985; VORLOP and LEHMANN, 1988). For an impeller-stirred reactor with a stationary tubular stator concentric with the impeller, only the tip velocity of the impeller, $u = u_i = \pi d n$, is known, and this is the value used. In a stirred reactor with stationary tubing, the anchor impeller is the best system by far since it produces an intense and uniform flow radially towards the tube. This is particularly true in comparison with low-pitch axial flow impellers

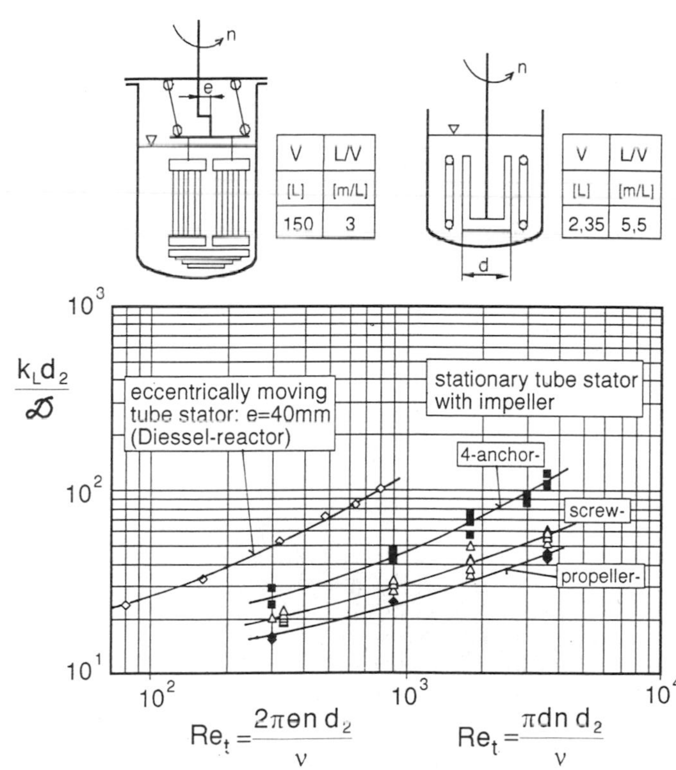

Fig. 13. Mass transfer characteristics for membrane oxygenation, Accurel membranes, water, $\vartheta = 37$ °C (from HENZLER and KAULING, 1993).

such as propellers, but it can be seen in Fig. 13 that this also applies to tall helical-blade impellers having the same axial length as the anchor impellers used.

For a reactor with stationary microporous tubing and a 2-blade anchor impeller, the data in Fig. 14 reduce to the following general relation for water in the range $20 < \vartheta < 37\,°C$:

$$Sh_t = (3.2 + 0.00088\, Re_t^{1.2})\, Sc^{1/3} \quad \text{for } 250 < Re_t < 6000 \text{ and } 200 < Sc < 500 \tag{33}$$

This relation was also confirmed by further results with diffusion membranes. The observed dependence on the Schmidt number is similar to that indicated in the literature for circular cylinders with cross flow (BRAUER and MEWES, 1971; PERKINS and LEPPART, 1962). In the case of microporous Accurel® membranes, lower $k_L a$ values were found in media containing albumin than in water (see Fig. 22), which is analogous to bubble aeration. This can be explained by additional diffusion resistance at the bubble surface due to the presence of protein. For diffusion membranes, no negative effect on mass transfer is observed with protein addition.

5.2.2 Diffusion Membrane Aeration

For diffusion membranes, the membrane resistance contributes significantly to the overall transfer resistance. Hence, the overall mass transfer coefficient k is required for predicting oxygenation with diffusion membranes. The membrane resistance can, in principle, be determined mathematically by means of Eqs. (25, 28 and 33) or Figs. 13 and 14, along with a knowledge of the membrane properties. However, when the membrane is pressurized with gas internally, there can be additional effects resulting from liquid supersaturation at the membrane–liquid interface. The condition for supersaturation is given by

$$\frac{C_{\text{membrane}}}{C_R^*} = \frac{Hy\, C_L}{(P_{\text{gas}})_R}\left(1 - \frac{k}{k_L}\right) + \frac{P_{\text{gas}}}{(P_{\text{gas}})_R}\,\frac{k}{k_L} > 1 \tag{34}$$

This is particularly likely when the gas concentration C_L in the liquid phase is high, when the gas pressure in the membrane, P_{gas}, is higher than the pressure in the reactor $(P)_R$, and when the ratio k/k_L is near one, i.e., when the membrane is thin or when there is little movement

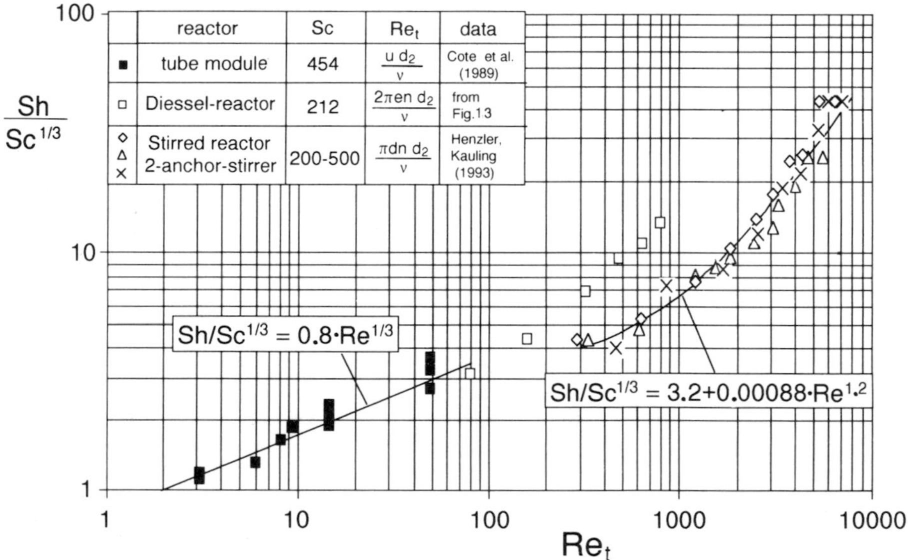

Fig. 14. Dimensionless mass transfer characteristics for membrane oxygenation.

at the membrane. Supersaturation is readily observed in mass transfer experiments (CÔTÉ et al., 1989; HENZLER and KAULING, 1990, 1993), and can lead to the formation of microbubbles, with the result that mass transfer is adversely affected relative to the theoretical maximum that would occur without bubbling. This phenomenon is seen in the upper diagram of Fig. 15, where for reinforced tubing, the apparent mass transfer coefficient decreases with increasing P_{O_2} in the gas phase. In this figure, the mass transfer coefficient is defined using the gas phase driving force, $ka=(G/V)/(P_{O_2}/Hy-C_L)$. When the membrane face reaches saturation, further increases in gas flux through the membrane create bubbles in the liquid. The oxygen mass in the bubble is not entirely transferred to the medium during its rise and disengagement; hence k decreases relative to the theoretical flux through the membrane for the given partial pressure driving force. Oxygen bubble formation and locally high levels of dissolved O_2 are potentially detrimental to culture health; it may be necessary to hold $C_{membrane}$ below a maximum tolerable value.

Mass transfer can be distinctly improved by a suitable impeller system, as shown in Figs. 13 and 16. The close-proximity anchor impeller developed for membrane aeration by HENZLER and KAULING (1990, 1993) is clearly superior to axial flow impellers frequently described in the literature and recommended by manufacturers, such as propellers and pitched-blade impellers. With these other impellers, k decreases with increasing coiling tightness, owing a stagnation in the reactor at and outside the stator. Conventional impellers require much higher shaft speeds and power inputs to give equivalent $k_L a$, with the result that shear on the cells is increased, defeating the purpose of the method. For coils with $w/d_2<1.3$, a

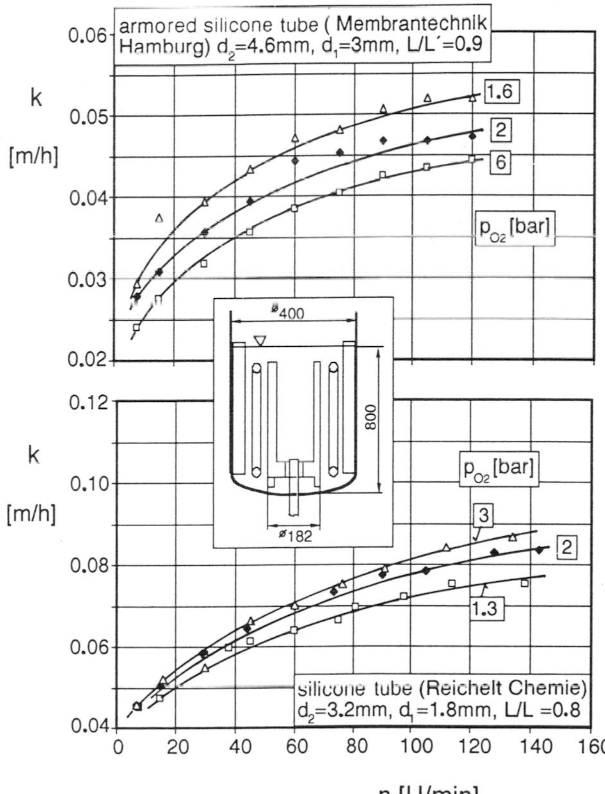

Fig. 15. Mass transfer coefficients for diffusion membranes, 2-blade anchor impeller, water, $\vartheta=35\ °C$, $(P_{O_2})_R=0.21$ bar, $C_L\approx 3$ mg/L (from HENZLER and KAULING, 1993).

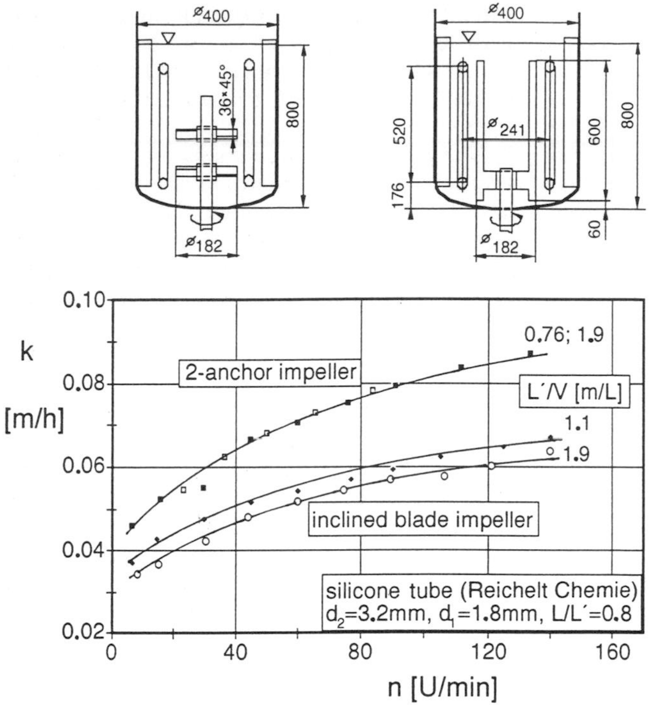

Fig. 16. Influence of impeller system on mass transfer coefficients, water, $\vartheta = 35\,°C$, $P_{O_2} = 3$ bar, $(P_{O_2})_R = 0.21$ bar, $C_L \approx 3$ mg/L (from HENZLER and KAULING, 1993).

pitched-blade impeller, e.g., can require more than 15 times as much power as the recommended anchor impeller to produce the same mass transfer!

5.2.3 *In situ* Membrane Aeration Design

Membrane aeration of cell culture reactors suffers from scaling law limitations like surface aeration; only a finite amount of tubing can be placed into a reactor without destroying the mixing properties of the reactor, unless ingenious ways to place more membranes in a given volume can be devised (e.g., the Diessel reactor). Fortunately, the additional $k_L a$ provided by the membrane, combined with tricks such as tubing pressurization for diffusion membranes, can give an absolute transfer rate large enough to devise membrane reactors for low *OUR* processes near the 1000 L scale. The tubing area per volume is given by

$$a_m = \frac{\pi z d_2 d_h h_h}{w D^2 H} = \frac{\pi z}{D}\left(\frac{d_2}{w}\right)\left(\frac{d_h}{D}\right)\left(\frac{h_h}{H}\right) = \pi z \frac{(H/D)^{1/3}}{V^{1/3}}\left(\frac{d_2}{w}\right)\left(\frac{d_h}{D}\right)\left(\frac{h_h}{H}\right) \quad (35)$$

where d_h is the helix diameter, h_h the helix height, and w the helix pitch, or spacing per turn. For helical coiling of the tubing, $z=4$, and for a toroidal winding as in Fig. 13, $z=8$. Winding the tubing toroidally as shown in Fig. 13 is preferable for the area advantage, and to prevent cell and particle deposition on the tubing surface. It also allows a simple support structure to achieve a cylindrical cage; helical windings sacrifice proximity to the impeller as the number of struts is decreased. In practice, the amount of tubing is limited by the need to mix the fluid in the region outside the tubing stator. This will be dependent on the impeller, process, and characteristics of any particles to be suspended. The winding pitch, or tube spacing may be constrained for good liquid circulation if particle suspension is involved; however, no experimental information is currently available. For mass transfer, Fig.16 demonstrates no decrease in efficiency at $w = 1.3\, d_2$.

Proper system selection and operation is constrained by commercially available tubing, and realizable tank configuration. Eq. (35) shows that there is no effect on area for different tubing sizes if d_2/w is constant. Small tubing is thinner and provides less membrane resistance; it also often has a higher burst pressure than large tubing since d_1/d_2 tends to be smaller (see Fig. 12). However, small tubes are more fragile under tensile strain, which is undesirable for a robust manufacturing process. Optimal tubing dimensions, as far as mass transfer and tubing strength are concerned, are discussed further in Sect. 7.1.1.

For microporous tubing aeration at constant power input, using Eq. (33) in the high Re_t limit (constant ratio. impeller Newton number to H/D and $Sh \propto Re_t^{1.2}$) gives a scale rule for $Re_t > 2000$:

$$\frac{(k_L a)_l}{(k_L a)_s} = \left(\frac{(H/D)_l}{(H/D)_s}\right)^{0.2} \left(\frac{V_l}{V_s}\right)^{-0.2} \tag{36}$$

Maximum possible ka values for diffusion membranes can be calculated using the tubing area and Eq. (28) for the membrane resistance. It is clear that where possible an anchor impeller type should be used to maximize transfer with minimal energy input. If an anchor impeller configuration is not possible and hence Eq. (33) is not valid, Figs. 13, 14 and 16 can be used to estimate reactor performance.

If the tank geometry must be varied as well, literature data on the Nusselt number for coils in stirred tanks can be used to advantage. Taking the average of available literature correlations (AUNINS et al., 1986), the heat–mass transfer analogy leads to the form

$$Sh = \frac{k_L d_2}{\mathscr{D}} = C \left(\frac{\upsilon}{\mathscr{D}}\right)^{0.34} \left(\frac{n d^2}{\upsilon}\right)^{0.63} \left(\frac{d_h}{D}\right)^{-0.26} \cdot \left(\frac{d}{D}\right)^{0.14} \left(\frac{d_2}{D}\right)^{0.53} \tag{37}$$

This compilation is for tubing coiled in a helix concentric around a central impeller shaft; d_h is the diameter of the helix. For radial flow impellers in close proximity to the tubing, a more appropriate correlator than the groups d_h/D and d/D might be the impeller–tubing spacing, $s = (d_h - d)/2$. For very small tubing, the exponent of d_2/D is probably negligible. Since the data of the authors cited cannot be converted to tubing Reynolds numbers, the quality of the impeller Reynolds number term is also unknown, although the exponent is well established to lie between 0.50 and 0.75. In truth, this equation expresses the effects of geometry on the circulation characteristics of a reactor–impeller–coil system, and hence the characteristic velocity at the tube face (DESPLANCHES et al., 1983), as mentioned above. The geometric factors largely disappear when agitation is provided by a jet that gives a constant mean circulation velocity for a given jet Reynolds number (HSU and SHIH, 1984). Although Eq. (37) is probably imperfect for these reasons, the dependences have been derived from a variety of reactors with different impeller types and baffling arrangements; it should serve as a first approximation for deviations from the reactor geometries investigated above.

Increasing stator diameter while keeping the impeller–cage spacing s constant results in the lowest film resistance, due to increased tip speeds and circulation velocities. In practice, the largest stator and impeller diameters possible which accomodate the necessary tank internals should be employed. In the above studies $0.46 < d/D < 0.5$, although in large tanks d/D can be increased to and beyond 0.6. Eq. (35) shows that the area is inversely proportional to D, favoring high aspect ratio tanks. Impeller tip speed is slightly decreased for high aspect ratios at constant power input. However, tank geometry is complicated by the intended usage. Membrane aeration is most appropriate for those cultures, usually microcarrier cultures, which can least stand high power inputs and impeller tip speeds. For calculating energy input tolerable in the culture, the cage volume is the appropriate dissipation volume. High aspect ratios also are not typically used for solids suspension, although the minimum suspension condition is approximately correlated with impeller tip speed (ZWEITERING, 1958). As an all-around compromise, tank aspect ratios $H/D \leq 2$ are favored for typical membrane aeration applications; for the above studies $0.85 < H/D < 2$.

5.3 *Ex situ* Membrane Aeration

For *ex situ* aeration, the membrane is in a module situated outside the fermenter in a circulation system. Culture fluid is enriched with O_2, stripped off CO_2 and returned to the reactor, as shown in Fig. 17. This configuration is common to many specialty loop reactors such as hollow fibers membranes. The logarithmic ΔC given by Eq. (3) should be used to calculate transfer rates. For a small gas concentration change, $C_1^* \approx C_2^* \approx C^*$ and so the gas mass flow is:

$$\frac{G}{V} = k\frac{A}{V}\frac{(C_2 - C_1)}{\ln\left(\frac{C^* - C_1}{C^* - C_2}\right)} \tag{38}$$

Recognizing that $G = q_L(C_2 - C_1)$, the unknown outlet concentration C_2 can be ascertained from the circulation system flow rate q_L and the operating concentration difference $(C^* - C_1)$ to give

$$\frac{G}{V} = \frac{q_L}{V}\left[1 - \exp\left(\frac{-kA}{q_L}\right)\right](C^* - C_1) \tag{39}$$

The turnover rate q_L/V necessary to provide a sufficient oxygen input $G/V = q_{O_2}X$ at $C_L = C_1$ can also be determined:

$$\frac{q_L}{V} = \frac{q_{O_2}X}{C_2 - C_1} \tag{40}$$

Eq. (40) shows that even at the relatively low oxygen demand of $q_{O_2}X = 6$ mg/L/h the entire contents of the fermenter must be circulated *at least* once per hour if $C_2 = C_{max} \approx 6$ mg/L. *Ex situ* oxygenation is therefore usually suitable only for low intensity cultures, or microcarrier, packed- or fluidized-bed cultures where the cells do not enter the aeration loop.

Figs. 14 and 18 show mass transfer measurements in membrane modules. The findings shown in Fig. 14 include values obtained by Côté et al. (1989) for a silicone tube module with gas on the inside and a transverse flow of water. This study agrees very well with the values for the Diessel reactor. Mass transfer behavior is less favorable if the liquid is fed into the tubular membrane lumen (cf. Figs. 14 and 18). The results in Fig. 18 for Accurel® membranes follow the mass transfer relationship given by Leveque (1928):

$$Sh_t' = \frac{k_L d_1}{\mathcal{D}} = 0.8(Re\,Sc(d_1/L))^{1/3} \tag{41}$$

which was derived theoretically for tubular flow with a fully developed laminar velocity profile. For transverse flow on the outside of the fibers, the data of Côté et al. (1989) in Fig. 14 reduce to the equation

$$Sh_t = \frac{k_L d_2}{\mathcal{D}} = 0.6\,Sc^{1/3} Re_t^{1/3} \tag{42}$$

Since the term d_1/L in Eq. (41) is always much less than one, it is seen that higher mass transfer results when the liquid is run outside the tubing in transverse flow. This can be especially advantageous to minimize recirculation

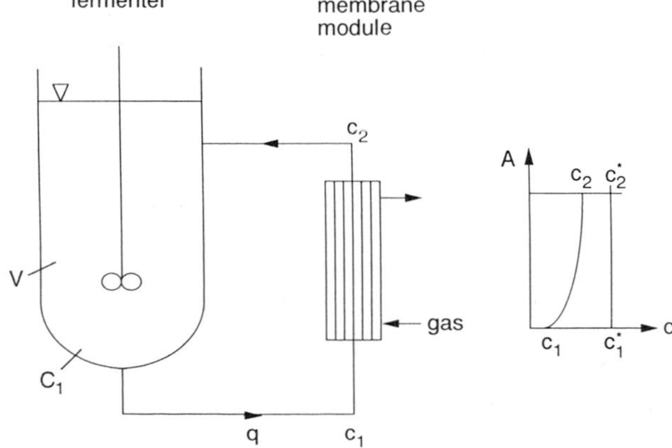

Fig. 17. Bioreactor with external oxygen supply.

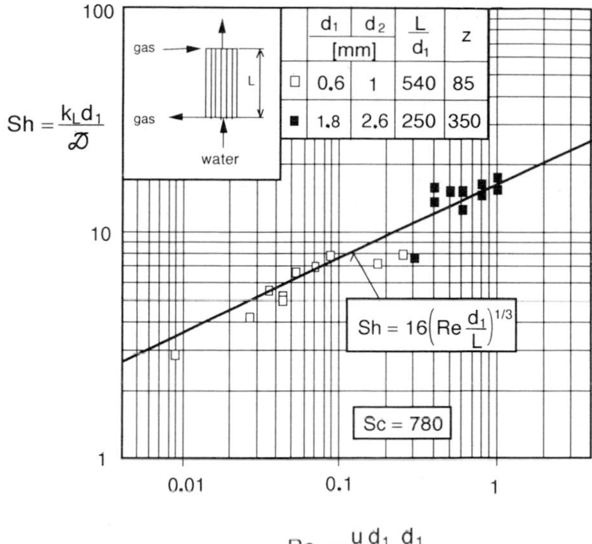

Fig. 18. Mass transfer characteristics for Enka Accurel module: LM2P06; MD070CP2L.

rates in systems where membrane fouling and particulate sedimentation are not issues.

Design of external aeration is straightforward when using commercial modules constrained in area, fiber dimensions, and number. Specifying the process constraints of q_{O_2}, X, C_2, and C_1 in Eq. (40) gives the minimum liquid circulation rate. This flow determines the superficial velocity through the module, and hence the mass transfer coefficient via Eq. (41) or (42). Using Eq. (39), the necessary gas phase partial pressure can be calculated, after substituting $q_{O_2}X = G/V$. If the partial pressure is beyond the limits of the membrane, the calculation must be repeated for the next larger module. For the microporous membranes, this will be the membrane bubble point. For silicone membranes, this will be somewhat below the burst pressure of the membrane (see Sect. 7.1.1 for further comments on optimization). An additional constraint which may enter is that the membrane face concentration per Eq. (34) is below a determined tolerable value.

6 Sparged Aeration

In sparged aeration, gas–liquid mass transfer depends not only on the operating conditions – superficial gas velocity v_s and volume- or mass-related impeller power, P/V or ε – but also on the material parameters – density ρ, viscosity υ, surface tension σ, the diffusion coefficient $\mathscr{D}$, interfacial film effects, and the coalescence behavior of the liquid. Apart from the type of sparger, which is of particular importance in bubble columns and airlift reactors at low gas loadings, and in stirred reactors at low power input, the geometry of the apparatus is unimportant unless it influences bubble coalescence, which it does in reactors of small diameter and low height. For accurate prediction and scale-up, it is therefore important to use values that were obtained in sufficiently large apparatus. When the coalescence behavior of the system is constant, the mass transfer relation can be represented purely formally by the following dimensionless numbers:

$$Y = k_L a \left(\frac{\upsilon^2}{g^2}\right)^{1/3} = f\left[\left(\frac{v_s}{(g\upsilon)^{1/3}}\right), \left(\frac{\varepsilon}{(g^4\upsilon)^{1/3}}\right), \left(\frac{\upsilon}{\mathscr{D}}\right), \left(\frac{\sigma}{\rho(\upsilon^4 g)^{1/3}}\right)\right] \quad (43)$$

As HENZLER (1982) has shown, the results for a given gas–liquid system can be correlated without referring to the dimensionless numbers $Sc = \upsilon/\mathscr{D}$ and $\sigma^* = \sigma/\rho(\upsilon^4 g)^{1/3}$. However, to adapt data to culture systems, material properties must be taken into account. Viscosity effects can be reproduced with sufficient accuracy by the sorption coefficient Y and the two process coefficients $[v_s/(g\upsilon)^{1/3}]$ and $[\varepsilon/(g^4\upsilon)^{1/3}]$. The Schmidt number Sc should be considered when gases other than oxygen are treated; the sorption coefficient Y is a power law function, $Y \propto Sc^c$. For a mobile phase interface, which is applicable for clean interfaces or large bubbles with $d_b > 2.5$ mm, the exponent c is 0.5; for surfactant-laden interfaces or bubbles smaller than 2.5 mm, the exponent c is 0.67 (CALDERBANK, 1959; CALDERBANK and MOO-YOUNG, 1961). For carbon dioxide and oxygen, steady-state sorption coefficients will differ by a factor of approximately 0.9, regardless of the exponent.

6.1 Gas Agitated Reactors

In theory, knowledge of bubble size, gas superficial velocity, and material properties should suffice to predict mass transfer in airlifts and bubble columns. However, the relations derived from physics are hampered by material specific characteristics. The simpler correlation approach above is thus usually applied to pneumatically agitated reactors.

6.1.1 Bubble Column Aeration

For a specified material system, reactor dimensions (D and H) and sparger design, the mass transfer coefficient $k_L a$ in bubble columns and loop reactors depends only on the superficial gas velocity v. The power law relation

$$k_L a = B v^b \tag{44}$$

is valid for this case over a limited range of gas velocities. The exponent b is a function of the gas velocity when undergoing the transition from the low gas holdup and low coalescence regime to the coalescing regime. Placing this equation into dimensionless form yields the relation

$$Y = \frac{k_L a}{v}\left(\frac{\upsilon^2}{g}\right)^{1/3} = A\left(\frac{v}{(\upsilon g)^{1/3}}\right)^{-a} \tag{45}$$

This correlation follows from Eq. (43) for $\varepsilon = 0$. It also gives an accurate representation of the influence of the gas velocity and medium viscosity, when taken over the regime of reactor operation for which it was determined. The results of experiments at the 50 to 150 L scale are shown in Fig. 19. The influence of the sparger hole diameter is very noticeable at these gas loadings, in the range desired for cell cultures. This is because the coalescence rate is low at low superficial velocities. That the coalescence rate is small is shown by the two curves for the 0.2 mm diameter sparger holes; a negligible decrease in $k_L a$ is observed with a three-fold increase in reactor height. With increasing v, the ratio $k_L a/v$, which is a measure

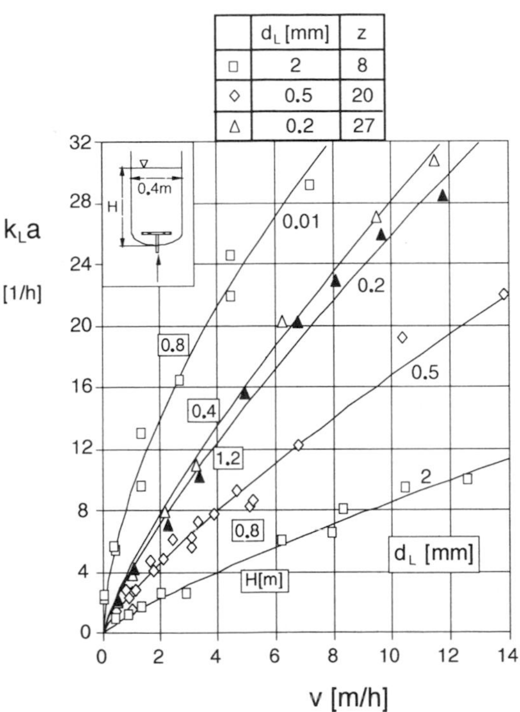

Fig. 19. Mass transfer coefficients for bubble columns, water, $\vartheta = 35\,°C$ (from HENZLER and KAULING, 1993).

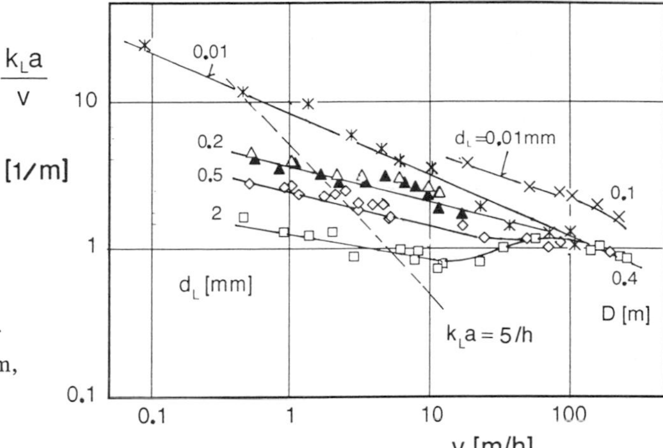

Fig. 20. Related mass transfer coefficients for bubble columns: influence of sparger and column diameter; $H > 0.8$ m, water, $\vartheta = 35\,°C$ (from HENZLER and KAULING, 1993).

of the aeration efficiency, i.e., of the utilization of the gas, decreases due to increased coalescence and bubble birth size at the sparger. This is shown in Fig. 20, in the range $v < 10$ m/h. In this range the sorption characteristic may be correlated by Eq. (45), and the resulting correlations are shown in Tab. 6.

A transition from the correlation regime can be seen at $v \geq 10$ m/h in Fig. 20. At this point the dynamic equilibrium caused by gas coalescence determines the bubble size and gas holdup, and k_La starts to become independent of sparger design. To achieve mass transfer without increased cell death rates from coalescence events and altered bubble size, it may be advisable to operate with increasing gas phase oxygen content rather than further increasing gas flow. The large bubble, air–water system used here is probably more coalescent than normal culture medium, so this is a conservative estimate of maximum gas flow rate.

As mentioned in Sects. 2 and 5, mass transfer is clearly hindered in medium with protein present. This is shown in Figs. 21 and 22 for sparging into a 1:1 DMEM/F12 mixture containing varying amounts of albumin. The achievable k_La with albumin present falls to less than 50% of the water value. This is due to increased viscosity, and the adsorbed protein layer at the interface. Sparged ($d_L = 2$ mm) and surface mass transfer into DMEM containing 5 vol% serum is decreased to 33% of phosphate buffered saline rates at $\vartheta = 37\,°C$ (AUNINS et al., 1991a). Although the protein content of such a mixture is only 1.8 g/L, the larger decrease compared to the values shown in Fig. 22 is probably due to the presence of multiple proteins and lipids which adsorb more avidly to the interface. Only about 10% of the observed difference from water values can be explained by viscosity, according to the dimensionless correlation of Eq. (45), and the data in Fig. 5 and Tab. 3. From these results and Fig. 21, where the intensity of the effect differs with the diameter of the sparger holes, it may be assumed that the remainder of the

Tab. 6. Sorption Characteristics for Bubble Columns (see Eq. 45)

Gas-sparger	d_L (mm)	A	a	$\frac{v}{(\upsilon g)^{1/3}}$
Frit	0.01	$6 \cdot 10^{-5}$	0.4	0.001–0.1
	0.2	$6.3 \cdot 10^{-5}$	0.2	
Double ring	0.5	$4.2 \cdot 10^{-5}$	0.2	0.01–0.2
	2	$2.2 \cdot 10^{-5}$	0.17	

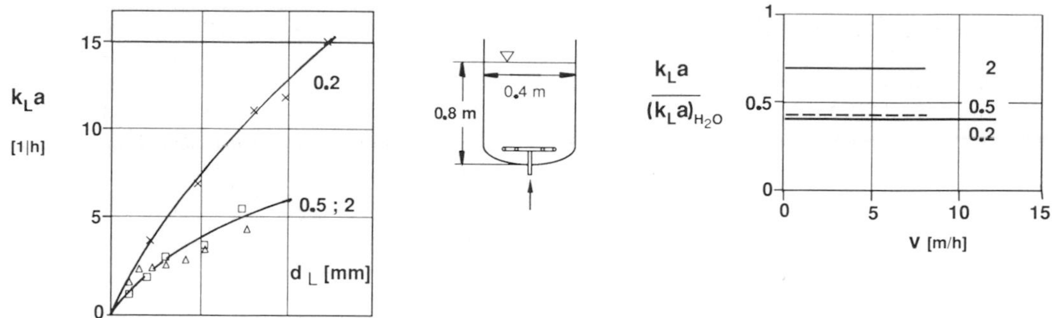

Fig. 21. Mass transfer coefficients for bubble columns, nutrient solution (left diagram) and the comparison with k_La values in water (right diagram); medium: DMEM/Ham's F12: 1:1, 3 g/L albumin; $t = 35$ °C (from HENZLER and KAULING, 1993).

mass transfer hindrance is caused by two surfactant-related phenomena. First, the protein solution may wet the sparger orifice better, causing undetermined changes in the initial bubble size due to changed contact angle. Second, the presence of an adsorbed layer at the interface decreases transfer. The greater reductions found with small sparger hole diameters ($d_L = 0.2$ and 0.5 mm) indicate that in the case of smaller bubbles, owing to the rigid interface, the inhibition of mass transfer by protein on the bubble surface is more pronounced than for large bubbles. Increased coalescence of bubbles may be ruled out, because a tendency to foam is observed in protein solutions, and the gas holdup is low.

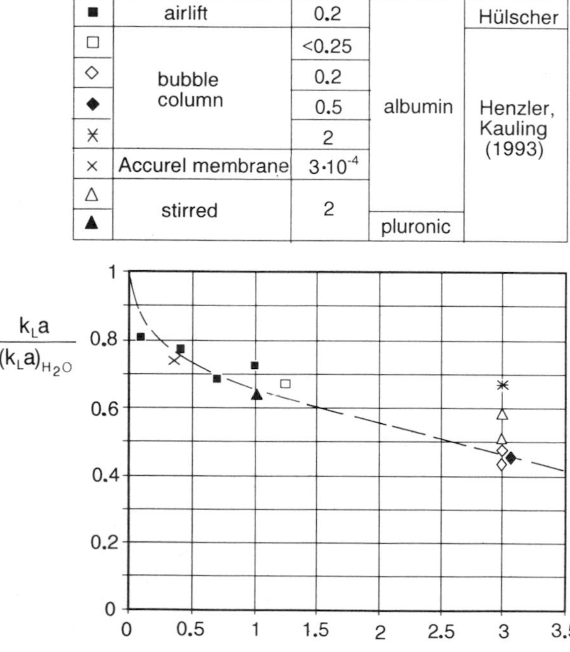

Fig. 22. Influence of polymer content on mass transfer.

6.1.2 Airlift Reactor Aeration

For the same superficial gas velocity per total cross section of the reactor, mass transfer is lower in a loop reactor than in a bubble column. This is because only part of the column cross-section is aerated, and because $k_L a$ is less than directly proportional to gas velocity. This is confirmed by the experimental results in Fig. 23. Assuming that bubble size is the same in

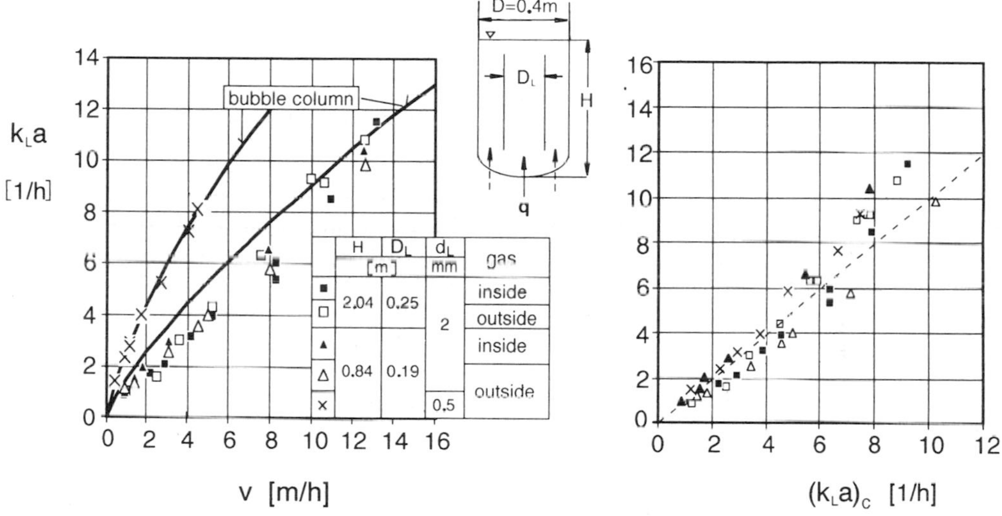

Fig. 23. Mass transfer coefficients $k_L a$ for airlift reactors as a function of superficial gas velocity $v = 4q/\pi D^2$ (left diagram) and comparison of measured $k_L a$ with $k_L a_c$ calculated by Eq. (46) (right diagram), water, $\vartheta = 34\,°C$ (from HENZLER and KAULING, 1993).

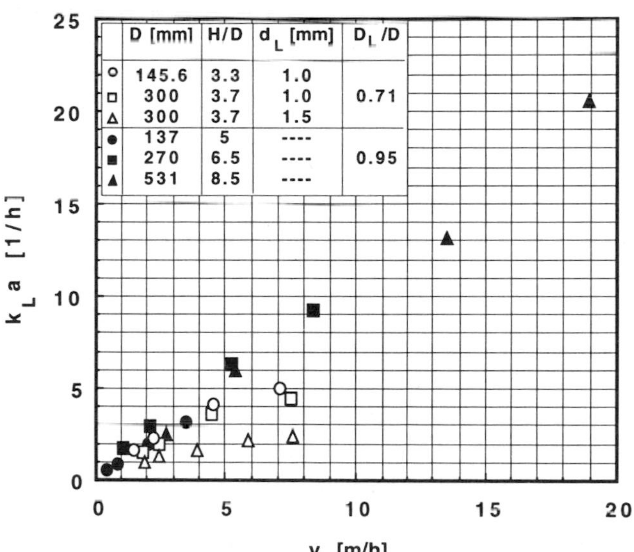

Fig. 24. Aeration rate versus draught tube velocity. Data of KATINGER et al. (1979): open symbols; BIRCH et al. (1987): closed symbols.

both reactor types and that no bubbles are entrained in the downcomer, it should be possible to calculate the mass transfer coefficients for loop reactors, $(k_L a)_c$, from those for the bubble columns $k_L a$. The actual superficial velocity v_s in the aerated volume is used here in the bubble column characteristic

$$(k_L a)_c = k_L a(v_s) \left(\frac{4A_u}{\pi D^2}\right)^2 \tag{46}$$

In this equation, A_u is the cross-sectional area of the upriser section. In the region of interest, there is no difference between reactors sparged in the annulus or the draught tube. From the right panel in Fig. 23, Eq. (46) allows a rough calculation up to $(k_L a)_c \approx 6$/h. Beyond this point, bubble circulation in the airlift downcomer increases the transfer rates for these columns.

Fig. 24 shows data taken from literature sources using water at $\vartheta = 37\,°C$ as the aerated fluid. The data from BIRCH et al. (1987) (also THOMPSON et al., 1985) were obtained with serum-free medium in 10, 100, and 1000 L production airlift reactors at Celltech Ltd., an eminent cell culture firm employing airlift reactor technology. Details on reactor and sparger construction are not readily available; however, from ancillary data presented by THOMPSON et al. (1985), it is possible to calculate $D_L/D = 0.95$ for these vessels, very close to a bubble column design. The data are plotted as $k_L a(\pi D^2/4A_u)^2/v_s$ versus v_s in Fig. 25. This compares well with the bubble column data in Fig. 20. For the series of reactors, gas velocity effects decrease with increasing scale, and for the 1000 L reactor there is no apparent effect. This may indicate that coalescence dominates bubble size for this reactor and medium, possibly due to the reactor height, ca. 4.5 meters.

Fig. 24 also shows data from KATINGER et al. (1979) taken in pilot-scale airlift reactors. These data are of interest since they were taken in spent, filtered medium containing 10% calf serum at $\vartheta = 37\,°C$. Probably they are the most representative data available as far as true broth properties are concerned. It is striking that the $k_L a$ values are similar to the data of BIRCH et al. and to Fig. 20, suggesting no apparent surfactant hindrance to mass transfer. Lacking data on the material properties of the broths, it is not possible to recommend a sorption characteristic for either set of data in Fig. 25.

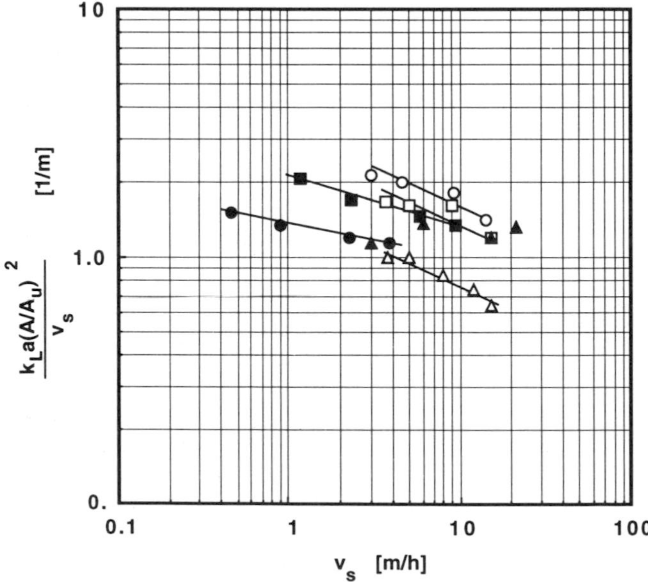

Fig. 25. Aeration efficiency for airlift reactors, data of KATINGER et al. (1979): open symbols; BIRCH et al. (1987): closed symbols.

6.1.3 Sparger and Pneumatic Reactor Design

Design for pneumatically agitated reactors has received little attention in the literature. The main objective is to provide the necessary aeration and ventilation capacity. This must be done within the context of providing adequate liquid circulation, and minimizing bubble-induced cell damage. Minimizing bubble damage involves controlling bubble size distribution, as discussed in Sect. 2.4.1. Either very large or very small bubbles are desirable for aeration of animal cell cultures. For small bubbles, the specific interfacial area is high and bubble rise velocity is low, so low sparge rates satisfy *OUR*. Although the liquid entrained (aerosolized) during the bubble burst is high per amount of gas sparged, the low sparge rates mean that the overall burst rate, and hence death rate, is low on a liquid volumetric basis. For large bubbles, very little liquid is ejected per amount of gas and bursting shear rates are lower, so although larger sparge rates are necessary, they can be tolerated due to the low death rate per bubble burst. Reactor design and operation should thus be aimed at producing either of these conditions and avoiding the intermediate bubble size range. ORTON (personal communication to J. AUNINS, 1992) has calculated that desired bubble sizes are roughly $d_b < 1$ mm or $d_b > 6$ mm; practically, it is much easier to achieve the larger size range for large-scale situations, hence the recommendation for using large-bubble systems in Sect. 2.

Bubble birth size is controlled by the forces acting on it during its growth and detachment from the aerator. Basically, two regimes of bubble formation exist for orifice spargers. These are individual bubble formation, and gas jet formation, with subsequent jet breakup into individual bubbles. The transition to the gas jet regime occurs in the vicinity of critical Weber or Froude numbers, *We* and *Fr*, through the orifice (RUFF et al., 1976):

$$We_{crit} = \frac{v_L^2 d_L \rho_g}{\sigma} = 2 \text{ for } d_L < d^* \qquad (47a)$$

$$Fr_{crit} = \frac{v_L^2}{g d_L} = 0.37 \left(\frac{\rho - \rho_g}{\rho_g}\right)^{5/4} \text{ for } d_L > d^* \qquad (47b)$$

where

$$d^* = 2.3 \left(\frac{\rho_g}{\rho - \rho_g}\right)^{5/8} \sqrt{\frac{\sigma}{\rho_g g}} \qquad (47c)$$

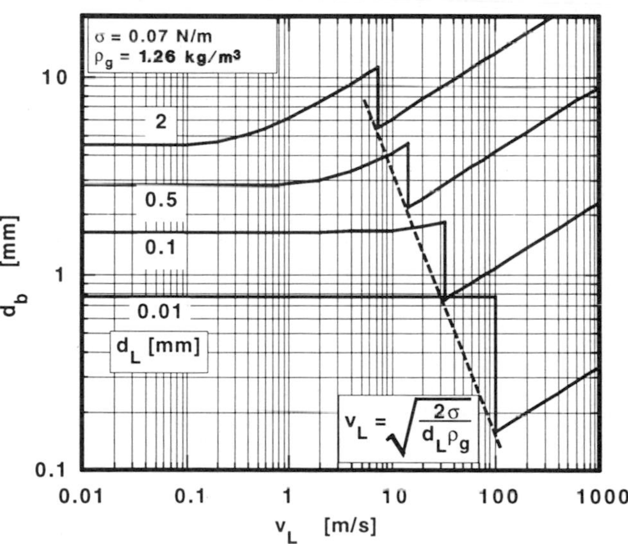

Fig. 26. Theoretical bubble birth size versus sparger velocity for varying orifice size, from Eqs. (48) and (49).

Here v_L is the superficial velocity through the sparger orifice. BRAUER (1971) correlated bubble birth size in the jetting regime by

$$\frac{d_b}{d_L} = 0.72 \left(\frac{v_L}{\sqrt{g d_L}}\right)^{1/3} \quad (48)$$

For very large diameter orifice spargers, $1 < d_L < 8$ mm, in the jet regime, AKITA and YOSHIDA (1974) found an identical correlation with a lead coefficient of 1.88 rather than 0.72. Their correlation was obtained for $1 < v_L < 280$ m/s, and the authors found no dependence of bubble size on material properties in this regime. The AKITA and YOSHIDA coefficient seems less realistic, as it results in bubble sizes larger than in the individual bubble regime. Interestingly, the gas velocities through the sparger in the reactors of KATINGER et al. (1979), $3 < v_L < 16$ m/s, are high enough to be in the jetting region according to Eq. (47), although no transitions are apparent. Similarly, some of the data in Fig. 19 are in the jetting region, but also show no evident transition for $k_L a$. It is not clear why transitions are not manifest. It is partly due to non-idealities in applying the correlations to multiple-orifice spargers with a common plenum. It may also indicate subtleties in sparger design and performance which are not revealed by Eqs. (47) and (48), and which result in differences between authors.

For bubble size prediction for orifice spargers below the jet threshold, GEARY and RICE (1991) present complete equations describing the force balance on a bubble through its growth. These require iterative solution; an analytical expression is available for a limited range of Froude and Weber numbers from MERSMANN (1962):

$$d_b = \left(\frac{3 \sigma d_L}{g(\rho - \rho_g)}\right)^{1/3} \left(1 + \sqrt{1 + \frac{We^2}{Fr}}\right)^{1/3}$$

$$\text{for } \sqrt{1 + \frac{We^2}{Fr}} = 1\text{–}26 \quad (49)$$

The predicted orifice sparger bubble size versus sparger velocity from Eqs. (48) and (49) is shown in Fig. 26 for material properties representative of basal medium and medium containing albumin and/or Pluronic F68. For porous frit spargers, the pores are usually so close to one another that coalescence occurs upon bubble formation, and Eqs. (48) and (49) do not apply. Gas channeling through a few of the pores also increases the local v_L. These phenomena are suggested in Fig. 20 and Tab. 6, where gas velocity has a much stronger effect on $k_L a/v$ for the 0.01 mm porous sparger than for the ring spargers, even at low v_s. Coalescence during rise should be negligible for these very small bubbles, and this is suggested by the lack of a transition at $v_s \approx 10$ m/h for this sparger.

From Eqs. (47–49), the prime variables to control in sparger design are d_L and v_L. Enough orifice area should be provided to place the sparger in the proper operating regime, and the design should preferably arrange it such that transitions which complicate bubble size control do not occur over the range of anticipated flow rates. Ring and perforated plate spargers are generally preferred over large open pipe spargers. PETROSSIAN et al. (1990) reported a practice where multiple single orifice spargers are opened sequentially at increasing oxygen uptake rates, thereby maintaining v_L constant. Although porous plate spargers avoid creating large bubbles, and the small bubbles produced are generally stable to colaescence, these spargers have practical drawbacks in certain instances. These are the birth coalescence and channeling mentioned above, the potential for poor CO_2 ventilation at low gas velocities mentioned in Sect. 2.2, and the difficulty of cleaning these spargers in a cGMP manufacturing situation.

Coalescence is a characteristic of large bubble systems, and its rate is practically unchangeable for a given v_s, d_b, and culture. For an infinite rise time, a dynamic equilibrium is established between turbulent forces that break bubbles, and surface tension holding the bubble together. Bubble size tends toward the maximum stable value. When coalescence equilibrium is established in turbulent liquid flow, maximum diameter is predicted by a critical Weber number for break-up, $We = \tau_{crit} d_{max}/(\sigma + E_s)$. Here E_s is the surface elasticity, a property of the material system, temperature, and a characteristic rate of bubble extension. Assuming that turbulent eddies in the inertial subrange couple with bubbles to cause rupture, several authors have proposed a correlation of the form:

$$d_{b,\max}=A\,\frac{(E_s+\sigma)^{3/5}}{\varepsilon^{2/5}\rho^{3/5}}\left(\frac{\eta}{\eta_g}\right)^b \tag{50}$$

WALTER and BLANCH (1986) reported $A=1.12$, $E_s=0.02$ N/m, and $b=0.1$ for a variety of fluids, with and without surfactants, and yeast and *Lactobacillus* broths. CALDERBANK (1977) reported $b=0.25$ and $E_s=0$, but offerd no value for A. For pneumatic power input,

$$\varepsilon_{\text{pneumatic}}=g\,v_s \tag{51}$$

Using Eq. (50) with the Walter and Blanch constants at cell culture sparge rates (and agitator power inputs), bubbles much larger than experimentally seen are predicted. The correlation is thus only useful for local bubble equilibria, at the high power dissipation rates under which it was developed.

Unfortunately, for culture properties and power inputs, the stable bubble diameter is in the undesirable range $1<d_b<6$ mm. Fortunately, this equilibrium is not always obtained. OOLMAN and BLANCH (1986) found the coalescence frequency to be $\Delta f\approx 0.36\,v_s$ for 0.25 M NaCl salt solutions, where v_s is in [cm/s] and Δf is defined as [no. coalescences/m rise height/bubble]. For $v_s=10$ m/h, $\Delta f H=1$ [coalescence/bubble] for $H=10$ m. Thus, it is only in very tall columns and at high gas velocities that coalescence alters birth size.

6.2 Stirred Reactor Aeration

Gas-liquid mass transfer can be intensified by stirring. A distincion is made between direct sparging, in which the gas bubbles are distributed throughout the bioreactor, and indirect sparging, in which sparging is confined to a bioreactor section separated from the cells by a screen.

6.2.1 Direct Sparging

The low range of impeller powers, i.e., the transition region between bubble column and stirred tank operation, is of interest for cell culture. It is commonly known, and shown in Fig. 27, that $k_L a$ values beyond a certain power input in stirred tanks can be correlated in the dimensional form

$$k_L a=K\,v_s^b\,\varepsilon^c \tag{52}$$

For sparging low-viscosity Newtonian cultures, the $k_L a$ enhancement from agitation is generally independent of the nature and geometry of the impeller. Since $k_L a$ by Eq. (52), and cell death rates by Eq. (9), do not overtly depend on impeller geometry, a wide range of impellers can potentially satisfy the oxygenation and mixing requirements of cell cultures at a given power input. This is shown in Fig. 27 for three different impeller systems, including the standard disc turbine with $d/D=0.33$. From the definition of the impeller Newton number Ne,

$$\varepsilon=Ne\,n^3 d^5 \tag{53}$$

a given power input per unit mass, ε, is achieved at lower impeller speeds with large radial-flow impellers than with small impellers, or propeller and pitched-blade impellers. This is not due strictly to geometry, but also to the lower Newton numbers for these impellers. Because power inputs and gas holdup are low, the Newton numbers for the unaerated state (e.g., HENZLER, 1978) can be used.

A dimensionless correlation of Fig. 27 for higher agitation powers is shown in Fig. 28. After rearrangement of the abscissal grouping, a relationship similar to that reported by HENZLER (1982) for the sorption of oxygen in water is obtained

$$Y_1=\frac{k_L a}{v_s}\left(\frac{\upsilon^2}{g}\right)^{1/3}=4.5\cdot 10^{-5}\left(\frac{\varepsilon}{\upsilon g}\right)^{1/2} \tag{54}$$

Tab. 7. Range for Validity of Sorption Characteristics of Eq. (54): $P/V<300$ W/m^3

Gas-sparger	d_L (mm)	P (W/m^3)
Frit	0.01	>100
Double ring	0.5	> 30
	2	> 10

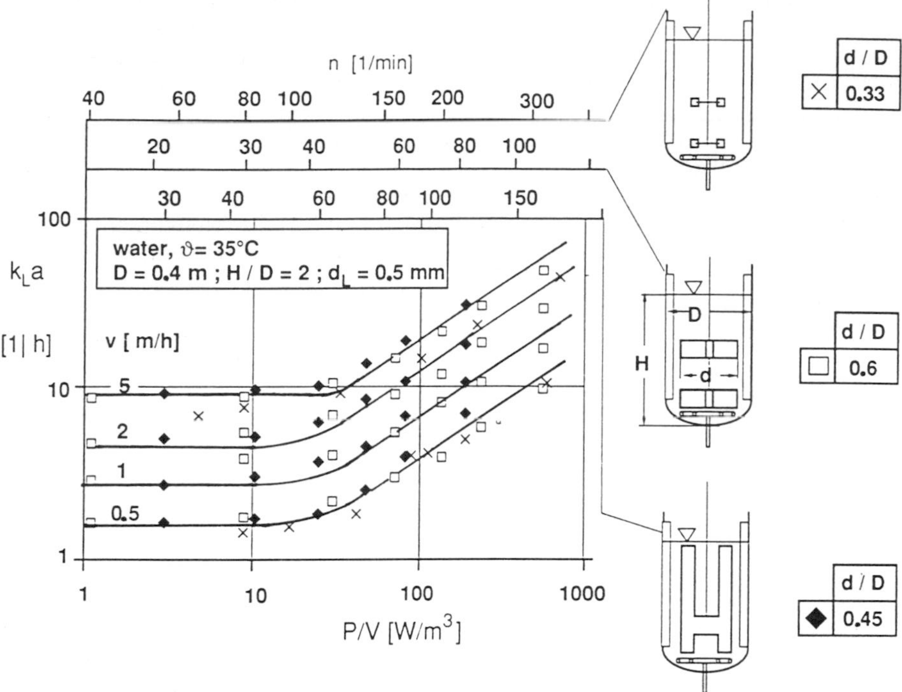

Fig. 27. Mass transfer coefficients $k_L a$ for various superficial gas velocities v and specific impeller power P/V or shaft speed n in stirred vessels (from HENZLER and KAULING, 1993).

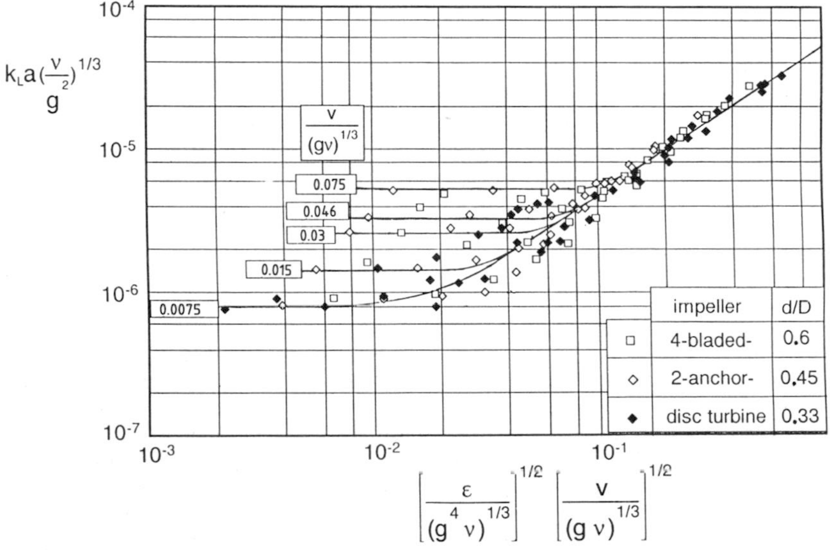

Fig. 28. Sorption characteristics for stirred reactors: dimensionless representation of data from Fig. 27.

This relationship is limited to the sparger operation regions described in Tab. 7. If the agitation power is less than the aeration power, i.e., if $(Ne\, n^3\, d^5)/(g\, \nu\, V) \leq 1$, the system operates purely like a bubble column, for which the sorption characteristics given in Tab. 6 are valid. The bubble column regime is clearly seen at low power in Fig. 28. As in bubble columns and loop reactors, surfactant presence leads to a decrease in the sorption parameter Y_1 per Fig. 22 and the discussion in Sect. 6.1.1. It should be noted that in the low power input regime of cell culture, the enhancement of $k_L a$ by agitation is not necessarily due to bubble size reduction and ensuing surface area increase, although this is an unknown. Enhancement can also occur due to higher bubble residence times, and higher bubble velocities relative to the liquid phase.

6.2.2 Impeller Bubble Dispersion

Although existing research lacks concrete proof of bubble dispersion and coalescence phenomena as effectors of cell viability, it is nevertheless strongly suspected that these events can kill cells, and should be avoided where possible. At a minimum, dispersion and coalescence cause bubble reduction into the lethal size regime. Most research has been focussed on the minimum agitation rates necessary to produce acceptable dispersion (see, e.g., HUDCOVA et al., 1989), however, these studies are outside the range of agitation and aeration that are needed in cell culture. Few data have been gathered on the specific conditions which avoid dispersion, but the former research often yields the conditions at which incomplete dispersion of bubbles occurs. WESTERTERP et al. (1963) correlated the onset of dispersion with the impeller tip speed, and found differences between turbine and paddle impellers. However, Fig. 27 suggests that for $k_L a$ to be affected, power input is the governing factor independent of tip speed. The results of WALTER and BLANCH (1986) in Eq. (50) also extend to include impeller-induced dispersion, and tend to corroborate the results of Fig. 27. Here ε would be taken as the local energy dissipation rate, and would be based on the volume of the impeller discharge region. In general, it is advisible to place the sparger in a position such that the locally high energy region of the impeller does not control bubble size distribution. The benefits of such an arrangement have been shown by OH et al. (1989) for impeller power inputs that cause bubble break up-and cell death. Spargers which control d_b at birth are by extension preferable to spargers relying on agitation dispersion, such as the open pipe spargers typically found in microbial culture vessels. It has been shown in microbial fermentation vessels that ring spargers with diameters greater than 1.2 d are useful to avoid impeller interactions (NIENOW, 1990).

6.2.3 Indirect Sparging

Indirect sparging has been proposed to overcome sparged gas damage to cells, especially microcarrier cultures. Here gas is sparged into a compartment separated from the cell suspension by a screen (WHITESIDE et al., 1985; JOHNSON et al., 1990; REITER et al., 1990) of a mesh size that does not allow either cells or bubbles to cross. From this compartment the oxygenated liquid enters the fermenter by a convective exchange flow F through the screen surface A_v. Since the oxygen is carried by fluid exchange and not simply down a diffusion gradient, the aeration rate expression is a two-equation network for sparged and cell compartments, which must be solved simultaneously. For sparged compartment transport $k_L a_s \gg k a$, and a small sparged compartment, $V_s/V \ll 1$, the overall mass transfer coefficient $k a$ to the cell compartment reduces to the form for series mass transfer resistance

$$\frac{1}{k a} = \frac{1}{k_v \dfrac{A_v}{V}} + \frac{V/V_s}{k_L a_s} \tag{55}$$

Here $k_v A_v/V$ is the screen transport coefficient, and this equals F/V, the rate of liquid exchange across the screen per unit volume of the cell compartment. $k_L a_s$ is the bubble transfer coefficient based on the sparged volume V_s.

The simplest indirect sparging is a spin filter with a sparging dip tube inside. In this case,

$k_L a_s$ can be determined from the bubble column situation in Fig. 20, or more conventional correlations for higher superficial velocities (CHISTI, 1989). To complete the calculation, a knowledge of the liquid exchange rate F is needed, and unfortunately, accurate data are not available. Only limited aeration data are available for spin filter systems. AUNINS et al. (1991a) examined aeration rate for an impeller-shaft mounted spin filter in a 75 L vessel, using dissolved oxygen probes placed in both compartments of the system. Their results are shown in Fig. 29 versus the filter velocity, $u_{\text{filter}} = \pi n d_s$ for dynamic ka determinations. The filter velocity does not affect bubble behavior within the sparged cage, so $k_L a_s$ remains constant. Filter velocity does affect the liquid exchange rate between compartments, so the cell compartment ka increases towards the sparged section $k_L a_s$. This indicates that the screen resistance controls ka, and it can be inferred that large-scale spin filter sparged systems might be limited in gas transfer rate by the screen area that can be accomodated within the vessel. Eq. (55) cannot be readily applied to these data to calculate F/V since $V_s/V = 0.17$ is relatively large, and ka and $k_L a_s$ are fairly close.

In contrast to systems where the screen rotates, the Chemap Chemcell system intensifies bubble aeration and transport through the screen by putting the entire sparged compartment and screen into a Vibromixer (REITER et al., 1990). Fig. 30 shows the results obtained by GEGENBACH (1989) for the Chemcell system in a dimensionless form. With parameters defined in the same way as for a stirred vessel per Eq. (43), the following correlation is obtained

$$Y_2 = \frac{ka}{v_s}\left(\frac{\upsilon^2}{g}\right)^{1/3} = 7.5 \cdot 10^{-5}\left(\frac{f^3 E^5}{\rho V_s}\frac{1}{g v_s}\right)^{0.23} \quad (56)$$

For Vibromixers, the power input is given from the frequency f and amplitude E of the vibration by the relation $P \sim f^3 E^5$. The reference volume for power input in this correlation is the sparged volume V_s, and the superficial velocity v_s is obtained using the sparged compartment cross-sectional area A_s. The slight dependence on Vibromixer power in comparison to Eq. (54) shows clearly that convective transfer has a decisive influence on the mass transfer coefficient.

The effect of vibrating the unit is to create bubble dispersion in the chamber and convective flow through the screen in this system. The convective flow should be proportional to the pumping rate of the Vibromixer, and dependent on the screen mesh. Although impel-

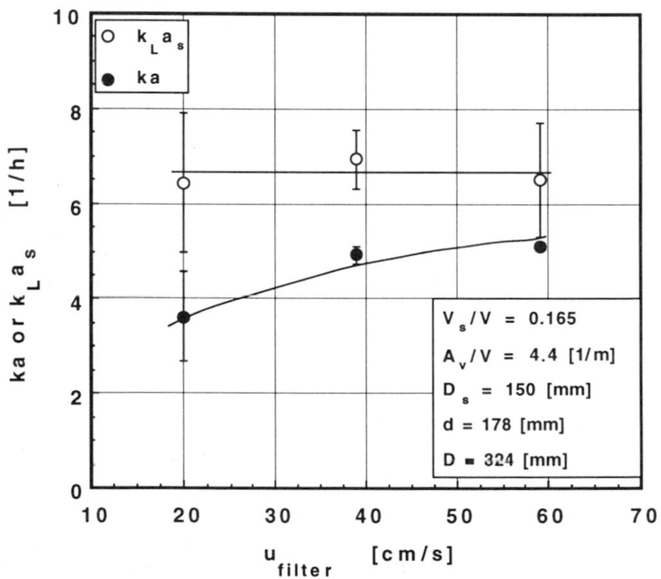

Fig. 29. Indirect sparging in a spin filter system (from AUNINS et al., 1991a).

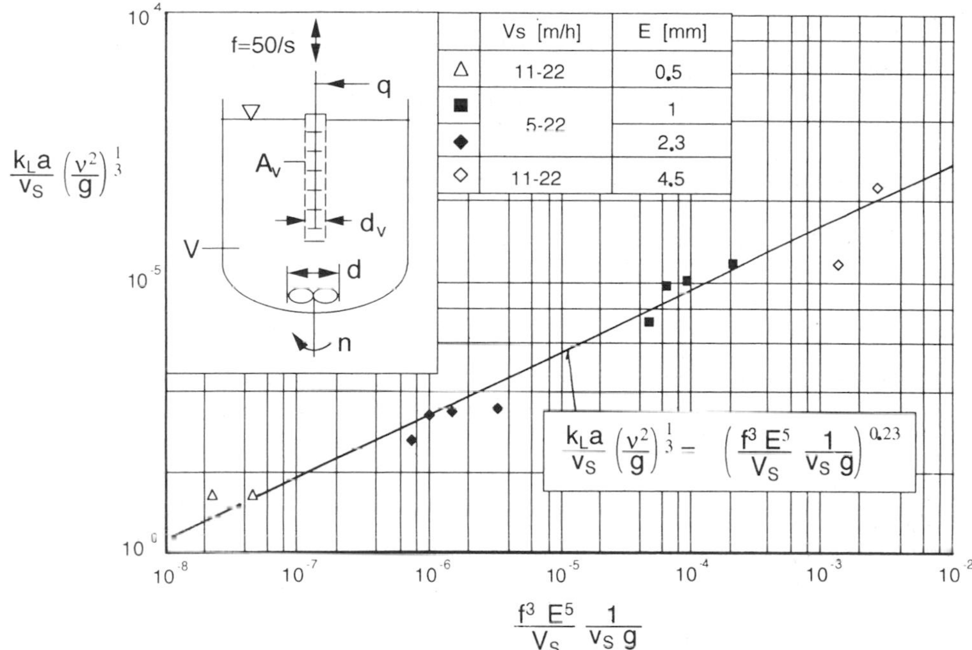

Fig. 30. Sorption characteristics for bioreactors with the Chemcell Vibromixer system (Chemap); $A_v/V = 3/$m, $d_s = 3.7$ cm, $V = 10$ L, $H/D = 1$, propeller stirrer, $d/D = 0.3$, water.

ler power input is not considered in this correlation, it would enter into a more general correlation as an additional power term. It should be noted that the data in Fig. 30 do not provide a general correlation, but are applicable only to GEGENBACH's reactor and Chemcell geometry with $A_v/V = 3/$m, a 50–80 μm screen mesh size, and water as the aerated fluid. Eq. (56) provides a general idea of the relevant parameters, and the lead constant can be modified according to the working fluid properties. No factually based recommendation can be made regarding impeller speed or power input for these gassing devices, although it is obvious that it would need to be placed relative to the screen surface in a manner that modifies F/V.

6.2.4 Stirred Tank Agitation Design

Only the impeller power input, superficial gas velocity, and sparger design affect mass transfer efficiency for sparged aeration. From the presentation above in Sects. 4.3 and 5.2 for surface and membrane aeration, not only the power input of the impeller, but also the impeller and tank design, determine the mass transfer rate and the characteristics of the reactor. Although the engineer is no exempt from considering the special features of his or her process to arrive at a vessel design, it is appropriate to recap general design considerations.

For a given scale, power input and Newton number, tip speed decreases by $d^{2/3}$. Although it was related in Sect. 2.3 that no good correlations of cell health versus impeller tip speed have appeared, larger impeller diameters offer the lowest tip speed available, and are thus the safest design. For a sparged vessel, a mild contradiction occurs in that it is difficult to have a large impeller, and a sparger which avoids the impeller discharge (and hence bubble dispersion) at the same time. This can be solved somewhat by placing the sparger above the (bottom) impeller.

Impeller flow characteristics, either radial or axial, are highly important to tank operation. For membrane aeration, the large anchor impellers achieve the highest transfer at the lowest power due to the fact that they direct flow radially outward against the gassing cage. It is therefore to be expected that they would also be best to provide flow for indirect sparging systems when arranged around the compartment screen. As mentioned above, the radial flow impellers also achieve the same power input at a lower tip speed due to their higher Newton numbers. VARECKA and BLIEM (1990) investigated power consumption of pitched-blade fan and propeller impellers versus the bulk flow velocity in internal loop reactors, to obtain a "pumping modulus" indicative of efficiency. Efficiency was maximized for fans with 20° pitch, and propellers with 0.57 pitch ratio (pitch/diameter). These specifications fairly correspond to commercially available hydrofoil impellers such as the Lightnin A310 or A315, Prochem Maxflo T, or Chemineer HE3. The question of which impeller types are truly best for cell culture can only be answered by more elegant biological experiments. Although it is commonly postulated that radial flow impellers are superior for shear-sensitive situations, recent data (HENZLER, unpublished results) suggest otherwise.

Baffles and other tank internals obviously affect reactor flow patterns, and as pointed out in Sect. 4.3, they affect surface aeration via bubble entrainment. Baffles have generally been looked upon unfavorably for cell culture applications, as many workers measure effect of tank design on cells at a constant impeller speed, and do not quote energy dissipation rates. Baffles raise the impeller power input for constant impeller speed, appearing unfavorable in such comparisons. However, they blunt the azimuthal component of impeller discharge velocity, and homogenize power dissipation throughout the vessel. If impeller speed is adjusted, so the same global power input is achieved, lower tip speed and impeller region power result, and sparged mass transfer is unaffected. It remains to be seen for microcarrier cultures whether the additional surface area for bead collision is more detrimental than the power homogenizing effects of the baffles. Internal loops, whether for airlift reactors or impeller agitated reactors, structure the liquid flow and minimize vessel turbulence for a given power input. It should be noted that industrial cell culture bubble columns are virtually unheard-of, and airlift reactors are the dominant design. For stirred reactors, they can help compartmentalize liquid flow to ensure that gas recirculation into impeller regions does not occur, and lower the impeller power needed to achieve desired liquid circulation velocities. These positive effects must be balanced against the loss in the aerated reactor volume, and the higher gas velocities necessary to achieve the same mass transfer.

As was shown in Sect. 6.2.1, the nature of the impeller system is of no importance for mass transfer. Limited answers can be given at present to the question of which impellers produce particularly low shear for a given power input, and are therefore particularly suitable. However, the following considerations do provide information on the influence of stirrer size, reactor geometry, and reactor size.

7 System Comparison

7.1 Aeration Performance Comparison

In a comparison of mass transfer systems, the attainable mass transfer coefficient $k_L a$, or the oxygen input per unit volume G/V, are the appropriate parameters for study. To constrain the problem, comparison of the various reactors can be carried out as a function of the maximum liquid velocity u, or the specific power input ε or P/V. The results calculated below have been obtained with the sorption characteristics reported above, a gas of pure oxygen, a typical nutrient solution containing 1 g/L albumin (according to the data of Fig. 22, a 35% impairment of mass transfer is to be expected when O_2 is supplied via Accurel membranes or by sparging) and a temperature of $\vartheta = 37\,°C$ with the assumed material parameters $Hy = 3.2 \cdot 10^6$ N m/kg, $\upsilon = 0.8 \cdot 10^{-6}$ m²/s, $\mathscr{D} = 3.2 \cdot 10^{-9}$ m²/s. In the case of membrane oxygenation the calculations were performed for a total membrane area of $a_m = 25$ m^{-1}. For open pore membranes only 75% and for reinforced membranes only 90% of this area is taken as effective. Turtuosity and other factors were taken as negligible contributors for simplicity. The total membrane area assumed requires the installation of 1.7–2.5 m tubing per L reactor volume, depending on the diameter of the tubing (see the table in Fig. 31). This is near the limit of technical feasibility for stirred vessels.

7.1.1 Bubble Free Aeration

Fig. 31 shows the attainable mass transfer coefficient ka and the oxygen input G/V as functions of the maximum velocity in the reactor u for some of the systems investigated above. Oxygenation via the surface of the liquid does not provide large rates of oxygen transfer, even with low height-to-diameter ratios. From the standpoint of the supply of O_2 the open-pore Accurel® membrane combined with the Diessel reactor offers the best solution. However, this reactor has not been accepted on the large scale owing to its complicated technology and the phenomenon of hydrophilization of the microporous membranes, causing wetting out and possible cell growth on the membrane surface.

The efficiency of aeration techniques without bubbles can be improved by the use of higher gas-side total pressures, as is shown in Fig. 32. Oxygen inputs G/V up to 180 mg/L h can then also be achieved with silicone membranes having tube lengths $L'/V > 1.75$ m/L. The comparison in Fig. 32 was conducted with stable diffusion membranes having relatively high wall thicknesses. By reduction of the wall thickness, it is possible to reduce the diffusion resistance of the membrane in accordance with Eqs. (25) and (28), but only a low gas pressure can then be achieved in the tube; there is thus an optimum tube geometry for the maximum oxygen transfer per unit area of membrane G/A. Stress tests show that commercially obtain-

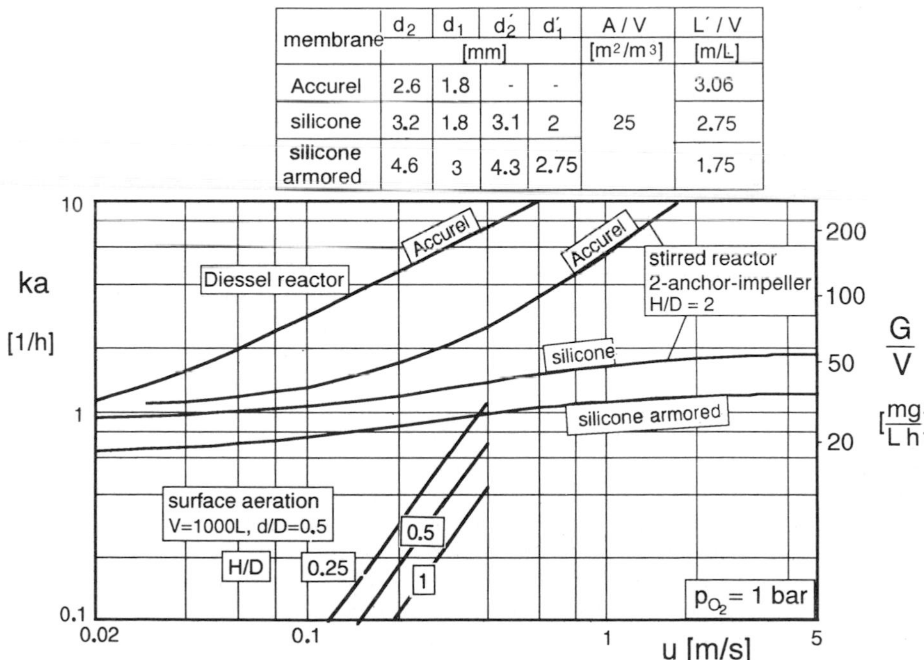

Fig. 31. Comparison of different methods of bubble free oxygen supply mass transfer coefficients ka and possible oxygen supply G/V for $Hy = 3.2 \cdot 10^6$ N m/kg, $C_L = 3$ mg/L, $\vartheta = 37$ °C.

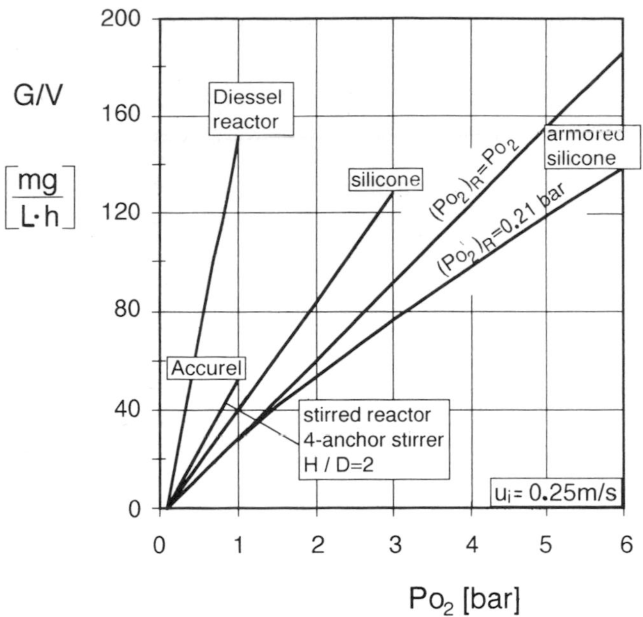

Fig. 32. Oxygen uptake in different membrane reactors: $Hy = 3.2 \cdot 10^6$ N m/kg, $C_L = 3$ mg/L, $\vartheta = 37\,°C$; see Fig. 31 for tube dimensions.

able non-reinforced silicone tubes can be exposed to a continuous, internal pressure stress of $P_{adm} \approx 2\sigma_{adm}\ (1 - d_1/d_2)$ in which $\sigma_{adm} = 4 \cdot 10^5$ N/m^2. The flux per unit area G/A with the highest possible pressure inside the tube is thus found to depend on the geometry of the tube as follows:

$$\frac{G}{A} = k(C^* - C_L) = k\left(\frac{P_{O_2}}{Hy} - C_L\right)$$

$$= \frac{2\,\sigma_{adm}(1 - d_2/d_1) - C_L}{Hy\left(\frac{1}{k_L} + \frac{d_2 \ln d_2/d_1}{2\,\mathscr{D}\,Hy/Hy_s}\right)} \qquad (57)$$

The mass transfer coefficient for the liquid boundary layer on the membrane k_L also depends on the outside diameter d_2 of the membrane, according to Figs. 13 and 14, and Eqs. (33), (41), and (42). Illustrative results are presented in Fig. 33; they show flat optima for the mass flux G/A, which move towards smaller diameter ratios n with increasing outside diameter d_2.

The ultimate choice of tubing size and wall thickness is a compromise between the mass transfer attainable and practicality. Mass transfer can be distinctly improved by the use of small bore micro-silicone tubes. However, a disadvantage of these tubes is the greater pressure loss as the gas passes through, which makes it necessary to divide the tube into several parallel segments. Not only is this technically more complicated, but it also leads to reduced reliability because of the great number of gas connections, and the fragility of the membranes. As shown by the calculation results of HENZLER and KAULING (1993), micro-silicone tubes are impractical even for semi-industrial plants.

7.1.2 Direct Sparging

Figs. 34 and 35 show a comparison of direct aeration with the bubble-free oxygenation methods for 25 and 1000 L bioreactors. Mass transfer in the sparged systems is not influenced by scale, while mass transfer increases for the membrane system, and decreases for a surface aerated system with increasing bioreactor volume. Even at atmospheric pressure, bubble columns and stirred tanks allow sufficient oxygen transfer through

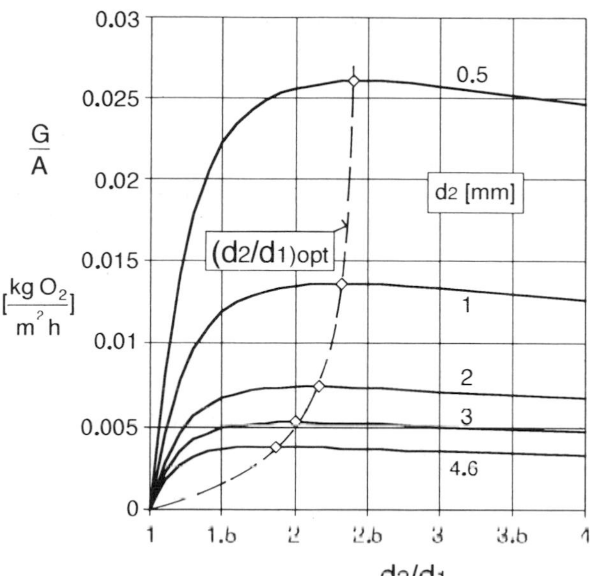

Fig. 33. Influence of silicone tube geometry on O_2 transport corresponding to Eq. (57), $Hy = 3.2 \cdot 10^6$ N m/kg, $C_L = 3$ mg/L, $\vartheta = 37$ °C.

aeration and/or stirring. Properly operated, the large-bubble system produced by the 0.5 mm orifice sparger should give no problem in connection with the elimination of CO_2, although the bubble sizes produced by this sparger d_L are potentially in the most lethal range, per Figs. 4 and 26. With very low gas throughputs of, e.g., $v = 2$ m/h, an oxygen input G/V of 200 mg/L/h can be achieved with a low impeller power $P/V \approx 80$ W/m^3. As was shown in Sect. 6.2.1, the nature of the impeller system is of no importance for mass transfer. Limited answers can be given at present to the question of which impellers produce particularly low shear for a given power input, and are therefore particularly suitable. However, the following considerations do provide information on the influence of stirrer size, reactor geometry, and reactor size.

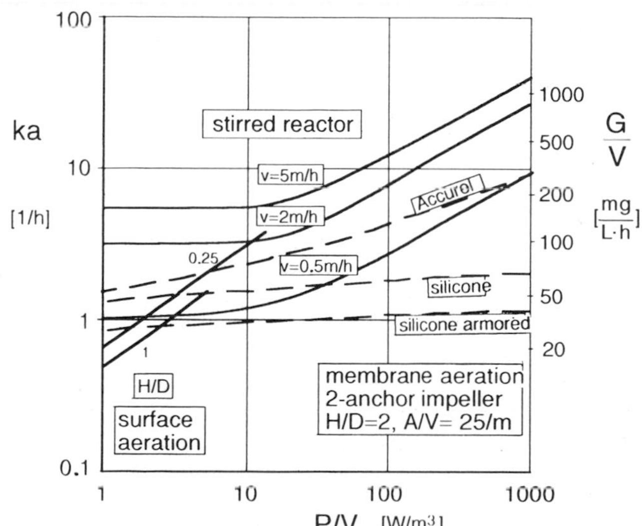

Fig. 34. Comparison of bubble-free oxygen supply and sparged oxygen supply: mass transfer coefficients $k\,a$ and possible oxygen supply G/V at 25 L volume; for conditions, see Fig. 31.

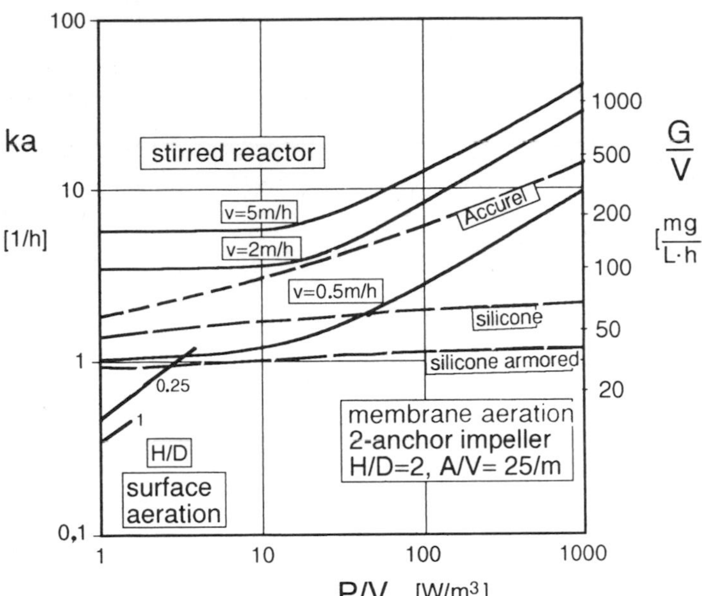

Fig. 35. Comparison of bubble-free oxygen supply and sparged oxygen supply: mass transfer coefficients $k\,a$ and possible oxygen supply G/V at 1000 L volume; for conditions, see Fig. 31.

7.2 Mechanical Stress Comparison

7.2.1 Notes on Mechanical Stress

Mechanical stress arises from the movement of the liquid in the reactor, which is essential for mixing, suspension, and gas/liquid mass transfer, and which is produced by an impeller and/or by ascending gas bubbles. Since densities of microorganisms, animal cells and even carriers with cells growing on them are only very slightly different from the density of the nutrient solution, these particles follow the mean movement of the liquid with practically no slip. Mechanical stress τ can therefore be assumed to result mainly from the frequent turbulent velocity fluctuations $\sqrt{\overline{u'^2}}$

$$\tau \sim \rho \sqrt{\overline{u'^2}} \tag{58}$$

Because of the low power inputs usually employed in reactors ($P/V \approx 1$–$100\ \text{W/m}^3$), the cells or the cell agglomerates and microcarriers are either smaller than or of the same order as the smallest turbulence elements, so that these are the only elements that can be said to affect cells or microcarriers. For isotropic turbulence these are the eddies in the dissipation range of the turbulence spectrum, for which $\sqrt{u'^2} \approx 0.26\, d_p \sqrt{\varepsilon/\upsilon}$ (see, e.g., HINZE, 1975; LIEPE, 1988). Thus Eq. (58) yields the following relation for shear stress on a particle surface

$$\tau_1 \sim \rho \sqrt{u'^2} = 0.068\, \rho\, d_p^2 \left(\frac{\varepsilon}{\upsilon}\right) \tag{59}$$

Because of the non-uniform energy distribution in the reactors, the particles experience different stresses as they circulate within the reactor. The maximum value of the energy dissipation ε is of interest for the determination of the maximum stress. If the cells are not destroyed during a single transit through this region, the viability of a culture will also depend on the frequency with which the maximum stress occurs. The maximum energy density differs very considerably from the mean energy dissipation $\varepsilon = P/(\rho V)$. Besides obtaining contradictory data (see discussion in HENZLER and KAULING, 1993), many authors (MÖCKEL, 1978; LAUFHÜTTE, 1986; GEISSLER, 1991) obtain the relation for impeller systems

$$\varepsilon \sim \frac{\bar{\varepsilon}}{(d/D)^3} \qquad (60)$$

which is to be used in calculating the maximum stress τ_1. In view of the contradictions between the results obtained so far together with the unanswered question whether they apply to all impeller systems, and also the lack of information on the energy dissipation distribution in bubble columns it is impossible at present to make any generally valid assessment of the various bioreactors on the basis of the shear hypothesis of Eq. (58). The relation $\sqrt{u'^2} \approx u$, which is generally valid for a given type of reactor should therefore be used for the identification of fundamental dependences on the aspect ratio and size of the reactor. The shear stress according to Eq. (58) can thus be estimated from the maximum velocity u present in the reactor:

$$\tau_2 \sim \rho \sqrt{u'^2} \sim \rho u^2 \qquad (61)$$

According to the available experimental results (see, e.g., HANDA-CORRIGAN et al., 1989; TRAMPER 1987; HÜLSCHER, 1990) an additional and particularly undesirable effect must be attributed to aeration as far as shear on animal cells is concerned. It is assumed that damage by gas bubbles is caused by the forces that occur during the formation or bursting of the bubbles at the surface of the liquid and by direct contact between the cell membrane and the bubble. Regardless of the mechanism responsible, the major shear resulting from the aeration should be proportional to the number of bubbles formed, and hence to the gas throughput per unit volume of the reactor q/V, which we shall consider as a third shear hypothesis:

$$\tau_3 \sim \frac{q}{V} \sim \frac{v}{H} \qquad (62)$$

7.2.2 Comparison of Methods

The relationships yielded by the three stress hypotheses for the individual types of reactor are given in Tab. 8 (HENZLER and KAULING, 1993). If the only effects considered are those of the geometric ratios d/D and H/D, the term $H/D/Ne$ is assumed to be constant in the case of the stirred tank, since the number of impellers is generally proportional to H/D. The relative changes in the mechanical stresses for the case of a constant superficial gas velocity v or a constant specific impeller power P/V, i.e., constant gas transfer, which follow from the relations in Tab. 8 and are presented graphically in Fig. 36, show that it is preferable to use impellers with high ratios of impeller diameter to reactor diameter d/D, and slender reactors. The loop reactor is an exception with regard to the motion of the liquid, which increases with the slenderness ratio H/D of the vessel.

Tab. 8. Influence of Geometry and Reactor Size on Stress of Cells

Reactor	$\tau_1 \sim \frac{\varepsilon}{\upsilon}$	$\tau_2 \sim \rho u^2$	$\tau_3 \sim \frac{q}{V}$
Stirred	$\frac{1}{(d/D)^3}\left[\frac{P}{V}\right]$	$\frac{V^{2/9}}{\left(\frac{H}{D}\right)^{2/9}\left(\frac{d}{D}\right)^{4/3}}\left[\frac{H/D}{Ne}\frac{P}{V}\right]^{2/3}$	
Bubble column		$\frac{V^{1/3}}{\left(\frac{H}{D}\right)^{1/3}}[v]$	$\frac{[v]^{3/4}}{\left(\frac{H}{D}\right)^{2/3} V^{1/3}}$
Airlift		$\left(\frac{H}{D}\right)^{2/3} V^{1/3}\,[v]$	

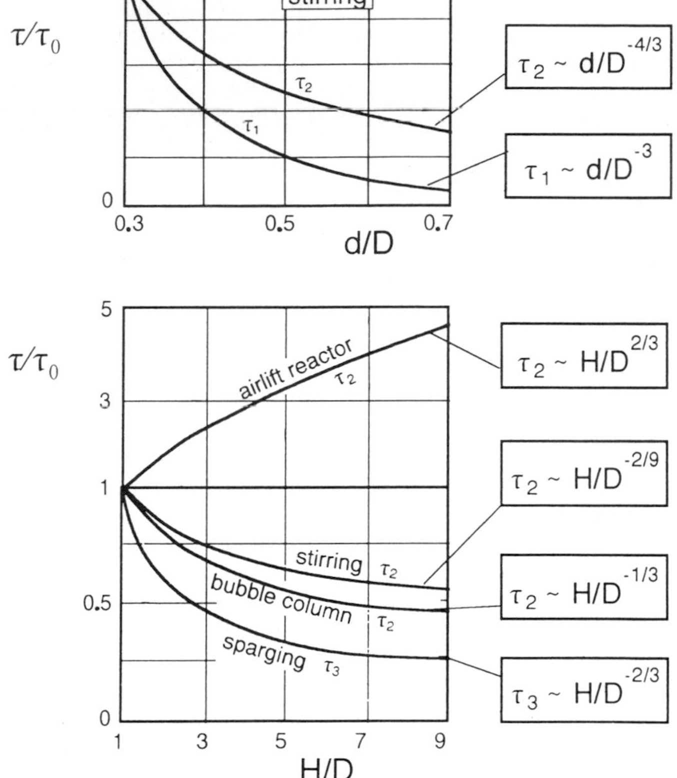

Fig. 36. Influence of impeller and reactor geometry on mechanical stress for v and $\varepsilon = \text{constant}$.

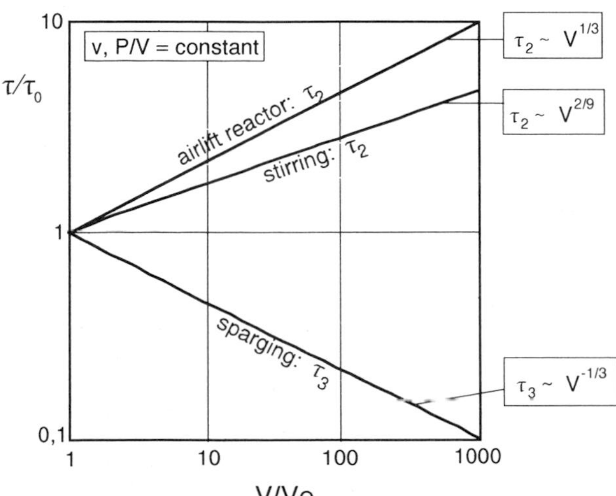

Fig. 37. Relative change of mechanical stress with increasing reactor volume.

Finally, Fig. 37 shows the scale-up relationships. Whereas the velocity of the liquid increases with the size of the reactor for ka = constant, the shear produced by the bubbles simultaneously decreases. If the size of the reactor is increased, e.g., from 1 L to 1000 L, the shear due to the aeration decreases by a factor of 10. Moreover, the slenderness ratios in the case of small fermenters are always low and hence unfavorable with regard to shear because of the small reactor diameter. This is a very important observation, since the information obtained so far indicates that the shear is mainly due to the aeration. It suggests the possibility that problems observed on the laboratory scale are of minor importance in larger reactors so that sparged aeration can in many cases be used for oxygenation even in production processes with animal cells.

8 References

AKITA, K., YOSHIDA, F. (1974), Bubble size, interfacial area, and liquid-phase mass transfer coefficient in bubble columns, *Ing. Eng. Chem. Process Des. Dev.* **13**, 84.

ALBERTY, R. A., DANIELS, F. (1980), *Physical Chemistry,* 5th Ed., pp. 194–196, New York: John Wiley & Sons.

AUGENSTEIN, D. C., SINSKEY, A. J., WANG, D. I. C. (1971), Effect of shear on the death of two strains of mammalian tissue cells, *Biotechnol. Bioeng.* **13**, 409.

AUNINS, J. G., GOLDSTEIN, J. M., CROUGHAN, M. S., WANG, D. I. C. (1986), Engineering developments in the homogeneous culture of animal cells: oxygenation of reactors and scale-up, *Biotechnol Bioeng. Symp.* **17**, 699.

AUNINS, J. G., WOODSON, B. A., HALE, T. K., WANG, D. I. C. (1989), Effects of paddle impeller geometry on power input and mass transfer in small-scale animal cell culture vessels, *Biotechnol. Bioeng.* **34**, 1127.

AUNINS, J. G., GLAZOMITSKY, K., BUCKLAND, B. C. (1991a), Aeration in pilot-scale vessels for animal cell culture, paper presented at *AIChE Annual Meeting,* Los Angeles, CA, Nov. 17–22.

AUNINS, J. G., WU, F., BUCKLAND, B. C. (1991b), Beneficial effects of agitation in porous microcarrier culture, paper presented at *AIChE Annual Meeting,* Los Angeles, CA, Nov. 17–22, *Biotechnol. Prog.,* submitted.

BAIJOT, B., DUCHENE, M., STEPHENNE, J. (1987), Production of Aujeszky vaccine by the microcarrier technology – From the ampoule to the 500 litre fermentor, *Dev. Biol. Stand.* **66,** 523.

BALIN, A. K., GOODMAN, D., RASMUSSEN, H., CRISTOFOLO, V. J. (1976), The effect of oxygen tension on the growth and metabolism of WI-38 cells, *J. Cell Physiol.* **89,** 235–250.

BAVARIAN, F., FAN, L. S., CHALMERS, J. J. (1991), Microscopic visualization of insect cell-bubble interactions. I: Rising bubbles, air–medium interface, and the foam layer, *Biotechnol. Prog.* **7**, 140.

BENTLEY, P. K., GATES, R. M. C., LOWE, K. C., DEPOMERAI, D. I., LUCY WALKER, J. A. (1989), *In vitro* cellular responses to a non-ionic surfactant. Pluronic F-68, *Biotechnol. Lett.* **11**, 111.

BIN, K. A. (1984), Mass transfer to the free interface in stirred vessels, *Chem. Eng. Commun.* **31,** 155.

BIRCH, J. R., THOMPSON, P. W., BORASTON, R., OLIVER, S., LAMBERT, K. (1987), The large-scale production of monoclonal antibodies in airlift fermenters, in: *Plant and Animal Cells: Process Possibilities* (WEBB, C., MAVITUNA, F., Eds.), pp. 162–171, Chichester: Ellis Horwood.

BLAISDELL, C. T., KAMMERMEYER, K. (1983), Counter-current and co-current gas separation, *Chem. Eng. Sci.* **28,** 1249.

BORASTON, R., THOMPSON, P. W., GARLAND, S., BIRCH, J. R. (1984), Growth and oxygen requirements of antibody producing mouse hybridoma cells in suspension culture, *Dev. Biol. Stand.* **55,** 103.

BRAUER, H. (1971), Grundlagen der Einphasen- und Mehrphasenströmungen, in: *Grundlagen der Chemischen Verfahrenstechnik,* Aarau–Frankfurt am Main: Verlag Sauerländer.

BRAUER, H., MEWES, D. (1971), Stoffaustausch einschließlich chemischer Reaktion, in : *Grundlagen der Chemischen Verfahrenstechnik,* Aarau–Frankfurt am Main: Verlag Sauerländer.

BRÄUTIGAM, H.-J. (1985), Untersuchungen zum Einsatz von nicht porösen Kunststoffmembranen als Sauerstoffeintragssystem, *Ph. D. Thesis,* Technische Hochschule Hamburg-Harburg.

BRÄUTIGAM, H.-J., SEKULOV, I. (1986), *Membranfließbettechnologie* **9**, 269–272.

BROWN, M. E. (1992), Process development for the production of recombinant antibodies using the glutamine synthetase (GS) system, *Cytotechnology* **9,** 231.

BÜNTEMEYER H., BÖDECKER, B. G. D., LEHMANN, J. (1987), Membrane-stirrer-reactor for bubble free aeration and perfusion, in: *Modern Approaches to Animal Cell Technology* (SPIER, R. E., GRIFFITHS, J.B., Eds.) p. 411, Frome, Somerset, UK: Butterworth.

CALDERBANK, P. H. (1959), Physical rate processes in industrial fermentation: 2, *Trans. Inst. Chem. Eng.* **37,** 173.

CALDERBANK, P. H., MOO-YOUNG, M. B. (1961), *Chem. Eng. Sci.* **16,** 39.

CALDERBANK, P. H., PEREIRA, J. (1977), *Chem. Eng. Sci.* **16,** 1427.

CANTOR, C. R., SCHIMMEL, P. R. (1979), *Biophysical Chemistry,* San Francisco, CA: W. H. Freeman & Co.

CHALMERS, J. J. (1993), The effect of hydrodynamic forces on insect cells, in: *Insect Cell Cultures: Production of Viral Biopesticides and Protein from Recombinant DNA* (SHULER, M., Ed.), München: Hanser, in press.

CHALMERS, J. J., BAVARIAN, F. (1991), Microscopic visualization of insect cell–bubble interactions. II: The bubble film and bubble rupture, *Biotechnol Prog.* **7,** 151.

CHERRY, R. S., KWON, K.-Y. (1990), Transient shear stresses on a suspension cell in turbulence, *Biotechnol. Bioeng.* **36,** 563.

CHERRY, R. S., PAPOUTSAKIS, E. T. (1989), Growth and death rates of bovine embryonic kidney cells in turbulent microcarrier bioreactors, *Bioprocess Eng.* **4,** 81.

CHERRY, R. S., PAPOUTSAKIS, E. T. (1986), Hydrodynamic effects on cells in agitated tissue culture reactors, *Bioprocess Eng.* **1,** 29.

CHISTI, M. Y. (1989), *Airlift Bioreactors,* pp. 59–62, London: Elsevier Applied Science.

CÔTÉ, P., BERSILON, J. L., HUYARD, A. (1989), Bubble free aeration using membranes: Mass transfer analysis, *J. Membr. Sci.* **47,** 91.

CRISTOFOLO, V. J., KRITCHEVSKY, D. (1966), Respiration and glycolysis in the human diploid cell strain WI-38, *J. Cell Physiol.* **67,** 125.

CROUCH, C. F., FOWLER, H. W., SPIER, R. E. (1985), The adhesion of animal-cells to surfaces – the measurement of critical surface shear-stress permitting attachment of causing detachment, *J. Chem. Technol. Biotechnol.* **35,** 273.

CROUGHAN, M. S., WANG, D. I. C. (1989), Growth and death in microcarrier cultures, *Biotechnol. Bioeng.* **33,** 731.

CROUGHAN, M. S., WANG, D. I. C. (1990), Reversible removal and hydrodynamic phenomena in CHO microcarrier cultures, *Biotechnol. Bioeng.* **36,** 316.

CROUGHAN, M. S., HAMEL, J.-F., WANG, D. I. C. (1985), Hydrodynamic effects on animal cells grown in microcarrier culture, paper presented at the *Worcester Colloquium of the New England Biotechnology Association,* Worcester, MA, March 21–22. Authors at M. I. T., Cambridge, MA.

CROUGHAN, M. S., HAMEL, J.-F., WANG, D. I. C. (1987), Hydrodynamic effects on animal cells grown in microcarrier culture, *Biotechnol. Bioeng.* **29,** 130.

CROUGHAN, M. S., HAMEL, J. F., WANG, D. I. C. (1988), Effects of microcarrier concentration in animal cell cultures, *Biotechnol. Bioeng.* **32,** 975.

DARNELL, J., LODISH, H., BALTIMORE, D. (1986), *Molecular Cell Biology,* p. 137, New York: Scientific American Books.

DEAN, R. C. JR., KARKARE, S. B., PHILLIPS, P. G., RAY, N. G., RUNSTADLER, P. W. JR. (1987), Continuous cell culture with fluidized sponge beads, in: *Large Scale Cell Culture Technology* (LYDERSEN, B. K., Ed.), pp. 145–167, München: Hanser.

DESPLANCHES, H., BRUXELMANE, M., CHEVALIER, J. L., DUCLA, J. (1983), Characteristic variable prediction and scale-up for heat transfer to coils in agitated vessels, *Chem. Eng. Res. Des.* **61,** 3.

DIAMOND, S. L., ESKIN, S. G., MCINTYRE, L. V. (1989), Fluid flow stimulates tissue plasminogen activator secretion by cultured human endothelial cells, *Science* **243,** 1483.

DRAPEAU, D., LUAN, Y.-T., WHITEFORD, J. C., LAVIN, D. P., ADAMSON, S. R. (1990), Cell culture scale-up in stirred tank bioreactors, paper presented at *SIM Annual Meeting,* Orlando, FL, August 1. Authors at Genetics Institute, Cambridge, MA.

FLEISCHAKER, R. J., SINSKEY, A. J. (1981), Oxygen demand and supply in cell culture, *Eur. J. Appl. Microbiol. Biotechnol.* **12,** 193.

GARCIA-BRIONES, M., CHATTOPADHYAY, D., CHALMERS, J. J. (1992), Mechanisms of cell damage as a result of sparging and rational criteria for designing medium to prevent this damage, paper presented at *Cell Culture Engineering III,* Palm Coast, FL, Feb. 2–7. Authors at The Ohio State University, Columbus, OH.

GARDNER, A. R., GAINER, J. L., KIRWAN, D. J. (1990), Effects of stirring and sparging on cultured hybridoma cells, *Biotechnol. Bioeng.* **35,** 940.

GARNER, F. H., ELLIS, S. C. M., LACEY, J. A. (1954), The size distribution and entrainment of droplets, *Trans. Inst. Chem. Eng.* **32,** 222.

GEARY, N. W., RICE, R. G. (1991), Bubble size prediction for rigid and flexible spargers, *AIChE J.* **37,** 161.

GEGENBACH, R. (1989), Preprints of *1. Taunustagung Membrantechnologie/Millipore.*

GEISSLER, R. K. (1991), Fluiddynamik und Leistungseintrag in turbulent gerührten Suspensionen, *Dissertation,* Technische Universität München.

GLACKEN, M. W., ADEMA, E., SINSKEY, A. J. (1989), Mathematical descriptions of hybridoma culture kinetics: II. The relationship between thiol chemistry and the degradation of serum activity, *Biotechnol. Bioeng.* **33**, 440.

GOLDBLUM, S., BAE Y.-K., HINK, W. F., CHALMERS, J. (1990), Protective effect of methylcellulose and other polymers on insect cells subjected to laminar shear stress, *Biotechnol. Prog.* **6**, 383.

GOLDSTEIN, J. M. (1986), Scale up of membrane oxygenated recombinant animal cell bioreactors, *M. S. Thesis,* M. I. T., Cambridge, MA, p. 140.

GREGG, C. T., MACHINIST, J. M., CURRIE, W. D. (1968), Glycolytic and respiratory properties of intact mammalian cells. Inhibitor studies, *Arch. Biochem. Biophys.* **127**, 101.

GRIFFITHS, J. B., CAMERON, D. R., LOOBY, D. (1987), A comparison of unit process systems for anchorage dependent cells, *Dev. Biol. Stand.* **66**, 331.

GUYTON, A. C. (1966), *Textbook of Medical Physiology,* 3rd Ed., p. 44, Philadelphia, PA: Saunders.

HANDA, A., EMERY, A. N., SPIER, R. E. (1987), On the evaluation of gas–liquid interfacial effects on hydridoma viability in bubble column bioreactors, *Dev. Biol. Stand.* **66**, 241.

HANDA-CORRIGAN, A., EMERY, A. N., SPIER, R. E. (1989), Effects of gas–liquid interfaces on the growth of suspended mammalian cells: Mechanisms of cell damage by bubbles, *Enzyme Microb. Technol.* **11**, 230.

HARRIS, A. K., WILD, P., STOPAK, D. (1980), Silicone rubber substrata: A new wrinkle in the study of cell locomotion, *Science* **208**, 177.

HENZLER, H.-J. (1978), Untersuchungen zum Homogenisieren von Flüssigkeiten und Gasen, VDI Forschungsh. **587**.

HENZLER, H.-J. (1980), Begasen höherviskoser Flüssigkeiten, *Chem. Ing. Tech.* **52**, 643.

HENZLER, H.-J. (1982), Auslegungsunterlagen für Rührbehälter als Fermenter, *Chem. Ing. Tech.* **45**, 461.

HENZLER, H.-J., KAULING, J. (1985), Scale-up of Mass Transfer in Highly Viscous Liquids, Paper 30, pp. 303–312, *5th Eur. Conference on Mixing*, Cranfield: BHRA.

HENZLER, H.-J., KAULING, J. (1990), Sauerstoffversorgung von Zellkulturen, Preprints of *GVC-Tagung Bioverfahrenstechnik,* Würzburg.

HENZLER, H.-J., KAULING, J. (1993), Oxygen supply to cell cultures, *Bioprocess Eng.* **9**, 61–75.

HENZLER, H.-J., SCHEDEL, M. (1991), Suitability of the shaking flask for oxygen supply to microbiological cultures, *Bioprocess Eng.* **7**, 123.

HENZLER, H.-J., SCHEDEL, M., MÜLLER, P. F. (1986), Nicht äquimolare Diffusion im Schüttelkolben, *Chem. Ing. Tech.* **58**, 234.

HINZE, J. O. (1971), Turbulent fluid and particle interactions, *Prog. Heat Mass Transfer,* **6**, 433.

HO, C. S., SMITH, M. D., SHANAHAN, J. F. (1987), Carbon dioxide transfer in biochemical reactors, *Adv. Biochem. Eng.* **35**, 83–125.

HSU, Y.-C., SHIH, R.-F. (1984), Heat transfer from a heating helical coil immersed in a jet-stirred vessel, *Can. J. Chem. Eng.* **62**, 474.

HU, W.-S., MEIER, J., WANG, D. I. C. (1986), Use of surface aerator to improve oxygen transfer in cell culture, *Biotechnol. Bioeng.* **28**, 122.

HU, W.-S., OBERG, M. G. (1990), Monitoring and control of animal cell bioreactors: Biochemical engineering considerations, in: *Large Scale Mammalian Cell Culture Technology* (LUBINIECKI, A. S., Ed.), p. 451, New York: Marcel Dekker.

HUDCOVA, V., MACHON, V., NIENOW, A. W. (1989), Gas–liquid dispersion with dual Rushton turbine impellers, *Biotechnol. Bioeng.* **34**, 617.

HÜLSCHER, M. (1990), *Fortschr. Ber. VDI Verfahrenstech.,* Reihe 3, **229**.

ISHIZAKI, A., SHIBAI, H., HIROSE, Y., SHIRO, T. (1971), Studies on ventilation in submerged fermentations. Part I. Dissolution and dissociation of carbon dioxide in the model system, *Agric. Biol. Chem.* **35**, 1733.

JOHANSSON, A. (1988), *Large Scale Animal Cell Cultivation for Production of Cellular Biologicals,* A/S Nunc, Roskilde, Denmark.

JOHNSON, M., ANDRÉ, G., CHAVARIE, C., ARCHAMBAULT, J. (1990), Oxygen transfer rates in a mammalian cell culture bioreactor equipped with a cell-lift impeller, *Biotechnol. Bioeng.* **35**, 43.

JUNKER, B. H., AUNINS, J. G., HUNT, G., GLAZOMITSKY, K., BUCKLAND, B. C. (1990), Retrofit and use of large scale microbial fermentors for animal cell culture, paper presented at *SIM Annual Meeting,* Orlando, FL, July 29–August 3. Authors at Merck, Inc., Rahway, NJ.

JUNKER, B. H., WU, F., WANG, S., WATERBURY, J., HUNT, G., AUNINS, J., JAIN, D., LEWIS, J., SILBERKLANG, M., BUCKLAND, B. (1992), Evaluation of microcarriers for large-scale cultivation of a viral vaccine, *Cytotechnology* **9**, 173.

KAMEN, A. A., TOM, R. L., CARON, A. W., CHAVARIE, C., MASSIE, B., ARCHAMBAULT, J. (1991), Culture of insect cells in a helical ribbon impeller bioreactor, *Biotechnol. Bioeng.* **38**, 619.

KATINGER, H. W. D., SCHEIRER, W., KROMER, E. (1979), Bubble column reactor for mass propagation of animal cells in suspension culture, *Ger. Chem. Eng.* **2**, 31.

KAWASE, Y., MOO-YOUNG, M. (1989), Mass transfer at a free surface in stirred tank bioreactors, *Trans. Inst. Chem. Eng. Part A,* **68,** 189.

KILBURN, D. G., WEBB, F. C. (1968), The cultivation of animal cells at controlled dissolved oxygen partial pressure, *Biotechnol Bioeng.* **10,** 801.

KING, A. T., LOWE, K. C., MULLIGAN, B. J. (1988), Microbial cell responses to a non-ionic surfactant. II. Effects as assessed by fluorescein diacetate uptake, *Biotechnol. Lett.* **10,** 873.

KRAMER, P., JEWETT, J., CROUGHAN, M. S. (1991), Adaptation of animal cells to high agitation, paper presented at *AIChE Annual Meeting,* Los Angeles, CA, Nov. 18–22. Authors at Genentech, Inc., S. San Francisco, CA.

KUNAS, K. T., PAPOUTSAKIS, E. T. (1990), Damage mechanisms of suspended animal cells in agitated bioreactors with and without bubble entrainment, *Biotechnol. Bioeng.* **36,** 476.

LAMONT, J. C., SCOTT, D. S. (1990), An eddy cell model of mass transfer into the surface of a turbulent liquid, *AIChE J.* **16,** 513.

LAUFHÜTTE, H.-D. (1986), Turbulenzparameter in gerührten Fluiden, *Dissertation,* Technische Universität München.

LAVERY, M., NIENOW, A. (1987), Oxygen transfer in animal cell culture medium, *Biotechnol. Bioeng.* **30,** 368.

LEHMANN, J., PIEHL, W., SCHULZ, R. (1985), Blasenfreie Zellkulturbegasung mit bewegten, porösen Membranen, *Biotech-Forum* **2,** 112.

LEVEQUE, M. A. (1928), Les lois de la transmission de chaleur par convection, *Ann. Mines,* **12,** 202.

LIEPE, F. (1988), Stoffvereinigungen in Fluid-Phasen, in: *Verfahrenstechnische Berechnungsunterlagen,* Weinheim: VCH.

LOVRECZ, G., GEBERT, C., GRAY, P. (1992), On-line qO_2 and qCO_2 determination for recombinant mammalian cell fermentations, paper presented at *Engineering Foundation Conference "Cell Culture III",* Palm Coast, FL, Feb. 2–7. Authors at University of New South Wales, Australia.

LUBINIECKI, A., AARATHOON, R., POLASTRI, G., THOMAS, J. WIEBE, M., GARNICK, R., JONES, A., VANREIS, R., BUILDER, S. (1989), Selected strategies for manufacture and control of recombinant tissue plasminogen, activator prepared from cell cultures, in: *Advances in Animal Cell Biology and Technology for Bioprocesses* (SPIER, R. E., GRIFFITHS, J. B., STEPHENNE, J., CROOY, P. J., Eds.), p. 442, Frome, Somerset, UK: Butterworth.

LUDWIG, A., KRETZMER, G., SCHÜGERL, K. (1992), Determination of a "critical shear stress level" applied to adherent mammalian cells, *Enzyme Microb. Technol.* **14,** 209.

MAIORELLA, B., INLOW, D., SHAUGER, A., HARANO, D. (1988), Large-scale insect cell culture for recombinant protein production, *Bio/Technology* **6,** 1406.

MATHER, J. P. (1990), Optimizing cell and culture environment for production of recombinant proteins, *Methods Enzymol.* **185,** 567.

MATHEWS, C. T., ROESING, T. G., MEEHAN, B. S. (1992), Robotic processing of culture vessels for the poduction of varicella virus vaccine in MRC-5 cells, paper presented at *Enginering Foundation Conference "Cell Culture III",* Palm Coast, FL, Feb. 2–7. Authors at Merck, Inc., West Point, PA.

MATSUMURA, M., MASUNAGA, H., KOBAYASHI, J. (1977), Correlation for flow rate of gas entrained from free liquid surface of surface aerated tank, *J. Ferment. Technol.* **55,** 388.

MATSUOKA, H., FUKADA, S., TODA, K (1992), High oxygen transfer rate in a new aeration system using hollow fiber membrane, *Biotechnol. Bioeng.* **40,** 346.

MCLIMANS, W. F. (1972), The gaseous environment of mammalian cells in culture, in: *Growth Nutrition and Metabolism of Cells in Culture* (ROTHBLAT, G. H., CRISTOFALO, V. J., Eds.), pp. 137–170, New York: Academic Press.

MCLIMANS, W. F., BLUMENSON, L. E., TUNNATH, K. V. (1968), Kinetics of gas diffusion in mammalian cell culture systems. II. Theory, *Biotechnol. Bioeng.* **10,** 741.

MCQUEEN, A., BAILEY, J. E. (1989), Influence of serum level, cell line, flow type and viscosity and flow-induced lysis of suspended mammalian cells, *Biotechnol. Lett.* **11,** 531.

MERSMANN, A. (1962), Druckverlust und Schaumhöhen von gasdurchströmten Flüssigkeitsschichten auf Siebböden, *VDI Forschungsh.* **491.**

MICHAELS, J. D., PAPOUTSAKIS, E. T. (1991), Polyvinyl alcohol and polyethylene glycol as protectants against fluid-mechanical injury of freely-suspended animal cells (CRL 8018), *J. Biotechnol.* **19,** 241.

MIELHOC, E., WITTRUP, K. D., BAILEY, J. E. (1990), Influence of dissolved oxygen concentration on growth, mitochondrial function, and antibody production of hybridoma cells in batch culture, *Bioprocess Eng.* **5,** 263.

MILLER, W. M., WILKE, C. R., BLANCH, H. W. (1987), Effect of dissolved oxygen concentration on hybridoma growth and metabolism in continuous culture, *J. Cell Physiol.* **132,** 524.

MILLER, W. M., WILKE, C. R., BLANCH, H. W. (1988), Transient responses of hybridoma metabolism to changes in the oxygen supply rate in continuous culture, *Bioprocess Eng.* **3,** 103.

MILTENBURGER, H. G., DAVID, P. (1980), Mass

production of insect cells in suspension, *Dev. Biol. Stand.* **46,** 183.

MITAKA, T., SATTLER, G. L., PITOT, H. C. (1991), The bicarbonate ion is essential for efficient DNA synthesis by primary cultured rat hepatocytes, *In Vitro* **27A,** 549.

MIZRAHI, A. (1984), Oxygen in human lymphoblastoid cell line cultures and effect of polymers in agitated and aerated cultures, *Dev. Biol. Stand.* **55,** 93.

MÖCKEL, H.-O. (1978), Hydrodynamische Untersuchungen in Rührmaschinen, *Dissertation,* Ingenieurhochschule Köthen.

MONTAGNON, B., VINCENT-FALQUET, J. C., FANGERT, B. (1984), Thousand litre scale microcarrier culture of vero cells for killed Polio virus vaccine. Promising results, *Dev. Biol. Stand.* **55,** 37.

MOREIRA, J.-L., ALVES, P. M., AUNINS, J. G., CARRONDO, M. J. T. (1992), Aggregate suspension cultures of BHK cells, in: *Animal Cell Technology: Developments, Processes, and Products* (SPIER, R. E., GRIFFITHS, J. B., MACDONALD, C., Eds.), p. 411, Jordan Hill, Oxford: Butterworth-Heinemann.

MURDIN, A. D., WILSON, R., KIRKBY, N. F., SPIER, R. E. (1987), Examination of a simple model for the duffusion of oxygen into dense masses of animal cells, in: *Modern Approaches to Animal Cell Technology* (SPIER, R. E., GRIFFITHS, J. B. Eds.), p. 353, Frome, Somerset, UK: Butterworth.

MURHAMMER, D. W., GOOCHEE, C. F. (1988), Scale-up of insect cell cultures: Proective effects of Pluronic F68, *Bio/Technology* **6,** 1411.

MURHAMMER, D. W., GOOCHEE, C. F. (1990), Structural features of nonionic polyglycol polymer molecules responsible for the protective effect in sparged animal cell cultures, *Biotechnol. Prog.* **6,** 142.

NEWITT, D. M., DOMBROWSKI, N., KNELMAN, S. H. (1954), 1. The mechanism of drop formation from gas or vapour bubbles, *Trans. Inst. Chem. Eng.* **32,** 244.

NIENOW, A. W. (1990), Gas dispersion performance in fermenter operation, *Chem. Eng. Prog.* **86,** 61.

OH, S. K. W., NIENOW, A. W., AL-RUBEAI, M., EMERY, A. N. (1989), The effects of agitation intensity with and without continuous sparging on the growth and antibody production of hybridoma cells, *J. Biotechnol.* **12,** 45.

OOLMAN, T. O., BLANCH, H. W. (1986), Bubble coalescence in air-sparged bioreactors, *Biotechnol. Bioeng.* **28,** 578.

ORTON, D. R., WANG, D. I. C. (1990), Effects of gas interfaces on animal cells in bubble aerated bioreactors, paper presented at *AIChE Annual Meeting,* Chicago, IL, Nov. 11–16. Authors at M. I. T., Cambridge, MA.

ORTON, D. R., WANG, D. I. C. (1991), Quantitative and mechanistic effects of gas sparging in animal cell bioreactors, paper presented at *202nd ACS National Meeting,* New York, NY, August 25–30. Authors at M. I. T., Cambridge, MA.

OZTURK, S. S., PALSSON, B. Ø. (1990), Effects of dissolved oxygen on hybridoma cell growth, metabolism, and antibody production kinetics in continuous culture, *Biotechnol. Prog.* **6,** 437.

PAPOUTSAKIS, E. T. (1991), Fluid-mechanical damage of animal cells in bioreactors, *Trends Biotechnol.* **9,** 427.

PERKINS, H. C., LEPPERT, G. (1982), Forced convection heat transfer for a uniformly heated cylinder, *Trans. ASME C, J. Heat Transfer* **84,** 257.

PETROSSIAN, A., BAKER, R. A., SINKULE, J. A. (1990), Application of airlift bioreactors (ALBs) for the large-scale manufacturing of bio-pharmaceuticals, paper presented at the *SIM Annual Meeting,* Orlando, FL, July 29–August 3. Authors at Techniclone, Inc., Tustin, CA.

PHILLIPS, A. W., BALL, G. D., FAUTES, K. H., FINTER, N. B., JOHNSTON, M. D. (1985), Experience in the cultivation of mammalian cells on the 8000 L scale, in: *Large-Scale Mammalian Cell Culture* (FEDER, J., TOLBERT, W. R., Eds.), pp. 87–96, New York: Academic Press.

PHILLIPS, H. A., SCHARER, J. M., BOLS, N. C., MOO-YOUNG, M. (1987), Effect of oxygen on antibody productivity in hybridoma culture, *Biotechnol. Lett.* **9,** 745.

PROKOP, A., ROSENBERG, M. Z. (1989), Bioreactor for mammalian cell culture, *Adv. Biochem. Eng.* **39,** 29.

QI, Z., CUSSLER, E. L. (1985), Microporous hollow fibers for gas absorption. II. Mass transfer across the membrane, *J. Membr. Sci.* **23,** 333.

RADLETT, P. J., TELLING, R. C., WHITSIDE, J. P., MASKELL, M. A. (1972), The supply of oxygen to submerged cultures of HBK 21 cells, *Biotechnol. Bioeng.* **14,** 437.

RAMÍREZ, O. T., MUTHARASAN, R. (1990), The role of plasma membrane fluidity on the shear sensitivity of hybridomas grown under hydrodynamics stress, *Biotechnol. Bioeng.* **36,** 911.

REID, R. C., PRAUSNITZ, J. M., SHERWOOD, T. K. (1977), *The Properties of Liquids and Gases,* 3rd. Ed., New York: McGraw-Hill.

REITER, M., WEIGANG, F., ERNST, W., KATINGER, H. W. D. (1990), High density microcarrier culture with a new device which allows oxygenation and perfusion of microcarrier cultures, *Cytotechnology* **3,** 39.

REUVENY, S., VELEZ, D., MACMILLAN, J. D., MILLER, L. (1986), Factors affecting cell growth and monoclonal antibody production in stirred reactors, *J. Immunol. Methods* **86**, 53.

RISSE, F. U., WEILAND, P., ONKEN, U. (1985), Messung des Stoffübergangs von Luftsauerstoff in turbulent strömende Flüssigkeiten, *Chem. Ing. Tech.* **57**, 628–629.

ROBB, W. L. (1968), Thin silicone membranes – their permeation properties and some applications, *Ann. NY Acad. Sci.* **146**, 119.

ROYCE, P. N. C., THORNHILL, N. F. (1991), Estimation of dissolved carbon dioxide concentrations in aerobic fermentations, *AIChE J.* **37**, 1680.

RUFF, K., PILHOFER, T., MERSMANN, A. (1976), *Chem. Ing. Tech.* **48**, 9.

SCHMID, G., HUBER, F., KERSCHBAUMER, R. (1992), Adaptation of hybridoma cells to hydrodynamic stress under continuous culture conditions, poster presented at *Cell Culture Engineering III,* Palm Coast, FL, Feb. 2–7. Authors at Hoffmann-La Roche, Basel, Switzerland.

SCHUMPE, A., QUICKER, G., DECKWER, W.-D. (1982), Gas solubilities in microbial culture media, *Adv. Biochem. Eng.* **24**, 1–38.

SIEGL, G., DE CHASTONAY, J., KRONAUER, G. (1984), Propagation and assay of hepatitis A virus *in vitro, J. Virol. Methods* **9**, 53.

SILBERKLANG, M. (1987), Investigations of the influence of physical environment on the cultivation of animal cells, in: *Modern Approaches to Animal Cell Technology* (SPIER, R. E., GRIFFITHS, J. B., Eds.), pp. 297–315, Frome, Somerset, UK: Butterworth.

SILBERKLANG, M., JAIN, D., GOULD, S., DISTEFANO, D., CUCA, G., BENINCASA, D., KRIPASHANKAR RAMASUBRAMANYAN, LENNY, A., MARK, G. E. (1992), A nutrient homeostasis fedbatch process design for recombinant antibody production, paper presented at *Cell Culture Engineering III,* Palm Coast, FL, Feb. 2–7. Authors at Merck, Inc., Rahway, NJ.

SINSKEY, A. J., FLEISCHAKER, R. J., TYO, M.A., GIARD, D. J., WANG, D. I. C. (1981), Production of cell-derived products: Virus and interferon, *Ann. NY Acad. Sci.* **369**, 47.

SLIWKOWSKI, M. B., GUNSON, J. V., WARNER, T. G. (1982), Effect of culture conditions on carbohydrate charge heterogeneity of recombinant human deoxyribonuclease produced in CHO cells, paper presented at *203rd ACS National Meeting,* San Francisco, CA, April 5–10. Authors at Genentech, Inc., S. San Francisco, CA.

SMITH, C. G., GREENFIELD, P. F., RANDERSON, D. H. (1987), A technique for determining the shear sensitivity of mammalian cells in suspension culture, *Biotechnol. Tech.* **1**, 39.

SMITH, C. M. II, HEBBEL, R. P., TUKEY, D. P., CLAWSON, C. C., WHITE, J. G., VERCELLOTTI, G. M. (1987), Pluronic F-68 reduces the endothelial adherence and improves the rheology of liganded sickle erythrocytes, *Blood* **69**, 1631.

SMITH, J. M., DAVISON, S. W., PAYNE, G. F. (1990), Development of a strategy to control the dissolved concentrations of oxygen and carbon dioxide at constant shear in a plant cell bioreactor, *Biotechnol. Bioeng.* **35**, 1088.

SPIER, R., E., GRIFFITHS, J. B. (1984), An examination of the data and concepts germane to the oxygenation of cultured animal cells, *Dev. Biol. Stand.* **55**, 81.

STATHOPOULOS, N. A., HELLUMS, J. D. (1985), Shear stress effects on human embryonic kidney cells *in vitro, Biotechnol. Bioeng.* **27**, 1021.

STERN, S. A., ONORATO, F. J., LIBOVE, C. (1977), The permeation of gases through hollow silicone rubber fibers: Effect of fiber elasticity on gas permeability, *AIChE J.* **23**, 567.

TELLING, R. C., ELSWORTH, R. (1965), Submerged culture of hamster kidney cells in a stainless steel vessel, *Biotechnol. Bioeng.* **7**, 417.

THEOFANOUS, T. G., HOUZE, R. N., BRUMFIELD, L. K. (1976), Turbulent mass transfer at free gas–liquid interfaces with applications to open channel, bubble, and jet flow, *Int. J. Heat Mass Transfer* **19**, 613.

THOMPSON, P. W., WOOD, L. A., BIRCH, J. R., LAMBERT, K., BORASTON, R. (1985), Antibody production in airlift fermenters, paper presented at *AIChE National Meeting,* Nov. 10–14, Chicago, IL. Authors at Celltech, Ltd, Slough, Berkshire, UK.

THORMAN, J. M., RHIN, H., HWANG, S.-T. (1975), *Chem. Eng. Sci.* **30**, 751.

TOLBERT, W., HITT, M. M., FEDER, J. (1980), Cell aggregate suspension culture for large-scale production of biomolecules, *In Vitro* **16**, 486.

TOLBERT, W. R., SRIGLEY, W. R. (1987), Manufacture of pharmacologically active proteins by mammalian cell culture, *BioPharm Manuf.,* Sept., U. S. Patent 4537860.

TOPIWALA, H. H. (1972), Surface aeration in a laboratory fermenter at high power inputs, *J. Ferment. Technol.* **50**, 668.

TRAMPER, J., SMIT, D., STRAATMAN, J., VLAK, J. M. (1988), Bubble column design for growth of fragile insect cells, *Bioprocess Eng.* **3**, 37.

TRAMPER, J., VLAK, J. M. (1987), Preprints of *GVC-Vortragstagung,* Tübingen 1987, p. 189.

TRAMPER, J., WILLIAMS, J. B., JOUSTRA, D., VLAK, J. M. (1986) Shear sensitivity of insect cells in suspension, *Enzyme Microbiol. Technol.* **8**, 33.

TYO, M., WANG, D. I. C. (1981), Engineering characterization of animal cell and virus production using controlled charge microcarriers, in: *Advances in Biotechnology* (MOO-YOUNG, M., Ed.), p. 142, Oxford: Pergamon Press.

VARECKA, R. M., BLIEM, R. F. (1990), Investigation into the pumping characteristics of axial flow impellers in an internal loop reactor for animal cell culture, *Chem. Eng. Commun.* **96,** 81.

VARGA, O. H. (1966), *Stress-Strain Behavior of Elastic Materials,* pp. 127–136, New York: J. Wiley & Sons.

VOGELMANN, H. (1981), Aspects of scale-up and mass cultivation of plant tissue culture, in: *Advances in Biotechnology,* Vol. 1 (MOO-YOUNG, M., Ed.), pp. 17–122, Elmsford, NY: Pergamon Press.

VORLOP, J., LEHMANN, J. (1988), Scale-up of bioreactors for fermentation of mammalian cell culture, *Chem. Eng. Technol.* **11,** 157.

WALTER, J. F., BLANCH, H. W. (1986), Bubble break-up in gas–liquid bioreactors: Break-up in turbulent flows, *Chem. Eng. J.* **32,** B7.

WESTERTERP, K. R., VANDIERENDONCK, L. L., DEKRAA, K. A. (1963), Interfacial areas in agitated gas-liquid contactors, *Chem. Eng. Sci.* **18,** 157.

WHITESIDE, J. P., FARMER, S., SPIER, R. E. (1985), The use of caged aeration for the growth of animal cells on microcarriers, *Dev. Biol. Stand.* **60,** 283.

WOHLPART, D., KIRWAN, D., GAINER, J. (1991), Oxygen uptake by entrapped hybridoma cells, *Biotechnol. Bioeng.* **37,** 1050.

YAGI, H., YOSHIDA, F. (1974), *J. Ferment. Technol.* **52,** 905.

YAMADA, K., FURUSHOU, S., SUGAHARA, T., SHIRAHATA, S., MURAKAMI, H. (1990), Relationship between oxygen consumption rate and cellular activity of mammalian cells cultured in serum-free media, *Biotechnol. Bioeng.* **36,** 759.

YANG, J.-D., WANG, N. S. (1992), Cell inactivation in the presence of sparging and agitation, *Biotechnol. Bioeng.* **40,** 806.

ZLOKARNIK, M. (1979), Scale up of surface aerators for waste water treatment, *Adv. Biochem. Eng.* **11,** 157.

ZWEITERING, T. N. (1958), Suspending of solid particles in liquid by agitators, *Chem. Eng. Sci.* **8,** 244.

12 Strategies for Fermentation with Recombinant Organisms

TADAYUKI IMANAKA

Osaka, Japan

1 Introduction

In the bioindustry, mutagenesis has been used as a means for strain improvement of microorganisms. However, only a few nucleotide base pairs are changed by mutation. In contrast, a DNA fragment of more than 1000 base pairs can be manipulated through genetic engineering techniques. Accordingly, the use of recombinant DNA allows approaches in which useful genetic information can be inserted directly into microorganisms. Qualitative success might involve the cloning of foreign DNA such as human genes in microorganisms, and quantitative success includes the engagement of products originating from the cloned genes due to the gene dosage effect. Thus, plasmids may serve as powerful tools in producing peptides and/or metabolites in the cultivation of microorganisms. However, unless cloned genes are fully expressed and are also stably kept *in situ* in vector plasmids during the growth of host cells, it is not possible to employ the recombinant plasmid as an agent to enhance production of specific materials in industry. This chapter describes how to select a suitable host–vector system and also pays attention to the subject of expression and stability of cloned genes in the cultivation of host cells.

2 Selection of Host Cells

(IMANAKA, 1986)

Despite the many examples of gene manipulation that are theoretically possible, some technical limitations for applications in industry still exist as follows: (1) Genetic maps do not always exist. (2) Gene exchange systems for industrially useful microorganisms, such as useful vectors and transformation procedures, are at an early stage of development. (3) Metabolic pathways leading from a raw material to the desired product, such as antibiotics, are not made clear in many cases. Because of these limitations, a suitable host strain should be carefully selected to make good use of the characteristics of the host.

The most popular organisms and their characteristics as host cells are summarized below.

Escherichia coli
This is the best understood bacterium. More than 1000 genes have been identified. It was shown that *E. coli* cells treated with cold calcium chloride became competent, and the competent cells could take up plasmid DNA as well as phage DNA.

Bacillus subtilis
This bacterium is genetically and biochemically well characterized. More than 300 genes have been identified on its circular chromosome. *B. subtilis* is a non-pathogenic soil microorganism which grows strictly under aerobic conditions and, thcrcfore, represents a safe host. *B. subtilis* does not contain pyrogenic lipopolysaccharides as does *E. coli*. *B. subtilis* is Gram-positive and has a rather simple cell envelope structure which consists of a single layer of membrane. Therefore, secretory proteins such as amylase and protease are released directly into the culture medium. This process obviates the necessity of disrupting cells and makes recovery and purification of secreted products simpler.

Bacillus stearothermophilus
This is a typical thermophile and produces thermostable enzymes. The amount of cooling water required for a specific cultivation of a thermophile in a large-scale fermentor would be reduced. Since thermophilic bacteria can generally grow faster than mesophiles, the cultivation time can be shortened. Because of high cultivation temperature ($>55\,°C$), the number of possible contaminating organisms would be reduced.

Streptomyces spp.
Streptomyces species are well known as producers of several thousand antibiotics. Many of them are constantly used in human and veterinary medicine and in agriculture. Application of recombinant DNA techniques to *Streptomyces* plays an important role in strain improvement aimed at increased antibiotic yields and the generation of novel antibiotics by incorporating parts of different antibiotic biosynthetic pathways into a strain.

Corynebacterium and the related bacteria
These include amino acid producer and biosurfactant producer.

Saccharomyces cerevisiae
This yeast is a commercially important strain and is genetically well characterized. Yeast is a suitable eukaryotic host, because it can be cultivated like bacteria on defined media and grows in colonies. The generation time is much shorter than those of other eukaryotic cells.
Animal cells
By and large, current methods of gene cloning in animal cells rely on the integration of the foreign DNA into the genome, using the genomes of special viruses (e.g., SV40) as cloning vehicles. Attempts are being made to develop vectors which can be maintained in an extrachromosomal state. Mass culture of animal cells is an important method to be improved for industrial applications.
Plant cells
Cloning in plant cells is currently being developed. It will open the way to the direct genetic modification of agricultural plants for breeding. Plant viruses and the Ti plasmid of the bacterium *Agrobacterium tumefaciens* are usually used as vectors.

3 Instability of Recombinant Plasmids

As mentioned above, plasmid is maintained under a subtle quasi-equilibrium condition in the host cell, and it is easily supposed that the recombinant plasmid carrying cloned genes would behave in a different manner compared to the original vector plasmid. Reasons for a general inclination towards instability of recombinant plasmids are summarized as follows;

- The higher the gene expression of a plasmid, the more segregants (plasmid-free cells) tend to appear.
- Recombinant plasmid is quite unstable when the products of cloned genes are inhibitory to host cells.
- Phenotypic instability of plasmid is due either to the disappearance of entire plasmid from host cells or to the deletion of a specific region from the recombinant plasmid.
- Transposable elements such as transposon and insertion sequence promote deletion of DNA from recombinant plasmids.
- The instability of the plasmid-carrying cells can be explained by the fact that their growth rate is lower than that of host cells without plasmid and thus, by the enrichment of plasmid-free cells during the culture.
- Plasmids also exist which do not change the phenotype, but the expression level decreases due to point mutation on the specific gene.
- The stability of plasmid cannot be guaranteed in chemostat culture, because plasmid loss from host cells is generally irreversible.

4 Assessment of Stability of a Plasmid in Host Cells

(IMANAKA and AIBA, 1981)

It is worthwhile examining the stability of plasmid in proliferating cells as a basis of genetic engineering. We considered that a mixed culture of plasmid-carrying (P) and plasmid-free (N) cells could be represented by the following reactions.

$$P \xrightarrow{\mu_1} (2-p)P$$
$$\searrow^{\mu_1} N \qquad (N=pP) \qquad (1)$$
$$N \xrightarrow{\mu_2} 2N$$

where p is the probability of plasmid loss and/or deletion per division of host cells, μ is the specific growth rate, and subscripts 1 and 2 are cells with and without plasmid, respectively.

Under the assumption of exponential cell growth, we noted an important consequence for the mixed culture. With $N_0=0$ for the initial condition, fraction F of cells carrying plasmid in the total population decreases continuously with generation number n, according to the form

$$F_n = \frac{1-\alpha-p}{1-\alpha-p\cdot 2^{n(\alpha+p-1)}} \tag{2}$$

where $\alpha = \mu_2/\mu_1$.

Taking $n=25$, for instance, F_{25} vs. α is shown in Fig. 1, the parameter being p. In general, α ranged nearly from 1.0 to 2.0. It is evident from Fig. 1 that F_{25} deteriorates drastically, as α increases from 1.0 to 2.0. An extreme case, $\mu_2=0$ ($\alpha=0$), may correspond to an auxotroph in a minimal medium; i.e., plasmid, on which biosynthetic enzymes for nutrients required are coded, is lost, the cells cannot grow. Another case, $\mu_2<0$ ($\alpha<0$), corresponds to the loss of drug-resistance plasmid. In other words, once the plasmid is lost, host cells are killed. It is clear from Eq. (2) that in batch system with any $\alpha>1.0$ and $p>0$, $F_\infty=0$.

A typical series of steps in the scale-up is as follows: (1) inoculum from slant into conical flask (20 mL of liquid medium), (2) large flask (300 mL, (3) jar fermentor (10 L), (4) pilot fermentor (3000 L), and (5) production size (100000 L). If the inoculum size in each step is 3%, $n=5$ is needed from step to step. Then the total amount of n required for the scale-up becomes $n=25$. Consequently, if correlated with the production performance, the F_{25} value might serve as a factor to judge whether or not the plasmid carriers in question are of practical use in industry. It is advisable to inoculate only a few cells (less than 50 cells per mL) into a fresh medium, because this small inoculum permits an assessment of F_{25}, i.e., the plasmid stability only in one step of the flask test.

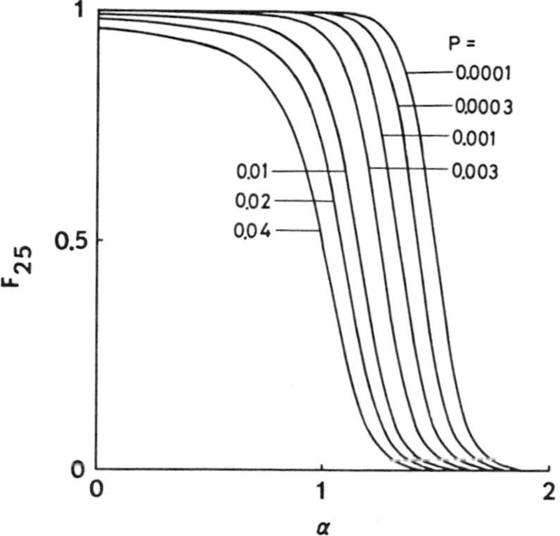

Fig. 1. F_{25} vs α. Parameter is p.

5 Some Proposals to Ensure the Stability of Recombinant Plasmids

To ensure the stability of recombinant plasmid, it would be of crucial importance that the probability of plasmid loss p and α ($=\mu_2/\mu_1$) be lowered and the genetic structure remains unchanged. Some proposals are presented as follows:

- Plasmids can be stabilized by joining with a particular DNA segment that provides the partition function.
- Plasmids can also be stabilized by joining with particular genes that couple host cell division to plasmid proliferation.
- Environmental selective pressure is effective. For instance, only the cells carrying drug resistance plasmid can grow in the presence of a specific drug and the plasmid stability is secured.
- Runaway-replication plasmids increase their copy number after the cultures are shifted from a low to a high temperature. This temperature increase would decrease plasmid-free cells, thus guaranteeing plasmid stability.
- It was confirmed experimentally that the higher the gene expression, the more segregants tended to appear. Conversely, plasmid is most likely to be kept unchanged when the gene expression is repressed. It is advisable to use a thermoinducible expression system of plasmid vector and/or host chromosome. Cells are grown at normal temperatures

to repress the gene expression in an early phase, followed by a temperature increase for the gene to be fully de-repressed.

- Plasmids can be stabilized by joining with a particular DNA fragment or gene that improves growth rate of the host cells.
- Transposable elements promote insertion and deletion. The use of plasmid with such a transposable element should be avoided.
- It is recommendable to eliminate unnecessary DNA fragment from plasmids, because the redundant DNA becomes a burden on the host cell and also increases the probability of rearrangement of DNA *in vivo*.
- In the case of self-cloning, a recombination-deficient mutant strain is required to avoid integration of a plasmid into the chromosome.
- Depending on the final products, a plasmid with optimal copy number should be used.
- Host cells should be improved so that a plasmid could become more stable.
- Immobilization of plasmid-carrying cells may be effective to maintain the plasmid stably in the reaction system.
- Plasmid copy number is sometimes influenced by the cultivation conditions of host cells. Therefore, the optimum condition should be adopted.

One concrete example of how plasmid was stabilized in host cells will be explained later.

6 Examples of Genetically Engineered Microorganisms

Recent advances in the technology of molecular cloning and the development of expression vectors allow efficient production of clinically important mammalian peptides, or industrially useful enzymes by microorganisms carrying the appropriate genes. Peptides can be directly synthesized as translation products by microorganisms. In contrast, non-peptide products are synthesized as the metabolites catalyzed by the specific enzymes originating from the cloned genes. For this reason, our work on amino acid production by a recombinant organism is briefly explained as an example.

6.1 Tryptophan Production by Recombinant *Escherichia coli*

(AIBA et al., 1982)

The main pathway of tryptophan biosynthesis is shown in Fig. 2. For the purpose of studying the production of L-tryptophan by *E. coli*, the deletion mutants of the *trp* operon (*trpAE1*) were transformed with mutant plasmid (pSC101trpI15·14) carrying the whole *trp* operon whose anthranilate synthase had been desensitized to tryptophan inhibition (Tab. 1). In addition, the deficiency of both tryptophan repressor (*trpR*) and tryptophanase (*tnaA*) was indispensable for host strains for the enhancement of L-tryptophan production (Tab. 2). The gene dosage effect on activities of specific enzymes of the *trp* operon was assessed. As-

Tab. 1. Bacterial Strains and Recombinant Plasmids

Strain/Plasmid	Relevant Properties/Phenotype
E. coli	
W3110	*trpAE1*[a]
W3110	*trpAE1 trpR27*(Am)[b]
W3110	*trpAE1 trpR tnaA*[c]
Plasmid	
pSC101-trp	Tc^rTrp^+
pSC101-trp·I15	$Tc^rTrp^+I^-$[d]
pSC101-trp·I15·14	$Tc^rTrp^+I^-$
RP4-trp·I15	$Ap^rKm^rTc^rTrp^+I^-$
RSF1010-trp·I15	$Sm^rTrp^+I^-$
pBR322-trp·I15	$Ap^rTc^rTrp^+I^-$

[a] Deletion mutant of *trpA-E*, designated AE1
[b] trp repressor amber mutant, designated Ram
[c] Tryptophanase-deficient mutant, designated Tna
[d] I^-, Insensitivity to feedback inhibition by tryptophan

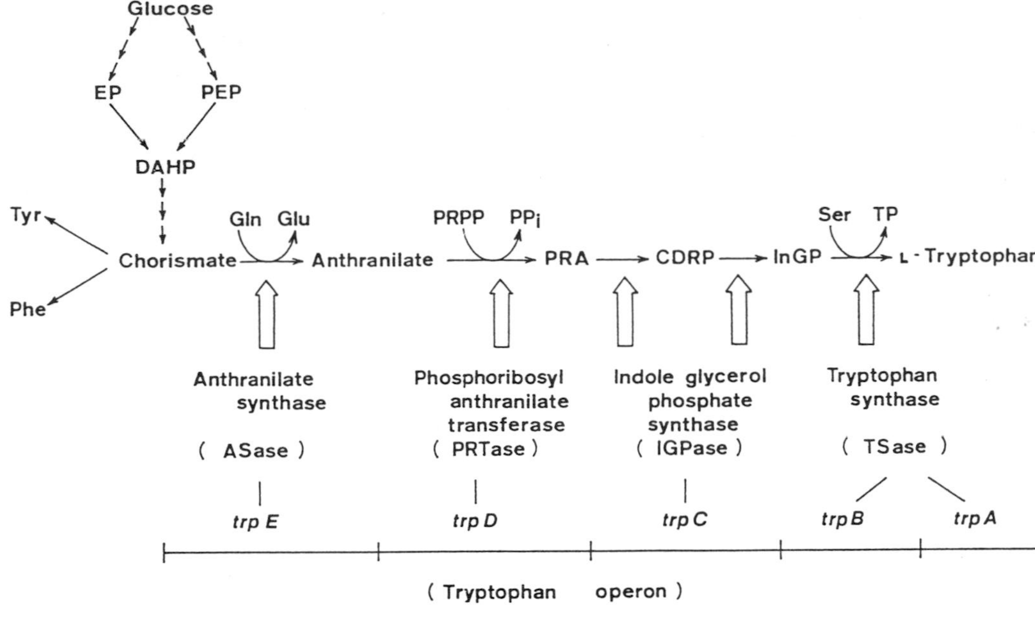

Fig. 2. Main pathway of L-tryptophan biosynthesis. Abbreviations: EP, D-erythrose-4-phosphate; PEP, phosphoenolpyruvate; DAHP, 3-deoxy-D-arabinoheptulosonate-7-phosphate; Tyr, tyrosine; Phe, phenylalanine; Glu, glutamate; PRPP, 5-phosphoribosyl-1-pyrophosphate; PP_i, pyrophosphate; PRA, 5-phosphoribosyl anthranilate; CDRP, 1-(*o*-carboxyphenylamino)-1-deoxyribulose-5-phosphate; InGP, indole-3-glycerol phosphate; Ser, serine; TP, triose phosphate.

Tab. 2. Tryptophan Production in Shake Flask Cultures[a]

Strain	Plasmid	Repression	Inhibition	Tryptophanase	Tryptophan (mg/L)	Plasmid Stability
AE1	pSC101-trp	+	+	+	7	~100 (%)
AE1	pSC101-trp·I15	+	–	+	11	~100
Ram	pSC101-trp	–	+	+	7	~95
Ram	pSC101-trp·I15	–	–	+	70	~95
Tna	pSC101-trp	–	+	–	8	~85
Tna	pSC101-trp·I15	–	–	–	360	~85

[a] Cells were grown in MTI medium (without anthranilic acid) at 37 °C for 36 h.

suming that tryptophan synthase (TSase) activity represents those of the *trp* operon enzymes, TSase activities were measured for high and low copy number plasmids (Tab. 3). The gene dosage effect was manifested in strain AE1 (*trpAE1*). Although the gene dosage effect could also be noted in strain Tna (*trpAE1 trpR tnaA*), the activity of TSase in Tna (pBR322-trp·I15) could not be measured because of the instability of plasmid.

We also examined the relationship between the copy number of the plasmids and trypto-

Tab. 3. Tryptophan Synthase Activities and Tryptophan Production[a]

Strain	Plasmid		TSase (U/mg of protein)	Tryptophan (g/L)
	Copy no./ Chromosome	Plasmid Stability (%)		
AE1 (RP4-trp·I15)	1–3	~100	4	
AE1 (pSC101-trp·I15)	~5	~100	25	
AE1 (RSF1010-trp·I15)	10–50	~100	47	
AE1 (pBR322-trp·I15)	60–80	~80	114	
Tna (RP4-trp·I15)	1–3	~100	36	1.7
Tna (pSC101-trp·I15)	~5	~35	107	3.1
Tna (RSF1010-trp·I15)	10–50	~35	215	2.6
Tna (pBR322-trp·I15)	60–80	–	–[b]	–

[a] Cells were grown in MM at 37 °C until the late log phase.
[b] Stable transformants difficult to obtain

phan production (Tab. 3). The gene dosage effect observed and the limitation suggest the optimum plasmid copy number for tryptophan production.

It is interesting that tryptophan production was enhanced in proportion to the release from feedback inhibition by tryptophan. When the release became more advanced, tryptophan production was adversely affected. Therefore, the release from feedback inhibition for anthranilate synthase (ASase) activity should be optimized for tryptophan production.

The newly constructed strain *E. coli* W3110 *trpAE1 trpR tnaA* (pSC101trpI15·14) exhibited high productivity of tryptophan (6.2 g L^{-1} 27 h^{-1} = 0.23 g L^{-1} h^{-1}) which exceeded any other data ever published. In this case, tetracycline was added to the medium to ensure the stability of the plasmid, and deletion mutants (*trpAE1*) were used as host strains to avoid the recombination of cloned *trp* operon with chromosomal DNA.

Although the batch culture of the above-mentioned strain was remarkable in the production of tryptophan, the continuous culture was not successful in establishing steady state. The failure of continuous culture was solely due to the instability of plasmid. Accordingly, the stability and/or instability of the plasmid becomes the most important factor concerning the actual use of a strain in industry.

6.2 Isolation of a Stable Plasmid Carrier Mutant

(Sakoda and Imanaka, 1990)

A host–plasmid system of *E. coli* that would warrant an industrial production of L-tryptophan by fermentation has been established. It was considered in the process that plasmid stability was one of the most important factors. The stability of a plasmid may be affected by the genetic characteristics of host cells, the copy number of the plasmid, culture conditions, and the genes carried on the plasmid. Some proposals to ensure plasmid stability have also been reported, and a new category is introduced here. As a model system we took the recombinant plasmid pSC101-trpI15·14 having a whole *trp* operon in *E. coli* cells.

E. coli Tna(pSC101-trpI15·14) was mutagenized with N-methyl-N′-nitro-N-nitrosoguanidine, and the mutants that were resistant to 6-fluoro-tryptophan (1000 μg/mL) were selected. Eighty mutant strains were obtained, a frequency of around 10^{-9}. Plasmid stability was examined on the basis of Tc^r and Trp^+ for all the mutant strains. Among them, an extremely stable plasmid carrier was found. This strain, 6F484, stably maintained the recombinant plasmid even for more than 100 generations.

It has been reported that Tc^r bacteria produce TET proteins to prevent tetracycline permeation across the membrane. TET proteins bind to metal ions, and therefore Tc^r cells are hypersensitive to lipophilic chelating agents such as fusaric acid. Only Tc^s colonies can grow well on the agar plates containing fusaric acid. This simple technique allows direct plate selection of Tc^s clones from a predominant Tc^r population. Thus, the strain FA14 cured of pSC101trpI15·14 was obtained from strain 6F484, and the phenotype was confirmed to be Tc^s and a tryptophan auxotroph (Trp^-). A Southern transfer experiment proved that strain FA14 did not have any homologous segment with pSC101-trpI15·14.

To examine which factor (host cell or plasmid) is responsible for the plasmid stability, *E. coli* strains Tna and FA14 were transformed with pSC101-trpI15·14. The transformants were cultured and their phenotypic stability was investigated (Fig. 3). Strain Tna(pSC101-trpI15·14) gradually lost the characteristics of Tc^r and Trp^+. After 100 generations, only 28% of the total bacteria had Tc^r and could grow on minimal medium. In contrast, the phenotype of strain FA14(pSC101-trpI15·14) was stable. After 100 generations, all the populations tested were Tc^r and Trp^+. When pSC101-trpI15·14 which had been prepared from strain 6F484 was used to transform strain Tna, the transformant showed nearly the same plasmid stability as that of Tna(pSC101-trpI15·14). It was, therefore, concluded that the extremely high plasmid stability was due to mutation in host strain Fa14 and not to the recombinant plasmid.

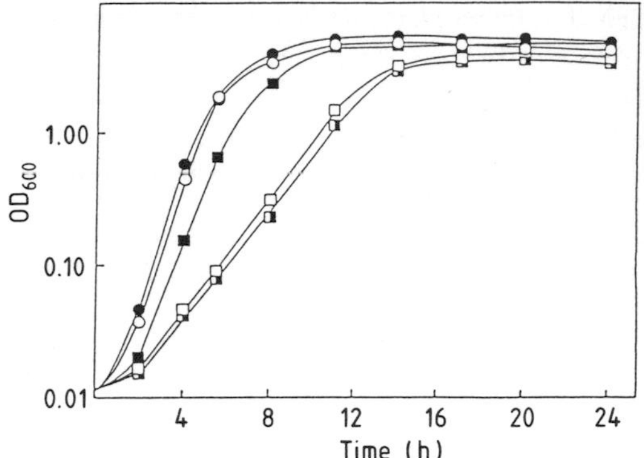

Fig. 4. Growth curves of *E. coli* cells with and without plasmid in a complete medium at 37 °C. Symbols: ○ Tna; ● FA14(pSC101-trpI15·14); □ FA14; ◨ FA14(pSC101); ■ FA14(pSC101-trpI15·14).

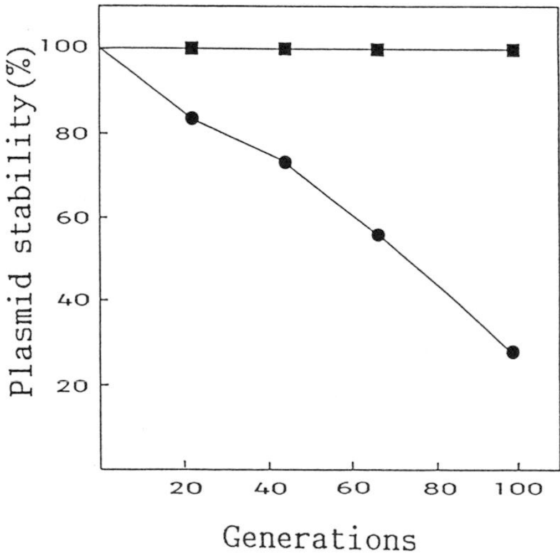

Fig. 3. Stability of plasmid pSC101-trpI15·14 during cell growth. Phenotype of the plasmid is TC^r Trp^+. Symbols: ● Tna(pSC101-trpI15·14); ■ FA14(pSC101-trpI15·14).

6.3 Reasons Why Recombinant Plasmid is Stably Maintained

(Sakoda and Imanaka, 1982)

We examined the growth characteristics in L broth at 37 °C of the strains Tna, FA14, and their transformants with pSC101 or pSC101-trpI15·14 (Fig. 4). FA14 grew more slowly than Tna; their specific growth rates were 0.50 h^{-1} and 1.31 h^{-1}, respectively. The growth rates of FA14 and FA14(pSC101) were nearly the same. However, the growth of strain FA14 was stimulated by the transformation with pSC101-trpI15·14, and the specific growth rate of the transformant was very similar to those of Tna cells with and without pSC101-trpI15·14 (Fig. 4). These results indicate that growth of strain FA14 was enhanced by the addition of the whole *trp* operon, in other words, the ability of tryptophan biosynthesis, even in a complete medium. It was, therefore, inferred that host strain FA14 could not efficiently take up tryptophan from the medium.

Tryptophan uptake by both strains, Tna and FA14, was examined using radioactive tryptophan. Since tryptophan uptake at 37 °C was too fast to measure accurately, the experiment was done at 20 °C (Fig. 5A). Tryptophan uptake, including both active transport and adsorption to the cell surface, by strain Tna continued for 2 min and leveled off at a higher value than that of strain FA14. The level of ^{3}H-tryptophan for FA14 was constant after 15 s, and the incorporation of radioactive substance might be considered as the adsorption to the cells. To eliminate the effects of active transport, an uncoupler, CCCP (inhibitor of ATP biosynthesis), was added to the reaction system. Consequently only the cell-bound tryptophan could be measured (Fig. 5B). In fact, both strains Tna and FA14 showed constant values from 15 s to 6 min. The difference between the values shown in Figs. 5A and 5B is concluded to correspond to the active transport of tryptophan. *E. coli* Tna had typical active transport of tryptophan, but strain FA14 showed little evidence of the active transport system. It is concluded by these results that strain FA14 lacks the active transport system for tryptophan, resulting in a low growth rate even in a complete medium. The slow growth of FA14 in a rich medium might be mainly supported by the non-specific incorporation of tryptophan.

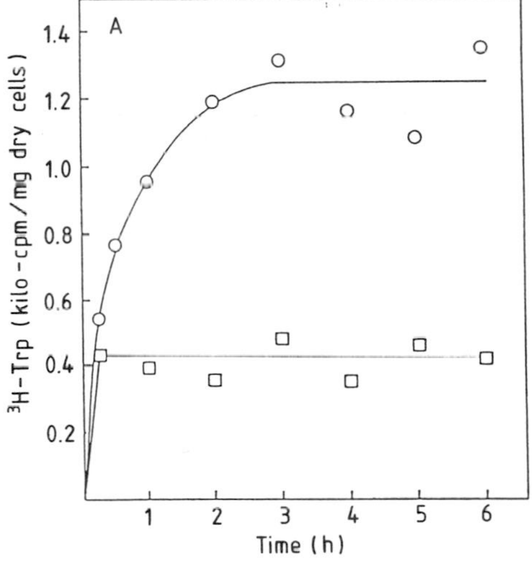

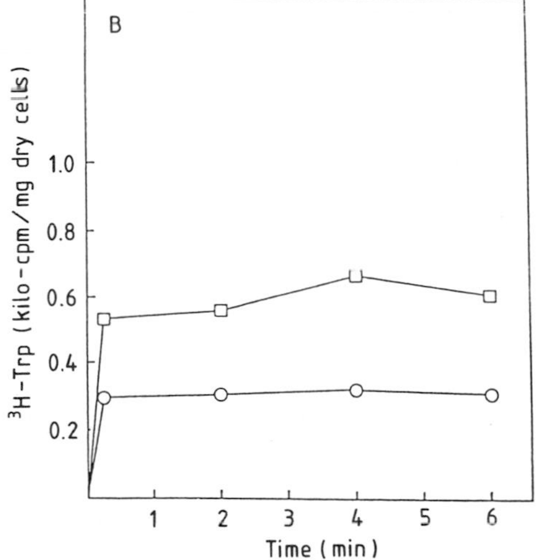

Fig. 5. (A) Uptake of [^{3}H]-tryptophan by *E. coli* cells at 20 °C. (B) Adsorption of [^{3}H]-tryptophan to *E. coli* cells at 20 °C. Symbols: ○ Tna; □ FA14.

6.4 A New Way of Stabilizing Recombinant Plasmids

(SAKODA and IMANAKA, 1990)

We established an extremely stable host-recombinant plasmid system.

This system has the following characteristics; (1) deletion of tryptophan operon in the host chromosome, (2) deficiency of active transport of tryptophan in the host cell, (3) transformation of the host strain with a recombinant plasmid carrying the tryptophan operon.

The tryptophan auxotrophic mutant Tna of *E. coli* cannot grow in a minimal medium. When it is transformed with pSC101-trpI15·14, the transformant can grow normally in the minimal medium and the plasmid is fairly stable, because the cell growth depends on the recombinant plasmid. However, in a complete medium the plasmid is not so stable, because the host cell growth does not depend on the tryptophan operon of the plasmid.

If the host strain cannot take up tryptophan efficiently, the *trp* operon recombinant plas-

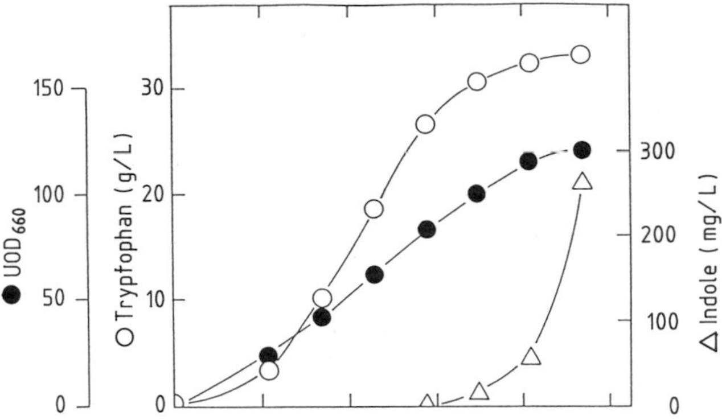

Fig. 6. Time course of L-tryptophan production. Feeding of glucose and anthranilate was started at 22 h.

mid carrier can preferentially grow even in a complete medium, thus plasmid could be stably maintained. Therefore, the three characteristics mentioned above are simultaneously required for the stability of the plasmid.

6.5 Hyperproduction of Tryptophan by Recombinant *Escherichia coli*

(TSUNEKAWA et al., 1989)

To produce a large amount of tryptophan, fed-batch culture was carried out by using the stable recombinant plasmid carrier. Production medium contains glucose, ammonium citrate, and hydrolysate of soybean meal as main carbon and nitrogen sources. Salts and anthranilate are also contained in the medium. Some nutrients and anthranilate were added during the fermentation. Time course of cell growth and tryptophan production in a 30 L fermentor is shown in Fig. 6. More than 30 g/L of tryptophan was accumulated at 70 h. Thereafter, indole concentration was increased quickly, and tryptophan production was deteriorated.

Three factors for indole accumulation were considered; (1) decrease of tryptophan synthase, (2) shortage of L-serine supply, and (3) product inhibition of tryptophan synthase by L-tryptophan. These possibilities were examined one by one. Intracellular concentration of the enzyme was nearly constant during the fermentation. Even if L-serine was supplied into the culture medium, indole accumulation was observed. However, it was found that tryptophan synthase activity was inhibited by tryptophan (Fig. 7). Therefore, to avoid the accumulation of indole, we have to keep the high activity of tryptophan synthase.

Since tryptophan is a hydrophobic amino acid, the solubility is low (about 15 g/L in water). The real solubility in the fermentation broth after 80 h was around 32 g/L, which corresponded to the inhibitory concentration for tryptophan synthase. To reduce the tryptophan solubility in the fermentation broth, a wide variety of habit modifiers were tested, and a good antifoam pluronic L-61 (non-ionic detergent, polyoxyethylene-polyoxypropylene ether) was screened. In fact, solubility of tryptophan was about 20 g/L in the presence of 1.0% pluronic L-61.

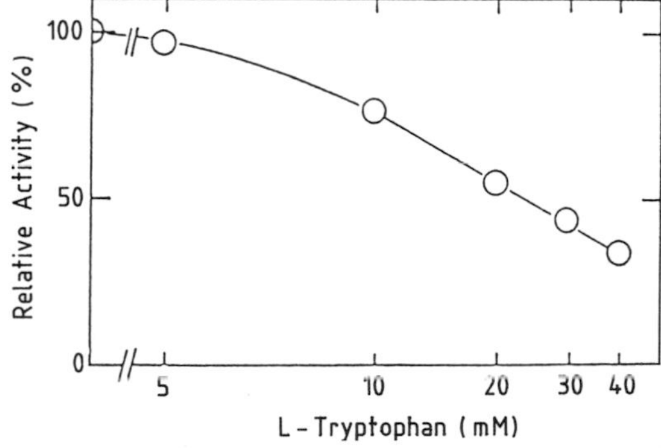

Fig. 7. Inhibition of tryptophan synthase activity by tryptophan.

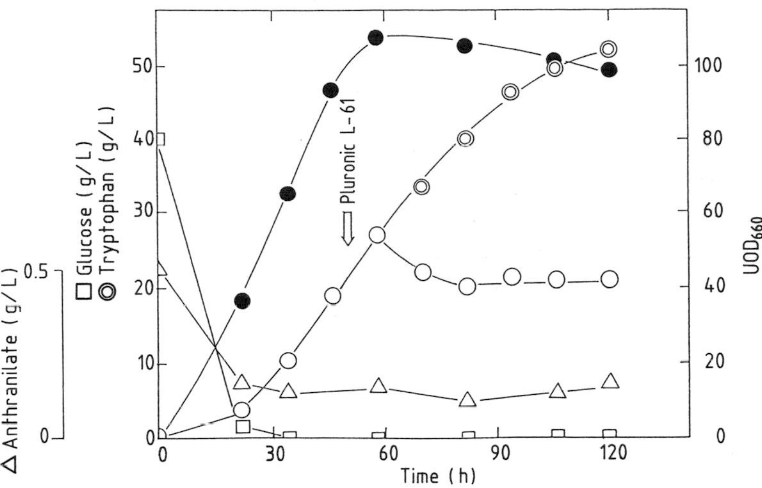

Fig. 8. Hyperproduction of tryptophan. Feeding of glucose and anthranilate was started at 22 h. Pluronic L-61 was added at 48 h

Tab. 4. Characteristics of a Recombinant Microorganism for Tryptophan Production and the Cultivation Condition

Regulation/Function	Host Chromosome	Plasmid	Effect	Total Effect
Repression	Deficiency of trp repressor (*trpR*)	–	Enhancement of *trp* operon enzymes	
Inhibition	–	Insensitive to feedback inhibition (I^- for ASase)	High rate of trp synthesis	High accumulation of trp
Degradation of trp	Deficiency of tryptophanase (*tnaA*)	–	No degradation of trp	
Recombination	Deletion of *trp* operon (*trpAE1*)	–	No recombination between chromosome and plasmid Trp auxotroph	
Trp synthesis	–	Whole *trp* operon	Trp synthesis	Extreme stability of recombinant plasmid
Active Transport	Deficiency of active transport of trp	–	Slow growth even in the presence of trp	
Cultivation Conditions		**Objective**	**Effect**	**Total Effect**
Feeding of anthranilate		Supply of intermediate without inhibition	High efficiency of trp synthesis	High productivity of trp
Addition of habit modifier		Decrease of trp solubility	Reduction of feedback inhibition for trp synthase	

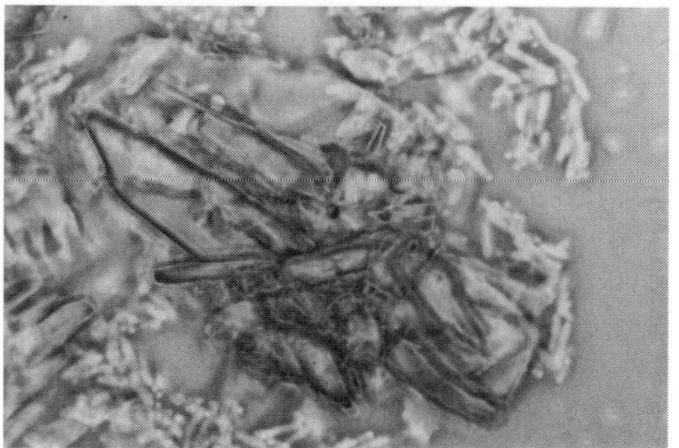

Fig. 9. Fermentation with crystallization. Small rods are *E. coli* cells. Large crystals are tryptophan.

Fig. 8 shows the time course of tryptophan production. Cell growth reached more than 100 UOD_{660} at 60 h. When tryptophan concentration became about 20 g/L, pluronic L-61 was added to the culture broth, and then the soluble concentration was kept at nearly the same level, resulting in no indole accumulation. Since tryptophan production continued up to more than 50 g/L, crystalline tryptophan appeared as an insoluble part. This is a typical crystalline fermentation for the production of tryptophan (Fig. 9). The features are summarized in Tab. 4. This fermentation method is feasible in industrial production of tryptophan and comparable to other methods such as chemical synthesis or enzymatic synthesis.

Acknowledgement
I thank Dr. SHUICHI AIBA for his support of this work, Drs. HIROSHI TSUNEKAWA and ROKURO OKAMOTO of Mercian Co. for large-scale fermentation.

7 References

AIBA, S., TSUNEKAWA, H., IMANAKA, T. (1982), *Appl. Environ. Microbiol.* **43,** 289–297

IMANAKA, T. (1986), *Adv. Biochem. Eng. Biotechnol.* **33,** 1–27.

IMANAKA, T., AIBA, S. (1981), *Ann. N.Y. Acad. Sci.* **369,** 1–14.

SAKODA, H., IMANAKA, T. (1990), *J. Ferment. Bioeng.* **69,** 75–78.

TSUNEKAWA, H., AZUMA, S., OKABE, M., OKAMOTO, R., IMANAKA, T., AIBA, S. (1989), in: *Abstract of the 1989 International Chemical Congress of Pacific Basin Societies, Bioscience and Technology*, 04–1.

13 Anaerobic Fermentations

LARRY E. ERICKSON
DANIEL Y. C. FUNG
PRAVATE TUITEMWONG
Manhattan, KS 66506, USA

List of Symbols

d moles of carbon dioxide per quantity of organic substrate containing one gram atom carbon (g mol/g atom carbon)
K_s saturation kinetic constant (g/L)
m_e rate of organic substrate consumption for maintenance (equivalents of available electrons per equivalent of available electrons in biomass $\cdot h^{-1}$)
m_s grams of substrate expended per gram of cells per hour for maintenance (h^{-1})
P product concentration (g/L)
S substrate concentration (g/L)
t time (h)
X biomass concentration (g/L)
Y_{ATP}^{max} true biomass yield based on ATP (g dry biomass/g mol ATP)
$Y_{x/s}^{max}$ true growth yield corrected for maintenance (g biomass/g substrate)
y_c biomass carbon yield, fraction of organic substrate carbon in biomass (dimensionless)
z fraction of organic substrate carbon in products (dimensionless)
α growth-associated kinetic parameter (dimensionless)
β maintenance-associated kinetic parameter (h^{-1})
γ_b reductance degree of biomass (equivalents of available electrons/g atom carbon)
γ_p reductance degree of products (equivalents of available electrons/g atom carbon)
γ_s reductance degree of substrate (equivalents of available electrons/g atom carbon)
δ number of equivalents of available electrons transferred to product to produce 1 g mol of ATP from ADP
η fraction of available electrons in organic substrate converted to biomass (dimensionless)
η_{max} "true" biomass available electron yield (dimensionless)
μ specific growth rate (h^{-1})
μ_{max} maximum specific growth rate (h^{-1})
ξ_p fraction of available electrons in organic substrate converted to products (dimensionless)
σ_b weight fraction carbon in dry biomass (dimensionless)
σ_p weight fraction carbon in products (dimensionless)
σ_s weight fraction carbon in organic substrate (dimensionless)

1 Introduction

Anaerobic fermentations are of great industrial and environmental significance. Microorganisms have been performing useful functions for man for many years; e.g., early recorded history reports wine in wineskins. As man learned to live in cities, waste management increased in importance and anaerobic processes for the digestion of wastes were developed. This work focuses on modern developments in anaerobic fermentations; the emphasis is on the science and technology of anaerobic bioprocessing.

The subject of anaerobic fermentations is treated in much greater detail elsewhere (ERICKSON and FUNG, 1988). The literature prior to 1988 has been reviewed comprehensively by the many authors who contributed to this handbook in which various aspects of anaerobic bioprocessing are considered (stoichiometry, process analysis, kinetics, bioenergetics, data collection and analysis, parameter estimation, control, and reactor design).

The form of the reactants and products must be considered in the design of anaerobic bioreactors. In this chapter, there are separate sections on gas phase reactants and products, liquid phase reactants and products, and solid phase reactants and products. The yogurt fermentation and several other examples of commercial anaerobic processes that are found throughout the world are considered because of their importance and because they are presently the subject of considerable research.

2 Strain Development and Management

2.1 Pure Cultures

Microorganisms have performed anaerobic fermentation in the natural environment since complex life forms began on this planet. They are involved with all types of bioconversion and various cycles of matter on earth. In nature, microorganisms compete with each other and the most suitable species predominate in certain niches. At any one time one would expect to find a mixed population interacting in the environment in antagonistic or symbiotic relationships.

In order to maximize the performance of desirable microorganisms it is imperative to isolate pure cultures from the environment and study their cultural characteristics in great detail. Certainly the development of pure culture technique was the cornerstone of the development of microbiology as a science. The work of LOUIS PASTEUR opened up the entire field of anaerobic fermentation as we understand the topic today (FUNG, 1992).

Microorganisms have complex requirement for molecular oxygen. Although in the study of anaerobic fermentation "anaerobes" are commonly used, many facultative anaerobes and aerobic organisms are also encountered in fermentation processes.

Isolation of pure cultures from the environment can be achieved by spreading mixed cultures on suitable solid surfaces such as agar, gelatin, or as ROBERT KOCH did, even on sterile potato surfaces and let each individual organism grow to a visible "colony". Further purification of this colony is required to assure that only one type of organism is in the culture. Another method is by dilution to extinction of a mixed culture in liquid. Eventually only one organism is diluted into a single tube. From growth within this tube one then can purify the culture and obtain a pure culture. Yet another method is to actually pass microbes into a capillary tube under the microscope and literally capture the target microbe in the tube and later cultivate this single organism into a large population for use in fermentation.

Once the pure culture is obtained it is necessary to maintain viability of the culture for future use. Methods in maintaining "stock" cultures are described by SUTTER et al. (1980), BORRIELLO and HARDI (1987), and SHAPTON and BOARD (1971). Basically cultures can be maintained on agar, in liquid, in frozen form, dried form, lyophilized state, in oil, paper, or other support bases. Different genera and species of organisms have different survival abilities in various storage conditions. Lyophilization seems to be the best method for long-term storage and maintenance of cultures.

Once the pure culture has been established to perform well in fermentation one may like to make the culture do special tasks. Through mutation and selection new strains can be developed to produce different compounds or increase production of desirable compounds. In the past 15 years, many efforts have been directed to biotechnology and bioengineering to insert desirable genes from foreign DNA into carrier microbes such that the recombinant DNA can express the foreign gene product in the new host. The field of biotechnology indeed has revolutionized pure and applied biological sciences and will continue to have great impact on the well being of humanity (BILLS and KUNG, 1990).

2.2 Mixed Cultures

The use of mixed cultures in fermentation can occur in two ways. "Natural" mixed cultures are used when one "seeds" substrates with portions of a successful batch of product. In sewage treatment, e.g., a portion of an ongoing batch of activated sludge can be pumped into the incoming raw sewage to assist in the development of a balanced population for effective bioconversion of organic matter. "Back inoculation" has been used for centuries in food fermentation when some successful dough or sausages are used to start another batch of product. Although this practice has been successful in many instances, the outcome is not entirely predictable. The utilization of "mixed-pure cultures" is another way to conduct mixed culture fermentation. In this method, known pure cultures are mixed in appropriate proportions to start the fermentation. A typical example of mixed-pure culture fermentation is the yogurt fermentation. *Streptococcus thermophilus* and *Lactobacillus bulgaricus* are mixed in appropriate ratios (one to one by weight) and then used in making good yogurt from milk.

Pure culture, natural mixed culture and mixed-pure culture are all useful in anaerobic fermentation depending on the processes involved. An excellent review of starter culture technologies was given by GILLILAND (1985).

2.3 Oxyrase as a Growth Stimulus

In studying the physiology of anaerobes and facultative anaerobes many growth factors and ingredients have been tested. One of the newest compounds with unusual properties is an enzyme commercially named "Oxyrase" (Oxyrase Inc., Ashland, OH). The enzyme was first reported by ADLER and CROW (1981) to have ability to allow growth of anaerobic organisms even under aerobic conditions. Oxyrase is an oxygen scavenger and rapidly reduces molecular O_2 to H_2O in the presence of an H_2 donor. Subsequently ADLER (1989, 1990) reported the use of this enzyme for the growth of a variety of anaerobes. At Kansas State University the enzyme has recently been found to stimulate growth of a large group of facultative anaerobes (YU and FUNG, 1991a, b, 1992). To date the enzyme has been successfully used to grow the following strict anaerobes and facultative anaerobes: *Clostridium difficile, C. tetani, C. kluyveri, C. sporogenes, C. sordelli, C. butyricum, C. bifermentans, C. acetobutyricum, C. perfringens, Bacteroides fragilis, B. ovatus, B. thetaiotamicron, B. vulgaris, Bifidobacterium* sp., *Peptostreptococcus anaerobius, P. magnus, P. asaccharolyticus, Desulfovibrio vulgaris, Fusobacterium nucleatum, Eubacterium lentum, E. limosum, Propionibacterium acnes, Veillonella parvula, Listeria monocytogenes, L. innocua, L. welshimeri, Streptococcus faecalis,* and *Escherichia coli* (HOSKINS et al., 1986; HOSKINS and DAVIDSON, 1988; OXYRASE TECHNICAL BULLETIN, 1990; YU and FUNG 1991a, b, 1992).

It is highly probable that this enzyme can stimulate the growth of starter cultures and can be used in fermented food such as bread and dairy products once the enzyme is determined safe for human consumption.

In terms of identification of anaerobes a variety of analytical tools such as GC, GLC, and GC/MS can be used. Authoritative information of instrumental analyses of a large variety of anaerobes can be found in HOLDEMAN et al. (1977). Recently, instrumentation for analysis of all types of microorganisms, including anaerobes, has been reviewed in a book edited by FUNG and MATTHEWS (1991). In FOX et al. (1990) several chapters are related to rapid identification of anaerobes by sophisticated instruments.

3 Kinetics, Bioenergetics, and Product Yields

The kinetics and bioenergetics of growth and product formation, microbial biomass and product yields are considered together because of the relationships which exist among the kinetic and yield parameters. The energy and composition regularities used by MINKEVICH and EROSHIN (1973), ERICKSON et al. (1978), ERICKSON and FUNG (1988), and others are employed because this allows a unified treatment in which comparisons may be made of yields on different substrates. The presentation in this section is based on earlier work which has been reported comprehensively in ERICKSON and FUNG (1988).

3.1 Kinetic and Yield Expressions

Anaerobic growth and product formation in which the adenosine triphosphate (ATP) formed is the result of substrate phosphorylation has been extensively investigated. Biomass growth, substrate consumption, and product formation are very closely coupled (ROELS, 1983; ONER et al., 1984; ERICKSON and FUNG, 1988). When no electron acceptors are present, the available electrons initially present in the substrate are distributed to the product and the biomass, with the allocation controlled by the bioenergetics of the process. Consider the anaerobic production of a simple product, such as ethanol, by fermentation. The chemical balance equation is

$$CH_mO_l + aNH_3 = y_cCH_pO_nN_q + zCH_rO_sN_t + cH_2O + dCO_2 \tag{1}$$

where CH_mO_l, $CH_pO_nN_q$, and $CH_rO_SN_t$ give the elemental compositions of carbon, hydrogen, oxygen, and nitrogen in the substrate, biomass, and extracellular product, respectively. Using the valences C=4, H=1, O=−2, and N=−3, the available electron balance is (MINKEVICH and EROSHIN, 1973; ERICKSON et al., 1978; ERICKSON and FUNG, 1988)

$$y_c \frac{\gamma_b}{\gamma_s} + z \frac{\gamma_p}{\gamma_s} = 1.0 \tag{2}$$

or

$$\eta + \xi_p = 1.0 \tag{3}$$

where γ is the reductance degree (MINKEVICH and EROSHIN, 1973; ERICKSON et al., 1978; ERICKSON and FUNG, 1988). The subscripts s, b, and p refer to substrate, biomass, and product, respectively. In Eq. (3), η is the biomass yield and ξ_p is the product yield based on the fractions of available electrons of substrate incorporated into biomass and product, respectively.

LUEDEKING and PIRET (1959a, b) found that product formation kinetics and growth kinetics are closely related; product formation was modeled with a growth-associated term and a maintenance-associated term, i.e.

$$\frac{dP}{dt} = \alpha \frac{dX}{dt} + \beta X \tag{4}$$

where X is biomass concentration and P is product concentration. Growth and product formation are dependent upon the substrate concentration; the Monod model

$$\frac{1}{X}\frac{dX}{dt} = \mu = \frac{\mu_{max} S}{K_s + S} \tag{5}$$

is commonly used to relate growth rate to substrate concentration. Here S is the substrate concentration, μ the specific growth rate, μ_{max} the maximum specific growth rate, and K_s the saturation constant. The organism uses substrate for growth and maintenance according to the linear model for substrate consumption (PIRT, 1975; ROELS, 1983; ONER et al., 1984; ERICKSON and FUNG, 1988), i.e.

$$-\frac{dS}{dt} = \frac{1}{Y_{x/s}^{max}}\frac{dX}{dt} + m_s X \tag{6}$$

where $Y_{x/s}^{max}$ is the true growth yield corrected

for maintenance and m_s is the maintenance coefficent. Both growth and product formation occur. The specific rate of substrate consumption can be written as

$$-\frac{1}{X}\frac{dS}{dt}=\frac{\mu}{Y_{x/s}}=\frac{\sigma_b\gamma_b\mu}{\sigma_s\gamma_s\eta}$$

where σ_b and σ_s are the weight fraction carbon in microbial biomass and substrate, respectively. Similarly,

$$\frac{1}{XY_{x/s}^{max}}\frac{dX}{dt}=\frac{\mu}{Y_{x/s}^{max}}=\frac{\sigma_b\gamma_b\mu}{\sigma_s\gamma_s\eta_{max}}$$

where η_{max} is the true growth yield corrected for maintenance in available electron units. The above equations may be substituted into Eq. (6) to obtain (ONER et al., 1984; ERICKSON and FUNG, 1988)

$$\frac{\mu}{\eta}=\frac{\mu}{\eta_{max}}+m_e \tag{7}$$

where m_e is the maintenance coefficient in available electron units. The specific rate of product formation can be written as

$$\frac{1}{X}\frac{dP}{dt}=\frac{\mu}{Y_{x/s}}Y_{p/s}=\frac{\sigma_b\gamma_b\mu}{\sigma_p\gamma_p\eta}\xi_p$$

where $Y_{p/s}$ is the product yield based on substrate. The above equation and Eq. (3) may be used to write Eq. (4) in the form (ERICKSON and FUNG, 1988)

$$\frac{\mu}{\eta}\xi_p=\frac{\mu}{\eta}(1-\eta)=\alpha_e\mu+\beta_e \tag{8}$$

where

$$m_e=\beta_e=m_s\frac{\sigma_s\gamma_s}{\sigma_b\gamma_b}=\beta\frac{\sigma_p\gamma_p}{\sigma_b\gamma_b} \tag{9}$$

$$\frac{1}{\eta_{max}}=\alpha_e+1=\frac{\sigma_s\gamma_s}{\sigma_b\gamma_b Y_{x/s}^{max}}=\alpha\frac{\sigma_p\gamma_p}{\sigma_b\gamma_b}+1 \tag{10}$$

The product formation kinetic parameters and the yield and maintenance parameters are related as shown in Eqs. (9) and (10). The above relationships exist when ATP generation is coupled to extracellular product formation. In these fermentation processes, microorganisms can meet their ATP requirements for growth and maintenance only by producing extracellular products.

If the available electrons that are transferred from organic substrate to products are viewed as being expended to provide energy for growth and maintenance, a true growth yield with available electron units η_{max} can be defined as follows (ERICKSON, 1980; ERICKSON and ONER, 1983; ERICKSON and FUNG, 1988)

$$\eta_{max}=\frac{\text{equivalents of available electrons of biomass formed per mol ATP}}{\text{equivalents of available electrons of biomass formed per mol ATP}+\text{equivalents of available electrons transferred to products to produce 1 mol ATP from ADP}}$$

$$\eta_{max}=\frac{(\sigma_b\gamma_b/12)\,Y_{ATP}^{max}}{(\sigma_b\gamma_b/12)\,Y_{ATP}^{max}+\delta} \tag{11}$$

where Y_{ATP}^{max} is the true growth yield based on ATP, δ is the equivalents of available electrons transferred to products to produce 1 mol ATP from ADP, σ_b is the weight fraction carbon in biomass, and γ_b is the reductance degree of biomass. Knowledge of the theoretical maximum yield with respect to ATP and the value of δ for a particular fermentation allows one to estimate the range of values of the growth-associated kinetic term in the product formation model. Eqs. (10) and (11) can be combined to obtain

$$\alpha_e=\frac{1}{\eta_{max}}-1=\frac{12\delta}{\sigma_b\gamma_b Y_{ATP}^{max}}$$

For $\delta=12$ equivalents of available electrons per mol ATP generated, $Y_{ATP}^{max}\leq 28.8$ g cells per mol ATP, $\sigma_b = 0.462$, and $\gamma_b = 4.291$, $\alpha_e \geq 2.52$ (ERICKSON and FUNG, 1988).

In summary, there are four independent parameters (μ_{max}, K_s, η_{max}, and m_e) needed in Eqs. (3)–(10) to describe growth and product formation in anaerobic fermentations with one chemical balance equation and one product that contains available electrons. The values of the product formation parameters, α_e and β_e may be obtained from Eqs. (9) and (10). Eq. (11) relates Y_{ATP}^{max}, δ and η_{max}. Experimental observations of ATP yields are needed to fix the values of δ and Y_{ATP}^{max} in Eq. (11).

3.2 Yield and Maintenance

The maintenance coefficient m_e in equivalents of available electrons expended per equivalent of available electrons of biomass per hour is related to m_{ATP}, the moles of ATP expended per gram of cells per hour, according to the relationship (ERICKSON and ONER, 1983; ERICKSON and FUNG, 1988)

$$m_e = \frac{12\delta m_{ATP}}{\sigma_b \gamma_b} \tag{12}$$

where the product δm_{ATP} is the equivalents of available electrons of substrate consumed for maintenance per gram of cells per hour, and $\sigma_b \gamma_b/12$ is the equivalents of available electrons of biomass per gram of cells. Eq. (7) can be used to estimate the maintenance parameter m_e and true growth yield η_{max} (ERICKSON and ONER, 1983). Inserting Eq. (3) into the numerator on the left-hand side of Eq. (7) gives (ERICKSON and FUNG, 1988)

$$\frac{\mu(\eta + \xi_p)}{\eta} = \frac{\mu}{\eta_{max}} + m_e \tag{13}$$

The carbon balance associated with Eq. (1) can be used to obtain

$$\frac{\mu(y_c + z + d)}{\eta} = \frac{\mu}{\eta_{max}} + m_e \tag{14}$$

where μ is the specific growth rate. Eqs. (7), (13), and (14) are similar and follow Pirt's model for growth and maintenance. Eq. (7) utilizes organic substrate, biomass, and specific growth rate measurements; Eq. (13) needs biomass, product, and specific growth rate measurements; Eq. (14) requires biomass, carbon dioxide, product, and specific growth rate measurements. Eqs. (13) and (14) would be identical to Eq. (7) if the available electron and carbon balances were satisfied at every data point.

Two forms of Eqs. (7), (13), and (14) have been used for parameter estimation; to obtain form I, each term in these equations should be divided by μ; form II is the form of these equations as presented above.

One can use any one of Eqs. (7), (13), and (14) to estimate m_e and η_{max}. Since different sets of experimental data are used with each of these equations, different estimates of m_e and η_{max} are obtained. This problem was solved by using the covariate adjustment method (SOLOMON et al., 1982, 1983, 1984; ONER et al., 1983, 1984, 1986c) to estimate η_{max} and m_e when sufficient data are available to use more than one of Eqs. (7), (13), and (14). This method allows all data to be used to obtain better estimates of m_e and η_{max}. The estimated values of the product formation kinetic parameters may then be found from Eqs. (9) and (10).

From Eqs. (3) and (7), one can derive the equations

$$\xi_p = 1 - \eta \tag{15}$$

$$\eta = \frac{\mu}{\mu/\eta_{max} + m_e} \tag{16}$$

$$\xi_p = 1 - \frac{\mu}{\mu/\eta_{max} + m_e} \tag{17}$$

Eq. (17) indicates that product yield ξ_p depends on the parameters m_e and η_{max} as well as on μ. To have a large product yield, the maintenance coefficient m_e should be large, and true growth yield η_{max} should be small. Eq. (17) shows that ξ_p approaches one as μ approaches zero. Thus, the desired goal of a product yield of one can be approached by using immobilized or membrane retained cells which have a large maintenance energy requirement. In ERICKSON and FUNG (1988), the estimated values of true growth yield, maintenance coefficient, and product formation kinetic parameters are summarized for several anaerobic fermentations. The statistical methods are presented as well.

There are several important conclusions for ethanol production based on the extensive results of many investigators and the above analysis of anaerobic processes. Organisms such as *Zymomonas mobilis,* which have a high maintenance coefficient and produce only 1 mol ATP/mol glucose consumed ($\delta = 24$), have the potential to have a high product yield. Operation at a very low specific growth rate also enhances product yield; however, economical production rates can only be obtained with high biomass concentrations. The active and productive cells must be retained in the bioreactor for further use when the product is separated from the fermentation broth. Research on processes with immobilized and/or membrane retained cells is discussed in a number of books (CHERYAN, 1986; TAMPION and TAMPION, 1987; ERICKSON and FUNG, 1988; MOO-YOUNG, 1988; GHOSE, 1989) and papers (CHEN et al., 1990a, b; DALE et al., 1990; GALAZZO and BAILEY, 1990; KANG et al., 1990; PARK et al., 1990).

3.3 Growth Kinetics

Methods to estimate the growth kinetic parameters μ_{max} and K_s in Eq. (5) are presented in ERICKSON and FUNG (1988). The covariate adjustment method may be used together with substrate, biomass, and product data to estimate the maximum specific growth rate μ_{max} (ONER et al., 1986a). The maximum specific growth rate may be estimated from the expression

$$\mu = \frac{1}{X}\frac{\mathrm{d}X}{\mathrm{d}t} \tag{18}$$

under exponential growth conditions. For exponential growth, $\mu = \mu_{max}$ and the growth yield η is constant; let $\eta = \eta_1$ and $\xi_p = \xi_{p1}$ be the yields during exponential growth where

$$\eta = \frac{\sigma_b \gamma_b}{\sigma_s \gamma_s} Y_{x/s}$$

and

$$\xi_p = \frac{\sigma_p \gamma_p}{\sigma_s \gamma_s} Y_{p/s}$$

The yield expressions (ERICKSON and FUNG, 1988)

$$\eta_1 = -\frac{\sigma_b \gamma_b}{\sigma_s \gamma_s}\frac{\mathrm{d}X}{\mathrm{d}S} \tag{19}$$

$$\eta_1 = \frac{\sigma_b \gamma_b (X - X_0)}{\sigma_s \gamma_s (S_0 - S)} \tag{20}$$

may be used with Eq. (18) to obtain

$$\mu = \frac{-(\sigma_s \gamma_s / \sigma_b \gamma_b)\eta_1 \,\mathrm{d}S/\mathrm{d}t}{X_0 + \eta_1 (\sigma_s \gamma_s / \sigma_b \gamma_b)(S_0 - S)} \tag{21}$$

If we introduce

$$Z = \frac{X_0 \sigma_b \gamma_b}{\eta_1 \gamma_s} + (S_0 - S)\sigma_s \tag{22}$$

the derivative is

$$\frac{\mathrm{d}Z}{\mathrm{d}t} = -\sigma_s \frac{\mathrm{d}S}{\mathrm{d}t} \tag{23}$$

and the expression

$$\mu = \frac{1}{Z}\frac{\mathrm{d}Z}{\mathrm{d}t} \tag{24}$$

can be obtained from Eq. (21).

Similarly, to utilize product data P, substitute the yields (ERICKSON and FUNG, 1988)

$$\frac{\eta_1}{\xi_{p_1}} = \frac{\sigma_b \gamma_b}{\sigma_p \gamma_p}\frac{\mathrm{d}X}{\mathrm{d}P} \tag{25}$$

and

$$\frac{\eta_1}{\xi_{p_1}} = \frac{\sigma_b \gamma_b}{\sigma_p \gamma_p}\frac{(X - X_0)}{(P - P_0)} \tag{26}$$

into Eq. (18) for X and $\mathrm{d}X$ to obtain the expression

$$\mu = \frac{1}{Y}\frac{\mathrm{d}Y}{\mathrm{d}t} \tag{27}$$

where

$$Y=\frac{X_0\sigma_b\gamma_b\xi_{p_1}}{\sigma_p\gamma_p\eta_1}+P-P_0 \quad (28)$$

For carbon dioxide data, one can write (ERICKSON and FUNG, 1988)

$$\mu=\frac{1}{W}\frac{dW}{dt} \quad (29)$$

where

$$W=\frac{12y_{c1}}{\sigma_b d_1}\int_{t_0}^{t} Q_c X dt+X_0 \quad (30)$$

where Q_cX is the rate of CO_2 evolution per unit volume of broth, y_{c1} is the biomass carbon yield, and d_1 is the carbon dioxide yield. The biomass carbon yield is related to the biomass available electron yield

$$y_{c1}=\frac{\gamma_s}{\gamma_b}\eta_1 \quad (31)$$

where the subscript 1 refers to exponential growth.

Eqs. (18)–(31) allow the maximum specific growth rate to be estimated based on all of the available data that are collected during the period of exponential growth. ONER et al. (1986a, b, c) and BUONO et al. (1990b) have obtained estimates of μ_{max} for yogurt cultures growing in complex mixtures where measurements to follow the progress of the fermentation are difficult to make. The procedures are described in detail in ONER et al. (1986a, b, c), ERICKSON and FUNG (1988), and BUONO et al. (1990b). Some of the results are shown in Tab. 1. There is good agreement among the values of maximum specific growth rate which are reported by the investigators.

Tab. 1. Point and 95% Confidence Interval Estimates of μ_{max} from ONER et al. (1986a) and BUONO et al. (1990b)

Point Estimate (h^{-1})	95% Confidence Interval (h^{-1})	Reference
0.4131	0.3953–0.4309	ONER et al.
0.4106	0.3755–0.4475	ONER et al.
0.341	0.300 – 0.383	BUONO et al.
0.373	0.308 – 0.439	BUONO et al.
0.296	0.182 – 0.411	BUONO et al.
0.478	0.474 – 0.481	BUONO et al.

4 Types of Anaerobic Processes

Several methods have been used to classify anaerobic processes. These include the feeding of media and gases, and the withdrawal of products; the mode of operation may be batch, fed-batch, or continuous. The classification may be based on the electron acceptor; the design may be for anaerobic or microaerobic conditions. The sterility requirements of pure and mixed culture processes with developed strains differ from those of environmental mixed culture processes which are based on natural selection. There are processes in which the fermentor is made by man and natural bioreactors such as the stomachs of the cow. In this work, the classification of anaerobic processes is according to the physical form of the reactants and products because of the emphasis on bioprocessing. Typical gas, liquid, and solid reactants and products are listed in Tab. 2.

Tab. 2. Major Anaerobic Fermentation Processes and Products

Gas Phase	Liquid Phase	Solid Phase
Methane	Ethanol	Silage
Carbon dioxide	Acetone	Sauerkraut
Nitrogen	Butanol	Pickle
Hydrogen	Lactic acid	Sausage
Carbon monoxide		Soybean products
Hydrogen sulfide		Sausage
Oxygen		Bread
		Fish

4.1 Processes with Gas Phase Products

Oxygen and carbon dioxide are the most common gas phase reactants and products.

Others include hydrogen, hydrogen sulfide, carbon monoxide, and methane. Generally, the concentration of the reactants and products in the liquid phase of the cellular microenvironment influences the kinetics of the cellular reactions. Mass transfer to and from the gas phase affects bioreactor performance in most processes with gas phase reactants or products. Most anaerobic processes are designed to limit or exclude oxygen. In some cases inert gases are bubbled into the anaerobic reactor to provide a gas–liquid interfacial area to remove the product gases. The phase equilibrium relationship is based on thermodynamic data while the rate of mass transfer depends on the gas–liquid interfacial area and the concentration driving force. Mechanical agitation increases the gas–liquid interfacial area.

Most large-scale anaerobic fermentors have carbon dioxide as one of the reactants or products. JONES and GREENFIELD (1982) and HO and SHANAHAN (1986) have reviewed the effects of carbon dioxide on growth and product formation. Carbon dioxide concentration can adversely influence membrane function and metabolic pathways at partial pressures above about 0.2 atm (ERICKSON and FUNG, 1988). LACOURSIERE et al. (1986) measured the specific growth rate for *Escherichia coli* growing on glucose under anaerobic conditions at several different carbon dioxide concentrations; the maximum specific growth rate was observed at 5 % carbon dioxide in the gas phase. HO and SMITH (1986) report that carbon dioxide affects the biosynthesis of membranes.

The dynamic equilibria of carbon dioxide in aqueous fermentations is affected by pH and the partial pressure of CO_2 which depends on gas phase composition and total pressure. YEGNESWARAN et al. (1990) have developed a model for carbon dioxide concentration which gives good agreement with experimental data for variations in pH and total pressure. ROYCE and THORNHILL (1991) have measured carbon dioxide transport rates and examined the departure from equilibrium.

Methane is produced through anaerobic degradation of waste products. It is a product of microbial action in landfills, bogs, and the stomachs of the cow. In anaerobic digestion both carbon dioxide and methane are produced as gas phase products. In their review of product inhibition in anaerobic processes, BAJPAI and IANNOTTI (1988) report that sparging nitrogen gas into anaerobic digestors improves process performance. The free energy driving force for methane production is greater for small values of methane and carbon dioxide concentration. As shown in Tab. 3, inhibition is found in many anaerobic fermentations.

Tab. 3. Product Inhibition of Anaerobic Fermentations

Product	Inhibitor
Ethanol	Ethanol, CO_2
Methane	CO_2/CH_4, propionic acid, acetic acid, H_2
Acetone/butanol	Butyric acid, high solvent concentration
Lactic acid	Lactic acid (1–4%), oxygen
Anaerobic digestion	NH_3, acetic acid, CO_2
Silage/lactic acid	NH_3, oxygen

Hydrogen is often found as both a reactant and a product in anaerobic processes. Its effects on anaerobic processes are considered in severeal chapters of ERICKSON and FUNG (1988).

Gas–liquid mass transfer must be considered in the design of anaerobic fermentations in which a fermentation product is removed from the fermentor through the gas phase or a reactant is supplied through the gas phase.

4.2 Processes with Liquid Phase Products

Many fermentors have liquid phase reactants and products. Ethanol, acetone, and butanol are examples of liquid products which can be produced by fermentation. The kinetics of biochemical reactions depend on the liquid phase concentrations of the reactants and in some cases the products. The Monod kinetic model and the Michaelis-Menten kinetic model show that many biochemical reactions have first order dependence on reactant (substrate) concentration at low concentrations and zero order dependence at higher concentrations. Rates are directly proportional to concentrations below 10 mg/L for many reactants under natural environmental conditions. At very high concentrations, inhibition is often observed (ERICKSON and FUNG, 1988).

Hydrocarbons such as hexadecane which are relatively insoluble in the water phase may also be reactants or substrates for biochemical reactions. Microbial growth on hydrocarbons has been observed to occur at the liquid–liquid interface as well as in the water phase.

4.2.1 Ethanol

One of the oldest and most widely practiced fermentations is the microbial production of ethanol and alcoholic beverages such as beer and wine. Since ethanol inhibits the fermentation at high concentrations, the process of inhibition has been extensively studied for this fermentation (BAJPAI and IANNOTTI, 1988; AMARTEY et al. 1991; LAPLACE et al., 1991). Ethanol affects the cell membrane and the activities of anabolic and catabolic enzymes. This inhibition limits the concentration of ethanol that can be obtained in a fermentor. Since ethanol is also produced for use as a motor fuel, there is still considerable research on ethanol production (DOELLE et al., 1991; LOPES et al., 1991). Because the cost of the substrate is a major expense, low cost raw materials such as wastes containing cellulose have been investigated (ABE and TAKAGI, 1991; LAWFORD and ROUSSEAU, 1991; TANAKA et al., 1991).

Alcohol fermentation depends on initial substrates, type of product, and organisms. Molasses is extensively used to produce alcohol for distilled beverages and alcohol for other purposes such as medical use. Grapes and malted cereal grains are used for both non-distilled beverages, such as beer and wine, and distilled beverages, such as rum and gin. Most of the time, yeasts, especially *Saccharomyces cerevisiae,* are used as the fermenting organisms. In a few cases, bacteria, e.g., *Zymomonas mobilis,* are used.

The alcoholic fermentation is an anaerobic process; it does not require air or oxygen. The stoichiometric mass yields on glucose are 51.1% alcohol and 48.9 % carbon dioxide and the actual yields are 90–95 % of the stoichiometric yield. Thus one can roughly expect to have half the amount of alcohol from an appropriate amount of sugar added. High alcohol concentration has a negative effect on the growth and product formation of the organism.

The alcoholic fermentation often follows the Embden-Meyerhof pathway which is shown elsewhere (STRYER, 1988). There are several types of yeast that can ferment alcohol and, at the same time, generate desirable aromatic compounds. European wine makers often use natural yeast on the skin of grapes. American wine makers, on the other hand, often select a pure culture of yeast, especially *Saccharomyces cerevisiae* var. *ellipsoides* montrachet strain. Specific strains can produce distinctive flavors and aroma.

The major influencing factors on fermenting yeast are: water, substrates such as carbohydrates (sucrose, dextrose, etc.), nitrogenous compounds, acid, minerals, and vitamins (AMERINE and CRUESS, 1980; BENDA, 1982). Sugar and nitrogenous compounds are the principal substrates for the fermentation. Normally, the desired initial sugar concentration is between 18–25 %. Sugar concentrations higher than 30 % (w/w) can kill the yeast due to the osmotic effect (ROSE, 1977). The simpler the form of the sugar, the faster the fermentation rate will be; e.g., most yeast can ferment glucose faster than sucrose or higher sugars.

Temperature is critical in fermentation processes because yeast will not produce the specific flavor characteristics away from the optimum. Therefore, for the fermentation the temperature should be kept constant or in an acceptable narrow range, such as 27–30 °C (BELL, 1989). Since alcoholic fermentation is an exothermic reaction, cooling might be needed during the early stage of fermentation.

Several methods for product removal and/or cell retention which can be used to improve the efficiency of anaerobic fermentations are listed in Tab. 4. Methods to remove ethanol from the fermentation broth are reviewed in the book edited by MATTIASSON and HOLST (1991). These include pervaporation, perstraction, adsorption, and distillation. In the

BIOSTIL process, the solids are separated from the liquid using a centrifugal separator and returned to the fermentor. The ethanol is separated from the liquid by distillation; the aqueous liquid stream is returned to the process. SHABTAI et al. (1991) have demonstrated improved ethanol production in which pervaporation is utilized for ethanol separation from the broth.

Methods to optimize product formation include cell retention in the fermentor (ERICKSON and FUNG, 1988; SCOTT, 1983). CHEN (1991) has obtained high ethanol productivities in an up-flow attached-bed continuous fermentor. LOPES et al. (1991) have investigated the effects of temperature and pH on the settling behavior of *Zymomonas mobilis;* sedimentation of flocculent strains can be used to recover and retain biomass. CIANI and ROSINI (1991) have used centrifugation to recover and recycle *Saccharomyces cerevisiae* in sparkling-wine production. TZENG et al. (1991) have immobilized yeast cells on calcium alginate particles for the production of ethanol in an eight-stage fluidized bed bioreactor.

4.2.2 Lactic Acid

Lactic acid can be produced chemically and biologically; this work confines the topic to the production through fermentation. Lactic acid fermentation is anaerobic; however, the bacteria responsible for the process are micro-aerophilic, e.g., *Lactobacillus* sp. and *Pediococcus* sp., the most common strain is the homofermentative *L. delbruckii* (MIALL, 1978). Strain selection depends on the choice of substrate, e.g., starch, sucrose, corn sugar, or lactose. Normally, the yield on glucose approaches 100% on a weight basis and product concentration is as high as 12% (WEIMER, 1991).

Efficient fermentation processes for lactic acid production continue to be investigated. YABANNAVAR and WANG (1991b) have shown higher productivities of lactic acid in an extractive fermentation system in which lactic acid was extracted using Alamine 336, a tertiary amine, and oleyl alcohol. NOMURA et al. (1991) have demonstrated that electrodialysis can be used to remove lactic acid and reduce product inhibition. BIBAL et al. (1991) have

Tab. 4. Methods to Remove Products or to Retain Cells in Anaerobic Fermentation

Processes	Products	References
Reactor containing flocs	Ethanol	ATKINSON and MAVITUNA (1983)
Slant tube	Wine	GODIA et al. (1987)
Tower fermentor	Ethanol	GODIA et al. (1987)
Horizontal parallel flow reactors in series	Ethanol	GODIA et al. (1987)
CSTR/settler/cell recycle	Ethanol in soy sauce	IWASAKI et al. (1991)
	Ethanol	GODIA et al. (1987)
Anaerobic upflow sludge blanket	Digested waste	ROZZI et al. (1988)
Reactor with microbial films	Methane	ATKINSON and MAVITUNA (1983)
Fixed bed reactor anaerobic filter	Treated distillate	SILVERIO et al. (1986)
Fluidized bed reactor with activated carbon	Digested coke waste water	EDELINE et al. (1986)
		GODIA et al. (1987)
Pervaporation	Acetone/butanol	ENNIS et al. (1986)
	Ethanol	SHABTAI et al. (1991)
Membrane, hollow fiber	Ethanol	GODIA et al. (1987)
Immobilization	Acetone/butanol	ENNIS et al. (1986)
Algenate entrapment	Ethanol	MARWAHA et al. (1988)
		TZENG et al. (1991)

applied cross-flow filtration to obtain high cell concentrations in a lactic acid fermentation.

A structured cell model for the lactic acid fermentation has been developed (NIELSEN et al., 1991a, b; NIKOLAJSEN et al., 1991). The model includes an A compartment which contains the active part of the cell and a G compartment in which the genetic and structural part of the cell is found. The model can simulate transient conditions and it has been extended to mixed substrates.

Lactic acid bacteria have their greatest importance in the food industries. Yogurt, acidophilus milk, and cheese are products of this fermentation. Lactic acid is retained in the foods to act as a preservative and flavor enhancer. The fermentation is usually a mixed culture process. The yogurt fermentation may use *Lactobacillus bulgaricus* and *Streptococcus thermophilus* which grow together symbiotically. Both cultures have lactic acid as the final catabolic product and use lactose as a substrate in a manner such that 1 mol lactose provides 4 mol lactic acid. The symbiotic growth of mixed cultures of lactic acid bacteria has a unique relationship which has been reported by a number of workers (e.g., TAMIME and ROBINSON, 1985). The details have been reviewed by TAMIME and ROBINSON (1985) and FUNG et al. (1988). Lactic acid is important for yogurt production because:

(1) Lactic acid destabilizes the casein micelles by converting the colloidal calcium-phosphate complex in the micelle to a soluble calcium-phosphate fraction (calcium lactate and calcium phosphate), resulting in depletion of calcium from the micelles. This leads to the agglutination of casein at pH 4.6–4.7, and the yogurt gel is formed.

(2) Lactic acid gives unique flavor characteristics to the yogurt; i.e., sharp, tart, and acidic (TAMIME and ROBINSON, 1985).

Lactic acid fermentations have been conducted to increase the acceptability of soybeans. Soybean products have been popular among Oriental people for more than one thousand years. Details on the production and fermentation of soy milk, tofu, natto, hamanatto, miso, soy sauces, tempeh have been reviewed elsewhere (STEINKRAUS, 1983). Research on the fermentation of soy milk has been reviewed and reported by BUONO (1988), BUONO et al. (1990a, b, c), and FUNG et al. (1988). The soy milk has a general composition which is similar in many ways to cow milk (United States Department of Agriculture, 1986). The yogurt bacteria can grow in soy milk; however, *Lactobacillus bulgaricus* cannot utilize sucrose, which is the main carbohydrate source in soy milk. The metabolites released from *Streptococcus thermophilus* appear to be enough for growth of the lactobacilli. The other sugar present in a similar amount to sucrose is stachyose, the tetraose which is believed to be a sugar causing flatulence. Raffinose, the flatulence causing triose sugar, is present at a very low concentration, 0.4–1 g/L, in soy milk (BUONO, 1988; TUITEMWONG et al., 1991). Due to the distinctive sugars in the soy milk, the fermentation is different from that found for regular cow milk (FUNG et al., 1988). Methods of lactic acid fermentation of soy milk to utilize and reduce the flatulence causing sugars in soy milk have been extensively studied (PATEL et al., 1980; BUONO et al., 1990a, b, c; TUITEMWONG et al., 1991). Lactic acid bacteria usually do not utilize either raffinose or stachyose effectively (BUONO et al., 1990b; TUITEMWONG et al., 1991).

The anaerobic kinetic and yield models for lactic acid fermentation of soy milk have been investigated by BUONO et al. (1990b). Sucrose and stachyose were taken into account as co-substrates; the biomass available electron yield on substrate during exponential growth, η, the maximum specific growth rate, μ_{max}, and product available electron yield on substrate during exponential growth, ξ_p, were estimated (BUONO, 1988; BUONO et al., 1990b).

An improved process of making soy milk was established by JOHNSON et al. (1981). Soybean flour was mixed with water and fed to the Rapid Hydration Hydrothermal Cooker for about 30–120 s. The soy slurry was cooked by the steam in the holding tube at 270–310 °F (132–154 °C) for 20–200 s. Soy milk produced with this method is sterile and has good characteristics; it is an ideal substrate for lactic acid fermentation (TUITEMWONG et al., 1991). This integrated process to produce soy yogurt is very promising because the use of soybean products as human foods is growing rapidly. Soy yogurts have no cholesterol, contain significant amounts of good proteins and amino acids, and are less expensive than animal protein. The application of lactic acid fermentation technology to produce these food products is not only challenging but also beneficial.

4.2.3 Acetone-Butanol

The anaerobic fermentation process for acetone and butanol has been well known since the first World War. Though it was first reported by FERNBACH in 1910, WEIZMANN, a chemist, turned this finding to a commercial reality. WEIZMANN worked with the production of acetone, which was needed for the production of cordite for small gun ammunition and as a propellant for heavy artillery (HASTINGS, 1978). He discovered that *Clostridium acetobutylicum* can ferment 4% corn starch and simple inorganic nutrients (ammonium phosphate and soluble phosphate) to acetone (30%) and ethanol (60%). The development of this sterile fermentation on a big scale was reviewed by HASTINGS (1978) and GIBBS (1983).

For pure culture fermentation, strains of *C. acetobutylicum* are used. The bacteria are strictly anaerobic spore formers with a typical cigar shape. The spore is bigger than the vegetative cell itself having a swollen portion at the subterminal region of the cell. They produce spores rapidly when nutrients are depleted. Cells isolated from liquid medium appear as single cells, chains, and boat-shaped clusters (HASTINGS, 1978).

Similar to methane production, the acetone-butanol anaerobic fermentation has two phases. One is acetogenic, a process that produces organic acids and H_2 (60%); the other is solventogenic, a process that converts those organic acids to solvents (acetone, butanol, and ethanol) and produces a smaller fraction of H_2 (30–40%) without growth (DATTA, 1988). However, if the starting substrate is cellulose, the fermentation would involve three phases. In this case, since the rate of hydrolysis is slow, the rate of acid production is dominant over that of solvent formation (WEIMER, 1991). The typical fermentation time course is as follows: after inoculation, the bacteria start growing and produce acids such as acetic and butyric acid. This is the cause of the decrease of pH of the broth from 5.8 (molasses) down to about 5.0–5.1. When the growth stops, the bacteria start producing solvents and the pH goes up to about 6.0, then the second phase takes over (HASTINGS, 1978).

Acetone-butanol ethanol fermentation can be carried out using sequential mixed cultures of *C. acetobutylicum* and *Kluyveromyces fragilis* (BALLERINI et al., 1985; WEIMER, 1991). Sugar beet juice (11% sucrose) with corn steep liquor, acetic acid, and NH_4OH were used as a fermenting broth for *C. acetobutylicum* at 35 °C for 10 h, then *K. fragilis* was inoculated into the medium. The broth was incubated at 30 °C for 30 h giving 1.2% butanol, 0.4% acetone, and 0.75% ethanol. The advantages of using mixed cultures are: (1) elimination of a multi-step process needed for this multi-phase fermentation when pure culture is applied, (2) a higher yield and better product quality since the desired reaction is more favorable and the side reaction is reduced (WEIMER, 1991). However, the microorganisms have limited tolerance to the concentration of butanol in the fermentation broth. The level of 1.2% by volume butanol is used to determine the maximum sugar concentration in the starting medium. DATTA (1988) applied the carbon and available electron balances using the method elaborated by ERICKSON, MINKEVICH and EROSHIN (ERICKSON et al., 1978; MINKEVICH and EROSHIN, 1973). This allows one to appreciate the efficiency and data consistency of the fermentation. The authors reported that 75–78% of available electrons from the substrate (carbohydrate) go to acetone, butanol, and ethanol. Butanol accounts for 50–55% of the available electrons. The non-solvent products such as organic acids, cell mass, and H_2 gas account for about 20–25% of the available electrons. Since this is a heterofermentative fermentation, the process is less efficient than that of homofermentative processes in which all available electrons go to solvent products. The complete details of the balances, metabolic pathways, and control have been described by DATTA (1988).

4.3 Processes with Solid Phase Reactants or Products

There are many examples of bioreactors with solid phase reactants. The cow may be viewed as a mobile bioreactor system which converts solid substrates to methane, carbon dioxide, milk, and body protein. While the cow is a commercial success, many efforts to transform low cost cellulosic solid wastes to commercial products in man-made bioprocesses have not achieved the same level of success.

Solid substrates such as soybean meal are commonly fed into commercial fermentations. Through the action of enzymes in the fermentation broth, the biopolymers are hydrolyzed and soluble reactants are obtained.

Many food fermentations involve the preservation of solid or semi-solid foods such as in the conversion of cabbage to sauerkraut and meats to sausage products. Cereals, legumes, vegetables, tubers, fruits, meats, and fish products have been fermented. Some fermented milk processes result in solid products such as cheeses and yogurts.

Other examples include the composting of yard wastes, silage production, biodegradation of crop residues in soil, microbial action in landfills, and the remediation of contaminated soil.

In many of these fermentations, physical mixing is difficult or expensive. Transport of essential reactants may depend on diffusion; the concentrations of reactants and products vary with position. Rates may be limited by the transport of essential reactants to the microorganisms.

Most compounds which are present as solids in bioreactors are somewhat soluble in the water phase. For reactants which are relatively insoluble, biochemical reaction rates may be directly proportional to the available interfacial area. The surface of the solid may be the location of the biochemical transformation. An example of microorganisms growing on the surface of a solid substrate is mold growing on bread. To design bioreactors for solid substrates and solid products, the solubility and the transport processes should be addressed as well as the kinetics of the process. Mixing prior to batch fermentation is often essential to distribute reactants and inoculum.

HUANG and CHOU (1990) have modeled growth on insoluble solid substrates. They have assumed that microorganisms assimilate solid substrates at the point of cell contact with the substrate on the solid surface. The surface area available for microbial adsorption onto the solid surface is an important variable in their model. They show good agreement of the model and the experimental data. MITCHELL et al. (1991) have developed a semimechanistic mathematical model for growth of biomass, starch hydrolysis, and glucose formation and consumption in a model solid-state fermentation system.

The rate of reaction in bioreactors is often directly proportional to the concentration of microbial biomass. In anaerobic bioprocesses, the biomass yield is low and the quantity of microbial biomass that can be produced from the substrate is limited. The economy of the operation and the rate of fermentation are enhanced by retaining the biomass in the bioreactor. Several types of immobilized cell reactors have been designed and operated because of the economy associated with reuse of cells and enzymes (TAMPION and TAMPION, 1987). In the anaerobic production of ethanol, butanol, lactic acid, and other fermentation products, the product yield is greatest when the organisms are not growing and all of the substrate is being converted to products (ERICKSON and FUNG, 1988). Continuous processes can be designed in which most of the cells are retained and the limiting maximum product yield is approached (DALE et al., 1990). Ultrafiltration membrane bioreactors have been used to retain cells, enzymes, and insoluble substrates (CHERYAN, 1986).

YABANNAVAR and WANG (1991a) have modeled the transport of substrate into and product out of k-carrageenan beads for a lactic acid fermentation in which the biomass is immobilized in the beads. Inhibitory product accumulation was found to be a problem for large beads. Thus, bead size is an important design variable; inhibition can be avoided by using smaller beads. This example shows the importance of diffusion in fermentations in which a solid phase is present.

4.3.1 Silage Production

Silage is important to farming because it overcomes the difficulty associated with season dependency, and it allows harvesting the crops at any stage of growth (WOOLFORD, 1985). The problem of ensilaging depends on the quality of raw materials; dry material with high solids content is desirable. The dryness and pH of the silage can prevent the growth of clostridia and enterobacteria already present in the ensilaging grasses. The recommended dry solids content ranges from 200–700 g/kg. Generally, the higher the dry solids, the greater the opportunity to achieve the desirable low pH. Lactic acid bacteria can tolerate low moisture conditions and outgrow the clostridia. They produce lactic acid, under anaerobic conditions, causing the rapid drop of pH to lower than 4.0 since plant cells have low buffering capacity. Some clostridia can survive and grow at a pH of about 4.0 (WOOLFORD, 1985). Thus, at the intermediate range of dry solids, the combination of low available water (dryness) and low pH is more effective to inhibit clostridia. Low dry solids, such as 150 g/kg, is considered undesirable for the fermentation due to the lack of the inhibitory conditions for clostridia. The moisture reduces the effectiveness of the acids produced from the fermentation. A higher solids content can be achieved by drying if the moisture content is high. Drying also helps reduce the activity of plant enzymes and the loss of active ingredients in crops during storage (MOSER, 1980). The anaerobic fermentation occurs under mixed culture conditions; thus, the other products are acetic acid, ethanol, and other minor organic compounds. The final pH may be as low as 3.8 (MUCK, 1991). Anaerobic conditions and good substrate selection are very important for high-quality silage. Maintaining anaerobic conditions requires keeping oxygen out of the silo. To increase the effectiveness and to control silage fermentation, other ingredients may be added. The objectives are to either increase the lactic acid fermentation rate, increase the starting number of the desired bacteria, or adjust the environment to favor lactic acid production. The "additives" are listed in Tab. 5.

Tab. 5. Additives in Silage Fermentation

Additive	Mode of Action	Reference
Sulfuric acid Hydrochloric acid Orthophosphoric acid Formic acid	Lowering the pH to the level unsuitable for Clostridia and to favor growth of lactic acid bacteria	WOOLFORD (1978, 1985), WALDO (1978)
Fatty acids Propionic acid Acrylic acid	Inhibition of growth of undesirable bacilli, and yeasts	COOK (1973), WILSON et al. (1979)
Formaldehyde 350 mg/L	Inhibition of growth of all types of organisms in silage fermentation	WOOLFORD (1975)
30–50 g/kg crude protein	Protection of protein from degradation	
Sodium metabisulfide	Sterilization of the whole system	WOOLFORD (1985)
Sodium nitrite	Inhibition of Clostridia	WOOLFORD (1985)
Antibiotics	Inhibition of Clostridia	WOOLFORD (1985), BOLSEN and HEIDKER (1985)
Enzyme antioxidant mixture	Preservation of dry matter, energy, and protein in wheat-rye grass silage and corn silage	BOLSEN (1978), BOLSEN and HEIDKER (1985), MCCULLOUGH (1975)

4.3.2 Sausage Fermentation

Fermented sausage is made from meat products containing sugar and other additives. Bacteria utilize the added sugar producing lactic acid. The acid can extend the shelflife of meat products by retarding growth of spoilage bacteria already present in the meat raw material, especially *Clostridium botulinum, Staphylococcus aureus,* and *Pseudomonas* sp. It also gives a unique tangy flavor to the products (BUEGE and CASSENS, 1980). The bacteria responsible for lactic acid production and nitrate reduction in sausage were discovered early in the 20th century.

Commercially, *Pediococcus cerevisiae* and *Lactobacillus* strains are used for the processing of fermented sausage. The optimum temperature for *P. cerevisiae* is 43 °C; it can grow in 5–7% saline medium. Though most starter organisms are capable of growing and producing lactic acid at a wide range of temperatures, generally the optimum growth is obtained in the range of 32–43 °C. The higher the temperature, the faster the acid is produced. High temperature combined with high acid or low pH in fermented sausage can inhibit growth of *S. aureus* (BUEGE and CASSENS, 1980); however, beyond 49 °C, most bacteria will die. The starter culture is usually prepared and supplied to consumers from starter culture companies in frozen or freeze dried form.

An excellent review on fermented sausage production is authored by BUEGE and CASSENS (1980). The major ingredients in sausage fermentation are bacterial starter, meat, salt, sugar, spice, nitrite or nitrate, ascorbate and erythorbate. The quality of meat should be good; minimal bacterial contamination is desirable to prevent off-flavors, gas production, and soft texture due to protein breakdown. Addition of lactic acid bacteria gives a safe and consistent product compared to a natural fermentation. A massive dose of lactic acid bacteria ensures the dominance of desirable fermentation and reduces the chance of growth of pathogenic and spoilage organisms. The starter culture is usually added during the final stages of mixing. Salt (2.5–3.5%) contributes to good texture by binding the meat particles together; it also provides a palatable flavor and restricts the growth of salt sensitive *Pseudomonas* normally present in meat (BUEGE and CASSENS, 1980). Sugar (dextrose or sucrose) serves as a carbon and energy source for the starter organism(s) to produce lactic acid under anaerobic conditions, and lower the pH of the sausage to the level of 4.7–5.3, which, in combination with salt, can preserve the meat product. JOHNSON (1991) and HAUSCHILD (1989) present the following guidelines for the control of bacterial pathogens in shelf-stable fermented sausage: (1) $pH < 5.0$; (2) water activity < 0.91; or (3) $pH \leq 5.2$ and water activity ≤ 0.95. Adding salt reduces the water activity; 1% salt based on the mix weight is recommended.

Spice can be added to enhance a specific flavor in the final product. However, spice might be a source of undesirable bacteria. Pretreated spice (by gassing or radiation) may be used to reduce bacterial contamination.

Nitrite and nitrate can fix the red pigment of meat. They also inhibit growth of *Clostridium botulinum,* the organism that produces botulin, the deadly neurotoxin. Federal regulations permit a quarter ounce of nitrite per 100 pounds of meat in the sausage formulation. Nitrite can be toxic at high concentrations. Ascorbic acid or erythorbate can preserve and maintain the pink color in fermented sausage during storage and also prevent undesirable oxidation and off-flavors. Moreover, these chemicals also prevent the formation of nitrosamine, the carcinogenic substance derived from nitrate or nitrite, in meat products.

Generally, the process starts with the mixing of ground or chopped meats, sugar, salt and spices; starter culture is added near the end of mixing cycle. The batter is stuffed into casings to ensure anaerobic conditions and incubated (fermented) at 32–43 °C for 12–24 h. In case of smoked sausage and summer sausage, it is smoked until the final inside temperature reaches about 60–63 °C.

4.4 Processes that Contain a Solid Phase

There are some anaerobic processes in which the reactants and products are soluble, but the fermentation environment contains a solid phase. Many of these fermentations have features which are similar to fermentations in which the reactant or product is a solid.

4.4.1 Bread Fermentation

Bread has been coined "the staff of life". Indeed, in many parts of the world bread is the most important food for daily maintenance of the population. Bread is fermented by yeast and is an excellent example of a solid phase fermentation. By definition bread is the product of moistening, kneading, and baking flour with addition of yeast or leaven. Although the biochemical pathway for bread fermentation and alcoholic fermentation is the same, utilization of the end product of fermentation is dramatically different. In alcoholic fermentation CO_2 is allowed to escape while in bread fermentation, the reverse is true, i.e., CO_2 is retained during the dough rising process by the elastic gluten, and alcohol is evaporated during the baking process. The key to a successful bread fermentation is the presence of yeast, fermentable sugar, and the effective development of gluten which is a mixture of wheat proteins (gliadin and glutenin), lipid, starch, ash, and water. As CO_2 is generated by yeast (*Saccharomyces cerevisiae*) it enters the existing air pockets and expands them to appropriate size. Excessive formation of large air pockets indicates faulty fermentation probably by contaminants or poor development of the dough matrix.

Current research in bread fermentation includes identification of compounds responsible for staling, effective frozen dough preparation, and acceleration of bread fermentation by improvement of yeast strains and stimulatory compounds for fermentation.

4.4.2 Sauerkraut Fermentation

Sauerkraut fermentation is a good example of a natural mixed culture fermentation in a solid state matrix. The process starts with using shredded cabbage supplemented with 2.5% salt and placed into a container with a small opening such that anaerobiosis can be achieved easily by sealing the opening. The microorganisms in the mixture consist mainly of enteric and to a lesser extent of lactic acid bacteria. Anaerobic condition is achieved by respiration of the plant materials themselves and consumption of oxygen by aerobic microorganisms. The microbial flora changes to enteric bacteria which are facultative anaerobic organisms. Since cabbage has poor buffering capacity, the drop of pH due to fermentation of the lactic acid bacteria kills the enteric and stimulates the further growth of lactic acid bacteria such as *Leuconostoc mesenteroides*. As this heterofermentative organism grows along with other microbes, lactic acid, acetic acid, ethanol, mannitol, dextran, esters, CO_2, etc. are formed. These compounds on one hand stimulate the growth of *Lactobacillus plantarum* and on the other hand inhibit the growth of undesirable organisms such as yeast in the developing sauerkraut. The last stage of sauerkraut fermentation invariably involves the growth of *Lactobacillus plantarum*, a homofermentative lactic acid bacterium which produces the desired final acidity of 1.7%. *L. plantarum* also utilizes mannitol and removes its bitter flavor from the sauerkraut. Sauerkraut fermentation is an example of microbial succession of a naturally fermented food product utilizing a solid state fermentation system (Erickson and Fung, 1988).

Starter culture is not necessary for this fermentation because the predominant bacterium at the end of the fermentation is always *L. plantarum* regardless of the presence or absence of starter culture.

4.4.3 Bioremediation

Bioremediation of contaminated soil may be carried out aerobically or anaerobically. Since maintaining aerobic conditions in soil is difficult, anaerobic biodegradation of organic contaminants offers some advantages. Erickson and Fan (1988) have reviewed the literature on the anaerobic degradation of toxic and hazardous waste. Bhatnager and Fathepure (1991) have summarized research on mixed cultures which consume hazardous organic compounds. The reductive dehalogenation of chlorinated organics is often best carried out under anaerobic conditions. This can be done in bioreactors containing soil slurries or *in situ* at contaminated sites. Maintaining the optimum environment for biodegradation is more easily achieved in a bioreactor in which pH, temperature, and redox reactions can be controlled.

One of the challenges for bioengineers and environmental engineers is to develop the necessary bioremediation technology for cleaning up spilled halogenated organic compounds which are more dense than water and form a second liquid phase below the aqueous phase in a ground water reservoir (MERCER and COHEN, 1990). Denser-than-water non-aqueous phase liquids (DNAPL) will continue to migrate downward until a barrier layer is found. BEEMAN and SUFLITA (1990) have shown that temperature and pH influence bioremediation rates significantly; this must be considered in the design of *in situ* bioremediation processes. Diffusion in soil can also limit the rate of bioremediation (DHAWAN et al., 1991).

Because of the difficulty of providing oxygen for aerobic bioremediation of petroleum hydrocarbons in soil, several studies have been conducted with nitrate and nitrous oxide as the terminal electron acceptor (HUTCHINS, 1991; HUTCHINS et al., 1991). In a field study where jet fuel had been spilled, nitrate was utilized by the organisms and a variety of organic contaminants were degraded (HUTCHINS et al., 1991). While the results are encouraging, more research and development are required because of the complexity of the processes and the high cost of remediation.

The research needs associated with bioremediation include an understanding of the degree of degradation that can be achieved; often the final concentration of the contaminant is unknown until treatability studies are carried out. The bioenergetics, kinetics, and optimum environmental conditions for biodegradation must be investigated if they are not already known. Knowledge of the adsorption of the contaminant to the soil, of the transport within the soil, and the character of the soil are needed as well.

5 Instrumentation and Measurement

The methodology of anaerobic fermentation has been reviewed by FUNG (1988). Measurement methods are summarized by SAN et al. (1988). Redox potential has been used to monitor oxygen-free conditions in anaerobic fermentations; an excellent review of this topic is available elsewhere (SRINIVAS et al., 1988).

DONLON and COLLERAN (1991) have reported on their work in which fermentation progress is monitored with a pressure transducer which measures changes in pressure associated with gas production. The pressure transducer signal can be sent to a computer for automated data collection and record keeping.

KURODA et al. (1991) have measured the dissolved hydrogen with a membrane covered electrode. The ability to measure the dissolved hydrogen in anaerobic fermentations represents an important advance which will have many applications. MARACHEL and GERVAIS (1991) have developed a sensor to measure osmotic pressure which will have many applications in anaerobic fermentations. The osmotic flux through a membrane is measured; it is directly proportional to the osmotic pressure gradient.

Fuzzy control has been applied to Japanese sake brewing by OISHI et al. (1991). Temperature and specific gravity were measured and signals were transmitted to a microcomputer where artificial intelligence and fuzzy control theory were utilized for management of the sake brewing process.

6 Conclusions

Anaerobic fermentations are of significant industrial importance. Much of the early work and many of the anaerobic processes are described in considerable detail in the *Handbook on Anaerobic Fermentations* (ERICKSON and FUNG, 1988). The present treatment is not as extensive. Anaerobic fermentation research is being conducted in many laboratories throughout the world because of numerous new commercial opportunities. Anaerobic bioremediation and biodegradation processes are being investigated because of the opportunities for low cost environmental applications of the technology.

Acknowledgements
Although the research described in this chapter has been funded in part by the U. S. Environmental Protection Agency under assistance agreement R-815709 to the Hazardous Substance Research Center for U. S. EPA Regions 7 and 8 with headquarters at Kansas State University, it has not been subjected to the Agency's peer and administrative review and therefore may not necessarily reflect the views of the Agency and no official endorsement should be inferred. This research was partially supported by the Kansas State University Center for Hazardous Substance Research.

7 References

ABE, S., TAKAGI, M. (1991), Simultaneous saccharification and fermentation of cellulose to lactic acid, *Biotechnol. Bioeng.* **37**, 93–96.

ADLER, H. I. (1989), Oxygen sensitivity of an *Escherichia coli* mutant, in: *Proc. 89th Meeting Am. Soc. Microbiol.*, pp. 14–18, New Orleans, LA.

ADLER, H. I. (1990), The use of microbial membranes to achieve anaerobiosis, *CRC Crit. Rev. Biotechnol.* **10**, 119–127.

ADLER, H. I., CROW, W. D. (1981), A novel approach to the growth of anaerobic microorganisms. *Biotech. Bioeng. Symp. No. 11*, pp. 533–540, New York: John Wiley & Sons.

AMARTEY, S. A., LEAK D. J., HARTLEY, B. S. (1991), Effects of temperature and medium composition on the ethanol tolerance of *Bacillus stearothermophilus* LLD-15, *Biotechnol. Lett.* **13**, 627–632.

AMERINE, M. A., CRUESS, W. V. (1980), *The Technology of Wine Making,* Westport, CT: AVI Publishing Co.

ATKINSON, B., MAVITUNA, F. (1983), *Biochemical Engineering and Biotechnology Handbook*. New York: The Nature Press.

BAJPAI, R. K., IANNOTTI, E. L. (1988), Product inhibition, in: *Handbook on Anaerobic Fermentations* (ERICKSON, L. E., FUNG, D. Y. C., Eds.), pp. 207–241. New York: Marcel Dekker.

BALLERINI, D., MAECHAL, D. R., HERMANN, M., BLANCHETTE, D., VANDECASTEELE, J. P. (1985), *French Patent* 2 550 222.

BEEMAN, R. E., SUFLITA, J. M. (1990), Environmental factors influencing methanogenesis in a shallow anoxic aquifer: a field and laboratory study. *J. Ind. Microbiol.* **5**, 45–57.

BELL, D. A. (1989), *Wine and Beverage Standards.* New York: Van Nostrand Reinhold.

BENDA, I. (1982), Wine and brandy, in: *Industrial Microbiology,* (REED, G., Ed.), 4th Ed., Westport, CT: AVI Publishing Co.

BHATNAGER, L., FATHEPURE, B. Z. (1991), Mixed cultures in detoxification of harzardous waste, in: *Mixed Cultures in Biotechnology* (ZEIKUS, G., JOHNSON, E. A., Eds.), pp. 293–340, New York: McGraw Hill.

BIBAL, B., VAYSSIER, Y., GOMA, G., PAREILLEUX, A. (1991), High concentration cultivation of *Lactococcus cremoris* in a cell recycle reactor, *Biotechnol. Bioeng.* **37**, 746–754.

BILLS, D. D., KUNG, S. D. (1990), *Biotechnology and Food Safety,* Stoneham, MA: Butterworth-Heinemann.

BOLSEN, K. K. (1978), The use of aids to fermentation in silage production, in: *Fermentation of Silage: A Review* (MCCULLOUGH, M. E., Ed.), pp. 181–231. West Des Moines, IA: National Feed Ingredients Association.

BOLSEN, K., HEIDKER, J. I. (Eds.) (1985), *Silage Additives USA,* Marlow: Chalcombe.

BORRIELLO, S. P., HARDI, J. M. (1987), *Recent Advances in Anaerobic Bacteriology,* Boston, MA: Martinus Nijhoff.

BUEGE, D. R., CASSENS, R. G. (1980), *Manufacturing of Summer Sausage,* Cooperative Extension Program, University of Wisconsin-Extension, Madison.

BUONO, M. A. (1988), An engineering, microbiology and sensory study of yogurt from soy milk, *Ph. D. Dissertation,* Kansas State University, Manhattan, KS.

BUONO, M. A., ERICKSON, L. E. FUNG, D. Y. C., JEON, I. J. (1990a), Carbohydrate utilization and growth kinetics in the production of yogurt from soymilk. Part I: Experimental methods. *J. Food Process. Preserv.* **14**, 135–153.

BUONO, M. A., ERICKSON, L. E., FUNG, D. Y. C. (1990b), Carbohydrate utilization and growth kinetics in the production of yogurt from soymilk. Part II. Experimental and parameter estimation results. *J. Food Process. Preserv.* **14**, 179–204.

BUONO, M. A., SETSER, C. S., ERICKSON, L. E., FUNG, D. Y. C. (1990c), Soymilk yogurt: sensory evaluation and chemical measurement. *J.*

Food Sci. **55,** 528–531.

CHEN, H. C. (1991), Up-flow attached-bed bioreactor for continuous ethanol fermentation, *Biotechnol. Prog.* **7,** 311–314.

CHEN, C., DALE, M. C., OKOS, M. R. (1990a), The long-term effects of ethanol on immobilized cell reactor performance using *K. fragilis, Biotechnol. Bioeng.* **36,** 975–982.

CHEN, C., DALE, M. C., OKOS, M. R. (1990b), Minimal nutritional requirements for immobilized yeasts, *Biotechnol. Bioeng.* **36,** 993–1001.

CHERYAN, M. (1986), *Ultrafiltration Handbook.* Lancaster, PA: Technomic.

CIANI, M., ROSINI, G. (1991), Sparkling-wine production by cell-recycle fermentation process (CRBF), *Biotechnol. Lett.* **13,** 533–536.

COOK, J. E. (1973), The use of additives to improve the stability of maize silage in aerobic conditions, *Bulletin of Maize Development Assoc.* **54** (8), 13–16.

DALE, M. C., CHEN, C., OKOS, M. R. (1990), Cell growth and death rates as factors in the long-term performance, modeling, and design of immobilized cell reactors, *Biotechnol. Bioeng.* **36,** 983–992.

DATTA, R. (1988), Control of carbon and electron flows in acetone-butanol fermentations, in: *Handbook on Anaerobic Fermentations* (ERICKSON, L. E., FUNG, D. Y. C., Eds.), pp. 269–289. New York: Marcel Dekker.

DHAWAN, S., FAN, L. T., ERICKSON, L. E., TUITEMWONG, P. (1991), Modeling, analysis and simulation of bioremediation in soil aggregates, *Environ. Prog.* **10,** 251–260.

DOELLE, H. W., KENNEDY, L. D., DOELLE, M. B. (1991), Scale-up of ethanol production from sugercane using *Zymomonas mobilis, Biotechnol. Lett.* **13,** 131–136.

DONLON, B., COLLERAN, E. (1991), Applications of pressure transducers in anaerobic systems, *Biotechnol. Lett.* **13,** 661–666.

EDELINE, F., LAMBERT, G., FATTICCIONO, H. (1986), Anaerobic treatment of coke plant waste water, *Process Biochem.* **26,** 58–60.

ENNIS, B. M., GUTIEREZ, N. A., MADDOX, I. S. (1986), The acetone-butanol-ethanol fermentation: a current assessment, *Process Biochem.* **21,** 131–135.

ERICKSON, L. E. (1980), Growth and product energetic yields of *Rhodopseudomonas sphaeroides* S in dark and aerobic chemostat cultures, *J. Ferment. Technol.* **58,** 53–59.

ERICKSON, L. E., FAN, L. T. (1988), Anaerobic degradation of toxic and hazardous wastes, in: *Handbook on Anaerobic Fermentations* (ERICKSON, L. E., FUNG, D. Y. C., Eds.), pp. 695–732. New York: Marcel Dekker.

ERICKSON, L. E., FUNG, D. Y. C. (Eds.) (1988), *Handbook on Anaerobic Fermentations,* New York: Marcel Dekker.

ERICKSON, L. E., ONER, M. D. (1983), Available electron and energetic yields in fermentation processes, *Ann. N. Y. Acad. Sci.* **413,** 99–113.

ERICKSON, L. E., MINKEVICH, I. G., EROSHIN, V. K. (1978), Application of mass and energy balance regularities in fermentation, *Biotechnol. Bioeng.* **20,** 1595–1621.

FOX, A., MORGAN, S. L., LARSON, L., ODHAM, G. (1990), *Analytical Microbiology Methods: Chromatography and Mass Spectrometry,* New York: Plenum Press.

FUNG, D. Y. C. (1988), Methodology of anaerobic cultivation, in: *Handbook on Anaerobic Fermentations* (ERICKSON, L. E., FUNG, D. Y. C., Eds.), pp. 3–25, New York: Marcel Dekker.

FUNG, D. Y. C. (1992), Food fermentation, in: *Encyclopedia of Food Science and Technology,* pp. 1034–1041, New York: John Wiley & Sons.

FUNG, D. Y. C., MATTHEWS, R. F. (1991), *Instrumental Methods for Quality Assurance in Foods,* New York: Marcel Dekker.

FUNG, D. Y. C., BUONO, M. A., ERICKSON, L. E. (1988), Mixed culture interaction in anaerobic fermentation, in: *Handbook on Anaerobic Fermentations* (ERICKSON, L. E., FUNG, D. Y. C., Eds.), pp. 501–536. New York: Marcel Dekker.

GALAZZO, J. L., BAILEY, J. E. (1990), Growing *Saccharomyces cerevisiae* in calcium-alginate beads induces cell alterations which accelerate glucose conversion to ethanol, *Biotechnol. Bioeng.* **36,** 417–426.

GHOSE, T. K. (Ed.) (1989), *Bioprocess Engineering.* Chichester, UK: Ellis Horwood.

GIBBS, D. F. (1983), The rise and fall (... and rise?) of acetone/butanol fermentations, *Trends Biotechnol.* **1,** 12–15.

GILLILAND, S. E. (1985), *Bacterial Starter Cultures for Foods,* Boca Raton, FL: CRC Press.

GODIA, F., CASAS, C., SOLA, C. (1987), A survey of continuous ethanol fermentation systems using immobilized cells, *Process Biochem.* **22,** 43–48.

HASTINGS, J. J. H. (1978), Acetone-butanol alcohol fermentation, in: *Primary Products of Metabolism* (ROSE, A. H., Ed.), New York: Academic Press.

HAUSCHILD, A. H. W. (1989), *Clostridium botulinum,* in: *Foodborne Bacterial Pathogens* (DOYLE, M. P., Ed.), pp. 111–189, New York: Marcel Dekker.

HO, C. S., SHANAHAN, J. F. (1986), Carbon dioxide transfer in bioreactors, *CRC Crit. Rev. Biotechnol.* **4,** 185–252.

HO, C. S., SMITH, M. D. (1986), Morphological alterations of *Penicillium chrysogenum* caused by carbon dioxide, *J. Gen. Microbiol.* **132,** 3479–3484.

HOLDEMAN, L. V., CATO, E. P., MOORE, W. F. C. (1977), *Anaerobic Laboratory Manual,* 4th ed., Virginia Polytechnic Institute and State University, Blacksberg, VA.

HOSKINS, C. B., DAVIDSON, P. M. (1988), Recovery of *Clostridium perfringens* from food samples using an oxygen-reducing membrane fraction. *J. Food Prot.* **51,** 187–191.

HOSKINS, C. B., RICO-MUNOZ, E., DAVIDSON, P. M. (1986), Aerobic incubation of *C. perfringens* in the presence of an oxygen reducing membrane fraction, *J. Food Sci.* **51,** 1585–1586.

HUANG, S. Y., CHOU, M. S. (1990), Kinetic model for microbial uptake of insoluble solid-state substrate, *Biotechnol. Bioeng.* **35,** 547–558.

HUTCHINS, S. R. (1991), Biodegradation of monoaromatic hydrocarbons by aquifer microorganisms using oxygen, nitrate, or nitrous oxide as the terminal electron acceptor, *Appl. Environ. Microbiol.* **57,** 2403–2407.

HUTCHINS, S. R., DOWNS, W. C., WILSON, T. T., SMITH, G. B., KOVACS, D. A., FINE, D. D., DOUGLASS, R. H., HENDRIX, D. J. (1991), Effect of nitrate addition on biorestoration of fuel-contaminated aquifer: field demonstration, *Ground Water* **29,** 571–580.

IWASAKI, K., NAKAJIMA, M., SASAHARA, H. (1991), Rapid ethanol fermentation for soy sauce production by immobilized yeast cells, *Agri. Biol. Chem.* **55,** 2201–2207.

JOHNSON, E. A. (1991), Microbiological safety of fermented foods, in: *Mixed Cultures in Biotechnology* (ZEIKUS, G., JOHNSON, E. A., Eds.) pp. 135–169, New York: McGraw-Hill.

JOHNSON, L. A., DEYOE, C. W., HOOVER, W. J. (1981), Yield and quality of soy milk processed by steam infusion cooking. *J. Food. Sci.* **43,** 239–243.

JONES, R. P., GREENFIELD, P. F. (1982), Effect of carbon dioxide on yeast growth and fermentation, *Enzyme Microb. Technol.* **4,** 210–223.

KANG, W., SHUKLA, R., SIRKAR, K. K. (1990), Ethanol production in a microporous hollow-fiber-based extractive fermentor with immobilized yeast, *Biotechnol. Bioeng.* **36,** 826–833.

KURODA, K., SILVEIRA, R. G., NISHIO, N., SUNAHARA, H., NAGAI, S. (1991), Measurement of dissolved hydrogen in an anaerobic digestion process by a membrane-covered electrode, *J. Ferment. Bioeng.* **71,** 418–423.

LACOURSIERE, A., THOMPSON, B. G., KOLE, M. M., WARD, D., GERSON, D. F. (1986), Effects of carbon dioxide concentration on anaerobic fermentations of *Escherichia coli, Appl. Microbiol. Biotechnol.* **23,** 404–406.

LAPLACE, J. M., DELGENES, J. P., MOLETTA, R., NAVARRO, J. M. (1991), Combined alcoholic fermentation of D-xylose and D-glucose by four selected microbial strains: process considerations in relation to ethanol tolerance, *Biotechnol. Lett.* **13,** 445–450.

LAWFORD, H. G., ROUSSEAU, J. D. (1991), Fuel ethanol from hardwood hemicellulose hydrolysate by genetically engineered *Escherichia coli* B carrying genes from *Zymomonas mobilis, Biotechnol. Lett.* **13,** 191–196.

LOPES, C. E., CALAZANS, G. M. T., RIOS, E. M. M. M., CARLOS, T. F. (1991), On the effect of temperature and pH on the settling behavior of a flocculent strain of *Zymomonas mobilis, Biotechnol. Lett.* **13,** 43–46.

LUEDEKING, R., PIRET, E. L. (1959a), A kinetic study of thc lactic acid fermentation. Batch process at controlled pH, *Biotechnol. Bioeng.* **1,** 393–412.

LUEDEKDING, R., PIRET, E. L. (1959b), Transient and steady states in continuous fermentation. Theory and experiment, *Biotechnol. Bioeng.* **1,** 431–459.

MARACHEL, P. A., GERVAIS, P. (1991), Development of a sensor allowing the evaluation of the osmotic pressure of liquid media, *Biotechnol. Bioeng.* **38,** 797–801.

MARWAHA, S. S., KENNEDY, J. F., SEHGAL, V. K. (1988), Simulation of process conditions of continuous ethanol fermentation of whey permeate using alginate entrapped *Kluyveromyces marxianus* NCYC 179 cells in a packed-bed reactor system, *Process Biochem.* **23,** 17–22.

MATTIASSON, B., HOLST, O. (Eds.) (1991), *Extractive Bioconversions*, p. 328, New York: Marcel Dekker.

MCCULLOUGH, M. E. (1975), The influence of silage additives on silage losses and feeding value, *Georgia Agric. Res.* **17,** (2).

MERCER, J. W., COHEN, R. M. (1990), A review of immiscible fluids in the subsurface: properties, models, characterization and remediation, *J. Contam. Hydrol.* **6,** 107–163.

MIALL, L. M. (1978), Organic acids, in: *Primary Products of Metabolism* (ROSE, A. H., Ed.), New York; Academic Press.

MINKEVICH, I. G., EROSHIN, V. K. (1973), Productivity and heat generation of fermentation under oxygen limitation, *Folia Microbiol.* (Praha) **18,** 376–385.

MITCHELL, D. A., DO, D. D., GREENFIELD, P. F., DOELLE, H. W. (1991), A semimechanistic mathematical model for growth of *Rhizopus oligosporus* in a model solid-state fermentation system, *Biotechnol. Bioeng.* **38,** 353–362.

MOO-YOUNG, M. (Ed.) (1988), *Bioreactor Immobil-*

ized Enzymes and Cells: Fundamentals and Applications, New York: Elsevier.

MOSER, L. E. (1980), Quality of forage as affected by post-harvest storage and processing, in: *Crop Quality, Storage, and Utilization,* (HOVELAND, C. S., Ed.), pp. 227–260. Madison, WI: American Society of Agronomy.

MUCK, R. E. (1991), Silage fermentation, in: *Mixed Cultures in Biotechnology* (ZEIKUS, G., JOHNSON, E. A., Eds.), pp. 171–204, New York: McGraw-Hill.

NIELSEN, J., NIKOLAJSEN, K., VILLADSEN, J. (1991 a), Structured modeling of a microbial system: I. A theoretical study of lactic acid fermentation, *Biotechnol. Bioeng.* **38**, 1–10.

NIELSEN, J., NIKOLAJSEN, K., VILLADSEN, J. (1991 b), Structured modeling of a microbial system: II. Experimental verification of a structured lactic acid fermentation model, *Biotechnol. Bioeng.* **38**, 11–23.

NIKOLAJSEN, K., NIELSEN, J., VILLADSEN, J. (1991), Structured modelding of a microbial system: III. Growth on mixed substrates, *Biotechnol. Bioeng.* **38**, 24–29.

NORMURA, Y., YAMAMOTO, K., ISHIZAKI, A. (1991), Factors affecting lactic acid production rate in the build-in electrodialysis fermentation, an approach to high speed batch culture, *J. Ferment. Bioeng.* **71**, 450–452.

OISHI, K., TAOMINAGA, M., KAWATO, A., ABE, Y., IMAYASU, S., NANBE, A. (1991), Application of fuzzy control theory to sake brewing process, *J. Ferment. Bioeng.* **72**, 115–121.

ONER, M. D., ERICKSON, L. E., YANG, S. S. (1983), Estimation of true growth and product yields in aerobic cultures, *Biotechnol. Bioeng.* **25**, 631–646.

ONER, M. D., ERICKSON, L. E., YANG, S. S. (1984), Estimation of yield, maintenance, and product formation kinetic parameters in anaerobic fermentations, *Biotechnol. Bioeng.* **26**, 1436–1444.

ONER, M. D., ERICKSON, L. E., YANG, S. S. (1986 a), Analysis of exponential growth data for yoghurt cultures, *Biotechnol. Bioeng.* **28**, 895–901.

ONER, M. D., ERICKSON, L. E., YANG, S. S. (1986 b), Utilization of spline functions for smoothing fermentation data and for estimation of specific rates, *Biotechnol. Bioeng.* **28**, 902–918.

ONER, M. D., ERICKSON, L. E., YANG, S. S. (1986 c), Estimation of the true growth yield and maintenance coefficient for yoghurt cultures, *Biotechnol. Bioeng.* **28**, 919–926.

OXYRASE TECHNICAL BULLETIN (1990), Properties of the Oxyrase™ enzyme system used to isolate and cultivate anaerobic microorganisms, Oxyrase Inc.

PARK, C. H., OKOS, M. R., WANKAT, P. C. (1990), Characterization of an immobilized cell trickle bed reactor during long term butanol (ABE) fermentation, *Biotechnol. Bioeng.* **36**, 207–217.

PATEL, A. A., WAGHMARE, W. M., GUPTA, S. K. (1980), Lactic fermentation of soymilk – a review, *Process Biochem.,* Oct./Nov. 9–13.

PIRT, S. J. (1975), *Principles of Microbe and Cell Cultivation.* New York: Wiley.

ROELS, J. A. (1983), *Energetics and Kinetics in Biotechnology.* New York: Elsevier Biomedical Press.

ROSE, A. H. (1977), History and scientific basis of beverage production, in: *Alcoholic Beverages* (ROSE, A. H., Ed.), New York: Academic Press.

ROYCE, P. N. C., THORNHILL, N. F. (1991), Estimation of dissolved carbon dioxide concentrations in aerobic fermentations, *AIChE J.* **37**, 1680–1686.

ROZZI, A., LIMONI, N., MENEGATTI, S., BOARI, G., LIBERTI, L., PASSINO, R. (1988), Influence of Na and Ca alkalinity on UASB treatment of olive mill effluents. Part I. Preliminary results, *Process Biochem.* **23**, 86–90.

SAN, K. Y., PAPOUTSAKIS, E. T., STEPHANOPOULOS, G. (1988), Measurements, data analysis and control, in: *Handbook on Anaerobic Fermentations* (ERICKSON, L. E., FUNG, D. Y. C., Eds.), pp. 447–462. New York: Marcel Dekker.

SCOTT, C. D. (1983), Fluidized-bed bioreactors using a flocculating strain of *Zymomonas mobilis* for ethanol production, *Ann. N. Y. Acad. Sci.* **413**, 448–456.

SHABTAI, Y., CHAIMOVITZ, S., FREEMAN, A., KATCHALSKI-KATZIR, E., LINDER, G., NEMAS, M., KEDOM, O. (1991), Continuous ethanol production by immobilized yeast reactor coupled with membrane pervaporation unit, *Biotechnol. Bioeng.* **38**, 869–876.

SHAPTON, D. A., BOARD, R. G. (1971), *Isolation of Anaerobes,* New York: Academic Press.

SILVERIO, C. M., ANGLO, P. G., MONTERO, G. V., PACHECO, M. F., LUIS, Jr., V. S. (1986), Anaerobic treatment of distillery slops using upflow anaerobic filter reactor, *Process Biochem.* **21**, 192–195.

SOLOMON, B. O., ERICKSON, L. E., HESS, J. L., YANG, S. S. (1982), Maximum likelihood estimation of growth yields, *Biotechnol. Bioeng.* **24**, 633–649.

SOLOMON, B. O., ERICKSON, L. E., YANG, S. S. (1983), Utilization of statistics and experimental design in data collection and analysis, *Biotechnol. Bioeng.* **25**, 2683–2705.

SOLOMON, B. O., ONER, M. D., ERICKSON, L. E., YANG, S. S. (1984), Estimation of parameters where dependent observations are related by equality constraints, *AIChE J.* **30**, 747–757.

SRINIVAS, S. P., RAO, G., MUTHARASAN, R. (1988), Redox potential in anaerobic and microaerobic fermentation, in: *Handbook on Anaerobic Fermentations* (ERICKSON, L. E., FUNG, D. Y. C., Eds.), pp. 147–186. New York: Marcel Dekker.

STEINKRAUS, K. H. (Ed.) (1983), *Indigenous Fermentation,* New York: Marcel Dekker.

STRYER, L. (1988), *Biochemistry,* 3rd ed. New York: Freeman.

SUTTER, V. L., CITRON, D. M., FINEGOLD, S. M. (1980), *Anaerobic Bacteriology Manual,* St. Louis, MO: C. V. Mosby.

TAMIME, A. Y., ROBINSON, R. K. (1985), *Yogurt: Science and Technology,* London: Pergamon.

TAMPION, J., TAMPION, M. D. (1987), *Immobilized Cells: Principles and Applications,* Cambridge: Cambridge University Press.

TANAKA, M., ASAI, M., AOKI, N., TANIMOTO, M., MATSUNO, R. (1991), Alcohol fermentation using lignin-related compounds as a carbon source, *J. Ferment. Bioeng.* **71**, 436–438.

TUITEMWONG, P., ERICKSON, L. E., FUNG, D. Y. C. (1991), Soy yogurt fermentation of rapid hydration hydrothermal cooked soy milk, *Proc. 21th Biochemical Engineering Symposium*, pp. 182–194, Colorado State University, Fort Collins, CO.

TZENG, J. W., FAN, L. S., GAN, Y. R., HU, T. T. (1991), Ethanol fermentation using immobilized cells in a multistage fluidized bed bioreactor, *Biotechnol. Bioeng.* **38**, 1253–1258.

UNITED STATES DEPARTMENT OF AGRICULTURE (1986), Composition of Foods: Legumes and Legume Products, *Agriculture Handbook* 8–16, Washington D.C.

WALDO, D. R. (1978), The use of direct acidification in silage production, in: *Fermentation of Silage: A Review* (MCCULLOUGH, M. E., Ed.), pp. 117–179, West Des Moines, IA: National Feed Ingredients Assoc.

WEIMER, P. J. (1991), Use of mixed cultures for the production of commercial chemicals, in: *Mixed Cultures in Biotechnology* (ZEIKUS, G., JOHNSON, E. A., Eds.), New York: McGraw-Hill.

WILSON, R. F., WOOLFORD, M. K., COOK, J. E., WILKINSON, J. M. (1979), Acrylic acid and sodium acrylate as additives for silage, *J. Agric. Sci.* (Cambridge) **92**, 409–415.

WOOLFORD, M. K. (1975), Microbiological screening of the straight chain fatty acid (C_1-C_{12}) as potential silage additives, *J. Sci. Food Agric.* **26**, 219–228.

WOOLFORD, M. K. (1978), Antimicrobial effects of mineral acids, organic acids, salts, and sterilizing agents in relation to their potential as silage additives. *J. Brit. Grassland Soc.* **33**, 131–136.

WOOLFORD, M. K. (1985), The silage fermentation, in: *Microbiology of Fermented Foods,* (WOOD, B. J. B., Ed.), vol. 2, London: Elsevier Applied Science.

YABANNAVAR, V. M., WANG, D. I. C. (1991a), Analysis of mass transfer for immobilized cells in an extractive lactic acid fermentation, *Biotechnol. Bioeng.* **37**, 544–550.

YABANNAVAR, V. M., WANG, D. I. C. (1991b), Extractive fermentation for lactic acid production, *Biotechnol. Bioeng.* **37**, 1095–1100.

YEGNESWARAN, P. K., GRAY, M. R., THOMPSON, B. G. (1990), Kinetics of CO_2 hydration in fermentors: pH and pressure effects, *Biotechnol. Bioeng.* **36**, 92–96.

YU, L. S. L., FUNG, D. Y. C. (1991a), Oxyrase^{TM} enzyme and motility enrichment Fung-Yu tube for rapid detection of *Listeria monocytogenes* and *Listeria* spp., *J. Food Safety* **11**, 149–162.

YU, L. S. L., FUNG, D. Y. C. (1991b), Effect of Oxyrase^{TM} enzyme on *Listeria monocytogenes* and other facultative anaerobes, *J. Food Safety,* **11**, 163–175.

YU, L. S. L., FUNG, D. Y. C. (1992), Growth kinetics of *Listeria* spp. in the presence of Oxyrase^{TM} in a broth model system, *J. Rapid Methods Automat. Microbiol.* **1**, 15–28.

14 Fermentation Monitoring and Control

THOMAS CHATTAWAY

Billingham, Cleveland, UK

GARY A. MONTAGUE
A. JULIAN MORRIS

Newcastle upon Tyne, UK

1 Introduction

One of the primary objectives of industrial fermentation research and development is the establishment of viable processes through increasing product yields and reduced operating costs. Historically, the most important means of achieving this have been strain improvement using a variety of techniques, growth medium development, and improvements in nutrient feeding. In recent years, however, tremendous progress has been made in the measurement of biotechnical parameters, bioprocess instrumentation and bioprocess modeling and control. The 'control' of biotechnological plants is a complex problem, since there are a great variety of processes. These can be considered not only from a biological but also an engineering view point. From the biological perspective - different substrates can be used (e. g., pure or synthetic substrates, natural substrates including waste). These are converted by a variety of microorganisms under anaerobic or aerobic conditions to the desired products (e. g., pharmaceuticals, amino acids, organic acids, enzymes, proteins, biodegradable products, biogas, alcohol, etc.). The engineering, or technological, aspects include different reactor types: stirred tank, tube, tower, cascaded reactors, etc. In addition there are different modes of reactor operation: batch, fed-batch, continuous, and all pose different problems in terms of control system structure. Despite such complexity, progress has been made over the past decade. A complex overview of the state of the art is given in Volume 4 of this series "Measuring, Modelling, and Control", edited by K. SCHÜGERL.

Rather than attempting to cover the broad area of biotechnological processes, this chapter will concentrate on the modeling and control of fermentation processes. It should be recognized, however, that the techniques developed for, and used in, fermentation are readily applicable, for example, to wastewater treatment and other biotechnological processes.

2 Current Practice in the Fermentation Industries

It is convenient to categorize common fermenter instrumentation as follows: *in situ* probes, other on-line instruments, gas analysis, and instruments for the off-line analysis of culture broth samples. With the exception of the latter, all these provide on-line measurements of variables of interest, generally through the amplification of an electrical signal. These signals may then be used for

- operator information in fermentation status monitoring
- direct closed loop control of the fermentation
- calculation of derived variables.

2.1 On-Line Fermentation Instrumentation

The performance of a fermentation process depends entirely on the maintenance of a well defined and controlled environment for biomass and product formation. The most straightforward and effective way to achieve this objective is to regulate the bioprocess by making a direct measurement of the fermentation conditions. Effective on-line instrumentation is therefore a prerequisite for efficient process operation. In specifying instrumentation, it is not just the ability to perform the measurement task which must be considered. A major consideration is that the measurements must provide demonstrable benefits without compromising the process. In particular, for most fermentation processes measurement must be achieved without increased risk of contamination. These considerations often lead to well instrumented research and development fermenters. However, in production-scale fermenters while replication of key measurement instruments is commonplace, there is a reluctance to utilize any other instrumentation. This is primarily due to the increased financial consequences of contamination.

Present on-line fermentation process measurement, and as a result most on-line control, is based upon a few robust commercial devices which are very similar to those found in chemical plants. However, it is worth noting that the most important sensors for fermentation control are some of the least reliable of the sensors widely used in chemical process control (FLYNN, 1982; LEES, 1976). The reliability problems are further compounded by the fact that fermentation probes must be capable of withstanding sterilization if they are inside the sterile boundary. This means not only elevated temperatures and wet heat, but also repeated heating and cooling cycles, all of which affect probe reliability and performance. In addition, probes are subject to surface fouling by the microorganisms. Not surprisingly, in addition to probe failure, significant drift problems are experienced, in particular with pH and DO_2 probes. To overcome drift, potential probe failure, and also account for non-homogeneities at large scale, several probes of each category are usually fitted on large production vessels. In addition, retractable probes that can be taken out for re-calibration and re-sterilized without compromising the vessel's sterile integrity are also used.

2.1.1 Instrumentation for Status Monitoring

In production fermentation a crucial role of instrumentation is to generate alarms or to automatically initiate corrective actions if certain threshold values are exceeded (e. g., if sterilization temperature is below a critical value). Here, for example, information is supplied to process operators regarding the performance of the fermentation control systems. Operator intervention may well be required for excursions of process variables outside the band of acceptable operation, or automatic alarm sequences can be triggered by such events. A vital instrumentation role is also to monitor the sterile integrity of the vessel and its ancillaries. For example, temperature probes are used extensively on steam-backed valves. To some extent trips are used to ensure process safety, though biological processes are not intrinsically hazardous, in the sense that they operate at near-ambient temperature and pressure and are not subject to run-away reactions. Indeed the major hazard area concerns sterility and containment, which is primarily dealt with by good design and manufacturing practice.

2.1.2 Instrumentation for Closed Loop Control

Present on-line fermentation process control is based upon a few robust, commercial instruments (FLYNN, 1982; CARLEYSMITH and FOX, 1984). Of the instruments listed in Tab. 1, the temperature, pH, and DO_2 probes are probably the most crucial and widespread, at all scales. Most microorganisms operate within a fairly narrow optimal temperature and pH range, hence the importance of these two measurements. However, closed loop control of pH is not always necessary. While most fermentations are acidifying, the media's buffer capacity is sometimes sufficient. Dissolved oxygen is a critical parameter for most aerobic fermentations and its control in practice often consists of maintaining it above a critical value rather than at a given set point.

Depending on the complexity of fermenter peripherals, a number of flows (gaseous or liquid), pressure, levels and temperatures may be measured on-line. As for any chemical plant some of these may be used to implement control actions. A key example is control of feeds to follow set product profiles in fed-batch fermentations.

2.1.3 Derived Variables

Some of the above measurements can be combined to provide valuable information on the process state, the reaction rates, or on equipment performance. For example, the process control action taken to maintain a measured environmental variable constant (acid/alkali addition, bioreactor heating/cooling, antifoam addition, etc.) is often growth and/or product formation related, although it is also subject to the effects of process disturbances, shifts in metabolism, and other con-

Tab. 1. Fermenter Instrumentation: Common Industrial Practice

Category	Measurement	Sensor	Control Mechanism	Comments
In situ probes	Temperature	Pt resistance thermometer	Circulation of cool water in coils. Heat addition by steam injection	Thermistors also used. Heating elements used at small scale
	pH	Glass and reference electrode	Addition of acid or alkali (liquid or gaseous)	
	Dissolved oxygen (DO_2)	Pt and Ag/AgCl, Ag and Pb probes, polarographic or galvanic type	Action on agitator speed, air rate, gas composition and overpressure; in isolation or combination	O_2 diffuses through membrane to be reduced at cathode. Polarographic probes generally more expensive and more robust
	Foam	Conductance probe/capacitance probes	On/off addition of aliquots of antifoam	Foam breaking device also used
Other on-line instruments	Agitation	Tachometer (speed) Watt-meter (power draw)	Variable speed drives	Power not normally measured at small scale
	Air flow	Mass flow meters Rotameters	Flow control valve	
	Level	Strain gauges/pressure Pads/load cells	Over-flow or in-flow of liquid	Load cells at small scale
	Pressure	Spring diaphragm	Pressure control valve	Not usual at small scale
	Feed flows	Electromagnetic flow meters	Flow control valve	Used for monitoring feeds and cooling water
Gas analysis	O_2 content	Paramagnetic analyzer/mass spectrometer		Used mainly to compute respiration data
	CO_2 content	Infra-red analyzer/ mass spectrometer		

trol actions. If pH is under feedback control, the control action taken in regulation provides an indication of metabolic rate. Such rates may be integrated with time to estimate reaction advancement. Total heat loads can be derived, to good accuracy at scales above hundred liters from coolant flow, and temperature measurements and heat transfer coefficients can be computed. Heat transfer coefficients are a key design variable, and their monitoring indicates problems such as high viscosity or fouling.

Probably some of the most valuable measurements are provided by the off-gas analysis and air flow measurements (MEYER et al., 1985). These are sometimes available continuously, e. g., with infra-red, paramagnetic and zirconia analyzers. More popular, and becoming more cost effective, is mass spectrometry. Indeed, it is now becoming an industrial standard, at least in research laboratories and pilot plants. This is inherently a fast, discrete analysis which is usually multiplexed (BUCKLAND et al., 1985; COPPELLA and DHURJATI, 1987). Although such measurements provide discrete data, they are relatively fast for control purposes. FLYNN (1982) has commented upon the accuracy and precision of calculations (giving derived variables) based on gas analysis (SPRIET et al., 1982), indicating that care must be taken in their use. The rates of O_2 and CO_2 transfer can be calculated by a mass balance over the fermenter and are usually a good enough approximation of the O_2 uptake rate (*OUR*) and CO_2 production rate (*CPR*), respectively. These provide access to the metabolic reaction rates. The ratio of *CPR* to *OUR,* known as respiratory quotient (*RQ*) provides insight into the metabolic state of the organism. In particular, it can provide indications of the metabolism shifts from growth to product formation or switches between primary substrates. Finally, volumetric oxygen transfer coefficients (k_La) are quite often computed on-line, since they can provide an indication of culture viscosity. These indirect measurements form much of the basis of estimation techniques, inferential control, and other advanced control methods for bioreactors.

2.2 Off-Line Fermentation Analysis

It is patent from Tab. 1 that, with the exception of pH and DO_2, no on-line measurements of the fermenter broth composition are routinely available. This is due to the difficulty of developing sterilizable probes or constructing a sampling system that does not compromise the fermenter's sterility or the containment of hazardous microbes when such constraints exist. Therefore, monitoring of many of the compounds of interest relies on manual sampling and off-line analysis. The fermenter broth is analyzed for substrates (carbohydrate, lipids, salts, amino acids) precursors, products and metabolites (organic or amino acids), as well as for the microorganisms themselves. A whole gamut of analytical methods is used here, ranging from simple wet chemistry to NMR. A characteristic of the off-line sample approach whether the analysis is performed with auto-analyzers or a variety of other specific analytical measurement techniques (spectrophotometry, atomic absorption, HPLC, GC, GCMS) is that process information is infrequent and delayed. It is often not the speed of analysis which causes the problems, the techniques can give quite rapid results, but in practice samples are usually taken relatively infrequently, with 'returned' results taking one or two hours. Even so, in some cases this approach can provide reasonable levels of process information. While the methods adopted for measurement of liquid phase components are relatively standard, the approaches taken for the microorganisms themselves require special mention since they are peculiar to the fermentation industry. The more common approaches are listed in Tab. 2. It is noteworthy that none of these methods is entirely satisfactory or universal; in particular, most give no information on the state of the microorganism. For this, various system-specific techniques are used: viability assessment by plate counts or staining and microscopic analysis; image analysis for assessing the morphology of filamentous microbes; measurement of intracellular enzyme activities, etc.

2.3 Traditional Control Policies

Some features of fermentation processes actually make them easier to control than chemical process systems. First, bioreactions are very largely self-regulating; through evolutionary pressure microorganisms have evolved the capacity to adapt to their environment. Thus, for instance, run-away reactions do not exist in bioprocesses (they will naturally slow down as conditions depart from the optimal). Indeed, operation of many bioprocesses is not crucially dependent on process control – this is not to say bioprocess performance cannot be considerably enhanced by better control. Where the probe and the actuator exist (as in temperature or pH measurement), control is achievable across a range of scales, processes, and states of processes. For instance, in practice it is usually possible to maintain conditions reasonably close to desired values (e. g., ± 0.5 °C, ± 0.2 pH) with well tuned control loops. Similarly, though much is sometimes made of the possibility of control-loop interactions, these are not manifest in practice. For instance, one could envisage an interaction between temperature and DO_2 control through agitation speed because of the power dissipated by the agitator; this does not in fact lead to difficulties, and such control loops may safely be treated as decoupled, obviating the need for sophisticated multivariable approaches.

Tab. 2. Off-Line Methods for Assessing Microorganisms

Method	Principle	Comments
Packed cell volume	Height of centrifuged pellet	Crude but rapid
Dry weight	Weight of suspended solids after drying	Difficult to interpret if medium contains solids
Optical density	Turbidity	Dilution necessary for linearity (same drawbacks as above)
Microscopic observation	Cell counts on hematocytometer	Labor-intensive; automation possible via image analysis
Coulter counter	Number of particles passing a micro-orifice	Gives size distribution; problems with small cells; not applicable to fungal organisms
Fluorescence or other chemical methods	Analysis of biomass related compounds (ATP, NAD, DNA, protein, etc.)	Indirect measurements only; calibration difficulties
Plate counts	Counting colonies formed on a plate after suitable dilution	Measures viability; requires long incubation period

The second "facilitating" feature of fermentation processes is their comparatively long time constants (though certain intracellular events probably have much shorter time scales). A convenient characteristic time is a microorganism's doubling time which for the fastest growing species is about twenty minutes, but for other organisms can be as long as twenty hours. The long dominant time constants probably explain why the potential problems resulting from time varying, non-linear dynamics and control–loop interactions do not manifest themselves in practice. The features that make the operation of bioprocesses more straightforward, have arguably resulted in slow advances in bioprocess control. Control is, however, recognized as a vital component of fermentation operation. For instance, regulatory authorities regard consistency of operation as key to good manufacturing practice. Despite the favorable bioprocess characteristics, achieving such consistency of operation can be a major problem, especially when the operating window has been tightly defined. Traditional feedback control of fermentation processes is limited to implementing a few simple and fairly robust control loops (as described in Tab. 1). For certain processes it is necessary or beneficial to regulate substrate or metabolic concentrations. As explained previously, the relatively long time constants often allow this through the use of manual sampling and off-line analysis, usually by manipulating feed rates.

In order to reap the benefits of process development, open-loop or feed forward control strategies are often used. These are mainly applied to feed profiles for fed-batch fermentations. Such strategies have enabled considerable optimization of secondary metabolite fermentations, in particular antibiotic production. Thus, not only is the potential of improved strains better realized, but the processes are best tailored to the heat and oxygen transfer capabilities of the fermentation vessels. Modern computer technology also permits set-point profiles to be tracked, as discussed in a following section. This not only applies to variables controlling aeration and agitation, so as to reduce the substantial electricity consumption of industrial fermentations, but also potentially to pH or temperature profiles where enhancements of microbial activity are possible. The use of inducible promotors (by temperature shifts or chemicals) in recombinant products can be placed into this category. Such strategies are obviously highly process-specific and, therefore, very proprietary;

not surprisingly, little has been published on their use. They are presumably fairly widespread, especially for the more mature processes where considerable process knowledge has been accumulated.

An emerging trend with respect to such feed-forward control policies seems to be to use defined fermentation events as 'triggers'. This introduces a degree of feedback, in a loose sense, to the strategies and makes use of the fermentation monitoring tools described above. For instance, feeds could be triggered by peaks in oxygen uptake, in alkali demand (or in pH if not regulated) or by discernible shifts in metabolism (as evidenced, for instance, by the respiratory coefficient) or in morphology (as detected by changes in viscosity and hence oxygen transfer coefficient) for filamentous organisms. Similarly, it is common practice to transfer the various inoculum stages on the basis of some measurable criteria, rather than after a fixed growth time. There is no doubt that the use of such 'event-based' control is driven as much by the quest for consistency as by that for optimality. Again, this type of control is highly process-specific. These strategies have inevitably been developed from considerable process experience and are subject to gradual improvement.

A description of current industrial practice would not be complete without considering the additional steps of the batch cycle. For the majority of fermentation processes being batch or fed-batch, capital productivity dictates that turn around time be minimized. Hence, the operations such as medium preparation, sterilization, vessel filling and emptying, etc. are automated, when justified by process economics. Automating these operations arguably also increases the reproducibility of the production process.

2.4 Problems Experienced with Current Control Policies

The incentive for bioprocess control improvement has not been as clear as for ordinary chemical processes (many of which could not operate safely without control). The potential benefits to be gained from, say, strain selection and process development have traditionally outweighed control scheme updating. Control has thus never been considered an enabling technology, more a means of achieving cost reduction. Thus, relatively little effort has been devoted to control system improvement. However, as bioprocesses become more mature and regulatory authority conditions become more strict, the incentive to control bioprocesses more effectively increases. In many bioprocess operations attaining 'competitive productivities' was once the target. Such process 'optimization' has to date been achieved through operating experience implemented via feedforward (and event triggered) control.

Though achieving maximum productivities and yields still remains an important goal, others such as minimizing waste or securing consistent product quality are gaining importance through regulatory and market pressure. The difficulty of controlling final product quality is best illustrated by a few examples of what defines it: organism viability for inoculants, enzyme activities, and performance for biocatalysts, flavor and texture in food products, impurities resulting from metabolic branch products for secondary metabolites, amino acid sequence for pharmaceutical proteins, etc. In all these cases quality is affected by both fermentation and downstream processing, and it is not straightforward to decide where control should best be exerted. The application of statistical data analysis procedures, such as control charts of product titers, may be useful for detecting gross non-compliance from set procedures; however, the appropriate corrective action to take when a deviation occurs is not always clear. Thus, the short falls of current practice are now becoming more apparent. While there exists obvious incentive for improving control (i. e., consistency) of biotechnological processes, the way forward is not so clear. Before outlining the contribution to be made by novel control developments such as seen in the abundant literature, it is worth specifying the difficulties they attempt to address.

The difficulties in measuring variables of interest have been described above, and the paucity of readily available probes is patent from Tab. 1. Progress here is limited by the complexity of many fermentation broths and more severely by the peculiar constraints posed by sterility (and containment). It is also fair to say that advances have been far slower because the cost benefit of specific probes has not always been clear. In spite of a consensus that enhanced instrumentation is desirable, the gains to be achieved from the availability of specific probes on specific processes are rarely obvious. This stems less from the potential not being real, than from the gross lack of understanding of the fundamental phenomena in bioprocesses, a point to be returned to.

While there are a number of bioprocess characteristics which assist control developments, there are also many problems impeding improvement. Most fermentations are batch (or fed-batch) processes. Thus, traditional control techniques developed for continuous processes (including many advanced control concepts) are not directly applicable. Moreover, bioprocess non-linearities are likely to invalidate control design based upon linear theory. Much of modern control theory has been devoted to solving problems of control–loop interactions (which is of lesser relevance to bioprocesses) or has dealt with systems for which either an accurate dynamic model exists or whose behavior could be represented by a simplified model. An accurate process description permits design of an effective control system. Indeed, the potential benefits of model-based control design have been demonstrated in recent years in the chemical industry where major financial rewards have been attained. Thus, if chemical process parallels are to be drawn, the root to improved bioprocess control lies in the development of dynamic system models.

Bioprocess systems do not readily lend themselves to mathematical description. The observable microbial reactions result from a complex and highly regulated network of parallel-series intracellular reactions, where kinetics are likely to be non-linear functions of the species involved. Thus, a complete mechanistic description of the system is likely to be impractical. Certain essential and process-specific features can, however, be modeled but in doing so, assumptions must very often be made to arrive at a tractable solution. Furthermore, even if simplifications were not necessary, the level of understanding is insufficient to fully account for all the reactions. Such complexity leads to intricate system behavior. This, coupled with the diversity in environmental conditions experienced within large bioreactors, results in a high level of variability. The ability to represent such complex behavior is far from industrial reality.

More often than not, the reality is that little is known of what determines the precise state of the microbial catalysts. The difficulty in responding to focused production objectives has been described above. Specifically defining the optimal environment for the microorganisms at any given time is a task that is extremely elusive. Thus, advances in understanding at a fundamental level are a prerequisite for significant process improvements. If such understanding were available, it could undoubtedly be efficiently captured in appropriate models, and very substantial gains through the application of control techniques could be made.

2.5 Emerging Industrial Trends

The present evolution in fermentation monitoring and control philosophy comes about as a result of the availability of low-cost and powerful computer systems. This has provided the opportunity for computer based fermentation monitoring and control from laboratory scale through to production plants.

2.5.1 Emerging On-Line Fermentation Instrumentation

Solutions to the problems induced by reliance upon off-line analyses are being found through new *in situ* sensors. The specificity and sensitivity required indicates significant potential for on-line enzyme-based sensors and biosensors (Brooks and Turner, 1987; Cleland and Enfors, 1983, 1984; Karube, 1984). Although there are problems with sterilization, stability and robustness, developments in continuous flow-line sampling, (Omstead and Greasham, 1988; Clarke et al., 1982; Mandenius et al., 1984), or in off-line discrete sample measurements (see references in MEYER et al., 1985; and the clinical laboratory techniques described by TRUCHAUD et al., 1980), may overcome these problems. However, many other options are being investigated, for example, in the field of biomass determination alone a wide range of measurement techniques are being applied (RAMSAY et al., 1985; HARRIS and KELL, 1985). They are based upon such diverse principles as: acoustics (CLARKE et al., 1982); piezo-electric membranes (ISHIMORI et al., 1981); bioelectrochemistry (RAMSAY et al., 1985); laser light scattering (LATIMER, 1982; CARR et al., 1987); electrical admittance spectroscopy (KELL, 1987; HARRIS and KELL, 1985); fluorescence (ZABRISKIE and HUMPHREY, 1978; SRINIVAS and MUTHARASAN, 1987); calorimetry (BIROU et al., 1987; RANDOLPH et al., 1990); and viscosity (PICQUE and CORRIEU, 1986).

In general, the measurements supplied by sensors are not simple linear correlations to the fermentation process variable of interest. Significant correlations can be made between these measurements and the state variables required for control, e. g., ATP or NAD(P)H for biomass. However, analysis of the measurement and the state variable usually indicates that the measured variable is a complex function of many factors. Under calibration conditions most of these factors vary little and/or cancel each other out to leave the required correlation. In practice, however, these factors can vary significantly, and as a result the calibrated correlation may not be valid. This may well be significant for any new sensors introduced (e. g., CLARKE et al., 1982; SRINIVAS and MUTHARASAN, 1987). For example, new techniques for biomass determination, such as admittance spectroscopy, IR optical fiber light scattering detection, on-line fluorescence probes for NAD(P)H, may all exhibit good direct correlations under suitable calibration conditions but are complex functions of biological and physicochemical effects. Similarly the development of a whole range of ion selective electrodes allows the direct measurement of a wide range of important medium constituents, but the measured values are activities, and correction is required for a whole range of interfering ions, ionic effects, and chelation (CLARKE et al., 1982).

Many of these devices are now commercially available, however, difficulties in sterilization and/or problems in interpreting the probe's response probably explain their uncommon use. Nevertheless, devices such as these are beginning to be used within an industrial environment, at least at laboratory and pilot scale.

2.5.2 Environmental Feedback Control

While the feedback of established environmental measurements is commonplace, the consequences of fermenter engineering design can affect the quality and applicability of the measurements made. For example, by the very nature of the operation of an air-lift fermenter, an organism is subjected to cyclic variations in dissolved oxygen levels. In this case, a number of dissolved oxygen probes may well be more appropriate in order to gain some insight into the cyclic nature of the variation. Even a relatively high frequency of exposure, compared with overall organism growth dynamics, can have serious consequences in terms of fermentation performance. REUSS and BRAMMER (1985) demonstrated that although control techniques may work well on the laboratory scale, the engineering implications of scale-up can significantly degrade the quality of performance obtainable from the fermentation. Cyclic variation of substrate in a large bioreactor due to inefficient mixing was demonstrated to markedly reduce the yield of biomass.

The developments in sensor technology which are currently taking place have been outlined earlier. It is apparent, however, that at present there are many measurements of environmental conditions (e. g., precursor, substrate concentrations) which are not routinely available on-line for direct feedback control. The usual method adopted for environmental fermentation regulation is, therefore, based upon the use of a combination of off-line and on-line measurements for single loop feedback control. Resorting to the use of off-line measurements in the feedback loop has important consequences on the quality of control which can be obtained. MONTAGUE et al. (1988) discussed in detail the resulting problems and the considerations necessary before reasonable control can be achieved. In summary, they conclude that it is the ability to obtain off-line measurements, with the minimum process/measurement delay and at a rate suitable for feedback control, which usually proves to be a problem. This said, the use of off-line measurement still forms a major component of many fermentation control schemes. More advanced techniques, aimed at overcoming the problems of off-line measurement and the resulting process delay, have been developed which utilize both on- and off-line measurements in on-line measurement-based feedback control schemes. These have been shown to be successful, as will be discussed in a later section.

2.5.3 Derived Variables in Fermentation Control

The successful maintenance of fermentation environment using feedback control, however, does not necessarily result in the fermentation being operated under optimal conditions. In order to improve the performance of a fermentation system, it is necessary to consider variables (or states) which give some indication of the way in which an organism is behaving and not just the environment to which it is exposed. It is worthwhile at this stage considering what the aims of a fermentation control scheme should be. Ideally it is desirable to be able to specify the condition (i. e., all the states) of the fermenter at any time. It is the present state coupled with the system inputs which completely specifies the future fermenter condition. The important consequences are that knowledge of all the critical variables (states) and inputs which affect the bioprocess are required to specify future fermentation conditions. In this context, environmental regulation is the control of only a few of the many system states and, therefore, will not achieve the quality of control often demanded for improved operation. To move towards this improved operation requires the control of further states of the system, while bearing in mind the problems of process measurement.

Off-gas (exit gas) measurements have proved to be a popular method of gaining insight into the performance of an organism since they are available on-line, without significant delay, and give some indication of growth. The behavior of *OUR*, *CPR*, and *RQ* varies with the type of fermentation under investigation and the particular operating conditions which prevail. JOHNSON (1987) with reference to the baker's yeast fermentation, discussed the behavior of *RQ*, *OUR*, and *CPR* during various operating regimes. In a fed-batch baker's yeast fermentation, two major causes for ethanol production can be identified. If the substrate concentration in the broth is too high, then ethanol is produced (negative Pasteur effect). In addition, if insufficient oxygen is supplied to combust the substrate, ethanol will again be produced. In this case, the production of large quantities of ethanol is undesirable, since it represents wasted sugar. When ethanol is produced, the CPR is observed to rise while the *OUR* remains constant. Hence, an increase in *RQ* is a good indication that ethanol is being produced. If a mass spectrometer is available, then it can be used to confirm the presence of ethanol.

The use of off-gas analysis-based control schemes has been adopted by many authors. AIBA et al. (1976), and WANG et al. (1977), for example, applied the method to the regulation of a fed-batch baker's yeast fermentation and demonstrated good performance. WILLIAMS et al. (1984) went further in the combination of *RQ* and dissolved oxygen control in an adaptive multivariable control scheme for regulation of a yeast fermentation. They showed that a high biomass and good conversion from nutrient to yeast could be obtained with their control strategy (see below).

The more straightforward control of *OUR* has also been widely studied. An early reference to *OUR* control can be found in HUMPHREY and JEFFREYS (1973), who investigated the control of organism growth through variation of the substrate addition rate. SQUIRES (1972) utilized the dissolved oxygen as an indication of *OUR* in an attempt to control the substrate addition to a fed-batch penicillin fermentation.

Heat production, from energy balances on the fermentation, have also been used in an attempt to define some measure of the metabolic activity. MOU and COONEY (1976) utilized heat evolution from a novobiocin fermentation to regulate the specific growth rate through variation of the substrate feed.

An alternative to energy balancing is the use of mass balancing techniques for on-line estimation. The conservation approach (whether it be mass or energy) avoids the necessity to specify yields and rate constants. The technique has been demonstrated in both fed-batch and continuous yeast fermentations (COONEY et al., 1977; COONEY and SWARTZ, 1982). MOU and COONEY (1983) extended the mass balancing technique to cover secondary metabolite fermentation. The balancing technique is more suited to fermentations which utilize defined/semi-defined media, although even in these cases a considerable proportion of carbon can be unaccounted for.

2.5.4 Computer-Based Supervision Developments

Three key tasks in a system can be identified: logging of process data, process data analysis, and bioprocess control. Data logging consists of sequentially, and/or on an interrupt demand, scanning of the sensor signals, applying signal conditioning and filtering and storage of the data in an orderly and easily accessible form. Data analysis is performed by algorithms written to extract information from the measured data to determine 'derived variables' to indicate the state and performance of the fermentation. This information is made available to the process operators and management through visual display units, printers, and plotters, as well as being used for process control. The process control function performs three important duties: event-based control and control loop set point variation, sequence control of process valves for sterilization, reactor charging, emptying, etc., and conventional closed loop control of reactor environmental variables. In addition, provision is made for alarm analysis and display. Sophisticated use of computer monitoring and control systems has tended to be used mainly at the pilot-plant scale where fermenters are more likely to be well instrumented. In established production plants computer-based monitoring and control is generally more limited in application sophistication. On the production scale, they tend to be used primarily for monitoring and sequence control. Some newer plants do, however, make full use of the facilities that computer control systems can provide.

The most advanced form of control – optimizing control to provide for a maximization of production – is still in its infancy even in pilot-plant laboratories. Recently, knowledge-based system approaches have also been used to improve the quality of information presented to operators and increase the level of automatic process supervision, although very few industrial applications have yet been reported. These approaches are discussed later.

3 The Future: Advanced Methodologies

3.1 Bioprocess Modeling for Control

From a control engineering point of view, an essential prerequisite for good supervision and operation of bioreactors is an understanding of the process (fermentation) behavior (for example, in the application of mass balancing techniques discussed previously). To progress beyond the purely physical and environmental control of the fermenter into the area of biological control requires comprehensive process understanding, allowing a process model that is sufficiently accurate to relate all important process inputs (strains, medium, feeds, environmental conditions) and outputs (biomass, product, pH, temperature, dissolved oxygen, off-gases, etc.). The availability of a sufficiently descriptive mathematical model would also provide insight into the behavior of the fermentation state variables which could be used for improved control. In general, there are four different forms of process models that might be considered for control purposes.

(1) In the *physiological model* knowledge of the physiology of the growth process can be expressed in causal/consequent, and usually non-mathematical, terminology. These models are useful in initial control strategy synthesis and may play an even more important role as knowledge-based systems (expert systems) are developed (see later on).
(2) In the *structured model* partial differential and algebraic equations are used to describe the dynamic behavior of the growth process. The basic idea here is that the biomass is structured, or classified, by some proper intracellular characteristic which describes growth, activity, metabolism, etc. of the biomass or cells. The classification basis might be any chemical species content of the cell (DNA, RNA, etc.), the cell's mass, volume, or chronological age (i. e., age distribution of the cell population or biomass). Sometimes the equations are simplified to provide a lumped parameter model. Such a simplification, however, will not explain the effects of the intracellular state on the process operation and dynamic behavior.
(3) In an *unstructured model* of the process the fermentation behavior is assumed to be represented by a single, homogeneously growing organism. In spite of its limitations, this type of model has been most frequently used for the development of fermentation control strategies.
(4) Finally, some of the newer methods being developed for fermentation control utilize *'black-box'* or *'input-output' models* of the process. Here, the primary model variable (output variable) is specified in terms of a function of the relevant process inputs (control or manipulated variables). The parameters of such a model do not necessarily have any biological relevance.

Once a mechanistic understanding of the process has been obtained the next step is to try to efficiently represent such mechanisms in a model that is appropriate for the particular problem being studied. The resulting model should be balanced with respect to its mathematical complexity and its ability to capture all the essential features for control purposes. It should also be simple enough to permit direct determination of its key parameters through feasible experimental procedures. Presently available structured models and lumped parameter models are not generally applicable due to the large numbers of parameters involved in the case of structured representations, and in contrast the inability of the lumped parameter models to accurately predict dynamic behavior.

The following models concentrate upon the fermentation phase. Nevertheless, the philosophies considered could equally well be applied to other phases of the fermentation cycle, such as sterilization. BOECK et al. (1988) discussed one such application.

3.1.1 Traditional Process Models

The most commonly adopted model in the design of model-based control strategies for fermentation processes is the homogeneous single organism form. Additionally, it is also commonly assumed that there is a single growth-limiting substrate. A typical example of such a model is highlighted by the work of BAJPAI and REUSS (1980) in their studies on penicillin fermentation. Although the traditional MONOD (1950) model has been found to perform well at low cell densities, a structurally similar model developed by CONTOIS (1959) which has the additional ability to account for diffusional limitations experienced at high cell densities was utilized. This can be particularly important in high cell density fermentations such as industrial penicillin fermentation. The model for the fed-batch fermentation set out below (BAJPAI and REUSS, 1980; MONTAGUE et al., 1986a, b) has been found to give good agreement with the practical results of PIRT and RIGHELATO (1967).

Growth of biomass is related to substrate concentration by the relationship

$$\frac{\mathrm{d}X}{\mathrm{d}t}=\frac{\mu_x SX}{K_x X+S}-\frac{X\mathrm{d}V}{V\mathrm{d}t} \tag{1}$$

where S represents substrate concentration, μ_x the maximum specific growth rate, X the biomass, V the fermenter volume, and K_x the Contois saturation constant for substrate limitation of biomass production.

Substrate inhibition kinetics, which other workers have used to good effect with the inclusion of a term to account for hydrolysis, has been used to model penicillin production. Penicillin production is related to substrate concentration and biomass by

$$\frac{dP}{dt} = \frac{\mu_p S X}{K_p + S(1 + S/K_I)} - KP - \frac{P dV}{V dt} \tag{2}$$

where μ_p represents the maximum specific rate of product formation, K_p is the Monod saturation constant for one substrate limitation of product formation, K_I is the substrate inhibition constant for product formation, and K is the first-order rate constant for penicillin hydrolysis. Substrate concentration is modeled by assuming constant yields and maintenance requirements:

$$\frac{dS}{dt} = -\frac{1}{Y_{x/s}} \cdot \frac{dX}{dt} - \frac{1}{Y_{p/s}} \cdot \frac{dP}{dt} - m_x X + F - \frac{S}{V} \cdot \frac{dV}{dt} \tag{3}$$

where $Y_{x/s}$ represents the yield of biomass on substrate, $Y_{p/s}$ the yield of product on substrate, m_x the maintenance requirement and F the term which accounts for the fermentation feed rate.

A term for the production of carbon dioxide, which the original BAJPAI and REUSS model lacked, was adopted from the work of CALAM and ISMAIL (1980). The carbon dioxide production relationship assumes that evolution is due to three processes: growth, maintenance, and penicillin biosynthesis:

$$CER = k_4 \frac{dX}{dt} + m_c X + k_5 \tag{4}$$

where k_4 relates CO_2 production to growth, m_c relates CO_2 production to maintenance and k_5 relates CO_2 production to penicillin formation. *CER* is the rate of CO_2 production per unit volume of broth.

An alternative approach by NESTAAS and WANG (1983) considered the various stages through which the penicillin mold develops. They constructed a segregated model for which good agreement was again obtained with experimental data. The segregated nature of this model, however, (biomass is considered to consist of a mixture of three different states) causes some difficulty in the measurement of kinetic parameters. This highlights the importance of selecting, or developing, a model the complexity of which balances parameter measurement difficulties and the ability to represent dynamic behavior.

A number of investigators have used the penicillin fermentation for their modeling, state estimation, optimization and control studies; for example, FISHMAN and BIRYUKOV (1974), MOU (1979), BIRYUKOV (1982), KISHIMOTO et al. (1982), NAKAMURA and CALAM (1983), THOMPSON (1984), MONTAGUE et al. (1986a, b). Recently, NICOLAI et al. (1991) have proposed an improved model for penicillin G fed-batch fermentation which allows for a smooth transition between maintenance and endogenous metabolisms. It also provides physically acceptable values for all the variables involved under various fermentation operating conditions.

3.1.2 Linear 'Black-Box' Models

To the microbiologist the identification of the 'important' process variables (i. e., the states) may be relatively straightforward, however, the specification of the model structure can be extremely difficult. The question as to what mathematical structure is most appropriate to fit observed behavior is often a major stumbling block. An alternative philosophy, which to some extent circumvents the structure specification problem, is to adopt 'general' structure type models and allow the model to 'learn' the process structural characteristics. The most straightforward description of a process assumes a linear representation. The approach of using a general linear input-output model of the process is similar to that adopted for the development of well known adaptive control laws.

It is usually assumed that the model is of the Autoregressive Moving Average (ARMA) or Controlled Autoregressive Moving Average (CARMA) form.

$$A(z^{-1})\, y(t) = z^{-k} B(z^{-1})\, u(t) + z^{-m} L(z^{-1}) \cdot v(t) + C(z^{-1})\, e(t) \tag{5}$$

where A, B, and L are polynomials in the backward shift operator z^{-1} [the backshift operator essentially delays information, for example, $z^{-1}y(t) = y(t\text{-}1)$, $z^{-2}y(t) = y(t\text{-}2)$]. $y(t)$, $u(t)$, and $v(t)$ are the system's output, manipulative input, and measurable load disturbance, respectively. k and m are the process time delays (expressed as integer multiples of the sample time) exhibited by the output to manipulative control input and load disturbance input, respectively. $e(t)$ is a random, zero mean disturbance with finite variance and for simplicity $C(z^{-1})$ is often considered to be unity.

An alternative, the Controlled Autoregressive Integrated Moving Average (CARIMA) form, can provide additional benefits in that it inherently includes integral action. Here, the process disturbances are described by the term $e(t)/\Delta$ where Δ is the differencing operator $(1-z^{-1})$. The disturbance term $e(t)/\Delta$, can be thought of as being Brownian motion and for controller design purposes realistically represents the form of disturbances affecting the process.

While a linear representation can be effective in certain situations, in reality an actual process can seldom be described by such a model. Thus, the linear representation is only used to reflect the pertinent system dynamics around the expected operating region. If the system dynamics are relatively linear, the use of a linear model may provide acceptable performance. However, in situations where the process is highly non-linear, the linearity assumption could be significantly detrimental to performance. In such cases, adaptation of linear model parameters could be considered but this tends to lower the robustness of the model. A desirable objective would be to develop a technique which possesses generality of model structure (facilitating rapid and cheap development) but which could also be capable of learning and expressing the process non-linearities and complexities.

3.1.3 Non-Linear 'Black-Box' Models

An approach that has recently received an immense amount of attention is that of artificial neural networks. This type of model requires little prior structural knowledge of the process, and unlike other commonly used black-box approaches, a neural network-based process model has the potential to describe more concisely the behavior of complex process dynamics.

The basic feedforward network is shown diagrammatically in Fig. 1. Scaled data (usually

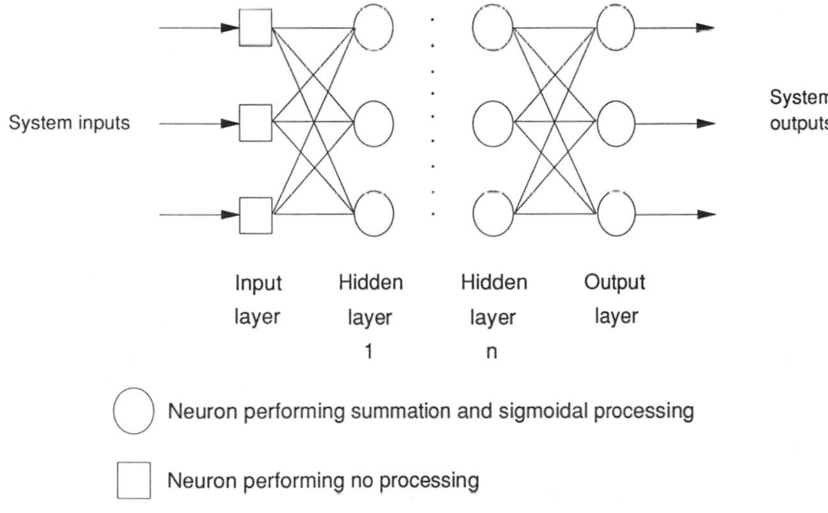

Fig. 1. Feedforward artificial neural network.

within the range 0–1) enter the network at the input nodes. The data are propagated forward through the network via the connections between hidden layers to the output layer. The network is fully connected (i. e., every neuron in a layer is connected to every neuron in adjacent layers). Each connection acts to modify the strength of the signal being carried. In the basic network the signals are modified by scalar multiplication (the connection weight). At neurons the input strengths are summed with a bias term (Eq. 6) and the resulting value passed through a sigmoidal non-linearity usually taking the form of Eq. (7) or a hyperbolic tangent.

$$N_s = wt_b + \sum_{i=1}^{N_{in}} wt_i \cdot s_i \tag{6}$$

$$N_{out} = 1/(1 + e^{-N_s}) \tag{7}$$

Here N_{in} is the number of neuron inputs, wt the scalar weight associated with each connection, s the signal strength and wt_b the bias weight. Data flows forward through successive layers of the network, with the outputs of neurons in one layer (N_{out}) becoming the signal strengths (s) of connections to the following layer. Ultimately, data arrives at the output layer, following which it is rescaled to engineering units.

The procedure for constructing an artificial neural network model given process data relies to a considerable degree on a general understanding of the process problem. Given the architecture shown in Fig. 1, it is necessary to determine:

a) the number of network inputs and outputs. This choice is primarily based upon an engineering appreciation of the problem. As with linear identification techniques, highly correlated inputs can degrade the quality of the resulting model. It is, therefore, necessary to attempt to minimize redundancy in network information.
b) number of hidden layers. The literature to date provides conflicting information regarding the choice of the number of hidden layers. CYBENKO (1989) postulated that two were sufficient to model any given continuous non-linearity, while other papers (HORNIK et al., 1989) claimed that one hidden layer was adequate. Experience to date suggests that this decision is best made heuristically; if one hidden layer is insufficient, then move to the use of two.
c) number of neurons per hidden layer. While trial and error determination of the 'optimal' number of neurons per layer usually produces acceptable results, more considered methods have been suggested. For instance, WANG et al. (1992) utilized an approach which assesses redundancy through analysis of hidden layer unit outputs.

Once the topology of the network has been specified, the network weights and biases must be obtained. Since essentially the problem is one of non-linear function minimization, a variety of methods exist for this task. For example, the simplex search procedure, a quasi-Newton approach or non-linear least squares will all perform acceptably. Further details of these techniques can be found in PRESS et al. (1985). WILLIS et al. (1991) outlined and compared alternative approaches when applied to neural network training.

3.2 Estimation Techniques in Fermentation

The sensors available today for bioreactor measurements do not cover all the necessary and important variables. The important 'internal' variables, such as biomass, substrate, and secondary product concentrations, that characterize the state and progress of a fermentation, are very difficult to measure reliably and fast enough for fermentation supervision and control purposes. For example, biomass is difficult to measure, but it may be estimated in proportion to the amount of oxygen consumed or carbon dioxide produced during the fermentation. Considering that biomass can multiply by a thousand-fold or more during a fermentation, the method relies on an accurate biomass measurement to start the integration of the off-gas analysis. It is important to stress here that the use of an inaccurate estimate of biomass, for control purposes, can result in the fermentation being driven away from its *a priori* defined optimum trajectory, thus defeating one of the main purposes of the control strategy. Periodic correction of estimated biomass by some form of direct measurement or assay is therefore essential. In spite of these difficulties, it is now relatively common to use measurements of related variables (secondary variables), such as gaseous oxygen and carbon dioxide, to estimate or infer unobtainable, or difficult to measure, (primary) variables such as biomass, products, etc.

The easiest, but potentially least accurate, way of obtaining an estimate of a primary variable with measurement difficulties is to establish, by experiment, a correlation between the primary and secondary related variables while neglecting all measurement errors, noise, etc. A more realistic approach is to take account of those errors and adopt a numerical estimation technique. Estimation methods can be used primarily for two different purposes: for estimating the parameters of a pre-defined model structure (parameter estimation or identification), or for estimating the actual process variables (state estimation). In comprehensive fermentation process models, the model parameters are usually physical or biochemical in nature, and the model structure is non-linear. The corresponding identification problems are therefore also non-linear. If estimation is carried out by off-line computation, then many well known numerical optimization algorithms can be used. On-line recursive estimation can also be carried out using any of the many existing recursive algorithms. Most of the methods used, to a greater or lesser extent, employ least squares-based ideas. When linearity and additivity assumptions on the process model are justified, then the optimal estimator is the Kalman filter. When they are not, then recourse must be made to a suboptimal estimator. The basic philosophy of an estimation scheme is shown in Fig. 2. Here, a process model is used to predict some of the more easily made measurements (e. g., off-gas analysis) and the error in the prediction is used to correct the future predictions of all the process measurements (including difficult to measure variables such as biomass concentration). The estimator gain is chosen either with knowledge of the process conditions and/or with consideration as to the speed of estimator response.

Clearly, the form of the estimation algorithm depends significantly upon the process model adopted. Unfortunately, the most useful and reliable algorithms have been developed for linear systems, whereas the biotechnical estimation problem is inherently non-linear.

3.2.1 Traditional Model-Based Estimation

Standard non-linear estimators such as the Extended Kalman Filter (EKF) (ANDERSON and MOORE, 1979), can suffer from some numerical problems and convergence difficulties

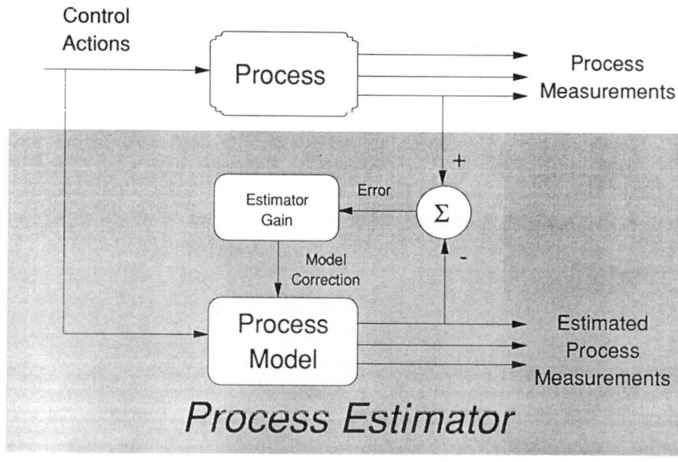

Fig. 2. Scheme of an estimator.

especially when the process noise characteristics are not well known (LJUNG, 1979). The use of extended Kalman filtering also requires the existence, or development, of a sufficiently accurate biochemical model of the fermentation being controlled. In spite of these problems, the EKF is an attractive means of tackling bioreactor identification and estimation. There have been a number of successful developments and applications of state estimation, and advanced fermentation control, reported by several authors (e. g., DEKKERS, 1982; HALME and SELKAINAHO, 1982; STEPHANOPOULOS and SAN, 1981, 1984; SWINIARSKI et al., 1982; NIHTILA et al., 1984; SAN and STEPHANOPOULOS, 1984; HALME et al., 1985; SHIOYA et al., 1985; TARBUCK et al., 1985; MONTAGUE et al., 1986a, b; LAKRORI and CHERUY, 1988).

The use of an adaptive filter based upon the techniques developed by JAZWINSKI (1970) has been recommended to overcome some of the problems encountered due to filter robustness in the presence of process and measurement uncertainties and growth model inaccuracy. A non-linear filter that has some useful innovations is that due to HALME and SELKAINAHO (1982). This filter basically provides a Bayesian maximum likelihood estimate for the state variables. A further problem commonly faced in practice is the variation of the model parameters, e. g., yield coefficients, maintenance, etc., either with time, point in the fermentation cycle, or environmental conditions. In such cases estimation of the model parameters and process state variables simultaneously is often necessary in order to obtain reliable results. This can also be achieved using well known filtering methods, or modifications of them, as well as, in principle, non-linear filtering procedures (HOLMBERG and RANTA, 1982; HOLMBERG and OLSSON, 1985; HALME et al., 1985).

When estimating parameters, either for the purposes of model identification or in adaptive state estimation, choosing the number of parameters which should be estimated can present potential difficulties. There are no standard ways of determining *a priori* the number of parameters that can be successfully estimated, and probably the best way is to rely on experience. It almost goes without saying, however, that the maximum number of parameters that might be successfully estimated depends upon the extent and quality of information made available to the estimator. For example, the more extensive the information pattern, the more reliable the estimation and the larger the number of parameters that might be estimated. It is, therefore, extremely important to utilize all the relevant information that is available for estimation. In biotechnological processes some of this information comes from laboratory assays and off-line analyses. The incorporation of these data into the estimation algorithms requires account to be taken of their irregular sample intervals and associated delay times.

SWINIARSKI et al. (1982) studied Kalman filtering methods for biomass estimation. The stationary extended adaptive Kalman filter and adaptive extended Kalman filter were derived and shown to give satisfactory results for biomass estimation. The algorithms, however, were demonstrated to be sensitive to errors in the initial estimate of substrate and very dependent upon an initial estimate of the noise covariances. LEIGH and NG (1984) reported significant batch to batch variations in what were initially expected to be nominally identical fermentations. Such unmodeled variations were shown to seriously degrade the biomass estimates when using a fixed parameter model for extended Kalman filter derivation. With this problem in mind, LEIGH and co-workers suggested a combination of approaches based on adaptive state estimation, improved modeling, and rigorous quality control to overcome possible batch to batch variations. The adaptation and extension of this work to represent a secondary metabolite antibiotic production system, *Streptomyces clavuligerus*, was later presented by TARBUCK et al. (1985). They showed that the application of an extended Kalman filter could provide reliable estimates of biomass. SHIOYA et al. (1982, 1985) investigated the application of an extended Kalman filter to the estimation of specific growth rate for control purposes. Modifications to the filter were proposed in order to use moving averages and dynamic mass balancing, as well as adaptively changing the noise covariance matrix according to the prediction error. An improvement in estimator performance was demonstrated. The estimator was then used in a 'profile control scheme' in a fed-batch baker's yeast fermentation.

An adaptive non-linear observer for the estimation of cell concentration and specific growth rate has been proposed by DOCHAIN and BASTIN (1985). Of the four different systems studied, stability and convergence proofs were given for three, and simulation studies demonstrated the performance of the simplified algorithms. However, for systems where the number of fermentation measurements might be more limited than those studied, a more rigorous mathematical model might well be required and the use of an extended Kalman filter approach might be at least as appropriate.

3.2.2 Linear Black-Box Model-Based Estimation

An alternative starting point for the development of estimation algorithms is to adopt the philosophy that the bioprocess dynamics can be represented by a quite general linear model (either in observer canonical form or input–output form). The approach using a general linear input–output model of the process is similar to that adopted for the development of well known adaptive control laws, except that an additional term representing a secondary process measurement is included (GUILANDOUST et al., 1988). In this technique the estimation of a 'primary' controlled output (e. g., biomass) is achieved using measurements of other 'secondary' outputs (e. g., CO_2 concentration in outlet gas, fermenter feed, etc.).

The original derivation of an adaptive linear model-based observer, using a state space approach was presented by GUILANDOUST et al. (1987) and discussed further by THAM et al. (1988). A diagrammatic representation of the estimation scheme is shown in Fig. 3.

The algorithm secondary measured variable, v (e. g., CO_2 in bioreactor off-gas), the infrequently measured, controlled variable, y (e. g., biomass assay), and manipulated variable or variables, u (e. g., bioreactor feeds), are those which are used in the development of the adaptive linear estimator model. The algorithm thus provides a means by which frequent estimates of biomass can be obtained (at the rate of measurement of the secondary variable) without measurement delay, hence improving the information available to the process operator.

The estimator has been applied so that, given infrequent, and irregular, laboratory dry weight data, and frequent CO_2 off-gas analysis and dilution rate data, it is able to predict hourly biomass concentrations. Results from a industrial fermentation are shown in Figs. 4

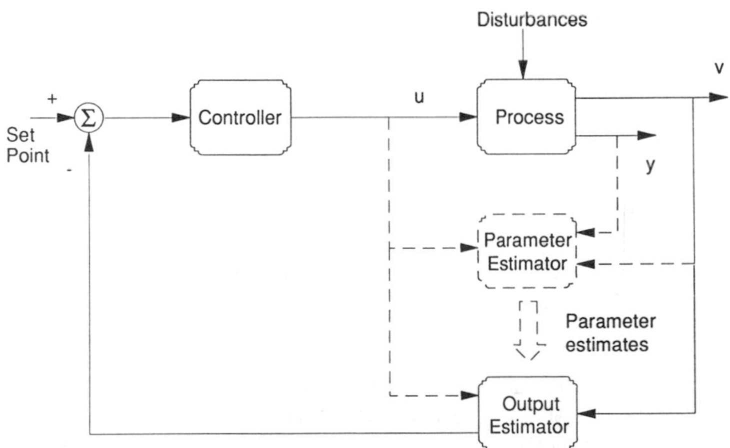

Fig. 3. Adaptive linear estimator structure.

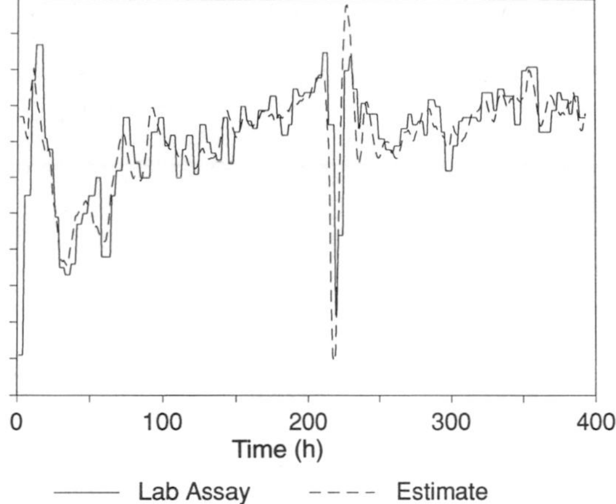

Fig. 4. Biomass estimates with linear 'black-box' model (off-line validation).

and 5. It should be noted that the lower point on the ordinate scale does not correspond to a zero data value. Fig. 4 shows results from off-line verification of the technique when the process is subjected to a significant disturbance. At around 210 hours into the fermentation process problems necessitate a reduction in the biomass concentration. It can be seen that the estimator performs reasonably well during this period. The 'undershoot' of the estimate as the biomass decreases and overshoot as normal production levels are re-instated is due to the process model within the estimator adapting to the changing process conditions. Following successful off-line verification, the estimator has been applied on-line as a process operator advisor. Fig. 5 shows results from a recent fermentation. Again, the estimator is seen to perform well.

3.2.3 Non-Linear Black-Box Model-Based Estimation

Rather than rely upon adaptation of the estimator model to track the changing process dynamics, it is possible to utilize the non-linear black-box modeling techniques discussed previously. Adopting an estimator model whose 'structure' is more closely related to the process reduces (if not eliminates) the need for adaptation, hence increases estimator robustness. Furthermore, the demands for off-line samples are significantly reduced. An example highlighting the benefits of this approach can be seen in the following. An artificial neural network has been developed to provide estimates of biomass concentration during the course of a fed-batch penicillin fermentation from readily available on-line process variables. The on-line measurements were selected as the network inputs, while biomass concentration is the network output. Previous studies (MONTAGUE, 1987) have shown that three process var-

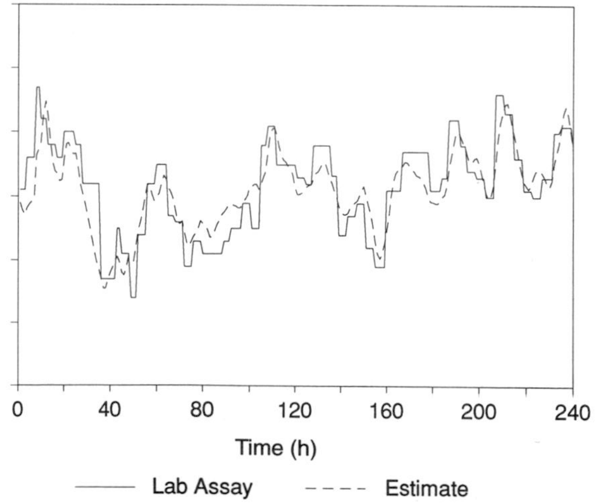

Fig. 5. Biomass estimates with linear 'black-box' model (on-line operator advisor).

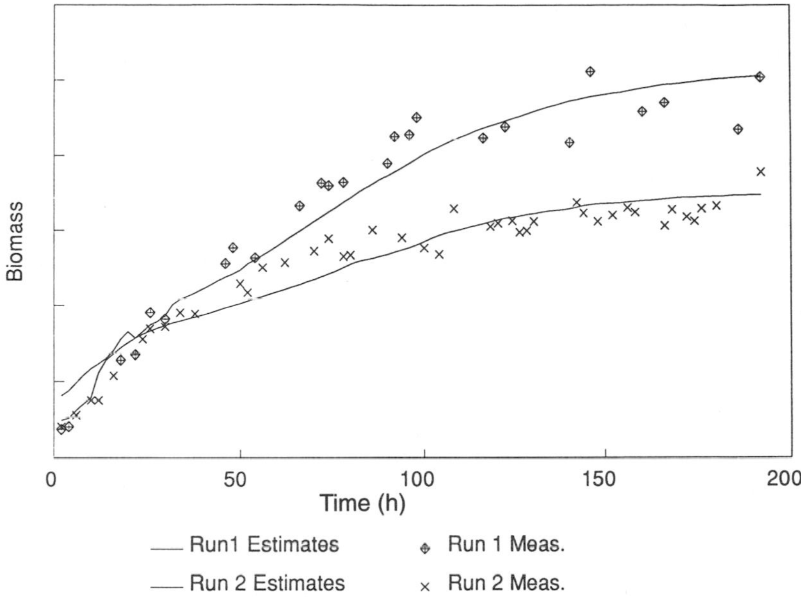

Fig. 6. On-line estimation of biomass.

iables provided the most pertinent information: carbon dioxide production (CPR), batch age, and substrate addition. Thus, these variables were used as inputs to the neural network. Since there are, in this case, two primary substrate feeds, utilized in a varying ratio, it is useful to separate the feed into its two components. Hence, the network topology specified is four inputs, two hidden layers, and one output. The network was then trained on selected fermentation data. Fig. 6 shows the biomass estimation results obtained solely with the on-line information. Network estimates are compared against off-line biomass measurements. Although the biomass assay results are corrupted with a high level of noise, it can be seen that the network estimate is representative of the underlying process behavior at two different operating conditions. This suggests that the artificial neural network model has captured the essential non-linear process characteristics.

3.3 Control Strategies for Fermentation Processes

Fermentation process control has become an active area of pursuit, especially in recent years. Most interest has centered on fed-batch processes, primarily because of their prevalence in the industry. Continuous fermenter operation is, however, sometimes preferred when the production of biomass or product needs to be optimized. Practical bioprocesses operating under industrial conditions exhibit non-linear and time-varying properties and will always be subject to unexplained disturbances. The control strategies developed must be able to efficiently handle these process disturbances. There are essentially four different approaches that can be adopted in control strategy design:

- linear three-term PID (Proportional-Integral-Derivative) control
- inferential control
- adaptive (predictive) control
- non-linear control.

3.3.1 Proportional-Integral-Derivative Control of Fermentation Processes

Here, a linear time-invariant model is assumed, and the parameters of a fixed structure Proportional + Integral + Derivative regulator are determined by well known controller tuning rules. Unfortunately, the time-varying properties of fermentation processes can make it difficult to set 'optimum' values for standard PID controllers in all but the simple standard bioreactor environment control loops (e. g., pH, level, temperature control).

One approach that addresses the non-linear and time-varying problems of fermentation processes is to adopt an autotuning PID control strategy. A number of techniques exist. For instance, one is based upon the relay-autotuning developments of ASTROM and colleagues (e. g., ASTROM and HAGGLUND, 1984). Here, the PID controller settings can be tuned on-line to provide for improved control performance over the whole fermentation cycle. JONES et al. (1991) have applied such an approach to the control of dissolved oxygen in a fed-batch baker's yeast fermentation. A second approach relies upon the adaptation of the parameters of a simplified process model followed by calculation of the PID settings, or by direct adaptation of the controller PID settings.

O'CONNOR et al. (1992) have developed and evaluated an approach to the design of control strategies for high cell density fermentations with *Saccharomyces cerevisiae* being used as the experimental system. Proportional (P), Proportional + Integral (PI), Feedforward and Estimated Feedforward, Oxygen Uptake Rate-based, and Exponential Schedule control strategies were evaluated. The individual control strategies were designed and 'tuned' to allow close regulation and control of a wide range of possible fermentation operations, for example, lag phase, growth phase, oxygen limitation, sensor failure, and the absence of off-gas measurements. The experiments showed the importance of designing the control strategy to cope with both low and high cell concentrations where different controller characteristics are required.

3.3.2 Inferential Control of Fermentation Processes

If good estimates of the key process variables can be obtained, then they can be utilized within the closed loop schemes discussed above to control the system to a pre-specified profile. The ultimate objective, of course, is to regulate all the process states in order to optimize the amount of product. However, 'optimal' profile specification is often extremely difficult. A more feasible aim is generally to regulate the most critical states. The appropriate profiles can be specified with the aid of historical fermentation records. In the following example, a PI controller was designed to make the biomass, and therefore the growth rate, in a penicillin fermentation follow a pre-specified profile. The PI controller made use of the error between biomass estimates (from a neural network-based estimator) and desired values to adjust the amount of carbon feed. The inferential control scheme is shown in Fig. 7.

Results from the implementation of the scheme in the pilot plant can be found in Fig. 8. Here, closed loop control is initiated at 40 hours into the fermentation. The controller increases sugar addition (control action, u) during this early closed loop stage to bring biomass concentration up to set-point. The profile then follows the set-point till around 150 hours when a disturbance occurs. The controller can be seen to return the process to set-point following the disturbance transients.

3.3.3 Adaptive (Predictive) Control of Fermentation Processes

A major problem in closed loop fermentation control is the tuning of the controller parameters. It is for this reason that adaptive controllers, the parameters of which can be estimated (identified) and varied on-line as the fermentation proceeds, have been studied quite extensively. Nevertheless, few adaptive controllers have found their way on to major production processes. The basis of the control designed is the use of an identified process

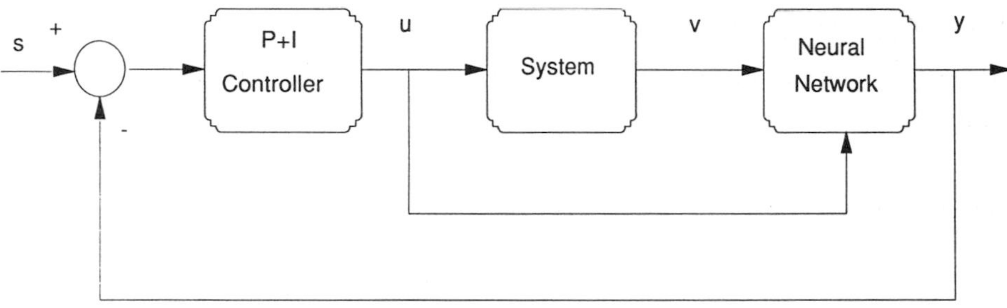

Fig. 7. Inferential control scheme.

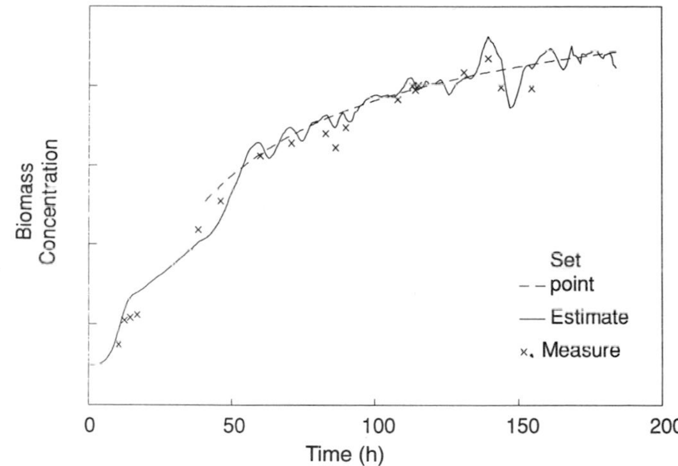

Fig. 8. Inferential control of biomass concentration.

model to predict the controlled process output at a pre-determined point of time in the future. Initial algorithm developments considered fixed-step-ahead prediction where the horizon is the effective time delay of the system. More recent, alternative approaches have adopted a multi-step-ahead prediction philosophy and are sometimes referred to as predictive controllers (e. g., long-range predictive control). Both fixed and multi-step-ahead controllers can be cast in adaptive or non-adaptive forms. Such control schemes have been shown to be capable of coping with and, to a greater or lesser degree, overcoming the complexities and control difficulties seen in biotechnological processes.

There are no clear guidelines available as to which algorithm is the most suitable for the particular type of process being studied, although most interest now centers around long-range predictive or linear quadratic forms. ÅSTROM and WITTENMARK (1984) gave a good overview of computer control systems and adaptive control, while BASTIN and DOCHAIN (1990) provided a good analysis and study of estimation and adaptive control for bioreactors.

There have been a significant number of studies of adaptive and self-tuning approaches to bioreactor control, and it is not possible to mention them all. The following studies, however, serve to highlight some of the early approaches: PERINGER and BLACHÈRE (1978), TAKAMATSU et al. (1979), DOCHAIN and BASTIN (1982, 1985), DAIRAKU et al. (1982), WU et al. (1985), POULISSE and VAN HELDEN (1985), FRUEH et al. (1985), DEKKERS and VOETTER (1985), MONTGOMERY et al. (1985), VERBRUGGEN et al. (1985), DOCHAIN (1986), and MONTAGUE et al. (1986a, b). Almost all the studies are concerned with yeast fermentations, although FRUEH et al. (1985) and MONTAGUE et al. (1986a, b) have investigated penicillin fermentation. DOCHAIN and BASTIN (1984) initially tried to use a minimum variance control policy before taking account of control effort by adopting a self-tuning GMV control law of the CLARKE and GAWTHROP (1979) type; while POULISSE and VAN HELDEN (1985), DEKKERS and VOETTER (1985) and MONTGOMERY et al. (1985) utilized LQ (Linear Quadratic) and LQG (Linear Quadratic Gaussian) methods of adaptive control. FRUEH et al. (1985) used a minimum variance type of self-tuning control law with a cascaded PI compensator, while VERBRUGGEN et al. (1985) applied a pole placement self-tuning controller with an outer loop integrator. Almost without exception, the models used for control law derivation have been of the Controlled Autoregressive Moving Average (CARMA) form leading to the problem of including integral action in the resulting control law. In the work of MONTAGUE et al. (1986a, b) integral action is inherently included in the controller by use of a CARIMA representation of the process with a Brownian motion type disturbance model.

Two examples of multivariable adaptive control are provided by the simulation and pilot plant studies of WILLIAMS et al. (1984, 1986), and the simulation studies of ANDERSEN and JORGENSEN (1988), in order to maximize productivity in a yeast fermentation. The studies of MONTAGUE et al. (1986a, b) on the control of biomass in a fed-batch penicillin fermentation serve to demonstrate the importance of the correct choice of control algorithm. Here, a Generalized Predictive Controller (GPC) was used to control biomass estimated by an extended Kalman filter. Of special interest in bioreactor control are the profile Model Reference Adaptive Control (MRAC) studies of SHIOYA et al. (1985). They suggest a programmed controller to follow the desired (biomass concentration or specific growth rate) profile, with a feedback compensator to regulate the system against disturbances.

Recent work has further studied and extended adaptive-predictive control approaches and applied the techniques to a wider range of bioprocesses. For example, VIGIE et al. (1990) presented theoretical and experimental studies examining the application of adaptive techniques to the control of effluent substrate concentration in a continuous multistage bioreactor fermenting glucose by yeast. LANDAU et al. (1990) provided a useful 'tutorial paper' on the development and design of robust performance-oriented adaptive control for biotechnological processes. The control methodology presented made use of control laws based upon the minimization of quadratic performance criteria combined with pole placement and robust parameter identification. Both continuous and fed-batch fermentations were addressed. SAMAAN et al. (1990) have experimentally studied the application of this performance-oriented robust adaptive control algorithm to a continuous alcohol fermentation (*Saccharomyces cerevisiae*). These two studies, together with their references to earlier developmental studies of the identification and control philosophies, provide a valuable contribution to the design of adaptive control methods for bioprocesses.

An interesting application of linear quadratic control to the continuous fermentation of *Saccharomyces cerevisiae* was presented by ANDERSEN et al. (1991). They also considered an alternative approach of robust controller design. Here a linear time-invariant process model is assumed and any model–process mismatch (model uncertainty) is accounted for either as part of the design process or as part of an on-line tuning exercise (see MORARI and ZAFIRIOU, 1989). A potential problem with the approach is that a knowledge of the process uncertainties is required and these will only tend to reflect the expected variations in process behavior, otherwise either a too conservative or too optimistic controller design will result. The well known Internal Model Control (IMC) approach was compared with both non-adaptive and adaptive linear quadratic control on a simulation of the continuous yeast fermentation pilot plant. Similar controlled performances were obtained for the PI, non-adaptive LQ and IMC control except for disturbance rejection where the IMC and PI controllers were observed to result in a more sluggish response. The study indicated that an improved PI controller performance could be obtained by using on-line model-based tuning. However, in order to be able to cope with changes in fermentation dynamics, the controller gain had to be reduced.

3.3.4 Non-linear Control of Fermentation Processes

The adaptive control approaches discussed above primarily utilize a linear time series model of the fermentation. It is, however, feasible to use knowledge of the non-linear behavior of a bioprocess to derive a non-linear controller. Since the underlying properties of the fermentation process are non-linear, improved control performance might be expected by explicitly taking into account known process non-linearities in the controller design. BASTIN and DOCHAIN (1988, 1990) have developed input–output linearizing control laws for the SISO bioreactor case and extended the method to allow for control adaptation to cope with bioprocess non-linearity and parameter uncertainty. Theoretical stability and convergence aspects were also extensively studied. A number of successful related control experiments were later reported by other workers. The adaptive linearizing control approach has recently been extended to include MIMO stirred tank bioreactors with equal number of inputs and outputs (DOCHAIN, 1991). The application of adaptive linearizing control, however, is limited to non-minimum phase processes. DOCHAIN and PERRIER (1991) considered this problem and proposed two possible solutions. One approach extends the continuous time form of the Generalized Minimum Variance controller (GMV), while the second proposes a linearizing controller designed using singular perturbation arguments. The two approaches are compared by simulation studies of a simple microbial growth process and an anaerobic digester process. POMERLLEAU et al. (1989) and RAMSEIER (1991) described studies of SISO and MIMO adaptive versions of non-linear Generic Model Control (GMC) (LEE and SULLIVAN, 1988). Experimental studies are restricted to the SISO case using the continuous fermentation of baker's yeast. An *a priori* non-linear representation of the bioprocess is combined with a simple adaptive scheme in order to help handle the bioprocess non-linearities. The incorporation of *a priori* bioprocess information allows reliance on adaptation to be reduced. The superior performance of the adaptive GMC algorithm over that using PI control was demonstrated. In addition, the disturbance rejection properties of the controller were studied and shown to be satisfactory.

An interesting approach to non-linear control of bioreactors was proposed by LAKRORI and CHERUY (1988) which addressed the non-linearity problem by taking into account the positivity constraint on control and state variables by a non-linear transformation (the logarithmic transformation). This allows the controller to be designed for the 're-scaled' process and for it to take a conventional PID structure. The controller is termed an L/A controller. This idea was further evaluated by BEATEAU et al. (1991) by simulation studies and application to an anaerobic digester pilot plant. It has been shown that L/A control design is straightforward and provides performances similar to those achieved by other more advanced non-linear controller design strategies.

Most recently, there has been an increasing interest in identifiability and observability studies for bioprocesses. CHEN and BASTIN (1991) described an approach which develops a general state space model of a bioreactor and goes on to study the identifiability problems, and their solution, associated with the yield parameters and kinetic parameters. The major identifiability problems arise in the modeling of the kinetics and the impact of the non-linear bioprocess characteristics.

Although a number of successful simulation and pilot plant studies of non-linear controllers have been reported, as exemplified by those referenced above, it is fair to observe that in general the algorithms still need to be 'engineered' to the point where robust consistent control can be achieved without operator intervention or supervision.

3.4 Optimization and Optimizing Control of Fermentations

The high costs associated with many fermentation processes make optimization of bioreactor performance very desirable. Some bioprocesses make use of expensive raw materials and involve capitally intensive plants. To a lesser extent the minimization of production costs of bulk production, low value added fermentations also provide impetus to optimize bioreactor performance. In both cases small improvements in performance can yield potentially high dividends. For example, a fermentation such as yeast might be optimized in order to maximize a yield of biomass, whereas a secondary metabolite fermentation can be controlled to maximize antibiotic production. Optimization is not only concerned with product maximization; other process considerations are essential, for example, production costs and fermentation time span. Optimization of present industrial fermentations has been predominantly carried out on an empirical basis. Although this has resulted in considerable improvements in operability, the application of modern mathematical optimization theory offers the potential of even greater benefits. Two good overviews setting out the background and principles involved can be found in CONSTANTINIDES (1979) and ARKUN and STEPHANOPOULOS (1980).

The success of optimizing the performance of any system depends upon the availability of a representative mathematical model of the process and the selection of an appropriate optimization routine. Optimization is essentially a mathematical procedure, the complexity of which depends upon the nature of the problem formulation and the process requirements such as speed of computation of solution. In order to obtain optimum results, a very accurate description of the process is necessary. Models that describe the process faithfully are often quite complex. In addition, process constraints must be taken into account such as oxygen limitations, temperature limits, and raw material limitations. All this increases the complexity of the problem, and relatively sophisticated algorithms are required to provide a solution. If, however, relatively simple models are used, it is possible that sub-optimal results will be obtained. A further difficulty arises in that external uncertainties which affect the process performance cannot be accurately modeled. Any optimization routine should embody as simple a mathematical representation as possible, include a procedure to take into account uncertainties and unforeseen process disturbances, and provide for relatively simple implementation procedures.

A number of the mathematical models discussed previously for estimation and control purposes can be, and have been, applied in fermentation optimization studies. For example, CONSTANTINIDES (1979) and CONSTANTINIDES and RAI (1974) used Pontryagin's Maximum Principle to predict the optimum temperature profile for the maximization of penicillin production. CONSTANTINIDES has shown that by using model optimization, increases in yield of 16% above the highest yield obtained with the optimal constant temperature can be achieved. BAJPAI and REUSS (1981) optimized penicillin production by varying different feeding profiles and concluded that productivity is relatively insensitive to feeding variations. The importance of rigorous optimization was stressed by SAN and STEPHANOPOULOS (1989). In their reassessment of the same problem using Pontryagin's Maximum Principle they showed significant increases in penicillin productivity. FISHMAN and BIRYUKOV (1982) applied the continuous maximum principle approach to the optimization of a fed-batch penicillin fermentation, and TAKAMATSU et al. (1975) applied the continuous maximum principle and Green's theorem to maximize the production of amino acid production and minimize transient time in a continuous fermentation. PROELL et al. (1991) compared different optimization and control schemes by application to an industrial-scale microalgae fermentation. The methods of Pontryagin's Maximum Principle, Feedback Sub Optimum, and Direct Search were used to maximize cellular productivity. Adaptive globally linearizing control and adaptive control techniques were then used to control the multivariable continuous fermentation. All the optimization algorithms produced similar results with the adaptive control scheme producing superior controlled performances.

BELLGARDT et al. (1991) provided an interesting approach to the optimum quality control of the fed-batch baker's yeast process using optimal substrate feeding strategies. A cell cycling model was proposed, and verified using experimental data, and used to form the basis of the optimal control scheme. A profit function was defined which incorporated economic indices such as quality, productivity, and yield. The profit function was maximized using the simplex method of NELDER and MEAD to provide an optimal time profile manipulation for feed flow rate, together with optimal values for the total sugar concentration in the substrate inflow, the gas volumetric flow rate, and the fermentation period. VAN BREUSEGEM and BASTIN (1990) have developed an optimal temperature profile scheme to maximize final biomass for a mixed culture bioprocess. They have shown that, under realistic modeling assumptions, the optimal temperature profile is constant whatever the structure of the specific growth rates providing that the admissible control is not bounded. KURTANJEK (1991) has studied the optimal, non-singular, control for a fed-batch fermentation using the orthogonal collocation approach. Optimal profiles for fermentation temperature, substrate concentration in the feed, feeding profile, and fermentation duration were generated. It was shown that the multivariable optimization approach adopted could be implemented in low cost hardware, is capable of handling constraints on both control and state variables, and can be incorporated into feedforward/feedback control strategies.

An alternative optimization approach, closely linked with the adaptive estimation and control ideas discussed previously, is the use of a generally structured linear model which adapts as changes in the process occur. This adaptation permits the tracking of process dynamics and hence provides a means by which on-line optimization can be achieved. GOLDEN and YDSTIE (1987) demonstrated the methodology in studies aimed at optimizing the yield of yeast in a continuous fermentation. PETERSEN and WHYATT (1990) used a dynamic optimization procedure, based upon the adaptive identification of a second-order Hammerstein model to optimize the production rate of a continuous bioreactor. KAMBHAMPATI et al. (1992) used an adaptive time series model to continuously optimize a yeast fermentation. This particular paper also analyzed the convergence conditions for an optimal solution.

3.5 Knowledge-Based Systems in Fermentation Supervision and Control

Although the algorithmic approaches described above can offer much insight and provide for improved process operability, the comprehensive management of a fermentation process requires more than simply an algorithmic approach at every level. 'Expert' and 'Knowledge-Based' Systems (KBS), as replacement experts, may not be the complete answer, but the artificial Intelligence (AI) techniques developed in the last twenty years do offer a number of solution strategies. Applications have been reported in, for example, wastewater treatment, antibiotic, enzyme, amino acid, yeast, and bacterial biomass production. An important early contribution was made by STEPHANOPOULOS and STEPHANOPOULOS (1986), who described the development of AI for biochemical process design. Quite a large number of approaches have since been proposed aimed at using knowledge-based system approaches to improve the quality of information presented to operators (e. g., KARIM and HALME, 1988; FOWLER et al., 1992) and to increase the level of automatic process supervision. COONEY et al. (1988) described an expert system for the supervisory control (equipment and metabolic fault detection and diagnosis) of a fermentation system. HALME and VISALA (1992) described a combined symbolic and numeric approach to the modeling of a bioprocess which builds upon the concept of functional state (conventional bioreactor state plus functional state describing phase of operation, etc.) set out by HALME (1988). ALFORD et al. (1992) outlined an intelligent remote alarming system for an industrial fermentation pilot plant which performs automated data analysis, validity checking, process troubleshooting, and harvest time determination. Knowledge-based systems can be designed to cope with uncertainty and allow the coupling of quantitative information with the qualitative or symbolic expressions (in the form of heuristics) so as to reproduce the actions of an experienced process operator. They might, therefore, be used as an intelligent, on-line assistant to the process operator, or, with extra knowledge as a supervisor in running and maintaining bioprocess operations within optimal operating conditions. An interesting paper by STEYER et al. (1992) describes a real-time application of AI for fermentation processes which exploits the features and advantages of qualitative physics to aid the on-line monitoring and understanding of bioprocess operations. AI may also have a wider role in the design and control of the biological processes actually taking place within the fermentation (SERESSIOTIS and BAILEY, 1988).

Real-time expert systems software environments are now becoming available, an excellent example being Gensym's G2 (MOORE and KRAMER, 1986; MOORE et al., 1987). An example of an integrated system for improved fermentation supervisory control, using a KBS approach, has been proposed by AYNSLEY et al. (1989, 1990). The system, Bio-SCAN (Bioprocess Supervision Control and Analysis), is a real-time KBS developed using the G2 Real-Time Expert System (MOORE, 1988). It has been designed to provide a general purpose framework for the monitoring and supervisory control of fermentation processes (MORRIS et al., 1991). The system is shown schematically in Fig. 9. Internally, the system incorporates knowledge of operating strategies (e. g., feeding regimes, set-points) and expected behaviors to enable rapid detection of any process faults.

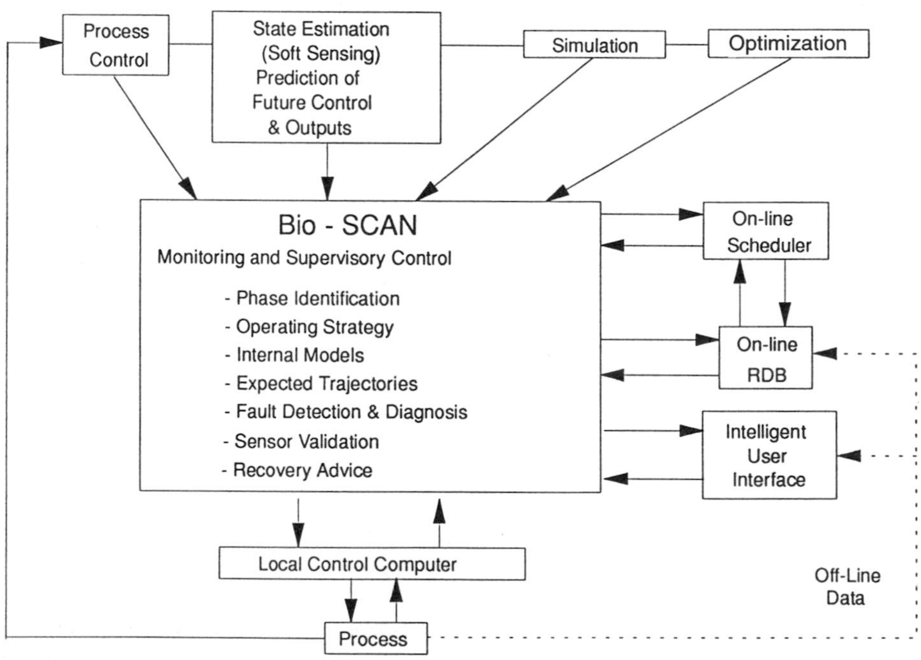

Fig. 9. Scheme of a knowledge-based system.

When faults do occur, the system attempts to diagnose the cause of the problem and offers recovery advice to process operators.

Since different fermentation processes are defined by different operating philosophies, specific knowledge (rule sets) are clearly required for optimal supervisory control. These "supervisory rule sets" are used to govern the general operation of the fermentation by identifying growth or production phases and invoking other specific rule sets accordingly (i. e., supervisory rule sets function as meta-rules). The supervisor is responsible for temperature/pH profiling, feed scheduling and on-line modifications to the schedule in response to changes in the process (i. e., faults). While specific rule sets are necessary in order to assess and respond to the metabolic state of a process organism in any given fermentation, for the diagnosis of hardware faults it is possible to write more generic rules, applicable across the range of fermentations under Bio-SCAN control. The diagnostic knowledge has also been structured into a number of discrete rule sets, and meta-rules again determine which rule set should be accessed next. The decision to invoke a particular rule set is based upon a comparison of the current raw, or processed data with that permitted or expected in the current phase of the fermentation. This ensures that knowledge can be focused in response to the occurrence of specific events.

Once a fault has been detected, the operator can be informed of the problem on a message board, and a list of possible causes is then presented together with advice on how to correct the fault, or confirm a diagnosis, if Bio-SCAN cannot solve the problem automatically. This is possible if the fermentation is sufficiently overdetermined, since the KBS capitalizes on redundancies to test for data consistencies and error identification. The source of these redundancies are usually balances or load cells, although estimation techniques can be used to reduce the need for hardware redundancy.

While a knowledge-based system, such as that described above, is expected to be capable of adequately supervising bioprocesses under most circumstances, the utility of the system can be greatly extended by interfacing to conventional real-time algorithmic methods of analyzing and controlling bioreactions. For example, the following features are important:

- Measured signal conditioning, statistical screening and interrogation to place confidence bounds around data
- Simulation using process models in imported subroutines
- Identification of process model parameters using on-line recursive parameter estimation algorithms
- Estimation of hidden states using 'soft sensors'
- Prediction of future process outputs given pre-specified open-loop feeding regimes. Use of adaptive–predictive control schemes to achieve closed-loop profiling of bioprocess states
- On-line constrained optimization using on-line identified process models that capture the essential non-linear structure of the bioprocess
- Continuous statistical monitoring and process control to provide interactive assessment of overall bioreaction performance against prespecified specifications
- Neural network modeling of non-linear bioprocess models for hidden state estimation, fault detection, and predictive control.

A useful minireview on knowledge-based control of fermentation processes has recently been published by Konstantinov and Yoshida (1992). However, in spite of the increasing number of publications on the whole area of KBS in bioprocesses, the number of industrial-scale applications remains quite small. Integration of lower-level control algorithms and the utilization of qualitative and quantitative biological data still remains a major goal in KBS bioprocess supervision and control.

4 Discussion

As the importance of biotechnologically based industries increases in the market place, there is a growing need to operate bioprocesses in a cost-effective manner. While the original route to process improvement was through strain development, an alternative complementary approach discussed above is that of improved process supervision and control. As outlined, these techniques provide quite significant potential for improvements in process operability and hence increased profitability. An important issue which this chapter has not addressed is the validation of process supervision and control techniques in order to satisfy regulatory authorities. ALFORD (1991) provided an interesting review of automatic control systems with reference to FDA regulations. Although the techniques and methodologies have been discussed with respect to their application to fermentation processes, the underlying fundamental theories, and to a greater or lesser extent their practice, are directly applicable to a wider spectrum of bioprocess systems, for example, wastewater treatment. The discussions on fermentation system control have made wide reference to four important IFAC meetings (HALME, 1982; JOHNSON, 1985; FISH and FOX, 1988; KARIM and STEPHANOPOULOS, 1992). These conference proceedings provide a valuable source of reference for modeling and control studies on the whole range of biotechnological processes.

Acknowledgements
The authors would like to acknowledge the support of Zeneca BioProducts, SmithKline Beecham, the U.K. Science and Engineering Research Council and the Department of Chemical and Process Engineering, University of Newcastle upon Tyne.

5 References

AIBA, S., NAGAI, S., NISHIZAVO, Y. (1976), Fed-batch culture of *Saccharomyces cerevisiae*: a perspective of computer control to enhance the productivity of baker's yeast cultivation, *Biotechnol. Bioeng.* **18**, 1001–1016.

ALFORD, J. (1991), Automatic control systems, in: *Drug Biotechnology Regulation, Scientific Basis and Practices* (CHIU, Y. GUERIGUIAN, J., Eds.), Washington, DC: US Food and Drug Administration.

ALFORD, J., FOWLER, G., HIGGS, R., CLAPP, D., HUBER, F. (1992), Development of real-time expert system applications for the on-line analysis of fermentation data, *Proc. 9th Int. Biotechnology Symposium and Exposition*, August, Washington, DC: American Chemical Society.

ANDERSEN, M., JORGENSEN, S. B. (1988), Multivariable adaptive control of a simulated continuous yeast fermentation, *IFAC Adaptive Control of Chemical Processes Conference*, Copenhagen.

ANDERSEN, M. Y., BRABRAND, H., JORGENSEN, S. B. (1991), Model based control of a continuous yeast fermentation, *Proc. Am. Control Conf.*, Boston, June, pp. 1329–1334.

ANDERSON, B. D. O., MOORE, J. L. (1979), *Optimal Filtering*, Englewood Cliffs, NJ: Prentice Hall.

ARKUN, Y., STEPHANOPOULOS, G. (1980), Optimizing control of industrial chemical processes: state of the art review, *Proc. Joint Atom. Control Conf.*, Paper WP5-A.

ASTROM, K. J., HAGGLUND, T. (1984), Automatic tuning of simple regulators with specifications on phase and amplitude margins, *Automatica* **20**, 645–651.

ASTROM, K. J., WITTENMARK, B. (1984), *Computer Controlled Systems*, Englewood Cliffs, NJ: Prentice Hall.

AYNSLEY, M., PEEL, D., MORRIS, A. J., MONTAGUE, G. A. (1989), A real time knowledge based system for fermentation control, *Proc. Am. Control Conf.*, Pittsburgh, pp. 2239–2244.

AYNSLEY, M., HOFLAND, A. G., MONTAGUE, G. A., PEEL, D., MORRIS, A. J. (1990), A real-time knowledge based system for the operation and control of a fermentation plant, *Proc. Am. Control Conf.*, San Diego, pp. 1992–1997.

BAJPAI, R. K., REUSS, M. (1980), A mechanistic model for penicillin production, *J. Chem. Tech. Biotechnol.* **30**, 332–344.

BAJPAI, R. K., REUSS, M. (1981), *Biotechnol. Bioeng.* **23**, 717.

BASTIN, G., DOCHAIN, D. (1988), Nonlinear adaptive control algorithms for fermentation processes, *Proc. Am. Control Conf.*, pp. 1124–1129.

BASTIN, G., DOCHAIN, D. (1990), *On-line Estimation and Adaptive Control of Bioreactors*, Amsterdam: Elsevier.

BEATEAU, J. F., FERRET, E. V., LAKRORI, M. L., CHERUY, A. C. (1991), Bioprocess control: An original approach taking into account some bioprocess constraints, *Proc. Am. Control Conf.*, Boston, pp. 1335–1340.

BELLGARDT, K. H., YUAN, J. Q., JIANG, W. S., DECKWER, W. D. (1991), Optimum quality control of a baker's yeast production process, *Proc. Eur. Control Conf.*, Grenoble, pp. 230–235.

BIROU, B., MARISON, I. W., VON STOCKAR, U. (1987), Calorimetric investigations of aerobic fermentations, *Biotechnol. Bioeng.* **30**, 650–660.

BIRYUKOV, V. V. (1982), Computer control and optimization of microbial metabolite production, *Proc. 1st IFAC Workshop on Modelling and Control of Biotechnical Processes* (HALME, A., Ed.), August, Helsinki, Finland, pp. 135–144, Oxford: Pergamon Press.

BOECK, L., WETZL, R., BURT, S., HUBER, F., FOWLER, G., ALFORD, J. (1988), Sterilization of bioreactor media on the basis of computer calculated thermal input designated as Fo, *J. Ind. Microbiol.* **3**, 305–310.

BROOKS, S. L., TURNER, A. P. F. (1987), Biosensors for measurement and control, *Trans. Inst. Meas. Control* **20**, 37–43.

BUCKLAND, B., BRIX, T., FASTERT, H., GBEWONYO, K., HUNT, G., JAIN, D. (1985), Fermentation exhaust gas analysis using mass spectrometry, *Biotechnology* **3**, 982–992.

CALAM, C. T., ISMAIL, B. A. K. (1980), Investigation of factors in the optimization of penicillin production, *J. Chem. Technol. Biotechnol.* **30**, 249–262.

CARLEYSMITH, S. W., FOX, R. I. (1984), Fermenter instrumentation and control, *Adv. Biotechnol. Processes* **3**, 1–51.

CARR, R. J. G., BROWN, R. G. W., RARITY, J. G., CLARKE, D. J. (1987), Laser light scattering and related techniques, in: *Biosensors: Fundamentals and Applications*, (TURNER, A. P. F., KARUBE, I., WILSON, G. S., Eds.), pp. 679–701, Oxford: Oxford University Press.

CHEN, L., BASTIN, G. (1991), On the model identifiability of stirred tank bioreactors, *Proc. Eur. Control Conf.*, Grenoble, pp. 242–247.

CLARKE, D. W., GAWTHROP, P. J. (1979), Self-tuning control, *Proc. Inst. Electr. Eng. Part D* **126**(6), 633–640.

CLARKE, D. J., KELL, D. B., MORRIS, J. G., BURNS, A. (1982), The role of ion-selective electrodes in microbial process control, *Ion-selective Electrodes Rev.* **4**, 75–131.

CLELAND, N., ENFORS, S. O. (1983), Control of glucose-fed batch cultivations of *E. coli* by means of an oxygen stabilized electrode, *Eur. J. Appl. Microbiol. Biotechnol.* **18**, 141–147.

CLELAND, N., ENFORS, S. O. (1984), Externally buffered enzyme electrode for determination of glucose, *Anal. Chim. Acta* **163**, 281–285.

CONSTANTINIDES, A. (1979), Application of rigorous optimization methods to the control and operation of fermentation processes, *Ann. NY Acad. Sci.* **326**, 193–221.

CONSTANTINIDES, A., RAI, V. R. (1974), Application of the continuous maximum principle to fermentation processes, *Biotechnol. Bioeng. Symp.* **4**, 663–680.

CONTOIS, D. E. (1959), Kinetics of bacterial growth: Relationship between population density and specific growth rate of continuous cultures, *J. Gen. Microbiol.* **21**, 40–50.

COONEY, C. L., SWARTZ, J. R. (1982), Application of computer control to yeast fermentation, *Proc. 1st IFAC Workshop on Modelling and Control of Biotechnical Processes* (HALME, A., Ed.), August, Helsinki, Finland, pp. 243–252, Oxford: Pergamon Press.

COONEY, C. L., WANG, Y. W., WANG, D. I. C. (1977), Computer aided material balancing for prediction of fermentation parameters, *Biotechnol. Bioeng.* **19**, 55–67.

COONEY, C. L., O'CONNOR, G. M, SANCHEZ-RIERA, F. (1988), An expert system for intelligent supervisory control of fermentation processes, Preprints of *8th Int. Biotechnol. Symp.*, Paris, July.

COPELLA, S. J., DHURJATI, P. (1987), Low cost computer-coupled fermentor off-gas analysis via quadrupole mass spectrometer, *Biotechnol. Bioeng.* **29**, 679–689.

CYBENKO, G. (1989), Approximation by superpositions of a sigmoidal function, *Mathematics of Control, Signals and Systems* **2**, 303–314.

DAIRAKU, K., IZUMOTO, E., MORIKAWA, H., SHIOYA, S., TAKAMATSU, T. (1982), Optimal quality control of baker's yeast fed-batch culture using population dynamics, *Biotechnol. Bioeng.* **24**, 2661–2674.

DEKKERS, R. M. (1982), State estimation of a fed-batch baker's yeast fermentation, *Proc. 1st IFAC Workshop on Modelling and Control of Biotechnical Processes* (HALME, A., Ed.), August, Helsinki, Finland, pp. 201–211, Oxford: Pergamon Press.

DEKKERS, R. M., VOETTER, M. (1985), Adaptive control of a fed-batch baker's yeast fermentation, *Proc. 1st IFAC Symposium on Modelling and Control of Biotechnological Processes* (JOHNSON, A., Ed.), December, Noordwijkerhout, The Netherlands, pp. 103–110, Oxford: Pergamon Press.

DOCHAIN, D. (1986), On-line parameter estimation, adaptive state estimation and adaptive control of fermentation processes, *Ph. D. Thesis*, University of Louvain, Belgium.

DOCHAIN, D. (1991), Design of adaptive controllers for nonlinear stirred tank bioreactors: extension to the MIMO situation, *J. Proc. Control* **1**, 41–48.

DOCHAIN, D., BASTIN, G. (1982), Adaptive identification and control algorithms for nonlinear bacterial growth systems, *Automatica* **20**, 621–634.

DOCHAIN, D., BASTIN, G. (1985), Stable adaptive algorithms for estimation and control of fermentation processes, *Proc. 1st IFAC Symposium on Modelling and Control of Biotechnological Processes* (JOHNSON, A., Ed.), December, Noordwijkerhout, The Netherlands, pp. 1–6, Oxford: Pergamon Press.

DOCHAIN, D., PERRIER, M. (1991), Nonlinear adaptive controllers for nonminimum phase bioreactors, *Proc. Am. Control Conf.*, Boston, pp. 1311–1316.

FISH, N., FOX, R. I. (Eds.) (1988), *Proceedings of the 4th International Congress on Computer Applications in Fermentation Technology - Modelling and Control of Biotechnical Processes*, SCI/IFAC, September, Cambridge, England.

FISHMAN, V. M., BIRYUKOV, V. V. (1974), Kinetic model of secondary metabolite production and its use in computation of optimal conditions, *Biotechnol. Bioeng.* **4**, 647–662.

FLYNN, D. S. (1982), Instrumentation for fermentation processes, *Proc. 1st IFAC Workshop on Modelling and Control of Biotechnical Processes* (HALME, A., Ed.), August, Helsinki, Finland, pp. 5–12, Oxford: Pergamon Press.

FOWLER, G., ALFORD, J., HIGGS, R. (1992), Development of real-time expert systems approach for the on-line analysis of fermentation respiration data, *Proc. 2nd IFAC Symposium on Modelling and Control of Biotechnical Processes*, Keystone, Colorado, April.

FRUEH, K., LORENZ, TH., NIEHOFF, J., DIEKMANN, J., HIDDESSEN, R., SCHÜGERL, K. (1985), On-line measurement and control of penicillin V production, *Proc. 1st IFAC Conference on Modelling and Control of Biotechnological Processes* (JOHNSON, A., Ed.), December, Noordwijkerhout, The Netherlands, pp. 45–48, Oxford: Pergamon Press.

GOLDEN, M. P., YDSTIE, B. E. (1987), Non-linear adaptive optimization of continuous bioreactors, *Proc. AIChE Meeting*, Miami, pp. 356–361.

GUILANDOUST, M. T., MORRIS, A. J., THAM, M. T. (1987), Adaptive inferential control, *Proc. Inst. Electr. Eng. Part D* **134**(3), 171–179.

GUILANDOUST, M. T., MORRIS, A. J., THAM, M. T. (1988), An adaptive estimation algorithm for inferential control, *Ind. Eng. Chem. Res.* **27**, 1658–1664.

HALME, A. (Ed.) (1982), *Proceedings of the 1st IFAC Workshop on Modelling and Control of Biotechnical Processes*, August, Helsinki, Finland, Oxford: Pergamon Press.

HALME, A. (1988), Expert system approach to recognize the state of fermentation to diagnose faults in bioreactors, *Proc. 4th Int. Cong. on Computer Applications in Fermentation Technology - Modelling and Control of Biotechnical Processes* (FISH, N., FOX, R., Eds.), SCI/IFAC, September, Cambridge, England.

HALME, A., SELKAINAHO, A. (1982), Application of a non-linear filter to multivariable parameter adaptive control in a distributed micro computer, *Proc. 6th IFAC Symp. on Identification and System Parameter Estimation*, Washington, DC.

HALME, A., VISALA, A. (1991), Combining symbolic and numerical information in modelling the state of biotechnological processes, *Proc. Eur. Control Conf.*, Grenoble, pp. 218–223.

HALME, A., KUISMIN, R., KORTENIEMI, M. (1985), A method to consider delayed laboratory analysis in state and parameter estimation of bioreactors, *Proc. 1st IFAC Symp. on Modelling and Control of Biotechnological Processes* (JOHNSON, A., Ed.), December, Noordwijkerhout, The Netherlands, pp. 179–184, Oxford: Pergamon Press.

HARRIS, C. M., KELL, D. B. (1985), The estimation of microbial biomass, *Biosensors* **1**, 17–84.

HOLMBERG, U., OLSSON, G. (1985), Simultaneous on-line estimation of oxygen transfer rate and respiration rate, *Proc. 1st IFAC Symp. on Modelling and Control of Biotechnological Processes* (JOHNSON, A., Ed.), December, Noordwijkerhout, The Netherlands, pp. 185–189, Oxford: Pergamon Press.

HOLMBERG, A., RANTA, J. (1982), Procedures for parameter and state estimation of microbial growth process models, *Automatica* **18**, 181–193.

HORNIK, K., STINCHCOMBE, M., WHITE, H. (1989), 'Multilayer feedforward networks are universal approximators', in: *Neural Networks*, Vol. 2, pp. 359–366.

HUMPHREY, A. E., JEFFREYS, P. (1973), Invited lecture presented at the *IV GIAM Meeting*, Sao Paulo, Brazil.

ISHIMORI, Y., KARUBE, I., SUZUKI, S. (1981), Determination of microbial populations with piezoelectric membranes, *Appl. Environ. Microbiol.* **42**, 632–637.

JAZWINSKI, A. H. (1970), *Stochastic Processes and Filtering Theory*, New York: Academic Press.

JOHNSON, A. (Ed.) (1985), *Proc. 1st IFAC Symp. on Modelling and Control of Biotechnological Processes*, December, Noordwijkerhout, The Netherlands, Oxford: Pergamon Press.

JOHNSON, A. (1987), The control of fed-batch fermentation processes - a survey, *Automatica* **23**(6), 691–705.

JONES, K. O., WILLIAMS, D., PHIPPS, D., MONTGOMERY, P. A. (1991), On-line tuning PID control of dissolved oxygen, *Proc. IFAC Symp. on*

Advanced Control of Chemical Processes, ADCHEM '91, Toulouse, France, pp. 41–46.

KAMBHAMPATI, C., THAM, M. T., MONTAGUE, G. A., MORRIS, A. J. (1992). Optimizing control of fermentation processes, *Proc. Inst. Electr. Eng. Part D* 60–66.

KARIM, M. N., HALME, A. (1988), Reconciliation of measurement data in fermentation using on-line expert system, *Proc. 4th Int. Cong. on Computer Applications in Fermentation Technology – Modelling and Control of Biotechnical Processes* (FISH, N., FOX, R. I., Eds.), SCI/IFAC, September, Cambridge, England.

KARIM, N., STEPHANOPOULOS, G. (Eds.) (1992), *2nd IFAC Symposium on Modelling and Control of Biotechnical Processes*, Keystone, Colorado, April.

KARUBE, I. (1984), Possible developments in microbial and other sensors for fermentation control, *Biotechnol. Genet. Eng. Rev.* **2**, 313–339.

KELL, D. B. (1987), The principles and potential of electrical admittance spectroscopy, in: *Biosensors: Fundamentals and Applications* (TURNER, A. P. F., KARUBE, I., WILSON, G. S., Eds.), Oxford: Oxford University Press.

KISHIMOTO, M., SAWANO, T., YOSHIDA, T., TAGUCHI, H. (1982), Optimization of a fed-batch culture by statistical data analysis, *Proc. 1st IFAC Workshop on Modelling and Control of Biotechnical Processes* (HALME, A., Ed.), August, Helsinki, Finland, pp. 161–168, Oxford: Pergamon Press.

KONSTANTINOV, K. B., YOSHIDA, T. (1992), Knowledge-based control of fermentation processes, Mini Review, *Biotechnol. Bioeng.* **39**, 479–486.

KURTANJEK, Z. (1991), Optimal nonsingular control of fed-batch fermentation, *Biotechnol. Bioeng.* **37**, 814–823.

LAKRORI, M., CHERUY, A. (1988), A new nonlinear adaptive approach to automatic control of bioprocesses, *Proc. 4th Int. Cong. on Computer Applications in Fermentation Technology – Modelling and Control of Biotechnical Processes* (FISH, N., FOX, R. I., Eds.), SCI/IFAC, September, Cambridge, England.

LANDAU, I. D., SAMAAN, M., M'SAAD, M. (1990), Robust performance oriented adaptive control for biotechnological processes, *Proc. Am. Control Conf.*, San Diego, pp. 2684–2687.

LATIMER, P. (1982), Light scattering and absorption methods of studying cell population parameters, *Annu. Rev. Biophys. Bioeng.* **11**, 129–150.

LEE, P., SULLIVAN, G. (1988), Generic model control, *Comput. Chem. Eng.* **12**, 573–580.

LEES, F. P. (1976), The reliability of instrumentation, *Chem. Ind.* **5**, 195–205.

LEIGH, J. R., NG, M. H. (1984), Estimation of biomass and secondary product in batch fermentation, *Proc. 6th Int. Conf. on Analysis and Optimization of Systems*, Nice, France, pp. 19–22.

LJUNG, L. (1979), Asymptotic behavior of the extended Kalman filter as a parameter estimator for linear systems, *IEEE Trans. Autom. Control* AC-**24**, 36–50.

MANDENIUS, C. F., DANIELSSON, B., MATTIASSON, B. (1984), Evaluation of a dialysis probe for continuous sampling in fermenters and in complex media, *Anal. Chim. Acta* **163**, 135–141.

MEYER, H.-P., KÄPPELI, O., FIECHTER, A. (1985), Growth control in microbial cultures, *Annu. Rev. Microbiol.* **39**, 299–319.

MONOD, J. (1950), La technique de culture continue, théorie et applications, *Ann. Inst. Pasteur* **79**, 390–410.

MONTAGUE, G. A. (1987), Inferential self-tuning control of the fed-batch penicillin fermentation. *Ph. D. Thesis*, University of Newcastle-upon-Tyne, England.

MONTAGUE, G. A., MORRIS, A. J., WRIGHT, A. R., AYNSLEY, M., WARD, A. C. (1986a), On-line estimation and adaptive control of penicillin fermentation, *Proc. Inst. Electr. Eng. Part D* **133**(5), 240–246.

MONTAGUE, G. A., MORRIS, A. J., WRIGHT, A. R., AYNSLEY, M., WARD, A. C. (1986b), Modelling and adaptive control of fed batch penicillin fermentation, *Can. J. Chem. Eng.* **64**, 567–580.

MONTAGUE, G. A., MORRIS, A. J., BUSH, J. R. (1988), Considerations in control scheme development for fermentation process control, *IEEE Control Systems Mag.*, 44–48.

MONTGOMERY, P. A., WILLIAMS, D., SWANICK, B. H. (1985), Control of a fermentation process by an on-line adaptive technique, *Proc. 1st IFAC Symp. on Modelling and Control of Biotechnological Processes* (JOHNSON, A., Ed.), December, Noordwijkerhout, The Netherlands, pp. 81–89, Oxford: Pergamon Press.

MOORE, R. L. (1988), The G2 real-time expert system for process control, *Int. Symp. on Advanced Process Supervision and Real-time Knowledge Based Control*, Newcastle-upon-Tyne, UK, November.

MOORE, R. L., KRAMER, M. A. (1986), Expert system in on-line process control, *Proc. 3rd Int. Conf. on Chemical Process Control*, Asilomar, California.

MOORE, R. L., HAWKINSON, L. B., LEVIN, M., HOFMAN, A. G., MATTHEWS, B. L., DAVID, M. H. (1987), Expert systems methodology for real-time process control, *Proc. 10th IFAC World Cong.*, July, Munich, FRG.

MORARI, M., ZAFIRIOU, E. (1989), *Robust Process*

Control, Englewood Cliffs, NJ: Prentice Hall.

MORRIS, A. J., MONTAGUE, G. A., THAM, M. T., AYNSLEY, M., DI MASSIMO, C., LANT, P. (1991), Towards improved process supervision – algorithms and knowledge based systems, *Proc. 4th Int. Conf. on Chemical Process Control, CPC IV*, Texas.

MOU, D. G. (1979), Toward an optimum penicillin fermentation by monitoring and controlling growth through computer aided mass balancing, *Ph. D. Thesis*, MIT, Cambridge, Massachusetts.

MOU, D. G., COONEY, C. L. (1976), Application of dynamic calorimetry for monitoring fermentation processes, *Biotechnol. Bioeng.* **18**, 1371–1392.

MOU, D. G., COONEY, C. L. (1983), Growth monitoring and control through computer-aided online mass balancing in a fed-batch penicillin fermentation, *Biotechnol. Bioeng.* **25**, 225–255.

NAKAMURA, I., CALAM, C. T. (1983), Optimal control of penicillin production using a mini-computer, *Biotechnol. Lett.* **5**, 561–566.

NESTAAS, E., WANG, D. I. C. (1983), Computer control of the penicillin fermentation using the filtration probe in conjunction with a structured process model, *Biotechnol. Bioeng.* **25**, 781–796.

NICOLAI, B. M., VAN IMPE, J. F., VANROLLEGHEM, P. A., VANDEWALLE, J. (1991), A modified unstructured mathematical model for the penicillin G fedbatch fermentation, *Biotechnol. Lett.* **13**(7), 489–494.

NIHTILA, M., HARMO, P., PERTTULA, M. (1984), Real-time growth estimation in batch fermentation, *Proc. 9th IFAC World Congress*, July, Budapest, Hungary, pp. 225–230.

O'CONNOR, G. M., SANCHEZ-RIERA, F., COONEY, C. L. (1992), Design and evaluation of control strategies for high cell density fermentations, *Biotechnol. Bioeng.* **39**, 293–304.

OMSTEAD, D. R., GREASHAM, R. L. (1988), Integrated fermentor sampling and analysis, *Proc. 4th Int. Cong. on Computer Applications in Fermentation Technology – Modelling and Control of Biotechnical Processes*, (FISH, N., FOX, R. I., Eds.), SCI/IFAC, September, Cambridge, England.

PERINGER, P., BLACHÈRE, H. T. (1978), Modelling and optimal control of baker's yeast production in repeated fed-batch culture, *Proc. 2nd Int. Conf. on Computer Applications in Fermentation Technology*, pp. 205–214, University of Pennsylvania, Philadelphia.

PETERSEN, J. N., WHYATT, G. A. (1990), Dynamic on-line optimization of a bioreactor, *Biotechnol. Bioeng.* **35**, 712–718.

PICQUE, D., CORRIEU, G. (1986), New instrument for on-line viscosity measurement of fermentation media, *Biotechnol. Bioeng.* **31**, 19–23.

PIRT, S. J., RIGHELATO, R. C. (1967), Effects of growth rate on the synthesis of penicillin by *Penicillium chrysogenum* on batch and chemostat cultures, *Appl. Microbiol.* **15**, 1284–1290.

POMERLLEAU, Y., PERRIER, M., DOCHAIN, D. (1989), Adaptive nonlinear control of the baker's yeast fed-batch fermentation, *Proc. Am. Control Conf.*, 2424–2429.

POULISSE, H. N. J., VAN HELDEN, C. (1985), Adaptive LQ control of fermentation processes, *Proc. 1st IFAC Symp. on Modelling and Control of Biotechnological Processes*, December, Noordwijkerhout, The Netherlands, pp. 7–11, Oxford: Pergamon Press.

PRESS, W. H., FLANNERY, B. P., TEUKOLSKY, S. A., VETTERLING W. T. (1985), *Numerical Recipes: The Art of Scientific Computing*, Cambridge: Cambridge University Press.

PROELL, T., HILALY, A., KARIM, M. N., GUYRE, D. (1991), Comparison of different optimization and control schemes in an industrial scale microalgae fermentation. *Proc. Am. Control Conf.*, Boston, pp. 1323–1328.

RAMSAY, G., TURNER, A. P. F., FRANKLIN, A., HIGGINS, I. J. (1985), Rapid bioelectrochemical methods for the detection of living organisms, *Proc. 1st IFAC Symp. on Modelling and Control of Biotechnological Processes*, (JOHNSON, A., Ed.), December, Noordwijkerhout, The Netherlands, pp. 65–71, Oxford: Pergamon Press.

RAMSEIER, M. (1991), Nonlinear adaptive control of fermentation processes utilizing *a priori* process knowledge, *PhD Thesis*, University of California, Santa Barbara.

RANDOLPH, T. W., MARISON, I. W., MARTENS, D. E., VON STOCKAR, U. (1990), Calorimetric control of fed-batch fermentations, *Biotechnol. Bioeng.* **36**, 678–684.

REUSS, M., BRAMMER, U. (1985), Influence of substrate distribution on productivities in computer controlled baker's yeast production, *Proc. 1st IFAC Symp. on Modelling and Control of Biotechnological Processes* (JOHNSON, A., Ed.), December, Noordwijkerhout, The Netherlands, pp. 119–124, Oxford: Pergamon Press.

SAMAAN, M., DAHHOU, T., QUEINNEC, J., M'SAAD, M. (1990), Experimental results in adaptive control of biotechnological processes, *Proc. Am. Control Conf.*, San Diego, pp. 2679–2683.

SAN, K. Y., STEPHANOPOULOS, G. (1984), Studies on on-line bioreactor identification. II, Numerical and experimental results, *Biotechnol. Bioeng.* **26**, 1189–1197.

SAN, K.-Y., STEPHANOPOULOS, G. (1989), *Biotechnol. Bioeng.* **34**, 72.

SERESSIOTIS, A., BAILEY, J. E. (1988), MPS: An artificial intelligence software system for the analysis and synthesis of metabolic pathways, *Biotechnol. Bioeng.* **31**, 587–602.

SHIOYA, S., TAKAMATSU, T., DAIRAKU, K. (1982), Measurement of state variables and controlling biochemical reaction processes, *Proc. 1st IFAC Workshop on Modelling and Control of Biotechnical Processes* (HALME, A., Ed.), August, Helsinki, Finland, pp. 13–25, Oxford: Pergamon Press.

SHIOYA, S., SHIMIZU, H., OGATA, M., TAKAMATSU, T. (1985), Simulation and experimental studies of the profile control of the specific growth rate in a fed-batch culture, *Proc. 1st IFAC Symp. on Modelling and Control of Biotechnological Processes* (JOHNSON, A., Ed.), December, Noordwijkerhout, The Netherlands, pp. 49–54, Oxford: Pergamon Press.

SPRIET, J. A., BOTTERMAN, J., DE BUYSER, D. R., DE VISCHER, P. L., VANDAMME, E. J. (1982), A computer-aided non-interfering on-line technique for monitoring oxygen-transfer characteristics during fermentation processes, *Biotechnol. Bioeng.* **24**, 1605–1621.

SQUIRES, R. W. (1972), Regulation of the penicillin fermentation by means of a submerged oxygen-sensitive electrode, *Dev. Ind. Microbiol.* **13**, 128–135.

SRINIVAS, S. P., MUTHARASAN, R. (1987), Inner filter effects and their interferences in the interpretation of culture fluorescence, *Biotechnol. Bioeng.* **30**, 769–774.

STEPHANOPOULOS, G., SAN, K. Y. (1981), State estimation for computer control of biochemical reactors, *Adv. Biotechnol.* **1**, 399–403.

STEPHANOPOULOS, G., SAN, K. Y. (1984), Studies on on-line bioreactor identification, I, Theory. *Biotechnol. Bioeng.* **26**, 1176–1188.

STEPHANOPOULOS, G., STEPHANOPOULOS, G. (1986), Artificial Intelligence in the development and design of biochemical processes, *Trends Biotechnol.*, September, 241–249.

STEYER, J. P., QUEINNEC, I., SIMOES, D. (1992), Biotech: a real time application of artificial intelligence for fermentation processes, *Proc. IFAC/IFIP/IMACS Int. Symp. on Artificial Intelligence in Real Time Control*, Delft, The Netherlands, pp. 353–358.

SWINIARSKI, R., LESNIEWSKI, A., DEWSKI, M. A. M., NG, M. H., LEIGH, J. R. (1982), Progress towards estimation of biomass in a batch fermentation process, *Proc. 1st IFAC Workshop on Modelling and Control of Biotechnical Processes* (HALME, A., Ed.), August, Helsinki, Finland, pp. 231–241, Oxford: Pergamon Press.

TAKAMATSU, T., HASHIMOTO, I., SHIOYA, S., MIZUHARA, K., KOIKE, T., OHNO, H. (1975), Theory and practice of optimal control in a continuous fermentation process, *Automatica* **11**, 141–148.

TAKAMATSU, T., SHIOYA, S., SHIOTA, M., KITABATA, K. (1979), Application of modern control theories to a fermentation process, *Biotechnol. Bioeng. Symp.* **9**, 283–302.

TARBUCK, L. A., NG, M. H., LEIGH, J. R., TAMPION, J. (1985), Estimation of the progress of *Streptomyces clavuligerus* fermentation for improved on-line control of antibiotic production, *Proc. 1st IFAC Symp. Modelling and Control of Biotechnological Processes* (JOHNSON, A., Ed.), December, Noordwijkerhout, The Netherlands, pp. 171–178, Oxford: Pergamon Press.

THAM, M. T., MONTAGUE, G. A., MORRIS, A. J. (1988), Application of on-line estimation techniques to fermentation processes, *Proc. ACC*, Atlanta, pp. 1129–1134.

THOMPSON, M. L. (1984), System analysis, simulation, control and optimization of the fed-batch penicillin fermentation, *M. S. Thesis*, MIT, Cambridge, Massachusetts.

TRUCHAUD, A., HERSANT, J., GLIKMANAS, G., FIEVET, P., DUBOIS, O. (1980), Parallel evaluation of Astra8 and Astra4 multichannel analyzers in two hospital laboratories, *Clin. Chem.* **26**, 139–141.

VAN BREUSEGEM, V., BASTIN, G. (1990), Optimal control of biomass growth in a mixed culture, *Biotechnol. Bioeng.* **35**, 349–355.

VERRBRUGGEN, H. B., EELDERINK, G. H. B., VAN DEN BROECKE, V. D. (1985), Multiloop controlled fed-batch fermentation process using a selftuning controller, *Proc. 1st IFAC Conf. on Modelling and Control of Biotechnological Processes* (JOHNSON, A., Ed.), December, Noordwijkerhout, The Netherlands, pp. 91–100, Oxford: Pergamon Press.

VIGIE, P. B., RENAUD, P. Y., CHAMILOTHORIS, G., DAHHOU, B., POURCEIL, J. B., GOMA, G. (1990), Adaptive predictive control of a multistage fermentation process, *Biotechnol. Bioeng.* **35**, 217–233.

WANG, H. Y., COONEY, C. L., WANG, D. I. C. (1977), Computer aided baker's yeast fermentations, *Biotechnol. Bioeng.* **19**, 69–86.

WANG, H. Y., COONEY, C. L., WANG, D. I. C. (1979), Computer control of baker's yeast production, *Biotechnol. Bioeng.* **21**, 975–995.

WANG, Z., THAM, M. T., MORRIS, A. J. (1992), Multilayer feedforward neural networks: Approximated canonical decomposition of nonlinearity, accepted for publication by *Int. J. Control.*

WILLIAMS, D., YOUSEFPOUR, P., SWANICK, B. H.

(1984), On-line adaptive control of a fermentation process, *Proc. Inst. Electr. Eng. Part D* **131**(4), 117–124.

WILLIAMS, D., YOUSEFPOUR, P., WELLINGTON, E. M. H. (1986), On-line adaptive control of a fed-batch fermentation of *Saccharomyces cerevisiae*, *Biotechnol. Bioeng.* **28**, 631–645.

WILLIS, M. J., DI MASSIMO, C., MONTAGUE, G. A., THAM, M. T., MORRIS, A. J. (1991), Artificial neural networks in process engineering, *Proc. Inst. Electr. Eng. Part D* **138**(3), 256–266.

WU, W. T., CHEN, K. C., CHIOU, H. W. (1985), On-line optimal control for fed-batch culture of baker's yeast production, *Biotechnol. Bioeng.* **27**, 756–760.

ZABRISKIE, D. W., HUMPHREY, A. E. (1978), Continuous dialysis for the on-line analysis of diffusible components in fermentation broth, *Biotechnol. Bioeng.* **20**, 1295–1301.

15 Fermentation Data Analysis for Diagnosis and Control

GREGORY STEPHANOPOULOS

Cambridge, MA 02139, USA

KONSTANTIN KONSTANTINOV

Newark, DE 19716, USA

URS SANER

Cambridge, MA 02139, USA

TOSHIOMI YOSHIDA

Osaka, Japan

List of Symbols and Abbreviations

General

p.u.	physical units according to the context where the symbol is used	
$\hat{x}$	estimated value of x	(p.u.)
$\tilde{x}$	predicted value of x	(p.u.)

Roman

a	node activity	(p.u.)
b	bias, offset	(p.u.)
BR	bioreactor	
c	concentration	(mol L^{-1}) or (g L^{-1})
CER	carbon dioxide evolution rate	(mol L^{-1} s^{-1}) or (g L^{-1} s^{-1})
CP	cell population	
CTR	carbon dioxide transfer rate	(mol L^{-1} s^{-1}) or (g L^{-1} s^{-1})
D	dilution rate	(s^{-1})
DO	dissolved oxygen concentration	(mol L^{-1}) or (g L^{-1})
EV	environmental variable	(p.u.)
f	feature	(p.u.)
F	convective flow rate	(mol s^{-1}) or (L s^{-1})
h	number of hidden nodes	(–)
I	cumulative uptake or production, integral variables	(mol) or (g)
IFB	internal feedback	
k	discrete time	(–)
$k_{L}a$	mass transfer coefficient	(s^{-1})
K	saturation constant	(mol L^{-1}) or (g L^{-1})
or	observer gain	(p.u.)
m	number of measured variables	(–)
MV	manipulated variable	(p.u.)
n	amount	(mol) or (g)
ORP	oxidation-reduction potential	(J mol^{-1})
OTR	oxygen transfer rate	(mol L^{-1} s^{-1}) or (g L^{-1} s^{-1})
OUR	oxygen uptake rate	(mol L^{-1} s^{-1}) or (g L^{-1} s^{-1})
p	parameter	(p.u.)
P	productivity	(g L^{-1} s^{-1}) or (mol L^{-1} s^{-1})
or	pressure	(bar)
PCA	principal component analysis	
PcUR	potential uptake rate of compound c	(mol g^{-1} s^{-1}) or (g g^{-1} s^{-1})
PSV	physiological state variable	(–) or (mol g^{-1} s^{-1}) or (g g^{-1} s^{-1})
PV	physiological variable	(mol L^{-1} s^{-1}) or (g L^{-1} s^{-1})
R	volume specific rate	(mol L^{-1} s^{-1}) or (g L^{-1} s^{-1})
or	gas constant	(J mol^{-1} K^{-1})
SGR	specific growth rate	(s^{-1})
SUR	substrate uptake rate	(mol L^{-1} s^{-1}) or (g L^{-1} s^{-1})
T	temperature	(K)
u	number of unmeasurable variables	(–)
v	total number of variables	(–)
V	volume	(m^3) or (L)

w	weights (in linear combinations)	(p.u.)
y	measured variable, output	(p.u.)
Y	averaged yield	($mol\ mol^{-1}$) or ($g\ g^{-1}$)

Greek

α	learning rate	(p.u.)
β	momentum rate	(p.u.)
∂	differential operator	(–)
ε	threshold	(p.u.)
or	hold-up	(–)
γ	headspace to liquid volume ratio	(–)
μ	specific growth rate	(s^{-1})
ρ	rate ratio, differential yield	($mol\ mol^{-1}$) or ($g\ g^{-1}$)

Indices

i	general compound
me	on-line measurable
um	not measurable on-line
s	substrate
x	biomass

1 Introduction

Efficiency in bioprocess development and the effectiveness of bioprocess control are directly dependent upon the quality of the information available about the biotechnical system. The continuing strong interest in development and application of new sensors probably attests to this fact. Presently, most modern laboratory bioreactors are equipped with a multitude of probes providing valuable information about the process under investigation. Although measurements remain primarily extracellular, they provide useful insight into cell physiology. Optical density sensors, based on the newest laser technology, provide biomass measurements with almost perfect accuracy. Mass spectrometers are widely applied as universal sensors for gases and volatiles. Commercial flow injection analyzers, capable of monitoring complex compounds, are gaining popularity in biotechnology laboratories. Furthermore, powerful computer systems linked to the bioreactors, are responsible for scanning these sensors and collecting data into data bases in formats and platforms that make them ready for further use. It is not uncommon to store tens, or even hundreds, of kilobytes of process data, just from a single cultivation.

Despite the growing sophistication in bioprocess monitoring equipment, the fact remains that the behavior of the biological system is still difficult to interpret and control. There are countless physiological phenomena which remain invisible to common sensors and control systems, although most of them leave their unique mark on the measured signals. Since such phenomena are often of crucial importance of the bioprocess, they must be considered in process development as well as control system design. The main concern here is about efficient utilization of the bioprocess information. Daily biotechnology practice shows that this valuable resource remains only partially utilized, either for on-line control purposes or for off-line design and optimization.

The problem of utilizing fermentation data is complex and multi-faceted. It can be basically described as the need to extract physiologically relevant information from the measured bioprocess signals. Derived quantities such as metabolic consumption and production rates often provide much more insight into the process characteristics than raw measurements such as concentrations. Therefore, transformation and combination of raw measurements to new variables offer great benefits. This chapter presents a summary of the basic concepts and techniques. Methods which are suited for online process monitoring, signal interpretation and control, as well as methods meant for offline application are discussed.

The chapter is organized as follows: Sect. 2 provides an overview of the measured and estimated variables in the fermentation system. The bioprocess variables are classified on the basis of the input–output relationships of the process. Attention is focused on the variables that are clearly related to cell physiology. Conventional methods for the calculation of metabolic consumption and production rates from material balances are discussed in Sect. 3. Various cultivation modes (batch, fed-batch and continuous), and diverse substances (gaseous, liquid, solid) are considered. Sect. 4 discusses the estimation of unknown bioprocess variables. A group of variables which can be elucidating in their integral or average form is considered in Sect. 5. Although they are not frequently used in process control, they are important for process evaluation and economic calculations. Sect. 6 focuses on the calculation of the so-called physiological state variables, which are most informative with respect to the current state of the cell culture. Several groups of physiological state variables, e.g., specific rates, ratios, degrees of limitation, and others, are presented. Finally, Sect. 7 summarizes recent applications of data-based techniques, such as pattern recognition methods for online supervision and process evaluation. In particular, on-line analysis of the trends of the bioprocess variables is discussed which, until recently, has not found the attention it deserves. Admittedly, many of the techniques discussed are not applied on a routine basis in industrial operations. In particular, this is true for Sects. 4, 6, and 7. We are thoroughly convinced, however, that they are useful and deserve more attention, not only in academia.

2 Classification of Fermentation Measurements and Estimated Quantities

Bioprocess variables are often classified in the literature on the basis of their nature: physical, chemical and biochemical. Although this grouping is also indicative of the origin of the measurement method for the corresponding variable, it is not very useful for control design and process development applications because it ignores the basic cause-effect relationships in bioprocess variables. As such relationships are fundamentally important, they are used here in a different classification scheme based on a structured input-output representation of the bioprocess.

2.1 Input-Output Representation of the Bioprocess

Intrinsic input-output relationships between bioprocess variables form the basis of every control system. These relationships are described below in the context of the bioreactor system structure shown in Fig. 1 (KONSTANTINOV et al., 1992a). The two major components of the system are the bioreactor (BR) and the cell population (CP). It must be emphasized that the cell population or biomass is not taken to be part of the bioreactor as opposed to more traditional system descriptions. Therefore, the bioreactor consists of the liquid phase without the biomass and the gas phase (entrapped bubbles and headspace).

In order to analyze the input-output relationships of the complete system, we first assume that no cell population is present. The input variables to the bioreactor are material and energy fluxes, represented quantitatively as flow rates (typically, volume-specific substrate feed rate, acid/base feed rates, air flow rate, etc.). In the BR, these inputs are transformed into either physical (e.g., temperature, pressure) or chemical (e.g., concentrations) variables. Jointly, these variables define the outputs of the BR and the environment of the culture and, thus, can be classified as *environmental variables* (*EV*s). In order to control the *EV*s, the input flow rates to the BR are manipulated. Therefore, they are called *manipulated variables* (*MV*s). To this end, each modern bioreactor is equipped with several (local) controllers, as shown in Fig. 1, that close the feedback loop from the *EV*s to the *MV*s.

Next, let us imagine that a bioreactor system, which is maintaining proper *EV*s, has been inoculated (cf. Fig. 2). The inoculation can be seen as exposing the cell population (CP) to the *EV*s. Apparently, for the CP the *EV*s play the role of input variables. The cells start to consume the substrates in the bioreactor, thereby reducing the corresponding concentrations, and to produce some other substances, such as end products and biomass. These activities are quantitatively represented by the output variables of the CP, which are typically (volume-specific) consumption or production rates. Examples are the glucose up-

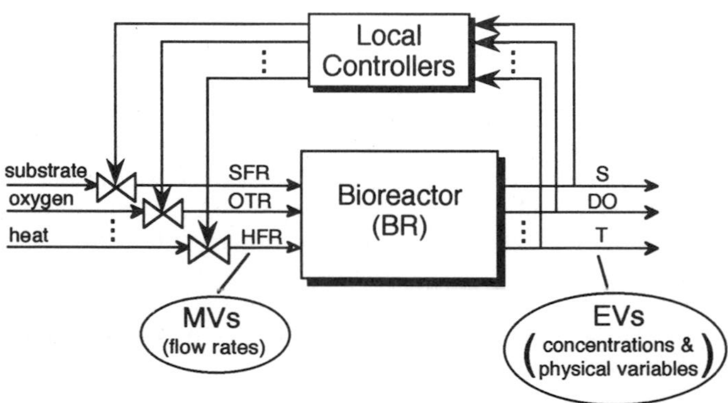

Fig. 1. Conventional input-output representation of a bioreactor with local controllers, FR, flow rate; TR, transfer rate.

take rate, the O_2 uptake rate, the growth rate, the CO_2 evolution rate, etc. Physically, they describe flows that either enter or leave the cell. However, as all of them are dependent on and can be controlled by the *EV*s, they are viewed as outputs of the CP. These variables are closely related to the physiological activities of the CP, and will be jointly referred to as *physiological variables* (*PV*s).

Since the cultivation takes place in the BR, all *PV*s influence the environment provided by the BR. The CP not only depends on its environment, but also affects it continuously. For example, the *PV* substrate uptake rate causes the *EV* substrate concentration to decrease. At the same time, the *MV* substrate feed rate typically increases the concentration. In Fig. 2, this is shown by the so-called internal feedback (IFB), which represents the effect of the CP on the *EV*s. Formally, this is expressed as a subtraction (or addition) of the *PV*s from (to) the corresponding input flow rates. Of course, this link is abstract and does not really exist, but it helps to understand the main interactions in the process. The IFB is negative for some variables, because the metabolic activity of the CP lowers the corresponding *EV* (e.g., substrate and *DO* concentrations), and positive for others (e.g., temperature increase due to the heat generated by the CP). Indeed, the main task of the local controllers is to counteract the IFB effect. As the cells produce some substances that did not exist in the BR before inoculation (biomass, products), these can be represented just as additional flows to the BR resulting in the appearance of new *EV*s. Usually, these cannot be controlled directly by the *MV*s. If those *EV*s have a significant influence on the CP, it might be useful to distinguish between two categories of *EV*s depending on whether they can be manipulated directly or not.

From the proposed scheme one can make the following observations:

1. Based on the input–output relationships, the bioprocess can be represented as two interconnected parts, BR and CP. Each of these is described by its own inputs and outputs, and thus, has its own state. The BR is a relatively simple physical system whose state can be completely described by the BR output, i.e., the *EV*s. The CP, on the other hand, is much more complicated, and its state can only partially be described by the *PV*s. The states of the BR and the CP depend on each other, but they are fundamentally different, and should not be mixed. In other words, the bioprocess

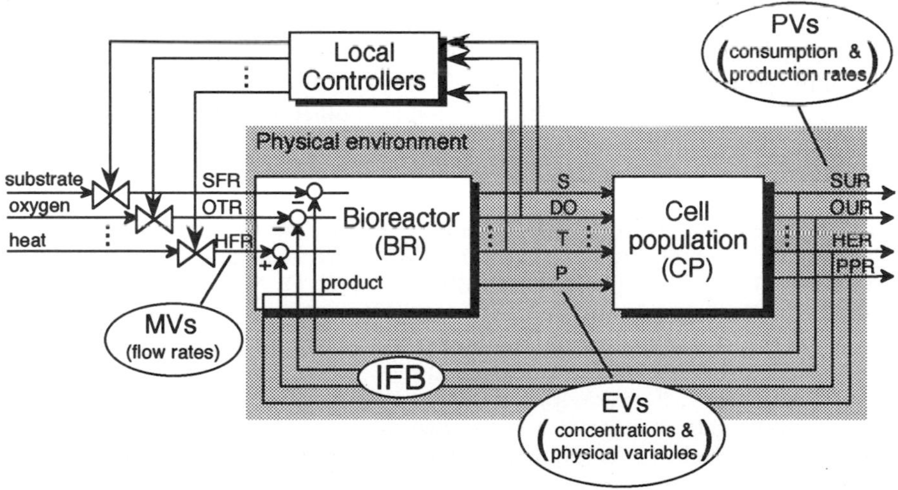

Fig. 2. Advanced input–output representation of a bioprocess based on manipulated (*MV*), environmental (*EV*) and physiological variables (*PV*). The role of the internal feedback (IFB) is clearly visible. FR, flow rate; TR, transfer rate; PR, production rate; ER, evolution rate.

cannot be adequately represented by a single state vector that mixes the *EV*s and the *PV*s, as often practiced. This would result in misleading unification of both parts of the system, neglecting the actual dependencies among the two types of output variables and complicating the analysis.

2. From the viewpoint of process control, there are three well-defined groups of variables in the bioprocess: *MV*s, *EV*s, and *PV*s. These are linked by natural input–output relationships, and each of them has a clear position in the above scheme. Below, the *PV*s will be identified with the metabolic state of the culture, though other types of variables can also be used for the description of the CP (e.g., intracellular concentrations, variables related to cell morphology) if they are available.
3. The conventional (local) controllers close the loop only over the BR (between the *EV*s and *MV*s), thus utilizing the available data only partially. The CP remains excluded from the control loop; its presence is accounted for only indirectly, through the IFB. This is why, at the level of the local controllers, the effect of the CP is considered and treated as a disturbance; its relationship with the *EV*s remains invisible. From a system viewpoint, better results can be expected if the control loop is closed over the entire system.

As can be seen in Fig. 2, *PV*s are the outputs of the CP, thus carrying explicit physiological information. They can be used to describe the physiological state (PS) of the culture, but there is still one more step that must be taken to improve this representation. The problem is that the individual *PV*s are somewhat ambiguous, as they are related to both physiological processes and other (non-physiological: transport) phenomena. For example, a low substrate uptake rate may have a physiological origin, but it might be simply caused by low biomass concentration. Similarly, a reduced CO_2 evolution rate in continuous culture may be caused by changes in physiology, or just by a lower glucose feed rate. It is for these reasons important to transform the *PV*s into variables whose values and trends relate primarily to cell physiology. Such *PV* derivatives will be referred to as *physiological state variables* (*PSV*s), and are summarized in Fig. 3 and discussed in more detail in Sect. 6. As it has been described elsewhere (KONSTANTINOV and YOSHIDA, 1989), the *PSV*s can be combined to form a *physiological state vector* that provides a most complete and transparent representation of the state of the cell population.

2.2 Computational Relationships

The sensors available today are primarily for the monitoring of *MV*s and *EV*s. With rare exceptions, *PV*s are not directly measurable. However, due to the IFB, the effects of *PV*s are propagated over virtually all other variables, which permits the on-line calculation of *PV*s, a well as of some unmeasurable *EV*s (such as concentrations), from the available measurements.

Fig. 3 outlines the typical flow of calculations required for the estimation of the variables representing the bioprocess. The major target is the calculation of the metabolic rates (i.e., the *PV*s). This is usually achieved using mass balance equations whose structure depends on the characteristics of the particular compound and on the cultivation mode (cf. Sect. 3). While such calculations are relatively simple for gaseous components (e.g., O_2 and CO_2), lack of sensors for the concentrations of other substances complicates the problem considerably. Therefore, more sophisticated estimation techniques are required (cf. Sect. 4). Often the calculation of the required quantities can be facilitated by application of stoichiometric equations balancing particular elements (C, O, H, or N) in the biological reaction.

Once the metabolic rates are available, a large group of *PSV*s can be calculated from them. They are classified in several groups, most important of which are the specific rates, rate ratios (including differential yields), and degrees of limitation. With some exceptions, the calculation of the *PSV*s is straightforward; most of them represent simple combinations of two or more metabolic rates. Usually, the determination of the rates is the most difficult

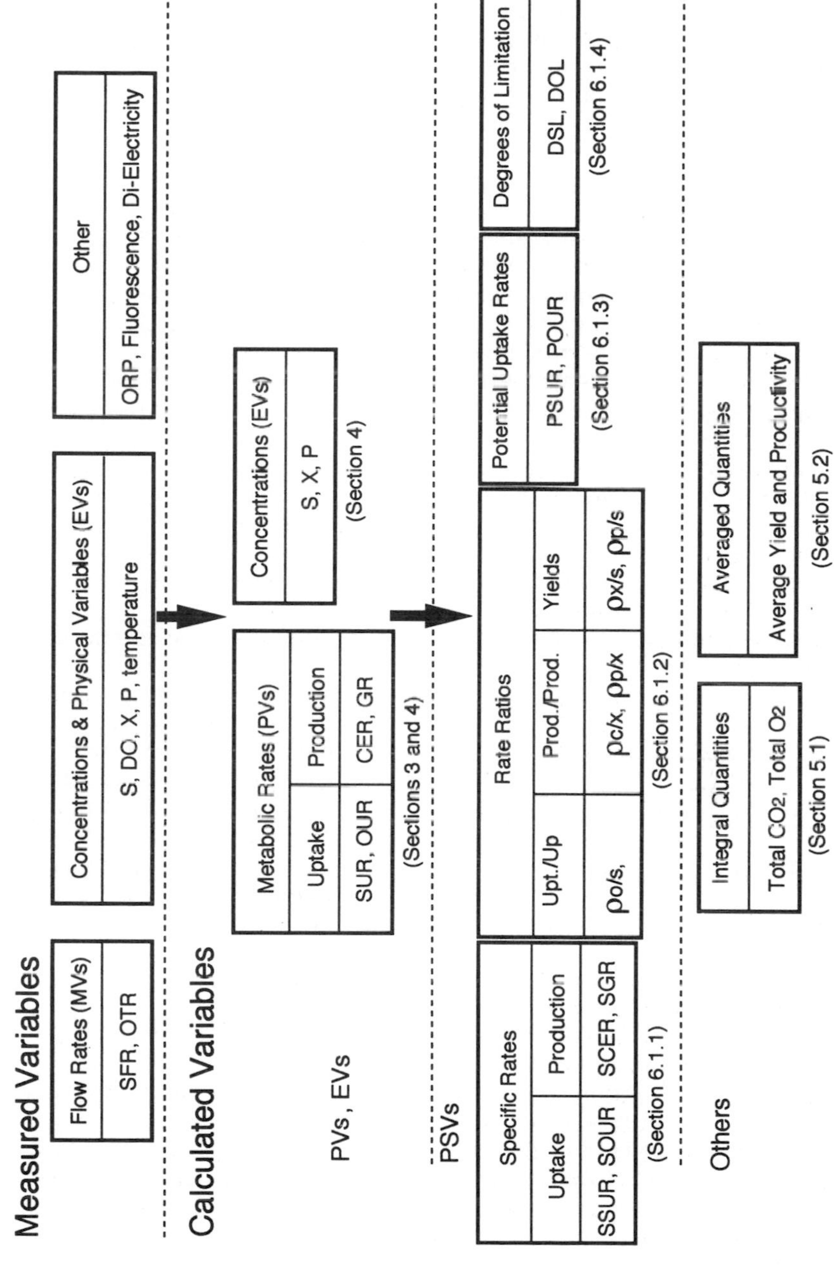

Fig. 3. Flow of calculation for variables used to characterize bioprocesses.

part of the calculations (cf. Sect. 3). In addition to the well-known *PSVs*, there are some other possibilities of constructing new variables representing the state of the cell culture based on non-conventional sensors, such as oxidation–reduction potential (ORP), fluorescence, capacitance–conductance, viscosity, and others. However, in spite of intensive recent research, the physiological interpretation of these signals is still an open question.

All of the variables presented so far have instantaneous character. This makes them suitable for control applications. There is one more group of variables which, though on-line calculable, do not share this feature: these are integral and averaged quantities. Integral variables show how much of a particular substance has been fed, consumed or produced up to the current moment. Examples are the total O_2 uptake, the total CO_2 production, the cumulative ammonia feed, etc. Averaged variables represent the averaged value of a particular instantaneous variable, such as yield or productivity. Both groups are important for the economic evaluation of the process – they represent the overall efficiency of the cultivation. Although these variables are not meant for process control, they can be used to trigger specific control actions in some cases.

3 Calculation of Metabolic Rates (Physiological Variables, *PVs*)

3.1 The General Balance Equation

The calculation of the metabolic rates makes use of the mass balances of the corresponding substrate or product. These balances appear in different forms depending on the nature of the compounds and cultivation conditions. A general structure, nevertheless, exists, representing the balance of compound i in the liquid phase, and is described by the following equation:

$$\frac{\partial n_i}{\partial t} = \frac{\partial (c_i V)}{\partial t} = V \frac{\partial c_i}{\partial t} + c_i \frac{\partial V}{\partial t} = \tag{1}$$

$$(R_{i,\text{flow}} + R_{i,\text{reaction}} + R_{i,\text{phase transfer}}) V$$

with

$$R_{i,\text{flow}} = R_{i,\text{in}} - R_{i,\text{out}}$$

(liquid flow entering and leaving the reactor)

$$R_{i,\text{reaction}} = R_{i,\text{production}} - R_{i,\text{uptake}}$$

$$R_{i,\text{phase transfer}} =$$

$$R_{i,\text{absorption}} - R_{i,\text{evaporation}} - R_{i,\text{crystallization}}$$

Here n_i and V are the amount of i in the liquid phase (total moles) and the volume of the liquid phase, respectively; c_i and R_i represent the liquid phase concentration of compound i and volume-specific rates of various processes which contribute to the change of the total amount of i in the bioreactor. $R_{i,\text{in}}$ and $R_{i,\text{out}}$ refer to convective flows in and out of the bioreactor confines whereas $R_{i,\text{absorption}}$ and $R_{i,\text{evaporation}}$ take transport from and to the gas phase into account. The left-hand side of Eq. (1) is generally referred to as the accumulation term and is set equal to zero for a process at steady state.

In order to assess the biological activity of the process, the uptake- and production rates $R_{i,\text{uptake}}$ and $R_{i,\text{production}}$ must be calculated. Using Eq. (1), several different cases can be described assuming that the liquid volume V is known. In most of these cases, the sole requirement for determining the particular metabolic rate is the concentration of the corresponding compound in the bioreactor. Depending on the cultivation mode and the characteristics of the compound of interest, this can present a serious constraint. Estimation techniques that can be quite useful in solving this problem, are discussed in Sect. 4.

3.2 Uptake Rates

Here, the calculation of the uptake rates for three different types of substrates will be presented, beginning with the balance of non-volatile substances followed by balances for volatile and gaseous substrates.

3.2.1 Non-Volatile Substrates, such as Glucose

Glucose is neither produced nor does it evaporate and, under normal conditions, it does not crystallize either. Therefore, its balance simplifies to:

$$\frac{\partial n_{\text{glucose}}}{\partial t} = V\frac{\partial c_{\text{glucose}}}{\partial t} + c_{\text{glucose}}\frac{\partial V}{\partial t} = (R_{\text{glucose, in}} - R_{\text{glucose, out}} - R_{\text{glucose, uptake}})\,V \tag{2}$$

Further reduction is possible depending on the cultivation mode. In batch processes, for example, $R_{\text{glucose, in}}$ and $R_{\text{glucose, out}}$ are both zero and the volume V is invariant, so that:

$$\frac{1}{V}\frac{\partial n_{\text{glucose}}}{\partial t} = \frac{\partial c_{\text{glucose}}}{\partial t} = -R_{\text{glucose, uptake}} \tag{3}$$

where c_{glucose} is the concentration in the liquid phase. Apparently, in order to determine the uptake rate, the glucose concentration in the medium must be known. In the case of fed-batch cultivations only $R_{\text{glucose, out}}$ is zero, and again the glucose concentration and the volume are required to calculate $R_{\text{glucose, uptake}}$. However, in some glucose-limited fed-batch cultures the glucose concentration is negligible (i.e., $c_{\text{glucose}}\ \partial V/\partial t \approx 0$) and constant (i.e., $V\ \partial c_{\text{glucose}}/\partial t \approx 0$), so that the glucose uptake rate can be equated to the glucose feed rate:

$$R_{\text{glucose, uptake}} \approx R_{\text{glucose, in}} = \frac{F\,c_{\text{feed}}}{V} \tag{4}$$

where F and c_{feed} are the total volumetric feed rate and the glucose concentration in the feed, respectively. Eq. (4) provides an easy and reliable means for determining the glucose uptake rate from the straightforward measurement of the two parameters F and c_{feed}. The two assumptions of constant and negligible glucose concentration leading to Eq. (4) need some further elaboration.

First, although the liquid volume of the fermenter is changing and the feed rate may vary with time in a fed-batch fermentation, the concentration of glucose, as well as other fermentation compounds, can be constant under pseudo-steady state conditions. This requires controlling the feed rate to the fermenter, so that the rate of glucose consumption is always balanced by the rate of glucose addition. The feed flow rate that yields pseudo-steady state conditions is determined from Eq. (2) which for fed-batch operation becomes:

$$\frac{\partial c_{\text{glucose}}}{\partial t} = D(c_{\text{feed}} - c_{\text{glucose}}) - \frac{1}{\rho_{\text{x/glucose}}}\,\mu\,c_{\text{x}} \tag{5}$$

where D is the dilution rate equal to F/V, μ the specific growth rate of biomass (cf. Sect. 6.1.1), c_{x} the biomass concentration and $\rho_{\text{x/glucose}}$ the rate ratio of biomass and glucose (cf. Sect. 6.1.2). In contrast to a continuous bioreactor where a steady state is attained for a set dilution rate smaller than the maximum specific growth rate, in fed batch operation the dilution rate needs to be controlled on a profile obtained from Eq. (5) by setting $\partial c_{\text{glucose}}/\partial t = 0$ for pseudo-steady state operation:

$$D = \frac{\mu\,c_{\text{x}}}{\rho_{\text{x/glucose}}(c_{\text{feed}} - c_{\text{glucose}})} \tag{6}$$

If accurate kinetic rate expressions for μ are available, Eq. (6) provides the basis for determining the dilution and feed rate profile; otherwise, more elaborate control structures are needed.

In order to determine the pseudo-steady state glucose concentration, Eq. (6) is solved together with its biomass balance counterpart:

$$\frac{\partial c_{\text{x}}}{\partial t} = -D\,c_{\text{x}} + \mu\,c_{\text{x}} = 0 \tag{7a}$$

with the accumulation term of the left-hand side set equal to zero. However, explicit kinetics for the specific growth rate are required. Assuming Monod kinetics for μ, Eq. (7a) can be rewritten:

$$\frac{\partial c_{\text{x}}}{\partial t} = -D\,c_{\text{x}} + \frac{\mu_{\max}\,c_{\text{glucose}}}{K_{\text{glucose}} + c_{\text{glucose}}}\,c_{\text{x}} = 0 \tag{7b}$$

Solving for the pseudo-steady state concentration of glucose yields:

$$c_{glucose} = \frac{D\,K_{glucose}}{\mu_{max} - D} \tag{8}$$

where $K_{glucose}$ and μ_{max} are the saturation constant and maximum specific growth rate, respectively, of a Monod model for biomass growth kinetics. As can be seen from Eq. (8), the glucose concentration at pseudo-steady state is proportional to $K_{glucose}$: a low saturation constant is a necessary condition for a small glucose concentration and applicability of Eq. (6).

3.2.2 Volatile Substrates, such as Ethanol

Assuming that the cells only consume, but do not produce ethanol, Eq. (1) is transformed to:

$$\frac{\partial n_{ethanol}}{\partial t} = (R_{ethanol,in} - R_{ethanol,uptake} - \\ - R_{ethanol,evaporation} - R_{ethanol,out})\,V \tag{9}$$

There is one additional term here accounting for the evaporation of ethanol into the gas phase. In the case of high ethanol concentrations and intensive aeration and agitation, the evaporation rate can be high and should not be neglected. Evaporation is, however, a physical phenomenon that depends on several variables (such as temperature, pressure, aeration, agitation, bioreactor geometry, etc.) which can be measured on-line, and on the ethanol concentration. Therefore, $R_{ethanol,evaporation}$ can be determined provided that the ethanol concentration is known. In fact, this is the sole requirement for the calculation of $R_{ethanol,uptake}$ in the case of batch, or non-limited fed-batch and continuous cultures. As with glucose above (cf. Sect. 3.2.1), substrate limitation (with low $K_{ethanol}$) results in great simplification of the calculations. A very low ethanol concentration and no accumulation eliminate both

$R_{ethanol,evaporation}$ and $R_{ethanol,out}$,

so that

$R_{ethanol,uptake}$ becomes equal to $R_{ethanol,in}$.

3.2.3 Gaseous Substrates, such as O_2

The oxygen uptake rate *OUR* is an important physiological variable. Due to its significance, ease of measurement and straightforward interpretation, it has become an almost standard component of fermentation monitoring. The O_2 balance in the liquid phase is given by:

$$\frac{\partial n_{O_2}}{\partial t} = \\ (R_{O_2,in} - R_{O_2,uptake} + R_{O_2,absorption} - R_{O_2,out})\,V \tag{10}$$

Since the liquid phase concentration of oxygen is very small, the convective and the accumulation terms can be neglected ($\partial n_{O_2}/\partial t$, $R_{O_2,in}$ and $R_{O_2,out}$ are set equal to zero). Therefore, *OUR* can be set equal to the oxygen transfer rate *OTR:*

$$R_{O_2,uptake} = OUR = R_{O_2,absorption} \\ = OTR = (k_L a)\,(DO^* - DO) \tag{11}$$

This equation can be used to calculate *OUR* from the measured dissolved oxygen concentration *DO*, the saturation concentration *DO** (Henry's law) and the previously determined mass transfer coefficient $k_L a$. It is preferable and more accurate, however, to calculate *OUR* from an overall O_2 balance in the gas phase, avoiding the tedious and unreliable estimation of $k_L a$.

In a typical aerated bioreactor, there are two distinct gas phases:

- The hold-up, i.e., the gas bubbles entrapped in the liquid phase.
- The reactor headspace, i.e. the gas phase above the liquid phase.

It can be assumed that all the gas bubbles leaving the hold-up and entering the headspace have the same oxygen concentration $y_{O_2,out}$. Assuming quasi-steady state, i.e., neglecting the accumulation term in the hold-up as well, *OTR* and consequently *OUR* can be described by the following balance of the gas phase entrapped in the liquid phase:

$$OUR = OTR = \frac{F_{in}\, y_{O_2,in} - F_{out}\, y_{O_2,out}}{V} \tag{12}$$

where F is the molar flow rate of air forced through the reactor. Gas analyzers or mass spectrometers are used to measure the O_2 mole fraction $y_{O_2,me}$ in the reactor headspace or the exhaust gas. Assuming that $y_{O_2,out}$ is equal to $y_{O_2,me}$, i.e., neglecting the headspace accumulation, the oxygen uptake rate can be calculated from Eq. (12). If F_{out} is not equal to F_{in}, it is usually calculated from an inert gas balance (nitrogen or argon) provided that the inert gas concentration is measured (e.g., by a mass spectrometer):

$$F_{out} = F_{in}\frac{y_{N_2,in}}{y_{N_2,out}} \quad \text{or} \quad F_{out} = F_{in}\frac{y_{Ar,in}}{y_{Ar,out}} \tag{13}$$

Before applying Eq. (12) one should check whether the accumulation terms are in fact negligible. For this purpose, a relationship between $y_{O_2,out}$ and $y_{O_2,me}$ is derived first assuming that $F_{in} \approx F_{out}$:

$$\frac{P\, V_H}{R\, T}\left(\frac{\partial y_{O_2,me}}{\partial t}\right) = F_{in}\,(y_{O_2,out} - y_{O_2,me}) \tag{14a}$$

or, after rearrangement:

$$\left(\frac{\partial y_{O_2,me}}{\partial t}\right) = \frac{y_{O_2,out} - y_{O_2,me}}{\dfrac{P\, V_H}{F_{in}\, R\, T}} = \frac{y_{O_2,out} - y_{O_2,me}}{\tau} \tag{14b}$$

Essentially, Eq. (14) represents a balance of the reactor headspace where V_H is the volume of the headspace. τ is the time constant of the first-order delay caused by the headspace. The larger τ is, the more the true value of OUR deviates from the value calculated by Eq. (12).

Second, the importance of the accumulation term in the gas phase is investigated by writing the complete O_2 balance as follows:

$$OTR = \frac{F_{in}\, y_{O_2,in} - F_{out}\, y_{O_2,out}}{V} - \frac{\varepsilon\, P}{(1-\varepsilon)\, R\, T}\left(\frac{\partial y_{O_2,out}}{\partial t}\right) \tag{15a}$$

Assuming that $F_{in} \approx F_{out}$, Eq. (15a) can be rewritten:

$$OTR = \frac{F_{in}\, \Delta y_{O_2}}{V} - \frac{\varepsilon\, P}{(1-\varepsilon)\, R\, T}\left(\frac{\partial y_{O_2,out}}{\partial t}\right) \tag{15b}$$

Tab. 1 explains the symbols and suggests typical values for bacterial fermentations. Assuming the most conservative values of Tab. 1 (boldface) and using Eq. (14), the time constant of the headspace is found to be 1.8 min. This implies that Eq. (12) is an accurate approximation ($\pm 2\%$) as long as the oxygen concentration does not change significantly during about 7 minutes ($\approx 4\,\tau$). In faster processes, $y_{O_2,me}$ lags behind $y_{O_2,out}$. In order to get an order of magnitude estimate of the error introduced by neglecting the hold-up dynamics, Eq. (15b) is evaluated for the most conservative case:

$$OTR = 16 \text{ mol m}^{-3}\text{ h}^{-1} + 0.2 \text{ mol m}^{-3}\text{ h}^{-1}$$

Here the error is 1.25%.

Following a similar approach for the liquid phase, Eq. (10) is rewritten for a batch reactor with the accumulation term:

$$OUR = OTR - \left(\frac{\partial DO}{\partial t}\right) \tag{16}$$

Although Eq. (16) holds for batch reactors only, very similar results are obtained for fed-batch and continuous reactors. Assuming that the dissolved oxygen concentration drops from air saturation at 2 bar reactor pressure to zero in 2 hours, which corresponds to a very active culture, the rate of change of the dissolved oxygen concentration at 30 °C is calculated:

$$\left(\frac{\partial DO}{\partial t}\right) = \frac{DO^*}{\Delta t} = \frac{-4.6\cdot 10^{-4}}{2} \text{ mol L}^{-1}\text{ h}^{-1} = -0.23 \text{ mol m}^{-3}\text{ h}^{-1}$$

At a higher temperature and/or a lower pressure, the above value will be smaller due to the decreasing saturation concentration DO^*. By comparing the rate of change to OTR from Eq. (15b), it can be seen that the accumulation term of the liquid phase is negligible as well in most cases. Due to the approximations, the

Tab. 1. Typical Values for Gas Balance Parameters in a Bacterial Fermentation. The Rate of Change of the Oxygen Mole Fraction is Negative Because the Mole Fraction Usually Decreases

Symbol	Description	From	To	Units
$\gamma = \frac{V_H}{V}$	Ratio of the volume of the reactor headspace to the volume of the liquid phase	0.2	**0.3**	–
$\frac{F_{in}}{V}$	Specific gassing rate	**800**[a]	2700[b]	mol m^{-3} h^{-1}
Δy_{O_2}	Difference in the oxygen mole fraction between inlet and outlet	**0.020** (2.0)	0.050 (5.0)	– (%)
$\varepsilon = \frac{V_G}{V + V_G}$	Hold-up = ratio of gas volume entrapped in liquid to total volume (hold-up and liquid)	0.1 (10)	**0.2** (20)	– (%)
P	Reactor pressure	10^5 (1)	$\mathbf{2 \cdot 10^5}$ (2)	N m^{-2} (bar)
R	Universal gas constant		8.314	J mol^{-1} K^{-1}
T	Reactor temperature	**305** (32)	310 (37)	K (°C)
$\frac{\partial y_{O_2,out}}{\partial t}$	Rate of change of the oxygen mole fraction	−0.005 (−0.5)	**−0.010** (−1.0)	h^{-1} (% h^{-1})

[a] corresponds to approx. 0.3 vvm
[b] corresponds to approx. 1.0 vvm

calculated value for *OUR* is usually smaller than the true value because Eq. (12) takes only the oxygen supplied from the outside of the reactor into account. In addition, some of the oxygen present in the liquid and gas phase at the beginning of the fermentation is consumed which results in a decrease of the dissolved oxygen concentration and the mole fraction. By neglecting this change, Eq. (12) tends to underestimate the consumption.

It must be emphasized that all the variables in Tab. 1 and Eq. (16) are usually available, possibly with the exception of the hold-up which can be estimated. Therefore, the accumulation terms can be included if the need arises. However, care should be taken when the rates of change are calculated from noisy measurements because this involves subtracting two numbers of the same order of magnitude thus greatly amplifying the relative error. A different approach to these calculations and other issues related to gas analysis can be found in HEINZLE (1987) or HEINZLE and DUNN (1991).

3.3 Production Rates

Determination of the production rates is of similar importance. In contrast to substrate concentrations, product concentrations cannot be considered zero regardless of the mode of cultivation. Thus, in order to calculate $R_{i,\text{production}}$ from Eq. (1), the product concentration c_i should always be available.

Often, the cells produce and degrade the same product, so that its uptake rate cannot be neglected. Such situations are complicated, and the actual production rate is difficult to determine. If the concentration c_i is known, the only possibility is to calculate the apparent production rate, i.e., the difference $R_{i,\text{production}} - R_{i,\text{uptake}}$. The equations derived in the following sections will yield the apparent production rate if both consumption and production are present.

3.3.1 Non-Volatile Products, such as Amino Acids

If there is no recycle stream, the inlet flow will not contain the product i, and the term $R_{i,in}$ equals zero. Then, Eq. (1) yields:

$$\frac{\partial n_i}{\partial t} = (R_{i,production} - R_{i,out})\ V \tag{17}$$

from which $R_{i,production}$ can be calculated from the concentration c_i.

3.3.2 Volatile Products, such as Ethanol

Product evaporation is a problem in many industrial processes. The most popular example is ethanol production, where large amounts of ethanol might be lost, if no special precautions are taken. Product removal by evaporation, however, becomes desirable in the case of product inhibition, provided that the product can be easily recovered from the exhaust gas. Taking evaporation into account, Eq. (1) becomes:

$$\frac{\partial n_{ethanol}}{\partial t} = (R_{ethanol,production} - R_{ethanol,evaporation} - R_{ethanol,out})\ V \tag{18}$$

As with the volatile substrates, $R_{ethanol,evaporation}$ can be determined as a function of the product concentration, $c_{ethanol}$, and several operating parameters. $R_{ethanol,production}$ can then be calculated from $c_{ethanol}$ whose value must be known. The determination of $R_{ethanol,evaporation}$ is simplified by the availability of mass spectrometers which allow for very accurate monitoring of the ethanol concentration in the gas phase. In this case, a gas phase balance can be used to calculate $R_{ethanol,evaporation}$ as shown for CO_2 in the subsequent section.

3.3.3 Gaseous Products, such as CO_2

A widely used variable, both in aerobic and anaerobic cultivations, is the CO_2 evolution rate *CER*. The liquid phase balance for CO_2 becomes:

$$\frac{\partial n_{CO_2}}{\partial t} = (R_{CO_2,in} + R_{CO_2,production} - R_{CO_2,evaporation} - R_{CO_2,out})\ V \tag{19}$$

where the accumulation term, as well as $R_{CO_2,in}$ and $R_{CO_2,out}$ can be neglected unless the fermentation is carried out at a pH above 5–6 (cf. Tab. 2). Therefore, Eq. (19) can be rewritten (cf. Eq. 11):

$$R_{CO_2,production} = CER = R_{CO_2,evaporation} = CTR = (k_L a)_{CO_2}(c_{CO_2} - c^*_{CO_2}) \tag{20}$$

where *CTR* is the CO_2 transfer rate. Although a reliable sensor for the dissolved CO_2 concentration c_{CO_2} has long been commercially available, it has not gained popularity. *CER* is determined instead from a gas phase balance and the measurement of the CO_2 mole fraction $y_{CO_2,out}$ in the exhaust gas (cf. Sect. 3.2.3):

$$CER = CTR = \frac{F_{out}\ y_{CO_2,out} - F_{in}\ y_{CO_2,in}}{V} \tag{21}$$

Due to the small concentration of carbon dioxide in air, the second term in Eq. (21) is usually negligible except when the biological activity is very low. The calculation of F_{out} has been described in Sect. 3.2.3. A similar analysis as for oxygen shows that the accumulation

Tab. 2. Distribution of Carbonic Species Depending on the pH in Water at 25 °C (in percent)

pH	CO_2 (aq)	HCO_3^-	CO_3^{2-}	Total Inorganic Carbon
5	96	4	0	100
6	70	30	0	100
7	19	81	0	100
8	2	97	1	100
9	0	95	5	100
10	0	64	36	100

of CO_2 in the gas phase is in fact negligible. In the liquid phase, the situation is more complicated because of the formation of bicarbonate and carbonate:

$$CO_2\,(\text{dissolved}) + H_2O \Leftrightarrow HCO_3^- + \\ + H^+ \Leftrightarrow CO_3^{2-} + 2H^+$$

Depending on the pH of the fermentation broth, a significant amount of the CO_2 is dissociated as shown in Tab. 2.

Although the dissociation of CO_2 is usually fast compared to mass transfer, the corresponding accumulation cannot always be neglected. To study the dynamic error, the rate of change of the total inorganic carbon concentration of the broth should be estimated and compared to the transfer rate determined from a gas balance (cf. the similar approach for O_2 in Sect. 3.2.3). If a sensor for the dissolved carbon dioxide concentration is available, it is important to find out whether it measures the total inorganic carbon or the dissolved CO_2 only.

If no sensor is available, the gas solubility can be used to obtain a rough approximation of the total inorganic carbon concentration. However, this estimate will usually be too low. The situation becomes even more complicated when changes in reactor pressure or aeration rate cause stripping of CO_2. Furthermore, presence of ions such as Ca^{2+} might increase the amount of CO_2 accumulated drastically due to precipitation of $CaCO_3$. Some corrective strategies to counter these problems and a different approach to the calculations related to gas analysis can be found in HEINZLE (1987) or HEINZLE and DUNN (1991).

3.3.4 Crystallizing Products

Some products crystallize when their concentration becomes high. Examples are certain amino acids, such as tryptophan and phenylalanine, and antibiotics. Crystallization may begin naturally, or it may be induced deliberately by addition of some chemicals or temperature changes. Crystallization reduces high product concentrations in the liquid phase, and eliminates product inhibition. However, this phenomenon adds an additional level of complexity in the determination of $R_{i,\text{production}}$. The balance is given by:

$$\frac{\partial n_i}{\partial t} = (R_{i,\text{production}} - R_{i,\text{crystallization}} - R_{i,\text{out}})\,V \tag{22}$$

where $R_{i,\text{crystallization}}$ depends on the product concentration, and a number of other physicochemical factors. If this relationship is quantitatively known, $R_{i,\text{production}}$ can be calculated provided that c_i is known. For this purpose, samples of the broth including the precipitate are analyzed off-line after the precipitate has been dissolved by shifting the pH or temperature.

4 Estimation of Unmeasurable Bioprocess Variables

4.1 Concepts and an Introductory Example

The lack of fast, reliable and accurate on-line measurement devices for key physiological variables *PVs* (e.g., metabolite production rate $R_{i,\text{production}}$) and environmental variables *EVs* (e.g., biomass concentration c_x) has prompted the development of methodologies for the calculation of such variables from generally available variables (e.g., carbon dioxide evolution rate *CER* or the dissolved oxygen concentration) by means of a mathematical model of the process. These methodologies are generally referred to as on-line state estimators or "soft sensors".

The following simplified example illustrates the concept: Assume a batch process where the biomass concentration $c_x(t)$ can be measured off-line only (e.g., by dry weight or cell count). However, on-line measurements of the carbon dioxide evolution rate $CER(t)$ are available (e.g., by means of a gas analyzer, cf.

Sect. 3.3.3). The following "model" was found to be a good description of the process:

$$CER(t) = \frac{\mu\, c_x(t)}{\rho_{x/CO_2}} \quad (23)$$

where μ and ρ_{x/CO_2} are two time-invariant parameters: μ, the specific growth rate (cf. Sect. 6.1.1), and ρ_{x/CO_2}, the rate ratio, i.e., the amount of biomass produced per amount of carbon dioxide produced (cf. Sect. 6.1.2).

This relationship was derived by comparison of on-line carbon dioxide evolution rate and off-line biomass concentration measurements. By combining ρ_{x/CO_2} and μ in a single time-invariant parameter p, Eq. (23) can be rewritten:

$$CER(t) - p\, c_x(t) \quad (24)$$

Once the parameter p is determined from historical data, (e.g., collected during process development or during a previous production run), the following estimate $\hat{c}_x(t)$ of the biomass concentration can be calculated on-line, as measurements of the carbon dioxide evolution rate $CER(t)$ become available:

$$\hat{c}_x(t) = \frac{CER(t)}{p} \quad (25a)$$

Obviously, the calculation of $\hat{c}_x(t)$ is affected by errors in the carbon dioxide evolution rate measurement and/or the value of p. Therefore, the value of $\hat{c}_x(t)$ computed by Eq. (25a) is only an estimate of the true value as symbolized by the "hat". Since digital computers are characterized by discrete operation with respect to time, the discrete notation for time will be used subsequently, i.e., time t equals $k\,T$ where k and T are the dimensionless discrete time and the sampling period, respectively. Furthermore, the presentation of observer-based state estimation is greatly simplified in discrete notation. Eq. (25a) can be rewritten as follows:

$$\hat{c}_x(k) = \frac{CER(k)}{p} \quad (25b)$$

In contrast to the examples presented in Sect. 3, where each species is analyzed independently, state estimation is based on relations between different compounds (e.g., biomass and CO_2 in the example above). Most of the techniques described in the literature can be assigned to either of the two following basic schemes to be presented in the two subsequent sections:

- Direct reconstruction including elemental balancing, empirical stoichiometric models and artificial neural networks (Sect. 4.2)
- State estimation with observers (Sect. 4.3).

Elemental balancing and empirical stoichiometric models both provide estimates of the unmeasurable *PV*s, whereas observers typically provide concentration estimates of the unmeasurable compounds, i.e., *EV*s. The latter can be transformed into the corresponding rates or *PV*s by means of a balance equation (cf. Sect. 3). Of course, the reverse transformation (*PV*→*EV*) is feasible as well. An example of the latter will be given at the end of Sect. 4.2.1. Artificial neural networks can estimate both types of variables. The number of required measurable variables to estimate a certain number of unmeasurable variables depends on the process and the estimation technique used. However, the latter is often larger than the former.

With the exception of some rare cases of elemental balancing, all of the techniques involve a "training step" or calibration procedure where the relationship between the unmeasurable and the measurable variables is derived from experimental data, i.e., the "training set", collected previously (e.g., during process development). Obviously, values for the unmeasurable variables must be measured by an appropriate off-line procedure at this stage (e.g., dry weight in the case of biomass). If possible, the training set should be split into a larger development set and a smaller cross-validation set. As the names imply, the development set is used to design the estimator whose predictive capabilities are then assessed with the crossvalidation set.

As will be shown in Sect. 4.4, each of the techniques has its own merits and the final decision should be based on the process at hand.

Selected references for all the presented techniques are given in Sect. 4.5.

4.2 Direct Reconstruction

In direct reconstruction, the estimates of the unmeasurable *EV*s and *PV*s at time k are computed from the measurable *EV*s and *PV*s at the same instant in time through straightforward calculations. Previous estimates and measurements have no influence on the current estimate. The biomass concentration estimator presented in Sect. 4.1 is a typical example of a direct reconstruction scheme. Several widely applied techniques will be presented in the following Sects. 4.2.1 to 4.2.3.

4.2.1 Elemental Balancing

This technique uses the generic conservation of elements in all (bio)chemical reactions to derive constraints between measurable and unmeasurable *PV*s. For this purpose a single reaction with all the compounds i present in the process and time-variant stoichiometric coefficients $\nu_i(k)$ is assumed:

$$\underset{\text{substrate}}{C_{sc}H_{sh}O_{so}} + \nu_{NH_3}(k)\,\underset{\text{nitrogen source}}{NH_3} + \nu_{O_2}(k)\,O_2 \rightarrow \nu_{pi}(k)\,\underset{\text{product i (metabolites, biomass)}}{C_{pic}H_{pih}O_{pio}N_{pin}} + \nu_{CO_2}(k)\,CO_2 + \nu_{water}(k)\,H_2O \tag{26}$$

For example, all the carbon in the substrate taken up must be balanced by the carbon in the biomass, in the carbon dioxide evolved and in the product(s) formed during the fermentation. The stoichiometric coefficient of the substrate is arbitrarily set to unity ($\nu_s = 1$) without loss of generality. Biomass is considered to be just another product, and ammonia represents any nitrogen source.

In the case of only one product (e.g., biomass) there are five unknown stoichiometric coefficients $\nu_j(k)$ which can be calculated from four elemental balances (C, H, O, N) and two measured *PV*s, $OUR(k)$ and $CER(k)$ (cf. Sects. 3.2.3 and 3.3.3, respectively), which put an additional constraint on the corresponding stoichiometric coefficients:

$$\frac{CER(k)}{OUR(k)} = \frac{\nu_{CO_2}(k)}{\nu_{O_2}(k)} \tag{27}$$

or after rearrangement:

$$\frac{\nu_{O_2}(k)}{OUR(k)} - \frac{\nu_{CO_2}(k)}{CER(k)} = 0 \tag{28}$$

Combining the elemental balances of Eq. (26) and the available measurements, Eq. (28) in matrix notation yields:

$$\begin{array}{c} \\ C \\ H \\ O \\ N \\ \text{rates} \end{array}
\begin{array}{c}
\begin{array}{ccccc} NH_3 & O_2 & P_i & CO_2 & H_2O \end{array} \\
\begin{bmatrix} 0 & 0 & pic & 1 & 0 \\ -3 & 0 & pih & 0 & 2 \\ 0 & -2 & pio & 2 & 1 \\ -1 & 0 & pin & 0 & 0 \\ 0 & (OUR(k))^{-1} & 0 & -(CER(k))^{-1} & 0 \end{bmatrix}
\end{array}
\begin{bmatrix} \nu_{NH_3}(k) \\ \nu_{O_2}(k) \\ \nu_{P_i}(k) \\ \nu_{CO_2}(k) \\ \nu_{water}(k) \end{bmatrix}
=
\begin{array}{c} S \\ \begin{bmatrix} sc \\ sh \\ so \\ 0 \\ 0 \end{bmatrix} \end{array} \tag{29}$$

The first 4 rows of the (5×5) matrix on the left-hand side of Eq. (29) are generally referred to as composition matrix, because each column represents the elemental composition of one compound. Reactants (e.g., NH_3 and O_2) are characterized by negative signs. In practice, the oxygen uptake rate $OUR(k)$ and the carbon dioxide evolution rate $CER(k)$ are measured and the system of 5 linear equations is solved to calculate the 5 unknown stoichiometric coefficients. Then, the consumption or production rate of any of the substances in Eq. (26) can be calculated.

The substrate uptake rate $R_{s,\text{uptake}}(k)$ for example is given by:

$$R_{s,\mathrm{uptake}}(k) = \nu_s \frac{CER(k)}{\nu_{CO_2}(k)} = \nu_s \frac{OUR(k)}{\nu_{O_2}(k)} \tag{30}$$

where ν_s is time-invariant and equal to 1 (see discussion above). Assuming that biomass is the only product besides CO_2 and water, its production rate is given by:

$$R_{x,\mathrm{production}}(k) = \nu_x(k) \frac{CER(k)}{\nu_{CO_2}(k)} = \nu_x(k) \frac{OUR(k)}{\nu_{O_2}(k)} \tag{31}$$

In a batch reactor, the substrate and the biomass concentration $c_s(k)$ and $c_x(k)$, i.e., the corresponding *EV*s, are then calculated as follows:

$$c_s(k) = c_s(0) - \int_0^{kT} R_{s,\mathrm{uptake}}(t)\,\mathrm{d}t \approx c_s(0) - T \sum_{j=1}^{k} R_{s,\mathrm{uptake}}(j) \tag{32}$$

$$c_x(k) = c_x(0) + \int_0^{kT} R_{x,\mathrm{production}}(t)\,\mathrm{d}t \approx c_x(0) + T \sum_{j=1}^{k} R_{x,\mathrm{production}}(i) \tag{33}$$

where $c(0)$ and T are the initial concentrations and the sampling period, respectively, which are assumed to be known. For different bioreactor operations involving flows in and/or out of the system, the expressions become slightly more difficult, but can be derived from the balance equations of Sect. 3. If more than two rates are measured (e.g., the ammonia uptake rate), a similar constraint as Eq. (28) is added to the system of Eqs. (29). If the resulting system of equations is overdetermined, the coefficients can be calculated by linear regression or the redundancy can be used to check for sensor failures (cf. Sect. 4.5 for references).

Due to its simplicity and generality, elemental balancing has drawn considerable attention in state estimation and fault detection. Shifts from one metabolic pathway to another caused by changes in the *EV*s are automatically taken into account by the time-variant stoichiometry. If all the compounds and their exact composition are known in advance (e.g., from the literature), the training procedure reduces to crossvalidation (cf. Sect. 4.1). However, applicability of elemental balancing to complex bioprocesses is limited, because the very same requirements cannot always be met even with training data. The efforts and costs to characterize industrial media containing complex components such as corn syrup or yeast hydrolysate, or to take all significant byproducts into account are often prohibitive. Furthermore, the customary measurements (typically carbon dioxide evolution and oxygen and ammonia uptake rates) are not sufficient to solve the resulting system of linear equations, if more than two products are formed (e.g., biomass and two or more metabolites besides water and carbon dioxide).

4.2.2 Empirical Stoichiometric Models

This technique is ideally suited for complex systems where neither the number of (bio)chemical reactions nor all the reactants and products nor their composition are identified. As in most techniques described in this section, a training and an application step can be distinguished:

Training

Using the data from the training set, values of all *PV*s accounted for with either on- or off-line measurements are calculated using the balance equations presented in Sect. 3. Assuming that ν different compounds were measured at n instances in time, all the available data are combined in a large, single $(n \times \nu)$ matrix D composed of two submatrices D_{me} and D_{um}:

$$D = \begin{bmatrix} PV_{me,1}(0) & PV_{me,2}(0) & \dots & PV_{um,1}(0) & PV_{um,2}(0) & \dots \\ PV_{me,1}(1) & PV_{me,2}(1) & \dots & PV_{um,1}(1) & PV_{um,2}(1) & \dots \\ \vdots & \vdots & \vdots & \vdots & \vdots & \vdots \\ PV_{me,1}(k) & PV_{me,2}(k) & \dots & PV_{um,1}(k) & PV_{um,2}(k) & \dots \\ \vdots & \vdots & \vdots & \vdots & \vdots & \vdots \\ PV_{me,1}(n-1) & PV_{me,2}(n-1) & \dots & PV_{um,1}(n-1) & PV_{um,2}(n-1) & \dots \end{bmatrix} = [D_{me}\, D_{um}] \quad (34)$$

with:

$$D_{me} = \begin{bmatrix} PV_{me,1}(0) & PV_{me,2}(0) & \dots \\ PV_{me,1}(1) & PV_{me,2}(1) & \dots \\ \vdots & \vdots & \vdots \\ PV_{me,1}(k) & PV_{me,2}(k) & \dots \\ \vdots & \vdots & \vdots \\ PV_{me,1}(n-1) & PV_{me,2}(n-1) & \dots \end{bmatrix} \quad (35)$$

$$D_{um} = \begin{bmatrix} PV_{um,1}(0) & PV_{um,2}(0) & \dots \\ PV_{um,1}(1) & PV_{um,2}(1) & \dots \\ \vdots & \vdots & \vdots \\ PV_{um,1}(k) & PV_{um,2}(k) & \dots \\ \vdots & \vdots & \vdots \\ PV_{um,1}(n-1) & PV_{um,2}(n-1) & \dots \end{bmatrix} \quad (36)$$

[dimensions: D: $(n \times v)$; D_{me}: $(n \times m)$; D_{um}: $(n \times u)$; $u + m = v$]

where the subscripts "me, 1" and "um, 2" refer to the first on-line measurable and the second unmeasurable component, respectively. Each row of D represents the data collected at a particular instance in time. m and u are the number of measurable and unmeasurable compounds, respectively. PVs of reactants should be multiplied by -1 before they are incorporated in D. Data from different cultivation modes (batch, fed-batch and continuous culture) can and should be incorporated in a single D-matrix to cover the largest possible range of operating conditions and process variations.

Applying *singular value decomposition*, matrix D can be rewritten as the product of three matrices U, S_v and V:

$$D = U \begin{bmatrix} S_v \\ \theta_{n-v,v} \end{bmatrix} V^T \quad (37)$$

[matrix dimensions:
$(n \times v) = (n \times n)\,(n \times v)\,(v \times v)$]

where S_v is a diagonal matrix with v elements. $\theta_{n-v,v}$ is a null matrix of dimension $(n-v, v)$. It can be shown that only the first d elements of S_v are significant whereas the remaining $v-d$ elements are zero unless D is corrupted by measurement errors. Since D consists of measured data the error-free case is hardly found in practical applications. Therefore, d which is equivalent to the number of independent reactions and, by itself, valuable information, must be determined first. Various ways and criteria are described in the references listed in Sect. 4.5. When d is determined, D is approximated by $\hat{D}$:

$$D \approx \hat{D} = Z\,N = D\,N^T N \quad (38)$$

where Z and N are $(n \times d)$ and $(d \times v)$ matrices, respectively $(d < v \leq n)$. N consists of the first d columns of V, and Z is the product of the first d columns of U with a diagonal matrix containing the first d elements of S_v. Each row of N contains the abstract stoichiometric coefficients for each one of the d reactions. They are referred to as abstract coefficients, because they have no chemical or biological meaning. Nevertheless, each row of N represents a linear constraint on the columns in $\hat{D}$, which is subsequently used for state estimation.

Since the difference between D and $\hat{D}$ consists of experimental error and possibly the effects of minor reactions, the entries of $\hat{D}$ are "closer" to the "true" values of the PVs than those of matrix D. As d is increased, the difference between D and $\hat{D}$ becomes smaller and smaller. Therefore, one way to determine d is to increase d until the difference between D and $\hat{D}$ is of the same order of magnitude as the experimental error.

Taking the partition of D, Eqs. (34)–(36), and the dimensions of Z and N into account, it can easily be shown that the two following equations hold:

$$D_{\rm me} \approx \hat{D}_{\rm me} = Z\, N_{\rm me} \tag{39}$$

$$D_{\rm um} \approx \hat{D}_{\rm um} = Z\, N_{\rm um} \tag{40}$$

where the same matrix Z appears in both equations. The last two equations can be rewritten with the data at a particular instance k in time (cf. the definition of D above):

$$\begin{bmatrix} PV_{\rm me,1}(k) \\ PV_{\rm me,2}(k) \\ \vdots \\ PV_{\rm me,m}(k) \end{bmatrix} = \boldsymbol{PV}_{\rm me}(k) \approx (N_{\rm me})^T z(k) \tag{41}$$

$$\begin{bmatrix} PV_{\rm um,1}(k) \\ PV_{\rm um,2}(k) \\ \vdots \\ PV_{\rm um,u}(k) \end{bmatrix} = \boldsymbol{PV}_{\rm um}(k) \approx (N_{\rm um})^T z(k) \tag{42}$$

where $z(k)^T$ is the $(k+1)^{\rm th}$ row of the matrix Z. Again $z(k)$ is the same in both equations. $\boldsymbol{PV}_{\rm me}$ and $\boldsymbol{PV}_{\rm um}$ are m- and u-dimensional column vectors and their transpose is equal to the $(k+1)^{\rm th}$ row of $D_{\rm me}$ and $D_{\rm um}$, respectively (cf. the definition of D above). The training step is completed when the two matrices $N_{\rm um}$ and $N_{\rm me}$ have been calculated by splitting N appropriately.

Application

During on-line operation, $\boldsymbol{PV}_{\rm um}(k)$ is to be calculated from $\boldsymbol{PV}_{\rm me}(k)$, $N_{\rm um}$ and $N_{\rm me}$. The two latter matrices were calculated during the training stage. Eq. (41) represents a system of linear equations for $z(k)$ which can be solved with the available data provided that the inverse (for $m=d$) or the pseudoinverse (for $m>d$) of $N_{\rm me}$ exist:

$m=d$:

$$z(k) = ((N_{\rm me})^T)^{-1}\, \boldsymbol{PV}_{\rm me}(k) \tag{43a}$$

$m>d$:

$$z(k) = (N_{\rm me}(N_{\rm me})^T)^{-1}\, N_{\rm me}\, \boldsymbol{PV}_{\rm me}(k) \tag{43b}$$

The existence of these matrices can be checked during the training stage, as soon as $N_{\rm me}$ is available. $z(k)$ is then used in Eq. (42) to calculate estimates of the unmeasurable PVs. By combining the two steps (calculation of $z(k)$ and evaluation) of Eq. (43), a direct relationship between $\boldsymbol{PV}_{\rm me}(k)$ and $\boldsymbol{PV}_{\rm um}(k)$ can be derived:

$m=d$:

$$\boldsymbol{PV}_{\rm um}(k) = ((N_{\rm me})^{-1}\, N_{\rm um})^T\, \boldsymbol{PV}_{\rm me}(k) \tag{44a}$$

$m>d$:

$$\boldsymbol{PV}_{\rm um}(k) = (N_{\rm um})^T (N_{\rm me}(N_{\rm me})^T)^{-1} N_{\rm me}\, \boldsymbol{PV}_{\rm me}(k) \tag{44b}$$

If the number of on-line measurable variables m is larger than the number of independent reactions d, the system of linear equations Eq. (41) is overdetermined. The resulting redundancy can be used to detect sensor failures or to reduce noise. The latter is achieved by Eq. (44b) which in fact is a linear regression. Failure detection techniques are described in the references.

Two independent reactions are often sufficient to describe the data in the training set, even if five to eight compounds were measured. Thus, a wide variety of processes is amenable to monitoring with on-line measurements of OUR and CER only provided that OUR and CER are independent (i.e., $RQ \neq 1$, cf. Sect. 6.1.2). This reduction in dimension is a consequence of the metabolic regulations synchronizing the many reactions in a given pathway or different pathways as a whole (e.g., energy production and growth). Since this technique is based on "interpolation" of historical data, it is important to include fermentations covering a wide variety of operating conditions in the training set.

As in elemental balancing, the EVs can be calculated from the PVs (cf. Sect. 4.2.1). For use in a mechanistic dynamic model, the abstract stoichiometry can be transformed to have (bio)chemical meaning. The references in Sect. 4.5 provide a more in-depth discussion of all the relevant aspects of this straightforward and robust technique (e.g., determination of the number of reactions and the transformation aspect).

Finally, it is worthwhile mentioning that elemental balancing and the empirical stoichiometry approach have a similar underlying structure, because both rely on identification of linear constraints between the *PV*s.

4.2.3 Artificial Neural Networks

As opposed to elemental balancing and empirical stochiometric models, artificial neural networks are non-linear mathematical structures without physical significance which map a set of input variables (i.e., the measurable *PV*s and *EV*s in this context) on a set of output variables (i.e., the unmeasurable *PV*s and *EV*s).

Fig. 4 shows the structure of a typical multilayer feed-forward neural network. It consists of an input layer, a hidden layer and an output layer. Each layer has a fixed number of nodes which are responsible for the mapping from the input to the output.

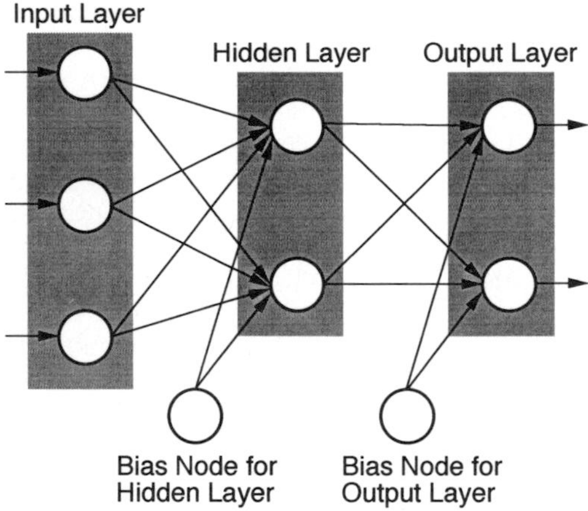

Fig. 4. Basic scheme of a feed-forward neural network with one hidden layer. The input, hidden and output layer contain 3, 2 and 2 nodes, respectively.

For the purpose of state estimation, the measurable and the unmeasurable *PV*s are the inputs and outputs of the network, respectively. Therefore, the number of nodes in the input and output layer equals the number of measurable and unmeasurable variables m and u, respectively (3 and 2 in Fig. 4). The number of hidden nodes h is determined from experimental data (see below, there are 2 hidden nodes in Fig. 4). Each connection from the ith to the jth node is characterized by a scalar weight $w_{i,j}$. In order to calculate estimates of unmeasurable *PV*s at time k, the following steps are required:

(1) The m output variables $y_{j,k}^{(in)}$ of the input layer are determined by normalizing the measurable *PV*s at time k:

$$y_{j,k}^{(in)} = \frac{PV_{me,j}(k) - \min_{me,j}}{\max_{me,j} - \min_{me,j}} \quad (j = 1, \ldots, m) \tag{45}$$

where $\min_{me,j}$ and $\max_{me,j}$ are the minimum and maximum conceivable values of the jth measurable *PV* for the process under consideration. This step essentially maps each *PV* to the interval [0, 1].

(2) The activity $a_{j,k}^{(hid)}$ of the jth hidden node is calculated as the weighted sum of all the outputs $y_{i,k}^{(in)}$ of the input layer and the constant bias $b_j^{(hid)}$:

$$a_{j,k}^{(hid)} = b_j^{(hid)} + \sum_{i=1}^{m} w_{i,j}^{(in \to hid)} \, y_{i,k}^{(in)} \tag{46}$$

$(j = 1, \ldots, h)$

where $w_{i,j}^{(in \to hid)}$ is the weight of the connection between the ith input and the jth hidden node.

(3) The output $y_{j,k}^{(hid)}$ of the jth hidden node is then calculated by filtering the activity $a_{j,k}^{(hid)}$ with a sigmoid function f (cf. Fig. 5):

$$y_{j,k}^{(hid)} = f(a_{j,k}^{(hid)}) = \frac{1}{1 + e^{-a_{j,k}^{(hid)}}} \tag{47}$$

$(j = 1, \ldots, h)$

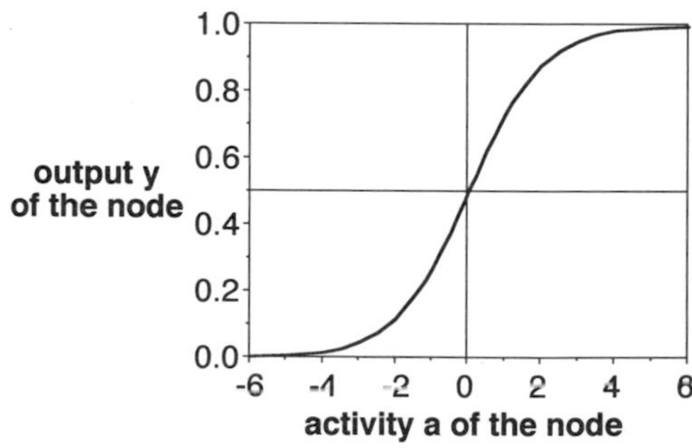

Fig. 5. Sigmoidal activation function of the hidden and output nodes, Eqs. (47) and (49): Node output y vs. node activity a.

(4) The activity $a_{j,k}^{(out)}$ of the jth output node is calculated as the weighted sum of all the outputs $y_{i,k}^{(hid)}$ of the preceding layer and the constant bias $b_j^{(out)}$:

$$a_{j,k}^{(out)} = b_j^{(out)} + \sum_{i=1}^{h} w_{i,j}^{(hid \to out)} y_{i,k}^{(hid)} \quad (j=1, \ldots, u) \tag{48}$$

In principle, this step is very similar to step (2).

(5) The output $y_{j,k}^{(out)}$ of the jth output node is then calculated by filtering the activity $a_{j,k}^{(out)}$ with a sigmoid function f (cf. Fig. 5):

$$y_{j,k}^{(out)} = f(a_{j,k}^{(out)}) = \frac{1}{1+e^{-a_{j,k}^{(out)}}} \quad (j=1, \ldots, u) \tag{49}$$

In principle, this step is very similar to step (3).

(6) Finally, estimates of the unmeasurable PVs are determined from $y_{j,k}^{(out)}$:

$$PV_{um,j}(k) = \min_{um,j} + (\max_{um,j} - \min_{um,j})\, y_{j,k}^{(out)} \quad (j=1, \ldots, u) \tag{50}$$

where $\min_{um,j}$ and $\max_{um,j}$ are the minimum and maximum conceivable values of the jth unmeasurable PV for the process under consideration. This step is necessary because the value of the sigmoid function in Eq. (49) is limited to the interval [0, 1].

These feed-forward networks have generated widespread interest because any non-linear input–output relationship can be modeled with one hidden layer and a sigmoid activation function. Furthermore, all the unknown weights w and biases b can be determined from a set of training data by means of an error back-propagation algorithm, the so-called *generalized delta rule*, possibly modified with a momentum term, which is summarized subsequently.

Assuming that n data samples are available in the training set, error back-propagation minimizes the following quadratic objective function $J(w)$ of all the weights:

$$J(w) = \sum_{k=1}^{n} J_k(w) = \frac{1}{2} \sum_{k=1}^{n} \sum_{j=1}^{u} (y_{j,k}^{(out,m)} - y_{j,k}^{(out)})^2 \tag{51}$$

where $J_k(w)$ is the sum of squared errors between the network prediction and the measurements of the kth element in the training set. As opposed to $y_{j,k}^{(out)}$ which represents the prediction of the network, Eq. (49), $y_{j,k}^{(out,m)}$ is a measured variable, possibly corrupted by measurement errors (cf. step (1) below). At each iteration step p new weights $w_{i,j}(p)$ and bias terms $b_j(p)$ are calculated from the previous values recursively by a steepest-descent algorithm:

$$w_{i,j}(p) = w_{i,j}(p-1) - \alpha \frac{\partial J(w)}{\partial w_{i,j}} + \beta(w_{i,j}(p-1) - w_{i,j}(p-2)) \tag{52}$$

$$b_j(p) = b_j(p-1) - \alpha \frac{\partial J(w)}{\partial b_j} + \beta(b_j(p-1) - b_j(p-2)) \tag{53}$$

where the positive constant α is the learning rate. The third term is the so-called momentum term ($0<\beta<1$). Although the latter is not part of the pure steepest-descent algorithm it is often included to speed up convergence. As can be readily seen from Eqs. (51), (52) and (53), the contribution of each element of the training set to the gradient is independent of all the other elements. The network is trained by the following algorithm:

(1) Normalize the input and output data in the training set according to Eqs. (45) and (50):

$$y_{i,k}^{(in)} = \frac{PV_{me,i}(k) - \min_{me,i}}{\max_{me,i} - \min_{me,i}} \quad (i=1, \ldots, m;\ k=0, \ldots, n-1) \tag{54}$$

$$y_{j,k}^{(out,m)} = \frac{PV_{um,j}(k) - \min_{um,j}}{\max_{um,j} - \min_{um,j}} \quad (j=1, \ldots, u;\ k=0, \ldots, n-1) \tag{55}$$

(2) Set the number of iterations p to zero.

(3) Set all the weights and bias terms equal to random numbers between -0.5 and 0.5.

(4) Set $k=0$, i.e., pick the first element of the training set.
(5) Using $y_{i,k}^{(in)}$ (i = 1, ..., m) and Eqs. (46) and (47), calculate all the predicted outputs $y_{j,k}^{(hid)}$ for each hidden node (j = 1, ..., h).
(6) Using $y_{j,k}^{(hid)}$ (i = 1, ..., h) and Eqs. (48) and (49), calculate all the predicted outputs $y_{j,k}^{(out)}$ for each output node (j = 1, ..., u).
(7) Calculate $J_k(w)$, i.e., the contribution of the kth element of the training set to the objective function with the current weights and bias terms:

$$J_k(w) = \frac{1}{2} \sum_{j=1}^{u} (y_{j,k}^{(out,m)} - y_{j,k}^{(out)})^2 \tag{56}$$

(8) Determine the contribution of the kth element of the training set to the weight and bias term adjustments in Eqs. (52) and (53):

$$\frac{\partial J_k(w)}{\partial w_{i,j}} = -\delta_{j,k}\, y_{i,k} \tag{57}$$

$$\frac{\partial J_k(w)}{\partial b_j} = -\delta_{j,k} \tag{58}$$

Eqs. (57) and (58) hold for all the weights and bias terms regardless of the layer they are associated with. $y_{i,k}$ is the output of the node where the connection weighted by $w_{i,j}$ starts. The error term $\delta_{j,k}$ is calculated as follows:
a) For all the weights between the hidden and the output layer and the bias terms of the output layer (i = 1, ..., h; j = 1, ..., u):

$$\delta_{j,k} = (y_{j,k}^{(out,m)} - y_{j,k}^{(out)})\, y_{j,k}^{(out)} (1 - y_{j,k}^{(out)}) \tag{59}$$

b) For all the weights between the input and the hidden layer and the bias terms of the hidden layer (i = 1, ..., m; j = 1, ..., h):

$$\delta_{j,k} = y_{j,k}^{(hid)} (1 - y_{j,k}^{(hid)}) \sum_{i=1}^{u} \delta_{i,k}^{(hid \to out)}\, w_{j,i} \tag{60}$$

Eqs. (57)–(60) are the core of the error back-propagation algorithm. The name is derived from the fact that the error used to correct the weights between the hidden and output layer, Eq. (59), is back-propagated through the summation of Eq. (60) to determine the error in the preceding layer. It can be shown easily that Eq. (60) is still applicable, if the network has more than one hidden layer.
(9) Set $k=k+1$, i.e., pick the next sample in the training set.
(10) If k is smaller than n continue with step (5).
(11) Determine the value of the objective function and the gradient term for the current set of weights and bias terms:

$$J(w) = \sum_{k=1}^{n} J_k(w) \tag{61}$$

$$\frac{\partial J(w)}{\partial w_{i,j}} = \sum_{k=1}^{n} \frac{\partial J_k(w)}{\partial w_{i,j}} \tag{62}$$

$$\frac{\partial J(w)}{\partial b_j} = \sum_{k=1}^{n} \frac{\partial J_k(w)}{\partial b_j} \tag{63}$$

(12) If the objective function still decreases significantly, set p to $p+1$ and adjust the network parameters according to Eqs. (52) and (53). Continue with step (4).
(13) The training of the network is completed.

Since its introduction in 1986, several modifications to the basic error back-propagation algorithm have been suggested, e.g., to circumvent local minima or to speed up convergence. Often, the network parameters are adjusted after each element of the training set instead of waiting until the whole set has been evaluated with a given set of parameters. However, there is still no generic procedure to determine the number of hidden nodes. Often, their number is chosen to minimize the prediction error in the cross-validation set (not the training set).

Of course, the network is not limited to *PV*s but can be set up for *EV*s or a combination of both types of variables as well. Besides state estimation, neural networks are used for fault detection and diagnosis, classification and development of predictive dynamic process models. For dynamic modeling, first-order recurrent networks where the outputs $y_{j,k}^{(out)}$ of the output layer at time k form additional inputs to the hidden layer at time $k+1$, exhibit better performance than the feed-forward networks presented above. In order to prevent the network from making unreliable predictions in cases for which no or only few training data were available, the sigmoid activation function has sometimes been replaced by a radial basis function.

Since neural networks are still a very active field of research, the reader is referred to the references in Sect. 4.5 and Chapter 14 by CHATTAWAY et al.) in this volume for further information. Some questions such as the efficient and systematic determination of the number of hidden nodes or the number of hidden layers are still debated.

4.3 State Estimation with Observers

In control system theory where the concept of state observer originated approximately thirty years ago, the state is a set of variables whose knowledge at a certain moment in time is sufficient to predict the future behavior of the process, provided that a process model is available and the future inputs are known. In the framework outlined in Sect. 2.1, this role is played by the *EVs* which is consistent with the empirical fact that observers are typically used to obtain concentration estimates.

In oder to apply observers, two mathematical models of the process are required as can be seen in Fig. 6. The first model represents the dynamics of the bioreactor (BR, usually mass balance equations, cf. Sect. 2.1) and predicts future *EVs* as a function of previous manipulated variables (*MVs*), *EVs* and *PVs*. As has been pointed out in Sect. 2.1, interaction of the *PVs* with the BR and, consequently, the *EVs* plays a key role through the internal feedback IFB. Although not shown explicitly in Fig. 6, the BR model takes this interaction into account. The second model essentially describes the dependencies of the *PVs* on the *EVs* at a given time, i.e., the behavior of the cell population CP.

These models are run in parallel to the real process subject to the same *MVs* (e.g., feed addition rate) and predict three types of variables at time k:

(1) Measurable *EVs* (e.g., dissolved oxygen concentration): $\widetilde{EV}_{\mathrm{me}}(k)$
(2) Measurable *PVs* (e.g., carbon dioxide production rate): $\widetilde{PV}_{\mathrm{me}}(k)$
(3) Unmeasurable *EVs* (e.g., biomass concentration): $\widehat{EV}_{\mathrm{um}}(k)$

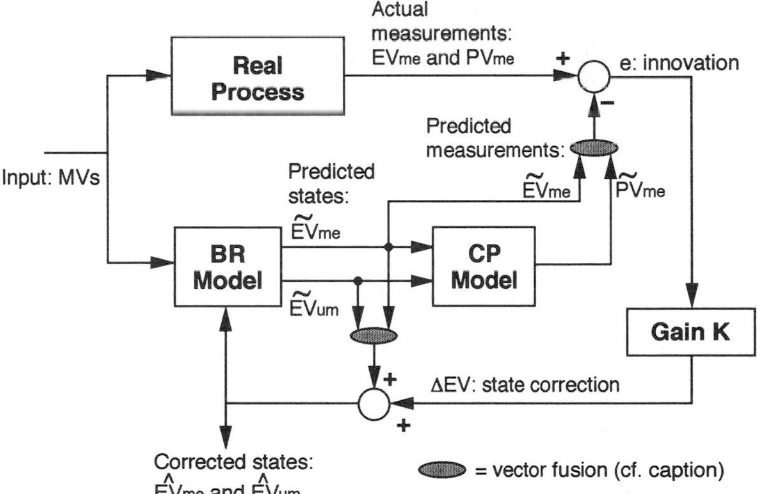

Fig. 6. Observer-based state-estimation within the framework of the bioreactor system structure of Sect. 2.1. For the sake of clarity, the internal feedback IFB is not shown. Vector fusion: The elements of two column vectors are combined in one new column vector, e.g., $\boldsymbol{x}=[x_1\ x_2]^T$ and $\boldsymbol{y}=[y_1\ y_2\ y_3\ y_4]^T$ are transformed into $\boldsymbol{z}=[x_1\ x_2\ y_1\ y_2\ y_3\ y_4]^T$.

Then, the first two types are compared to the actual measurements $EV_{me}(k)$ and $PV_{me}(k)$ of the process. In the next step, the predictions of both types of EVs are corrected according to the difference between the predicted and the actual measurements weighted by a gain variable $K(k)$ in order to yield the estimates $\widehat{EV}_{um}(k)$ and $\widehat{EV}_{me}(k)$. This two-step procedure is often referred to as predictor–corrector method and is widely used in on-line system identification and adaptive control. At a first glance, it might seem unnecessary or even misleading to correct $\widetilde{EV}_{me}(k)$. However, it must be taken into account that $\widetilde{EV}_{me}(k)$ is corrupted by model errors. The choice of $K(k)$ often represents an optimal compromise between model accuracy as represented by the predictions, and the measurement accuracy. Therefore, $\widehat{EV}_{me}(k)$ is closer to the true value than $\widetilde{EV}_{me}(k)$ in a statistical sense. The larger the value of the gain $K(k)$, the more weight is placed on the measurements and the less weight is put on the model predictions (cf. Eq. (68) below). The estimates $\widehat{EV}_{um}(k)$ and $\widehat{EV}_{me}(k)$ are in turn used as basis for prediction in the next step.

Without doubt, the Kalman filter is the most popular state estimation algorithm due to its strong theoretical foundation, although other techniques exist which differ in the way $K(k)$ is calculated. Despite the fact that state observers primarily estimate unmeasurable EVs, the corresponding PVs can be determined by applying the methodologies of Sect. 3 or by using the model of the CP to derive the PVs from the EV estimates.

It is important to note that the number of measured variables (EVs and PVs) is not necessarily equal to or larger than the number of variables in the state. In fact, in most practical applications, the full state of a system can be in essence reconstructed from a rather small number of measurements. Besides having an accurate model, a necessary condition for such state reconstruction is that all state variables have a direct effect on at least one of the measurements used in the corrector part of the observer algorithm. For linear systems, strict observability criteria exist to test the extent of state variable–measurement interaction. Although there is no formal theory, some insight can be gained into non-linear systems as well.

Since control system theory uses different terms and symbols than the framework developed in Sect. 2.1, Tab. 3 presents a comparison to facilitate understanding of the literature recommended in Sect. 4.5.

Before proceeding with the Kalman filter algorithm, an observer for the example described in the introduction is developed in order to illustrate the concept. In the context of state estimation with observers, the model represented by Eq. (24) is not sufficient, because it is not able to predict future values of the EVs, i.e., it does not contain any dynamics. Therefore, another model equation (the so-called state equation, cf. Tab. 3) is required to predict the time course of the EV $c_x(k)$:

$$c_x(k) = (1 + \mu T)\, c_x(k-1) \tag{64}$$

Eq. (64) is an approximate discrete equivalent of exponential growth in a batch reactor. In this very simple case, no MVs or PVs affect the process. The following algorithm lists all the necessary steps to compute on-line estimates $\hat{c}_x(k)$ of the biomass concentration:

(1) Predictor

(1a) Calculate a prediction of the EVs or state variables at time k using the estimates at time $k-1$ and the model of the BR or state equation, Eq. (64):

$$\tilde{c}_x(k) = (1 + \mu T)\, \hat{c}_x(k-1) \tag{65}$$

(1b) Predict the measurable PVs at time k using the model of the CP or measurement equation, Eq. (24), and the predicted EV $\tilde{c}_x(k)$:

$$C\tilde{E}R(k) = p\, \tilde{c}_x(k) \tag{66}$$

(2) Corrector

(2a) Compare the actual measurements to the predicted values:

$$\mathrm{e}(k) = CER(k) - C\tilde{E}R(k) \tag{67}$$

In this example, no EVs were measurable. Otherwise, the difference between the measured and the predicted EVs would have been calculated as well.

(2b) Calculate the observer gain $K(k)$ according to the particular algorithm chosen (e.g.,

Tab. 3. List of Frequently Used Terms and Symbols in Control System Theory and Their Interpretation in the Framework of Section 2.1

System Theory	Symbol	Framework of Section 2.1
State vector at time k	$\boldsymbol{x}(k)$	Column vector with all the *EVs* of interest at time k
Measurement vector at time k	$\boldsymbol{y}(k)=\begin{bmatrix} y_{PV}(k) \\ y_{EV}(k) \end{bmatrix}$	Column vector with all the on-line measurable *PVs* and *EVs* at time k
Input vector at time k	$\boldsymbol{u}(k)$	Column vector with all the *MVs* at time k
State equation	$\boldsymbol{x}(k)=\boldsymbol{f}(\boldsymbol{x}(k-1), \boldsymbol{u}(k-1))$	Model of the BR describing the dynamics of the *EVs*
Measurement equation	$\boldsymbol{y}(k)=\boldsymbol{g}(\boldsymbol{x}(k), \boldsymbol{u}(k))$	Model of the CP describing the relationship between all the *EVs* and the measurable *EVs* and *PVs*

for a Kalman filter). Since the various techniques (deterministic and stochastic observers) differ substantially in the way $K(k)$ is calculated, no specific details are given here and the reader is referred to the description of the Kalman filter below and the extensive literature cited in Section 4.5.

(2c) Calculate the final estimate $\hat{c}_x(k)$ by correcting the predicted value $\tilde{c}_x(k)$:

$$\hat{c}_x(k)=\tilde{c}_x(k)+K(k)\,e(k) \tag{68}$$

Eqs. (65)–(68) can be combined to give one recursive equation for the calculation of the estimates $\hat{c}_x(k)$:

$$\hat{c}_x(k)=(1-K(k)\,p)\,(1+\mu\,T)\,\hat{c}_x(k-1) + K(k)\,CER(k) \tag{69}$$

Most models of the CP are rather crude approximations and, therefore, only valid in a very limited range of operating conditions. In order to extend the applicability of observers, the list of estimated variables has often been extended to include not only environmental variables but also unknown, lumped or changing model parameters (e.g., the specific growth rate μ in the example above). Instead of trying to determine and model all the significant influences of the various *EVs* on μ (e.g., from substrate and product concentration), μ is estimated along with the states. For prediction (step 1a, above), these parameters are usually assumed to be time-invariant. Thus Eq. (64) for μ would read:

$$\mu(k)=\mu(k-1) \tag{70}$$

If some information about the dependencies is known, it can of course be included in the model.

We close this section with the basic equations for the extended Kalman filter (EKF) that apply to a system with the non-linear state and measurement equations shown in Tab. 3:

state equation:

$$\boldsymbol{x}(k)=\boldsymbol{f}(\boldsymbol{x}(k-1), \boldsymbol{u}(k-1))+\boldsymbol{w}(k-1) \tag{71}$$

measurement equation:

$$\boldsymbol{y}(k)=\boldsymbol{g}(\boldsymbol{x}(k), \boldsymbol{u}(k))+\boldsymbol{v}(k) \tag{72}$$

Both $w(k-1)$ and $v(k)$ are sequences of normally distributed random numbers with zero mean and covariance matrices $Q(k-1)$ and $R(k)$, respectively. $\boldsymbol{w}$ is a measure for the model error of $\boldsymbol{f}$, whereas $\boldsymbol{v}$ represents the measurement error associated with $\boldsymbol{y}$. The algorithm is an extension of the theory developed for linear systems. It is presented here in its most basic form, and the steps correspond to those of the simple example above. The reader is referred to Sect. 4.5 for more references on this topic.

(1) Predictor

(1a) Calculate a prediction $\tilde{\boldsymbol{x}}(k)$ of the state at time k using the estimates at time $k-1$ and the state equation Eq. (71):

$$\tilde{x}(k)=f(\hat{x}(k-1),\ u(k-1)) \tag{73}$$

(1b) Predict the measurement at time k using the predicted state $\tilde{x}(k)$ and the measurement equation Eq. (72):

$$\tilde{y}(k)=g(\tilde{x}(k),\ u(k)) \tag{74}$$

(1c) Predict the covariance $\tilde{P}(k)$ of the state prediction $\tilde{x}(k)$:

$$\tilde{P}(k)= \Phi(k-1\rightarrow k)\,\hat{P}(k-1)\,\Phi(k-1\rightarrow k)^T+Q(k) \tag{75}$$

where Φ is the linearized, discrete state transition matrix. The (i, j)-element of Φ is given by:

$$\Phi_{i,j}=\left.\frac{\partial f_i}{\partial x_j}\right|_{\hat{x}(k-1)} \tag{76}$$

Eq. (76) stipulates that the derivative of f_i with respect to x_j is evaluated using the state estimate at time $k-1$.

(2) Corrector
(2a) Compare the actual measurements to the predicted values:

$$e(k)=y(k)-\tilde{y}(k) \tag{77}$$

(2b) Calculate the Kalman filter gain $K(k)$:

$$K(k)=\tilde{P}(k)\,C(k)^T\,[C(k)\,\tilde{P}(k)\,C(k)^T+R(k)]^{-1} \tag{78}$$

where $C(k)$ is the linearized measurement equation matrix given by:

$$C_{i,j}(k)=\left.\frac{\partial g_i}{\partial x_j}\right|_{\tilde{x}(k)} \tag{79}$$

Eq. (78) strikes an optimum balance between the accuracy of the model prediction and the measurement uncertainty.
(2c) Calculate the final estimate $\hat{x}(k)$ by correcting the predicted value $\tilde{x}(k)$:

$$\hat{x}(k)=\tilde{x}(k)+K(k)\,e(k) \tag{80}$$

(2d) Calculate the corrected state covariance matrix $\hat{P}(k)$:

$$\hat{P}(k)=[I_n-K(k)\,M(k)]\,\tilde{P}(k)\,[I_n-K(k)\,M(k)]^T+K(k)\,R(k)\,K(k)^T \tag{81}$$

The square roots of the diagonal elements of the corrected state covariance matrix $\hat{P}(k)$ can serve as rough indicators of how accurate a state estimate is.

4.4 Assessment of the Different Techniques

As opposed to direct reconstruction which has no inherent feedback characteristics, the comparison of actual and predicted measurements in observers yield some information about the validity of the model used and, consequently, the accuracy of the state estimates. Depending on the algorithm used to calculate the gain K, unknown disturbances and model errors are to a certain extent taken into account (e.g., measurement noise, wrong initial concentrations, approximate kinetics). This feature becomes very important, if the estimates are to be used in feedback control. Biased estimates will result in a permanent offset from the setpoint even with a perfect controller.

All the techniques except for artificial neural networks provide some criteria to check *a priori* whether the unmeasurable variables can in fact be estimated. This is generally referred to as observability analysis. In elemental balancing and empirical stoichiometries it is sufficient to show that a certain matrix has full rank. A similar but slightly more involved analysis can be done for observers.

Given the recursive nature of observers, previous state estimates influence future estimates through the prediction step. This feature often filters out the effects of outliers in the measurements, but slows down the tracking ability of the state estimator when drastic changes take place. Neither occurs with direct reconstruction. The dynamic model necessary for prediction in observers can be used to predict the effects of different control strategies on-line by simulation (stochastic optimal control) whereas direct reconstruction represents strictly static relationships or constraints.

Although observers are superior from a purely technical point of view, their advantages are balanced by higher costs for design and development and larger computational requirements during operation. The application of state estimators for control ("inferential control") is discussed in more detail in Chapter 14 in this volume.

4.5 State Estimation References

The following compilation presents selected literature references for the different state estimation techniques described above:

Direct reconstruction

Elemental balancing for state estimation: COONEY et al. (1977), ERICKSON (1979), WANG et al. (1979a, b), PARK and STEPHANOPOULOS (1991).
Elemental balancing for fault detection: ROMAGNOLI and STEPHANOPOULOS (1980, 1981), WANG and STEPHANOPOULOS (1983), VAN DER HEIJDEN (1991).
Empirical stoichiometric models: State estimation: LIAO (1989), SANER (1991); Modeling: HAMER (1989), SANER (1991).
Artificial neural networks: RUMELHART et al. (1986a, b), HOSKINS and HIMMELBLAU (1988); BHAT and MCAVOY (1990); KRAMER and LEONARD (1990); THIBAULT et al. (1990); WILLIS et al. (1991); THIBAULT and GRANDJEAN (1992); CHATTAWAY et al. (1993).
Continuing journals: *IEEE Transactions on Neural Networks* since 1990, edited by the IEEE Neural Networks Council, and *Neural Networks*, the official journal of the International Neural Network Society, since 1988, Pergamon Press, New York.

Observers

Deterministic observers: DOCHAIN and BASTIN (1985), BASTIN and DOCHAIN (1986, 1990), DOCHAIN et al. (1988).
Theory of stochastic observers (Kalman filters); JAZWINSKI (1970), GELB (1974), ANDERSON and MOORE (1979), FRIEDLAND (1987), OGATA (1987).
Applications of stochastic observers (Kalman filters): STEPHANOPOULOS and SAN (1984), SAN and STEPHANOPOULOS (1984a, b), THAM et al. (1988), AGARWAL and BONVIN (1989), CHATTAWAY and STEPHANOPOULOS (1989a, b).

5 Calculation of Integral and Averaged Quantities

5.1 Integral Variables

These variables are calculated by integrating the feed, consumption or production rates of batch, fed-batch or continuous processes with respect to time t. For example, the amount of substrate i consumed up to the current moment is given by the integral:

$$I_i = \int_0^t R_{i,\mathrm{uptake}}(t)\, V(t)\, \mathrm{d}t \tag{82}$$

where $t=0$ corresponds to the beginning of the fermentation (i.e., $I_i(0)=0$). At the end of the fermentation, they represent the total turnover of some key substance. Typical variables of this class are the total amount of CO_2 produced, the total amount of glucose or O_2 consumed, etc. Although integral variables do not have physiological meaning, they do provide important quantitative information about the overall progress of the cultivation. This is why they are sometimes used in process control, particularly for triggering supervisory control actions. For example, the total amount of sodium hydroxide fed to control pH in animal cell cultivations can be used to activate feeding of fresh medium after the batch phase. This ad-hoc strategy is justified because of the lack of more suitable measurements representing the actual state of the cell culture.

5.2 Averaged Variables

Typical variables of this class are the averaged process yield and productivity. These are calculated from the corresponding integral variables, cf. Eq. (82), as follows:

$$\text{Yield:} \quad Y_{i/j}(t) = \frac{I_i(t)}{I_j(t)} \tag{83}$$

$$\text{Productivity:} \quad P_i(t) = \frac{I_i(t)}{t\,V(t)} \tag{84}$$

where i and j are product and substrate, respectively. Usually, these variables are used for off-line process evaluation and optimization, rather than for on-line process control. It is common practice to represent the overall process efficiency by its yield and productivity, calculated at the point of process termination. It should be emphasized that these two values have only a static sense. Practically, there is no reason to restrict their calculation to the final point. This can be done throughout the cultivation, showing at any time instant t the values of the yield and productivity averaged up to the current moment.

Time profiles of the averaged quantities have already much higher value. For example, they show exactly at which point the particular variable passes its maximum, i.e., when the process has to be terminated. This is important either for dynamic evaluation of the termination time (resulting in event-based instead of time-based triggering), or for post-mortem process evaluation and planning. On-line calculated averaged quantities can be combined into multi-variable criteria, providing an optimal compromise between yield, productivity, and product concentration.

6 Calculation of Physiological State Variables (*PSVs*)

It was mentioned earlier (Sect. 2) that traditional bioreactor controls focus exclusively on the *EV–MV* interaction thus reducing the effect of the culture on the bioreactor environment to that of a mere disturbance. A more explicit accounting of the cause–effect relationships between CP and *EV* will enhance control performance significantly. To this end, a necessary first step is the identification of variables characteristic of the state of the cell population. Such variables, previously referred to as physiological state variables (*PSV*), are presented in this section along with the means for calculating them from presently available measurements.

The main groups of *PSVs* are shown in Fig. 3 according to the classification proposed elsewhere (KONSTANTINOV et al., 1992a). Most *PSVs* are ratios of a metabolic rate (*PV*) and the biomass concentration (an *EV*) or ratios of two metabolic rates. Application of some new sensors, such as mass spectrometers equipped with membrane inlet systems, or FIAs and HPLCs for the monitoring of various complex substances, would significantly broaden the spectra of this type of variables and at the same time enhance the usefulness of these rather expensive devices.

6.1 Classification of Physiological State Variables

6.1.1 Specific Rates

The specific rates represent ratios of consumption or production rates to the biomass concentration (the amount of active catalyst). Their significance derives from the assumption that the total metabolic activity in a reactor is proportional to the amount of biomass. Changes in the metabolic rates *per se* are ambiguous, because they could be the result of a change in the biomass concentration or a metabolic shift. In contrast, specific rates do not change merely because of a change in the amount of biomass. Therefore, they provide a better indicator of the process behavior. However, specific rates calculated in this way do not always have purely "physiological" meaning. For example, in glucose-limited cultivations, low values of the specific O_2 uptake rate may have no physiological origin, but may be caused by too low a feed rate. These ambiguities can be eliminated by analyzing ratios (cf. Sect. 6.1.2). In any case, due to its role in the calculation of specific rates, the biomass concentration should be monitored continuously. With the recent advances in laser optical density sensor technology (ISHIHARA et al., 1989; YAMANE et al., 1992; KONSTANTINOV et al., 1992c), this requirement can now be satisfied in many cultivations.

Depending on the metabolic rate considered (i.e., the numerator), two types of specific rates can be defined:

(a) Specific uptake rates (*SiUR*) defined by:

$$SiUR = \frac{\text{uptake rate of compound i}}{\text{biomass concentration}} \tag{85}$$

Typical variables of this type are the specific glucose uptake rate (*SGUR*) and the specific O_2 uptake rate (*SOUR*).

(b) Specific production rates (*SiPR*) defined by:

$$SiPR = \frac{\text{production rate of compound i}}{\text{biomass concentration}} \tag{86}$$

Probably, the most popular variables of this type are the specific CO_2 production rate (known as the specific CO_2 evolution rate, *SCER*) and the specific biomass production rate (known as the specific growth rate, *SGR* or μ). For an economic process evaluation, the specific production rate of the end product is very often a key variable in any fermentation.

6.1.2 Rate Ratios ($\rho_{i/j}$)

The ratios are a large group of variables represented by the general formula:

$$\rho_{i/j} = \frac{\begin{array}{c}\text{uptake (production) rate}\\ \text{of compound i}\end{array}}{\begin{array}{c}\text{uptake (production) rate}\\ \text{of compound j}\end{array}} \tag{87}$$

Practically, it is possible to combine any two rates (*PVs*) in defining ratios. Some are among the most popular bioprocess variables; others are less popular – yet they carry significant information in the proper context. According to the numerator and the denominator of Eq. (87), three types of ratios can be distinguished:

(a) Uptake/Uptake Ratios

In this case, both i and j are substrates. For example, in aerobic, glucose-limited cultivations, the ratio of the O_2 uptake rate to the glucose uptake rate ($\rho_{O_2/\text{glucose}}$) can easily be calculated on-line and has been found very informative (KONSTANTINOV et al., 1990). Nevertheless, it appears that it has not been widely applied yet. The ratio of the ammonia uptake rate to the glucose uptake rate ($\rho_{\text{ammonia/glucose}}$) might be used to characterize the efficiency of biomass formation. Unless the ammonia concentration is negligible, the calculation of this variable requires an ammonia sensor which, though available today, is still not proven technology.

(b) Production/Production Ratios

A typical representative of this type is the ratio of the cell growth rate to the CO_2 production rate (ρ_{x/CO_2}) which shows how much biomass is synthesized per unit of CO_2 evolved. Similarly, the ratio of the ethanol production rate to the CO_2 production rate ($\rho_{\text{ethanol}/CO_2}$) can contribute to reliable monitoring of the physiological transitions in ethanol fermentations.

(c) Production/Uptake Ratios

These ratios are commonly known as differential, instantaneous or momentary yields. Examples are the ratio of the cell growth rate to the glucose uptake rate (known as the yield of biomass from glucose: $\rho_{x/\text{glucose}}$), the product yield ($\rho_{\text{product/substrate}}$) and the ratio of the cell growth rate to the ammonia uptake rate (yield of biomass from ammonia). The respiratory quotient (*RQ*), defined as the ratio of the CO_2 evolution rate to the O_2 uptake rate, is most popular because of the availability of *OUR* and *CER* measurements and the distinct shifts in *RQ* observed in some cultures (e.g., yeast). The differential yields should not be confused with the average yields (cf. Sect. 5.2) which are cumulative markers usually calculated after the cultivation to quantify the overall efficiency of the process.

6.1.3 Potential Uptake Rates (*PiUR*)

A common feature of most current applications is that they are run under single-substrate limitation. The substrate uptake rate is then equal to its feed rate, and any variations of the latter will also induce changes in the former. In this sense, the current uptake rate and the specific uptake rate by themselves are not very informative; they simply reflect external control actions rather than the actual culture state. Under these circumstances, the maximum specific uptake rate of the cell culture, called for brevity potential uptake rate (*PiUR*), is of more interest because it is closely related to the state of the cells, and in some cultivations undergoes dramatic changes.

Unfortunately, there is no direct way to calculate the *PiUR*s under limitation by compound i. Potential uptake rates are determined instead by introducing artificial perturbations that bypass temporarily such substrate limitations. Thus, if for a short period of time (usually a few minutes), the concentration of substrate i is increased beyond the saturation concentration by a pulse addition, the specific uptake rate *SiUR* will reach its maximum (assuming fast adaptation), and provide a measure of *PiUR*. This technique, called induced perturbation, has been found useful for determining the potential oxygen uptake rate in oxygen-limited amino acid production (AKASHI et al., 1979), and the potential methanol uptake rate in methanol-limited biomass production (KONSTANTINOV and YOSHIDA, 1989). It is easy to implement and has been used to determine other bioprocess variables as well (KONSTANTINOV and YOSHIDA, 1990). However, great care is required if high substrate concentrations have a toxic effect on the cells. Furthermore, induced perturbation is not suitable for analyzing production rates. The specific production rates of particular substances should, nevertheless, be measured simultaneously during the *PiUR* estimation for a substrate in order to calculate the corresponding rate ratios (cf. Sect. 6.1.2).

6.1.4 Degrees of Limitation (*DiL*)

These variables are derived from the *PiUR*s (cf. Sect. 6.1.3), and can provide convenient setpoints for *PSV* control (cf. Sect. 6.2). They are represented by the formula

$$DiL = \frac{\text{specific uptake rate of compound i}}{\text{potential uptake rate of compound i}} \tag{88}$$

The *DiL*s take their value from the interval [0, 1]; total limitation corresponds to a value of zero, while a lack of any limitation by the given substance will result in a value of one. A controller for the degree of limitation will be more effective than a controller for the specific uptake rate, because changes in the potential uptake rate are taken into account without having to change the setpoint. This of course assumes that *PiUR* is determined regularly. The usefulness of these variables was shown by AKASHI et al. (1979).

6.1.5 *PSV*s of the Future

Undoubtedly, with development and widespread application of novel measurement devices, new informative *PSV*s will emerge. Even at present, there are some sensors which offer intriguing, but not yet investigated opportunities. An example is the ratio of the electric capacitance of the culture medium to the biomass concentration. The capacitance sensor manufacturers claim that the response of this device is proportional to the *viable cell number*, while the optical density measured by a laser OD sensor is related to the *total cell concentration*. It follows that their ratio (biomass-specific capacitance) should represent the *viability* of the culture which is particularly important in animal cell cultivations. However, the soundness of this approach has not yet been corroborated and is presently under investigation.

Other candidates for the derivation of biomass-specific quantities include fluorescence and bioluminescence measurements. The latter might be useful for investigating various physiological phenomena in recombinant bioluminescent strains. When physiological changes are accompanied by changes in shape and size of the cells, on-line monitoring of these quantities becomes desirable. Unfortunately, measurement techniques are not up to this task yet. Assuming that it is related to the size and shape of the microorganisms, the ratio of the fermentation broth viscosity to the biomass concentration has been used to characterize and control a fungal fermentation (OLSVIK and KRISTIANSEN, 1992).

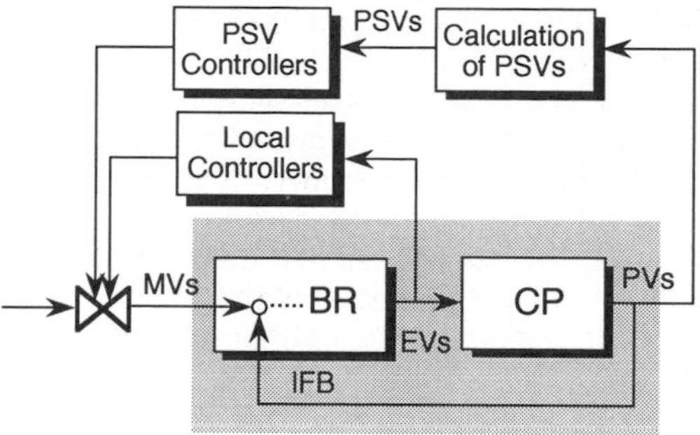

Fig. 7. Direct physiological state control. Local controllers and *PSV* controllers work independently of each other to control environmental variables (*EV*s) and physiological state variables (*PSV*s), respectively.

6.2 Physiological State (PS) Control Structures

Fig. 2 (Sect. 2.1) shows that conventional controllers close the loop only over the BR, excluding the CP. Since the *PSV*s provide useful information, it is possible to improve the performance of the control system by including the *PSV*s into the control loop. Three different approaches will be presented in the subsequent Sects. 6.2.1 to 6.2.3 although more structures are possible (KONSTANTINOV et al., 1992a). Control structures based on modern control theory are also discussed in Chapter 14 in this volume.

6.2.1 Direct PS Control

In this case, additional control loops are added, from the *PSV*s directly to the *MV*s that are not involved in the local loops (Fig. 7). To date, there have been some attempts to control isolated *PSV*s in this way, without any explicit intention to handle the integral PS of the culture. Examples described in the literature include the control of the specific growth rate by the glucose feed rate (*GFR*), and of the *RQ* by the *GFR* (SHIMIZU et al., 1989; AIBA et al., 1976; MEYER and BEYELER, 1984). The *PSV*s are usually used as isolated variables, and thus the resulting controllers closely resemble local controllers, though they close the loop over both parts of the plant.

Systems capable of monitoring and controlling several *PSV*s simultaneously, and thereby aiming at the explicit handling of the PS of the culture, are not yet available. Their development is hampered by the requirement for an accurate mathematical model of the plant, which, due to the great complexity of the CP, is hard to obtain. Also, the required complex logic of operation of the controllers is difficult to express only in the form of conventional algorithms.

6.2.2 Cascaded PS Control

Cascade control as shown in Fig. 8 might greatly facilitate tight control of a *PSV*, provided that certain requirements regarding the process dynamics are met. The technique based on two controllers is applicable, if one

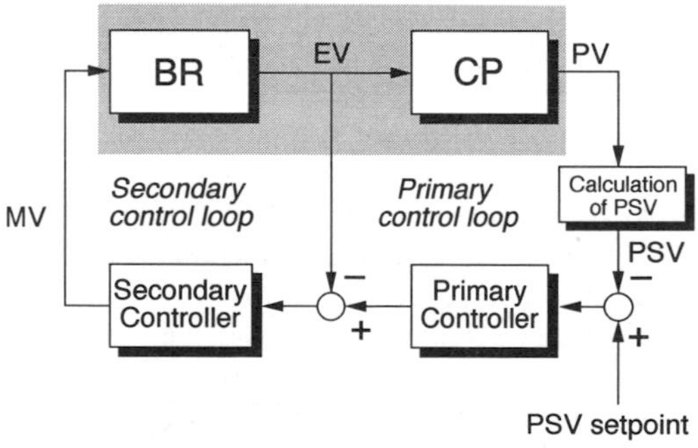

Fig. 8. Cascade control of a physiological state variable (*PSV*). The output of the primary controller provides the setpoint of the secondary controller which controls the environmental variable (*EV*) on which the *PSV* of interest depends.

MV is used to control one *PSV* which in turn mainly depends on one *EV*. The controller for the *PSV*, i.e., the primary controller, does not directly set the *MV* but provides the setpoint for the secondary controller. The latter compares this setpoint to the current value of the *EV* and then adjusts the *MV* accordingly.

As in direct PS control above (Sect. 6.2.1), the *PSV* setpoint is the only parameter a user has to specify. Apart from using two measurements to guide one *MV*, thus making better use of available information, this configuration is able to reject disturbances in the *EV* before they can affect the *PSV*. However, this is only feasible, if the dynamics of the secondary loop are much faster than those of the primary loop. Before implementing cascade control, this requirement has to be checked carefully, because the system is prone to instability if it is not met. Further reference for this widely applied technique (e.g., temperature control of chemical reactors) can be found in most textbooks on process control (e.g., STEPHANOPOULOS, 1984).

6.2.3 Supervisory PS Control

Fig. 9 shows an alternative structure of a control system, meant to monitor and handle the PS of the culture. In contrast to the first scheme, the system is now *hierarchically* organized in two levels. The higher level explicitly identifies the PS (using the current values and trends of the *PSV*s) and, based on this, supervises the activities of the lower control level. Typically, the supervisory commands are for activation/deactivation of control algorithms (change of control strategy), change of the setpoints and/or modification of the control algorithm parameters.

Because of the complexity and the informal, knowledge-intensive nature of the procedures involved in PS identification, knowledge-based and/or pattern recognition techniques seem particularly promising for the design of the higher level. The lower level, on the other hand, performs only standard control functions, and can be designed using the traditional control techniques. It consists of standard local controllers (for temperature, pH, *DO*, etc.), and some other case-dependent controllers, possibly computer-implemented. All of them have to be able to interpret the supervisory commands mentioned above. This indirect PS control structure offers new possibilities for the implementation of high-performance and versatile control systems for fermentation processes.

7 New Methods Based on Pattern Recognition Techniques

7.1 Why Pattern Recognition Now?

Although most pattern recognition techniques are rather old and well established in other fields such as image analysis, they have

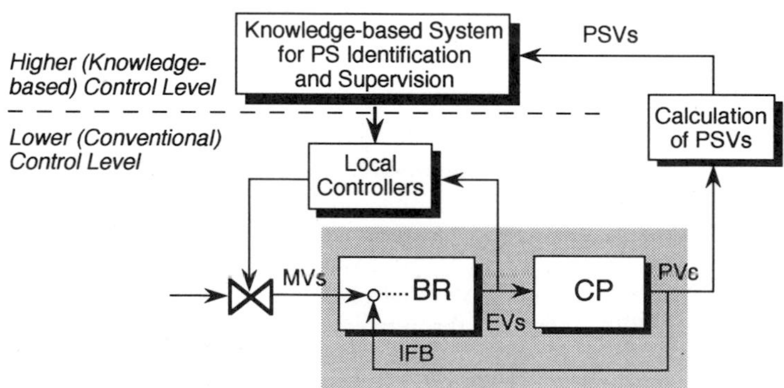

Fig. 9. Supervisory physiological state control. The physiological state variables (*PSV*s) are the input to a knowledge-based system on the highest hierarchy level. After PS identification, this system adjusts setpoints and parameters of the local controllers responsible for control of the environmental variables (*EV*s) on the lower level.

not been applied widely to fermentation data analysis. The reason is that until recently the emphasis was on data interpretation through modeling. However, despite extensive efforts to develop accurate and transparent mechanistic models of bioprocesses based on first principles, the inherent complexity of the systems under study has prevented a major breakthrough in the last twenty years. At the same time, computers and advanced analytical equipment have dramatically increased the amount and quality of data collected on a routine basis. Therefore, pattern recognition techniques have come into focus, because they promise to derive useful information from these data at lower cost than mechanistic models.

It must be emphasized that the whole notion of data analysis by pattern recognition is based on the assumption that the effects of the dominant but hidden intracellular processes are somehow visible in the measured variables. In system theory, this requirement is formalized by the concept of observability. Loosely speaking, pattern recognition helps to detect what is already there in the data. Nevertheless, its benefits should not be dismissed too easily, because it provides a consistent, rational and quantitative approach to process diagnosis and data interpretation that can handle a large amount of data.

In this section, some promising new methodologies based on shape or trend analysis, spectral representation, and identification of constraints are first outlined in a rather general way. Then, real-time shape analysis for process supervision and control is presented in more detail in order to convey a sense of how pattern recognition approaches actually work. This particular example was chosen for its intuitiveness and for the fact that one form of shape analysis or another is at the core of many of the methodologies proposed in the literature.

Admittedly, these methodologies still await the ultimate test of effectiveness in an industrial context. This fact does not diminish their inherent qualities, but might rather be due to the lack of accessible, user-friendly computer software, a point which is valid for many of the advanced methodologies described in this chapter.

7.2 Overview of Pattern Recognition Approaches to Data Analysis

7.2.1 Trend Analysis

It is not uncommon to observe experienced process operators and engineers use the characteristic shape of time profiles of measured variables in order to evaluate process behavior or performance. The former might look for a particular peak in the oxygen uptake rate *OUR* to initiate feed addition, while the latter might draw conclusions for process optimization based on the presence or absence of a feature in the dissolved oxygen concentration (e.g., whether depletion occurs depending on the substrate feed addition rate). They use the "pattern" represented by the time profiles or parts thereof to diagnose, analyze, compare and discriminate between several processes (typically different batch or fed-batch runs) or the effects of variations in the operating strategy (e.g., high or low substrate addition rate). If one wants to automate or simulate this process as in pattern recognition, two interconnected issues arise:

(1) Trend detection
The general underlying "characteristic" trend must be extracted from the noisy signal where noise encompasses both stochastic measurement uncertainties originating in the measurement process as well as minor, though real, variations of the signal without particular significance (step 1, Fig. 10). Wavelet decomposition provides a very consistent and rigorous solution to this problem (BAKSHI and STEPHANOPOULOS, 1992a, b, c). Polynomial curve fit is a less complex and more pragmatic, though largely heuristic, approach (cf. Sect. 7.3).

(2) Feature extraction
Features describing the detected trend in a convenient and unique way must be defined and then extracted (step 2, Fig. 10). One approach is to divide the signal into episodes during which the signs of the first and the second derivative with respect to time (>0, $=0$, <0) do not change (CHEUNG and STEPHANOPOULOS, 1990a, b). Taking into account that the second derivative is always zero if the first derivative

Tab. 4. Definition of Qualitative Features for Feature Extraction Depending on the Sign of the First and Second Derivative of Signal y with Respect to Time

Derivatives	Features A	B	C	D	E	F	G
$\frac{\partial y}{\partial t}$	>0	<0	<0	>0	>0	<0	$=0$
$\frac{\partial^2 y}{\partial t^2}$	<0	<0	>0	>0	$=0$	$=0$	$=0$

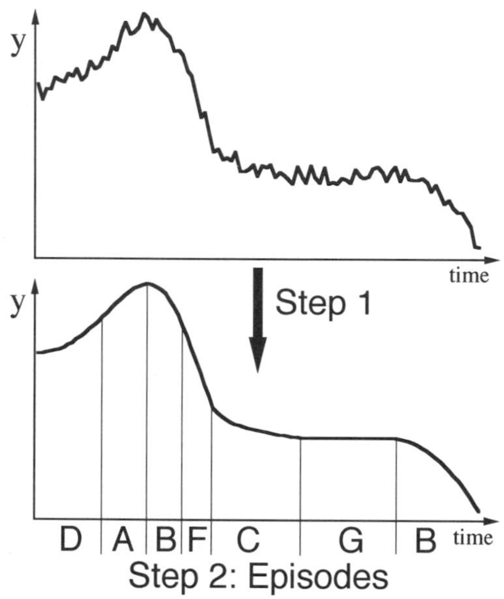

Fig. 10. Trend analysis and feature extraction. Step 1: determination of stable trends, step 2: feature extraction (episodes).

is zero, there are 7 distinct combinations shown in Tab. 4. Episodes with the same combination of signs are assigned the same label or feature (e.g., the letters A, B, ... in Tab. 4 and Fig. 10). Each run is thus characterized by a string of letters indicative of a series of episodes with distinct features, e.g., {A, B, C, F, D, E, C}.

It must be emphasized that this approach is not limited to qualitative features, since the description of the episodes might include quantitative information as well, such as the mean slope during and the duration of an episode. The methodology presented in Sect. 7.3 uses a library of so-called characteristic shapes, i.e., the basic episodes of Tab. 4 and combinations of them (e.g., "Passed Over Maximum" which corresponds to {B, C}), to characterize and describe events.

For the purpose of on-line process supervision and control, the qualitative and quantitative features are determined in real-time and compared to those of a historical database with particular properties (e.g., runs with specific faults). The objective is to find the best or sufficient match from which conclusions can be drawn about the state of the current process. For *a-posteriori* analysis or process evaluation, the features are extracted from several batch or fed-batch runs which differ with respect to certain properties of interest (e.g., operating strategy and final yield or productivity; cf. Sect. 5.2). Then, the features that have the greatest discriminating power among runs with different properties are identified using decision trees. The implicit assumption is that these features were in fact responsible for the different observed outcome of the processes.

A very simple example is shown in Fig. 11. There are 5 "good" fermentations (filled circles; e.g., with high yield) and three "bad" fermentations (shaded circles; e.g., with low yield). Each fermentation is characterized by two features f_1 and f_2 (e.g., vitamin concentration and feed addition rate). If f_1 is used to discriminate between "good" and "bad", the separation is not as good as if f_2 was used. Therefore, f_2 is assumed to be more typical or responsible for differences in the observed outcome than f_1, i.e., the feed addition rate is a more important parameter.

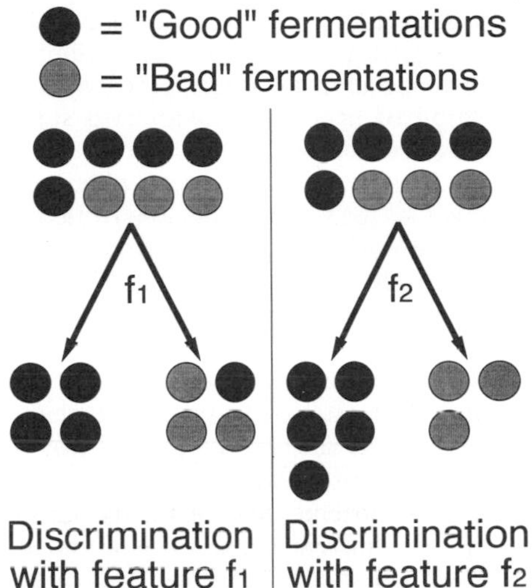

Fig. 11. Comparison of discrimination power of features f_1 and f_2 using decision trees.

For real problems with many more features where the conclusions are not so clear-cut and obvious, several layers of decisions are used and a performance measure is calculated for each decision based on its discriminating power. By carefully selecting the type of variables analyzed, different objectives can be met. For process evaluation, it is obviously preferable to include the *MV*s in order to derive an optimum operating strategy. This procedure has been successfully applied to derive explicit non-obvious rules for a complex industrial fed-batch fermentation (BAKSHI and STEPHANOPOULOS, 1992a, c).

Fourier transformation provides an alternative approach to describing time profiles of fixed length (LOCHER et al., 1990). Here, instead of quantitative and qualitative shape features, the signals are characterized by their spectral coefficients which depend of the corresponding frequency in the signals. Using historical data, a set of spectral coefficients is calculated for each type of fermentation run available in the database. These sets represent the so-called "prototypes". When new data become available, their spectral coefficients are determined and compared to the "prototypes". The new data are assigned to the type of fermentation providing the best match between the spectral coefficients. The performance of the methodology was studied using a continuous culture of *Saccharomyces cerevisiae* growing on glucose. The algorithm could recognize correctly some complicated states in the yeast cultivation, associated with the natural oscillations of the cells. It was also possible to detect sensor failures, isolating them from the overall process picture.

7.2.2 Principal Component Analysis

An altogether different approach is the application of Principal Component Analysis (PCA) which is a multivariate data analysis technique to identify linear relationships between many measured variables. In a first step, the number of principal components d, which is usually smaller than the number of measured variables, is determined from available experimental data. This step is very similar to the determination of the number of reactions for empirical stoichiometric models mentioned in Sect. 4.2.2. Then, a new set of fewer variables (d) is determined which can replace the original measured variables with a minimum loss of information.

To illustrate this method assume that *OUR*, *CPR*, the substrate uptake rate *SUR* and the dissolved oxygen concentration *DO* were measured and that the number of principal components or factors was found to be two ($d=2$). Then, each measurement can be represented by a linear combination of two principal components z_1 and z_2, e.g.:

$$OUR(k) \approx \alpha_1 z_1(k) + \alpha_2 z_2(k) \tag{89}$$

$$CPR(k) \approx \beta_1 z_1(k) + \beta_2 z_2(k) \tag{90}$$

$$SUR(k) \approx \gamma_1 z_1(k) + \gamma_2 z_2(k) \tag{91}$$

$$DO(k) \approx \delta_1 z_1(k) + \delta_2 z_2(k) \tag{92}$$

where α, β, γ and δ are constant coefficients determined by PCA. Although z_1 and z_2 have no physical meaning, they preserve most of the variability of the original variables. Therefore, they can be used to replace the original variables for many purposes such as process supervision (WISE and RICKER, 1989; KRESTA et al., 1991) or simply to visualize the fermentation data (ROUSSET and SCHLICH, 1989). Instead of monitoring all the original variables, it is sufficient to focus on the fewer principal components and examine whether any clustering of runs with similar performance occurs in the projection space (z_1, z_2). The presence of clusters gives rise to patterns of performance that can form the basis of process diagnosis.

It should be noted that the measured variables are not completely independent because they must satisfy Eqs. (89) to (92). For given values of $OUR(k)$ and $CPR(k)$ at any time k, the approximate values of $SUR(k)$ and $DO(k)$ are determined and *vice versa*. These constraints can be used to develop process supervision schemes by assuming that the process follows standard behavior as long as the constraints are not violated (SANER and STEPHANOPOULOS, 1992).

7.3 A Specific Example: Real-Time Trend Analysis of Temporal Profiles for Supervision and Control

The temporal scope of a typical bioprocess control system is restricted to a narrow interval around the current time (cf. Fig. 12). In most cases only the current values of the process variables are used. Such limited representation is probably sufficient at the level of local controllers maintaining pH, temperature and DO at fixed setpoints. It is, however, virtually impossible to develop adequate schemes for interpretation of process behavior and supervision using only current values. To detect and interpret complex physiological phenomena, the control systems should be able to take into account temporal profiles or trends of the process variables over sufficiently long episodes from the (recent) process history. Although during manual control process operators rely heavily on such characteristic shapes as reliable expressions of underlying phenomena, their use in control design has not been considered seriously hitherto. Present bioprocess control systems are relatively good at working with current data; they are weak at processing trends or historical information. Therefore, the historical data remain almost completely unutilized. The availability of a consistent set of features, which allow control systems to utilize historical data, will significantly broaden their scope, providing more reliable and complete interpretation of the plant behavior. In fact, such capabilities are considered to be a requisite in any highly dynamic, complex environment (MOORE et al., 1990; CHEUNG and STEPHANOPOULOS, 1990a, b).

A generic approach for *qualitative* representation of temporal shapes of bioprocess variables has been reported recently (KONSTANTINOV and YOSHIDA, 1992). The proposed concept is at the core of a flexible algorithm designed especially for use in real-time knowledge-based systems ("expert systems", cf. Chapter 14. It acts as an independent front-

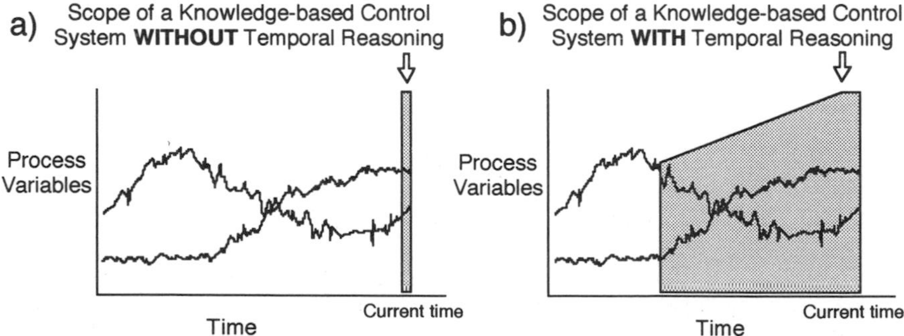

Fig. 12. Comparison of the temporal scopes of knowledge-based control systems without (a) and with (b) temporal reasoning.

end procedure, supplying the inference engine with information about the temporal shapes of specified variables. Due to its efficiency, the algorithm can be used even on low-performance personal computers.

The procedure enhances expert systems by the capability of handling rules of the following general form:

IF
(TimeInterval Variable ShapeDescriptor),
THEN
(Conclusion) (93)

where *TimeInterval* defines a certain period $[t_1, t_2]$ of the process history, *Variable* is a particular process variable, and *ShapeDescriptor* specifies the expected pattern of the temporal profile. Such a rule might look as follows:

IF
(DuringTheLast2h RQ has PassedOverMaximum),
(TimeInterval: DuringTheLast2h; Variable: RQ; ShapeDescriptor: PassedOverMaximum)
THEN
(Conclusion)

Usually, the right boundary t_2 coincides with the current time, i.e., the reasoning is in respect to the most recent process history. The left time boundary t_1 can be specified either explicitly or in respect to a certain event in the past. The *ShapeDescriptor* might specify simple trends, or more complicated forms. Each rule can combine several facts, providing sophisticated logic for capturing and handling complex phenomena. Each fact is finally characterized by its degree of certainty *dc* ($\varepsilon [0, 1]$), which shows to what extent the given fact is true at time t_2.

The calculation of the *dc* requires the introduction of an appropriate procedure *P* that assigns a *dc* to each fact:

$$dc = P(\text{fact}) \tag{94}$$

The estimated *dc* is further used by the inference mechanism to calculate the certainty of the condition of the corresponding rule(s). In order to derive *dc*, the proposed approach evaluates the similarity between the time profile of a given process variable $x_j(t)$ over the time interval $[t_1, t_2]$, and the expected temporal pattern represented linguistically by the *ShapeDescriptor*. The *ShapeDescriptor* takes its value from a library of template shapes (e.g., *"Increasing"*, *"DecreasingConvexly"*, *"PassedOverMaximum"*, etc.) stored in computer memory (cf. Fig. 13).

The similarity between the library and the real-time profiles is searched for qualitatively, by representing and comparing the real and the library shapes in a symbolic form as ordered compositions of elementary shape features. As in the qualitative process theory (FORBUS, 1984; KUIPERS, 1986), the signs of the first and second derivatives of the variables over the interval $[t_1, t_2]$ are used. The sequential combination of these features provides a rich set of profiles (cf. Fig. 13), which can cover a large number of real bioprocess situations. Formally, the extraction of the sequence of the derivative signs from the real-time profiles is described by the operators *SD1* and *SD2* for the first and the second derivative, respectively:

$$SD1\,[x_j(t)] = sd1 = (+, -, \ldots) \quad t\,\varepsilon\,[t_1, t_2] \tag{95}$$

$$SD2\,[x_j(t)] = sd2 = (+, -, \ldots) \quad t\,\varepsilon\,[t_1, t_2] \tag{96}$$

which transform the continuous variable $x_j(t)$ into discrete symbolic strings *sd1* and *sd2*, respectively. The *qualitative shape* of $x_j(t)$ is represented by the combination of these strings (Fig. 14):

$$qshape\,[x_j(t)] = \{SD1\,[x_j(t)];\ SD2\,[x_j(t)]\} = \{(+, -, \ldots);\ (+, -, \ldots)\}\ t\,\varepsilon\,[t_1, \quad t_2] \tag{97}$$

Hence, two temporal shapes are considered qualitatively equivalent if their *qshapes* coincide. The analyzing procedure first extracts *sd1* and *sd2* of $x_j(t)$ over $[t_1, t_2]$ in real-time and then compares them with those in the shape library. The feature strings $sd1^L$ and $sd2^L$ of the library shapes are stored in computer memory in the same symbolic form (cf. Fig. 13).

The degree of certainty *dc* represents a measure of the similarity between the qualitative representation of the real and library shapes. As shown in Fig. 15, evaluation of *dc* is composed of three modules – approxima-

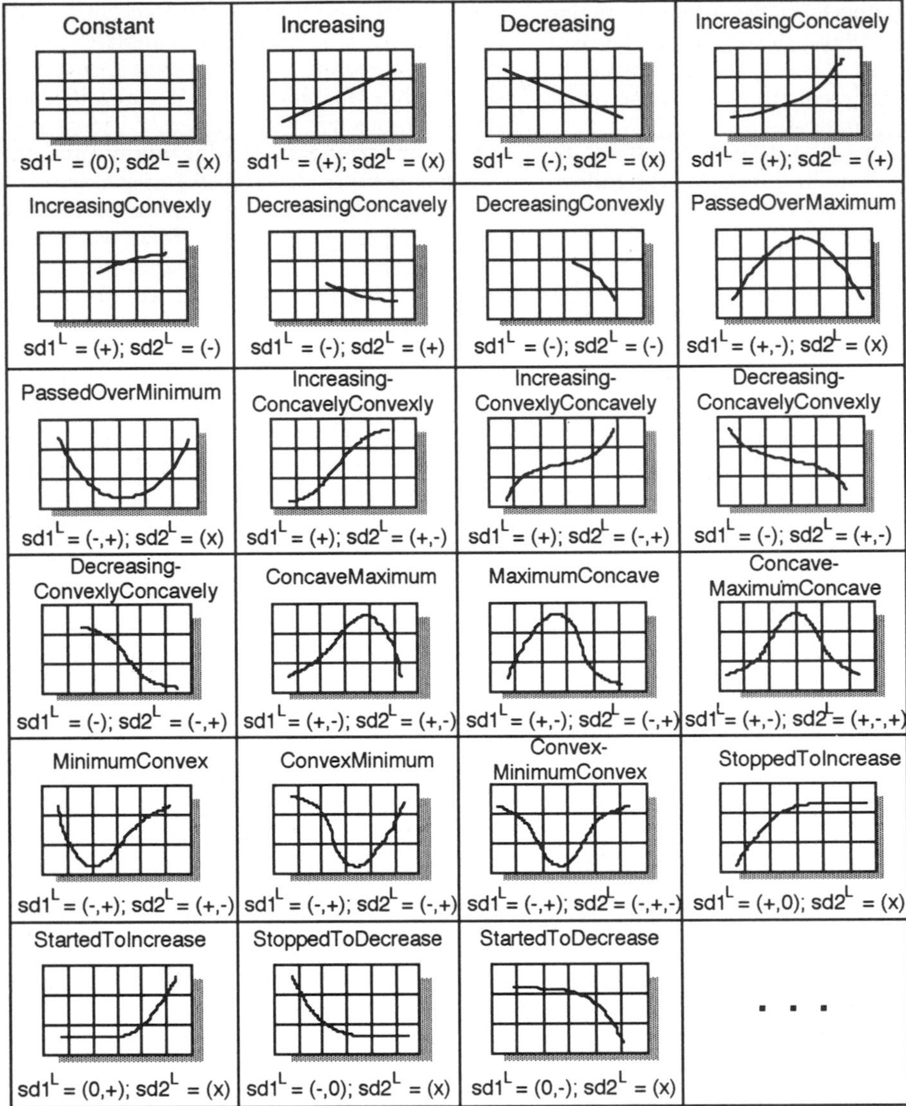

Fig. 13. Complete library of template shapes used to identify trends.

tion, transformation into qualitative form, and *dc* calculation – invoked sequentially.

As the response time of the inference engine is always a major concern in on-line applications, the authors have proposed a fast algorithm providing adequate response times even in heavily loaded multi-task environments. The first step of the procedure consists of an approximation of the variable $x_j(t)$ by a proper analytical function $x_j^*(t)$ over the time interval $[t_1, t_2]$. This is necessary to: (1) provide a convenient model for the subsequent analysis; (2) reduce the noise; (3) eliminate non-essential (superimposed) details from the real profile. Then, the feature strings *sd1* and *sd2* comprising the qualitative model *qshape* of $x_i(t)$ are

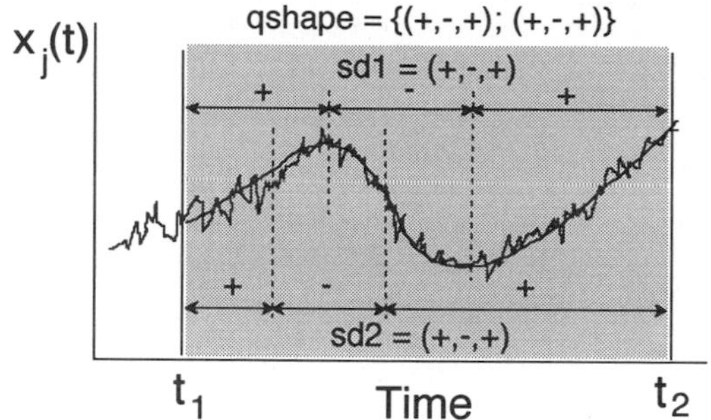

Fig. 14. An example of sequence extraction. The continuous temporal profile between t_1 and t_2 is transformed into the qualitative shape characterized by *sd1* and *sd2*.

extracted analytically from $x_j^*(t)$. Finally, the corresponding degree of certainty *dc* of the fact is calculated as follows:

a) If the strings *sd1* and $sd1^L$ do not coincide, *dc* will be set to zero.
b) If the strings *sd1* and $sd1^L$ do coincide, *dc* is given by:

$$dc = \min\left[1 - w_1 \frac{p}{p_{\max}} - w_2 \frac{\sum_{i=1}^{n} (x_j^*(i) - x_j(i))^2}{SSE_{\max}}, 0\right] \quad (98)$$

with *p*, number of elements in *sd2* that do not match $sd2^L$; $p_{\max}$, number of elements in *sd2*; *n*, number of data points in the time-interval $[t_1, t_2]$; w_1, w_2, positive weight constants (<1); $SSE_{\max}$, maximum tolerable deviation (sum of squared errors).

Obviously, this scheme puts much more weight on the first derivative. Lack of coincidence between *sd2* and $sd2^L$ despite of agreement between *sd1* and $sd1^L$ results in a reduction of *dc*. The procedure accounts for the difference between $x_j^*(t)$ and $x_j(t)$ by lowering *dc* proportionally to the sum of squared errors. Consequently, *dc* will not reach its maximum value of 1 unless $x_j^*(t)$ and $x_j(t)$ are identical. The min operator prevents *dc* from becoming negative. The shape analyzing procedure terminates after passing the calculated value of *dc* to the inference engine.

The shape analyzer has been embedded into a compact knowledge-based system for bioprocess control on a personal computer equipped with an 80286 processor and the QNX operating system. This modest computer has successfully controlled fed-batch cultivations of recombinant *Escherichia coli* for phenylalanine production (KONSTANTINOV et al., 1992b). The shape analysis procedure was found particularly useful in this process to monitor the large number of physiological phenomena and to trigger adequate control actions. Essentially, feedforward control according to a fixed predetermined schedule was substituted by feedback or event-triggered control capable of correcting for the influence of unknown disturbances. Two examples of typical situations are shown below.

Handling of process phase transitions: A typical example of the application of the shape analysis procedure is the automatic detection and handling of the transition from the first (batch) to the second (fed-batch) cultivation phase. The most informative variable, closely related to the underlying physiological

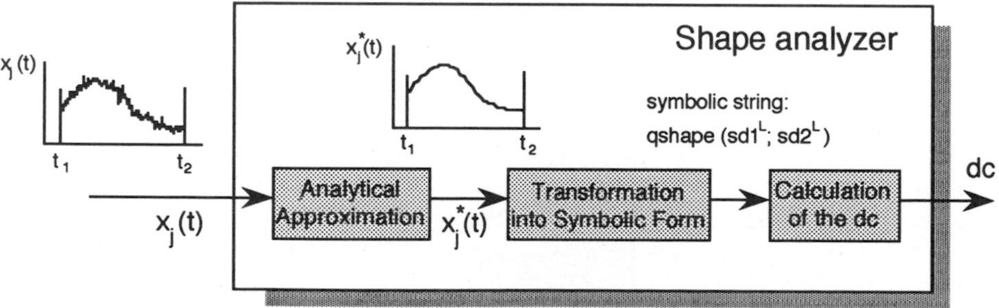

Fig. 15. The three steps in the evaluation of the degree of certainty *dc*.

changes, is the dissolved oxygen concentration *DO*, because it exhibits a characteristic shape during the transiton phase (Fig. 16). Initially, the following rule was used to detect the glucose depletion, i.e., the end of the batch phase:

IF (DOincrement > 5%),
THEN (Report: Glucose depletion) and
(Activate glucose feeding)

However, this rule, which is based only on the momentary *DO* increment, does not always work reliably due to unpredictable disturbances. It was therefore replaced by the rule:

IF (DOincrement > 5%) and (During
TheLast30secDO has been Increasing),
THEN (Report: Glucose depletion) and
(Activate glucose feeding)

This correction practically eliminated the errors observed previously. It is worth noting, that the above rule represents an effective combination of qualitative and quantitative reasoning. Such constructions were found to provide high reliability and expressive power.

To confirm that the transition has passed smoothly, the inference engine invokes another rule which checks for the expected *DO* shape a few minutes after starting the feed:

IF (SinceActivationOfTheGlucoseFeeding DO
hasPassedOverMaximum),
THEN (Report: Normal transition to Phase 2)

If this rule does not fire, the system will call other rules to identify the problem and its possible cause.

Detection of foaming: During the first stage of the cultivation, the culture produces foam which is eliminated by the addition of an antifoam agent. Since it is impossible to use a special foaming sensor, occurrence of foam has to be detected using other signals. A detailed analysis of the *DO* showed that its value decreases following a convex pattern during this stage of the process. In the case of foaming, however, this pattern changes to concave (Fig. 17), which can be efficiently detected by a shape analyzing rule

IF (DuringTheLast1hr DO
has beenDecreasingConvexlyConcavely),
THEN (Report: Foaming) and
(Feed antifoam agent)

These examples illustrate how shape analysis algorithms can help to incorporate operator experience or qualitative knowledge of cause–effect relationships in an advanced control scheme. However, all the rules were known *a priori* in this application. Some of the methodologies described earlier in Sect. 7.2 extract those rules themselves from collected data without *a priori* information. Subsequently, these rules can be used on-line as shown above or for process optimization.

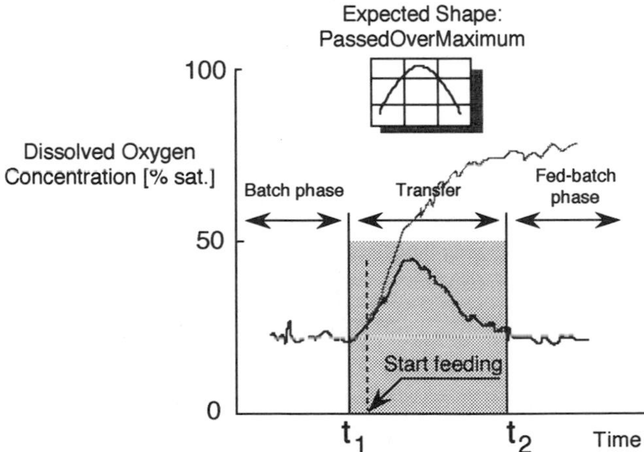

Fig. 16. An example how process phase transitions are handled: feed initiation.

8 Conclusions

In the preceding sections, various methodologies were presented to transform the raw measurements into variables which are more informative with respect to the underlying biological phenomena. We conclude this chapter by making a few remarks concerning their practical application and their significance within a wider range of common data analysis and representation techniques.

In order to get the highest benefit possible from all the measurements collected at considerable cost (e.g., analytical equipment, laboratory personnel) the methodologies must be applied on a routine basis. It must be emphasized that most of the required calculations are not very demanding in terms of computing power and complexity. Therefore, their implementation on today's personal computers using a spreadsheet or similar data manipulation software is straightforward even for the scientist or engineer with no programming experience at all. Some of the more involved techniques are accessible through commercially available software packages. Developing data analysis software from scratch is very time-consuming and only justified in a few rare cases.

Taking all possible combinations into account, the number of derived variables will be rather large. In most cases, however, a subset is sufficient and preferable to characterize the process at hand, because some of the variables are correlated or have a low information content. Elimination of unimportant and selection of the most discriminating variables reduces the number of degrees of freedom, thus facilitating process analysis and experimental design at the development stage. For the purpose of process supervision and control, redundancy is instrumental in detecting sensor malfunctioning, while variables with a low information content should be discarded to arrive at a simple and robust control scheme. Unfortunately, no general guidelines for variable selection have been developed hitherto. It is still the noble task of the scientist or engineer to make a choice using process-specific knowledge and his or her best judgment.

Often, dynamic models with a varying degree of detail are developed from experimental data for process simulation and optimization, or simply to get a more compact and transparent representation of a large volume of data. Most dynamic models have three components, although this is not always explicitly stated:

1. The mass balance equations for each species in the system under consideration,

2. the stoichiometry which describes how each species participates in a reaction, i.e., the relationships between different *PV*s,

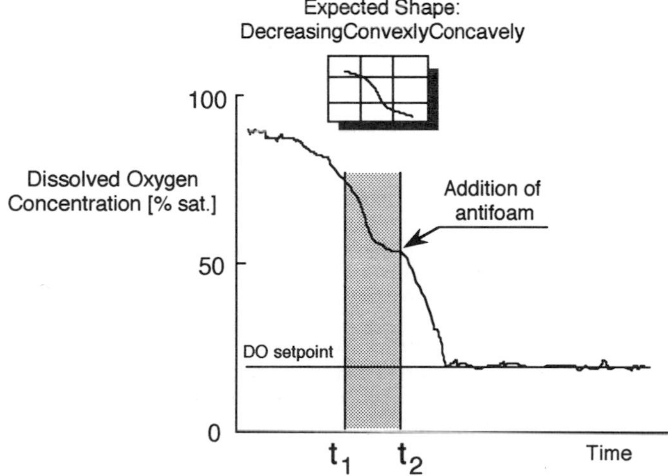

Fig. 17. An example how specific events are detected: foaming.

3. the kinetic rate expressions which describe the dependencies of the *PV*s on the *EV*s, i.e., a mathematical description of the CP.

Despite the popularity of dynamic models we have proposed a different, more data-oriented approach in this chapter. In Sects. 3–6, mainly balance equations, (1) above, and to a smaller extent stoichiometric relationships, (2) above, were exploited to derive the *PV*s. Instead of going one step further towards a dynamic model by deriving kinetic rate expressions, (3) above, *PSV*s were defined to characterize the CP. Even more information can be extracted from the *PSV*s without any mechanistic knowledge of the CP by pattern recognition techniques some of which were discussed in Sect. 7.

The list of applicable techniques is certainly not exhaustive, and there is no doubt that future research will provide additional useful algorithms and exemplary case studies. We believe that this data-oriented approach is well justified for the analysis of complex processes or when not enough time is available to develop a full-fledged mechanistic process model. Both situations are very common in the development stage of industrial bioprocesses.

Acknowledgement
The writing of this chapter was supported in part by the MIT BioProcess Engineering Center and a fellowship by the Swiss National Science Foundation. The authors are very grateful to the reviewers for their most valuable comments and suggestions.

References

AGARWAL, M., BONVIN, D. (1989), Limitations of the extended Kalman filter, *Proc. "IFAC DYCORD"*, Maastricht, The Netherlands, pp. 299–306.

AIBA, S., NAGAI, S., NISHIZAWA, Y. (1976), Fed-batch culture of *Saccharomyces cerevisiae:* a perspective of computer control to enhance the productivity of baker's yeast fermentation, *Biotechnol. Bioeng.* **18,** 1001–1016.

AKASHI, K., SHIBAI, H., HIROSE, Y. (1979), Effect of oxygen supply on L-phenylalanine, L-proline, L-glutamine and L-argenine fermentations, *J. Ferment. Technol.* **57,** 321–327.

ANDERSON, B. D. O., MOORE, J. B. (1979), *Optimal Filtering*, Englewood Cliffs, NJ: Prentice Hall.

BAKSHI, B. R., STEPHANOPOULOS, G. (1992a), Temporal representation of process trends for diagnosis and control; *Preprints IFAC Symp. "On-Line Fault Detection and Supervision in the Chemical Process Industries"*, Newark, pp. 69–74.

BAKSHI, B. R., STEPHANOPOULOS, G. (1992b), Representation of process trends – Part III. Multi-scale extraction of trends from process data, *Comput. Chem. Eng.*, submitted.

BAKSHI, B. R., STEPHANOPOULOS, G. (1992c), Representation of process trends – Part IV. Induction of real-time patterns from operating data for diagnosis and supervisory control, *Comput. Chem. Eng.*, submitted.

BASTIN, G., DOCHAIN, D. (1986). On-line estimation of microbial specific growth rates, *Automatica* **22,** 705–709.

BASTIN, G., DOCHAIN, D. (1990), *On-Line Estimation and Adaptive Control of Bioreactors*, Amsterdam: Elsevier.

BHAT, N., MCAVOY, T. J. (1990), Use of neural nets for dynamic modeling and control of chemical process systems, *Comput. Chem. Eng.* **14,** 573–583.

CHATTAWAY, T., STEPHANOPOULOS, G. (1989a), An adaptive state estimator for detecting contaminants in bioreactors, *Biotechnol. Bioeng.* **34,** 647–659.

CHATTAWAY, T., STEPHANOPOULOS, G. (1989b), Adaptive estimation of bioreactors: Monitoring plasmid instability, *Chem. Eng. Sci.* **44,** 41–48.

CHEUNG, J. T.-Y., STEPHANOPOULOS, G. (1990a), Representation of process trends – Part I. A formal representation framework, *Comput. Chem. Eng.* **14,** 495–510.

CHEUNG, J. T.-Y., STEPHANOPOULOS, G. (1990b), Representation of process trends – Part II. The problem of scale and qualitative scaling, *Comput. Chem. Eng.* **14,** 511–539.

COONEY, C. L., WANG, H. Y., WANG, D. I. C. (1977), Computer-aided material balancing for prediction of fermentation parameters, *Biotechnol. Bioeng.* **19,** 55–67.

DOCHAIN, D., BASTIN, G. (1985), Stable adaptive algorithms for estimation and control of fermentation processes, in: *Modelling and Control of Biotechnological Processes* (JOHNSON, A., Ed.), pp. 37–42, Oxford: Pergamon Press.

DOCHAIN, D., DE BUYL, E., BASTIN, G. (1988),

Experimental validation of a methodology for on-line state estimation, in: *Computer in Fermentation Technology* (FISH, N. M., FOX, R. I., THORNHILL, N. F., Eds.), pp. 187–194, New York: Elsevier Applied Science.

ERICKSON, L. E. (1979), Applications of mass-energy balances in on-line data analysis, *Biotechnol. Bioeng. Symp.* **9,** 49–59.

FORBUS, K. D. (1984), Qualitative process theory, *Artif. Intell.* **24,** 85–168.

FRIEDLAND, B. (1987), *Control System Design*, New York: McGraw-Hill.

GELB, A. (1974), *Applied Optimal Estimation*, Cambridge, MA: MIT Press.

HAMER, J. W. (1989), Stoichiometric interpretation of multireaction data: Application to fed-batch fermentation data; *Chem. Eng. Sci.* **44,** 2363–2374.

HEINZLE, E. (1987), Mass spectrometry for on-line monitoring of biotechnological processes, *Adv. Biochem. Eng. Biotechnol.* **35,** 1–45.

HEINZLE, E., DUNN, I. J. (1991), Methods and instruments in fermentation gas analysis, in: *Biotechnology Second Edition* (REHM, H.-J., REED, G., Eds.), Vol. 4, pp. 27–74, Weinheim–New York–Basel–Cambridge: VCH.

HOSKINS, J. C., HIMMELBLAU, D. M. (1988), Artificial neural network models of knowledge representation in chemical engineering, *Comput Chem. Eng.* **12,** 881–890.

ISHIHARA, K., SUZUKI, T., YAMANE, T., SHIMIZU, S. (1989), Effective production of *Pseudomonas fluorescens* lipase by semi-batch culture with turbidity-dependent automatic feeding of both olive oil and iron ion, *Appl. Microbiol. Biotechnol.* **31,** 45–48.

JAZWINSKI, A. H. (1970), *Stochastic Processes and Filtering Theory*, New York: Academic Press.

KONSTANTINOV, K. B., YOSHIDA, T. (1989), Physiological state control of fermentation processes, *Biotechnol. Bioeng.* **33,** 1145–1156.

KONSTANTINOV, K. B., YOSHIDA, T. (1990), On-line monitoring of representative structural variables in fed-batch cultivation of recombinant *Escherichia coli* for phenylalanine production; *J. Ferment. Bioeng.* **70,** 421–427.

KONSTANTINOV, K. B., YOSHIDA, T. (1992), Qualitative analysis of the temporal shapes of bioprocess variables, *AIChE J.* **38,** 1703–1715.

KONSTANTINOV, K. B., AARTS, R., YOSHIDA, T. (1992a), On the selection of variables representing the physiological state of the cell cultures, *Proc. IFAC Symp. Modeling and Control of Biotechnical Processes*, Keystone, Oxford: Pergamon Press.

KONSTANTINOV, K. B., MATANGUIHAN, R., YOSHIDA, T. (1992b), Physiological state control of recombinant amino acid production using micro expert system with modular, embedded architecture, *Proc. IFAC Symp. Modeling and Control of Biotechnical Processes*, Keystone, Oxford: Pergamon Press.

KONSTANTINOV, K. B., PAMBAYUN, R., MATANGUIHAN, R., YOSHIDA, T., PERUSICH, C., HU, W.-S. (1992c), On-line monitoring of hybridoma cell growth using a laser turbidity sensor; *Biotechnol. Bioeng.*

KRAMER, M. A., LEONARD, J. A. (1990), Diagnosis using backpropagation neural networks – analysis and criticism, *Comput. Chem. Eng.* **14,** 1323–1338.

KRESTA, J. V., MAC GREGOR, J. F., MARLIN, T. E. (1991), Multivariate statistical monitoring of process operating performance, *Can. J. Chem.* **69,** 35–47.

KUIPERS, B. (1986), Qualitative simulation, *Artif. Intell.* **29,** 289–338.

LIAO, J. C. (1989), Fermentation data analysis and state estimation in the presence of incomplete mass balance, *Biotechnol. Bioeng.* **33,** 613–622.

LOCHER, G., SONNLEITNER, B., FIECHTER, A. (1990), Pattern recognition: a useful tool in technological processes, *Bioprocess Eng.* **5,** 181–187.

MEYER, C., BEYELER, W. (1984), Control strategies for continuous bioprocess based on biological activities, *Biotechnol. Bioeng.* **26,** 916–925.

MOORE, R., ROSENOF, H., STANLEY, G. (1990), Process control using real-time expert system; *Preprints 11th IFAC World Congress*, Tallin, Vol. 7, pp. 234–239.

OGATA, K. (1987), *Discrete-Time Control Systems*, Englewood Cliffs, NJ: Prentice-Hall.

OLSVIK, E. S., KRISTIANSEN, B. (1992), On-line rheological measurements and control in fungal fermentations, *Biotechnol. Bioeng.* **40,** 375–387.

ROMAGNOLI, J. A., STEPHANOPOULOS, G. (1980), On the rectification of measurement errors for complex chemical plants, *Chem. Eng. Sci.* **35,** 1067–1081.

ROMAGNOLI, J. A., STEPHANOPOULOS, G. (1981), Rectification of process measurement data in the presence of gross errors, *Chem. Eng. Sci* **36,** 1849–1863.

ROUSSET, S., SCHLICH, P. (1989), Amylase production in submerged culture using principal component analysis, *J. Ferment. Bioeng.* **68,** 339–343.

RUMELHART, D. E., HINTON, G. E., WILLIAMS, R. J. (1986a), Learning representations by back-propagating errors, *Nature* **323,** 533–536.

RUMELHART, D. E., HINTON, G. E., WILLIAMS, R. J. (1986b), Learning internal representations by error propagation, in: *Parallel Distributed Processes* (RUMELHART, D. E., MCCLELLAND, J. L., Eds.), pp. 318–362, Cambridge, MA: MIT Press.

SAN, K.-Y., STEPHANOPOULOS, G. (1984a), Studies on on-line bioreactor identification. II. Numerical and experimental results, *Biotechnol. Bioeng.* **26**, 1189–1197.

SAN, K.-Y., STEPHANOPOULOS, G. (1984b), Studies on on-line bioreactor identification. IV. Utilization of pH measurements for product estimation, *Biotechnol. Bioeng.* **26**, 1209–1218.

SANER, U. (1991), Modelling and On-Line Estimation in a Batch Culture of *Bacillus subtilis, Ph. D. Thesis* No. 9481, ETH Zürich.

SANER, U., STEPHANOPOULOS, G. (1992), Application of pattern recognition techniques to fermentation data analysis, *Proc. IFAC Symp. on Modeling and Control of Biotechnical Processes*, Keystone, Oxfored: Pergamon Press.

SHIMIZU, H., SHIOYA, S., SUGA, K., TAKAMATSU, T. (1989), Profile control of the specific growth rate in fed-batch experiments, *Appl. Microbiol. Biotechnol.* **30**, 276–282.

STEPHANOPOULOS, G. (1984), *Chemical Process Control*, Englewood Cliffs, NJ: Prentice-Hall.

STEPHANOPOULOS, G., SAN, K.-Y. (1984), Studies on on-line bioreactor identification. I. Theory, *Biotechnol. Bioeng.* **26**, 1176–1188.

THAM, M. T., MONTAGUE, G. A., MORRIS, A. J. (1988), Application of on-line estimation techniques to fermentation processes, *Proc. Am. Control Conf.* **7**, 1129–1133.

THIBAULT, J., GRANDJEAN, B. P. A. (1992), Neural networks in process control – a survey, in: *Advanced Control of Chemical Processes (ADCHEM'91)* (NAJIM, K., DUFOUR, E., Eds.), Oxford: Pergamon Press.

THIBAULT, J., VAN BREUSEGEM, V., CHÉRUY, A. (1990), On-line prediction of fermentation variables using neural networks, *Biotechnol. Bioeng.* **36**, 1041–1048.

VAN DER HEIJDEN, R. T. J. M. (1991), State estimation and error diagnosis for biotechnological processes, *Ph. D. Thesis*, University of Technology, Delft, The Netherlands.

WANG, N. S., STEPHANOPOULOS, G. (1983), Application of macroscopic balances to the identification of gross measurement errors, *Biotechnol. Bioeng.* **25**, 2177–2208.

WANG, H. Y., COONEY, C. L., WANG, D. I. C. (1979a), Computer control of baker's yeast production, *Biotechnol. Bioeng.* **21**, 975–995.

WANG, H. Y., COONEY, C. L., WANG, D. I. C. (1979b), On-line gas analysis for material balances and control, *Biotechnol. Bioeng. Symp.* **9**, 13–23.

WILLIS, M. J., DI MASSIMO, C., MONTAGUE, G. A., THAM, M. T., MORRIS, A. J. (1991), Artificial neural networks in process engineering, *IEE Proc. D.* **138**, 256–266.

WISE, B. M., RICKER, N. L. (1989), Feedback strategies in multiple sensor systems, *AIChE Symp. Ser.* **267**, 19–23.

YAMANE, T., HIBINO, W., ISHIHARA, K., KADOTANI, Y., KOMINAMI, M. (1992), Fed-batch culture automated by use of continuously measured cell concentration and cell volume, *Biotechnol. Bioeng.* **39**, 550–555.

16 Design of Aseptic, Aerated Fermentors

MARVIN CHARLES

Bethlehem, PA 18015, USA

JACK WILSON

Allentown, PA 18106, USA

List of Symbols

A_c	vessel cross-sectional area (m^2)
A_s	jacket surface area (m^2)
d_i	impeller diameter (m)
D_t	tank diameter (m)
F	gas flow (mol min^{-1} or mol h^{-1})
g	gravitational constant (m s^{-2})
g_c	Newton's law conversion (m^3 kg^{-1} s^{-2})
h_l	liquid height (m)
H_g	gas holdup fraction
$K_g a$	oxygen transfer coefficient (mmol L^{-1} h^{-1} bar^{-1})
N	impeller speed (rps or rpm)
N_a	aeration number
N_p	power number
N_{Re}	Reynolds number
OTR	oxygen transfer rate (mmol L^{-1} h^{-1})
P^*	equilibrium oxygen partial pressure (bar)
P_g	gassed mechanical power input (kW)
P_i	inlet gas oxygen partial pressure (bar)
P_o	outlet gas oxygen partial pressure (bar)
P_t	average pressure in the vessel (bar)
$P_{t,i}$	inlet gas total pressure (bar)
$P_{t,o}$	outlet gas total pressure (bar)
P_{ug}	ungassed power draw (kW)
$(P)_{lm}$	log mean pressure driving force (bar)
Q	gas flow rate (m^3 min^{-1}, L min^{-1}, m^3 h^{-1}, or L h^{-1})
Q_g	heat generation rate (W)
T_f	fermentation temperature (°C)
T_i	cooling water inlet temperature (°C)
T_o	cooling water outlet temperature (°C)
$(\Delta T)_{lm}$	log mean temperature driving force (K)
U	overall heat transfer coefficient (W m^{-2} K^{-1})
V_l	liquid volume (L or m^3)
V_s	linear gas flow rate (cm min^{-1})
X	cell mass concentration (dry basis) (g L^{-1})
y_i	inlet gas oxygen mole fraction
y_o	outlet gas oxygen mole fraction
y^*	equilibrium oxygen mole fraction
Y_o	cell yield coefficient on oxygen (g g^{-1})
η	viscosity (kg m^{-1} s^{-1})
μ	specific growth rate (h^{-1})
ρ	fluid density (kg m^{-3})
Φ	oxygen transfer efficiency

1 Scope

The scope of this chapter is limited to major factors of the design and construction of stirred tank fermentors for aerobic, aseptic, microbial fermentations. Applications of basic principles and actual practices are illustrated by means of a design example for a 12000 liter (working volume) fermentor.

We assume that the reader has at least some background in fermentation; therefore, we present briefly only those basic principles which are pertinent to the design philosophies and practices discussed.

2 Introduction

A fermentor is only one part of a fermentation process. Its design must derive from a rational plant design basis (see below) which includes factors such as annual production, efficiencies, scheduling considerations, containment and sterility requirements, etc. The designer also must consider safety, reliability, cost, architectural constraints, labor requirements, and must cope effectively with conflicting specifications, incomplete information, unreliable correlations, and other factors, many of which arise from realities of fabrication, shipping, vendors, personal perceptions, and a host of other interrelated factors which are difficult or impossible to quantify in any rigorous sense.

It should be clear that fermentor design and construction are not simple tasks which can yield to generic, off-the-shelf solutions. Each case must be considered on its own merits so that the inevitable compromises can be made in the most intelligent way possible. The bottom line is that really good design and construction require a lot of experience and insight applied intelligently, adaptively and with a great deal of attention paid to detail.

We do not have the space here to consider in detail all of the factors mentioned above. We will, however, attempt to give a sense of the nature of the process of rational design – at least in so far as the fermentor itself is concerned.

3 Basic Design Calculations

The primary driving forces for basic fermentor design calculations are the aeration, agitation, and heat transfer required to satisfy production specifications. In theory, these calculations appear to be relatively simple and straightforward. This is not so in practice because of the interrelationships not only between oxygen and heat transfer, but also because of the interaction of these with practical constraints on vessel geometry, impellers, etc., and of course those imposed by the laws of physics.

3.1 Oxygen Transfer

The oxygen transfer rate (*OTR*) required depends on the cell mass concentration (X), the specific growth rate (μ) and the yield coefficient on oxygen (Y_o). For the "usual" case in which the *OTR* is just about equal to the oxygen requirement of the organisms,

$$OTR = \mu X / Y_o \tag{1}$$

Y_o and μ vary during most fermentations. The variation in Y_o is not large in most practical fermentations: Y_o usually may be taken constant for design purpose. Variations in μ often can be large enough to have very significant effects on the design calculations, and these should be evaluated quantitatively. More often than not, however, complete, reliable data are not available, and reasonable assumptions must be made.

The rate at which oxygen *can* be transferred depends on the oxygen transfer coefficient ($K_g a$) and the partial pressure driving force. If the liquid phase is mixed perfectly and the gas phase is in plug flow, then (WANG et al., 1979a)

$$OTR = K_g a (\Delta P)_{lm} \tag{2}$$

where $(\Delta P)_{lm}$ is the log mean partial pressure driving force defined as

$$(\Delta P)_{lm} = (P_i - P_o) / \ln[(P_i - P^*)/(P_o - P^*)] \tag{3}$$

P_o and P_i are the outlet and inlet oxygen partial pressures in the gas, respectively, and P^* is the oxygen partial pressure that would be in equilibrium with the actual dissolved oxygen concentration in the liquid. None of the assumptions noted is strictly true in any practical fermentor; however, in most cases they are "close enough" for practical calculations.

The oxygen partial pressures can be expressed in terms of the corresponding oxygen mole fractions (y) and total pressures (P_t)

$$P_i = P_{t,i}\, y_i \tag{4}$$

$$P_o = P_{t,o}\, y_o \tag{5}$$

where $P_{t,o}$ is the headspace pressure and $P_{t,i}$ is the pressure at the bottom of the tank:

$$P_{t,i} = P_{t,o} + \rho (g/g_c) h \tag{6}$$

where ρ is the fluid density and h is the fluid height. $P_{t,i}$ obviously depends on the height of the broth which, in turn, depends on the vessel geometry.

The inlet mole fraction (y_i) can be set independently. For air, the value is approximately 0.21. This can be increased by oxygen enrichment which is a typical lab approach, but which can be quite costly on a commercial scale. The outlet mole fraction (y_o) is determined by material balance:

$$OTR\ V_l = F(y_i - y_o) \tag{7}$$

or

$$y_o = y_i - (V_l\ OTR)/F \tag{8}$$

where F is the total molar flow of gas and V_l is the liquid volume. For most practical fermentations, F may be assumed to be reasonably constant. y_o also can be expressed in terms of the oxygen transfer efficiency (Φ):

$$y_o = (1 - \Phi) y_i \tag{9}$$

Φ usually is in the range of 0.15 to 0.35 for most practical yeast and bacterial fermentations. For mycelial fermentations, the range usually is 0.1 to 0.2.

The total gas molar flow, F, can, in theory, be set independently. There are, however, practical considerations which usually set a limit of about 180 cm/min on the actual linear gas flow rate (V_s) through the vessel. This will be discussed later.

The mass transfer coefficient ($K_g a$) is related in a complex way to the input power under gassed conditions (P_g), the gas linear velocity and the geometries of the tank and the impellers. The literature is replete with correlations for $K_g a$ which usually are expressed in the form (WANG et al., 1979b):

$$K_g a = a (P_g / V_l)^b (V_s)^c \tag{10}$$

where a, b, and c are "constants", and V_l is the unaerated liquid volume. Unfortunately, not only is the reliability of such correlations restricted to the specific fermentor geometries and the fermentations on which the correlations are based, but also to the scales used to obtain them (most of which are for small lab fermentors under conditions which all-too-often are not meaningful practically). Indeed, different published correlations often predict widely divergent results for the same proposed conditions, and more often than not, the range of predicted values is nowhere near what is found in practice on commercial scales. If at all possible, then, one should use experimental values or at least temper the calculated results with experience.

The problem is exacerbated by the fact that it is difficult to predict accurately P_g, and that P_g and V_s are related intimately with all other aspects of the fermentor design (e.g., vessel geometry) and operation (e.g., bulk mixing quality).

3.2 Power

P_g depends in a complex way on vessel geometry, gas flow, impeller characteristics and speed, etc. Because of this complex dependence of P_g on so many factors, most of the published correlations for calculating it are cast in terms of the ratio P_g/P_{ug}, where P_{ug} is the power draw of the same impeller operated under the same conditions but without aeration. P_{ug} for a given impeller is a function of impeller diameter and speed, and of the density and viscosity of the unaerated fluid. Most correlations (WANG et al., 1979c) for P_{ug} usually are cast in terms of the impeller power number, N_p, and the impeller Reynolds number N_{Re}:

$$N_p = P_{ug}(N^3 d_i^5 \rho \tag{11}$$

and

$$N_{Re} = N\, d_i^2 \cdot \rho/\eta \tag{12}$$

where P_{ug} is the ungassed power (W), N is the impeller speed (rps, revolutions per second), and d_i is the impeller diameter (m), ρ is the fluid density (kg m^{-3}), and η is viscosity (kg m^{-1} s^{-1}). Such correlations usually are reliable enough for most design calculations so long as the designer accounts for the fact that each such correlation is valid only for a specific geometry and for a single impeller.

The total P_{ug} for multiple impellers is not a simple multiple of P_{ug} for a single impeller, but depends on impeller spacing unless the impeller flow patterns do not interact strongly with each other (OLDSHUE, 1983a).

It is also important to note that when N_{Re} is greater than 10000, the fluid flow is "fully turbulent", and N_p becomes practically constant. Most highly aerated fermentors are operated under fully turbulent conditions; hence, it is common practice to speak of a single N_p for a given impeller, and to infer from this the relative power draw. Some care should be exercised here because what is really important for oxygen transfer is P_g, not P_{ug}. It is not necessarily true that if N_p for impeller A is higher than N_p for impeller B, then impeller A will have the higher P_g under similar aerated operating conditions. This brings us back to the relationship between P_g and P_{ug}. In the absence of any experimental data (which is by far the most usual case), the P_g/P_{ug} correlation we use most frequently as a preliminary guide is one based on the dimensionless aeration number (N_a) (NAGATA, 1975a):

$$P_g/P_{ug} = f(N_a) \tag{13}$$

where

$$N_a = Q/N\, d_i^3 \tag{14}$$

Q is the gas volumetric flow rate, N is the impeller speed, and d_i is the impeller diameter. This correlation usually is presented in graphical form (HICKS and KIME, 1976) with different curves for each type of impeller. Unfortunately, it turns out in practice that the power ratio does not behave so simply. For a fixed N_a, P_g/P_{ug} depends on the specific values of Q, N, and d_i, on the vessel geometry, and on the scale. So, once again, the calculated results must be tempered with a healthy dose of experience, the dosage depending on the specific case.

P_g/P_{ug} also depends very strongly on the impeller type. This is one reason to avoid jumping to conclusions based simply on N_p. The most widely used impeller in fermentation is the Rushton turbine which has a high N_p ($N_p = 6$ for a 6-blade turbine and $N_p = 9.6$ for a 12-blade turbine (RUSHTON et al., 1950). The Rushton impeller does a good job on gas dispersion and oxygen transfer, but it has a relatively low (0.4–0.5) P_g/P_{ug} under practical fermentation conditions. It also produces primarily radial flow which does little to contribute to good bulk mixing (see below). Some newer impellers (e.g., the A-315) have lower N_ps but have considerably higher P_g/P_{ug} values than the Rushton impeller (OLDSHUE et al., 1988). Some, such as the Lightnin A-315 (Mixco Equipment Company), also produce good bulk mixing. It also has been reported (BALMER et al., 1988; GBEWONYO et al., 1986) that some of the newer designs have a higher *OTR*/P_g under some conditions than do Rushtons; however, not enough practical information is yet available about them to allow reliable predictions. They do, nevertheless, warrant serious consideration now for some applications (see the discussion of bulk mixing, below).

Cases involving multiple impellers are even more complicated, primarily because in practical circumstances each of the impellers operates under somewhat different conditions (e.g., effective fluid density). Unfortunately, there are no really reliable methods available to predict quantitatively the power distribution among the impellers or what would be the "best" distribution. There are several different rules of thumb and calculational procedures used commonly, but we do not have the space here to discuss them. The reader should consult OLDSHUE (1983b) and CHARLES and WILSON (1992) for details.

We conclude this section with some additional practical factors one should consider when determining impeller diameter, number, speed and spacing:

(1) Geometric constraints (e.g., liquid height to tank diameter ratio) imposed by mixing quality considerations, vessel internal diameter, etc. (see below).

(2) The fact that the power input is an extremely strong function of impeller diameter (almost fifth power).

(3) The fact that shaft torque varies directly with d_i which must be considered in the light of standard gearbox specifications and costs – particularly for larger vessels. This involves a simultaneous consideration of impeller speed in that one must take into account standard gearbox ratios and motor specifications including realistic turn-down ratios.

(4) Impeller spacing and number do not appear to have much effect on $K_g a$ at fixed P_g, but spacing does have a considerable effect on power input itself (OLDSHUE, 1983a) and on the bulk mixing characteristics (see below). Guidelines that we have found to work well call for the bottom impeller to be mounted one impeller diameter from the bottom of the vessel, and additional impellers to be mounted 1 to 1.5 impeller diameters apart. The top impeller should be mounted such that the liquid height above it is approximately 1 to 1.5 impeller diameters. There are other published guidelines (CHARLES and WILSON, 1992), but most of them are based primarily on studies in which all the impellers were of the same type (not always the case in fermentation) and were operated in non-aerated vessels. Also, one must factor into the compromise equation the fact that during a fed-batch fermentation, the volume (hence liquid height) changes continually. This is also true for batch fermentations in which the gas holdup changes appreciably.

(5) The potential for vibration must be minimized to avoid jeopardy to workers and serious damage to equipment. The potential for vibration is determined by

- the impeller sizes, types, weights and locations,
- the shaft length, weight and diameter,
- the weights and locations of any fittings on the shaft,
- the nature of the shaft suspension, and
- the shaft speed.

These are all interrelated; therefore, vibration considerations can impose important constraints of the selection of impeller diameter, shaft speed, etc. which must be reconciled with the requirements of the fermentation. The bottom line is that the design must insure that the maximum impeller speed will be at least 25% below the natural frequency of the shaft/impeller assembly. See CHARLES and WILSON (1992) for a detailed discussion of calculations for vibration analysis.

(6) There are organisms which are sensitive to the fluid mechanical forces developed in a fermentor. Bacteria are at one end of this sensitivity scale, and mammalian cells are at the other (most sensitive). A host of theories and rules of thumb about fluid mechanical forces and their effects have been published. These fall into the general categories of tip speed, fluid shear, and fluid turbulence theories. None of these yet provides enough reliability for general design applications, and some have led to the unwarranted acceptance of "magic numbers" (e.g., the 1500 fpm rule) which usually are based on only very limited experience but are all-too-often inferred to be widely applicable. Needless to say, such "magic numbers" are not the answer for rational design, and once again the designer must bring to the problem a lot of practical experience. See CHARLES and WILSON (1992) and HU and WANG (1986) for further information.

(7) The selection of impeller diameter must be guided by all of the factors discussed above (e.g., power required, mixing quality, tank geometry). This often presents the designer with some difficult compromises which call for considerable experience and judgement. There are no absolute rules. All we can offer are a few guidelines which we have found to be helpful:

- The ratio of impeller diameter to tank diameter (d_i/D_t) usually falls in the range of 0.33 to 0.5. Note: there is a popular mis-conception that requires d_i/D_t to be approximately 0.33. Other than the fact that most power correlations are based on this ratio, there is no rational basis for selecting it. The simple fact is that in all-too-many cases it requires unreasonable compromises for other factors. It is one more example of the "magic number" syndrome which does little other than to discourage rational design and the exercise of good judgement.

- For very viscous broths, the best compromises usually result in a d_i/D_t at the high end of the range. The requirements for low viscosity broths usually can be met with ratios in the lower end of the range; however, we have found a fairly large number of cases for which it has been necessary to specify high ratios (0.45 to 0.48). This usually turns out to be the case when the power requirement is very high.
- Mammalian cell cultures require very special attention. The combined requirements of low power, very low speed, and good bulk mixing usually are satisfied best by use of a very large impeller. Indeed, we have found that a very satisfactory solution is to use as large an impeller as will fit into the vessel and which will distribute the power as uniformly as possible throughout the culture (see CHARLES and WILSON, 1992, for a description of the "elephant ear" impeller).

3.3 Linear Gas Velocity

Linear gas velocity, V_s, affects OTR not only through its effect on $K_g a$, but also by its influence on the oxygen partial pressures (via the oxygen material balance). The overall theoretical effects of V_s can be shown by combining Eqs. (2)–(5), (7), (9), and (10):

$$OTR = \frac{26786\, V_s A_c P_t}{V_l} \cdot \frac{298}{T_f + 273} [1 - \exp(-Z \cdot V_s^{c-1})] \quad (15)$$

where

$$Z = 3.73 \cdot 10^{-5} \left(\frac{T+273}{298}\right) V_l\, a \left(\frac{P_g}{V_o}\right)^b / A_c \quad (16)$$

T_f is the fermentation temperature, P_t the average pressure in the tank, and A_c is the tank cross-sectional area.

Eq. (15) is plotted in Fig. 1 for the following conditions: $D_t = 2.44$ m; $P_t = 2$ bar; $T_f = 30$ °C; $y_i = 0.21$; $V_l = 10000$ L; $P_g/V_l = 4.98$ kW m^{-3};

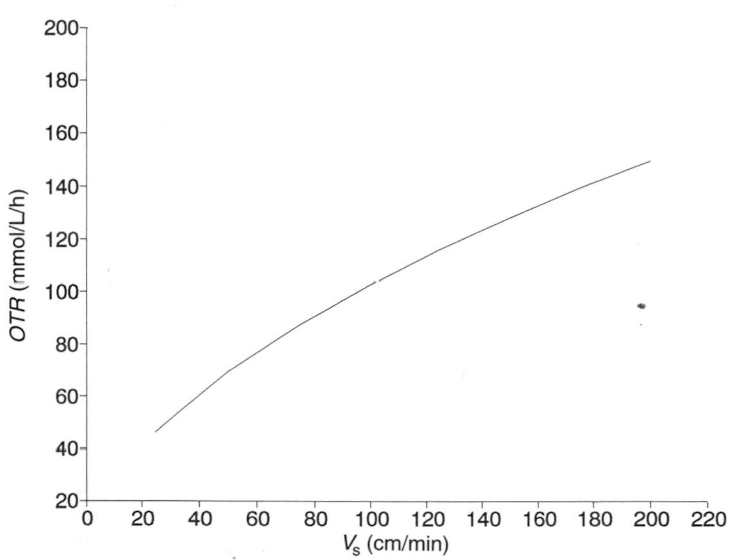

Fig. 1. Oxygen transfer rate as a function of gas linear velocity.

$y^* = 0.0$. The $K_g a$ correlation used was:

$$K_g a = a(P_g/V_l)^b V_s^c$$

with $a = 12.18$, $b = 0.5$, and $c = 0.5$.

It is important to note that the intent here is to give only a semiquantitative sense of the effect because of the uncertainty associated with the $K_g a$ correlation.

The net theoretical effect of increasing V_s alone is to increase *OTR*. This also is observed in practice, but only up to the point of impeller flooding (the impeller becomes enshrouded in air) or the practical limitations imposed by gas holdup, foaming and/or broth blow-out. At the flooding point, the gassed power draw decreases precipitously which causes a marked drop in *OTR*. There are published correlations (DICKEY, 1979) for flooding, but in applying them one should take into account that the physical properties of the fluid used to develop the correlations probably were not the same as those of the subject fermentation broth. In addition, such correlations tend to depend on the specifics of the equipment used and on scale.

Gas holdup also increases with increasing V_s. One correlation for this effect is given by NAGATA (1975b):

$$H_g = \alpha(P_g/V_l)^{0.47} V_s^{0.53} \tag{17}$$

The same caveats noted for flooding correlations apply here.

Obviously, useful volume and productivity decrease as holdup increases. Also, very high gas holdups will decrease heat transfer rates (see below) to unacceptably low levels. For these reasons, holdup should not exceed 20–25% of the unaerated liquid volume.

Foaming also is exacerbated by high levels of V_s. While the effect varies considerably from one fermentation to another, a fairly safe rule of thumb is that conditions which do not cause excessive holdup also will be acceptable with regard to foaming.

The same rule of thumb usually applies to broth blow-out.

V_s also must be considered with regard to its interrelationship with P_t, F, and D_t:

$$V_s/D_t^2 P_t/F = 4 \cdot 22.4\,[(T_f + 273)/298]/10\,\pi \tag{18}$$

where P_t is the total pressure in bar, T_f is the fermentation temperature in °C, and V_s is the actual linear velocity at fermentation conditions. (P_t may vary considerably from the top to the bottom of tall fermentors, but for most practical design calculations, it is adequate to use the average pressure in the vessel.)

Eq. (18) reveals some important design considerations:

- The limits imposed on V_s constrain the vessel geometry for a fixed operating pressure.
- For a fixed V_s and tank diameter, we can increase the molar flow rate of gas, F, if we increase the operating pressure. The increased pressure alone will tend to increase *OTR* by increasing directly the oxygen partial pressures. Increasing F also will tend to increase *OTR* due to oxygen material balance considerations. On the other hand, if we keep the *OTR* constant and increase P_t and/or F, we will decrease the $K_g a$; therefore, less power will be required to obtain the *OTR* specified.

But there are limits here also, particularly with regard to P_t. The higher we make P_t, the thicker we must make the vessel walls. Very thick vessel walls can be negative for several reasons:

- The vessel cost will increase.
- The thicker the walls, the greater the resistance to heat transfer (stainless steel has a very low thermal conductivity).
- If the walls get thick enough, the vessel must be fabricated from plate rather than from gauge metal. This results in a poorer quality vessel and a poorer finish.

There also are practical limits on the tank diameter imposed by mixing quality requirements, vessel fabrication practices and by several other practical considerations. More will be said later about these.

We conclude this section with a comment concerning the widely used term "gas volume flow per liquid volume per minute" (vvm). We have found all-too-often people discuss scale-up in terms of the vvm they have used in lab size fermentors operated at atmospheric pressure. This can be very misleading, and can cause serious problems depending on the size and operating conditions of the larger-scale vessel. The reason for this is quite simply that at a fixed vvm, V_s increases very rapidly with increasing tank diameter; therefore, the 2 vvm used in the 14-liter lab fermentor could lead easily to a dangerously high V_s in a 10000 liter production fermentor. The extent of the danger will depend on the vessel operating pressure. We urge, therefore, that scale-up be done on the basis of V_s, and that operating conditions be chosen on the basis of material balance and transfer rate estimates as discussed above. This is clearly a more rational way to proceed.

3.4 Bulk Mixing

Good bulk mixing (broth homogeneity) is required to insure meaningful measurements, good control and that only one fermentation is occurring. Mixing quality is affected by agitation, aeration, vessel geometry, and broth physical properties the most important of which are the rheological properties. Quantification of mixing quality is a real art. There are some correlations which describe mixing quality in terms of "mixing time", but these are at best very rough guides.

We have space here to discuss only a few of the important design considerations required to insure adequate bulk mixing. The reader is directed to the references CHARLES and WILSON (1992), CHARLES (1978), MOO-YOUNG et al. (1972), BLAKEBROUGH and SAMBAMURTHY (1966) for more detailed discussions.

3.4.1 Effects of Impellers

High power radial flow impellers (e.g., Rushton turbines) do not provide adequate bulk mixing, particularly in larger vessels. There are several ways to overcome this problem:

(1) Use radial flow impellers at the lower positions and an axial flow impeller at the upper position. Also, use in this hybrid the largest diameter radial flow impellers consistent with all the constraints discussed previously.

This approach also requires that the tank be completely baffled (as does any design employing internal impellers) to prevent broth swirling and vortex formation (sudden vortex collapse can cause serious mechanical damage). In practice, this requires four baffles each having a width equal to about 10% of the tank diameter, and spaced 90° apart.

(2) Use new impellers which have been designed specifically to provide good oxygen transfer and good bulk mixing. As noted previously, one of the most promising of these newer designs is the A-315. But until more practical information becomes available about them (see above), we recommend that they be used only as the axial flow impeller in the hybrid system discussed above.

(3) Use broth recirculation in concert with special aeration devices (e.g., the "venturi" in the IZ deep jet fermentor). This method also has the advantage of improved heat transfer (see below), because it allows for an external heat exchanger. Unfortunately, this approach introduces what we feel are unacceptably high contamination and containment risks. Also, it requires very special and very expensive pumps (to handle the entrained gas). It obviously is not appropriate at all for sensitive organisms.

3.4.2 Vessel Geometry

The liquid height to vessel diameter ratio is also an important factor in determining mixing quality, even for impellers or impeller combinations that give good vertical mixing. Experience has shown that the ratio should be kept within the range of 1 to 3 (preferably 1.4–2.0). Smaller ratios require very large diameter impellers (very high torque), and limit severely the number of impellers that can be included (this also limits the power input). Larger ratios result in liquid heights that are too large for even good vertical mixers to "turn over" the broth effectively. In either case, the result usually is poor bulk mixing (see CHARLES and WILSON, 1992; CHARLES, 1978; MOO-YOUNG et al., 1972; BLAKEBROUGH and SAMBAMURTHY, 1966, for further details).

3.4.3 Broth Rheological Properties

Highly viscous and/or non-Newtonian broths (e.g., mycelial broths, broths containing polysaccharides) must be treated as special cases. There is even less reliable information about these than about rheologically "simple" broths. What is known with certainty is that it is much more difficult to provide them with good oxygen transfer (and heat transfer) and good bulk mixing. Also, the literature contains many misconceptions and erroneous information concerning the rheology of fermentation broths (see CHARLES, 1978, for several examples). See CHARLES and WILSON (1992), CHARLES (1978), MOO-YOUNG et al. (1972), and BLAKEBROUGH and SAMBAMURTHY (1966) for more detailed discussions.

3.5 Heat Transfer

Coping with heat transfer requirements for highly aerated fermentors can be extremely difficult. Indeed, in many cases heat transfer can become the limiting design factor.

The rate of heat generation in a fermentor is given approximately by the sum of the metabolic heat generated by the organisms and the heat generated by agitation:

$$Q_g = 0.502\, OTR\; V_l + P_g \tag{19}$$

where Q_g is the rate of heat generation in kW, OTR is the oxygen transfer rate in mmol^{-1} h^{-1}, V_l is the liquid volume, and P_g is the gassed mechanical input in kW.

The rate at which heat can be transferred from the fermentor is

$$Q_r = U A_s (\Delta T)_{lm} \tag{20}$$

where U is the overall heat transfer coefficient in W m^{-2} K^{-1}, A_s is the heat transfer surface in m^2 and $(\Delta T)_{lm}$ is the log mean temperature driving force defined as

$$(\Delta T)_{lm} = (T_o - T_i)/\ln[(T_o - T_f)/(T_i - T_f)] \tag{21}$$

where T_f, T_t, and T_o are the fermentation, inlet coolant and outlet coolant temperatures, respectively. There are some cases for which the log mean does not apply as presented (CHARLES and WILSON, 1992), but for fermentation these usually occur only for heat-up, cool-down, and sterilization.

The biggest practical problem in using Eq. (20) is obtaining reliable values of U. U depends on the nature of the broth (e.g., rheological properties), agitation and aeration levels, the thermal properties and thickness of the vessel wall, the thermal nature and flow characteristics of the coolant, and the nature of the heat transfer surfaces (e.g., jacket, coils). Suffice it to say for present purposes that there is not a lot of reliable information available for aerated fermentors – particularly in the form of generally applicable correlations. Once again, experience is the order of the day. For single cell organisms in relatively low viscosity, Newtonian broths under strong agitation and high aeration, U can be taken as approximately 550–850 W m^{-2} K^{-1} for new, clean vessels. The values for coils tend to be somewhat higher than those for jackets. It is most important to recognize that these values are only estimates based on experience, and that surface area and/or coolant temperature requirements depend strongly on them; therefore, prudent, conservative design is called for. More will be said about this when we discuss the design example.

Unfortunately, this is not the end of the story. High oxygen transfer rates and the concomitant high power levels required to get them generate a lot of heat. But this heat is generated at a low temperature level which, in turn, calls for large surface areas and/or low temperature coolant flowing at fairly high rates. The most desirable circumstance is to require only jacket cooling and unchilled cooling water. The lowest temperature of natural cooling water is ca. 12 °C (ground water), but there are many cases in which this is not available. Even if it is, it is often not low enough to make the jacket area alone sufficient. This problem begins to occur around an OTR of about 200 mmol L^{-1} h^{-1}, the exact value depending on the liquid volume and the nature of the fermentation (e.g., the problems start at lower OTRs and volumes for highly viscous, non-Newtonian broths than they do for "simple" single cell cultures).

The options available for handling problem cases are internal coils, refrigerated coolant, combinations of coils and refrigerated coolant, and recirculation through external exchangers. None of these is desirable. External recirculation is the least desirable because of reasons cited previously with regard to oxygen transfer. Coils are not only expensive, but also present a constant danger of contamination as a result of leakage of non-sterile coolant through pinholes which often develop around welds (see below for more details). Refrigerated coolant obviously requires a refrigeration system which is not only expensive, but also is a considerable maintenance problem. Subfreezing coolant presents not only additional refrigeration costs but also the real and present danger of freeze-ups in lines, valves, and fittings.

When faced with such options, one should give very serious thought to solving the problem by slowing down the fermentation and/or decreasing the final cell mass so as to decrease the maximum *OTR*. In many cases, the cost of productivity loss resulting from this option will be much less than the costs associated with a design that is based on conditions which occur for only a short period during the cycle, and which can be the source of considerable risk to the success of the fermentation.

In addition to all the above, there are problems associated with vessel construction (by qualified vendors), cleanability, architectural constraints, validation, transportation, standard sizes, and several others. We will discuss some of these (those associated with the mechanical construction of the fermentor) after we have done the basic calculations for the design example.

4 Design Example

4.1 Design Basis

A fermentor is to be built for the production of a protein from a recombinant organism. Some of the major specifications for the design basis are given in Tab. 1. All calculations are based on equations and considerations discussed previously; however, the reader should note that the following specific forms and conditions have been used throughout the detailed calculations:

(1) The linear gas velocity, V_s, has been taken to be 180 cm min^{-1} under specified conditions of pressure and temperature in the vessel. It also has been assumed that 180 cm min^{-1} is the arithmetic average V_s with the understanding that the velocity would be somewhat lower at the bottom of the tank, and would be somewhat higher at the top. We have found this approach to be quite adequate for design calculations.

(2) The correlation between $K_g a$ and gassed power draw used is illustrated graphically in Fig. 2. This is a composite based on our experience with similar fermentations at similar scales to that of the example. It is important to note that the correlation is valid over only a relatively narrow range, and should not be viewed as having general applicability.

(3) No oxygen transfer "credit" has been taken for the power draw of the A-315 impeller. The reasons for this were discussed previously. Also, the power draw of this impeller will be less than 10% of the draw of the Rushton's specified (under the specific conditions of the example).

(4) The aeration number (N_a) correlation which has been used to obtain the ratio of P_g/P_{ug} is:

$$P_g/P_{ug} = 1.0 - 18.56\,N_a + 215.6\,N_a^2 - 1082.7\,N_a^3 + 1859.7\,N_a^4 \tag{22}$$

The same caveats apply here as for the $K_g a$ correlations.

(5) The correlation used here for gas holdup (H_g) is:

$$H_g = 0.0114(P_g/V_l)^{0.35}(0.6\,V_l)^{0.5} - 0.03 \tag{23}$$

Again, the caveats apply.

(6) A bottom jacket has been assumed for all cases discussed. The area of the bottom jacket has been taken to be the projected area for the bottom dish. This takes into account the space required for the bottom drain valve and the fact that the bottom jacket is somewhat less effective than the straight side jacket.

Tab. 1. Partial Design Basis

Product: Extracellular protein from recombinant organisms
Annual protein production (kg, dry basis): 6000
Operating days/year: 330
Operating hours/day: 24
Fermentation:

1. Cell mass concentration (g L^{-1}): 25
2. Specific growth rate (h^{-1}): 0.32
3. Yield on glucose (g cells dw/g glucose): 0.4
4. Yield on oxygen (g cells dw/g oxygen): 1.0
5. Protein (g protein dw/g cells dw): 0.05
6. Temperature (°C): 30
7. pH: 6.8
8. Medium: Defined medium with glucose, salts, yeast extract, and a proprietary ingredient. Maximum viscosity of $6 \cdot 10^{-3}$ Pa s
9. Mode: Fed-batch; glucose concentration <0.5 g L^{-1}
10. Total cycle time (h): 24
11. Turn around time (h): 6
12. Product induction time (h): 4
13. Sterilization: Continuous sterilization for batching medium and feed. Fermentor sterilized empty
14. Seed culture: 5% of initial broth volume
15. Cool down: The broth must be cooled to below 10 °C in no more than 1 hour after the fermentation ends

Recovery Efficiency: 60%
Recovery time: The product must be recovered and purified in no more than 12 hours after it leaves the fermentor

dw, dry weight

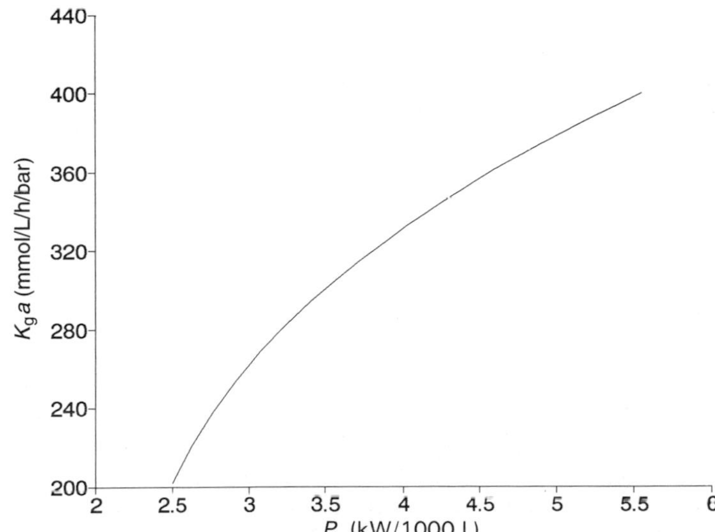

Fig. 2. K_ga correlation used in an illustrative example.

(7) The overall heat transfer coefficient has been assumed to be 851.7 W m^{-2} K^{-1} for all cases.
(8) We have specified that the full straight side of the vessel will be jacketed; however, the heat transfer area used in the calculations has been based on the actual area contacted by the broth under design conditions. This has been based on the aerated liquid height calculated from the gas holdup correlation.

4.2 Calculations and Results

From material balances based on the concentrations and timing factors given in the basis, we find that we must produce daily 24000 liters of broth, and that the maximum oxygen transfer rate will be 250 mmol L^{-1} h^{-1}. We can minimize the fermentor capital cost by choosing one 24000 liter (wv) fermentor to do

Tab. 2. Initial Sizing Results for 24000 Liter Vessels

	Case 1	Case 2	Case 3[a]
Total volume (m^3)	34.13	34.13	34.13
Liquid volume (m^3)	24.00	24.00	24.00
Tank outer dia. (m)	2.74	2.74	2.74
Tangent ht. (m)	5.57	5.57	5.57
Unaer. liq. ht. (m)	4.31	4.31	4.31
Unaer. liq. ht./dia.	1.58	1.58	1.58
Total ht./dia.	2.29	2.29	2.29
Percent fill	70.3	70.3	70.3
OTR (mmol L^{-1} h^{-1})	250	250	250
V_s (cm min^{-1})	180	180	180
Pressure (bar)	3.42	3.08	2.05
O_2 mole frac., inlet (%)	21	40	40
O_2 transfer eff. (%)	29.0	16.8	24.5
$K_g a$ (mmol L^{-1} h^{-1} bar^{-1})	265	179	253
Gassed power req'd. (kW)	133	61	121
Air flow (std. m^3 min^{-1})	36.8	25.3	17.3
O_2 flow (std. m^3 min^{-1})	0	8.0	5.5
No. of turbines	2	2	2
No. of A-315 impellers	1	1	1
Shaft speed (rpm)	205	129	142
Turbine dia. (m)	1.14	1.14	1.14
Axial imp. dia. (m)	0.91	0.91	0.91
Turbine D_l/D_t ratio	0.42	0.42	0.42
Aeration number	0.059	0.074	0.060
P_g/P_{ug}	0.46	0.42	0.45
Gassed power (turb.) (kW)	179	82	164
Heat load ($W \cdot 10^{-6}$)	0.971	0.898	0.958
U (W m^{-2} K^{-1})	852	852	852
Coolant temp. (°C)	−5.56	−3.89	4.44
Coolant flow (L min^{-1})	681	681	681
Avail. jkt. (m^2)[b]	48.2	45.5	47.3
Area req'd. (m^2)	48.2	45.5	88.9

[a] Each case is for a different combination of total pressure and inlet oxygen mole percent.
[b] Total of bottom jacket and straight side jacket contacted by aerated broth (may be less than total straight side jacket).
Abbreviations: dia., diameter; unaer. liq., unaerated liquid; std., standard; ht., height; frac., fraction; eff., efficiency; req'd., required; temp., temperature; avail. jkt., available jacket; imp., impeller; turb., turbine

the job, but this choice has important disadvantages:

- It maximizes the risk of large losses resulting from ruined fermentations.
- It maximizes the size and hence the cost of the recovery system which may well exceed any capital savings for the fermentor. This is because the recovery must be done within 12 hours after completion of the fermentation; therefore, the use factor for the recovery/purification train will be only about 50% (based on the 24 hour fermentation cycle time).
- Preliminary sizing calculations for three sets of operating conditions for a 24000 liter (wv) vessel (see Tab. 2) show that we will have to use sub-freezing coolant and/or internal coils to handle the heat transfer requirements at the maximum oxygen transfer rate. As discussed previously, this will cause increased cost and operating problems.

For these and other reasons, we should consider a larger number of smaller vessels. Based on the process timing, two 12000 liter (wv) fermentors appear to be a rational choice to investigate.

Tab. 3 gives the results for initial sizing calculations using three different vessel diameters for a 12000 liter (wv) fermentor. The 2.29 m diameter vessel would give us a fairly low ratio of liquid height to vessel diameter. This would limit us to one large radial flow impeller plus one axial flow impeller which would have to be operated at fairly high speeds. The result would be very high torque and considerable difficulties in avoiding vibration. The 1.98 m diameter vessel would allow at least two reasonably sized radial flow impellers plus one axial flow impeller. Also, of the cases considered in Tab. 3, the 1.98 m diameter vessel would have the highest ratio of jacket surface area to culture volume; therefore, it would seem to be the best choice with regard to heat transfer. But it has a rather large liquid height to tank diameter ratio, and this might cause significant problems in achieving good top-to-bottom mixing (as noted previously). The 2.13 m diameter vessel appears to give the best balance, and hence it will be studied further.

Tab. 4 gives the results of sizing calculations for operating pressures of 2.047 and 3.423 bar, and for inlet oxygen concentrations of 21, 30, and 40 mol%. Pressure and oxygen were selected as the primary variables because they have the most direct, the greatest and the most reliably predictable effects on oxygen transfer rate. For all six cases V_s was fixed at 180 cm/min (maximum "safe" value as explained previously). The impellers chosen were two 94 cm, 12-blade Rushton turbines combined with an A-315. This combination should give good bulk mixing (for reasons stated above) and in five of the six cases should give adequate power input to provide a 250 mmol/L/h oxygen transfer rate when operated at maximum speeds between 100 and 150 rpm. This speed range can be achieved easily with standard mo-

Tab. 3. Initial Sizing Results for 12000 Liter Vessels

	Case 1[a]	Case 2	Case 3
Total volume (m^3)	17.04	17.04	17.04
Liquid volume (m^3)	12.00	12.00	12.00
Tank outer dia. (m)	2.29	2.13	1.98
Tangent ht. (m)	3.96	4.66	5.51
Unaer. liq. ht. (m)	3.13	3.56	4.10
Unaer. liq. ht./dia.	1.38	1.68	2.09
Total ht./dia.	2.00	2.43	3.00
Jacket area (m^2)[b]	30.38	32.52	34.93
Percent fill	70.3	70.4	70.4

[a] Each case is for a different tank diameter
[b] Straight side jacket plus full bottom jacket
Abbreviations see Tab. 2

Tab. 4. Sizing for 2.13 m Diameter Vessels

	Case 1	Case 2	Case 3	Case 4	Case 5	Case 6[a]
Total volume (m^3)	17.04	17.04	17.04	17.04	17.04	17.04
Liquid volume (m^3)	12.00	12.00	12.00	12.00	12.00	12.00
Tank outer dia. (m)	2.13	2.13	2.13	2.13	2.13	2.13
Tangent ht. (m)	4.41	4.41	4.41	4.41	4.41	4.41
Unaer. liq. ht. (m)	3.56	3.56	3.56	3.56	3.56	3.56
Unaer. liq. ht./dia.	1.68	1.68	1.68	1.68	1.68	1.68
Total ht./dia.	2.43	2.43	2.43	2.43	2.43	2.43
Percent fill	70.4	70.4	70.4	70.4	70.4	70.4
OTR (mmol L^{-1} h^{-1})	250	250	250	250	250	250
V_s (cm min^{-1})	180	180	180	180	180	180
Pressure (bar)	2.05	3.42	2.05	3.42	2.05	3.42
O_2 mole frac., inlet (%)	21	21	30	30	40	40
O_2 transfer eff. (%)	39.4	24.3	27.6	17.0	20.7	12.8
$K_g a$ (mmol L^{-1} h^{-1} bar^{-1})	731	396	462	262	328	192
Gassed power req'd. (kW)	298	64	86	36	46	29
Air flow (std. m^3 min^{-1})	13.5	21.9	12.0	19.4	10.3	16.7
O_2 flow (std. m^3 min^{-1})	0	0	1.5	2.5	3.3	5.6
No. of turbines	2	2	2	2	2	2
No. of A-315 impellers	1	1	1	1	1	1
Shaft speed (rpm)	205	129	142	108	118	102
Turbine dia. (m)	0.94	0.94	0.94	0.94	0.94	0.94
Axial imp. dia. (m)	0.76	0.76	0.76	0.76	0.76	0.76
Turbine D_l/D_t ratio	0.44	0.44	0.44	0.44	0.44	0.44
Aeration number	0.037	0.059	0.053	0.070	0.064	0.074
P_g/P_{ug}	0.55	0.46	0.47	0.43	0.44	0.42
Gassed power (turb.) (kW)	297	64	86	36	47	30
Heat load ($W \cdot 10^{-6}$)	0.717	0.482	0.508	0.454	0.465	0.449
U (W m^{-2} K^{-1})	852	852	852	852	852	852
Coolant temp. (°C)	−10.6	2.22	2.22	2.78	2.78	3.33
Coolant flow (L min^{-1})	378	454	492	435	454	454
Avail. jkt. (m^2)[b]	34.8	30.2	31.0	29.0	29.5	28.4
Area req'd. (m^2)	34.8	29.8	30.9	28.6	29.1	28.4

[a] Each case is based on a different combination of total pressure and inlet oxygen mole percent
[b] Total bottom jacket area and straight side area contacted by aerated broth (may be less than total straight side area of the jacket)
Abbreviations see Tab. 2

tors and gear boxes, and will not cause vibration problem which are difficult to design out. For these reasons we will assume here that the maximum speed will be 150 rpm. Note: This effectively eliminates Case 1 (Tab. 4). The elimination of this case will be discussed below in greater detail.

We also specifiy at this point that the vessel will have a full bottom jacket (all 6 cases) because the heat load will be very high (at least $4.4 \cdot 10^6$ W), and we will need every bit of jacket area we can manage to get if we are going to avoid internal coils and/or subfreezing coolant.

The combination of operating variables selected above is not unique: it is one example among several that will give results that can be engineered in a practical, reliable fashion. We now consider narrowing down the ranges so that we arrive at a final, detailed set of design specifications.

We rule out Case 1 because it will require an impractically high power input (298 kW). (Note: This is equivalent to 25.1 kW m^{-3}

which should not be viewed as a reliable number, because it is outside the range of data on which the correlations we used are based; nevertheless, even if it is "high" by a factor of two, it still gives an unacceptably high power. Furthermore, the very large uncertainty alone would speak against using this set of conditions for the design case.) We also rule out this case because it has an extremely high heat load (a result of the very high power requirement) which will require internal coils and/or subfreezing coolant. In addition, the speed required could cause difficult-to-solve vibration problems. (Note: The speed problem could be overcome by using larger impellers, but this would not solve the other problems, and could drive the shaft torque up to impractical values.)

Choices from among the remaining cases will depend on whether the user wants oxygen enrichment and the maximum pressure with which he/she is comfortable for whatever reason(s). Based on Cases 4–6, we would recommend that the design be based on oxygen enrichment up to 40 mol% and a maximum operating pressure of 3.42 bar (the vessel would be rated for at least 3.77 bar). The reason is that we then could specify reliably a 56 kW drive (includes consideration of drive efficiency), and be confident that the user would have plenty of "operating room" within which to achieve the desired results of 250 mmol/L/h oxygen transfer rate. Without oxygen enrichment and/or with a pressure maximum of 2.047 bar, we would have to specify a 75–93 kW drive (includes drive efficiency) as can be seen from Cases 2 and 3. We assume here that the user chooses to agree with our recommendations.

On the basis of the preceding considerations we would specify the vessel dimensions given in Tab. 4 (same for all cases) along with the following:

Impellers: Two 94 cm, 12-blade Rushtons; one 76 cm A-315

Spacing: The first Rushton will be approximately 94 cm above the bottom of the tank.
The second Rushton will be 94–107 cm above the first.
The A-315 will be 94–107 cm above the second Rushton.
(Note: These spacings are based on calculated aerated liquid heights which will be between 4.0 m and 4.8 m above the tank bottom depending on operating conditions and gas holdup.)

Speed: 125 rpm (revolutions min^{-1}), maximum

Air flow: 25 scmm (standard m^3 min^{-1}), maximum

O_2 flow: 5 scmm (standard m^3 min^{-1}), maximum

A very important additional point to consider in the final mechanical design is that the shaft dimensions be specified such that the shaft's natural frequency is at least 25% higher than the 125 rpm maximum impeller speed.

The specifications above obviously are not intended to be complete (e.g., many of the details such as vessel finish, wall thickness, etc., are not included). They should, however, give the reader a reasonable, general sense of what is required. Also, the reader should be able to see that the specifications allow for fairly broad operating ranges thereby affording a reasonable degree of insurance. Such insurance usually can be had at relatively small cost; however, the extent to which it really will be valuable will depend considerably on how well the instrumentation and control systems are designed, constructed and maintained. Discussion of these points, however, is beyond the scope of this chapter.

Before concluding this section, we find it important to address in greater detail the heat transfer requirements. The calculations show that the cooling load can be handled with above freezing coolant (2.8–3.3 °C) and without an internal coil. This assumes the use of a full bottom jacket, and requires a fairly high (30.4–31.7 L min^{-1}) coolant flow for this size of vessel. We also are concerned that the design is rather "tight". The calculations are based on an assumed value of 852 W m^{-2} K^{-1} for the overall heat transfer coefficient. This value is based on our experience with similar fermentations in similar vessels. But it is obvious that the actual value might well be lower (e.g., the jacket might get fouled), and that it probably will decrease as the vessel ages. Consider Case 6 in Tab. 4 which has the lowest heat load ($4.49 \cdot 10^5$ W) of the three cases on which our design is based (Cases 4–6). If the heat transfer coefficient were to drop to 710 W $m^{-2} K^{-1}$ for this case (only a 17% drop), subfreezing coolant would be required to handle the heat load; hence our concern. The user could opt to design in a subfreezing coolant system to cope with this contingency, but this would add considerably to the capital cost and to the operating problems and costs. An alternative is to slow down the fermentation and/or to lower the final cell mass so as to reduce the oxygen transfer rate to 223 mmol $L^{-1} h^{-1}$ which would give a heat load that could be handled with 2.8 °C coolant even for the lower heat transfer coefficient. This could be achieved by lowering the specific growth rate about 10% near the end of the fermentation which would add at most an hour to the fermentation time. This approach would have very little (if any) practical effect on the overall plant productivity. It (and/or any others which "shave" the high peaks) should be at least considered whenever the kind of situation being considered (and/or similar situations) arises which happens frequently.

We now go on to consider several important aspects of the details of the mechanical design.

5 Mechanical Design Details

5.1 Vessel Codes

Fermentation vessels for aseptic operations must be pressure vessels if for no other reason than that they must withstand steam sterilization. For purposes of safety, legality, and insurability, all such vessels must be tested by licenced inspectors as per ASME (American Society of Mechanical Engineers) code for nonfired pressure vessels (Yokel, 1986). Vessels which pass inspection are ASME stamped (directly and permanently on the vessel). Anyone purchasing such a vessel should make certain that the stamp is affixed, and should obtain from the vendor all of the relevant documentation. The owner of the vessel also must be aware that if certain modifications (e.g., any modification requiring a new hole greater than 5 cm in diameter in the vessel wall) are made to the vessel, it must be retested as above.

5.2 Vessel Dimensions

5.2.1 Total Volume

Total volume should be chosen to give about 30% freeboard to allow for holdup and foaming. The primary considerations are usually those related to architectural constraints and shipping considerations. Cost usually is not a significant factor, because providing some additional freeboard requires only some additional straight side for which the cost is quite small.

5.2.2 Wall Thickness

This is determined primarily by the design pressure and the vessel diameter, and was discussed above.

5.2.3 Vessel Diameter

The diameter is selected not only on the basis of considerations discussed previously, but also on the following:

(1) The height of the building into which the vessel will go. One must consider not only the full height of the vessel itself, but also the manner in which the vessel will be supported (e.g., legs, skirt), the dimensions of the drive system and its supports, the location of the drive, and the space required for operation and for servicing.

(2) Standard diameters of heads. In most instances these will be standard flanged and dished heads the dimensions of which are available in standard tables (Precision Stainless Inc., 1989). There are some cases of mammalian cell culture for which a hemispherical bottom head is specified. These must be treated as special cases. It should be mentioned here that the only material effect we have noted for

hemispherical heads is that they increase the cost of the vessel considerably. There are, however, those who believe that such heads give better mixing quality for the gentle agitation required by mammalian cells. We have yet to see many convincing evidence of this.
(3) One should try to keep the diameter below 3.7 m. One reason is that vessels with larger diameters may be difficult to ship. Indeed, very large-diameter vessels must be field-erected. This should be avoided if at all possible not only because of the markedly increased cost, but also because field-erected vessels usually do not have as high a quality as shop-fabricated vessels. In addition, as the size of the vessel increases, it becomes more difficult to get a high quality interior finish (more later). Finally: As the diameter goes up, the length of the qualified vendor list goes down. This will tend to drive up the cost, and will limit the purchaser's clout.

5.3 Cleanability

A fermentor for aseptic service must be designed and built to be easily cleanable. This requires great attention to detail which should be guided by the following principles:

- All surfaces must be smooth (see below) and free-draining.
- Ideally, there should be no crevices in which solids can accumulate.
- Heating surfaces should be designed and operated to minimize solids bake on.
- All interior joints should be welded, if at all possible.

The simple fact is that the easier the vessel is to clean, the simpler and less arduous will be the cleaning protocols; hence, the greater is the likelihood that cleaning will be done properly.

Finally: The same considerations should be given to all equipment contacting sterile medium and/or pure cultures.

5.4 Surface Finishes

The primary purpose of a high quality finish is to facilitate cleaning and sterilization. The following are some general guidelines concerning finishes:

- All surfaces which contact sterile medium and/or pure cultures should have the same quality finish. This is nothing more than a restatement of the "weakest link in the chain" concept. This thinking should be applied not only to the fermentor but to all peripherals.
- The better the finish, the easier it will be to clean.
- Quality of surface finish and cleaning/sterilization protocols are related intimately: the better the finish, the simpler will be the protocols.

Surface finishes usually are specified in terms of "grit number": The higher the number, the smoother the finish. Finishes commonly used for aseptic service are 180, 220, and 320 grit. Surfaces for very demanding service (e.g., mammalian cell culture) should be electropolished (EP) after being polished to 320 grit.

Demanding service usually requires surface passivation to restore the stainless protective film which is at least partially destroyed during polishing operations. This entails surface treatment with sodium hydroxide and citric acid followed by nitric acid and a complete water rinse (BJURSTROM and COLEMAN, 1987). It should be noted, however, that electropolishing is itself among the best means of passivating a vessel: Usually, an electropolished vessel does not require chemical passivation. This is one important factor that can justify the additional cost of electropolishing.

Finally, the better the finish, the higher the cost; therefore, one should choose the finish appropriate to the service required.

5.5 Materials

The materials used most for fermentors designed for aseptic service are SS304, SS304L, SS316, and SS316L (SS stands for stainless steel). The primary differences among these are carbon and alloying contents which affect machining and welding characteristics, corrosion resistance and fabrication characteristics. The choice should be dictated by the service required.

The "L" designation implies low carbon content, and is frequently inferred to mean that the steel is "better". One should keep in mind that there is considerable difference of opinion on what "better" means, but there is no question that "L" will add about 15% to the fermentor cost. The fact is that "L" is not always the best choice (e.g., it is not better for cases in which single-pass welding will be used).

5.6 Welds

Aside from basic concerns about the safety and strength of welds, one should focus on corrosion and cleanability. Some basic guidelines are:

- Do all welding under an inert gas atmosphere.
- Use the TIG (tungsten inert gas) method.
- Completely polish all welds to the same standards as specified for all other vessel surfaces.
- All welds should be done by a validated, documented procedure whether or not the initial intent is to put the fermentor into a GMP (good manufacturing practice) facility. This will insure that the procedures used are always the same, and that a fermentor built for a GMP facility is not built with "special" procedures for the manufacturer.

5.7 Jackets and Coils

Any of several commonly used jacket types (e.g., half pipe, dimple) (MARKOVITZ, 1971) usually will be satisfactory for fermentors. Regardless of the jacket type, however, one should pay careful attention to the following:
(1) The jacket should be designed to minimize the possibility of solids bake-on during sterilization. This usually can be accomplished through proper zoning of the jacket. Proper zoning also is important for limiting jacket pressure drop. This is particularly important for larger vessels requiring very large coolant flows. It also affects the sizes of external piping, pumps, and valves which can be important cost factors.
(2) It usually pays to jacket the full straight side of the vessel to allow for maximum heat transfer at maximum gas holdup. It is important, however, to recognize that in calculating coolant flows and temperatures for design conditions, the actual area used probably will be less than the total jacket area.
(3) Bottom jackets do not contribute much (other than to cost) for vessels less than about 5000 L.
(4) Stainless jackets can increase vessel life appreciably. SS304 is adequate for this purpose. The additional cost usually is more than compensated for by the increased life.
(5) Jacket insulation should be non-chloride bearing. Chloride bearing insulation can lead to rapid vessel corrosion.
(6) Piping connections to the jacket (or coil) should be made via flanges or sanitary clamps to minimize stress on the welds.
(7) The jacket should be free draining to avoid solids buildup.
(8) Coolant should be filtered.

If one must use coils (although everything within reason should be done to avoid them), the coils should be designed, built, and mounted in such a way that
(1) they can handle the thermal and mechanical stresses developed during fermentation (see Fig. 3 for an example),
(2) they disrupt the bulk mixing minimally,
(3) they can be cleaned easily. It is important to understand that cleaning coils is extremely difficult, and that it may be necessary to do part of the cleaning manually – at least after several fermentations (depending on the nature of the broth, etc.),
(4) the spacing between turns should be at least 15 cm for fluid mechanical and cleaning considerations. Also, joints should be butt welded, and the welds should be "sealed" by welding a concentric tube around each butt weld. Joints should be tested according to the method described by PERKOWSKI (1984).

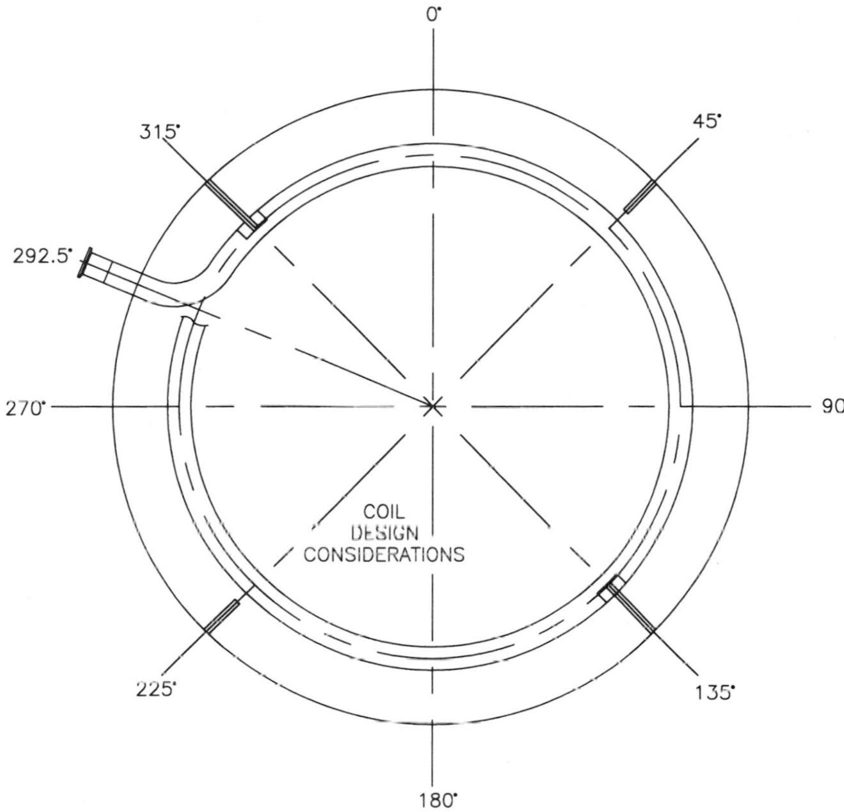

Fig. 3. Recommended method for mounting fermentor cooling coils.

5.8 Addition Ports

Ports should be designed to be free draining and to minimize any nooks and crannies in which solids can "hide". One design which works quite well is illustrated in Fig. 4. Note that:

- the port is mounted at an angle between 5° and 15° from the horizontal to assist free draining,
- it extends slightly beyond the interior wall to prevent additions from running down the wall. This extension should be made as short as possible,
- the external piping also is kept as short as possible to minimize sterilization problems,
- the steam lock assembly is designed so that all valves may be "steamed through" to insure that all surfaces which contact sterile additions and/or pure cultures are exposed directly to steam. In general, it is poor practice to rely on thermal conduction to insure sterility.

These principles should be applied to all sterile piping in the system.

5.9 Static Seals

Ethylene propylene diene monomer (EPDM) O-rings are the best choice for most fermentor static seals (e.g., nozzle seals as shown in

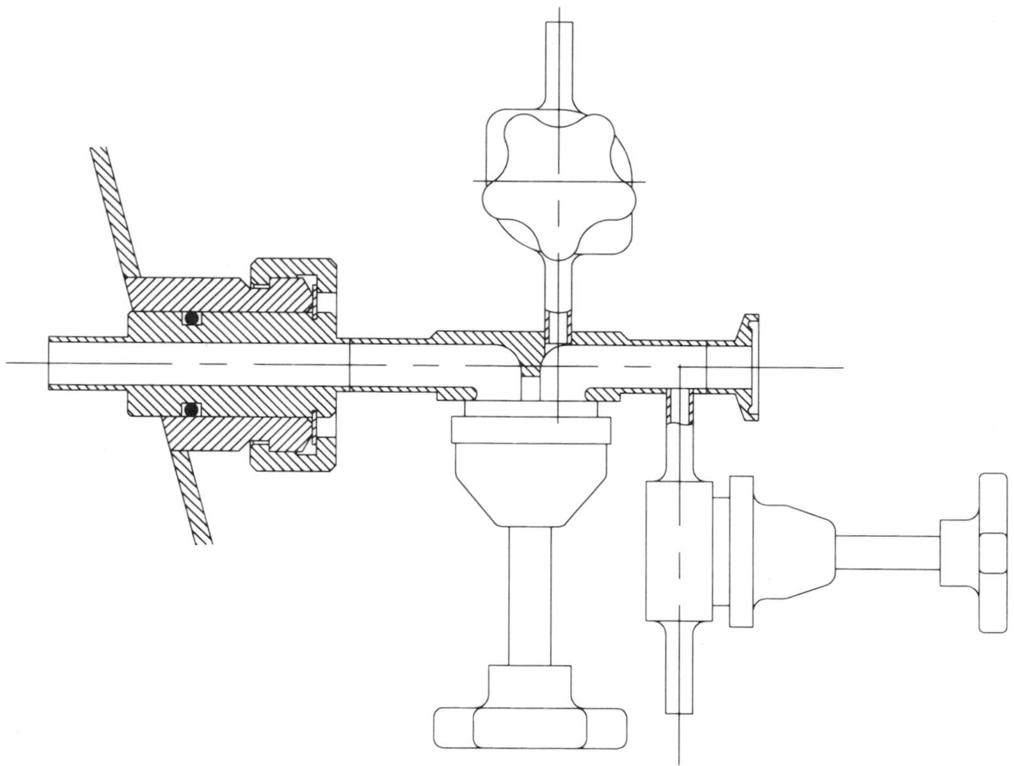

Fig. 4. Typical addition port/steam lock assembly.

Fig. 4). Other materials have superior properties of one kind or another (e.g., teflon has a higher temperature resistance), but these usually are out-weighed by their disadvantages (e.g., teflon does not stretch, viton hardens). One major exception to this generalization is the preferred use of silicone O-rings for head plate seals.

5.10 Air Exhaust Line

The air exhaust line must be designed to insure that

- organisms do not escape from the vessel,
- foreign organisms do not enter the vessel,
- mist and/or water vapor condensate do not accumulate on filters or in the back pressure regulator,
- foam cannot cause serious damage (e.g., damaging the back pressure regulator or exhaust filters).

It also is important that back pressure regulation effected in the exhaust line be uncoupled to the greatest extent possible from the air flow control system in the air inlet line.

5.11 Foam Control

Virtually all aerated fermentation broths foam: It is only a question of degree. This means that some means for foam control should always be included. The two methods used almost universally for commercial fermentations are mechanical foam breakers and addition of chemical antifoam agents. Some designs employ both.

A mechanical foam breaker is a high-speed

impeller designed very much like a centrifugal pump impeller. Foam is drawn into the impeller where it is collapsed by strong mechanical forces (obviously not suitable for delicate organisms). The impeller must be mounted on its own shaft, and be driven independently from the main agitator. Obviously, this requires a separate agitator seal which is another source of potential problems.

Chemical antifoamers collapse foams by altering their surface tension characteristics. Sterile antifoamers usually are pumped into the fermentor (air pressure pumping is best) automatically from dedicated addition vessels. The choice of antifoamer should be made not only on the basis of its compatibility with the fermentation, but also on the basis of its compatibility with the recovery equipment (e.g., silicone antifoamers can decrease dramatically the permeate fluxes for certain types of membrane filters even at very low concentrations).

Finally, a good foam control system also will include the means for cutting down automatically the air flow and the agitation speed when the foam starts to overwhelm the system. This must be done with great care to avoid damaging the drive system.

5.12 Bottom Drain Valve

Drain valves should be designed so as to avoid any pockets in which solids can accumulate. They also must be capable of independent steam sterilization. The best choice available to accomplish these objectives is a diaphragm valve flush mounted with the bottom of the vessel, and fitted with appropriate steam and condensate lines (see Fig. 5 for an example). Careful attention also must be paid to the valve internals.

Finally, it is important to avoid any design which employs a coupling of any kind between the valve and the bottom of the tank.

5.13 Agitation Drive System

Among the major factors in the design of the drive system are sizing, orientation, seals, and vibration. We do not have the space here to consider all the major factors in detail, but we will try to give the reader a reasonable sense of them and the kinds of thinking that go into the design of a reliable drive.

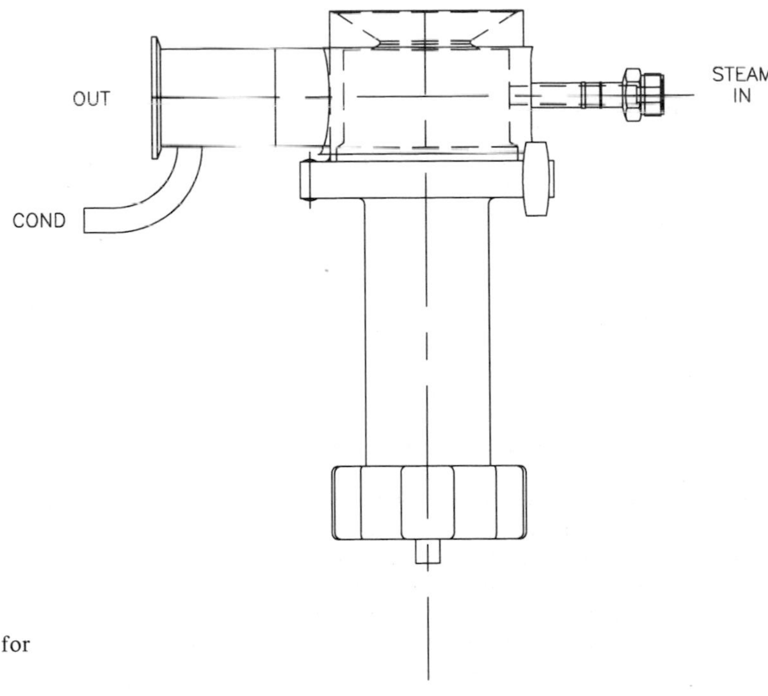

Fig. 5. Bottom drain valve for aseptic operation.

5.13.1 Sizing

Sizing of the drive is based not only on the calculations discussed previously, but also on practical considerations such as standard motor speeds, standard gearbox ratios, and torque ratings, the potential of motor overheating, desired agitation turndown ratio, and torques exerted on the shaft. In sizing motors and gearboxes, it usually is wise to be as conservative (oversizing) as cost, space, and weight considerations will allow: the design should be as "bullet proof" as possible. One of the major reasons for urging a very conservative approach is that the calculations for power required and power delivered under aerated conditions are not very reliable (see previous comments). It also is desirable to build in a safety factor to protect against some of the inadvertant operating errors which can occur even in well run plants (e.g., turning down the air flow too much when the drive is running at full speed), and which can result in severe damage to the drive and to the vessel.

The extent to which this approach will influence cost materially varies with fermentor size and power level. For small vessels (e.g., a few hundred liters), oversizing by as much as a factor of two will not increase the price materially in comparison to the "insurance" it buys. For very large vessels (greater than 5000–10000 liters), cost increases more rapidly with increasing power, and the design usually must be considerably "tighter"; nevertheless, it usually is worth at least considering a somewhat more powerful drive than is calculated.

5.13.2 Mounting

The drive can be top mounted or bottom mounted, top mounted being the more commonly used. The major advantages (WILSON and ANDREWS, 1983) of bottom drive are a shorter shaft, elimination (in most cases) of the steady bearing usually required for top drives, greater accessibility for servicing, and simpler, less expensive supporting structures. The disadvantages most frequently cited are the potentials for seal grinding by particulates and greater danger and consequences of leaks. Our experience is that there is no firm basis for such assertions – so long as the bottom seals are designed and maintained properly.

The bottom line is that the choice usually turns out to be a matter of personal preference. There is no real difference in terms of aseptic operation so long as design and maintenance are sound.

5.13.3 Seals

Proper seal design and maintenance are critical to achieving aseptic operation. The most reliable design employs a double mechanical seal. In the classic double seal, the seals are arranged back-to-back, and the seal housing is operated at a pressure higher than in the vessel (WILSON and ANDREWS, 1983). This design has the disadvantages of a fairly complex seal lubricant system, and the potential of leakage of seal lubricant into the vessel. There is also the danger of accumulation of solids in the pockets inherent in this design. These problems can be overcome by reversing the top seal (WILSON and ANDREWS, 1983). This Arrangement along with its sterile lubricant system is illustrated in Fig. 6.

5.13.4 Vibration

The need for a vibration analysis has been discussed previously. We will simply reemphasize here the importance of the analysis, and the general rule of designing in such a way that the maximum speed of the agitator is at least 25% below the natural frequency of the shaft/impeller assembly. It is also important that a great deal of attention be paid to construction details to insure that the shaft is straight, that the shaft/impeller assembly is balanced, etc. The reader should consult CHARLES and WILSON (1992) for additional information about drive systems.

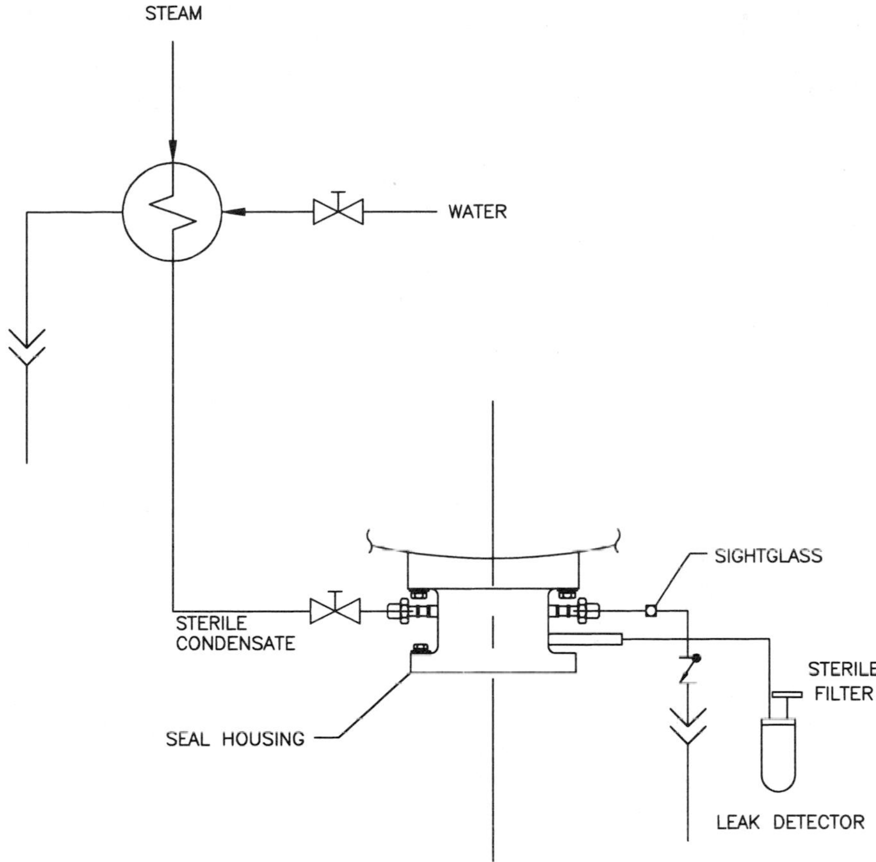

Fig. 6. Agitator seal system recommended for aseptic operation.

6 Conclusion

We have attempted to present to the reader a reasonable sense of the thinking and some of the details involved in the design of a particular class of fermentors. We trust that the reader appreciates the fact that there is much that we have not been able to include, and that much of the art of fermentor design and construction does not lend itself readily to adequate written description (why should this be different from any other art?). There also are whole areas that we have had to leave out because of space limitations. One of the most important among these is validation. With regard to the fermentor itself, we will only note here that observance of the principles discussed in this chapter will go a long way toward achievement of validation. But the reader must recognize that there are many other considerations, and that compliance with cGMP (current good manufacturing practice) is a demanding task which should not be undertaken alone by the uninitiated – regardless of how much he or she has read or knows about the "theory". That is an extremely important fact not only for validation but also for fermentor design and construction. If you take no other message than that, we will have achieved one of our major purposes in writing this chapter.

7 References

BALMER, G. J., MOORE, I. P. T., NIENOW, A. W. (1988), Aerated and unaerated power and mass transfer characteristics of Prochem agitators, in: *Biotechnology Processes, Scale-up and Mixing* (HO, C. S., OLDSHUE, J. Y., Eds.), pp. 116-127, New York: American Institute of Chemical Engineers.

BJURSTROM, E. COLEMAN, D. (1987), Water for injection system design, *Biopharm* **1** (12), 42-47.

BLAKEBROUGH, N., SAMBAMURTHY, K. (1966), Mass transfer and mixing rates in fermentation vessels, *Biotechnol. Bioeng.* **8**, 25-42.

CHARLES, M. (1978), Technical aspects of the rheological properties, of microbial cultures, *Adv. Biochem. Eng.* **8**, 1-62.

CHARLES, M., WILSON, J. D. (1993), Fermentor design, in: *Engineering Design Handbook for Biotechnology* (LEYDERSEN, K., NELSON, K., Eds.), pp. 1-63, New York: Marcel Dekker.

DICKEY, D. S. (1979), Turbine agitated gas dispersion, Paper presented at the *72nd AIChE Annual Meeting*, San Francisco. Washington, DC: American Institute of Chemical Engineers.

GBEWONYO, K., MASE, D., BUCKLAND, B. C. (1986), The use of hydrofoil impellers to improve oxygen transfer efficiency in viscous mycelia fermentation broths, in: Papers presented at the *International Conference on Bioreactor Fluid Dynamics*, Cambridge, England, April 15-17, Bedford, UK: BHRA The Fluids Engineering Centre.

HICKS, R., GATES, L. E. (1976), How to select turbine agitators for dispersing gas into liquids. *Chem. Eng.* **83** (15), 141-150.

HU, W.-S., WANG, D. I. C. (1986), in: *Mammalian Cell Technology* (THILLY, W. G., Ed.), pp. 167-198, Boston: Butterworth.

MARKOVITZ, R. E. (1971), Picking the best vessel jacket, *Chem. Eng.* **78** (11), 156-163.

MOO-YOUNG, M., TICHAR, K., DULLEN, F. A. (1972), The blending efficiencies of some impellers in batch mixing, *AIChE J.* **18**, 178-182.

NAGATA, S. (1975a), *Mixing: Principles and Applications*, pp. 59-62, New York: Halsted Press.

NAGATA, S. (1975b), *Mixing: Principles and Applications*, pp. 341-350, New York: Halsted Press.

OLDSHUE, J. Y., POST, T. A., WEETMAN, R. J., COYLE, C. (1988), Comparison of mass transfer characteristics of radial and axial flow impellers. Paper presented at the *6th European Conference of Mixing*, Pavia, Italy, May 24-26.

OLDSHUE, J. Y. (1983a), *Fluid Mixing Technology*, pp. 64-68, 178, New York: McGraw-Hill.

OLDSHUE, J. Y. (1983b), *Fluid Mixing Technology*, pp. 264-266, New York: McGraw-Hill.

PERKOWSKI, K. (1984), Detection of microscopic leaks in fermentor cooling coils, *Biotechnol. Bioeng.* **26**, 857-859.

PRECISION STAINLESS, INC. (1989), *Tank Head Components Division, Publication TH 5/89.*

RUSHTON, J. H., COSTICH, E. W., EVERETT, H. J. (1950), Power characteristics of some mixing impellers, *Chem. Eng. Prog.* **46**, 395-404

WANG, D. I. C., COONEY, C. L., DEMAIN, A. L., DUNNILL, P., HUMPHREY, A. E., LILLY, M. D. (1979a), *Fermentation and Enzyme Technology*, pp. 173-174, New York: John Wiley & Sons.

WANG, D. I. C., COONEY, C. L., DEMAIN, A. L., DUNNILL, P., HUMPHREY, A. E., LILLY, M. D. (1979b), *Fermentation and Enzyme Technology*, pp. 178-192, New York: John Wiley & Sons.

WANG, D. I. C., COONEY, C. L., DEMAIN, A. L., DUNNILL, P., HUMPHREY, A. E., LILLY, M. D. (1979c), *Fermentation and Enzyme Technology*, pp. 166-172, New York: John Wiley & Sons.

WILSON, J. D., ANDREWS, T. E. (1983), Current practice in designing bottom entering fermentor drives, *Biotechnol. Bioeng.* **25**, 1205-1214.

YOKEL, S. (1986), Understanding the pressure vessel codes, *Chem. Eng.* **93** (9), 75-85.

17 Biotransformations and Enzyme Reactors

ANDREAS S. BOMMARIUS

Hanau (Wolfgang), Federal Republic of Germany

List of Symbols

c	constant (−)
Δc	concentration difference (mol/L)
d_p	diameter of particle (m)
D	diffusion coefficient (m^2/s)
D_{eff}	effective diffusion coefficient (m^2/s)
$[E]$	enzyme concentration (mol/L)
$[E]^*$	enzyme concentration (g/L)
ee	enantiomeric excess (%)
k_{cat}	catalytic constant (s^{-1})
k_s	mass transfer coefficient (m/s)
K_{eq}	equilibrium constant (changing units)
K_i	inhibition constant (mol/L)
K_M	Michaelis constant (mol/L)
L	characteristic length (m)
m_E	total mass of enzyme (g)
N	flux ($mol/(m^2 \cdot h)$)
Q	volumetric flux (m^3/h)
Δp	pressure drop (bar)
$[P]$	product concentration (mol/L)
r	radius (m)
r	reaction rate ($mol/(g \cdot s)$)
r'	volumetric reaction rate ($mol/(L \cdot s)$)
$[S]$	substrate concentration (mol/L)
$[S_0]$	initial substrate concentration (mol/L)
t_{rct}	reaction time (h)
T	temperature (K)
v	velocity (m/s)
V	volume (m^3)
V_m	maximum rate ($mol/(L \cdot s)$)
x	degree of conversion (–)
δ	distance (m)
ν	kinematic viscosity ($kg/(m \cdot s)$)
ρ	density (kg/m^3)
τ	residence time (s, h)
Re	Reynolds number (–)
Sc	Schmidt number (–)
Sh	Sherwood number (–)
η	effectiveness factor (–)
Φ	Thiele modulus (–)

1 General

1.1 Introduction

During the last decade, the role of biocatalysis in organic synthesis, biochemical and biomedical processes has increased dramatically (CROSBY, 1992). The improved availability of cellular and enzyme catalysts paralleling an increasing willingness and ability of organic chemists to use them was reflected in a rising number of applications in bioorganic synthesis (for reviews of transformations, see: WHITESIDES and WONG, 1985a, b; TRAMPER et al., 1985; CROUT and CHRISTEN, 1989; ELFERINK et al., 1991; SERVI, 1992). Biocatalysts feature superior properties in comparison to other catalysts:

- They allow transformations under mild conditions of temperature and pH, which permits consecutive biocatalytic reactions to be run in one vessel with products sensitive to harsh reaction conditions.
- They catalyze a variety of reaction types (Diels–Alder reactions and Cope rearrangements are notable exceptions, see HAYASHI et al., 1983), according to well worked-out kinetics (WHITESIDES and WONG, 1985a, b), and with high turnover numbers.
- They possess nearly unsurpassed selectivity. While chemical and regioselectivity are important qualities of many biocatalysts, it is enantioselectivity that is most desired.

Primarily, the need for enantiomerically pure compounds in the pharmaceutical, food, and crop-protection industries owing to consumer and regulatory demands will continue to fuel the interest in biocatalysts and associated processes.

Biotransformation reactions to a desired product often are in competition with alternative routes, such as extraction from natural sources, fermentations, or asymmetric chemical synthesis. The extraction route depends on the natural availability of a product and often cannot satisfy demand or purity constraints. Efficient fermentations must be based on a controllable biosynthetic pathway to the product and need a large-volume process to be economical. Asymmetric synthesis processes often are cumbersome, less enantioselective than biocatalytic routes, and are characterized by low turnover numbers. Fig. 1 illustrates that the proportion of biocatalytic uses to generate asymmetric molecules has risen considerably in the last years. Tab. 1 lists industrial-scale biocatalytic processes in use today.

As biotechnology advances, the emphasis will shift from generation of superior products to design of improved processes because substrate utilization and downstream processing will account for an increasing fraction of the total transformation cost. The optimal process features a high-productivity reactor containing a stable biocatalyst with an efficient, rapid separation scheme of biocatalyst, substrates, and products. Several reactor designs have been developed for biocatalytic processes. For many such processes, especially on small scale, a batch reactor suffices. For continuous processes, the two most general types are (1) the

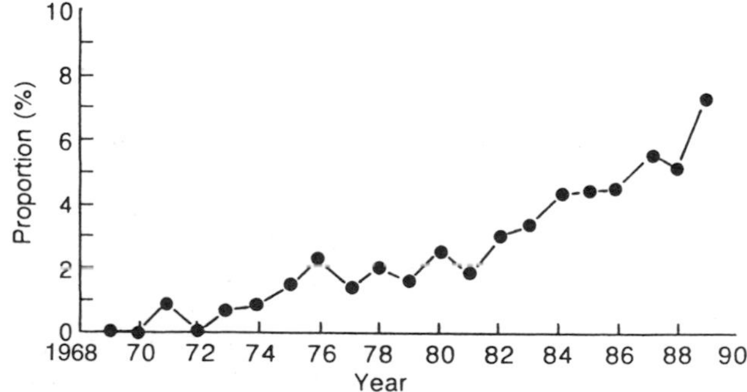

Fig. 1. Significance of biocatalysis in asymmetric synthesis (from CROSBY, 1992, with permission of Wiley & Sons).

Tab. 1. Selected Industrial-Scale Biotransformations

Scale, Enzyme	Product	Reactor	Company	Reference
>1000000 t/a				
Glucose isomerase	Fructose	Fixed-bed, IME	various	Jensen and Rugh, 1987
>10000 t/a				
Nitrile hydratase	Acrylamide	Batch	Nitto Co	Nagasawa and Yamada, 1990
>1000 t/a				
Penicillin acylase	6-APA	Fixed-bed, IME	Eli Lilly	Baldaro et al., 1992
Aspartase	L-Asp	Fixed-bed, IME	Tanabe	Chibata et al., 1986
Thermolysin	Aspartame		Tosoh/DSM	Homandberg et al., 1978; Oyama et al., 1987
Hydantoinase/carbamoylase	D-*p*-OH-Phg	Resting cells	Recordati	Olivieri et al., 1981; Degussa AG, 1992; Drauz et al., 1991a and b; Bommarius et al., 1992
>100 t/a				
Fumarase	L-Malic acid	Fixed-bed, IME	Tanabe	Chibata et al., 1983; Takata et al., 1983
Aminoacylase	L-Met, L-Val	EMR	Degussa	Wandrey and Flaschel, 1979; Bommarius et al., 1992a, c
Aminoacylase	L-Met, L-Phe L-Val	Fixed-bed, IME	Tanabe	Tosa et al., 1969; Chibata et al., 1976
>1 t/a				
Lipase	(*R*)-Glycidyl butyrate	Batch	BASF, DSM	Crosby, 1992

IME, immobilized-enzyme reactor; EMR, enzyme-membrane reactor; 6-APA, 6-aminopenicillanic acid; D-*p*-OH-Phg, D-*p*-OH-phenylglycine; L-Asp, L-aspartic acid; L-Met, L-methionine; L-Val, L-valine; L-Phe, L-phenylalanine

immobilized enzyme reactor, usually run as a fixed-bed plug-flow reactor (PFR), and (2) the enzyme membrane reactor (EMR), commonly run as a continuous stirred-tank reactor (CSTR).

1.2 Scope and Limitations of the Review

This chapter covers transformations, most of them on advanced scale, that are catalyzed by either isolated enzymes or resting cells, either in soluble or immobilized form. The vast majority of these transformations are conversions from a well-defined substrate to a well-defined product. Not covered are systems needing an oxygen supply to the biocatalyst, e.g., production of high-value pharmaceuticals with growing cells or fermentations in membrane reactors. While some of the most spectacular successes of reactions are described that can be carried out in any reactor scheme, emphasis will be laid on conversions and products where reactor design is an integral part of the transformation.

The chapter consists of four parts: after an introduction, Sect. 2 treats reactor design aspects, enzyme reactor modeling, and the use of membrane reactors for enzymatic kinetic investigations. Sect. 3 covers biotransformations, classified according to reaction type; most of the examples are carried out in immobilized enzyme or in membrane reactors. The bulk of featured reactions is catalyzed by hydrolases (Sect. 3.2), but oxidoreductases (Sect. 3.3) and lyases (Sect. 3.5) are also well represented. In the last part, some of the recent developments both in the reactor design area

(Sect. 4) as well as in the biotransformation field (Sect. 5) are illustrated.

1.3 Classification of Biocatalytic Reactors

As in chemical processing, biocatalytic reactions are carried out most simply in a batch reactor (Fig. 2a). If scale is small and biocatalyst is inexpensive or deactivates fast, this might be the best solution. If one wishes to recover the biocatalyst, but keep to a batch reactor design, a batch reactor with subsequent ultrafiltration, batch-UF reactor (Fig. 2b), might be optimal. In all batch reactor schemes, residence time of reactants and catalyst are identical.

As biocatalysis in organic synthesis and large-scale processing grew more sophisticated, methods had to be devised to retain the catalyst efficiently within the reactor. Fig. 3 provides an overview of the options. Attachment of enzymes to solid surfaces, immobilization, was the first development. Instead of a planar surface, immobilized enzymes are usually configured in beads. This allows packing of the beads in a fixed-bed reactor (Fig. 2c) which is usually run as a plug-flow reactor. Undesired radial or axial gradients of pH or temperature can be eliminated by increasing fluid velocity into the column, so that the reactor is operated as a fluidized-bed reactor (Fig. 2d).

Membrane reactors are another technique to separate biocatalysts from substrates or products and retain the catalysts in a given reactor space. Size-specific pores allow substrate and product molecules to cross the membrane but not the biocatalyst. Membrane reactors can be operated as CSTRs with dead-end filtration (Fig. 2e) or as loop or recycle reactor (Fig. 2f) with tangential (crossflow) filtration.

Both immobilized enzyme and enzyme membrane reactors are simultaneously a reaction and a downstream processing device because both possess inherent capability for *in situ* separation of substrates and products from the biocatalyst. If run continuously, as both of them mostly are, residence times of reactants and the biocatalyst are uncoupled. Encapsulation by membranes and immobilization on solid beads are often viewed as the two main separation and retention techniques of biocatalysts. However, especially if one considers that the surface of immobilization can very well be a membrane, the lines between enzyme membrane and immobilized enzyme reactors are not clearly drawn; often membrane reactors are referred to as a special case of immobilized enzyme reactors.

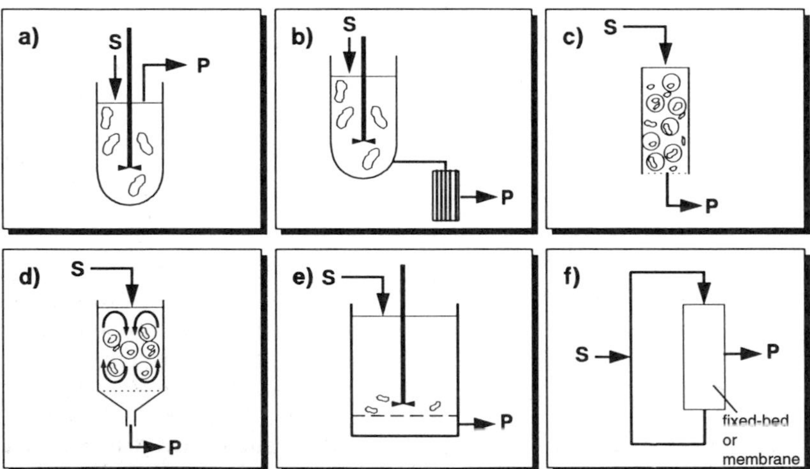

Fig. 2. Enzyme reactor designs. a) batch reactor, b) batch-UF reactor, c) fixed-bed reactor, d) fluidized-bed reactor, e) continuous stirred-tank membrane reactor, f) recycle reactor.

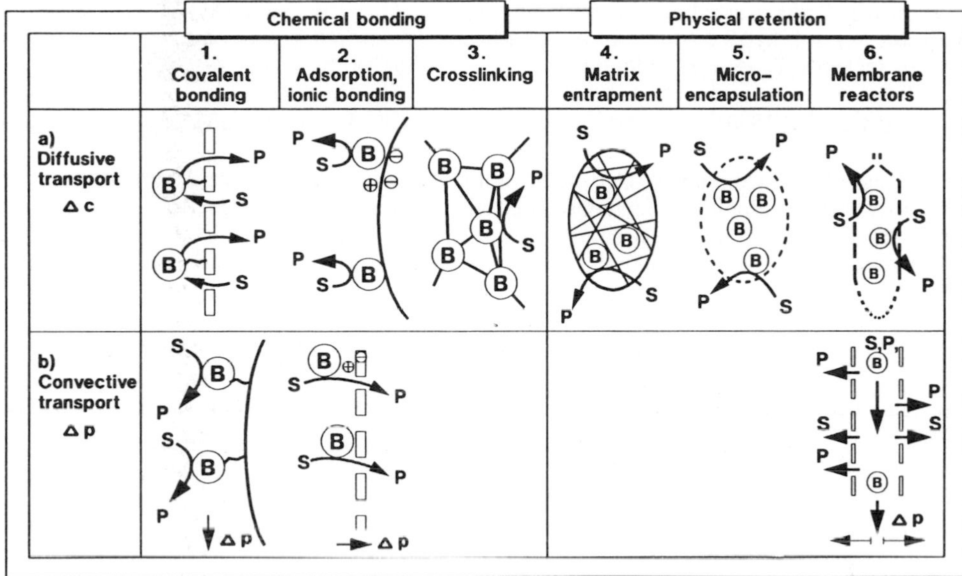

Fig. 3. Configurations for biocatalyst retention. B, biocatalyst; S, substrate; P, product. Diffusive transport: driving force Δc, concentration gradient; convective transport: driving force Δp, pressure gradient.

Tab. 2. Advantages and Disadvantages of Membrane- and Immobilization Reactors

	Advantages	Disadvantages
Membrane reactors	No mass transfer limitations Pyrogen-free products Scale-up simple	Pre-filtration necessary No polymeric products No product precipitation
Immobilization reactors	Potentially more stable enzyme Co-immobilization possible	Difficult to sterilize Mass transfer limitations Immobilization adds extra cost

Immobilized enzyme and enzyme membrane reactors both have advantages and disadvantages. They are listed in Tab. 2.

Immobilization changes activity and stability of biocatalysts much more profoundly than membrane encapsulation; increased activity (rarely seen) or stability (more common) upon immobilization can be a major consideration to choose this technique.

With membrane reactors, the absence of mass transfer limitations, even at high catalyst concentrations, convenient scale-up by adapting membrane area, and the ease to counteract enzyme deactivation at constant throughput (see Sect. 2.5) are major advantages. Furthermore, product streams past the membrane filtration step are usually already free of microorganisms and pyrogens, an increasingly attractive feature for the production of pharmaceutical products. However, product isolation and purification by precipitation cannot be employed owing to membrane blockage.

1.4 History of Biocatalytic Reactors

Three years after the first publication on the use of aminoacylase for splitting of acetylamino acid racemates in a large-scale continuous process in 1966 (Tosa et al.), the first industrial process was started by Tanabe Seiyaku with an immobilized acylase in a fixed-bed reactor in 1969 (TOSA et al., 1969; CHIBATA et al., 1976). The process had been run in a batch mode since the 1950s, but with continuous processing significant cost savings were realized. Tanabe started up more fixed-bed reactor processes with immobilized enzymes: L-aspartate with aspartase in 1973 (CHIBATA et al., 1986) and L-malic acid with fumarase a year after (CHIBATA et al., 1983; TAKATA et al., 1983). The penicillin acylase-catalyzed hydrolysis of penicillin G (BALDARO et al., 1992) and the isomerization of glucose to fructose with glucose isomerase (JENSEN and RUGH, 1987) helped further to establish immobilized enzyme processes towards the end of the 1970s.

In 1970, BUTTERWORTH et al. used a dead-end filtration system to hydrolyze starch to glucose with the help of α-amylase. After some time, however, due to concentration polarization, flux through the membrane decreased dramatically and enzyme leaked through the membrane. Instead of further developing this mode of filtration, e.g., as a new method of immobilization (DRIOLI et al., 1975), tangential flow filtration soon became the method of choice for operating membrane reactors. Already in 1972, the first report by KNAZEK et al. on a hollow fiber reactor for mammalian cells appeared. Based on the work by WANDREY et al. (1979), Degussa introduced a continuous acylase process employing an enzyme membrane reactor (EMR) configuration in 1981 (DEGUSSA AG/GBF, 1981; LEUCHTENBERGER et al., 1984).

1.5 Reviews of Enzyme Reactors

The topic of immobilized enzymes has been reviewed extensively. Already in 1971, ZABORSKY reviewed the subject, while a summary on the use of enzymes in industrial reactors took until 1975 (MESSING, 1975). Several updates were published (see, for instance, BUCHHOLZ and FIECHTER, 1982; SHARMA et al., 1982a, b), and several volumes of the series *Methods in Enzymology* dealt with immobilized enzymes and cells (MOSBACH, 1976, 1987a, b).

On the subject of membrane reactors, groundwork was covered by FLASCHEL et al. (1983). BELFORT (1989) described the use of synthetic membranes with biological catalysts, such as enzymes and cells, including those altered by genetic engineering. Recently, KRAGL et al. (1992) reviewed membrane bioreactors from a point of reaction engineering performance with the focus on soluble enzymes. The often favorable economics of running biocatalytic processes with sophisticated continuous reactor design has been reviewed for a case of an immobilized enzyme reactor (DANIELS, 1987) as well as for enzyme membrane reactors (FLASCHEL and WANDREY, 1978; WANDREY and FLASCHEL, 1979).

2 Enzyme Reactors

2.1 Immobilized Enzyme Reactors

Without innovative process designs, biocatalysis would have stayed confined to a few batch processes. It was with immobilization of biocatalysts that their potential could be tapped in large-scale processes. Today, immobilization does not just have importance for enzymes in bioorganic synthesis, but also has importance in the fields of biomedical reactors (LANGER et al., 1990), and, as suitably coated enzyme pharmaceuticals, in medical therapy (CHANG, 1976).

2.1.1 Classification of Immobilized Enzymes

Immobilized enzyme systems can be classified according to the mode of immobilization, carrier properties, and rate-controlling step.

Mode of Immobilization: Immobilization can be achieved chemically, either by covalent bonding of the biocatalyst to the surface (Fig. 3, option 1; Fig. 4), adsorption or ionic interactions between catalyst and surface (option 2), or crosslinking of biocatalyst molecules to enhance size (option 3), or encapsulation within a matrix (option 4).

Carrier Properties: Carriers come in a variety of configurations and shapes such as films, fibers, planar surfaces, particles or beads. A major influence on the reaction is exerted by the morphology of the carrier, i.e., surface structure and porosity, and the carrier material; most important are (1) inorganic materials such as ceramic or glass, (2) man-made polymers such as nylon or polystyrene, and (3) polysaccharide materials (cellulose, agarose or dextran).

Rate-Controlling Step: In immobilized systems, especially porous ones, the biocatalytic step is only one within a sequence of rate processes:

(1) film diffusion of the substrate molecule through the stagnant layer close to a macroscopic surface,
(2) pore diffusion from the surface to the active site,
(3) biocatalytic step(s),
(4) pore diffusion of product molecule back to the surface, and
(5) film diffusion of product back to the bulk medium.

Depending on reaction conditions, one of the transport steps (1), (2), (4), or (5) can become rate-limiting (see below) instead of the biocatalytic steps.

2.1.2 Influence of Immobilization on Mass Transfer

Immobilization potentially influences the biocatalytic reaction in two major ways: by altering reaction parameters (pH, charge density, and hydrophobicity) and by changing the rate-controlling step from the (bio)chemical reaction to the preceding or subsequent transport step.

On ionic or inhomogeneous surfaces, local charge density is changed in comparison to conditions in the bulk medium. Local charge density affects H^+ ion concentration and thus the local pH. Close to positively charged surfaces, local concentration of H^+ ions is reduced, the apparent pH value is shifted to the *alkaline region* with respect to bulk conditions; accordingly, negatively charged surfaces enhance local concentration of hydrogen ions in the vicinity causing an apparent pH shift to *more acidic* values. Similarly, hydrophobic surfaces cause an apparent *broadening* of the pH scale.

With catalysts within beads, external mass transfer (film diffusion) and internal mass transfer (pore diffusion) has to be investigated. External mass transfer limitations occur if the rate of diffusional transport through the stagnant layer close to a macroscopic surface is rate-limiting. Internal diffusional limitations within porous carriers indicate that the slowest step is the transport from the surface of the particle to the active site buried in the particle interior.

If dominant, external diffusion reduces observed enzyme activity. The flux N through the stagnant film at the surface can be written as:

$$\begin{aligned} N &= (k_s/\delta)\cdot\Delta[S] \\ &= (k_s/\delta)\cdot([S]_{\text{bulk}} - [S]_{\text{surface}}) \end{aligned} \tag{1}$$

where δ is the thickness of the stagnant layer and k_s is the mass transfer coefficient of the solute. k_s can be estimated by the following simple relationship:

$$\begin{aligned} &Sh = 2 + c\cdot Re^{1/2} Sc^{1/3} \quad \text{or} \\ &(k_s\cdot d_p/D) = 2 + c(v\cdot d_p/\nu)^{1/2}(\nu/D)^{1/3} \end{aligned} \tag{2}$$

Sh, *Re*, and *Sc* are the dimensionless Sherwood, Reynolds, and Schmidt numbers, d_p is the relevant length scale such as the diameter of a particle; for fixed beds and laminar flow, the constant $c = 4.7$.

Internal diffusion in porous catalysts also reduces observed biocatalyst activity if important. To check whether internal diffusion is important, the effectiveness factor η is determined. It is defined as

$$\eta = \frac{\begin{array}{c}\text{reaction rate}\\ \text{with internal diffusion effects}\end{array}}{\begin{array}{c}\text{intrinsic reaction rate}\\ \text{without diffusion}\end{array}} \tag{3}$$

If $\eta \sim 1$, the reaction is not hindered by diffusion, if $\eta \ll 1$, pore diffusion is the sole rate-

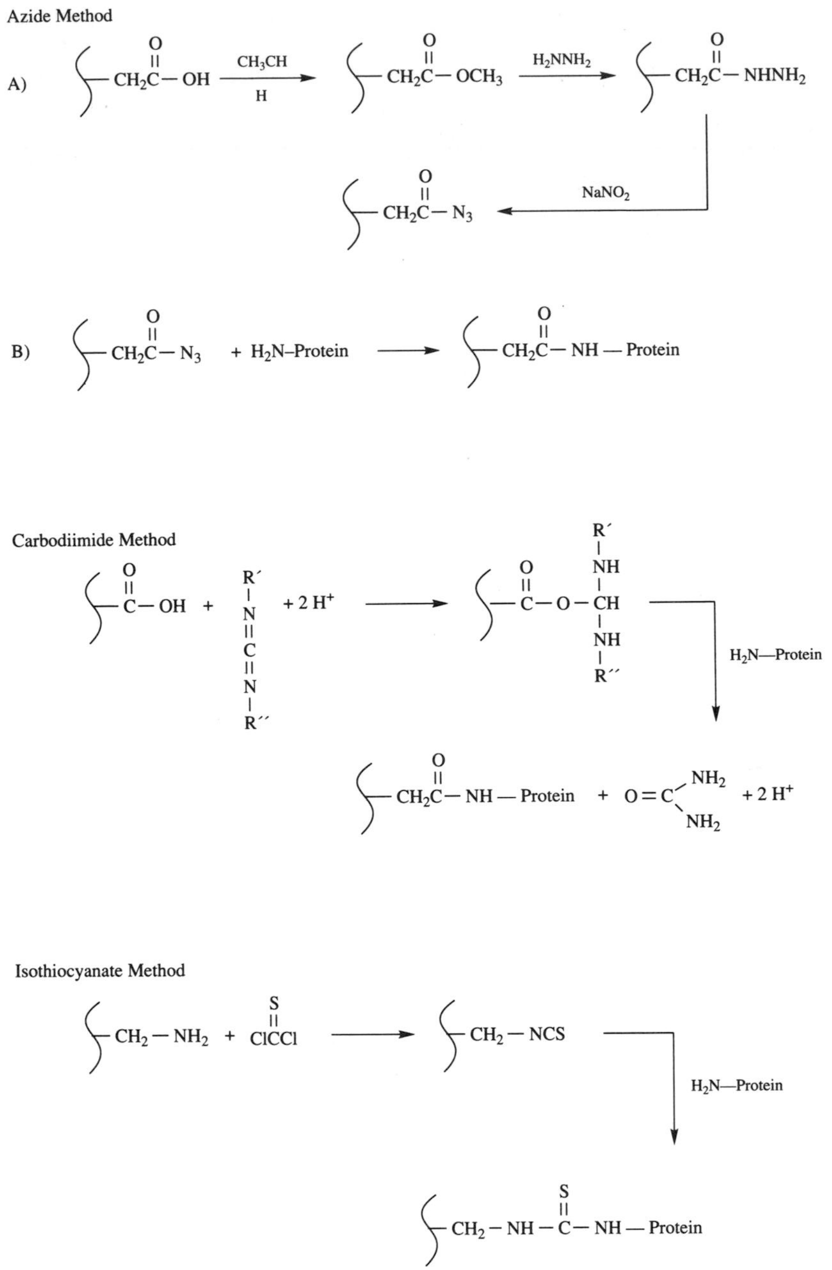

Azide Method

A)

$$\text{—}CH_2\overset{O}{\overset{\|}{C}}\text{—}OH \xrightarrow[H]{CH_3CH} \text{—}CH_2\overset{O}{\overset{\|}{C}}\text{—}OCH_3 \xrightarrow{H_2NNH_2} \text{—}CH_2\overset{O}{\overset{\|}{C}}\text{—}NHNH_2 \xrightarrow{NaNO_2} \text{—}CH_2\overset{O}{\overset{\|}{C}}\text{—}N_3$$

B)

$$\text{—}CH_2\overset{O}{\overset{\|}{C}}\text{—}N_3 + H_2N\text{–Protein} \longrightarrow \text{—}CH_2\overset{O}{\overset{\|}{C}}\text{—}NH\text{—Protein}$$

Carbodiimide Method

$$\text{—}\overset{O}{\overset{\|}{C}}\text{—}OH + R'\text{—}N{=}C{=}N\text{—}R'' + 2\,H^+ \longrightarrow \text{—}\overset{O}{\overset{\|}{C}}\text{—}O\text{—}CH(NH\text{—}R')(NH\text{—}R'') \xrightarrow{H_2N\text{—Protein}} \text{—}CH_2\overset{O}{\overset{\|}{C}}\text{—}NH\text{—Protein} + O{=}C(NH_2)_2 + 2\,H^+$$

Isothiocyanate Method

$$\text{—}CH_2\text{—}NH_2 + Cl\overset{S}{\overset{\|}{C}}Cl \longrightarrow \text{—}CH_2\text{—}NCS \xrightarrow{H_2N\text{—Protein}} \text{—}CH_2\text{—}NH\text{—}\overset{S}{\overset{\|}{C}}\text{—}NH\text{—Protein}$$

Fig. 4. Most common methods of immobilization of enzymes by covalent bonding (SHARMA et al., 1982a, b).

Cyanogen Bromide Method

A)

$$\left\{\begin{array}{l}-CH_2OH \\ -CH_2OH\end{array}\right. + BrCN \longrightarrow \left\{\begin{array}{l}-CH_2O{-}CN \\ -CH_2OH\end{array}\right. \longrightarrow \left\{\begin{array}{l}-CH_2-O \\ -CH_2-O\end{array}\right\rangle C{=}NH$$

B)

$$\left\{\begin{array}{l}-CH_2-O \\ -CH_2-O\end{array}\right\rangle C{=}NH + H_2N{-}Protein \quad \left\{\begin{array}{l}-CH_2-O \\ -CH_2-O\end{array}\right\rangle C{=}N{-}Protein$$

$$\xrightarrow{H_2O} \left\{\begin{array}{l}-CH_2{-}O{-}\overset{O}{\overset{\|}{C}}{-}NH{-}Protein \\ -CH_2OH\end{array}\right.$$

Azo Method

$$\{-CH_2-C_6H_4-NH_2 \xrightarrow[HCl]{NaNO_2} \{-CH_2-C_6H_4-\overset{\oplus}{N_2}\overset{\ominus}{Cl}$$

$$\xrightarrow{HO-C_6H_4-CH_2-Protein} \{-CH_2-C_6H_4-N{=}N-C_6H_3(OH)-CH_2{-}Protein$$

limiting step. To calculate η, the combined diffusion-reaction equation is solved. Modeled as sequential events, the equation reads for spherical geometry and Michaelis–Menten kinetics, $r=k_{cat}[E]\cdot[S]/(K_M+[S])$:

$$d^2[S]/dr^2+2/r\cdot d[S]/dr - k_{cat}[E][S]/D_{eff}(K_M+[S])=0 \tag{4}$$

with the solution:

$$\eta=3/(\Phi\cdot\tan\Phi)-1/\Phi \tag{5}$$

where $\Phi=L\cdot(k_{cat}[E]/(K_M D_{eff}))^{1/2}$ is the dimensionless Thiele modulus, L is the characteristic length, i.e., the half width of a flat plate, $r/2$ of a long cylindrical pellet, and $r/3$ of a sphere. If $\Phi<0.5$, $\eta\sim 1$. If $\Phi>5$, $\eta=1/\Phi$, the substrate concentration drops to zero in the pore very quickly; this case is called *strong pore diffusion*. The dependence of η on Φ was calculated for different geometries and kinetics of arbitrary order (ARIS, 1957).

How is a biocatalyst pellet checked for transport effects? External diffusion is reduced by stronger stirring which compresses the stagnant layer at the surface; if the observed reaction rate is unchanged, external diffusion is not important. Fixed-bed reactors can be checked by comparing rates at different bed heights and constant residence time or more simply by comparing rates at constant bed height and different flow rates. Once the absence of film diffusion effects is verified and if the reaction order n is known, the expression for the rate law $r=\eta\cdot k_{cat}[E][S]_{bulk}/K_M$ (assumed first order) can be inserted into the definition for Φ and the unknown k_{cat} eliminated (WEISZ and PRATER, 1954):

$$-r\cdot L^2/(D_{eff}[S]_{bulk})=\eta\cdot\Phi^2 \tag{6}$$

If pore diffusion is unimportant, $\eta\sim 1$, $\Phi<1$, and

$$-r\cdot L^2/(D_{eff}[S]_{bulk})<1 \tag{7}$$

with strong pore diffusion, $\eta=1/\Phi$ and $\Phi>1$, so Eq. (6) becomes

$$-r\cdot L^2/(D_{eff}[S]_{bulk})>1 \tag{8}$$

Eqs. (7) and (8) only contain observables. A quick test for internal diffusion can be conducted by changing L and observing the effect on r; if there is no effect, pore diffusion cannot be important.

2.1.3 Influence of Mass Transfer on Reaction Parameters

Mass transfer affects the observed kinetic parameters of enzyme reactions. Indications are non-linear Lineweaver–Burk or similar diagrams, non-linear Arrhenius plots, or vastly different K_M values for native and immobilized enzyme. Several expressions have been developed for the apparent Michaelis constant under the influence of external mass transfer limitations:

$$K_{M(app)}=K_M+k_{cat}[E]/k_s \tag{9}$$

(HORNBY et al., 1968)

$$K_{M(app)}=K_M+3/4\,k_{cat}[E]/k_s \tag{10}$$

(KOBAYASHI and MOO-YOUNG, 1971)

$$K_{M(app)}=K_M+k_{cat}[E]\cdot K_M/k_s(K_M+[S_0]) \tag{11}$$

(SCHULER et al., 1972)

Pore diffusion causes the apparent activation energy to decrease by a factor of two (LEVENSPIEL, 1972). It has been pointed out that, owing to broadening of the temperature and pH ranges optimal for activity, pore diffusion apparently increases temperature and pH stability (KARANTH and BAILEY, 1978). Hindered pore diffusion of a substrate might reduce the effects of substrate inhibition (DICKENSHEETS et al., 1977; TRAMPER et al., 1978).

2.1.4 Optimal Conditions for an Immobilized Enzyme Reactor

Biocatalysts are advantageously used in immobilized form for two reasons: (1) the immobilized catalyst is more active or more stable than the native one, or (2) the reactor design of a fixed-bed plug-flow reactor (PFR) produces higher space-time yields.

- Owing to deceleration of the unfolding process, immobilized enzymes are usually more temperature-stable than native ones, an example is penicillin acylase (Sect. 3.2.2).
- Suitable carrier surfaces often accelerate an enzymatic reaction.
- Co-immobilization of two or more enzymes catalyzing consecutive reactions often results in enhanced reaction rates and improved pH or temperature stability.
- While a CSTR is suitable for reactions with substrate inhibition because mixing causes dilution of substrate at the reactor inlet, a PFR carries away product fast and is best for product inhibition situations.
- Since very high conversions can be achieved with much smaller volume in a PFR compared to a CSTR (LILLY and DUNNILL, 1976; VIETH et al., 1976), an immobilized enzyme plug-flow reactor is suited best if conversion has to be complete, such as in the case of the isomerization of glucose (Sect. 3.4) or the decarboxylation of DL-aspartate with L-aspartate β-decarboxylase (Sect. 3.5.3).

2.2 Membrane Reactors

Membrane reactors became available as option for retaining biocatalysts and as a reaction engineering tool when membrane processing technology advanced to a point where precise control of thickness and pore structure became possible. The primary function of the membrane is to retain the biocatalyst which can be an enzyme, an organelle, a resting microorganism or a living one, or a plant or animal cell. For the latter two, the membrane also serves an important function in the transfer of oxygen. In multiphase membrane reactors, the membrane often helps to stabilize the phase boundary. In fermentations, membranes also serve important functions in the cleaning and sterilization of air or other gases.

A membrane can be generated by polymerization around a few biocatalyst molecules enclosing a small space of dimension of hundreds of micrometers or smaller, microencapsulation (Fig. 3, option 5), or be macroscopic (Fig. 3, option 6). Membrane reactors can be classified according to (1) driving force, (2) pore structure, (3) pore size, and (4) direction of flux.

2.2.1 Classification by Driving Force

Membrane reactor devices can be classified according to the main driving forces effecting separation. The membrane can be a barrier for diffusive flux only, thus basing separation on the concentration difference Δc (part a in Fig. 3); the process of diffusing against a reservoir of water is referred to as dialysis. If diffusion of charged species, mostly small ions, is aided by a charge gradient, the process is called electrodialysis. If a pressure difference ΔP is imposed across the membrane, the main separation power stems from convective forces (part b in Fig. 3). From the point of reaching high flux, any design with enforced flow across the membrane is superior to an approach with diffusive mass transfer.

2.2.2 Classification by Pore Structure

Conventional membranes consist of a network of fibers and retain solutes by a stochastic adsorption mechanism; they are termed depth filters (Fig. 5a); most sterilization applications are performed with this type of filter. In contrast, the vast majority of membranes relevant for retention of biocatalysts contain holes or pores of a fairly narrow size range and thus separate on the basis of size and shape of the solute only. Membranes operating by this principle are called screen filters (Fig. 5b). Membrane processing technology has advanced to a point today where the cutoff curves with respect to solute size are very sharp. Protein biocatalysts usually are approximately globularly shaped hard spheres and thus easy to filter; coiled polymers uncoil

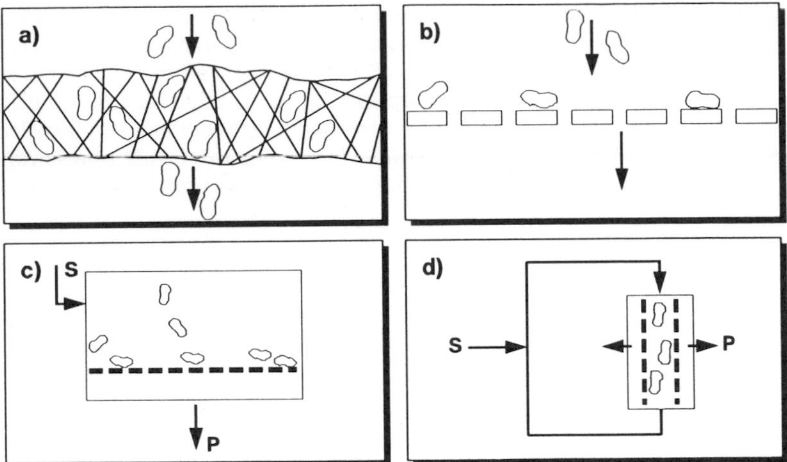

Fig. 5. Membrane reactor configurations. a) depth filtration, b) screen or membrane filtration, c) dead-end filtration, d) tangential or crossflow filtration.

Tab. 3. Advantages and Disadvantages of Depth and Screen Filters

	Advantages	Disadvantages
Depth Filters	High flux Inexpensive device	No absolute retention Grow-through of bacteria Channeling observed Leakage of filter material
Screen Filters	Absolute retention Pyrogen-free product Stable phase boundary	Easier blockage

much more easily and thus still slip through nominally smaller pores. Tab. 3 lists advantages and disadvantages of both depth and screen filters.

2.2.3 Classification by Pore Size

Membranes are classified into microfiltration (MF) membranes with pore sizes in the range of 0.1 to 10 μm (commonly 0.1 or 0.22 μm), ultrafiltration (UF) membranes with a molecular weight cut-off (MWCO) between 0.5 and 500 kD (commonly 10 to 100 kD), and reverse osmosis (RO) membranes suitable for low-molecular weight solutes such as salts. The diameter of most cells is around 1 μm, so they are retained by MF membranes. Enzymes, with sizes between 10 and 500 kD, are not retained by MF but by UF membranes with a typical MWCO of 10 kD of commercial membranes. Membrane materials are commonly either hydrophobic polysulfone or hydrophilic regenerated cellulose or cellulose acetate; other materials are nylon, polytetrafluoroethylene (PTFE, teflon), polyether-ether ketone (PEEK), or polydivinylfluoride (PDVF).

2.2.4 Classification by Direction of Flux

Originally, membrane reactors were run by dead-end filtration (Fig. 5c), with the direction of flow perpendicular to the membrane surface. In this setup, concentration polarization occurs because rejected material is not removed from the upstream part of the membrane and blocks the settling of oncoming particles. Although such problems can be minimized by a stirrer (on the lab scale), nowadays mostly tangential or crossflow filtration is employed within a loop reactor (Fig. 5d). The fluid travels tangentially to the membrane surface, rejected material is swept away and thus can no longer block the membrane surface.

Today, membranes are a routine processing option. Owing to excellent process stability, the cost of membranes is not dominating, so instead of minimizing membrane area at given flux sufficient area is provided to enhance total operation time of a membrane bioreactor. Sterile operation and removal of pyrogens by the UF membrane reduce fouling of catalyst and product validation efforts. The main challenge to membrane stability is fouling by high total protein concentration stemming from repeated addition of enzyme to counter deactivation. Sufficient membrane area and occasional cleaning prevent any limitations from concentration polarization by protein or other debris.

2.3 Membrane Reactor Configurations

2.3.1 Biocatalysis with Microencapsulation

In work pioneered by T. M. S. Chang (1964, 1987), enzymes were encapsulated within spherical semipermeable membranes of about 100 μm diameter ('artificial cells', Fig. 3a, 5). Encapsulation was effected by cellulose nitrate or polyamide membranes, the latter is produced by interfacial polymerization and is most efficient. A variety of proteins were protected from diffusing out of the capsule. Since the enzymes and reactants inside the capsule are in solution, multi-enzyme reactions and cofactor regeneration reactions are possible.

2.3.2 Dialysis Reactor

One of the simplest membrane reactors has been pioneered by Bednarski et al. (1987) (membrane-enclosed enzymatic catalysis, MEEC). A bag from dialysis tubing filled with biocatalyst was suspended in aqueous medium (Fig. 3a, 6). This simple setup was probed with a variety of enzymes both crude and purified, both single and combined. It is especially advantageous for crude enzyme preparations which cause problems during immobilization on beads. The small membrane area for the diffusive flux seems to be the limiting quantity of this reactor configuration.

2.3.3 Immobilization on Membranes

A membrane can serve as a surface for immobilization of a biocatalyst. Often, an enzyme is inadvertently immobilized onto a membrane surface by ionic or adsorptive forces (Fig. 3a and b) but this mode can be dictated by design as in multiphase reactors (see below). Biocatalysts immobilized onto membranes have to be treated as other immobilized enzyme reactors rather than recycle membrane reactors (see next section).

2.3.4 Recycle Reactor

A recycle reactor (Fig. 2f) is a valuable reaction engineering tool to gather kinetic and transport data for process development and scale-up. Furthermore, a membrane reactor with soluble biocatalysts is run preferentially as a recycle reactor to achieve tangential flow along the membrane. The most characteristic quantity of a recycle reactor is the recycle ratio, the ratio of substrate flow (and hence permeate flow through the membrane) to recycle flow. A recycle reactor can be operated as a PFR if the recycle ratio approaches zero, or as a CSTR if the recycle ratio approaches infinity. In membrane reactors run as CSTRs, multiple stable states and unsteady operating regions have been observed, just as in conventional catalytic reaction engineering (Schmidt et al., 1986). The recycle reactor design has

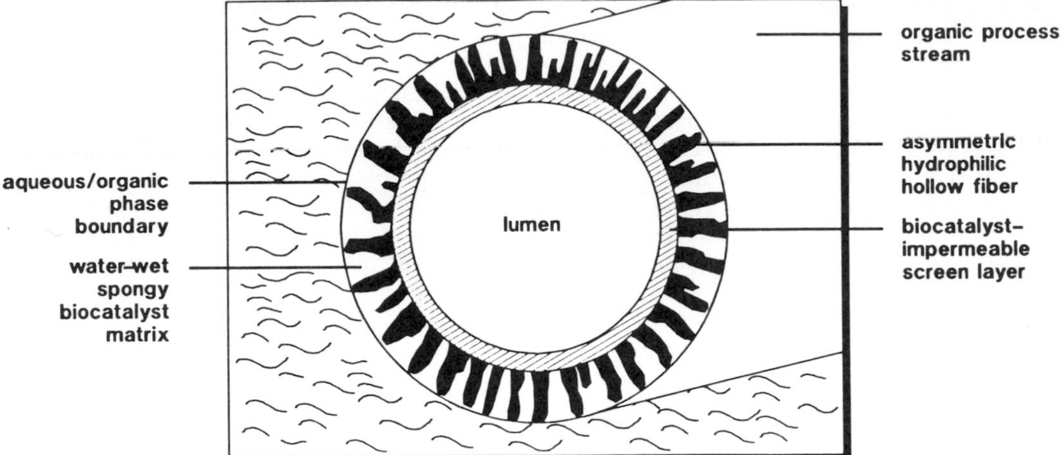

Fig. 6. Entrapment of a biocatalyst in a hydrophilic membrane (YOUNG, 1990).

been incorporated into most membrane reactor processes today, examples are discussed in Sects. 3.2.1 and 3.3.1.

2.3.5 Multiphase Reactors

Conversion of hydrophobic substrates sparsely soluble in aqueous systems presents a particular challenge to reaction engineering. Among other attempts to attack this challenge, liquid membranes (SCHEPER et al., 1989; for modeling, see: MARR and KOPP, 1980; LORBACH and HATTON, 1988) are an equilibrium-driven process by necessity reaching equilibrium, and microemulsions (LUISI, 1985a, b; MARTINEK et al., 1986; LUISI et al., 1988; SHIELD et al., 1986), though not limited by mass transfer for most systems, present difficulties in separating products. Multiphase membrane reactors present a solution (MC CONVILLE et al., 1990): the membrane acts as a phase contactor, phase separator, and interfacial catalyst. Enzymes are immobilized in the interstices of an asymmetric hydrophilic membrane (Fig. 6).

The enzyme is charged from the shell side. It is entrapped because on the lumen side the cutoff is chosen so that no leakage occurs; the shell side is contacted with the organic solvent, the enzyme's low distribution coefficient prevents partitioning into the organic phase. Deactivated enzyme can be flushed by immersing the membrane in water on both sides and setting a gradient towards the shell side. Control of pressure gradient (positive towards the lumen) is decisive not to push organic solvent into the membrane. For the reactions carried out in this biphasic reactor, see Sect. 4.1.1.

2.4 Enzyme Reaction Engineering

The modeling procedure has been published several times (KRAGL et al., 1992; BOMMARIUS et al., 1992a), so it is recapitulated here only briefly. The main goal of enzyme reaction engineering is to achieve high space-time yield at high substrate conversion with high chemical selectivity, regioselectivity, and/or enantioselectivity. Figs. 7a and 7b show the schematic plan of procedure.

Starting with the *reaction system*, values for pH and temperature are set according to reactant, product and enzyme properties. Fortunately, in biocatalysis most reactions proceed optimally within very narrow ranges of temperature (30–50 °C) and pH (7–10). Initial substrate concentration and the ratio of the two substrate concentrations in a bimolecular reaction have to be fixed. Biocatalyst concentration is important for achieving the desired space-time yield and for influencing selectivity in case of undesired parallel or consecutive

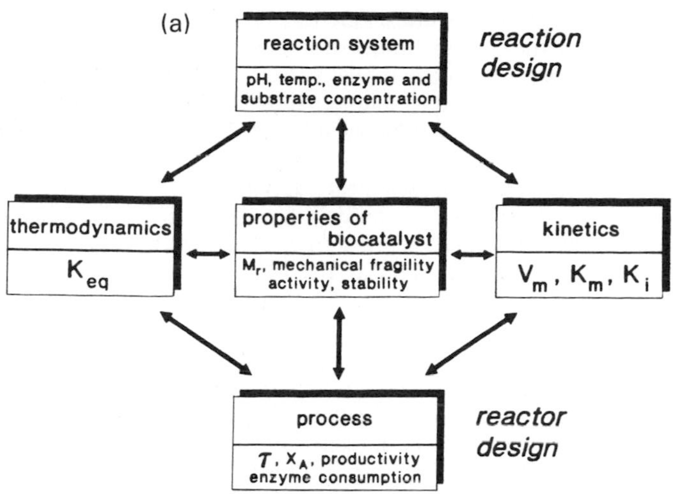

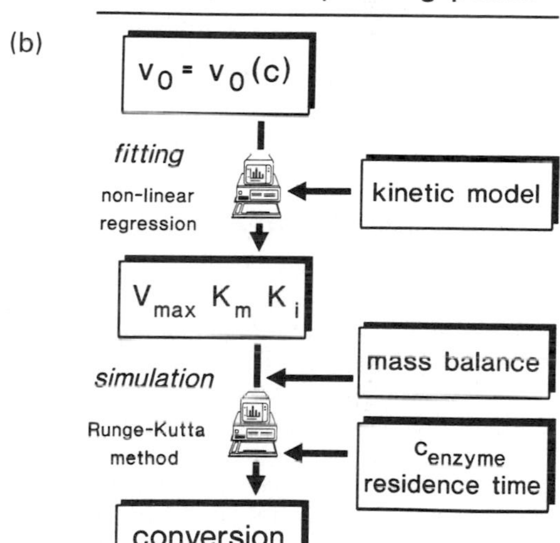

Fig. 7. Procedure of kinetic modeling of enzyme reactions (BOMMARIUS et al., 1992a). (a) Relationship between different aspects of enzyme reaction engineering. (b) Parameter estimation and determination of operating points.

side reactions. In multi-enzyme systems, the optimal activity ratio of all enzymes has to be determined. Enzyme activity and stability have to be checked as functions of reaction conditions prior to kinetic measurements. Enzyme stability is an important aspect in biocatalytic processes and is expressed as biocatalyst consumption per unit weight of product (see Sect. 2.5). In multi-enzyme systems, the stability of all enzymes has to be balanced for optimal overall rate and space-time yield.

The *reaction kinetics* have to be determined at process-relevant conditions, whereas most kinetic measurements found in the literature refer to optimum enzyme pH and temperature conditions and to very dilute solutions. The influence of all reactants and products has to be measured as function of conversion instead of just measuring initial rate kinetics. *Thermodynamics* of the reaction in question has to be integrated into the kinetic model because there is no "forward reaction" and separate "reverse reaction" enzyme, so the forward reaction cannot be treated separately from the back reaction in the kinetic model. Preferably, parameter estimation for kinetic models should be obtained by non-linear regression routines and not by linearization such as the Lineweaver–Burk method (Fig. 7b). The best proof for a kinetic model is a proper fit of concentration versus time along the entire range of conversion in a batch experiment (integral method). Sometimes it may be difficult to integrate the equations of the kinetic models, in which numerical methods such as the well-known Runge–Kutta method are required (Fig. 7b). Kinetic modeling has been extended to multienzyme systems, for instance to redox reactions with cofactor regeneration (VASIC-RACHI et al., 1989). While a membrane reactor is advantageous for reactions with substrate inhibition, it is a disadvantage in case of severe product inhibition because product cannot be removed immediately.

2.5 Enzyme Deactivation in Continuous Reactors

The topic of operational stability of enzymes has not received proper attention in enzyme reaction engineering. Even when studying biocatalytic reactions instead of stability of a protein molecule, enzyme stability is often expressed as a 'half-life'; the time after which enzyme activity has decreased to 50% of initial activity. The half-life is determined either by measuring initial rates after resting the enzyme for a specified time at standard conditions ('temperature stability') or by comparing initial rates in a series of successive batch reactions at ever-changing conversion of substrate.

Both methods are inadequate for characterizing enzyme stability under operating conditions. Temperature stability refers to the stability of the protein molecule; since deactivation occurs in the absence of reaction, it is usually not a measure of operational stability. Stability results at varying conversions are not reproducible unless the same conversion profile is followed. Defined measurements of operational enzyme stability are carried out at constant conversion of substrate, which implies constant turnover number in case of stable enzymes and necessitates the use of a continuous reactor. With immobilized enzymes, a fixed-bed plug-flow reactor is most appropriate. A continuously stirred tank is more advantageous for soluble enzymes (TOSA et al., 1966), so membrane reactors are excellent tools to determine operational stability of soluble biocatalysts.

The relevant parameter for operational stability studies of enzymes is the product of active enzyme concentration $[E]^* = m_E/V$ (g/L) and residence time τ, $[E]^* \cdot \tau$. The ensuing derivation highlights the utility of this parameter. The reaction rate is defined as

$$-\mathrm{d}[S]/\mathrm{d}t = r' = r \cdot [E]^* \tag{12}$$

with volumetric reaction rate r' (mol/L/min) and specific (mass-based) reaction rate r (mol/g/min). Introducing the degree of conversion x:

$$[S_0] \cdot \mathrm{d}x/\mathrm{d}t = r \cdot [E]^* \tag{13}$$

Rearranged and integrated over the degree of conversion from time 0 to residence time τ:

$$\begin{aligned} \int_0^x (1/r)\mathrm{d}x &= [E]^*/[S_0] \int_0^\tau \mathrm{d}t \\ &= ([E]^* \cdot \tau)/[S_0] \end{aligned} \tag{14}$$

In a CSTR, the integral $\int (1/r)\mathrm{d}x$ reduces to an algebraic equation because r is constant throughout the reactor:

$$([E]^* \cdot \tau)/[S_0] = x/r(x) \tag{15}$$

A very useful feature of Eq. (15) is its *independence of the enzyme kinetics*, however, r still depends on x. At constant mean residence time τ as well as substrate concentration $[S_0]$ at the reactor inlet, the degree of conversion x is constant as long as the concentration of active enzyme $[E]^*$ is constant. If no wash-out of the enzyme through the membrane is assumed, $[E]^*$ can only decrease through enzyme deactivation. A lower value of $[E]^*$ from enzyme deactivation results in a decreased degree of conversion x. To reach the previous value of $[E]^* \cdot \tau$ and thus the present level of conversion, two strategies are possible:

- either the residence time τ is increased by lowering the flow rate Q to compensate for the decrease in $[E]^*$, or
- fresh enzyme is added to bring the concentration of active enzyme $[E]^*$ back to previous levels.

Raising τ is convenient to use for immobilized enzyme in a plug-flow reactor, because replacement of the decaying enzyme is difficult while maintaining the same enzyme concentration across the length and breadth of a column. Soluble enzymes are most easily investigated with the second method in a CSTR. The amount of enzyme added over time in either of the two strategies is scaled to the amount of product formation ($[P]=x[S]$) to yield an accurate measure of operational stability.

2.6 Enantioselectivity of Enzymes

While many enzymes feature excellent enantioselectivity, not all of them do. Most lipases are notable exceptions, which has been linked to their flexible structure (KLYOSOV et al., 1975; CHEN and SIH, 1989a, b). Low enantioselectivity is most acute in kinetic resolutions with both enantiomers present in similar quantities. The enantiomeric ratio, E, characterizes the enantioselectivity of an enzyme and is defined as (ln = natural logarithm to the base e)

$$E=\ln[(1-x)(1-ee_R)]/\ln[(1-x)(1+ee_R)] \quad (16)$$

or

$$E=\ln[(1-x(1+ee_P))]/\ln[(1-x)(1-ee_P)] \quad (17)$$

or

$$E=\ln([A]/[A_0])/\ln([B]/[B_0]) \quad (18)$$

or

$$E=\ln(k_{cat}/K_M)_A/\ln(k_{cat}/K_M)_B \quad (19)$$

with the degree of conversion x, and the enantiomeric excess of a compound A over B defined as $ee=([A]-[B])/([A]+[B])$. For derivations of these relationships, see CHEN et al. (1982); CHEN and SIH (1989a, b). Useful graphical representation of the dependence of ee_R on x and E has been provided by MARTIN et al. (1981). An E value of 1 means no enantioselectivity at all, whereas a value of infinity indicates perfect selectivity. In general, enzymes with an E value of 20 to 50 are regarded as sufficiently selective, and with a value of greater than 100 as completely selective.

3 Biotransformations in Enzyme Reactors

3.1 Overview

Many different classes of enzymes have been utilized for large-scale transformations. No longer are hydrolases the only game in town; with large-scale cofactor regeneration of NAD/NADH, dehydrogenases have made considerable progress (Sect. 3.3) and lyases are used in three established processes (Sect. 3.5). However, the largest biotransformation process is conducted with an isomerase (Sect. 3.4). From the number of commercialized processes, amino acids are by far the most important class of compounds, starting with the aminoaclase process (Sect. 3.2.1) and finishing with the aforementioned lyases.

Several important biotransformation processes are not covered in this chapter. Instead, attention is focused on conversions where process development was a significant amount of the total effort. Achievements such as the amidase process from DSM (KAMPHUIS et al., 1990, 1992) to enantiomerically pure α-amino acids or the nitrilase process from Nitto Co converting nitriles to amides (NAGASAWA and YAMADA, 1990) not covered extensively here, should, nevertheless, be paid attention to.

The number of reactions conducted with immobilized enzymes has become too numerous to follow. For membrane reactors, the number is not quite as large. Taken from a recent re-

Tab. 4. List of Single-Enzyme Reactions in an Enzyme Membrane Reactor (KRAGL et al., 1992)

Enzyme	Biocatalytic Reaction	pH	T (°C)	Substrate Concentration (mol/L)	Reactor Volume (L)	Residence Time (h)	Conversion (%)	Space-Time Yield (g/L/d)	Enzyme Consumption (U/kg)	Applications
Aminoacylase	Hydrolysis	7.0	37	0.8 *N*-Acetyl-DL-methionine	1	1.8	85	1200	1000	Separation of enantiomeric amino acids
PEN-G-acylase	Hydrolysis	7.8	25	0.002 Phenacetyl-L-tyrosyl-L-alanine	0.01	0.25	98	29	7200	Deacylation of dipeptide
Fumarase	Dehydration	7.0	37	1.0 Fumaric acid	1.2	1.5	42	1000		Production of L-malic acid
(*R*)-Oxinitrilase	Formation of a C-C bond	3.75	20	0.059 Benzaldehyde, 0.2 HCN	0.01	0.06	80–90	24000	17000	Production of mandelic acid via (*R*)-mandelic acid nitrile
Aldolase	Formation of a C-C bond	6.0	25	0.03 (*R*,*S*)-3-Acido-2-hydroxypropanol, 0.019 dihydroxyacetone phosphate	0.005	2–3	72	40	512000	Production of non-natural sugar analogs
Aspartase	Amination	8.5	20	0.4 Fumaric acid, 0.44 NH_3	0.2	5.0	80	400	5000	Production of L-aspartic acid
β-Decarboxylase	Decarboxylation	5.7	30	0.2 L-Aspartic acid, 0.005 pyruvate	0.18	4.5	74	53	144000	Production of L-alanine
Aspartate-phenylpyruvate transaminase	Transamination			0.090 L-Aspartic acid, 0.061 phenylpyruvate	0.01	0.5	75			Production of L-phenylalanine
Formate dehydrogenase (FDH)	Reduction	8.0	25	0.2 Fumaric acid, 0.0194 NAD^+	0.01	2.0	90			Production of NADH
α-Chymotrypsin	Formation of a C-N bond	9.5	25	0.01 ForTyr-OProp, 0.02 Arg-OProp	0.2	0.5	>98	91	750 mg/kg	Production of kyotorphin
α-Chymotrypsin	Formation of a C-N bond	9.5	25	0.48 MalTyr-OEt, 0.8 Arg-NH_2	0.0021	0.083	80	25600	175 mg/kg	Production of kyotorphin
Papain	Formation of a C-N bond	9.0	25	0.005 MalTyrArg-OEt, 0.05 Ala-OEt	0.01	0.25	62	154.4	2.1 g/kg	Production of tripeptide
Carboxypeptidase Y	Formation of a C-N bond	8.4	25	0.002 PhenacTyr-OMe, 1.0 L-alanine	0.01	0.25	>98	33.7	1060	Production of dipeptide

Tab. 5. List of Dual-Enzyme Reactions in a Membrane Reactor (KRAGL et al., 1992)

Enzymes	Biocatalytic Reaction	pH	T (°C)	Substrate Concentration (mol/L)	Reactor Volume (L)	Residence Time (h)	Conversion (%)	Space-Time Yield (g/L/d)	Enzyme Consumption (U/kg)	Number of Cycles	Applications
LeuDH/FDH	Reductive amination	8.0	25	0.1 α-Ketoisocaproate, 0.4 NH_4-formate	0.01	1.2	80	250	260 LeuDH, 300 FDH	75000	Production of L-leucine
LeuDH/FDH	Reductive amination	8.0	25	0.1 α-Ketoisocaproate, 1.0 NH_4-formate	0.01	1.0	87	239	653 LeuDH, 412 FDH	50000	Production of L-leucine
LeuDH/FDH	Reductive amination	8.0	25	0.5 Trimethylpyruvate, 1.0 NH_4-formate	0.01	2.0	85	640	930 LeuDH, 2280 FDH	125000	Production of L-*tert*-leucine
PheDH/FDH	Reductive amination	8.5	25	0.12 Phenylpyruvate, 1.0 NH_4-formate	0.01	1.0	95	456	1500 PheDH, 150 FDH	600000	Production of L-phenylalanine
L-HicDH/FDH	Reduction	7.0	25	0.1 2-Ketoisocaproate, 0.4 Na-formate	0.01	0.7	90	411	730 L-HicDH, 2280 FDH	69400	Production of hydroxy acids
MaDH/FDH	Reduction	8.0	25	0.5 Phenylglyoxylic acid, 1.0 NH_4-formate	0.01	2–3	95	698	1070 MaDH, 660 FDH	24000	Production of (*R*)-mandelic acid
D-HicDH/FDH	Reduction	7.5	28	0.08 3,4-Dihydroxy-phenylpyruvate, 0.2 NH_4-formate	0.01	1.7	80	205	5125 HicDH	5000	(*R*)-DHPL (constituent of rosmarinic acid production)
HK/AK	Phosphorylation	7.6	25	0.1 Glucose, 0.06 acetylphosphate	0.01	1.33	87	10.2	14 HK, 6 AK	20000	Production of glucose-6-phosphate
D-RK/PK	Phosphorylation	8.3	25	0.025 D-Ribulose, 0.03 phosphoenolpyruvate	0.01	1.0	93	117	7300 D-RK, 25800 PK		Production of D-(−)-ribulose-5-phosphate
EPI/ALD	Formation of a C-C bond	7.5	25	0.2 *N*-Acetylglucosamine, 0.1 pyruvate	0.014	2.85		109			Production of *N*-acetyl-neuraminic acid
D-AO/CAT	Oxidative deamination	7.5	25	0.0005 D-Methionine, 0.1 pyruvate	0.01		100	99			Production of L-methionine

Abbreviations: ALD, aldolase; AK, acetate kinase; D-AO, D-amino acid oxidase; CAT, catalase; EPI, epimerase; FDH, formate dehydrogenase; HicDH, hydroxyisocaproate dehydrogenase; HK, hexokinase; LeuDH, leucine dehydrogenase; MaDH, mandelate dehydrogenase; PheDH, phenylalanine dehydrogenase; PK, pyruvate kinase; D-RK, D-ribulokinase

view (KRAGL et al., 1992), a list of reactions carried out in membrane reactors is shown for single-enzyme reactions (Tab. 4) and dual-enzyme reactions (Tab. 5).

3.2 Hydrolases

3.2.1 Aminoacylase

The most established method for enzymatic L-amino acid synthesis is the resolution of racemates of *N*-acetyl-DL-amino acids by the enzyme acylase I (aminoacylase; E.C. 3.5.1.14.). *N*-acetyl-L-amino acid is cleaved to yield L-amino acid whereas the *N*-acetyl-D-amino acid does not react (Eq. (20)).

$$\text{R-CH(COOH)-NH-CO-CH}_3 \text{ (DL)} + H_2O \xrightarrow{\text{aminoacylase}} \text{R-CH(COOH)-NH}_2 \text{ (L)} + \text{R-CH(COOH)-NH-CO-CH}_3 \text{ (D)} \quad (20)$$

After separation of the L-amino acid through ion exchange or crystallization, the remaining *N*-acetyl-D-amino acid can be racemized by acetic anhydride in alkaline solution or by adding a racemase (TAKEDA CHEMICAL INDUSTRIES LTD., 1989) to achieve very high overall conversions to the L-amino acid. *N*-acetyl-DL-amino acids are conveniently accessible on laboratory as well as industrial scale through acetylation of DL-amino acids with acetyl chloride or acetic anhydride in a Schotten–Baumann reaction (for a review, see SONNTAG, 1953).

With the aminoacylase process, Tanabe Seiyaku commercialized the first immobilized enzyme reactor system ever in 1969 after running the process in batch mode since 1954 (TOSA et al., 1969; CHIBATA et al., 1976). Enzyme from *Aspergillus oryzae* fungus was immobilized by ionic binding to DEAE-Sephadex (TOSA et al., 1966). In a fixed-bed reactor, the reaction is carried out at elevated temperature to produce L-methionine, L-valine, and L-phenylalanine. Costs are significantly lower than in a batch process with native enzyme.

At Degussa, several enzyme membrane reactor (EMR) set-ups are in operation covering six orders of magnitude from laboratory via pilot stage to full production scale; the process has been scaled up to an annual production level of several 100 tons of enantiomerically pure α-amino acids, mostly L-methionine and L-valine. The reactor design is a recycle reactor operated as a CSTR (recycle ratio up to 200; Fig. 2f). For both pilot and large-scale operation, the necessary membrane areas are configured into hollow-fiber modules resulting in a nearly quantitative rejection rate of the aminoacylase.

Aminoacylase has also been immobilized on a nylon membrane (IBORRA, 1992). While the half-life as measured by thermal stability of 161 days is superior to the data for immobilized acylase (65 days, CHIBATA et al., 1976) or soluble enzyme in an EMR (WANDREY and FLASCHEL, 1979), reactor productivity at 0.186 L-valine kg/L/d is lower than that for DEAE-Sephadex-immobilized acylase (0.5 kg/L/d, CHIBATA et al., 1976) or that for a membrane reactor (0.35 kg/L/d, WANDREY and FLASCHEL, 1979).

Acylase I is available commercially from porcine kidney and *Aspergillus oryzae* (GENTZEN et al., 1979, 1980; CHENAULT et al., 1989). Stability tests in repeated-batch mode (GENTZEN et al., 1979; CHENAULT et al., 1989; WANDREY, 1977) revealed that resistance to deactivation and oxidation is superior for the fungal enzyme. Results on operational stability of both acylases in a recycle reactor at constant conversion (WANDREY, 1977) with reaction conditions close to intended large-scale conditions demonstrated much better stability of the *Aspergillus* enzyme, while renal enzyme is not stable enough for long-term operation (WANDREY, 1977; BOMMARIUS et al., 1992c). Moreover, on the process scale achieved today, the supply of renal acylase is insufficient so that fungal acylase is now used almost exclusively, especially since the price per unit is comparable.

The substrate specificity of aminoacylase I is unusually wide (CHENAULT et al., 1989): enzymes from both sources prefer long straight-chain, hydrophobic substrates but fungal acylase also readily accepts aromatic and branched substrates. Fig. 8 lists both proteinogenic and non-proteinogenic amino acids prepared at Degussa in bulk quantities by the resolution of the respective *N*-acetyl amino acids.

3.2.2 Penicillin Acylase

Penicillins represent the most important class of antibiotics; they consist of a common core, 6-aminopenicillanic acid (6-APA), and varying side chains. Only two examples, penicillin G (Pen G), with phenylacetic acid side chain, and penicillin V (Pen V), with phenoxyacetic acid side chain, are fermented from *Penicillium chrysogenum*; all others are made from 6-APA which is produced almost exclusively by enzymatic hydrolysis today, mostly from Pen G with the help of penicillin G acylase (E.C. 3.5.1.11.; Eq. 21).

carbef combines this synthetic capability with the complete enantiospecificity at the amino group to be acylated (ZMIJEWSKI et al., 1991). For an overview of synthetic applications of Pen G acylase, see BALDARO et al. (1992).

3.2.3 Enzymatic Resolution Reactions with Complete Conversion

Hydantoins, cyclically protected α-amino acids, can be converted to α-amino acids in a two-step reaction sequence catalyzed by the enzyme hydantoinase (E.C. 3.5.2.2.) and carbamoylase. Depending on the enantioselectivity of the enzymes, L- or D-amino acids can be obtained (Fig. 9).

The ability of hydantoins to racemize, either enzymatically or base-catalyzed, above pH 8 opens the possibility of attaining up to 100% yield of the desired product enantiomer. Several processes, both for the L-series (SYLDATK et al., 1990; WAGNER et al., 1990; SYLDATK et al., 1992; RÜTGERSWERKE AG, 1988; SYLDATK and WAGNER, 1990) as well as for the D-products (OLIVIERI et al., 1981; DEGUSSA AG, 1992; DRAUZ et al., 1991a, b; BOMMARIUS et al., 1992b), have been developed. Today, more than 1000 tons of D-*p*-hydroxyphenylglycine, the side chain of the antibiotic ampicillin, are manufactured via this route, mostly by Recordati.

PEN G $+H_2O$ —PGA→ 6-APA + phenylacetic acid (21)

Pen G acylase represents one of the first and greatest successes of immobilized enzyme technology and is run on a scale of 6000 tons per year (BALDARO et al., 1992). It has been immobilized by a wide variety of methods, mostly to enhance operational stability. A particular challenge of Pen G to 6-APA processes is precise pH control during hydrolysis because the β-lactam ring is very sensitive to chemical hydrolysis at alkaline and acidic pH values.

Pen G acylase has characteristics and mechanisms similar to serine proteases, so it has been used for kinetically controlled peptide syntheses (FUGANTI et al., 1986; STOINEVA et al., 1992) and deprotection of peptides (DIDZIAPETRIS et al., 1991; HERRMANN, 1991; WALDMANN, 1991). Its use for the resolution of an intermediate of the new antibiotic Loracarbef combines this synthetic capability with the complete enantiospecificity at the amino group to be acylated.

In the Toray process to L-lysine, DL-α-amino-ε-caprolactam is hydrolyzed by L-α-amino-ε-caprolactam hydrolase from *Candida humicola*, while the D-enantiomer is racemized, optionally enzymatically with the help of D-α-amino-ε-caprolactam racemase from *Alcaligenes faecalis* (Eq. 22) (FUFKUMURA, 1976).

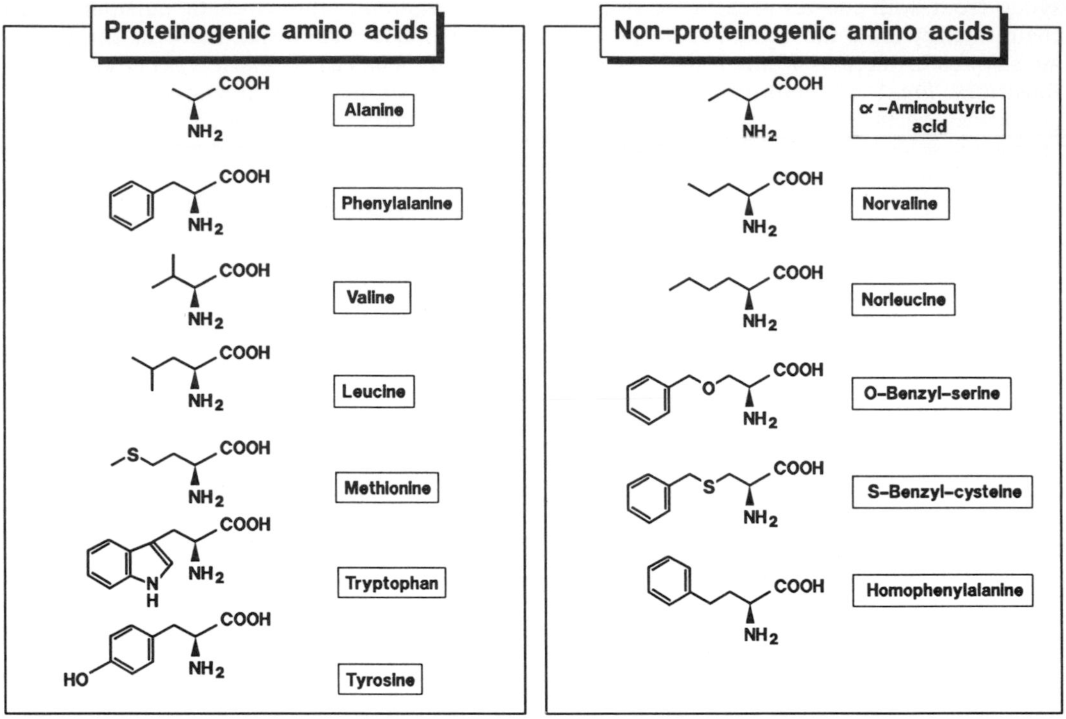

Proteinogenic amino acids
Alanine
Phenylalanine
Valine
Leucine
Methionine
Tryptophan
Tyrosine
Non-proteinogenic amino acids
α-Aminobutyric acid
Norvaline
Norleucine
O-Benzyl-serine
S-Benzyl-cysteine
Homophenylalanine

Fig. 8. L-Amino acids prepared in bulk quantities by acylase I resolution of *N*-acetyl-DL-amino acids.

L-Carbamoylic acid
L-Hydantoin
D-Hydantoin
D-Carbamoylic acid
D-Hydantoinase
D-Carbamoylase
L-Amino acid
D-Amino acid

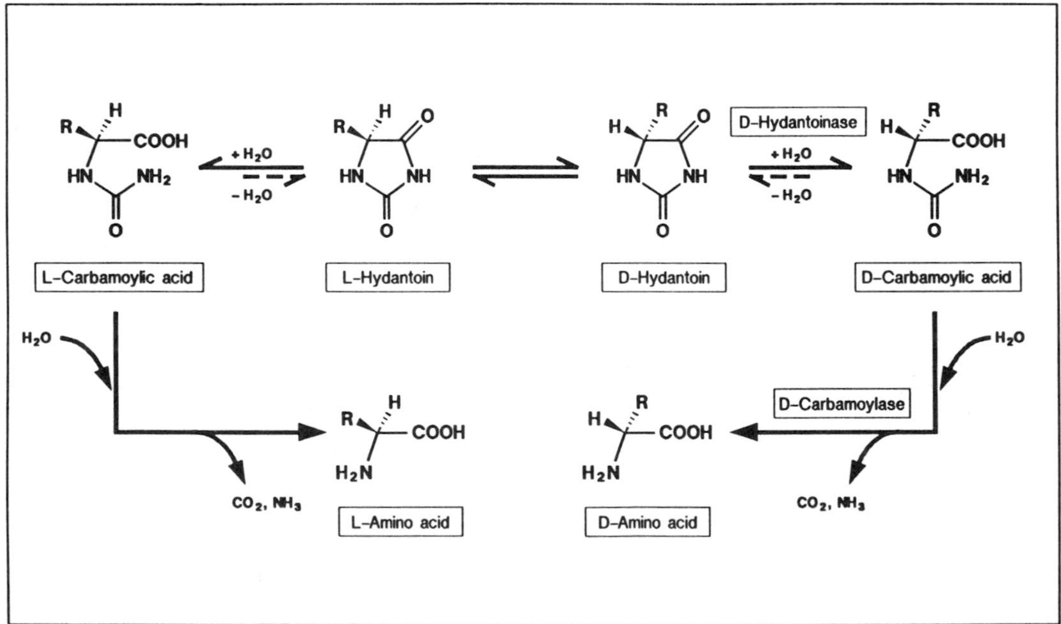

Fig. 9. Reaction path from hydantoins to L- and D-amino acids.

$$\text{L-}\alpha\text{-amino-}\varepsilon\text{-caprolactam} + H_2O \xrightarrow{\text{L-}\alpha\text{-amino-}\varepsilon\text{-caprolactam hydrolase}} \text{L-Lys} + \text{D-}\alpha\text{-amino-}\varepsilon\text{-caprolactam} \quad (22)$$

(racemase: D-α-amino-ε-caprolactam → L-α-amino-ε-caprolactam)

A similar enzymatic method for preparing 2-substituted L-ornithine was developed by Merrell Dow. The method is based on the enantioselective hydrolysis of racemic α-amino-δ-valerolactams such as DL-α-amino-α-difluoromethyl-ornithine catalyzed by L-α-amino-ε-caprolactam hydrolase (MERRELL DOW PHARMACEUTICALS, 1990a, b) (Eq. 23).

$$F_2HC\text{-}\alpha\text{-amino-}\delta\text{-valerolactam} + H_2O \xrightarrow{\text{L-}\alpha\text{-amino-}\varepsilon\text{-caprolactam hydrolase}} H_2N(CH_2)_3C(CHF_2)(NH_2)COOH \quad (23)$$

3.2.4 Arginase: Transformation with a Batch-UF Reactor

L-Ornithine and its salts are of growing importance in parenteral nutrition (L-Orn·HCl and L-Orn·acetate), in the treatment of hepatic diseases (L-Orn·L-Asp and L-Orn·α-ketoglutarate) or as starting material for the chemoenzymatic synthesis of citrulline (DRAUZ et al., 1991a, b). In large-scale synthesis, L-Orn can advantageously be produced by fermentation processes (AJINOMOTO, 1990) or by enzymatic hydrolysis of L-arginine using L-arginase (L-arginine amidinohydrolase, E. C. 3.5.3.1.) applied by Degussa (DEGUSSA AG, 1992a, b) (Eq. 24).

Native soluble enzyme superior in activity to any immobilized enzyme as well as high solubility of reactants and products point to an enzyme membrane reactor as the process design of choice. However, since arginase from calf or bovine liver is sensitive to mechanical agitation, the usual recycle reactor process with a pumped cycle leads to rapid inactivation. To enhance enzyme stability, a batch-UF reactor concept was developed using a quiescent medium with hydraulic conveying of substrate and enzyme solution (DEGUSSA AG, 1992c; BOMMARIUS et al., 1991) (Fig. 10) decreasing arginase deactivation by a factor of 20.

3.2.5 Enzymatic Peptide Synthesis: Aspartame

Aspartame is a dipeptide ester, α-L-Asp-L-Phe-OMe, 200 times sweeter than sucrose whose sweetness was discovered accidentally by scientists from G. D. Searle working on the synthesis of the digestive hormone gastrin (MAZUR et al., 1969). It has found wide-

$$\underset{\text{L-Arg}}{H_2N\text{-}C(=NH)\text{-}NH(CH_2)_3CH(NH_2)COOH} + H_2O \longrightarrow \underset{\text{L-Orn}}{H_2N(CH_2)_3CH(NH_2)COOH} + \underset{\text{urea}}{H_2N\text{-}CO\text{-}NH_2} \quad (24)$$

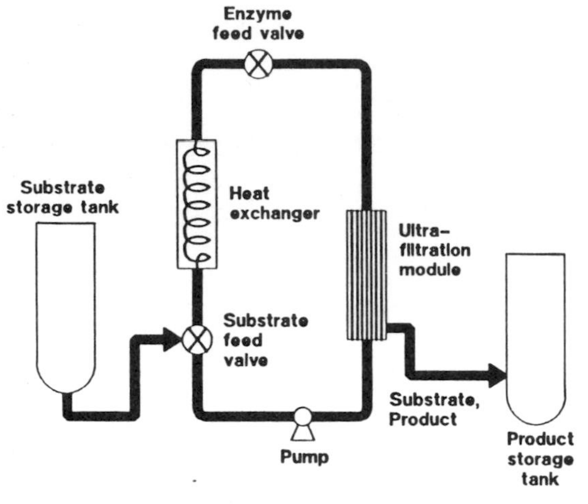

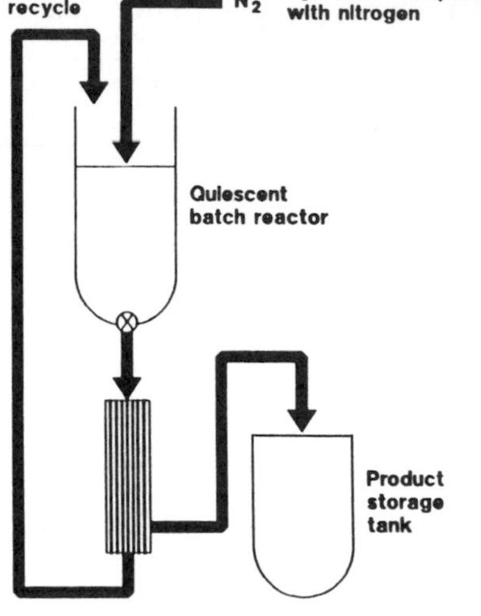

Fig. 10. Alternative process design of the enzyme membrane reactor without mechanical agitation (BOMMARIUS et al., 1991).

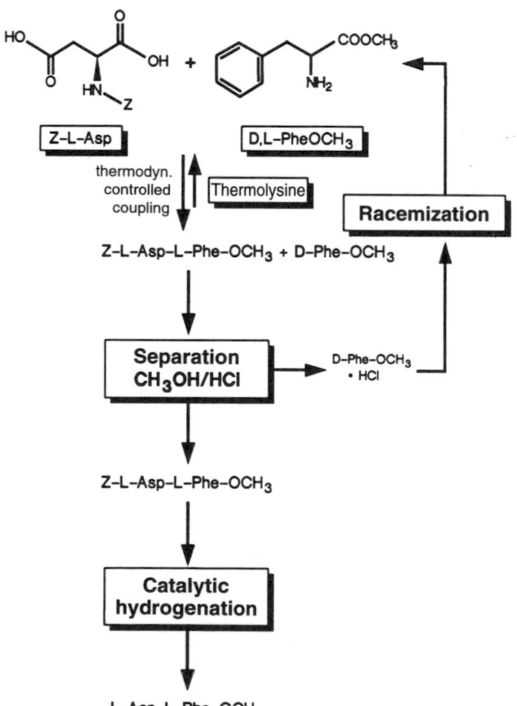

Fig. 11. Tosoh process for the enzymatic manufacture of aspartame (OYAMA and KIHARA, 1984).

spread use as a low-caloric sweetener in soft drinks, instant mixes, dressings, table-top sweeteners, and pharmaceuticals with a market volume of several thousand tons in 1992 envisaged to grow to more than 10000 t by the year 2000. As of 1993, its patents have expired in most countries. Several syntheses have been developed, however, the focus of this section is on the enzymatic process developed by Tosoh and operated by Holland Sweetener Company (joint venture between Tosoh and DSM) in the Netherlands (OYAMA and KIHARA, 1984) (Fig. 11).

The process employs thermolysin, a neutral Zn-protease (E.C. 3.4.24.4.) stable up to 80 °C. The advantages of this enzyme are:

- its complete enantiospecificity: L-Phe-OMe or DL-Phe-OMe can be used, depending on price advantage,
- its complete regiospecificity: no β-Asp-bonds are formed, as are to significant degrees in most other syntheses,
- its thermal stability: optimal at 50–60 °C, and
- as neutral protease, its lack of esterase activity.

The kinetics (OYAMA et al., 1981) of this equilibrium controlled peptide synthesis process have been worked out in detail. From the several possible approaches to shift the equilibrium towards the synthesis of the dipeptide, addition of water-soluble solvents (HOMANDBERG et al., 1978; OYAMA et al., 1987), or of water-insoluble organic solvents (HOMANDBERG et al., 1978; OYAMA et al., 1981, 1987), or precipitation of the product, the latter approach was adopted (OYAMA et al., 1984). With a solubility of the addition compounds of Z-L-Asp-L-Phe-OMe and D-Phe-OMe of 0.005 g/L, the equilibrium ($K=1.5$ L/mol) can be shifted far towards the synthesis.

3.2.6 Enzymatic Peptide Synthesis: Tyrosyl-Arginine

An EMR-based process has been developed and partially scaled up for the continuous α-chymotrypsin (α-CT; E.C. 3.4.4.5.) catalyzed, kinetically controlled synthesis of the dipeptide kyotorphin (L-Tyr-L-Arg, a strong analgesic, FLÖRSHEIMER et al., 1989; HERRMANN et al., 1991) (Eq. 25).

$$\text{Mal-L-Tyr-OEt} + \text{L-Arg-OEt} + H_2O \xrightarrow{\alpha\text{-CT}} \text{Mal-L-Tyr-L-Arg-OEt} + \text{EtOH} \qquad (25)$$

$$\text{glycidyl butyrate} + H_2O \xrightarrow{\text{lipase}} (R)\text{-glycidyl butyrate} + (R)\text{-glycidol} \qquad (26)$$

(R)-glycidyl butyrate

(R)

This reaction performed in an EMR system was improved over conventional enzymatic peptide synthesis by the introduction of new, highly water-soluble protecting groups (Mal = maleyl) resulting in one of the highest space-time yields ever seen in continuously operated enzymatic peptide synthesis; with esters as nucleophiles, 13.4 kg/L/d (FORSCHUNGSZENTRUM JÜLICH GmbH/DEGUSSA AG, 1992; FISCHER et al., 1991, 1993), with L-Arg-NH_2 25.0 kg/L/d have been achieved, albeit on a laboratory scale. Mal-L-Tyr-L-Arg-OEt precipitates from aqueous solution when cooled, and the process has been scaled up to the 72 mol scale (400 L) in a batch reactor yielding 50.4% of pure Tyr-Arg acetate product after deprotection and purification (FISCHER et al., 1993).

3.2.7 Lipase-Catalyzed Reactions

There is much industrial interest in (*R*)-glycidyl butyrate as an intermediate for β-blockers (CROSBY, 1992). Lipases have been found to catalyze the hydrolysis of the racemate, albeit at modest enantioselectivity ($E=18$) (Eq. 26) (LADNER and WHITESIDES, 1984). Conversion has to exceed 50% significantly to produce a product with sufficient enantiomeric purity.

Fatty acids are important raw materials which can be obtained from naturally occurring triglycerides. The thermal process, however, produces colored products, so a mild enzymatic route is sought. Lipases (E.C. 3.1.1.3.) catalyze the hydrolysis of lipids at the oil–water interface. In a membrane reactor, the enzyme has been immobilized both at the water-phase side of a hydrophobic membrane (YAMANE et al., 1988) and at the hydrophobic side of a hydrophilic one (PRONK et al., 1988). In both cases, use of a membrane obviated other stabilization measures of the emulsion. The synthesis reaction to *n*-butyl oleate was achieved with lipase from *Mucor miehei* immobilized at a hollow-fiber membrane wall. Conversion reached was 88% but the substrate butanol caused the membrane system to decay before the enzyme deactivated (HABULIN and KNEZ, 1991).

Phosphatidylglycerol (PG), useful as a lung surfactant to treat the respiratory distress syndrome, can be obtained from phosphatidylcholine (PC) in a transphosphatidylation reaction catalyzed by phospholipase D (PLD, E.C. 3.1.4.4.), with hydrolysis to phosphatidic acid (PA) as side reaction:

$$\text{PC} + \text{glycerol} \rightarrow \text{PG} + \text{choline} \quad (27)$$

$$\text{PC} + H_2O \rightarrow \text{PA} + \text{choline} \quad (28)$$

Usually, the reaction is carried out in an emulsion stabilized by surfactants containing the PC substrate. In a microporous membrane reactor, PLD stability could be enhanced several-fold by the presence of ethers on the shell side (LEE et al., 1985). Moreover, no surfactants were needed, the product could be separated readily, and 20% of PG was formed instead of at most 4% in the emulsion system.

3.2.8 Hydrolysis of Polymeric Substrates

Membrane reactors can be used for hydrolysis of proteins (DEESLIE and CHERYLAN, 1982a, b; BRESSOLIER et al., 1988), starch (BUTTERWORTH et al., 1970; UTTAPAP et al., 1989; DARNOKO et al., 1989), and cellulose (HONG et al., 1981; PIZZICHINI et al., 1989). The first account on the use of membrane reactors for enzyme reactions was on the hydrolysis of starch with α-amylase (E.C. 3.2.1.1.) (BUTTERWORTH et al., 1970). In such a case, the reaction will proceed until the substrate is hydrolyzed far enough to pass through the pores of the membrane. MATSUMURA and HIRATA successfully modeled simultaneous saccharification and fermentation (SSF) with raw sweet potato starch by means of amyloglycosidase (E.C. 3.2.1.3.) in a membrane reactor by a two-site deactivation model, one for ethanol and the other for starch through adsorption onto the substrate (MATSUMURA and HIRATA, 1989).

In biomedical engineering, the removal from the blood of the anticoagulant heparin, a glycoaminoglycan polymer of 6 to 30 kD, by heparinase (E.C. 4.2.2.7.) immobilized on an agarose matrix (YANG et al., 1988) has been modeled and verified experimentally (LANGER et al., 1990; BERNSTEIN and LANGER, 1988). The model was based on a CSTR design without adjustable parameters. The enzyme has also been immobilized on cellulose hollow fibers for use as a clinical device for heparin removal from blood (COMFORT et al., 1989a, b, c).

3.3 Oxidoreductases

Racemization of non-reacted enantiomer when producing enantiomerically pure α-amino acids can be avoided by starting from prochiral compounds. Reductive amination of α-keto acids by means of L-amino acid dehydrogenases (E.C. 1.4.1.) yields the corresponding α-amino acids. Reductive amination reactions depend on NADH, so for proper utilization of NADH cofactor regeneration has to be considered. An elegant solution for the regeneration of NADH from NAD^+ is the use of the system formate/formate dehydrogenase (FDH; E.C. 1.2.1.2.) (SCHÜTTE et al., 1976; WEDY, 1992) commonly available nowadays (Fig. 12).

The FDH-catalyzed reaction of formate oxidation to CO_2 serves two purposes: it recycles NAD^+ to NADH and causes the coupled equilibrium-limited reductive amination reaction to go to completion because CO_2 is removed from the reaction system.

While the enzymes are retained by a UF membrane, native cofactor permeates through the membrane because UF membranes cannot be built size-specifically enough to retain cofactor (MW 789) but not amino acid molecules (MW ~150). NAD^+ covalently bound to a water-soluble polymer such as polyethyleneglycol (PEG, MW 20000) (BÜCKMANN et al., 1981), however, enlarged the size sufficiently to achieve retention of PEG-NAD^+ of greater than 99.9% while still functioning as cofactor.

The synthesis of L-*tert*-leucine from trimethyl pyruvate has been studied in detail

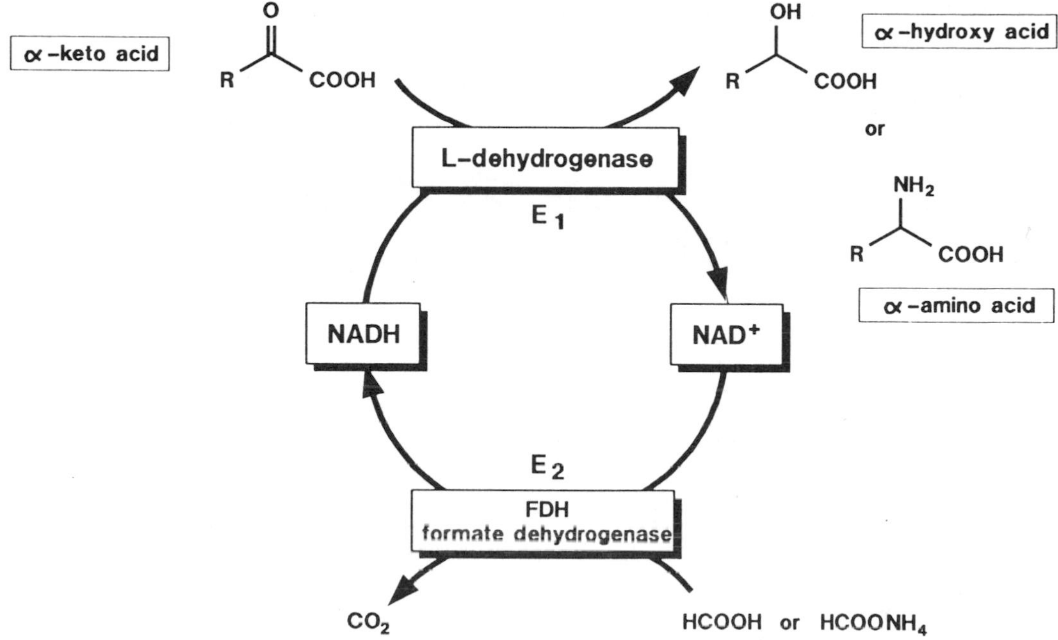

Fig. 12. Cofactor regeneration of NAD^+/NADH with FDH/formate (BOMMARIUS et al., 1992a). E_1, L-dehydrogenase; E_2, FDH (formate dehydrogenase).

(KRAGL et al., 1992) (Eq. 29). L-*tert*-Leucine as an unnatural amino acid cannot be made by a fermentative route, but there is a strong need owing to its use as chiral auxiliary in a number of syntheses (SCHÖLLKOPF and NEUBAUER, 1982; SCHÖLLKOPF, 1983; SCHÖLLKOPF and SCHREVER, 1984).

trimethyl-pyruvate $\xrightarrow[\text{NADH + H}^+ \;\rightarrow\; \text{NAD}^+]{\text{LeuDH}}$ L-Tle (29)

In Fig. 13, the reaction rate of the coupled system FDH/LeuDH (E.C. 1.4.1.9) is displayed as a function of both initial substrate concentration and cofactor concentration. The rate increases monotonously with cofactor concentration, but shows a maximum with increasing substrate concentration owing to severe product inhibition by L-*tert*-leucine.

With a substrate concentration of 0.5 M, L-*tert*-leucine was produced continuously in a membrane reactor with an average conversion of 85% and a space-time yield of 0.64 kg/L/d. Total turnover number over a period of two months was 125000. This result should help to overcome the notion that cofactor cost is still a limiting factor for enzymatic redox reactions.

In an analogous reaction, phenylpyruvate can be reductively aminated to L-phenylalanine with PheDH. Other substrates are 2-keto-4-phenylbutyric acid to L-homophenylalanine (BRADSHAW et al., 1991; HUMMEL et al., 1987a, b). However, despite an impressive total turnover number of 600000 (WANDREY, 1987), such a system cannot compete with the fermentative route to L-phenylalanine starting from inexpensive glucose and ammonia. For this reason, commercialization of continuous

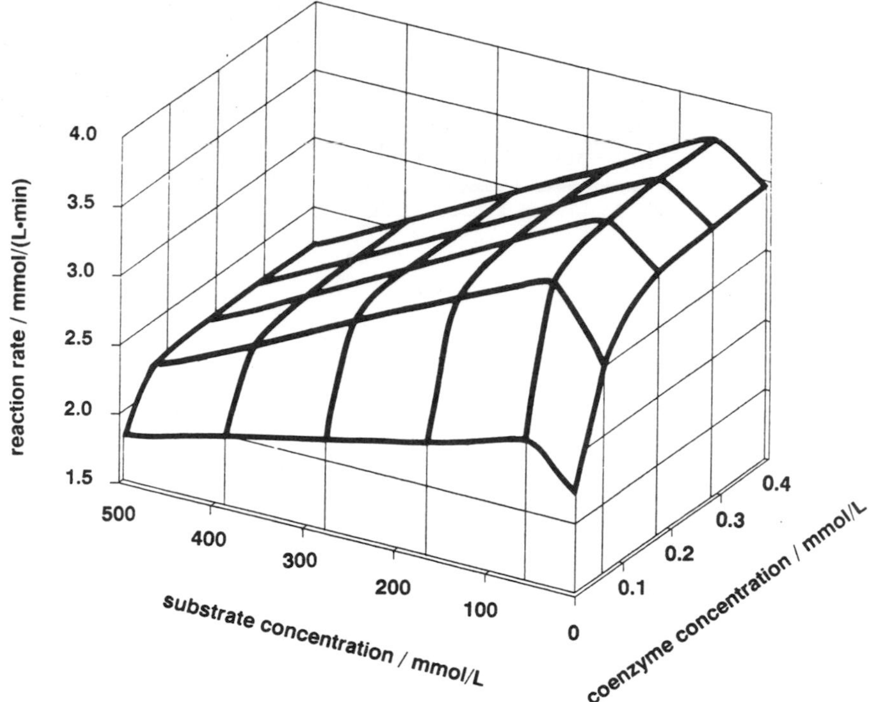

Fig. 13. Calculated reaction rate of LeuDH as a function of substrate and cofactor (PEG-NADH) concentration in a continuous process (BOMMARIUS et al., 1992a). Conditions: LeuDH, 30 U/mL; FDH, 30 U/mL; conversion: 90%, controlled by variation of residence time; $[HCOO^- NH_4^+] = 1$ mol/L.

cofactor regeneration was achieved with a non-proteinogenic amino acid such as L-*tert*-leucine. Degussa has produced multi-kilogram amounts of enantiomerically pure non-proteinogenic α-amino acids by the dehydrogenase route.

The cofactor regeneration scheme using FDH has also been used to synthesize enantiomerically pure α-hydroxy acids: D-Hydroxyisocaproate dehydrogenase (D-Hic-DH) reduces 3,4-dihydroxypyruvic acid to (*R*)-(+)-3-(3,4-dihydroxyphenyl)lactic acid (caffeic acid useful against inflammatory symptoms (Eq. 30) (PABSCH et al., 1991). Space-time yields of more than 1 kg/L/d have been attained.

3.4 Isomerases: Glucose Isomerase

The isomerization of glucose to fructose is desired in the food industry because of the threefold sweetness of fructose compared to sucrose which is achieved at reduced caloric intake. The chemical process using sulfuric acid causes degradation and coloring of the glucose syrup. A much milder isomerization has been developed with glucose isomerase (E.C. 5.3.1.5.) which in nature isomerizes xylose to xylulose. Known and carried out in batches for decades by the corn syrup industry, the process vastly increased in scale in the 1970s (1984: $5 \cdot 10^6$ tons, JENSEN and RUGH, 1987) owing to

3,4-dihydroxypyruvic acid —(D-HicDH; NADH + H⁺ → NAD⁺)→ (*R*)-(+)-3-(3,4-dihydroxyphenyl)-lactic acid (30)

high sugar prices and profound process development resulting in cost advantages.

Nowadays, the process starts with a syrup at approximately 50% dry weight of glucose. The high concentration with the concomitant viscosity necessitates processing at high temperature. Temperature and age of the enzyme are adjusted to control production capacity: high capacity is reached at 63–65 °C and new enzyme, whereas low throughput is obtained at 55–57 °C and, on average, partially deactivated enzyme. The process is run in several parallel PFRs at varying degrees of enzyme deactivation.

3.5 Lyases

3.5.1 Aspartase

Owing to the commercialization of Aspartame, demand for L-aspartic acid has been rising steeply. L-Aspartic acid can be synthesized by enantioselective ammonia addition to fumaric acid catalyzed by aspartase (E.C. 4.3.1.1.), Eq. (31).

$$\text{HOOC-CH=CH-COOH} + NH_3 \xrightarrow{\text{aspartase}} \text{HOOC-CH}_2\text{-CH(NH}_2\text{)-COOH} \quad (31)$$

fumaric acid → L-Asp

$$\text{HOOC-CH=CH-COOH} + H_2O \xrightarrow{\text{fumarase}} \text{HOOC-CH}_2\text{-CH(OH)-COOH} \quad (32)$$

fumaric acid → L-malic acid

Since 1960, Tanabe has been producing L-aspartic acid using intact *Escherichia coli* cells in a batch reactor, but switched to an immobilized cell system in 1973. Aspartase extracted from *E. coli* cells was found to be inactivated by 1 M ammonium fumarate substrate solution (50% after 30 days at 30 °C); in contrast, entrapment of the cells in polyacrylamide gels increased the half-life to 120 days at 37 °C (CHIBATA et al., 1986). The reaction is run in a fixed-bed reactor with external water cooling owing to the exothermicity of reaction (31) (NISHIMURA and KISUMI, 1988; NISHIMURA et al., 1989). Further improvements were achieved upon immobilization with κ-carrageenan gel hardened by glutaraldehyde and hexamethylenediamine treatment (1978; half-life 2 years) and by switching to a genetically engineered *E. coli* strain, EAPc7, with much higher activity in 1982 (CHIBATA et al., 1992). The process is run on the 4000 tons/year scale.

3.5.2 Fumarase

L-Malic acid is used in salts of basic amino acids in infusion solutions and as acidulant in the food industry on the scale of 500 tons annually; it can be produced in analogy to Eq. (31) from fumaric acid with the help of fumarase (E.C. 4.2.1.2), Eq. (32):

Similar process development to the L-aspartic acid synthesis was performed by Tanabe so that the process based on *Brevibacterium ammoniagenes* introduced in 1974 could be improved 25-fold by immobilizing a *Brevibacterium flavum* strain on κ-carrageenan gel with polyethyleneimine to reach a half-life of 310 days at 37 °C (CHIBATA et al., 1983; TAKATA et al., 1983).

3.5.3 L-Aspartate β-Decarboxylase

Tanabe manufactures L-alanine from L-aspartate (see Sect. 3.5.2) by decarboxylation with L-aspartate β-decarboxylase (E.C. 4.1.1.12.) from *Pseudomonas dacunhae* (YAMAMOTO et al., 1980), Eq. (33):

HOOC–CH₂–CH(NH₂)–COOH (L-Asp) —L-Asp-β-decarboxylase→ CH₃–CH(NH₂)–COOH (L-Ala) $+ CO_2$ (33)

Since it was difficult to maintain pH at the optimum of 6.0 for the organism due to CO_2 evolution, a closed-loop reactor was developed which kept the CO_2 dissolved at 10 bar and thus the pH stable. Co-immobilization of both *E. coli* and *P. dacunhae* cells to produce L-alanine directly from fumaric acid was not successful, because the *E. coli* cells operate best at 8.5 versus an optimum of 6.0 for the decarboxylase; the sequential process has been run since 1982 (FURUI and YAMASHITA, 1983). The very good enantioselectivity of the L-aspartate β-decarboxylase led to a process started up in 1989, in which inexpensive DL-aspartate was converted to both L-alanine (L-Ala) and D-aspartate (D-Asp) products, the latter being useful for a synthetic penicillin (Eq. 34, SENUMA et al., 1989).

NH_4-fumarate —aspartase→ DL-Asp —L-Asp-β-decarb.→ D-Asp + L-Ala + CO_2 (34)

4 New Reactor Designs for Enzyme Reactors

4.1 Multiphase Reactors

This setup has been used, notably by Sepracor, to resolve racemates of hydrophobic esters (YOUNG, 1990). The analgesic (*S*)-ibuprofen was produced by resolution of racemic esters with *Candida cylindracea* lipase enantiomeric excess (*ee*) (MCCONVILLE et al., 1990). An enantiomeric purity of 96% was obtained with the methoxyethyl ester, at a productivity of 6.8 kg acid per year per m^2 (or 0.7 µmol per hour per mg enzyme). Neat ibuprofen ester was circulated in the shell compartment. (*R*)-glycidyl butyrate, useful as an intermediate for β-blocker cardiovascular drugs (LADNER and WHITESIDES, 1984), was obtained in good enantiomeric purity by reacting neat racemic ester with porcine pancreatic lipase (PPL). The imperfect enantioselectivity of PPL ($E = 18$, CHEN et al., 1982) was compensated for by reacting the system to 67% conversion. Productivity was 17.6 g/h/m^2 or 28.4 µmol/h/mg enzyme. The main drawback of the multiphase reactor system is the low effectiveness factor of the enzyme, commonly between 30 and 50%.

4.2 Novel Cofactor Regeneration Schemes

Recently, STECKHAN et al (1990a, b) performed cofactor regeneration with an organometallic rhodium complex substituting for FDH:

[(Cp*)Rh(H_2O)(bipyridine–R)]$^{2+}$ (Rh $3+$); R = H; R = CH_2OPEG; R = CH_2OEt (35)

Hydride donor is formate or electrons from a cathode. Similar to size enhancement of NAD^+ (BÜCKMANN et al., 1981), the authors increased the size of the complex by binding it to polyethyleneglycol (PEG 20000) to render it impassable through the membrane. This arrangement might offer advantages over FDH, because the complex is more stable than FDH, does not suffer from product inhibition, and is insensitive to O_2. However, turnover number (16.9 h^{-1}), cycle number (1000), and space-time yield (44.4 mmol/L/d) of the system formate/PEG-Rh/NAD^+ were still very low.

ATP (adenosine triphosphate) is a cofactor for a wide variety of syntheses and has to be recycled to be economical (CRANS et al., 1987; Fig. 14).

Two regeneration schemes are currently most promising: one uses acetyl phosphate or, with higher phosphoryl donor strength, meth-

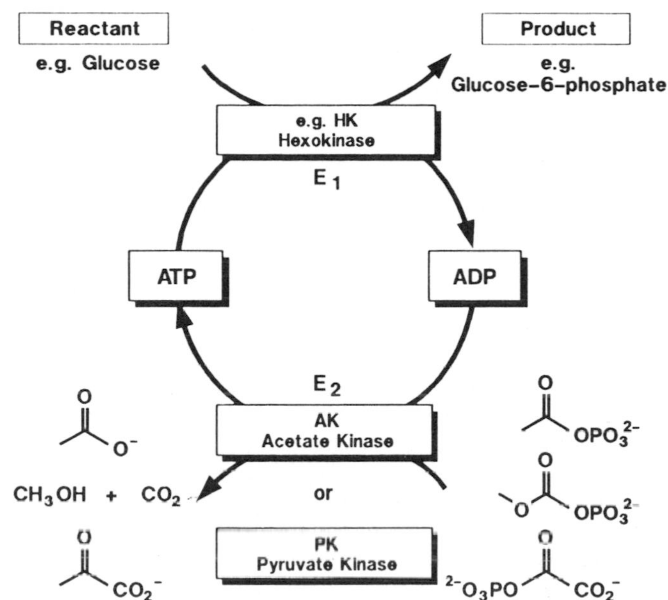

Fig. 14. ATP regeneration scheme. E_1, synthesis enzyme; E_2, regeneration enzyme. (For explanations of substrates and products, see text.)

oxycarbonyl phosphate as phosphorylating agents with acetate kinase (E.C. 2.7.2.1.) as catalyst (CRANS and WHITESIDES, 1983; KAZLAUSKAS and WHITESIDES, 1985), the other is based on the reaction of phosphoenolpyruvate (PEP) to pyruvate with the help of pyruvate kinase (E.C. 2.7.1.40.) (HIRSCHBEIN et al., 1982). Both enzymes are commercially available, highly active, and possess sufficient stability. ATP is necessary for the synthesis of compounds such as 5-phospho-α-D-ribose-pyrophosphate (PRPP), an important intermediate in nucleoside and nucleotide biosynthesis (GROSS et al., 1983; HARTMANN and BUCHANAN, 1958) or *sn*-glycerol 3-phosphate, a building block for the synthesis of enantiomerically pure phospholipids (RIOS-MERCADILLO and WHITESIDES, 1979; RADHAKRISHNAN et al., 1981). In a hollow-fiber EMR, ATP recycle numbers of 2130 could be achieved at space-time yields of 2 $mol/m^3/h$ and conversions of 92.8% (ISHIKAWA et al., 1989).

NADPH cofactor regeneration has been investigated with glucose dehydrogenase (GDH, E.C. 1.1.99.10.) for sorbitol production (IKEMI et al., 1990). Continuous production of L-menthol was carried out with NAD^+ regeneration in a two-phase membrane reactor system (KISE and HAYASHIDA, 1990). Synthesis of S-sulcatol from sulcatone was achieved with a thermophilic ADH from *Thermoanaerobicum brockii*, while recycling $NADP^+$ within a charged UF-membrane reactor (RÖTHIG et al., 1990).

5 New Biotransformations

5.1 Toluene Dioxygenase

The finding of an intermediate oxidation product between benzene and CO_2, benzene-*cis*-dihydrodiol in a *Pseudomonas putida* mutant in 1968 (GIBSON et al., 1968) has since spawned a large effort to find more transformations and to utilize the diols for further synthesis (BROWN, 1993; SHELDRAKE, 1992).

The reaction, catalyzed by toluene dioxygenase (E.C. 1.13.11.) has no simple equivalent in nature, but opens a route to enantiomerically pure, stereochemically predictable intermediates from a variety of benzene derivatives which can then be further converted to carbohydrates, prostaglandins, alkaloids, and other

$$\text{C}_6\text{H}_5\text{R} \xrightarrow{\text{dioxygenase}} \text{cis-dihydrodiol} \xrightarrow[\times]{\text{dehydrogenase}} \text{catechol} \rightarrow \rightarrow CO_2 \quad (36)$$

R = H, Me, Cl, Br, CN

products (LEE, 1990; HUDLICKY et al., 1988, 1989, 1990; CARLESS et al., 1989). Several reactor designs have been proposed (BRAZIER et al., 1990; WOODLEY et al., 1991), as has a process based on continuous extraction of the diols (VAN DE TWEEL et al., 1988). Typically, resting cells of a *Pseudomonas* strain are used as carrier of the enzymatic activity in a batch process. Several companies such as ICI, Enzymatix, and many others are offering several of the dihydrodiols in pilot quantities, and increased use of these compounds for further synthesis can be expected.

5.2 Oxinitrilase

In the 1960s, the potential of (*R*)-oxinitrilase (E.C. 4.1.2.10.) isolated from bitter almonds (*Prunus amygdalus*) was recognized to catalyze the (*R*)-enantiospecific addition of HCN to a variety of aldehydes (BECKER et al., 1965a, b) utilizing the reverse reaction used by plants for protection against enemies. In the last decade, it was shown (EFFENBERGER et al., 1987a, b, 1990, 1991; BRUSSEE et al., 1988, 1990; NIEDERMEYER and KULA, 1990a, b) that a variety of aldehyde substrates yield cyanohydrins of high *ee*-values (Eq. 37; Tab. 6). Enantiomerically pure cyanohydrins can be converted to α-amino and α-hydroxy acids, α-hydroxy aldehydes, β-amino alcohols, and many other compounds.

$$\text{R–CHO} + \text{HCN} \xrightarrow{\text{R-oxinitrilase}} \text{R–CH(OH)CN (R-cyanohydrin)}$$
$$\text{R–CHO} + \text{HCN} \xrightarrow{\text{chemical addition}} \text{R–CH(OH)CN (racemate)} \quad (37)$$

Reaching high *ee*-values in the cyanohydrin formation reaction is not easy, because the chemical addition to aldehydes yielding racemic cyanohydrins is in competition to the enzymatic addition. Since the chemical addition is fast above the p*K* of cyanide at pH 4.5, reac-

Tab. 6. Enzymatic Formation of Cyanohydrins from Aldehydes (EFFENBERGER et al., 1987a, b, 1990, 1991)

Aldehyde	Cyanohydrins in H_2O/EtOH			Cyanohydrins in Ethylether/Cellulose		
	t_{rct} (h)	Yield (%)	ee[a] (%)	t_{rct} (h)	Yield (%)	ee[a] (%)
Benzaldehyde	1	99	86	2.5	95	99
3-Phenoxybenzaldehyde	5	99	10.5	192	99	98
Furfural	2	86	69	4	88	98.5
Nicotinaldehyde	2.5	78	6.7	4.5	88	14
Crotonaldehyde	1.5	68	76	3	68	97
Phenylacetaldehyde	4	82	27	4.5	89	14
3-Methylthiopropionaldehyde	3	87	60	6.5	97	80
Pivalaldehyde	2.5	56	45	4.5	78	73
Butyraldehyde	2	75	69	4.5	75	96

[a] as (*R*)-(+)-MPTA derivatives (MPTA = α-methoxy-α-(trifluoromethyl)phenylacetic acid; Mosher's reagent)

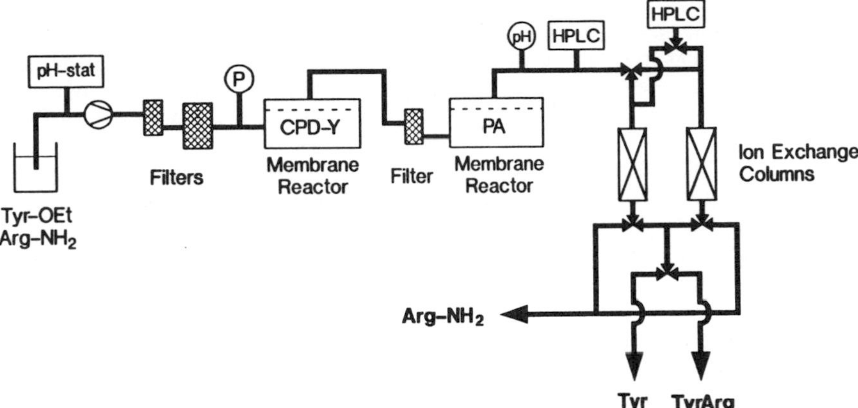

Fig. 15. Two-stage enzyme membrane reactor for continuously operated enzymatic peptide synthesis with integrated product recovery (SCHWARZ, 1991).

tor and medium designs have been proposed to suppress the unspecific chemical addition: (1) reaction with immobilized enzyme in organic solvents (EFFENBERGER et al., 1987a, b, 1990, 1991), (2) use of acetone cyanohydrin as a donor of HCN (OGNYANOV et al., 1991), (3) batch reaction at low pH in water (NIEDERMEYER and KULA, 1990a, b), (4) an enzyme-membrane reactor using high enzyme concentration and very short residence times at a pH of 3–5 (DEGUSSA AG, 1989a, b), and (5) a two-phase system with a small amount of organic solvent (SOLVAY DUPHAR N. V., 1991).

5.3 Peptide Amidase

Recently, a new amidase was detected in the flavedo of oranges by STEINKE and KULA which selectively hydrolyzes the amide bond of peptide amides or *N*-acylated amino acid amides without cleaving the peptide bond or acyl amino acid itself (STEINKE and KULA, 1990a, b, c; FORSCHUNGSZENTRUM JÜLICH GmbH/DEGUSSA AG, 1991) (Eq. 38).

R_1 is an amino acid or peptide residue or *N*-terminal protecting group of an amino acid amide, R_2 is an amino acid side chain.

Whereas the substrate specificity towards the C-terminal amino acid is broad, D-amino acid amides in this position or non-*N*-acylated L-amino acid amides are not accepted (Tab. 7). Thus, this highly selective amidase can be used for the resolution of diastereomeric peptides and the selective C-terminal deprotection of peptide amides (FORSCHUNGSZENTRUM JÜLICH GmbH/DEGUSSA AG, 1991; STEINKE and KULA, 1991).

Furthermore, in thermodynamically controlled peptide syntheses, the peptide amide as the primary reaction product can be continuously removed from the reaction equilibrium by the action of peptide amidase, thus enabling high yields of the peptides. Based on these findings, a continuously operated process for enzymatic peptide synthesis catalyzed by subsequent action of carboxypeptidase Y (CPD-Y; E.C. 3.4.16.1.) and peptide amidase (PA) has been developed using a two-stage enzyme membrane reactor (SCHWARZ, 1991).

$$R_1\text{-CO-NH-CH}(R_2)\text{-CO-NH}_2 \xrightarrow[\text{H}_2\text{O}]{\text{peptide amidase}} R_1\text{-CO-NH-CH}(R_2)\text{-COO}^- + NH_4^+ \quad (38)$$

peptide amide → peptide

Tab. 7. Substrate Specificity of Peptide Amidase of Orange Flavedo (STEINKE and KULA, 1990)

Substrate[a]	Conversion after 6 h (%)
Ac-L-Trp-NH_2	100
Bz-L-Arg-NH_2	100
Bz-L-Tyr-NH_2	100
H-L-Val-L-Phe-NH_2	100
H-L-Asp-L-Phe-NH_2	100
H-L-Ala-L-Phe-NH_2	100
H-L-Arg-L-Met-NH_2	100
H-L-Phe-L-Leu-NH_2	100
Z-Gly-L-Tyr-NH_2	100
Bz-L-Tyr-L-Thr-NH_2	100
Bz-L-Tyr-L-Ser-NH_2	100
Bz-L-Tyr-L-Ala-NH_2	100
Boc-L-Leu-L-Val-NH_2	20
Trt-Gly-L-Leu-L-Val-NH_2	80
Z-L-Pro-L-Leu-Gly-NH_2	100
Z-Gly-Gly-L-Leu-NH_2	100
H-Gly-D-Phe-L-Tyr-NH_2	100
H-Gly-L-Phe-D-Phe-NH_2	0[b]
H-L-Arg-L-Pro-D-Ala-NH_2	0[b]
Z-L-Arg-L-Arg-pNa	0[b]

[a] Ac, acetyl; Bz, benzoyl; Boc, *tert*-butyloxycarbonyl; Trt, trityl; Z, benzyloxycarbonyl; pNa, *p*-nitroanilide; amino acid amides are not hydrolyzed.
[b] No conversion in 12 h

This process includes integrated recovery of the peptide and by-products as well as separation and recycling of the nucleophilic amino acid amide by ion exchange chromatography (Fig. 15).

6 Summary

Biocatalytic processes have steadily gained ground during the last years. The selectivity of biocatalysts fills the demand facing new processes in the light of more sophisticated customers and of the need for environmentally benign transformations. In many cases, biocatalytic processes have been demonstrated to deliver better products at a lower price than processes based on conventional technology.

Besides simple hydrolyses and reactions with low steric demands on substrates and products, highly enantioselective and stable enzymes, cofactor regeneration schemes, and novel reactions through altered pathways are increasingly dominating the scene in biotransformations.

The two basic reaction engineering tools of biocatalysis, the immobilized-enzyme and the enzyme-membrane reactor already cover many current processing needs. Reactions at elevated temperatures, proteins with biologically altered reactivity profile, reactions on very large and very small scale as well as multi-phase reactors can be envisaged to broaden biocatalytic processing options even further.

7 References

AJINOMOTO (1990), *Eur. Pat. Appl.* 0393708.

ARIS, R. (1957), *Chem. Eng. Sci.* **6**, 262–268.

BALDARO, E., FUGANTI, C., SERRI, S., TAGLIANI, A., TERRENI, M. (1992), in: *Microbial Reagents in Organic Synthesis* (SERVI, S., Ed.), *NATO ASI Ser.* **C381**, 175–188, Dordrecht-London-Boston: Kluwer Acad. Publ.

BECKER, W., FREUND, H., PFEIL, E. (1965a), *Angew. Chem.* **77**, 1139.

BECKER, W., FREUND, H., PFEIL, E. (1965b), *Angew.Chem. Int. Ed. Engl.* **4**, 1079.

BEDNARSKI, M. D., CHENAULT, H. K., SIMON, E. S., WHITESIDES, G. M. (1987), *J. Am. Chem. Soc.* **109**, 1283–1285.

BELFORT, G. (1989), *Biotechnol. Bioeng.* **33**, 1047–1066.

BERNSTEIN, H., LANGER, R. (1988), *Proc. Natl. Acad. Sci. USA* **85**, 8751–8755.

BOMMARIUS, A. S., MAKRYALEAS, K., DRAUZ, K. (1991), *Biomed. Biochim. Acta* **50**, 249–255.

BOMMARIUS, A. S., DRAUZ, K., GROEGER, U., WANDREY, C. (1992a), in: *Chirality in Industry* (COLLINS, A. N., SHELDRAKE, G. N., CROSBY, J., Eds.), pp. 371–398, Chichester-New York: Wiley & Sons.

BOMMARIUS, A. S., KOTTENHAHN, M., DRAUZ, K. (1992b), in: *Microbial Reagents in Organic Synthesis* (SERVI, S., Ed.), *NATO ASI Ser.* **C381**, 161–174, Dordrecht-London-Boston: Kluwer Acad. Publ.

BOMMARIUS, A. S., DRAUZ, K., KLENK, H., WANDREY, C. (1992c), *Ann. N. Y. Acad. Sci. (Enzyme Eng.* **11)** **929**, 126–136.

BUCHHOLZ, K., FIECHTER, A. (Eds.), (1982), *Adv. Biochem. Eng.* **24**, 39–71.

BÜCKMANN, A. F., KULA, M.-R., WICHMANN, R., WANDREY, C. (1981), *J. Appl. Biochem.* **3**, 301-315.

BUTTERWORTH, T. A., WANG, D. I. C., SINSKEY, A. J. (1970), *Biotechnol. Bioeng.* **12**, 615-631.

BRADSHAW, C. W., WONG, C.-H., HUMMEL, W., KULA, M.-R. (1991), *Bioorg. Chem.* **19**, 29-39.

BRAZIER, A. J., LILLY, M. D., HERBERT, A. B. (1990), *Enzyme Microb. Technol.* **12**, 90.

BRESSOLLIER, P., PETIT, J. M., JULIEN, R. (1988), *Biotechnol. Bioeng.* **31**, 650.

BROWN, S. M. (1993), in: *Organic Synthesis; Theory and Applications* (HUDLICKY, T., Ed.), JAI Press, in press.

BRUSSEE, J., ROOS, E. C., VAN DER GEEN, A. (1988), *Tetrahedron Lett.* **29**, 4485-4488.

BRUSSEE, J., LOOS, W. T., KRUSE, C. G., VAN DER GEEN, A. (1990), *Tetrahedron* **46**, 979-986.

CARLESS, H. A. J., BILLINGE, J. R., OAK, O. Z. (1989), *Tetrahedron Lett.* **30**, 3113-3116.

CHANG, T. M. S. (1964), *Science* **146**, 524.

CHANG, T. M. S. (1976), *Methods Enzymol.* **44**, 676-698.

CHANG, T. M. S. (1987), *Methods Enzymol.* **136**, 67-82.

CHEN, C. S., SIH, C. J. (1989a), *Angew. Chem.* **101**, 711-724.

CHEN, C. S., SIH, C. J. (1989b), *Angew. Chem. Int. Ed. Engl.* **28**, 695-708.

CHEN, C. S., FUJIMOTO, Y., GIRDAUKAS, G., SIH, C. J. (1982), *J. Am. Chem. Soc.* **104**, 7294-7299.

CHENAULT, H. K., DAHMER, J., WHITESIDES, G. M. (1989), *J. Am. Chem. Soc.* **111**, 6354-6364.

CHIBATA, I., TOSA, T., SATO, T., MORI, T. (1976), *Methods Enzymol.* **44**, 746-759.

CHIBATA, I., TOSA, T., TAKATA, I. (1983), *Trends Biotechnol.* **1**, 9.

CHIBATA, I., TOSA, T., SATO, T. (1986), *Appl. Biochem. Biotechnol.* **13**, 231.

CHIBATA, I., TOSA, T., SHIBATANI, T. (1992), in: *Chirality in Industry* (COLLINS, A. N., SHELDRAKE, G. N., CROSBY, J., Eds.), pp. 351-370, Chichester-New York: Wiley & Sons.

COMFORT, A. R., ALBERT, E. C., LANGER, R. (1989a), *Biotechnol. Bioeng.* **34**, 1366-1373.

COMFORT, A. R., ALBERT, E. C., LANGER, R. (1989b), *Biotechnol. Bioeng.* **34**, 1374-1382.

COMFORT, A. R., BERSKOWITZ, S., ALBERT, E. C., LANGER, R. (1989c), *Biotechnol. Bioeng.* **34**, 1383-1390.

CRANS, D. C., WHITESIDES, G. M. (1983), *J. Org. Chem.* **48**, 3130-3132.

CRANS, D. C., KAZLAUSKAS, R. J., HIRSCHBEIN, B. L., WONG, C.-H., ABRIL, O., WHITESIDES, G. M. (1987), *Methods Enzymol.* **136**, 263-280.

CROSBY, J. (1992), in: *Chirality in Industry* (COLLINS, A. N., SHELDRAKE, G. N., CROSBY, J., Eds.), pp. 1-68, Chichester-New York: Wiley & Sons.

CROUT, D. H. G., CHRISTEN, M. (1989), in: *Modern Synthetic Methods* (SCHEFFOLD, R., Ed.), pp. 1-114, Berlin-Heidelberg-New York: Springer.

DANIELS, M. J. (1987), *Methods Enzymol.* **136**, 371-379.

DARNOKO, D., CHERYLAN, M., ARTZ, W. E. (1989), *Enzyme Microb. Technol.* **11**, 154.

DEESLIE, W. D., CHERYLAN, M. (1982a), *Biotechnol. Bioeng.* **23**, 2257.

DEESLIE, W. D., CHERYLAN, M. (1982b), *Biotechnol. Bioeng.* **24**, 69.

DEGUSSA AG (1989), *Eur. Pat.* 326063.

DEGUSSA AG (1989), *Eur. Pat.* 350908.

DEGUSSA AG (1992a), *Ger. Pat. Appl.* 4020980.

DEGUSSA AG (1992b), *Ger. Pat. Appl.* 4119029.

DEGUSSA AG (1992c), *Ger. Pat. Appl.* 3917057.

DEGUSSA AG/GBF (1981), *U.S. Pat.* 4304858.

DICKENSHEETS, P. A., CHEN, L. F., TSAO, G. T. (1977), *Biotechnol. Bioeng.* **19**, 365.

DIDZIAPETRIS, R., DRABNIG, B., SCHELLENBERGER, V., JAKUBKE, H.-D., SVEDAS, V. (1991), *FEBS Lett.* **287**, 31-33.

DRAUZ, K., KOTTENHAHN, M., MAKRYALEAS, K., KLENK, H., BERND, M. (1991a), *Angew. Chem.* **103**, 704-706.

DRAUZ, K., KOTTENHAHN, M., MAKRYALEAS, K., KLENK, H., BERND, M. (1991b), *Angew. Chem. Int. Ed. Engl.* **30**, 712-714.

DRIOLI, E., GIANFREDA, L., PALESCANDOLO, R., SCARDI, V. (1975), *Biotechnol. Bioeng.* **17**, 1365-1367.

EFFENBERGER, F., ZIEGLER, T., FÖRSTER, S. (1987a), *Angew. Chem.* **99**, 491-492.

EFFENBERGER, F., ZIEGLER, T., FÖRSTER, S. (1987b), *Angew. Chem. Int. Ed. Engl.* **26**, 458-459.

EFFENBERGER, F., HÖRSCH, B., FÖRSTER, S., ZIEGLER, T. (1990), *Tetrahedron Lett.* **31**, 1249-1252.

EFFENBERGER, F., HÖRSCH, B., WEINGART, F., ZIEGLER, T., KÜHNER, S. (1991), *Tetrahedron Lett.* **32**, 2605-2608.

ELFERINK, V. H. M., BREITGOFF, D., KLOOSTERMAN, M., KAMPHUIS, J., VAN DEN TWEEL, W. J. J., MEIER, E. M. (1991), *Rec. Trav. Chim. Pays-Bas* **110**, 63-74.

FISCHER, A., SCHWARZ, A., WANDREY, C., BOMMARIUS, A. S., KNAUP, G., DRAUZ, K. (1991), *Biomed. Biochim. Acta* **50**, 169-174.

FISCHER, A., BOMMARIUS, A. S., DRAUZ, K., WANDREY, C. (1993), submitted.

FLASCHEL, E., WANDREY, C. (1978), in: *Enzyme Engineering*, Vol. 4 (BROUN, G. B., MANECKE,

G., WINGARD, L. B., Eds.), pp. 83–88, New York: Plenum Press.

FLASCHEL, E., WANDREY, C., KULA, M.-R. (1983), *Adv. Biochem. Eng.* **26**, 73–142.

FLÖRSHEIMER, A., KULA, M.-R., SCHÜTZ, H.-J., WANDREY, C. (1989), *Biotechnol. Bioeng.* **33**, 1400–1405.

FORSCHUNGSZENTRUM JÜLICH GMBH/DEGUSSA AG (1991), *Ger. Pat.* 4014564.

FORSCHUNGSZENTRUM JÜLICH GMBH/DEGUSSA AG (1992), *Ger. Pat. Appl.* 4101895.

FUFKUMURA, T. (1976), *Agric. Biol. Chem.* **40**, 1687.

FUGANTI, C., GRASSELLI, P., CASATI, P. (1986), *Tetrahedron Lett.* **27**, 3191–3194.

FURUI, M., YAMASHITA, K. (1983), *J. Ferment. Technol.* **61**, 587.

GENTZEN, I., LÖFFLER, H.-G., SCHNEIDER, F. (1979), in: *Metalloproteins* (ESER, U., Ed.), pp. 270–274, Stuttgart: Thieme.

GENTZEN, I., LÖFFLER, H.-G., SCHNEIDER, F. (1980), *Naturforsch.* **35c**, 544–550.

GIBSON, D. T., KOCH, J. R., KALLIO, R. E. (1968), *Biochemistry* **7**, 2653.

GROSS, A., ABRIL, O., LEWIS, J. M., GERESH, S., WHITESIDES, G. M. (1983), *J. Am. Chem. Soc.* **105**, 7428–7435.

HABULIN, M., KNEZ, Z. (1991), *J. Membrane Sci.* **61**, 315–324.

HARTMAN, S. C., BUCHANAN, J. M. (1958), *J. Biol. Chem.* **233**, 451.

HAYASHI, T., KONISHI, M., FUKUSHIMA, M., KANEHIRA, K., HIOKI, T., KUMADA, M. (1983), *J. Org. Chem.* **48**, 2195–2202.

HERRMANN, P. (1991), *Biomed. Biochim. Acta* **50**, 19–31.

HERRMANN, G., SCHWARZ, A., WANDREY, C., KULA, M.-R., KNAUP, G., DRAUZ, K., BERNDT, H. (1991), *Biotechnol. Appl. Biochem.* **13**, 346–353.

HIRSCHBEIN, B. L., MAZENOD, F. P., WHITESIDES, G. M. (1982), *J. Org. Chem.* **47**, 3765–3766.

HOMANDBERG, G. A., MATIS, J. A., LASKOWSKI, JR., M. (1978), *Biochemistry* **17**, 5220.

HONG, J., TSAO, G. T., WANKAT, P. C. (1981), *Biotechnol. Bioeng.* **23**, 1501.

HORNBY, W. E., LILLY, M. D., CROOK, E. M. (1968), *Biochem. J.* **107**, 669–674.

HUDLICKY, T., LUNA, H., BARBIERI, G. KWART, L. D. (1988), *J. Am. Chem. Soc.* **110**, 4735–4741.

HUDLICKY, T., LUNA, H., PRICE, J. D., RULIN, F. (1989), *Tetrahedron Lett.* **30**, 4053–4054.

HUDLICKY, T., LUNA, H., PRICE, J. D., RULIN, F. (1990), *J. Org. Chem.* **55**, 4683–4687.

HUMMEL, W., SCHÜTTE, H., SCHMIDT, E., WANDREY, C., KULA, M.-R. (1987a), *Appl. Microbiol. Biotechnol.* **26**, 409–416.

HUMMEL, W., SCHÜTTE, H., SCHMIDT, E., KULA, M.-R. (1987b), *Appl. Microbiol. Biotechnol.* **27**, 283–291.

IBORRA, J. L., OBON, J. M., MANJON, A., CANOVAS, M. (1992), *Biotechnol. Appl. Biochem.* **15**, 22–30.

IKEMI, M., ISHIMATSU, Y., KISE, S. (1990), *Biotechnol. Bioeng.* **36**, 155–165.

ISHIKAWA, H., TAKASE, S., TANAKA, T., HIKITA, H. (1989), *Biotechnol. Bioeng.* **34**, 369–379.

JENSEN, V. J., RUGH, S. (1987), *Methods Enzymol.* **136**, 356–370.

KAMPHUIS, J., BOESTEN, W. H. J., BROXTERMAN, Q. B., HERMES, H. F. M., VAN BALKEN, J. A. M., MEIJER, E. M., SCHOMAKER, H. E. (1990), *Adv. Biochem. Eng.* **42**, 133–186.

KAMPHUIS, J., BOESTEN, W. H. J., KAPTEIN, B., HERMES, H. F. M., SONKE, T., BROXTERMAN, Q. B., VAN DEN TWEEL, W. J. J., SHOEMAKER, H. E. (1992), in: *Chirality in Industry* (COLLINS, A. N., SHELDRAKE, G. N., CROSBY, J., Eds.), pp. 187–208, Chichester–New York: Wiley & Sons.

KARANTH, N. G., BAILEY, J. E. (1978), *Biotechnol. Bioeng.* **20**, 1817.

KARRENBAUER, M., PLÖCKER, U. (1984), *Ann. N.Y. Acad. Sci.* (*Enzyme Eng.* **7**) **434**, 78.

KAZLAUSKAS, R. J., WHITESIDES, G. M. (1985), *J. Org. Chem.* **50**, 1069–1076.

KISE, S., HAYASHIDA, M. (1990), *J. Biotechnol.* **14**, 221–228.

KLYOSOV, A. A., VAN VIET, N., BEREZIN, I. V. (1975), *Eur. J. Biochem.* **59**, 3.

KNAZEK, R. A., GULLINO, P. M., KOHLER, P. O., DEDRICK, R. L. (1972), *Science* **178**, 65–67.

KOBAYASHI, T., MOO-YOUNG, M. (1971), *Biotechnol. Bioeng.* **13**, 893–910.

KRAGL, U., VASIC-RACKI, D., WANDREY, C. (1992), *Chem. Ing. Tech.* **64**, 499–509.

LADNER, W., WHITESIDES, G. M. (1984), *J. Am. Chem. Soc.* **106**, 7250–7251.

LANGER, R., BERNSTEIN, H., BROWN, L., CIMA, L. (1990), *Chem. Eng. Sci.* **45**, 1967–1978.

LEE, S. V. (1990), *Pure Appl. Chem.* **62**, 2031.

LEE, S. Y., HIBI, N., YAMANE, T., SHIMIZU, S. (1985), *J. Ferment. Technol.* **63**, 37–44.

LEVENSPIEL, O. (1972), in: *Chemical Reaction Engineering*, 2nd Ed., Chap. 14, New York: John Wiley & Sons.

LILLY, M. D., DUNNILL, P. (1976), *Methods Enzymol.* **44**, 717–746.

LORBACH, D. M., HATTON, T. A. (1988), *Chem. Eng. Sci.* **43**, 405–418.

LUISI, P. L. (1985a), *Angew. Chem.* **97**, 449–460.

LUISI, P. L. (1985b), *Angew. Chem. Int. Ed. Engl.* **24**, 439–450.

LUISI, P. L., GIOMINI, M., PILENI, M.-P., ROBINSON, B. H. (1988), *Biochim. Biophys. Acta* **947,** 209–246.

MARR, R., KOPP, A. (1980), *Chem. Ing. Tech.* **52,** 399–410.

MARTIN, V. S., WOODARD, S. S., KATSUKI, T., YAMADA, Y., IKEDA, M., SHARPLESS, K. B. (1981), *J. Am. Chem. Soc.* **103,** 6237–6240.

MARTINEK, K., LEVASHOV, A. V., KHMELNITSKI, Y. L., KLYACHKO, N., BEREZIN, I. V. (1986), *Eur. J. Biochem.* **155,** 453–468.

MATSUMURA, M., HIRATA, J. (1989), *J. Chem. Tech. Biotechnol.* **46,** 313–326.

MAZUR, R. H., SCHLATTER, J. M., GOLDKAMP, A. H. (1969), *J. Am. Chem. Soc.* **91,** 2684–2691.

MCCONVILLE, F. X., LOPEZ, J. L., WALD, S. A. (1990), in: *Biocatalysis* (ABRAMOWICZ, D. A., Ed.), pp. 167–177, New York: van Nostrand Reinhold.

MERRELL DOW PHARMACEUTICALS (1990a), *U.S. Pat.* 4902719.

MERRELL DOW PHARMACEUTICALS (1990b), *Eur. Pat. Appl.* 0357029.

MESSING, R. A. (1975), *Immobilized Enzymes for Industrial Reactors*, New York: Academic Press.

MOSBACH, K. (Ed.), (1976), Immobilized enzymes, in: *Methods Enzymol.* **44.**

MOSBACH, K. (Ed.,) (1987a), Immobilized enzymes and cells, Part B, in: *Methods Enzymol.* **135.**

MOSBACH, K. (Ed.), (1987b), Immobilized enzymes and cells, Part C, in: *Methods Enzymol.* **136.**

NAGASAWA, T., YAMADA, H. (1990), in: *Biocatalysis* (ABRAMOWICZ, D. A., Ed.), pp. 277–318, New York: van Nostrand Reinhold.

NIEDERMEYER, U., KULA, M.-R. (1990a), *Angew. Chem.* **102,** 423–425.

NIEDERMEYER, U., KULA, M.-R. (1990b), *Angew. Chem. Int. Ed. Engl.* **29,** 386–388.

NISHIMURA, N., KISUMI, M. (1988), *J. Biotechnol.* **7,** 11.

NISHIMURA, N., TANIGUCHI, T., KOMATSUBARA, S. (1989), *Ferment. Biotechnol.* **67,** 107.

OGNYANOV, V. I., DATCHEVA, V. K., KYLER, K. S. (1991), *J. Am. Chem. Soc.* **113,** 6992–6996.

OLIVIERI, R., FASCETTI, E., ANGELINI, L., DEGEN, L. (1981), *Biotechnol. Bioeng.* **23,** 2173–2183.

OYAMA, K., KIHARA, K. (1984), *Chemtech* **14,** 100–105.

OYAMA, K., NISHIMURA, S., NONAKA, Y., KIHARA, K., HASIMOTO, T. (1981), *J. Org. Chem.* **46,** 5241-5244.

OYAMA, K., KIHARA, K., NONAKA, Y. (1981), *J. Chem. Soc. Perkin Trans.* **2,** 356–360.

OYAMA, K., IRINO, S., HARADA, T., HAGI, N. (1984), *Ann. N. Y. Acad. Sci.* (*Enzyme Eng.* **7**) **434,** 95.

OYAMA, K., IRINO, S., HAGI, N. (1987), *Methods Enzymol.* **136,** 503–516.

PABSCH, K., PETERSEN, M., RAO, N. N., ALFERMANN, A. W., WANDREY, C. (1991), *Rec. Trav. Chim. Pays-Bas* **110,** 199–205.

PIZZICHINI, M., FABIANI, C., ADAMI, A., CAVAZZONI, V. (1989), *Biotechnol. Bioeng.* **33,** 955.

PRONK, W., KERKHOF, P. J. A. M., VAN HELDEN, C., VAN'T RIET, K. (1988), *Biotechnol. Bioeng.* **32,** 512–518.

RADHAKRISHNAN, R., ROBSON, R. J., TAKAGAKI, Y., KHORANA, H. G. (1981), *Methods Enzymol.* **72,** 408.

RIOS-MERCADILLO, V. M., WHITESIDES, G. M. (1979), *J. Am. Chem. Soc.* **101,** 5828–5829.

RÖTHIG, T. R., KULBE, K. D., BÜCKMANN, F., CARREA, G. (1990), *Biotechnol. Lett.* **12,** 353–356.

RÜTGERSWERKE AG (1988), *Ger. Pat. Appl.* 3712539.

SCHEPER, T., BARENSCHEE, E. R., BARENSCHEE, T., HASLER, A., MAKRYALEAS, K., SCHÜGERL, K. (1989), *Ber. Bunsenges. Phys. Chem.* **93,** 1034–1038.

SCHMIDT, E., FIOLITAKIS, E., WANDREY, C. (1986), *GBF Monogr. Ser.* **9,** 213–220.

SCHÖLLKOPF, U. (1983), *Tetrahedron* **39,** 2085–2091.

SCHÖLLKOPF, U., NEUBAUER, H. J. (1982), *Synthesis,* 861–864.

SCHÖLLKOPF, U., SCHREVER, R. (1984), *Liebigs Ann. Chem.*, 939–950.

SCHÜTTE, H., FLOSSDORF, J., SAHM, H., KULA, M.-R. (1976), Eur. J. Biochem. **62,** 151–160.

SCHULER, M. L., ARIS, R., TSUCHIYA, H. M. (1972), *J. Theor. Biol.* **35,** 67–76.

SCHWARZ, A. (1991), *Dissertation*, Universität Bonn, Germany.

SENUMA, M., OTSUKI, O., SAKATA, N., FURUI, M., TOSA, T. (1989), *J. Ferment. Bioeng.* **67,** 233.

SERVI, S. (Ed.), (1992), *Microbial Reagents in Organic Synthesis, NATO ASI Ser.* **C381,** Dordrecht–London–Boston: Kluwer Acad. Publ.

SHARMA, B. P., BAILEY, L. F., MESSING, R. A. (1982a), *Angew. Chem.* **94,** 836–852.

SHARMA, B. P., BAILEY, L. F., MESSING, R. A. (1982b), *Angew. Chem. Int. Ed. Engl.* **21,** 837–853.

SHELDRAKE, G. N. (1992), in: *Chirality in Industry* (COLLINS, A. N., SHELDRAKE, G. N., CROSBY, J., Eds.), pp. 127–166, Chichester–New York: Wiley & Sons.

SHIELD, J. W., FERGUSON, H. D., BOMMARIUS, A. S., HATTON, T. A. (1986), *Ind. Eng. Chem. Fundam.* **25,** 603–612.

SOLVAY DUPHAR N. V. (1991), *Eur. Pat. Appl.* 203214.

SONNTAG, N. O. V. (1953), *Chem. Rev.* **52**, 237–416.

STECKHAN, E., HERRMANN, S., RUPPERT, R., THÖMMES, J., WANDREY, C. (1990a), *Angew. Chem.* **102**, 445–447.

STECKHAN, E., HERRMANN, S., RUPPERT, R., THÖMMES, J., WANDREY, C. (1990b), *Angew. Chem. Int. Ed. Engl.* **29**, 388–390.

STEINKE, D., KULA, M.-R. (1990a), *Enzyme Microb. Technol.* **12**, 836–840.

STEINKE, D., KULA, M.-R. (1990b), *Angew. Chem.* **102**, 1204–1206.

STEINKE, D., KULA, M.-R. (1990c), *Angew. Chem. Int. Ed. Engl.* **29**, 1139–1141.

STEINKE, D., KULA, M.-R. (1991), *Biomed. Biochim. Acta* **50**, 143–148.

STOINEVA, I. B., GALUNSKY, B. P., LOZANOV, S. V., IVANOV, I. P., PETKOV, D. D. (1992), *Tetrahedron* **48**, 115–122.

SYLDATK, C., WAGNER, F. (1990), *Food Biotechnol.* **4**, 87–95.

SYLDATK, C., LÄUFER, A., MÜLLER, R., HÖKE, H. (1990), *Adv. Biochem. Eng.* **41**, 29–75.

SYLDATK, C., LEHMENSIEK, V., ULRICHS, G., BILITEWSKI, U., KROHN, K., HÖKE, H., WAGNER, F. (1992), *Biotechnol. Lett.* **14**, 99–104.

TAKATA, J., TOSA, T., CHIBATA, J. (1983), *Appl. Biochem. Biotechnol.* **8**, 31–38.

TAKEDA CHEMICAL INDUSTRIES LTD. (1989), *Eur. Pat. Appl.* 0304021.

TOSA, T., MORI, T., FUSE, N., CHIBATA, I. (1966), *Enzymologia* **31**, 214.

TOSA, T., MORI, T., FUSE, N., CHIBATA, I. (1969), *Agric. Biol. Chem.* **33**, 1047–1052.

TRAMPER, J., MULLER, F., VAN DER PLAS, H. C. (1978), *Biotechnol. Bioeng.* **20**, 1507.

TRAMPER, J., VAN DER PLAS, J. C., LINKO, P. (Eds.), (1985), *Biocatalysis in Organic Synthesis*, Amsterdam: Elsevier.

UTTAPAP, D., KOBA, Y., ISHIZAKI, A. (1989), *Biotechnol. Bioeng.* **33**, 542.

VAN DE TWEEL, W. J. J., DE BONT, J. A. M., VORAGE, M. J. A. W., MARSMAN, E. H., TRAMPER, J., KOPPEJAN, J. (1988), *Enzyme Microb. Technol.* **10**, 134.

VASIC-RACKI, D., JONAS, M., WANDREY, C., HUMMEL, W., KULA, M.-R. (1989), *Appl. Microbiol. Biotechnol.* **31**, 215–222.

VIETH, W. R., VENKATASUBRAMANIAN, K., CONSTANTINIDES, A., DAVIDSON, B. (1976), *Appl. Biochem. Bioeng.* **1**, 221.

WAGNER, F., SYLDATK, C., LEHMENSIEK, V., KROHN, K., HÖKE, H., LÄUFER, A. (1990), *Eur. Pat. Appl.* 0377083.

WALDMANN, H. (1991), *Kontakte* **2**, 33–54.

WANDREY, C. (1977), *Habilitationsschrift*, Technische Universität Hannover, Germany.

WANDREY, C. (1987), *Proc. 4th Eur. Congr. Biotechnol.* **4**, 171–188 (NEIJSSEL, O. M., VAN DER MEER, R. R., LUYBEN, K. CH. A. M., Eds.), Amsterdam: Elsevier Science Publishers B.V.

WANDREY, C., FLASCHEL, E. (1979), *Adv. Biochem. Eng.* **12**, 147–218.

WANDREY, C., FLASCHEL, E., GHOSE, T. K., FIECHTER, A., BLAKEBROUGH, N. (1979), *Adv. Biochem. Eng.* **12**, 147–218.

WEDY, M. (1992), *Dissertation*, Rheinisch-Westfälische Technische Hochschule Aachen, Germany.

WEISZ, P. B., PRATER, C. D. (1954), *Adv. Catal.* **6**, 143.

WHITESIDES, G. M., WONG, C.-H. (1985a), *Angew. Chem.* **97**, 617–638.

WHITESIDES, G. M., WONG, C.-H. (1985b), *Angew. Chem. Int. Ed. Engl.* **24**, 617–638.

WOODLEY, J. M., BRAZIER, A. J., LILLY, M. D. (1991), *Biotechnol. Bioeng.* **37**, 133.

YAMAMOTO, K., TOSA, T., CHIBATA, I. (1980), *Biotechnol. Bioeng.* **22**, 2045.

YAMANE, T., HOQ, M. M., SHIMIZU, S. (1988), *Ann. N.Y. Acad. Sci.* (*Enzyme Eng.* **9**) **542**, 224–228.

YANG, V. C., BERNSTEIN, H., LANGER, R. (1988), *Ann. N.Y. Acad. Sci.* (*Enzyme Eng.* **9**) **542**, 515–520.

YOUNG, J. W. (1990), *Chemicals in Britain*. 18–21.

ZABORSKY, O. R. (1971), *Immobilized Enzymes*, Cleveland: CRC Press.

ZMIJEWSKI JR., M. J., BRIGGS, B. S., THOMPSON, A. R., WRIGHT, I. G. (1991), *Tetrahedron Lett.* **32**, 1621–1622.

III. Product Recovery and Purification

18 Cell and Cell Debris Removal: Centrifugation and Crossflow Filtration

RAJIV V. DATAR

Glen Cove, NY 11542, USA

CARL-GUSTAF ROSÉN

Björnlunda, Sweden

List of Symbols

a	$=21\ \mu/d_p \cdot \rho_l$
A	surface area
b	$=6\ (a/21)^{0.5}$ or blade width of impeller
B	Darcy's permeability constant
c	defined in Eq. (16.3)
c_d	drag coefficient, defined in Eq. (14)
C	volumetric concentration of solids
$C_{b,g,w}$	bulk-, gel- or wall concentration
$C_{e,f}$	effluent- or feed concentration
d, d_p	diameter of particle
D	diameter of cylinder
$\mathcal{D}$	diffusion coefficient
$F_{e,f}$	fraction of particles in effluent or feed
g	gravitational constant
g'	corrected gravity value for particle sedimentation
G	grade efficiency function
J	filtrate (or permeate) flux
k	mass transfer coefficient
k_1, k_2	constants in Eq. (19)
L	axial length of flow channel
$P_{i,o,p}$	inlet-, outlet- or permeate pressure
ΔP_{CF}	pressure drop along flow channel
ΔP_{TM}	transmembrane pressure drop
Q	throughput
R	hydraulic resistance
$R_{m,g,p,c}$	hydraulic resistance offered by membrane, gel, polarization or deposited solids layers
S	thickness of bed or deposited solids layer
v	tangential (or crossflow) velocity
V_s	sedimentation velocity
x	axial position
y	position normal to membrane surface
Y	yield or recovery
α	cone angle of disks in disk stack centrifuges
ε	bed porosity
γ, γ_w	local- or wall shear rate
ϕ	radius of particle
μ, η	apparent viscosity
$\rho_{l,s}$	density of liquid or solid phases
Σ	sigma coefficient or summation

1 Introduction

Over the last decade, advances in recombinant-DNA (rDNA) techniques have overshadowed efforts in understanding the specific needs and problems of downstream processing (DSP). As a result the operational limitations of DSP, along with its costs, are now widely accepted as a major obstacle in the ability to rapidly develop and place commercially viable biotechnological products in the market place. In particular, the recovery of miroorganisms and/or their debris from fermented broth presents the bioprocess engineer with a difficult solid/liquid separation, for which few practical solutions exist. The nature of the so-called 'solids' phase – typically characterized as being gelatinous, compressible and of low density, severely limits the number of unit operations that can be used. Ideally, cell recovery must be a cheap, simple and reliable operation, as relatively large volumes of whole broth containing product at dilute concentrations are handled at this stage of processing. In practice though, these ideal requirements are rarely met. Instead, most cell and debris removal operations impose both technical and financial restrictions.

This chapter outlines the present state of the art and identifies some of the technical and economic problems that need to be addressed.

2 Elements of a Recovery Strategy

In Fig. 1, which outlines a process for rDNA based products, we have indicated some of the most common applications of centrifugation and crossflow filtration, which are in the general order in which they appear in a process:

(1) cell/broth separation
(2) cell debris removal
(3) collection of protein precipitate
(4) washing/buffer exchange (diafiltration)
(5) protein concentration (ultrafiltration)

In the following we will concentrate only on the primary separation steps. The objectives of the initial solid/liquid separation steps are simple enough:

- a well clarified supernatant,
- solids of maximum dryness, and
- contained operation.

Depending upon the degree of separation required, a variety of mechanical, thermal, chemical or electrokinetic separation techniques can be employed (ATKINSON and SAINTER, 1982). However, from the aspect of cost, mechanical methods are preferable. These are basically centrifugal sedimentation,

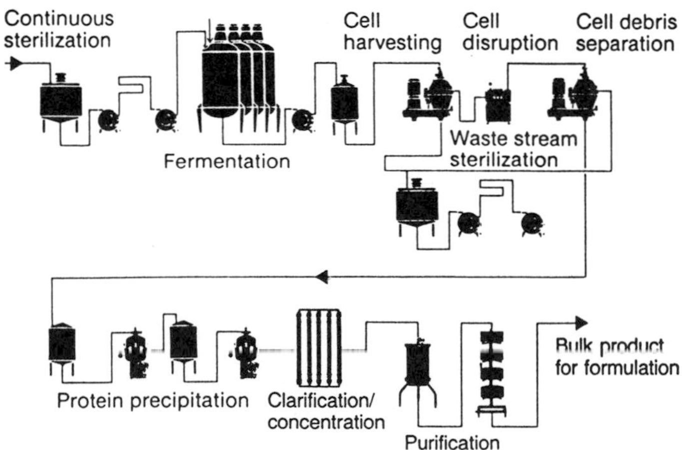

Fig. 1. Schematic representation of an rDNA-based production plant.

filtration and sedimentation/flotation. In this chapter, we discuss only centrifugation and crossflow filtration techniques.

It is important to note that while centrifugation is generally regarded as the classical technique for primary recovery operations, crossflow filtration is being increasingly considered as a viable alternative.

2.1 Process Characteristics

Product type, biomass concentrations and recovery costs for different products and organisms vary considerably as indicated in Tab. 1. Usually the lower the product concentration, the smaller the scale of operation and the higher the market price of the product. At one extreme are the large-scale processes which are governed primarily by the cost of the fermentation raw materials.

For medium-scale processes, the economics of production are finely balanced. Among the major factors determining product price are the costs associated with the cell recovery step. Here the total DSP costs can be as much as 40–60% of the total production costs.

Volumes of processing for small-scale fermentation plants fall into the 1–50 m^3 per batch range. Recovery costs dominate the overall economics. As is clearly seen, product concentration and method of recoveries of some human therapeutic products can easily push DSP costs beyond 80% of overall production costs.

2.2 Nature of Solids Phase

Tab. 2 lists a number of solid phases characteristic of today's biotechnological processes. Due to the different morphologies, whole broths exhibit a range of different viscosities and densities. For example, mycelial broths exhibit high viscosities and low specific gravities. In centrifugal separators, the maximum concentration factor achievable is highly dependent on the slurry's flow characteristics. The high viscosities of mycelial suspensions can also render crossflow membrane filtration uneconomic. Energy costs associated with the pump size, required to maintain adequate crossflow velocities, can be steep.

Plant and animal cells are not able to withstand the same degree of applied shear as microbial cells. Therefore, separation techniques which generate high mechanical shear, such as conventional crossflow filtration and classical centrifugation, may not be applicable. This is certainly problematic.

Tab. 1. Examples of Biotechnological Processes (ATKINSON and SAINTER, 1982; ROSÉN and DATAR, 1983; SPALDING, 1991)

Process/Product	Typical Volumes (m^3/batch)	Cell Type	Conc. of Product in Fermenter (kg/m^3)	Ratio of Recovery to Fermentation Cost (Estimates)	Average Ratio of Recovery Cost to Total Cost (%)
Large scale	>200			0.2	20
Single cell protein		Bact./yeast	30– 50		
Ethanol		Yeast	70–120		
Medium scale	50–200				50
Antibiotics		Fungi	10– 50	1	
Enzymes		Bacteria	5– 10	1–2	
Small scale	< 50				
Riboflavin		Bacteria	1	2	67
rec-human insulin		Bacteria	0.5	2–3	70
rec-tPA		Animal	0.05	3–4	78
Glucocerebrosidase		Extraction	0.02	4 5	82

Tab. 2. Some Solid Phase Characteristics

(a) Sizes and Specific Gravities of Representative Solids (BOWDEN, 1984; DATAR, 1984a)

Solids Type	Size (μm)	Density Difference Between Solids and Broth (kg/m^3)	Cost of Recovery
Cell debris	0.2×0.2	0–120[a]	Highest
Bacterial cells	1×2	70	∧
Yeast cells	7×10	90	∧
Mammalian cells	40×40	70	∧
Plant cells	100×100	50	∧
Fungal hyphae	1×10×(matted)	10	∧
Floccules	100×100	—	Lowest

[a] Cell debris densities depend on composition, e.g., lipid content.

(b) Estimates of Apparent Viscosities of Whole Broth

Biomass Concentration of Whole Broth (kg/m^3)	Apparent Viscosity of Broths (mPa s)		
	Yeast, Bacteria, Cell Debris	Mammalian, Plant Cells	Mycelial Fungi
2	1.2	1	30
10	1.3	2	900
20	1.5	3	8000
30	2	10	40000
70	4	600	≫40000

As stated earlier, in the preliminary design of a process, the bioprocess engineer has one simply defined objective: to recover the solids in the most successful (economic) manner. The objective is straightforward, but the factors necessary for a successful solids separation can be complex.

One such factor is 'broth conditioning'. This is best defined as the application of physicochemical as well as biological techniques to alter the separation characteristics of the broth. The profound importance of cell surface chemistries and cell–cell interactions is underscored by the observation that most large- and medium-scale processes (see Tab. 1) incorporate a broth conditioning stage prior to cell recovery. Along with rheology then, these two factors can be instrumental in determining whether a separation will be a success.

3 Characteristics of Centrifugal Separators

Centrifuges have been used for cell separation since the late 19th century, when the first separators were installed in a yeast factory in Sweden. The basic design, first developed for cream separation, is still valid, although many new principles and design variables have emerged over the years. Some of the common features of centrifuges are discussed below.

Continuous processing

Today's self-cleaning centrifuges can be run for hundreds of hours with no need for stopping and cleaning. Under favorable circumstances, performance remains constant throughout an extended run. Although of a mechanically advanced design, centrifuges are generally very reliable, and with proper maintenance they may be used for many years – in extreme cases several decades.

Short retention times

Passage of liquid through the bowl (rotor) of a continuous centrifuge takes the order of a fraction of a second to a few seconds. The time in which sensitive biological materials are under stress is thus very limited.

Small space requirements

Since retention times are very short, the active volume of a centrifuge is very small – a small fraction of a percent of the hourly throughput. Space requirements are therefore minimal compared to other types of separation equipment.

Adjustable separation efficiency

With liquid and solid streams relatively constant through the centrifuge, separation efficiency is easily adjustable to the optimum point for each particular product.

Closed systems

Centrifuges may be designed to comply with the most stringent demands of process containment. All running parts are enclosed in a pressure vessel with double or triple mechanical seals on moving parts, such as drive shafts. Valves and other functional parts are controlled by contained hydraulic or pneumatic systems.

No external materials needed

A major advantage of centrifuges over classical-type filters is the fact that no materials such as filter aid need to be added. Processed materials therefore come out uncontaminated.

4 Principles of Centrifugation

4.1 Sedimentation Theory

In sedimentation, particles move under the influence of gravity. Centrifugation is an analogous process, in which gravity is replaced by a centrifugal force. Sedimentation is basically described by Stokes' law, which strictly applies to spherical particles and Newtonian fluids:

In actual systems there are many causes for deviations from Stokes' law

$$V_s = \frac{d_p^2 \cdot (\rho_s - \rho_l) \cdot g}{18\,\mu} \tag{1}$$

- non-spherical particles
- non-Newtonian rheology
- hindered settling
- non-laminar flow.

Nevertheless, the formula gives many useful indications as to possible improvements:

- Sedimentation rate being proportional to the square of the particle diameter, increasing particle size, e.g., by flocculation, strongly enhances sedimentation.
- Denser particles or less dense medium improves settling. Dilution is sometimes a possible route; but a more common way to take advantage of an increased density difference is to work with denser particles such as inclusion bodies.
- In many industrial separations viscosity is reduced by heating, but this is rarely a possible route with heat-labile materials. An example of viscosity reduction in this context is, however, to remove or degrade nucleic acids prior to centrifugation.
- By substituting gravity by centrifugal force, the settling rate increases several thousandfold. This is the very basis of separation by means of centrifugation.

Fig. 2 illustrates the analogy between settling by means of sedimentation and by means of centrifugation. Fig. 2a is a lamella settler, in

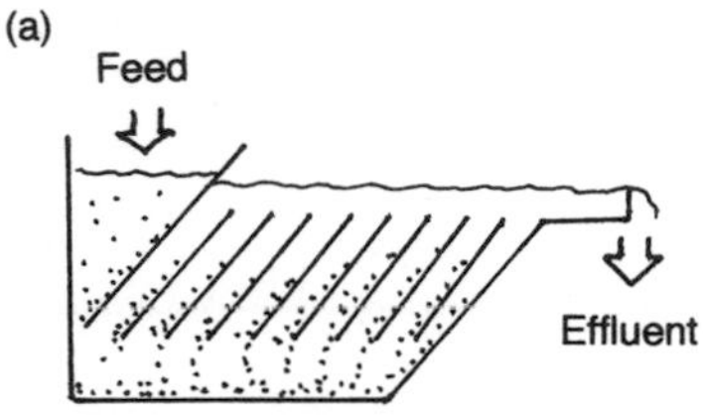

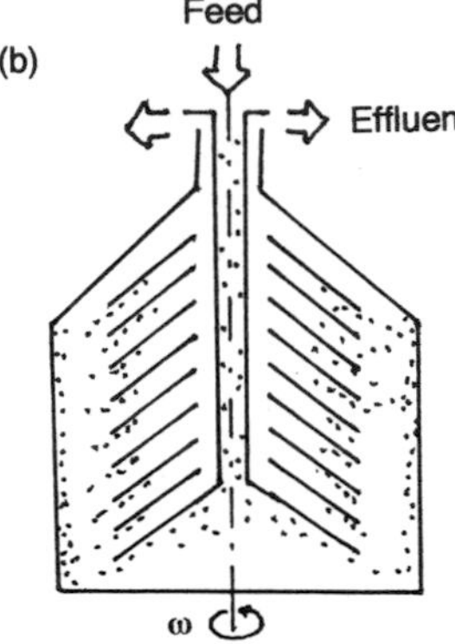

Fig. 2. Comparison of a lamellae thickener (a) with a disk stack centrifuge (b).

which separation is enhanced by means of lamellae, which reduce the settling distance, i.e., the distance which particles have to migrate before they reach a solid surface, where they are no longer exposed to the influence of liquid flow.

Fig. 2b shows in an analogous way a disk stack centrifuge, which has a similar flow, but where conical disks have replaced the flat lamellae of the settler. Settling distance there is reduced to a maximum of less than one millimeter, which is the space between two adjacent disks.

4.2 Separation in a Centrifuge

The migration of a particle within the disk stack of a centrifuge is shown in Fig. 3. Liquid flow between the disks is laminar, i.e., flow rate is maximum half-way between two disks and drops to zero at the surface of either disk. Particle A enters the space at the left side of the figure and moves upwards between the disks under the influence of the liquid flow (v_2), which is parallel to the plates, and the centrifugal force, which has a radial orientation (v_1). The resultant velocity (v) determines the path of the particle.

Particle A is sufficiently large (and/or dense) to reach the upper disk before leaving the disk stack even if it enters in the least favorable position as indicated in Fig. 3. Once it reaches the upper disk, by definition it is no longer influenced by the liquid flow (this being zero at the surface). Hence, all particles which have settled on the upper disk will slide down along its surface and leave the outer rim of the disk stack as aggregates, which eventually collect along the outer periphery of the rotating bowl (from where they are removed by different mechanisms to which we shall return).

Smaller and/or less dense particles will be less influenced by the centrifugal force. The resultant of the two forces which determine the path of these particles will consequently be at a smaller angle to the disks. As indicated in Fig. 3, some of the small particles (particle B) will nevertheless reach the surface of the upper disk before reaching the axial end of the space between the disks and will thus be removed from the fluid stream. Other particles, however, which enter the space between two adjacent disks closer to the lower disks (such as particle A), will escape at the axial end before they

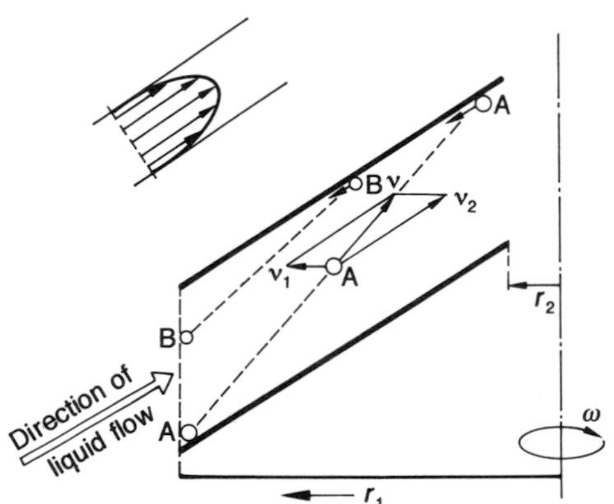

Fig. 3. Movement of particles between two adjacent disks in a disk stack centrifuge with outer radius r_1 and inner radius r_2 of the stack. Particle diameters $d_A > d_B$. The small picture shows (ideal) laminar flow pattern between disks. v_1, velocity vector due to centrifugal force; v_2, velocity vector due to liquid flow.

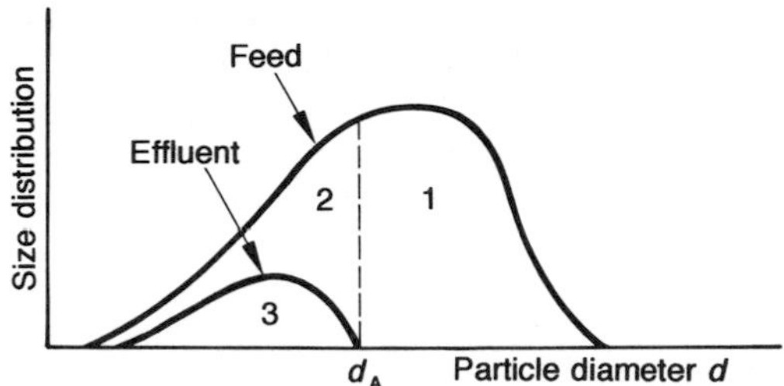

Fig. 4. Size distribution of particles before and after passage of a centrifuge. All particles with $d > d_A$ are removed. Some particles with $d < d_A$ are removed depending on where they enter the space between disks.

reach the upper disk. Such particles will not be removed from the liquid.

This simplified picture of centrifugation suggests an important aspect of centrifugation, namely that in polydisperse systems it is seldom an absolute method of separation. This is further illustrated in Fig. 4, which gives the particle size distribution in a slurry before and after passage through a centrifuge. The assumptions are the same as in Fig. 3, i.e., that all particles of $d > d_A$ are separated, whereas particles with $d < d_A$ will be only partly removed. The area marked 1 represents particles with $d > d_A$, which are all removed. Area 2 represents particles with $d < d_A$, which have nevertheless been removed, whereas area 3 represents those particles which have not been removed by the centrifugation.

4.3 Throughput and Area Equivalent

Throughput (Q) of a sedimentation tank may be expressed as

$$Q = V_{\text{lim}} \cdot A \tag{2}$$

where V_{lim} is the sedimentation velocity for a limit particle, and A is the surface area of the tank. An analogous expression may be derived for the throughput of a centrifuge

$$Q = \left[\frac{d_{\text{lim}}^2 \Delta\rho}{18\eta}\right]\left[\frac{2\pi}{3}\omega^2 N \cot\alpha (r_1^3 - r_2^3)\right] \tag{3}$$

where N is the number of disks in the stack, and the other new constants and variables are defined in Fig. 5. It can be seen that the first factor is media-dependent, whereas the second is an expression of design and size of the centrifuge. Multiplying the first factor by the gravitational constant g and dividing the second factor by g, leads to Eq. (4)

$$Q = \left[\frac{d_{\text{lim}}^2 \Delta\rho}{18\eta} g\right]\left[\frac{2\pi}{3g}\omega^2 N \cot\alpha (r_1^3 - r_2^3)\right] \tag{4}$$

One recognizes Stokes' law, Eq. (1), as the first factor, which thus expresses the sedimentation velocity of a limit particle under gravity. The second factor must then be equivalent to the second factor of Eq. (2), i.e., to the area of a sedimentation tank. It is therefore called the area equivalent and is widely used as a measure of the capacity of a given factor. In the literature it is often called the sigma value, Σ. Sigma values have been derived for a number of alternative centrifuge designs, such as tubular bowl and decanter centrifuges. It is a useful tool for comparison of different centrifuges and for calculation of throughputs. Manufacturers of centrifuges have additionally derived a number of variations from the sigma values, which take into account non-ideal flow conditions in the centrifuge. One such expression is the KQ value, which is used by Alfa-Laval as a standard value for comparison of centrifuge capacities.

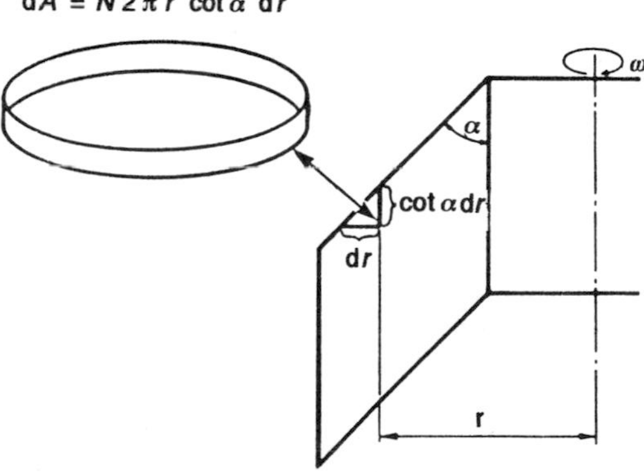

Fig. 5. Schematic representation of a disk stack centrifuge with disk angle α, radius r, and number of disks in the stack N.

4.4 Grade Efficiency Curve

The separating performance of a sedimenting centrifuge can be assessed by measuring the grade efficiency curve (SVAROVSKY, 1977). Fig. 6 shows the form of such a curve – the x-axis gives the particle size and the y-axis the fraction of particles of a given size which are sedimented. The grade efficiency function $G(d)$ has a value of 100% if all particles of that size are removed and a value of 0% if none are removed. The function possesses probability characteristics because the following factors influence the separation process:

- the unequal sedimentation conditions at different regions within the centrifuge
- the finite dimensions of the inlet and outlet regions
- the different surface properties of the particles.

Fig. 7 shows the effect of process variables on the theoretical grade efficiency curve of a disk stack centrifuge. In practice, improvements in separation efficiency will be less significant because of non-ideal flow conditions and other specific interactions between particles.

Solids recovery (Y) can be calculated from the particle size distribution

$$Y = 1 - (\Sigma F_e(d) \cdot C_e)/(\Sigma F_f(d) \cdot C_f) \quad (5)$$

where $F_f(d)$ is the fraction of particles of size d or smaller in the feed, $F_e(d)$ is the fraction of particles of size d or smaller in the effluent, and C_f, C_e are the solids contents of feed and effluent, respectively.

The recovery can also be related to the grade efficiency

$$Y = [\Sigma G(d) \cdot F_f(d) \cdot C_f]/[\Sigma F_f(d) \cdot C_f] \quad (6)$$

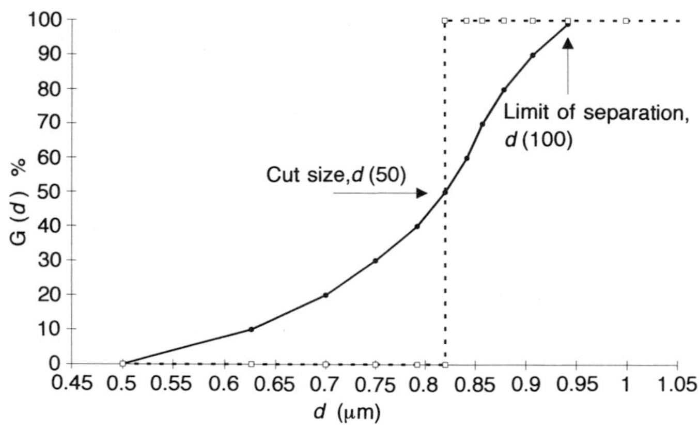

Fig. 6. Typical grade efficiency curve for a sedimenting (tubular) centrifuge with definitions of $d(50)$ and $d(100)$.

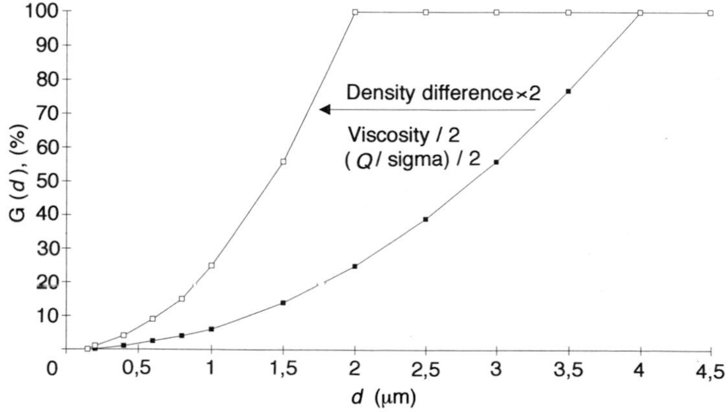

Fig. 7. Effect of process variables on theoretical grade efficiency of a disk stack centrifuge.

In practice, particle size distributions are determined using a method based on settling size and time.

4.5 Sample Calculation Based on *Escherichia coli* Separation

The following example has been given by Westfalia Separator (BRUNNER, 1983). Data pertain to their pilot-plant size disk stack centrifuge SB 7. The particle size distribution was determined for an *E. coli* suspension. Particles ranged in sizes from 0.8 to 1.7 μm nominal diameter. Throughput was calculated from the following product and machine data:
Bowl speed, $n = 8400$ rpm $\rightarrow \omega = 880\ s^{-1}$
number of disks, $N = 72$
outside radius of disks, $r_1 = 0.081$ m
inside radius of disks, $r_2 = 0.036$ m
half disk angle, $\alpha = 38°$
smallest particle diameter, $d_{lim} = 0.8$ μm
density of particles, $S = 1.05\ g/cm^3$
density of liquid, $S' = 1.02\ g/cm^3$
dynamic viscosity, $\eta = 1.02 \cdot 10^{-3}$ kg/m·s
$g = 9.81\ m/s^2$.
Using these values in Eq. (3), we find that $Q = 273$ L/h. This is the rate at which we may separate *E. coli* with nominal particle sizes down to 0.8 μm in an SB7 centrifuge. This value happens to be within the range of values experienced with this particular system, but if we examine the data more closely, we will find that such calculations are subject to large errors depending on small variations in some of the measured variables. Especially the density factor, which is obtained by subtraction of two values which differ only by three percent, can introduce a large error in the calculated capacity.

Actual selection of a centrifuge for a particular separation is therefore nearly always based on calculations from experimentally obtained data for each particular system.

5 Centrifuge Selection

5.1 Objectives in Separation

First of all it is necessary to decide the objectives of a given separation. The following main alternatives exist.

Concentration: the desired product is in the solid phase in the feed to the separator

(a) The heavy phase after centrifugation contains the product, and any remaining liquid may contribute to a deterioration of the quality.
(b) Any solids remaining in the light phase after centrifugation represent a loss of product yield.

Clarification: the product is dissolved in the feed to the separator

(a) Any solids remaining in the light phase after centrifugation represent a possible loss of quality (final impurity or disturbance of further downstream processing).
(b) Any liquid remaining in the heavy phase after centrifugation represents a loss of product yield.

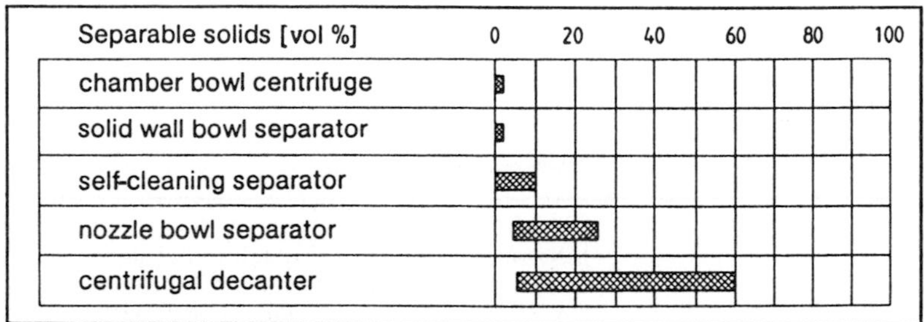

Fig. 8. Sludge concentrations handled by different types of centrifuges (BRUNNER, 1983).

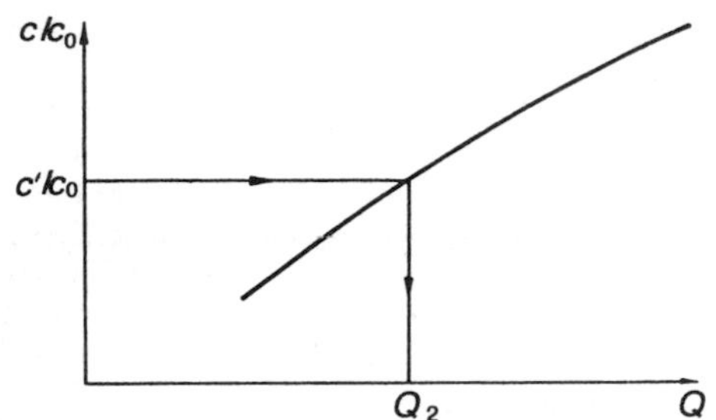

Fig. 9. Determination of machine size from sample run in a test machine. c' is the permissible concentration of solids in the effluent, cf. Eq. (8) (ALFA-LAVAL, 1983).

5.2 Primary Choice: Solids Content

The solids content of a bioslurry determines which type of centrifuge can be used. The choice of centrifuge is made according to Fig. 8 (BRUNNER, 1983), depending on the separated solids content of the suspension, measured in vol%.

5.3 Determination of Machine Size

As we have seen above, an estimate of machine size may be obtained by theoretical calculation. Furthermore, the major manufacturers of centrifuges have considerable experience and methods for predicting centrifuge performance from test tube centrifugation data and scale-up data for pilot-scale tests.

Fig. 9 shows relative concentrations of solids in the effluent as a function of throughput in a test centrifuge. In Fig. 9, c'/c_0 represents maximum permissible concentration of solids and Q_2 the corresponding throughput. This value is determined by gradually increasing the feed rate to the test centrifuge until c/c_0 is reached. A scale-up factor, E, is then used to determine maximum throughput in full-size centrifuges (ALFA-LAVAL, 1983):

$$Q_1 = E \cdot Q_2 \tag{7}$$

5.4 Final Choice of Centrifuge

The sludge handling capacity normally determines the final choice of centrifuge. As we shall see below, continuous centrifuges have various mechanisms for removal of solids from the bowl. In some, removal is continuous, in others intermittent, but in either case, the sludge removal mechanism sets a limit as to how much sludge the centrifuge can handle in a given time.

6 Centrifuge Types and Their Applications

6.1 Overview

Tab. 3 gives an overview of the most common types of industrial centrifuges used in bioprocesses.

Fig. 10 illustrates the main features of these centrifuges.

6.2 Solids-Retaining Centrifuges

This is the oldest type of centrifuge. It is used for separations where solids concentra-

Tab. 3. Some Common Industrial Centrifuge Types

Centrifuge Type	Transport of Sediment
Solids bowl separators	Stays in bowl
Solids-ejecting separator	Intermittent discharge through radial slots
Solids-ejecting nozzle separator	Intermittent discharge through axial channels
Nozzle separator	Continuous discharge through open nozzles
Decanter	Internal screw conveyor

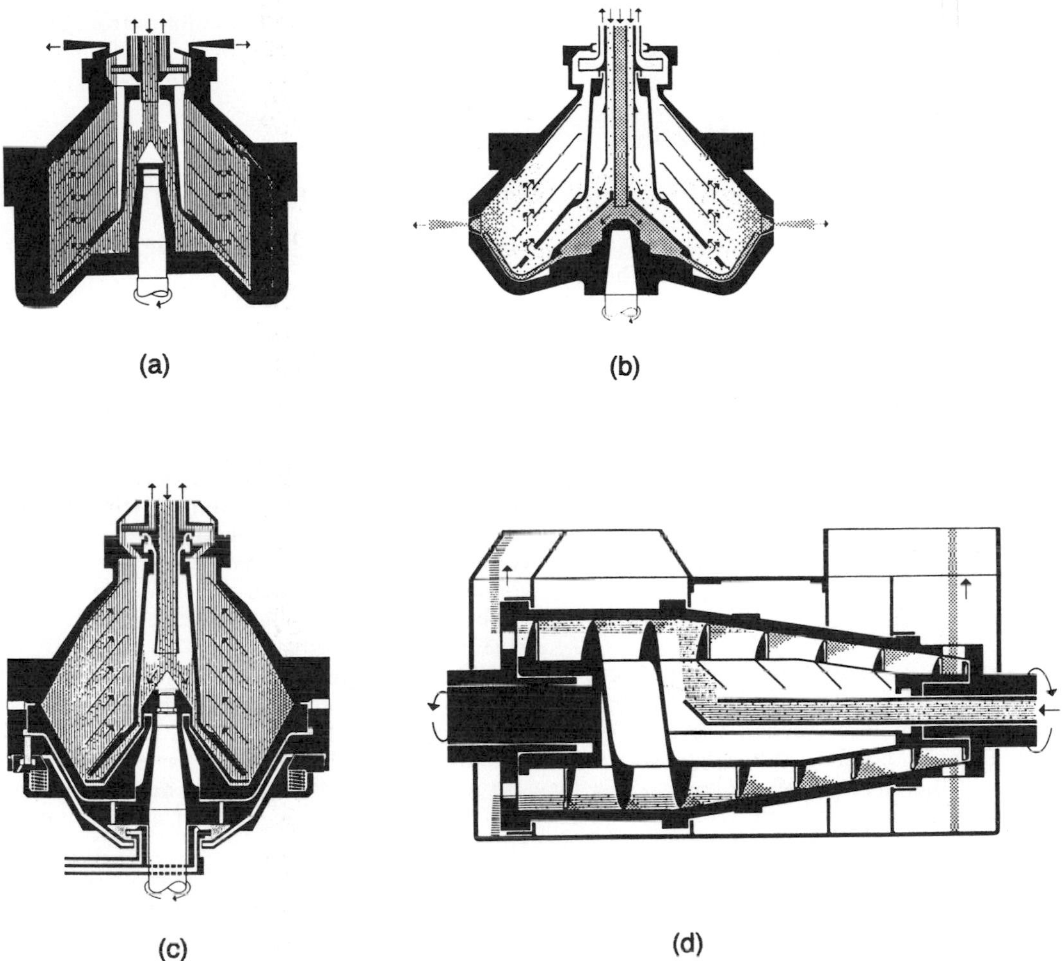

Fig. 10. Some common types of centrifuges: (a) solid wall, (b) open nozzles, (c) self-ejecting, (d) decanter with internal screw conveyor (ALFA-LAVAL, 1990).

tions are very low, e.g., in vaccine production. They can also be used for clarification and simultaneous separation of two liquids. The separated solids accumulate inside the separator bowl which must be stopped at certain intervals for manual removal of solids.

The cylinder bowl centrifuge is a variant of this machine. It has no disk stack but a series of concentric cylindrical solids retainers. It has an extended solids handling capacity and is suitable for the recovery of valuable solids, such as proteins or special cells, which are obtained as a very dry cake. They come in capacities up to about 65 m^3/h.

6.3 Solids-Ejecting Centrifuges

In these centrifuges, solids are removed intermittently while the machine is running at full speed. Discharge can be actuated by a timer or by a sensor indicating that the bowl has been filled to a certain level. It is therefore called a self-triggering separator and is suitable when the solids concentration in the feed fluctuates.

Solids-ejecting centrifuges are very versatile machines, used in the production of yeast, bacteria, mycelium, antibiotics, enzymes, amino acids and many more bioproducts. They come in capacities up to at least 60 m^3/h.

6.4 Solids-Discharging Nozzle Centrifuges

These have nozzles through which the solids are continuously fed either directly to the outside of the bowl or to an internal chamber, from which they are pumped to the outside by means of a paring tube.

Nozzle separators can handle sludge concentrations up to about 30 vol% and are used in many industrial applications, such as baker's yeast production, and in the antibiotics, enzymes and vitamin industries. They come in capacities up to more than 200 m^3/h.

6.5 Decanter Centrifuges

The basic decanter centrifuge consists of a drum, part of which is cylindrical and part of which has a conical shape. The feed enters through axial nozzles, and solids are transported in the direction of the conical part by means of an internal screw conveyor. Solids are continuously discharged at the conical end, while the liquid phase is discharged at the cylindrical end. Levels inside the rotor may be set by means of external nozzles. Decanters are capable of handling very high solids concentrations and may be used also for very troublesome materials, such as mycelia and gelatinous sludges.

Two interesting variants of the classical decanter centrifuge have emerged in recent years. Sharples have produced a machine, which combines the basic features of a decanter with some of the features of a disk stack centrifuge. Thanks to the screw conveyor it can handle high sludge concentrations, while inclined plates in the cylindrical part improve clarification in a way similar to a disk stack. Effluent quality is improved by 50 percent as compared to regular decanters.

The other development is Westphalia's extraction centrifuge, which combines the action of a decanter with that of a centrifugal countercurrent extractor. One example of its many uses is the extraction of antibiotics from mycelia.

7 Centrifugal Separation of Cells and Cell Debris in a Model System

7.1 *Escherichia coli* Cells and Debris

One of the best studied microorganisms and the first one to be used in an industrial process involving recombinant organisms is *Escherichia coli*. It has also been chosen as a model organism for separation and scale-up studies in bioprocess technology (DATAR and ROSÉN, 1987; SANCHEZ-RUIZ, 1989).

Most of the available data have been generated in experiments with a self-ejecting centrifuge designated BTPX 205 and manufactured by Alfa-Laval. This machine has a capacity of up to a few hundred liters of bacterial slurry per hour and a sludge space of approximately 1.2 liters.

Fig. 11 gives a schematic representation of a separation experiment with a BTPX 205 using *E. coli* K12 as a test organism. It is seen that cells are concentrated tenfold, giving a cell concentration of 50% (v/v) in the ejected sludge. The cell concentration in the effluent is

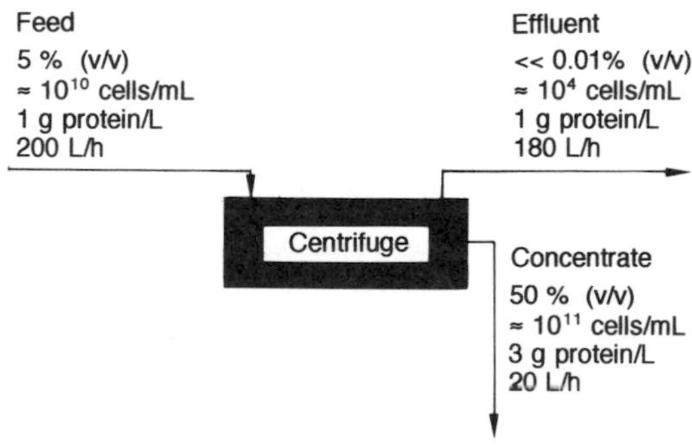

Fig. 11. Separation of *Escherichia coli* in a pilot-size centrifuge (Alfa-Laval BTPX 205).

reduced by a factor of 10^6, which gives a very low loss of cellular material in the effluent.

With cell debris, separation is less sharp. After homogenization and some dilution, the same cells were run through the separator once again. Optimum conditions in this case resulted in 98% recovery of solids (cell debris) at a throughput of only 50 L/h.

One more observation came out of the experiments summarized in Fig. 11. The extracellular protein concentration remains unchanged by passage through the centrifuge of the liquid phase, while there is a certain increase in the sludge phase, probably reflecting a leakage of periplasmic proteins due to the stress the cells undergo as a result of shear and pressure drop during sludge discharge.

Loss of solids into the effluent occurs for the following reasons:

- inhomogeneous particle population (variation of size and density)
- unavoidable disturbance of flow by intermittent discharges
- lack of fine-tuning of centrifuge operation.

A population of *E. coli* harvested under optimal conditions represents a fairly ideal case, where particle sizes vary only ±25%. Cell debris is a different matter. Recent studies of particle size distributions in homogenized *E. coli* slurries (SANCHEZ-RUIZ, 1989) showed two populations of particles: small fragments less than 0.2 μm and aggregates of nominal diameters of 0.5–3 μm. As we have seen above (see Sect. 3.2), separation in such a system can never be an absolute. The results reported above are to be expected in a polydisperse system such as fragmented cells.

During discharge of the sludge, the bowl (rotor) opens at the periphery for a fraction of a second. This totally upsets the flow pattern in the centrifuge. Even if pure water is injected at the moment of discharge to replace the contents of the bowl, product loss will occur until a normal (ideally laminar) flow pattern is resumed. From the flow rates in these particular experiments and centrifuge data (1.2 liters sludge space) we estimate a maximum time interval between discharges of 7 minutes. In this case, obviously the flow pattern is disturbed for a negligible fraction of the total run time, something which is borne out by the sharp separation which takes place in the case of whole cells. In other cases, however, this may be a serious problem. If the centrifuge is working close to its sludge handling capacity, separator performance may be impaired by too frequent discharges.

In the experiments reported above, centrifuge experts from Alfa-Laval were involved in the fine-tuning of the centrifuge. These results therefore correspond to optimum running conditions. In practice, we have found that many users of centrifuges in industry settle for less than optimum conditions. Instead of a yield >99.9%, which is practically feasible for a relatively homogeneous population of bacteria, some operators unnecessarily settle for 98% recovery and less.

7.2 Separation Efficiency versus Throughput

As we have seen above (Sect. 3.2), separation of particles by means of continuous centrifugation is rarely an absolute phenomenon. Working with biologically derived materials one often has to make a compromise between throughput and separation efficiency. This is true especially for cell debris, which is characterized by a wide variation in particle sizes and densities. The experiments reported below were designed so as to illustrate the relationship between throughput and separation efficiency for an industrial-type continuous centrifuge. They have previously been reported by DATAR (1985) and DATAR and ROSÉN (1987).

The experimental setup was that shown in Fig. 12. *E. coli* ATCC 15224, which expresses constitutively β-galactosidase, was cultivated at 37 °C in a 1000 L working volume fermentor and harvested in the late exponential phase when the cell concentration had reached 10.2 g/L dry weight (approximately 6% v/v). A continuous steam-sterilizable disk-bowl centrifuge (Alfa-Laval BTPX 205) was used for harvesting the cells and, again for clarification of the broth after disruption in a two-stage high-pressure homogenizer (Bran & Luebbe SHL 20).

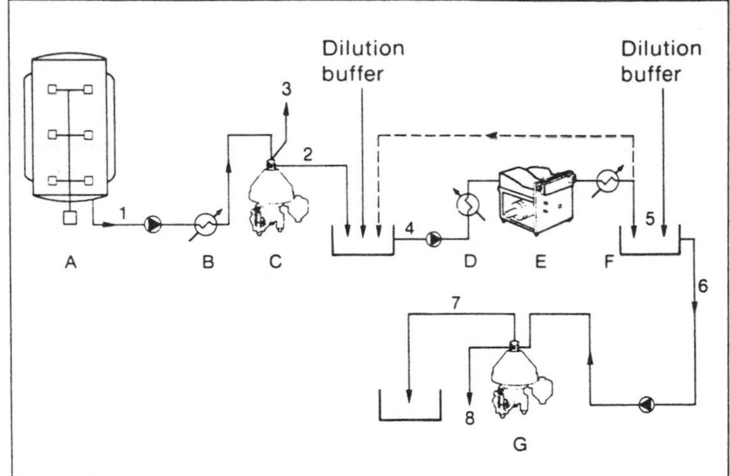

Fig. 12. Flow diagram showing the unit operations for centrifugal separation of *E. coli* cells and cell debris. **A,** 1500 L fermentor; **B, D, F,** heat exchangers; **C, G,** disk bowl centrifuge model BTPX 205 (Alfa-Laval); **E,** high-pressure homogenizer, model SHL 20 (Bran & Luebbe) (DATAR and ROSÉN, 1987).

In order to establish breakthrough curves, the 6 vol% bacterial broth was subjected to centrifugation at various throughputs. The timing for discharge of solids was controlled manually, and the escape of solids in the overflow was measured after various time intervals. Following cell harvest the concentrated slurry (approximately 50 vol%) was centrifuged at different flow rates. Temperature was maintained at 15 °C ± 1 °C.

Cells were disrupted by repeated discrete passes through the homogenizer at a flow rate of 200 L/h. One operating pressure (630 bar) and one- and two-stage homogenization were investigated. The inlet temperature was controlled at 5 °C, and the rise in temperature during homogenization was between 4 and 10 °C. Following the requisite number of passes through the homogenizer, the batches were centrifuged individually at various combinations of flow rates and solids concentrations. These experiments were carried out at 5 °C.

β-Galactosidase activity and protein concentration were assayed as described in the original publication. Complete removal of particles from suspensions prior to analysis was effected by means of an Alfa-Laval Cyrotester batch centrifuge run at 13000 g for 15 minutes. Viscosities were measured at 5 °C in a rotation viscometer (Gebrueder Haake Rotovisko RV2).

Fig. 13 shows the resulting homogenization as a function of the number of passes. This figure leads to an interesting observation. Residual packed solids decrease more or less linearly on a semi-logarithmic plot with the number of passes, while more than 80% of the β-galactosidase activity is released already in the first pass. Leakage of β-galactosidase from these cells is thus a sensitive indicator of cell damage. Cells are progressively fragmented with increasing numbers of passes.

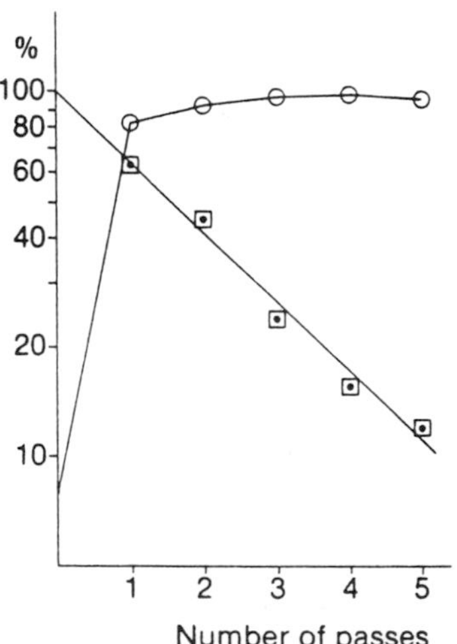

Fig. 13. Disruption of *Escherichia coli* in a high-pressure homogenizer, model SHL 20, using discrete passes at 630 bar. ○, enzyme release (%); ⊡, residual packed solids (% of original value).

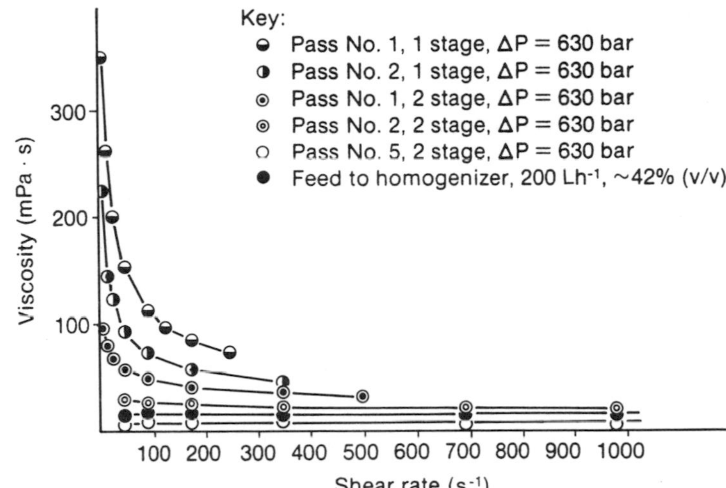

Fig. 14. The effect of one- and two-stage homogenization on the apparent viscosities of *E. coli* suspensions (DATAR and ROSÉN, 1987).

Viscosities are recorded in Fig. 14. The most interesting observation is the fact that shear obviously reduces viscosity. Five two-stage passes even reduce viscosity to below that of the feed to the homogenizer. This is in accordance with theory, which stipulates that DNA, which is released by homogenization, is fragmented by shear. Multiple passes through the homogenizer thus serve the dual purpose of releasing the intracellular contents and reducing the viscosity which, as we have seen above, Eq. (1), strongly influences the sedimentation rate of particles.

The effect of varying throughput on separation efficiency is illustrated in Fig. 15a, which shows the percentage of unsedimented solids in the overflow of the separator as a function of time following solids discharges at various throughputs. At flow rates up to 200 L/h, the breakthrough of solids occurs quite suddenly, making a sharp separation possible. However, with increasing flow rates, this breakthrough becomes more gradual. Plotted in a different way (Fig. 15b), data from the same set of ex-

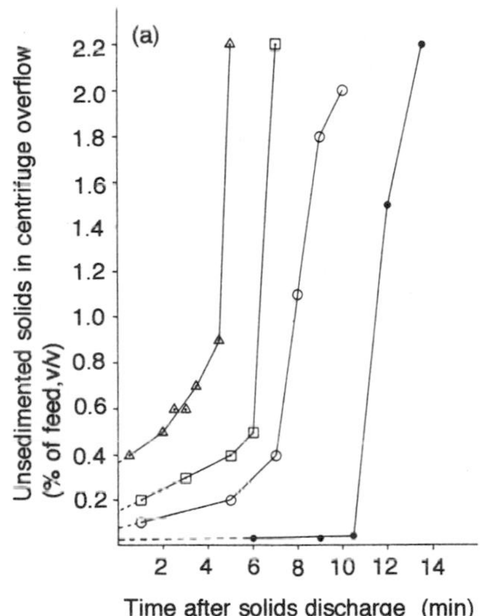

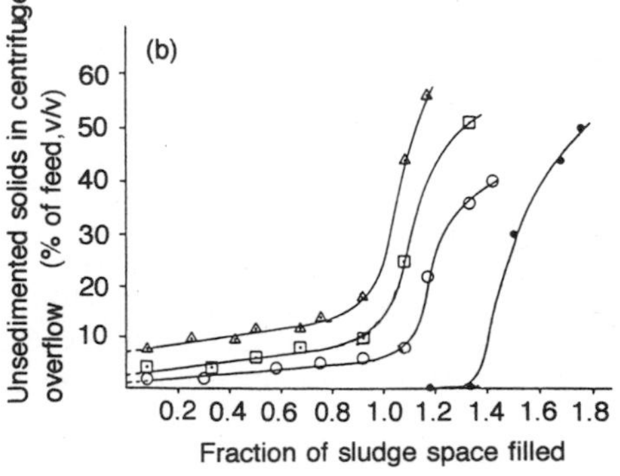

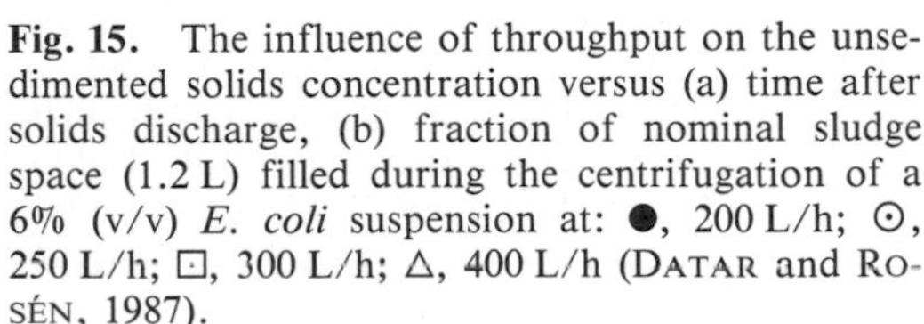

Fig. 15. The influence of throughput on the unsedimented solids concentration versus (a) time after solids discharge, (b) fraction of nominal sludge space (1.2 L) filled during the centrifugation of a 6% (v/v) *E. coli* suspension at: ●, 200 L/h; ⊙, 250 L/h; ⊡, 300 L/h; △, 400 L/h (DATAR and ROSÉN, 1987).

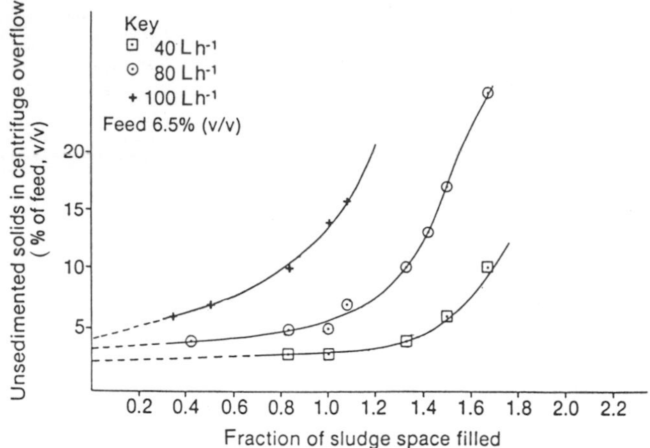

Fig. 16. The influence of throughput on the unsedimented solids concentration vs. fraction of nominal sludge space filled during a centrifugation in a BTPX 205 of a suspension of *E. coli* cell debris (DATAR and ROSÉN, 1987).

periments show unsedimented solids vs. fractions of nominal sludge space (1.2 L) filled during the centrifugation. This figure shows even more clearly that at flow rates up to 200 L/h, breakthrough occurs abruptly when the sludge space has become overfilled.

For cell debris the situation is different (Fig. 16). There, breakthrough is a gradual phenomenon even at a flow rate as low as 40 L/h, reflecting the wide variation in particle sizes and densities for that suspension. Even at very low flow rates some fraction of particles remains unsedimented. On theoretical grounds (DATAR and ROSÉN, 1987) it may be computed from these experiments that a flow rate as low as 10% of that sufficient for quantitative removal of cells will still leave 0.1% unsedimented solids in the overflow.

7.3 Scope for Enhanced Separation

Many different substances have been used to improve the efficiency of centrifugal separation of difficult materials such as cell debris (SANCHEZ-RUIZ, 1989). Some of the more common are polyethyleneimine, chitosan, EDTA, and calcium salts. Various synthetic polymers are also in use but are generally prohibitively expensive in large-scale operations. pH is often a determining factor in the flocculation behavior of protein-containing materials. Most procedures and results in this area are proprietary, but it seems that unsedimented solids as well as impurities such as nucleic acids may be reduced to a very small fraction of that obtained for a straight separation if flocculants are introduced prior to cell debris separation.

8 Industrial Application

Continuous cell separation by means of centrifugation is used in hundreds of different industrial applications.

Tab. 4 gives a survey of industrial separations using different types of centrifuges and the respective relative and maximum throughputs (partly after AXELSSON, 1985).

In pure bacterial systems, such as the genetically modified *E. coli* used as the model organisms in the experiments described above, with no pretreatment of the broth, throughputs larger than a few cubic meters per hour are rarely achieved. The most efficient specialized biotech centrifuges, such as Alfa-Laval's BTAX 215 and Westphalia's CSA 160, which have *g*-values of 15000 times gravity, generally give throughputs of 3000–4000 L/h.

Tab. 4. Cell Types and Throughputs for Some Industrial Separations (partly after AXELSSON, 1985)

Organism	Cell Size (nm)	Relative Throughput	Type of Centrifuge	Maximum Throughput (m^3/h)
Saccharomyces (baker's yeast)	7–10	100	Open nozzles	300
Saccharomyces (alcohol yeast)	5– 8	60	Open nozzles	300
			Self-ejecting	100
Candida	4– 7	50	Open nozzles	300
Aspergillus	(mycelium)	20–30	Self-ejecting	100
			Decanter	200
Actinomyces	10–20	7	Self-ejecting	100
Bacillus	1– 3	7	Self-ejecting	100
Clostridium	1– 3	5	Solid bowl	150

9 Final Remarks on Centrifugation in Cell and Cell Debris Separation

Centrifugation is a very efficient unit operation in most cell/broth separations, particularly when the cells are large and of high relative density. Flow rates can be as high as 200 m^3/h or more for yeast separation, and yields can be above 99.9%. However, with decreasing cell sizes and/or lower densities (relative to the surrounding medium), sedimentation may be too slow a process to be economically satisfactory.

Another difficult case is encountered with filamentous organisms, which may sediment too slowly due to their large cross-section (cf. Tab. 4). We have also seen that heterogeneous materials such as cell debris offer special difficulties during centrifugation. While there are clear-cut cases where centrifugation is the most practical and economic alternative in separation, there are also cases where filtration offers distinct advantages. Between these extremes there are cases for which there is no distinct advantage by either centrifugation or filtration.

This will become clearer after the description of theory and practice of tangential flow filtration (see below).

10 Crossflow Filtration

Crossflow filtration operates by applying a driving force across a semi-permeable membrane, while maintaining tangential flow of the feed stream parallel to the separation surface. The aim is to provide sufficient shear close to the membrane surface, thereby keeping solids and other particulate matter from settling on and within the membrane structure. In theory, recoveries of 100% are possible. Such tangential flow filtration is used for a variety of separation applications, and it is common to classify these membrane processes as follows:

- *microfiltration:* separation of particulates, typically 0.02–10 µm
- *ultrafiltration:* separation of particulates and polymeric solutes in the 0.001–0.02 µm range.
- *hyperfiltration:* also referred to as reverse osmosis, where ionic solutes, typically less than 0.001 µm, are separated.

Of these three membrane processes, microfiltration (MF) and ultrafiltration (UF) are most widely applicable in the primary recovery stages.

Crossflow microfiltration is commonly used to remove suspended particles from a process fluid and comprises operations such as the recovery of cells from fermentation broth and the clarification of cell debris homogenates. In this sense it is a unit operation that is comparable to centrifugation, and it provides the bioprocess engineer an interesting alternative. Both unit operations sclcctively remove particles from suspensions, albeit by different mechanisms. Centrifugation, as we have seen, relies on the exploitation of the differences in densities between the particles to be separated and the suspending medium. Membrane microfiltration, on the other hand, is based on the retention of particles by size, while allowing the passage of the solution through the membrane. In practice, both techniques can be used to effect separation on the basis of size and are to some extent interchangeable.

Ultrafiltration is an effective technique for concentrating or separating smaller particulate matter, as well as dissolved molecules of different sizes. As such, ultrafiltration has several applications in bioprocessing: (1) initial separation of cells from the fermentation or culture medium; (2) separation of cellular fragments from the medium following cell homogenization or lysis; (3) it may be used for order-of-magnitude fractionation of protein solutions; and (4) for recycling of biomass in continuous fermentation applications.

Theoretical aspects common to both microfiltration and ultrafiltration will be discussed first. Then the influence of process parameters on the performance of crossflow filtration, the types of membranes and hardware used, the scale-up of these unit operations and, finally, economics will be considered.

11 Background and Theory

11.1 The Nature of Secondary Membranes

During crossflow micro- and ultrafiltration invariably a 'secondary membrane' is formed, often also called the 'dynamic membrane' on top of the actual separating (or primary) membrane. The formation of this secondary membrane is due to the presence of colloidal and other macromolecular species in the feed.

The intrinsic properties of the primary membrane typically control the process only during the early stages of crossflow filtration. As filtration proceeds, formation of the secondary membrane continuously alters the flux and rejection characteristics of the primary membrane. This phenomenon has been known and is well documented for UF applications involving macromolecules.

Similar effects observed with microfiltration membranes are certainly more surprising. Intuitively, one would not expect this phenomenon to occur when a great difference exists between the molecular weight or size of a macromolecule and the pore size of the separating membrane. Crossflow microfiltration experiments (Fig. 17) with the enzyme β-galactosidase (MW ca. 500000) using a 0.2 μm polymeric membrane (Versapor® from Gelman Sciences, Ann Arbor, MI. 34) produced a de-

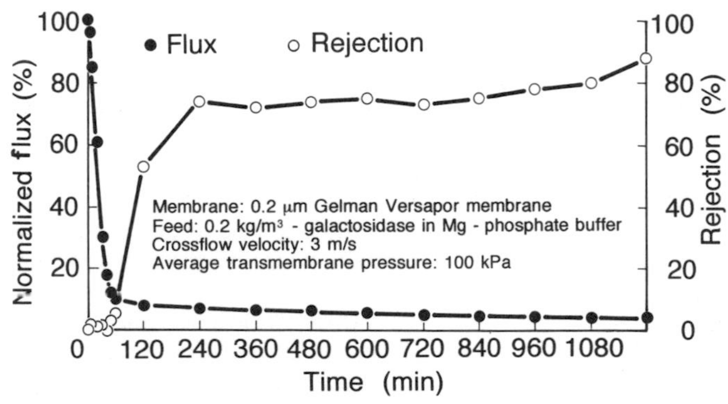

Fig. 17. Flux and rejection profiles during crossflow microfiltration (CFF) of β-galactosidase through a 0.2 μm membrane.

creasing flux and increasing rejection of the enzyme as filtration proceeded (DATAR, previously unpublished data from doctoral research). Eventually almost constant values of flux and rejection were obtained. The size difference between the enzyme and the membrane pore makes this result noteworthy. It is obvious that only the existence of a secondary membrane – formed by a protein layer of constant thickness – could cause this phenomenon. The formation of such a dynamic membrane is likely due to one or more of the following mechanisms: (a) protein deposition, (b) non-specific adsorption of denaturated enzyme, and/or (c) mechanical retention of aggregated denatured protein, probably caused by shear induced conformational changes during multiple pump passage.

TANNY (1978) has classified secondary membranes into three categories:

Class 1 formed by the rejection of macromolecules on the ultrafilter surface due to the formation of a gel layer (gel polarization).

Class 2 formed by colloidal suspensions whose size may not necessarily be close to that of the membrane. Pore blocking and bridging of pore entries are indicated (Standard law of filtration in combination with 'cake filtration' given by Ruth's law).

Class 3 formed when the pore size is very close to one of the dimensions of the molecule in solution. Flexible molecules can partly enter the pores and remain adsorbed to the walls, significantly changing flux/rejection characteristics.

Ultrafiltration is primarily concerned with Class 1-type dynamic membranes, whereas microfiltration can show characteristics of both Class 1- and Class 2-type dynamic membranes.

KRONER et al. (1984) found that the use of membranes with different pore sizes did not remarkably influence long-term filtration rates. DATAR (1984b) has reported that fluxes of *Escherichia coli* homogenates using different media – 0.45 and 0.2 μm – became the same after an initial filtration time. Until that point was reached, the quality of the filtrate through the more open media was hazy; this was also corroborated by considerably higher optical density readings. It is important, therefore, to fully characterize and understand the phenomenon of dynamic membrane formation so that system performance can be optimized, rather than be left to chance.

11.2 Models

11.2.1 Gel Polarization Theory

The modeling of crossflow filtration has proven to be a difficult task. Historically, the earliest models developed were those for reverse osmosis and ultrafiltration. These have been extensively dealt with by PORTER (1972). Modeling of the flux-transmembrane pressure relationship for UF has produced various approaches, including:

- *film theory* – applicable to the mass transfer controlled (MTC) region
- *resistance model* – applicable to both the pressure controlled and MTC regions
- *osmotic pressure model* – applicable to both regions

Over the years, considerable experimental data have been generated for ultrafiltration that appear to show a logarithmic relationship between the inlet concentration of the dispersed phase and the magnitude of the flux, J. The mechanism is illustrated in Fig. 18. Under all process conditions, the flux is usually several orders of magnitude smaller than the water flux through the membrane under identical conditions. To explain this behavior, it has been suggested that the flux is limited by the formation of a concentration polarization boundary layer consisting of a high concentration of solutes, which do not pass through the membrane. This layer – previously referred to as the 'dynamic' layer – provides an additional hydraulic resistance to flow. At steady state the boundary layer is in equilibrium; the rate of convection of the dispersed phase to the membrane is balanced by back diffusion into the bulk

$$J(x) \cdot C = \mathscr{D} \cdot (\mathrm{d}c/\mathrm{d}y) \tag{8}$$

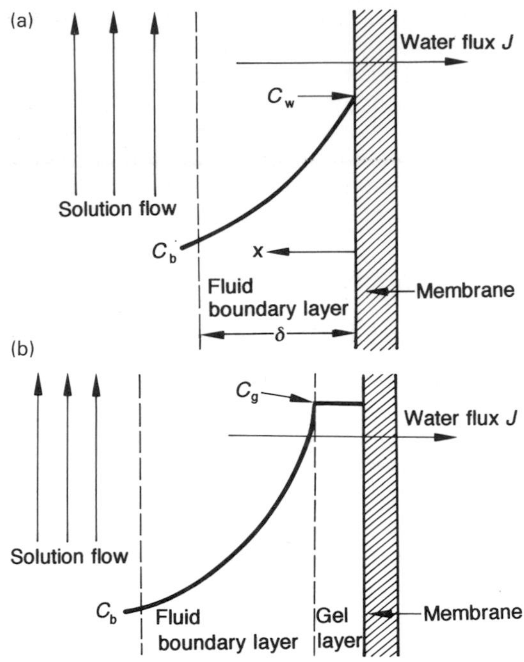

Fig. 18. Schematic of (a) macromolecular concentration polarization and (b) gel polarization at a membrane surface.

where $J(x)$ is the local filtrate flux at axial position x, C the concentration of dispersed phase, $\mathscr{D}$ the diffusion coefficient, and y the distance above the membrane. Eq. (8) can be integrated across the boundary layer to give

$$J = \frac{\mathscr{D}}{\delta(x)} \ln \frac{C_w}{C_b} = k(x) \cdot \ln \frac{C_w}{C_b} \quad (9)$$

where $\delta(x)$ is the local boundary layer thickness, $k(x)$ the local mass transfer coefficient, and C_w and C_b are the wall and bulk concentrations, respectively. As the flux increases, C_w also increases to its maximum value of C_g, the gel concentration. At this point, the flux attains its maximum, pressure-independent value, and the membrane is said to be 'gel-polarized'. The mass transfer coefficient can be evaluated from the solution of LEVEQUE (1938) for the case of laminar flow and constant C_w as

$$k(x) = 0.544 \left\{ \frac{\mathscr{D}^2 \cdot \gamma_w}{x} \right\}^{0.33} = \text{const } 1 \cdot \gamma_w^{0.33} \quad (9.1)$$

and for turbulent flow as

$$k = \text{const } 2 \cdot \gamma_w^{0.80} \quad (9.2)$$

where γ_w is the wall shear rate. The mathematical representation by Eq. (9) for boundary layer transport has proven very successful in the ultrafiltration of macromolecules (BLATT et al., 1970), but quantitative agreement with experimental results for the crossflow filtration of colloidal and particulate suspensions has been poor.

Relation to Darcy's Equation

At pressures below the region where the flux begins to level off, the flux increases roughly linearly with ΔP_{TM} for small pressure drops. Expressions of the form of Eq. (9) can be used to assess the relative magnitudes of the different contributions to the resistance to flow:

$$J = B \cdot \frac{\Delta P_{TM}}{\mu \cdot \Sigma R} \quad (10)$$

where B is Darcy's permeability constant, μ is the permeate viscosity, $R = R_m + R_g + R_p + R_c$, with R_m, the hydraulic resistance of the membrane, R_g, the hydraulic resistance of the gel layer, R_p, the hydraulic resistance of the polarization layer, and R_c, the hydraulic resistance of the deposited solids layer.
By combining Eqs. (8) and (9), one obtains

$$\ln \frac{C_w}{C_b} = \frac{B \cdot \Delta P_{TM}}{k \cdot \mu (R_m + R_g + R_p + R_c)} \quad (11)$$

Each of the resistances identified are functions of the operating conditions, feed stream properties, and the time (e.g., R_c is a function of the layer thickness, porosity as well as effective particle size). Not all of the resistances may be operative all the time. For instance, in the crossflow filtration of pure water, resistance to flow is only due to that of the membrane (R_m). With real process streams at moderate tangential velocities and transmembrane pressures, R_g may be negligible but R_p could be dominating.

As long as the concentration C_w is less than C_g, C_w increases with pressure. However, when C_w becomes equal to C_g, further increases in the transmembrane pressure can cause compaction of the gel layer and the layer of deposited solids which consequently increases the values of R_g and R_c. At that point the flux begins to level off and no longer varies with the pressure. If, however, the tangential velocity is sufficiently high, the thicknesses of the polarization and deposited layers will be minimized, thereby enabling the flux to remain at a high value. At low tangential velocities the flux is often not a strong function of pressure.

The debate on the validity of Eq. (9) continues, since it is argued that in crossflow filtration mass transfer does not only occur through diffusion, but shear effects need to be taken into account as well. Whatever the case, crossflow filtration is less sensitive to concentration and thus there is seldom a need for prefiltration of the feed. The ability to filter liquids containing a dispersed phase of almost any arbitrary fineness, and within a broad concentration interval, is a unique property of crossflow filtration that lends it to the selection of suitable applications.

11.2.2 Sedimentary Pipe Flow and Fluidization Theory

A more stringent approach to modeling combines sedimentary pipe flow expressions with those of Darcy's law. For a suspension flowing in a horizontal pipe, the settling force is gravity while the driving force for particle transport originates in the flow velocity due to momentum transfer. In crossflow filtration, the settling force is the drag experienced by the particles due to the flow of the filtrate through the filter membrane. Standard fluidization theory provides an estimate of the equivalent gravity from the following equation

$$\Delta P = (1-\varepsilon)\cdot(\rho_s - \rho_l)\cdot S \cdot g' \quad (12)$$

where ΔP is the pressure drop across the bed, ε the porosity of bed, ρ_s the density of solid, ρ_l the density of liquid, S the thickness of bed, and g' the corrected gravity value for particle sedimentation.

Re-arranging Eq. (12) provides

$$g' = (\Delta P)/[(1-\varepsilon)(\rho_s - \rho_l)\cdot S] \quad (12.1)$$

If the effects of real gravity are ignored due to the small particle sizes, the sedimentation force is an inverse function of the bed (or cake) thickness. Counteracting this settling force are the dispersion forces due to fluid turbulence, along with other factors. Turbulent eddies play a major role in constantly "lifting" the particles into the fluid stream (lift forces) and keeping it dispersed.

A fundamental problem of suspension flow in horizontal pipes is the evaluation of the forces exerted on the individual solid particle by the fluid and the relation to the actual particle behavior. Particles travel with a velocity only slightly lower than that of the liquid. At low linear velocities, a bed forms at the bottom of the pipe due to sedimentation; intermediate regions from the bottom of the bed to the center line are marked by a "sliding bed flow". NEWITT (1955) found that for each value of suspension concentration there is a critical velocity which marks the transition between flow with a stationary bed and fully suspended flow. DURAND (1953) developed an expression for the critical velocity v, below which a stationary bed forms. For a cylinder with diameter D and in gravity field without crossflow, this expression is

$$v = 2.43\,[2g'\cdot D\cdot(\rho_s/\rho_l - 1)]^{0.5}\cdot(C^{0.33}/c_d^{0.25}) \quad (13)$$

where C is the volumetric concentration of solids and c_d is the drag coefficient for a particle in free fall, and defined as

$$c_d = 21/Re_p + 6/Re_p^{0.5} + 0.28 \quad (14)$$

and

$$Re_p = (d_p\cdot v\cdot \rho_l)/\mu \quad (15)$$

where Re_p is the particle Reynolds number ($0.1 < Re < 4000$), μ the liquid viscosity, and d_p the particle diameter.

It is this transitional flow phenomenon which has been used in conjunction with Darcy's law to explain, at least partly, certain behavioral aspects of crossflow filtration. MURKES and CARLSSON (1988) have combined the equations for critical velocity, Eqs. (11–15) and solved the resulting equation for flux J in Darcy's equation, to produce the following relation:

$$J=\frac{B\cdot\Delta P_{\mathrm{TM}}}{\mu}\cdot\left\{\frac{(c^2\cdot\Delta P_{\mathrm{TM}})}{(a\cdot v^3+b\cdot v^{3.5}+0.28\,v^4)^{0.5}}+R_{\mathrm{m}}\right\}^{-1} \tag{16}$$

where

$$a=21\,\mu/d_{\mathrm{p}}\cdot\rho_{\mathrm{l}} \tag{16.1}$$
$$b=6(a/21)^{0.5} \tag{16.2}$$
$$c=2.43\,C^{0.33}\,[2D/(1-\varepsilon)\,\rho_{\mathrm{l}}]^{0.5} \tag{16.3}$$

Although Eq. (16) does not give quantitatively exact results, it seems to qualitatively describe the phenomena of crossflow filtration as will be seen in Sect. 12.

11.2.3 Incorporation of a Shear-Enhanced Diffusion Coefficient

It has been proposed (ZYDNEY and COLTON, 1982) that discrepancies between experimental data and theoretical predictions for the crossflow filtration of colloidal and/or particulate matter can be eliminated by incorporation of a shear-enhanced diffusion coefficient that accounts for the lateral migration of larger particles. Essentially, the diffusivity term in Eq. (9) is "adjusted" for the effect(s) of wall shear rate on the particle size moving in an axial direction. Microeddies generated in the wake of larger particles such as red blood cells tend to provide a lift force which acts in conjunction with the back-diffusion of the particle, keeping it from depositing on the surface of the wall. In an earlier experimental study of this effect, ECKSTEIN et al. (1972) had developed the following correlation for the enhanced diffusion coefficient (valid for $C_{\mathrm{b}}>0.20$)

$$\mathscr{D}=0.025\cdot\phi^2\cdot\gamma \tag{17}$$

where ϕ is the particle radius. ZYDNEY and COLTON (1982) substituted this diffusion coefficient into Eqs. (9) and (9.1) which, on integrating, lead to:

$$J=0.070\,\{(\phi^4/L)\}^{0.33}\cdot\gamma_{\mathrm{w}}\cdot\ln(C_{\mathrm{w}}/C_{\mathrm{b}}) \tag{18}$$

where J is the length-averaged flux with γ_{w} and C_{b} assumed constant. Experimental data for the plasma filtrate flux covering a wide range of operating conditions were found to be in good agreement with theoretical predictions. DATAR (unpublished results) tried to adapt the ZYDNEY and COLTON model, Eq. (18), to other systems with mixed results: The crossflow filtration of baker's yeast produced flux data that were in reasonable agreement with those obtained for the filtration of red blood cells. However, for considerably smaller, and nonspherical cells such as *Escherichia coli,* the model could not explain the results obtained.

A theory that combines the polarization effects of solutes (macromolecular and low molecular weight species) with the cake filtration effects of suspended particles (both depth and surface phenomena) and which incorporates the rheological properties of the retentate and deposited layer of solids appears to be the best approach.

12 Relevant Process Parameters

Unlike centrifugation, specific interactions play a significant role in membrane separation processes. These interactions may include particle/particle, particle/membrane, solute/membrane, particle/medium, and membrane/medium. The most important factors to consider are:

- the crossflow (or tangential) velocity, DPCF
- the driving pressure, DPTM
- the separation characteristics of the membrane in terms of permeability and pore size
- the size of particulates relative to the pore size of the membrane
- the hydrodynamic conditions within the flow module.

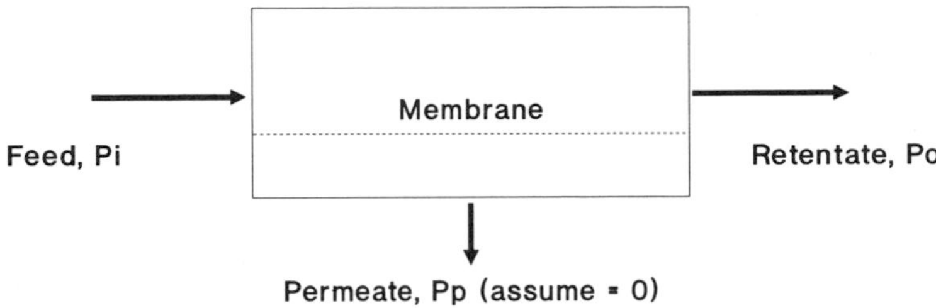

Fig. 19. Definition and principle of tangential flow filtration (TFF).

del Pcf = Pi - Po where del Pcf = pressure drop along channel

del Ptm = [(Pi + Po)/2] - Pp del Ptm = transmembrane pressure

del Ptm = Pi - (del Pcf/2)

Fig. 19 shows a schematic of a tangential flow filtration process and defines the important operating parameters. The permeate flux is a function of various factors:

- ΔP_{TM}, the transmembrane pressure drop
- crossflow (or tangential) velocity, related to ΔP_{CF}, the pressure drop along the flow channel
- hydrodynamic conditions – channel geometry and shear rate
- mass transfer coefficient and thickness of deposit
- solids concentration of retentate
- temperature
- rheological properties of the retentate
- membrane characteristics.

It is generally observed that the permeate flux during filtration of particulate suspensions is less concentration-dependent than during the ultrafiltration of molecules. Consequently, the microfiltration of particulates is likely to exhibit features of both ultrafiltration, e.g., gel polarization, and of "dead-end" filtration, e.g., cake build-up. The concentration polarization phenomenon associated with the TFF of macromolecules, as shown in Fig. 18, is superimposed on the deposition of particles over the surface of the membrane (Fig. 20).

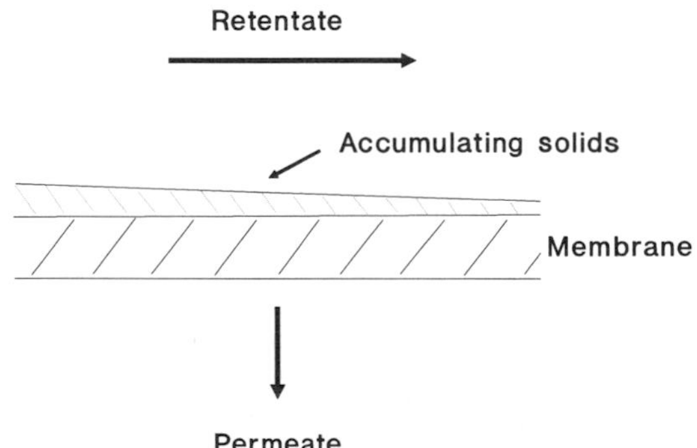

Fig. 20. Schematic of solids build-up during tangential flow filtration (TFF).

12.1 Filtrate Flux versus Transmembrane Pressure Drop

For small pressure drops, the flux increases linearly with the transmembrane pressure whereas for large transmembrane pressures the flux approaches a constant asymptote, due to the increased values of R_g and R_c (Fig. 21).

The actual relationship between J and ΔP_{TM} varies from one system to another and is usually affected by both the rheological characteristics of the retentate and the fouling properties of the suspension.

12.2 Filtrate Flux versus Tangential Flow Velocity

When flux versus flow velocity is plotted from Eq. (17), for small flow rates J increases as $v^{1.5}$, whereas for high flow rates the resistances offered by R_c, R_p and R_g become infin-

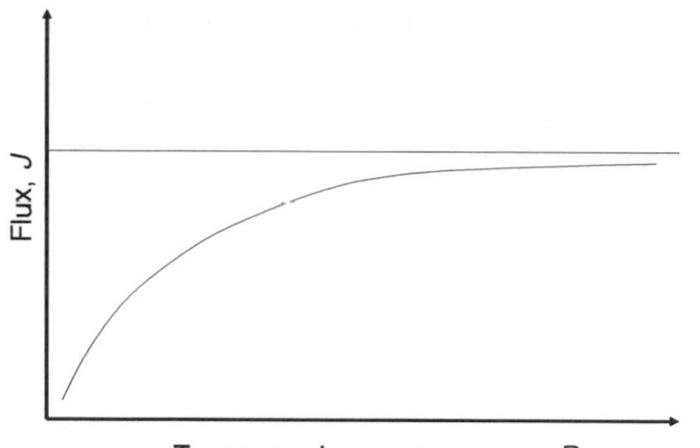

Fig. 21. Filtrate flux versus pressure drop in tangential flow filtration (TFF).

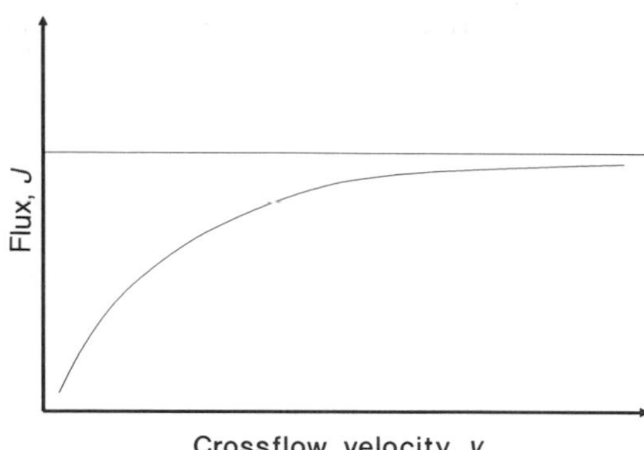

Fig. 22. Filtrate flux versus crossflow velocity in tangential flow filtration (TFF).

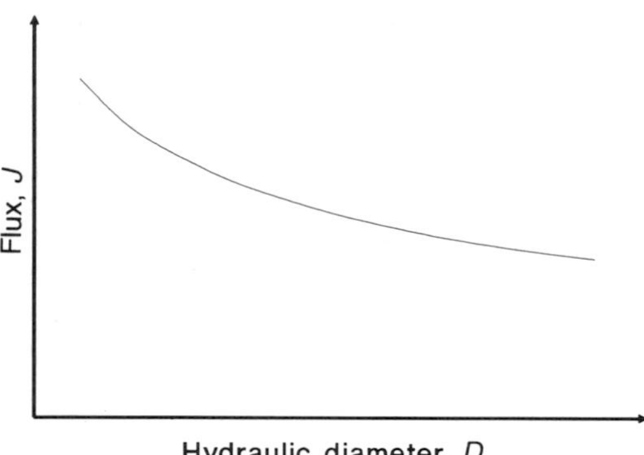

Fig. 23. Filtrate flux versus hydraulic diameter in tangential flow filtration (TFF).

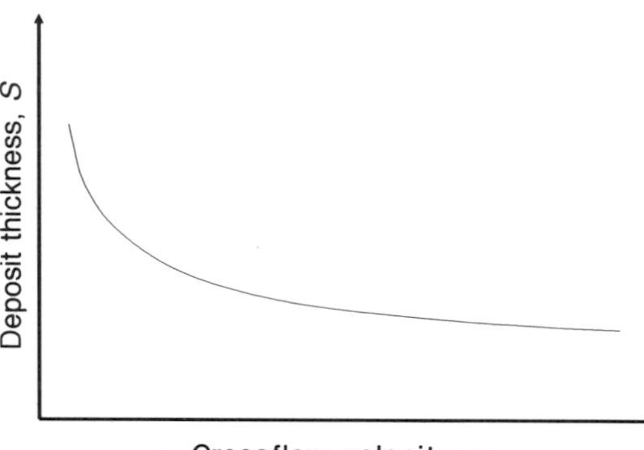

Fig. 24. Deposit thickness versus flow velocity in tangential flow filtration (TFF).

itely small and the flux should be limited by the resistance of the filter medium (R_m) alone, giving a constant asymptote which is independent of v (Fig. 22)

$$J = \frac{B \cdot \Delta P_{TM}}{\mu \cdot R_m} \tag{19}$$

12.3 Filtrate Flux versus Hydraulic Diameter of Flow Channel

Substituting the equation for c, Eq. (16.3), into Eq. (16) gives the dependence of flux on the tube diameter

$$J = \frac{k_1 \cdot \Delta P_{TM}}{k_2 \cdot \Delta P_{TM} \cdot D + R_m} \tag{20}$$

The relation: $J = f(D)$ is illustrated in Fig. 23.

12.4 Cake Deposit Thickness versus Flow Velocity

Substituting Eq. (11) into Eq. (13) and solving for deposit thickness, S, gives

$$S = \frac{c^2 \cdot \Delta P_{TM}}{(a v^3 + b v^{3.5} + 0.28 v^4)^{0.5}} \tag{21}$$

A generalized plot of S versus flow velocity, v, is illustrated in Fig. 24. For small velocities, the second and third terms in the denominator are negligible and the deposit thickness decreases with increasing velocity as

$$S = \text{const} \cdot \Delta P_{TM} / v^{1.5} \tag{21.1}$$

For $v \rightarrow$ infinity, the third term in the denominator dominates, and the deposit thickness decreases with increasing velocity as

$$S = \text{const} \cdot \Delta P_{TM}/v^2 \quad (21.2)$$

12.5 Filtrate Flux versus Solids Concentration of Retentate

The solids concentration influences the flux because of its effect on the rheological properties of the retentate. The flux decreases with increasing solids content because of an increase in the solids deposited at the membrane surface due to a reduction in both the back diffusion and the mass transfer rates. Fig. 25 shows that there can be a significant difference between the behavior of tangential flow ultrafiltration (TFUF) and tangential flow microfiltration (TFMF) processes – microfiltration often gives a sigmoidal type curve.

12.6 Effect of Temperature

All other things being constant, an increase in the temperature of the feed stream will lead to an increase in the permeate flux because of a reduction in the permeate viscosity. In general, biological products are sensitive to heat. Certain stronger enzymes such as β-galactosidase are able to withstand temperatures up to 60 °C; however, at these elevated temperatures other undesirable effects such as cell lysis and precipitation of solutes could severely foul the membrane.

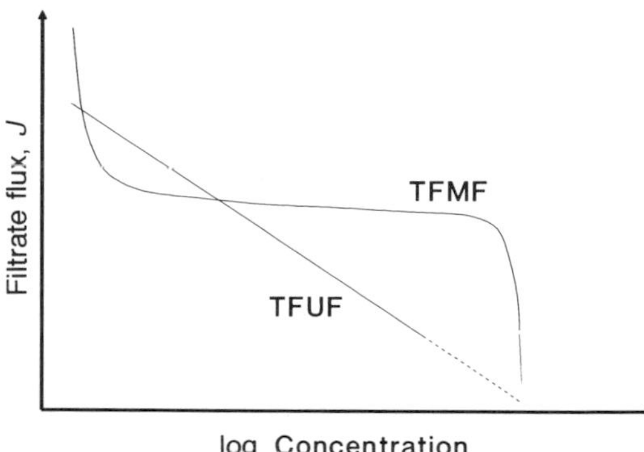

Fig. 25. Relationship between flux and concentration in tangential flow filtration (TFF). TFMF, tangential flow microfiltration; TFUF, tangential flow ultrafiltration.

12.7 Rheological Properties

Solutes which pass through the membrane directly affect the flux by virtue of their influence on the viscosity of the liquid. Solutes that fail to pass through but which increase the viscosity of the retentate will lower permeation rates either by reducing the back diffusion rate and/or by increasing the resistance of the gel layer. In addition, the viscosity/solids concentration relationship influences the performance of TFF by affecting pumping costs, as also the maximum attainable solids content. In biological fluids, increasing solids content typically changes the flow characteristics from Newtonian to non-Newtonian behavior, with concomitant changes in the tangential flow velocity, the Reynolds number, and hence the thicknesses of the resistance layers.

12.8 Membrane Characteristics

The following properties of the primary membrane significantly affect the performance of tangential flow filtration (TFF):

- *pore size distribution* – the sizes of the pores and their distribution relative to the particles determine whether the pores become blocked;
- *permeability* other things being equal, the higher the permeability the greater the flux. A high value reflects large pore sizes, a broad pore size distribution and/or a high porosity;
- *porosity* – a measure of the openness of the structure;
- *thickness* – the greater the thickness the greater the resistance to liquid flow;
- *surface chemistry* – the nature of the membrane surface has a major impact on the extent to which it interacts with solutes and particulates in the process stream.

13 Time-Dependent Flux Decline

With few exceptions the permeate flux will decrease with time. The loss of flux when other variables are held constant is called membrane fouling. This behavior is detrimental to TFF operation and is caused by effects such as convective deposition, specific and non-specific adsorption and cake build-up. Some of these effects are reversible while others are not. The main contributing factors to loss of flux are:

(1) Changes in the properties of the membrane-compaction due to applied pressure or chemical deterioration can irreversibly reduce the pore sizes and porosity of the membrane.
(2) Concentration polarization – reversible.
(3) Membrane fouling by specific species – effectively irreversible.

Appropriate selection of the membrane can minimize the effects due to time-dependent changes in the physical properties. The effects of concentration polarization are generally reversible through altering the transmembrane pressure, the feed concentration, or the mass transfer coefficient. Fouling by specific species, however, is effectively irreversible and is due to

- the deposition/accumulation of colloidal particles within the pores and on the surface of the membrane
- the precipitation of small solutes on the surface and within the pores of the membrane
- the adsorption of macromolecular solutes on the surface and within the pores of the membrane.

Effective cleaning procedures are able to maintain relatively higher flux rates (DATAR, 1985a) although a continuous and long-term decline is inevitable (Tab. 5). The majority of the constituents of a process stream will foul a membrane to some extent. These can be generally classified as:

- low molecular weight inorganic compounds, e.g., salts which can form complexes or precipitate as a result of changes in the chemistry of the medium;
- low molecular weight organic compounds, e.g., antifoam additives, especially those that are silicon-based, and other surfactants;
- natural macromolecular species, e.g., DNA/RNA, proteins, lipids and polysaccharides;
- polymeric species such as flocculants;
- colloidal and larger particulates, e.g., fatty oils, cellular fragments and other debris.

Tab. 5. Effect of Cleaning on Pure Water Fluxes (DATAR, 1985a)

Experimental Sequence[a] Expt. Number	Wash No.	Pure Water Flux[b] ($m^3/h \cdot m^2$)
New membrane	—	21.0 ± 19%
1	1	18.0 ± 14%
2	2	17.2 ± 10%
3	3	17.0 ± 7%
4	4	16.9 ± 6%

System 1: Fresh *E. coli* homogenate used for every experiment.
System 2: TFF, with 0.45 μm Gelman Versapor membrane, under turbulent flow conditions.
[a] Wash no. 1 is performed after except. no. 1, wash no. 2 after expt. no. 2, and so on.
[b] Pure water flux measured after following cleaning procedure:
(1) a 10-minute rinse with de-ionized water at 60 °C;
(2) a 30-minute recirculating wash with 2% NaOH solution at 60 °C;
(3) two 10-minute rinses with water at 60 °C and 25 °C, respectively.

Membrane fouling not only reduces the permeate flux but can also affect the rejection properties of the primary membrane. In the following section the operational parameters that need to be controlled in order to improve the overall performance of crossflow filtration will be described.

14 Improving Crossflow Filtration Performance

Fig. 26 provides a general guide to the relationship between operating parameters and permeate flux, *J*. The relationship shown is typical for UF. It applies also to MF operation although the mass transfer controlled region is less marked. From an operational perspective, the throughput and solute transmission performance of crossflow filtration can be improved by (1) operating at a transmembrane pressure in the region of the change from pressure control to mass transfer control; (2) operating under suitable flow conditions; (3) minimizing the causes of time-dependent flux decline; (4) correct selection of membrane; (5) correct selection of the configuration.

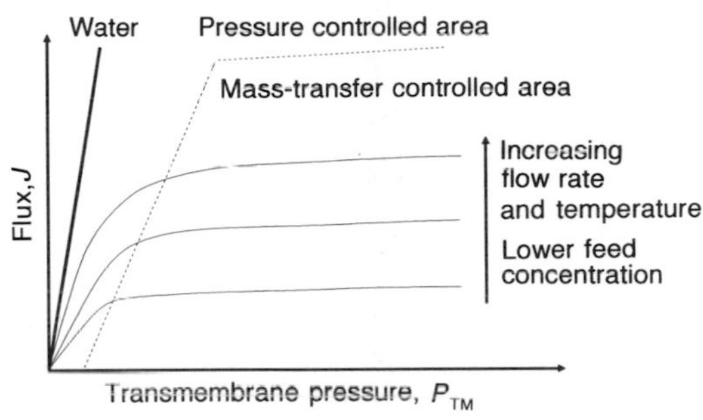

Fig. 26. Generalized correlation between flux and operating parameters indicating areas of control.

14.1 Transmembrane Pressure

Operating at too high a pressure can increase the detrimental effects due to concentration polarization and can cause compaction of the deposited layers of solids. The optimum pressure is often the minimum pressure within the mass transfer controlled region. However, as shown in Fig. 26, the tangential flow velocity will affect this value.

14.2 Flow Conditions

High tangential velocities reduce the thickness of the polarization layers thereby minimizing the accumulation of solids at the membrane surface. The maximum feasible velocity will depend on the channel size, the pumping costs, and the maximum pressure drop which the configuration can accept along the length of the flow channels.

Conventional TFF systems typically operate at tangential velocities between 2 and 6 m/s. As the retentate concentration builds up (see Sect. 11.5), it is inevitable that the changing viscosity will lower both the flux and product transmission rates. Recent developments in hardware technology (LOPEZ-LEIVA, 1979; GOLDINGER et al., 1986; KRONER, 1990) permit membrane face velocities upwards of 20 m/s. The differences in performance between the conventional (or low to moderate shear) TFF systems, and the newer developments, are illustrated in Tab. 6 (DATAR, 1985a; KRONER et al., 1987). It is expected that these newer filtration techniques will gain prominence in the coming years.

14.3 Minimization of Time-Dependent Flux Decline

The long-term loss of flux can be minimized in a number of ways as discussed below.

(1) Boundary layer losses associated with concentration polarization can be reduced by

- reducing the solids concentration in the feed, although this is likely to also reduce the throughput
- reducing the transmembrane pressure in accordance with Sect. 14.1
- reducing the concentration of solutes at the membrane surface mainly by
  - introducing turbulence promoters adjacent to the surface
  - increasing the mass transfer coefficient through increased shear rate and back diffusivity (e.g, by reducing channel size and length).

(2) Structural changes associated with the membrane can be controlled by

Tab. 6. Comparison of Enzyme Transmission Rates Between TFF and Rotary Shear Filters (DATAR, 1985b; KRONER et al., 1987)

Enzyme	Molecular Weight (×1000)	Microorganism	Source	Tangential Flow Filtration Transmission (%)	Tangential Flow Filtration Flux (L/h·m²)	Rotary Shear Filtration Transmission (%)	Rotary Shear Filtration Flux (L/h·m²)
Protease[a]	25	*Bacillus licheniformis*	Extracellular	80	33	94	70
Formate dehydrogenase	78	*Candida boidinii*	Cell lysate	40	15	80	60
Lactate dehydrogenase	168	*Lactobacillus confusus*	Cell lysate	40	21	70	60
Fumarase	180	*Brevibacterium ammoniagenes*	Cell lysate	25	11	85	35
Aspartase	190	*Escherichia coli*	Cell lysate	30	18	75	50
Leucine dehydrogenase	310	*Bacillus cereus*	Cell lysate	20	25	50	65
β-Galactosidase[b]	540	*Escherichia coli*	Cell lysate	14	20	95	77

Data summarized from different experiments as indicated in references. Biomass concentration 10–20% wet w/w. All experiments with 0.2 μm membrane except where stated.

[a] Whole broth concentration

[b] Experiment with 0.45 μm membrane

- suitable membrane selection
- appropriate pre-conditioning of the membrane
- use of cleaning agents and methods that do not cause excessive deterioration of polymeric membranes.

(3) Loss of flux associated with membrane fouling can be minimized by

- correct selection of the membrane and its properties, in particular the pore size distribution, surface porosity, roughness and chemistry
- modifying the temperature, pH and/or ionic strength of the process stream to reduce the degree of fouling by adsorptive phenomena, or by adding surface-active compounds that might remove a particular foulant
- chemical modification of the membrane's surface to increase the wettability so that hydrophobic foulants are less likely to adsorb by hydrophobic interactions
- selection of a suitable cleaning-in-place procedure
- use of optimum hydrodynamic conditions.

The best ways to minimize loss of flux is determined by the particular process, but operating at high tangential flow velocity and designing an appropriate cleaning procedure are the most common solutions.

14.4 Membrane Selection

As indicated earlier, the choice of the membrane is a significant factor in the performance of the process. The membrane not only governs the permeate flux and rejection characteristics, but also the degree of fouling and subsequent ease of cleaning.

14.5 Hardware Selection

A list of characteristics of membrane modules from selected manufacturers appears in Tab. 7, while the advantages and disadvantages of different configurations are indicated

Tab. 7. Characteristics of Some Popular TFF Modules

Manufacturer	Trade Name	Typical Configurations	Flow Characteristics Hydrodynamics	Channel Size	Maximum Temperature (°C)
A/G Technology	A/G Technology	Hollow fiber	Mixed-flow	Narrow	
Alfa-Laval	RO-module	Circular disks	Turbulence-promoted	Wide	
Amicon/Grace	Diaflo	Hollow fiber	Mixed-flow	Narrow	75
	Diaflo	Spiral wound	Turbulence-promoted	Narrow	Autoclavable
	Thin Channel	Spiral channel	Laminar	Narrow	—
DDS	Type 36	Flat sheet	Laminar	Narrow	100
Dorr Oliver	Ioplate	Flat sheet	Turbulent	Narrow	70
Enka	Microdyn	Tubular	Turbulent	Wide	Autoclavable
Filtron	Minisette	Flat sheet	Turbulence-promoted	Narrow	70–121
	Centrisette	Flat sheet	Mixed-flow	Narrow-wide	70–121
Koch	Abcor	Spiral wound	Turbulence-promoted	Narrow	85
	Abcor	Tubular	Turbulent	Wide	85
Millipore	Pellicon	Flat sheet	Turbulence-promoted	Narrow	Autoclavable
	Prostack	Flat sheet	Mixed-flow	Narrow-wide	121
	PUF	Spiral wound	Turbulence-promoted	Narrow	—
Rhone-Poulenc	Pleiade	Flat sheet	Mixed-flow	Narrow	—
Romicon	Romicon	Hollow fiber	Mixed-flow	Narrow	75
	Romicon	Tubular	Turbulent	Wide	140
Sartorius	Sartocon	Flat sheet	Mixed-flow	Narrow-wide	121
SFEC Carbosep		Tubular	Turbulent	Wide	121

in Tab. 8. Given the wide variety of options available, it is important to select the correct configuration during the design stage itself.

15 Scale-Up

The modular design of TFF configurations permits scale-up from laboratory- through pilot- to full-scale operation to be performed relatively easily. It is important, nevertheless, to ensure that the following variables be kept constant during scale-up

- inlet and outlet pressures
- tangential flow velocity
- flow channel sizes
- feed stream properties – test slurries should be representative of the actual process streams
- membrane type and configuration – test data from one design cannot directly be used to design another geometry.

It is also necessary to assess operating life times including cleaning cycles, so as to allow for the effects of membrane fouling. As indicated earlier, multiple passes through pumps (and pipework) can affect feed stream proper-

Tab. 8. Advantages and Disadvantages of TFF Module Configurations

Configuration	Pros	Cons
Tubular	Wide channels less prone to blockage Crossflow velocity between 2–6 m/s Reynolds number >10000 Easy to clean Membrane replacement simple	Low surface area to volume ratio High liquid hold-up High energy consumption
Hollow fiber	High wall shear rates (4000–14000 s^{-1}) High surface area to volume ratio Low liquid hold-up Low energy consumption Can be backflushed	Narrow channels more prone to blockage Crossflow velocity between 0.5–2.5 m/s Reynolds number 500–3000 Maximum operating pressure limited to approx. 2 bar Membrane replacement cost high
Spiral wound	High surface area to volume ratio Low capital cost Reasonably economical	Narrow channels more prone to blockage Design prone to collapse Requires clean feed streams
Flat sheet	Membrane replacement relatively simple Visual observation of permeate from each membrane pair Moderate energy consumption	Cannot be backflushed Low surface area to volume ratio Relatively high capital cost

ties; it is important, therefore, to take these into consideration during the pilot-scale trials.

16 Comparative Economics

The comparison of the economics of using different separation methods for a particular feed stream is rarely straightforward, because the evaluations of specific contributions to the overall cost are often not based on equivalent criteria. In addition, it is not advisable to apply conclusions from one costing comparison to a different process, as valid economic analysis can only be made given adequate laboratory and/or pilot data.

It is possible, nevertheless, to provide general guidelines on the relative contributions to the overall cost of various items. The following list provides an indication of the most relevant factors

(1) capital cost and depreciation
(2) labor costs
(3) maintenance costs
(4) utilities – energy, water, etc.
(5) overhead costs – related to operating, supervision and maintenance labor
(6) patents and royalties – where relevant
(7) taxes and insurance.

First, centrifugation costs alone shall be considered. An important factor in selecting centrifuges is the relationship between particle size, d^2 in Eq. (1), and ease of recovery. As a rule of thumb, the smaller the particles the higher the recovery costs. This is clearly seen in Fig. 27. A quick calculation shows when conventional (i.e., low to moderate shear) TFF may not be economically feasible. Large-volume applications are, for example, typically dominated by centrifuges. In the case of baker's yeast concentration, continuous nozzle centrifuges like Alfa-Laval's FEUX 512 model can process a batch at approximately 50 m^3/h. The capital cost of this centrifuge in 1990 would have been in the range of US $ 150000. A realistic long-term flux rate for this separation using a 0.45 μm-based TFF system would

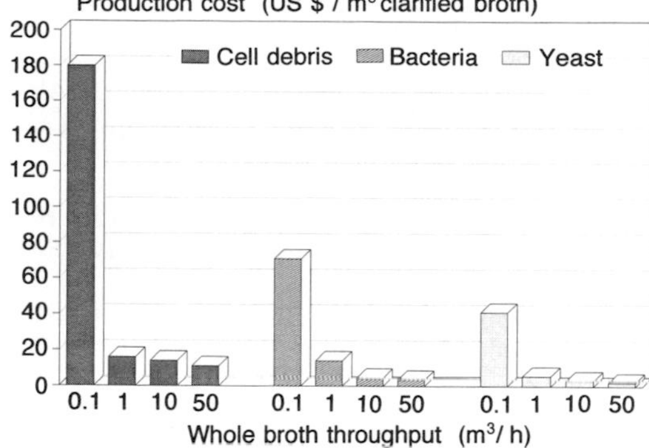

Fig. 27. Centrifugation costs for production of clarified broth (ca. 10% v/v in whole broth).

be about 100 L/(h·m^2). On a unit area basis, an average capital cost for tangential flow filtration (TFF) is $ 3500 per m^2, and the process will require 500 m^2 of membrane area. The capital cost of such a TFF installation would therefore be $ 1.75 million, or an order of magnitude higher than that of the centrifuge.

For volumes less than about 3000 L and/or smaller particle sizes (e.g., *Escherichia coli* bacteria, 1–2 μm), the economics can begin to shift in favor of TFF (DATAR, 1985a). Nevertheless, given the present flux rates achievable with TFF systems, the scale of operation will continue to be an important consideration. The picture changes, however, for bacterial cell debris processing (typically around 0.2 μm and hundreds of liters), where TFF is likely to be less expensive. Figs. 28 and 29 illustrate the relative costs for centrifugation and crossflow microfiltration for cell harvesting and cell debris separation, respectively.

Most current mammalian cell process volumes are of the order of a few hundred liters for batch operations; daily flow rates for continuous processes tend to be even lower. An important concern is the sensitivity of these cells to shear denaturation. Consequently, if the cells cannot be discarded, it is rare that a centrifuge will be used. Based on the authors' experience, TFF techniques dominate this application. Re-circulating pumps are normally over-sized and used at reduced capacities, so as not to destroy the cells by shear. While this procedure allows for lower tangential velocities, it can lead to lowered throughputs. The problem is easily countered by using additional membrane modules. Smaller process volumes, combined with relatively clean feed streams makes this an inexpensive option.

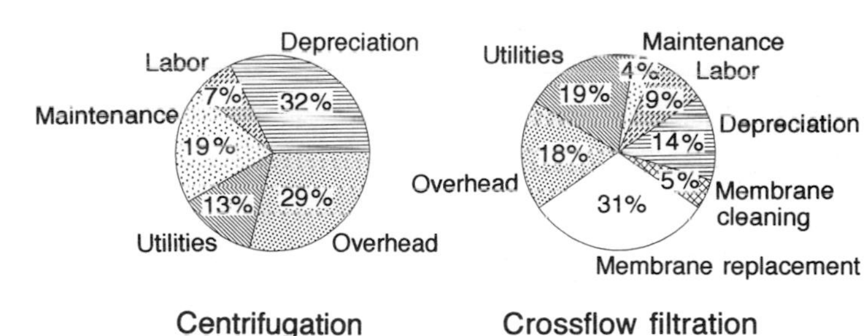

Fig. 28. Typical relative costs for harvest of *E. coli* cells (DATAR, 1985b).

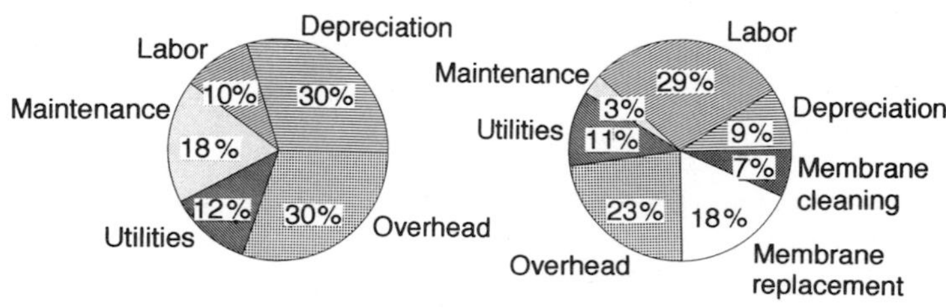

Fig. 29. Typical relative costs for separation of *E. coli* cell debris.

17 Conclusions

Both centrifugal sedimentation and crossflow filtration offer the bioprocess engineer a complementary choice for the purposes of the primary separation steps. Each technique has its own advantages and disadvantages, but through careful design considerations the most suitable separator for a given process can be identified.

For the processing of large-volume broths containing larger cells (>6 μm), it is likely that centrifugation will prove to be the optimum solution. This application is well proven in the harvesting of yeast cells.

For the processing of small (1–3 μm) cells the problems are significant, although not as acute as for the dewatering of cell debris. It appears that the advantages of centrifugation will outweigh those of crossflow as the scale of operation increases. At relatively small scales of operation (<3000 L) the advantages of crossflow may be more attractive, while on larger scales of operation (>20000 L) centrifugation may be better.

At any reasonable scale of operation (>5000 L batch) the use of centrifugal separation for the dewatering of cell debris is problematic, unless flocculation is allowed. However, acute fouling of membranes poses its own set of problems for crossflow filtration, and both techniques – centrifugation or crossflow filtration – present the bioengineer with severe cost penalties. Novel process solutions to the dewatering of cell debris suspensions are urgently required. A new generation of rotating membrane filters currently in development, and designed for tangential velocities an order of magnitude higher than those in conventional tangential flow filtration systems, appear to provide one solution. We can expect to see such developments enter the market place in the near future.

18 References

ALFA-LAVAL (1983), Laboratory separator LAPX 202 user's guide, *Technical Brochure* PM 40229 E3, Tumba, Sweden.

ALFA-LAVAL Separation AB (1990), Centrifuges for the Pharmaceutical and Fermentation Industries, *Technical Brochure* PB 41163E2, Tumba, Sweden.

ATKINSON, B., SAINTER, P. (1982), Downstream biological process engineering, *Chem. Eng.* (November), 410.

AXELSSON, H. A. C. (1985), Centrifugation, in: *Comprehensive Biotechnology* (MOO-YOUNG, M., Ed.), pp. 325–346, Oxford: Pergamon.

BLATT, W. F., DRAVID, A., MICHAELS, A. S., NELSON, L. (1970), Solute polarization and cake formation in membrane ultrafiltration – cause, consequences and control techniques, in: *Membrane Science and Technology* (FLINN, J. E., Ed.), pp. 47–97, New York: Plenum Press.

BOWDEN, C. P. (1984), Primary solid–liquid separation: The recovery of micro-organisms from fermentation broth, *Proc. Biotechnol. Conf.* **84,** 139, Washington DC, Online Confs.

BRUNNER, K.-H. (1983), Separators in biotechnology, *Chem. Tech.* **4,** 39–45.

DATAR, R. (1984a), Centrifugal and membrane filtration methods in biochemical separation, *Filtr. Sep.* (Nov./Dec.), 402.

DATAR, R. (1984b), Studies on the tangential flow membrane filtration of bacterial cell debris from crude homogenates, *Proc. Biotechnol. Conf.* **84,** A15–A30, Washington, DC, Online Confs.

DATAR, R. (1985a), The separation of intracellular soluble enzymes from bacterial cell debris by tangential flow membrane filtration, *Biotechnol. Lett.* **7** (7), 471–476.

DATAR, R. (1985b), A Comparative Study of Primary Separation Steps in Fermentation, *Ph. D. Thesis,* The Royal Institute of Technology, Stockholm, Sweden.

DATAR, R., ROSÉN, C. G. (1987), Centrifugal separation in the recovery of intracellular protein from *E. coli, Chem. Eng. J.* **34,** B49–B56.

DURAND, R. (1953), Basic relationships of the transportation of solids in pipes, *Proc. Minnesota Int. Hydraulic Convention,* St. Anthony Falls Hydraulic Laboratories, Minneapolis.

ECKSTEIN, E. C., BAILEY, D. G., SHAPIRO, A. H. (1972), Self-diffusion of particles in a shear flow of a suspension, *J. Fluid Mech.* **79,** 191–195.

GOLDINGER, W., REBSAMEN, E., BRAENDLI, E., ZIEGLER, H. (1986), Dynamic micro- and ultrafiltration in biotechnology, *Sulzer Tech. Rev.* No. **3,** Winterthur, Switzerland: Sulzer Brothers Ltd.

KRONER, K. H., SCHÜTTE, H., HUSTEDT, H., KULA, M. R. (1984), Cross-flow filtration in the downstream processing of enzymes, *Process Biochem.* (April), 67–74.

KRONER, K. H., NISSINEN, V., ZIEGLER, H. (1987), Improved dynamic filtration of microbial suspensions, *Bio/Technology* **5**, 921–926.

LEVEQUE, M. A. (1938), cited in: *Transport Phenomena* (1960) (BIRD, R. B., STEWART, W. E., LIGHTFOOT, E. N., Eds.), pp. 364, 369, New York: Wiley & Sons.

LOPEZ-LEIVA, M. (1979), Ultrafiltration in Rotary Annular Flow, *Ph. D. Thesis,* University of Lund, Sweden.

MURKES, J. (1967), Separering i centrifugalseparatorer. *Teknisk Tidskrift,* 347–349.

MURKES, J., CARLSSON, C. G. (1988), in: *Crossflow Filtration – Theory and Practice*, pp. 18–38, New York: John Wiley & Sons.

NEWITT, D. M. (1955), Hydraulic conveying of solids in horizontal pipes, *Trans. Inst. Chem. Eng.* **33**.

PORTER, M. C. (1972), Concentration polarization with membrane ultrafiltration, *Ind. Eng. Chem. Prod. Res. Dev.* **11**, 243–248.

ROSÉN, C.-G., DATAR, R. (1983), Primary separation steps in fermentation processes, *Proc. Biotechnol. Conf.* **83**, 201–223, London, Online Confs.

SPALDING, B. J. (1991), *Bio/Technology,* **9**, 600.

SVAROVSKY, L. (Ed.) (1977), *Efficiency of Separation of Particles from Fluids, Solid–Liquid Separation,* 2nd Ed., pp. 33–63, *Butterworth's Monographs in Chemistry and Chemcial Engineering,* London: Butterworth.

TANNY, G. (1978), Dynamic membranes, in: *Ultrafiltration and Reverse Osmosis, Separation and Purification Methods* **7** (2), 183–220.

ZYDNEY, A. L., COLTON, C. K. (1982), Continuous flow membrane plasmaphoresis – theoretical models for flux and hemolysis prediction, *Trans. Am. Soc. Artif. Intern. Organs* **28**, 408–412.

19 Cell Disruption and Isolation of Non-Secreted Products

HORST SCHÜTTE

Berlin, Federal Republic of Germany

MARIA-REGINA KULA

Jülich, Federal Republic of Germany

1 Introduction

Cells are separated from the surrounding medium by a composite structure of membranes and cell walls. Membranes provide the diffusion barrier and the means for selective transport, while the cell walls contribute elasticity together with mechanical strength, e.g., the strength necessary to withstand fast changes in osmotic pressure. Depending on the order, class and genus of the organism, cell walls are composed of various complex polysaccharides which may be crosslinked by peptides. A schematic illustration is presented in Fig. 1. The mechanical strength of a microorganism may be very high, e.g., small Gram-positive bacteria; however, this strength is not a constant, but depends on the species involved, growth conditions and the history of the cells (ENGLER and ROBINSON, 1981; HARRISON et al., 1990). Under conditions of growth limitation and starvation, microbial cells reinforce the cell wall to prepare for survival and become more difficult to disrupt. This physiological response has to be taken into account in upstream developments in order to ease further processing steps in recovery.

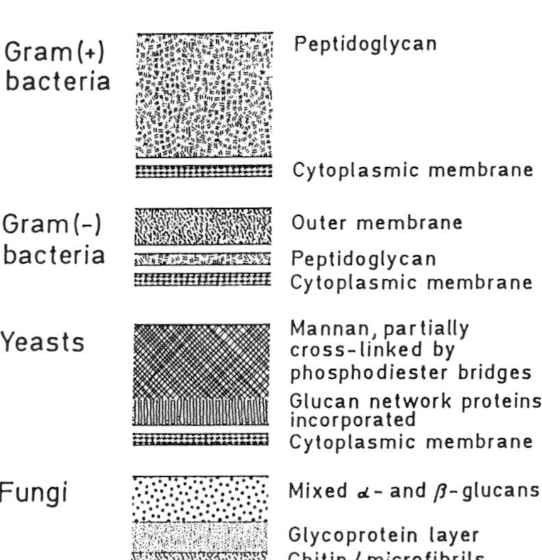

Fig. 1. Schematic illustration of cell envelopes of different classes of microorganisms.

A disintegration step is necessary in order to isolate proteins (or other products) that are accumulated within the cells during the production stage. In principle three different strategies are possible: to break the cell structure by mechanical forces, to damage preferentially the cell wall, e.g., by enzymatic lysis, or to damage primarily the membranes by treatment with chemicals. Each strategy has specific advantages and disadvantages, depending on the product and its application. Mechanical disintegration is generally applicable, while the other approaches need specific developments for each individual case. The goals of a cell disintegration step may be summarized as follows:

- to solubilize the maximal amount of product present in the cells with maximal biological activity,
- to avoid secondary alteration of product, e.g., denaturation, proteolysis, oxidation,
- to minimize the impact of the disintegration step on the performance of the following separation step(s), especially solid/liquid separation.

If the desired product is an inactive inclusion body, the disintegration step serves to facilitate removal of soluble cellular proteins prior to dissolution of the aggregated protein. Similarly, cells have to be disrupted to recover granules of polyhydroxybutyrate (HARRISON et al., 1990).

2 Analysis of Cell Disruption

The degree of disintegration may be judged by different methods. While microscopic observation allows one to distinguish intact and broken cells, it is advisable as a quick check only; quantitation is tedious and gives no information on the quality of the product. Sedimentation analysis has recently been described as a method to quantify cell disintegration for cells carrying inclusion bodies by measuring the volume frequency distribution in the *Escherichia coli* homogenate in the range 0.7–1.5 μm (MIDDELBERG et al., 1991). Indirect methods like determination of the amount of released activity or total protein require the definition of a 100% value. This is not without problems, since the activity per unit weight of cells may vary with cultivation conditions, time and conditions of storage, etc. Commonly employed protein assays do not yield absolute values, but are related to standard proteins and may be confounded by proteinaceous material which may be liberated upon prolonged treatment of membranes and cell walls. It is therefore necessary, especially during process development, to establish a standard protocol using the most efficient and gentle laboratory method of cell breakage and to frequently measure 100% values upon which the success and degree of a cell disintegration step can be based.

It is possible to follow the solubilization of product by measurement of activity or protein concentration on-line by flow-injection techniques adopting available assay procedures (RECKTENWALD et al., 1985a, b). Such data may also be useful for process documentation. In practice, until now, the disintegration step is generally carried out under well defined operating conditions while measuring on-line only physical parameters such as temperature, flow, agitator speed or pressure. The concentration of product is usually determined off-line after the process is completed.

The size of the insoluble fragments generated during cell disintegration is a very important parameter for the necessary solid/liquid separation. Unfortunately, it is a quantity most difficult to measure. Given the small size of microbial cells (1–10 μm), cell wall fragments may be expected more or less as a continuum from intact cells or even agglomerates reaching down to the colloidal range. Also, the number of fragments generated from each cell is not known and is not expected to be uniform. This adds to the difficulties because the dynamic range of different particle size analyzers does not span the entire range of interest (BARTH, 1984).

Counting devices such as a Coulter counter have intrinsic limits below 1 μm; particles with dimensions below 0.5–0.3 μm cannot be detected. Sedimentation analysis suffers from the unknown density of cell fragments and the small density difference between fragments and aqueous media; size analysis of fragments in the submicron range is quite slow and in our hands did not yield reproducible results. This finding may reflect also time-dependent changes in the sample, agglomeration, enzymatic lysis, etc., or slight differences in the redispersion of the sample prior to analysis. Under these circumstances size distributions are very difficult to access and did not correspond to microscopic observations (VOGELS, 1990). Results of sedimentation analysis are biased in favor of the larger particles. Similar difficulties were encountered using laser light scattering devices, which may be a consequence of the small difference in the refractive index between particle and medium or the mathematical analysis of the primary data. Certainly more collaborative efforts are necessary in the future to obtain reliable data for the size of fragments generated by different treatments.

Chromatographic techniques following the clarification of the homogenate require a particle-free feed. Therefore, the absolute size of cell wall fragments and the size distribution into the submicron range is important, since these factors will strongly influence the performance of solid/liquid separation devices. It should be remembered that in commonly employed unit operations, such as centrifugation or filtration, throughput is decreasing at least with the square of the particle diameter. However, such small fragments have to be removed, since they lead to fouling of adsorbents or membranes and to clogging of chromatography columns. In contrast, extraction of proteins from cell homogenates is not influenced by the method of cell disruption and size of fragments, since the separation depends on thermodynamic properties (KULA et al., 1982). Different techniques of cell disintegration have rarely been compared with regard to fragment size produced (VOGELS and KULA, 1992; KULA et al., 1990). With the improvement in mechanical as well as non-mechanical methods to solubilize intracellular products, such a comparison will prove necessary in order to optimize the initial step with regard to cost, productivity and integration into a recovery train.

3 Mechanical Methods of Cell Disruption

3.1 High-Pressure Homogenizers

High-pressure homogenizers have been produced for about 100 years for process steps such as emulsification, dispersion, disagglomeration and disintegration of microorganisms. These machines are employed predominantly in the dairy industries, for example, for homogenization of cream, or dispersion of fat in milk. In the pharmaceutical and food industries, homogenizers of various designs are used with pressures up to 30 MPa in order to obtain stable emulsions, to increase the viscosity or to reduce the addition of stabilizing agents. However, for cell disruption 30 MPa are insufficient to get significant disruption of microorganisms as discussed below.

3.1.1 Construction and Function

High-pressure homogenization is the most widely used method for large-scale cell disruption. The machine consists of a high-pressure positive displacement pump(s) coupled to an adjustable, restricted-orifice discharge valve. For disruption, cells are suspended in an aqueous medium and delivered by the piston pump into the homogenizing valve at a pressure higher than 50 MPa. The discharge pressure in the valve unit can be controlled by a spring-loaded valve rod or by an automatic hydraulic regulator, which positions the valve in relation to the valve seat. During discharge at a preset pressure the suspension passes between the valve and its seat at a very high velocity (≈ 350 m s^{-1}) and impinges on an impact ring as illustrated in Fig. 2. The impingement of the high-velocity jet of suspended cells on stationary surfaces and the magnitude of the pressure drop are the mechanisms contributing to cell rupture. High shear forces occurring during the high acceleration of the liquid in the gap may also play a role (BROOKMAN, 1974; KESHAVARZ et al., 1990).

3.1.2 Release of Intracellular Proteins and Enzymes

HETHERINGTON et al. (1971) studied the effect of some operating parameters such as homogenizing pressure, valve design, number of passes through the valve assembly, temperature and cell concentration. From this work it is also known that cell disruption in a high-pressure homogenizer follows first-order kinetics and can be described by the equation:

$$\ln \frac{R_\mathrm{m}}{R_\mathrm{m}-R} = k \cdot N \cdot p^a \qquad (1)$$

with

R_m maximal obtainable protein content per unit weight of cells (kg protein/kg cells),
R concentration of proteins released at time t (kg protein/kg cells),
k first-order rate constant (s^{-1}),
N number of passages of the cell suspension through the valve unit,
p operating pressure (MPa), and
a pressure exponent.

The pressure exponent a depends on the microorganism itself as well as on its history. Values

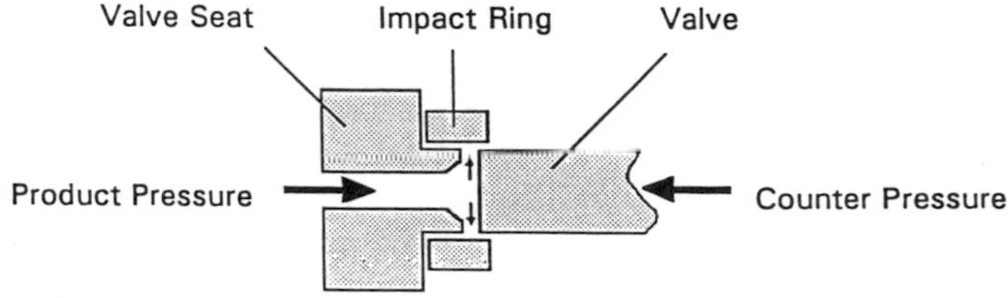

Fig. 2. Schematic view of the cell disruption valve in a high-pressure homogenizer.

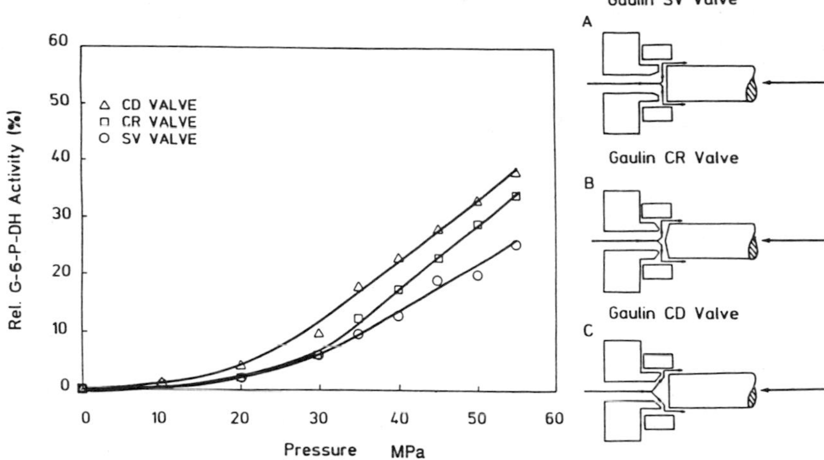

Fig. 3. Solubilization of glucose-6-phosphate dehydrogenase from *Saccharomyces cerevisiae* as a function of pressure and valve design.

of *a* between 1.86 and 2.9 have been reported for *Saccharomyces cerevisiae* (FOLLOWS et al., 1971; KULA et al., 1990), 2.2 for *Escherichia coli* (GRAY et al., 1972), and 1.59–3.08 for *Alcaligenes eutrophus* (HARRISON et al., 1990). The kinetics of disruption of filamentous fungi seems to be different from micellular organisms and shows only weak dependence on pressure (KESHAVARZ et al., 1990).

In comparison to a bead mill only a few operating parameters affect disruption in high-pressure homogenizers. The homogenizing pressure and the choice of the valve unit are the most significant parameters at present. The effect of both variables is shown in Fig. 3 where the disruption of *S. cerevisiae* for a single passage is shown as a function of the homogenizing pressure using three different valve designs. It is obvious that the degree of disruption is strongly influenced by the operating pressure and with all the valve designs tested (SCHÜTTE and KULA, 1990a; MADSEN and IBSEN, 1987; KESHAVARZ et al., 1987; MASUCCI, 1985; KESHAVARZ-MOORE et al., 1990). With the best valve unit – the cell disruption valve (CD-valve) – only about 40% of the total amount of enzymes is solubilized at 55 MPa in a single pass. Reference is made to a review by ENGLER (1990).

3.1.3 Changes in Design and Operation in Recent Years

In recent years, homogenizers were reconstructed for pressures up to 125 MPa and new valve designs investigated to meet the requirements of the expanding biotechnology industry. Using a homogenizing pressure of 100 MPa in a single passage, the same degree of disruption for *S. cerevisiae* can be obtained as with four passes at 55 MPa. Time and energy costs can be reduced in this way. However, it must be taken into account that the temperature in the liquid rises linearly with the operating pressure, approximately 2.5 °C per 10 MPa, because virtually all the energy is converted into thermal energy. This means, when starting at 10 °C the temperature in the cell suspension will rise to 40 °C at an operating pressure of 120 MPa. The heat generation may be deleterious for heat-labile enzymes and biologically active proteins. Therefore, rapid chilling of the processed cell suspension is necessary especially when a second or third pass is required. In Tab. 1 the results for the release of soluble intracellular enzymes from various microorganisms are listed, and disruption data at 55 MPa and 120 MPa are compared. The

Tab. 1. Effect of Operating Pressure on Disruption Efficiency During High-Pressure Homogenization

Microorganism	Enzyme	Percentage Disrupted	
		55 MPa	120 MPa
Escherichia coli	Fumarase	79	97
Bacillus cereus	L-Leucine DH	62	91
Lactobacillus confusus	L-2-Hydroxyisocaproate DH	52	94
Saccharomyces cerevisiae	D-Glucose-6-phosphate DH	39	89
Candida boidinii	Formate DH	32	77

DH, dehydrogenase

advantage of using 120 MPa is evident, and it is suggested that most organisms can be completely disrupted in two passes through the machine (BUESCHELBERGER and LONCIN, 1989).

At these high pressures, the stress and erosion of the valve unit by the cell debris or inclusion bodies in rDNA organisms is a crucial point in the construction. The impact increases with rising homogenizing pressure. Therefore, wear-resistant materials such as tungsten carbide, special ceramics or polycrystalline diamonds are used for the valve components to ameliorate the problem of erosion. The lifetime of valves made of high-density special ceramics is increased up to 10 times compared to valves made from rexalloy. In addition the surface quality of stainless steel parts in contact with the cell suspension, e.g., cylinder blocks and homogenizer valve bodies has been improved. By means of an electropolishing process the metal surfaces are made smooth and devoid of roughness down to the submicron range. The specific microsurface is reduced by more than 80% compared to surfaces after grinding. The low potential energy of the metal grain structure reduces adhesion and deposition of cell debris and released material resulting in decreased fouling.

On the other hand, the smooth and porefree surface can be cleaned more easily, thereby meeting the standard demands in the biotechnology industry. For aseptic operation the machines are equipped with plungers with special ceramic or chromium coatings and special temperings to make the surface resistant to aggressive cleaning agents and desinfectants. Homogenizers are available for aseptic operation also with steam-sterilizable surfaces and in addition with mechanical safety arrangements (see Sect. 6). High-pressure homogenizers are on the market with an operating pressure of 100 MPa and flow rates in the range of 10 to 6000 liters per hour, which means that in principle machines can be delivered by the manufacturers for laboratories, pilot- and process plants. It is expected that optimized impact rings (KESHAVARZ et al., 1990) will be commercially available soon.

Recently, a laboratory high-pressure homogenizer for small sample volumes (20 to 40 mL) was introduced, operating at pressures up to 160 MPa (APV Gaulin GmbH, Lübeck, FRG). The homogenizer Micron Lab 40 is a batch device (a continuous device is in preparation) and utilizes the principle of high-pressure expansion in a precisely controlled knife-edge gap. For operation the sample is poured into a cylinder, and the valve body is put in place. The complete stack is clamped hydraulically. The desired homogenizing pressure is selected, and the pressure stroke started. The sample is pressurized and flows through the valve unit into the collection chamber. The whole procedure requires about 20 s. The mechanism of disruption in this Micron Lab 40 is comparable to technical high-pressure homogenizers. First studies with the Micron Lab 40 and a pilot plant high pressure homogenizer (30 CD, 250 L/h) have shown that the degree of disruption is identical if the same pressure and cell suspension are used in both machines as demonstrated in Fig. 4.

3.2 High-Speed Agitator Bead Mills

Wet milling is a common unit operation in the processing of ore or clay minerals. High-

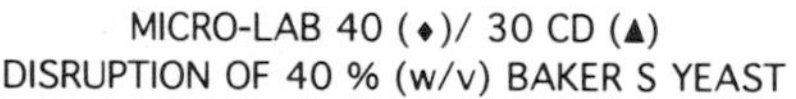

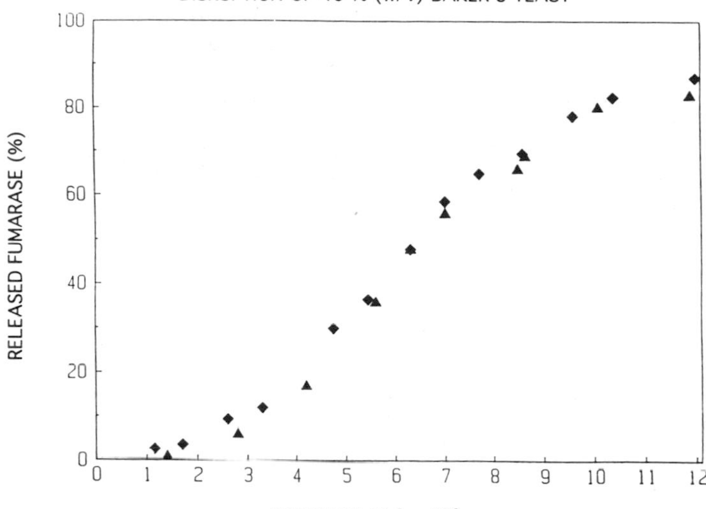

Fig. 4. Comparison between a small, batch-operated high-pressure homogenizer (Micro-Lab 40, ◆) and a continuously operated, pilot-scale high-pressure homogenizer (30 CD, ▲). Performance is measured as relative activity of fumarase released from *Saccharomyces cerevisiae* in a single pass

speed agitator bead mills have been developed for the production of pigments in the printing industry where micron-sized particles with a narrow size distribution are required. Such mills have been studied (CURRIE et al., 1972; VAN GAVER and HUYGHEBAERT, 1990; KULA and SCHÜTTE, 1987; LIMON-LASON et al., 1979; MARFFY and KULA, 1974; REHACEK and SCHAEFER, 1977; SCHÜTTE et al., 1983; SCHÜTTE and KULA, 1990a) and improved for the disruption of microorganisms (SCHÜTTE et al., 1986, 1988; KULA et al., 1990). The different constructions of the mills and the single parameters contributing to cell disintegration have been discussed in detail by SCHÜTTE and KULA (1990b).

3.2.1 Construction and Function

In Fig. 5 a schematic drawing of a mill with a horizontally oriented grinding chamber is presented; such mills are currently preferred for cell disintegration. The mill chamber is nearly completely filled with grinding beads during operation. The optimal size of the beads depends on the density and viscosity of the feed. For microbial suspensions the bead size varies between 0.2–0.5 mm for bacteria and 0.4–0.7 mm for yeast. The beads should be of rather uniform size in order to avoid crushing and extensive wear during processing. A special bead separator keeps the grinding elements in the mill chamber but allows the cell suspension to pass in continuous operation (SCHÜTTE and KULA, 1990b). At approximately 85% filling of the free volume with beads, heat generation and power consumption rise strongly during operation (SCHÜTTE et al., 1983), as does bead wear, thereby limiting the maximal bead loading to approximately 90%. Since the cell disintegration depends on the bead loading, the mills are usually operated close to the maximum. The agitator transfers energy to the grinding beads, its design is of crucial importance for successful product liberation from microbial cells by the milling process.

In principle, mills with three different types of agitators are commercially available for different purposes:

(1) disc agitators,
(2) pin agitators, and
(3) rotor/stator systems.

The energy density in the grinding zone increases from 1 to 3, which appears desirable, but may be detrimental for product quality, especially with feed dispersions of low viscosity

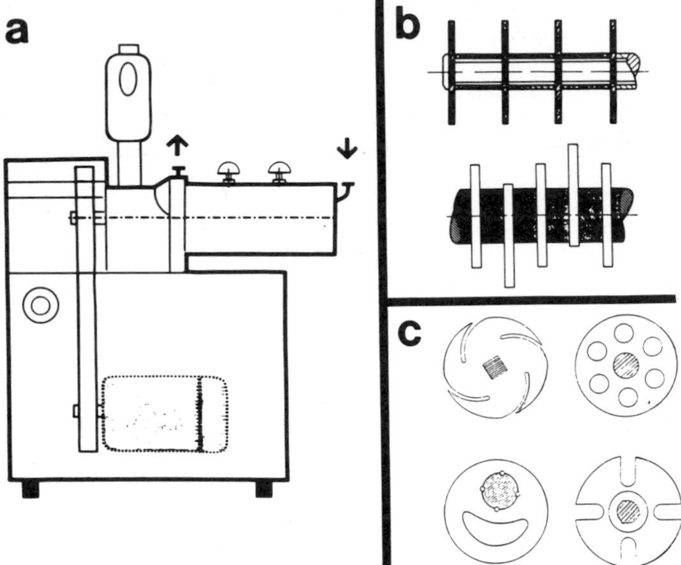

Fig. 5. (a) Schematic view of a high-speed agitator bead mill, (b) centric and eccentric arrangement of agitator discs on the drive shaft, (c) selection of discs used in wet milling.

such as microbial suspensions (SCHÜTTE et al., 1988). In general, pin agitators and rotor/stator systems are applied to process highly viscous media, while disc agitators, as shown in Fig. 5c, are preferred for cell disintegration. Disc agitators contain a variable number of discs mounted on the central shaft in a centric or eccentric position. The discs differ in design and may be solid, perforated or slitted in various ways (see Fig. 5) and are manufactured from metal or polymer material such as polyurethane. The transfer of kinetic energy from the turning shaft to the grinding beads depends on adhesion as well as displacement forces, which are related in a complex manner to the geometry and material properties of the agitator (KULA and SCHÜTTE, 1987). The beads are accelerated by the turning disc(s) in radial direction and oriented in streaming layers of different velocity. In this way, liquid shear forces are created leading to cell disintegration. In addition to such shear forces, the frequency and strength of the collision events during milling contribute also to the cell disruption process. The profile of the stream layers depends on the size, density and the load of the grinding beads as well as the distance between fixed and moving parts in the mill chamber. The energy of the beads is maximal at the periphery of the disc and minimal near the axis and the wall. Therefore, the peripheral speed should be used as criterion to compare different mills and to scale up processes. The number of parameters which influence cell disintegration by wet milling is quite large, as illustrated in Tab. 2 (MÖLLS and HÖRNLE, 1972).

Most, if not all, machine-related parameters of the milling process are fixed with the purchase of the mill. For a given product the biological and physical parameters of the feed suspension will vary with the producing organism. These parameters are fixed and controlled during the development of the biological production step. Fixing these parameters leaves the operator mainly the choice of processing conditions in order to optimize the disintegration step (Tab. 2). Since most of the energy introduced during milling is converted to heat, effective cooling has to be incorporated for cell disintegration. For small- and medium-sized mills, adequate cooling is provided by circulating brine through a jacket at the chamber wall and the bearing housing. If desired mill chambers can be sterilized chemically or by steam. Installation of a dual-action mechanical seal on the stirrer shaft is required to secure containment. Cleaning in place is possible and usually is carried out employing special wash cycles.

Tab. 2. Selection of Relevant Parameters in Wet Milling (According to MÖLLS and HÖRNLE, 1972)

Machine-Related Parameters[a]	Operating Conditions[b]	Biological and Physical Parameters of Cell Suspension[c]
Design of agitator	Tip speed	Type of organism
Material of construction	Feed rate	Cultivation conditions
Design of bead separator	Bead size	Storage conditions
Free volume	Bead loading	Growth phase at harvest
Cross-sectional area	Cell concentration	Viscosity and rheological properties of suspension
Ratio diameter/length		pH, buffer capacity
Cooling		
Containment		

[a] The most important variables of different machines are the stirrer design, the diameter and distance disc to disc, disc to wall, the form of the blades, etc., which will determine energy consumption, mixing behavior and efficiency of cell disintegration.
[b] Agitator tip speed and feed rate influence the residence time distribution in a complex manner; increasing feed rates (shorter residence times) reduce back mixing and improve performance of the mill.
[c] These parameters should be primarily addressed in the upstream process design, the selection of host and growth conditions. The known magnitude of change in the pressure exponent, which is a measure of mechanical strength, should alert the operator to changes in the feed stock with time.

3.2.2 Solubilization of Enzymes and Proteins

Despite the many parameters involved cell disruption in a bead mill can be described as a first-order process for batch operation by the equation

$$\ln \frac{R_m}{R_m - R} = \ln D = k \cdot t \qquad (2)$$

with D as the reciprocal fraction of unreleased protein.

The first-order rate constant k is a lumped parameter and depends on the operational, physical and biological factors summarized in Tab. 2.

The rate of release of enzymes or other products into the soluble fraction depends to some extent on their location within the cell. Enzymes located within the periplasmic space are liberated faster than total protein, while for release of intracellular enzymes the rates are nearly identical. Products from organelles such as mitochondria are released slower than total protein (FOLLOWS et al., 1971). Product release occurs parallel with product inactivation, e.g., by shear stress, proteolysis or thermal denaturation (MARFFY and KULA, 1974). It may be masked in the initial stages of a cell disruption process. The impact of operating conditions on product stability can be determined by plotting total activity in the soluble fraction as a function of time during extended batch processing. Batch disruption is usually carried out to determine the first-order rate constant for a given organism and product under specified conditions. With the optimization of the operational parameters the time necessary to disrupt the cells is minimized. This minimal time may be used to determine the feed rate for a continuous production process. In practice, the mean residence time is set three times the necessary time in batch processing. This empirical factor reflects the fact that the agitator speed as well as the feed rate influence the mixing behavior and, therefore, the residence time distribution in a given mill (SCHÜTTE et al., 1986, 1988).

Continuous disintegration can also be described as a first-order process; Eq. (2) must be modified, however, to take into account the residence time. Residence time distribution in a disc type bead mill may be described as a cascade of stirred tank reactors (CSTR) (LIMON-LASON et al., 1979; KULA et al., 1990;

Tab. 3. Disintegration Rate Constants of Microorganisms for Various Rotor Geometries in Horizontally Operated High-Speed Agitator Bead Mills

Bead Mill	Rotor Type	Diameter of Grinding Element (cm)		$K_{protein}$ (min^{-1})	
				Saccharomyces cerevisiae	*Brevibacterium ammoniagenes*
LMD 1	9 eccentric rings	4.4	120° [a]	0.15	0.039
	5 conical discs	6.6		1.20	0.091
	7 spiral discs	7.3		0.71	n.d.
LMD 4	12 eccentric rings	7.2	120° [a]	0.41	0.035
	6 conical discs	10.5	45°	1.00	0.140
	8 spiral discs	10.3		0.75	0.032
LMD 20	12 eccentric rings	13.0	90°	0.46	n.d.
	15 eccentric rings	13.0	90°	1.08	n.d.
	8 conical discs	19.4		1.41	n.d.

[a] disc to disc angular offset
n.d., not determined
Mills of similar geometry of 1, 4 and 20 L chamber volume, respectively, were tested at the same agitator tip speed (supplier: Netzsch Feinmahltechnik, Selb, FRG).

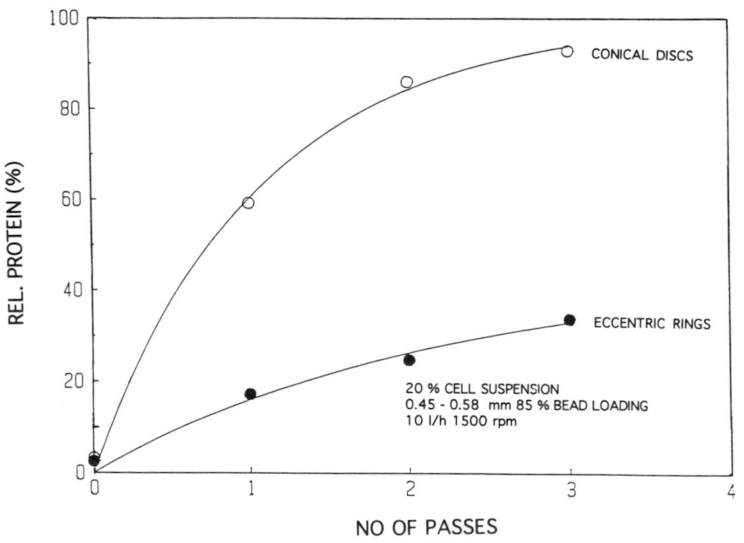

Fig. 6. Disruption of *Brevibacterium ammoniagenes* using different stirrer blades under identical operating conditions in the same high-speed mill (LMD-1). For stirrer blades see Fig. 5c (lower line).

SCHÜTTE et al., 1988). The fraction of intact cells D (unreleased protein) is then given by

$$D = \frac{R_m}{R_m - R} = \left(1 + \frac{k\,\tau}{j}\right)^j \tag{3}$$

where τ is the mean residence time (free volume divided by the flow rate), j is the number of CSTRs in the cascade; j is determined experimentally and may be represented by a non-integral number, it depends also on the geometry and the agitator design. It should be noted that the optimal agitator is different for batch or continuous operation.

For batch processing, the energy dissipation per unit volume is the decisive factor in cell disruption. In continuous processing the mixing behavior of the mill also becomes important, especially for feed streams of low viscosity as encountered in cell disintegration. Be-

Tab. 4. Preferred Operation Conditions for the Disruption of Yeast and Bacteria Using a High-Speed Agitator Bead Mill

Parameter	Yeast	Bacteria
Agitator tip speed	8–12 m s^{-1}	10–15 m s^{-1}
Flow rate of feed	10 to 15-fold[a]	2.5 to 7.5-fold[a]
Cell concentration	10–15%[b]	10–15%[b]
Diameter of glass beads	0.4–0.7 mm	0.2–0.5 mm
Bead loading	80–85%[c]	85%[c]
Capacity	4–10 kg[d]	2–6 kg[d]

[a] chamber volume per hour
[b] cell dry mass
[c] of free chamber volume
[d] per hour and liter chamber volume

cause of the complex nature of the mixing behavior, scale up of a high-speed agitator bead mill is non-linear, and optimal conditions should be experimentally verified during process development. Mills of different chamber volumes and different agitator design are compared in Tab. 3. High-speed agitator bead mills – most often constructed originally for other purposes – are commercially available in the range of 0.15–250 L chamber volume.

A study has been performed in one case to optimize the agitator design especially for cell disintegration (Schütte et al., 1988; Kula et al., 1990). Processing Gram-positive bacteria, a 4-fold increase in productivity has been achieved for a given mill, using different stirrer blades; at the same time the specific energy consumption was decreased. As an example the release of proteins from *Brevibacterium ammoniagenes* is shown in Fig. 6, which illustrates, that there is room for improvements in cell disintegration. The obvious targets are the biological system and the agitator design in its dual function for energy transfer and mixing in the mill.

Tab. 4 summarizes preferred operating conditions for the disruption of yeast and bacteria in high-speed agitator bead mills. In principle, cells of all origin may be efficiently disrupted by wet milling and a high degree of product release achieved. This holds also for small Gram-positive bacteria where high-pressure homogenizers fail (Schütte and Kula, 1990b).

4 Non-Mechanical Methods of Cell Disruption

4.1 Chemical Treatment

The use of chemicals to extract intracellular components from microorganisms by permeabilizing the cell membrane has considerable potential for application. This technique provides several advantages over complete mechanical disruption including selectivity and simplification of cell removal. Gram-negative and Gram-positive bacteria as well as yeasts and fungi can be permeabilized; the choice of method depends very much on the type of microorganism and the composition of the cell wall and cell membrane. Furthermore, the concentration of the agent, the time required, pH, temperature, etc., have to be optimized for each organism, so that the efficiency of the chemical treatment depends very much on the individual case and hence is difficult to generalize.

The most important chemical agents that have been employed for permeabilizing cells are organic solvents, detergents, chaotropic agents, chelating agents and alkali. The following part will discuss some examples. In every case the stability of the desired product under the specific operating conditions must be tested. The techniques for chemical lysis do not require expensive or specialized cell breakage equipment, rather, only commonly available mixing tanks and filtration equipment are used. There are hazards associated with the handling of inflammable compounds, however, necessitating the use of spark-proof equipment and special precautions with regard to fire safety.

4.1.1 Organic Solvents

One of the most frequently used organic solvents for cell permeabilization is toluene. Low concentrations of toluene ($\approx 0.5\%$) are sufficient only for *in situ* enzymatic assays, where the structure of the cells is left intact and the enzyme activity remains inside (MURAKAMI et al., 1980). Intracellular proteins may also be recovered from cells, but then the toluene concentration must be increased to 5–8%. In Gram-negative bacteria, toluene most likely acts by dissolving inner membrane phospholipids (DE SMET et al., 1978; HOBOT et al., 1982), and proteins with molecular weights less than 50000 are mainly released under these conditions (DEUTSCHER, 1974). After treatment of *Escherichia coli* cells with 5% toluene approximately 25 to 30% of the intracellular protein content was released (JACKSON and DE MOSS, 1965; TEUBER, 1970; DEUTSCHER, 1974; MURATA et al., 1980). Yeast cells tend to dissolve by autolysis rather than permeabilize under appropriate conditions for protein recovery. Acceleration of the autolysis of *Saccharomyces cerevisiae* by various chemical treatments has been studied by BREDDAM and BEENFELDT (1991) and optimized for the recovery of carboxypeptidase.

A selective liberation of enzymes from the periplasmic space is possible by treatment with water-miscible solvents such as ethanol, methanol, isopropanol or *t*-butanol. The procedure has been investigated in detail for the isolation of β-D-galactosidase from *Kluyveromyces fragilis* by FENTON (1982). Reference is made to a review by FELIX (1982) covering a wide range of organic solvents that effect the permeability of many microorganisms.

4.1.2 Detergents

The non-ionic detergent Triton X-100 has a specific solubilizing effect for membrane proteins. This method could be used as a first step in the purification of smaller proteins from *E. coli* because proteins with molecular weights below 70 kD are preferentially released from permeabilized cells (MIOZZARI et al., 1978). One of the few large-scale procedures using Triton X-100 involves recovery of cholesterol oxidase from *Nocardia* sp. (BUCKLAND et al., 1974). Besides Triton X-100 other detergents such as cholate (KRONER et al., 1978), sodium dodecyl sulfate (WOLDRINGH, 1970) or Brij (BIRDSELL and COTA-ROBLES, 1968) are able

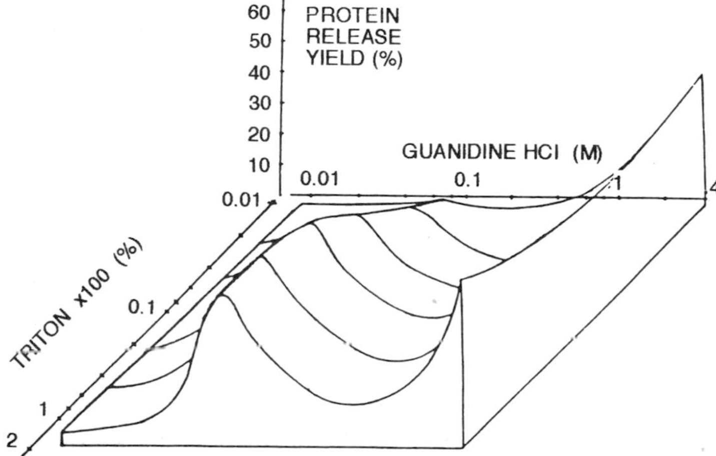

Fig. 7. Protein release from *Escherichia coli* as a function of guanidine-HCl and Triton X-100 concentrations (from HETTWER and WANG, 1989).

to act on the cytoplasmic and outer cell membranes. Synergistic effects of detergents with other chemicals for the release of intracellular enzymes from yeast have been reported by BREDDAM and BEENFELDT (1991).

4.1.3 Chaotropic Agents

Guanidine-HCl and urea are able to solubilize hydrophobic compounds from the *E. coli* membrane (HAMMES and SWANN, 1967; HATEFI and HANSTEIN, 1974; SCHNAITMAN, 1971). The effect of chaotropic agents on protein release from *E. coli* was studied in detail by INGRAM (1981). Chaotropic agents in combination with a detergent can interact synergistically with the *E. coli* cell structure, which can lead to an effective permeabilization method. HETTWER and WANG (1989) and NAGLAK and WANG (1990) have used the combination of 0.1 M guanidine-HCl and 0.5 to 2% Triton X-100 to release protein from *E. coli*, yielding 50–60% of the total protein content (Fig. 7). Chaotropic agents are quite expensive and contaminate the product stream; often they must be removed, e.g., by ultrafiltration or gel filtration prior to further processing. In addition there is an inherent danger of protein denaturation in the presence of chaotropic agents, especially for proteins exhibiting complex quaternary structures.

4.1.4 Alkali

This technique can be used only if the desired protein is stable at extreme alkaline pH for about 30 min at 20 °C, because such conditions tend to damage most of the proteins and enzymes along with the cell walls. Treatment with alkali was used for the isolation of the therapeutic enzyme L-asparaginase from *Erwinia carotovora* (WADE, 1971) and recombinant human growth hormone from *E. coli* (SHERWOOD et al., 1982).

4.2 Enzymatic Cell Lysis

Autolysis (plasmolysis) is an old procedure to prepare microbial extracts. The major disadvantages of this simple cell lysis procedure are the long reaction time required (24 to 48 h) and the low yield of protein (KNORR et al., 1979). The use of a lytic enzyme or of enzyme systems, which provide biological specificity to the process of cell disruption, shows high promise as a method of controlled operation and selective product release (ANDREWS and ASENJO, 1987; ANDREWS et al., 1990). A number of variables determine the success of a lysis method. The degree of cell lysis and activity yield may be influenced by strain differences, choice of growth conditions, the presence of proteases, the choice of buffers, cell density, osmolarity of the resuspension buffer, pH and temperature as well as the growth phase at cell harvest. A trial and error approach is often required to optimize lysis conditions (CULL and MCHENRY, 1990).

The best known lytic enzyme is lysozyme, which is widely distributed in nature. For commercial purposes chicken egg-white has become the major source of this enzyme, which catalyzes the hydrolysis of the β-1,4-glycosidic bonds in the peptidoglycan layer of bacterial cell walls (WITHOLT et al., 1976; HAMMOND et al., 1983; WECKE et al., 1982; KNORR et al., 1979). Gram-positive bacteria have a much thicker peptidoglycan layer as cell wall than Gram-negative bacteria and lack an outer membrane (see Fig. 1). Therefore, Gram-positive bacteria are more susceptible to the attack of lysozyme. After removal of the peptidoglycan layer the internal osmotic pressure of a cell will burst the periplasmic membrane releasing the intracellular content. Lysis of Gram-negative bacteria requires passage of lysozyme through the outer membrane, which is aided by the addition of EDTA to chelate divalent cations. Lysozyme has been used for several large-scale processes including enzyme extraction (SCOTT and SCHEKMAN, 1980; YANSHINSKI, 1984). Examples are lysis of *Erwinia aroideae* for penicillin V-acylase purification (VANDAMME and VOETS, 1974); *Pseudomonas putida* for alkane hydroxylase isolation (FISH et al., 1983) or enzyme production from *Staphylococcus aureus* (WECKE et al., 1982). Besides lysozyme other bacteriolytic enzyme systems have been described from *Cytophaga* sp. (LECORRE et al., 1985); *Staphylococcus* sp. (VALISENA et al., 1982); *Streptomyces globi-*

sporius (HAYASHI et al., 1981); *Streptomyces coelicolor* (BRÄU et al., 1991); *Micromonospora* sp. (SUZUKI et al., 1985) and *Bacillus subtilis* (BOROVIKOVA et al., 1980).

Yeasts differ from bacteria and have more complex cell walls composed of mannan, partially cross-linked by phosphodiester bridges and an inner β-glucan layer (see Fig. 1). Several lytic systems that digest yeast cells have been described. The major active components of these preparations are a β-1,3-glucanase for the degradation of the inner layer and a specific wall-lytic protease to degrade the outer layer of protein mannan. Yeast-lytic enzyme systems are described from *Arthrobacter luteus* (KITAMURA and YAMAMOTO, 1972); *Oerskovia* (SCOTT and SCHEKMAN, 1980; OBATA et al., 1977); *Rhizoctania* (KOBAYASHI et al., 1982); *Arthrobacter* (VRSANSKA et al., 1977) and *Cytophaga* (ANDREWS and ASENJO, 1987).

The enzyme-dependent release of invertase from *Saccharomyces cerevisiae* is shown in Fig. 8. In combination with the reducing agent L-cysteine, time for complete release of invertase can be shortened significantly. The enzyme system of *Cytophaga* sp. has also been studied for lysis of *Bacillus subtilis* and *Corynebacterium lilium* (LECORRE et al., 1985).

Present drawbacks for the application of lytic enzymes or enzyme systems are the lack of availability, high costs and the need for removal in subsequent purification steps. Potential advantages are selectivity in the release of cellular products and operation under mild conditions. References should be made to a current review covering the whole field of enzymatic lysis of microbial cells (ASENJO and ANDREWS, 1990).

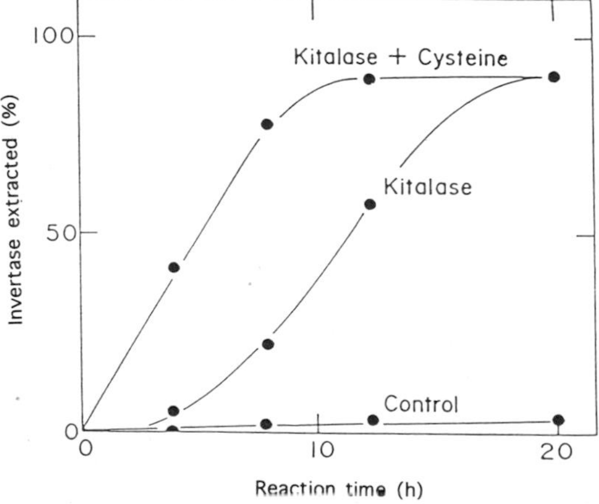

Fig. 8. Cell-wall degrading enzyme-induced release of invertase from *S. cerevisiae* cells (from KOBAYASHI et al., 1982).

4.3 Combination of Lysis with Mechanical Disintegration

The large-scale application of enzymes to cell lysis in a protein solubilization process requires either long incubation times or high enzyme concentrations in order to release the product in high yield. For bacterial proteins only a single enzyme is needed, lysozyme, which is commercially available in high quality at reasonable cost. It has been demonstrated using *Bacillus cereus* that a combination of enzymatic and/or thermal pretreatment - in itself not sufficient to yield extensive product release - enhances the efficiency of a subsequent mechanical disintegration step considerably (VOGELS and KULA, 1992). Such a pretreatment can easily be incorporated in established processing procedures and leads to savings in energy consumption and total process time. The latter is of importance, if the product is rather unstable in the crude homogenate. Energy savings result not only from the reduction in the number of passages through a mechanical disintegrator but also from improvements in further processing, since by the combined treatment larger fragments are generated from the cell walls (VOGELS and KULA, 1992). A combination of chemical pretreatment and mechanical disintegration was also recommended for the recovery of polyhydroxybutyrate granules from *Alcaligenes eutrophus* (HARRISON et al., 1990).

5 New Developments

5.1 Microfluidizer

5.1.1 Construction and Function

The Microfluidizer (Microfluidizer Corp., Newton, MA, USA) consists of an air-driven,

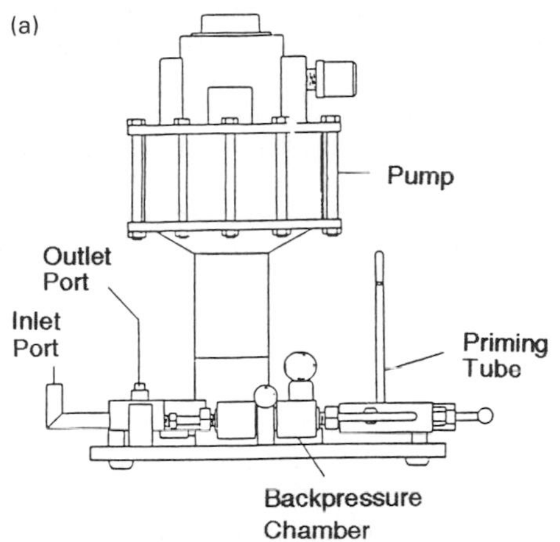

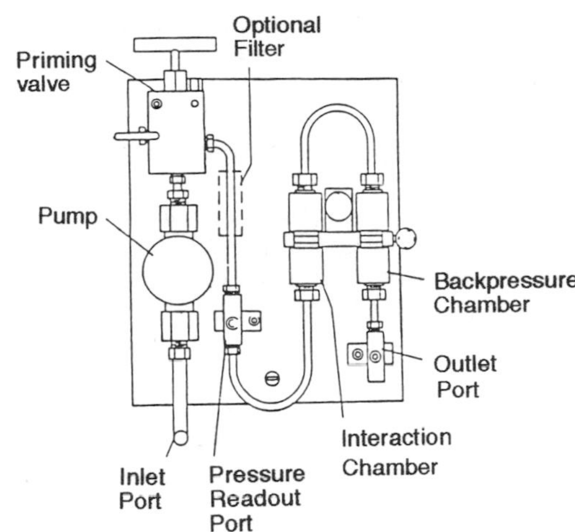

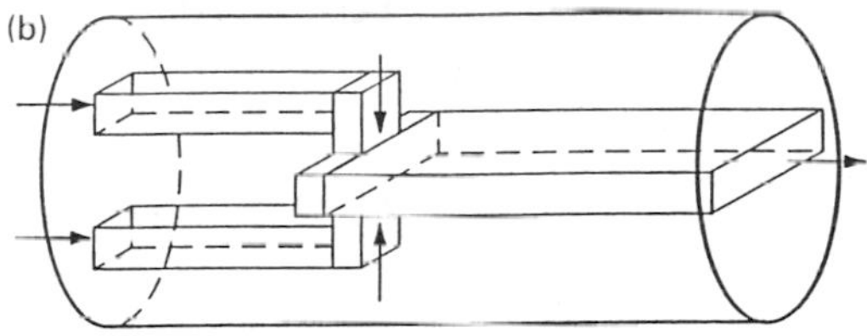

Fig. 9. (a) View of the Microfluidizer model M 110 T from the side and top, (b) device for impinging jet fluidizer. The process stream is divided into two jets that impinge against each other at high velocities (redrawn from Microfluidizer Corporation).

high-pressure pump (ratio 1:250; required air pressure 0.6–1 MPa) and a special disruption chamber with an additional back-pressure unit. The cell suspension is fed under pressure through an interaction chamber, where the process stream is split into two fluid streams (channel diameter 100 μm), which are then directed head-on at each other at high velocity. A schematic diagram of the Microfluidizer and a sectional drawing of the disruption chamber is given in Fig. 9a, b. The cell slurry enters the ceramic micro-channel of the interaction chamber under high pressure (maximum allowable working pressure 140 MPa) and emerges at atmospheric pressure. The mechanisms involved in microfluidization include laminar flow (shear), turbulence (inertial force) and vapor bubble implosion (cavitation) which results from the energy released in the pressure drop.

5.1.2 Release of Intracellular Proteins

The Microfluidizer can be operated either in a batch (minimum 30 mL) or a continuous flow mode. The achievable flow rate depends on the operating pressure and the viscosity of the cell suspension.

With the Microfluidizer model M 110 T, flow rates of 2–23 L/h with operating pressures of 10–95 MPa can be obtained, which corresponds to residence times between 0.045 to 0.025 s (30 MPa–95 MPa) in the disruption chamber. SAUER et al. (1989) have investigated the disruption of both native and recombinant strains of *Escherichia coli* over a wide range of operating conditions. Similar to the high-pressure homogenizer the efficiency of the Microfluidizer was found to be highly dependent on the operating pressure and the number of passes of the cell suspension through the disruption chamber.

Disruption was also dependent on the initial cell concentration and on the specific growth rate of the cells during the fermentation stage, as well as being strain-dependent. SEVA et al. (1986) have investigated the disruption of *Saccharomyces cerevisiae, E. coli* and *Bacillus subtilis* using a Microfluidizer at an operating pressure of 65 MPa. They observed that 30 passes are necessary to break 95% of the yeast cells, one pass to get 90% degree of disruption for *E. coli*, whereas three passes had to be carried out for *B. subtilis*.

6 Safety and Automation

6.1 Containment

The introduction of recombinant DNA technology has greatly extended the scope of biotechnology. Genetic engineering has opened up many new opportunities for industrial application, but has also prompted an intense and necessary debate over the anticipation and recognition of potential hazards to health or to the environment and the criteria for practical risk assessment. The microorganisms used in the biotechnology industry and generally accepted as safe (GRAS status) are not pathogenic in themselves, however, the compounds they produce or express may present a risk to health, if the microorganism is able to colonize the body after incorporation. In addition, there is the more obvious problem of direct exposure to proteins, cell debris and other products in the form of aerosols during downstream processing, where various toxic or allergic reactions have to be considered after inhalation (BENNET et al., 1990). In particular, allergic reactions can be induced in susceptible people. Following repeated exposure to an allergen, workers may become sensitized and will then react to extremely low challenge concentrations.

The most likely route of occupational exposure is via aerosols generated by breaches of containment during cell growth in bioreactors and subsequent processing to separate, concentrate and purify the desired product (STEWART and DEANS, 1990). In practice the various types of equipment used in downstream processing are by their nature often less well adapted to strict asepsis. The major consideration in a cell disruption step is to provide proper containment, so that aerosols containing organisms do not escape to the workplace. The simplest way to overcome potential hazard during the cell disruption process is to enclose the equipment used for this unit operation within a ventilated cabinet (HARRIS-SMITH and EVANS, 1974; DUNNILL, 1982; TUIJNENBURG-MUIJS, 1987). Ventilated safety cabinets of various design and efficiencies have been described ranging from simple boxes to a complete system of modular cabinets with integral disinfectant liquid locks, air locks, and autoclaves. The principle of enclosure is well known, but relatively few examples have been described for application with pilot and process-scale machines.

As outlined above, the most likely route of occupational exposure of cells, cell debris and intracellular components is via aerosols generated by breaches of containment, e.g., during cell disruption. In the following part cell disruption with high-pressure homogenizers will be taken as an example, because these machines operate at high pressures, and any leaks are likely to form aerosols of process media. Considering that these machines were originally developed for use in the dairy industry where containment is not critical, a new generation of high-pressure homogenizers was specifically developed to meet the needs of the biotechnology industry. All parts in contact with the cell suspension are electropolished to fulfill extreme requirements in hygiene and to tolerate aggressive cleaning agents and disinfection. The design of the pump valves – suction valve and discharge valve – meets sanitary demands, and the materials used assure a long service life. But the most important part of the new generation is the aseptic cylinder assembly (APV GAULIN INC., 1990; MEYER, 1988). Aseptic homogenizer construction confines the plungers in a sterilizable shell. The cylinder itself is constructed with a steam fitting, which supplies steam to each chamber through a common steam connection to all plunger bores. The steam is directed into each plunger bore and confined between sets of primary and secondary high-temperature plunger packing. The primary packing is spring-loaded and is used to seal the product being pumped into the pumping chamber. It also forms part of the steam seal of the sterilizing chamber. The secondary packing assembly is also spring-loaded and helps to complete the individual steam chamber around each plunger. The depth of the cylinder is designed to prevent any part of the plunger normally in contact with the cell suspension from being withdrawn from the sterile chamber.

In addition, it is possible to run the homogenizer during the process with steam or sterile media at all gaskets and seals along the product stream through the machine (Fig. 10). The double sealing allows a control system for all dynamic and static seals with controlled leakage draining, e.g., through a killing tank, and leakage detection by pressure, turbidity, flow or conductivity. To sterilize the cylinder, wet steam or superheated water at temperatures up to 148 °C is normally used. In operation, steam at 0.7 to 1 bar is fed into the top of the cylinder, and the condensate is removed through the built-in steam trap. Care must be taken that the steam pressure does not exceed the product inlet pressure during processing. In order to prolong packing life, steam feeding to plunger seals should be stopped during in-place cleaning cycles. In principle, safety control can be obtained during disruption of microbial cells in a high-pressure homogenizer by measuring

- flow
- feeding pressure
- homogenizing pressure
- product temperature
- sterilizing temperature and pressure
- leakage at dynamic seals (plunger packing system)
- leakage at static seals (e.g., homogenizing valve).

Nevertheless, it may be necessary for a fully contained process to operate the machine inside a cabinet, particularly if pathogens or other organisms suspected to be potentially hazardous are disrupted. Also the cooling water for the stuffing boxes should be contained in a closed system or passed to a kill tank.

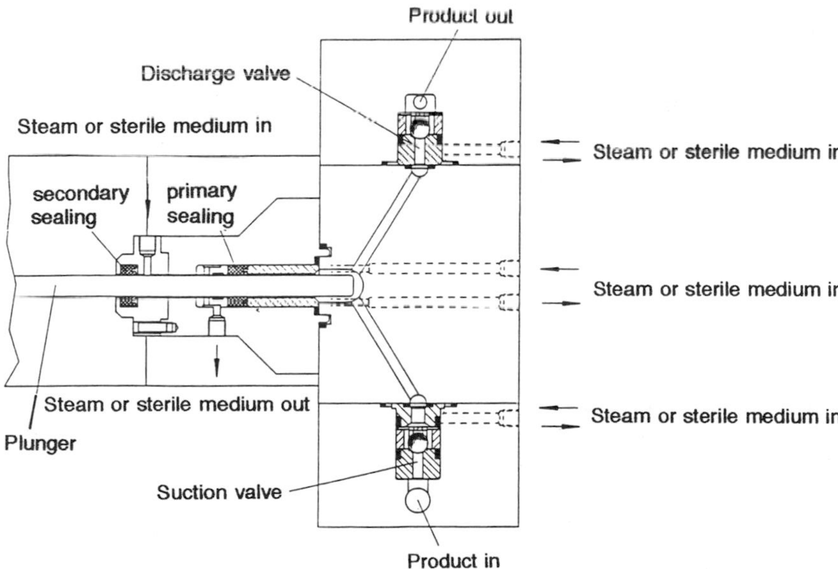

Fig. 10. Cylinder block assembly of a high-pressure homogenizer with double sealing for application in biotechnology.

6.2 Automation and Control

Besides the aseptic cylinder assembly the mechanical safety arrangements are very important when working on a production scale. In laboratory homogenizers the operating pressure, and thus the disintegration, is controlled by altering the pressure of the spring on the valve piston by means of a large handwheel, thus enabling operation between 0 and 125 MPa. For pilot- and process-scale machines a hydraulic valve actuator (HVA) allows automatic adjustment of homogenizing pressure ensuring uniform process conditions. The HVA system includes a hydraulic relief valve which should be carefully set to control the maximum desired homogenizing pressure, or the maximum safe operating pressure of the machine.

The operating pressure is controlled by a separate pressure-reducing valve permitting independent control. The electric pressure control (EPC) allows operation of the homogenizer with different throughputs at preset pressures. After starting the homogenizer, the desired operating pressure, indicated at the pressure transmitter, will be controlled by the EPC system. In case of failure, the servo valve in the HVA system will open and set the pressure to zero to avoid damage to the homogenizer.

Feeding control (FC) in combination with the HVA system prevents mechanical damage of the homogenizer by air or gas in the product line or loss of product. If no product is available at the suction side of the homogenizer or excessive air (gas) is in the feed, an electrical impuls from the inductive pressure gauge, which is set to a minimum of 2 bar (0.2 MPa), will attenuate a quick-unloading valve in the HVA system. The homogenizing pressure will be set to zero within fractions of a second, and the operator will be alarmed by a light or acoustic signal. A schematic view of such a mechanical safety system is shown in Fig. 11.

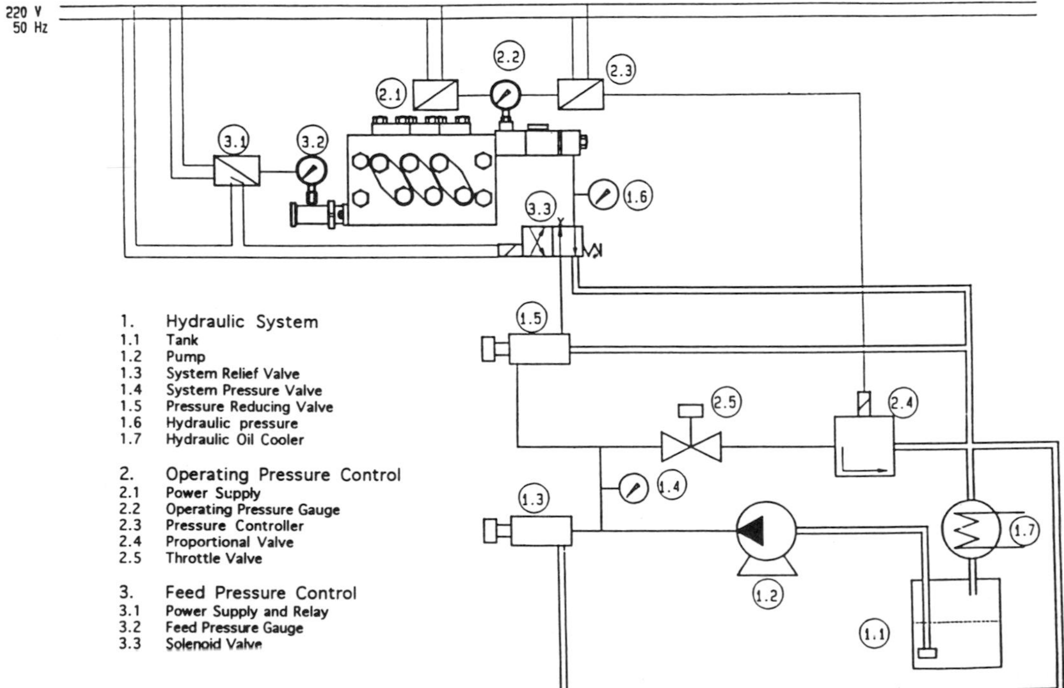

Fig. 11. Control and automation of high pressure homogenizer.

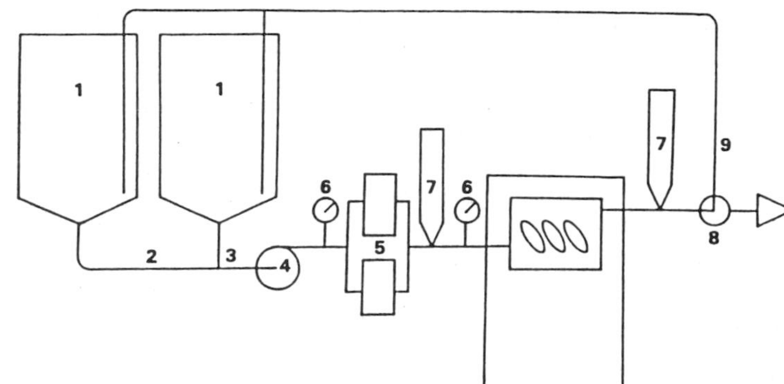

Fig. 12. Installation plan for cell disintegration, instead of the homogenizer depicted a bead mill may be placed. For designations see text.

6.3 Installation of a Mechanical Cell Disintegrator

Mainly two mechanical cell disintegrators are in common use in the biotechnology industry, the high-pressure homogenizers and the high-speed agitator bead mills. Microbial cells can be disrupted effectively with both types of machines; nevertheless, usually more than one pass is necessary through the machines to yield a high degree of product release. Therefore, installation should permit recycle of the product as shown in Fig. 12. It would be desirable to follow product release on-line (RECKTENWALD et al., 1985a). To improve performance the agitator in the feed tank (1 in Fig. 12) has to be selected to avoid early formation of a vortex, resulting in air entrapment. A low-level detector should be installed to sound alarm or automatically stop the machine and/or lower homogenizing pressure/agitator speed before the feed tank is completely emptied. Unnecessary length (2) and sharp turns in pipe lines should be avoided as far as practical, and diameters should be selected not less than those specified for the homogenizer/pump suction and discharge connections, respectively. For changes between feed tanks (3) the valve arrangement in the suction line should be such that the change-over will be gradual, allowing stagnant product in the line to accelerate to normal flow velocity before the feed from the first tank is cut completely. The feed pump (4) has to be dimensioned in order to ensure positive feed pressure immediately at the homogenizer inlet. The suction strainer (5) is recommended for protection of pump cylinders, valves and seals. Preferably duplex arrangement should be installed for easy maintenance without interruption of product flow. Suction pressure gauges (6) should be placed upstream of the filter and one immediately at the homogenizer inlet to register the inlet pressure and the pressure drop across the suction strainer to indicate any strainer problem. Hydraulic accumulators or vibration dampeners (7) are recommended for inlet and outlet line, to prevent vibrations produced by high flow velocities in the pipes. Under no circumstances should shut-off valves be installed in the discharge line; to divert product flow, a threeway valve (8) or similar designs have to be used which keep discharge open in at least one of several directions. Lastly, return line (9) must be installed with a discharge point preferably under the level of product in the feed tank to prevent incorporation of air. In any case the heat produced during the disintegration process must be removed by efficient and rapid chilling of the suspension in a direct (bead mills) or indirect (high-pressure homogenizers) manner via external heat exchangers to prevent thermal and/or proteolytic degradation of the desired product.

7 References

ANDREWS, B. A., ASENJO, J. A. (1987), Enzymatic lysis and disruption of microbial cells, *TIBTECH* **5**, 273–277.

ANDREWS, B. A., HUANG, R.-B., ASENJO, J. A. (1990), Differential product release from yeast cells by selective enzymatic lysis, in: *Separations for Biotechnology*, Vol. 2 (PYLE, D. L., Ed.), pp. 21-28, London: Elsevier.

APV GAULIN (1990), Homogenizers operation and service manual, *Technical Bulletin: Micron Lab 40*.

ASENJO, J., ANDREWS, B. (1990), Enzymatic cell lysis for product release, in: *Separation Processes in Biotechnology* (ASENJO, J., Ed.), pp. 143-175, New York: Marcel Dekker.

BARTH, H. G. (Ed.) (1984), *Modern Methods of Particle Size Analysis*, New York: John Wiley & Sons.

BENNET, A. M., BENBOUGH, J. E., HAMBLETON, P. (1990), Biosafety in downstream processing, in: *Separations for Biotechnology*, Vol. 2 (PYLE, D. L., Ed.), pp. 592-600, London: Elsevier.

BIRDSELL, D. C., COTA-ROBLES, E. H. (1968), Lysis of spheroplasts of *Escherichia coli* by a nonionic detergent, *Biochem. Biophys. Res. Commun.* **31**, 438-446.

BOROVIKOVA, V. E., ARSENOVSKAYA, V. E., LAURENOVA, G. I., KISLUKHINA, O. V., KALUNYANTS, K. A., STEPANOV, V. M. (1980), Isolation and characterization of lytic enzyme from *Bacillus subtilis, Biokimya* **45**, 1524-1533.

BRÄU, B., HILGENFELD, R., SCHLINGMANN, M., MARQUARDT, R., BIRR, E., WOHLLEBEN, W., AUFDERHEIDE, K., PÜHLER, A. (1991), Increased yield of a lysozyme after self-cloning of the gene in *Streptomyces coelicolor* "Müller", *Appl. Microb. Biotechnol.* **34**, 481-487.

BREDDAM, K., BEENFELDT, T. (1991), Acceleration of yeast autolysis by chemical methods for production of intracellular enzymes, *Appl. Microb. Biotechnol.* **35**, 323-329.

BROOKMAN, J. S. (1974), Mechanism of cell disintegration in a high pressure homogenizer, *Biotechnol. Bioeng.* **16**, 371-383.

BUCKLAND, B. C., RICHMOND, W., DUNNILL, P., LILLY, M. D. (1974), in: *Industrial Aspects of Biochemistry*, Vol. 30, Part I (SPENCER, B., Ed.), pp. 65-79, Amsterdam: North-Holland-American Elsevier.

BUESCHELBERGER, H. G., LONCIN, M. (1989), Untersuchungen zum mechanischen Aufschluß von Mikroorganismen in Hochdruck-Homogenisatoren, *Chem. Ing. Tech.* **61**, 420-421.

CULL, M., MCHENRY, C. S. (1990), Preparation of extracts from prokaryotes, *Methods Enzymol.* **182**, 147-153.

CURRIE, J. A., DUNNILL, P., LILLY, M. D. (1972), Release of protein from baker's yeast by disruption in an industrial agitator mill, *Biotechnol. Bioeng.* **14**, 725-736.

DE SMET, M. J., KINGMA, J., WITHOLT, B. (1978), The effect of toluene on the structure and permeability of the outer and cytoplasmic membranes of *Escherichia coli, Biochim. Biophys. Acta* **506**, 64-80.

DEUTSCHER, M. P. (1974), Preparation of cells permeable to macromolecules by treatment with toluene: studies of transfer ribonucleic acid nucleotidyltransferase, *J. Bacteriol.* **118**, 633-639.

DUNNILL, P. (1982), Biosafety in the large-scale isolation of intracellular microbial enzymes, *Chem. Ind.* (Nov.), 877-879.

ENGLER, C. R. (1990), Cell disruption by homogenizer, in: *Separation processes in Biotechnology* (ASENJO, J. A., Ed.), pp. 95-105, New York: Marcel Dekker.

ENGLER, C. R., ROBINSON, C. W. (1981), Effects of organism type and growth conditions on cell disruption by impingement, *Biotechnol. Lett.* **3**, 83-88.

FELIX, H. (1982), Permeabilized cells, *Anal. Biochem.* **120**, 211-234.

FENTON, D. M. (1982), Solvent treatment for β-D-galactosidase release from yeast cells, *Enzyme Microb. Technol.* **4**, 229-232.

FISH, N. M., HARBRON, S., ALLENBY, D. J., LILLY, M. D. (1983), Oxidation on *n*-alkanes: Isolation of alkane hydroxylase from *Pseudomonas putida, Appl. Microbiol. Biotechnol.* **17**, 57-63.

FOLLOWS, M., HETHERINGTON, P. J., DUNNILL, P., LILLY, M. D. (1971), Release of enzymes from baker's yeast by disruption in an industrial homogenizer, *Biotechnol. Bioeng.* **13**, 549-560.

GRAY, P. P., DUNNILL, P., LILLY, M. D. (1972), The continuous flow isolation of enzymes, in: *Fermentation Technology Today* (TERNI, G., Ed.), pp. 347-351, Tokyo: Society for Fermentation Technology Japan.

HAMMES, G. G., SWANN, J. C. (1967), Influence of denaturing agents on solvent structure, *Biochemistry* **6**, 1591-1596.

HAMMOND, P. M., PRICE, C. P., SCAWEN, M. D. (1983), Purification and properties of aryl acylamidase from *Pseudomonas fluorescens* ATCC 39004, *Eur. J. Biochem.* **132**, 651-655.

HARRIS-SMITH, R., EVANS, C. G. T. (1974), Bioengineering and protection during hazardous microbiological processes, *Biotechnol. Bioeng. Symp.* **4**, 837-855.

HARRISON, S. T., DENNIS, J. S., CHASE, H. A. (1990), The effect of culture history on the disruption of *Alcaligenes eutrophus* by high pressure homogenization, in: *Separations for Biotechnology*, Vol. 2 (PYLE, D. L., Ed.), pp. 38-47, London: Elsevier.

HATEFI, J., HANSTEIN, W. G. (1974), Destabilization of membranes with chaotropic ions, *Meth-*

ods Enzymol. **31**, 770–790.

HAYASHI, K., KASUMI, T., KUBO, N., TSUMURA, N. (1981), Purification and characterization of the lytic enzyme produced by *Streptomyces rutgersensis* H-46, *Agric. Biol. Chem.* **45**, 2289–2300.

HETHERINGTON, P. J., FOLLOWS, M., DUNNILL, P., LILLY, M. D. (1971), Release of protein from baker's yeast by disruption in an industrial homogenizer, *Trans. Inst. Chem. Eng.* **49**, 142–148.

HETTWER, D., WANG, H. (1989), Protein release from *E. coli* cells permeabilized with guanidine-HCl and Triton X-100, *Biotechnol. Bioeng.* **33**, 886–895.

HOBOT, J. A., FELIX, H. R., KELLENBERGER, E. (1982), Ultrastructure of permeabilised cells of *Escherichia coli* and *Cephalosporium acremonium, FEMS Microbiol. Lett.* **13**, 57–61.

INGRAM, L. O. (1981), Mechanism of lysis of *Escherichia coli* by ethanol and other chaotropic agents, *J. Bacteriol.* **146**, 331–336.

JACKSON, R. W., DE MOSS, J. A. (1965), Effects of toluene on *Escherichia coli, J. Bacteriol.* **90**, 1422–1425.

KESHAVARZ, E., HOARE, M., DUNNILL, P. (1987), Biochemical engineering aspects of cell disruption, in: *Separations for Biotechnology*, Vol. 1 (VERRALL, M. S., HUDSON, M. J., Eds.), pp. 62–79, New York: Halsted Press - John Wiley & Sons.

KESHAVARZ, E., BONNERJEA, J., HOARE, M., DUNNILL, P. (1990), Disruption of a fungal organism, *Rhizopus nigricans*, in a high-pressure homogenizer, *Enzyme Microb. Technol.* **12**, 494–498.

KESHAVARZ-MOORE, E., HOARE, M., DUNNILL, P. (1990), Disruption of baker's yeast in a high-pressure homogenizer: New evidence on mechanism, *Enzyme Microb. Technol.* **12**, 764–770.

KITAMURA, K., YAMAMOTO, Y. (1972), Purification and properties of an enzyme, zymolyase, which lyses viable yeast cells, *Arch. Biochem. Biophys.* **153**, 403–406.

KNORR, D., SHETTY, K. J., KINSELLA, J. E. (1979), Enzymatic lysis of yeast cell walls, *Biotechnol. Bioeng.* **21**, 2011–2021.

KOBAYASHI, R., MIWA, T., YAMAMOTO, S., NAGASAKI, S. (1982), Preparation and evaluation of an enzyme which degrades yeast cell walls, *Eur. J. Appl. Microbiol. Biotechnol.* **15**, 14–19.

KRONER, K. H., HUSTEDT, H., GRANDA, S., KULA, M.-R. (1978), Technical aspects of separation using aqueous two-phase systems in enzyme isolation process, *Biotechnol. Bioeng.* **20**, 1967–1988.

KULA, M.-R., SCHÜTTE, H. (1987), Purification of proteins and the disruption of microbial cells, *Biotechnol. Prog.* **3**, 31–42.

KULA, M.-R., KRONER, K. H., HUSTEDT, H. (1982), Purification of enzymes by liquid–liquid extraction, *Adv. Biochem. Eng.* **24**, 73–118.

KULA, M.-R., SCHÜTTE, H., VOGELS, G., FRANK, A. (1990), Cell disintegration and purification of intracellular proteins, *Food Biotechnol.* **4**, 169–183.

LECORRE, S., ANDREWS, B. A., ASENJO, J. A. (1985), Use of a lytic enzyme system from *Cytophaga* sp. in the lysis of Gram-positive bacteria, *Enzyme Microb. Technol.* **7**, 73–78.

LIMON-LASON, J., HOARE, M., ORSBORN, C. B., DOYLE, D. J., DUNNILL, P. (1979), Reactor properties of a high-speed bead mill for microbial cell rupture, *Biotechnol. Bioeng.* **21**, 745–774.

MADSEN, F. S., IBSEN, C. I. (1987), *Cell disruption by means of high pressure*, Dept. of Biotechnology, Engineering Academy of Denmark, Copenhagen, January.

MARFFY, F., KULA, M.-R. (1974), Enzyme yields from cells of brewer's yeast disrupted by treatment in a horizontal disintegrator, *Biotechnol. Bioeng.* **16**, 623–634.

MASUCCI, S. F. (1985), Improving the efficiency of high pressure homogenizers for cell disruption application, *B. S. Thesis*, MIT, Cambridge, MA, USA.

MEYER, L. D. (1988), Hochdruckhomogenisiermaschinen für den Zellaufschluß: Einsatzmöglichkeiten und steriltechnische Maßnahmen, *DECHEMA-Monogr.* **113**, 145–163.

MIDDELBERG, A. P. J., O'NEILL, B. K., BOGLE, I. D. L. (1991), A novel technique for the measurement of disruption in high-pressure homogenization: Studies on *E. coli* containing recombinant inclusion bodies, *Biotechnol. Bioeng.* **38**, 363–370.

MIOZARRI, G., NIEDERBERGER, P., HÜTTNER, R. (1978), Permeabilization of microorganisms by Triton X-100, *Anal. Biochem.* **90**, 220–233.

MÖLLS, H., HÖRNLE, R. (1972), Wirkungsmechanismus der Naßzerkleinerung in der Rührwerkskugelmühle, *DECHEMA-Monogr.* **69**(2), 631–661.

MURAKAMI, K., NAGURA, H., YOSHINO, M. (1980), Permeabilization of yeast cells: application to study on the regulation of AMP deaminase activity *in situ, Anal. Biochem.* **105**, 407–413.

MURATA, K., TANI, K., KATO, J., CHIBATA, I. (1980), Glutathione production coupled with an ATP regeneration system, *Eur. J. Appl. Microbiol. Biotechnol.* **10**, 11–21.

NAGLAK, T. J., WANG, H. Y. (1990), Protein release from the yeast *Pichia pastoris* by chemical permeabilization: comparison to mechanical disruption and enzymatic lysis, in: *Separations for Biotechnology*, Vol. 2 (PYLE, D. L., Ed.), pp. 55–64, London: Elsevier.

OBATA, T., IWATA, H., NAMBA, Y. (1977), Proteolytic enzyme from *Oerskovia* sp. CK lysing viable yeast cells, *Agric. Biol. Chem.* **41**, 2387-2394.

RECKTENWALD, A., KRONER, K. H., KULA, M.-R. (1985a), Rapid on-line protein detection in biotechnological processes by flow injection analysis (FIA), *Enzyme Microb. Technol.* **7**, 146-149.

RECKTENWALD, A., KRONER, K. H., KULA, M.-R. (1985b), On-line monitoring of enzymes in downstream-processing by flow-injection-analysis (FIA), *Enzyme Microb. Technol.* **7**, 607-612.

REHACEK, J., SCHAEFER, J. (1977), Disintegration of microorganisms in an industrial horizontal mill of novel design, *Biotechnol. Bioeng.* **19**, 1523-1534.

SAUER, T., ROBINSON, C. W., GLICK, B. R. (1989), Disruption of native and recombinant *Escherichia coli* in a high-pressure homogenizer, *Biotechnol. Bioeng.* **33**, 1330-1342.

SCHNAITMAN, C. A. (1971), Solubilization of the cytoplasmic membrane of *Escherichia coli* by Triton X-100, *J. Bacteriol.* **108**, 545-552.

SCHÜTTE, H., KULA, M.-R. (1990a), Pilot- and process scale techniques for cell disruption, *Biotechnol. Appl. Biochem.* **12**, 599-620.

SCHÜTTE, H., KULA, M.-R. (1990b), Bead mill disruption, in: *Downstream Processing in Biotechnology* (ASENJO, J. A., Ed.), pp. 107-142, New York: Marcel Dekker.

SCHÜTTE, H., KRONER, K. H., HUSTEDT, H., KULA, M.-R. (1983), Experiences with a 20 l industrial bead mill for the disruption of microorganisms, *Enzyme Microb. Technol.* **5**, 143-148.

SCHÜTTE, H., KRAUME-FLÜGEL, R., KULA, M.-R. (1986), Scale-up of mechanical cell disintegration - influence of the stirrer geometry on the residence time distribution and the cell disintegration in a 20 l high speed bead mill, *Germ. Chem. Eng.* **9**, 149-156.

SCHÜTTE, J., JÜRGING, B., PAPAMICHAEL, N., OTT, K., KULA, M.-R. (1988), Improvement of rotor design of high-speed agitator ball mills for continuous disruption of microorganisms, in: Enzyme Engineering 9, *Ann. NY Acad. Sci.* **542**, 121-125.

SCOTT, J. H., SCHEKMAN, R. (1980), Lyticase: Endoglucanase and protease activities that act together in yeast cell lysis, *J. Bacteriol.* **142**, 414-423.

SEVA, R., FIESCHKO, J., SACHDEV, R., MANN, M. (1986), Cell breakage using a microfluidizer, presented at: *Annual Meeting of the Society of Industrial Microbiology*, 1986, San Francisco.

SHERWOOD, R. F., COURT, J., MOTHERSHAW, A., KEEVIL, W., ELLWOOD, D., JACK, G., GILBERT, H., BLAZEK, R., WADE, J., ATTKINSON, T., HOHNSTROM, B. (1982), in: *From Gene to Protein: Translation into Biotechnology* (AHMAD, F., SCHULTZ, J., SMITH, E. E., WHELAN, W. J., Eds.), pp. 564-565, London: Academic Press.

STEWART, J. W., DEANS, J. S. (1990), Containment testing of cell disruptions, *Warren Spring Laboratory Report no. LR 767 (BT)*.

SUZUKI, K., UYEDA, M., SHIBATA, M. (1985), *Serratia marcescens*-lytic enzyme produced by *Micromonospora* sp. strain no. 152, *Agric. Biol. Chem.* **49**, 1719-1726.

TEUBER, M. (1970), Release of the periplasmic penicillinases from *Escherichia coli* by toluene, *Arch. Microbiol.* **73**, 61-64.

TUIJNENBURG-MUIJS, G. (1987), Air and surface contamination during microbiological processing, *Swiss Biotech.* **5** (2a), 43-49.

VALISENA, S., VARALDO, P. E., SATTA, G. (1982), Purification and characterization of three separate bacteriolytic enzymes excreted by *Staphylococcus aureus, Staphylococcus simulans*, and *Staphylococcus saprophyticus, J. Bacteriol.* **151**, 636-647.

VAN GAVER, D., HUYGHEBAERT, A. (1990), Optimization of yeast cell disruption with a newly designed bead mill, *Enzyme Microb. Technol.* **13**, 665-671.

VANDAMME, E. J., VOETS, J. P. (1974), Properties of the purified penicillin V-acylase of *Erwinia aroideae, Experientia* **31**, 140-143.

VOGELS, G. (1990), Untersuchungen kombinierter Zellaufschlußverfahren zur Verbesserung der Proteinfreisetzung aus Mikroorganismen, *Dissertation*, Heinrich-Heine-Universität, Düsseldorf, FRG.

VOGELS, G., KULA, M.-R. (1992), Combination of enzymatic and/or thermal pretreatment with mechanical cell disintegration, *Chem. Eng. Sci.* **47**, 123-131.

VRSANSKA, M., BIELY, P., KRATKY, Z. (1977), Enzymes of yeast lytic system produced by *Arthrobacter* GJM-1 bacterium and their role in lysis of yeast-cell walls, *Z. Allg. Mikrobiol.* **17**, 465-480.

WADE, H. E. (1971), Improvement in or relating to extraction processes, *Brit. Patent Specification B 125 8063*.

WECKE, J., LAHAV, M., GRINSBURG, I., GIESBRECHT, P. (1982), Cell wall degradation of *Staphylococcus aureus* by lysozyme, *Arch. Microbiol.* **131**, 116-123.

WITHOLT, B., VAN HEERIKHUIZEN, H., DE LEIJ, L. (1976), How does lysozyme penetrate through the bacterial outer membrane?, *Biochim. Biophys. Acta* **443**, 534-544.

WOLDRINGH, C. L. (1970), Lysis of the cell membrane of *Escherichia coli* K12 by ionic detergents, *Biochim. Biophys. Acta* **224**, 288-290.

YANSHINSKI, S. (1984), Protein engineering makes its mark on enzymes, *New Sci.* **103** (1420), 27.

20 *In vitro* Protein Refolding

JEFFREY L. CLELAND

South San Francisco, CA 94080, USA

DANIEL I. C. WANG

Cambridge, MA 02139, USA

1 Introduction

Before attempting to produce proteins as biologicals or pharmaceuticals, the producer must have an understanding of the biophysical characteristics of the protein. The composition and structure of the final protein product must be the same for every batch of material produced. In addition, the pharmaceutical company must convince the appropriate government regulatory agency that the physicochemical properties of the protein are well understood and that any variations in these properties can be controlled. The behavior of the protein in different physical environments such as elevated temperature must also be well documented.

To acquire the knowledge of the physicochemical properties of the protein, the bioprocess engineer must have a general understanding of protein structure and the effects of the solvent environment on this structure. An excellent review of protein structure has been presented by CREIGHTON (1984). In general, 20 different amino acids along with various amino acid analogs can be incorporated into a protein backbone. The proportion of each amino acid and their sequential order as well as the surrounding environment (pH, temperature, solvent, etc.) determine the secondary structure of the protein. Three secondary structures are common elements of a protein. These structures are *alpha*-helices, *beta*-sheets, and random coils as shown in Fig. 1. Additional elements of secondary structure include

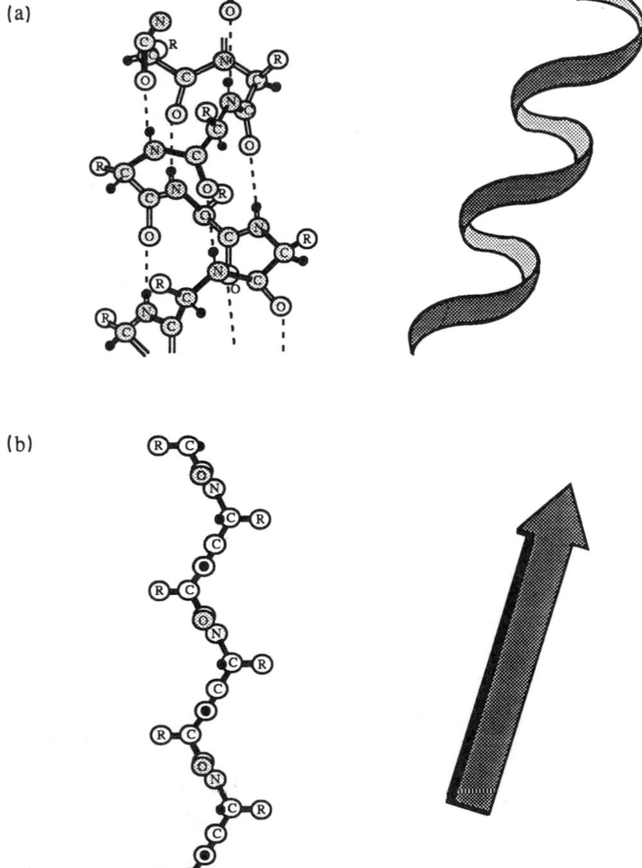

Fig. 1. Molecular models (CREIGHTON, 1984) and ribbon diagrams (RICHARDSON, 1981) of two major components of protein structure (a) *alpha*-helix and (b) *beta*-sheet.

external *omega*-loops and various conformational turns. These sections of secondary structure determine the possible tertiary structures which are thermodynamically stable. The protein may be held in a stable structural state by disulfide bonds between cysteine residues or by proline residues that have obtained a stable configuration (*cis* or *trans*). Since there are a multitude of unknown potential energy minima for a given protein, it is impossible with the current technology to determine the final conformation of the protein from its primary sequence. Therefore, X-ray crystallography or multi-dimensional nuclear magnetic resonance (NMR) must be performed to assess the final tertiary structure of the protein. Unfortunately, these techniques are cost- and labor-intensive and cannot be performed on a routine basis. The manufacturer must then rely on several different measurements to assure that the protein has the correct tertiary structure, biological activity, and chemical composition.

Tab. 1. Methods for Protein Structure and Chemical Analysis

Methods	Structure/Composition Detected	[Protein] (mg/mL)[a]
Spectroscopic Techniques		
Absorbance		0.1–1.0
Zero-order spectrum	Gross structural changes, sensitive to aromatic amino acids and disulfides	
Second derivative	Small perturbations in aromatic amino acid orientation	
Fluorescence		0.01–1.0
Anisotropy	Molecular size, fluorophore environment	
Polarization	Changes in fluorophore environment, usually tryptophans	
Quenching	Fluorophore accessibility, requires quenching agent (e.g. iodide)	
Hydrophobic probes	Exposure of hydrophobic sections, requires fluorophore probe	
Circular dichroism (CD)		0.1–1.0
Near UV	Tertiary structural changes, aromatic amino acid environment	
Far UV	Secondary structure changes, *alpha*-helix, *beta*-sheet, or random coil	
Light scattering		
Classical	Precipitates or large aggregates	>0.5
Quasi-elastic	Collapse of protein from random coil and formation of multimers	<1.0
Chromatography		>0.01
Size exclusion	Molecular size, hydrodynamic radius	
Ion exchange	Exposure of charged groups, deamidation	
Hydrophobic interaction	Proteolysis and exposure of hydrophobic regions	
Reverse phase	Oxidation of methionines, exposure of hydrophobic sections	
Affinity	Specific conformational epitopes	
Activity		Variable
Biological activity	Active site or binding site	
Antibodies	Multiple conformational epitopes	
Ligands	Specific binding conformation	
Inhibitors	Inhibitor binding site formation	
Miscellaneous		
Differential scanning calorimetry	Gross conformational changes, thermodynamics of folding	0.10–1.0
Electron spin resonance	Specific environment of labeled residue, size of labeled protein	0.01–1.0
Nuclear magnetic resonance (NMR)	Three-dimensional structure (liquid or solid)	>10 mg/mL
X-ray crystallography	Solid state three-dimensional structure	Crystals

[a] Range of protein concentrations usually required for analysis.

These techniques and their utility are outlined in Tab. 1. Each of these techniques has been used in studies of protein folding to measure formation of both native protein and various unwanted by-products. Assessment of protein stability in different environments is also performed with these techniques.

Once the final native conformation of the protein has been ascertained, the bioprocess engineer can consider different routes of production which will yield the greatest recovery of final product. A variety of different host organisms are currently used to produce proteins as biologicals or pharmaceuticals. The most common host organisms are *Escherichia coli, Saccharomyces cerevisiae,* and Chinese Hamster Ovary (CHO) cells. Since the majority of proteins expressed by recombinant techniques are originally derived from eukaryotic organisms, CHO cells provide the most natural production method. In most cases, CHO cells will produce the desired protein with the same composition and structure as the natural protein. Unfortunately, the use of CHO cells for protein production is often cost-intensive, since these cells require complex media for growth, have low productivities, and often require the use of expensive purification techniques. As an alternative, yeast can be used as an expression system. Yeast will usually synthesize and process the protein in the correct conformation, but it will not provide the proper carbohydrate structures on the protein. If the protein does not require glycosylation for biological activity, slower clearance time, or immune tolerance, it could be expressed and produced in *E. coli.* This host expression system provides the least expensive production method for many proteins. However, *E. coli* often does not produce the protein in its native state and, therefore, additional processing is required to refold the protein to its native conformation. If these additional processing costs can be reduced by increasing the yield of active protein during refolding, the production of therapeutic proteins in *E. coli* could be a very cost-effective method.

Several methods of protein refolding have been developed in an attempt to achieve high yields of active protein. As shown in Fig. 2, insoluble protein aggregates which form in *E. coli* must be isolated and solubilized with a

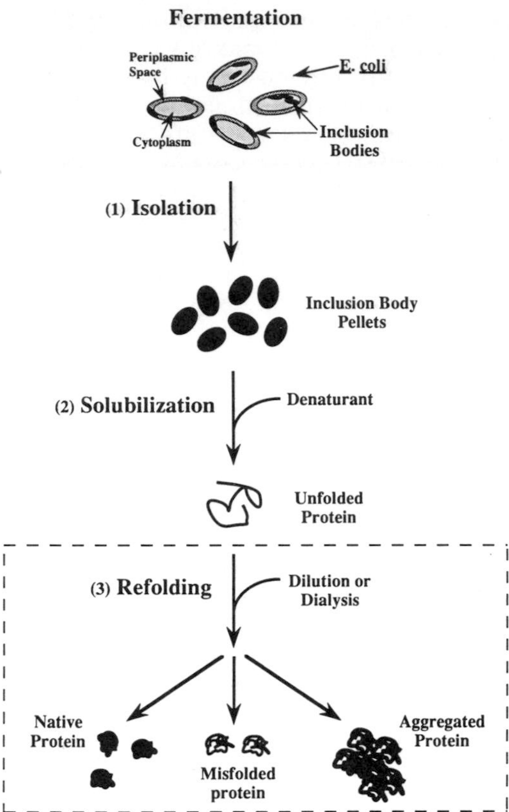

Fig. 2. Recovery of native protein from *Escherichia coli* inclusion bodies. Fermentation of *E. coli* overproducing a recombinant protein often results in the formation of inclusion bodies within the periplasm or cytosol of the cell (see Sects. 2.2 and 2.3). To recover the native protein from these insoluble aggregates, the inclusion bodies must be isolated from the cell (1). Isolation usually involves homogenization to disrupt the cell membrane and centrifugation to separate the resulting cell debris from the protein aggregates. The inclusion bodies are then solubilized with a denaturant or acidic or basic conditions (2). In addition, a reducing agent must also be used if the protein contains cysteine residues. The final step is then removal of the denaturant and reducing agent by dilution or dialysis (3). During this step, the protein may form aggregates or misfolded protein structures instead of refolding to the native state. The dashed box indicates the processing steps that are not well understood and are addressed in this chapter.

denaturant such as urea, guanidine hydrochloride (GuHCl), or sodium dodecyl sulfate (SDS), and a reducing agent such as glutathione, β-mercaptoethanol, or dithiothreitol (DTT). The denaturant and reducing agent must be removed to allow the protein to regain its native state. Denaturant and reducing agent removal has typically been performed by dilution or dialysis with a stabilizing buffer solution. The reducing agent is either exchanged with similar agents which provide oxidizing conditions, or molecular oxygen is used to promote disulfide bond formation. In general, refolding must be performed at low protein concentrations (μg/mL) and high denaturant concentrations to reduce the degree of aggregation.

To develop improved methods of protein refolding, it is necessary to understand the mechanisms which govern folding both *in vivo* and *in vitro*. The pathway of folding is dependent on the host cell processing as well as the protein composition. In addition, *in vitro* folding is affected by the solvent environment and physical conditions such as temperature. To understand protein folding, one must have a rudimentary knowledge of a few specific principles which relate the final protein structure to its processing and environment.

2 Folding in Eukaryotic and Prokaryotic Hosts

Most proteins can be expressed in their native form in eukaryotic hosts such as CHO cells. Therefore, a comparison between eukaryotic and prokaryotic protein expression could provide insight into folding and inclusion body formation in prokaryotic systems. Protein expression in eukaryotic hosts involves several processing steps which have been described previously (DARNELL et al., 1986; ROBINSON and AUSTIN, 1987). The cellular machinery involved in the production and export of proteins in eukaryotes diverges significantly from that used by *E. coli*. Several cellular compartments, post-translational modifications and folding aids such as molecular chaperones (discussed below) are present in eukaryotes. In particular, a majority of the protein folding reactions in animal cells occurs in the endoplasmic reticulum (ER).

2.1 Expression, Folding, and Secretion in Eukaryotes

The production of recombinant proteins in eukaryotes such as CHO cells is usually performed with cell lines which will secrete the desired protein in its native state. Therefore, the protein must be expressed with the appropriate precursors to assure complete processing and export. Proteins designated for secretion in eukaryotes are initially synthesized with precursors which are subsequently removed in the ER. The leader or signal sequence usually consists of three different domains (NOTHWEHR and GORDON, 1990; KIKUCHI and IKEHARA, 1991). The amino terminal region of the leader has a net positive charge and is followed by a hydrophobic core region which typically consists of 7 to 16 amino acids. Finally, the carboxy terminal region contains 4 to 6 relatively polar amino acids (NOTHWEHR and GORDON, 1990). The carboxy terminus of the leader peptide usually consists of proline and glycine residues which likely provide a structural turn away from the nascent protein. The last three residues at the carboxy terminus of the leader sequence may also be required for subsequent processing (KIKUCHI and IKEHARA, 1991). This signal sequence is acknowledged by the signal recognition particle (SRP) as it is translated from the ribosome into the cytoplasm (Fig. 3a). As the polypeptide chain emerges from the ribosome, SRP binds to the ribosome and the signal sequence, stops translation, and docks the ribosome on the ER membrane by binding to the SRP receptor which is anchored in the membrane. SRP is a very unique molecule with three amphipathic *alpha*-helices containing several methionine residues. The residues provide a flexible surface which can adapt to the variety of different residues on the leader sequence of the protein (GELLMAN, 1991). The leader sequence is removed by a signal peptidase as the protein is translocated into the ER as described previously (WALTER et al., 1981; MEYER, 1982).

Alternatively, translocation of the protein into the ER may occur through interactions with binding protein (BiP) which is a member of the heat shock family of proteins (Hsp 70) discussed below. BiP has been shown to bind to proteins which have begun to fold or have folded in the ER. By binding to the protein, BiP maintains the protein in an unfolded state which is competent for translocation across the ER membrane. However, once the protein

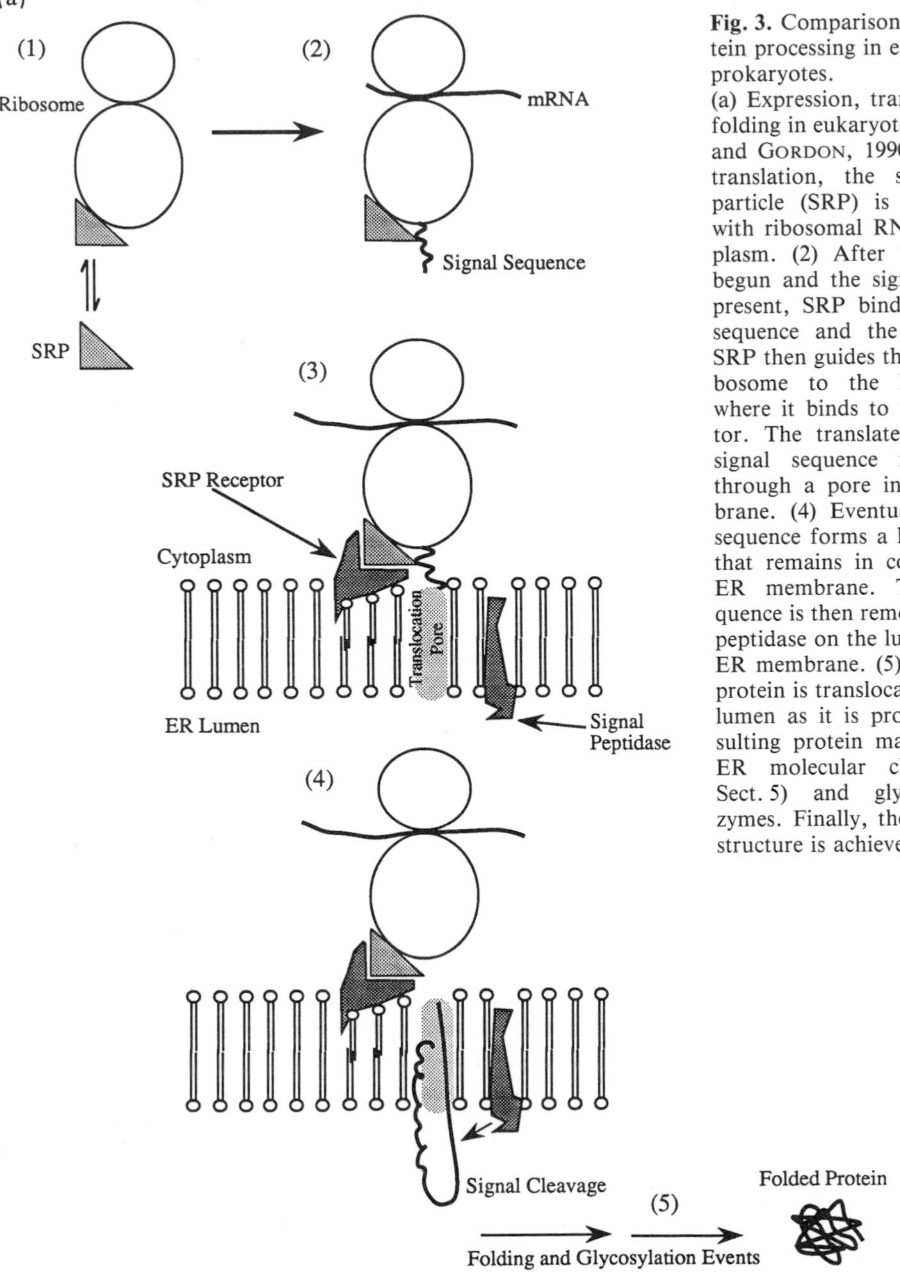

Fig. 3. Comparison between protein processing in eukaryotes and prokaryotes.
(a) Expression, translocation, and folding in eukaryotes (NOTHWEHR and GORDON, 1990). (1) Prior to translation, the signal receptor particle (SRP) is in equilibrium with ribosomal RNA in the cytoplasm. (2) After translation has begun and the signal sequence is present, SRP binds to the signal sequence and the ribosome. (3) SRP then guides the translating ribosome to the ER membrane where it binds to the SRP receptor. The translated protein with signal sequence is translocated through a pore in the ER membrane. (4) Eventually, the signal sequence forms a helical structure that remains in contact with the ER membrane. The signal sequence is then removed by a signal peptidase on the lumen side of the ER membrane. (5) The remaining protein is translocated into the ER lumen as it is produced. The resulting protein may interact with ER molecular chaperones (see Sect. 5) and glycosylation enzymes. Finally, the native protein structure is achieved.

has been translocated into the ER, BiP may remain bound until proper folding has been completed (HUBBARD and SANDER, 1991; NGUYEN et al., 1991). The action of BiP combined with overexpression of protein can result in protein aggregation in the ER. Overexpression can strain the cellular machinery required to process the protein and, thus, partially folded protein structures accumulate. For example, high levels of protein expression in

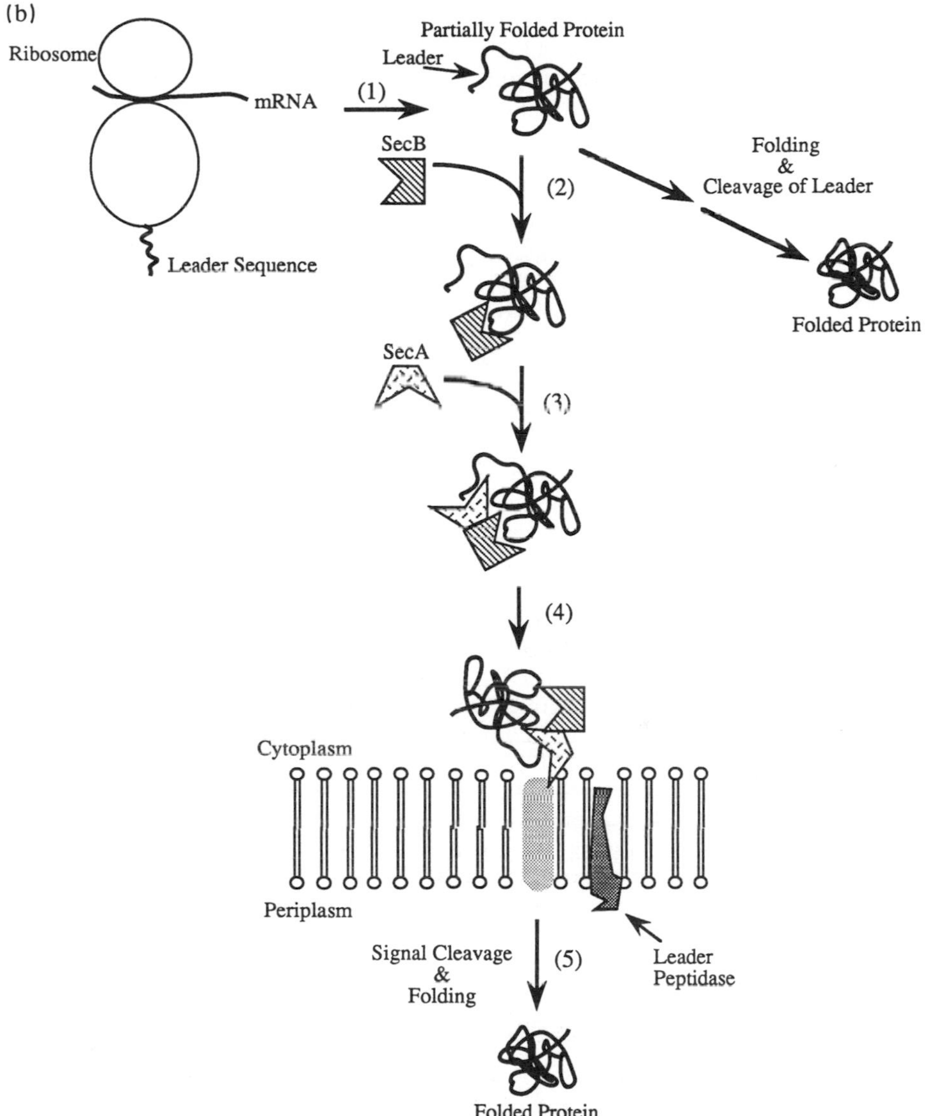

Fig. 3. (b) Expression, translocation, and folding in prokaryotes (WICKNER et al., 1991). (1) After the protein is translated by the cytoplasmic ribosomes, it slowly folds. (2) The leader sequence on the protein interferes with the folding allowing SecB to bind to a partially folded intermediate. Alternatively, the leader sequence can be removed allowing the protein to fold and not be exported. (3) For exported proteins, SecA binds to the leader sequence and SecB and (4) facilitates translocation of the protein into the periplasm. (5) During translocation, the leader sequence is cleaved by a leader peptidase. The protein may then fold or remain in a partially folded state.

yeast have resulted in the formation of intracellular aggregates (COUSENS et al., 1987).

After removal of the signal sequence or dissociation from BiP in the ER, the protein initiates folding into its native conformation. Several processes then occur during folding in the ER. First of all, the protein will collapse to form a compact structure and subsequently rearranges to form a native-like structure. The rate of folding to this first native-like protein structure can be catalyzed by a molecular chaperone, peptidyl-prolyl isomerase (PPI) or cyclophilin, which facilitates *cis-trans* isomerization of proline residues within the protein (LODISH and KONG, 1991). The isomerization of proline residues is the slow step in the folding of several proteins (PTITSYN et al., 1990). This catalytic step may then be necessary to prevent intracellular accumulation of folding intermediates.

After partially folding or immediately upon translocation, the protein can become glycosylated by interactions with several ER enzymes (DARNELL et al., 1986). Prior to glycosylation, the amino acid sequence required for glycosylation is recognized by the glycosylation site binding protein (GSBP). This protein has also been observed to catalyze folding and is sequentially identical to the molecular chaperone, protein disulfide isomerase (PDI), which catalyzes disulfide exchange during folding (LAMANTIA et al., 1991), GSBP/PDI may assist the protein in obtaining a native-like state prior to glycosylation and, thereby, prevent incorrect glycosylation of the protein. Proper glycosylation is also necessary for complete folding and inhibition of aggregation within the ER (RIEDERER and HINNEN, 1991). However, glycosylation is usually not required to achieve the native conformation as shown by the successful refolding of several unglycosylated eukaryotic proteins expressed in *E. coli* (MARSTON, 1986).

2.2 Folding and Secretion in Prokaryotes

In contrast to eukaryotic systems, protein expression in prokaryotes such as *E. coli* occurs by significantly different mechanisms as described in recent reviews (RANDALL and HARDY, 1984; RANDALL et al., 1987; KUMAMOTO, 1991). The export of folded protein from *E. coli* may also proceed as the result of a leader sequence on the protein (PARK et al., 1988). Randall and colleagues have proposed that export of protein from *E. coli* is limited by a kinetic competition between export and folding since the protein must be in a partially folded state to be competent for export (see Fig. 3b). This theory was demonstrated through mutations in both the leader sequence and the nascent protein. The defect in the leader sequence decreased export of the protein (RANDALL et al., 1990). An additional mutation in the nascent protein sequence of the mutant protein slowed the rate of folding and increased the export (TESCHKE et al., 1991). These results have also been explained by the mechanism of the molecular chaperone, SecB, in protein export from *E. coli*. SecB bound partially folded proteins with a high affinity and maintained them in a transport competent state (HARDY and RANDALL, 1991; TESCHKE et al., 1991). Thus, if the protein folds slowly, a partially folded state will be highly populated such that SecB binding will be favored resulting in protein export. Additional studies have also indicated that SecB may have a limited specificity and that some proteins may be transported without its assistance (LAMINET et al., 1991). However, these proteins may be held in an export competent state by other molecular chaperones (LAMINET et al., 1991).

Attempts to produce secreted and properly folded protein from *E. coli* have had only limited success. For example, bovine somatotropin was expressed with a signal sequence (LamB) which is a common signal sequence on proteins that have been observed to bind to SecB. The signal sequence was removed during export resulting in secretion of mature bovine somatotropin, but the yield of mature native protein (1–2 $\mu g/OD_{550}$) was not useful for commercial production (KLEIN et al., 1991). The low yield of native protein was likely the result of the limitations of the other cellular machinery (KLEIN et al., 1991). To overcome this problem, it has been postulated that the coexpression of eukaryotic folding aids such as PPI, PDI and those discussed below could be used to increase folding yields in bacterial systems (SCHEIN, 1989; GOLOUBINOFF et al., 1989). Unfortunately, this method may have limited utility and does not address the other environmental differences between folding in prokaryotes and eukaryotes (e.g., differences in redox potential). As observed in some cases for eukaryotes, overexpression of proteins in *E. coli* can result in the formation of inclusion bodies, and inclusion body formation is usually independent of protein composition or origin, prokaryotic or eukaryotic (MARSTON, 1986).

3 Inclusion Body Formation

Since prokaryotic organisms do not have the necessary machinery for complete protein processing and export, production of eukaryotic proteins in prokaryotes such as *E. coli* often results in the formation of inactive or aggregated protein. Proteins expressed in *E. coli* can achieve three different final forms (KANE and HARTLEY, 1988). The protein can be expressed as either a stable species which forms soluble native protein or an unstable species. The unstable intermediate species can then be degraded or it accumulates in the form of inclusion bodies (KANE and HARTLEY, 1988). A recent review of inclusion body formation suggested that proteins with low proline content and long sequences of acidic amino acids do not form insoluble aggregates (SCHEIN, 1989). However, this observation may not be generally valid as noted in reviews by MITRAKI and coworkers (MITRAKI and KING, 1989; MITRAKI et al., 1991a). They have studied the mechanisms of inclusion body formation and concluded that these aggregates are the result of partially folded intermediates (MITRAKI and KING, 1989). The aggregation of these intermediates was observed to be a function of the environment of the protein. In particular, increased temperatures were observed to enhance the formation of aggregates as discussed below (MITRAKI et al., 1991a). Since the environment of the intermediate affects its aggregation properties, manipulation of the growth conditions for *E. coli* could result in modifications of inclusion body formation. For example, β-lactamase inclusion body formation which occurred in the periplasmic space of *E. coli* was reduced by the addition of non-metabolized sugars such as sucrose and raffinose to the culture media (GEORGIOU et al., 1986; BOWDEN and GEORGIOU, 1988). Therefore, it should be possible to control the extent of inclusion body formation by modifications in the growth conditions. These modifications could include growth at low temperatures and addition of aggregation inhibitors such as sugars to the media. Higher concentrations of soluble protein could then be achieved with these modifications. In addition, the expression of the precursor portion of the protein has been observed to be essential for complete protein folding (OHTA et al., 1991). Before expressing the precursor protein, the extent to which the precursor affects folding should be assessed, since precursors have also been observed to facilitate inclusion body formation (TOKATLIDIS et al., 1991). By controlling the external environment (i.e., temperature) and expressing the precursor portion of the protein, higher recovery of soluble native protein may be achieved in *E. coli*.

4 Effect of Folding Mutations

In vivo folding and inclusion body formation have been altered by folding mutations which stabilize a folding intermediate. Reviews of the methodology for applying protein engineering to folding studies are useful in providing conceptual background (KING, 1986; GOLDENBERG, 1988). An excellent genetic study of protein refolding *in vivo* has been published by KING and coworkers. Initially, the trimeric tailspike protein from phage p22 was mutated such that the protein folding was temperature-sensitive (GOLDENBERG et al., 1983). As shown in Fig. 4, refolding at 40 °C prevented the formation of the native trimer and resulted in the accumulation of partially folded intermediates and aggregates (KING et al., 1987).

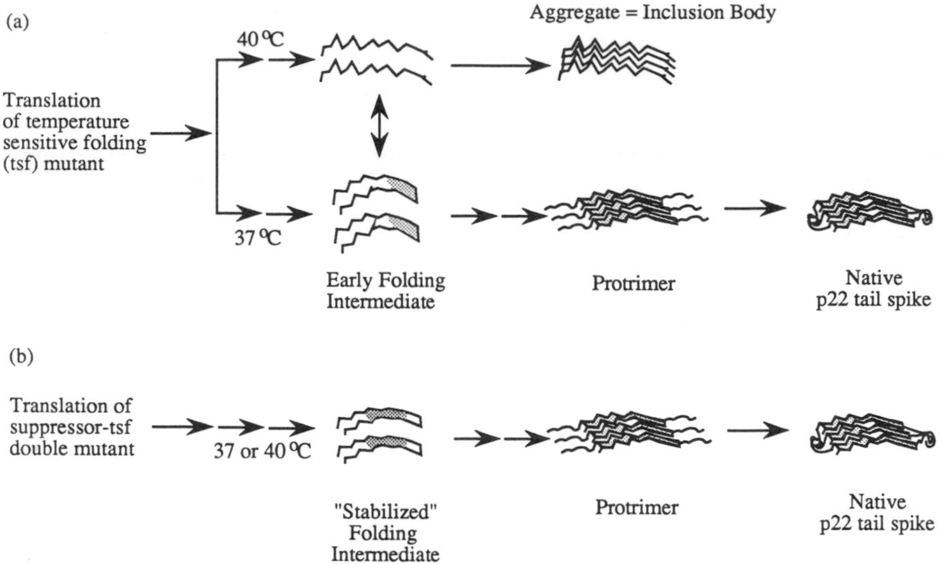

Fig. 4. *In vivo* folding pathways for temperature-sensitive folding (tsf) mutant (a) and suppressor-tsf double mutant (b) of p22 tailspike (MITRAKI et al., 1991b). At high temperature (40 °C), the tsf mutant will form unstable intermediates which aggregate and lead to inclusion body formation. The suppressor-tsf double mutant likely forms a more stable intermediate that will fold to form native tailspike at growth temperatures (37 °C) or elevated temperatures.

The temperature-sensitive mutant was shown to form the native trimer and did not result in a thermolabile native state (STURTEVANT et al., 1989). Therefore, this mutant provided a mechanism for studying the *in vivo* folding pathway and the formation of inclusion bodies. In particular, *in vivo* refolding at thermolabile conditions resulted in the stabilization of an intermediate which aggregated to form inclusion bodies (HAASE-PETTINGELL and KING, 1988). These studies revealed that aggregation occurred through a hydrophobic intermediate species which was kinetically trapped by the temperature-sensitive mutation. Recent work with this protein has shown that the temperature-sensitive mutations can be suppressed by additional modifications in the protein sequence (MITRAKI et al., 1991b; FANE et al., 1991). Therefore, it may be possible to make single amino acid substitutions in other proteins which will allow them to fold more rapidly and, thereby, avoid inclusion body formation.

Genetic engineering approaches to provide refolding analysis have also been applied to *in vitro* protein refolding. As in the *in vivo* studies, the approach primarily involved the stabilization of an intermediate on the refolding pathway. For example, an intermediate of bovine growth hormone was stabilized by a point mutation. This mutation was postulated to have caused enhanced hydrophobic attraction and increased aggregation (BREMS et al., 1988). On the other hand, a mutant of bovine growth hormone which did not aggregate was prepared by replacing the proposed aggregation region (residues 109–133) with the same sequence from the homologous protein, human growth hormone. The folding of this mutant was observed to occur by the same pathway as the wild-type protein (LEHRMAN et al., 1991). Therefore, it is possible to produce a mutant molecule which does not aggregate, but has the same folding pathway and native conformation as the wild type.

Other folding mutations have been performed on several proteins to elucidate their folding pathways. The structure of an intermediate in the refolding of barnase, ribonuclease from *Bacillus amyloliquefaciens,* was elucidated by nuclear magnetic resonance (NMR) techniques and mutants which resulted in a stable intermediate that folded to the native state (MATOUSCHEK et al., 1990). The comparison of mutant protein refolding results to the wild-type refolding pathway could be valid for many proteins, since it has been shown that substitutions can be made without altering the native structure (LIM and SAUER, 1991; BOWIE et al., 1990). On the other hand, compact proteins such as dihydrofolate reductase which have hydrophobic cores can be destabilized by substitution of hydrophilic residues for core hydrophobic residues (GARVEY and MATTHEWS, 1989). As pointed out by SHORTLE, single site mutations of proteins can be used to probe folding pathways, but the effect of the mutation on the thermodynamics of the protein must also be studied to validate the analysis (SHORTLE, 1989). In general, protein engineering can be used to assist in the determination of folding pathways and the structure of folding intermediates.

5 Molecular Chaperones

To avoid the aggregation of partially folded intermediates, eukaryotic organisms utilize proteins which bind to these intermediates. The binding proteins which have been observed in the *in vivo* folding pathway were initially believed to occur only when the cells were exposed to external stresses such as heat. Heat shock proteins such as the Hsp70 class of proteins have been shown to facilitate folding under normal conditions, and these proteins may require energy in the form of a nucleoside triphosphate for activity (BECKMAN et al., 1990). Molecular chaperones have been postulated to have many roles in cellular processes. These roles include catalysis of folding, prevention of aggregation, and removal of denatured non-host proteins. In addition, the onset of several diseases could be the result of a failure of these proteins to recognize foreign proteins (HORWICH et al., 1990). Recent reviews of chaperones as mediators of folding have provided an overview of the structure and function of these molecules (RANDALL et al., 1987; ROTHMAN, 1989; ELLIS and VAN DER VIES, 1991; HUBBARD and SANDERS, 1991; LANGER and NEUPERT, 1991). In general, it has been suggested that molecular chaperones operate primarily by binding to partially folded proteins in a stepwise manner which allows folding to occur while preventing aggregation (ROTHMAN, 1989). Structural and functional similarities between eukaryotic and prokaryotic binding proteins have been proposed from the results of a study which compared mitochondrial and prokaryotic protein processing (HARTL and NEUPERT, 1990). Tab. 2 summarizes the currently known chaperones, their typical cellular location, and their proposed function.

5.1 Non-Catalytic Chaperones

In both eukaryotes and prokaryotes, several proteins exist which bind to unfolded, partially folded, or aggregated proteins, but do not catalyze folding (Fig. 5). Molecular chaperones in eukaryotes were discovered in initial studies of heat shock effects on cells, since the chaperones were observed to associate with the thermally denatured proteins in the cell (LANGER and NEUPERT, 1991). From these early studies, the proteins were classified as heat shock proteins (Hsp). Three different groups of proteins were observed and were grouped by their approximate molecular weights: 60 kDa (Hsp60), 70 kDa (Hsp70), and 90 kDa (Hsp90). Hsp60 which is also referred to as chaperonin 60 has been observed to bind partially folded proteins in the mitochondria. The dissociation of these bound proteins has been observed to occur by interactions with chaperonin 10 (approximate molecular weight of 10 kDa) and hydrolysis of adenosine triphosphate (ATP) (ELLIS and VAN DER VIES, 1991). It has not yet been conclusively proven whether Hsp70 is the precursor for chaperonin 60 and chaperonin 10 in the mitochondria, but the similarity in its function would suggest this possibility. Hsp70 is usually found in the cytoplasm, and it has been shown to bind only unfolded proteins at physiological temperatures (PALLEROS et al., 1991). In addition, Hsp70 will bind ADP resulting in a complex which rapidly associates with unfolded proteins

Tab. 2. Molecular Chaperones (Ellis and van der Vies, 1991; Hubbard and Sander, 1991; Langer and Neupert, 1991)

Name	Location	Proposed Function	Cofactors
Non-catalytic Chaperones[a]			
Chaperonin 60		Binding to partially folded protein or aggregates	Chaperonin 10
Hsp 60	Mitochondria		
groEL	*E. coli*		
Chaperonin 10		Dissociation of proteins from Chaperonin 60	ATP
Hsp 10	Mitochondria		
groES	*E. coli*		
Heat shock protein 70		Binding to partially folded proteins and translocation	ATP or GTP
Hsp 70	Cytoplasm in eukaryotes		
DnaK	Cytoplasm in *E. coli*	Also, active in DNA replication	
BiP	Cytoplasm/ER in eukaryotes		
SecB	Cytoplasm in *E. coli*	Binding to partially folded protein and translocation	ATP, SecA
SRP	Cytoplasm in eukaryotes	Binds to leader sequence and facilitates translocation	GTP
Catalytic Chaperones			
Protein disulfide isomerase (PDI)	Endoplasmic reticulum	Facilitates disulfide bond exchange Possible role in glycosylation	
Peptidyl-prolyl *cis-trans* isomerase (PPI)	Endoplasmic reticulum	Facilitates proline isomerization	

[a] For a more complete list of non-catalytic chaperonins, see Ellis and van der Vies (1991).

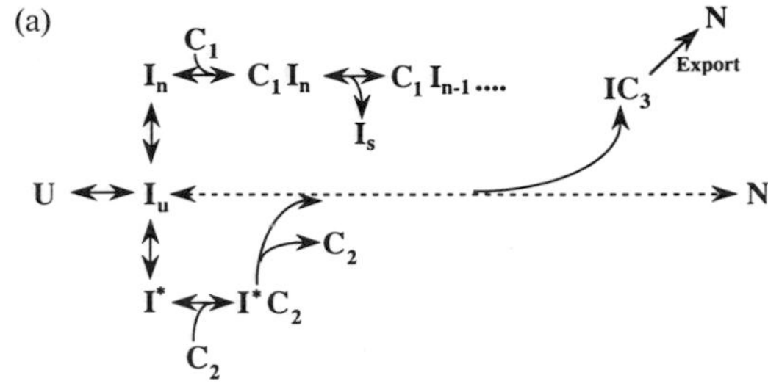

Fig. 5. Schematic representation of the roles of molecular chaperones in protein folding. (a) Non-catalytic chaperones can act by three different mechanisms. The unfolded protein (U) can fold to form an unstable intermediate (I_u) that can become misfolded (I*), aggregated (I_n), or folded to the native state (N). Molecular chaperonins (C_1) can bind to the aggregated protein to form a complex ($C_1 I_n$). These chaperones may then cause dissociation of the aggregates and partial refolding with concomitant release of a stable folding intermediate (I_s). Molecular chaperones can also bind to misfolded intermediates to form a complex (I^*C_2). They may correct the folding "error" and release a more stable folding intermediate that will eventually fold to the native state. Finally, non-catalytic chaperones have also been observed to bind to partially folded proteins (IC_3) and facilitate export.
(b) Catalytic chaperones such as PDI and PPI bind to partially folded proteins forming a complex (IC). These molecules will then catalyze the formation of native-like intermediates as well as the native protein.

(PALLEROS et al., 1991). As discussed previously, a member of the Hsp70 class of proteins is BiP which binds to folding intermediates. BiP maintains the protein in a partially folded state to facilitate translocation from the cytoplasm to the ER. BiP as well as the other Hsp family of proteins have been proposed to bind through recognition of the solvent exposed peptide backbone and not through interactions with specific hydrophobic patches which are normally buried (HUBBARD and SANDER, 1991). This hypothesis would explain the observation that these chaperones bind to several different proteins (ELLIS and VAN DER VIES, 1991; HUBBARD and SANDER, 1991; LANGER and NEUPERT, 1991).

Heat shock proteins and other chaperonins are also present in *Escherichia coli.* An Hsp70 homolog, Dnak, binds to partially folded proteins in the cytoplasm of *E. coli.* The chaperone releases the protein with concomitant hydrolysis of ATP which is mechanistically similar to Hsp70 function in eukaryotes (LIBEREK et al., 1991). Other molecular chaperones which have homology to those found in eukaryotes are chaperonin 60 and chaperonin 10, which are referred to as groEL and groES in *E. coli.* Many recent *in vitro* folding studies have begun to elucidate the structure and function of these chaperones. groEL forms a unique structure consisting of a double ring, and each ring contains seven subunits. The structure of groES is similar to groEL, since it also consists of a ring with seven subunits (HUBBARD and SANDER, 1991). However, groEL, unlike groES, has several exposed hydrophobic sections which may interact with partially folded proteins (MENDOZA et al., 1991a). It has been postulated that these chaperones provide a template for binding and release of partially folded or aggregated proteins (ELLIS and VAN DER VIES, 1991; HUBBARD and SANDER, 1991; LANGER and NEUPERT, 1991). The mechanistic relationship between these structures and chaperonin function have not been determined.

To determine the function of groEL and groES, the folding of several proteins has been studied. Aggregation during the refolding of citrate synthase was inhibited by groEL, and the rate of refolding was not increased. In this case, BiP, which actually inhibited folding, and Hsp70 did not prevent aggregation. groES and ATP were required to facilitate release of the partially folded protein from groEL (BUCHNER et al., 1991). The requirement of groES and ATP to release protein bound to groEL has been observed for other proteins (MARTIN et al., 1991). The specific interaction between groEL and folding intermediates has also been examined. groEL bound to a compact folding intermediate of both rhodanese and dihydrofolate reductase (MARTIN et al., 1991). The compact intermediate state is often referred to as a molten globule (see discussion of folding pathways) and similar structures have been observed during the refolding of several proteins (PTITSYN et al., 1990). Further analysis of this interaction has revealed that groEL binds to the *alpha*-helical amino terminus of rhodanese and promotes the formation of helical structure (LANDRY and GIERASCH, 1991). In addition, chemical synthesis of groES has suggested that these chaperonins can fold without assistance from other molecules (MASCAGNI et al., 1991). These results suggest a possible mechanism for these chaperonins. groEL self-assembles and may bind to folding intermediates or aggregates with a specificity for *alpha*-helical regions or, perhaps, hydrophobic regions. groES along with ATP interacts with the groEL-folding intermediate complex to correct misfolding or facilitate dissociation. These chaperones have not been observed to cause an increase in the rate of refolding and, therefore, they may only have a salvage function to prevent accumulation of misfolded or aggregated protein.

Overall, non-catalytic molecular chaperones are likely to have the primary function of binding to partially folded protein structures with high affinity and low selectivity. These chaperonins block the formation of aggregates by binding to hydrophobic misfolded or partially folded intermediates. It has been demonstrated that these proteins can be applied to *in vitro* refolding to inhibit aggregation (MENDOZA et al., 1991a; BUCHNER et al., 1991). Further studies should provide insight into the specific residues which modulate interactions between folding intermediates and molecular chaperones. Refolding aids which mimic the action of chaperonins could then be designed and could employ either synthetic molecules or existing chaperones which have been mutated to provide the desired properties.

5.2 Catalytic Chaperones

Although most chaperones have not been shown to increase the rate of folding, a few chaperones which catalyze specific folding reactions have been observed in eukaryotes. Two catalytic proteins have been studied in detail and discussed previously (see Sect. 2.1). Protein disulfide isomerase (PDI) catalyzes the formation of correct disulfide bonds during folding and may have several other functions (BASSUK and BERG, 1989). Recent studies have revealed that PDI must be in a dithiol state to be effective in catalysis of disulfide bond formation (LYLES and GILBERT, 1991a). Excess PDI can also inhibit folding be occupying exposed thiols on the protein (LYLES and GILBERT, 1991b). PDI has also been shown to act independently of the second observed catalytic protein, peptidyl-prolyl *cis-trans* isomerase (PPI) (LANG and SCHMID, 1988). PPI catalyzed the slow proline isomerization step in the refolding of the S protein of bovine RNAse A and can be effective in catalyzing the folding of other proteins (LANG et al., 1987). The slow step in folding of several proteins has been postulated to be the result of proline isomerization or disulfide bond formation (STELLWAGEN, 1979; SEMISOTNOV et al., 1990; PTITSYN et al., 1990; KIM and BALDWIN, 1990). Therefore, PDI and PPI could be utilized to enhance refolding. Unfortunately, both proteins must be able to access the substrate residues and, therefore, the application of these proteins *in vitro* may be limited (LANG et al., 1987). However, a better understanding of the mechanisms of these catalysts would assist in the design of improved folding processes both *in vivo* by protein engineering of the catalysts and *in vitro* with the development of methods to circumvent accessibility problems.

6 Synthetic Folding Aids

Many different methods of refolding have been developed to simulate the action of molecular chaperones on folding. Most of these methods operate on the principle of isolating each individual protein molecule and allowing it to refold before it can associate. For example, reversed micelles have been used to assist refolding by enclosing a single protein within each micelle and, then, removing the denaturant (HAGEN et al., 1990). Unfortunately, this system could not be used with hydrophobic proteins which interact with the surfactant. Another method for isolating denatured protein involved ion exchange chromatography. In this case, the denatured protein was bound to the column resin followed by denaturant removal (CREIGHTON, 1985). CREIGHTON has also suggested binding of the denatured protein to a solid support at a specific surface residue which would not inferfere with refolding (CREIGHTON, 1985). As an alternative to isolating the protein, attempts have been made to increase the rate of refolding. If the rate of refolding can be significantly increased, the aggregation of intermediates could be reduced.

6.1 Specific Ligands

To alter rate or extent of refolding, conformation-specific ligands have been used. Refolding studies have been performed with ligands and antibodies which are specific to the native structure. In particular, creatine kinase was refolded in the presence of antibodies to the native structure. The antibody was observed to sterically interfere with the interaction between the protein domains that are required for refolding (MORRIS et al., 1987). Other studies of the effect of antibodies on refolding have shown some success in stabilizing intermediate structures and increasing the refolding (CARLSON and YARMUSH, 1992). Antibodies to the native protein likely alter folding by binding to the native epitopes as they are formed and, thus, stabilize sections of the protein (Fig. 6a). The binding of an antibody to a native-like structure could prevent its further refolding. However, if the antibody is only in contact with the native epitope after complete folding of that segment, the additional folding reactions may not be inhibited. Additional studies on the effect of antibodies in refolding should elucidate the role of native structure formation in early folding steps as well as provide a possible mechanism for increasing the rate or extent of refolding.

In addition to antibodies, ligands and cofactors have been used to enhance refolding. These molecules would be more likely to interact with the folding protein after formation of the native protein. Therefore, the folding

(a)

AbI_n ⇢ AbN

Ab

U ↔ I_1 … I_n … I_a ↔ N

(b)

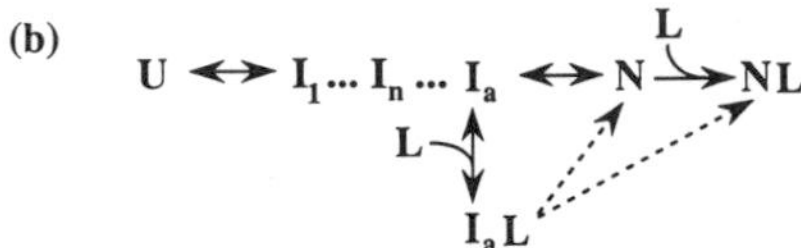

Fig. 6. Potential folding pathways in the presence of specific ligands. Initially, the unfolded protein folds to form a compact intermediate with little native structure (I_1). This intermediate slowly folds to form native-like intermediates (I_n). The active site may then form a native-like intermediate (I_a). The protein will eventually assume its native conformation (N).
(a) Antibodies (Ab) to the native protein may bind to the native-like intermediate forming a stable complex (AbI_n). The protein can then continue to fold and become released as a later folding intermediate or the native protein.
(b) Ligands (L) to the active site or inhibitor regions of the protein can bind to intermediates that have the correct conformation to form an inactive complex (I_aL). This complex may then fold to form the native protein. Alternatively, the ligand may bind to the native protein to form an inactive complex (NL) and drive the folding equilibrium toward the native state.

equilibrium could than be "driven" to the native state (Fig. 6b). For example, the rate of refolding of ferricytochrome c was enhanced by the extrinsic ligand for the axial position of the heme iron (BREMS and STELLWAGEN, 1983). Similar studies with hemoglobin revealed that the recovery of native protein could be increased with the addition of heme during refolding. Hemoglobin inclusion bodies were also more easily solubilized through the addition of heme indicating a partially formed native structure in the aggregate (HART and BAILEY, 1992). Studies of inhibitors and refolding revealed that an allosteric inhibitor stabilized pyruvate kinase against denaturation by guanidine hydrochloride (CONSLER and LEE, 1988). Other investigations into the use of substrate analogs and inhibitors to alter folding revealed that inhibitors may increase the rate of refolding, but assessment of native conformation is difficult due to reduced activity (CLELAND, 1991). Additional studies of cofactors, substrate analogs and inhibitors should indicate the utility of this approach for increasing the rate of refolding.

6.2 Cosolvents

Instead of specifically targeting native-like conformations with ligands, the solvent environment can be changed to alter the folding pathway. Some cosolvents create a change in the solvent environment which results in an increased hydration of the protein. For example, sugars such as sucrose and glycerol have been observed to cause the formation of more compact, hydrated proteins (ARAKAWA and TIMASHEFF, 1982). The use of sugars has been studied for the refolding of several proteins. Sucrose was successfully used to increase the recovery of β-lactamase during refolding (VALAX and GEORGIOU, 1991). However, other refolding studies with sugars have not been successful in improving the recovery of native protein. When sucrose was used in the refolding of porphyrin c, the refolding of the protein was not altered (BREMS et al., 1982). In contrast, the first step in ribonuclease (RNAse) refolding occurs more slowly in the presence of sugars (TSONG, 1982). Glycerol and glucose have been observed to decrease the rate of refolding of octopine dehydrogenase (TESCHNER et al., 1987). Additional studies have also shown that sugars reduce the rate of refolding (VAUCHERET et al., 1987). Recent refolding studies of bovine carbonic anhydrase B (CAB) indicated that sugars can inhibit precipitation during refolding, but they do not increase the recovery of active protein (CLELAND and Wang, 1992a). These results suggest that cosolvents which cause hydration of the protein may not be useful for improving the recovery of active protein.

A different class of cosolvents are molecules which bind to the protein through non-specific interactions such as hydrogen bonding. Several polymers and surfactants have been successful in preventing aggregation during refolding by altering the refolding pathway. Polyethylene glycol (PEG) and poly(vinylpyrrolidone) did not decrease the rate of refolding for octopine dehydrogenase, but did alter the refolding pathway (TESCHNER et al., 1987). Non-denaturing detergents have been successfully used to prevent aggregation during refolding of the membrane protein, rhodanese (HOROWITZ and CRISCIMAGNA, 1986). In particular, it has been proven that the detergent, lauryl maltoside, will interact with non-polar regions of a rhodanese refolding intermediate and, thereby, prevent aggregation of the intermediate (TANDON and HOROWITZ, 1987). The role of the detergent in rhodanese folding is very similar to that observed for chaperonins (MENDOZA et al., 1991a). Both the chaperonin, groEL, and the detergent bind to partially folded rhodanese and prevent aggregation (LANDRY and GIERASCH, 1991). However, the actual sites of action for these folding aids may be quite different. The application of less hydrophobic detergents or surfactants was also suggested to prevent aggregation of rhodanese (TANDON and HOROWITZ, 1986).

A less hydrophobic polymer, PEG, has been extensively studied for its effect on refolding and aggregation of CAB. PEG was found to weakly bind only to the molten globule first intermediate on the CAB folding pathway (CLELAND and RANDOLPH, 1992). In addition, the rate of folding was not increased in the presence of PEG (CLELAND et al., 1992a). However, aggregation during refolding was inhibited by the addition of PEG (CLELAND and WANG, 1990a). As shown in Fig. 7, the pathway for refolding of CAB in PEG is postulated to involve the formation of a PEG-intermediate complex which does not associate and folds to form the native protein (CLELAND et al., 1992a). Since PEG inhibits aggregation of the hydrophobic first intermediate, refolding at high protein concentrations (>1.0 mg/mL) and at low denaturant concentrations (<0.5 M guanidine hydrochloride) could be performed without the formation of aggregates (CLELAND and WANG, 1990a). Additional studies have shown that PEG may have general utility as a folding aid. PEG increased the recovery of active protein for the refolding of both recombinant human deoxyribonuclease and recombinant human tissue plasminogen activator (CLELAND et al., 1992b). These results showed a relationship between the amount of PEG required for enhanced refolding and the

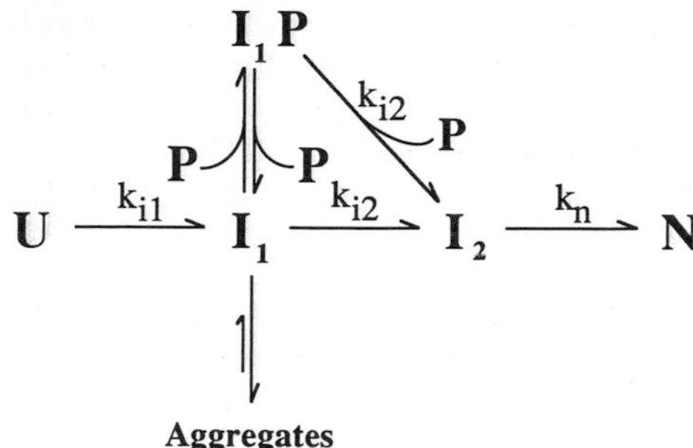

Fig. 7. Model for PEG enhanced refolding of CAB (CLELAND et al., 1992a). Upon rapid dilution from 5 M GuHCl, the unfolded protein (U) rapidly forms the first intermediate (I_1). This intermediate will then either proceed through the normal refolding to the second intermediate (I_2) and native protein (N), or it will bind to PEG, P, to form a non-associating complex (I_1P). The refolding of the PEG-intermediate complex (I_1P) to the second intermediate occurs at the same rate as that observed for refolding without PEG at low protein concentrations (<10 μM) and 1.0 M GuHCl.

hydrophobicity of the protein. More hydrophobic proteins require greater molar ratios of PEG to protein, but the overall PEG concentration required for enhancement is usually quite low (<1 mg/mL) (CLELAND et al., 1992b). By weakly binding to a folding intermediate and inhibiting aggregation, PEG may function as a molecular chaperone which does not require energy or other molecules for release of the protein. Investigations of PEG analogs indicated that the properties of PEG dictate its ability to facilitate refolding. More hydrophobic PEG analogs inhibited refolding of CAB (CLELAND and WANG, 1992a). Therefore, a molecule such as PEG which weakly binds to a folding intermediate without inhibiting folding will have greater utility in assisting refolding.

Further analyses of cosolvents and refolding lead to the development of a general rule for the desired physicochemical properties of the cosolvent. The desired properties include weak binding only to hydrophobic folding intermediates which aggregate, exclusion from the protein after a stable intermediate is formed, and ease of separation of the protein from the cosolvent (CLELAND and WANG, 1992a). Several cosolvents have been studied to assess their effect on native proteins (ARAKAWA and TIMASHEFF, 1982, 1984, 1985a, b; TIMASHEFF and ARAKAWA, 1988; ARAKAWA et al., 1990). The mechanism of various cosolvents is illustrated in Fig. 8. Cosolvents which lie in the center between hydration (steric exclusion) and

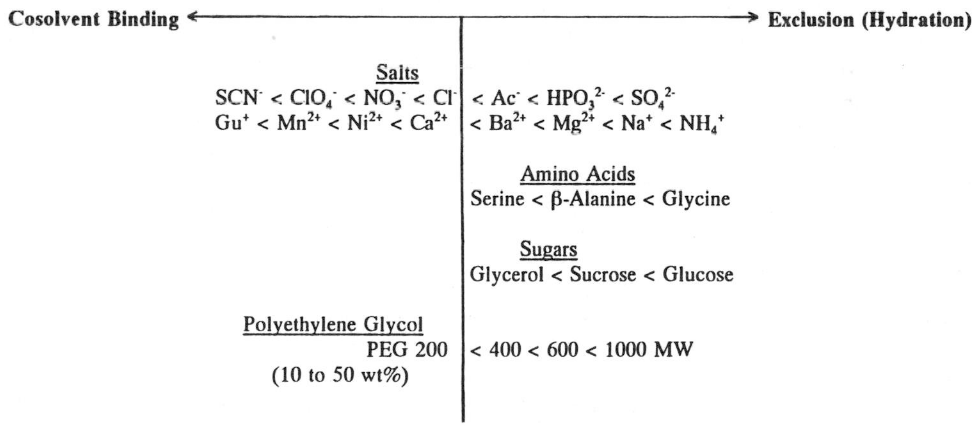

Fig. 8. Cosolvent interactions with native proteins. The cosolvents which are left of the midpoint have been shown to bind non-specifically to proteins. Cosolvents to the right of the midpoint have been observed to act by preferential hydration (exclusion) of the protein resulting in an increase in the concentration of water at the protein surface. The mechanism of action for polyethylene glycol (PEG) has been observed to depend upon the PEG concentration where high concentrations (10–50 wt%) result in exclusion and low concentrations (<10 wt%) can result in PEG binding to the protein (ARAKAWA and TIMASHEFF, 1985a; CLELAND and WANG, 1992a; CLELAND et al., 1992a).

binding have been postulated to facilitate folding, since they can weakly bind to exposed hydrophobic regions and then become excluded from the native structure (CLELAND and WANG, 1992a). Additional studies of these and other cosolvents should provide information on the physicochemical properties which will facilitate folding. Once these properties are understood, the development of synthetic folding aids which inhibit aggregation and catalyze refolding can be successfully performed.

7 Models of *in vitro* Protein Refolding

Studies of the *in vitro* refolding of denatured proteins has resulted in the formulation of several different models of refolding (KARPLUS and WEAVER, 1976; KIM and BALDWIN, 1982; KING, 1989). These models include biased random search, nucleation-growth, diffusion-collision-adhesion, and sequential folding. From an analysis of several reviews of potential models for refolding, a general theme can be derived. First of all, the formation of secondary structure in the denatured protein occurs very rapidly in most cases and can be described by the diffusion-collision-adhesion model where microdomains are formed from a few residues and then collide to form a stable structure. According to the diffusion-collision theory of protein folding developed by KARPLUS and WEAVER, microdomains are segments of unstable native secondary structure and consist of only a few amino acid residues (8–10) such that all of the conformational alternatives can be searched in a short time period (<1 ms) (KARPLUS and WEAVER, 1979). More recent development of this model has suggested that the diffusion-collision process occurs prior to the formation of an initial compact intermediate (BASHFORD et al., 1988). In addition, this model has been partially confirmed by NMR studies of the initiation of protein folding (WRIGHT et al., 1988). An investigation of folding pathways has been made by using the assumptions that the dominant free energy contribution to native conformation is contributed by hydrophobic interactions (DILL, 1990), the native protein contains a history of its folding, and denatured protein forms a random coil. Based on these assumptions, the folding pathways of 19 different proteins were determined through computational analysis. The resulting pathways confirmed the hypothesis of nucleation followed by propagation and diffusion-collision (MOULT and UNGER, 1991). If these pathways can be experimentally confirmed, they would provide strong support for the folding model developed by KARPLUS and WEAVER (1979).

The diffusion-collision model also complements the observations of a molten globule state which rapidly forms in the initial stages of refolding for many proteins (PTITSYN, 1987; KUWAJIMA, 1989; PTITSYN et al., 1990). The compact globular structure has several general features which include a native-like secondary structure and exposed hydrophobic surfaces (KUWAJIMA, 1989). In addition, this compact state has been observed to be general to several proteins independent of class or composition (PTITSYN et al., 1990). The driving force for the diffusion-collision process to form the compact molten globule was postulated to be hydrophobic interactions (BASHFORD et al., 1988; KUWAJIMA, 1989). This hypothesis is supported by the analysis of protein structural interactions in a recent review (DILL, 1990). DILL concluded that hydrophobic interactions are the strongest driving forces for the formation of native protein structure. Another recent review by KIM and BALDWIN led to the conclusion that the initial structural intermediates in refolding form quickly and that the rate-limiting step in folding then becomes the correct "adhesion" of different segments of structure (KIM and BALDWIN, 1990). In accordance with the diffusion-collision-adhesion model, the next reaction would then proceed with a pairing of the domains formed from microdomains to give a stable compact native structure (KARPLUS and WEAVER, 1979). Domain pairing does not occur in small proteins such as ribonuclease or lysozyme, since they do not contain multiple domains (KARPLUS and WEAVER, 1976). However, the slow step in folding of single domain proteins may be the result of inherent structural limitations such as proline isomerization or correct disulfide bond formation (KIM and BALDWIN, 1990).

With these overall hypotheses of refolding, a general structural model of protein refolding can be proposed as shown in Fig. 9. The two major events of folding are the rapid formation of a hydrophobic compact structure followed by the slow shuffling of the protein to its native conformation. The rapid collapse of the initial random coil involves initial nucleation sites of ordered structure which quickly propagate throughout the protein. The secondary structures then coalesce to form a somewhat rigid compact state. The sponge-like compact state slowly rearranges to form a native-like molecule with some correct disulfides and proline orientations. The native-like molecule can then form the native state or, in the case of multimeric proteins, it may self-associate. The general framework of this folding model should be applicable to studies of most globular proteins.

8 *In vitro* Refolding Dependence on Denaturant Concentration

The stabilizing interactions in the formation of the native protein structure are all affected by the environmental conditions during the refolding process. Therefore, any model or analysis must include these environmental parameters in the refolding process. These physical parameters include effects such as denaturant concentration, viscosity, temperature, ionic strength, and solvent properties (see also Sects. 6 and 9). The denaturant concentration has been observed to be one of the critical factors in protein refolding. Denaturants such as urea and GuHCl have been shown to bind to proteins (PRAKASH et al., 1981; ARAKAWA and TIMASHEFF, 1984). By preferentially binding or solvating interior hydrophobic residues,

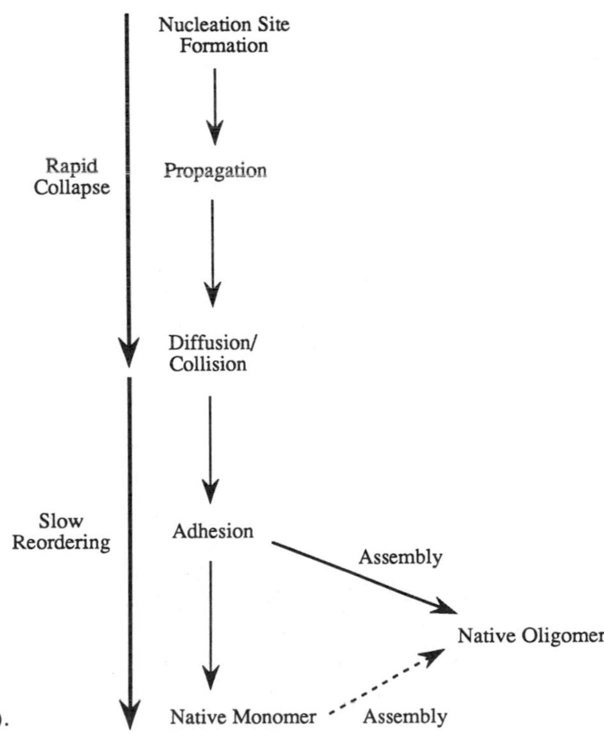

Fig. 9. Structural model of protein folding as proposed by KARPLUS and WEAVER (1976) (see text for description).

these denaturants unfold the protein to expose the hydrophobic core (DILL and SHORTLE, 1991). The removal of these denaturants from the protein surface is required to refold the protein. The denaturant bound to the protein is in equilibrium with the bulk denaturant. Therefore, as the bulk concentration of the denaturant is reduced, the equilibrium concentration of denaturant bound to the protein will also decrease allowing the protein to refold. During denaturant removal, the unfolded protein will form stable intermediates and their rates of formation depend on the structural properties of the protein, the denaturant removal rate, and the final denaturant concentration. Protein folding intermediates have been studied by incrementally reducing the denaturant concentration and measuring the equilibrium formation of protein structure (cf. ALONSO and DILL, 1991; RODIONOVA et al., 1989; BREMS, 1988). Additional analysis of the effect of denaturant on refolding has been detailed previously (DILL and SHORTLE, 1991; CREIGHTON, 1984).

Several relevant studies of denaturant effects on refolding and recovery of active protein have been performed. For example, the rate of refolding of bovine serum albumin was enhanced by decreasing the urea concentration below 2.0 M (DAMODARAN, 1987). The rate of refolding was also dependent on the final denaturant concentration for refolding of lactate dehydrogenase (LDH) (RUDOLPH et al., 1979). More recent refolding studies revealed that the rate of refolding and the rate of aggregation depend on the denaturant concentration as well as the protein concentration. Refolding of CAB at several protein and denaturant concentrations yielded an operating diagram as shown in Fig. 10 (CLELAND and WANG, 1990b). This diagram indicates the conditions at which CAB will form aggregates during refolding. In addition, the rates of aggregation were measured for each case. The rate of aggregation for refolding at a given protein concentration increased dramatically with decreasing denaturant concentration and increasing protein concentrations resulting in multimer

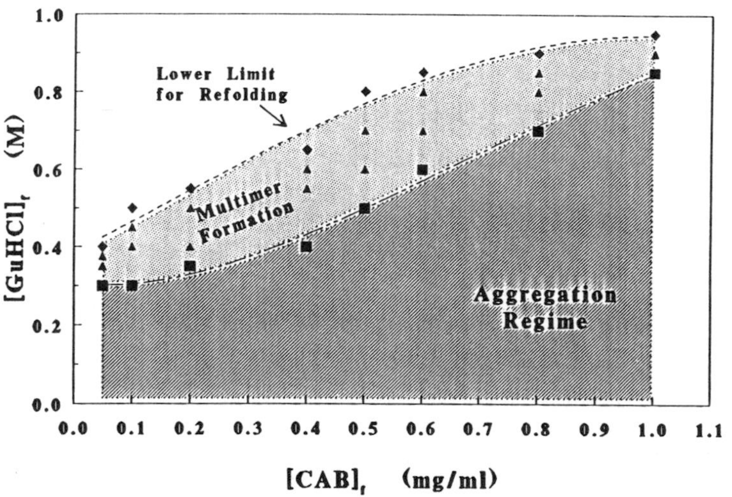

Fig. 10. Regimes of refolding and aggregation of CAB (CLELAND and WANG, 1990b). Each data point represents rapid dilution of CAB in 5 M GuHCl to a given final protein and GuHCl concentration. The aggregation regime is defined as the final solution conditions which result in the immediate formation of micron-sized aggregates. The upper boundary of the aggregation regime is defined by the lower data points (■). The cases where multimers are observed prior to micron-size aggregation constitute the multimer formation regime (▲). The lower limit of refolding is the distinct regime where multimers form, but do not proceed to form micron-sized aggregates (◆). From the lower limit of refolding to 1 M GuHCl, the protein refolds to form either a stable intermediate or the native structure in the absence of aggregation.

formation followed by precipitation (CLELAND an WANG, 1990b). In general, the optimum denaturant concentration and removal rate would be a balance between the formation of native structure and the reduction in misfolded intermediates or aggregates.

9 Temperature Effects on Refolding

In addition to the solvent environment, temperature can affect the rate and extent of refolding (see Sect. 3 and 4 for effect of temperature on *in vivo* folding). In particular, two proteins, phosphoglycerate kinase (PGK) and aspartokinase-homoserine dehydrogenase (AK-HDH), have been shown to depend on temperature for refolding (PORTER and CARDENAS, 1980; VAUCHERET et al., 1987). The extent of reactivation of PGK decreases with temperature above 32 °C and remains constant from 0 to 25 °C. The decrease in refolding of PGK at higher temperatures is due to protein aggregation and the formation of incorrectly folded monomers (PORTER and CARDENAS, 1980). The slow reaction step in the folding of AK-HDH monomers is also strongly dependent on temperature from 0 to 50 °C (VAUCHERET al., 1987). A hydrophobic membrane protein, rhodanese, could only be successfully refolded without detergents or chaperonins at low protein concentrations (< 10 μg/mL) and low temperature (10 °C) (MENDOZA et al., 1991b). The exact role of temperature in the kinetics of refolding and aggregation was not determined for these proteins. However, temperature directly affects the molecular interactions within the protein. Hydrophobic interactions which have been postulated as the major driving force for folding are endothermic (DILL, 1990). Therefore, increases in temperature should result in a faster rate of folding. A balance must be achieved, however, between the enhancement in the hydrophobic interactions and the thermal denaturation of the protein. The stabilization of denatured states at higher temperatures can also result in protein aggregation. Since protein aggregation may be driven by hydrophobic interactions between proteins, higher temperatures should also stabilize aggregates. In general, an optimum temperature probably exists for each set of solvent conditions such that the rate of refolding is much greater than the rate of aggregation for a given protein.

10 Aggregation During Refolding

As observed in the *in vivo* studies of inclusion body formation, protein aggregation of intermediates can also occur during *in vitro* folding. The similarity between *in vivo* and *in vitro* aggregation during folding has been assessed for the phage p22 tailspike protein (SECKLER et al., 1989). For this protein, the *in vitro* folding was observed to occur with different kinetics than in the *in vivo* studies. The difference in folding kinetics was postulated as the lack of molecular chaperonins for the *in vitro* systems (SECKLER et al., 1989). Early studies on protein aggregation revealed that the kinetics of aggregation competed with the folding kinetics (ZETTLMEISSL et al., 1979). The rate and extent of aggregation has also been shown to depend on the final protein and denaturant concentrations for CAB (CLELAND and WANG, 1990b) and LDH (ZETTLMEISSL et al., 1979; RUDOLPH et al., 1979). In addition, refolding of PGK at a "critical" concentration of denaturant resulted in the formation of aggregates (MITRAKI et al., 1987). The phenomenon of a critical denaturant concentration for irreversible aggregation has been observed in the refolding of several other proteins (MITRAKI et al., 1987). At the critical concentration of denaturant, the rate of aggregation for a folding intermediate may exceed the rate of refolding to a non-associating intermediate resulting in a low recovery of active protein. For example, extensive studies of the CAB refolding and aggregation pathways have revealed that the formation of the hydrophobic first intermediate occurs more rapidly at low denaturant concentrations. The refolding and aggregation pathway of CAB as shown in Fig. 11a includes

(a)

$$U \xrightarrow{k_{i1}} I_1 \xrightarrow{k_{i2}} I_2 \xrightarrow{k_n} N$$

$I_1 \xrightarrow{k_d} D$, $D \xrightarrow{k_t} T$; D and T - - → **Aggregates (micron size)**

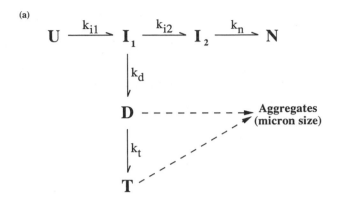

(b)

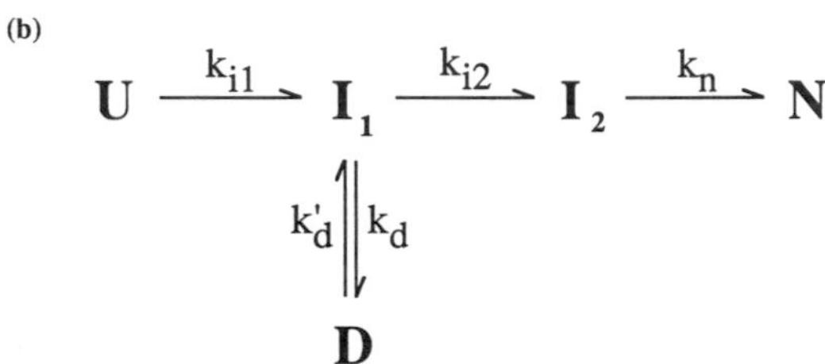

Fig. 11. Proposed pathways for refolding and aggregation of CAB (CLELAND and WANG, 1990b, 1992b). (a) The unfolded protein (U) collapses to form the molten globule first intermediate (I_1). The first intermediate can then proceed to form the second intermediate or the dimer species (D). The second intermediate continues to refold to form the native protein structure (N) as described previously (STEIN and HENKENS, 1978; SEMISOTNOV et al., 1987, 1990). The dimer may form the trimer species (T) with the addition of another intermediate protein (I_1) to the aggregate. The dimer, trimer, and first intermediate will then form the micron-sized aggregates under aggregating conditions (see Fig. 10). Each rate constant is dependent on the final protein and GuHCl concentration as discussed previously (CLELAND and WANG, 1990b).
(b) Refolding at 1.0 M GuHCl and high protein concentrations (>10 μM) has been observed to result in transient association of the molten globule first intermediate (CLELAND and WANG, 1992b). Complete recovery of active protein was successfully achieved under these conditions.

the formation of multimers prior to precipitation (CLELAND and WANG, 1990b). Other studies have only assessed the change in light scattering as the result of large aggregate formation, but these studies also conclude that aggregation occurs from an early folding intermediate (KIEFHABER et al., 1991). In addition to irreversible aggregation, transient association of the folding intermediates can occur without a reduction in the yield of native protein. Bovine growth hormone was observed to transiently associate during refolding and completely recovered activity (BREMS et al., 1987). The transient formation of a first intermediate homodimer was measured during the refolding of CAB (Fig. 11b). The rate of association correlated well with protein concentration. As the protein concentration increased, the rate of refolding decreased as the result of the slow dissociation of the dimer (CLELAND and WANG, 1992b).

Association of folding intermediates has also been observed at equilibrium by stabilizing the intermediate in high denaturant and protein concentrations. For example, equilibrium studies of CAB indicated that the molten globule first intermediate associated to form dimers and trimers in 2.0 M GuHCl (CLELAND and WANG, 1991). BREMS and coworkers have shown that a folding intermediate of bovine growth hormone can be stabilized at 3.7 M GuHCl or 8.5 M urea and will associate at high protein concentrations (BREMS et al., 1986; HAVEL et al., 1986). The association was reduced by the addition of a peptide fragment (37 amino acids) from the native protein (BREMS et al., 1986). These studies suggest that aggregation between specific structural units may occur as hypothesized previously (MITRAKI and KING, 1989). The intramolecular and intermolecular interactions which occur during refolding result in the formation of native protein as well as aggregates. Furthermore, the rate and extent of this aggregate formation is dependent on the final protein and denaturant concentrations.

Non-covalent interactions are the most likely driving force for aggregation as postulated by several researchers (IKAI et al., 1978; FISH et al., 1985; FUKE et al., 1985; TANDON and HOROWITZ, 1987; MITRAKI and KING, 1989). Hydrophobic interactions are the strongest non-polar forces and have also been postulated as the driving force for refolding (DILL, 1990). The evidence that hydrophobic interactions are responsible for aggregation includes the endothermic kinetics characteristic of hydrophobic reactions (MITRAKI et al., 1987; BREMS et al., 1986; STELLWAGEN and WILGUS, 1978), concomitant loss of entropy upon aggregation (GHELIS and YON, 1982; MITRAKI et al., 1987), and exposure of hydrophobic regions by chaotropic agents or solvents (MARSTON, 1986; MITRAKI et al., 1987; IKAI et al., 1978). A specific example of hydrophobic effects is the removal of the hydrophobic end group of a viral glycoprotein which resulted in higher protein solubility (MARSTON, 1986). In addition, rhodanese aggregates with concomitant exposure of hydrophobic surfaces. The domain interfaces of rhodanese are hydrophobic and cause protein–protein interactions leading to aggregation (TANDON and HOROWITZ, 1986, 1987). Another example is the partial denaturation of bovine growth hormone which results in the exposure of a hydrophobic face on an *alpha*-helix. The hydrophobic surfaces interact between molecules resulting in aggregation (BREMS et al., 1986). Several hydrophobic molten globule intermediates have been shown to aggregate (CAB, CLELAND and WANG, 1990b; bovine growth hormone, BREMS et al., 1986; rhodanese, TANDON and HOROWITZ, 1986). MITRAKI and coworkers have provided additional evidence to support the hypothesis that hydrophobic interactions are the driving force for aggregation (MITRAKI et al., 1991a). In conclusion, aggregation during refolding can occur when hydrophobic intermediates exist on the folding pathway and the extent of aggregation will depend on the final denaturant and protein concentrations both of which impact the kinetics of refolding and aggregation.

11 Summary

The ultimate objective of protein folding is the attainment of the native protein conformation without the production of by-products such as aggregates. The recovery of native biologically active protein can be achieved through both *in vivo* and *in vitro* folding in *E. coli*. For *in vivo* folding of proteins, the protein must be expressed with a signal sequence which will activate the folding and export machinery of *E. coli*. The protein may also require mutations which alter the folding pathway and prevent aggregation. If folding mutants cannot be used, it would be necessary to amplify the expression of chaperonin proteins such as groEL and groES to avoid the formation of aggregates. This approach would require the desired protein to be expressed with several other proteins and, therefore, the actual yield of desired protein from each cell would be reduced. It is also likely that the cellular machinery of *E. coli* would fail if forced to overproduce several proteins simultaneously. Thus, the large quantities of soluble protein can only be produced in *E. coli*, if folding mutants can be successfully engineered.

Alternatively, the protein can be expressed as an inclusion body in *E. coli* and subsequently folded *in vitro*. The insoluble protein aggregates are separated from the cell debris (see Fig. 1). Then, the inclusion bodies are solubilized with denaturant to cause dissociation and unfolding. The denaturant must then be removed to recover native protein. During denaturant removal, the protein may reassociate to form insoluble aggregates. To avoid aggregate formation, refolding must usually be performed at low protein concentrations (<0.1 mg/mL). However, refolding at high protein concentrations (>0.1 mg/mL) can be successfully achieved, if the folding conditions are optimized. Since refolding at high protein concentrations can result in aggregation, aggregation inhibitors must be added during denaturant removal. These inhibitors include chaperonins or cosolvents such as PEG. However, chaperonins may not be feasible for commercial protein production, since they could not be easily reused and would be costly to produce. Cosolvents such as PEG are more likely to provide increased recovery of native proteins at a lower cost. Further optimization of folding conditions would also include assessment of the optimal temperature for folding. In addition, a range of final protein, redox reagent (for proteins with disulfides), and denaturant concentrations should be explored to determine conditions which provide the greatest yield. Additional studies of protein folding pathways and aggregation should provide insight into other methods that would provide greater recovery of native protein. Eventually, general rules of protein folding will be developed that relate the protein sequence and physical environment of the protein to its attainment of native structure. These rules can then be applied to devise methods that provide complete recovery of native protein without the formation of misfolded intermediates or aggregates.

12 References

ALONSO, D. O. V., DILL, K. A. (1991), Solvent denaturation and stabilization of globular proteins, *Biochemistry* **30**, 5974–5985.

ARAKAWA, T., TIMASHEFF, S. N. (1982) Stabilization of protein structure by sugars, *Biochemistry* **21**, 6536–6544.

ARAKAWA, T., TIMASHEFF, S. N. (1984), Mechanism of protein salting in and out by divalent cation salts: balance between hydration and salt binding, *Biochemistry* **23**, 5924–5929.

ARAKAWA, T., TIMASHEFF, S. N. (1985a), Mechanism of poly(ethylene glycol) interaction with proteins, *Biochemistry* **24**, 6756–6762.

ARAKAWA, T., TIMASHEFF, S. N. (1985b), The stabilization of proteins by osmolytes, *Biophys. J.*, 411–414.

ARAKAWA, T., BHAT, R., TIMASHEFF, S. N. (1990), Why preferential hydration does not always stabilize the native structure of globular proteins, *Biochemistry* **29**, 1924–1931.

BASHFORD, D., COHEN, F. E., KARPLUS, M., KUNTZ, I. D., WEAVER, D. L. (1988), Diffusion-collision model for the folding kinetics of myoglobin, *Proteins Struct. Funct. Genet.* **4**, 211–227.

BASSUK, J. A., BERG, R. A. (1989), Protein disulphide isomerase, a multifunctional endoplasmic reticulum protein, *Matrix* **9**, 244–258.

BECKMAN, R. P., MIZZEN, L. A., WELCH, W. J. (1990), Interaction of Hsp70 with newly synthesized proteins: implications for protein folding and assembly, *Science* **248**, 850–854.

BOWDEN, G. A., GEORGIOU, G. (1988), The effect of sugars on β-lactamase aggregation in *Escherichia coli, Biotechnol. Prog.* **4**, 97–101.

BOWIE, J. U., REIDHAAR-OLSON, J. F., LIM, W. A., SAUER, R. T. (1990), Deciphering the message in protein sequences: tolerance to amino acid substitutions, *Science* **247**, 1306–1310.

BREMS, D. N. (1988), Solubility of different folding conformers of bovine growth hormone, *Biochemistry* **27**, 4541–4546.

BREMS, D. N., LIN, Y. C., STELLWAGEN, E. (1982), The conformational transition of horse heart porphyrin c, *J. Biol. Chem.* **257**, 3864–3869.

BREMS, D. N., STELLWAGEN, E. (1983), Manipulation of the observed kinetic phases in the refolding of denatured ferricytochromes c, *J. Biol. Chem.* **258**, 3655–3661.

BREMS, D. N., PLAISTED, S. M., KAUFFMAN, E. W., HAVEL, H. A. (1986), Characterization of an associated equilibrium folding intermediate of bovine growth hormone, *Biochemistry* **25**, 6539–6543.

BREMS, D. N., PLAISTED, S. M., DOUGHERTY, J. J., HOLZMAN, T. F. (1987), The kinetics of bovine growth hormone folding are consistent with a framework model, *J. Biol. Chem.* **262**, 2590–2596.

BREMS, D. N., PLAISTED, S. M., HAVEL, H. A., TOMICH, C.-S. C. (1988), Stabilization of an associated folding intermediate of bovine growth hormone by site-directed mutagenesis, *P.N.A.S.* **85**, 3367–3371.

BUCHNER, J., SCHMIDT, M., FUCHS, M., JAENICKE, R., RUDOLPH, R., SCHMID, F. X., KIEFHABER, T. (1991), GroE facilitates refolding of citrate synthase by suppressing aggregation, *Biochemistry* **30**, 1586–1591.

CARLSON, J. D., YARMUSH, M. L. (1992), Antibody assisted protein refolding, *Biotechnology* **10,** 86–91.

CLELAND, J. L. (1991), Mechanisms of protein aggregation and refolding, *PhD Thesis,* Department of Chemical Engineering, Massachusetts Institute of Technology, Cambridge, Massachusetts.

CLELAND, J. L., RANDOLPH, T. W. (1992), Mechanism of polyethylene glycol interaction with the molten globule folding intermediate of bovine carbonic anhydrase B, *J. Biol. Chem.*

CLELAND, J. L., WANG, D. I. C. (1990a), Cosolvent assisted protein refolding, *Biotechnology* **8,** 1274–1278.

CLELAND, J. L., WANG, D. I. C. (1990b), Refolding and aggregation of bovine carbonic anhydrase B: quasi-elastic light scattering analysis, *Biochemistry* **29,** 11072–11078.

CLELAND, J. L., WANG, D. I. C. (1991), Equilibrium association of a molten globule intermediate in the refolding of bovine carbonic anhydrase, in: *Protein Refolding* (GEORGIOU, G., DE BERNARDEZ-CLARK, E., Eds.) *A.C.S. Symp. Ser.* **470,** 169–179, Washington, DC: American Chemical Society.

CLELAND, J. L., WANG, D. I. C. (1992a), Cosolvent effects on refolding and aggregation, in: *Protein Folding* (GEORGIOU, G., HIMMEL, M., Eds.), *A.C.S. Symp. Ser.*

CLELAND, J. L., WANG, D. I. C. (1992b), Transient association of the first intermediate during the refolding of bovine carbonic anhydrase B, *Biotechnol. Prog.*

CLELAND, J. L., HEDGEPETH, C., WANG, D. I. C. (1992a), Polyethylene glycol enhanced refolding of bovine carbonic anhydrase B: reaction stoichiometry and refolding model, *J. Biol. Chem.*

CLELAND, J. L., BUILDER, S. E., SWARTZ, J. R., WINKLER, M., CHANG, J. Y., WANG, D. I. C. (1992b), Polyethylene glycol enhanced protein refolding, *Biotechnology.*

CONSLER, T. G., LEE, J. C. (1988), Domain interaction in rabbit pyruvate kinase: I. Effects of ligands on protein denaturation induced by guanidine hydrochloride, *J. Biol. Chem.* **263,** 2787–2793.

COUSENS, L. S., SHUSTER, J. R., GALLEGOS, C., KU, L., STEMPIEN, M. M., URDEA, M. S., SANCHEZ-PESCADOR, R., TAYLOR, A., TEKAMP-OLSON, P. (1987), High level expression of proinsulin in the yeast *Saccharomyces cerevisiae, Gene* **61,** 265–275.

CREIGHTON, T. E. (1984), *Proteins: Structures and Molecular Properties,* New York: W. H. Freeman & Company.

CREIGHTON, T. E. (1985), Folding of proteins adsorbed reversibly to ion-exchange resins, in: *Protein Structure Folding and Design* (OXENDER, D. L., Ed.), pp. 249–251, New York: Alan R. Liss, Inc.

DAMODARAN, S. (1987), Influence of solvent conditions on refolding of bovine serum albumin, *Biochim. Biophys. Acta* **914,** 114–121.

DARNELL, J. E., LODISH, H., BALTIMORE, D. (1986), *Molecular Cell Biology,* Chapter 4, *Scientific American Books.* New York: W. H. Freeman & Company.

DILL, K. (1990), Dominant forces in protein folding, *Biochemistry* **29,** 7133–7155.

DILL, K., SHORTLE, D. (1991), Denatured states of proteins, *Annu. Rev. Biochem.* **60,** 795–825.

ELLIS, R. J., VAN DER VIES, S. M. (1991), Molecular chaperones, *Annu. Rev. Biochem.* **60,** 321–347.

FANE, B., VILLAFANE, R., MITRAKI, A., KING, J. (1991), Indentification of global suppressors for temperature-sensitive folding mutations of the p22 tailspike protein, *J. Biol. Chem.* **261,** 11640–11648.

FISH, W. W., DANIELSSON, A., NORDLING, K., MILLER, S. H., LAM, C. F., BJORK, I. (1985), Denaturation behavior of antithrombin in guanidinium chloride. Irreversibility of unfolding caused by aggregation, *Biochemistry* **24,** 1510–1515.

FUKE, Y., SEKIGUCHI, M. MATSUOKA, H. (1985), Nature of stem bromelian treatment on aggregation and gelation of soybean proteins, *J. Food Sci.* **50,** 1283–1289.

GARVEY, E. P., MATTHEWS, C. R. (1989), Effects of multiple replacements at a single position on the folding and stability of dihydrofolate reductase from *Escherichia coli, Biochemistry* **28,** 2083–2093.

GELLMAN, S. H. (1991), On the role of methionine residues in the sequence-independent recognition of nonpolar protein surfaces, *Biochemistry* **30,** 6633–6636.

GEORGIOU, G., TELFORD, J. N., SHULER, M. L., WILSON, D. B. (1986), Localization of inclusion bodies in *Escherichia coli* overproducing β-lactamase or alkaline phosphatase, *Appl. Environ. Microbiol.* **52,** 1157–1161.

GHELIS, C., YON, J. (1982), *Protein Folding,* New York: Academic Press.

GOLDENBERG, D. P. (1988), *Annu. Rev. Biophys. Chem.* **17,** 481–507.

GOLDENBERG, D. P., SMITH, D. H., KING, J. (1983), Genetic analysis of the folding pathway for the tail spike protein of phage p22, *P.N.A.S.* **80,** 7060–7064.

GOLOUBINOFF, B., GATENBY, A. A., LORIMER, G. (1989), Reconstitution of active dimeric ribulose bisphosphate carboxylase from an unfolded state depends on two chaperonin proteins and Mg-ATP, *Nature* **337**, 44–47.

HAASE-PETTINGELL, C., KING, J. (1988), Formation of aggregates from a thermolabile *in vivo* folding intermediate in p22 tailspike maturation, *J. Biol. Chem.* **263**, 4977–4983.

HAGEN, A., HATTON, T. A., WANG, D. I. C. (1990), Protein refolding in reversed micelles: interactions of the protein with micelle components, *Biotechnol. Bioeng.* **35**, 966–975.

HARDY, S. J. S., RANDALL, L. L. (1991), A kinetic partitioning model of selective binding of nonnative proteins by the bacterial chaperone SecB, *Science* **251**, 439–443.

HART, R. A., BAILEY, J. E. (1992), Solubilization and regeneration of *Vitreoscilla* hemoglobin isolated from protein inclusion bodies, *Biotechnol. Bioeng.*

HARTL, F.-U., NEUPERT, W. (1990), Protein sorting to mitochondria: evolutionary conservations of folding and assembly, *Science* **247**, 930–938.

HAVEL, H. A., KAUFFMAN, E. W., PLAISTED, S. M., BREMS, D. N. (1986), Reversible self-association of bovine growth hormone during equilibrium unfolding, *Biochemistry* **25**, 6533–6538.

HOROWITZ, P. M., CRISCIMAGNA, N. L. (1986), Low concentrations of guanidinium chloride expose apolar surfaces and cause differential perturbation in catalytic intermediates of rhodanese, *J. Biol. Chem.* **261**, 15652–15658.

HORWICH, A. L., NEUPORT, W., HARTL, F.-U. (1990), Protein-catalysed protein folding, *TIBTECH* **8**, 126–131.

HUBBARD, T. J. P., SANDER, C. (1991), The role of heat-shock and chaperone proteins in protein folding: possible molecular mechanisms, *Protein Eng.* **4**, 711–717.

IKAI, A., TANAKA, S., NODA, H. (1978), Reactivation kinetics of guanidine denatured bovine carbonic anhydrase B, *Arch. Biochem. Biophys.* **190**, 39–45.

KANE, J. F., HARTLEY, D. L. (1988), Formation of recombinant protein inclusion bodies in *Escherichia coli, TIBTECH* **6**, 95–101.

KARPLUS, M., WEAVER, D. L. (1976), Protein-folding dynamics, *Nature* **260**, 104–111.

KARPLUS, M., WEAVER, D. L. (1979) Diffusion-collision model for protein folding, *Biopolymers* **18**, 1421–1426.

KIEFHABER, T., RUDOLPH, R., KOHLER, H.-H., BUCHNER, J. (1991), Protein aggregation *in vitro* and *in vivo*: a quantitative model of the kinetic competition between folding and aggregation, *Biotechnology* **9**, 825–829.

KIKUCHI, M., IKEHARA, M. (1991), Conformational features of signal sequences and folding of secretory proteins in yeasts, *TIBTECH* **9**, 208–211.

KIM, P. S., BALDWIN, R. L. (1982), Specific intermediates in the folding reactions of small proteins and the mechanism of protein folding, *Annu. Rev. Biochem.* **51**, 459–472.

KIM, P. S., BALDWIN, R. L. (1990), Intermediates in the folding reactions of small proteins, *Annu. Rev. Biochem.* **59**, 631–660.

KING, J. (1986), Genetic analysis of protein folding pathways, *Biotechnology* **4**, 297–303.

KING, J. (1989), Deciphering the rules of protein folding, *Chem. Eng. News* **67**, 32–54.

KING, J., HAASE, C., YU, M. (1987), Temperature-sensitive mutations affecting kinetic steps in protein-folding pathways, *Protein Eng.* **1**, 109–121.

KLEIN, B. K., HILL, S. R., DEVINE, C. S., ROWOLD, E., SMITH, C. E., GALOSY, S., OLINS, P. O. (1991), Secretion of active bovine somatotropin in *Escherichia coli, Biotechnology* **9**, 869–872.

KUMAMOTO, C. A. (1991), Molecular chaperones and protein translocation across the *Escherichia coli* inner membrane, *Mol. Microbiol.* **5**, 19–22.

KUWAJIMA, K. (1989), The molten globule state as a clue for understanding the folding, and cooperativity of globular-protein structure, *Proteins Struct. Funct. Genet.* **6**, 87–103.

LAMANTIA, M., MIURA, T., TACHIKAWA, H., KAPLAN, H. A., LENNARZ, W. J., MIZUNAGA, T. (1991), Glycosylation site binding protein and protein disulfide isomerase are identical and essential for cell viability in yeast, *P.N.A.S.* **88**, 4453–4457.

LAMINET, A. A., KUMAMOTO, C. A., PLÜCKTHUN, A. (1991), Folding *in vitro* and transport *in vivo* of pre-β-lactamase are SecB independent, *Mol. Microbiol.* **5**, 117–122.

LANDRY, S. J., GIERASCH, L. M. (1991), The chaperonin GroEL binds a polypeptide in an α-helical conformation, *Biochemistry* **30**, 7359–7362.

LANG, K., SCHMIDT, F. X. (1988), Catalysis of protein folding by prolyl isomerase, *Nature* **331**, 453–455.

LANG, K., SCHMIDT, F. X., FISCHER, G. (1987), Protein-disulphide isomerase and prolyl isomerase act differently and independently, *Nature* **329**, 268–270.

LANGER, T., NEUPERT, W. (1991), Heat shock proteins hsp60 and hsp70: their roles in folding, assembly and membrane translocation of proteins, *Curr. Top. Microbiol. Immunol.* **167**, 2–30.

LEHRMAN, S. R., TUIS, J. L., HAVEL, H. A., HASKELL, R. J., PUTNAM, S. D., TOMICH, C.-S. C. (1991), Site-directed mutagenesis to probe protein folding: evidence that the formation and aggregation of a bovine growth hormone folding intermediate are dissociable processes, *Biochemistry* **30**, 5777–5784.

Liberek, K., Skowyra, D., Zylicz, M., Johnson, C., Georgopoulos, C. (1991), The *Escherichia coli* DnaK chaperone, the 70-kDa heat shock protein eukaryotic equivalent, changes conformation upon ATP hydrolysis, thus triggering its dissociation from a bound target protein, *J. Biol. Chem.* **266**, 14491–14496.

Lim, W. A., Sauer, R. T. (1991), The role of internal packing interactions in determining the structure and stability of a protein, *J. Mol. Biol.* **219**, 359–376.

Lodish, H., Kong, N. (1991), Cyclosporin A inhibits an initial step in folding of transferrin within the endoplasmic reticulum, *J. Biol. Chem.* **266**, 14835–14838.

Lyles, M. M., Gilbert, H. F. (1991a), Catalysis of the oxidative folding of ribonuclease A by protein disulfide isomerase: dependence of the rate on the composition of the redox buffer, *Biochemistry* **30**, 613–619.

Lyles, M. M., Gilbert, H. F. (1991b), Catalysis of the oxidative folding of ribonuclease A by protein disulfide isomerase: pre-steady-state kinetics and the utilization of the oxidizing equivalents of the isomerase, *Biochemistry* **30**, 619–625.

Marston, F. A. O. (1986), The purification of eukaryotic polypeptides synthesized in *Escherichia coli, Biochem. J.* **240**, 1–12.

Martin, J., Langer, T., Boteva, R., Schramel, A., Horwich, A. L., Hartl, F.-U. (1991), Chaperonin-mediated protein folding at the surface of groEL through a 'molten globule'-like intermediate, *Nature* **352**, 36–42.

Mascagni, P., Tonolo, M., Ball, H., Lim, M., Ellis, J., Coates, A. (1991), Chemical synthesis of 10 kDa chaperonin, *FEBS Lett.* **286**, 201–203.

Matouschek, A., Kellis, J. T., Serrano, L., Bycroft, M., Fersht, A. R. (1990), Transient folding intermediates characterized by protein engineering, *Nature* **346**, 440–445.

Mendoza, J. A., Rogers, E., Lorimer, G. H., Horowitz, P. M. (1991a), Chaperonins facilitate the *in vitro* folding of monomeric mitochondrial rhodanese, *J. Biol. Chem.* **266**, 13044–13049.

Mendoza, J. A., Rogers, E., Lorimer, G. H., Horowitz, P. M. (1991b), Unassisted refolding of urea unfolded rhodanese, *J. Biol. Chem.* **266**, 13587-13591.

Meyer, D. I. (1982), The signal hypothesis - a working model, *TIBS* **2**, 320–321.

Mitraki, A., King, J. (1989), Protein folding intermediates and inclusion body formation, *Biotechnology* **7**, 690–697.

Mitraki, A., Betton, J.-M., Desmadril, M., Yon, J. M. (1987), Quasi-irreversibility in the unfolding-refolding transition of phosphoglycerate kinase induced by guanidine hydrochloride, *Eur. J. Biochem.* **163**, 29–34.

Mitraki, A., Haase-Pettingell, C., King, J. (1991a), Mechanisms of inclusion body formation, in: *Protein Refolding* (Georgiou, G., De-Bernardez-Clark, E., Eds.); *A.C.S. Symp. Ser.* **470**, 35–49, Washington DC: American Chemical Society.

Mitraki, A., Fane, B., Haase-Pettingell, C., Sturtevant, J., King, J. (1991b), Global suppression of protein folding defects and inclusion body formation, *Science* **253**, 54–58.

Morris, G. E., Frost, L. C., Newport, P. A., Hudson, N. (1987), Monoclonal antibody studies of creatine kinase: antibody binding sites in the N-terminal region of creatine kinase and effects of antibody on enzyme refolding, *Biochem. J.* **248**, 53–57.

Moult, J., Unger, R. (1991), An analysis of protein folding pathways, *Biochemistry* **30**, 3816–3824.

Nguyen, T., Law, D. T. S., Williams, D. B. (1991), Binding protein Bip is required for translocation of secretory proteins into the endoplasmic reticulum in *Saccharomyces cerevisiae, P.N.A.S.* **88**, 1565–1569.

Nothwehr, S. F., Gordon, J. I. (1990), Targeting of proteins into the eukaryotic secretory pathway: signal peptide structure/function relationships, *BioEssays* **12**, 479–484.

Ohta, Y., Hojo, H., Almoto, S., Kobayashi, T., Zhu, X., Jordan, F., Inouye, M. (1991), Pro-peptide as an intermolecular chaperone: renaturation of denatured subtilisin E with a synthetic pro-peptide, *Mol. Microbiol.* **5**, 1507–1510.

Palleros, D. R., Welch, W. J., Fink, A. L. (1991), Interaction of hsp70 with unfolded proteins: effects of temperature and nucleotides on the kinetics of binding, *P.N.A.S.* **88**, 5719–5723.

Park, S., Liu, G., Topping, T. B., Cover, W. H., Randall, L. L. (1988), Modulation of folding pathways of exported proteins by the leader sequence, *Science* **239**, 1033–1035.

Porter, D. H., Cardenas, J. M. (1980), Analysis of the renaturation kinetics of bovine muscle pyruvate kinase, *Biochemistry* **19**, 3447–3452.

Prakash, V., Loucheux, C., Scheufele, S., Gorbunoff, M. J., Timasheff, S. N. (1981), Interactions of proteins with solvent components in 8 M urea, *Arch. Biochem. Biophys.* **210**, 455–464.

Ptitsyn, O. B. (1987), Protein folding: hypotheses and experiments, *J. Protein Chem.* **6**, 273–277.

PTITSYN, O. B., PAIN, R. H., SEMISOTNOV, G. V., ZEROVNIK, E., RAZGULYAEV, O. I. (1990), Evidence for a molten globule state as a general intermediate in protein folding, *FEBS Lett.* **262,** 20–24.

RANDALL, L. L., HARDY, S. J. S. (1984), Export of protein in bacteria: dogma and data, in: *Modern Cell Biology,* Vol. 3, pp. 1–20, New York: A. R. Liss, Inc.

RANDALL, L. L., HARDY, S. J. S., THOM, J. R. (1987), Export of protein: a biochemical view, *Annu. Rev. Microbiol.* **41,** 507–541.

RANDALL, L. L., TOPPING, T. B., HARDY, S. J. S. (1990), No specific recognition of leader peptide by SecB, a chaperone involved in protein export, *Science* **248,** 860–863.

RICHARDSON, J. S. (1981), The anatomy and taxonomy of protein structure, *Adv. Protein Chem.* **34,** 167–339.

RIEDERER, M. A., HINNEN, A. (1991), Removal of N-glycosylation sites of the yeast acid phosphatase severely affects protein folding, *J. Bacteriol.* **173,** 3539–3546.

ROBINSON, A., AUSTIN, B. (1987), The role of topogenic sequence in the movement of proteins through membranes, *Biochem. J.* **246,** 249–261.

RODIONOVA, N. A., SEMISOTNOV, G. V., KUTYSHENKO, V. P., UVERSKII, V. N., BOLOTINA, I. A., BYCHKOVA, V. E., PTITSYN, O. B. (1989), Two-stage equilibrium unfolding of carboanhydrase B by strong denaturing agents, *Mol. Biol.* (Moscow), **23,** 683–692.

ROTHMAN, J. E. (1989), Polypeptide chain binding proteins: catalysts of protein folding and related processes in cells, *Cell* **59,** 591–601.

RUDOLPH, R., ZETTLMEISSL, G., JAENICKE, R. (1979), Reconstitution of lactic dehydrogenase, noncovalent aggregation vs. reactivation. 2. Reactivation of irreversibly denatured aggregates, *Biochemistry* **18,** 5572–5575.

SCHEIN, C. H. (1989), Production of soluble recombinant proteins in bacteria, *Biotechnology* **7,** 1141–1149.

SECKLER, R., FUCHS, A., KING, J., JAENICKE, R. (1989), Reconstitution of the thermostable trimeric phage p22 tailspike protein from denatured chains *in vitro, J. Biol. Chem.* **264,** 11750–11753.

SEMISOTNOV, G. V., RODIONOVA, N. A., KUTYSHENKO, V. P., EBERT, B., BLANCK, J. PTITSYN, O. B. (1987), Sequential mechanism of refolding carbonic anhydrase B, *FEBS Lett.* **224,** 9–13.

SEMISOTNOV, G. V., UVERSKY, V. N., SOKOLOVSKY, I. V., GUTIN, A. M., RAZGULYAEV, O. I., RODIONOVA, N. A. (1990), Two slow stages in refolding of bovine carbonic anhydraase B are due to proline isomerization, *J. Mol. Biol.* **213,** 561–568.

SHORTLE, D. (1989), Probing the determinants of protein folding and stability with amino acid substitutions, *J. Biol Chem.* **264,** 5315–5318.

STEIN, P. J., HENKENS, R. W. (1978), Detection of intermediates in protein folding of carbonic anhydrase with fluorescence emission and polarization, *J. Biol. Chem.* **253,** 8016–8018.

STELLWAGEN, E. (1979), Proline peptide isomerization and the reactivation of denatured enzymes, *J. Mol. Biol.* **135,** 217–229.

STELLWAGEN, E., WILGUS, H. (1978), Relationship of protein thermostability to accessible surface area, *Nature* **275,** 342–347.

STURTEVANT, J. M., YU, M., HAASE-PETTINGELL, C., KING, J. (1989), Thermostability of temperature-sensitive folding mutants of the p22 tailspike protein, *J. Biol. Chem.* **264,** 10693–10698.

TANDON, S., HOROWITZ, P. M. (1986), Detergent-assisted refolding of guanidinium chloride-denatured rhodanese: the effect of lauryl maltoside, *J. Biol. Chem.* **261,** 15615–15618.

TANDON, S., HOROWITZ, P. M. (1987), Detergent-assisted refolding of guanidinium chloride-denatured rhodanese: the effects of the concentration and type of detergent, *J. Biol. Chem.* **262,** 4486–4491.

TESCHKE, C. M., KIM, J., SONG, T., PARK, S., PARK, C., RANDALL, L. L. (1991), Mutations that affect the folding of ribose-binding protein selected as suppresors of a defect in export in *Escherichia coli, J. Biol. Chem.* **226,** 11789–11796.

TESCHNER, W., RUDOLPH, R., GAREL, J.-R. (1987), Intermediates on the folding pathway of octopine dehydrogenase from *Pecten jacobaeus, Biochemistry* **26,** 2791–2796.

TIMASHEFF, S. N., ARAKAWA, T. (1988), Mechanism of protein precipitation and stabilization by co-solvents, *J. Crystal Growth* **90,** 39–46.

TOKATLIDIS, K., DHURJATI, P., MILLET, J., BEGUIN, P., AUBERT, J.-P. (1991), High activity of inclusion bodies formed in *Escherichia coli* overproducing *Clostridium thermocellum* endoglucanase D, *FEBS Lett.* **282,** 205–208.

TSONG, T. Y. (1982), Viscosity-dependent conformational relaxation of ribonuclease A in the thermal unfolding zone, *Biochemistry* **21,** 1493–1497.

VALAX, P., GEORGIOU, G. (1991), Folding and aggregation of RTEM β-lactamase, in: *Protein Refolding* (GEORGIOU, G., DEBERNARDEZ-CLARK, E., Eds.), *A.C.S. Symp. Ser.* **470,** 97–109, Washington, DC: American Chemical Society.

VAUCHERET, H., SIGNON, L., LE BRAS, G., GAREL, J.-R. (1987), Mechanism of renaturation of a large protein, aspartokinase-homoserine dehydrogenase, *Biochemistry* **26,** 2785–2789.

WALTER, P., IBRANIMI, I., BLOBEL, G. (1981), Translocation of proteins across the endoplasmic reticulum. I. Signal recognition particle (SRP) binds to *in-vitro*-assembled polysomes synthesizing secretory proteins, *J. Cell. Biol.* **91,** 545–550.

WICKNER, W., DRIESSEN, A. J. M., HARTL, F.-U. (1991), The enzymology of protein translocation across the *Escherichia coli* plasma membrane, *Annu. Rev. Biochem.* **60,** 101–124.

WRIGHT, P. E., DYSON, H. J., LERNER, R. A. (1988), Conformation of peptide fragments of proteins in aqueous solution: implications for initiation of protein folding, *Biochemistry* **27,** 7167–7175.

ZETTLMEISSL, G., RUDOLPH, R., JAENICKE, R. (1979), Reconstitution of lactic dehydrogenase, noncovalent aggregation vs. reactivation. 1. Physical properties and kinetics of aggregation, *Biochemistry* **18,** 5567–5571.

21 Liquid–Liquid Extraction (Small Molecules)

KARL SCHÜGERL

Hannover, Federal Republic of Germany

1 Introduction

The formation of several low-molecular weight compounds by microorganisms can be more economical than their chemical synthesis, if their inexpensive recovery from the cultivation medium is possible. Because of the complex composition of the cultivation media, the high protein contents, and the low primary and secondary metabolite concentrations, recovery and purification of the desired products can be rather difficult. Several common separation techniques can be used for recovery: precipitation, adsorption, membrane processes, extraction, and in case of volatile compounds, fractional distillation.

- Precipitation is usually not selective enough. Sometimes a poorly soluble salt of the product (e.g., carboxylic acid) is a suitable form for recovery. However, during purification, an inorganic salt is formed as by-product or waste (because of its low purity) which has to be disposed.
- Adsorption often is not efficient enough on account of the low loading capacity of the adsorbents.
- Membrane processes suffer from low selectivity and fouling (usually caused by proteins).
- Fractional distillation is restricted to volatile and heat stable products.
- Extraction often is not specific enough, although specificity can be improved by selection of proper carriers.

Extraction often is combined with other separation methods, e.g., adsorption or distillation for recovery of the product from the culture medium. However, extraction is seldom used for fine purification.

This chapter examines common extraction techniques and their different modern variants: carrier, supercritical, liquid membrane, and reversed micelle extractions. Finally, the *in situ* extraction from the cultivation medium during product formation is considered.

The prerequisite for extraction is the presence of two immiscible phases. The partition coefficient P of the desired product (solute) between the phases is an indicator of the separation performance. Definitions are:

$$P_x = \frac{x_\mathrm{o}}{x_\mathrm{a}}, \quad P_c = \frac{c_\mathrm{o}}{c_\mathrm{a}} \text{ and } P_w = \frac{w_\mathrm{o}}{w_\mathrm{a}}$$

where P_x, P_c and P_w are the partition coefficients based on mole fraction x, mole concentration c and mass w, respectively. Index o means organic phase, index a aqueous phase.

Since the mutual solubilities of the aqueous and organic phases are not negligible, the two phases have to be mutually saturated for the determination of the partition coefficients.

The distribution coefficent K_D is often used in practice:

$$K_\mathrm{D} = \frac{c_\mathrm{ot}}{c_\mathrm{at}}$$

where c_ot, c_at (mol L^{-1}) are the measured total solute concentrations in the organic and the aqueous phase without any correction.

The distribution coefficient can be influenced by the extracting solvent (hydrophilic or hydrophobic character), the pH value (especially for weak acids and bases), temperature, salts (counter ions for ionic solutes), and other additives (modifiers). It also depends on the solute concentration.

The solvent should be inexpensive, non-toxic, non-volatile, non-flammable, slightly soluble in the aqueous phase and should not form stable emulsions.

Since several primary and secondary metabolites are weak acids or bases, the pH value can frequently be used to control the distribution coefficient. The quantitative relationships will be given by practical examples in Sect. 3.

As long as the distribution coefficient does not depend on the solute concentration, the calculation of the concentrations in the aqueous phase c_a^e and the organic phase c_o^e at equilibrium with the concentrations in the feed c_a^0 and c_o^0 in a batch process is possible by means of a simple mass balance equation:

$$c_\mathrm{a}^0 A + c_\mathrm{o}^0 O = c_\mathrm{a}^\mathrm{e} A + c_\mathrm{a}^\mathrm{e} O$$

$$c_\mathrm{o}^\mathrm{e} = \frac{K_\mathrm{D} c_\mathrm{a}^0}{1+E} \quad c_\mathrm{a}^\mathrm{e} = \frac{c_\mathrm{a}^0}{1+E} \quad E = \frac{K_\mathrm{D} O}{A}$$

where E is the extraction factor,

c_a^0 and c_o^0 are the initial solute concentrations in the aqueous and the organic phase,

c_a^e and c_o^e the equilibrium solute concentrations in the aqueous and the organic phase,

A and O the volumes of the aqueous and the organic phase.

In continuous countercurrent extraction with

$$c_o = K c_a^e$$

the mass balance is given by

$$F_a c_a + c_o^0 F_o = F_a c_a^0 + c_o F_o$$

Here, F_a and F_o are the throughputs of the aqueous and the organic phase,

c_a^0 and c_o^0 the concentrations of the solute in the aqueous and organic feed,

c_a and c_o the actual concentrations of the solute in the aqueous and the organic phase.

$$c_o = \frac{F_a}{F_o}(c_a^0 - c_a)$$

if, as usual, $c_o^0 = 0$.

The variation of the actual concentration along axis z of the column is

$$\frac{dc_a}{dz} = \left(\frac{ka}{F_a/F_o}\right)(c_a - c_a^e)$$

where k is the mass transfer coefficient,

a the specific interfacial area between the aqueous and the organic phase.

Extraction can be carried out in various equipment. The most common equipment type is the mixer settler consisting of a stirred vessel in which the extraction takes place and an unstirred vessel in which the phase separation (droplet coalescence and settling) occurs. For better separation, a cascade of several mixer settlers is operated in a countercurrent mode. The number of necessary stages for the extraction degree can be calculated by means of the equilibrium curve and the mass balance.

In chemical industry, various types of columns are used (plate, packed, stirred, pulsed columns; LO et al., 1983). They are usually operated in a countercurrent mode. Different types of extraction columns and their modeling are considered in Sect. 3.

In biotechnology, however, centrifugal extractors are applied almost exclusively. On account of their high throughput, the contact time between the phases and the extraction time is very short. Sensitive solutes, therefore, can be extracted under rather harsh conditions. Furthermore, fairly stable emulsions, common in biotechnology, can also be separated.

Some performance data of centrifugal extractors are presented in Sect. 3.

Extraction is followed by re-extraction. The solvent is regenerated and reused. From an economical and ecological point of view, it is important that the solvent loss caused by entrainment, vaporization, or high solubility in the aqueous phase is kept low.

The advantages of the extraction processes are low temperature (vs. distillation), liquid handling (vs. precipitation, adsorption), slight sensitivity to fouling (vs. membrane separation), high separation capacity, and option for continuous operation. They are mainly used for the recovery of bulk products.

2 Extraction with Organic Solvents

Low molecular biotechnological products usually are primary or secondary metabolites. Most of them are hydrophilic compounds. Organic solvents are not suited for their recovery.

Extraction of hydrophobic products from the aqueous cultivation medium or from the cells by organic solvents is a common technique. Hydrophilic compounds, however, have a rather low solubility in organic solvents with a low solubility in the aqueous phase. Therefore, organic solvents with a fairly high solubility in the aqueous phase are used for the extraction of hydrophilic compounds.

Typical hydrophilic compounds are low-molecular weight alcohols: methanol, ethanol, propanol, and butanol. Extraction from their aqueous solutions into water-immiscible organic solvents increases with increasing molecular weight of the alcohol (Tab. 1), concentration in the aqueous feed, and temperature. Obviously, this hydrophobicity governs the partition coefficient of aliphatic alcohols in organic solvents. With increasing polarity of the solvents, the partition coefficient of alcohols increases, but with additional coextraction of water. The overall water content of the organic phase increases alcohol concentration, as does the solvent content of the aqueous phase.

To select the suitable solvent, energy and capital costs should be minimized. In Tab. 2, the ethanol recovery from cultivation media by different solvents is shown. Because of the low energy and capital costs, heptanal seems to be a good choice. In Tab. 3 the comparative costs of solvent extraction of ethanol with heptanal and its distillation are compared. Obviously distillation is more economical. This holds also truc for other low-molecular weight alcohols, esters, ethers, and ketones.

Several aliphatic carboxylic acids are formed during aerobic cultivation as intermediates of the metabolic pathways of the microorganisms. Their purification with carbon-bonded oxygen bearing extractants (CBO) is not economical, because of both their low partition and distribution coefficients: acetic acid (Tab. 4), lactic acid (Tab. 5), and citric acid (Tab. 6). The distribution coefficients are

Tab. 1. Partition Coefficients P_c of Methanol MeOH, Ethanol EtOH, *n*-Propanol *n*-PrOH, and *n*-Butanol *n*-BuOH between Water and Organic Solvents at 25 °C (KERTES and KING, 1986)

Solvent	MeOH	EtOH	*n*-PrOH	*n*-BuOH
n-Hexane	0.0024	0.0083	0.03	0.25
n-Heptane		0.0074		
n-Octane	0.0024	0.0068	0.03	0.25
n-Nonane		0.0064		
n-Decane	0.0024	0.0060	0.03	0.025
n-Undecane		0.0057		
n-Dodecane		0.0055		
n-Tridecane		0.0053		
n-Tetradecane		0.0052		
n-Hexadecane		0.051		
Cyclohexane	0.0016	0.0079	0.028	0.13
Benzene	0.013	0.031	0.23	0.66
Toluene	0.007	0.03	0.15	0.50
Chlorobenzene	0.011	0.021	0.18	0.50
Chloroform	0.055	0.14	0.40	2.2
Carbon tetrachloride	0.008	0.018	0.12	0.51
Diethyl ether	0.14	0.32	0.40	4.1
n-Octanol	0.18	0.50	1.8	7.6
Tri-*n*-butyl phosphate	0.20	0.54	2.7	14.5

Tab. 2. Data for Solvent Selection for the Recovery of Ethanol from Cultivation Medium (ESSIEN and PYLE, 1987)

Solvent	Solvent Inventory	Solvent Make-up	Energy Costs	Other
Alcohols	Low	Low	High	
Heptanal	Moderate	High	Low	Lowest inventory
Esters	Moderate	Moderate	High	
Ketones	High	Low	Moderate	
Butyl ether	Very high	Very high	Moderate	

Tab. 3. Comparative Cost of Solvent Extraction and Straight Distillation of Ethanol (ESSIEN and PYLE, 1987)

Ethanol in Aqueous Feed	Distillation 2 mol% 5 wt%	4 mol% 9.6 wt%	6 mol% 14 wt%	Extraction 2 mol% 5 wt%	4 mol% 9.6 wt%	6 mol% 14 wt%
			$/h			
Utilities	75	59	55	66	51	49
Capital-related costs	60	42	38	95	69	59
Labor and overheads	60	57	56	65	61	60
Solvent make-up	—	—	—	406	199	131
Total						
Cost ($/h)	182	143	139	635	380	300
($/L prod.)	0.036	0.029	0.028	0.127	0.076	0.060

Tab. 4. Distribution Coefficients of Acetic Acid in Solvent/Water Systems (KING, 1983)

Solvent		Range of K_D
n-Alcohols	(C_4–C_8)	1.68–0.64
Ketones	(C_4–C_{10})	1.20–0.61
Acetates	(C_4–C_{10})	0.89–0.17
Ethers	(C_4–C_8)	0.63–0.14

Tab. 5. P_c of Lactic Acid in Organic Solvent/Water Systems (KERTES and KING, 1986)

Solvent	P_c
Diethyl ether	0.1
Diisopropyl ether	0.04
Methyl-isobutyl ketone	0.14
n-Octanol	0.32
Isobutanol	0.66
n-Pentanol	0.40
n-Hexanol	0.26

Tab. 6. Distribution Coefficients of Citric Acid in Solvent/Water Systems (KERTES and KING, 1986)

Solvent	Range of K_D
n-Butanol	0.29
Ethyl acetate	0.1
Ethyl ether	0.1
Methyl-isobutyl ketone	0.1
Methyl-ethyl ketone	0.33
Cyclohexane	0.21

somewhat higher in the presence of salts (in cultivation media) due to the *salting out effect* (SHAH and TOWARI, 1981), but they are still not high enough for an economical separation.

Acetic acid is usually purified by distillation (at high concentration) or extractive distillation (at low concentration). The concentration of diluted acetic acid (20–40 wt%) is increased by extraction up to 50–70 wt% and, finally, the acid is recovered with distillation. The economics of acetic acid purification is determined by the cost of the solvent recovery.

If a high-boiling point solvent is used, the product is distilled away from the solvent, thereby minimizing the thermal load on the system. However, in this case it may be difficult to prevent contaminating the aqueous raffinate with solvent while recycling the aqueous layer back to the fermentation operation (BUSCHE, 1983).

If a low-boiling point solvent is used, the solvent is distilled away from the product. In effect, this scheme treats the vaporization of solvent like vaporization of water at usually only one sixth of the latent heat of the vaporization of water, and hence, at an equivalent reduction in heat load on the still. Overall, extraction reduces the operating costs associated with simple distillation, but at the expense of added investment for an extractor and solvent stripper (BUSCHE, 1983).

Diethyl ether, ethylene dichloride and ethyl acetate are used as solvents for acetic acid recovery. Ketones and ethers are generally superior extractants, but overall system costs with ketones are usually higher as a result of mutual solubility with water. After the acid extraction from aqueous solution with, e.g., ethyl acetate an ethyl acetate/acetic acid/water mixture is formed. Acetic acid is withdrawn at the bottom of the distillation column and a binary azeotropic mixture of water and ethyl acetate with a boiling point of 70.4 °C is formed at the head, from which ethyl acetate is recovered by fractional distillation (Fig. 1).

During the production of lactic acid by Lactobacilli the pH has to be controlled. The pH value is maintained by the addition of $CaCO_3$ or $Ca(OH)_2$. The poorly soluble calcium lactate is precipitated from the cultivation medium. After its separation from the medium, the precipitate is treated with sulfuric acid. The insoluble $CaSO_4$ is separated and the free acid is recovered.

For citric acid production, a low pH is necessary for citric acid formation. At the end of the cultivation, the culture solution is filtered to remove the mycelium. A polishing filtration may be necessary to remove residual antifoam, mycelium, or oxalate. Calcium citrate is precipitated from the clear solution by addition of calcium hydroxide. Calcium citrate is then filtered off and the filter cake is transferred to a tank where it is treated with sulfuric acid to precipitate calcium sulfate. The dilute filtrate is purified by activated carbon, demineralized by ion exchangers, and the purified solution is evaporated yielding crystals of citric acid.

In the case of well-soluble calcium salts of the acids, sodium hydroxide is used for pH control. The salt solution has to be concentrated, e.g., in a falling film evaporator, after its separation from the cells. With increasing salt concentrations, the boiling point increases. Aqueous solutions of sodium acetate exhibit boiling elevations from 1 °C at 7.5 wt% to 25 °C at the saturation point of 64.4 wt%. Electrical demands for evaporating such solutions increase from 3×10^3 to 2×10^4 BTU/LB salt for 1 wt% feed concentration and from 1.5×10^2 to 5×10^2 BTU/LB salt solution for 20 wt% feed concentration (BUSCHE et al., 1982). The salt must be acidified subsequently to recover the free acid by conventional distillation or extraction.

Several amino acids are produced commercially by enzymatic biotransformation or microbial biosynthesis (SCHMIDT-KASTNER and EGERER, 1984). Because of their amino and carboxyl groups they behave as cations at low pH, as anions at high pH, and they are of zwitterionic character at intermediate pH. Their extraction with CBO extractants is not possible or economical because of their low solubility. Their extraction with alcohols, ethers, esters, and ketones is inefficient.

Antibiotics are secondary metabolites, usually secreted by cells into the medium. The

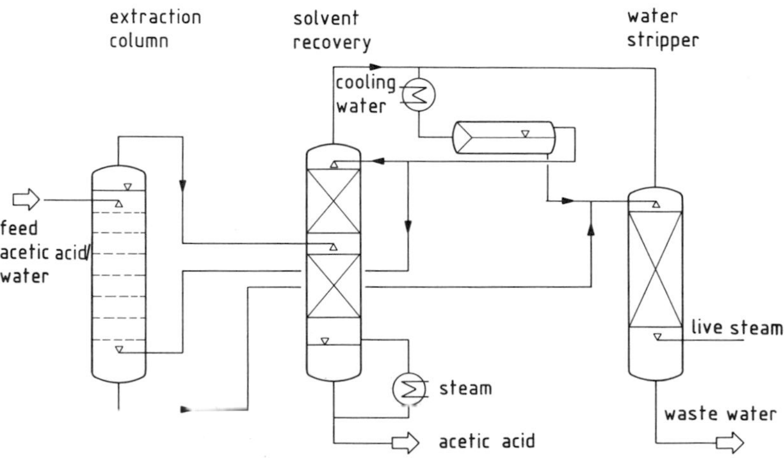

Fig. 1. Flowsheet of acetic acid recovery by extraction (QVF, Corning Co).

Tab. 7. Solvent Extraction of Several Antibiotics (SCHÜGERL et al., 1992)

Products	Medium	Solvent	pH
Actinomycin	cake	1 MeOH + 2 $MeCl_2$	2.5
Adrianimycin	cake	acetone	acid.
Bacitracin	broth	*n*-BuOH	7.0
Chloramphenicol	broth	EtAc	n-alk.
Clavulanic acid	broth	*n*-BuOH	2.0
Cycloheximide	broth	acetone/chloroform	acid.
Erythromycin	broth	AmAc	alk.
Fusidic acid	broth	MIBK	6.8
Griseofulvin	cake	BuAc	n.
Macrolides	broth	MIBK, EtAc	alk.
Nisin	broth	$CHCl_3$ + sec. oct. alc.	4.5
	mash	$CHCl_3$	2.0
Oxytetracycline	broth	BuOH	
Penicillin G	broth	BuAc, AmAc	2.0
Salomycin	cake	BuAc	9.0
Tetracycline	broth	BuOH	
Tylosin	broth	AmAc, EtAc	
Virginiamycin	broth	MIBK	acid.

MeOH (methanol), AmAc (amyl acetate), BuOH (1-butanol), BuAc (butyl acetate), EtAc (ethyl acetate), $MeCl_2$ (methylene dichloride), MIBK (methyl-isobutyl ketone), $CHCl_3$ (chloroform), n. (neutral), acid. (acidic), alk. (alkaline), broth (cell-free medium), cake (medium-free cells), mash (broth with cells)

cells are withdrawn from the medium by using a filter press or a rotary filter and the product is recovered from the medium by solvent extraction or adsorption. In case of adsorption, the desorption from the sorbent is an extraction process. If some of the product remains in the cells, the solid filter cake is also extracted. In Tab. 7 several examples for extraction of antibiotics with CBO extractants are shown.

Some of them are weak acids or bases. Because only the free acid or the free base can be extracted by the organic solvent it is necessary to reduce the pH value (for acids) of the medium for the extraction below the pK_a value, or increase it (for bases) above their pK_b value. However, some of them are very unstable at these extreme pH values. Therefore, the extraction is carried out at low temperature as quickly as possible, to avoid the decomposition of the product.

A typical representative of such antibiotics is penicillin G, a weak acid ($pK_a = 2.75$), which can only be extracted by organic solvents at pH values below its pK_a value.

The recovery of penicillin G is the best investigated antibiotic extraction process. In Tab. 8 the distribution coefficients for penicillin G at pH 4 between aqueous solutions and different CBO extractants are compiled. Most of these solvents are more effective than butyl acetate and amyl acetate, which are the most widely used solvents. However, they are not as economical as butyl acetate and amyl acetate due to their higher price and/or difficulties

Tab. 8. Distribution Coefficients for Penicillin G at pH 4 between Aqueous Solutions and the Listed Solvents (RIDGWAY and THORPE, 1983)

Solvent	K_D
Methyl cyclohexanone	180
Dimethyl cyclohexanone	160
Methyl cyclohexanol	80
2-Chloro-2'-methoxy diethyl ether	57
Cyclohexyl acetate	62
Furfuryl acetate	44
Methyl-isobutyl ketone	33
Dimethyl phthalate	30
2-Ethyl hexanol	26
Amyl acetate	20
Diethyl oxalate	20

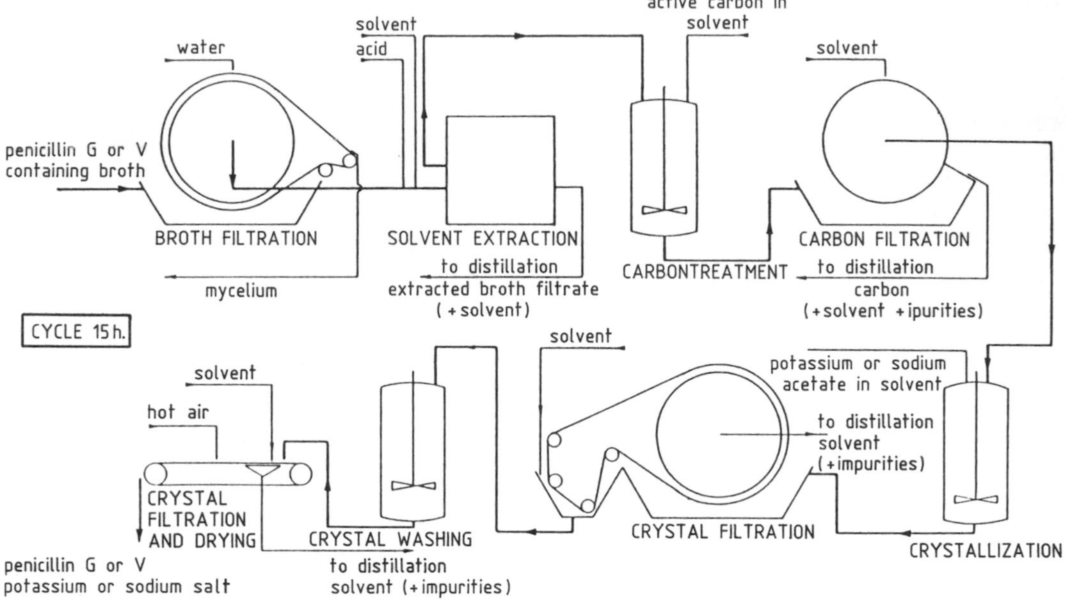

Fig. 2. Penicillin recovery according to Gist Brocades (HERSBACH et al., 1984).

in recovering penicillin from these organic phases.

Fig. 2 shows the classical penicillin recovery according to Gist Brocades, the largest penicillin producer. The mycelium free medium is cooled down to low temperatures (<3 °C), acidified with sulfuric acid (pH 2.0–3.0), and usually extracted using amyl acetate or butyl acetate in centrifugal extractors (Podbielnak, Luwesta, Westfalia) with a solvent-to-medium ratio of 0.1. Fig. 3 shows a Podbielnak extractor, which is the most popular one for penicillin extraction. In order to repress emulsification, a de-emulsifier is added to the medium. Typically a 3:1 or 4:1 medium-to-solvent ratio is used. The extract is treated with activated carbon, filtered, and contacted with an aqueous buffer solution at pH 6. The resulting crystals are separated and washed. The largest fraction of the 18% loss during recovery and purification occurs in the first three steps.

According to other companies (Biogal, Beecham, Biochemie, Toyo-Jazo), the cell-containing mash may be extracted in a countercurrent decanter (Type CA 366-290 of Westfalia; Fig. 4) in two stages using butyl acetate. Typical operation data are

- mash throughput (L h^{-1}): 2000–5000
- solvent throughput (L h^{-1}): 700–4000
- phase ratio (aqueous phase/organic phase): 1:1–4:1
- acid feed rate (L h^{-1}): 20–100

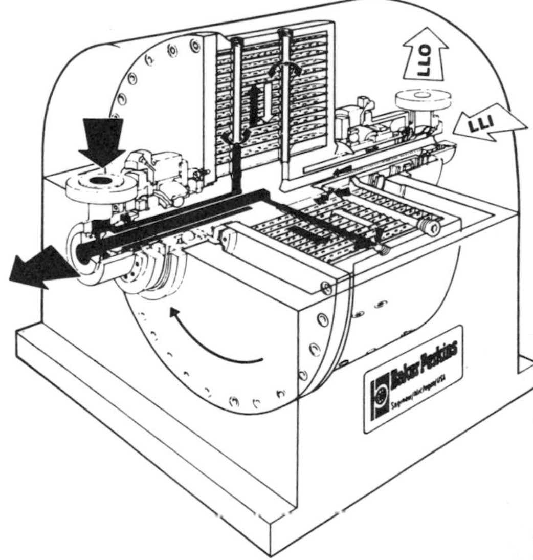

Fig. 3. Podbielnak centrifugal extractor (Baker Perkins, Michigan).

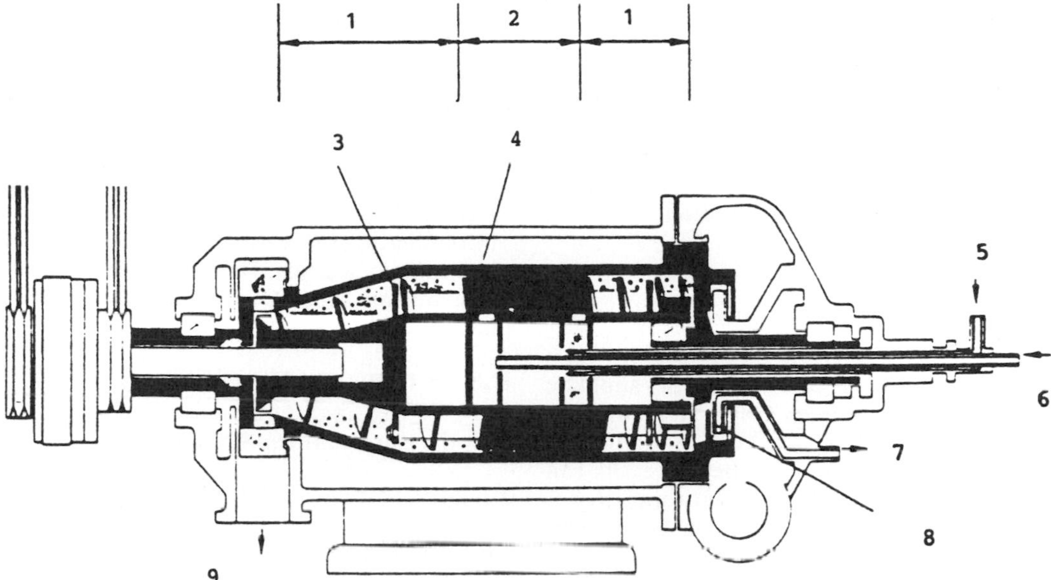

Fig. 4. Schematic view and operation mode of countercurrent extraction decanter CA 226-290 of Westfalia Co. Heavy phase: cultivation medium, light phase: *n*-butyl acetate.
1 clarifier zone, **2** countercurrent extraction, **3** separation disc, **4** scroll, **5** medium inlet, **6** solvent inlet, **7** extract outlet, **8** control disc, **9** medium outlet.

By avoiding filtration, a 5–8% yield improvement can be achieved by direct extraction of the mash.

According to KATINGER et al. (1981), mashes with a 50 vol% wet cell mass content can be extracted, and the solvent content of the cell mass leaving the decanter is about 2%. The yield improvement is 15%, and 50% less solvent is used because the culture filtrate is not diluted with the washing water used for cake elution.

For fat-soluble products such as β-carotene or steroids volatile apolar solvents are used (e.g., petroleum ether). In case of steroids, filtration prior to the extraction markedly reduces the amount of steroids that can be extracted by methylene chloride, ethylene chloride, or chloroform. It is thus desirable to conduct the extraction in the presence of the cells. By proper pretreatment conditioning, the fibrous cell material can be transformed into a more suitable material which makes a centrifugal extraction possible (TODD and DAVIES, 1973).

In this chapter, it has been shown that extraction of alcohols with CBO extractants cannot compete with their fractional distillation (Tabs. 1–3). Therefore, research has concentrated on extractive fermentation, i.e., recovery of metabolites by solvent extraction from cultivation media during product formation. This will be discussed in Sect. 7.

3 Carrier Extraction

Several metabolites have a high solubility in the aqueous cultivation medium. Their extraction with organic solvents is not economical due to their unfavorable distribution coefficients and the high solubility of the organic solvents in the aqueous phase. They are able to react with phosphorous and amine compounds insoluble in the aqueous culture medium and form selective solvation bonds or stoichiometric complexes which are also insoluble in the aqueous phase. These phosphorous and amine compounds are called carriers, because they "carry" the metabolites into the organic phase. This extraction is also called "reactive extraction" because a reaction – the formation of the solvation bond or the complex – is the prerequisite for the extractive separation.

3.1 Carboxylic Acids

The distribution coefficients of aliphatic carboxylic acids between CBO extractants (alcohols, esters, ethers, and ketones) and water are rather low (Tabs. 4–6). Better results can be obtained by extractants which form strong, selective, and reversible solvation bonds with the solute.

Examples of such compounds include phosphorous-bonded oxygen donor extractants (PBO). They contain phosphoryl groups which are strong Lewis bases. Furthermore, these extractants coextract low amounts of water, and their solubility in water is low. Thus, in the equilibirum system of 30 wt% trioctyl phosphine oxide (TOPO) in Chevron 25 and 10% citric acid solution, the solubility of TOPO in the aqueous phase is less than 1 ppm (KING, 1983).

The basicity of phosphoryl oxygen increases in the order: trialkyl phosphate ($(RO)_3P{=}O$) < dialkyl phosphonate ($(R{=})_2RP{=}O$) < alkyl dialkyl phosphinate ($(RO)R_2P{=}O$) < trialkyl phosphine oxide ($R_3P{=}O$).

The distribution coefficients of the acids are substantially higher with PBO-extractants than with CBO-extractants.

According to WARDELL and KING (1978) the distribution coefficient of acetic acid varies with the number of butoxy groups ($-OCH_2CH_2CH_2CH_3$; Tab. 9). The higher the acid concentration, the lower the distribution coefficient (Tab. 10). Also, the diluents have been shown to influence the distribution coefficients.

Considering the extraction equilibrium of weak carboxylic acids, H_nS, by a strong solvating extractant, TOPO, in the organic phase (index o)

$$H_nS + m\text{TOPO}_o \rightleftharpoons H_nS(\text{TOPO})_{m,o}$$

with the extraction equilibrium constant K

$$K = [H_nS(\text{TOPO})_m]_o / ([H_nS]\,[\text{TOPO}]_o^m)$$

where m is the solvation number of the acid, n is the number of carboxylic groups of the acid molecule,
the following relationship can be obtained in a non-polar diluent (hexane) (Tab. 11; HANO et al., 1990):

$$n = m \quad \text{and} \quad K = K_D \beta_n$$

Tab. 10. Influence of the Acetic Acid Concentration on the K_D Value with 22 wt% TOPO in Chevron 25 (RICKER et al., 1979)

Acetic Acid Concentration (wt%)	K_D
0.189	3.12
1.27	1.33
3.20	0.766
7.54	0.450

Tab. 9. Effect of Butoxy Groups on the Distribution of Acetic Acid from 0.5 wt% Water Solution (WARDELL and KING, 1978)

Extractant	Number of Butoxy Groups in Extractant	Diluent	K_D
Tributyl phosphate	3	—	2.3
Dibutyl phosphate	2	—	2.7
Butyl-dibutyl phosphinate	1	—	?
Tributyl phosphine oxide	0 37.3 wt%	Chevron 25	4.4

Tab. 11. Dissociation Constants of Different Carboxylic Acids and their Equilibrium and Distribution Constants with TOPO in Hexane (HANO et al., 1990)

Acid	Structure	pK_a	K	log K_D
Monocarboxylic Acids				
Acetic acid	CH_3COOH	4.56	5.32	−3.03
Glycolic acid	CH_2OH COOH	3.63	0.67	−4.02
Propionic acid	CH_3CH_2COOH	4.67	37.9	−2.50
Lactic acid	CH_3CHOH COOH	3.66	1.19	−3.51
Pyruvic acid	CH_3CO COOH	2.26	1.67	−3.09
Butyric acid	$CH_3(CH_2)_2COOH$	4.63	51.9	−1.98
Dicarboxylic Acids				
Succinic acid	HOOC CH_2CH_2COOH	4.00; 5.24	21.2	−3.47
Fumaric acid	HOOC CH:CH COOH	2.85; 4.10	174.5	−2.58
Maleic acid	HOOC CH:CH COOH	1.75; 5.83	33.5	−3.41
L-Maleic acid	HOOC CH_2CHOH COOH	3.24; 4.71	3.52	−4.18
Itaconic acid	HOOC $(CH_2(C{=}CH_2)COOH$	3.65; 5.13	80.8	−3.12
Tartaric acid	HOOC CHOH CHOH COOH	2.99; 4.44	2.19	−4.60
Tricarboxylic Acids				
Citric acid	CH_2COOH – C(OH)COOH – CH_2COOH	2.87; 4.35; 5.69	19.61	−4.67
Isocitric acid	CH_2COOH – CHCOOH – CH(OH)COOH	3.29; 4.71; 6.4	32.7	n.d.

where β_n is the association constant of acid with TOPO in the organic phase:

$$\beta_n = [H_nS \cdot TOPO]_o / ([H_nS]_o [TOPO]_o^n)$$

The association constant of each individual acid with TOPO in the organic phase is the same for acids having the same number of carboxyl groups. The extraction equilibrium constant is controlled by the hydrophobicity of the acids, not by the pK_a value.

SMITH and PAGE (1948) reported on the acid binding properties of long chain aliphatic amines (A) which are based on their insolubility in aqueous solutions and solubility in organic solvents. When using these high-molecular weight aliphatic amines, the extraction occurs by proton transfer or by ion pair formation often according to a well defined stoichiometric relation:

$$HS_a + mA_o \rightleftharpoons [HS(A)_m]_o$$

where index a means aqueous and index o organic phase, with

$$K = [HS(A)_m]_o / ([HS]_a [A]_o^m)$$

The acid extracted into the amine containing organic phase is regarded as an ammonium salt. At higher amine concentrations in inert diluents, molecular association of the salt AHS can be formed in the organic phase:

$$q\,AHS_o \rightleftharpoons [AHS]_{q,o}$$

with

$$K_q = [(AHS)_q] / [AHS]^q$$

q depends on the chemical nature of the salt, its concentration, the nature of the diluent, and on temperature (KERTES and KING, 1986).

Primary amines have a high mutual solubility with water. Secondary amines yield high values of K_D for acetic acid (Tab. 12), however, they are subject to amide formation during regeneration by distillation. Also, tertiary amines have high K_D values for acetic acid (Tab. 13). Long-chain tertiary amines are practically insoluble in the aqueous phase at acid concentrations below 10 wt%.

Diluents are used to reduce the viscosity of the organic phase and change the density difference between the phases. Alcohol diluents

Tab. 12. Extraction of Acetic Acid by Several Secondary Amines (RICKER et al., 1979)

Amine	vol% of Amine	Diluent	Equil. Acid Conc. (wt%)	K_D
Amberlite LA-1	50	Chevron 25	1.16	1.27
			3.78	2.26
	30	chloroform	0.047	4.48
	100	none	2.52	4.22
Amberlite LA-2	50	Chevron 25	0.53	3.82
			2.19	4.48
	30	chloroform	0.0218	9.86
	100	none	1.83	6.49
Adogen 283-D	50	Chevron 25	0.983	9.65
			2.63	4.55
			3.85	3.46
	30	chloroform	0.0133	32.11
	100	none	3.83	33.4

Tab. 13. Extraction of Acetic Acid by Several Tertiary Amines (RICKER et al., 1979)

Amine	vol% of Amine	Diluent	Equil. Acid Conc. (wt%)	K_D
Adogen 381	20	2-heptanone	3.23	1.72
	30	chloroform	0.0448	7.79
Adogen 364	20	2-heptanone	2.90	1.98
	30	chloroform	0.038	9.69
Adogen 368	20	2-heptanone	3.11	1.83
	30	chloroform	0.0447	8.23
Adogen 363	20	2-heptanone	3.44	1.58
	30	chloroform	0.0623	7.82
Alamine 366	20	2-heptanone	2.99	2.24
	30	chloroform	0.0465	9.68
Methyl ditridecyl	50	methyl-isoamyl ketone	2.91	3.54

yield the highest K_D values but are subject to esterification with the acid. In general, the K_D values diminish in the following succession: alcohols, ketones, esters, hydrocarbons.

In non-polar diluents, the solubility of the amine–acid complexes is low. At high acid concentrations, a third phase with high concentrations of the complex can form. To increase the solubility of the complex and eliminate the third phase, so-called *modifiers* are used. Usually high-molecular weight alcohols (e.g., isodecanol) are employed as modifiers. They prevent the formation of a third phase and improve the distribution coefficient. For instance, the addition of isodecanol to amine extractants, tri-*n*-decyl/*n*-octyl amine (Hostarex A 327, Hoechst) and tri-*n*-isooctyl amine (Hostarex A 321, Hoechst), in a hydrocarbon diluent (Shellsol T) improved the K_D value at low acetic acid concentrations considerably (SIEBENHOFER and MARR, 1983).

However, if polar diluents are used, no differentiation between diluent and modifier is possible. Only in the case of non-polar diluents, diluent and modifier roles can be distinguished.

There is an optimal molecular weight of extractants (340–400) which is a compromise between a high K_D value (low molecular weight) and low solubility in the aqueous phase (high molecular weight). For the diluent, low solubility in the aqueous phase is important. The modifier should increase the solubility of the acid–base complex but only slightly influence the solubility of the extractant and the diluent in the aqueous phase.

A popular extractant is Alamin 336 which is sometimes diluted with diisobutyl ketone (DIBK), as the relative volatility of acetic acid to DIBK is about 3.3 for an organic phase consisting of 40 vol% Alamine 336 in DIBK. With this system, $K_D = 2.5$ was attained with 1 wt% acetic acid in the aqueous phase and 50 vol% Alamine 336 in DIBK as the organic phase (RICKER et al., 1980). On the basis of costs in 1978, RICKER et al. (1980) estimated total operating costs of US \$ 1.9/m^3 water for extraction of acetic acid from 5 wt% aqueous solution into ethyl acetate, a solvent mixture of 50% Alamine 336 in DIBK and 40 wt% TOPO in 2-heptanone, respectively (Tab. 14). According to this calculation, the Alamine 336/DIBK extraction would lower the economically recoverable feed concentration by a factor of 2 in comparison with the use of ethyl acetate as a solvent.

Tab. 14. Estimated Costs (1978 Basis) for Extraction of Acetic Acid from 5 wt% Solution (22700 kg h^{-1}) Using Various Solvents (RICKER et al., 1980)

Make-up Solvent	Ethyl Acetate	40% TOPO in 2-Heptanone	50% Alamine 336 in DIBK
Fixed capital costs (\$)	1732	1342	1215
Operating costs [k \$/a (\$/m^3 water)]			
Steam	300 (1.65)	32 (0.18)	42 (0.26)
Cooling water	90 (0.50)	9 (0.05)	13 (0.08)
Make-up solvent	10 (0.05)	221 (1.22)	34 (0.21)
depreciation, Maintenance, etc.	311 (1.71)	242 (1.33)	219 (1.35)
	711 (3.91)	504 (2.78)	308 (1.90)
Value of acetic acid recovered (k \$/a)	2846	2846	2846
Return on investment before taxes	123%	175%	220%

k \$/a, 1000 US dollars per year

By combination of different extractants and polar diluents the acetic acid extraction performance can be increased considerably.

WOJTECH and MAYER (1985) combined several alcohols and phenols as polar diluents for a 50:50 mixture of tri-*n*-octyl/tri-*n*-decyl amine (Hostarex A 327, Hoechst). They found the highest distribution coefficient with 90% *p*- and 10% *o*-nonyl phenol and in the following succession: *n*-butanol > *n*-pentanol > 3-methyl butanol (isoamyl alcohol) > *n*-octanol > 2-ethyl hexanol > isodecanol, with decreasing effect. The advantage of the acidic nonyl phenol over alcohols is that it does not form an ester with acetic acid during distillation. A sharp maximum in the distribution coefficient ($K_D = 300$) appeared as a function of the extractant content at 30 wt% Hostarex A 327 and 70 wt% nonyl phenol (mole ratio 4:1) with 1.4 wt% acetic acid and a phase ratio V_a/V_o of 2:1 at 25 °C.

An alternative system of 60 wt% *n*-pentanol and 3-methyl butanol, respectively, with 40 wt% Hostarex A 327 (mole ratio of trioctyl/tridecyl amine of 7:1) at 1.4 wt% acid gave a lower distribution coefficient ($K_D = 30$) but a higher loading capacity. This system was recommended for removal of acetic acid from industrial effluents.

With the same extractant/diluent combination, the K_D values of other acids (butyric, malonic, lactic, and glycolic acid) were also improved by a factor of 10–100 (WOJTECH and MAYER, 1985).

Also PBO-extractants and long-chain amines were combined by SIEBENHOFER and MARR (1985) employing dioctyl phosphonate and trioctyl amine in a hydrocarbon diluent (Shellsol T) for the extraction of acetic acid. By changing the composition from 20 vol% triisooctyl amine and 80 vol% Shellsol to 20 vol% triisooctyl amine, 40 vol% dioctyl phosphonate and 40 vol% Shellsol T the K_D value increased by a factor of 10.

Lactic acid recovery from the cultivation medium is performed presently by precipitation with $Ca(OH)_2$ or $CaCO_3$ and acidification by H_2SO_3. The disadvantage of this process is the by-product formation of $CaSO_4$ and its disposal. Lately, several investigations have been carried out for its extractive recovery by different carriers and diluents/modifiers. Two different diluents were used: one with a very low K_D value and low solubility in the aqueous phase (kerosene) and another with high a K_D value but rather high solubility in the aqueous phase (butyl acetate) (VON FRIELING, 1991).

In the latter (butyl acetate), the physical extraction of the undissociated acid and the carrier extraction of dissociated acid have to be taken into account. The lactic acid (HL) is distributed between the aqueous (a) and the organic (o) phase, dissociates to anion L^- and proton H_3O^+ and reacts with the carrier (CA):

(1) Physical extraction:

$$HL_{a,p} \underset{}{\overset{k_P}{\rightleftharpoons}} HL_{o,p}$$

$$k_P = [HL]_{o,p}\, V_o / [HL]_{a,p}\, V_a$$

(2) Dissociation:

$$HL_a + H_2O \overset{k_a}{\rightleftharpoons} H_3O + L_a^-$$

$$[HL]_a = \frac{(CA)_o (HL)}{1 + 10^{(pH - pK_s)}}$$

(3) Complex formation ($V_a : V_o = 1:1$):

$$HL_a + (CA)_o \overset{k_1}{\rightleftharpoons} (HL(CA))_o$$

$$k_1 = \frac{[(HL(CA))]_o}{[HL]_a\, [CA]_o}$$

$$HL_a + (HL(CA))_o \rightleftharpoons ((HL)_2(CA))_o$$

$$k_2 = \frac{[(HL)_2(CA)]_o}{[HL]_a^2\, [HL(CA)]_o}$$

where k_1 and k_2 are the complex formation constants with a single and two lactic acids (VON FRIELING, 1991). With increasing modifier concentration in presence of Hostarex A 327, k_2 increases. With Cyanex 923, no $(HL)_2$-complex is formed.

VON FRIELING (1991) developed a relationship for the calculation of k_1 and k_2 for these systems. There is a good agreement between the calculated and the measured complex formation constants.

In an extractant–butyl acetate diluent, the distribution coefficients are mainly controlled by the partition coefficient of lactic acid between the aqueous and the butyl acetate phase. The partition coefficient at 23 °C $k_p = 0.28$ differs from the coefficient at 20 °C $k_p = 0.11$ evaluated by HOLTEN (1971).

For the Hostarex A 327/butyl acetate system, the complex formation constants are: $k_1 = 8.82$ L mol^{-1} and $k_2 = 2.38$ L mol^{-1} (VON FRIELING, 1991).

WENNERSTEN (1983) investigated citric acid extraction with three tertiary amines: tribenzyl amine, tri-dodecyl amine and Alamine 223 (tri-*n*-octyl amine) in different diluents. In Tabs. 15 and 16, the effect of the diluents on the K_D value of citric acid with Alamine 336 and tri-dodecyl amine as extractants are shown at different temperatures and acid concentrations. At low acid concentrations, a strong diluent effect prevails. K_D is a function of the polarity of the diluent and its ability to form hydrogen bonds. At higher acid concentrations, however, this diluent effect decreases. With increasing temperature K_D drops considerably. This behavior can be exploited to re-extract the citric acid into the aqueous phase (BANIEL, 1982; BANIEL et al., 1981; ALTER and BLUMBERG, 1981).

Tab. 15. Effect of Diluents on the Distribution Coefficient of Citric Acid with 50 vol% Alamine 336 at Different Acid Concentrations and Temperatures (WENNERSTEN, 1983)

Diluent	Acid Conc. (mol L^{-1})	K_D	T (°C)
n-Hexane	14.3×10^{-3}	2.9	25
	0.382	1.9	25
	27.9×10^{-3}	0.3	60
Toluene	3.37×10^{-3}	10.8	25
	0.258	2.1	25
	22.1×10^{-3}	1.4	60
Ethyl acetate	2.45×10^{-3}	23.0	25
	0.275	3.1	25
	15.7×10^{-3}	2.5	60
MIBK	2.07×10^{-3}	27.5	25
	0.276	3.1	25
	5.17×10^{-3}	11.5	60
2-Ethyl hexanol	0.83×10^{-3}	70.0	25
	0.351	2.1	25
	1.97×10^{-3}	28.9	60
Isoamyl alcohol	0.67×10^{-3}	87.8	25
	0.305	2.7	25
	1.37×10^{-3}	42.2	60
n-Butanol	0.35×10^{-3}	168.3	25
	0.274	3.1	25
	0.8×10^{-3}	72.9	60

Tab. 16. Effect of the Diluent on the Distribution Coefficient of Citric Acid with Tri-Dodecyl Amine at Different Acid Concentrations and Temperatures (WENNERSTEN, 1983)

Diluent	Acid Conc. (mol L^{-1} 10^3)	K_D	T (°C)
MIBK	1.7	34.0	25
	9.87	4.8	60
2-Ethyl hexanol	0.33	177	25
	1.8	31.7	60
Toluene	8.20	6.0	25
	329	0.55	60
Chloroform	0.42	136	25
	3.13	17.7	60

In Tabs. 17 and 18 the distribution coefficients are shown for ISOPAR H (paraffine/kerosene mixture) and methyl-isobutyl ketone (MIBK) as diluents and Alamine 336 as extractant at different acid concentrations. With increasing acid concentration K_D decreases. This holds true at 25 °C as well as at 60 °C. According to WENNERSTEN (1983) the optimum citric acid-to-Alamine 336 mole ratio is one mole amine to one mole carboxylic group. At higher acid concentrations a third phase is formed.

RÜCKL et al. (1986) found an optimum with a blend of 30% Hostarex A 324, 30% isodecanol and 40% alkanes.

All of these investigations dealt with pure acid solutions. However, cultivation media consist of several components which can influence the distribution coefficient of the solute. Little data on the extraction of carboxylic acids from their mixture have been published.

In lactic acid production with several Lactobacilli the supplementation of the medium with acetic and citric acid increases productivity (SIEBOLD, 1991). However, lactic acid must then be separated from the other carboxylic acids.

In Fig. 5, the degree of coextractions of acetic, lactic, and citric acid with 30 wt% Host-

Tab. 17. Effect of Diluent and Acid Concentration on the Distribution Coefficient of Citric Acid with Alamine 336 at 25 °C (WENNERSTEN, 1983)

Extractant/ Diluent (vol/vol)	Equil. Acid Conc. (mmol L^{-1}) in Aqueous Phase	K_D
Alamine 336/ Isopar H (1:1)	28	9.6
	67	8.0
	128	4.75
	193	3.48
	295	2.34
	390	1.85
	621	1.19
	1000	0.76
Alamine 336/ Isopar H/ MIBK (2:1:1)	7.7	38.4
	22.7	26.3
	62	11.1
	128	5.9
	230	3.36
	323	2.5
	565	1.44
	954	0.86

Tab. 18. Effect of Diluent and Acid Concentration on the Distribution Coefficient of Citric Acid with Alamine 336 at 60 °C (WENNERSTEN, 1983)

Extractant/ Diluent (vol/vol)	Equil. Acid Conc. (mmol L^{-1}) in Aqueous Phase	K_D
Alamine 336/ Isopar H (1:1)	201	1.86
	272	1.57
	331	1.51
	415	1.30
	497	1.18
	700	0.92
	1067	0.63
Alamine 336/ Isopar H/ MIBK (2:1:1)	28	5.22
	79.3	6.63
	186	3.65
	369	2.03
	600	1.28
	982	0.8

arex A 327 and 10 wt% isodecanol at different temperatures is shown. The degree of coextraction of the same acids with 30 wt% Cyanex 923 at different temperatures is shown in Fig. 6. It is not possible to separate lactic acid from acetic and citric acid by extraction with Hostarex A 327 and isodecanol. The degree of reextraction of lactate with 0.1 N NaOH is high (90%), and amounts only to 55% with water.

With Cyanex 923, acetic acid is coextracted with a high degree of extraction, lactic acid only with a moderate degree, and citric acid is not extracted at all. Therefore, their separation is possible. Re-extraction of lactate with NaOH is not possible because a stable emulsion is formed.

Re-extraction of the free acids from the Hostarex A 327 and isodecanol containing organic phase by 2 M HCl yields the highest degree of re-extraction of citric acid, a high degree of re-extraction of lactic acid and a low degree of re-extraction of acetic acid (Fig. 7).

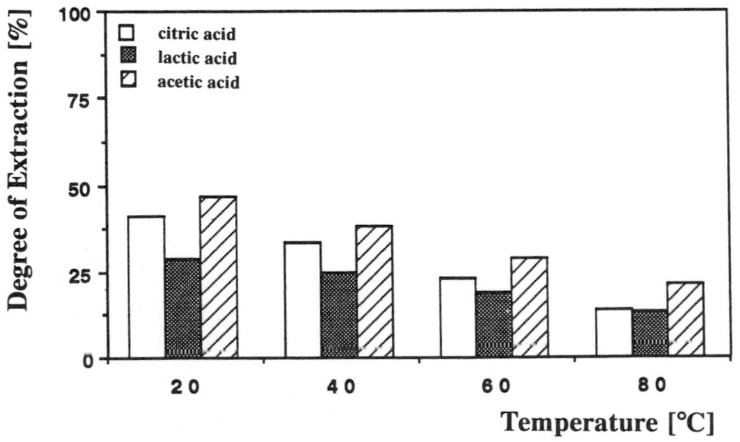

Fig. 5. Temperature dependence of extraction degrees of citric, lactic, and acetic acid with the Hostarex A 327-decanol system (VON FRIELING, 1991).

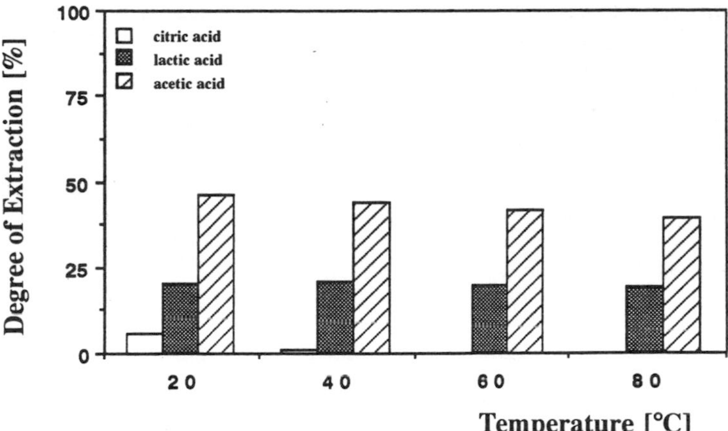

Fig. 6. Temperature dependence of extraction degrees of citric, lactic, and acetic acid with the Cyanex 923 system (VON FRIELING, 1991).

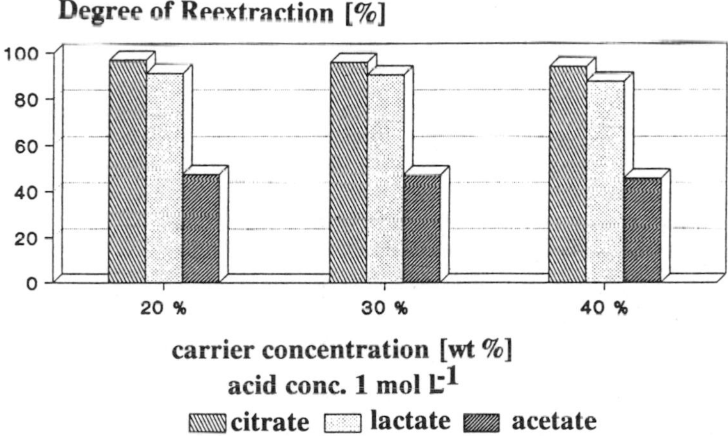

Fig. 7. Selectivities of the re-extraction of citric, lactic, and acetic acid from the Hostarex A 327 system (VON FRIELING, 1991).

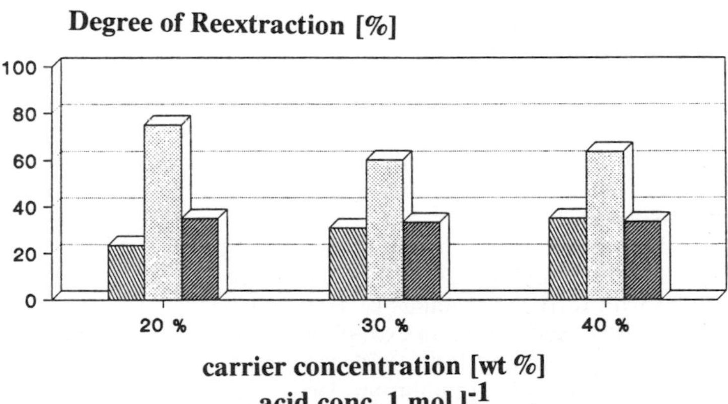

Fig. 8. Selectivities of the re-extraction of citric, lactic and acetic acid from the Cyanex 923 system (VON FRIELING, 1991).

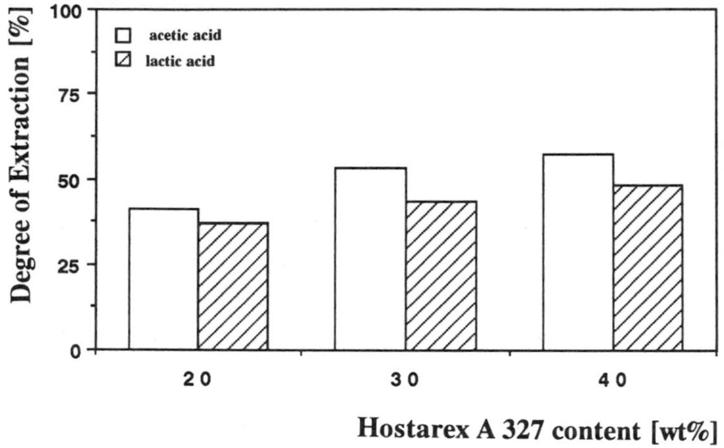

Fig. 9. Extraction degrees of acetic and lactic acid with different concentrations of Hostarex A 327 (VON FRIELING, 1991).

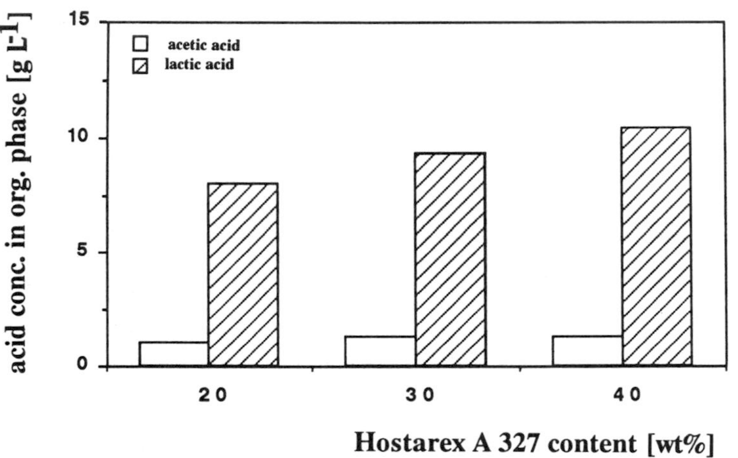

Fig. 10. Concentrations of acetic and lactic acid in the organic phase with different concentrations of Hostarex A 327 (VON FRIELING, 1991).

However, the total yield of lactic acid is rather low (0.45). The total yield with 1 M H_2SO_4 is somewhat lower. The re-extraction of the free acids from the Cyanex 923 containing organic phase with 2 M HCl yields the highest degree of re-extraction of lactic acid, and low degrees of re-extraction of acetic and citric acid (Fig. 8). However, the total yield of the lactic acid is still low (max. 0.20).

If the ultrafiltered cultivation medium of *Lactobacillus salivarius* ssp. *salivarius* is extracted with different amounts of Hostarex A 327 and 10% isodecanol, an extraction degree of 50% for acetic and lactic acid can be obtained (Fig. 9). The equilibrium lactic acid concentration in the organic phase amounts to about 10%, and because of the low acetate content of the cultivation medium, the equilibrium acetic acid concentration in the organic phase is low (2.5%; Fig. 10).

With Cyanex 923 a lower extraction degree of acetate and lactate is obtained from the ultrafiltered cultivation medium of *Lactobacillus salivarius* ssp. *salivarius* (Fig. 11). The equilibrium lactic acid concentration in the organic phase amounts to about 8% and the equilibrium acetic acid concentration to about 2.5–3.0% (Fig. 12). The re-extraction of lactic acid with 2 M HCl from the carrier-containing organic phase yields about a 75% degree of extraction with Hostarex A 327, and 55% with Cyanex 923 (Fig. 13; VON FRIELING, 1991). Again, the total yield coefficients are low (0.37 for Hostarex and 0.22 for Cyanex).

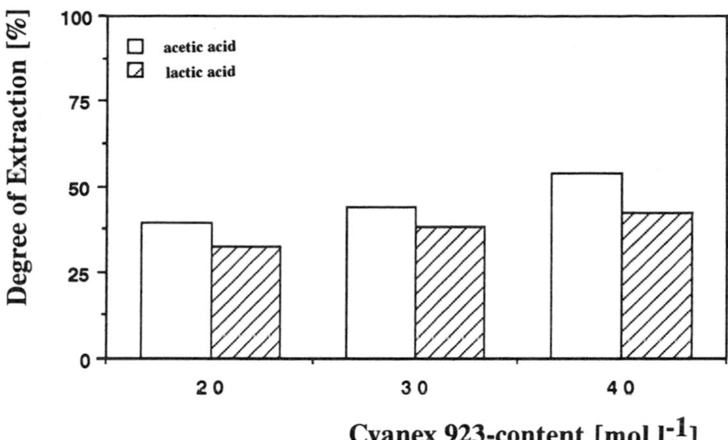

Fig. 11. Extraction degrees of acetic and lactic acid with different concentrations of Cyanex 923 (VON FRIELING, 1991).

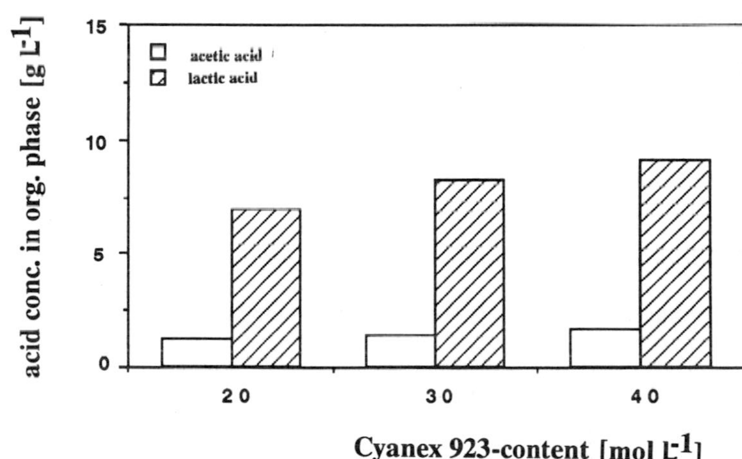

Fig. 12. Concentrations of acetic and lactic acid in the organic phase with different concentrations of Cyanex 923 (VON FRIELING, 1991).

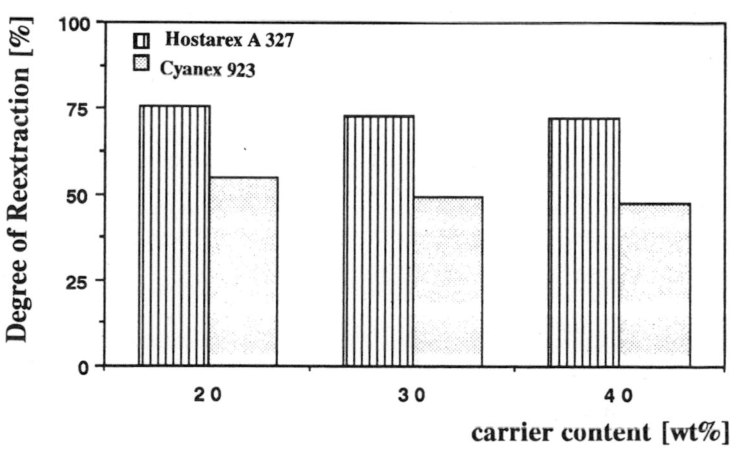

Fig. 13. Re-extraction degrees of lactic acid from the organic phase with different concentrations of Hostarex A 327 and Cyanex 923 (VON FRIELING, 1991).

3.2 Amino Acids

Amino acids have two or three charged groups, form zwitterions at neutral pH and, except for threonine, all of the bifunctional acids have pK_1 (COO^-) values between 2 and 3, and pK_2(NH_4^+) values between 9 and 10.

This holds also true for trifunctional amino acids, except cysteine with $pK_1 = 1.71$ and $pK_2 = 10.78$, and histidine with $pK_1 = 1.82$.

The trifunctional amino acids differ only in their pK_3 (side group) values: 3.36 (aspartic acid), 4.25 (glutamic acid), 6.0 (histidine), 8.33 (cysteine), 10.07 (tyrosine), 10.53 (lysine), and 12.40 (arginine).

For the extraction of these compounds, quaternary ammonium ions, quaternary phosphonium ions, as well as dialkyl or trialkyl phosphoric acids can be used. The extraction is performed at very low or very high pH values in order to eliminate the zwitterionic form.

The separation of amino acids of basic character (lysine, arginine, histidine) from those of acidic character (aspartic acid, glutamic acid) by extraction is fairly easy. Likewise, the separation of amino acids with polar side-groups (glycine, serine, threonine, cysteine, tyrosine, glutamine) from those with a non-polar side group (alanine, valine, leucine, isoleucine, proline, phenylalanine, tryptophan, methionine) causes no difficulties.

The following equilibrium reactions of the amino acid (HS) system exist in the alkaline (0.01 N NaOH) aqueous phase (a) and of the quaternary ammonium chloride (Q^+Cl^-) in the organic phase (o):

(1) Dissociation

$$HS_a + H_2O \rightleftharpoons H_3O + S_a^-$$

(2) Reaction of the amino acid with the extractant

$$S_a^- + (Q^+Cl^-)_o \overset{K}{\rightleftharpoons} (Q^+S^-)_o + Cl_a^-$$

$$K = [Q^+S^-]_o\,[Cl^-]_a/([S_a^-]_a\,[Q^+Cl^-]_o)$$

(3) Coextraction of HO^- ions with the carrier

$$OH_a^- + Q^+Cl_o^- \rightleftharpoons Q^+OH_o^- + Cl_a^-$$

$$K_{OH} = [Q^+OH^-]_o\,[Cl^-]_a/([OH^-]_a\,[Q^+Cl^-]_o)$$

(4) Exchange of coextracted OH^- anions with amino acid anions

$$S_a^- + Q^+OH_o^- \rightleftharpoons Q^+S_o^- + OH_a^-$$

In order to furnish enough amino acid anions for reaction (2), the pH value of the aqueous phase must be at least two units above the pK value of the N-terminal proton (tryptophan: pK (NH_2) = 9.39). In the case of tyrosine, it is necessary to consider the formation of dianions (S_a^{2-}) by the proteolysis of a phenolic (ph) proton (tyrosine: pK (ph-OH) = 10.07) at pH values higher than 12.

In Tab. 19, the extraction equilibrium constants K in the presence of tri-*n*-octyl-methylammonium chloride (TOMAC) in the high pH range and the isoelectric points of various amino acids are compiled.

Except for tyrosine, TOMAC reacted with the anion of amino acids and OH^- according to the stoichiometric ratio determined by ionic valency. Tyrosine dissociates into divalent anions in the pH range higher than 13. Therefore, the equilibrium constant is given by

$$K = [Q_2{+}S^-]_o\,[Cl^-]_a^2/[S^{2-}]_a\,[Q^+Cl^-]_o$$

Tab. 19. Extraction Equilibrium Constants K and Isoelectric Points pI of Various Amino Acids (HANO et al., 1991)

Amino Acid	pI	K
Glycine (Gly)	6.0	0.036
Alanine (Ala)	6.0	0.038
Valine (Val)	6.0	0.089
Leucine (Leu)	6.0	0.29
Isoleucine (Ile)	6.0	0.24
Methionine (Met)	5.7	0.21
Phenylalanine (Phe)	5.5	0.97
Tryptophan (Trp)	5.9	0.89
Tyrosine (Tyr)	5.7	0.40
Histidine (His)	7.6	0.083
Arginine (Arg)	10.8	0.062
Serine (Ser)	5.7	0.049
Threonine (Thr)	6.2	0.071

K increases as the number of carbon atoms increases. The highest K value obtained (for tryptophan) is 260 times higher than the lowest value (for glycine). K values of aromatic amino acids are rather high compared to those of aliphatic amino acids. There is a clear relationship between the K value and the hydrophobicity scale of the amino acids, which is defined as the free energy change with each amino acid transfer from water to the ethanol phase (NOZAKI and TANFORD, 1971). With increasing hydrophobicity, the K value increases (HANO et al., 1991).

During the extraction of amino acids with TOMAC at high pH values, all types of inorganic anions (SO_4^{2-}, CO_3^{2-}, PO_4^{3-}) are coextracted. Therefore, it is recommended that these be removed from the cultivation medium before extraction. However, in absence of these anions, OH^- is still coextracted.

3.3 Antibiotics

Since the carrier extraction of penicillin G has been investigated in detail, this antibiotic will be considered here.

Penicillin G is a weak acid ($pK_a = 2.75$). Only the free undissociated acid can be extracted by CBO-extractants. Therefore, extraction is performed at pH 2.0–2.5 where the free acid is unstable. In spite of the low extraction temperature, considerable losses occur during extraction.

By extracting penicillin in the neutral pH range, the losses could be reduced. A patent by BEECHAM GROUP LTD. (1975) has pointed out that the solubility of penicillin can be increased by using a phase transfer catalyst, a quaternary ammonium salt (Q^+X^-), to increase the solubility of the penicillin anion (P^-) in apolar extractants by forming a salt (Q^+P^-). This technique was investigated later by LIKIDIS (1986), LIKIDIS et al. (1989) and HARRIS et al. (1990).

Better results were obtained with secondary and tertiary amines (A) (RESCHKE and SCHÜGERL, 1984a, b, c, 1985, 1986; MÜLLER et al., 1987; LIKIDIS, 1986; LIKIDIS and SCHÜGERL, 1987a, b; LIKIDIS et al., 1989). They react with the penicillin acid anion (P^-) and the proton (H^+) according to the following relationship:

$$A_o + P_a^- + H_a^+ \overset{K}{\rightleftharpoons} AHP_o$$

with the equilibrium constant K:

$$K = [AHP]/([A][P^-][H^+])$$

which depends on the pH value:

$$K = P_c[1 + 10^{pH - pK_a}]^{-1}$$

where

$$P_c = [HP]_o/[HP]_a$$

is the partition coefficient of the free acid on a molar basis.

The equilibrium degree of extraction E is defined as the fraction of the extracted solute with regard to its overall concentration

$$E = \frac{[HP]_o}{[HP_o + HP_a + P_a^-]} = \frac{100}{1 + (1 + 10^{pH - pK_a})/P_c}$$

Fig. 14 shows the measured and the calculated equilibrium degrees of extraction of penicillin as a function of the pH value with different concentrations of Amberlite LA-2.

The reaction of the acid with quaternary ammonium chloride (Q^+Cl^-) corresponds to an ion exchange:

$$(Q^+Cl^-)_o + P_a^- \rightleftharpoons (Q^+P^-)_o + Cl_a^-$$

with the equilibrium constant:

$$K = [Q^+P^-]\,[Cl^-]/([Q^+Cl^-]\,[P^-])$$

In the neutral range, this reaction is independent of the pH value. In the acidic range, the dissociation of penicillin must be taken into account. This pH independence has disadvantages, since at high equilibrium constants the re-extraction is incomplete. Therefore, the sodium salt of p-toluene sulfonic acid was used as a counter ion, and it was regenerated by NaOH or chloride ions (LIKIDIS, 1986; LIKI-

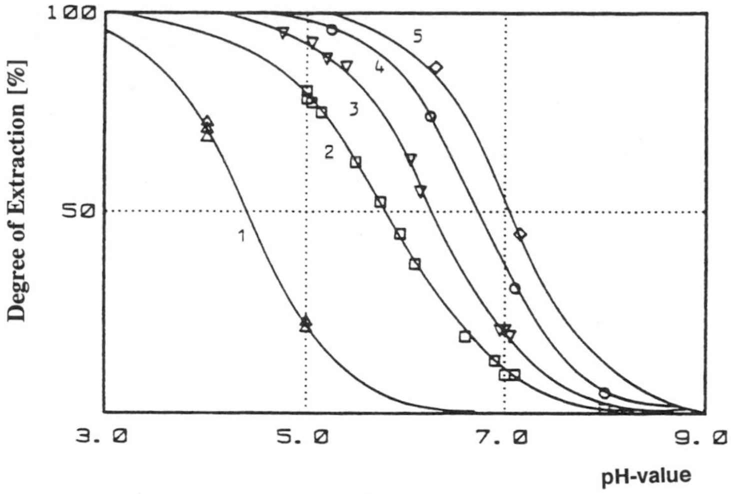

Fig. 14. Degree of extraction as a function of the medium pH value for the extraction of penicillin G with Amberlite LA-2 and *n*-butyl acetate. Initial penicillin concentration $C_P(0) = 10$ mM, at different LA-2 concentrations C_A (RESCHKE and SCHÜGERL, 1984b).
Curve 1, $C_A = 0$;
curve 2, $C_A = 10$ mM;
curve 3, $C_A = 20$ mM;
curve 4, $C_A = 50$ mM;
curve 5, $C_A = 100$ mM.
Symbols are measured values, — calculated.

DIS and SCHÜGERL, 1987a). Nitrate ions were also employed for re-extraction of penicillin (HARRIS et al., 1990).

With the secondary amine Amberlite LA-2, several investigations were carried out in a stirred cell (RESCHKE and SCHÜGERL, 1984c), in a laboratory extraction column (RESCHKE and SCHÜGERL, 1985; LIKIDIS and SCHÜGERL, 1988a), in a pilot plant column (MÜLLER et al., 1987), in a laboratory centrifugal extractor (LIKIDIS and SCHÜGERL, 1987b), in a pilot plant centrifugal extractor (LIKIDIS and SCHÜGERL, 1987b) and in a pilot plant countercurrent decanter (LIKIDIS et al., 1989). The extraction process in the columns was mathematically modeled and simulated on a computer (RESCHKE and SCHÜGERL, 1986; LIKIDIS and SCHÜGERL, 1988; MÜLLER et al., 1988). The agreement between the calculated and measured data in a stirred tank (Fig. 15), a laboratory (Fig. 16) and a pilot plant (Fig. 17) extraction column is satisfactory.

The precursors for penicillin G (phenyl acetic acid) and penicillin V (phenoxy acetic acid) can also be recovered with a high degree of extraction (RESCHKE and SCHÜGERL, 1984b).

One can conclude that the extraction of penicillin G and penicillin V, respectively, by

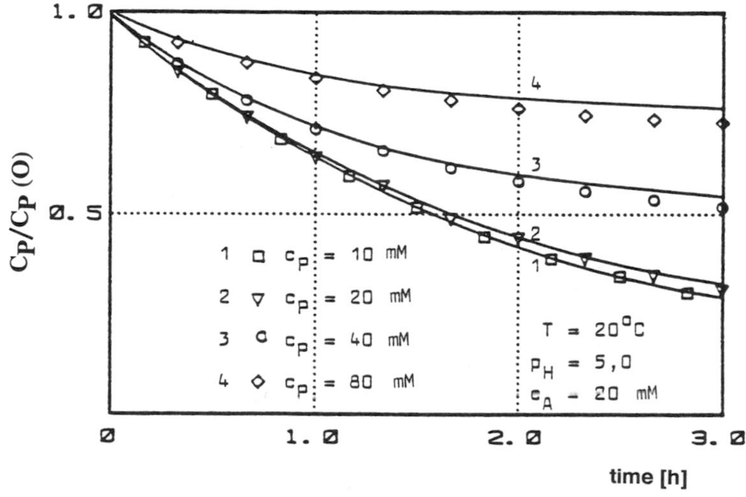

Fig. 15. Dimensionless penicillin concentration in the aqueous phase as a function of the extraction time at pH 5 for various penicillin concentrations showing the extraction kinetics of penicillin G in an Amberlite LA-2-*n*-butyl acetate system ($T = 20$ °C, $C_A(0) = 20$ mM) at different penicillin concentrations C_P (RESCHKE and SCHÜGERL, 1984c).
Symbols are measured values, — calculated.

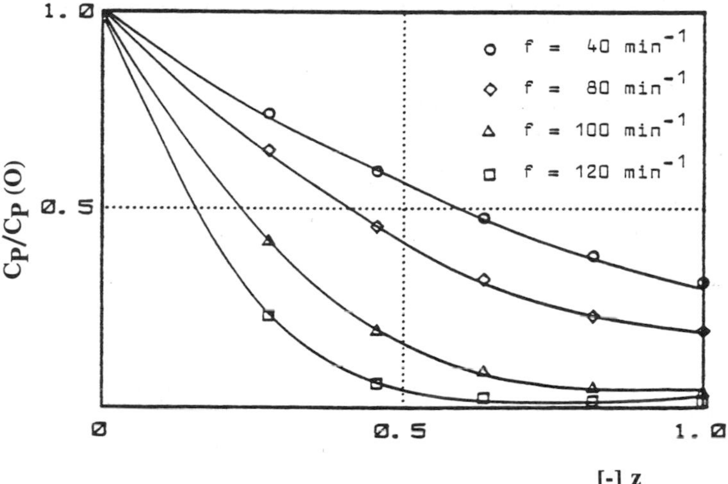

Fig. 16. Longitudinal profiles of penicillin concentration in the aqueous phase with regard to the feed concentration, $C_P/C_P(0)$ for various stroke frequencies f (RESCHKE and SCHÜGERL, 1985). Symbols are measured values, — calculated.

the secondary amine Amberlite LA-2 and the tertiary amine DITA (diiso-tridecyl amine) in butyl acetate at pH 5 is amenable (LIKIDIS and SCHÜGERL, 1987a). Its re-extraction at pH 7–8 with buffer solutions is also possible. It yields high-purity penicillin G and penicillin V, respectively, with high degrees of extraction in common extraction columns at low-phase separation times and at low losses of the carrier. These extractant–diluents are also suitable for the penicillin recovery in centrifugal extractors (LIKIDIS and SCHÜGERL, 1987b) as well as in a countercurrent decanter (LIKIDIS et al., 1989).

4 Supercritical Fluid Extraction

Since the early 1980s supercritical fluid (SCF) extraction has been investigated as an alternative to liquid–liquid extraction or distillation. The main attraction of SCF extraction is that the solvent properties are highly sensitive to changes in both pressure and temperature, allowing the tailoring of the solvent strength to a given application (Tab. 20). Other advantages of the SCF extraction include

- high diffusivity
- low surface tension
- easy recovery of the solute from the solution by evaporating the solvent
- low-temperature processing
- high capacity for solutes
- low reactivity and toxicity of the common solvents
- extraction may be coupled with further purification by fractionation with changing pressure and temperature (RANDOLPH, 1990).

However, SCF extraction has some serious disadvantages as well:

- Solubilities are much lower than those achievable in many liquid solvents also for compounds of low polarity.
- Esters, ethers and lactones can be extracted below 70–100 bar, benzene derivatives up to three —OH groups, or two —OH groups and one —COOH group, can still be extracted, but above 100 bar. Benzene derivatives with three —OH groups and one —COOH group, as well as sugars and amino acids, cannot be extracted below 500 bar.
- Capital costs are high for high-pressure equipment.
- Insufficient data exist on the physical properties of many biomolecules, mak-

ing prediction of phase behavior difficult.

- Selectivity can be achieved by the use of cosolvents, but this may obviate the advantage of minimal solvent residues in the final product.

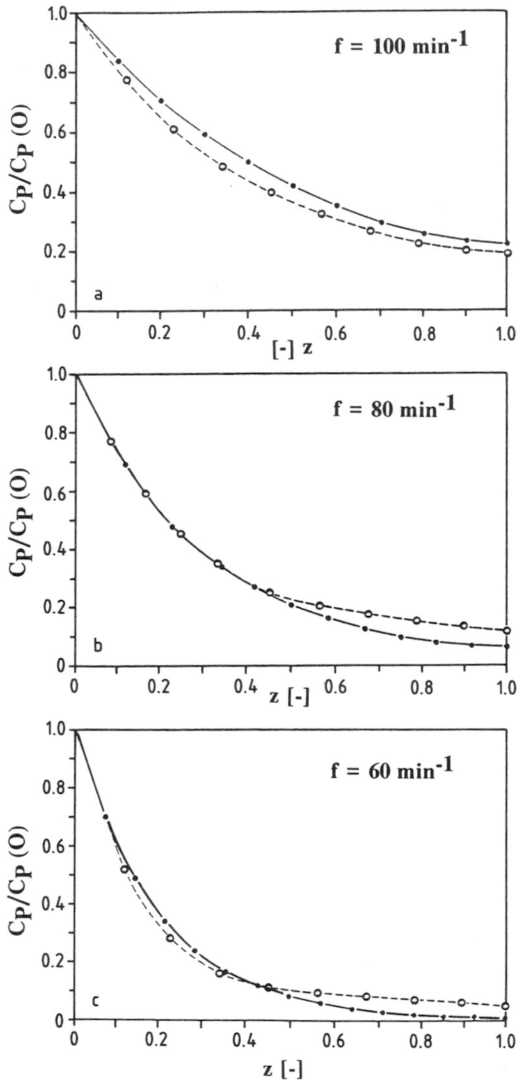

Fig. 17. Comparison of the measured and calculated longitudinal penicillin concentration profiles in the 8 m high pilot plant Karr-column at different stroke frequencies with model media (MÜLLER et al., 1988).
(a) $f=60\ min^{-1}$, (b) $f=80\ min^{-1}$, (c) $f=100\ min^{-1}$, --○-- measured, —●— calculated.

Tab. 20. Critical Properties of Fluids Used for SCF Extraction (RANDOLPH, 1990)

Compound	Critical Temperature (°C)	Critical Pressure (bar)	Critical Density ($g\ cm^{-3}$)
Butane	135.0	37.5	0.228
Carbon dioxide	31.3	72.9	0.443
Ethane	32.2	48.1	0.203
Ethylene	9.2	49.7	0.218
Dinitrogen oxide	36.5	71.7	0.45
Pentane	196.6	37.5	0.232
Propane	96.6	41.9	0.217
Water	374.2	217.6	0.322

In industry, SCF extraction is used in the food industry as well as in the mineral oil industry (STAHL and SCHILZ, 1976; KING et al., 1987).

For the extraction of ethanol from aqueous solutions supercritical carbon dioxide, ethane, and propene were used (BRUNNER, 1987).

The solubility of ethanol in CO_2 and ethane increases with an increasing ethanol concentration in the aqueous phase. At a 10% ethanol concentration in the aqueous phase, the solubility in CO_2 amounts to 1.5 wt%. The solubility of ethanol in ethane is about the same as that in CO_2. Therefore, a large amount of solvent is needed for ethanol recovery.

Ethanol was extracted by supercritical CO_2 (BRUNZENBERGER and MARR, 1987), propene (VICTOR, 1983), CO_2, and ethane, respectively (BRUNNER and KREIM, 1985). The recovery of ethanol from aqueous solutions with supercritical CO_2 extraction was performed in a 3.4 m high bubble column, 17 mm in diameter (BRUNNER and KREIM, 1985). After 60 min, the ethanol concentration decreased from 10 wt% to 2.3 wt% in the aqueous phase and increased to 0.8% in the CO_2 phase at 45 °C, $p=155$ bar, and a superficial solvent velocity of $0.12\ cm\ s^{-1}$.

In the countercurrent flow of the two phases in this column with a 13 wt% ethanol feed, the head product contained 51 wt% ethanol (in the CO_2 phase), and the bottom product 0.1 wt% ethanol (in the aqueous phase) at a superficial gas velocity of $7.2\ mm\ s^{-1}$ (BRUNNER and KREIM, 1985).

According to the estimation of these authors, the separation costs in a 50000 t/a 90% ethanol plant are of the same order of magnitude as the separation costs in a commercial 30000 t/a ethanol plant. No exact data, however, are available on the investment and operating costs of the SCF ethanol extraction plant.

5 Liquid Membrane Techniques

Liquid membrane (LM) systems introduced by NORMAN LI (1966) consist of three immiscible phases: usually two aqueous phases and one organic phase which separates the aqueous phases. Two types of liquid membrane systems are used: the emulsion LM (ELM) and the solid supported LM (SLM).

The ELM consists of small aqueous droplets (diameters in the μm range) within globules of the organic phase (diameters in the mm range), which are suspended in an external aqueous phase (Fig. 18). The internal aqueous phase is dispersed in the organic phase at a high stirrer speed (5000–10000 rpm) and the emulsion is dispersed in the external aqueous phase at a low stirrer speed (300–400 mm). The mass transfer is directed from the external aqueous phase through the organic membrane into the internal aqueous phase. If this mass transfer is selective, it may be used for separation of components in the outer aqueous feed phase.

The SLM consists of a membrane, the pores of which are filled with the organic phase. The two aqueous phases are separated by this membrane. The mass transfer is directed from one of the aqueous phases into the other. The

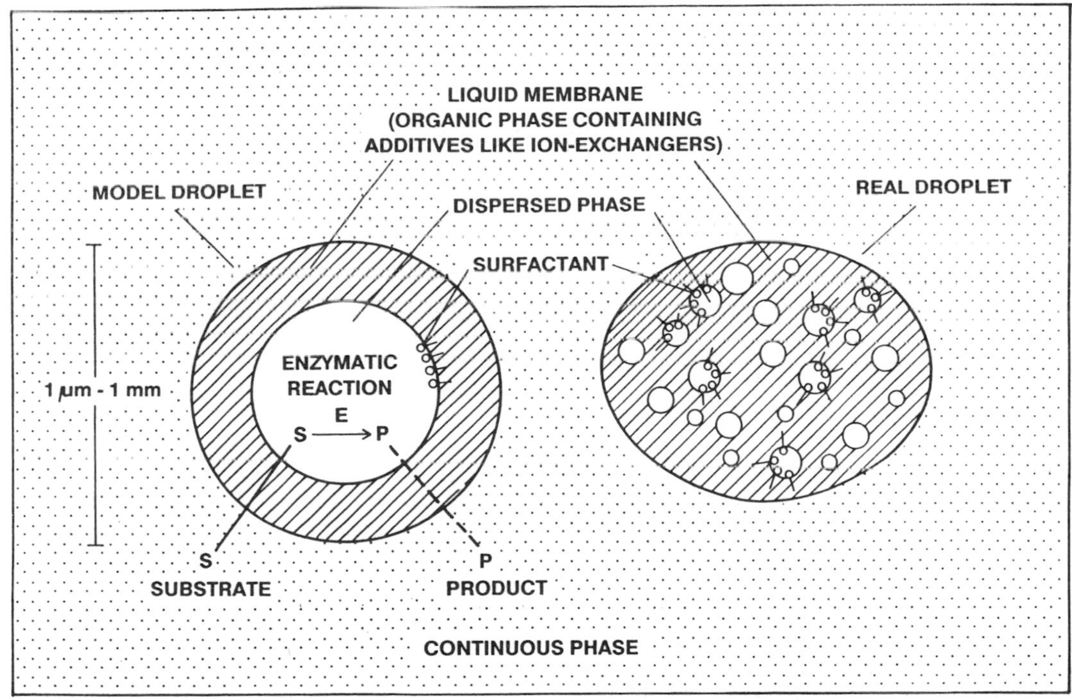

Fig. 18. Idealized model of emulsion liquid membrane (ELM) (left) and real droplet of ELM (right). In this specific case the inner aqueous phase contains an enzyme which converts the substrate S into the product P.

separation is based again on the selective transfer through the membrane.

The LM technique corresponds to an extraction and re-extraction process in a single step. Thus, the extraction is not limited by the distribution coefficient between the external aqueous phase and the organic membrane phase. The most important parameter is the permeation coefficient across the membrane, which is a product of the solubility and diffusivity of the solute in the organic phase. Also, the loading capacity of the LM is much higher than that of a common solvent phase. It is mainly applied in the low concentration range where a considerable enrichment of the solute in the internal aqueous phase is possible.

Extraction of **carboxylic acids** (formic, acetic, propionic, and *n*-butyric acid) from the aqueous phase were carried out at high pH values, e.g., the extraction of these acids by the SLM using TOPO as extractant and kerosene as diluent into alkaline solution by NUCHNOI et al. (1987), the batch extraction of citric and lactic acid (CHAUDHURI and PYLE, 1987), and of lactic acid (CHAUDHURI and PYLE, 1990) by the ELM with Alamine 336/Span 80 into the internal alkaline solution.

BEHR and LEHN (1973) were the first to report on the separation of **amino acids** that permeated through a liquid membrane with a highly lipophilic, positively charged carrier from an alkaline to an acidic solution, or through a liquid membrane with a highly lipophilic, negatively charged carrier from an acidic to an alkaline solution. According to that, amino acids are either extracted at low pH values with a cation exchanger, or at high pH values with an anion exchanger.

Most extractions were carried out at a high pH, e.g., amino acid extractions at pH 11 by Aliquat 336 as extractant, with decyl alcohol as modifier and Paranox 100 as surfactant in a kerosene diluent (THIEN et al., 1986, 1988), and the extraction of valine by SLM with Aliquat 336 and decyl alcohol (DEBLAY et al., 1990).

Extraction at a low pH was seldom used, e.g., for the extraction of lysine from an acidic solution (1 N HCl) by D2 EHPA/Span 80 into the alkaline solution (YAGODIN et al., 1986).

Enantiomeric separation of amino acids is based on the different transmembrane permeabilities of the free amino acid and its ester: SCHEPER et al. (1983, 1984) carried out the separation of D- and L-phenylalanine enantiomers by enzymatic conversion of their L-ester in a liquid membrane system.

Emulsion liquid membranes were also used for the production of L-amino acids from precursors by biotransformation, e.g., the production of L-leucine from α-keto-isocaproate and NH_3 by MAKRYALEAS et al. (1985) with enzymatic cofactor regeneration in a continuous stirred tank reactor.

Extraction of **antibiotics** (penicillin G) by the ELM was investigated by LIKIDIS (1983). Due to the low solubility of the penicillin G–Amberlite LA-2 complex in kerosene and its low permeation rate across the kerosene LM, the performance of an ELM extraction with kerosene as diluent was very low. On the other hand, the ELM with *n*-butyl acetate was unstable.

HANO et al. (1990) used dioctyl amine as extractant in an *n*-butyl acetate-kerosene mixture as diluent and ECA4360J (Exxon Chem. Co.) as surfactant. Penicillin G was extracted by this ELM from the aqueous external phase (at pH 6, controlled by citrated buffer) across the organic liquid membrane into the internal aqueous Na_2CO_3 solution. These investigations indicated that penicillin G can be extracted with the ELM rapidly from aqueous phases at pH 6.

In order to accelerate the extraction of penicillin G by ELM using Amberlite LA-2 as extractant, kerosene as diluent, and Span 80 as surfactant, the driving force across the membrane was increased by converting penicillin G enzymatically to 6-amino penicillanic acid (6-APA) in the internal aqueous phase (SCHEPER et al., 1989). This ELM was used for *in situ* extraction of penicillin G and V and the conversion to 6-APA and ampicillin. Therefore, it will be discussed in Sect. 7.

6 Reversed Micelles

Reversed micelles are formed by surfactants in organic solvents. The polar groups (heads) of the surfactant molecules are directed towards the interior of the spheroidal aggregate forming a polar core, and the aliphatic chains are directed towards the organic solvent. This is reverse to the situation in normal micelles in water (Fig. 19; LUISI and LAANE, 1986). They are mainly used for solubilization of enzymes in organic solvents (LUISI and LAANE, 1986; LESER et al., 1989).

The extraction of amino acids with reversed micelles was investigated by HATTON and coworkers (LEODIDIS and HATTON, 1990a, b, 1991; LEODIDIS et al., 1991). They used Aerosol-OT (bis(2-ethylhexyl)sodium sulfo-succinate; AOT) as surfactant in isooctane as well as dodecyl trimethylammonium chloride (DTAC) as surfactant in heptane/hexanol and contacted these organic phases with amino acids in a phosphate buffer solution at pH 6–6.5. These two phases were vigorously shaken and emulsified for 15 s in a vibrating shaker, and the phase separation was performed at 25 °C between 2 and 6 days.

Of the two phases formed, the lower (denser) phase was an electrolyte solution containing the buffer, a low amount of AOT and DTAC, and the residual amino acid. The upper phase was a reversed micellar solution. They evaluated the partition coefficient of the amino acids between the aqueous and the organic phase. They also considered the physicochemical basis of this phenomenon and found that the free energy of transfer of a large number of amino acids from water to the surfactant interfaces of AOT/isooctane and DTAC/heptane/hexanol W/O microemulsions correlate well with the existing hydrophobicity scales (LEODIDIS and HATTON, 1990b). It is possible to calculate the partition coefficient P_c^S of the amino acids between water and the surfactant interface of AOT and DTAC-reversed micelles by using simple phase-equilibrium experiments:

$$P_c^S = \frac{55.5\ V_{a,in}}{N_{S,tot}(1+R_C)}\left(\frac{c_{a,in}-c_{a,f}}{c_{a,f}}\right)$$

where $V_{a,in}$ is the initial volume of the aqueous phase before the contact,
$N_{s,tot}$ the total number of moles of the surfactant,
R_C the ratio of moles of alcohol (octanol, heptanol) to the moles of surfactant (AOT, DTAC) in the reversed micellar interface: 2.88, and
$c_{a,in}$ and $c_{a,f}$ are initial molar concentrations of the solute (amino acid) in the aqueous phase before the contact (in) and for the final aqueous phase (f).

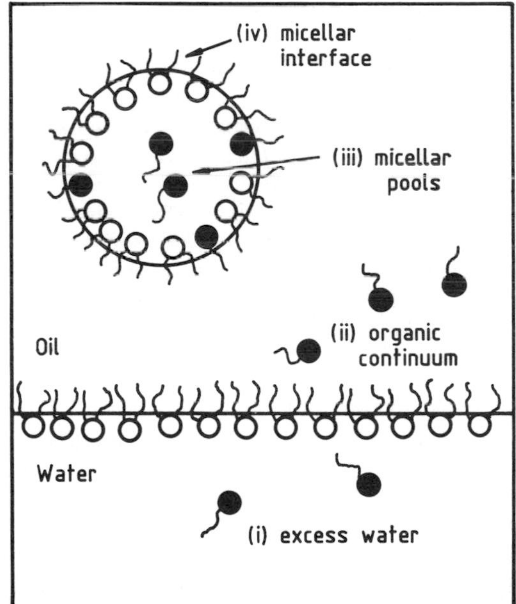

Fig. 19. Schematic picture of a reversed micelle with four solubilization environments in a two-phase system at equilibrium (LEODIDIS and HATTON, 1990a).

Tab. 21. Interfacial Partition Coefficients of Some Amino Acids (LEODIDIS and HATTON, 1990b)

Compound	P_c^S (Water → AOT)
Valine	4.6
Leucine	26.1
Isoleucine	17.0
Proline	2.2
Phenylglycine	16.8
Phenylalanine	89.3
Tyrosine	9.0
Tryptophan	283.1
Methionine	19.2

In Tab. 21, several partition coefficients of amino acids are compiled. In spite of the considerable amount of information available on reversed micelles, no practical application of this technique for amino acid separation is known.

7 *In situ* Extraction

Several biotechnological production processes are impaired by product inhibition. This holds especially true for batch processes in which the product concentration attains high values at the end of the batch process. Therefore, *in situ* product recovery from the cultivation medium during its formation can considerably increase the productivity of the processes.

In situ product recovery can be performed by adsorption (e.g., PITT et al., 1983), distillation (e.g., ROYCHOUDHURY et al., 1986), precipitation (e.g., PODOJIL et al., 1984), electrophoresis (e.g., LEE and HONG, 1988), pervaporation (e.g., PONS and TILLIER-DORION, 1987), gas stripping (e.g., GROOT et al., 1989), dialysis (e.g., LECHNER et al., 1988), reverse osmosis (e.g., CHOUDHURY et al., 1985), and also by extraction.

Most of the investigations concerning extraction deal with the *in situ* recovery of ethanol, acetone-butanol, carboxylic acids, and few of them with the separation of antibiotics from the cultivation medium.

The main problem of *in situ* extraction is the availability of suitable solvents. Therefore, systematic investigations were carried out with regard to reducing solvent toxicity, increasing the distribution coefficient of the solute between solvent and cultivation medium, reducing the emulsion stability of the solvent/cultivation medium system, and increasing the easy recovery of the solvent after the extraction.

For **ethanol** recovery, several solvents were tested regarding these criteria (KOLLERUP and DAUGULIS, 1985a, 1986). These investigations show that dodecyl alcohol, higher iso-alcohols, higher *n*-alcohols, tributyl phosphate, dibutylphthalate dodecane, fluorocarbons, etc. are preferred. For instance, dodecyl alcohol is fairly biocompatible, however, it tends to form stable emulsions with cultivation media, it has a relatively high solubility in water, and its melting point is 26 °C, which is too high for some cultivation processes. In addition, the ethanol distribution coefficient in the dodecanol/water system is only about 0.35 on the mass basis. An ethanol distribution coefficient of 0.4 in the water/solvent system could not yet be achieved. In addition, the distribution coefficient of ethanol is lower in the solvent/culture medium than in the solvent/water system. Also the mass transfer coefficient is considerably reduced if cells are present in the two-phase system (CRABBE et al., 1986).

To avoid the direct contact of the culture medium with the solvent, they may be separated by a thin gas layer in the pores of a membrane module. This separation is called transmembrane distillation (BANDINI and GOSTOLI, 1990).

Other groups investigated extraction in hollow fiber modules (VATAI and TEKIC, 1991). According to the former authors, ethanol production with membrane extraction at least is as competitive as the conventional process (the ethanol costs can be reduced by 3–4%).

Cell immobilization has several problems: low mechanical stability, leakage of viable cells, non-negligible internal mass transfer resistance, and decrease of the viability of entrapped microorganisms in the presence of solvents (GIANETTO et al., 1988). However, MATSUMURA and MÄRKL (1984) showed that Poropack Q protected the immobilized cells from tributyl phosphate, 3-phenyl-propanol, 2-octanol, and HONDA et al. (1986) indicated that castor oil protected immobilized cells from *o-tert*-butyl-phenol. However, the other problems of immobilized cells remain.

Another method was recommended by CHRISTEN et al. (1990) to separate the cultivation medium from the solvent during the extraction of ethanol. A teflon sheet was soaked with iso-tridecyl alcohol and used as solid supported liquid membrane (SLM). This assembly increased biocompatibility, permeation efficiency, and stability. By removal of ethanol from the cultures, the ethanol volumetric productivity was increased by a factor of 2.5, and the cell viability was maintained above 95%.

At the same time, ethanol was purified from the medium component and removed from the extractant by aqueous and gaseous stripping, respectively. It seems that the former has a higher performance.

Oliveira and Cabral (1991) used a chemical conversion of ethanol to increase its distribution coefficient. Oleic acid was used as extractant, allowing the esterification of ethanol catalyzed by lipase. The distribution coefficient of ethanol between the aqueous and the organic phase was improved tenfold by esterification. A better performance was obtained when co-immobilized *Saccharomyces bayanus* cells and lipase in microemulsion of phosphatidyl choline were used.

Kollerup and Daugulis (1985a, b) and Daugulis et al. (1987, 1991) investigated the *in situ* extraction of ethanol. They found that conventional and extractive ethanol production plants have an almost identical design. It should be possible to retrofit a conventional plant with an *in situ* extractive unit at substantial cost savings over the grass roots plant. Optimal conditions were evaluated by a computer program. Three different strategies were investigated: for the first two, operating conditions were based on the optimal conditions (medium dilution rate $D=0.15\ h^{-1}$ and solvent dilution rate $D_S=1.25\ h^{-1}$ and $D=0.18\ h^{-1}$, $D_S=1.50\ h^{-1}$, respectively). They provided ethanol selling prices of 32.70 cents per liter and 31.01 c/L, respectively, compared to 45.06 c/L for the conventional plant. The third strategy fitted the capacity of the extraction unit better to the product formation unit. This caused an increase of D_S to $2.0\ h^{-1}$. This yielded an ethanol selling price of 29.44 c/L.

However, as expected, the largest component (60%) of the overall production costs of ethanol continues to be the raw material cost (mainly substrate cost) followed by utilities (largely energy) at 26%. Equipment depreciation accounts for only 3% of the production costs of ethanol indicating the relative unimportance of numerous efforts aimed at developing better novel reactors for the ethanol process. Relative to the conventional ethanol production, the *in situ* extractive unit has a major saving in energy because it uses significantly higher substrate concentrations, i.e., less water (Daugulis et al., 1991).

Similar to ethanol formation, the production of **acetone-butanol** by *Clostridium acetobutylicum* is impaired by product inhibition. Therefore, several researchers investigated the *in situ* extraction of butanol (acetone inhibition can be neglected in comparison to that of butanol). The extraction may be carried out with the whole of cell-free medium.

Wayman and Parekh (1987) extracted acetone and butanol from the cell-containing medium with dibutyl phthalate, whereby the sugar concentration could be increased to 80–100 $g\ L^{-1}$ and the solvent (acetone-butanol) concentration from 18–20 $g\ L^{-1}$ to 28–30 $g\ L^{-1}$.

Roffler et al. (1987, 1988a) extracted the solvent produced by *Clostridium acetobutylicum* from molasses by oleyl alcohol in a Karr-column. In batch culture, the productivity was increased by 20%, which lowered the capital costs by 20% (Roffler et al., 1987). In continuous culture, the productivity was 70% higher than in batch culture (Roffler et al., 1988a).

Other research groups recovered the solvent by extraction from membrane permeate. Jeon and Lee (1989) and Groot et al. (1987) used silicon tubing as the membrane and oleyl alcohol and isopropyl myristate as extractants. Jeon and Lee (1989) succeeded in increasing the productivity in the batch culture fourfold at high glucose concentrations. Groot et al. (1987) doubled the productivity of the batch cultivation by extraction.

Eckert and Schügerl (1987) used a hydrophilized polypropylene membrane for cell retention. The permeate of the continuous culture was extracted by *n*-dodecanol in a four-stage mixer settler. The productivity was increased by a factor of four. After 600 h of continuous operation, the cell viability decreased. This was caused by a contaminant of the technical *n*-dodecanol: 1,3-hexanediol.

Shukla et al. (1989) used a hollow fiber module (Cellgard X-20, Hoechst Celanese) for extracting the product with 2-ethyl-1-hexanol and for its re-extraction. The productivity was increased by 40%.

The productivity of 2,3-butanediol by *Klebsiella oxytoca* with D-xylose as substrate did not change, if the product was extracted by dodecanol in an external column (Eiteman and GAINER, 1989). The cells did not grow in the reactor because of their contact with dodecanol.

Several mathematical models have been developed for *in situ* alcohol extraction (SHI et al., 1990; ROFFLER et al., 1988b; HONDA et al., 1987; FOURNIER, 1988) which indicate that significant improvement of the production performance can be attained with *in situ* extraction.

Since cell growth and product formation are reduced at high **carboxylic acid** concentrations, several authors have dealt with the *in situ* extraction of acids. Three approaches were investigated:

- the extractant contacted with the immobilized cells (BAR and GAINER, 1987; YABANNAVAR and WANG, 1987, 1991),
- separation of the cells from the extractant by the membrane (BARTELS et al., 1987; PRASAD and SIRKAR, 1987),
- conversion of the acids into a hydrophobic derivative of high solubility in the extractant (AIRES-BARROS et al., 1989).

YABANNAVAR and WANG (1987) produced lactate by immobilized *Lactobacillus delbrückii,* and the acid was extracted at a low pH by TOPO + dodecane in Isopar M and Alamine 336 in oleyl alcohol. The latter increased the productivity from 7 to 12 g L^{-1} (gel) h^{-1} (YABANNAVAR and WANG, 1991). BAR and GAINER (1987) produced citric acid by immobilized *Aspergillus niger,* acetic acid by immobilized *Acetobacter aceti,* and lactic acid by immobilized *L. delbrückii.* The acids were extracted with different extractants (long-chain hydrocarbons, perfluoro decalin, methyl oleate). The main problem besides the toxicity of the extractants is the low pH necessary for the acid extraction.

BARTELS et al. (1987) used microporous membrane modules to extract lactic acid by Alamine 336 + MIBK. PRASAD and SIRKAR (1987) applied hydrophilic, hydrophobic, and composite membranes to extract succinic acid and acetic acid from aqueous solutions by *n*-butanol.

The biocompatibility of different extractants with regard to the extraction of lactic acid produced by Lactobacilli (SIEBOLD, 1991) and their use for *in situ* extraction (v. FRIELING, 1991) have been discussed in Sect. 3.1.

In order to increase the solubility of the acids in extractants, they were converted into hydrophobic esters by lipase. Esterification with oleyl alcohol increased the apparent distribution coefficient in CBO extractants 4- to 15-fold (AIRES-BARROS et al., 1989); however, thereby the back-extraction becomes more difficult. It is doubtful whether the economy of the acid recovery is improved by this technique.

Only few investigations have been published on the *in situ* extraction of **secondary metabolites.** ENGLAND et al. (1988) investigated the extraction of codeine/norcodeine with functionalized polysiloxanes, and MAVITUNA et al. (1987) extracted capsaicin, with sunflower oil. Cycloheximide and gibbelinic acid were extracted from the permeate by substituted polyglycols, Genapol 2822 and nonyl phenol-8, 1-polyethylene (HOLLMANN et al., 1990; MÜLLER et al., 1989). Other antibiotics, penicillin G and V, were extracted *in situ* and converted into 6-amino-penicillanic acid (6-APA) by penicillin G-amidase, immobilized in ELM (SCHEPER et al., 1989; SCHÜGERL et al., 1992). Fig. 20 shows the transfer of penicillin from the cultivation medium (outer phase) across the liquid membrane by carrier extraction into the inner phase, its enzymatic conversion into 6-APA and phenyl acetic acid (PhA), and the back-transfer of PhA into the outer aqueous phase. The complete apparatus for the integrated process is shown in Fig. 21. By this integrated process, the production of ampicillin can be reduced to seven process stages.

It is expected that in the future integrated product formation and product recovery processes with *in situ* extraction will gain importance.

8 Future Developments

Extraction with organic solvents is a major method of recovering biological products from the culture medium. Due to volatility, flamma-

outer phase | membrane phase | inner phase | membrane phase | outer phase

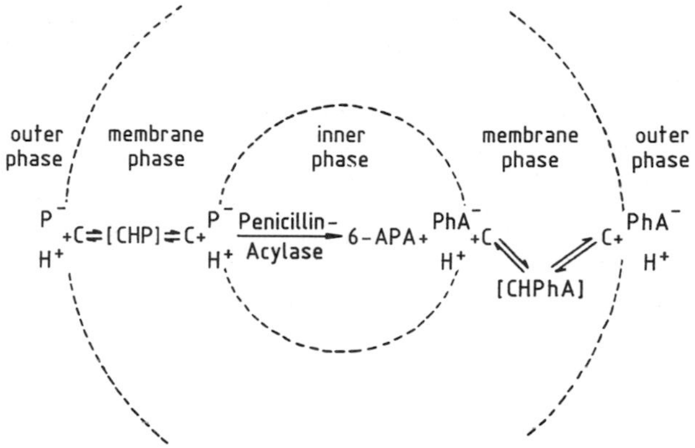

$P^- + C \rightleftharpoons [CHP] \rightleftharpoons C + P^-$ Penicillin-Acylase → 6-APA + PhA^- + C ⇌ [CHPhA] ⇌ C + PhA^-; H^+

Fig. 20. Extraction of penicillin by ELM across the carrier (LA-2) containing membrane phase into the enzyme (penicillin G acylase) containing internal aqueous phase, in which penicillin is converted into 6-APA, which is enriched here, and PhA, which returns into the external aqueous phase (SCHEPER et al., 1989).

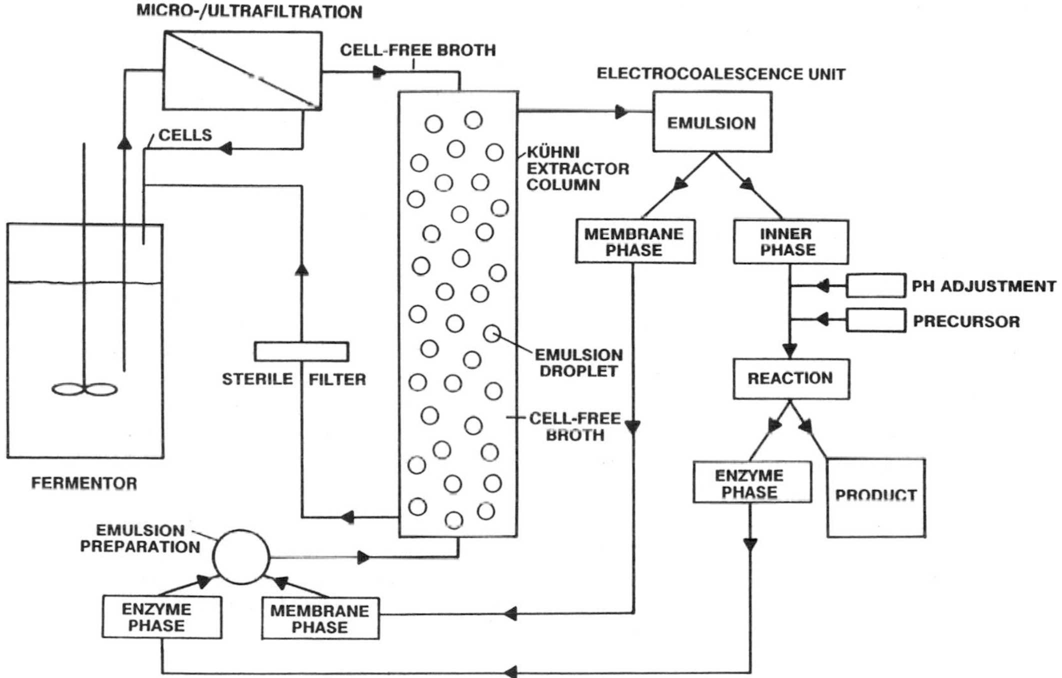

Fig. 21. Flowsheet of the integrated process penicillin *in situ* extraction and *in situ* conversion into 6-APA for production of ampicillin (product) in seven process stages (BARENSCHEE et al., 1992).

bility, and solubility in the aqueous cultivation medium and the toxicity of various solvents used in industry, their use will gradually be restricted. They are progressively replaced by solvents of low volatility, flammability, and solubility in the aqueous culture and non-toxic properties. The distribution coefficients in these solvents, however, are rather low. Thus, they are combined with "carriers" which chemically interact with the solute and increase their loading capacity, distribution coefficient, and selectivity. The carriers on the market, however, have been developed for hydrometallurgical applications, where the market volume is high enough. These carriers are not optimal for metabolite recovery. New inexpensive carriers with higher loading capacities, distribution coefficients, and selectivity are imperative. They will be employed preferably for highvalue products. The combination of carrier extraction with the emulsion liquid membrane technique will gain importance in environmental biotechnology, where low concentration solutes must be recovered from wastewater. It will compete with the ion exchange and flotation techniques.

Contrary to food technology, the supercritical extraction in biotechnology will be restricted to few special cases with hydrophobic solutes owing to the hydrophilic character of most metabolites.

Extraction with reversed micelles is not yet well-adapted for metabolite recovery, it remains a technique for the recovery of special products.

In situ extraction is becoming more and more important, because product concentration in a number of processes is increasing gradually, and product inhibition is becoming a more serious problem with screening, mutation, and recombinant techniques. It competes with electrodialysis, ion exchange, and adsorption.

9 References

AIRES-BARROS, M. R., CABRAL, J. M. S., WILLSON, R. C., HAMEL, J.-F. P., COONEY, C. L. (1989), Esterification-coupled extraction of acids: partition enhancement and underlying reaction and distribution equilibria, *Biotechnol. Bioeng.* **34,** 909–915.

ALTER, J. E., BLUMENBERG, R. (1981), *U.S. Patent* 4251671.

BANDINI, S., GOSTOLI, C. (1990), Ethanol production by extractive fermentation: a novel membrane extraction, in: *Separations for Biotechnology* (PYLE, D. L., Ed.), pp. 539–548. Amsterdam: Elsevier Science Publishers.

BANIEL, A. M. (1982), *Eur. Patent* EP 49429.

BANIEL, A. M., BLUMENBERG, R., HAJDN, K. (1981), *U.S. Patent* 4275234.

BAR, R., GAINER, J. L. (1987), Acid fermentation in water-organic solvent two-liquid phase systems, *Biotechnol. Progr.* **3,** 109–114.

BARENSCHEE, T., SCHÜGERL, K., SCHEPER, T. (1992), An integrated process for the production and biotransformation of penicillin, submitted.

BARTELS, P. V., DROST, J. C. G., DE GRAAUW, J. (1987), Solvent extraction using a porous membrane module, *Proc. 4th Eur. Congr. Biotechnology,* Vol. 2, pp. 558–559 (NEIJSSEL, O. M., VAN DER MEER, R. R., LUYBEN, K. Ch. A. M., Eds.). Amsterdam: Elsevier Science Publishers B.V.

BEECHAM GROUP LTD. (1975), *British Patent* 1565656 (Dec. 13).

BEHR, J. P., LEHN, J. M. (1973), Transport of amino acids through organic liquid membranes, *J. Am. Chem. Soc.* **95** (18), 6108–6110.

BRUNNER, G. (1987), Stofftrennung mit überkritischen Gasen (Gasextraktion), *Chem. Ing. Tech.* **59,** 12–22.

BRUNNER, G., KREIM, K. (1985), Abtrennung von Ethanol aus wäßrigen Lösungen mittels Gasextraktion, *Chem. Ing. Tech.* **57,** 550–551.

BRUNZENBERGER, G., MARR, R. (1987), Application of supercritical extraction by use of a countercurrent extraction column, Poster Presentation at the *Engineering Foundation Conference on Separation Technology,* April 26–May 1, 1987, at Schloss Elmau, Germany.

BUSCHE, R. M. (1983), Recovering chemical products from dilute fermentation broths, *Biotechnol. Bioeng.* **13,** 597–615.

BUSCHE, R. M., SHIMSHICK, E. J., YATES, R. A. (1982), Recovery of acetic acid from dilute acetate solution, *Biotechnol. Bioeng.* **12,** 249–262.

CHAUDHURI, J., PYLE, D. L. (1987), Liquid membrane extraction, in: *Separations in Biotechnology* (VERRALL, M. S., HUDSON, M. J., Eds.), pp. 241–259. Chichester, U.K.: Ellis Horwood Ltd.

CHAUDHURI, J. B., PYLE, D. L. (1990), A model for emulsion liquid membrane extraction of organic acids, in: *Separations for Biotechnology* (PYLE, D. L., Ed.), pp. 112–121. Amsterdam: Elsevier Scientific Publishers.

CHOUDHURY, J. P., GHOSH, P., GUHA, B. K. (1985), Separation of ethanol from ethanol-water mixture by reverse osmosis, *Biotechnol. Bioeng.* **27,** 1081–1084.

CHRISTEN, P., MINIER, M., RENON, H. (1990), Ethanol extraction by supported liquid membrane during fermentation, *Biotechnol. Bioeng.* **36,** 116–123.

CRABBE, P. G., TSE, C. W., MUNRO, P. A. (1986), Effect of microorganisms on rate of liquid extraction of ethanol from fermentation broths, *Biotechnol Bioeng.* **28,** 939–943.

DAUGULIS, A. J., SWAINE, D. E., KOLLERUP, F., GROOM, C. A. (1987), Extractive fermentation-integrated reaction and product recovery, *Biotechnol. Lett.* **9,** 425–430.

DAUGULIS, A. J., AXFORD, D. B., MCLELLAN, P. J. (1991), The economics of ethanol production by extractive fermentation, *Can. J. Chem. Eng.* **69**, 488–497.

DEBLAY, P., MINNIER, M., RENON, H. (1990), Separation of L-valine from fermentation broths using supported liquid membrane, *Biotechnol. Bioeng.* **35**, 123–131.

ECKERT, G., SCHÜGERL, K. (1987), Continuous acetone–butanol production with direct product removal, *Appl. Microbiol. Biotechnol.* **27**, 211–228.

EITEMAN, M. A., GAINER, J. L. (1989), *In situ* extraction versus the use of an external column in fermentation, *Appl. Microbiol. Biotechnol.* **30**, 614–618.

ENGLAND, R., ABED-ALI, S. S., BRISDOM, B. J. (1988), Direct extraction of biotransformation products using functionalized polysiloxanes, *DECHEMA Biotechnology Conferences*, Vol. 2, pp. 71–81. Weinheim: VCH Verlagsgesellschaft.

ESSIEN, D. E., PYLE, D. L. (1987), Fermentation ethanol recovery by solvent extraction, in: *Separations for Biotechnology* (VERRALL, M. S., HUDSON, M. J., Eds.), pp. 320–332. Chichester, U.K.: Ellis Horwood Ltd.

FOURNIER, R. L. (1988), Mathematical model of microporous hollow-fiber membrane extractive fermentor, *Biotechnol. Bioeng.* **31**, 235–239.

GIANETTO, A., RIGGERI, B., SPECCHIA, V., SASSI, G., FORNA, R. (1988), Continuous extraction loop reactor: alcoholic fermentation by fluidized entrapped biomass, *Chem. Eng. Sci.* **43**, 1891–1896.

GROOT, W. J., TIMMER, J. M. K., LUYBEN, K. Ch. A. M. (1987), Membrane solvent extraction for *in situ* butanol recovery in fermentations, *Proc. 4th Eur. Congr. Biotechnology*, Vol. 1, pp. 564–566 (NEIJSSEL, O. M., VAN DER MEER, R. R., LUYBEN, K. Ch. A. M., Eds.). Amsterdam: Elsevier Science Publishers B.V.

GROOT, W. J., VAN DER LAANS, R. G. J. M., LUYBEN, K. Ch. A. M. (1989), Batch and continuous butanol fermentations with free cells: integration with product recovery by gas stripping, *Appl. Microbiol. Biotechnol.* **32**, 305–308.

HANO, T., MATSUMOTO, M., OHTAKE, T., SASAKI, K., HORI, F., KAWANO, Y. (1990), Extraction equilibria of organic acids with tri-*n*-octylphosphineoxide, *J. Chem Eng. Jpn.* **23**, 734–738.

HANO, T., OHTAKE, T., MATSUMOTO, M., OGAWA, S., HORI, F. (1990), Extraction of penicillin with liquid surfactant membrane, *J. Chem. Eng. Jpn.* **23**, 772–775.

HANO, T., OHTAKE, T., MATSUMOTO, M., KITAYAMA, D., HORI, F., NAKSHIO, F. (1991), Extraction equilibria of amino acids with quaternary ammonium salt, *J. Chem. Eng. Jpn.* **24**, 20–24.

HARRIS, T. A. J., KHAN, S., REUBEN, B. G., SHOKOYA, T., VERRALL, M. S. (1990), Reactive solvent extraction of beta-lactam antibiotics, in: *Separations for Biotechnology* (PYLE, D. L., Ed.), pp. 172–180. Amsterdam: Elsevier Science Publishers.

HERSBACH, G. J. M., VAN DER BEEK, C. P., VAN DIJK, P. W. M. (1984), The penicillins: properties, biosynthesis and fermentation, in: *Biotechnology of Industrial Antibiotics* (VANDAMME, E. J., Ed.), pp. 45–140. New York, Basel: Marcel Dekker.

HOLLMANN, D., MERRETTIG-BRUNS, U., MÜLLER, U., ONKEN, U. (1990), Secondary metabolites by extractive fermentation, in: *Separations for Biotechnology* (PYLE, D. L., Ed.), pp. 567–576. Amsterdam: Elsevier Science Publishers B.V.

HOLTEN, C. H. (1971), *Lactic Acid.* Weinheim: Verlag Chemie.

HONDA, H., TAYA, M., KOBAYASHI, T. (1986), Ethanol fermentation associated with solvent extraction using immobilized growing cells of *Saccharomyces cerevisiae* and its lactose-fermentable fusant, *J. Chem. Eng. Jpn.* **19**, 268–273.

HONDA, H., MANO, T., TAYA, M., SHIMIZU, K., MATSUBARA, M., KOBAYASHI, T. (1987), A general framework for the assessment of extractive fermentations, *Chem. Eng. Sci.* **42**, 493–498.

JEON, Y. J., LEE, Y. Y. (1989), *In situ* product separation in butanol fermentation by membrane-assisted extraction, *Enzyme Microb. Technol.* **11**, 575–582.

KATINGER, H., WIBBELT, F., SCHERFLER, H. (1981), Kontinuierliche direkte Extraktion von Antibiotika mit Separatoren und Dekantern, *VT "Verfahrenstechnik"* **15**, 179–182.

KERTES, A. S., KING, C. J. (1986), Extraction chemistry of fermentation product carboxylic acids, *Biotechnol. Bioeng.* **28**, 269–282.

KING, C. J. (1983), Acetic acid extraction, in: *Handbook of Solvent Extraction* (LO, T. C., BAIRD, M. H. I., HANSON, C., Eds.), pp. 567–573. New York: John Wiley & Sons.

KING, M. B., BOTT, T. R., CHAMI, J. H. (1987), Extraction of biomaterials with compressed carbon dioxide and other solvents under near critical conditions, in: *Separations for Biotechnology* (VERRALL, M. S., HUDSON, M. J., Eds.), pp. 293–319. Chichester, U.K.: Ellis Horwood Ltd.

KOLLERUP, F., DAUGULIS, A. J. (1985a), Screening and identification of extractive fermentation solvents using a database, *Can. J. Chem. Eng.* **63**, 919–927.

KOLLERUP, F., DAUGULIS, A. J. (1985b), A mathematical model for ethanol production by extractive fermentation in continuous stirred tank fermentation, *Biotechnol. Bioeng.* **27,** 1335–1346.

KOLLERUP, F., DAUGULIS, A. J. (1986), Ethanol production by extractive fermentation – solvent identification and prototype development, *Can. J. Chem. Eng.* **64,** 598–606.

LECHNER, M., MÄRKL, H., GÖTZ, F. (1988), Lipase production of *Staphylococcus carnosus* in a dialysis fermentor, *Appl. Microbiol. Biotechnol.* **28,** 345–349.

LEE, C. K., HONG, J. (1988), Membrane reactor coupled with electrophoresis for enzymatic production of aspartic acid, *Biotechnol. Bioeng.* **32,** 647–654.

LEODIDIS, E., HATTON, T. A. (1990a), Amino acids in AOT reversed micelles. 1. Determination of interfacial partition coefficients using the phase-transfer method, *J. Phys. Chem.* **94,** 6400–6411.

LEODIDIS, E., HATTON, T. A. (1990b), Amino acids in AOT reversed micelles. 2. The hydrophobic effect and hydrogen bonding as driving forces for interfacial solubilization, *J. Phys. Chem.* **94,** 6411–6420.

LEODIDIS, E., HATTON, T. A. (1991), Amino acids in reversed micelles. 4. Amino acids as co-surfactants, *J. Phys. Chem.* **95,** 5957–5965.

LEODIDIS, E., BOMMARIUS, A. S., HATTON, T. A. (1991), Amino acids in reversed micelles. 3. Dependence of the interfacial partition coefficient on excess phase salinity and interfacial curvature, *J. Phys. Chem.* **95,** 5943–5956.

LESER, M. E., LUISI, P. L., PALMIERI, S. (1989), The use of reverse micelles for the simultaneous extraction of oil and proteins from vegetable meal, *Biotechnol. Bioeng.* **34,** 1140–1146.

LI, N. N. (1966), Separating hydrocarbons with liquid membranes, *U.S. Patent* 3410794, Ser. No. 533,933 (Cl. 208–308).

LIKIDIS, Z. (1983), *Master Thesis,* University of Hannover, FRG.

LIKIDIS, Z. (1986), Reaktivextraktion und Rückextraktion von Penicillin aus Fermentationsmedien in Zentrifugal-Extraktoren und Extraktionskolonnen im Technikumsmaßstab, *PhD Thesis,* University of Hannover, FRG.

LIKIDIS, Z., SCHÜGERL, K. (1987a), Reactive extraction and re-extraction of penicillin with different carriers, *J. Biotechnol.* **5,** 293–303.

LIKIDIS, Z., SCHÜGERL, K. (1987b), Recovery of penicillin by reactive extraction in centrifugal extractors, *Biotechnol. Bioeng.* **30,** 1032–1040.

LIKIDIS, Z., SCHÜGERL, K. (1988a), Continuous extraction of penicillin G and its re-extraction in three different column types – a comparison, *Chem. Eng. Sci.* **43,** 27–32.

LIKIDIS, Z., SCHÜGERL, K. (1988b), Simulation of the continuous re-extraction of penicillin G from the solution of its ion-pair complex in three different types of bench-scale columns, *Chem. Eng. Sci.* **43,** 1243–1246.

LIKIDIS, Z., SCHLICHTING, E., BISHOFF, L., SCHÜGERL, K. (1989), Reactive extraction of penicillin G from mycel-containing broth in a countercurrent extraction decanter, *Biotechnol. Bioeng.* **33,** 1385–1392.

LO, T. C., BAIRD, M. H. I., HANSON, C., Eds. (1983), *Handbook of Solvent Extraction.* New York: John Wiley & Sons.

LUISI, P. L., LAANE, C. (1986), Solubilisation of enzymes in apolar solvents via reverse micelles, *TIBTECH,* June, 153–161.

MAKRYALEAS, K., SCHEPER, T., SCHÜGERL, K., KULA, M. R. (1985), Enzymatic production of L-amino acid with continuous coenzyme regeneration by liquid membrane technique, *Ger. Chem. Eng.* **8,** 345–350.

MATSUMURA, M., MÄRKL, H. (1984), Application of solvent extraction to ethanol fermentation, *Appl. Microbiol. Biotechnol.* **20,** 371–377.

MAVITUNA, F., WIKINSON, A. K., WILLIAMS, P. D. (1987), Liquid–liquid extraction of plant metabolites as an integrated stage with bioreactor operation, in: *Separations for Biotechnology* (VERRALL, M. S., HUDSON, M. J., Eds.), pp. 333–339. Chichester, U.K.: Ellis Horwood Ltd.

MÜLLER, B., SCHLICHTING, E., BISHOFF, L., SCHÜGERL, K. (1987), Reactive extraction of penicillin G in a pilot plant Karr-column. II. Fermentation broths, *Appl. Microbiol. Biotechnol.* **26,** 206–210.

MÜLLER, B., SCHLICHTING, E., SCHÜGERL, K. (1988), Reactive extraction of penicillin G in a pilot plant Karr-column. III. Modelling and simulation of the extraction process, *Appl. Microbiol. Biotechnol.* **27,** 484–486.

MÜLLER, U., TRÄGER, M., ONKEN, U. (1989), Selection of solvents for extractive fermentations, *Ber. Bunsen-Ges. Phys. Chem.* **93,** 1001–1004.

NOZAKI, Y., TANFORD, C. (1971), The solubility of amino acids and two glycine peptides in aqueous ethanol and dioxane solutions, *J. Biol. Chem.* **246,** 2211–2217.

NUCHNOI, P., YANO, T., NISHIO, N., NAGAI, S. (1987), Extraction of volatile fatty acids from diluted aqueous solutions using a supported liquid membrane, *J. Ferment. Technol.* **65,** 301–310.

OLIVEIRA, A. C., CABRAL, J. M. S. (1991), Production and extractive biocatalysis of ethanol using microencapsulated yeast cells and lipase system, *J. Chem. Technol. Biotechnol.* **52,** 219–225.

PITT JR., W. W., HAAG, G. L., LEE, D. D. (1983), Recovery of ethanol from fermentation broths using selective sorption–desorption, *Biotechnol. Bioeng.* **25**, 123–131.

PODOJIL, M., BLUMAUEROVA, M., VANEK, Z., CULIK, K. (1984), The tetracyclines: properties, biosynthesis, and fermentation, in: *Biotechnology of Industrial Antibiotics* (VANDAMME, E. J., Ed.), pp. 259–279. New York, Basel: Marcel Dekker, Inc.

PONS, M. N., TILLIER-DORION, F. (1987), Gas membrane removal of volatile inhibitors, *Proc. 4th Eur. Congr. Biotechnology*, Vol. 1, pp. 149–152 (NEIJSSEL, O. M., VAN DER MEER, R. R., LUYBEN, K. Ch. A. M., Eds.). Amsterdam: Elsevier Science Publishers B.V.

PRASAD, R., SIRKAR, K. K. (1987), Solvent extraction with microporous hydrophilic and composite membranes, *AIChE J.* **33**, 1057–1066.

RANDOLPH, T. W. (1990), Supercritical fluid extractions in biotechnology, *TIBTECH* **8**, 78–82.

RESCHKE, M., SCHÜGERL, K. (1984a), Reactive extraction of penicillin I: Stability of penicillin G in presence of carriers and relationships for distribution coefficients and degrees of extraction, *Chem. Eng. J.* **28**, B1–B9.

RESCHKE, M., SCHÜGERL, K. (1984b), Reactive extraction of penicillin II: Distribution coefficients and degrees of extraction, *Chem. Eng. J.* **28**, B11–B20.

RESCHKE, M., SCHÜGERL, K. (1984c), Reactive extraction of penicillin III: Kinetics, *Chem. Eng. J.* **29**, B25–B29.

RESCHKE, M., SCHÜGERL, K. (1985), Continuous reactive extraction of penicillin G in a Karr column, *Chem. Eng. J.* **31**, B19–B26.

RESCHKE, M., SCHÜGERL, K. (1986), Simulation of the continuous reactive extraction of penicillin G in a Karr column, *Chem. Eng. J.* **32**, B1–B5.

RICKER, N. L., MICHAELS, J. N., KING, C. J. (1979), Solvent properties of organic bases for extraction of acetic acid from water, *J. Separ. Process Technol.* **1**, 36–41.

RICKER, N. L., PITTMANN, E. F., KING, J. C. (1980), Solvent extraction with amines for recovery of acetic acid from dilute industrial streams, *J. Separ. Process Technol.* **1**, 23–30.

RIDGWAY, K., THORPE, E. E. (1983), Use of solvent extraction in pharmaceutical manufacturing processes, in: *Handbook of Solvent Extraction* (LO, T. C., BAIRD, M. H. I., HANSON, C., Eds.), pp. 583–591. New York: Wiley Interscience.

ROFFLER, S. R., BLANCH, H. W., WILKE, C. R. (1987), Extractive fermentation of acetone and butanol: Process design and economic evaluation, *Biotechnol. Progr.* **3**, 131–140.

ROFFLER, S. R., BLANCH, H. W., WILKE, C. R. (1988a), *In situ* extractive fermentation of acetone and butanol. *Biotechnol. Bioeng.* **31**, 135–143.

ROFFLER, S. R., WILKE, C. R., BLANCH, H. W. (1988b), Design and mathematical description of differential contactors used in extractive fermentations, *Biotechnol. Bioeng.* **32**, 192–204.

ROYCHOUDHURY, P. K., GHOSE, T. K., GHOSH, P., CHOTANI, G. K. (1986), Vapor liquid equilibrium behavior of aqueous ethanol solution during vacuum coupled simultaneous saccharification and fermentation, *Biotechnol. Bioeng.* **28**, 927–976.

RÜCKL, W., SIEBENHOFER, M., MARR, R. (1986), Separation of citric acid from aqueous fermentation solutions by extraction re-extraction process, *International Solvent Extraction Conference ISEC '86,* III-653–658.

SCHEPER, T., HALWACHS, W., SCHÜGERL, K. (1983), Preparation of L-amino acids by means of continuous enzyme catalysed D,L-amino acid ester hydrolysis inside liquid surfactant membranes, *International Solvent Extraction Conference ISEC '83,* 389–390.

SCHEPER, T., HALWACHS, W., SCHÜGERL, K. (1984), Production of L-amino acids by continuous enzymatic hydrolysis of D,L-amino acid methyl ester by liquid membrane technique, *Chem. Eng. J.* **29**, B31–B37.

SCHEPER, T., BARENSCHEE, E. R., BARENSCHEE, T., HASLER, A., MAKRYALEAS, K., SCHÜGERL, K. (1989), A combination of selective mass transport and enzymatic reaction: enzyme immobilization in liquid surfactant membranes, *Ber. Bunsen-Ges. Phys. Chem.* **93**, 1034–1038.

SCHMIDT-KASTNER, G., EGERER, P. (1984), Amino acids and peptides, in: *Biotechnology,* 1st Ed. (REHM, H. J., REED, G., Eds.) Vol. 6a: *Biotransformations* (KIESLICH, K., Vol. Ed.) pp. 387–419. Weinheim: Verlag Chemie.

SCHÜGERL, K., MATTIASSON, B., HATTON, A. (1992), *Extraction in Biotechnology*. Heidelberg: Springer Verlag.

SHAH, D. J., TOWARI, K. K. (1981), Effect of salt on the distribution of acetic acid between water and organic solvent, *J. Chem. Eng. Data* **26**, 375–378.

SHI, Z., SHIMIZU, K., IIJIMA, S., MORISUE, T., KOBAYASHI, T. (1990), Theoretical development and evaluation for extractive fermentation using multiple extractants, *Biotechnol. Bioeng.* **36**, 520–529.

SHUKLA, R., KANG, W., SIRKAR, K. K. (1989), Acetone–butanol–ethanol (ABE) production in a novel hollow fiber fermentor–extractor, *Biotechnol. Bioeng.* **34**, 1158–1166.

SIEBENHOFER, M., MARR, R. (1983), Acid extraction by amines, *International Solvent Extraction Conference ISEC '83,* 219.

SIEBENHOFER, M., MARR, R. (1985), Auswirkungen der Extraktionsmittelzusammensetzung auf den Stoffaustausch und Apparateauswahl am Beispiel der Essigsäure-Extraktion. *Chem. Ing. Tech.* **57**, 558.

SIEBOLD, M. (1991), Konversion von Glucose in Milchsäure, *PhD Thesis*, University of Hannover, FRG.

SMITH, E. L., PAGE, J. E. (1948), The acid binding properties of long-chain aliphatic amines, *J. Soc. Chem. Eng.* (London) **67**, 48–51.

STAHL, E., SCHILZ, W. (1976), Extraktion mit überkritischen Gasen in direkter Kopplung mit der Dünnschicht-Chromatographie, *Chem. Ing. Tech.* **48**, 773–778.

THIEN, M. P., HATTON, T. A., WANG, D. I. C. (1986), Separation of amino acids from fermentation broths using liquid emulsion membranes, *ISEC '86*, III-685–693.

THIEN, M. P., HATTON, T. A., WANG, D. I. C. (1988), Separation and concentration of amino acids using liquid emulsion membranes, *Biotechnol. Bioeng.* **32**, 604–615.

TODD, D. B., DAVIES, G. R. (1973), Centrifugal pharmaceutical extractions, *Filtr. Sep.*, November/December, 663–666.

VATAI, GY., TEKIC, M. N. (1991), Membrane-based ethanol extraction with hollow fiber module, *Sep. Sci. Technol.* **26**, 1005–1011.

VICTOR, J. (1983), Extraction of ethanol from water with liquid propylene, *International Solvent Extraction Conference ISEC '83*, 511–512.

VON FRIELING, P. (1991), Entwicklung und Charakterisierung von Systemen für die Reaktivextraktion von Milchsäure, *PhD Thesis*, University of Hannover, FRG.

WARDELL, J. M., KING, C. J. (1978), Solvent equilibria for extraction of carboxylic acids from water, *J. Chem. Eng. Data* **23**, 144–148.

WAYMAN, M., PAREKH, R. (1987), Production of acetone–butanol by extractive fermentation using dibutylphthalate as extractant, *J. Ferment. Technol.* **65**, 295–300.

WENNERSTEN, R. (1983), The extraction of citric acid from fermentation broth using a solution of a tertiary amine, *J. Chem. Techn. Biotechnol.* **33B**, 85–94.

WOJTECH, B., MAYER, M. (1985), Synergistische Effekte bei der Extraktion mit tertiären Aminen, *Chem. Ing. Techn.* **57**, 134–136.

YABANNAVAR, V. M., WANG, D. I. C. (1987), Bioreactor system with solvent extraction for organic acid production, *Ann. N. Y. Acad. Sci.* **506**, 523–535.

YABANNAVAR, V. M., WANG, D. I. C. (1991), Extractive fermentation for lactic acid production, *Biotechnol. Bioeng.* **37**, 1095–1100.

YAGODIN, G. A., YURTOV, E. V., GOLUBKOV, A. S. (1986), Liquid membrane extraction of amino acids, *International Solvent Extraction Conference ISEC '86*, III-677–683.

22 Protein Purification by Liquid–Liquid Extraction

BRIAN D. KELLEY

Andover, MA 01810, USA

T. ALAN HATTON

Cambridge, MA 02139, USA

1 Introduction

The initial stages of the downstream processing of protein products reflect engineering problems common to many biologically derived materials. The protein is in dilute solution, often present with contaminating solids such as cells or cell debris, and with other contaminating macromolecules in solution. The initial goals of the first stages, then, are meant to yield a clarified liquid stream concentrated in protein and with major classes of soluble contaminants removed (LEE, 1989; KELLEY and HATTON, 1991). Traditional unit operations employed are rotary vacuum filtration or continuous discharge centrifugation to remove biomass, ultrafiltration to concentrate the product, and product precipitation by salt or solvent addition to reduce soluble contaminants. Process flowsheets (Fig. 1) often have each of these three processes listed, and product recoveries reflect losses incurred for each of the steps. A single unit operation which effects biomass removal, concentration, and partial purification could replace these conventional processing stages and their associated equipment, and perhaps increase the efficiency and recovery of the protein product. A clear candidate to perform this concentration, clarification and purification is liquid–liquid extraction.

Liquid–liquid extraction is a technique by which a solute of interest is transferred from a liquid feed stream into a second, immiscible liquid phase. By the selective partitioning of the solute relative to the contaminating species, a purification is achieved. Biomass re-

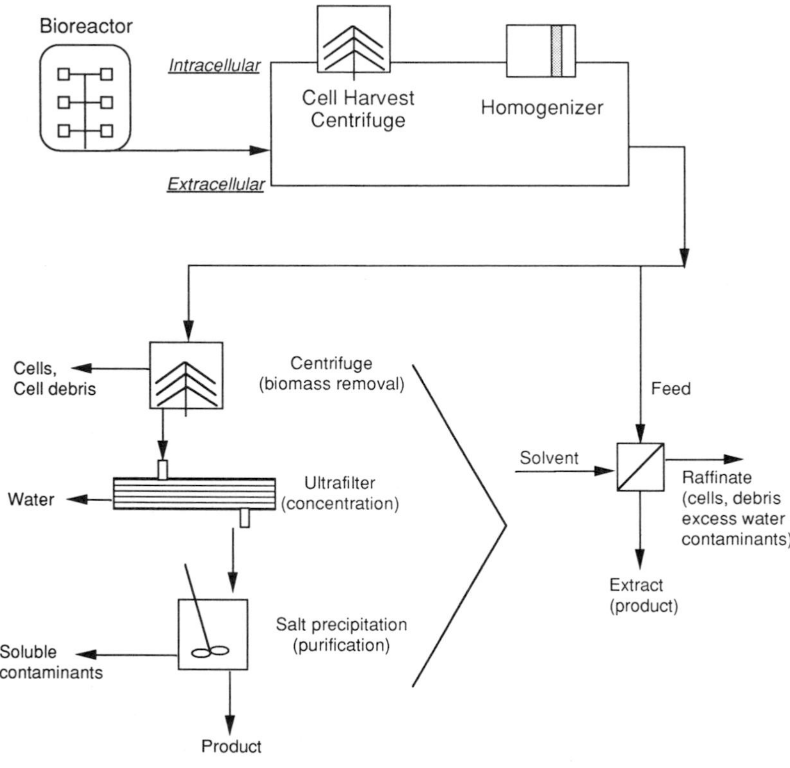

Fig. 1. Process flowsheet for extracellular and intracellular products, with liquid–liquid extraction replacing several unit operations in the initial downstream processing.

mains in the stream, and is generally not taken up by the product-containing extract phase. With large partition coefficients and the correct choice of phase volume ratios, concentration is also possible.

Liquid–liquid extraction is a widely used unit operation in the chemical processing industry (TREYBAL, 1980), and it has been used in the pharmaceutical industry for the initial separation of fermentation products from either a whole or a clarified broth. Despite a wealth of knowledge concerning liquid–liquid extraction, however, the technique has only recently been applied to the large-scale separation of biologically derived macromolecules such as proteins (KULA, 1985; ABBOTT and HATTON, 1988). Typical extracting solvents used for processing chemicals are immiscible with water due to their extremely nonpolar nature; ketones, for instance, are commonly used for penicillin extraction. Such solvents would not be suitable for protein purification, and so biocompatible liquids which phase separate with water needed to be identified before significant progress could be made in this field.

Several types of extraction systems have been developed in which protein structure and function are preserved despite the formation of two phases during extraction. The phenomenon of phase separation by the addition of polymers or salts has been well established as a means of achieving non-denaturing conditions for the partitioning of proteins. The use of water-in-oil microemulsions (reversed micelles) relies on the hosting of active proteins within water pools stabilized by surfactants in an organic phase. New technologies, such as phase-separated micellar solutions and partitioning with salts and aliphatic alcohols, are being explored.

This chapter reviews the use of these liquid–liquid extraction techniques as applied to protein purification. The bases of the separations are presented, the controlling factors governing protein uptake are examined, and applications are discussed.

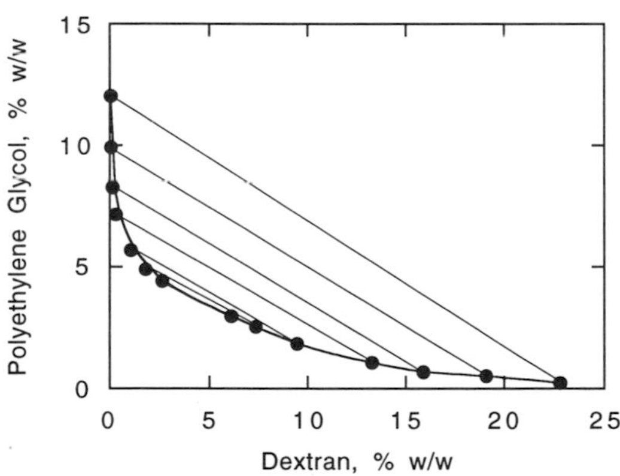

Fig. 2. Phase diagram for a PEG 6000-dextran D48 two-phase system at 20 °C (redrawn from Albertsson, 1986).

2 Two-Phase Aqueous Polymer Systems

2.1 Basis of Phase Separation

The addition of a mixture of water-soluble polymers to an aqueous phase can result in the formation of two immiscible phases for certain regions of the ternary phase diagram (Fig. 2). A single polymer and a salt may show similar phase separation behavior. The co-existing phases are each enriched in one of the two polymers or salts. Phase diagrams delineate polymer and salt concentrations in the top and bottom phases, respectively, which is important information for consideration of recycle and reuse of components. Lists of published phase diagrams for both polymer–polymer and polymer–salt systems are available (ALBERTSSON, 1986, KULA, 1985; TJERNELD and JOHANSSON, 1990).

Subtle differences in the two chemical environments of the separate phases arise because of the polymer or salt partitioning, and proteins, macromolecules, cell debris or whole cells may then partition unequally between the phases. This is the basis for the use of two-phase aqueous polymers in liquid–liquid extraction. Because of the high concentration of water in both phases (typically over 85%), the protein is often quite stable in this environment. Contaminants such as cells and cell debris generally partition to the bottom phase,

Tab. 1. Two-Phase Aqueous Polymer Systems

Polymer/Salt Combination	References
Polyethylene glycol (PEG)-dextran	ALBERTSSON, 1986
PEG-phosphate	ALBERTSSON, 1986
PEG-citrate	VERNAU and KULA, 1990
PEG-hydroxypropyl starch	TJERNELD and JOHANSSON, 1990
Ethylhydroxyethyl cellulose-hydroxypropyl starch	TJERNELD, 1989
PEG-polyvinyl alcohol	TJERNELD, 1989
PEG-pullulan	NGUYEN et al., 1988
PEG-maltodextrins	SIKDAR et al., 1991

although there are examples of cells collecting at the interface or even populating the top phase due to differences in the surface properties of the bacteria (ANDERSSON and HAHN-HÅGERDAL, 1988).

Tab. 1 lists some of the many polymer/polymer and polymer/salt systems. The most common are dextran and polyethylene glycol (PEG) or PEG and potassium phosphate, although novel polymers are being examined (STURESSON et al., 1990; NGUYEN et al., 1988; SIKDAR et al., 1991; PATRICKIOS et al., 1992). The PEG-rich phase is less dense than either the dextran-rich or salt-rich phase, and is referred to as the top phase. Typical concentrations of PEG–dextran systems are 10% (w/w) PEG and 15% dextran, or with salt-rich bottom phases, 15% PEG and 15% salt. The novel polymer systems may have several advantages when compared to the PEG–dextran system. The ethylhydroxyethyl cellulose-hydroxypropyl starch combination phase separates at very low polymer concentrations (as low as 1 and 4%, respectively), considerably lower than found with PEG–dextran or PEG–salt systems (TJERNELD, 1989). The additional cost of biological remediation of phosphate salts has motivated the development of PEG-citrate systems, where the citrate is easily degraded (VERNAU and KULA, 1990). Less expensive polymers like starches, maltodextrins or pullulans, greatly reduce processing costs. Other efforts to use crude polymers are motivated by the high expense of size-fractionated materials with narrow molecular weight distributions, and so polydisperse polymers still capable of forming two-phase systems would be advantageous (KRONER et al., 1982). The drive to eliminate PEG is not so great, as it is available at low cost, and forms two-phase systems with other neutral polymers as well as salts. The use of novel polyampholytes as one of the phase-forming polymers has some attraction, both because charge interactions can be modified readily using pH swings, and because the polymers can be recovered easily using isoelectric precipitation (PATRICKIOS et al., 1992).

Partition coefficients are commonly used to quantify the performance of the extraction, and are defined as the product concentration in the top phase divided by that in the bottom phase:

$$K=[\text{protein}]_{\text{T}}/[\text{protein}]_{\text{B}}$$

Partition coefficients near unity provide for poor separation, while very large or very small values provide an opportunity to exploit this phase preference, if contaminating species do not partition in a similar fashion. The phase volume ratio, $R=V_{\text{T}}/V_{\text{B}}$, impacts the recovery also, as shown by the following relations describing the percent recovery in a single theoretical stage for both the top and bottom phases:

$$\text{Top phase recovery (\%)} = 100\ KR/(1+KR)$$
$$\text{Bottom phase recovery (\%)} = 100/(1+KR)$$

For continuous flow operations, the relative flow rates will substitute for the phase volumes; because of inefficiencies in mixing under flow conditions, the full equilibrium partitioning may be approached, but is not usually achieved.

Tab. 2. Factors Affecting Protein Partitioning in Two-Phase Aqueous Polymer Systems

Protein molecular weight
Protein charge, surface properties
Polymer molecular weight
Phase composition, tie-line length
Salt effects
Affinity ligands attached to polymers

2.2 Factors Influencing Protein Partitioning

The partitioning of the protein is affected by many system parameters. These variables are often impossible to separate and analyze individually, making *a priori* prediction of partition coefficients impractical. Several of these controlling factors are listed in Tab. 2. Very general effects result from the size and charge of the protein extracted, specific salt types in solution, and the polymer molecular weight and concentration; quite specific interactions can be exploited in affinity partitioning processes using polymer-attached ligands. Genetic engineering has been used to increase protein partitioning in both derivatized and "native" PEG–salt systems, pointing out the ability to influence partitioning strongly with only a few amino acid changes or additions.

The size of the protein has been found to have a general influence on protein partitioning for many proteins (SASAKAWA and WALTER, 1972). As shown in Fig. 3, an increase in protein size decreases the partition coefficient, from a situation weakly favoring top-phase partitioning for small proteins to strong bottom-phase partitioning for the largest proteins. It is important to note that the charge properties of the proteins were isolated as best possible by partitioning them at solution pHs equal to their isoelectric points. This molecular weight correlation to partitioning is only a generalization, with several reports of anomalous behavior, pointing out the importance of more complex interactions with the phase-forming polymers and the chemical and physical environments of both phases.

Protein partitioning can be very sensitive to the salt type and concentration in two-phase aqueous polymer systems, as evidenced by the results shown in Fig. 4 (WALTER et al., 1972).

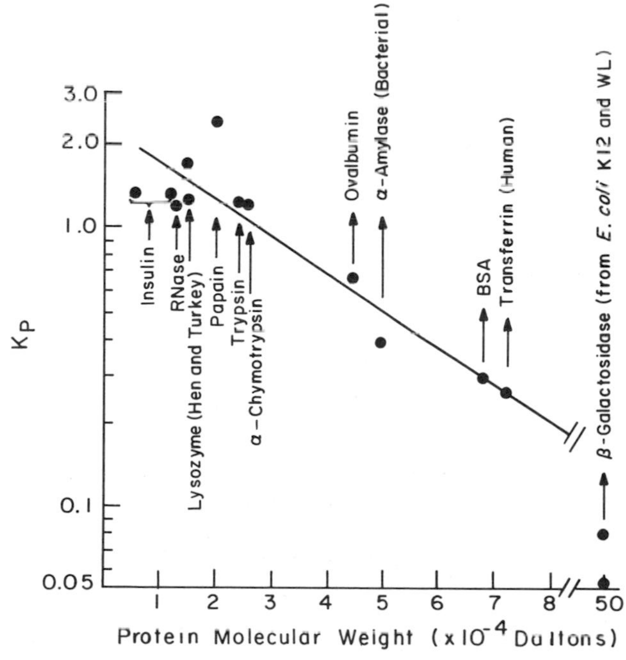

Fig. 3. Effect of protein molecular weight on partitioning in a PEG–dextran system at the protein *pI*-values. System: PEG 4000 (4.4%)/Dx 500 (7.0%) adjusted to 0.1 M NaCl or 0.05 M Na_2SO_4, 10 mM phosphate or glycine; 20 °C (from SASAKAWA and WALTER, 1972).

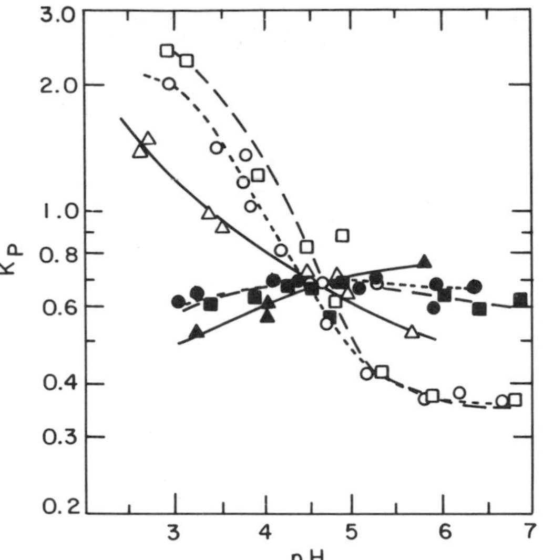

Fig. 4. Dependence of ovalbumin partition coefficient K_p on solution pH and salt type. Hollow symbols denote chloride salts, filled symbols sulfates (squares, potassium; circles, sodium; triangles, lithium). Note that all curves intersect at the isoelectric point of the protein (from WALTER et al., 1972).

The partition coefficient depends both on whether chloride or sulfate salts are used, and on the particular cation in solution. The strong pH dependence of the partitioning behavior illustrates that this is an electrostatic effect; this argument is supported by the fact that the common intersection point of the curves corresponds closely to the isoelectric pH, or point of zero net charge, of the protein (ALBERTSSON et al., 1970). This effect has been interpreted in terms of an apparent electrical potential difference between the two coexisting phases which will influence the partitioning of charged proteins (JOHANSSON, 1974; BROOKS et al., 1984). This potential difference, which is caused by the differing chemical affinities of the different ions for the two polymer phases, coupled with the requirements for electrical neutrality in each phase, can be of the order of 2–10 mV.

Polymer molecular weight can affect phase compositions and phase diagrams. It is not surprising, then, that both molecular weight and concentration influence protein partitioning. Increasing molecular weight of a polymer tends to decrease the protein partitioning to that phase. In one study, the $\log K$ of the proteins partitioned was inversely proportional to the PEG molecular weight for a PEG–dextran system (ALBERTSSON et al., 1987). In other studies, an increasing weight fraction of PEG increased the preference of proteins for the bottom phase (BASKIR et al., 1989). Differences in polymer molecular weight and concentration should always be interpreted in relation to the phase diagram and the resulting tie-line length, a measure of the changes in the polymer and salt concentrations in the two phases. Correlations of the partition coefficient with tie-line length indicate that it becomes further removed from unity as the tie-line length increases (CARLSON, 1988).

Increased selectivity for the enzyme of interest can be achieved by the coupling of a biospecific ligand to one of the polymeric species (Fig. 5), usually PEG. The strong partitioning of the PEG set up by the two-phase equilibrium results in most of the ligand reporting to the top phase. With this additional mechanism for interaction between the targeted protein and one of the polymer phases, the enzyme transfer to the top phase may be enhanced under conditions favoring the retention of contaminating species in the bottom phase, thereby increasing the separation efficiency. Dramatic increases in the partition coefficient relative to the control study of partitioning in the absence of ligand have been obtained by this addition of a biospecific ligand to the top-phase polymer. The convention is to report increases in $\Delta \log K$, the difference between $\log K$ values obtained with and without the ligand. Conditions are chosen to minimize the partitioning of contaminating proteins to the upper phase, with the best results obtained when the partition coefficient can be kept below 0.1. Affinity partitioning has been modelled by several groups, with good agreement with experimental data (FLANAGAN and BARONDES, 1975; BROOKS et al., 1985).

Many types of affinity ligands have been employed for directed purifications (KOPPERSCHLAGER and BIRKENMEIER, 1990). The most common strategy is to attach the ligand to the PEG polymer via derivatization of the free hydroxyls at the termini of the polymer chain. Chemistries have included initial con-

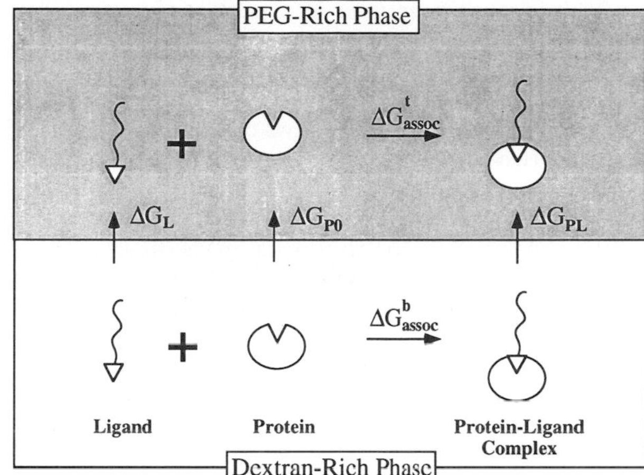

Fig. 5. Affinity partitioning of proteins in PEG–dextran systems, indicating the various free energies of transfer between the phases, and of association between the ligand and the protein (adapted from FLANAGAN and BARONDES, 1975).

version of the hydroxyl to halides, sulfonate esters, or epoxide derivatives (HARRIS and YALPANI, 1985). These reactive intermediates may then couple to the ligand to form a polymer capped with two protein binding sites. The derivatization of dextran is less common, as a result of less efficient chemistry and the general preference of many contaminating species to partition to the dextran-rich phase, reducing the efficacy of directed partitioning to this phase. Ligands from the reactive dye families including Cibacron blue, Procion red and Procion yellow have been used for group-specific isolations of many intracellular enzymes. Typical $\Delta \log K$ values are between 2 and 3, as shown in Fig. 6 (JOHANSSON and ANDERSSON, 1984), reflecting a strong interaction between protein and ligand, although differences were noted for the ligand concentration needed for half-maximal increases in protein partitioning (KIRCHBERGER et al., 1989). These differences in binding strength in the polymer phase were mirrored by measured dissociation constants with protein in free solution, suggesting that the protein–ligand interaction is similar in both environments.

After partitioning to the top phase and removal of the contaminated bottom phase, the proteins must be dissociated from the ligand-polymer to yield a homogeneous solution. Ul-

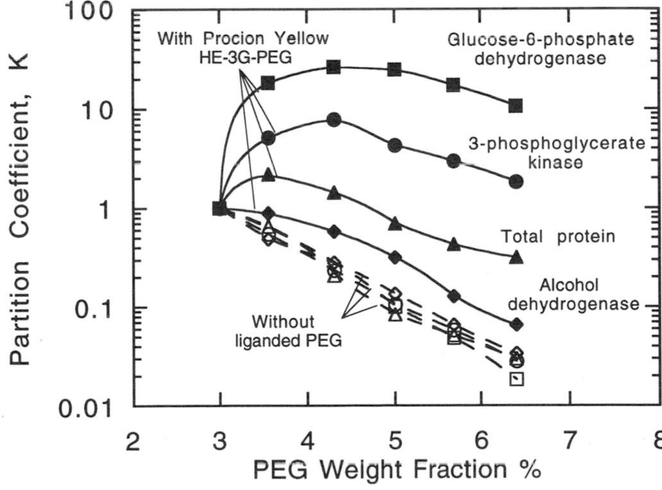

Fig. 6. Affinity partitioning of enzymes in PEG–dextran systems, with and without Procion yellow HE-3G-PEG. System: PEG 6000/Dx 500, 25 mM sodium phosphate buffer, pH 7.0, and 10% yeast extract (from JOHANSSON and ANDERSON, 1984).

trafiltration has been used in a pilot study to separate the dissociated protein from the polymer-ligand, permitting recycle of the polymer-ligand (CORDES and KULA, 1986). In another approach, soluble effectors can be added to a new two-phase system to compete with the bound ligand for the protein's binding site, thus reversing the partition coefficient and shifting the product to the bottom phase (KOPPERSCHLAGER and BIRKENMEIER, 1990).

One processing scheme employs PEG as a "shuttle" polymer which phase separates with both dextran and salt in sequential partitioning steps (Fig. 7). In an initial extraction, contaminating proteins are collected in the top PEG phase. The dextran phase, which now contains the product, is contacted with a second PEG phase, this time one containing ligand; the product is recovered in the top phase, and is then washed. If the contaminating proteins are more likely to partition to the dextran solution, then the first step should be bypassed, and the extract contacted directly with the ligand-bearing PEG phase. The loaded PEG solution is separated from the dextran-rich bottom phase, and salt is added resulting in a PEG–salt biphasic system. If the high salt concentration is sufficient to dissociate the product from the ligand-polymer, the product can be recovered in the second bottom phase. When increased ionic strength does not cause dissociation, inhibitors can be added to cause protein dissociation and recovery in the bottom phase.

It should be emphasized that affinity-based polymer extractions such as these are unlikely to provide the resolution and purification shown by solid-phase affinity adsorption, followed by column chromatography. The use of the attached ligand serves merely to enhance the selectivity of the process, and not to transform this front-end unit operation for protein purification into a high resolution step. An example of the use of affinity partitioning in just this scenario is given for the purification of lactate dehydrogenase from muscle (TJERNELD et al., 1987). Up to 5 kg of muscle per hour was processed using a Procion yellow derivatized PEG, with 68% yield and relative purification factor of 10. The importance of this extraction as a front-end purification arises from the rapid processing without stream pretreatment, reducing proteolytic losses. Also, as the separation is a rapid process, the need to operate at reduced temperatures to minimize protein degradation losses, as re-

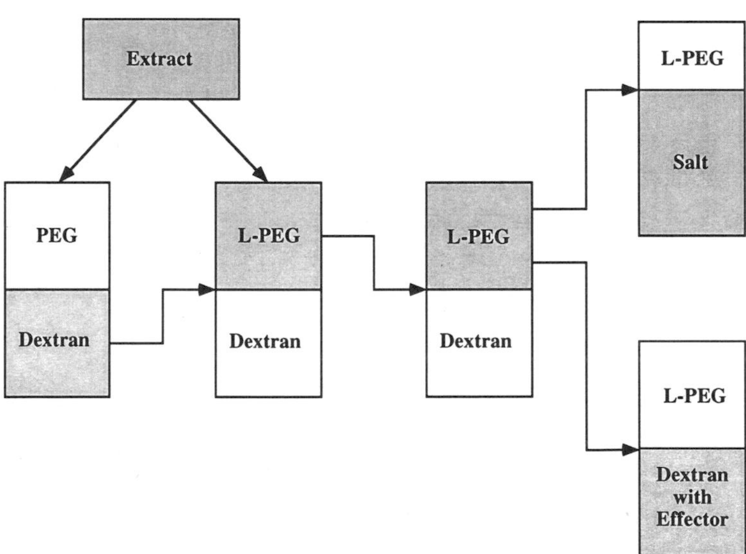

Fig. 7. Affinity partitioning using PEG as a shuttle between two biphasic systems. See text for details. The product-containing phase is the shaded one in each step of the process (KOPPERSCHLAGER and BIRKENmeier, 1990).

quired in many conventional processes, is minimized.

Just as affinity partitioning uses modified polymers to extract proteins, modified proteins can be designed and produced by genetic engineering to make them more amenable to extraction into the top phase. In one example, the strong partitioning of *beta*-galactosidase into the upper phases of PEG–salt systems was exploited. Fusion of this protein to targeted peptides resulted in a five-fold increase in the peptide partitioning coefficient (KOHLER et al., 1991b). Partitioning studies of free amino acids and dipeptides showed tryptophan partitioning strongly to the top phase, and it was hypothesized that proteins engineered to express a tryptophan fusion may have improved partitioning. The fusion was a tetrapeptide repeat of Ala-Trp-Trp-Pro, with the alanine residue designed as a spacer and the proline used to increase the fusion tail accessibility by virtue of its constrained conformational geometry (KOHLER et al., 1991a). A single tetrapeptide fusion increased the partition coefficient in a PEG–salt system 8-fold, while addition of three repeating tetrapeptides resulted in a 16-fold increase. Fig. 8 summarizes the partitioning changes, both as a function of the number of fusions and the PEG concentration difference between phases. The strong dependence of the partition coefficient of the fusion proteins on the PEG concentration confirms the influence of the tryptophan residues in increasing protein partitioning. Such a protein fusion eliminates the need to derivatize polymers, and allows the use of inexpensive biphasic systems based on PEG and salt.

Another general form of affinity partitioning uses immobilized metal ions on the polymers to extract proteins capable of forming charge-transfer complexes with the metal ion. Immobilized metal ion chromatography has been used for many years for purification of proteins, and the principles of protein–metal

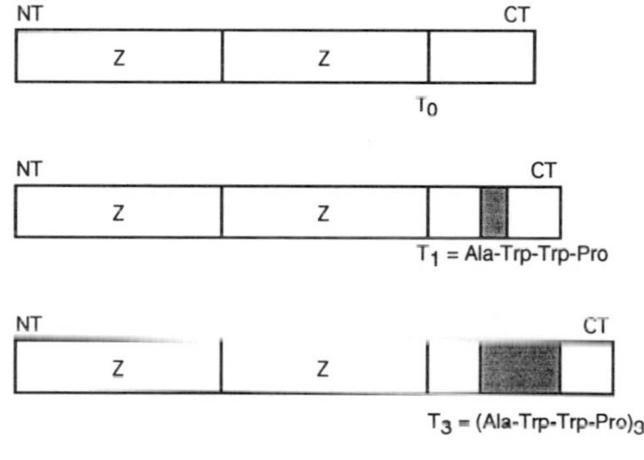

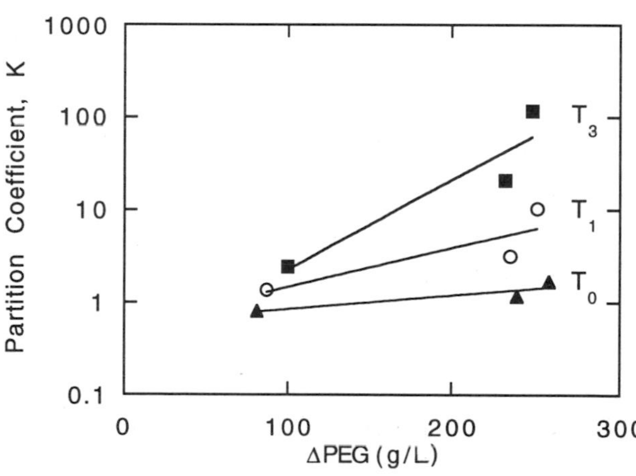

Fig. 8. Effect of tryptophan-rich fusions on protein partitioning in PEG–salt systems as a function of PEG concentration difference between phases, for three different fusion lengths (HUDDLESTON et al., 1991).

ion interactions have been well-studied (SULKOWSKI, 1985; PORATH, 1988; ARNOLD, 1991). Fusion proteins have also been designed to interact strongly with the immobilized metal ion (SMITH et al., 1988). The use of immobilized metal ions attached to polymers is a natural extension of the principles established for chromatographic operation, because of the known stability of the protein–metal ion interaction at high salt concentration.

In this technique, the metal ion is attached to PEG through coordination with a chelating agent such as iminodiacetic acid (IDA) present on the end of the PEG polymer. Two different synthetic routes have been used to produce IDA–PEG (CHUNG and ARNOLD, 1991; BIRKENMEIER et al., 1991). By including copper or ferric ions in the extraction buffers, up to 40-fold increases in the protein partition coefficient have been achieved. Fig. 9 shows the increase in partitioning of various hemoglobins as the number of accessible histidines in the protein increases. The pH dependence is a result of the protonation of the imidazole ring at low pH, preventing complexation with the copper–IDA ligand. This reversal of partitioning through pH shifts would provide an excellent means of dissociating the protein-polymer complex. Other metals can also be used for the extraction of proteins. For instance, the partitioning of a phosphoprotein was strongly influenced by the presence of Fe–IDA in the extracting phase (WUENSCHELL et al., 1990).

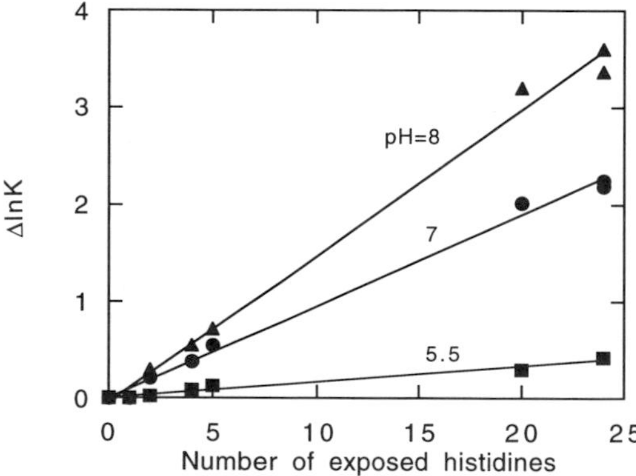

Fig. 9. Dependence of protein partitioning in IDA–PEG systems on pH and accessible histidine residues (WUENSCHELL et al., 1990).

The influence of surface histidines has been demonstrated clearly for both chromatographic and extractive separations; genetic engineering and site-directed mutagenesis have resulted in proteins expressing two histidine residues within an alpha-helix that have remarkably strong interactions with bound copper (TODD et al., 1991).

2.3 Applications

Applications of two-phase aqueous polymer systems for protein purification must use one or more of the above mechanisms for protein partitioning, while also addressing problems arising from the increased scale of operation, complexities of poorly-defined feed streams, or build-up of contaminants during recycle of phase-forming components. Economic analyses of liquid–liquid extraction by two-phase polymer systems indicate successful competition with existing technology, if these problems can be solved (KRONER et al., 1984; HUSTEDT et al., 1985). These issues will be addressed by reviewing examples of industrially-relevant studies.

An early example of large-scale partitioning involved the recovery of pullulanase and glucan phosphorylase from *Klebsiella pneumoniae* using a PEG–dextran system (HUSTEDT et al., 1978). Over five kg of cells were lysed and added to PEG and dextran solutions to yield a final volume of 19 liters. The upper phase was just over 13 liters and contained 91% of the pullulanase activity. Excellent purity and recovery were demonstrated, and traditional processing operations like ultrafiltration and ion-exchange adsorption were integrated into the separation.

In a recent pilot plant study, superoxide dismutase was extracted from 24 kg of bovine liver using PEG–phosphate (BOLAND et al., 1991). High recovery was demonstrated (83%) with a four-fold increase in the purification factor, and the clarified solution was suitable for further processing by chromatographic methods. Computer process control was used to set flow rates to two pilot-scale disk stack separators. Mixing of the initial two-phase system was conducted in a stirred tank, while the second two-phase system was mixed with static in-line mixers. Biomass concentration was maintained at 20%, higher than the level of PEG (15%) or salt (8%). Flow rates of up to 100 L/h were achieved in the first centrifuge, with only a 6% loss of the enzyme in the bottom phase. This demonstration stresses the ability of two-phase aqueous polymer systems to process large volumes of material with conventional pilot-plant equipment.

Two-phase systems may be formed through the addition of phase-forming polymers and salts directly to the bioreactor used to cultivate the cells. In a recent study by KOHLER (1991) (see also HUDDLESTON et al., 1991), the protein of interest was secreted by an *Escherichia coli* culture, and PEG and phosphate salts added after the fermentation was complete. The product partitioned to the top phase while over 99.9% of the cells remained in the bottom phase or collected at the interface. This replaced a centrifugation step and a microfiltration step in the conventional process; a 90% recovery was achieved, while a threefold increase in concentration factor resulted in a substantial reduction in the liquid handling volume. The time necessary to allow phase separation under normal gravity was not addressed, however, as the phase separation was conducted with batch centrifugation.

The use of expensive polymers results in the need to recover them for reuse. Ultrafiltration has been employed to separate PEG and salt, with the PEG recycled back for further extractions (HUSTEDT, 1986). Ninety five percent of the PEG was returned to the first extracting phase, and no negative effects on yield and purity were seen. The loss of salt was rather high, however, and would result in increased costs through the need both for replacement and for effluent processing.

Recycling of salts will also improve process economics for those systems which rely on high concentrations of phosphates or sulfates for phase separation. Operating costs and environmental impact could be minimized by effective recycle of the phase-forming salts as well as the polymers. Countercurrent extraction of the bottom phase containing phosphate by contacting with aliphatic alcohols resulted in 95% recovery of phosphate (GREVE and KULA, 1991a, b). A flowsheet suggesting one possible recycling scheme is given in Fig. 10, with both PEG and salt reuse. Economic analysis showed this to be a feasible solution, with

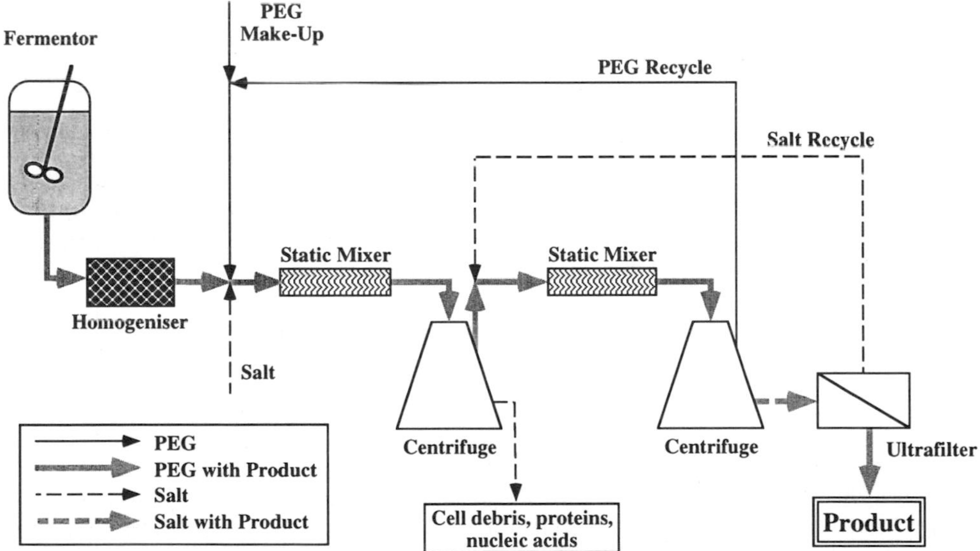

Fig. 10. Process scheme for protein extraction in aqueous two-phase systems for the downstream processing of intracellular proteins, incorporating PEG and salt recycling (adapted from GREVE and KULA, 1991b).

the cost of recovery only slightly higher than the purchase price of the phosphate, and certainly lower after disposal costs are accounted for (GREVE and KULA, 1991a, b).

Problems with polymer recycle can be addressed through utilization of derivatized Sepharose particles, which partition into the PEG-rich upper phase (KU et al., 1989). The large size of the beads allows rapid settling and removal from the polymer phase. A great diversity of ligands are available as commercial affinity chromatographic resins, suitable for immediate application. Purification of pyruvate kinase from a recombinant yeast fermentation resulted in a yield of up to 95%, with a two-fold increase in specific activity (the recombinant yeast strain overproduced pyruvate kinase to 15–20% of total intracellular protein, explaining the small increase in specific activity). Other traditional liquid chromatographic resins were also tested for their compatibility with these phase-forming systems; one third partitioned into the PEG-rich top phase.

The integration of fermentation and separation allows production of organic solvents or acids without inhibition of growth by the build-up of toxic components. These compounds can be partitioned from the broth with suitable organic solvents, or, if volatile, removed by a stripping gas. Proteins, however, cannot be extracted by these methods. Fortunately, the two-phase aqueous polymer systems are favorable both for enzyme stabilization and bacterial fermentation (KAUL and MATTIASSON, 1991). The fermentation is carried out in the phase separated polymer mixture, and both macromolecules and cells partition between the phases, as shown in Fig. 11.

Alpha-amylase production by *Bacillus subtilis,* with simultaneous partitioning of the product between the phases of a two-phase aqueous polymer system as it was produced, has been reported (ANDERSSON et al., 1985). Batch fermentations were conducted with PEG and dextran present; the cells partitioned completely to the bottom phase, while 73% of the enzyme was present in the top phase, resulting in a two-fold increase in culture productivity. Withdrawal of the enzyme-rich top phase and replacement with fresh solution allowed continual removal of the product as it was

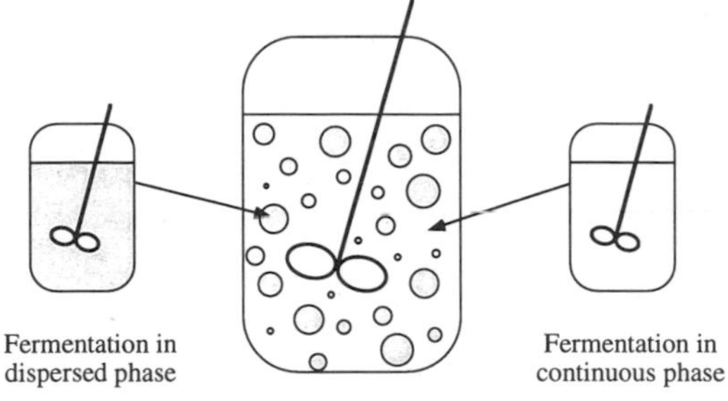

Fig. 11. Integration of fermentation and separation achieved by *in situ* extraction of enzyme as it is produced (ALAM et al., 1989).

formed, and replenishment of nutrients to the stationary culture. Advantages of the two-phase cultivation process over immobilized cell systems include the reduction in diffusional transport resistance, the ability to use particulate-containing media, and a higher *alpha*-amylase concentration in the product stream.

The processing of the product phase is usually carried out in a semi-continuous manner, as demonstrated by investigations with cellulase produced by *Trichoderma reesei* (PERSSON et al., 1991). Alkaline protease produced by *Bacillus licheniformis* has been extracted with PEG and dextran phase-forming systems (LEE and CHANG, 1990). The protease was found to partition almost equally between the two phases, yet a 30% increase in total activity was achieved by this extraction when semi-continuous changing of the top phase was performed. Other workers have investigated molecular weight effects for amylase purification, and have documented enzyme instability and deactivation with high levels of phosphate (KIM and YOO, 1991).

Cells in such systems often partition to the bottom phase. When the product enzyme prefers the top phase, this is an effective means of separating biomass and product, as well as integrating production and extraction. If the enzyme remains in the bottom dextran phase, however, it may be possible to induce cells to grow in the PEG-rich upper phase. This has been shown to work with *Bacillus amyloliquefaciens* cells by suitable adjustment of the polymer compositions (ALAM et al., 1989).

2.4 Equipment for Processing

Industrial separations with two-phase aqueous polymer systems require relatively simple processing equipment. The feed stream must be throroughly mixed with the phase-forming components, equilibration time must be allowed and the phases separated. Complications arise from high viscosities of cell homogenate streams, or the low interfacial tensions of these systems. Phase separation investigations have been performed with traditional phase separating equipment such as disk-stack centrifuges and simple gravity settlers. Hollow-fiber contactors can be used to obviate the need to disperse one phase in another and, therefore, the need for subsequent phase disengagement, while still providing the intimate contact between the phases required for respectable overall interphase mass transfer rates.

The biphasic interface in two-phase polymer systems is characterized by extremely low interfacial tensions. This will contribute to rapid mass transfer through the formation of many small droplets upon agitation, although it may lead to difficulties in phase disengagement. Phase equilibrium in agitated vessels has been examined for scales up to 160 liters, with a PEG–phosphate system (FAUQUEX et al., 1985). Less than two minutes were required for attainment of equilibrium at all scales for a modest power input from the stirrer. If the PEG was added as a solid instead of a solution, the time increased by one minute. These mass transfer studies confirm that the two-phase mixing can be scaled up on a power-per-unit volume basis to facilitate scale-up and continuous processing.

Traditional equipment has been used for conducting the phase separation after mixing, including disk-stack centrifuges (CORDES and KULA, 1986). They may be operated with both continuous or intermittent discharge, with the intermittent nozzle-type discharge recommended for operation with extremely viscous lower phases (as found for the processing of cell homogenate). Balancing of the outgoing streams to maintain the bulk phase interface within the disks is important to ensure phase separation and prevent redispersion of one phase in the other.

Some equipment employed and tested in handling two-phase aqueous polymer systems includes mixer–settler trains, extraction columns, Graesser contactors, and centrifugal separators (KULA et al., 1982). Rotating disk contactors have been used for bench-scale contacting of two-phase polymeric solutions to generate the data necessary for scale-up of these separators (KULA, 1985). The behavior of two-phase polymer systems in Alfa Laval gyrotesters and separators is also well understood (KULA et al., 1981).

Multi-stage, countercurrent extraction performed in column contactors will provide increase in recovery and product concentration over single-stage or batch systems. Spray columns may be suitable for two-phase aqueous polymer systems, and the wealth of data for equipment design and operation would speed application, if the problems of low interfacial tensions can be overcome and an analysis of drop formation can be performed (JOSHI et al., 1990).

Hollow fiber modules show promise as effective phase-contacting devices. In contrast to ultra- or microfiltration operations, where the purpose of the membrane is to provide the size-selective sieving usually associated with these processes, in extraction operations the purpose of the membrane is simply to stabilize the biphasic interface. The feed stream can be introduced to the shell side of the module, and the extractant fed countercurrently through the fiber lumens (or *vice versa*), and interfacial transport occurs within the membrane pores themselves. Mass transfer rates are reasonably high for such systems, comparable to those found in more traditional liquid–liquid extracton applications, as the very high surface to volume ratios achieved in hollow fiber modules (at least 50 cm^2/cm^3) combine with the modest mass transfer coefficient to provide excellent mass transfer (YANG and CUSSLER, 1988). An advantage of these contactors is that it is not necessary to disperse one phase as droplets in the other to obtain the required high mass transfer areas, and thus the complications associated with phase disengagement are avoided. The units are relatively easy to scale up, as correlations for the mass transfer coefficients as functions of shell side and lumen flow rates, fluid properties and protein partition coeffcients are readily available (DAHURON and CUSSLER, 1988).

2.5 Summary and Potentials

Aqueous two-phase polymer systems are the most advanced technology available for the liquid-liquid extraction of proteins. The processing is proven to work on crude preparations without pretreatment of the feed stream, and allows rapid processing with minimal material loss. Recoveries are consistently high, emphasizing the gentle processing conditions. Their industrial use is on the rise, resulting from continued publication of new research results, refinements of older techniques, and repeated demonstration of pilot-scale separations.

The processing cost is being reduced by the development of effective recycle schemes for both polymer and salts, as well as the use of less expensive polymers capable of phase-separating at lower concentrations. Affinity separations have demonstrated extremely high selectivity, with both group-specific dye ligands and low molecular weight enzyme inhibitors; several generic chemistries allow derivatization of the phase-forming polymers for coupling of specific ligands. Genetic engineering has emerged as a means of increasing partitioning, either by alteration of the histidine content and placement for interaction with metal chelating polymers, or by the addition of tryptophan residues which increase top phase partitioning in PEG–salt systems. The operation of appropriate equipment for mixing and phase separation is well understood, and innovative operations such as the use of hollow-fiber contactors may allow the whole extraction-phase separation operation to take place in a single unit.

3 Reversed Micellar Extractions

3.1 Basis of Phase Separation

Reversed micelles are nanometer-scale water droplets stabilized within an organic solvent by a surrounding monolayer of surfactant molecules, and can be formed by contacting an aqueous phase with an immiscible organic phase containing these surfactants. The amount of water taken up by the organic phase can be quite large; it is not uncommon that over 20% of the organic phase volume is water-solubilized in reversed micelles (KITAHARA, 1980); these solutions are more properly termed water-in-oil microemulsions. The water transfer is controlled by the ionic strength of the aqueous phase and the concentration of surfactant in the organic phase (KITAHARA and KON-NO, 1966). These effects are well-characterized; the ionic strength primarily controls the size of the individual reversed micelles, while the surfactant concentration determines their number.

The polar environment of the interior water pool of the reversed micelle allows the uptake by the organic solvent phase of various hydrophilic species originally found in the aqueous phase, including proteins, as shown in Fig. 12 (LUISI and MAGID, 1986; LUISI et al., 1988; LEODIDIS and HATTON, 1989b). Spectroscopic examination of the proteins hosted within these reversed micelles indicates no gross differences in their UV and CD spectra relative to their spectra in their natural aqueous environment (LUISI et al., 1979; BRUNO et al., 1990; MARZOLA, et al., 1991). Many retain enzymatic activity both within reversed micelles and after transfer back into an aqueous environment. The shell of water around the protein generally protects it from denaturation in the organic phase, although in some cases the protein may be associated with the surfactant interface.

The surfactant systems employed as reversed micellar media fall into two categories: those that require a cosurfactant, and those that do not. The commonly used AOT (sodium bis-(2-ethylhexyl)-sulfosuccinate) and isooctane system requires no cosurfactant, and displays phase stability over a wide range of surfactant and water concentrations. As an example of the former category, the surfactant CTAB (cetyltrimethylammonium bromide) does not usually form reversed micelles in linear or branched alkanes unless a medium-length alcohol such as hexanol is added to the solution. There are disadvantages in the use of such cosurfactant-requiring systems for sepa-

Tab. 3. Surfactant/Solvent Systems Capable of Forming Reversed Micelles

Surfactant	Solvent
Aerosol OT (AOT)	Isooctane or linear alkanes
Cetyltrimethylammonium bromide (CTAB)	Linear alkanes with alcohols as cosurfactants
Phospholipids	Benzene, alcohols, or ethers
Dioleyl phosphoric acid	Isooctane
Ethoxylated nonylphenols	Cyclohexane

rations, as the cosurfactant often partitions between the organic and aqueous phases, requiring constant addition of cosurfactant, if a continuous processing format is used. Tab. 3 lists several surfactant/organic solvent systems used for reversed micellar investigations on protein extractions and enzymatic catalysis.

3.2 Factors Influencing Protein Partitioning

The success of the extraction will depend on the driving forces which cause proteins to solubilize within a reversed micellar phase, and on the product's properties relative to those of the contaminating species. These forces have been determined from investigations of several reversed micellar extractions (GOKLEN and HATTON, 1985, 1987; DEKKER et al., 1986; FLETCHER and PARROTT, 1988; LESER and LUISI, 1990; JOLIVART et al., 1990). Almost all surfactants used for reversed micellar extractions are ionic in character. The electrostatic interactions between these groups and polar solutes influence the transfer of certain species initially in the aqueous phase to the reversed micelles within the organic continuum. For example, a protein below its *pI* value has a net positive charge. An anionic surfactant would have an attraction for the protein, and the protein would be likely to leave the aqueous phase and enter the reversed micellar phase (GOKLEN and HATTON, 1987). The opposite is true for quaternary ammonium surfactants with a positively charged head group (JOLIVART et al., 1990). The transfer of protein is not simply controlled by the pH of the solution, however. It has also been shown that

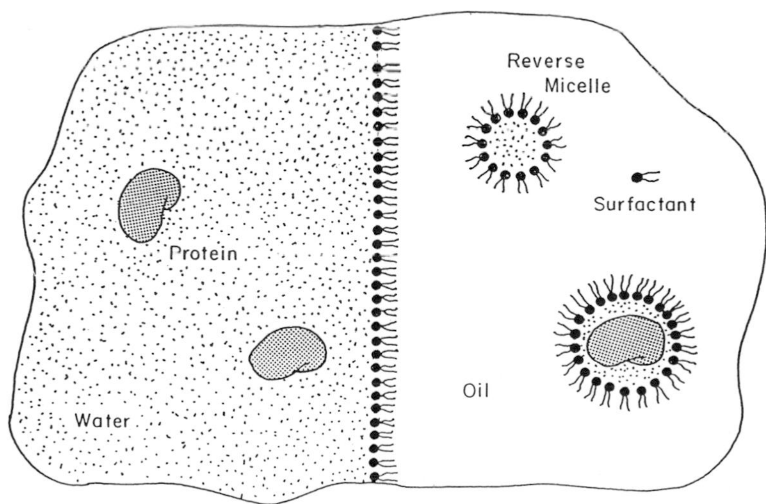

Fig. 12. Partitioning of proteins between an aqueous phase and a reversed micellar solution can be used for protein recovery by liquid–liquid extraction (from GOKLEN and HATTON, 1987).

the addition of a small amount of non-ionic surfactant to a reversed micellar phase consisting of charged surfactants can increase the capacity of the system for transferred protein (DEKKER et al., 1986).

The salt type as well as the concentration will determine the size of the reversed micelles formed, and directly influence the partitioning of proteins. For AOT reversed micelles, the cationic species of the salt used as the electrolyte or buffering species controls the water transfer, with sodium and ammonium supporting larger reversed micelles than potassium or cesium (LEODIDIS and HATTON, 1989a, b). As a result of these larger reversed micelles, protein extraction in AOT–isooctane reversed micelles is favored when sodium is the only salt present in the extracting buffer. Specific salt effects have been shown to arise at the lowest pHs investigated, where surface charges are numerous and the electrostatic interaction with the surfactant is strongest (KELLEY et al., 1992; MARCOZZI et al., 1991). Surfactant concentration was also shown to influence protein transfer, but for AOT–isooctane reversed micelles, this effect could be normalized by plotting protein transfer as a function of the water transfer to the reversed micellar phase. This resulted in a single universal curve which was independent of the surfactant concentration, as shown in Fig. 13. The position of this curve and final equilibrium protein transfer at the highest levels of water transfer are functions of the pH of the solution.

With current surfactant systems, very large proteins (molecular weights >100 kD) cannot be solubilized within reversed micelles under phase transfer conditions, where there is an aqueous phase in equilibrium with the reversed micellar phase. Experiments employing these large proteins in enzymatic catalysis invariably use the injection method to force solubilization without a supporting aqueous phase. This upper limit on the size of solubilized protein can perhaps be explained from correlations of protein transfer and molecular weight. In one study, it was found that the pH of maximum solubilization could be correlated with molecular weight for 15 proteins transferred into TOMAC (trioethylmethylammonium chloride) reversed micelles, as shown in Fig. 14 (WOLBERT et al., 1989). The results show that for the larger proteins, a greater deviation from the isoelectric point is required to effect an uptake into the reversed micelle. In most cases, the transfer measured was above 50%, unlike the studies using AOT where 100% transfer was achieved for nearly all proteins. The AOT data, however, were slightly more poorly correlated with molecular weight. One interesting analysis was made of the effect of surface charge asymmetry, as determined by the three-dimensional structures of 12 different proteins. The most asymmetrical proteins showed the highest protein transfer, while those having nearly complete symmetry were only weakly transferred. These studies were conducted at low surfactant concentrations, where protein

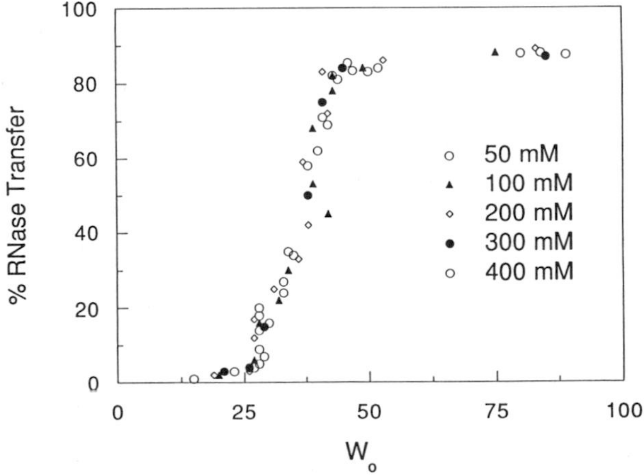

Fig. 13. Effect of surfactant concentration on partitioning in AOT/isooctane reversed micelles (from KELLY et al., 1992).

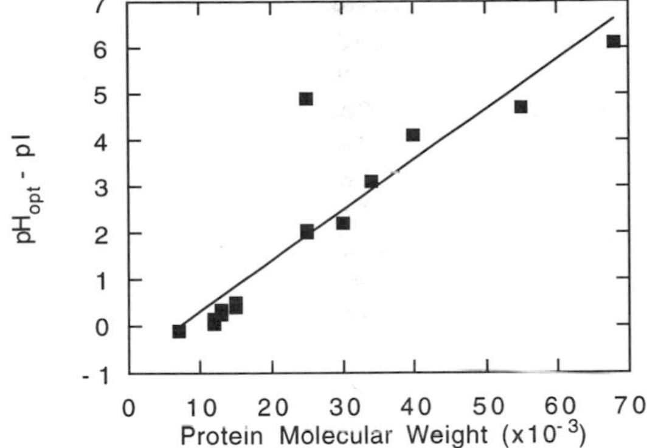

Fig. 14. The pH at which maximal solubilization occurs relative to the protein isoelectric point correlates well with the protein molecular weight. System: TOMAC reversed micelles with Rewopal HV5 and octanol as cosurfactants (from WOLBERT et al., 1989).

transfer is usually not complete, and illuminated the fundamentals of the interactions responsible for transferring protein into the reversed micelles; optimized conditions will employ higher surfactant concentrations resulting in better protein transfer.

Affinity cosurfactants have also been used to increase the selectivity of protein transfer. As shown in Fig. 15, affinity cosurfactants are amphiphilic molecules with hydrophilic ligands as head groups, which the protein is able to recognize and bind. Attached to the ligand is a hydrophobic tail, which imparts surface activity to the protein-ligand complex. This acts as an additional driving force for protein solubilization, in addition to the electrostatic interactions described above. In the first reported use of affinity cosurfactants, the molecule *beta*-D-glucopyranoside was added to the normal AOT–isooctane reversed micellar phase (WOLL et al., 1989). This molecule has a hydrophobic *N*-alkyl tail which interacts with the AOT tails in the reversed micelle, while the glucopyranosyl head group most probably resides in the water pool. Concanavalin A was transferred with much greater selectivity over ribonuclease under identical conditions of pH and ionic strength, presumably because of the biospecific interaction between the protein and the cosurfactant molecule. Control experiments emphasized the biospecificity of the extraction; they included the addition of soluble, competing ligands which reduced concanavalin A transfer by lowering the number of affinity cosurfactant–protein complexes.

Recent results show this technique to apply to more general ligand-protein pairs. The novel surfactant *N*-laurylbiotinamide was synthesized through *N*-hydroxysuccinimide condensation of biotin and laurylamide, and used to extract avidin (COUGHLIN and BACLASKI, 1990). The highest increases in transfer were reported at pHs above 9.5, where very few contaminant proteins are expected to be transferred in the AOT–isooctane reversed micelles

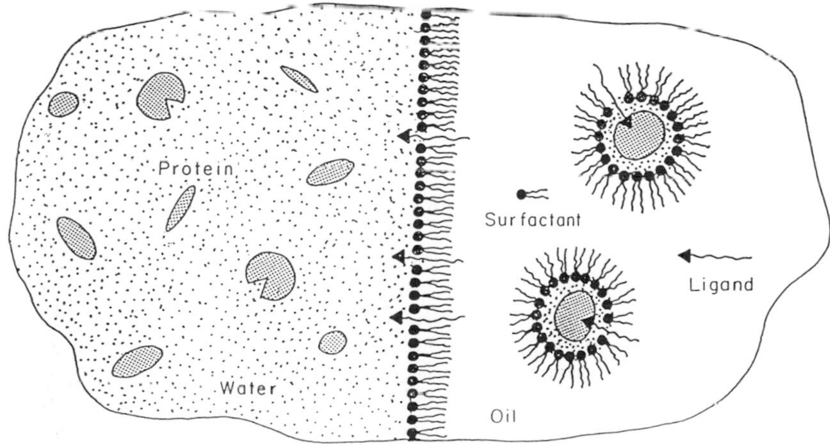

Fig. 15. Principle of affinity-based reversed micellar protein extraction (from WOLL et al., 1990).

used. This demonstrates clearly the ability of affinity cosurfactants to increase the operating range of an extraction, and should result in greatly increased specific activities after back extraction. The work also demonstrates an application of well-established coupling chemistries used for synthesis of affinity chromatographic resins as one means of producing affinity cosurfactants.

Some affinity interactions between two proteins can facilitate protein transfer without the use of amphiphilic cosurfactants. The interaction between concanavalin A and horse radish peroxidase allowed uptake of the peroxidase under conditions precluding transfer without the lectin (PARADKAR and DORDICK, 1991). Interpretation of electron spin resonance spectra suggested that the very large protein complex may be present in the surfactant interface due to some motional restriction of the attached spin label; however, it is not clear what motional restrictions would be present for proteins known to reside in the reversed micellar water pool.

3.3 Applications

The feasibility of using reversed micelles to extract industrially relevant proteins is being tested by several groups. The proteins are both intracellular and extracellular products, usually from bacterial sources. Each of these works also examines the problems associated with back-transfer of the product, a process which is known to occur more slowly than forward transfer (DEKKER et al., 1990; DUNGAN et al., 1991).

The recovery of an extracellular alkaline protease from a *Bacillus* fermentation broth demonstrated the robustness of the AOT–isooctane system, despite the unknown variety of contaminants present from the growth on an undefined, complex medium (RAHAMAN et al., 1989). The pH optimum for protease transfer was between 5.4–6.0.Increased pH resulted in adequate back-transfer. There were some problems with protein denaturation at low pH values (≤ 5.2), and some instability in protease activity during extraction at higher pHs. Increases in specific activity of between 2.2 and 8 were achieved, limited perhaps by the broad spectrum of contaminating proteins with different isoelectric points.

Continuous flow extraction of extracellular *alpha*-amylase produced by *Bacillus amyloliquefaciens* into reversed micelles and subsequent stripping by a second aqueous phase in a mixer–settler train was used by DEKKER et al. (1986) to concentrate the protein about eightfold with 45% enzyme recovery. The feed solution contained purified enzyme in a buffer solution, and had few complex components present from the fermentation broth. The reversed micellar phase consisted of 0.4% TOMAC plus 0.1% octanol (a cosurfactant) in isooctane. Back-transfer was achieved by contacting with a high salt low pH buffer. For this system, the partition coefficient of the amylase between the aqueous feed solution and the reversed micellar extracting phase was approximately 10. The extraction suffered from slow loss of surfactant and the low yield for the process. This latter problem may have been due to a denaturation of the protein by the cationic surfactant, and manifested itself as a precipitate formation at the phase interface. Later improvements in the process included the addition of a non-ionic surfactant (Rewopal HV5, nonylphenolpentaethoxylate), and resulted in an increase in the partition coefficient from 10 to 100, and better recovery of the amylase (75%) (DEKKER et al., 1989).

More recently, the same surfactant system was subjected to temperature increase up to 35 °C, which resulted in a partial dewatering of the organic phase (DEKKER et al., 1991). Good recovery of the amylase was achieved in the low-volume aqueous phase present after heating, with an amazing 2000-fold concentration increase. The stability of the amylase could be demonstrated for temperatures up to 45 °C, slightly lower than the measured stability in the absence of surfactants, suggesting some cooperativity to the denaturation. Two continuous flow centrifuges were used to process 60 L/h aqueous feed, and the reversed micellar phase was recycled during the extraction (see Fig. 16). It is not clear, however, that this temperature-induced back-extraction would act as a universal strategy with all reversed micellar systems.

Intracellular enzymes have been purified successfully with reversed micelles, using a

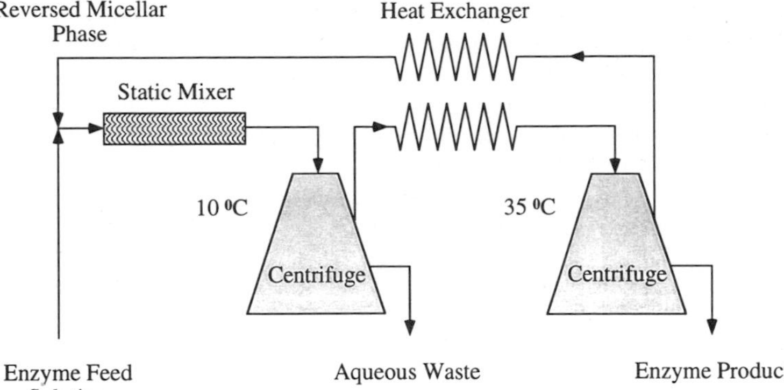

Fig. 16. Extraction of amylase using twin centrifuges and high temperature back-extraction (from DEKKER et al., 1991).

unique combination of lysis and extraction (GIOVENCO et al., 1987). Injection of cell paste of *Azetobacter vinelandii* cells into a reversed micellar solution of CTAB and hexanol in octane resulted in rapid cell lysis, in accordance with the known ability of CTAB to disrupt cell membranes. Back-extraction was achieved with the addition of tertiary ammonium salts at high concentrations. Recovery of two different dehydrogenases showed some sensitivity to the extraction, as one was recovered with 100% activity, the other with only 30%. A third dehydrogenase could not be recovered in an active form at all. The recovered enzymes were purified six-fold by the reversed micellar treatment.

Lipases are becoming important enzymes in the processing of foods and production of fine chemicals, and reversed micelles are natural environments for the synthesis of certain monoglycerides because of their ability to support high concentrations of the reacting long-chain fatty acids without phase separation (HAYES and GULARI, 1991). The separation of lipases from crude *Chromobacterium viscosum* preparations has been achieved with a single reversed micellar extraction (CAMARINHA-VINCENTE et al., 1990). Two distinct lipases were present in the crude powder, but only lipase B could be extracted with AOT–isooctane reversed micelles. By manipulating the phase ratios, a two-fold concentration of the extracted lipase was achieved with an 85% recovery of activity (AIRES-BARROS and CABRAL, 1991). The back-extraction employed 2.5% ethanol to disrupt protein–surfactant interactions.

3.4 Equipment for Processing

In most cases the equipment used to perform reversed micellar extraction has been simple bench-scale beakers and test tubes, with few studies employing the liquid–liquid contactors required for large-scale applications. The notable exceptions used membrane separators as a means to immobilize the interface between the organic and aqueous phases, which served to stabilize the interfacial surface area and prevent foaming or emulsions. Cytochrome-c and *alpha*-chymotrypsin were extracted by the same hollow-fiber contactor used for the two-phase aqueous polymer systems described above (DAHURON and CUSSLER, 1988). Some protein was lost due to denaturation, but this could be avoided by removal of entrapped air in the apparatus; upon reduction of the air–liquid interface, denaturation was avoided. The surfactant used to form the reversed micellar phase also adsorbed to the hydrophobic membrane surface, decreasing the breakthrough transmembrane pressure. Precise liquid flow control allowed effective operation despite this additional constraint.

In another example, *alpha*-amylase was extracted using a polypropylene hollow-fiber membrane with 0.2 mm pores to contact reversed micelles formed by TOMAC and octanol in isooctane with the feed solution (DEK-

KER et al., 1987). Mass transfer analysis was unable to determine the amount of resistance offered by the membrane, as the pressure across the membrane was found to influence the rate of enzyme transfer. A low Reynolds number for fluid flow in the membrane suggested that the boundary layer resistance was likely to be the dominant resistance in the system.

3.5 Summary and Potentials

Reversed micelles can be used for the extractive recovery of proteins produced as intracellular or extracellular products, and have demonstrated robust response to adverse conditions including whole fermentation broths and concentrated cell homogenate. Increases in protein partitioning can be achieved by adding affinity cosurfactants to target the desired protein through biospecific recognition and binding. Back-extraction of the protein from the reversed micelle to a stabilizing buffer, although slow relative to protein uptake, is being addressed in several studies. Novel recovery techniques include the addition of cosolvents capable of disrupting surfactant interactions and causing protein expulsion to the aqueous phase (WOLL et al., 1989; CARLSON and NAGARAJAN, 1992). New topics being addressed include novel surfactants which respond to pH changes by decreasing the concentration of solubilized water (GOTO et al., 1990), 2-D NMR and amide exchange of protein within the reversed micelle to detect structural perturbations (VENKATARAMAN et al., 1991), and natural surfactant systems based on phospholipids (WALDE et al., 1990).

Problems still to be solved include occasional low recoveries after back-extraction, examination of continued recycle of the reversed micellar phase for all common surfactant systems, and the lack of knowledge about processing large volumes of aqueous feed streams by reversed micellar extraction. Models describing protein uptake are not universal, and fundamentals of protein–surfactant interactions need further study. In many ways, the problems addressed and solved by the two-phase aqueous polymer research community a decade ago must be tackled for reversed micellar extraction, to justify further investigation of reversed micelles as a tool for protein purification.

4 Other Extractions

Other biphasic aqueous systems have been examined for protein partitioning, and could thus be used for liquid–liquid extraction of enzymes.

Under certain conditions of surfactant concentration and temperature, members of the Triton surfactant family (polyoxyethylene octyl phenols) phase separate and come out of solution. This is a reflection of the formation of micellar aggregates present at the cloud point (BORDIER, 1981). Hydrophilic proteins are found exclusively in the aqueous phase, while membrane proteins, which are more hydrophobic, partition to the detergent-rich phase. By altering the length of the ethylene oxide units in the surfactant, the temperature at which phase separation occurs can be controlled. Further separations can be made on whole cell extracts, as the glycoprotein species present in the aqueous phase can be precipitated by exhaustive dialysis and removal of the surfactant (PRYDE and PHILLIPS, 1986). PRYDE (1986) summarized other aspects of this phase partitioning technique.

Aliphatic alcohols and certain salts have been shown to phase separate under certain conditions, and phase diagrams have been published recently (KULA, 1991). Whether these systems will be biocompatible with proteins remains to be seen, however, as there may be denaturation of the protein from high concentrations of the alcohol.

5 Overall Summary

Liquid–liquid extractions allow purification of protein from complex feed streams typical of fermentation broths or cell homogenates. Clarification, concentration, and purification can all be achieved with a single unit operation. Economic analysis suggests that some of these systems will compete favorably with established technologies, and the added advantages of readily scalable operation, rapid purification of labile enzymes, and high selectivities due to affinity interactions will result in further acceptance in conventional scale protein purification operations.

6 References

Abbott, N. L., Hatton, T. A. (1988), Liquid-liquid extraction for protein separations, *Chem. Eng. Prog.,* August, 31–41.

Aires-Barros, M. R., Cabral, J. M. S. (1991), Selective separation and purification of two lipases from *Chromobacterium viscosum* using AOT reversed micelles, *Biotechnol. Bioeng.* **38,** 1302–1307.

Alam, S., Weigand, W. A., Hong, J. (1989), On the mechanism of growth of cells in the mixed aqueous two-phase system, *Appl. Biochem. Biotechnol.* **20,** 421–436.

Albertsson, P.-A. (1986), *Partition of Cell Particles and Macromolecules,* 3rd Ed. New York: Wiley.

Albertsson, P.-A., Sasakawa, S., Walter, H. (1970), Cross-partition and isoelectric points of proteins, *Nature,* **228,** 1329–1330.

Albertsson, P.-A., Cajarville, A., Brooks, D. E., Tjerneld, F. (1987), Partition of proteins in aqueous polymer two-phase systems and the effect of molecular weight of the polymer, *Biochim. Biophys. Acta* **926,** 87–93.

Andersson, E., Hahn-Hägerdal, B. (1988), High concentrations of PEG as a possible uncoupler of the proton motive force: Amylase production with *Bacillus amyloliquefaciens* in aqueous two-phase systems and PEG solutions, *Appl. Microbiol. Biotechnol.* **29,** 329–336.

Andersson, E., Johansson, A.-C., Hahn-Hägerdal, B. (1985), Amylase production in aqueous two-phase systems with *Bacillus subtilis, Enzyme Microb. Technol.* **7,** 333–338.

Arnold, F. H. (1991), Metal-affinity separations: A new dimension in protein processing, *Bio/Technology* **9,** 151–156.

Baskir, J. N., Hatton, T. A., Suter, U. W. (1989), Thermodynamics of the partitioning of biomaterials in two-phase aqueous polymer systems: Comparison of lattice model to experimental data, *J. Phys. Chem.* **93,** 2111–2122.

Birkenmeier, G., Vijayalakshmi, M. A., Stigbrand, T., Kopperschlager, G. (1991), Immobilized metal ion affinity partitioning, a method combining metal-protein interactions and partitioning of proteins in aqueous two-phase systems, *J. Chromatogr.* **539,** 267–277.

Boland, J. G., Hesseling, P. G. M., Papamichael, N., Hustedt, H. (1991), Extractive purification of enzymes from animal tissue using aqueous two phase systems: Pilot scale studies *J. Biotechnol.* **19,** 19–34.

Bordier, C. (1981), Phase separation of integral membrane proteins in Triton X-114 solution, *J. Biol. Chem.* **256,** 1604–1607.

Brooks, D. E., Sharp, K. A., Bamberger, S., Tamblyn, C. H., Seaman, G. V. F., Walter, H. J. (1984), Electrostatic and electrokinetic potentials in two phase aqueous polymer systems, *J. Colloid Interface Sci.* **102,** 1–13.

Brooks, D. E., Sharp, K. A., Fisher, D. (1985), Theoretical aspects of partitioning, in: *Partitioning in Aqueous Two-phase Systems. Theory, Methods, Uses and Applications in Biotechnology* (Walter, H., Brooks, D. E., Fisher, D. Eds.), pp. 11–84, Orlando: Academic Press.

Bruno, P., Caselli, M., Luisi, P. L., Maestro, M., Traini, A. (1990), A simplified thermodynamic model for protein uptake into reverse micelles. Theoretical and experimental results, *J. Phys. Chem.* **94,** 5908–5917.

Camarinha-Vincente, M. L., Aires-Barros, M. R., Cabral, J. M. S. (1990), Purification of *Chromobacterium viscosum* lipases using reversed micelles, *Biotechnol. Tech.* **4,** 137–142.

Carlson, A. (1988), Factors influencing the use of aqueous two phase partition for protein purification, *Sep. Sci. Technol.* **23,** 785–817.

Carlson, A., Nagarajan, (1992), *Biotechnol. Prog.* **8,** 85–90.

Chung, B. H., Arnold, F. H. (1991), Metal-affinity partitioning of phosphoproteins in PEG/dextran two-phase systems. *Biotechnol. Lett.* **13,** 615–620.

Cordes, A., Kula, M.-R. (1986), Process design for large-scale purification of formate dehydrogenase from *Candida boidinii* by affinity partition, *J. Chromatogr.* **376,** 375–384.

Coughlin, R. W., Baclaski, J. B. (1990), *N*-Laurylbiotinamide as affinity cosurfactant, *Biotechnol. Prog.* **6,** 307–309.

Dahuron, L., Cussler, E. L. (1988), Protein extractions with hollow fibers, *AlChE J.* **32,** 130–136.

Dekker, M., van't Riet, K., Weijers, S. R. (1986), Enzyme recovery by liquid–liquid extraction using reversed micelles, *Chem. Eng. J.* **33B,** 27–33.

DEKKER, M., VAN'T RIET, K., WINJNANS, J. M. G. M., BALTUSSEN, J. W. A., BIJSTERBOSCH, B. H., LAANE, C. (1987), Membrane based liquid-liquid extraction of enzymes using reversed micelles, *Congr. on Membranes and Membrane Processes*, Tokyo.

DEKKER, M., VAN'T RIET, K., BIJSTERBOSCH, B. H., WOLBERT, R. B. G., HILHORST, R. (1989), Modeling and optimization of the reversed micellar extraction of amylase, *AIChE J.* **35**, 321-324.

DEKKER, M., VAN'T RIET, K., BIJSTERBOSCH, B. H., FIJNEMAN, P., HILHORST, R. (1990), Mass transfer rate of protein extraction with reversed micelles, *Chem. Eng. Sci.* **45**, 2949-2957.

DEKKER, M., VAN'T RIET, K., VAN DER POL, J. J., BALTUSSEN, J. W. A., HILHORST, R., BIJSTERBOSCH, B. H. (1991), Effect of temperature on the reversed micellar extraction of enzymes, *Chem. Eng. J.* **46**, B69-B74.

DUNGAN, S. R., BAUSCH, T., HATTON, T. A., PLUCINSKI, P., NITSCH, W. (1991), Interfacial transport processes in the reversed micellar extraction of proteins,*J. Colloid Interface Sci.* **145**, 33-50.

FAUQUEX, P.-F., HUSTEDT, H., KULA, M.-R. (1985), Phase equilibration in agitated vessels during extractive enzyme recovery, *J. Chem. Tech. Biotechnol.* **35B**, 51-59.

FLANAGAN, S. D., BARONDES, S. (1975), Affinity partitioning. A method for purification of proteins using specific polymer ligands in aqueous polymer systems, *J. Biol. Chem.* **250**, 1484-1489.

FLETCHER, P. D. I., PARROTT, D. (1988), The partitioning of proteins between water-in-oil microemulsions and conjugate aqueous phases, *J. Chem. Soc.Trans. 1* **84**, 1131-1144.

GIOVENCO, S., VERHEGGEN, F., LAANE, C. (1987), Purification of intracellular enzymes from whole bacterial cells using reversed micelles, *Enzyme Microb. Technol.* **9**, 470-473.

GOKLEN, K. E., HATTON, T. A. (1985), Protein extraction using reverse micelles, *Biotechnol. Prog.* **1**, 69-74.

GOKLEN, K. E., HATTON, T. A. (1987), Liquid-liquid extraction of low molecular weight proteins by selective solubilization in reversed micelles, *Sep. Sci. Technol.* **22**, 831-841.

GOTO, M., KONDO, K., NAKASHIO, F. (1990), Protein extraction by reversed micelles using dioleyl phosphoric acid, *J. Chem. Eng. Jpn.* **23**, 513-515.

GREVE, A., KULA, M.-R. (1991a), Cost structure and estimation for the recycling of salt in a protein extraction process, *Bioprocess Eng.* **6**, 173-177.

GREVE, A., KULA, M.-R. (1991b), Recycling of salts in partition protein extraction processes, *J. Chem. Tech. Biotechnol.* **50**, 27-42.

HARRIS, J. M., YALPANI, M. (1985), Polymer-ligands used in affinity partitioning and their synthesis; in: *Partition in Aqueous Two-Phase Systems. Theory, Methods, Uses, and Applications to Biotechnology* (WALTER, H., BROOKS, D. E., FISHER, D., Eds.), pp. 589-626, Orlando: Academic Press.

HAYES, D. G., GULARI, E. (1991), 1-Monoglyceride production from lipase-catalyzed esterification of glycerol and fatty acids in reverse micelles, *Biotechnol. Bioeng.* **38**, 507-517.

HUDDLESTON, J., VEIDE, A., KOHLER, K., FLANAGAN, J., ENFORS, S.-O., LYDDIAT, A. (1991), The molecular basis of partitioning in aqueous two phase systems, *TIBTECH* **9**, 381-388.

HUSTEDT, H. (1986), Extractive enzyme recovery with simple recycling of phase forming chemicals, *Biotechnol. Lett.* **8**, 791-796.

HUSTEDT, H., KRONER, K. H., STACH, W., KULA, M.-R. (1978), Procedure for the simultaneous large-scale isolation of pullulanase and 1,4-glucan phosphorylase from *Klebsiella pneumoniae* involving liquid-liquid separations, *Biotechnol. Bioeng.* **20**, 1989-2005.

HUSTEDT, H., KRONER, K. H., MENGE, U., KULA, M.-R. (1985), Protein recovery using two-phase systems, *Trends Biotechnol.* **3**, 139-143.

JOHANSSON, G. (1974), Partition of proteins and microorganisms in aqueous biphasic systems, *Mol. Cell. Biochem.* **4**, 169-180.

JOHANSSON, G., ANDERSSON, M. (1984), Parameters determining affinity partitioning of yeast enzymes using polymer-bound triazine dye ligands, *J. Chromatogr.* **303**, 39-51.

JOLIVALT, C., MINIER, M., RENON, H. (1990), Extraction of *alpha*-chymotrypsin using reversed micelles, *J. Colloid Interface Sci.* **135**, 85-96.

JOSHI, J. B., SAWANT, S. B., RAGHAVA RAO, K. S. M. S., PATIL, T. A., ROSTAMI, K. M., SIKDAR, S. K. (1990), Continuous counter-current two-phase aqueous extraction, *Bioseparation* **1**, 311-324.

KAUL, R., MATTIASSON, B. (1991), Extractive bioconversions in aqueous two-phase polymer systems, in: *Extractive Bioconversion* (MATTIASSON, B., HOLST, O., Eds.), pp. 173-188, New York: Marcel Dekker.

KELLEY, B. D., HATTON, T. A. (1991), The fermentation/downstream processing interface, *Bioseparation* **1**, 303-349.

KELLEY, B. D., RAHAMAN, R. S., HATTON, T. A. (1992), Salt and surfactant effects on protein transfer in AOT-isooctane reversed micelles, in: *Analytical Chemistry and Organized Media* (HINZE, W., Ed.), Greenwich, CT: Jai Press.

KIM, S. H., YOO, Y. J. (1991), Characteristics of an aqueous two-phase system for *alpha*-amylase production, *J. Ferment. Bioeng.* **71**, 373–375.

KIRCHBERGER, J., CADELIS, F., KOPPERSCHLAGER, G., VIJAYALAKSHMI, M. A. (1989), Interaction of lactate dehydrogenase with structurally related triazine dyes using affinity partitioning and affinity chromatography, *J. Chromatogr.* **483**, 289–299.

KITAHARA, A. (1980), Solubilization and catalysis in reversed micelles, *Adv. Colloid Interface Sci.* **12**, 109–140.

KITAHARA, A., KON-NO, K. (1966), Mechanism of solubilization of water in nonpolar solution of oil-soluble surfactants: Effects of electrolytes, *J. Phys. Chem.* **70**, 3394–3399.

KOHLER, K. (1991), Aqueous two-phase extraction of fusion proteins, *PhD Thesis,* Royal Institute of Technology, Stockholm, Sweden.

KOHLER, K., LJUNGQUIST, C., KONDO, A., VEIDE, A., NILSSON, B. (1991a), Engineering proteins to enhance their partition coefficients in aqueous two-phase systems, *Bio/Technology,* **9**, 642–646.

KOHLER, K., VEIDE, A., ENFORS, S.-O. (1991b), Partitioning of *beta*-galactosidase fusion proteins in PEG/potassium phosphate aqueous two-phase systems, *Enzyme Microb. Technol.* **13**, 204–209.

KOPPERSCHLAGER, G., BIRKENMEIER, G. (1990), Affinity partitioning and extraction of proteins, *Bioseparation* **1**, 235–254.

KRONER, K. H., HUSTEDT, H., KULA, M.-R. (1982), Evaluation of crude dextran as a phase-forming polymer for the extraction of enzymes in aqueous two-phase systems in large scale, *Biotechnol. Bioeng.* **24**, 1015–1045.

KRONER, K. H., HUSTEDT, H., KULA, M.-R. (1984), Extractive enzyme recovery: Economic considerations, *Process Biochem.,* October, 170–179.

KU, C. A., HENRY, J. D., BLAIR, J. B. (1989), Affinity-specific protein separations using ligand-coupled particles in aqueous two-phase systems: 1. Process concept and enzyme binding studies for pyruvate kinase and alcohol dehydrogenase from *Saccharomyces cerevisiae, Biotechnol. Bioeng.* **33**, 1081–1097.

KULA, M.-R. (1985), Liquid–liquid extraction of biopolymers, in: *Comprehensive Biotechnology,* (COONEY, C. L., HUMPHREY, A. E., Eds.) pp. 451–471, New York: Pergamon Press.

KULA, M.-R. (1991), Phase diagrams of new aqueous phase systems composed of aliphatic alcohols, salts, and water, *Fluid Phase Equilibria* **62**, 53–63.

KULA, M.-R., KRONER, K. H., HUSTEDT, H., SCHÜTTE, H. (1981), Technical aspects of extractive enzyme purification, *Ann. N. Y. Acad Sci.* **369**, 341–354.

KULA, M.-R., KRONER, K. H., HUSTEDT, H. (1982), Purification of enzymes by liquid–liquid extraction, *Adv. Biochem. Eng.* **24**, 73–118.

LEE, S. M. (1989), The primary stages of protein recovery, *J. Biotechnol.* **11**, 103–118.

LEE, Y. H., CHANG, H. N. (1990), Production of alkaline protease by *Bacillus licheniformis* in an aqueous two-phase system, *J. Ferment. Bioeng.* **69**, 89–92.

LEODIDIS, E. B., HATTON, T. A. (1989a), Specific ion effects in electrical double layers: selective solubilization of cations in Aerosol-OT reversed micelles, *Langmuir* **5**, 741–753.

LEODIDIS, E. B., HATTON, T. A. (1989b), Interphase transfer for selective solubilization of ions, amino acids, and proteins in reversed micelles, in: *Structure and Reactivity in Reversed Micelles* (PILENI, M. P. Ed.), pp. 270–302, Amsterdam: Elsevier Publ. Co.

LESER, M. E., LUISI, P. L. (1990), Application of reverse micelles for the extraction of amino acids and proteins, *Chimia* **4**, 270–282.

LUISI, P. L., MAGID, L. J. (1986), Solubilization of enzymes and nucleic acids in hydrocarbon micellar solutions, *Crit. Rev. Biochem.* **20**, 409–474.

LUISI, P. L., BONNER, F. J., PELLEGRINI, A., WIGET, P., WOLF, R. (1979), Micellar solubilization of proteins in aprotic solvents and their spectroscopic characterisation, *Helv. Chim. Acta* **62**, 740–753.

LUISI, P. L., GIOMINI, M., PILENI, M. P., ROBINSON, B. H. (1988), Reverse micelles as hosts for proteins and small molecules, *Biochim. Biophys. Acta* **947**, 209–246.

MARCOZZI, G., CORREA, N., LUISI, P. L., CASELLI, M. (1991), Protein extraction by reverse micelles: A study of the factors affecting the forward and backward transfer of alpha-chymotrypsin and its activity, *Biotechnol. Bioeng.* **3**, 1239–1246.

MARZOLA, P., FORTE, C., PINZINO, C., VERACINI, C. A. (1991), Activity and conformation changes of *alpha*-chymotrypsin in reverse micelles studied by spin labelling, *FEBS Lett.* **289**, 29–32.

NGUYEN, A. L., GROTHE, S., LUONG, J. (1988), Applications of pullulan in aqueous two-phase systems for enzyme production *Appl. Microbiol. Biotechnol.* **2**, 341–346.

PARADKAR, V. M., DORDICK, J. S. (1991), Purification of glycoproteins by selective transport using concanavalin-mediated reverse micellar extraction, *Biotechnol. Prog.* **7**, 330–334.

PATRICKIOS, C., ABBOTT, N. L., FOSS, R., HATTON, T. A. (1992), Synthetic polyampholytes for protein partitioning in two phase aqueous polymer systems, *AIChE Symp. Ser.* **88**, 80–88.

PERSSON, M, STALBRAND, H., TJERNELD, F.,

HAHN-HÅGERDAL, B. (1991), Semicontinuous production of cellulolytic enzymes with *Trichoderma reesei* Rutgers C30 in an aqueous two-phase system, *Appl. Biochem. Biotechnol.* **27,** 37–43.

PORATH, J. (1988), IMAC-immobilized metal ion affinity based chromatography, *Trends Anal. Chem.* **7,** 254–259.

PRYDE, J. G. (1986), Triton X-114: a detergent that has come in from the cold, *Trends Biochem. Sci.* **11,** 160–163.

PRYDE, J. G., PHILLIPS, J. H. (1986), Fractionation of membrane-proteins by temperature-induced phase-separation in triton X-114: Application to subcellular fractions of the adrenal medulla, *Biochem. J.* **233,** 525–533.

RAHAMAN, R. S., CHEE, Y. J., CABRAL, J. M. S., HATTON, T. A. (1989), Recovery of extracellular alkaline protease from whole fermentation broth, *Biotechnol. Prog.* **5,** 218–224.

SASAKAWA, S., WALTER, H. (1972), Partition behavior of amino acids and small peptides in aqueous dextran-PEG phase systems, *Biochemistry* **11,** 2760–2765.

SIKDAR, S. K., COLE, K. D., STEWART, R. M., SZLAG, D. C., TODD, P., CABEZAS, JR., H. (1991), Aqueous two-phase extraction in bioseparations: An assessment, *Bio/Technology* **9,** 253–256.

SMITH, M. C., FURMAN, T. C., INGOLIA, T. C., PIDGEON, C. (1988), Chelating peptide-immobilized metal ion affinity chromatography, *J. Biol. Chem.* **263,** 7211–7215.

STURESSON, S., TJERNELD, F., JOHANSSON, G. (1990), Partition of macromolecules and cell particles in aqueous two-phase systems based on hydroxypropyl starch and PEG, *Appl. Biochem. Biotechnol.* **26,** 281–295.

SULKOWSKI, E. (1985), Purification of proteins by IMAC, *Trends Biotechnol.* **3,** 1–7.

TJERNELD, F. (1989), New polymers for aqueous two-phase systems, in: *Separations Using Aqueous Phase Systems* (FISHER, D., SUTHERLAND, I. A., Eds.) pp. 429–438, London: Plenum Press.

TJERNELD, F., JOHANSSON, G. (1990), Aqueous two-phase systems for biotechnical use, *Bioseparation* **1,** 255–263.

TJERNELD, F., JOHANSSON, G., JOELSSON, M. (1987), Affinity liquid–liquid extraction of lactate dehydrogenase on a large scale, *Biotechnol. Bioeng.* **30,** 809–816.

TODD, R. J., VAN DAM, M. E., CASIMIRO, D., HAYMORE, B. L., ARNOLD, F. H. (1991), Cu(II)-binding properties of a cytochrome c with a synthetic metal-binding site: His-X_3-His in an alpha-helix, *Proteins* **10,** 156–161.

TREYBAL, R. E. (1980), *Mass-Transfer Operations,* New York: McGraw-Hill.

VENKATARAMAN, G., GLEASON, K., KIM, P., HATTON, T. A. (1991), Protein conformation at charged interfaces, *Abstract #212e, AIChE Meeting,* Los Angeles.

VERNAU, J., KULA, M.-R. (1990), Extraction of proteins from biological raw material using aqueous polyethylene glycol–citrate phase systems, *Biotechnol. Appl. Biochem.* **12,** 397–404.

WALDE, P., GIULIANI, A. M., BOICELLI, C. A., LUISI, P. L. (1990), Phospholipid-based reverse micelles, *Chem. Phys. Lipids* **53,** 265–288.

WALTER, H., SASAKAWA, S., ALBERTSSON, P.-A. (1972), Cross-partition of proteins: Effect of ionic composition and concentration, *Biochemistry* **11,** 3880–3883.

WOLBERT, R. B. G., HILHORST, R., VOSKUILEN, G., NACHTEGAAL, H., DEKKER, M., VAN'T RIET, K., BIJSTERBOSCH, B. H. (1989), Protein transfer from an aqueous phase into reversed micelles: The effect of protein size and change distribution, *Eur. J. Biochem.* **184,** 627–633.

WOLL, J. M., HATTON, T. A., YARMUSH, M. L. (1989), Bioaffinity separations using reversed micellar extraction, *Biotechnol. Prog.* **5,** 57–62.

WUENSCHELL, G. E., NARANJO, E., ARNOLD, F. H. (1990), Aqueous two-phase metal affinity extraction of heme proteins, *Bioprocess Eng.* **5,** 199–202.

YANG, M.-C., CUSSLER, E. L. (1988), Designing hollow-fiber contactors, *AIChE J.* **32,** 1910–1916.

23 Protein Separation and Purification

JAN-CHRISTER JANSON

LARS RYDÉN

Uppsala, Sweden

1 Introduction – From Art to Rational Protein Management

Proteins are the answer of the living cell to all kinds of functional requirements. Regulation, catalysis, control, etc. are all carried out by a countless number of proteins manufactured by the cell, with structures perfectioned after millions of years of evolution. In biotechnology this potential of the living cell is tapped for a multitude of applications. In many cases this can be done by using the cell as such or a simple extract. Bacteria and yeast have been put into the service of man since the beginning of civilization, and their capacities in various industrial processes such as production of acetic acid, acetone and butanol have been added this century. Still this use depends on a single or a few of the hundreds or thousands of enzymes present in the cell. Likewise the use of antisera for clinical diagnosis depends on a specified antibody present, although the whole serum is used. Specificity alone is enough to allow these applications.

However, in many instances mixtures of hundreds of proteins are not the answer to the problem at hand. A purified protein is required to allow the problem to be solved. Competing catalytic activities may hinder the reaction needed or competing antibodies disturb the reaction measured. If a protein is used as a pharmaceutical, impurities may be dangerous. In all these instances it is vital that the protein asked for is prepared in pure form. The original objective of protein purification was, however, investigation of protein structure and function. These studies required that proteins were homogeneous to allow proper conclusions from measurements. The slogan "There is no point in wasting clean thinking on dirty enzymes" (KORNBERG, 1980) is still valid in the research laboratory. The tremendous achievement of protein chemistry and, to no lesser extent, of molecular biology over the last couple of decades is to a significant degree due to improved techniques for protein purification. Today these techniques allow biotechnologists to regard protein purification as just another step in their laboratory routine, a step that has a rather good chance of being successful.

The history of modern protein purification begins around 1950. Before then, the separation and purification of enzymes and other proteins was primarily performed by precipitation (COHN et al., 1946) and adsorption methods (ZECHMEISTER and ROHDEWALD, 1951) using a variety of inorganic as well as organic materials. The first attempts to separate proteins by elution chromatography were based on the weakly acidic resin cation exchanger Amberlite IRC-50 (a co-polymer of methacrylic acid and divinylbenzene) (PALEUS and NEILANDS, 1950) and on calcium phosphate (SWINGLE and TISELIUS, 1951) (see Tab. 1). The early resin ion exchangers suffered primarily from low porosity and too high hydrophobicity, restricting their application area to low-molecular weight basic proteins. The calcium phosphate crystallites were both chemically and physically unstable. These pioneering efforts were followed by two major breakthroughs: the introduction of cellulose ion exchangers (SOBER and PETERSON, 1954; PETERSON and SOBER, 1956) and of hydroxyapatite (TISELIUS et al., 1956). The hydrophilic nature of these materials made them especially suited for purification of high-molecular weight proteins with a wide variety of isoelectric points and polarities.

The possibility of separating water-soluble molecules according to their size by molecular sieving in neutral gel materials was discussed already in the late 1940s (SYNGE, 1981). Some years later, preliminary reports on starch (LINDQVIST and STORGÅRDS, 1955; LATHE and RUTHVEN, 1956) were followed by the development of more suitable materials such as cross-linked dextran (PORATH and FLODIN, 1959), cross-linked polyacrylamide (HJERTÉN and MOSBACH, 1962) and agarose (HJERTÉN, 1962). An account of the history of the development of cross-linked dextran (Sephadex®) has been published by JANSON (1987). The cross-linked dextran gels were also derivatized to ion exchangers which were rapidly adopted also to industrial applications (BJÖRLING, 1972).

Agarose is a spontaneously gel-forming polysaccharide, capable of forming rigid gels also at low matrix concentrations (2–6%). Agarose

Tab. 1. Some Milestones in the Development of Matrices and Media for Elution Chromatography of Proteins

Year	Authors	Medium
1950	PALEUS and NEILANDS	Acrylic ion exchanger (Amberlite)
1954	SOBER and PETERSON	Cellulose ion exchangers
1956	TISELIUS et al.	Hydroxyapatite
1959	PORATH and FLODIN	Cross-linked dextran (Sephadex)
1960	FLODIN	Sephadex ion exchangers
1962	HJERTÉN and MOSBACH	Cross-linked polyacrylamide
1962	HJERTÉN	Agarose (Sepharose, BioGel A)
1965	HALLER	Controlled pore glass
1966	URIEL	Agarose/polyacrylamide (Ultrogel AcA)
1969	DETERMANN et al.	Spherical cellulose
1971	PORATH et al.	Cross-linked agarose
1977	JOHANSSON	Polyacrylamide/dextran (Sephacryl)
1978	HASHIMOTO et al.	Vinyl polymer (Toyopearl)
1983	UGELSTAD et al.	Monosized PS-DVB ion exchangers
1988	KÅGEDAL et al.	Agarose/dextran (Superdex)
1991	AFEYAN et al.	Perfusion gel media

gels are characterized by their macroporosity (Fig. 1) (ARNOTT et al., 1974) and neutral hydrophilicity based on alcohol hydroxyl groups. In the late 1960s the CNBr method for activation and coupling of water-soluble, primary amino group-containing molecules to agarose gels was developed (AXÉN et al., 1967). By the utilization of this technology for the synthesis of biospecific adsorbents for the purification of enzymes and other proteins (CUATRECASAS et al., 1968) the technique of affinity chromatography was introduced. Its rapid acceptance was facilitated by the commercial availability of CNBr preactivated 4% agarose (PHARMACIA, 1971).

The polysaccharide matrix of agarose gels is held together primarily by hydrogen bonds. Such gels are thermoreversible and, furthermore, substitution with, e.g., ionizable groups would lead to gel instability and eventually leakage. To overcome these weaknesses, cross-linked bead-shaped agarose was introduced by PORATH et al. (1971). Later work showed that it is possible to accomplish considerably higher rigidity using improved cross-linking chemistries for agarose enabling the synthesis of 10 μm particles for HPLC of proteins (HJERTÉN and YAO, 1981; ANDERSSON et al., 1984) and also the manufacturing of media suitable for industrial applications (JANSON and PETTERSSON, 1992).

Other important chromatographic materials developed over the last decades include various porous silica-bonded phases (UNGER et al., 1974; REGNIER and NOEL, 1976) and several synthetic organic polymer-based materials (HASHIMOTO et al., 1978). Particular interest has been focused on monosized polystyrene-divinylbenzene particles (UGELSTAD et al., 1983, 1992) and the macroporous perfusion gel media developed by AFEYAN et al. (1991).

An excellent complement to size exclusion chromatography and ion exchange chromatography, hydrophobic interaction chromatogra-

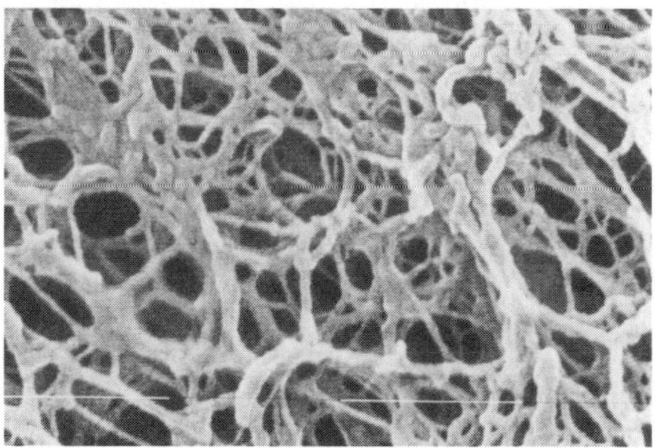

Fig. 1. Scanning electron micrograph of 2% agarose gel. The white bar represents 500 nm. Preparation and photo: A. MEDIN, Biochemical Separation Centre, Uppsala University, Uppsala.

phy was introduced in the early 1970s by YON (1972), ER-EL et al. (1972), HOFSTEE (1973), PORATH et al. (1973), and HJERTÉN (1973).

Along with the development of various media for protein preparation and chromatography, the chemistry for surface design of the matrices has developed. This is most apparent in affinity procedures where various ligands are introduced on to the gels which then specifically bind to a single protein or group of proteins. The affinity methods range from practically monospecific ligands, such as biotin for biotin-binding proteins, to ligands which bind a whole category of proteins such as the cofactor NAD.

A corresponding development has occurred with regard to the understanding of the structure of the proteins to be purified. The surface properties of globular proteins may allow the design of purification strategies suited to a particular protein with precision. To this end, LOWE et al. (1992) have developed new highly selective "biomimetic" dye ligands by molecular modeling directed against target proteins such as certain dehydrogenases. However, it should be stressed that this "rational protein purification" or "rational protein management" is at a very early stage of development and often more a goal than reality. It shows that protein purification today is moving from the field of skillful handicraft or art to rational science.

Another approach is the use of genetic engineering for the design of a fusion tail, i.e., an extra domain with well-defined binding properties that allow purification of recombinant proteins in a predetermined way. Examples of fusion tails are Staphylococcal protein A (MOKS et al., 1987) and histidine oligopeptides (HOCHULI et al., 1988).

2 Strategies for Protein Separation

Before starting the design of the purification process, there are a few crucial considerations to be made. First of all it is important to make the best possible choice of raw material. Different sources of the same protein may differ widely with regard to concentration in the starting material, both regarding specific activity and stability, as well as other properties. In addition, raw materials may be very different as to their availability and cost. Furthermore, the handling of various raw materials may differ considerably with regard to the optimum extraction procedure and the presence of critical impurities such as pyrogens, DNA, difficult-to-remove contaminating proteins and proteases. In general, extracellular liquids – clarified cell culture media and fermentation broths, serum, milk, etc. – are by far the best choice. Animal tissues are more expensive and more complex, although the absence of cell walls makes homogenization easy. Fat present in such tissues is, however, often a source of problems. Yeast and bacteria both have a more or less tough cell wall and are thus more difficult to homogenize. Plants contain large amounts of fibrous bulk material and often polyphenols and pigments, that may interfere in the purification process. Bacteria have a relatively high content of nucleic acids which have to be removed.

Today many proteins are purified from genetically engineered bacterial strains, in most cases *Escherichia coli*. Here it is an advantage if the protein is excreted into the surrounding medium (Fig. 2). In *E. coli* this is mostly achieved by exporting the protein into the periplasmic space followed by selective lysis of the outer membrane. Often already the cloning of the gene is the starting point of the design of a purification process. Then, when several alternative expression systems are at hand, the advantages of overexpression, stability to proteases, and excretion into the medium are factors that should be considered. An interesting new possibility lies in cloning of enzymes from thermostable bacteria. Many intracellularly expressed recombinant proteins accumulate as insoluble aggregates, so-called inclusion bodies. Substantial benefit can be gained from developing efficient washing procedures, often involving non-ionic detergents, before solubilization and refolding to their native, biologically active state (MARSTON, 1986; TITCHENER-HOOKER et al., 1991; FISHER et al., 1993).

For soluble systems, the design of the extraction procedure is important for the final

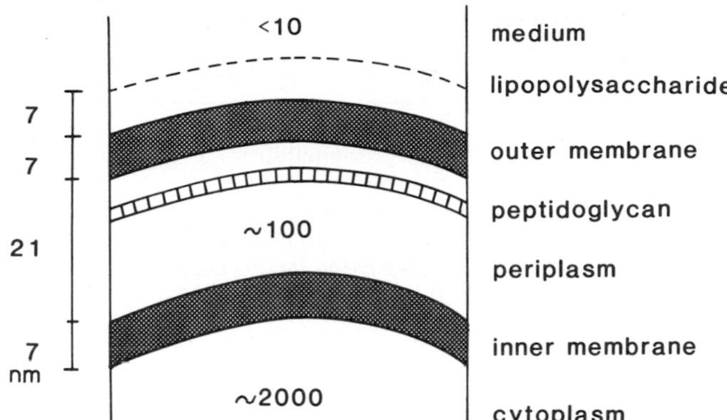

Fig. 2. Location and approximate numbers of the proteins in *Escherichia coli*.
Courtesy: M. UHLÉN, Royal Institute of Technology, Stockholm.

outcome of a preparation scheme. Time and conditions of extraction have to be limited to prevent the release of unwanted components, when a reasonable recovery of the desired activity has been achieved. The dangers here are denaturation, proteolysis, bacterial contamination, and contamination with nucleic acids and lipopolysaccharides (pyrogens), especially relevant for Gram-negative bacteria. By proper choice of extraction medium some of these problems can be handled satisfactorily. The pH value, buffer, and additives, such as detergents, metal ions and chelators or protease inhibitors should be considered when designing a good extraction medium.

The aim of a protein purification process is three-fold: to remove unwanted contaminants, to concentrate the desired component, and to transfer the protein to an environment where it is stable and in a form ready for the intended application. The ultimate goal for the ambitious process engineer is to meet all these objectives in a single, integrated procedure. This is, however, possible very seldom. Instead most protein purification schemes consist of on an average of 3–6 separate steps: an initial capturing step that results in a clarified and concentrated protein solution, one or several intermediary purification steps, mostly chromatographic, where the majority of the contaminating proteins and other unwanted substances are removed, and finally a polishing step, where the remaining trace impurities, such as aggregates or partially degraded or modified forms of the protein itself, are removed. The polishing is often achieved by a single gel filtration step, sometimes, however, combined with, or replaced by, high resolution reverse phase chromatography.

A wide variety of protein purification techniques are available today to the biochemical engineer (Tab. 2). Precipitation, two-phase extraction, membrane techniques, adsorption and various forms of chromatography are most frequently applied. In practice the chromatographic techniques have become dominant due to their high resolving power. However, combinations with membrane techniques are gaining increased popularity because of their relative simplicity, and the often easily handled equipment required. The fouling problem, however, is still waiting for its final solution. The electrophoretic techniques are used primarily for analytical separations, and the extraction techniques for initial steps as described below.

For more detailed discussions of protein purification strategies, techniques and methodologies, as well as case studies, the reader is referred to recent handbooks on the subject (SCOPES, 1987; JANSON and RYDÉN, 1989; DEUTSCHER, 1990).

3 Techniques for Initial Fractionation of Crude Extracts

There are several alternative means available to achieve preliminary purification, clarification, and concentration of a crude protein ex-

Tab. 2. Separation Principles and Techniques in Biotechnology

Separation Principle	Separation Technique
Temperature stability	Heat denaturation
Solubility	Salt precipitation
	Solvent precipitation
	Polymer precipitation
	Isoelectric precipitation
	Partitioning in aqueous two-phase systems
	Partition chromatography
Size and shape	Size exclusion chromatography (gel filtration)
	Ultrafiltration
	Sieving gel electrophoresis
Net charge	Non-sieving electrophoresis
	Isotachophoresis
	Ion exchange chromatography
Isoelectric point	Isoelectric focusing
	Chromatofocusing
Hydrophobicity	Hydrophobic interaction chromatography
	Reversed phase chromatography
Biological function	Biospecific affinity chromatography
Antigenicity	Immunosorption
Carbohydrate content	Lectin affinity chromatography
Content of free −SH	Covalent chromatography (chemisorption)
Metal binding	Immobilized metal ion affinity chromatography
Miscellaneous	Hydroxylapatite chromatography
	Dye ligand affinity chromatography

tract. The aim is to enrich the desired protein in one phase, either as a precipitate, partitioned to an aqueous polymer phase, or adsorbed to a solid phase medium (e.g., chromatographic particles). The classical approach is the use of precipitants such as ammonium sulfate, ethanol or polyethylene glycol (FOSTER, 1993). Precipitations can also be done when the extract still contains particles and lipids which is often the case after low-speed centrifugation of a homogenate. Certain substances should be removed that have a tendency to disturb the protein purification procedures. Thus, plant pigments may be removed by adsorption to silica (e.g., Celite™) and nucleic acids present in bacterial homogenates may be precipitated by the addition of streptomycin sulfate. Precipitation methods are relatively easy to adapt to large volumes. Often two sequential precipitations are carried out, the first to remove bulk impurities, the second to precipitate and thus concentrate the protein under study. After dissolution in a chosen buffer, the second precipitate is then used to continue the preparation scheme. A circumstance that is often neglected is that precipitation depends on protein concentration in a critical way, due to the simple fact that there is a finite solubility of a protein also after the addition of the precipitating agent. If the protein under study is present in the extract in very low concentrations, precipitation to concentrate the protein, the second of the two suggested steps, is not possible then with acceptable recovery.

An often advantageous alternative to precipitation is using an aqueous two-phase polymer system, such as polyethylene glycol, PEG. In this procedure a considerable concentration of a protein in one of the phases, often the upper PEG phase, can be achieved if the right conditions are found. Only few transfer steps are normally required to achieve adequate separation, and the transfers can either be done manually or by using simple equipment. Particles are normally collected at the interphase, and this is why two-phase extraction often can be applied directly to unclarified extracts. One disadvantage is the presence of viscosity-increasing phase polymers in the fraction collected for the next step in the purification procedure.

A third strategy for concentrating a particular protein from a crude extract is capturing high-capacity chromatographic particles either in stirred batch, by fixed-bed adsorption or by expanded-bed adsorption. The latter, newly introduced technology, is based on high-capacity derivatized cross-linked agarose beads filled with inert, density-increasing particles that are allowed to expand by upward buffer liquid flow in a purpose-designed column. The concept of expanded-bed adsorption differs from traditional fluidized-bed adsorption mainly by its bed stability, enabling the use of plug flow mode and thus a more efficient utilization of the binding capacity of the adsorbent. When the expanded bed has stabilized in the buffer liquid flow, the crude, unclarified sample is applied. The conditions are chosen such that the desired protein is efficiently captured by the adsorbent. The non-adsorbed proteins, cells and cell debris are washed out of the column in expanded mode. The adsorbed proteins are finally eluted, concentrated and partially purified, downward in packed-bed mode (PHARMACIA, 1993).

Most often adsorbents containing ion exchange groups or hydrophobic groups are used for initial capturing of proteins from crude extracts. The main reasons are their high binding capacity, their applicability to harsh CIP protocols and their relatively low cost (SOFER and NYSTRÖM, 1989, 1991).

Membrane separation techniques are sometimes a viable alternative. Thus, crossflow microfiltration is often used instead of, or in combination with, conventional centrifugation. By diafiltration, salts and smaller proteins can easily be separated from larger proteins, and this is often combined with concen-

Tab. 3. Naturally Occurring Amino Acids Classified According to their Relative Hydrophobicity and Chemical Nature. Courtesy of FOSTER (1993)

Hydrophobic Index[a]	Classification of Amino Acids			
	Aromatic	Aliphatic (and heterocyclic[b])	Polar Aliphatic	Ionizable Aliphatic (and heterocyclic[b])
1	Phenylalanine			
2		Leucine		
3		Isoleucine		
4	Tyrosine			
5	Tryptophan			
6		Valine		
7		Methionine		
8		(Proline)		
9			Cysteine	
10		Alanine		
11		Glycine		
12			Threonine	
13			Serine	
14				Lysine
15			Glutamine	
16			Asparagine	
17				(Histidine)
18				Glutamic acid
19				Aspartic acid
20				Arginine

[a] BLACK and MOULD (1991)
[b] in parentheses

tration of the protein solution as well. Newly developed, selective affinity membranes show promising results (BRIEFS and KULA, 1992; CHAMPLUVIER and KULA, 1992).

4 Protein Chromatography – General Aspects

Chromatography is by far the most important technique available for protein purification and is often the only technique used in modern downstream processes. This situation is the result of the development of new media over the last decade that combine high selectivity and efficiency with adequate scaling-up capability.

Chromatography of proteins differs markedly from chromatography of low-molecular weight substances, primarily because of the inherently lower diffusivity of the larger molecules, but also due to the large number of possible chemical interactions arising from variations in the frequency and distribution of the amino acid side chains on the surface of proteins (Tab. 3). Furthermore, the delicate, tertiary structures of globular proteins impose restrictions on both the surface chemistry of the chromatographic media used and on superimposed conditions such as buffer composition, pH, and temperature (Tab. 4).

Adsorption chromatography of proteins is based on a very complex series of interactions composed of thermodynamic, mass transfer and kinetic elements (JANSON and HEDMAN, 1987) that gives rise to differential retardation of the components in the protein mixture. It is important to emphasize that there are no simple rules that describe these phenomena, but that adsorption chromatography of proteins still has to be regarded largely an empirical science.

In various forms of adsorption chromatography the protein molecules bind reversibly to chemical groups on the surface of the gel matrix. This is the result of the formation of several secondary chemical bonds, i.e., ion–ion bonds or salt bridges, ion–dipole, dipole–dipole, hydrogen bonds and van der Waals bonds in addition to hydrophobic interactions. The hydrophobic interaction differs from other adsorption phenomena in that it has a large entropic component. Since each of these bonds are rather weak, with binding enthalpies in the order of a few up to 20 kJoules per mole, several bonds are necessary to give rise to dissociation constants of $K_D = 10^{-3}$, the lower limit for retardation, up to 10^{-5}–10^{-6} which give rise to complete binding. Even if these bonds account for the strength of the binding, they do not alone explain the selectivity in protein chromatography. Here steric factors are important. The binding groups on the surface of a protein need to be distributed in such a way that complementarity between the immobilized ligand and the protein surface arises. Increased freedom of movement of the ligand will increase the probability of optimal binding and thus of higher selectivity and efficiency. This is why ligands separated from the rigid matrix backbone by a polymer chain have shown promising results. The concept of "fit" explains why proteins with minor differences, in the extreme a single amino acid replacement,

Tab. 4. Options for Exploiting Adsorbability Parameters for Proteins

Variable Environment	Variable Protein	Fixed Protein
Ionic strength	Charge	Molecular size and conformation
pH	Ion binding	Amino acid composition
Dielectric constant		Amino acid sequence
Temperature		Ratio, distribution and chemical nature of polar/non-polar residues
Buffer composition		Number of ionizable residues
Specific ion effects		Dissociation constants

sometimes also can be separated on media of comparatively low selectivity such as ion exchangers (MÜLLER, 1990). However, the concept of fit is of course maximally exploited in affinity chromatography and especially in immunosorption.

The different interactions are each influenced by the solvent in some particular way. Increased ionic strength thus decreases ionic interactions, whereas hydrophobic interactions are favored. The ways used to elute adsorbed proteins differ for each type of chromatography. It should, however, be kept in mind that, since a single type of interaction is seldom responsible for the binding, non-conventional approaches are sometimes worth trying.

Two forms of protein adsorption enlarge the concept of binding beyond the ionic and hydrophobic forces. In chemisorption, also called covalent chromatography, the binding between protein and immobilized ligand is still reversible, but now due to covalent interaction between thiol groups, using a cysteine moiety on the protein to form a disulfide link that can be reduced allowing elution of the bound protein. Metal chelate chromatography depends on metal–ligand interactions between a metal fixed to a gel surface by a chelating group and metal binding amino acid side chains, mostly histidine, on the protein.

As a general rule, highly specific adsorbents, such as immunosorbents based on immobilized monoclonal antibodies, however tempting, should not be subjected directly to crude extracts. Unless carefully equilibrated, crude protein solutions have a tendency to precipitate when introduced to a chromatographic column, they may contain proteases that degrade the immobilized protein and reduce the life span of a column considerably. In practice, in most purification schemes an initial fractionation step such as precipitation or adsorption to an expanded or fluidized bed is applied before the chromatographic steps.

In different forms of partition chromatography the components remain in solution during the entire separation process. This is based on partition between the mobile buffer phase and the liquid portion of the stationary phase. An interesting variety of this technique was developed by MÜLLER (1988) utilizing the ability of certain gel media to selectively attract one of the phases in aqueous polymer two-phase systems. Protein mixtures equilibrated in the mobile moiety of the two-phase system and applied to the likewise equilibrated column, will partition according to differences in size and surface characteristics that are complementary to those utilized in adsorption chromatography. A special case of partition chromatography and the simplest type of protein chromatography, thus theoretically best developed, is size exclusion chromatography (gel filtration) which, by definition, is independent of any interactions with the medium, other than purely steric.

Many different base matrices have been proposed in attempts to find the ideal medium for protein chromatography. Three main types may be distinguished based on the structure of the gel-forming matrix. In the first, the gel is formed by point cross linking of individual polymer chains to give a three-dimensional network. This gives rise to so-called xerogels that swell and shrink in the presence and absence of solvent. At porosities suitable for protein–protein separation, these gels are impracticably soft and this is why they are primarily used for desalting operations and group separations. Typical examples are Sephadex® from Pharmacia and BioGel P® from BioRad.

The second type is based on polymers that aggregate during gel formation leaving large voids inside the often bead-shaped particles for easy access to the large internal surface also for high-molecular weight proteins. To this category, the so-called aerogels, belong agarose gels, macroporous silica, and media based on macroporous synthetic organic polymer gels such as polystyrene-divinylbenzene and polyacrylates. For agarose gels the matrix volume varies between 2 and 6%, whereas it usually lies between 20 and 40% for synthetic polymer gels. For the latter, as for macroporous silica, there is a correlation between the internal pore size distribution, pore volume, and available surface area for protein interaction. To a significant extent the binding capacity can also be controlled by proper surface modification. By increasing the internal pore diameter of the particles, perfusion, i.e., intraparticle convection, will give a significant contribution to the mass transport efficiency (AFEYAN et al., 1991; RODRIGUES et al., 1992).

The third matrix type is represented by a variety of composite gels where aggregated matrix elements provide a mechanically strong support, whereas the desired pore-size distribution is obtained by polymer chains grafted to the walls of the supporting matrix structure. Examples of this gel type are Superdex® (agarose/dextran) from Pharmacia and Ultrogel® AcA (agarose/polyacrylamide) from Sepracor/IBF.

During the last decade, stationary phase technology has improved dramatically allowing the production of media that meet all kinds of user demands, from microanalytical to industrial process scale. Traditional high performance liquid chromatography of proteins utilizes beads of sizes between 3 and 15 μm with flow rates normally in the order of 200 to 400 cm/h. Perfusion chromatography (AFEYAN et al., 1991) has extended this range by one order of magnitude. In standard elution chromatography flow rates are normally in the range of 30–300 cm/h, allowing preparative chromatographic separation cycles lasting a couple of hours down to a few minutes.

Size exclusion chromatography (gel filtration) has a special position. Here the separation of two proteins is primarily governed by the steepness of the selectivity curve of the gel filtration medium, the column length, the sample band width, the flow rate, and the relative sample viscosity. For optimum performance, the sample viscosity relative to the eluent buffer should not exceed 1.5, which corresponds to a protein concentration of approximately 70 mg/mL. The flow rates used for protein separations seldom exceed 30 cm/h (HAGEL, 1989; HAGEL and JANSON, 1992).

5 Protein Size Methods – Gel Filtration

In size exclusion chromatography (gel filtration), the separation mechanism is based on partitioning of the proteins between two liquid phases, one stationary inside the gel particles and one mobile making up the void volume between the particles. The degree of separation is governed by the relative proportion of each protein which at every moment is present in the mobile phase. The separation mechanism is thus only dependent on differences in the sizes and shapes of the proteins and the pore size distribution of the three-dimensional network of the gel materials used for the chromatography. The molecules in the sample solution mixture, which is pumped into the packed bed as a narrow zone, are distributed practically momentarily between the stationary gel phase and the flowing aqueous buffer outside the gel particles. At any particular moment during this process, a certain mass fraction of each molecular species, the size of which depends on the fraction of the gel phase sterically available to that molecule, is moving down the column with the speed of the flowing liquid. The total speed of axial mass transport of each molecular species down the column is inversely proportional to the fraction of the gel phase available to this particular molecule.

All molecules which enter the gel phase are distributed through the whole gel particle cross-section occupying only that fraction which is available to them by pure steric restriction. The experimentally determined relationship between solute size (Stokes' radius or molecular weight) and the distribution coefficient is a fundamental characteristic of any gel filtration medium known as the selectivity curve.

The retardation of a protein in gel filtration depends on the Stokes' radius of the protein. This in turn depends on two factors: shape which usually is expressed as the frictional ratio and solvatization expressed in % bound solvent. For the few proteins where both parameters have been determined, solvatization normally is about 20% for typical globular proteins in aqueous buffers, while the frictional ratio varies from about 1.10 to 1.25 also for proteins that by crystallographic studies appear to be practically spherical. The value increases to 1.46 for gamma globulin with its well-known asymmetry. The frictional ratios can be interpreted in terms of axial ratios of ellipsoids of rotation. However, these figures, often given for proteins earlier, do not have a meaning for other shapes than ellipsoids. For globular proteins with their rough surfaces, often characterized by crevices for active sites or binding sites, more or less elongated as several domains of subunits associate, the interpretation of frictional ratios can only be done by comparison with known protein structures.

The Stokes' ratio for a protein is also influenced by the presence of larger prosthetic groups. Thus, carbohydrates in glycosylated proteins increase Stokes' radius both due to solvatization of the carbohydrate chains and increased asymmetry.

Despite these limitations selectivity curves are often useful for estimating molecular weights. The prerequisite is that a gel filtration column is calibrated with proteins with the same general solvatization and asymmetry as the sample protein, an assumption that is reasonably valid for globular proteins as noted. However, large deviation from this pattern can occur. Thus, a large increase in asymmetry, most accentuated for rod-shaped molecules results in gross overestimations of molecular weights. For random coils, the solvatization is far greater than for compact globular proteins, about 98%, and thus again an extreme overestimation of molecular weights will be done when compared to globular proteins. One way of avoiding the problem is to run the chromatography in conditions where all proteins assume the same shape. 6 M guanidine HCl has proven to be a medium where all peptide chains assume a random coil conformation (FISH et al., 1969). Crosslinks, such as disulfides, should be removed. Shorter side chains, such as oligosaccharides, can be shown to be without any important influence on the size of the random coil, unless they are found close to the end of the longer peptide chain and thus in practice elongate it.

Since the proteins are eluted according to the available pore volume fraction, the total separation volume cannot be larger than the total volume of the liquid in the gel phase. The relative pore volumes vary between approximately 52 and 97% for different gel filtration media, with the lower value being typical for porous glass and silica materials and the higher for certain polysaccharide-based media. Due to the restricted separation volume, the maximum number of components that can be separated in gel filtration is very small as compared with other liquid chromatography techniques (HAGEL, 1992). Even in HPLC mode, no more than approximately 12 peaks can be separated by gel filtration if a complete separation is required (i.e., if $R_s = 1.5$). In this context it should be remembered that the only parameter which governs the separation selectivity in gel filtration is the steepness and shape of the selectivity curve. The size of the chromatographic particles only shortens the time required for the molecules to achieve diffusive equilibrium in the gel medium. In other words, one can in principle get the same resolution in gel filtration irrespective of particle size merely by optimizing the operating flow rate for the eluting buffer.

Since ideal gel filtration means separation without any interaction with the gel matrix itself, the factors which govern the chromatographic separation capacity are those connected to parameters such as sample zone width versus particle diameter and column length and the concentration of the sample components. The effect of sample concentration in gel filtration is normally small and makes itself felt through its effect on the viscosity of the sample relative to the eluent rather than the dry weight content as such.

The structure of the gel matrix is also of decisive importance. In some respect, it is misleading to refer to 'pores' in gel filtration media, since this often implies the existence of well-defined spaces in a matrix composed of stationary elements. At the molecular level, the gel-forming elements of many of the most useful media have a mobility not far from the mobility they would have in free solution. Any individual space in such a gel is continually changing both its size and shape, just those properties which determine its ability of steric exclusion of other molecules. The dynamic nature of the spaces in these gels means that the pore-size distribution must be defined operationally in terms of its exclusion properties (HAGEL, 1988). It almost certainly explains the observation that pore-size distributions for gels where the exclusion properties are defined by flexible polymer chains are smoother than those for gels were the excluding elements are expected to be stiffer.

In general, the resolution in gel filtration is more affected by increasing the sample zone width than increasing the flow rate. This fact can be used for a simple strategy for the optimization of column productivity. In the first optimization step, one adjusts the column length and flow rate to obtain a reasonable resolution, but primarily satisfying the desired chromatographic cycle time. In the next step, the desired resolution is obtained by adjusting the relative sample volume. Finally, the desired column productivity is achieved by adjusting the diameter of the gel filtration column keeping the linear flow rate and the relative sample volume (i.e., the sample zone width) constant. This simple strategy requires that the columns are packed to the same density and homogeneity irrespective of the column diameter. There are many examples in industry where this is possible, but it requires experience and skill in packing large-diameter columns. In cases when this is not possible to achieve, one may compensate for less than optimal column packing quality by increasing the column length as discussed by NAVEH (1990).

One characteristic feature of size exclusion chromatography is that the influence of the composition of the eluent on the separation can usually be completely neglected. Within wide limits set by the stability of the separation medium, the eluent can thus be chosen to suit the properties of the sample, in particular the stability of the biological activity of the target protein. Special components may be added almost without restriction to solve special problems of solubility or to meet other specific requirements. For example, detergents at concentrations below the critical micelle concentration may be added to improve solubility of membrane-derived proteins, pH and ionic strength may be chosen to suit the requirements of a subsequent step like ion exchange chromatography or product formulation. This is often a way to minimize dialysis. However, gel filtration always results in dilution of a sample which may be inconvenient unless a subseqeuent concentration step is applied. It should be noted that a number of additives which improve protein solubility can also affect the shape of the protein molecule and thus its elution position.

6 Protein Charge Methods – Ion Exchange Chromatography and Chromatofocusing

Ion exchange chromatography is one of the most important and most general liquid chromatography methods for purifying proteins: It is easy to apply, it is very widely applicable, it can give very good resolution, it gives high yields of active material, it is a concentrating

Tab. 5. Charged Amino Acid Side Chains and Other Groups in Proteins. Reproduced with permission from KARLSSON et al. (1989)

Group	Structure	pK_a	Average Occurrence in Proteins (%)
Arginine	Guanido	12	4.7
Aspartic acid	Carboxylate	4.5	5.5
Cysteine	Thiol	9.1–9.5	2.8
Glutamic acid	Carboxylate	4.6	6.2
Histidine	Imidazole	6.2	2.1
Lysine	ε-Amino	10.4	7.0
Tyrosine	Phenol	9.7	3.5
α-Amino	Amino	6.8–7.9	
α-Carboxyl	Carboxylate	3.5–4.3	
Sialic acid	Carboxylate		Sialoglycoproteins
γ-Carboxyglutamate	Carboxylate		Blood coagulation factors
Phosphoserine etc.	Phosphate		Phosphoproteins

technique and, finally, it can be easily controlled to meet specific needs.

In ion exchange chromatography proteins are separated according to differences in their surface charges. Charges on the proteins bind by coulombic forces to charges of the opposite sign on the chromatographic medium, an ion exchanger. The strength of the binding depends on the surface charge density of the protein and on the ion exchanger, the more highly charged the protein and the ion exchanger, the stronger the interaction. Ionically bound proteins are eluted differentially from the ion exchanger, either by increasing the concentration of ions which compete for the same binding sites on the ion exchanger by the addition of a simple salt, or by changing the pH of the eluent so that the proteins lose their charges by titration.

Since the pH of the eluent influences the charge on the proteins, it must be carefully controlled. At an unsuitable pH, the charge on the protein may be of the wrong sign for binding or may be so small that the ion exchanger can only bind a small proportion of the sample. Separations performed at different pHs may give vastly different results, and changing the pH is a powerful way of modifying the resolution obtained. Likewise, the ionic strength of the eluent must also be controlled. Other ions, coming for example from added salt, will compete for binding to the charged groups on the ion exchanger and thus suppress protein binding.

Interactions other than charge–charge interactions, for example hydrogen-bonding or hydrophobic interactions, may modify the binding of a protein to an ion exchanger, but these are seldom of practical significance when well-designed ion exchange media are used. Where necessary, for example during chromatography of hydrophobic membrane proteins, nonionic detergents or organic solvents may be added to suppress hydrophobic interactions. In actual practice, hydrophobic interactions between the protein molecules are likely to be more important than hydrophobic interactions between the proteins and the ion exchanger.

Practically all proteins have both positive and negative charges on their surface, by which they bind to ionic groups on a matrix. Five of the twenty amino acid side chains are charged at neutral pH values. Aspartic acid and glutamic acid, in addition to the C-terminal carboxyl group, are negatively charged. Lysine, arginine and histidine in addition to the N-terminal alpha-amino group are positively charged (Tab. 5). In addition, modifications of proteins can confer new groups which also contribute to the overall charge. Thus sialoglycoproteins contain sialic acid carboxylates and phosphoprotein phosphate groups. Other post-translational modifications that provide new charged groups or change the acid–base properties of existing charged groups are: methylation of lysine and histidine, which turns their side chains into much stronger bases (CREIGHTON, 1983), acylation of the N-terminus, cyclization of an N-terminal glutamate to pyroglutamate, and the amidation or esterification of the C-terminal alpha-carboxylate which remove the charges on these groups. It should be noted that many post-translational modifications only affect part of a population of molecules and give rise to charge heterogeneity. Such charge variants may often be resolved by high performance ion exchange chromatography.

At neutral pH, the typical globular protein has a slight excess of negatively charged groups and net negative charge. As the pH decreases, these groups are protonized and the net charge will turn positive. At high pH values deprotonation of histidines and alpha-amino groups and later on lysine will lead to an increased net negative charge. The detailed titration curve of a protein depends on the environment of the individual charge of the amino acid side chains, and thus on the conformation of the protein. The relationship between total charge and pH can be determined as the titration curve, e.g., by electrophoresis (ROSENGREN et al., 1977; HAFF et al., 1983).

Although the overall charge on a protein is usually taken as a good indicator of its behavior in ion exchange chromatography, it is the distribution of charges on the surface that affects binding, particularly close to the isoelectric point. This may cause a protein to bind to a cation exchanger with a "patch" of positive residues also when the overall charge is slightly negative, and *vice versa* to an anion exchanger when the overall charge is slightly positive. The surface charge is also a function of the conformation of the protein, and agents that affect conformation may also affect binding to the ion exchanger.

Protein size is of some importance in ion exchange chromatography. An increase in molecular weight will cause stronger binding to an ion exchanger, especially evident when a protein forms aggregates. For very large proteins, there is a reduction of binding capacity due to size exclusion on the matrix itself.

Ion exchangers consist of an inert gel matrix that carries the charged groups which take part in the ion exchange process. Although the properties of the matrix strongly affect the practical use of an ion exchanger, the charged groups are most important for the actual process of ion exchange and the separation result which can be obtained. Most ion exchangers are either anion exchangers or cation exchangers. Zwitter-ionic exchangers (PORATH and FRYKLUND, 1970; PORATH and FORNSTEDT, 1970) have generally proved to be less useful.

Anion exchangers carry positively charged groups and bind anions. Typical anion exchange groups are: diethylaminoethyl (DEAE) and quaternary amino (Q). Similarly, cation exchangers carry negatively charged groups and bind cations. Typical cation exchange groups are: carboxymethyl (CM) and sulfonate (S). See Fig. 3 for structures of common ion exchange groups.

Q and S ion exchangers are termed strong ion exchangers, since they have extreme pK_a values and therefore are charged at most pHs (all pHs in the case of Q). DEAE and CM ion exchangers are termed weak ion exchangers, since they are weakly dissociated, and therefore uncharged, outside a relatively narrow pH range. While, in principle, it should be possible to demonstrate significant differences between two different anion or cation exchange groups in terms of their ability to bind proteins, under pH conditions where they both are charged, it is difficult to give any hard and fast recommendations as to which will give the most useful result. Since an ion exchanger must be charged and should ideally be applicable at a wide range of pH values, strong ion exchangers, e.g., Q and S, are most generally used.

The total ionic capacity of an ion exchanger depends on its degree of substitution with charged groups, typical values being ca. 20–40 µmol/mL. However, most of the charged groups are in the interior of the gel. The gel matrix must, therefore, be porous enough to allow proteins to diffuse inside to bind.

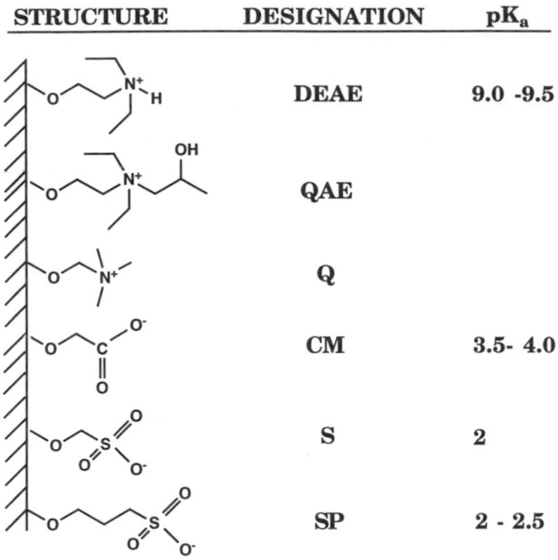

STRUCTURE	DESIGNATION	pK_a
	DEAE	9.0 -9.5
	QAE	
	Q	
	CM	3.5- 4.0
	S	2
	SP	2 - 2.5

Fig. 3. The chemical structures of common functional groups of ion exchangers for proteins.

Apart from the surface charge and porosity of the ion exchanger, the amount of protein which an ion exchanger can bind depends on many factors which vary widely from case to case. These include the pH and salt concentration of the eluent, the eluent flow rate, and the size and surface charge of the proteins. Typical values are 100 mg/mL, but the actual capacity for a particular sample can only be determined by experiment under relevant conditions.

Anion and cation exchangers based on several different matrices are available, and these differences are often of practical significance, since it is often the matrix which determines properties such as mechanical and chemical stability as well as cost.

In contrast to the situation described above for gel filtration, the composition of the mobile phase is of prime importance and must be controlled within the limits set by the requirements of the technique. As described above, the pH and ionic composition of the eluent is critical.

Elution of bound proteins is most often achieved by superimposing a gradient of ionic strength on a constant concentration of buffer salt at constant pH by adding an increasing concentration of a non-buffering, salt, e.g., NaCl, to the eluent. Different salts have different elution powers for different proteins (SÖDERBERG, 1983). However, these effects are difficult to predict and seldom of practical significance. Similarly, effects of using specific buffer salts are only seldom noticed. Elution by changing the pH of the eluent is not as common, partly due to the difficulty of finding suitable buffer combinations.

Various additives may be used, typically to improve protein solubility (detergents) or to modify protein conformation, naturally with the restriction that they do not contribute to the ionic strength of the eluent. Note that anionic detergents will bind extremely strongly to an anion exchanger (and cationic detergents to a cation exchanger), thereby turning it into a highly hydrophobic medium and rendering it useless for its original purpose.

Chromatofocusing is a special case of ion exchange chromatography in which the solid phase is substituted with charged groups exhibiting a high buffering capacity over a wide pH interval. The presence of quaternary ammonium groups secures high protein binding capacity also at high pH values. The elution of the adsorbed proteins takes place using a pH gradient. The ion exchanger is equilibrated with a suitable buffer salt solution at the higher pH value of the selected pH interval. The sample solution, containing the proteins to be separated having isoelectric points within the selected pH interval, is usually equilibrated in the same buffer. After sample application, a special elution buffer called "polybuffer" is pumped through the column with constant speed. The pH of the polybuffer is adjusted to the lowest pH of the selected interval. Characteristically, the polybuffer has an even and high buffering capacity over the entire pH interval which leads to the development of an even and stable pH gradient in the mobile phase as a consequence of its slow titration by the ion exchanger.

At the beginning of the elution, the proteins of the sample are adsorbed as a relatively narrow band at the top of the column. When the polybuffer is introduced, the pH value will slowly decresae in the mobile phase around the adsorbed proteins. When the pH gradient passes the isoelectric point of a particular protein, its charge will change sign and it will be repelled from the ion exchanger. It will then be transported down the column by the flowing liquid. The pH of the polybuffer surrounding the moving protein is, however, not constant but changes by titration due to the high buffering capacity of the charged groups bound to the ion exchanger particles. Suddenly the pH has passed the isoelectric point of the protein, but now from the other direction. The protein becomes negatively charged again and will be adsorbed to the positive charges of the ion exchanger. However, as the gradient moves constantly down the column, the described procedure will be repeated for each protein in the mixture over and over again. The result is that each protein will be enriched, focused, to a very narrow moving band leaving the outlet of the column at a pH value in the gradient close to its isoelectric point.

Every separated protein in the mixture will move at the same speed, the speed of the gradient. Of crucial importance is the fact that due to the constant titration taking place in the column between the polybuffer and the ion exchanger, the moving speed of the gradient will be approximately one tenth of that of the liquid flow. Thus, proteins lagging behind in the gradient will be quickly transported back to the immediate neighborhood of their isoelectric points. After a while, an equilibration situation will prevail in which each protein occupies a band the width of which is a complex function of the steepness of the pH gradient, the gross kinetics of adsorption–desorption, the longitudinal diffusion rate, and the degree

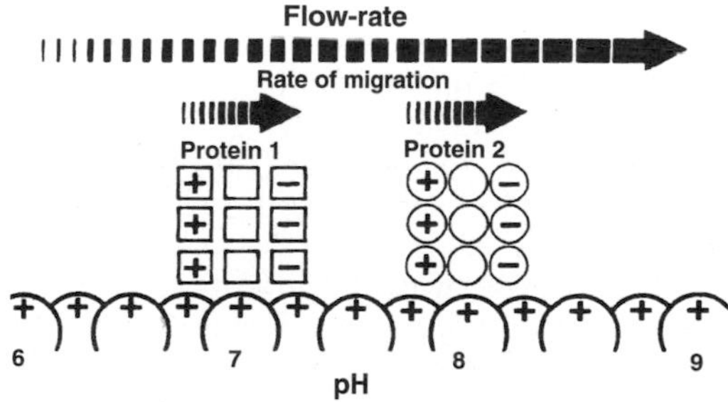

Fig. 4. Schematic representation of the principle for chromatofocusing of proteins. See text for details.

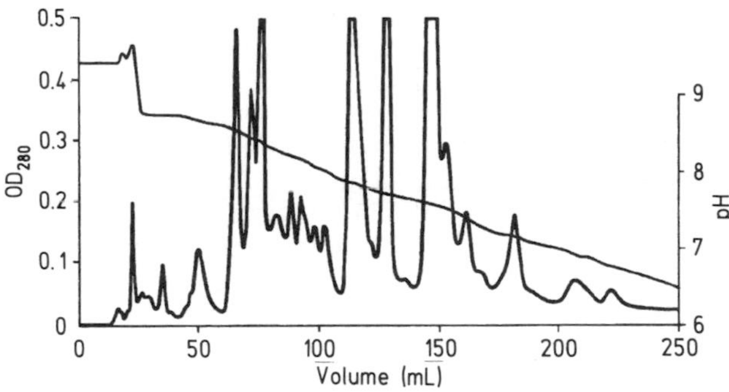

Fig. 5. Elk muscle proteins separated by chromatofocusing. 10×35 cm column packed with PBE 94. Applied sample: 5 mL elk meat extract in 25 mM ethanolamine-HCl buffer pH 9.4. Eluted with Polybuffer 96, pH 6, 7.5 mM/pH-unit/mL at 20 cm/h. (Work from Pharmacia LKB Biotechnology AB, Uppsala)

of flow inhomogeneity as a consequence of the packing quality of the column. Fig. 4 shows schematically the principle of chromatofocusing, and Fig. 5 presents the result of a separation of a complex protein mixture.

7 Protein Hydrophobicity Methods – Hydrophobic Interaction Chromatography and Reversed Phase Chromatography

Hydrophobic interaction chromatography (HIC) has become a popular technique for purifying proteins primarily because it displays binding characteristics complementary to other protein chromatography techniques such as ion exchange chromatography. HIC is a versatile technique for purifying proteins, exploiting the often small differences in surface-located hydrophobic patches of the proteins. It is closely related to the classical "salting-out chromatography" introduced by TISELIUS (1948). It has many desirable characteristics, of which the following are important when purifying a given protein in a biological sample, e.g., recombinant proteins in cell culture supernatants or whole cell extracts. It is most effective at an early stage of a purification strategy and also for concentration. Sample pretreatment such as dialysis or desalting after ammonium sulfate precipitation are usually not necessary. It generally functions by the principle of group separation and normally 50% or more of the extraneous impurities are removed, including the major portion of the proteolytic enzymes. In addition, HIC offers relatively high adsorption capacity combined with good selectivity and satisfactory yield of active material.

In HIC proteins are separated based on differences in their content of hydrophobic amino acid side chains on their surface, or in clefts and crevices at or near the protein surface. The surface of globular proteins typically has around 45% hydrophobic residues and 55% charged or hydrophilic uncharged residues accessible to the surrounding water. These are more or less unevenly distributed on the surface. The separation takes place by differential interaction with individual alkyl or aryl substituents on neutral, hydrophilic carriers such as agarose gels. The binding to the HIC adsorbent is primarily driven by the weak hydrophobic interaction within the aqueous solvent and to a lesser extent by the creation of van der Waals interactions. The strength of the binding depends on the density of hydrophobic groups on the surface, or near the surface of the protein and on the type and degree of substitution of the hydrophobic ligand coupled to the polymer matrix. As a rule, HIC requires the presence of moderately high concentrations of salt (1–2 M), and the term salt-promoted adsorption is often used for this type of chromatography. The adsorbed proteins are typically eluted differentially either by decreasing the salt concentration and/or by increasing

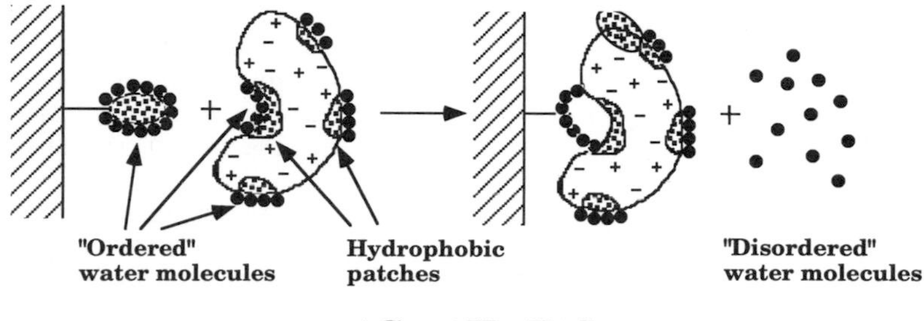

Fig. 6. Thermodynamic interpretation of the principle of hydrophobic interaction chromatography. See text for a discussion.

the eluent concentration of polarity perturbants such as ethylene glycol.

According to the second law of thermodynamics, a process which is accompanied by a decrease in free energy tends to occur spontaneously. For an isothermal process the change in free energy is obtained from the well-known equation $\Delta G = \Delta H - T\Delta S$. The underlying principle of HIC is thus assumed to be the gain in entropy resulting from the disruption of the ordered water molecules surrounding the interacting hydrophobic groups into the bulk of disordered water molecules (Fig. 6). This will increase the entropy of the system ($\Delta S > 0$) thus leading to a decrease in free energy, ΔG.

The protein binding mechanism of HIC is superficially similar to that of reversed phase chromatography (RPC), i.e., the adsorption involves interaction with hydrophobic groups attached to a chromatographic carrier material. However, as the retarding process in HIC does not normally involve those hydrophobic amino acid residues that are buried in the protein's interior, and takes place in neutrally buffered aqueous salt solutions, it does not significantly change the protein conformation and thus conserves the native, biologically active state of most native globular proteins. In RPC, on the other hand, the separation mechanism is based on a total coverage of the solid phase surface with hydrocarbon groups (HORVATH and MELANDER, 1977), and elution takes place by the application of concentration gradients of organic solvents, which in most cases leads to protein denaturation and irreversible inactivation of biological activity.

The most widely used ligands for HIC and RPC are straight chain alkanes ($-C_n$) or simple aromatic compounds, notably phenyl. The latter results in HIC adsorbents of a mixed-mode type where aromatic interactions are superimposed on the hydrophobic character of the phenyl group. This is reflected in the adsorption selectivity of these two types of immobilized ligands. The immobilized *n*-alkanes form a homologous series of adsorbents where their adsorption selectivity decreases with increase in the alkyl chain length. Furthermore, the strength of interaction increases with increase in the *n*-alkyl chain length. This can be summarized by the following hydrophobicity scale of HIC ligands, arranged in increasing order of hydrophobicity or decreasing selectivity:

methyl < ethyl < propyl < butyl < phenyl <
pentyl < hexyl < heptyl < octyl

In recent years the term thiophilic affinity has been used for a salt-promoted protein adsorption technique based on thioether containing alkyl ligands and related to hydrophobic interaction chromatography (PORATH et al., 1985; OSCARSSON and PORATH, 1990). Obviously, there seems to be an additional selectivity involved when a sulfur atom is substituted for the oxygen. Thus butyl-S showed higher selectivity than butyl-O for the purification of recombinant hepatitis B surface antigen produced by transformed CHO-cells (BELEW et al., 1991). In Fig. 7 the structures of common ligands for HIC of proteins are shown.

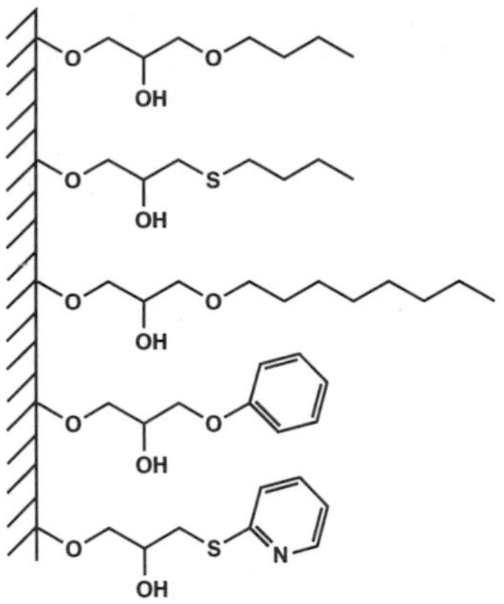

Fig. 7. The chemical structures of some ligands for hydrophobic interaction chromatography, HIC.

The higher the concentration of the immobilized hydrophobic ligand, usually the stronger the interaction. The main reason is the higher probability of forming multi-point interactions between the protein and the attached ligands. Sometimes, and especially in RPC, this may lead to the denaturation of the protein arising from structural changes that can occur during such strong interactions. Elution of bound protein can also be difficult requiring the use of aqueous–organic mixtures for desorption, thus leading, in some instances, to the denaturation of the eluted protein. Ligand density is therefore an important optimization parameter in the design of HIC adsorbents, to such an extent that manufacturers, in some cases, offer products based on the same ligand but displaying two or more different ligand densities to satisfy the demand of selectivity and binding capacity for different categories of proteins.

Proteins are hydrophobic chiefly because they contain amino acids with hydrophobic side chains (alanine, leucine, isoleucine, valine, phenylalanine, tryptophan, tyrosine, proline, methionine) exposed on the surface or in clefts or crevices close to the surface of the protein. See Tab. 6 for a comparison of two hydrophobicity scales for amino acids normally found in proteins. Those buried in the inner structure of the protein normally do not participate in

Tab. 6. Hydrophobicity Scale for Amino Acids. Scale 1 is based on the different amino acids, expressed as the free energy of transfer from ethanol or dioxane to water. Scale 2 is based on the fraction of the number of amino acids buried within proteins. Reproduced with permission from ERIKSSON (1989)

Scale 1	(kJ/mol)	(kcal/mol)	Scale 2 (Fraction buried)	
Trp	15.83	3.77	Phe	0.87
Ile	13.23	3.15	Trp	0.86
Phe	12.05	2.87	Cys	0.83
Pro	11.63	2.77	Ile	0.79
Tyr	11.21	2.67	Leu	0.77
Leu	9.11	2.17	Met	0.76
Val	7.85	1.87	Val	0.72
Met	7.01	1.67	His	0.70
Lys	6.89	1.64	Tyr	0.64
Cys	6.38	1.52	Ala	0.52
Ala	3.65	0.87	Ser	0.49
His	3.65	0.87	Arg	0.49
Arg	3.57	0.85	Asn	0.42
Glu	2.84	0.67	Gly	0.41
Asp	2.77	0.66	Thr	0.38
Gly	0.42	0.10	Glu	0.38
Asn	0.38	0.09	Asp	0.37
Ser	0.29	0.07	Pro	0.35
Thr	0.29	0.07	Gln	0.35
Gln	0.00	0.00	Lys	0.31

binding to HIC adsorbents. As discussed above, this is an important distinction between HIC and reversed phase chromatography (RPC). The hydrophobicity is not only affected by the number and the type of exposed hydrophobic amino acids, but also by their relative position in the protein. In particular many active sites and binding sites in globular proteins are covered by hydrophobic groups – they often function as "the organic solvent in an aqueous environment". These sites may also constitute the binding site to a hydrophobic group on an HIC gel. The structure of these sites also depends on conformation, and conformation perturbing agents also influence the behavior of a protein in hydrophobic interaction chromatography.

Changing the pH primarily affects the charge of the protein; however, this in turn may affect its conformation. Surface charge based expansion or contraction of a protein may affect the number of exposed hydrophobic sites and consequently its overall hydrophobicity. Thus, acid expansion increases the hydrophobicity of albumin and thereby causes it to be eluted later in HIC. It is generally thought that increasing the pH decreases the hydrophobic interaction between proteins and the immobilized hydrophobic ligands, but this is not always the case. Although many molecules bind stronger to the matrix at lower pH, some show weaker attachment. In some cases it is even possible to apply the sample at a low pH with a relatively low salt concentration. Each protein will behave differently at different pH values. Some proteins, like lysozyme, do not easily bind to HIC adsorbents at neutral pH and moderate salt concentrations, although their 3-D structure clearly shows the presence of several hydrophobic patches on sections of their surface exposed to the solvent.

The size of the protein is also a contributing factor: the larger the protein, the higher the probability for multi-point attachment leading to an apparently stronger adsorption. However, very large proteins may have limited access to hydrophobic groups in the interior of the adsorbent resulting in reduction of binding capacity. Finally, glycoproteins appear to be less hydrophobic than non-glycosylated proteins.

Non-buffering ions, coming, for example, from added neutral anti-chaotropic salt, promote binding of proteins to the alkyl or aryl groups on the HIC adsorbent. The salt effect in HIC is often discussed in relation to the Hofmeister (lyotropic) series: decreasing salting-out effect and increasing chaotropic effect:

Anions: SO_4^{2-}, HPO_4^{2-}, CH_3COO^-, Cl^-, Br^-, NO_3^-, ClO_4^-, I^-, SCN^-

Cations: NH_4^+, Rb^+, K^+, Na^+, Cs^+, Li^+, Mg^{2+}, Ca^{2+}, Ba^{2+}

Too high concentrations of salt will, however, reduce the binding selectivity by causing salting-out effects to the solid phase rather than just promoting the hydrophobic interaction. This phenomenon was described by TISELIUS (1948). A brief general discussion of salting-out adsorption techniques for protein purification has been published by PORATH (1987).

Increasing the temperature enhances hydrophobic interactions; conversely, decreasing the temperature decreases hydrophobic interactions. Lowering the temperature generally causes a protein to be eluted earlier, and as a sharper peak. However, this may not always be true since hydrophobic interactions within the protein itself may also be affected, changing both conformation and surface hydrophobicity. If the samples adhere too strongly to the HIC adsorbent, a lower temperature will decrease the interactions and sometimes improve peak symmetry. A 20–30% reduction in binding strength has been observed when the temperature was reduced from 25 to 4 °C.

Various additives may be used to improve protein solubility or to modify protein conformation. Detergents are often added to increase solubility of proteins and particularly for keeping integral membrane proteins in solution. All HIC adsorbents are stable in the presence of detergents, but since detergents bind to the hydrocarbon chains, they will change. When ionic detergents are bound to the hydrophobic gel, a mixed mode separation of ion exchange and HIC will be promoted due to the presence of the charged group of the detergent. Detergents can be removed from the HIC media using cleaning instructions provided by the supplier.

The binding of proteins to hydrocarbon-containing gels is influenced by the hydrophobicity of the ligand such that short-chain alkyl is less hydrophobic than aryl which is less hydrophobic than long-chain alkyl such as octyl. The ionic strength of the buffer is an important parameter in HIC, and those salts which cause salting out, e.g., $(NH_4)_2SO_4$, also promote the binding of proteins to hydrophobic ligands. Binding to butyl octyl and phenyl ligands in standard HIC media is generally negligible unless high salt buffer solutions are used. A salt concentration just below that used for salting out of the protein is normally adequate. Increasing the salting-out effect strengthens the hydrophobic interactions, whereas increasing the chaotropic effect weakens them.

Elution is effected by reducing hydrophobic interactions, either by reducing the concentration of salting-out ions in the buffer with a negative salt gradient, or by increasing the concentration of chaotropic ions in the buffer in a positive gradient.

If the protein of interest is very hydrophobic and is eluted at a very low ionic strength (i.e., at the end of the gradient), lowering the initial concentration of the added salt may save time. Another advantage is that many of the contaminants that would normally bind to the matrix do not bind to the column.

8 Protein Affinity Methods – Affinity Chromatography

Traditionally, affinity methods are based on the formation of biospecific pairs of the type listed in Tab. 7. In practice, however, the term is also used for separation methods based on more or less specific interactions between proteins and a variety of ligand molecules covalently attached to a solid phase. Classical biospecific affinity chromatography (CUATRECASAS et al., 1968; MOSBACH, 1974), is primarily based on the use of immobilized enzyme inhibitors such as substrate analogs or cofactor analogs. Thus, WILCHEK et al. (1984) listed the conditions for affinity chromatography of 293 enzymes. Later, other methods were adopted that are based on, e.g., the formation of chelated metal-ion complexes (see below) and charge-transfer complexes (PORATH, 1989). Dyes such as Cibacron Blue F3G-A have been popular ligands not only for group separation of a variety of enzymes (dehydrogenases, kinases, etc., KOPPERSCHLÄGER et al., 1982), but also for the purification of proteins like human serum albumin (HARVEY, 1980) and human interferons (JANKOWSKI et al, 1976; PESTKA, 1983). By chemical modification of the dye aromatic ring system, it is possible to alter the binding specificity and to change the dye-protein affinity by several orders of magnitude (LOWE et al., 1986). A strategy for enzyme isolation using dye ligand affinity chromatography has been reported by SCOPES (1986).

Tab. 7. Examples of Biospecific Pairs Utilized in Classical Affinity Chromatography

Enzyme	Substrate, substrate analog Cofactor, cofactor analog Inhibitor
Hormone	Carrier protein Receptor
Glycoprotein	Lectin
Antibody	Antigen Hapten
Nucleic acid	Complementary polynucleotide Polynucleotide-binding protein

Introduced by PORATH et al. (1975), immobilized metal affinity chromatography (IMAC) offers a selectivity factor apparently mainly dependent on the number of available histidine residues on the surface of a protein (SULKOWSKI, 1985). In the synthesis of an IMAC adsorbent, a metal chelate complex forming group such as iminodiacetic acid is covalently attached to a suitable support such as bead-shaped agarose. Before use, the chelating groups are saturated with the appropriate metal ion (most often Me^{2+}). The most useful metal ions can be found among the first series transition metals (Cu^{2+} and Zn^{2+}, and to some extent Co^{2+} and Ni^{2+}; SULKOWSKI, 1985). In one example, a Ni^{2+} chelate adsorbent was used for the purification of recombinant human interferon gamma (ZHANG et al., 1992). Phosphoproteins have been isolated with Fe^{3+} gels (ANDERSSON and PORATH, 1986).

Other important affinity purification methods utilize protein ligands such as lectins for glycoprotein purification (KRISTIANSEN, 1974), immuno ligands such as monoclonal antibodies for isolation of a variety of proteins (SECHER and BURKE, 1980; GOODING, 1986) and "pseudoimmunoaffinity" ligands such as staphylococcal protein A (HJELM and SJÖQUIST, 1975; LANGONE, 1982) and streptococcal protein G (ÅKERSTRÖM et al., 1985) for the purification of IgG from various mammalian sera.

Heparin is a sulfated glycosaminoglycan with anticoagulant properties which after immobilization to an appropriate carrier has been shown to selectively bind several different enzymes and other proteins (see *Handbook on Heparin Sepharose® CL-6B* from Pharmacia for a comprehensive list of references). Attached to Sepharose® Fast Flow, heparin is used for the industrial isolation of antithrombin III from human plasma, packed in a 120 cm (i.d.) × 10 cm column (113 liters), according to a method developed by EKETORP and described by JANSON and PETTERSSON (1992). Finally, chemisorption of proteins containing free-SH groups to activated mixed disulfides (covalent chromatography) (BROCKLEHURST et al., 1973) finds occasional use.

Proteolytic degradation, causing microheterogeneity and the possible formation of neodeterminant immunogenic structures, is one of the major problems in biotechnology downstream processing. The inhibition, or quick removal, of proteases is therefore of considerable interest. Affinity adsorbents based on immobilized protease inhibitors have primarily been used for the isolation of serine proteases. Examples of such inhibitors are soybean trypsin inhibitor, aprotinin (Trasylol), pepstatin, D-tryptophan methyl ester, glycyl-L-tyrosylazobenzylsuccinic acid, L-alanyl-L-alanyl-L-alanine, *p*-aminobenzamidine and phosphoryl difluoride (MILLQVIST et al., 1988). Thiol proteases can be isolated by mixed disulfide formation with 2-pyridyldisulfide-activated thiol groups (covalent chromatography). Immobilized bacitracin has been shown to efficiently bind serine-, aspartyl-, and metallo-proteases from various sources (STEPANOV and RUDENSKAYA, 1983). The application of crude extracts from homogenized baker's yeast and *E. coli* to HIC adsorbents has been shown to considerably reduce their protease content (HEDMAN and GUSTAFSSON, 1985).

In the design of an affinity separation experiment, it is important to consider, in addition to the choice of ligand, the choice of support and chemistry for the immobilization of the ligand. Still, 25 years after the birth of modern affinity chromatography, 4% agarose and the CNBr activation method (AXÉN et al., 1967; WILCHEK et al., 1984) are most widely used. Other useful coupling techniques are based on activation with bisoxiranes (SUNDBERG and PORATH, 1974), organic sulfonyl chlorides such as tresyl chloride (NILSSON and MOSBACH, 1984) and N-hydroxysuccinimide (NHS). NHS-activated agarose reacts readily with primary or secondary amino group-containing ligands at pH 5 – 8. The main advantage over the CNBr method is the high stability of the resulting amide bond over a wide pH range and in the presence of nucleophilic reagents.

By attaching low-molecular weight ligands used in low-affinity systems (dissociation constant K_D of 10^{-3}–10^{-5}) at some distance from the matrix using spacers or "leashes", their functional concentration and availability to binding will be increased significantly (CUATRECASAS et al., 1968; STEERS et al., 1971). The spacer concept pinpoints the importance of high degrees of substitution in affinity chromatography using low-molecular weight ligands. One way to achieve this, and also to reduce the ligand leakage, is to use polymer spacers such as polyacryl hydrazide, as was suggested by WILCHEK and MIRON (1974).

Other important issues to address are: which elution agents should be investigated and which are the optimum operating conditions, e.g., regarding temperature, buffers, flow rate, and column size? For a more detailed discussion of these and the other affinity separation topics reviewed in this chapter, the reader is referred to a number of excellent monographs and handbooks on the subject (JACOBY and WILCHEK, 1974; LOWE, 1979; SCOUTEN, 1981; DEAN et al., 1985). Also it is often profitable to search for original work published in the proceeding volumes of the latest affinity chromatography symposia (e.g., ISHII, 1992; WILCHEK, 1990; JENNISSEN and MÜLLER, 1988; TURKOVA et al., 1986; CHAIKEN et al., 1983; GRIBNAU et al., 1982).

9 Concluding Remarks

The often extreme binding equilibria and slow interaction kinetics encountered in the chromatographic separation and purification of proteins and large peptides differ markedly from what one experiences with low-molecular weight substances. This is primarily due to the large number of possible chemical interactions arising from variations in the frequency and distribution of the amino acid side chains on the surface of these molecules, but also from the inherently lower diffusivity of the larger molecules.

Many biospecific affinity systems have been exploited in the laboratory because of their high purification qualities in a single step. However, for planning large-scale routine protein purification, the best strategy is to design separation procedures which best fit into the overall process to give the best long-term recovery and economy. This is often achieved by a carefully controlled combination of simple, sturdy separation principles such as those based on differences in ionic charge, hydrophobic interaction, and size exclusion. An enormous amount of empirical knowledge of these separation principles has accumulated in the literature, ready to be tapped by the progressive investigator, and much of it has already been converted into detailed advice in logical expert systems that are useful to both the novice and the expert.

The wide range of chromatographic media now available reduces the need for the biochemical engineer to modify the basic support material or to manipulate more or less hazardous activating reagents. Furthermore, media suppliers nowadays are forced of Regulatory Support Files to disclose commercially sensitive, but important, details about composition, manufacture, and critical properties of the separation tools. This, and the demand for ISO 9001 certification, means that many industrially applied chromatographic media are extensively characterized and documented with respect to operational parameters, and their production processes are made highly reproducible. Different outcomes of allegedly similar separation procedures can therefore be ascribed with greater confidence than before to sample variation rather than to more or less untraceable differences in media characteristics.

10 References

AFEYAN, N. B., FULTON, S. P., REGNIER, F. (1991), Perfusion chromatography packing materials for proteins and peptides, *J. Chromatogr.* **544**, 267–279.

ÅKERSTRÖM, B., BRODIN, T., REIS, K., BJÖRK, L. (1985), Protein G: A powerful tool for binding and detection of monoclonal and polyclonal antibodies, *J. Immunol.* **135**, 2589–2592.

ANDERSSON, L., PORATH, J. (1986), Isolation of phosphoproteins by immobilized metal (Fe^{3+}) affinity chromatography, *Anal. Biochem.* **154**, 250–254.

ANDERSSON, T., CARLSSON, M., HAGEL, L., PERNEMALM, P.-Å., JANSON, J.-C. (1984), Agarose based media for high-resolution gel filtration of polymers, *J. Chromatogr.* **326**, 33–44.

ARNOTT, S., FULMER, A., SCOTT, W. E., DEA, I. C. M., MOORHOUSE, R., REES, D. A. (1974), The double agarose helix and its function in agarose gel structure, *J. Mol. Biol.* **90**, 269–284.

AXÉN, R., PORATH, J., ERNBACK, S. (1967), Chemical coupling of peptides and proteins to polysaccharides by means of cyanohalides, *Nature* **214**, 1302–1304.

BELEW, M., MEI, Y., LI, B., BERGLÖF, J., JANSON, J.-C. (1991), Purification of recombinant hepatitis B surface antigen produced by transformed Chinese hamster ovary (CHO) cell line grown in culture, *Bioseparation* **1**, 397–408.

BJÖRLING, H. (1972), Plasma fractionation methods used in Sweden, *Vox Sang.* **23**, 18–25.

BLACK, S. D., MOULD, D. R. (1991), Development of hydrophobicity parameters to analyze proteins which bear post- and cotranslational modifications, *Anal. Biochem.* **193**, 72–82.

BRIEFS, K.-G., KULA, M.-R. (1992), Fast protein chromatography on analytical and preparative scale using modified microporous membranes, *Chem. Eng. Sci.* **47** (1), 141–149.

BROCKLEHURST, K., CARLSSON, J., KIERSTAN, M. P. J., CROOK, E. M. (1973), Covalent chromatography – preparation of fully active papain from dried papaya latex, *Biochem. J.* **133,** 573–584.

CHAIKEN, I. M., WILCHEK, M., PARIKH, I. (Eds.), (1983), *Affinity Chromatography and Biological Recognition*, pp. 1–515, Orlando: Academic Press.

CHAMPLUVIER, B., KULA, M.-R. (1992), Dye-ligand membranes as selective adsorbents for rapid purification of enzymes: A case study, *Biotechnol. Bioeng.* **40,** 33–40.

COHN, E. J., STRONG, L. E., HUGHES, W. L., JR., MULFORD, D. J., ASHWORTH, J. N., MELIN, M., TAYLOR, H. L. (1946), Preparation of components of human plasma, *J. Am. Chem. Soc.* **68,** 459–475.

CREIGHTON, R. E. (1983), *Proteins. Structure and Molecular Principles*, New York: Freeman.

CUATRECASAS, P., WILCHEK, M., ANFINSEN, C. B. (1968), Selective enzyme purification by affinity chromatography, *Proc. Natl. Acad. Sci. USA* **61,** 636–643.

DEAN, P. D. G., JOHNSON, W. S., MIDDLE, F. A. (1985), *Affinity Chromatography, A Practical Approach*, Oxford: IRL Press.

DETERMAN, H., MEYER, N., WIELAND, T. (1969), Ion exchanger from pearl-shaped cellulose gel, *Nature* **223,** 499–500.

DEUTSCHER, M. P. (1990), Guide to protein purification, *Methods Enzymol.* **182,** 1–894.

ER-EL, Z., ZAIDENZAIG, Y., SHALTIEL, S. (1972), Hydrocarbon coated Sepharoses. Use in the purification of glycogen phosphorylase, *Biochem. Biophys. Res. Commun.* **49,** 383–390.

ERIKSSON, K.-O. (1989), Hydrophobic interaction chromatography, in: *Protein Purification, Principles, High Resolution Methods and Applications* (JANSON, J.-C., RYDÉN, L., Eds.), pp. 207–226, New York: VCH Publishers Inc.

FISH, W. W., MANN, K. G., TANFORD, C. (1969), The estimation of polypeptide chain molecular weights by gel filtration in 6 M guanidine hydrochloride, *J. Biol. Chem.* **244,** 4989–4994.

FISHER, B., SUMNER, I., GOODENOUGH, P. (1993), Isolation, renaturation and formation of disulfide bonds of eukaryotic proteins expressed in *Escherichia coli* as inclusion bodies, *Biotechnol. Bioeng.* **41,** 3–13.

FLODIN, P., (1960), Internal research report, Pharmacia AB, Uppsala, Sweden.

FOSTER, P. R. (1993), Protein precipitation, in: *Engineering Processes for Bioseparations* (WEATHERLEY, L. R., Ed.), Oxford: Butterworth-Heinemann.

GOODING, J. W. (1986), *Monoclonal Antibodies: Principles and Practice,* pp. 108–125, London: Academic Press.

GRIBNAU, T. C. J., VISSER, J., NIVARD, R. J. F. (Eds.) (1982) *Affinity Chromatography and Related Techniques, Analytical Chemistry Series,* Vol. 9, pp. 1–584, Amsterdam: Elsevier.

HAFF, L. A., FÄGERSTAM, L. G., BARRY, A. R. (1983), Use of electrophoretic titration curves for predicting optimal chromatographic conditions for fast ion-exchange chromatography of proteins, *J. Chromatogr.* **266,** 409–425.

HAGEL, L. (1988), Pore size distributions of chromatography media, in: *Aqueous Size-Exclusion Chromatography* (DUBIN, P., Ed.), p. 146, Amsterdam: Elsevier.

HAGEL, L. (1989), Gel filtration, in: *Protein Purification, Principles, High Resolution Methods and Applications* (JANSON, J.-C., RYDÉN, L., Eds.), pp. 63–106, New York: VCH Publishers Inc.

HAGEL, L. (1992), Peak capacity of columns for size exclusion chromatography, *J. Chromatogr.* **591,** 47–54.

HAGEL, L., JANSON, J.-C. (1992), Size exclusion chromatography, in: *Chromatography* (HEFTMAN, E., Ed.), 5th Ed., pp. A267–A307, Amsterdam: Elsevier.

HALLER, W. (1965), Chromatography on glass of controlled pore size, *Nature* **206,** 693–694.

HARVEY, M. J. (1980), The application of affinity chromatography and hydrophobic chromatography of the purification of human serum albumin, in: *Methods in Plasma Protein Fractionation* (CURLING, J., Ed.), pp. 189–200, New York: Academic Press.

HASHIMOTO, T., SASAKI, M., AIURA, M., KATO, Y. (1978), *J. Polym. Sci. Polym. Phys. Ed.* **16,** 1789.

HEDMAN, P., GUSTAFSSON, J.-G. (1985), Reduction of proteolytic break-down in microbial homogenates, *Dev. Biol. Stand.* **59,** 31.

HJELM, H., SJÖQUIST, J. (1975), *Scand. J. Immunol.* **4** (Suppl. 3), 51–57.

HJERTÉN, S. (1962), Chromatographic separation according to size of macromolecules and cell particles on columns of agarose suspensions, *Arch. Biochem. Biophys.* **99,** 466–475.

HJERTÉN, S. (1973), Some general aspects of hydrophobic interaction chromatography, *J. Chromatogr.* **87,** 325–331.

HJERTÉN, S., MOSBACH, R. (1962), "Molecular-sieve" chromatography of proteins on columns of cross-linked polyacrylamide, *Anal. Biochem.* **3,** 109–118.

HJERTÉN, S., YAO, K. (1981), High performance liquid chromatography of macromolecules on agarose and its derivates, *J. Chromatogr.* **215,** 317–322.

HOCHULI, E., BANNWARTH, W., DÖBELI, H., GENTZ, R., STÜBER, D. (1988), Genetic approach to facilitate purification of recombinant proteins with a novel metal chelate adsorbent, *Bio/Technology* **6**, 1321–1325.

HOFSTEE, B. H. J. (1973), Hydrophobic affinity chromatography of proteins, *Anal. Biochem.* **52**, 430–448.

HORVATH, C., MELANDER, W. (1977), Liquid chromatography with hydrocarbonaceous bonded phases; theory and practice of reversed-phase chromatography, *J. Chromatogr. Sci.* **15**, 393–404.

ISHII, S. (Ed.) (1992), *J. Chromatogr. Symp.* **597**, 1–452.

JACOBY, W. B., WILCHEK, M. (1974), *Methods Enzymol.* **39**, 1–80.

JANKOWSKI, W. J., VON MÜNCHHAUSEN, W., SULKOWSKI, E., CARTER, W. A. (1976), Binding of human interferons to immobilized Cibacron Blue F3G-A: The nature of molecular interaction, *Biochemistry* **15**, 5182–5187.

JANSON, J.-C. (1987), On the history of the development of Sephadex, *Chromatographia* **23**, 361–369.

JANSON, J.-C., HEDMAN, P. (1987), On the optimization of process chromatography of proteins, *Biotechnol. Progr.* **3** (1), 9–13.

JANSON, J.-C., PETTERSSON, T. (1992), Large scale chromatography of proteins, in: *Preparative and Production Scale Chromatography* (GANETSOS, G., BARKER, P. E., Eds.), pp. 559–590, New York: Marcel Dekker, Inc.

JANSON, J.-C., RYDÉN, L. (1989), *Protein Purification; Principles, High Resolution Methods and Applications,* New York: VDH Publishers, Inc.

JENNISSEN, H. P., MÜLLER, W. (1988), *Makromol. Chem., Macromol. Symp.*, Vol. 17, pp. 1–497, Heidelberg: Hüthig & Wepf Verlag.

JOHANSSON, I. (1977), *Sephacryl S-200 Superfine,* Pharmacia Fine Chemicals AB, Uppsala, Sweden.

KÅGEDAL, L., SÖDERBERG, L., ENGSTRÖM, B. (1988), Internal research report, Pharmacia LKB Biotechnology AB, Uppsala, Sweden.

KÅGEDAL, L,. ENGSTRÖM, B., ELLEGREN, H., LIEBER, A.-K., LUNDSTRÖM, H., SKÖLD, A., SCHENNING, M. (1991), Chemical, physical and chromatograpic properties of Superdex 75 prep grade and Superdex 200 prep grade gel filtration media, *J. Chromatogr.* **537**, 17–32.

KARLSSON, E., RYDÉN, L., BREWER, J. (1989), Ion exchange chromatography, in: *Protein Purification, Principles, High Resolution Methods and Applications* (JANSON, J.-C., RYDÉN, L., Eds.), pp. 107–148, New York: VCH Publishers Inc.

KOPPERSCHLÄGER, G., BÖHME, H. J., HOFMANN, E. (1982), Cibacron Blue F3G-A and related dyes as ligands in affinity chromatography, *Adv. Biochem. Eng.* **25**, 102–138.

KORNBERG, A. (1980), *Statement at a seminar* in Uppsala, Sweden.

KRISTIANSEN, T. (1974), Group specific separation of glycoproteins, *Methods Enzymol.* **34**, 331–341.

LANGONE, J. J. (1982), Applications of immobilized protein A in immunochemical techniques, *J. Immunol. Methods* **55**, 277–296.

LATHE, G. H., RUTHVEN, C. R. J. (1956), The separation of substances and estimation of their relative molecular sizes by the use of columns of starch in water, *Biochem. J.* **62**, 665–674.

LINDQVIST, B., STORGÅRDS, T. (1955), *Nature* **175**, 511.

LOWE, C. R. (1979), An introduction to affinity chromatography, in: *Laboratory Techniques in Biochemistry and Molecular Biology Series* (WORK, T. S., WORK, E., Eds.), Amsterdam: North Holland.

LOWE, C. R., BURTON, S. J., PEARSON, J. C., CLONIS, Y. D., STEAD, V. (1986), Design and application of bio-mimetic dyes in biotechnology, *J. Chromatogr.* **376**, 121–130.

LOWE, C. R., BURTON, S. J., BURTON, N. P., ALDERTON, W. K., PITTS, J. M., THOMAS, J. A. (1992), Designer dyes: "Biomimetic" ligands for the purification of pharmaceutical proteins by affinity chromatography, *Trends Biotechnol* **10**, 442–448.

MARSTON, F. A. O. (1986), The purification of eucaryotic polypeptides synthesized in *Escherichia coli, Biochem. J.* **240**, 1–12.

MILLQVIST, E., PETERSSON, H., RÅNBY, M. (1988), An affinity gel for the inhibition, binding and isolation of serine proteases, *Anal. Biochem.* **170**, 289–292.

MOKS, T., ABRAHAMSÉN, L., HOLMGREN, E., BILICH, M., OLSSON, A., POHL, G., SERKY, C., HULTBERG, H., JOSEPHSON, S., HOLMGREN, A., JÖRNVALL, H., UHLÉN, M., NILSSON, B. (1987), Expression of human insulin-like growth factor I in bacteria: Use of optimized gene fusion vectors to facilitate protein purification, *Biochemistry* **26**, 5244–5250.

MOSBACH, K. (1974), General ligand affinity chromatography, *Biochem. Soc. Trans.* **2**, 1294–1296.

MÜLLER, W. (1988), *Liquid-Liquid Partition Chromatography of Biopolymers*, pp. 1–60, Darmstadt: GIT Verlag GmbH.

MÜLLER, W. (1990), New ion exchangers for the chromatography of biopolymers, *J. Chromatogr.* **510**, 133–140.

NAVEH, D. (1990), Industrial-scale downstream

processing of biotechnology products, *BioPharm* (May), 28–36.

NILSSON, K., MOSBACH, K. (1984), Immobilization of ligands with organic sulfonyl chlorides, *Methods Enzymol.* **104,** 56–69.

OSCARSSON, S., PORATH, J. (1990), Protein chromatography with pyridine and alkyl thioether based agarose adsorbents, *J. Chromatogr.* **499,** 235–247.

PALEUS, S., NEILANDS, J. B. (1950), Preparation of cytochrome C with the aid of ion exchange resin, *Acta Chem. Scand.* **4,** 1024–1030.

PESTKA, S. (1983), The human interferons – from protein purification and sequence to cloning and expression in bacteria: Before, between and beyond, *Arch. Biochem. Biophys.* **221,** 1–37.

PETERSON, E. A., TORRES, A. R. (1984), Displacement chromatography of proteins, *Methods Enzymol.* **104,** 113–133.

PHARMACIA (1971), CNBr Activated Sepharose was introduced by Pharmacia Fine Chemicals AB, Uppsala, Sweden.

PHARMACIA (1993), Streamline™ media for expanded bed adsorption of proteins, Pharmacia BioProcess Technology AB, Uppsala, Sweden.

PORATH, J. (1987), Salting-out adsorption techniques for protein purification, *Biopolymers* **26,** S193–S204.

PORATH, J. (1989), Electron-donor-acceptor chromatography (EDAC) for biomolecules in aqueous solutions, in: *Protein Recognition of Immobilized Ligands*, pp. 101–122, New York: Alan R. Liss, Inc.

PORATH, J., FLODIN, P. (1959), Gel filtration: A method for desalting and group separation, *Nature* **183,** 1657–1659.

PORATH, J., FORNSTEDT, N. (1970), Group fractionation of plasma proteins on dipolar ion exchangers, *J. Chromatogr.* **51,** 479–489.

PORATH, J., FRYKLUND, L. (1970), Chromatography of proteins on dipolar ion exchangers, *Nature* **226,** 1169–1170.

PORATH, J., JANSON, J.-C., LÅÅS, T. (1971), Agar derivatives for chromatography, electrophoresis and gel-bound enzymes. 1. Desulphated and reduced cross-linked agar and agarose in spherical form, *J. Chromatogr.* **60,** 167–177.

PORATH, J., SUNDBERG, L., FORNSTEDT, N., OLSSON, I. (1973), A new approach to hydrophobic adsorption. Salting-out in amphipatic gels, *Nature* **245,** 465–466.

PORATH, J., CARLSSON, J., OLSSON, I., BELFRAGE, G. (1975), Metal chelate chromatography, a new approach to protein fractionation, *Nature* **258,** 598–599.

PORATH, J., MAISANO, F., BELEW, M. (1985), Thiophilic adsorption – a new method for protein fractionation, *FEBS Lett.* **185** (2), 306–310.

REGNIER, F. E., NOEL, R. J. (1976) *J. Chromatogr. Sci.* **14,** 316–320.

RODRIGUES, A. E., LOPES, J. C., LU, Z. P., LOUREIRO, J. M., DIAS, M. M. (1992), Importance of intraparticle convection in the performance of chromatographic processes, *J. Chromatogr.* **590,** 93–100.

ROSENGREN, Å., BJELLQVIST, B., GASPARIC, V. (1977), in: *Electrofocusing and Isotachophoresis* (RADOLA, B. S., GRAESSLIN, D., Eds.), p. 165, Berlin: Walter de Gruyter.

SCOPES, R. K. (1986), Strategies for enzyme isolation using dye-ligand and related adsorbents, *J. Chromatogr.* **376,** 131–140.

SCOPES, R. K. (1987), *Protein Purification, Principles and Practice*, pp. 1–282, 2nd Ed., New York: Springer Verlag.

SCOUTEN, W. H. (1981), Affinity chromatography, bioselective adsorption on inert matrices, *Chem. Anal.* (NY) **59,** 1–348.

SECHER, D. S., BURKE, D. C. (1980), A monoclonal antibody for large-scale purification of human leucocyte interferon, *Nature* **285,** 446–450.

SOBER, H. A., PETERSON, E. A. (1954), Chromatography of proteins on cellulose ion-exchangers, *J. Am. Chem. Soc.* **76,** 1711–1712.

SÖDERBERG, L. (1983), Physicochemical considerations in the use of Monobeads for the separation of biological molecules, *Protides Biol. Fluids* **30,** 629–634.

SOFER, G. K., NYSTRÖM, L.-E. (1989), *Process Chromatography, a Practical Guide*, pp. 1–145, London: Academic Press.

SOFER, G. K., NYSTRÖM, L.-E. (1991), *Process Chromatography, a Guide to Validation*, pp. 1–80, London: Academic Press.

STEERS, E. J. M., CUATRECASAS, P., POLLARD, H. B. (1971), The purification of beta-galactosidase from *Escherichia coli* by affinity chromatography, *J. Biol. Chem.* **246,** 196–200.

STEPANOV, V. M., RUDENSKAYA, G. N. (1983), Proteinase affinity chromatography on Bacitracin-Sepharose, *J. Appl. Biochem.* **5,** 420–428.

SULKOWSKI, E. (1985), Purification of proteins by IMAC, *Trends Biotechnol.* **3** (1), 1–7.

SUNDBERG, L., PORATH, J. (1974), Preparation of adsorbents for biospecific affinity chromatography. I. Attachment of group-containing ligands to insoluble polymers by means of bifunctional oxiranes. *J. Chromatogr.* **90,** 87–98.

SWINGLE, S. M., TISELIUS, A. (1951), Tricalcium phosphate as an adsorbent in the chromatography of proteins, *Biochem. J.* **48,** 171–174.

SYNGE, R. L. M. (1981), The Faraday Society's discussion at Reading in 1949 and the exploitation of molecular-sieve effects for chemical separations, *J. Chromatogr.* **215,** 1–6.

TISELIUS, A. (1948), Adsorption separation by salting out, *Arkiv Kemi, Mineral. Geol.* **26B** (1), 1–5.

TISELIUS, A., HJERTÉN, S., LEVIN, Ö. (1956), Protein chromatography on calcium phosphate columns, *Arch. Biochem. Biophys.* **65,** 132–155.

TITCHENER-HOOKER, N. J., GRITSIS, D., MANNWEILER, K., OLBRICH, R., GARDINER, S. A. M., FISH, N. M., HOARE, M. (1991), Integrated process design for producing and recovering proteins from inclusion bodies, *Pharm. Technol.* **3** (9), 42–48.

TURKOVA, J., CHAIKEN, I. M., HEARN, M. T. W. (Eds.) (1986) *J. Chromatogr. Symp. Ser.* Vol. **376,** 1-451.

UGELSTAD, J., SÖDERBERG, L., BERGE, A., BERGSTRÖM, J. (1983), Monodisperse polymer particles - a step forward for chromatography, *Nature* **303,** 95–96.

UGELSTAD, J., BERGE, A., ELLINGSEN, T., SCHMID, R., NILSEN, T.-N., MÖRK, P. C., STENSTAD, P., HORNES, E., OLSVIK, Ö. (1992), Preparation and application of new monosized polymer particles, *Prog. Polym. Sci.* **17,** 87–161.

UNGER, K. K., KERN, R., MINOU, M. C., KREBS, K.-F. (1974), *J. Chromatogr.* **99,** 435.

URIEL, J. (1966), *Bull. Soc. Chim. Biol.* **48,** 969.

WILCHEK, M. (Ed.) (1990), *J. Chromatogr. Symp.* **510,** 1–375.

WILCHEK, M., MIRON, T. (1974), Polymers coupled to agarose as stable and high capacity spacers, *Methods Enzymol.* **34,** 72–76.

WILCHEK, M., MIRON, T., KOHN, J. (1984), Affinity chromatography, *Methods Enzymol.* **104,** 3–55.

YON, R. J. (1972), Chromatography of lipophilic proteins on adsorbents containing mixed hydrophobic and ionic groups, *Biochem. J.* **126,** 765–767.

ZECHMEISTER, L., ROHDEWALD, M. (1951), Some aspects of enzyme chromatography, *Fortschr. Chem. Org. Naturst.* **8,** 341–364.

ZHANG, Z., TONG, K.-T., BELEW, M., PETTERSSON, T., JANSON, J.-C. (1992), Production, purification and characterization of recombinant human interferon gamma, *J. Chromatogr.* **604,** 143–155.

24 Affinity Separations

Srikanth Sundaram
Martin L. Yarmush

Piscataway, NJ 08854 0909, USA

1 Introduction

Traditional methods of purification of biological molecules have relied on differences in physicochemical properties such as solubility (partitioning of a substance between two immiscible phases), size (gel permeation/size exclusion) and charge (ion exchange). Typically, these methods (which can be placed in the broad category of all variations in liquid chromatography) have found widespread use in the purification and characterization of many biological macromolecules both at the laboratory and industrial scales. The major limitation inherent in these methods is their nonspecific nature, which generally results in low yields, low purity, multiple steps, and high purification costs.

Affinity purification techniques, on the other hand, rely on specific, reversible molecular interactions between the substance to be purified (termed "ligate") and a binding molecule (termed "ligand"). In all ligate-ligand interactions, the buffer conditions such as pH, ionic strength, and chemical composition, along with other parameters such as temperature and pressure dictate the stability of the complex. Thus, the ligand-ligate complex can be disrupted by simply changing the environment either in a nonspecific manner (e.g., changing pH) or in a specific manner (e.g., adding specific competing molecules).

Affinity purification techniques, thus, enable the purification of almost any biological molecule on the basis of its individual structure. Purification is based on specific binding of the ligate to the ligand, and thus recoveries of active material are often high (i.e., more than 1000-fold higher than nonspecific techniques). During the early and mid 1980s, when affinity chromatography began to be widely accepted and used, there was great enthusiasm that affinity-based separations opened the possibility of "one-step" total purification. The technique has not, however, lived up to such expectations. Many affinity chromatographic purifications (especially those involving the use of group specific ligands) involve multiple steps. A 1986 survey of 100 papers on protein purification concluded that the average purification achieved with affinity procedures was just over 100-fold, while all other techniques achieved less than 12-fold purification (BONNERJEA et al., 1986). It was also reported in this study that the losses associated with each step varied from 39% (affinity and ion-exchange methods) to 19% (purification). The low "average" purification reported with affinity procedures may be attributed to the fact that a majority of affinity procedures used tend to be based on group specific ligands (such as dyes, coenzymes, NAD derivatives, lectins, etc.) that have lower average purifications (of about 20-fold) than specific-ligand based purifications (typically several 1000-fold). For example, the purification of human erythrocyte acyl phosphatase using a specific anti-acyl phosphatase antibody effected a 135000-fold purification in a single step with a 44% yield (DEGL'INNOCENTI et al., 1990). However, it must be noted that general comparisons of losses and average purification factors between techniques are limited because both of these numbers are going to depend on the purity of the starting feed material. Thus, average purification factors are going to be low in general for size exclusion chromatography since this is usually a final polishing step with the feed solution already at a high degree of purity. Consequently, the maximum theoretically attainable purification factor in this case is quite limited.

For practical purposes, in an affinity separation scheme, the strength (i.e., association constant) of the ligand-ligate interaction should be neither too low nor too high. Low association constants result in inferior selectivities, while very high binding constants can result in problems with ligate recovery and with reduced ligand stability due to essentially harsh schemes used to recover the ligate. Typically, binding constants in the range of 10^5–10^8 are well-suited for affinity purification processes.

Affinity purification techniques may be classified on the basis of the affinity ligand used for the separation (VIJAYALAKSHMI, 1989) as shown in Fig. 1. "Biospecific" or affinity separations are said to be based on reversible and specific interactions between biologically active substances. In these cases, the interaction between the ligand and ligate is based on a highly specific complementarity of

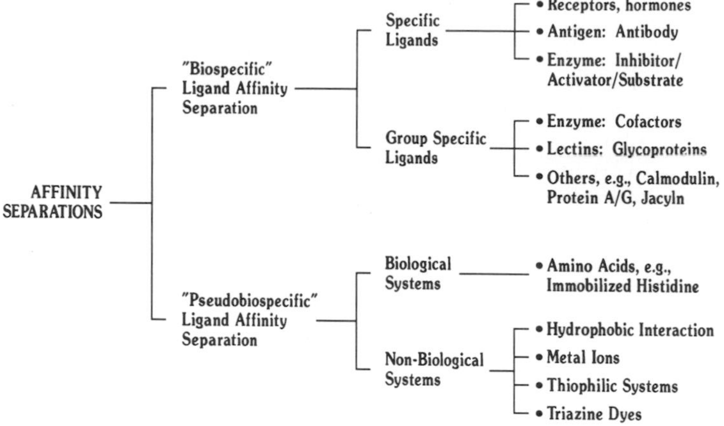

Fig. 1. Classification of affinity purification techniques based on the ligand used (adapted from VIJAYALAKSHMI, 1989).

charge, hydrophobicity, shape, etc. Examples of such "biospecific" interactions are highly specific interactions such as those between antigens and antibodies, hormones and receptors, enzymes and their inhibitors, and other group specific interactions such as those between enzymes and cofactors or lectins and glycoproteins. On the other hand, affinity separations can also be based on the interaction of biological molecules with simpler ligands such as hydrophobic ligands, dyes, metals, etc. These interactions are termed "pseudobiospecific" interactions. The same forces that govern biospecific interactions also play a role in pseudobiospecific interactions, but differ in their relative magnitudes.

The most traditional affinity purification scheme is affinity adsorption chromatography wherein the ligand is first covalently attached to an insoluble support and the "immobilized" ligand is packed into a chromatographic column. Affinity adsorption can also be carried out with microfiltration membranes as affinity supports which eliminates much of the kinetic limitations incurred with packed beds of beaded supports. Other less commonly used schemes which have been developed primarily to overcome some of the disadvantages of affinity adsorption chromatography include affinity partitioning, affinity precipitation and affinity cross-flow filtration. In affinity partitioning, the ligand is attached to a water soluble carrier (most often PEG) which participates as one of the components in an aqueous two-phase polymer system. Thus, the partition coefficient for a ligate interacting with the polymer-attached ligand is dramatically altered. In affinity precipitation, the ligand is attached to a soluble carrier, either forming bifunctional moieties capable of directly precipitating oligomeric proteins, or giving polyfunctional polymer derivatives which can be precipitated by, e.g., a change in pH (JANSON, 1984).

2 Affinity Adsorption Chromatography

2.1 Basic Principles

The general principles governing affinity adsorption chromatography are depicted in Fig. 2. For affinity adsorption (or for any other affinity-based technique) to work, a ligand that exhibits specific and reversible binding to the substance to be purified must be available. It should also have chemically modifiable groups which would enable the covalent attachment of the ligand to a chromatographic material (the matrix). This step is called "immobilization" and it is important that the immobilized ligand retains its specific activity to the ligate. In the first step, the source material containing the ligate and a variety of impurities and contaminants is passed through the chromato-

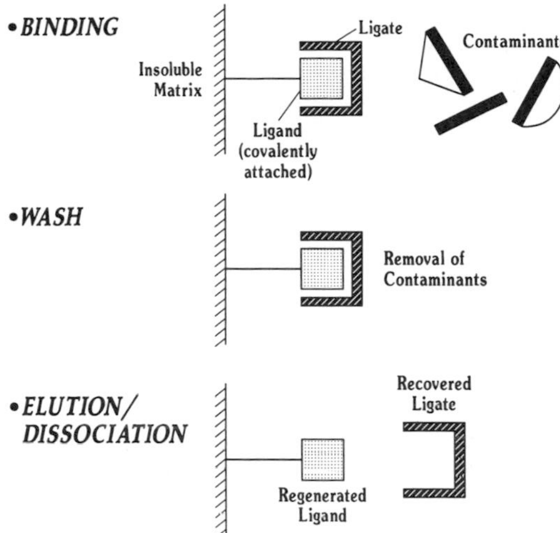

Fig. 2. Principles of affinity adsorption chromatography.

graphic bed ("loading"). Only the ligate which has appreciable affinity for the ligand is retained on the column while other impurities and contaminants pass through unretarded ("washing"). The specifically adsorbed ligate is then recovered from the column ("elution") by a variety of techniques which alter the physical environment or solvent composition to favor dissociation of the ligand-ligate complex.

2.2 Affinity Chromatography Supports

A key component in the development and widespread use of affinity chromatography has been the availability of increasing numbers of support materials and methods for attaching ligands to them. The basic requirements of an ideal support for affinity chromatography have been described and are listed in Tab. 1 (NARAYAN and CRANE, 1990; MOHR and POMMERENING, 1985). Briefly, the ideal matrix should exhibit: (1) good chemical, mechanical, biological, and thermal stability; (2) high binding capacity with high specificity and hydrophilicity (no surface charge and no hydrophobic sites); (3) good recovery, reproducibility, and ease of synthesis in terms of low cost and/or commercial availability; (4) flow characteristics; (5) kinetic characteristics.

Tab. 1. Characteristics of an Ideal Support for Affinity Adsorption Chromatography

Insolubility
Rigidity
Permeability
Macroporosity
Hydrophilicity
Potential for chemical derivatization
Low nonspecific adsorption
Biological inertness
Resistance to chemical degradation
Ease of synthesis/commercial availability
Zero/low surface charge content
High binding capacity
Low cost
Good reproducibility
Good recovery

The relative importance of each of these properties will vary depending on the scale and specifics of each affinity chromatography protocol. In general, for process-scale applications, stability may be the most significant factor in determining the overall economics of the separation, followed closely by specificity and capacity (ITO et al., 1987).

2.2.1 Macrosupports

Traditionally, nearly all affinity chromatography applications have been carried out at low pressure using beaded particles of a size range between 50–400 μm (macrosupports). A partial list of macrosupports (defined as particulate supports greater than 5 μm in diameter) that can serve in both low and high pressure affinity chromatography is shown in Tab. 2. By far, the most popular support is agarose. A variety of supports have been developed to replace agarose such as cross-linked agarose, polyacrylamide, cellulose, synthetic polymers, and inorganic materials such as silica.

2.2.1.1 Agarose

Agarose is a purified linear copolymer of 1,3-linked β-D-galactose and 2,4-linked 3,6-an-

Tab. 2. Macrosupports

Material	Trade Name	Supplier
Low/Medium Pressure		
Cross-linked dextran	Sephadex	Pharmacia
Cross-linked polyarcylamide	Biogel P	Bio-Rad
	Affi-gel	Bio-Rad
Polyacrylamide/dextran	Sephacryl	Pharmacia
Acrylates		
a) Polyacrylamide with monomers containing oxirane groups	Eupergit C	Röhm Pharma
b) N-acryloyl-amino-2-hydroxymethyl-1,3-propanediol	Trisacryl	IBF
c) N-acryloyl morpholine and methylene bisacrylamide	Enzacryl	IBF
Polyacrylamide/agarose	Ultrogel	IBF
Agarose	Sepharose	Pharmacia
	Biogel A	Bio-Rad
Controlled pore glass	CPG	Pierce
Polymer-clad silica	Bakerbond Wide-Pore	J. T. Baker
Hydroxyethyl methacrylate	Dynospheres	Dyno Particles
	Separon	Tessek
Vinyl polymer	TSK/Toyopearl	Merck
	Fractogel	Merck
Cross-linked cellulose	Cellufine	Amicon
Organic powders	Macrosorb	Sterling Organics
High Pressure		
Silica	Lichrosphere	Merck
	Ultrasphere	Beckman
	Hypersol	Shandon
	Spheron	Waters
	Zorbax	Dupont
Cross-linked agarose	Superose	Pharmacia

hydro-α-L-galactose. Agarose solidifies in the form of a porous network consisting of triple strands of carbohydrates, held together by hydrogen bonds. Consequently, any substance or treatment which disperses hydrogen bonds leads to a loss of structure of the agarose. The main advantages of agarose include: (1) its highly porous, hydrophilic structure, (2) its relative inertness and chemical stability, (3) its capacity for easy chemical derivatization. The major drawbacks are its modest rigidity, high cost, thermal lability, and instability with organic solvents (needed in many derivatization schemes) and denaturants and detergents (which may be used as eluents).

The stability of the agarose matrix to mechanical, chemical, and microbial degradation has been considerably increased with chemical cross-linking using epichlorohydrin, divinyl sulfone or 2,3-dibromopropanol (PORATH et al., 1971; PORATH, 1975; LAAS, 1975). Sepharose, e.g., has limited pH (4–9) and temperature stability (0–40 °C) and low stability to chaotropic ions, organics, and high concentrations of urea and detergents. Cross-linked agarose, Sepharose CL, on the other hand, has a wider pH and temperature stability (pH of 3–14, temperatures of up to 70 °C and can even withstand autoclaving at pH 7 and 110–120 °C for limited periods of time), and high stability to denaturants (up to 6 M guanidium and 8 M urea), chaotropics, and organic solvents. While cross-linking improves stability to extremes of pH, temperature, and high concentrations of organic solvents and chaotropic agents, it also may result in a decrease in porosity and ligand attachment capacity due to reduction in the number of hydroxyl groups available for chemical modification. MADDEN and THOM (1982), however, showed that cross-linking of Sepharose does not decrease the effective pore size. Thus, it appears that cross-links take place mainly between chains in a single gel fiber, probably between the oxygens at position 6 of galactose and position 2 of anhydrogalactose. Several types of well characterized agarose and cross-linked agarose with differing exclusion volumes are commercially available from a variety of suppliers under the trade names of Sepharose, Sepharose CL, Biogel A, etc.

Agarose can also be used in combination with synthetic polymers as a supporting material. Ultrogel, thus, consists of a three-dimensional polyacrylamide lattice and an interstitial agarose gel. The acrylamide–agarose ratio may vary from 1:1 to 10:1. Functionalization of these gels can be achieved through either the hydroxyl groups of agarose or the amide group of the polyacrylamide. Ultrogel AcA has excellent rigidity and low compressibility, thus permitting higher flow rates. Agarose-based supports are also available in magnetized forms for use in magnetic affinity chromatography. Magnogel A4R is a cross-linked agarose whose magnetic properties result from the incorporation of 7% Fe_3O_4 in the interior of the gel beads. Magnogel AcA is a magnetized support, derived from Ultragel AcA. Magnogel AcA is rendered magnetic in a similar manner to Magnogel A4R.

Even though other supports with considerably better quality are now available, agarose and cross-linked agarose continue to be the most widely used solid-phase matrix for affinity chromatography. The popularity of agarose as an affinity adsorbent is demonstrated quite convincingly (and has been propagated in part) by the wide selection of unactivated, activated, and ligand-coupled agaroses that are commercially available.

2.2.1.2 Cross-Linked Dextran

Dextran is a polysaccharide composed of 1,6-linked α-D-glucose units. Dextran cross-linked with epichlorohydrin in alkaline solution is one of the most widely used supports for size exclusion chromatography as well. Dextran gels with a low degree of cross-linking have been used successfully in affinity chromatography. The main representative of this type is available from Pharmacia under the trade name Sephadex. Sephadex is chemically very stable, with little effect observed even after two months in 0.25 M NaOH at 60 °C or six months in 0.02 M HCl. Activation of Sephadex is achieved through the –OH groups of the polysaccharide backbone; however, this usually results in further cross-linking of the gel resulting in a lowering of the porosity and permeability. Like agarose, Sephadex is hydrophilic and exhibits moderate to good stability in aqueous, organic and alkaline solvents. Another dextran-based product, Sephacryl (produced by cross-linking allyl-dextran with N,N′-methylene bisacrylamide), has enhanced rigidity and flow properties, when compared to Sephadex, but also shows slightly higher nonspecific adsorption.

2.2.1.3 Polyacrylamide and Other Acrylates

Polyacrylamide gels are obtained by the copolymerization of acrylamide and N,N′-methylene bisacrylamide resulting in a polymer with a hydrocarbon framework carrying carboxamide sidechains. Polyacrylamide, used widely as electrophoresis media and also for size permeation chromatography, can be easily functionalized for affinity chromatography. Its advantages are its stability to extremes of pH (1–10) and many organic solvents, lack of charged groups if initiated by free radicals, biological inertness, and resistance to degradation. However, the polymer is mechanically weak and cannot be used at high pressures because of compaction. Increasing the concentration ratio of the cross-linking agent N,N′-methylene bisacrylamide to the acrylamide monomer results in a more rigid polymer with decreased porosity. Commercial polyacrylamide beads in a wide range of porosities are available from Bio-Rad under the trade name Bio-Gel P. Hydroxyalkyl methacrylate supports are prepared by the copolymerization of hydroxyethyl methacrylate and ethylene dimethacrylate in aqueous solution in the presence of inert solvents (COUPEK, 1982). Two well known matrices of this type are available under the trade name Spheron and Separon. Both are neutral, hydrophilic gels with excellent mechanical stability available in a wide range of porosities ranging from heavily cross-linked microparticles to macroparticles with macroporous structures. These matrices retain most of the advantages of polyacrylamide supports such as excellent stability in a wide pH range and in organics, and excellent resistance to thermal and biological degradation. For example, they are able to withstand heating for 8 h in 1 M sodium glycolate solution at 150 °C or boiling in 20% (v/v) HCl for 24 h without changing their structures. However, nonspecific adsorption is greater compared to agarose largely due to hydrophobic interactions. The chemically reactive groups are hydroxyl groups that can be easily derivatized. Alternatively, the gels may be formed by copolymerization with monomers containing reactive groups such as epoxide or *p*-nitrophenyl ester groups in a manner similar to polyacrylamide gels that lack hydroxyl groups. The epoxide function, e.g., can be directly built into the backbone by copolymerizing methacrylamide, N,N′-methylene bisacrylamide and glycidyl methacrylate. This support is commercially available as Eupergit C. Eupergit C is hydrophilic and electrically neutral, with the epoxide functionality being stable for months when dry at −30 °C. The flow properties of Eupergit C compare favorably with that of agarose. As will be discussed later, the epoxide-based linkage is very stable and ligand leakage is minimal.

Another important acrylate is N-acryloyl-2-amino-2-hydroxymethyl-1,3-propanediol (available as Trisacryl) which has distinct advantages over acrylamide in that it is much more hydrophilic due to the three hydroxymethyl groups. The polyethylene backbone is buried beneath a layer of hydroxymethyl groups, thus creating a favorable microenvironment that promotes the adsorption of hydrophilic proteins to the polymer surface. Trisacryl supports are resistant to moderate pressures (up to 3 bar) and have excellent chemical, temperature (−20 °C–121 °C), and pH stability (0–13).

Enzacryl, a copolymer of N-acryloyl morpholine and N,N′-methylene bisacrylamide, is another modified acrylamide that has also been used for affinity chromatography. The Enzacryl series of pre-activated polyacrylamides consist of the following members: Enzacryl AH (which is a hydrazide derivative), Enzacryl AA (which contains aromatic acid residues) and Enzacryl Polyacetal (which is a polymer of N-acryloyl aminoacetaldehyde dimethylacetal with N,N′-methylene bisacrylamide).

2.2.1.4 Inorganic Supports

Controlled pore glass (CPG) consists of 97% SiO_2 and 3% B_2O_3 and is the most commonly used inorganic matrix for immobilization of biological macromolecules. A wide variety of organic functional groups have been applied to the support, and activation can then be achieved by the same techniques used for other supports. The advantages include high mechanical and chemical stability and biological inertness. Disadvantages include high nonspecific adsorption and the relatively high solubility of silica at the surface especially in alkaline solutions. Both these effects can be reduced by treating the surface with a cross-linked organic coating of dextran or with a hydrophilic silane such as γ-aminopropyltriethoxy silane or glycerolpropyl silane (WEETALL, 1973; REGNIER and NOEL, 1976). Further reduction in solubility can be achieved by using glass coated with other metal oxides such as zirconium or titanium oxides.

Silica is another inorganic support material that is being increasingly used in affinity chromatography, especially in high performance liquid affinity chromatography (HPLAC) applications. Like CPG, silica suffers from high nonspecific adsorption due to residual charge on the surface and solubility at alkaline pH. Both these problems can be minimized by extensive derivatization of the silanol groups on the surface of silica with functional groups that shield the surface from interacting with the hydroxyl ions in the solvent. Reaction with silanes such as 3-glycidoxypropyltrimethoxy silane is commonly used; anhydrous procedures form a monolayer of silane while aqueous procedures are used when a polymeric layer is desirable.

2.2.1.5 Cellulose

Cellulose is a 1,4-β-linked glucose polymer derived from both plant and bacterial sources. Most of the early work with this material was done with powdered semicrystalline cellulose of low porosity and heterogeneity, thereby reducing its utility as a substrate for immobilization of ligands. Native cellulose takes up only 30–40% (w/w) water with a correspondingly small degree of swelling; this results in a small available pore surface and relatively small pores. Highly swollen regenerated celluloses with larger pore sizes are prepared by dissolving cellulose either as a derivative or in its native form in DMSO/CH_2O followed by regeneration in the presence of additives (TURBAK et al., 1982). Since the early experience with fibers, special procedures necessary to produce cellulose beads, films, and other formats that permit adequate diffusion of large proteins have been described. Cellulose has several advantages as substrate for affinity chromatography such as its insolubility in almost all solvents, lack of ion-exchange properties, low hydrophobic character, and excellent mechanical and physical rigidity even in nonpolar solvents. Beaded cellulose is a highly porous spherical product which is composed of pure regenerated cellulose (15%) and water. During the formation of beaded cellulose, a microheterogeneous structure is formed consisting of regions of crystalline order interconnected by amorphous material. For this reason, beaded cellulose is mechanically more stable and rigid than even cross-linked polysaccharide gels of comparable porosity. Cellulose may also exceed other synthetic polymers in its chemical reactivity.

2.2.2 Microsupports

The major disadvantage of porous macrosupports is that their porous nature impedes diffusional mass transfer rates. These pores, however, are necessary to provide a large surface area for the coupling of ligands in a small column volume. One method of overcoming such diffusional limitations associated with macrosupports is to use non-porous microparticles (particles less than 5 µm in diameter) that have a surface area comparable to that of porous macrosupports. In any bead suspension, the interfacial surface area increases exponentially with decreasing particle diameter. Also, the distance that a molecule has to migrate before it reaches its adsorption site will decrease with particle size. Thus, one would expect a dramatic improvement in adsorption–desorption kinetics upon decreasing particle size. Decreasing the particle size is, however, disadvantageous in a packed chromatographic column, since the mechanical stress caused by drag forces is higher with smaller particles resulting in a higher risk of particle deformation and/or destruction. Consequently, conventional affinity separations involving microsupports are not likely to be based on usual chromatographic techniques, but rather on batch adsorption techniques. Microsupports that can be made magnetic have the most potential since they are easily recovered using high-gradient magnetic filtration. Commercially available microsupports are somewhat limited. Among these are polymer-based particles (Dyno Particles, 0.5 and 3 µm), magnetic particles (Advanced Magnetics, 0.5–2 µm) and silica particles (Merck, 1.5 µm). Notwithstanding this limitation, the synthesis of many microparticles on a lab scale is quite simple, with a variety of polymer types and particle sizes and functionalization schemes possible.

2.3 Activation and Coupling Procedures

The first step in covalently linking biologically active ligands to the support is the activation of the chemically inert support. The second step is the coupling of the ligand to the activated matrix, and the final step is the blocking of excess unreacted active groups. Many factors play a role in selecting an activation/coupling scheme: (1) type and site of the chemical linkage; (2) steric conditions in the vicinity of the immobilized ligand, Tab. 3 summarizes many features of commonly used activation/coupling schemes; (3) the stability of the covalent linkage formed between the support and the ligand.

Different supports require different activation schemes. Polysaccharides such as agarose,

Tab. 3. Activation/Coupling Schemes

Activating Agent	Reactive Matrix	Group Ligand	Type of Linkage	Toxicity of Reagent	Stability of Complex	Non-specific Interactions	Activation Time (h)	Ligand Coupling Time (h)	pH Coupling
Glutaraldehyde	$-NH_2$	$-NH_2$	Secondary amine	Moderate	Good	–	1–8	6–16	6.5–8.5
CNBr	$-OH$	$-NH_2$	Isourea, imido carbamate, N-substituted carbamate	High	Unstable at pH < 5, pH > 10	Ionic	0.2–0.4	2–4, RT[a] overnight, 4 °C	8–10.0
Epoxy activation (bisoxiranes)	$-OH$	$-NH_2$, $-OH$, $-SH$ (alkaline), $-COOH$ (acidic)	Alkylamide, ether, thioether	Moderate	Excellent	–	5–18	15–48	8.5–12.0
CDI	$-OH$, $-NH_2$	$-NH_2$	Urea linkage	Moderate	Good	–	0.2–0.4	Overnight to 6 days	8–9.5
Tresyl chloride	$-OH$	$-NH_2$	Alkylamine	Moderate	Good	–	0.5–0.8	Rapid	7.5–10.5
Diazonium	$-OH$, $-NH_2$	$-OH$	Azo	Moderate	Moderate	Hydrophobic		0.5–1.0	6–8
Periodate	$-OH$	$-NH_2$	Alkylamine	Good (non-toxic)	Good	–	14–20	Overnight	7.5–8.5
Triazine	$-NH_2$	$-NH_2$	Triazonyl	High	Good	Aromatic (hydrophobic)	0.5–2.0	4–16	7.5–9.5
DVS	$-OH$	$-OH$, $-SH$, $-NH_2$	Ether, thioether, secondary amine	High	Unstable at high pH	–	0.5–2.0	Rapid	8–10.0
p-Nitrophenyl formate	$-OH$	$-NH_2$	Carbamate	–	Hydrolysis, pH sensitive	–	0.5	Rapid	8.5–9.5

[a] RT, room temperature

dextran, and cellulose (and other affinity chromatography supports such as methacrylates) have reactive hydroxyl groups that can be used for the coupling of ligand to the support (CUATRECASAS, 1970). Polyacrylamides can be derivatized by the primary modification of the carboxamide side chains of the preformed polyacrylamide beads or by copolymerizing acrylamide and bisacrylamide with active esters (Fig. 3). In the former, primary modification is achieved by reacting the support with a large excess of ethylene diamine or hydrazine at 50 °C (INMAN and DINTZIS, 1969; INMAN, 1974). The amido group is displaced by the aminoethyl or hydrazide group resulting in primary amino groups available for further functionalization. Alkaline hydrolysis can also be used to modify polyacrylamide supports, generating carboxyl groups. Glutaraldehyde has also been used for derivatization of polyacrylamide supports (WESTON and AVRAMEAS, 1971).

Silanization of CPG represents the first step in its functionalization. The reaction between an amino functional trialkoxy silane such as 3-aminopropyltriethoxy silane and the hydroxyl

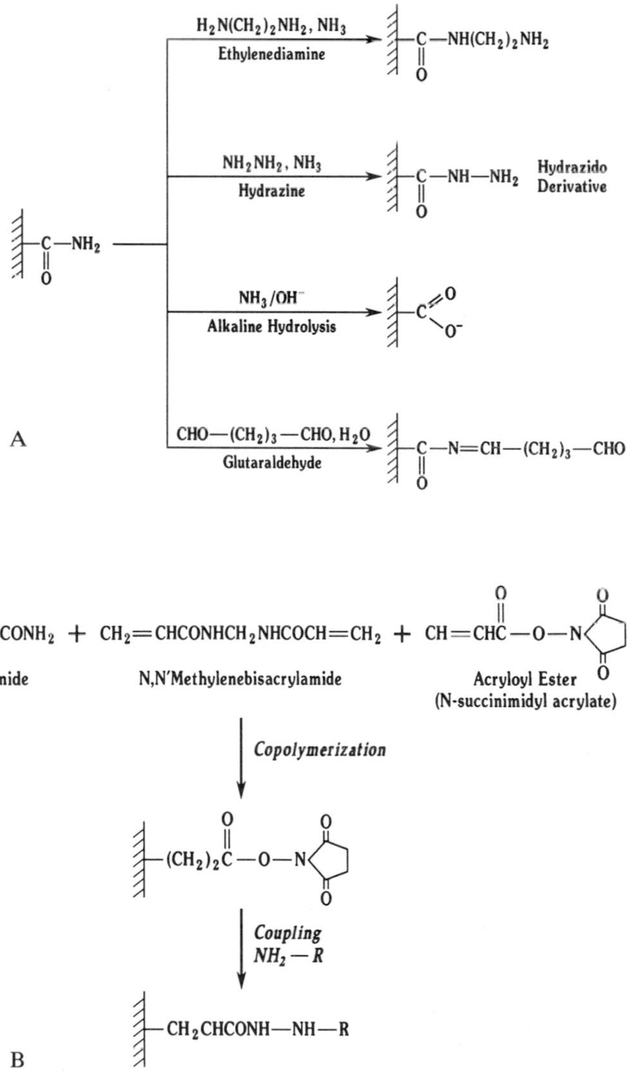

Fig. 3. (A) Primary modification of polyacrylamide supports; (B) polyacrylamide copolymerized

groups on the glass surface results in the incorporation of amino groups on the surface which is the reactive moiety for further functionalization (WEETALL and FILBERT, 1974). The following sections will briefly review some of the commonly used procedures available for the activation of affinity chromatography supports.

2.3.1 Cyanogen Bromide Activation

Cyanogen bromide (CNBr) activation, first introduced by AXEN et al. (1967) is the most commonly used protocol for activation of polysaccharide supports. Extensive studies on the mechanism of CNBr activation have resulted in the formulation of the reaction scheme shown in Fig. 4 (KOHN and WILCHECK, 1982). Thus, the products of the reaction include cyanate esters, cyclic and acyclic imidocarbonates, carbamates and carbonates. The predominant active species on agarose is the cyanate ester, and that on dextran and cellulose is the cyclic imidocarbonate. While the imidocarbonate is extremely labile at low pH, the cyanate ester moiety, on the other hand, is rapidly hydrolyzed at alkaline conditions. This difference in the pH sensitivity of the two different active species can be utilized to produce homogeneous activated supports in terms of the active groups involved prior to coupling.

CNBr activated supports are not very stable in aqueous solutions. Inactivation of the active groups to inert carbamates and cyclic carbonates occurs fairly rapidly under these conditions. Therefore, the coupling of the ligand to the activated support is usually made immediately after activation. Coupling occurs through the primary amino groups on the ligand resulting in the N-substituted isourea derivative. The isourea linkage is positively charged at neutral pH, which introduces an element of ion exchange to the adsorbent. Another problem with this method is the instability of the isourea linkage under alkaline conditions which may lead to slow leakage of the ligand from the support.

KOHN and WILCHEK (1984) have devised a method for the enhancement of the electrophilicity of CNBr, which is based on the formation of a "cyano-transfer" agent by CNBr and certain bases such as triethylamine (TEA) and dimethylamino-pyridine (DAP). Replacement of the bromide ion with non-nucleophilic anions such as tetrafluoroborate or perchlorate greatly enhances the stability of these agents. The cyano-transfer complex formed with TEA (N-cyanotriethylammonium tetrafluoroborate, CTEA) is not quite as stable as

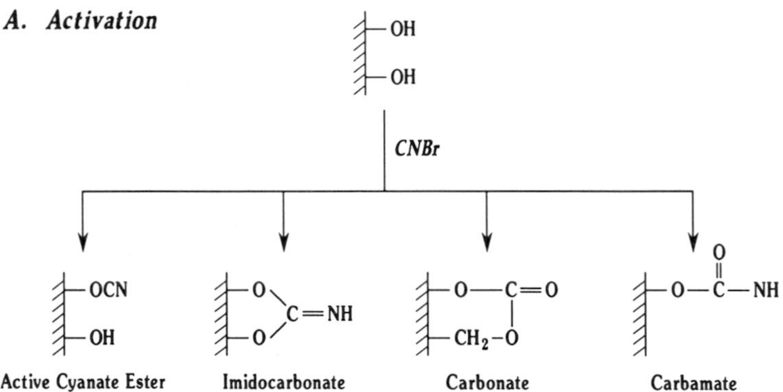

Fig. 4. (A) Activation of hydroxyl containing supports via cyanogen bromide; (B) coupling of active cyanate ester to the amino group of the ligand.

the aromatic N-cyano complex formed with DAP (1-cyano-4-(dimethylamino)-pyridinium tetrafluoroborate, CDAP). However, both are non-volatile and pose a reduced health hazard as compared to CNBr. In addition, CDAP is only slightly hygroscopic and therefore easy to handle.

The cyano-transfer reagents can, due to their increased electrophilicity, cyanalate matrix bound hydroxyl groups to cyanate esters at much lower pH than used in conventional procedures. The yield of active cyanate esters produced is also improved dramatically. However, the reaction has to be performed in a mixed organic solvent–water system which restricts the use of these agents to those supports that can be transferred to such solvents without shrinking or undergoing any degradation.

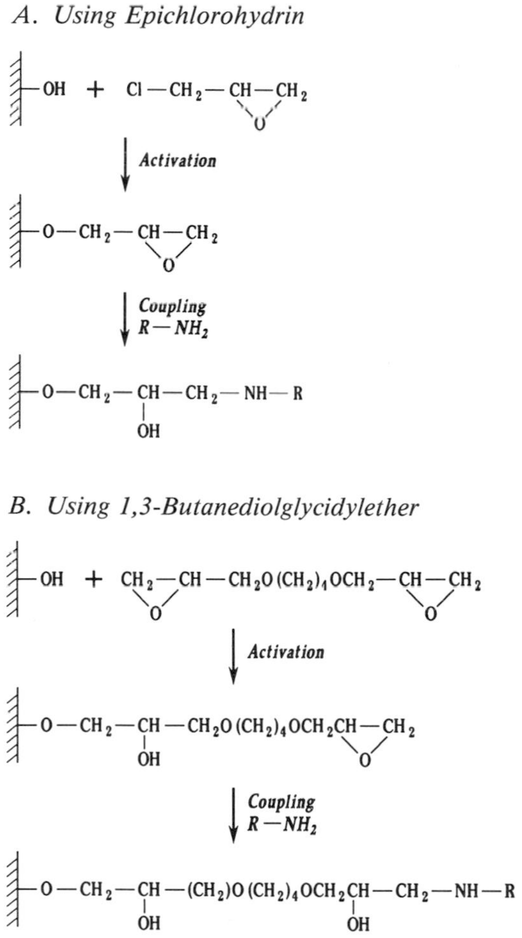

A. Using Epichlorohydrin

$-OH + Cl-CH_2-CH-CH_2$ (epoxide, O)

↓ *Activation*

$-O-CH_2-CH-CH_2$ (epoxide, O)

↓ *Coupling* $R-NH_2$

$-O-CH_2-CH(OH)-CH_2-NH-R$

B. Using 1,3-Butanediolglycidylether

$-OH + CH_2-CH-CH_2O(CH_2)_4OCH_2-CH-CH_2$ (two epoxides, O)

↓ *Activation*

$-O-CH_2-CH(OH)-CH_2O(CH_2)_4OCH_2CH-CH_2$ (epoxide, O)

↓ *Coupling* $R-NH_2$

$-O-CH_2-CH(OH)-(CH_2)O(CH_2)_4OCH_2CH(OH)-CH_2-NH-R$

Fig. 5. Activation of hydroxyl bearing supports via epoxide chemistry using (A) epichlorohydrin and (B) 1,3-butanediol-glycidyl ether.

2.3.2 Epoxide Activation

Epoxide activation of hydroxyl bearing supports (Fig. 5) can be achieved with epichlorohydrin or more commonly, a bisoxirane, 1,3-butanediolglycidyl ether (SUNDBERG and PORATH, 1974). In the latter case, a hydrophilic spacer arm containing 11 atoms is automatically introduced between the matrix and the reactive oxirane, and the simultaneous cross-linking which occurs may result in a more stable matrix (albeit at the cost of reduced permeability). Epoxy groups are somewhat less active than cyanate esters, but can react with several different groups on the ligand, depending on the pH of the coupling step. In alkaline media, both amino and thiol groups react, and under high temperatures, hydroxyl groups also react. In acidic conditions, the epoxy group reacts with carboxyl moieties on the ligand.

Unlike CNBr activation, the resulting linkage in this case is very stable and spacer arm is useful when immobilizing small ligands. Elimination of the excess epoxy groups may be problematic requiring either prolonged storage in alkaline conditions or prolonged treatment with small nucleophiles.

2.3.3 Carbonyldiimidazole Activation

1,1′-Carbonyldiimidazole (CDI) is a highly reactive reagent suitable for coupling a variety of functional groups (Fig. 6). CDI reacts quite rapidly with hydroxyl groups to form the corresponding carbonate ester that is comparable to cyanate esters in reactivity (BETHELL et al., 1979). The formation of the imidazoylcarbonate ester is catalyzed dramatically by sodium alcoholates at room temperature or by traces of alkoxide at higher temperatures. However, no catalyst appears to be needed for activation of agarose. The ester reacts easily with amines to produce a stable urethane linkage between

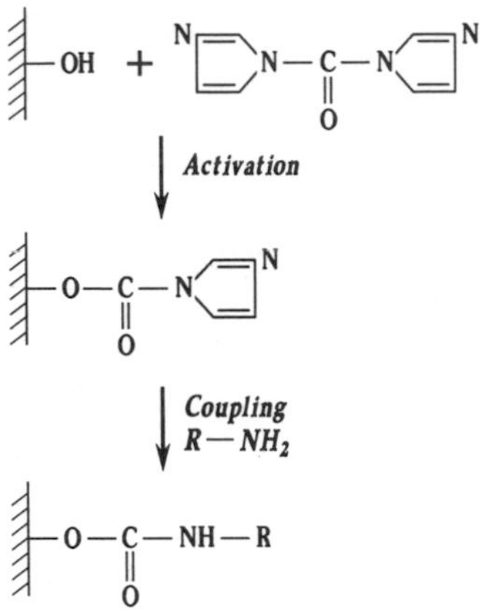

Fig. 6. Activation of hydroxyl bearing supports by 1,1′-carbonyldiimidazole (CDI).

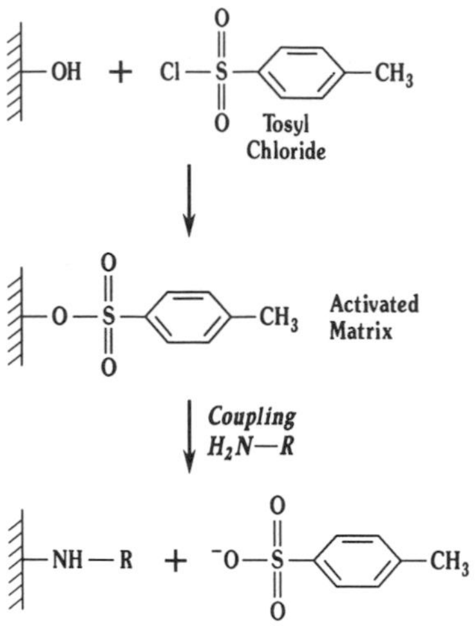

Fig. 7. Activation of hydroxyl bearing supports by sulfonyl chlorides.

the support and ligand. CDI can also be used to activate supports containing amino groups instead of hydroxyl groups. In this case the first intermediate is a carbonyldiamide which reacts readily with nucleophilic groups of the ligand to produce a stable urea linkage between the support and the ligand.

The major advantages of the CDI chemistry are the ease of reaction and stability of the resulting linkage. However, at high ratios of CDI to support and especially with small-pore-size supports, an increasing fraction of unreactive cyclic carbonates forms as a result of interchain cross-links.

2.3.4 Sulfonyl Chloride Activation

Alkyl and aryl sulfonates participate readily in nucleophilic substitution reactions resulting in direct bonding between the nucleophile and the ester group of the sulfonate. The presence of electrophilic substituents on the sulfonyl group facilitates these reactions. Suitable substituents include toluyl (tosyl), trifluoromethyl (mesyl) and trifluoroethyl (tresyl) groups (Nilsson and Mosbach, 1980).

The activation step (Fig. 7) between the sulfonyl chloride and the hydroxyl group is carried out in essentially non-aqueous media to minimize hydrolysis of the acid chloride, and results in the introduction of the corresponding ester derivative onto the support. The introduced tosyl, tresyl or mesyl group reacts rapidly to give a stable secondary amine. No titratable ion-exchange activity is found on derivatized agarose using this activation method (in contrast to CNBr), and there is no evidence of further cross-linking of the agarose (in contrast to the CDI method).

2.3.5 Periodate Oxidation

Sodium periodate reacts with vicinal hydroxyls on agarose (as well as dextran and cellulose) to produce aldehyde groups which can then react with primary amines to form Schiff's bases; these are subsequently stabilized by reduction with sodium borohydride or cyanoborohydride. The noncoupled dialdehydes which are unstable in alkaline solution are also stabilized by the borohydride reduction. The principal advantage of the periodate method is that the reduced Schiff's base does not contribute charge effects to the adsorbent while maintaining good ligand stability (Sanderson and Wilson, 1971).

2.3.6 Triazine Activation

The triazine method (Kay and Lilly, 1970) can be used to activate supports containing

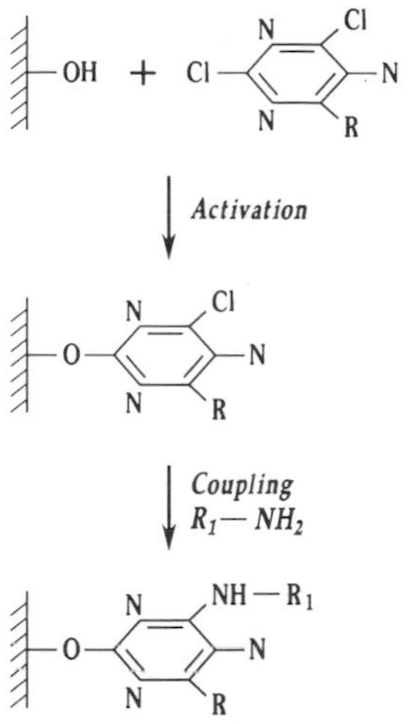

Fig. 8. Activation of hydroxyl bearing supports by triazine chemistry.

either hydroxyl or amine groups (Fig. 8). The first step is the activation of the support by cyanuric chloride or its derivatives (typically aminodichloro triazine) in which a chlorine has been replaced by a solubilizing group. The ligand is then coupled via a primary amino group by the nucleophilic substitution of another chlorine. Any residual chlorine on the triazine ring can be blocked by storing the immobilized protein in a solution of ethanolamine. The advantage of the triazine activation is the stability of the resulting linkage. The major disadvantage is that an aromatic ring is introduced between the support and the ligand which may give the adsorbent some hydrophobic character.

2.3.7 Diazonium Activation

This activation scheme can be used with supports bearing either hydroxyl or amino groups (Fig. 9). The first step in the diazotization procedure is the introduction of the *p*-nitrophenyl group into the matrix. The nitro group is then converted to its amino derivative by reduction. The polymer is treated with nitrous acid to prepare the diazonium derivatives. Coupling of the ligand to the support takes place mainly through the reaction of the diazonium derivatives with the aromatic rings of the ligand, such as the tyrosine or histidine. One feature of this activation scheme is that the bound ligand can be cleaved off by complete reduction which permits the isolation of ligand–ligate complexes under mild conditions.

2.3.8 Glutaraldehyde Activation

Glutaraldehyde has found widespread use in protein chemistry as an intermolecular cross-linking agent. Affinity matrices containing amino groups can be derivatized using glutaraldehyde as shown in Fig. 10. One aldehyde group reacts with the amino group of the support to form a Schiff's base while the second aldehyde group couples with an amino moiety of a ligand thus achieving linkage of the ligand to the support. Reduction with borohydride results in the formation of stable secondary amines. This method of coupling produces a stable link between the ligand and the support without introducing charged groups and/or other sources of nonspecific interactions. It also provides a means for increasing the spacing between the support and ligand. However, as with other bifunctional reagents used for coupling, cross-linking of the support, resulting in decreased porosity and permeability, may occur.

2.3.9 Succinylation and N-Hydroxysuccinimide Activation

Supports containing amino groups can be reacted with succinic anhydride to introduce free carboxylic groups. N,N′-disubstituted carbodiimides, which promote condensation between a free amino group and a free carboxyl group to form a peptide link by acid-catalyzed removal of water, are then used to couple the ligand to the support via their amino groups. A water soluble carbodiimide, such as N-ethyl-N′-(3-dimethylaminopropyl)carbodiimide hydrochloride (EDAC) or N-cyclohexyl-N′-2-(4′-methyl-morpholinium)ethyl carbodiimide-*p*-toluene sulfonate (CMC), is commonly used.

Succinylated matrices can also be activated with N-hydroxysuccinamide by reacting the

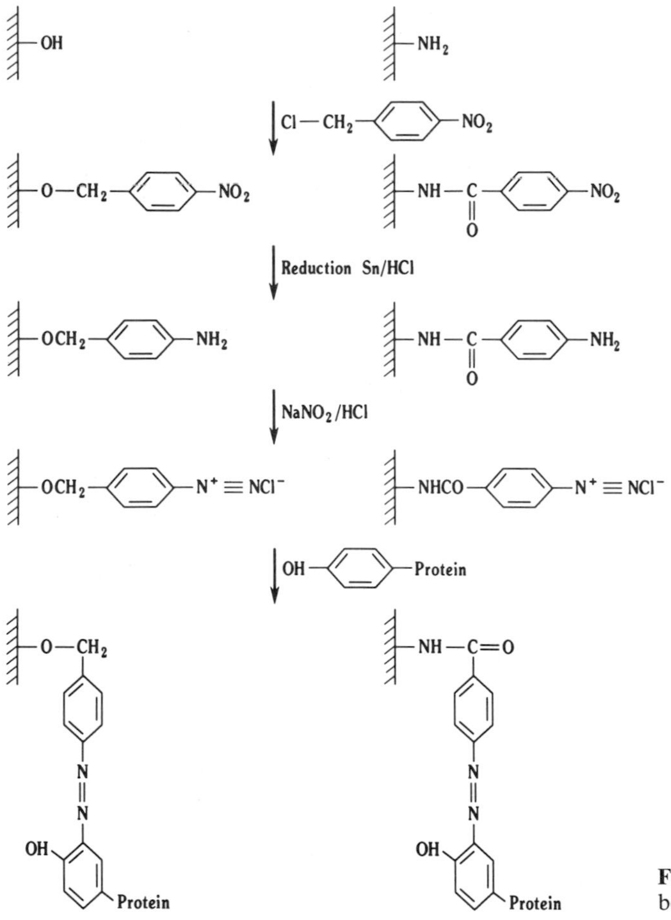

Fig. 9. Activation of hydroxyl and amino bearing supports by diazotization.

support in anhydrous dioxane with N-hydroxysuccinimide and N,N′-dicyclohexyl carbodiimide (CDC). After washing with methanol to remove the precipitated dicyclohexyl urea, the activated support can be stored as a 50% suspension in dioxane under anhydrous conditions or lyophilized. Coupling of the ligand to the support is performed with buffers at a pH of 5–8.5. The unprotonated form of the amino groups on the ligand reacts rapidly under very mild conditions to form stable amide bonds. Above pH 8.5, however, there is rapid hydrolysis of the active ester. Thiol groups may also be effectively coupled with the activated ester.

2.3.10 Divinyl Sulfone Activation

Divinyl sulfone (DVS), occasionally used for cross-linking synthetic polymers, is another bifunctional reagent that can be used for the attachment of ligands containing amino or hydroxyl groups to matrices containing amino or hydroxyl groups (PORATH and SUNDBERG, 1972). Treatment of the matrix with DVS results in the introduction of highly reactive vinylsulfonyl groups into the matrix, with cross-linking as a side reaction. Unlike epoxy activation, these vinyl groups are more reactive than the oxirane groups, and thus will react with amines and hydroxyl groups at lower tempera-

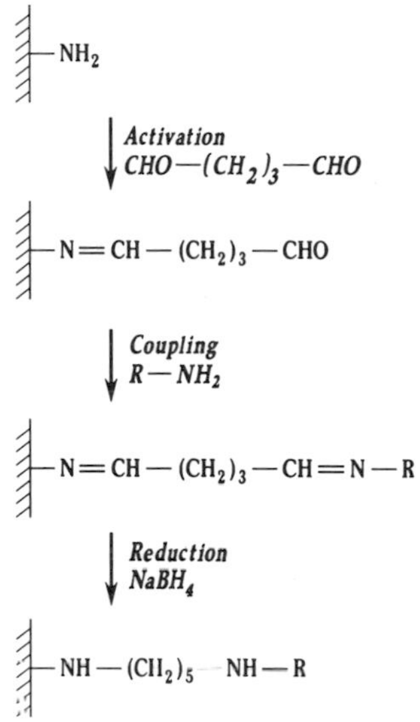

$$\text{—}NH_2 \xrightarrow[\text{CHO—(CH}_2)_3\text{—CHO}]{\text{Activation}} \text{—}N{=}CH\text{—}(CH_2)_3\text{—}CHO \xrightarrow[R\text{—}NH_2]{\text{Coupling}} \text{—}N{=}CH\text{—}(CH_2)_3\text{—}CH{=}N\text{—}R \xrightarrow[NaBH_4]{\text{Reduction}} \text{—}NH\text{—}(CH_2)_5\text{—}NH\text{—}R$$

Fig. 10. Activation of amino bearing supports by glutaraldehyde.

tures and lower pH than the oxirane intermediates. However, the resulting linkages are not stable in alkaline solution with the amino link becoming labile at about pH 8 and the hydroxyl link at about pH 9–10.

2.3.11 Nitrophenyl Chloroformate Activation

p-Nitrophenyl chloroformate has been used to activate affinity supports containing hydroxyl groups. The reaction results in the formation of the active *p*-nitrophenyl carbonate ester, which couples with amino groups of the ligand at pH 8.5–9.5 to yield a carbamate linkage and *p*-nitrophenol. The stability of the resulting complex is similar to that resulting from CNBr activation.

2.4 Spacers

In some instances, the steric conditions in the vicinity of the immobilized ligand may hinder the binding of the ligate to its ligand. This is often true in ligand–ligate systems where the ligand is a small molecule (e.g., enzyme–cofactor, antibody–hapten). Such steric limitations can be avoided by attaching the ligand to the matrix via a flexible arm or "spacer". These spacer molecules are generally linear aliphatic hydrocarbons with optional polar moieties such as secondary amino, hydroxyl, and peptide groups. The length of the spacer arm is often critical in determining the success of the affinity separation. If it is too short, then the arm is often ineffective with the immobilized ligand binding ligate rather poorly. If it is too long, it may contribute significantly to non-specific adsorption thereby compromising selectivity (O'Carra et al., 1973). In such cases, incorporation of additional polar moieties such as secondary amino or hydroxyl groups can reduce the non-specific binding effects. In addition, the spacers may themselves generate local steric hindrance while at the same time reducing steric interference by the matrix (O'Carra, 1981).

There is no general "universal" spacer that can be specified with respect to length and chemical nature. Four methylene groups, e.g., give a spacer length of 0.5 nm and eight methylene groups about 1 nm. In most cases, a C_6 chain is sufficient to provide an adequate spacer effect. The most commonly used spacer is based on substituted alkyl side chains with either ω,ω'-diaminoalkalines or ω-amino carbonic acids bound to the support matrix. Typically, hexamethylene diamine and ε-aminocaproic acid are used. Coupling of the ligand to the $-NH_2$ or $-COOH$ group is usually done using the carbodiimide method. These spacers, however, display some hydrophobic properties. Spacers with hydrophilic properties based on structural units such as 3,3-diamino-dipropylamine (Harris et al., 1973), 1,3-diamino-2-propanol (O'Carra et al., 1974), or oligopeptides (Lowe et al., 1973) have been described. Some supports possess an intrinsic spacer arm such as hydroxyalkyl methacrylate which has a side chain of two methylene groups. Some supports, such as porous glass, have a spacer incorporated during derivatization. Some activating procedures also introduce side chains between the matrix and ligand. Examples of such procedures are glutaraldehyde activation, diazotization, hydroxysuccinamide activation and epoxide activation (see above).

2.5 Ligand Types

As discussed earlier, affinity chromatography can be classified on the basis of the ligand used to effect the separation. Tab. 4 lists several examples of selective and reversible biospecific affinity complexes used in affinity purifications. Below we describe both conventional biospecific and pseudobiospecific separation procedures starting with the latter first.

2.5.1 Pseudobiospecific Affinity Chromatography

2.5.1.1 Immobilized Metal Affinity Chromatography

Immobilized metal affinity chromatography (IMAC) is based on the differences in the affinity of proteins for metal ions coordinated to immobilized chelates on a suitable support (PORATH et al., 1975). IMAC is a well established technique for the purification of proteins on a laboratory scale and is gradually finding acceptance as a versatile and useful technique for industrial applications.

In IMAC, a metal chelating ligand such as iminodiacetic acid is immobilized on a chromatographic support, preferably via a spacer group. Loading of the metal is such that some coordination sites are left free on the resulting complex for binding the ligand. This is not necessary if the complex can rearrange itself to allow for the ligate to participate in the formation of complexes with the metal ion. After elution of the bound ligate, the column is usually stripped of metal ions by washing with EDTA before reuse.

Beaded agarose is the most commonly used support for IMAC of proteins. Other supports such as cross-linked agarose, hydrophilized resins and silica are used especially in HPLC applications. The chelating groups most commonly used in IMAC include N-methyliminodiacetic acid and N-(hydroxyethyl)ethylenediaminetriacetic acid. These chelators are cap-

Tab. 4. Examples of Affinity Complexes Used in Affinity Separation Methods

Ligand	Ligate
Biospecific Complexes	
Group-specific ligands	
Cofactors/coenzymes	Enzyme
Lectins	Glycoproteins
Protein A/G	Immunoglobulins
Boronic acid derivatives	Nucleotides, tRNA
Carbohydrates	Lectins
Heparin	Coagulation factors, protein kinases, restriction endonucleases and RNA polymerases
Poly T	mRNA
Specific ligands	
Antibody	Antigen
Hapten	Antibody
Hormone	Receptor protein
Substrate/inhibitor	Enzyme
Pseudobiospecific Complexes	
Group-specific ligands	
Alkyl/aryl	Various proteins
Dye	Various proteins
Metal-chelate	Various proteins

able of forming 1:1 complexes with various metal ions such as copper, nickel, zinc, cobalt, calcium, magnesium, and iron with binding affinities varying from 10^4 to 10^{15} (KAGEDAL, 1989). Immobilization of the chelating ligand is usually done by coupling the ligand to epoxy-activated supports. This immobilization chemistry is ideally suited to IMAC since it provides a chemically stable ether linkage while at the same time introducing a 12-atom spacer between the ligand and the support.

The specificity of IMAC depends on the exposure of certain amino acid residues on the surface of the protein. Thus, molecules that have very similar properties with respect to size, charge, and amino acid composition can be separated. The amino acid residues that have been implicated in binding of proteins to IMAC supports include histidine and cysteine, and to a lesser extent, tyrosine and tryptophan (VIJAYALAKSHMI, 1989; SULKOWSKI, 1987). The chelating group used as well as the choice of metal ion can dictate the specificity of the IMAC support. Buffer conditions such as pH, ionic strength and composition can modulate the interaction of a given protein with the support. Several elution schemes are available for the desorption of the bound protein such as pH gradients, organic solvents, competitive ligands, and chelating agents. A major drawback of using IMAC for protein purifications is that there is no rational way to predict whether a given protein will be adsorbed onto a metal chelate support, much less the conditions under which this interaction is optimal.

2.5.1.2 Hydrophobic Interaction Chromatography

Some investigators also include hydrophobic interaction chromatography (HIC) in the group of pseudobiospecific techniques. In hydrophobic interaction chromatography, adsorbents with immobilized hydrophobic ligands are used to separate protein mixtures on the basis of the hydrophobicity of the individual components. It is well known that proteins have extensive hydrophobic patches on their surface in addition to the expected hydrophilic groups. A recent report studied 46 different proteins with known structures and found that between 50–68% of the surface of the protein was occupied by hydrophobic residues. In buffers favoring hydrophobic interactions, these regions on the protein can bind to hydrophobic ligands. Hydrophobic ligands of different structures and length, typically aminoalkanes, have been used successfully in HIC separations. A number of HIC gels containing phenyl, octyl, pentyl, and other alkyl groups are commercially available for both conventional and HPLC applications.

Since most proteins bind to HIC gels, this technique has wide utility in the purification of many different proteins. The adsorption is generally favored by a high salt concentration which apart from stabilizing the proteins, also suppresses any undesirable ionic interaction between the support and the protein. The effectiveness of different salts in promoting hydrophobic interactions depends on their contribution to the surface tension of the solution; salts that tend to increase the surface tension favor hydrophobic interactions (MELANDER and HORVATH, 1977). Specific interactions between the salt ions and the protein can also effect the strength of the hydrophobic interaction (FAUSNAUGH and REGNIER, 1986). The strength of the interaction as well as the capacity follows the Hofmeister series:

Anions $SO_4^{2-} > Cl^- > Br^- > NO_3^- > ClO_4^- > I^- > SCN^-$

Cations $Mg^{2+} > Li^+ > Na^+ > K^+ > NH_4^+$

Sodium sulfate, sodium chloride, and ammonium sulfate are the most widely used salts in HIC applications. As with other chromatographic techniques, the pH and temperature used to play an important role in determining the binding capacity and strength of interaction. Generally, a decrease in temperature decreases the interaction, while no general trend is observed with pH. The choice of the pH and temperature is usually dictated by the stability requirements of the protein and the matrix.

Elution is usually achieved by either changing the salt concentration, or changing the polarity of the solvent by adding organic solvents such as polyethylene glycol (PEG) or isopropanol or adding detergents. Changing the salt species to a chaotropic ion such as SCN^- also works well. The gels are regenerated by washing with 6 M urea or guanidine hydrochloride to remove the strongly adsorbed proteins.

2.5.1.3 Dye–Ligand Affinity Chromatography

Dye–ligand affinity chromatography uses reactive dyes, typically triazine dyes, as immobilized ligands for affinity separations. The reactive dye usually consists of the dye component (e.g., anthraquinone, azo dyes) and a reactive residue, such as triazine, epoxide, or ethylene imine residues, through which they are covalently bound to the affinity matrix.

The most important member of the family of reactive dyes is Cibacron Blue F3G-A. Cibacron Blue shows affinity for a wide variety of enzymes and proteins including dehydrogenases, kinases, transferases, cyclases, reductases, serum proteins, and interferons (KOPPERSCHLAGER et al., 1982). The success of Cibacron Blue as an affinity ligand has spurred the development of more reactive dyes, particularly of the Cibacron and Procion series, many of which are commercially available. The low cost of these dyes, ease of immobilization, and the high capacity and stability of the immobilized dyes have made this technique useful in both small and large-scale purifications.

The mechanism of dye–protein interaction is not clearly understood. From the wide spectrum of proteins and enzymes that have been purified using Cibacron Blue, it is clear that different binding mechanisms exist for different proteins. Thus, it is rather difficult to establish rules for selecting reactive dyes as affinity ligands to purify a given protein. An initial screening of the protein of interest against various dyes will have to be carried out using different conditions such as pH and ionic strength.

Elution of the bound protein from the immobilized dye supports is easily accomplished by changes in ionic strength or pH. However, the best results are obtained with specific elution protocols using cofactors or substrates as eluents.

2.5.2 Biospecific Ligand Affinity Chromatography

2.5.2.1 Group-Specific Ligand Affinity Chromatography

Group-specific ligands recognize a group of biological macromolecules that share some common property or determinant. Examples of such ligands are coenzymes and cofactors (e.g., adenosine triphosphate, ATP, and nicotinamide-adenine dinucleotide, NAD), lectins (proteins that bind glycoproteins), and protein A and protein G (which bind various immunoglobulins).

Protein A and G are cell surface proteins isolated from Staphylococcal bacteria that bind many immunoglobulins with high affinity and have been used extensively in the recovery of immunoglobulins from a variety of sources. The specificities of protein A and G depend upon the species and class or subclass of the immunoglobulin. Generally, the binding of protein G to immunoglobulins is stronger requiring harsher elution (low pH of about 2.5) than protein A. Pre-coupled protein A and protein G are commercially available from a variety of sources for both conventional and HPLC applications.

Lectins, in particular concanavalin A (Con A), are useful ligands for the purification of carbohydrates and glycoproteins. Elution is performed by nonspecific means or by washing with a competing monomeric sugar which results in the displacement of the bound ligate. Many glycoproteins such as immunoglobulins (WEINSTEIN et al., 1972), rhodopsin (STEINEMANN and STRYER, 1973), and many membrane glycoproteins (MOHR and POMMERENING, 1985) have been purified using lectins as affinity ligands.

Coenzymes, such as ATP and NAD and their analogs, interact with many enzymes such as dehydrogenases and kinases. Immobilization is usually done via a spacer group to avoid any steric problem. Given their broad specificity, these ligands do not provide high purification factors and the separations often have to be optimized or coupled with other steps to improve purification efficiency.

2.5.2.2 Specific Ligand Affinity Chromatography

Specific ligand–ligate pairs can be classified into two broad categories: (1) enzymes and their substrate analogs and/or inhibitors and (2) immunospecific reagents such as antigen–antibody pairs and receptor–modulator pairs (KLEIN, 1991).

The high specificity of antibodies for antigens has been exploited extensively in the purification of proteins. Before the development of monoclonal antibodies, several problems were encountered with the use of polyclonal antibodies which display a broad heterogeneity of affinities for antigen. The presence of high affinity antibodies necessitated harsh elution schemes which adversely affected both the product and the immunoadsorbent. In some cases, a substantial fraction of the product could become irreversibly bound to the support leading to losses in column capacity (EVELEIGH, 1982). The advent of hybridoma technology has led to a general availability of monoclonal antibodies with homogeneous binding affinities and specificities and has had a significant impact on protein purification and detection. Recent developments in the area of antibody engineering (WINTER and MILSTEIN, 1991) have considerably expanded the potential of immunospecific reagents in both industrial and biomedical applications.

2.6 Operational Methodologies

2.6.1 Adsorption

The adsorption step is generally carried out under conditions that maximize optimum binding between the ligand and ligate while minimizing the nonspecific binding of extraneous materials. Specific adsorption of ligate can be optimized using process parameters such as ionic strength, pH, and chemical composition of buffer, along with temperature and flow rate.

The adsorption step is carried out by applying the protein sample to the previously equilibrated affinity column. Equilibration is performed by washing the column with the loading buffer till a stable baseline is obtained. After loading the sample, an incubation period may be required depending upon the kinetics of the binding process. Material that is not bound by the ligand is then removed by washing the column with several column volumes of buffer. While it is theoretically possible to eventually recover the bound ligate by continued washing, this procedure is generally very time-consuming and results in dilute product. Typically, recovery of the ligate is effected by changing the pH, ionic strength, and chemical composition of the buffer (chemical elution) or by using non-chemical means (e.g., temperature, pressure, electric fields, etc.)

The degree to which a sample has to be processed before it can be applied to an affinity column depends upon the sample. Often, very little processing is required except for a filtration step to remove particulates that may clog the column. In some cases, a preliminary fractionation may be advantageous or even necessary. Examples of cases where preliminary purification prior to the affinity step may be needed are the crude samples containing substances that degrade the ligand, or the samples containing substances that otherwise interfere in the affinity purification step.

2.6.2 Elution

The most critical step in affinity purification schemes is the recovery or elution of the adsorbed ligate from the adsorbent. Elution is usually accomplished by using conditions that result in a significant drop in the affinity constant of the ligand–ligate binding. Optimal elution conditions are those that permit complete and effective recovery of the ligate in a relatively small volume while at the same time not compromising the biological activity of both the ligand and the ligate. Several general methods of elution have been described for a wide variety of affinity applications.

2.6.2.1 Chemical Elution

Chemical elution is usually effected by changing the pH, ionic strength, and chemical composition of the buffer. Chemical elution methods are simple and relatively inexpensive to carry out.

Reducing the pH to 2 or 3 is possibly the most practiced elution technique today and in the past (CHASE, 1984; EVELEIGH and LEVY, 1977). The most common choice of buffers is glycine/HCl or acetic acid, or occasionally propionic acid. Low pH is not effective, especially when hydrophobic interactions dominate the interaction between ligand and ligate. High pH solutions have been used less often because proteins are more susceptible to irreversible denaturation under those conditions. High pH has been achieved with the use of NaOH, ammonia, glycine/NaOH buffers, and various amines, such as ethanolamine.

Chaotropic salts have also been used as effective chemical eluents and are believed to affect hydrophobic interactions. 3 M thiocyanate (SCN^-) is a commonly used chaotrope (DE SAUSSURE and DANDLIKER, 1969). Other salts have been used as eluents, including iodide (AVRAMEAS and TERNYNCK, 1967) and perchlorate.

Raising or lowering the ionic strength of the solvent can be an extremely mild elution technique, although often too mild to be effective. Concentrations of NaCl that have been used were in the range 0.75–1.2 M. NaCl is probably effective in those cases where ionic interactions play a major role. In a few cases, salts of divalent cations, namely $MgCl_2$ and $CaCl_2$, have proven to be effective eluents (MAINS and EIPPER, 1976).

Denaturants such as urea and guanidine salts have also been used for chemical elution. Generally, they tend to be used when the ligand–ligate reaction is of extremely high affinity so that no other eluent was found to be effective or when the ligate is to be recovered in denatured form so that it could be gradually renatured later.

Over the years, a number of water-miscible organic solvents have been used as eluents. Ethylene glycol is potentially the most appealing because it denatures proteins only at very high concentrations (SINGER, 1962). It has been found effective in some cases at 50–60% when combined with a pH of 4 or 9.5 (VAN OSS et al., 1979) or a pH of 12 (ANDERSSON et al., 1978). Another useful solvent is DMSO at relatively lower concentrations of up to 10% (ANDERSSON et al., 1978). The chief disadvantages of organic solvents are their effects on polysaccharide matrices and problems with viscosity.

Biospecific elution is accomplished by addition of an excess of a low molecular weight compound that competes with the ligand for binding sites on the ligate. The eluent is subsequently removed from the product stream by dialysis or gel filtration. Affinity elution has been used extensively in enzymology with cofactors, inhibitors, cosubstrates, etc. and in immunoadsorption with haptens and peptides. Biospecific elution is, however, quite expensive, and thus is generally limited to lab-scale purifications.

2.6.2.2 Non-Chemical Elution

Electrophoretic elution has been successfully used to recover a broad range of biological products including steroids, enzymes, antibodies, and virus particles. Essentially, the technique involves placing an electric field across a biospecific adsorbent after the usual adsorption and wash steps of affinity chromatography are performed. The applied electric field acts to remove dissociated antigen without substantially altering ligand–ligate affinity. Recovery of the eluate can be accomplished by elution into a chamber bounded by ultrafiltration membranes. High or complete recoveries of bound material have typically been obtained in 2–5 h elution at a field of 1–20 V/cm (YARMUSH and OLSON, 1988). Thus, this technique offers 100% theoretical recoveries of bound material, yields a highly concentrated product, and requires no solvent exchange or column regeneration. On the other hand, electrophoretic elution is slower than chemical elution and the experimental setup is more intricate and difficult to control. Electrophoretic elution is carried out in non-denaturing solvents, typically aqueous solutions buffered at near neutral pH. The field strengths used in this techniques can only induce electrophoresis. Thus, one might expect both the ligand and the ligate to retain full biological activity. Although limited in scope, activity testing of electroeluted materials has largely corroborated these notions.

The oldest non-chemical elution technique involves changes in temperature. Le Chatelier's principle states that the more exothermic the binding process, the more susceptible the complex will be to dissociation upon an increase in temperature. In most protein systems, however, the practical operating window for temperature elution is fairly limited. In addition, it is not clear if the change in affinity produced by temperature variation is large enough to provide an effective elution scheme. In antigen–antibody systems, e.g., a phenomenon termed "enthalpy–entropy" compensation has been defined which results in the free energy (and hence affinity) being fairly independent of temperature.

An alternative to temperature as an elution process is high pressure. The rationale for pressure elution can be summed up as follows. A sizeable experimental literature demonstrates that pressures of about 1000–2000 bar can dissociate non-covalent protein complexes. Pressures required to denature monomeric proteins, however, appear to be much higher, 6000 bar or more (WEBER and DRICKAMER, 1983; HEREMANS, 1982, 1987). The response of a system to pressure is governed by Le Chatelier's principle, which states that pressure shifts an equilibrium in the direction of decreasing volume. The variation of the equilibrium association (or affinity) constant K with pressure P is given by the following expression:

$$[\partial(\ln K)/\partial P]_T = -\Delta V/RT$$

where ΔV is the reaction volume ($V_p - V_r$), V_p is the volume of the products, V_r is the volume of the reactants, R is the gas constant, and T is the absolute temperature. This means that for an order of magnitude decrease in K at 1000 bar, the required volume of reaction ΔV is about 60 mL/mol. It has been shown that pressure can provide an effective yet mild elution scheme for the recovery of protein molecules bound to some immunospecific supports (OLSON et al., 1989).

2.7 High Performance Liquid Affinity Chromatography

By combining the biospecificity of affinity chromatography with the robust support materials and hardware of high performance liquid chromatography (HPLC), one can obtain high specificity, speed, and resolution in chromatography both on analytical and preparative scales. High performance liquid affinity chromatography (HPLAC) has several advantages over conventional "soft-gel" affinity chromatography. The small rigid particles allow for high resolution and very high flow rates, while the wide selection of HPLC equipment including high speed pumps, sophisticated injection units, detectors, autosamplers, and data-handling capabilities provide sensitive detection and ease of operation. Method development and scale-up are accomplished with relative ease, promoting higher productivity especially in an industrial setting. However, the cost of such specialized equipment is not a trivial matter.

As with conventional affinity chromatography, there is no ideal matrix for HPLAC. The material that is most often used is silica (sometimes incorporating zirconium and titanium oxides), because of outstanding mechanical stability. Its major problems, limited alkaline stability and non-specific adsorption, have been reduced significantly by coating silica surfaces with hydrophilic layers. This process also introduces efficient groups for further derivatization and/or ligand immobilization. Synthetic polymers, with their high chemical stability and relatively good mechanical characteristics, are also gaining popularity as HPLC supports. Pre-activated HPLAC columns with a range of immobilization chemistries (e.g., tresyl, aldehyde, CDI, epoxy, hydrazine, and chloroformate groups) are commercially available. Pre-activated supports are extremely convenient since users do not have to be concerned about activation and packing of the support. Ligand-coupled HPLAC columns are also commercially available for a number of ligands such as albumin, concanavalin A, protein A/G, iminodiacetic acid, etc.

In analytical applications, HPLAC offers new ways of monitoring biomolecules in complex mixtures, providing on-line analysis of bioreactors and downstream processing. In preparative applications, HPLAC has been used to isolate vaccines, immunomodulators, monoclonal antibodies, and other important biological macromolecules cost-effectively at high purity, while retaining complete biological activity.

2.8 Large-Scale Affinity Chromatography

Although affinity chromatography has become a widely practiced technique for the purification and analysis of biotechnological products at the laboratory scale, there are several concerns associated with application of this technology to large-scale use. For economical operation of large-scale processes, the "throughput" has to be maximized (LOWE et al., 1973). Matrices that have high adsorption capacities, high mass transfer coefficients, and high liquid flow rates are desirable for large-scale affinity chromatography. The maximum effective flow rate for any affinity chromatography application is determined by the gel compressibility and effective diffusivity of the ligate. The adsorption capacity of the affinity chromatography support is dictated, at least in part, by the interfacial area between the solid and mobile phases. The smaller the bead radius, the larger the surface area available for ligand immobilization, and hence the adsorptive capacity of the support also increases. Smaller particle size also means shorter times are required for the protein to diffuse and reach the immobilized ligand in the interior of the particle; consequently, the adsorption–desorption kinetics are significantly enhanced by using smaller beads. However, as the particle size decreases, the drag forces on the gel increase proportionally. Thus, small diameter packings made from rigid materials are most ideal for large-scale affinity chromatography.

It is also important to consider the configuration of the affinity chromatography column while scaling-up. Generally, it is preferable to use a wider-diameter column than one with a longer length since the overall compressive force and pressure drop across the bed increase with bed height thereby limiting flow rates (and hence throughput). Increasing column diameter, however, decreases the supporting effect from the column wall since a smaller fraction of the particles is in contact with the wall. Thus, bed compression is a major problem in scaling-up affinity chromatography processes. Several approaches such as using stacked columns, mixing compressible and non-compressible particles, or inserting layers of mesh discs in the gel have been suggested to overcome these problems (CLONIS, 1987).

Supports best suited to large-scale affinity chromatography include cross-linked and improved agaroses, such as Sepharose-CL, Fast Flow Sepharose and Superose (Pharmacia), cellulose-based beaded gels such as Matrex Cellufine (Amicon), synthetic macroporous rigid matrices such as TSK-gel PW (Toyo Soda), Dynospheres (Dynoparticles), and several others. Rigid macroporous beaded gels of particle size less than 20 μm with narrow particle-size distribution and appropriate functional groups for ligand immobilization are available from several manufacturers.

The cost and stability of affinity ligands is of great importance in determining the economics of any large-scale process. Generally, pseudobiospecific ligands such as reactive dyes or immobilized metal chelators are both robust and cost-effective. Protein ligands such as lectins and monoclonal antibodies tend to be more expensive and much less stable. The activation chemistry used can also add significantly to the cost of the affinity adsorbent. The most efficient chemicals such as tresyl chloride and CDI are very expensive. Cyanogen bromide activation is most widely used, as described above, but suffers from ligand leakage and the introduction of undesirable charged groups. Epoxide activation is generally more appropriate to large-scale affinity chromatography in terms of both stability and economy (CLONIS, 1987). The choice of the elution scheme is also important in determining the overall cost of the purification process; harsh conditions that tend to shorten affinity adsorbent lifetime and degrade the product are obviously not suitable for large-scale affinity chromatography. Chemical elution schemes must also take into account disposal of the eluent which can be costly depending upon its composition.

3 Other Affinity Techniques

3.1 Affinity Adsorption with Membranes

Scale-up of affinity chromatography poses several problems, the principal being transport limitations. Membranes offer the potential for making significant improvements by providing high throughput capacities and low product losses.

Several microfiltration membranes made of materials such as cellulose acetate, nylon, polyvinyl chloride, polysulfone, etc. have been developed. All of these can be activated using some of the procedures described earlier, and affinity ligands can be coupled to them to produce microporous affinity membranes. Such microporous affinity membranes have nominal pore dimensions and surface areas similar to those found in traditional affinity chromatography supports with one critical difference; namely the diffusion length (characteristic length from the bulk solution to the ligand). The diffusion length in a membrane is generally about 0.2–1.5 µm, a distance which is several fold smaller than those of affinity chromatography supports (25–300 µm). Thus, affinity membranes can dramatically improve the kinetics of the affinity adsorption process while providing for a shallow bed configuration recommended for reduced pressure drops. These two features lend themselves to extremely rapid processing. In order to maintain small diffusion lengths, scaling-up entails preparation of large surface area microfilters. For example, to produce a membrane equivalent of a 200 mL gel column (with an interstitial volume of 30% of total bed volume), one would need a membrane unit of 10000 cm^2 area with a thickness of 0.02 cm (KLEIN, 1991). To date, the main technological limitation of affinity membrane process development has been the lack of suitable membranes. The ideal membrane must display high porosity and hydrophilicity with good mechanical, chemical, and thermal stability. Membranes with these characteristics are just now becoming available. Fouling of membranes is another major bottleneck in the development of affinity membrane technology; thus membranes with a high fouling resistance or operational procedures which limit fouling are required for future development.

3.2 Affinity Cross-Flow Filtration

Affinity cross-flow filtration is the combination of the principles of affinity interactions and cross-flow membrane separations (MATTIASSON and LING, 1986; MATTIASSON, 1991b). The membrane chosen for affinity cross-flow filtration has an MW cutoff which does not retain the ligate, but does retain a very high molecular-weight ligand (macroligand). In this process, the ligate is bound by the macroaffinity ligand and thus, is retained by the membrane (while the unbound components of the crude mixture pass through). The isolated ligand–ligate complex is then treated with a suitable eluent which dissociates the ligate from the macroligand. The free ligate is collected by filtration while the macroligand is recycled.

The binding of the ligate to the macroligand can be achieved in two ways. If the crude mixture contains no particulates, the macroligand is directly mixed with the sample. A more general approach is to deliver the crude sample and the macroligand to opposite sides of a suitable MW cutoff membrane. Binding occurs, in this case, when the ligate passes through the membrane to the side where the macroligand is contained. Affinity cross-flow filtration has been used to purify alcohol dehydrogenase from a crude extract of *Saccharomyces cerevisiae* using Cibacron Blue immobilized on starch granules as the macroligand (MATTIASSON and LING, 1986) and to purify β-galactosidase from cell homogenates of *Escherichia coli* using immobilized *p*-aminobenzyl-thio-β-D-galactopyranoside agarose (PUNGOR et al., 1987). A high molecular-weight water-soluble macroligand having an acrylamide backbone with *m*-aminobenzamidine groups was used to purify trypsin from pancreatic extract (MALE et al., 1987). Affinity cross-flow filtration seems to be limited by both the availability of suitable macroligands and membranes. However, the high resolution and recovery potential of this technique along with the ability to process unclarified viscous liquids provide the incentive for further development.

3.3 Affinity Partitioning

Affinity partitioning in an aqueous polymer two-phase system has been used successfully for the purification and analysis of a number of enzymes (MATTIASSON, 1991a). In affinity partitioning, the separation is based on the preferential distribution of a soluble ligand–product complex to one of the phases in an aqueous polymer two-phase system (FLANAGAN et al., 1976). Typically, this is achieved by covalently attaching the affinity ligand to a carrier such as PEG 6000, which constitutes the less polar polymer in the two-phase system. The polar phase is usually based on high-molecular weight dextran (Dextran T-500). Most proteins, nucleic acids, and cell debris prefer the polar phase; only proteins with binding affinity for the ligand will bind to the PEG phase, and will thus be selectively extracted and concentrated in the PEG phase. Addition of potassium phosphate to the separated PEG phase will, in general, dissociate the ligand–ligate complex as well as induce the formation of a new two-phase system. The product usually favors the salt-rich phase and can be recovered by ultrafiltration. The ligand-PEG polymer phase also has an appreciable amount of salt, which must be removed (by ultrafiltration or solvent extraction) to recycle the ligand bound polymer. The recycle step is sufficiently tedious to constitute a serious limitation in the widespread applicability of affinity partitioning. Affinity partitioning of trypsin with *p*-aminobenzamidine as the PEG-bound ligand was the first published attempt in this area (TAKERKART et al., 1974). In a dextran-PEG system lacking the affinity ligand, recovery of trypsin in the PEG phase was only 40%; attachment of the affinity ligand to the PEG phase boosted trypsin recovery to 92%. Extraction of serum albumin using PEG-bound palmitic acid is another example of affinity partitioning (SHANBHAG and JOHANSSON, 1974). Much of the work on affinity partitioning has dealt with PEG derivatives of general ligands such as cofactors or triazine dyes for the purification of glycolytic and other enzymes (KULA et al., 1979; JOHANSSON et al., 1985).

For affinity partitioning to reach its full potential, it is essential that effective and simple methods are developed for the synthesis and purification of the polymeric derivatives. A few dextran derivatives such as octadecylamino dextran, octadecylamide dextran and N-amino dextran have been described (HARRIS and YALPANI, 1985) that will permit attachment of amines to dextran or substrates that will react with dextran amine. Several PEG derivatives have also been described for use in affinity partitioning. Activated PEGs have been prepared via several methods used for activation of hydroxyl supports. PEG has been reacted with sulfonyl chlorides to yield sulfonate esters that are reactive towards nucleophiles. PEG epoxides (PITHA et al., 1979) have been prepared by reacting PEG with epichlorohydrin followed by reaction with NaOH. Other derivatives of PEG that have been used for attachment of protein ligands include the cyanuric chloride derivative (BOCCU' et al., 1983; HARRIS et al., 1984), the carbonyl diimidazole derivative (BEAUCHAMP et al., 1983) and the succinimidyl derivative (ABUCHOWSKI et al., 1984). An aldehyde-terminated PEG prepared by reacting PEG-chloride with the phenoxide of 4-hydroxy benzaldehyde has also been described. Other PEG derivatives such as PEG-monopalmitate, PEG-octadecyl ether, ethylenediamine PEG, trimethylamino PEG, PEG-Cibacron Blue, etc. have been described (HARRIS and YALPANI, 1985).

Affinity partitioning has several features which render it very suitable for large-scale purification work such as its direct applicability to crude homogenates, high binding capacity and rapid operation. Recently, FLYGARE et al. (1990) found that addition of magnetically susceptible material to an aqueous two-phase system induces rapid separation when the mixed system is placed into a magnetic field. The time required for separation was reduced by up to 240000-fold.

3.4 Affinity Precipitation

This technique involves coupling of the ligand to a soluble carrier followed by precipitation of the ligand–product complex. Two approaches for affinity precipitation have been described. The first approach, which is limited to ligands binding to oligomeric proteins, involves the derivatization of the affinity ligand to form bifunctional reagents, also called bis-ligands. If the molar ratio of the bifunctional bis-ligand and the oligomeric ligate is optimal, a 3-D network will be formed which precipitates from solution as an insoluble aggregate. The first investigation of affinity precipitation (LARSSON and MOSBACH, 1979) used this approach to affinity precipitate the tetrameric enzyme lactate dehydrogenase (LDH) with a bifunctional nucleotide derivative N_2,N_2-adipohydrazido-bis-(N^6-carbonylmethyl NAD) in the presence of pyruvate. Addition of NADH to the separated precipitate resulted in the dissociation of the enzyme from the affinity bis-ligand. Subsequently, gel filtration was used to remove the enzyme from the pyruvate and bis-NAD. Affinity precipitation is acknowledged as a useful technique for oligomeric enzymes (such as dehydrogenases), with bis-functionalized nucleotide derivatives (LARSSON et al., 1984). The alternative approach, which is of more general applicability, involves coupling of affinity ligands to polymers to form multifunctional soluble reagents. Upon complexation with the ligate of interest, these reagents are caused to aggregate and finally precipitate by changing the composition of the solution, e.g., by addition of salt or changing the pH. This approach was used by SCHNEIDER et al. (1978) for the purification of trypsin from bovine pancreas. A water-soluble polyacrylamide polymer was developed bearing two functionalities: (1) a ligand group (*p*-aminobenzamidine) and (2) a precipitation group (benzoic acid) which permitted a quantitative precipitation of the affinity polymer upon changing pH. The polymer was added directly to the crude extract unter conditions favoring the binding of trypsin to the ligand and then recovered by precipitation. After elution of trypsin, the polymer could be recycled for repeated use without losing its binding efficiency. The recovery yield was large (>90%), with the recovered trypsin retaining its activity. The amount of polymer added was low (0.1–0.5%) so that nonspecific adsorption was very low resulting in essentially pure product. Affinity precipitation has much potential as a large-scale separation technique since precipitation and recovery of precipitate are commonly used procedures in the separation industry that can be easily scaled up.

3.5 Engineering Proteins for Affinity Purification

Advances in recombinant DNA technology have permitted the modification of cloned genes to facilitate protein purification (SASSENFELD, 1990). This is commonly done by adding a DNA segment encoding a polypeptide or protein tag to either the 3′ or 5′ end of the gene coding for the protein of interest. Expression of this recombinant gene in a suitable host results in the production of a fusion protein consisting of the desired protein and an affinity tag. An enzymatic or chemical cleavage site is included at the junction between the desired protein and the affinity tag to facilitate the removal of the tag after purification.

The choice of an appropriate affinity tag is dictated by several factors such as the host organism, the state of the secreted protein (whether it is secreted as soluble functional fusion protein or accumulates as denatured aggregates known as inclusion bodies), the effect of the fusion on protein function, and the method of removal of the affinity tag. Affinity tags that have been used to facilitate purification of recombinant proteins include group-specific ligands such as protein A, specific ligands such as cellulose and maltose binding proteins, glutathione S-transferase, β-galactosidase, chloramphenicol acetyltransferase, immunoaffinity tag, and pseudobiospecific ligands such as metal chelate affinity tags, hydrophobic tags, and covalent chromatography tags. Removal of the tag which may be required either to restore native protein function or reduce immunogenicity is accomplished via enzymatic procedures using proteases such as enterokinase (recognition sequence = XDDDK†X), trypsin (XKR†X), factor Xa (XIEGR†X), thrombin (XGVRGPR†X), or chemical methods such as cyanogen bromide (XM†X) or hydroxylamine (XN†GX). Chemical methods are generally harsh and not selective while enzymatic methods, although specific, are not efficient and can result in undesirable protein degradation as a result of protease contamination.

4 New or Improved Concepts and Applications

4.1 Reversed Affinity Chromatography

In reversed affinity chromatography, a protein immobilized onto a stationary phase is used for both ligand and chiral molecule separation. Examples of reversed affinity chromatography include the use of immobilized bovine serum albumin (BSA) and α-1-acid glycoprotein for the enantiomeric separation of numerous compounds (ALLENMARK, 1986; HERMANSSON, 1989). A major problem with this technique is its low binding capacity which results from the large size of the protein molecule with respect to the molecule of interest. The use of proteolytic fragments instead of the intact protein may help alleviate this problem (ERLANDSSON and NILSSON, 1989). More recent applications include the use of immobilized cellulase for the separation of β-adrenergic antagonists, and immobilized inactivated trypsin for the separation of amino acids, amino acid derivatives and peptides (ERLANDSSON et al., 1990).

4.2 Receptor Affinity Chromatography

Traditionally, immunoaffinity chromatography has been the method of choice for the purification of hormones such as cytokines. Receptor affinity chromatography, which exploits the biochemical interaction between a soluble protein and its receptor, is becoming increasingly useful as an alternative to immunoaffinity purification. Recently, WEBER and BAILON (1990) have described the purification of recombinant human interleukin 2 (rIL-2) using immobilized IL-2 receptors (IL-2R). In this study, a soluble form of IL-2R was produced in genetically engineered Chinese Hamster Ovary (CHO) cells and purified by IL-2 affinity chromatography. The isolated receptor was then coupled to a hydroxysuccinimide activated support and used to purify rIL-2 produced in bacteria or CHO cells. After the receptor affinity step, both forms of rIL-2 had a specific activity of 1.8×10^7 U/mg, with final recoveries of 58% and 88%, respectively. In contrast, rIL-2 purified via an immunoaffinity procedure has a lower specific activity (0.8×10^7 U/mg) and contained significant amounts of oligomeric and aggregated forms of rIL-2. The receptor sorbent was stable for at least 500 runs and could be used to purify rIL-2 mutants and homologs and IL-2 fusion proteins.

4.3 Fluorocarbon-Based Immobilization Technology

A new immobilization technology based on the adsorption of suitably modified biomolecules onto fluorocarbon supports has been described (KOBOS et al., 1989; STEWART et al., 1990). In this technique, protein ligands are controllably modified at their amino, carboxyl, or carbohydrate moieties to contain a significant level of fluorine atoms by reacting them with perfluoroalkylating agents. This modification greatly enhances the avidity of the protein for the support resulting in essentially irreversible binding. Retention of biological activity and specificity as well as solubility is a function of the protein and the extent of substitution. For most enzymes, perfluoroalkylation of 10–20% of the detectable amino groups results in secure immobilization and retention of more than 70% of the native biological activity. In most cases, it is possible to optimize binding to the support and biological activity by altering the ratio of perfluoroalkylating agent to the protein.

Fluorocarbon supports have several advantages including excellent chemical inertness (to acids, bases, and organic solvents), high mechanical stability and excellent flow properties (which makes them ideal for high performance applications), low nonspecific adsorption and good temperature stability (can be autoclaved). The particles are essentially fused aggregates of submicron fluorocarbon latex spheres; thus, they behave essentially as nonporous supports which have no diffusional limitations. However, this also means that binding capacities are low, a disadvantage that is compensated by its other advantages such as high throughput, fast operational times and ease of scale-up.

4.4 Molecular Imprinting and Bio-Imprinting

Molecular imprinting refers to the process of preparing tailor-made adsorption materials that are selective for a particular compound. In this technique, polymerizable monomers are first allowed to arrange themselves around the molecule of interest (the "print molecule"). The print molecule can be either covalently but reversibly bound to the monomers or the interactions between the monomer and the print molecule can be entirely noncovalent. Following polymerization, the print molecule is removed by extraction. Since the residual cavity will recognize the original print molecule by its shape and the complementary interactions between the print molecule and the groups on the polymer, the polymer will selectively adsorb the print molecule from a crude mixture, and can thus be used for its purification (EKBERG and MOSBACH, 1989).

Using this approach, polymers capable of separating amino acid derivatives on the basis of both substrate- and enantio-selectivity have been made (ANDERSSON et al., 1990). Polymers that exhibit some catalytic activity have also been prepared using transition-state analogs as print molecules (ROBINSON and MOSBACH, 1989). For example, polymers that mimic the hydrolytic action of proteases on amino acid esters have been prepared from imidazole monomers using *N-tert*-butyloxy carbonylamino acid-2-picoylamides as print molecules. These catalytic polymers were also selective since there was a correlation between the amino acid used in the imprinting and the hydrolytic rate of the active amino acid ester.

Another technique known as "bio-imprinting" has been used to alter the selectivity of a protein such as chymotrypsin or BSA to a template or print molecule by either precipitating or lyophilizing the protein in the presence of the print molecule (STAHL et al., 1990; BRACO et al., 1990). BSA, e.g., was mixed with a print molecule such as L-tartaric acid or *p*-hydroxybenzoic acid in an acetone–water solution and lyophilized. Following this treatment, there was enhanced binding of the print molecules to BSA, both in an organic solvent and when immobilized to CPG.

Yet another approach to produce tailor-made ligands for affinity separations utilizes "antisense" peptides. Antisense peptides are sequences encoded in antisense DNA. Over the past five years, several studies have shown that sense–antisense peptide interactions are observable in many systems and are fairly selective. Some of the systems studied thus far include adrenocorticotropic hormone, insulin, β-endorphin, angiotensin II, luteinizing hormone releasing hormone, etc. (CHAIKEN, 1988). The observed selectivity of sense–antisense peptide interaction has prompted the use of antisense peptides as tools in affinity technology. The use of antisense peptides in affinity chromatography has been termed "pattern-recognition affinity chromatography". Recently, receptor–peptide complexes were isolated chromatographically using immobilized antisense peptide as the affinity sorbent (LU et al., 1991). This study indicates the vast potential of antisense technology in designing tailor-made affinity ligands for various proteins.

4.5 Perfusion Chromatography

In conventional chromatographic media, intraparticle transport occurs essentially by molecular diffusion, which is a slow process for biological macromolecules. In "perfusion chromatography", transport into the particles occurs by a combination of convection and diffusion (AFEYAN et al., 1990). This study reported development of macroporous HPLC media containing two distinct classes of pores: "throughpores" of 600–800 nm which are large enough to allow some convective flow through the particle, and "diffusive pores" of 50–150 nm that form a network between the larger pores and provide a large adsorptive surface area. The support exhibits transport characteristics of much smaller particles (1 μm), but the pressure drops are much smaller; thus, mobile phase velocities of 10 to 100 times higher can be used than is possible with conventional media. Both resolution and capacity are essentially independent of the flow rate due to the rapid perfusive transport within the throughpores and exceedingly small diffusion lengths. While no reports describing the use of these supports in affinity chromatographic applications have been published yet, this technique should be applicable to immobilized ligand systems and considerably advance the use of affinity chromatography.

4.6 Site-Directed Immobilization of Proteins

The immobilization chemistry discussed earlier in this chapter such as the CNBr activation protocol couples proteins randomly through various reactive moieties such as primary amino groups, scattered throughout the molecule. In such cases, the matrix has the potential to bind at or near those regions on the ligand molecule implicated in binding. This can lead to a loss of binding activity mainly due to steric hindrance effects but also due to the modification of key residues involved in the binding. The binding capacity of the immobilized ligand is thus significantly reduced. In the case of antibodies, e.g., which possess two binding sites per molecule, immobilization via CNBr procedures typically results in a 50% or greater reduction in binding capacity. Much of the work on oriented immobilization of affinity ligands has been performed with antibodies, given the immense popularity of immunoaffinity chromatography. Antibodies consist of three domains; two identical antigen-binding domains termed Fab (for "fragment antigen-binding") and a third domain not involved in antigen binding termed Fc (for "fragment crystallizable"). Antibodies are also glycoproteins with much of the carbohydrate content located in the Fc portion.

One approach to oriented immobilization of antibodies has been to use immobilized protein A as an antibody affinity matrix. Protein A binds specifically to antibodies near their Fc domains; thus, antibodies bound to immobilized protein A matrices are likely to have their Fab arms oriented away from the matrix. Use of protein A as a spacer and coupling agent for antibody immobilization has been shown to yield immunoadsorbents with twice the binding capacity of those obtained with conventional CNBr coupling (Avila et al., 1990). Another approach to site-directed immobilization of antibodies utilizes their carbohydrate moieties (Domen et al., 1990; O'Shannessy, 1990; Turkova et al., 1990). This method relies on the oxidation of the carbohydrate groups to yield aldehydes; these are then coupled to supports bearing primary amino groups or hydrazido-modified supports resulting in a Schiff's base and hydrazone derivative, respectively. Mild reduction with sodium cyanoborohydride results in the formation of stable secondary amine or hydrazine linkages. The oxidation of the carbohydrate groups can be performed by either enzymatic (galactose oxidase) or chemical (sodium *meta*-periodate) means. The coupling efficiency is a function of both carbohydrate content and the accessibility of these moieties for oxidation. Immunosorbents prepared by this technique consistently demonstrated the optimum molar ratio of immobilized antibody to antigen of 1:2; coupling efficiencies were about 70% which was attributed to the polyclonal nature of the antibodies used in the study (Domen et al., 1990). Site-directed immobilization of antibody molecules has also been accomplished via free sulfhydryl groups generated in the hinge region of the molecule via mild reduction. Incorporation of a free Cys by site-directed mutagenesis in recombinant antibody fragments such as Fab or the Fv has also been used as a handle for site-directed mutagenesis. Recently, a novel scheme for orienting $F(ab')_2$ fragments (bivalent antigen binding fragments of antibodies generated by pepsin digestion) on supports has been developed (Lu et al., 1992). The strategy involved first reversibly blocking all the accessible carboxyl groups on the molecule followed by generation of the $F(ab')_2$ fragment by proteolytic digestion. This fragment would have only two C-terminal carboxylic groups free for coupling to a support containing a primary amino group or a hydrazido-modified support. After coupling, the unreacted carboxylic groups were deblocked. Preliminary results indicated that while the binding capacity of immunoadsorbents generated in this manner was comparable to that obtained with carbohydrate-oriented immunoadsorbents, the coupling efficiencies were somewhat greater.

5 Summary

The commercial realization of biotechnology depends ultimately on the ready availability of highly purified biomolecules for industrial and biomedical applications. Traditional purification schemes utilize a combination of techniques that resolve substances based on physico-chemical properties add substantially to the cost of the product. Up to 90% of the overall processing cost of many bioproducts is incurred at the purification stage. Affinity-based separations (especially affinity chromatography) have been used in the laboratory for many years as a powerful tool for the purification of biological macromolecules. In the past decade, impressive progress has been made in almost every aspect of affinity purification technology and new and improved concepts and applications have been discovered, resulting in its increasing use on the process scale.

In the future, the combination of molecular biology, process engineering, and innovative products from material science will undoubtedly result in new generations of affinity systems with improved, robust purification procedures and high resolution and recovery.

6 References

ABUCHOWSKI, A., KAZO, G. M., VERHOEST JR., C. R., VAN ES, D., NUCCI, M. L., KAFKEWITZ, T., VIAU, A. T., DAVIS, F. F. (1984), Cancer therapy with chemically modified enzymes. I. Antitumor properties of PEG-asparaginase conjugates, *Cancer Biochem. Biophys.* **7**, 175–186.

AFEYAN, N. B., GORDON, N. F., MAZSAROFF, I., VARADY, L., FULTON, S. P., YANG, Y. B., REGNIER, F. E. (1990), Flow-through particles for the high-performance liquid chromatographic separation of biomolecules: Perfusion chromatography, *J. Chromatogr.* **519**, 1–29.

ALLENMARK, S. (1986), Optical resolution by liquid chromatography in immobilized bovine serum albumin, *J. Liq. Chromatogr.* **9**, 425–442.

ANDERSSON, K. K., BENYAMIN, Y., DOUZOU, P., BALNY, C. (1979), The effects of organic solvents and temperature on the desorption of yeast 3-phosphoglycerate kinase from immunoadsorbents, *J. Immunol. Methods* **25**, 375–381.

ANDERSSON, L. I., O'SHANNESSY, D. J., MOSBACH, K. (1990), Molecular recognition in synthetic polymers: preparation of chiral stationary phases by molecular imprinting of amino acid amides, *J. Chromatogr.* **513**, 167–179.

AVILA, D. M., KAUSHAL, V., BARNES, L. D. (1990), Immunoaffinity chromatography of diadenosine 5′,5′′-P^1,P^4-tetraphosphate phosphorylase from *Saccharomyces cerevisiae, Appl. Biochem. Biotechnol.* **12**, 276–283.

AVRAMEAS, S., TERNYNCK, T. (1967), Use of iodide salts in the isolation of antibodies and the dissolution of specific immune complexes, *Biochem. J.* **102**, 37C–39C.

AXEN, R., PORATH, J., ERNBACK, S. (1967), Chemical coupling of peptides and proteins to polysaccharides by means of cyanogen halides, *Nature* **214**, 1302–1304.

BEAUCHAMP, C. O., CRONIAS, S. L., MENAPACE, D. P., PIZZO, S. V. (1983), A new procedure for the synthesis of PEG-protein adducts, *Anal. Biochem.* **131**, 25–33.

BETHELL, G. S., AYERS, J. S., HANCOCK, W. S., HEARN, M. T. W. (1979), A novel method of activation of cross-linked agarose with 1,1′-carbonyldiimidazole which gives a matrix for affinity chromatography devoid of additional charged groups, *J. Biochem.* **254**, 2572–2574.

BOCCU', E., LARGAJOLLI, R., VERONESE, F. M. (1983), Coupling of monomethoxy polyethyleneglycols to proteins via active esters, *Bioscience* **38C**, 94–99.

BONNERJEA, J., OH, S., HOARE, M., DUNHILL, P. (1986), Protein purification: the right step at the right time, *Bio/Technology* **4**, 954–958.

BRACO, L., DABULIS, K., KLIBANOV, A. M. (1990), Production of abiotic receptor by molecular imprinting of proteins, *Proc. Natl. Acad. Sci. USA* **87**, 274–277.

CHAIKEN, I. (1988), The design of peptide and protein recognition mimics using ideas from sequence simplification and antisense peptides, in: *Molecular Mimicry in Health and Disease* (LERNMARK, A., DYRBERG, T., TERENIUS, L., HOKFELT, B., Eds.), pp. 351–367. Amsterdam: Elsevier Science (Biomedical Division).

CHASE, H. A. (1984), Affinity separations using immobilized monoclonal antibodies – a new tool for the biochemical engineer, *Chem. Eng. Sci.* **39,** 1099–1125.

CLONIS, Y. D. (1987), Large-scale affinity chromatography, *Bio/Technology* **5,** 1290–1293.

COUPEK, J. (1982), Macroporous spherical hydroxylethyl methacrylate copolymers, their properties, activation and use in high performance affinity chromatography, in: *Affinity Chromatography and Related Techniques* (GRIBNAU, T. C. J., VISSON, J., NIVARD, R. J. F., Eds.), pp. 165–179. Amsterdam: Elsevier.

CUATRECASAS, P. (1970), Protein purification by affinity chromatography: derivatization of agarose and polyacrylamide beads, *J. Biol. Chem.* **245,** 3059–3065.

DEGL'INNOCENTI, D., BERTI, A., STEFANI, M., LIGURI, G., RAMPONI, G. (1990), Immunoaffinity purification and immunoassay determination of human erythrocyte acyl phosphatase, *Appl. Biochem. Biotechnol.* **12,** 450–459.

DE SAUSSURE, V. A., DANDLIKER, W. B. (1969), Ultracentrifuge studies of the effects of thiocyanate ion on antigen–antibody systems, *Biochemistry* **6,** 77–83.

DOMEN, P. L., NEVENS, J. R., MALLIA, A. K., HERMANSSON, G. T., KLENK, D. C. (1990), Site-directed immobilization of proteins, *J. Chromatogr.* **510,** 293–302.

EKBERG, B., MOSBACH, K. (1989), Molecular imprinting: a technique for producing specific separation materials, *Trends Biotechnol.* **7,** 92–96.

ERLANDSSON, P., NILSSON, S. (1989), Use of fragment of bovine serum albumin as a chiral stationary phase in liquid chromatography, *J. Chromatogr.* **482,** 35–51.

ERLANDSSON, P., MARLE, I., HANSSON, L., ISAKSSON, R., PETERSSON, C., PETERSSON, G. (1990), Immobilized cellulase as a chiral stationary phase for direct resolution of enantiomers, *J. Am. Chem. Soc.* **112,** 4573–4574.

EVELEIGH, J. W. (1972), Practical considerations in the use of immunoadsorbents and associated instrumentation, in: *Affinity Chromatography and Related Techniques* (GRIBNAU, T. C. J., VISSON, J., NIVARD, R. J. F., Eds.), pp. 293–303. Amsterdam: Elsevier.

EVELEIGH, J. W., LEVY, D. E. (1977), Immunochemical characteristics and preparative applications of agarose-based immunoadsorbents, *J. Solid-Phase Biochem.* **2,** 45–78.

FAUSNAUGH, J. L., REGNIER, F. E. (1986), Solute and mobile phase contributions to retention in hydrophobic interaction chromatography of proteins, *J. Chromatogr.* **359,** 131–136.

FLANAGAN, S. D., BARONDES, S. H., TAYLOR, P. (1976), Affinity partitioning of membranes: cholinergic receptor-containing membranes from *Torpedo californica, J. Biol. Chem.* **251,** 858–865.

FLYGARE, S., WIKSTROM, P., JOHANSSON, G., LARSSON, P. O. (1990), Magnetic aqueous two-phase separation in preparative applications, *Enzyme Microb. Technol.* **12,** 95–103.

HARRIS, J. M., YALPANI, M. (1985), Polymer-ligands used in affinity partitioning and their synthesis, in: *Partitioning in Aqueous Two-Phase Systems* (WALTER, H., BROOKS, D. E., FISHER, D., Eds.), pp. 589–625. New York: Academic Press.

HARRIS, R. G., ROWE, J. J. M., STEWART, P. S., WILLIAMS, D. C. (1973), Affinity chromatography of glucuronidase, *FEBS Lett.* **29,** 189–192.

HARRIS, J. M., YALPANI, M., VAN ALSTINE, J. M., STRUCK, E. C., CASE, M. G., PALEY, M. C., BROOKS, D. E. (1984), Synthesis of PEG derivatives, *J. Polym. Sci.* **22,** 341–352.

HEREMANS, K. (1982), High pressure effects on proteins and other biomolecules, *Annu. Rev. Biophys. Bioeng.* **11,** 1–21.

HEREMANS, K. (1987), Pressure effects on the secondary and tertiary structure of biopolymers, in: *Current Perspectives in High Pressure Biology* (JANNASCH, H. W., MARQUIS, R. E., ZIMMERMAN, A. M., Eds.), pp. 225–234, New York: Academic Press.

HERMANSSON, J. (1989), Enantiomeric separation of drugs and related compounds based on their interaction with α1-acid glycoprotein, *Trends Anal. Chem.* **8,** 251–259.

INMAN, J. K. (1974), Covalent linkage of functional groups, ligands and proteins to polyacrylamide beads, *Methods Enzymol.* **34,** 30–58.

INMAN, J. K., DINTZIS, H. M. (1969), The derivatization of cross-linked polyacrylamide beads: controlled introduction of functional groups for the preparation of special purpose biochemical adsorbents, *Biochemistry* **8,** 4074–4082.

ITO, N., NOGUCHI, K., KAZAMA, M., KASAI, K. I. (1987), Analysis of human glutamyl- and lysyl-plasminogen by high-performance affinity chromatography, *J. Chromatogr.* **400,** 163–167.

JANSON, J.-C. (1984), Large-scale affinity purification – state of the art and future prospects, *Trends Biotechnol.* **2,** 31–38.

JOHANSSON, G., JOELSSON, M., AKERLUND, H. E. (1985), An affinity-ligand gradient technique for purification of enzymes by counter-current distribution, *J. Biotechnol.* **2,** 225–237.

KAGEDAL, L. (1989), Immobilized metal-ion chromatography, in: *Protein Purification: Principles, High Resolution Methods and Applications* (JANSON, J.-C., RYDEN, L., Eds.), New York: VCH Publishers.

KAY, G., LILLY, M. D. (1970), The chemical attachment of chymotrypsin to water-insoluble polymers using 2-amino-4,6-dichloro-*s*-triazine, *Biochim. Biophys. Acta* **198,** 276–285.

KLEIN, E. (1991), Applications for affinity microfiltration membranes, in: *Affinity Membranes: Their Chemistry and Performance in Adsorptive Separation Processes,* pp. 128–147. New York: John Wiley & Sons, Inc.

KOBOS, R. K., EVELEIGH, J. W., ARENTZEN, R. (1989), A novel fluorocarbon-based immobilization technology, *Trends Biotechnol.* **7,** 101–105.

KOHN, J., WILCHEK, M. (1982), The determination of active species on CNBr and trichloro-*s*-triazine activated polysaccharides, in: *Affinity Chromatography and Related Techniques* (GRIBNAU, T. C. J., VISSON, J., NIVARD, R. J. F., Eds.), pp. 235–244. Amsterdam: Elsevier.

KOHN, J., WILCHEK, M. (1984), The use of cyanogen bromide and other novel cyanylating agents for the activation of polysaccharide resins, *Appl. Biochem. Biotechnol.* **9,** 285–305.

KOPPERSCHLAGER, G., BOHME, H. J., HOFMANN, E. (1982), Cibacron Blue F3G-A and related dyes as ligands in affinity chromatography, *Adv. Biochem. Eng.* **25,** 101–138.

KULA, M. R., JOHANSSON, G., BUCKMANN, A. F. (1979), Large-scale isolation of enzymes, *Biochem. Soc. Trans.* **7,** 1–5.

LAAS, T. (1975), Agar derivatives for chromatography, electrophoresis and gel-bound enzymes, II. A benzylated dibromopropanol cross-linked Sepharose as an amphophilic gel for hydrophobic salting-out chromatography of enzymes with special emphasis on denaturing disks, *J. Chromatogr.* **111,** 373–387.

LARSSON, P. O., MOSBACH, K. (1979), Affinity precipitation of enzymes, *FEBS Lett.* **98,** 333–338.

LARSSON, P. O., FLYGARE, S., MOSBACH, K. (1984), Affinity precipitation, *Methods Enzymol.* **104,** 364–369.

LOWE, C. R., DEAN, P. D. G. (1974), *Affinity Chromatography.* London: John Wiley & Sons.

LOWE, C. R., HARVEY, M. J., CRAVEN, D. B., DEAN, P. D. G. (1973), Some parameters relevant to affinity chromatography on immobilized nucleotides, *Biochem. J.* **133,** 499–506.

LU, F. X., AIYAR, N., CHAIKEN, I. (1991), Affinity capture of [Arg^8]vasopressin-receptor complex using immobilized antisense peptide, *Proc. Natl. Acad. Sci. USA* **88,** 3642–3646.

LU, X.-M., YARMUSH, D. M., YARMUSH, M. L. (1992), Oriented coupling of antibody binding fragments to solid phase supports: site-directed binding of $F(ab')_2$ fragments, *J. Immunol.,* submitted.

MADDEN, J. K., THOM, D. (1982), Properties and interactions of polysaccharides underlying their use as chromatographic supports, in: *Affinity Chromatography and Related Techniques* (GRIBNAU, T. C. J., VISSON, J., NIVARD, R. J. F., Eds.), pp. 113–129. Amsterdam: Elsevier.

MAINS, R. E., EIPPER, B. A. (1976), Biosynthesis of adrenocorticotropic hormone in mouse pituitary tumor cells, *J. Biol. Chem.* **251,** 4115–4120.

MALE, K. B., LUONG, J. H. T., NGUYEN, A.-L. (1987), Studies on the application of a newly synthesized polymer for trypsin purification, *Enzyme Microb. Technol.* **9,** 374–378.

MATTIASSON, B. (1991a), Affinity Partitioning, in: *Chromatographic and Membrane Process in Biotechnology* (COSTA, C. A., CAHRAL, J. S., Eds.), pp. 309–322. Dordrecht: Kluwer Academic Publishers.

MATTIASSON, B. (1991b), Membrane affinity filtration, in: *Chromatographic and Membrane Processes in Biotechnology* (COSTA, C. A., CAHRAL, J. S., Eds.), pp. 335–350. Dordrecht: Kluwer Academic Publishers.

MATTIASSON, B., LING, T. G. I. (1986), Ultrafiltration affinity purification: a process for large-scale biospecific separations, in: *Membrane Separation in Biotechnology* (MCGREGOR, W. C., Ed.), pp. 99–114. New York: Marcel Dekker.

MELANDER, W., HORVATH, C. (1977), Salt effects on hydrophobic interactions in precipitation and chromatography of proteins: an interpretation of the lyotropic series, *Arch. Biochem. Biophys.* **183,** 200–215.

MOHR, P., POMMERENING, K. (1985), *Affinity Chromatography: Practical and Theoretical Aspects.* New York: Marcel Dekker, Inc.

NARAYAN, S. R., CRANE, L. J. (1990), Affinity chromatography supports: a look at performance requirements, *Trends Biotechnol.* **8,** 12–16.

NILSSON, K., MOSBACH, K. (1980), *p*-Toluene sulfonyl chloride as an activating agent of agarose for the preparation of immobilized affinity ligands and proteins, *Eur. J. Biochem.* **112,** 397–402.

O'CARRA, P. (1981), Biospecific binding to immobilized small ligands in affinity chromatography, *Biochem. Soc. Trans.* **9,** 283.

O'CARRA, P., BARRY, S., GRIFFIN, T. (1973), Spacer arms in affinity chromatography: the need for a more rigorous approach, *Biochem. Soc. Trans.* **1,** 289–290.

O'CARRA, P., BARRY, S., GRIFFIN, T. (1974),

Spacer arms in affinity chromatography: use of hydrophilic arms to control or eliminate non-specific adsorption effects, *FEBS Lett.* **43,** 169–175.

OLSON, W. C., LEUNG, S. K., YARMUSH, M. L. (1989), Recovery of antigens from immunoadsorbents using high pressure, *Bio/Technology* **7,** 369–373.

O'SHANNESSY, D. J. (1990), Hydrazido-derivatized supports in affinity chromatography, *J. Chromatogr.* **510,** 13–21.

PITHA, J., KOCIOLEK, K., CARON, M. G. (1979), Detergents linked to polysaccharides: preparation and effects on membranes and cells, *Eur. J. Biochem.* **94,** 11–18.

PORATH, J., SUNDBERG, L. (1972), High capacity chemisorbents for protein immobilization, *Nature New Biol.* **238,** 261–262.

PORATH, J., JANSON, J. C., LAAS, T. (1971), Agar derivatives for chromatography, electrophoresis and gel-bound enzymes. I. Desulfated and reduced cross-linked agar and agarose in spherical bead form, *J. Chromatogr.* **60,** 167–177.

PORATH, J., CARLSSON, J., OLSSON, I., BELFRAGE, G. (1975a), Metal chelate affinity chromatography: a new approach to protein fractionation, *Nature* **258,** 598–599.

PORATH, J., JANSON, J. C., LAAS, T. (1975b), Agar derivatives for chromatography, electrophoresis and gel-bound enzymes. III. Rigid agarose gel cross-linked with divinylsulfone (DVS), *J. Chromatogr.* **60,** 167–177.

PUNGOR, E., AFEYAN, N. B., GORDON, N. F., COONEY, C. L. (1987), Continuous affinity-recycle extraction: a novel protein separation technique, *Bio/Technology* **5,** 604–608.

REGNIER, F. E., NOEL, R. (1976), Glycerolpropylsilane bonded phases in the steric exclusion chromatography of biological macromolecules, *J. Chromatogr. Sci.* **14,** 316–320.

ROBINSON, D. K., MOSBACH, K. (1989), Molecular imprinting of a transition state analogue leads to a polymer exhibiting esterolytic activity, *J. Chem. Soc. Commun.* **14,** 969–970.

SANDERSON, C. J., WILSON, D. V. (1971), Methods for coupling proteins or polysaccharides to red cells by periodate oxidation, *Immunochemistry* **8,** 163–168.

SASSENFELD, H. M. (1990), Engineering proteins for purification, *Trends Biotechnol.* **8,** 88–93.

SCHNEIDER, M., GUILLOT, C., LARRY, B. (1978), *U.S. Patent* 4066505.

SHANBHAG, V. P., JOHANSSON, G. (1974), Specific extraction of human serum albumin by partition in aqueous biphasic systems containing poly(ethylene glycol) bound ligand, *Biochem. Biophys. Res. Commun.* **61,** 1141–1146.

SINGER, S. J. (1962), The properties of proteins in non-aqueous solvents, *Adv. Protein Chem.* **17,** 1–68.

STAHL, S., MÅNSSON, M. O., MOSBACH, K. (1990), The synthesis of a D-amino acid ester in an organic media with α-chymotrypsin modified by a bio-imprinting procedure, *Biotechnol. Lett.* **12,** 161–166.

STEWART, D. J., PURVIS, D. R., LOWE, C. R. (1990), Affinity chromatography on novel perfluorocarbon supports: immobilization of C.I. Reactive Blue on a polyvinyl alcohol-coated perfluoropolymer support and its application in affinity chromatography, *J. Chromatogr.* **510,** 177–187.

SULKOWSKI, E. (1987), Immobilized metal ion affinity chromatography, in: *Protein Purification: Micro to Macro* (BURGESS, R., LISS, A. K., Eds.), UCLA Symposia on Molecular and Cellular Biology, Vol. 68, pp. 149–162.

SUNDBERG, L., PORATH, J. (1974), Preparation of adsorbents for biospecific affinity chromatography: Attachment of group-containing ligands to insoluble polymers by means of bifunctional oxiranes, *J. Chromatogr.* **90,** 87–98.

TAKERKART, G., SEGARD, E., MONSIGNY, M. (1974), Partition of trypsin in two-phase systems containing a diamidino-a,w-diphenyl carbamyl poly(ethyleneglycol) as a competitive inhibitor of trypsin, *FEBS Lett.* **42,** 218–220.

TURBAK, A. F., EL-KAFRACY, A., SNYDER, W., AUERBACK, A. B. (1982), Process of forming shaped cellulosic products, *U.S. Patent* 4352770.

TURKOVA, J., PETKOV, L., SAJDOK, J., KAS, J., BENES, M. J. (1990), Carbohydrates as a tool for oriented immobilization of antigens and antibodies, *J. Chromatogr.* **500,** 585–593.

VAN OSS, C. J., ABSOLOM, D. R., GROSSBERG, A. L., NUEMANN, A. W. (1979), Repulsive van der Waals forces. I. Complete dissociation of antigen–antibody complexes by means of negative van der Waals forces, *Immunol. Commun.* **8,** 11–29.

VIJAYALAKSHMI, M. A. (1989), Pseudobiospecific ligand affinity chromatography, *Trends Biotechnol.* **7,** 71–76.

WEBER, D. V., BAILON, P. (1990), Application of receptor-affinity chromatography to bioaffinity purification, *J. Chromatogr.* **510,** 59–69.

WEBER, G., DRICKAMER, H. H. (1983), The effect of high pressure upon proteins and other biomolecules, *Quart. Rev. Biophys.* **16,** 89–112.

WEETALL, H. H. (1973), Affinity chromatography, *Sep. Purif. Methods* **2,** 199–229.

WEETALL, H. H., FILBERT, A. M. (1974), Porous glass for affinity chromatography applications, *Methods Enzymol.* **34,** 59–76.

WESTON, P. D., AVRAMEAS, S. (1971), Proteins coupled to polyacrylamide beads using glutaraldehyde, *Biochem. Biophys. Res. Commun.* **45**, 1574–1580.

WINTER, G., MILSTEIN, C. (1991), Man-made antibodies, *Nature* **349**, 293–299.

YARMUSH, M. L., OLSON, W. C. (1988), Electrophoretic elution from biospecific adsorbents: principles, methodology, and applications, *Electrophoresis* **9**, 111–120.

25 Electrokinetic Separations

ALAN J. GRODZINSKY

Cambridge, MA 02139, USA

MARTIN L. YARMUSH

Piscataway, NJ 08854, USA

1 Introduction

Electrokinetic separation techniques are widely used in biotechnology to purify and characterize a variety of biological macromolecules of commercial and research value. While the use of gel electrophoresis and related techniques for large-scale purifications is limited by many factors, its utility for analytical purposes is virtually unsurpassed. The advantages of electrokinetic methods as analytical tools include simplicity and ease of operation, good sensitivity (detection down to nanogram quantities), reproducibility, and accuracy. Today, a wealth of electrokinetic techniques and increasingly sophisticated instrumentation are available that render electrophoresis an easy and versatile tool to acquire even for a novice.

Electrokinetic phenomena involve the flow of a fluid electrolyte relative to an adjacent charged solid surface. An applied electric field can induce motion of fluid through a charged capillary or a charged porous medium (electroosmosis) or motion of charged molecules with respect to the fluid (electrophoresis). Conversely, pressure-induced motion of fluid past a charged surface can induce streaming potentials, while sedimentation charged particles in an electrolyte will produce a sedimentation potential. These classical electrokinetic effects were described over a century ago, and have been studied in great detail experimentally and theoretically (OBERBEEK, 1952b; DUKHIN and DERJAGUIN, 1974; LEVICH, 1962; HUNTER, 1989).

In this chapter, we will focus on recent advances in electrokinetic separations, including capillary electrophoresis and the use of alternating electric fields with slab gel electrophoresis. We first summarize the fundamental laws that describe fields, forces, and flows at charged interfaces (e.g., macromolecule–electrolyte and glass capillary–electrolyte interfaces). These laws constitute a complete description (within the confines of certain limiting assumptions) of the equilibrium electrical double layer and the non-equilibrium electrophoretic and electroosmotic flows that can occur at double layer interfaces. We then summarize some current methodologies based on these fundamental principles, and give examples of modern practice.

2 Electrokinetic Fundamentals

2.1 Basic Laws

The basic formulation of electromechanical and electrochemical transduction includes laws that describe the individual electrical, chemical and mechanical subsystems and their coupling. The electrical subsystem focuses on charge groups fixed to the solid medium, mobile ions in the electrolyte, and the electrical interactions between them. These interactions are described, classically, by Poisson's equation of electrostatics. It is useful and important to recall the fundamental electroquasistatic subset of Maxwell's equations (HAUS and MELCHER, 1989) from which Poisson's equation is derived, since these laws are the basis for defining the concept of "voltage" in electrochemical systems at all frequencies of interest to electrokinetics:

$$\nabla \times E = -\frac{\partial B}{\partial t} \simeq 0; \quad E = -\nabla \Phi \tag{1}$$

$$\nabla \cdot \varepsilon E = \rho = \sum_i z_i F c_i \tag{2}$$

Eq. (1) is the quasistatic form of Faraday's law, which states that the electric field E has no curl (no circulation) when there is negligible magnetic flux density B in the region, or when time rates of change of B are small (i.e., low frequencies); E is then said to be a "conservative" field. Only under these conditions can we define the concept of electrical potential (voltage), Φ, whose derivative (gradient) is then related to E as shown on the right-hand side of Eq. (1). Gauss' law, Eq. (2), relates E to the total space charge density ρ which, in an electrolyte, is further expressed in terms of the ionic concentrations, c_i, valence, z_i, and the Faraday constant, F; ε is the dielectric permittivity of the fluid, assumed to be a linear, isotropic medium. The chemical subsystem includes relations for the flux N_i and continuity of each ionic species, respectively:

$$N_i = \frac{z_i}{|z_i|} u_i c_i E - D_i \nabla c_i \tag{3}$$

$$\frac{\partial c_i}{\partial t} + v \cdot \nabla c_i = -\nabla \cdot N_i + (G_i - R_i) \tag{4}$$

where u_i and D_i are the mobility and diffusivity of each electrolyte species, respectively, v is the fluid velocity, and the generation and recombination rates G_i, R_i account for chemical reactions. We note that the total current density J is related to the ionic fluxes by $J = \sum z_i F N_i$, and that conservation of current within the electrolyte phase, $\nabla \cdot J = -(\partial \rho / \partial t)$, is obtained from appropriate summation of the continuity laws Eq. (4). Finally, the mechanical subsystem must account for the relative flow of a viscous, incompressible fluid past a charged, solid surface (e.g., electroosmotic flow through a charged glass capillary tube, or flow around a charged macromolecule as in electrophoresis). These fluid flows are described by the law for fluid continuity, Eq. (5), and the Navier–Stokes equation (6) for viscous dominated creeping flow, including the electrical force density ρE when the fluid contains net ionic space charge:

$$\nabla \cdot v = 0 \tag{5}$$

$$\rho_m \left(\frac{\partial v}{\partial t} + v \cdot \nabla v \right) = -\nabla p + \eta \nabla^2 v + \rho E \tag{6}$$

where ρ_m and η are the fluid mass density and kinematic viscosity, and p is the fluid pressure.

2.2 The Interfacial Electric Double Layer

Charge separation naturally occurs at phase boundaries in electrochemical and at biological interfaces, due to chemical or physical adsorption from the solution, the presence of ionizable fixed charge groups on the solid surface, or other discontinuities in properties at the interface. The charge on the solid phase leads to the formation of an electrical double layer, with one sign of charge fixed on the solid, an excess of mobile counter-ions in the adjacent electrolyte phase, and a corresponding equilibrium potential difference across the interface. This double layer is the site of electromechanical coupling inherent in electrokinetic effects, and has therefore been the subject of many detailed treatments (e.g., OVERBEEK, 1952a; LEVICH, 1962; HUNTER, 1989). Experimental and theoretical investigations have shown that the double layer can be divided into one or more compact regions of molecular dimensions immediately adjacent to the solid surface, along with a diffuse region that extends into the fluid phase over several electrical Debye lengths (Fig. 1) (OVERBEEK, 1952a).

Most importantly, the spatial distribution of the electrical potential and ionic species in the equilibrium double layer remains essentially undisturbed by the tangential electrical and viscous stresses that produce the non-equilibrium transport inherent in electrokinetic separations. In order to highlight the critical double layer parameters that permit such behavior, we briefly derive the equilibrium distribution of Φ and c_i from the basic laws, Eqs. (1)–(6). In thermal equilibrium, there is no net flux N_i of any species; Eq. (3) thus reduces to the Boltzmann distribution for ions in a potential field:

$$c_i(r) = c_{io} \exp(-z_i F \Phi(r)/RT) \tag{7}$$

Eq. (7) incorporates the Nernst–Einstein relation, $(D_i/u_i) = (RT/|z_i|F)$, and c_{io} are the bulk ion concentrations far away from the solid surface. Eqs. (1) and (2) combine to give Poisson's equation of electrostatics and, with Eq. (7), give the familiar Poisson–Boltzmann equation (for a $z{:}z$ electrolyte):

$$\nabla^2 \Phi = -\frac{\rho}{\varepsilon} = \kappa^2 \sinh(zF\Phi/RT). \tag{8}$$

When Eq. (8) can be linearized, solution gives a simple exponential decay for the potential Φ, governed by the characteristic decay length (Debye length) κ^{-1}:

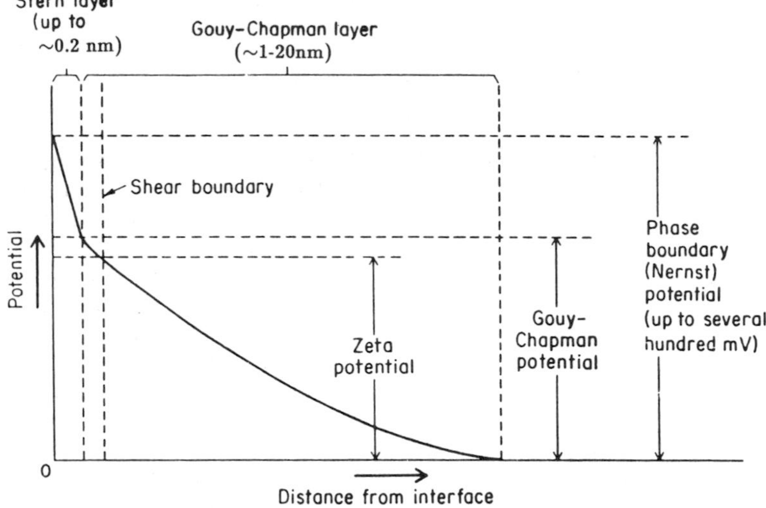

Fig. 1. Schematic of the electrical double layer (from JAIN, 1972).

$$\kappa^{-1} = \sqrt{\frac{\varepsilon R T}{2 z^2 F^2 c_o}}. \tag{9}$$

This distribution of Φ is pictured in Fig. 1 as the so-called Gouy–Chapman layer or diffuse double layer region. Also shown is the inner "Stern layer" demarking the region of closest approach of finite-size hydrated ions. During electrokinetic flows, a shear boundary exists at which tangential fluid velocities decay to zero; the electrical potential at this fluid slip-plane is defined as the ζ (zeta) potential of electrokinetics. For a 0.1 M NaCl solution, the Debye length characterizing the diffuse layer is about 1 nm; the slip plane is drawn within the diffuse layer in Fig. 1, but may also occur within the compact (Stern) layer.

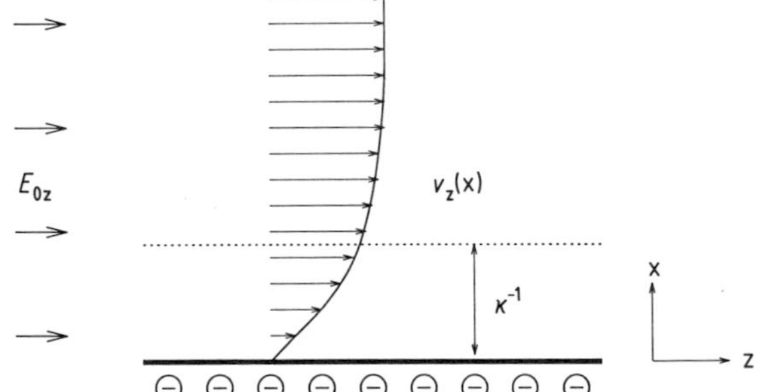

Fig. 2. Fluid velocity profile resulting from a balance of electrical and viscous shear stresses within the diffuse double layer, produced by an electric field applied tangentially to the interface.

2.3 Electrokinetic Transduction: Electroosmosis

We now extend the basic laws (Sect. 2.1) and the description of the double layer (Sect. 2.2) to demonstrate the critical parameters that govern capillary electrophoresis (Sect. 4) and electrophoresis of solutes in gel media (Sect. 3). In general, electrokinetic phenomena arise from a balance between viscous and electrical shear stresses in the electrical double layer. Fig. 2 shows this balance at a planar surface, an example that is useful for a physical understanding of electroosmotic flow in capillaries as well as "free" electrophoresis within such capillaries. In equilibrium, a double layer exists at the planar surface; in fused silica glass, for example, silanol groups contribute to a negative surface charge density at pH above 3. When an electric field E_{0z} is applied tangential to the interface, a ρE force is applied to the mobile portion of the double layer. Since the tangential electric field is continuous at the interface (a boundary condition consistent with Eq. (1)), there is also an electrical force on the glass wall surface charge. However, the solid surface cannot move; therefore, an electroosmotic flow of fluid past the solid is generated by the applied field E_{0z}. In the geometry of Fig. 2, assuming temporarily that the pressure drop along the direction of flow is negligible, the Navier–Stokes Eq. (6) for low Reynolds number fully developed creeping flow takes the form:

$$\eta \frac{\partial^2 v_z}{\partial x^2} = -\rho E_z = +\frac{\partial}{\partial x}\left[\left(\frac{\partial \Phi(x)}{\partial x}\right) E_{0z}\right] \tag{10}$$

The right-hand equality in Eq. (10) is based on the assumption that the double layer charge density ρ is unperturbed by the applied field E_{0z}. This is well justified for ionic concentration ranges in buffers of interest, which correspond to double layer thicknesses of 1–10 nm; ρ corresponds to an x-directed double layer electric field that is orders of magnitude greater than the tangential applied field (even in capillary electrophoresis). Two integrations of Eq. (10) give the velocity profile within the double layer and the associated uniform, bulk electroosmotic fluid velocity up to the edge of the double layer where $\Phi(x) \simeq 0$:

$$v_z(x) = -\frac{\varepsilon(\zeta - \Phi(x))}{\eta} E_{0z} \tag{11}$$

using the boundary conditions $(\partial v_z/\partial x) = (\partial \Phi/\partial x) = 0$ several Debye lengths away from the surface, and $\Phi = \zeta$, $v_z = 0$ at the slip plane near the solid surface (called the "shear boundary" in Fig. 1). The resulting bulk fluid velocity $v_z = (\varepsilon\zeta/\eta)E_{0z}$ (VON SMOLUCHOWSKI, 1921) is applicable even for curved surfaces (e.g., glass capillaries) whose radius of curvature is much larger than a Debye length. This bulk fluid velocity is that of interest in capillary electrophoresis (Sect. 4) when the pressure drop along the capillary is negligible.

More generally, the fluid velocity profile within a capillary, including axial pressure drops, is derived in a similar manner by integration of Eq. (6) (including the pressure term) in cylindrical coordinates with analogous boundary conditions (DUKHIN and DERJAGUIN, 1974; KOH and ANDERSON, 1975):

$$v_z(r) = \left[\frac{\varepsilon(\zeta - \Phi(r))}{\eta \ell}\right] \Delta\Psi + \left[\frac{r^2 - (R-\delta)^2}{4\eta\ell}\right] \Delta P \tag{12}$$

where $\Delta\Psi$ and ΔP are the electrical potential and pressure drops across the capillary of length ℓ and radius R, and the shear boundary is located at the position $(R-\delta)$. Fig. 3 shows the fluid velocity profile associated with each of the terms in Eq. (12), for the case where $R \sim \kappa^{-1}$. Note that without an applied electric field, an applied pressure drop ΔP will generate the usual parabolic Poiseuille flow (Fig. 3a). However, when $\Delta P = 0$, an applied electric field E_{0z} will generate an electroosmotic fluid velocity profile that is flat in the central region of the capillary where $\rho \simeq 0$ (Fig. 3b). Thus, in capillary electrophoresis, where $R \sim 20$–$200\ \mu\text{m} \gg \kappa^{-1}$, a plug-flow fluid velocity profile results. This is critical to understand the applications described in Sect. 4 below.

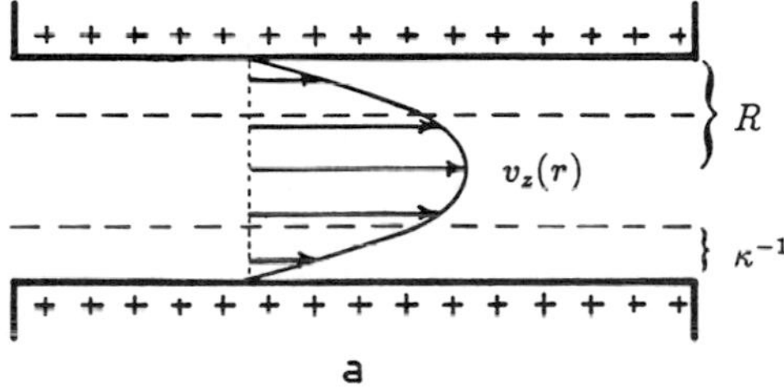

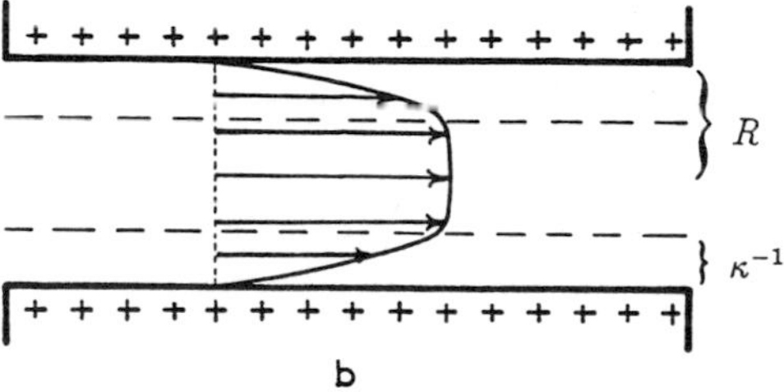

Fig. 3. Fluid velocity profile within a cylindrical capillary having positive surface charge density, corresponding to the analytical expression of Eq. (12). (a) Velocity distribution produced by an applied pressure drop with $\Delta\Psi = 0$; (b) velocity distribution produced by an applied electric field ($E_{0z} = \Delta\Psi/l$) with $\Delta P = 0$.

2.4 Electrokinetic Transduction: Free Electrophoresis

When an electric field is applied to a charged particle in free solution, the electrical and viscous stresses in the double layer interact to produce relative flow of fluid and solid just as in electroosmosis. The result is an electrophoretic migration of the particle through the fluid. VON SMOLUCHOWSKI (1921) showed that for a particle size much larger than a Debye length, but of otherwise arbitrary shape, the electrophoretic particle velocity U for a wide range of conditions is:

$$U = \left(\frac{\varepsilon\zeta}{\eta}\right) E_{0z} \tag{13}$$

From Eq. (13), the electrophoretic mobility is defined as $U/E_{0z} = (\varepsilon\zeta/\eta)$. We note that this result, Eq. (13), also follows directly from the physical argument in planar geometry leading to Eq. (11), in the limit where the particle radius is much larger than a Debye length. LEVICH (1962) extended this treatment to include the effects of convective charge transport in the double layer, which could slow the particle under conditions of very low ionic strength; for the case of buffers used in electrokinetic separations, this effect would be unimportant. Models accounting for the effects of smaller particle radii ($R \sim \kappa^{-1}$) have also been derived (DUKHIN and DERJAGUIN, 1974).

2.5 Zone Electrophoresis

TISELIUS first described the use of free electrophoresis in 1937 (as in Sect. 2.4) in the separation of globular proteins. Major disadvantages included Ohmic (Joule) heating produced by the high current densities, which often resulted in convective disruption of the protein bands. This led to the introduction of zone electrophoresis, in which a support medium is used to suppress convection and other mechanical disturbances arising from high temperature and sharp concentration gradients of macromolecules. Support media include filter paper, cellulose acetate membranes, and starch, agar, agarose and polyacrylamide gels (LASS, 1989). Materials such as paper, cellulose, and silica do not participate in the separation; in these cases the separation is based solely on the specific electrophoretic mobility of the molecules. On the other hand, polyacrylamide and agarose gels, whose pore sizes may be of the same order as the size of the molecules of interest, can interact with the molecules. This results in a separation based both on charge and molecular shape. The extent of such molecular sieving depends upon the exact relation between the pore size of the gel and the size of the molecules involved. For example, electrophoresis of most proteins in agarose gels results in minimal sieving due to the large agarose pore size. However, agarose gel electrophoresis of DNA results in significant molecular sieving effects, so that separation is dependent solely on DNA length and fairly independent of the base composition and sequence.

The problems associated with gel electrophoresis include inefficient heat removal and, in the case of charged support media, electroosmotic convection of buffer and macromolecules within the support medium. Convective transport due to electroosmosis can be significant enough to cause even negatively charged molecules to migrate towards the cathode, especially in low ionic strength buffers. While higher field strengths yield faster separation times, the above limitations are also proportional to field strength.

3 Gel Electrophoresis Techniques

Gel electrophoresis has gained widespread use during the past decades. We now summarize the most common methods in use today, including recent advances which incorporate alternating ("pulsed" or spatially "rotating") electric fields. Polyacrylamide gel electrophoresis (PAGE) is one of the oldest and most widely practiced techniques. Native polyacrylamide gel electrophoresis (native-PAGE) separates molecules based on both charge and size/shape, while denaturing gel electrophoresis such as sodium dodecyl sulfate–polyacrylamide gel electrophoresis (SDS-PAGE) separates molecules solely on the basis of molecular size. Isoelectric focusing techniques are used to separate molecules based strictly on their respective isoelectric points (defined as the pH at which the molecule exhibits no net charge). Two-dimensional gel electrophoresis (in which separation is first carried out in one dimension using native gel electrophoresis or electrofocusing followed by separation in the second dimension using SDS gel electrophoresis) results in enhanced separation of the components of a given mixture. Agarose gel electrophoresis (AGE) is widely used in the analysis of RNA and DNA molecules, but is limited in the size of the molecules it can resolve (less than 50000 bp). This limitation has been overcome with the advent of pulsed field gel electrophoresis (PFGE) which has been used to analyze DNA molecules up to 6000 kbp.

3.1 Polyacrylamide Gel Electrophoresis

Polyacrylamide (PA) gel is made by polymerization of acrylamide in the presence of a cross-linking agent such as N,N'-methylenebisacrylamide. The polymerization process is initiated by adding either ammonium persulfate or riboflavin (requires irradiation), with N,N,N',N'-tetramethylethylenediamine (TEMED) serving as a catalyst. The effective pore size of polyacrylamide gels depends on the concentration of the acrylamide monomer and the crosslinker. The pore size decreases as the acrylamide concentration is increased; for a given acrylamide concentration, increasing the concentration of crosslinker decreases pore size. PA gels have been used to separate small DNA fragments between 6 and 1000 bp and proteins between 6000 and 700000 daltons.

PAGE can be carried out in a rod or slab gel format. Slab gels permit the analysis of a large number of samples simultaneously, while rod gels are used mainly in two-dimensional electrophoresis to perform the separation in the first dimension. PAGE can be performed in either dissociating or non-dissociating buffer systems (HAMES, 1990). Typically, protein systems are denatured by heating at 100 °C in the presence of sodium dodecyl sulfate (SDS) and a thiol reagent such as β-mercaptoethanol to cleave disulfide bonds (SDS-PAGE). Under these conditions, most polypeptides bind SDS in a constant weight ratio of 1.4 g SDS per gram of polypeptide (REYNOLDS and TANFORD, 1970). Consequently, the intrinsic charge of the protein components is masked by the charges provided by the bound detergent. As a result, all components have identical charge densities. Separation thus occurs solely on the basis of molecular size as a result of molecular sieving effects. WEBER and OSBORN (1969) and MORRIS and MORRIS (1971) showed that the specific electrophoretic mobility of similarly shaped molecules under such conditions is linearly related to the logarithm of the molecular weight via a function of the form $[a - b \log(MW)]$. High concentrations of urea (up to 8 M) can also be used instead of SDS but may not be quite as effective for protein samples. Urea is commonly used as a denaturant system for DNA samples as is 98% formamide. Several buffer systems for use in SDS-PAGE have been evaluated (BURG, 1981). Electrophoresis of samples under non-dissociating buffer conditions (native-PAGE) is designed to separate a mixture on the basis of both molecular size and charge. Conditions used for native-PAGE permit retention of subunit structure, conformation, and biological activity of the individual components, thereby permitting detection using specific reagents such as antibodies.

PAGE can be carried out in either a continuous or discontinuous (multiphasic) buffer system (HAMES, 1990), In continuous buffer systems, the same buffer ions are present throughout the sample, gel, and electrode reservoirs, and the samples are loaded directly onto the resolving gel where separation occurs. Discontinuous buffer systems utilize different buffers (both buffer composition and pH) in the gel and the electrode reservoirs. Also, the samples are loaded onto a large-pore stacking gel where sieving effects are minimal. In the stacking gel, the components of the mixture are concentrated and stacked in the order of their mobilities before entering the resolving gel. In the resolving gel, separation occurs on the basis of molecular size alone or both charge and size depending upon whether a dissociating or non-dissociating buffer is used in the gel.

The principle of discontinuous buffer systems is as follows. In its simplest form, the resolving gel and the stacking gel both contain a fast moving ion, F (chloride, for example) while the buffer in the upper electrode reservoir contains a slowly moving ion S (glycine, for example). When the voltage is applied to the system, the fast moving ions leave the upper part of the stacking gel and are replaced by the slower moving ions entering the gel from the upper electrode reservoir. The differential migration of the fast and slow ions results in the creation of a zone of "ionic vacuum". The low ionic strength of this zone results in a higher electric field strength, since the current density is constant. Thus, the slowly moving ions entering this zone migrate faster such that both the slow and fast ions reach the same migration speed. A self-regulating moving boundary is thus set up with the fast ions in the front and the slow ions in the rear. The components in the applied sample usually have intermediate electrophoretic mobilities, and are trapped between the two zones, sorted in the order of their mobilities, and enter the resolving gel as a well-defined thin zone of high protein density. The most commonly used discontinuous buffer system is the one described by LAEMMLI (1970).

Two of the parameters that must be adjusted for optimizing the PAGE process are the buffer pH and gel porosity. Gradient gels (gels with a continually varying acrylamide concentration) are generally better than homogeneous gels for several reasons (ARCUS, 1970). First, gradient gels are able to fractionate proteins over a wider range of molecular weights. Second, the protein bands obtained on gradient gels are sharper, because the migration rate of any given component decreases and asymptotically reaches zero as it approaches the gel concentration whose porosity corresponds to its Stokes radius (or its "pore limit").

3.2 Isoelectric Focusing

Isoelectric focusing (IEF) is a technique that separates molecules on the basis of differences in their isoelectric (p*I*) values. Used mainly with proteins, this method is generally carried out under non-denaturing conditions and is a very high resolution technique (RIGHETTI, 1983; RIGHETTI et al., 1990). IEF is performed by carrying out electrophoresis in a pH gradient set up between two electrodes, with the cathode being at a higher pH than the anode. As the protein moves through the pH gradient, its net charge and, hence, mobility will decrease as it approaches the pH equalling its p*I*. At this pH, the protein has no net charge and will stop migrating. Diffusion of the protein away from this isoelectric position does not occur since this would impart a charge to the molecule and the imposed electric field would drive it back to the point where its charge is zero. Thus, the protein condenses or focuses into a sharp band in the pH gradient at its individual characteristic isoelectric point (hence the term "focusing").

The key to the development of IEF into a widely used analytical technique was the establishment of pH gradients which remained stable in the presence of electric fields. This is usually accomplished with commercially available synthetic carrier ampholytes which are essentially mixtures of relatively small, multicharged amphoteric molecules with closely spaced p*I* values and high conductivities. During electrophoresis, the ampholytes separate to form smooth pH gradients. Differences in p*I* values as small as 0.02 pH units can be quite easily resolved in such systems.

In contrast to these "natural" gradients that develop automatically during electrophoresis, "artificial" gradients can be created in advance by immobilizing certain charged species to the gel matrix. Special derivatives of acrylamide containing amine or carboxylic acid functionalities, available under the trade name of Immobiline, are used to form these immobilized gradients. Several types of Immobiline molecules with different p*K* values are currently available. Resolution is much higher in such systems, of the order of 0.001 pH units.

3.3 Two-Dimensional Gel Electrophoresis

In high resolution 2-D gel electrophoresis of proteins (DUNBAR, 1987; DUNBAR et al., 1990), the samples (after reduction) are separated in the first dimension according to their isoelectric points using IEF with carrier ampholytes. Following this separation, they are then subjected to a fractionation in the second dimension according to their molecular size using SDS-PAGE. Electrophoresis in the first dimension is carried out in a rod gel; after removal of the gel from the tube, it is introduced into the SDS-PAGE apparatus which is run in a slab gel format. The development and standardization of simplified equipment for reproducible and large-scale analysis and the availability of high quality reagents has made this powerful and versatile technique increasingly popular in the past decade. Two-dimensional gel electrophoresis is particularly useful in the analysis of complex protein mixtures and is especially important when the proteins of interest are present in low concentrations. 2-D gel electrophoresis is one of the most powerful methods available for protein mapping of cellular or subcellular homogenates.

2-D gel electrophoresis of complex RNA samples has been carried out in the following three ways. In the "urea shift" method, both the first- and second-dimension gels have the same gel composition and neutral pH. However, one of the two gels is run at a high concentration of urea, while the second gel is run under non-denaturing conditions. This method is generally used for the analysis of small RNA fragments (13–80 bases). In the "concentration shift" method, the only difference between the two gels is a change in the gel concentration, and it is mainly used for the analysis of medium-sized RNA fragments (80–400 bases). The third type of 2-D gel electrophoresis of RNA combines a pH shift with urea and concentration shifts and is used in the analysis of large viral RNA molecules.

2-D gel electrophoresis of DNA is performed by digesting large DNA molecules with one restriction enzyme and running out the fragments in the first dimension. The gel pattern is then transferred by blotting to a suitable membrane following which digestion with a second enzyme takes place. The second electrophoresis is then carried out with the electric field perpendicular to the direction of the first separation.

3.4 Agarose Gel Electrophoresis

Agarose gel electrophoresis (AGE) is widely used in the analysis of RNA and DNA samples. Agarose gels can be used to analyze double-stranded DNA fragments from 70–50000 bp. AGE is often carried out in a horizontal slab gel apparatus, with the gel completely immersed in buffer ("submarine" gels). A vertical slab gel apparatus generally provides better resolution but places more mechanical stress on the gel.

3.5 Analysis of Gels after Electrophoresis

The final step in any gel electrophoretic method is the visualization and quantitation of the separated protein or nucleic acid bands. Except for some naturally colored proteins such as myoglobin or hemoglobin which can be visualized directly as long as their chromophores are not destroyed during the process, most proteins require organic or metal-based stains for visualization. Many organic stains, such as Coomassie Blue, Bromophenol Blue, etc., have been used for detection of protein bands. Of these, Coomassie Blue has proved to be the most sensitive (BLAKESLY and BOEZI, 1977). Metal-based stains such as the widely used silver stain offer high sensitivity and are over a 100-fold more sensitive than Coomassie Blue staining (SAMMONS et al., 1981; PORRO et al., 1982). For detection using specific reagents such as radiolabeled antibodies, the gel pattern is usually first blotted onto a nitrocellulose or nylon filter, fixed, incubated with the labeled reagents, and then visualized by autoradiography. Nucleic acids can be visualized by staining with ethidium bromide, a fluorescent dye that binds DNA, by UV shadowing, or via hybridization using radiolabeled probes (MERRIL, 1990).

3.6 Pulsed Field Gel Electrophoresis

The recent development of pulsed field gel electrophoresis (PFGE) has permitted separation of DNA molecules up to 12000 kb or longer. In this technique, DNA is electrophoresed through an agarose gel under the influence of two electric fields aligned almost perpendicular to one another, which are alternatively switched or "pulsed". The principle of PFGE, first outlined by SCHWARTZ and CANTOR (1984), can be described qualitatively as follows. At high gel concentrations and voltage gradients, DNA molecules must be aligned in the direction of the field to enable them to enter the pores of the gel and migrate. By switching field directions, the DNA molecules are forced to reorient themselves in the direction of the newly imposed field before they can migrate in that direction (SOUTHERN et al., 1987). The longer the molecule, the longer it takes to attain this new orientation and consequently, the more the molecule is retarded in the gel. This permits separation of large fragments of DNA, which would otherwise co-migrate using the traditional steady, uniform electric field in gel electrophoresis. The separation achieved depends on several conditions such as gel concentration, field strength, orientation angle between the two pulsed fields, pulse duration, and temperature.

Several different electrode configurations have been devised for PFGE. In the single inhomogeneous field system, one homogeneous field and one inhomogeneous field (created by using an array of electrodes as cathodes and one electrode as anode) are used on the same gel (SCHWARTZ and CANTOR, 1984). In the double homogeneous field system, the electrodes are placed along the diagonal axes with the cathodes being long continuous wires and the anodes short electrodes (CARLE and OLSON, 1984). In field inversion gel electrophoresis (FIGE), a conventional gel apparatus is used but the polarity of the field is switched at the end of each cycle (CARLE et al., 1986). Overall forward movement is made possible by using a greater part of the switching cycle or a higher voltage in the forward direction. In the rotating gel system, the field is held in a fixed position but a circular gel is turned at each pulse (SOUTHERN et al., 1987). In the contour clamped homogeneous field system (CHEF), a hexagonal electrophoresis chamber with electrodes along four sides of the hexagon is used (CHU et al., 1986). Several of the aforementioned apparatuses have been compared in a recent review (DAWKINS, 1989) and are available commercially from a variety of sources. Continued advances and new methodologies can be expected as the fundamental principles are further clarified and quantified.

The applications ot these techniques in recombinant DNA technology are numerous. In lower eukaryotes, such as yeast, they have permitted the production of chromosome-specific libraries, assignment of cloned DNA to individual chromosomes, etc. In higher eukaryotes, such as mammals, the chromosomes are much too large to be separated as individual bands. Instead, restriction enzymes, such as Not I, whose recognition sequences are relatively rare in mammalian DNA, are used to generate fragments that can then be analyzed by PFGE.

4 Capillary Electrophoresis

Although traditional modes of gel electrophoretic separations are widely practiced today, they are slow and labor-intensive, have limited quantitative capability, suffer from poor reproducibility, and are not easily automated. Capillary electrophoresis (CE) is a generic term applied to electrophoretic separations in narrow bore plastic tubes, glass capillaries and thin films between parallel plates. This method offers the possibility of rapid analysis of complex mixtures with extremely high resolution and sensitivity. In describing this method, we will refer back to the fundamental Eq. (12) and Fig. 3b for electroosmotic fluid flow in the capillary, and Eq. (13) for free electrophoresis of a solute with respect to the fluid in the capillary.

As discussed earlier, faster separation times can be achieved by using high electric field strengths. One of the limits to increasing field strength is the Joule heating generated by the applied power. The result of Joule heating would be a temperature gradient from the center of the gel to the walls, leading to sample diffusion and band broadening. In CE, the use of thin-walled narrow bore glass or fused-silica capillary tubes in a temperature-controlled bath results in rapid heat dissipation, thereby permitting the separation to be run at much higher fields than in traditional electrophoresis. Typical capillaries used for CE are 20–200 μm in diameter and 25–100 cm in length, for which only small temperature effects have been estimated. Many commercial systems are now available which combine rapid, high resolution separation with a variety of in-line continuous detection schemes. These advances, reminiscent of the development of HPLC, have led to the name, "high performance capillary electrophoresis" (see KARGER et al., 1989, for a recent review).

CE techniques have been classified based on the mode of operation into three groups (THORMANN and FIRESTONE, 1989). In the first class, the applied electric field is parallel to the capillary axis, and the separation is usually carried out in narrow bore plastic, glass, or fused-silica capillaries under flow or non-flow conditions. Examples of this group of conventional CE are capillary zone electrophoresis (CZE), discontinuous capillary zone electrophoresis (DCZE) and capillary isoelectric focusing (CIEF).

In capillary zone electrophoresis, the sample is taken into the capillary initially by either electroosmotic flow from the sample reservoir or by hydrostatic means. Sample volumes are in the nanoliter range or smaller, and detection volumes can be as small as 30 pL (GORDON et al., 1988). Application of 20–30 kV axially along the capillary then produces electroosmotic flow of the component-containing buffer towards the cathode Eq. (12), and simultaneous free electrophoresis of the positively or negatively charged components with respect to the convecting buffer, Eq. (13). As pictured in Fig. 3b above, the plug-flow profile inherent in electroosmosis is critical to minimizing potential band spreading within the capillary. Methods for increasing the selectivity of separations using CZE include partitioning into micelles within the capillary (micellar electrokinetic capillary chromatography, TERABE et al., 1984), and the use of gel-filled capillaries to incorporate isoelectric focusing and SDS-PAGE within the capillary (see GORDON et al., 1988). Detection methods for CZE include UV absorbance, fluorescence, conductivity, radioactivity, and electrochemical techniques (GORDON et al., 1988). The major limitation of CZE involves solute–wall interactions (i. e., in the Stern layer of Fig. 1), which has led to studies of additives and wall coatings.

The second class of CE techniques involves those in which the applied electric field is perpendicular to the capillary axis. The electrophoretic separation is enhanced by using a suitable mobile phase and is carried out in thin ribbon-like channels or in hollow ultrafiltration fibers. Examples of such applications include electrical field flow fractionation (EFFF) and electrical hyper-layer field flow fractionation (EHFFF). EFFF couples an electrical field with a mobile phase profile to achieve differential migration. Under the electrical field, solute molecules accumulate in a layer of distinct thickness along one wall of the channel; the thickness of this solute layer then determines the rate of transport of the solute along the channel. In EHFFF, separation occurs as a result of an additionally imposed pH gradient. The third category of CE techniques includes those in which the electrical field is applied perpendicular to the direction of flow as in continuous flow electrophoresis (CFE). CFE is mainly a preparative technique (see below) and is conducted in a thin film of fluid flowing between two charged parallel plates. Different components in the applied sample migrate to different positions in the flow profile and are then fractionated by an outlet array. Thus, the discussion of free electrophoresis (Sect. 2.4) is important to this category of applications as well.

The advantages of CE are numerous: (1) high resolution, (2) high degree of automation, (3) fast separation times, (4) on-column detection of sample via a wide variety of detectors, and (5) easy adaptation to micropreparative work. Over the past two decades, CE has been used to separate a broad spectrum of molecules ranging from small organic and inorganic ions to large proteins and cells (GUZMAN et al., 1990; NIELSEN and RICKARD, 1990; KARGER et al., 1989).

5 Scale-Up of Electrokinetic Separations: Examples

New developments in the scale-up of electrophoretic techniques for preparative applications have been recently reviewed by RIGHETTI et al. (1991). Below, we briefly mention processes involving electrophoresis and membrane-based electrokinetic separations which embody the use of multiple electrokinetic mechanisms; the reader should consult RIGHETTI et al. (1991) for a more complete survey.

One promising recent example of a technique for scaling up electrophoresis separations is the development of recycle zone electrophoresis (RZE) (IVORY, 1990). RZE is usually carried out in a multiported thin-film chamber into which the solutes are introduced in a recycle port. In order to compensate for the lateral displacement of solutes in the electric field, the effluent from the chamber is "back shifted", i. e., recycled back to the inlet after being shifted laterally by one or more ports. The magnitude of this shift is roughly equal to the average of the displacement of the two "key" components. The electric field is adjusted so that solutes with low electrophoretic mobilities tend to move downstream with the flow, while those with high electrophoretic mobilities are carried upstream against the flow. Thus, one can in principle obtain a split of a continuous multicomponent feed into two fractions. RZE has been used successfully in the fractionation of synthetic protein mixtures such as bovine serum albumin (BSA) + hemoglobin and azocasein + hemoglobin. It has also been used in the purification of monoclonal antibodies from hybridoma supernatants. The processing rates have been about 20–30 mL/h (>1.0 g/h). Instrumentation of RZE as well as on-line UV-visible monitoring have now been developed which will permit implementation of process controls and eventual automation.

Another example suggesting the potential advantages of membrane-based electrokinetic separations for scale-up involves electrically controlled membranes. Application of an electric field across a charged polyelectrolyte gel membrane can simultaneously promote membrane charging and swelling, drive electroosmotic convection across the membrane, and produce electrophoresis of charged solutes within and across the membranc (GRIMSHAW et al., 1989, 1990a). The resulting selective changes in transmembrane solute flux in such a system offer a separation technique based on the size and charge of the solutes. This technique combines many of the electrokinetic processes described above with the scale-up provided by the use of a membrane geometry.

In the presence of a pH gradient across a membrane, an applied electric field can induce changes in membrane swelling. Fig. 4 shows measured changes in membrane thickness induced by electric fields applied across a poly(methacrylic acid) (PMAA) membrane supporting a pH gradient (pH 6 on one side,

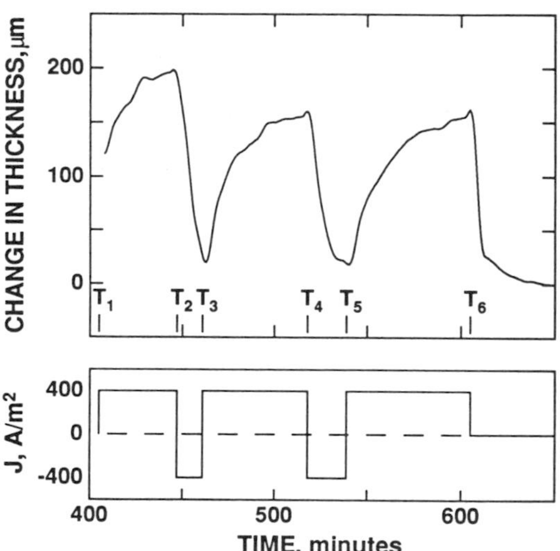

Fig. 4. Measured change in thickness vs. time (top) of a PMAA membrane for uniaxial confined swelling in 50 mM KCl with 5 mM malonic acid buffer. The top surface of the membrane was in contact with a pH 3 solution and the bottom surface with a pH 6 solution. A current density J of 400 A/m^2 was applied from bath A (pH 6) to bath B (pH 3) starting at time T_1 (bottom). The direction of the current was reversed at times T_2, T_3, T_4, and T_5; the current was turned off at time T_6. Thickness changes are relative to the zero-current thickness of 266 µm (from GRIMSHAW et al., 1990a).

and pH 3 on the other). An 80–150 μm increase in thickness over a period of 50–60 minutes resulted each time when a 400 A/m^2 current density was applied from the pH 6 to the pH 3 bath (T_1, T_3, T_5 in Fig. 4). Each time the current was applied in the opposite direction (T_2, T_4), a similar decrease in thickness occurred over a shorter time. The changes in membrane thickness observed in Fig. 4 were shown to be produced by an electrodiffusion mechanism (GRIMSHAW et al., 1990b), which involved the alteration of the intramembrane pH profile by the applied electric field. In deformable membranes with ionizable charge groups, such changes in the intramembrane pH can modulate electrostatic swelling forces, resulting in membrane swelling or shrinking as seen in Fig. 4.

The changes in swelling seen in Fig. 4 resulted in significant changes in the permeability of the membrane to neutral and charged solutes. Four distinct mechanisms were shown to control the permeability: (1) modulation of electrostatic swelling forces altered the effective pore size of the membrane matrix, resulting in permeability changes which are sensitive to solute size; (2) changes in membrane-fixed charge density altered the electrostatic (Donnan) partitioning of charged solutes within the membrane, adding to separation on the basis of charge; (3) electroosmotic flow of buffer across the charged membrane, induced by the applied electric field, altered the transport of both neutral and charged solutes across the membrane via convection, a permeability change that was sensitive to solute size; (4) transport of charged proteins was further enhanced (or suppressed) via intramembrane electrophoresis.

Mechanisms (3) and (4) above are directly analogous to the electroosmotic and electrophoretic processes that occur in capillary electrophoresis using open or gel-filled capillaries. Here, the pores in the gel membrane correspond to using an array of capillaries in parallel for scale-up.

A combination of the 4 transport mechanisms described above produced large, selective changes in the separation of proteins, as demonstrated by the 21-fold change in the relative flux of fluorescently labeled bovine serum albumin and ribonuclease shown in Fig. 5. A current density of 100 A/m^2 applied at time T_1 across a PMAA membrane supporting a pH gradient (pH 7 to 5.5), led to a dramatic increase in the flux of the positively charged ribonuclease, due to constructive superposition of all four mechanisms. In contrast, these mechanisms produced changes in the flux of the negatively charged BSA which essentially cancelled. (The current density was turned off at T_2 (Fig. 5).)

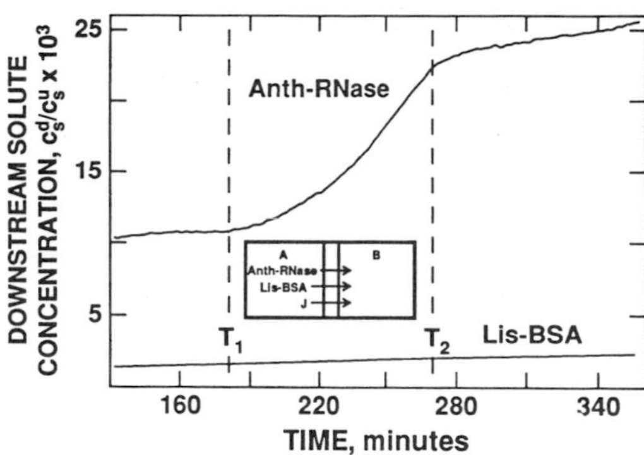

Fig. 5. Normalized downstream concentrations of fluorescently labeled ribonuclease (Anth-RNase = anthracene-labeled ribonuclease) and bovine serum albumin (Lis-BSA = lissamine-labeled BSA) versus time for transport across a poly(methacrylic acid) (PMAA) membrane in 100 mM KCl and 5 mM imidazole buffer. The flux of both proteins was from bath A (pH 7) to bath B (pH 5.5). A transmembrane current density of 100 A/m^2 was applied from bath A and B starting time T_1 and turned off at T_2 (from GRIMSHAW et al., 1990a).

Thus, the four mechanisms can be strategically superimposed to achieve maximal changes in separation under real-time control. In a separation process one could employ one or more of these mechanisms to maximize the transport of a desired product across the membrane while retaining the undesired components of the mixture; alternatively one could retain the desired products while passing the impurities. The concept of electrically controlled membrane separations may permit scale-up for production, where, for example, conventional electrophoresis and chromatography techniques may not be practical. Dynamic control of membrane permeability and selectivity could also be advantageous for separation processes requiring time-varying control.

Acknowledgements

The authors thank THOMAS M. QUINN for careful reading of the manuscript; the data of Figures 4 and 5 came from the Ph. D. thesis research of Dr. PAUL E. GRIMSHAW. This work was supported in part by NSF Grant CDR-8803014 to the Biotechnology Process Engineering Center, MIT.

6 References

ARCUS, A. C. (1970), Protein analysis by electrophoretic molecular sieving in a gel of graded porosity, *Anal. Biochem.* **37**, 53-63.

BLAKESLY, R. W., BOEZI, J. A. (1977), A new staining technique for proteins in polyacrylamide gels using Coomassie Brilliant Blue G 250, *Anal. Biochem.* **82**, 580-582.

BURG, A. F. (1981), Evaluation of different buffer systems for PAGE-SDS, *J. Chromatogr.* **213**, 491-500.

CARLE, G. F., OLSON, M. V. (1984), *Nucleic Acid Res.* **12**, 5647-5664.

CARLE, G. F., FRANK, M., OLSON, M. V. (1986), *Science* **232**, 65-68.

CHU, G., VOLLRATH, D., DAVIS, R. W. (1986), *Science* **234**, 1582-1585.

DAWKINS, H. J. S. (1989), Large DNA separation using field alternation agar gel electrophoresis, *J. Chromatogr.* **492**, 615-639.

DUKHIN, S. S., DERJAGUIN, B. V. (1974), Electrokinetic phenomena, in: *Surface and Colloid Science*, (MATIJEVIC, E., Ed.), Vol. 7, New York: John Wiley.

DUNBAR, B. S. (1987), *Two-Dimensional Gel Electrophoresis and Immunological Techniques*, New York: Plenum Press.

DUNBAR, B. S., KIMURA, H., TIMMONS, T. M. (1990), Protein analysis using high-resolution two-dimensional polyacrylamide gel electrophoresis, *Methods Enzymol.* **182**, 441-458.

GORDON, M. J., HUANG, X., PENTONEY, S. L., ZARE, R. N. (1988), Capillary electrophoresis, *Science* **242**, 224-228.

GRIMSHAW, P. E., GRODZINSKY, A. J., YARMUSH, M. L., YARMUSH, D. M. (1989), Dynamic membranes for protein transport: Chemical and electrical control, *Chem. Eng. Sci.* **44**, 827-840.

GRIMSHAW, P. E., GRODZINSKY, A. J., YARMUSH, M. L., YARMUSH, D. M. (1990a), Selective augmentation of macromolecular transport in gels by electrodiffusion and electrokinetics, *Chem. Eng. Sci.* **45**, 2917-2929.

GRIMSHAW, P. E., NUSSBAUM, J. H., GRODZINSKY, A. J., YARMUSH, M. L. (1990b), Kinetics of electrically and chemically induced swelling in polyelectrolyte gels, *J. Chem. Phys.* **93**, 4462-4472.

GUZMAN, N. A., HERNANDEZ, L., TERABE, S. (1990), High-resolution nanotechnique for separation, characterization, and quantitation of micro- and macromolecules, in: *Analytical Biotechnology: Capillary Electrophoresis and Chromatography* (HORVATH, C., NIKELLY, J. F., Eds.), pp. 1-36, Washington, DC: American Chemical Society.

HAMES, B. D. (1990), An introduction to polyacrylamide gel electrophoresis, in: *Gel Electrophoresis of Proteins: A Practical Approach* (HAMES, B. D., RICKWOOD, D., Eds.), pp. 1-148, Oxford: IRL Press.

HAUS, H. A., MELCHER, J. R. (1989), *Electromagnetic Fields and Energy,* Chap. 3, New York: Prentice Hall.

HUNTER, J. J. (1989), *Foundations of Colloid Science*, Chap. 13, Oxford Science Publications.

IVORY, C. F. (1990), The development of recycle zone electrophoresis, *Electrophoresis* **11**, 919-926.

JAIN, M. K. (1972), *The Bimolecular Lipid Membrane*, p. 38, New York: Van Nostrand Reinhold.

KARGER, B. L., COHEN, A. S., GUTTMAN, A. (1989), High-performance capillary electrophoresis in the biological sciences, *J. Chromatogr.* **492**, 585-614.

KOH, W. H., ANDERSON, J. L. (1975), Electroosmosis and electrolyte conductance in charged microcapillaries, *AIChE J.* **21**, 1176-1188.

LAAS, T. (1989), Electrophoresis in gels, in: *Protein Purification: Principles, High Resolution Methods and Applications* (JANSON, J.-C., RYDÉN, L., Eds.), pp. 349-375, New York: VCH Publishers.

LAEMMLI, U. K. (1970), Cleavage of structural proteins during the assembly of the head of bacteriophage T4, *Nature* **227**, 680-685.

LEVICH, V. G. (1962), *Physiochemical Hydrodynamics*, Chap. 9, Englewood Cliffs, NJ: Prentice Hall.

MERRIL, C. R. (1990), Gel staining techniques, *Methods Enzymol.* **182**, 477-487.

MORRIS, C. J. O. R., MORRIS, P. (1971), Molecular sieve chromatography and electrophoresis in polyacrylamide gels, *Biochem. J.* **124**, 517-528.

NIELSEN, R. G., RICKARD, E. C. (1990), Applications of capillary zone electrophoresis to quality control, in: *Analytical Biotechnology: Capillary Electrophoresis and Chromatography* (HORVATH, C., NIKELLY, J. F., Eds.), pp. 37-49. Washington, DC: American Chemical Society.

OVERBEEK, J. TH. G. (1952a), Electrochemistry of the double layer, in: *Colloid Science* (KRUYT, H. R., Ed.), pp. 115–193, Amsterdam: Elsevier.

OVERBEEK, J. TH. G. (1952b), Electrokinetic phenomena, in: *Colloid Science* (KRUYT, H. R., Ed.), pp. 194–243, Amsterdam: Elsevier.

PORRO, M., VITI, S., ANTONI, G., SALETTI, M. (1982), Ultrasensitive silver-stain method for the detection of protein in polyacrylamide gels and immunoprecipitates on agarose gels, *Anal. Biochem.* **127**, 316–321.

REYNOLDS, J. A., TANFORD, C. (1970), Binding of dodecyl sulfate to proteins at high binding ratios: Possible implications for the state of proteins in biological membranes, *Proc. Natl. Acad. Sci. USA* **66**, 1002–1007.

RIGHETTI, P. G. (1983), *Isoelectric Focusing: Theory, Methodology and Applications*, Amsterdam: Elsevier.

RIGHETTI, P. G., GIANAZZA, E., GELFI, C., CHIARI, M. (1990), Isoelectric focusing, in: *Gel Electrophoresis of Proteins: A Practical Approach* (HAMES, B. D., RICKWOOD, D., Eds.), pp. 149–216, Oxford: IRL Press.

RIGHETTI, P. G., FAUPEL, M., WENISCH, E. (1991), Preparative electrophoresis with and without immobilized pH gradients, in: *Advances in Electrophoresis* (CHRAMBACH, A., DUNN, M. J., RADOLA, B. J., Eds.), Vol. 5, pp. 161–200, Weinheim–New York–Basel–Cambridge: VCH.

SAMMONS, D. W., ADAMS, L. D., NISHIZAWA, E. E. (1981), Ultrasensitive silver-based colour staining of polypeptides in polyacrylamide gels, *Electrophoresis* **2**, 135–140.

SCHWARTZ, D. C., CANTOR, C. R. (1984), *Cell* **37**, 67–75.

SOUTHERN, E. M., ANAND, R., BROWN, W. R. A., FLETCHER, D. S. (1987), *Nucleic Acids Res.* **15**, 5925–5943.

TERABE, S., OTSUKA, K., ICHIKAWA, K., ANDO, T. (1984), *Anal. Chem.* **56**, 113.

THORMANN, W., FIRESTONE, M. A. (1989), Capillary electrophoretic separations, in: *Protein Purification: Principles, High Resolution Methods and Applications* (JANSON, J.-C., RYDÉN, L., Eds.), pp. 107–148, New York: VCH Publishers.

VON SMOLUCHOWSKI, M. (1921), in: *Handbuch der Electrizität und des Magnetismus* (GRAETZ, Ed.), Vol. II, p. 366, Leipzig: Barth.

WEBER, K., OSBORN, M. (1969), The reliability of molecular weight determination by dodecyl-sulfate polyacrylamide gel electrophoresis, *J. Biol. Chem.* **244**, 4406–4412.

26 Final Recovery Steps: Lyophilization, Spray-Drying

CHRISTIAN F. GÖLKER

Wuppertal, Federal Republic of Germany

1 Introduction

Following concentration and purification, the obtained product usually is converted into a "finished product" suitable for special applications. Enzymes for technical use, e.g., are best handled as dry powder for stability reasons and easy handling. If very high purity is required, compounds may have to be crystallized. For pharmaceutical preparations, in addition to high purity, stability during formulation and extended shelf life is mandatory. For these reasons the production of biotechnological products is not terminated after the last high-performance purification step but additional steps for product formulation have to be added. From the properties of biomolecules it follows that only very gentle methods can be used as finishing steps for biotechnological products, especially proteins. These final recovery steps are drying methods, particularly spray-drying and freeze-drying and crystallization.

Finished products should meet a number of characteristics:

- high product quality,
- long-term storage stability,
- acceptable appearance,
- sufficient bioavailability,
- easy reconstitution in case of dried products.

2 Freeze-Drying (Lyophilization)

Lyophilization is the method of choice for the production of temperature-sensitive materials like vitamins, enzymes, vaccines, microorganisms, or therapeutic proteins. Freeze-drying has many applications in the food industry and especially in the pharmaceutical industry (WILLIAMS and POLLI, 1984; WOOG, 1989; PIKAL, 1990a, b). The most frequently used method of stabilization of proteins for therapeutic or diagnostic applications is their conversion into the dry state.

Freeze-drying represents a substantial fraction of the whole production cost and should therefore attract special attention during process development. But, in many cases, freeze-drying is still operated on an empirical basis although increasing scientific information about properties of solutions at low temperatures, freezing behavior and freeze-drying has been published over the last years (e.g., MACKENZIE, 1977; FRANKS, 1985b, 1989, 1990).

The structural integrity as well as the functionality of biomolecules like membranes, nucleic acids, phospholipid bilayers, and proteins depend on the presence of water (TANFORD, 1980). For this reason, it is suitable to know the properties of biomolecules as well as the properties of the water molecule for developing a successful drying process.

2.1 Properties of Biomolecules

Most biological molecules, e.g., nucleic acids or proteins, are labile substances and sensitive to extreme pH values, high salt concentrations, extreme temperatures, or organic solvents. Very often, they exhibit a unique three-dimensional structure which enables them to function as biocatalysts, structural elements or antibodies. The specificity of these reactions depends on reversible molecular interactions. These interactions as well as the three-dimensional structure of the molecules are regulated by different kinds of non-covalent bonds: electrostatic bonds, hydrogen bonds, van der Waal's bonds, and hydrophobic bonds. Some structural elements necessary for the functionality of proteins are listed in Tab. 1.

The many different functions of proteins depend on their conformation, i.e., the spacial arrangement of polypeptide chains or different protein molecules. This three-dimensional structure of proteins is mainly determined by their primary structure, i.e., the linear arrangement of amino acids.

Numerous bonds contribute to the stability of proteins. The individual amount of free energy of each bond can be substantial, however, the difference in free energy between the native

Tab. 1. Structural Parameters of Proteins

Primary structure	Linear arrangement of amino acids
Secondary structure	Structured domains of the polypeptide chain due to hydrogen bonds between carbonyl-O and amine-N, e.g., α-helix, β-plated sheet
Tertiary structure	Spatial arrangement of the polypeptide chain due to interactions between different amino acid side chains
Quaternary structure	Spatial arrangement of different protein molecules

and the denatured state is only of the order of 50–100 kJ/mol (FRANKS, 1988). In many cases two denaturation temperatures exist with a temperature of maximal stability between the two temperature limits, i.e., denaturation of proteins might result from exposure to either high or low temperatures. Whereas the denaturation of proteins by high temperatures is a well known phenomenon, fewer examples are known for cold denaturation. Denaturation by cold was experimentally demonstrated for chymotryspinogen (FRANKS and HATLEY, 1985) and lactate dehydrogenase (HATLEY and FRANKS, 1986). Myoglobin undergoes a reversible conformational transition into a non-compact disordered state by heating as well as by cooling (PRIVALOV et al., 1986).

2.2 Properties of Water Molecules

Water has very special properties. Water is a polar molecule. The atoms in the water molecule are arranged in a triangular shape which leads to an unsymmetrical charge distribution. Water can react very easily with many polar groups. For this reason water is an excellent solvent for many compounds.

Water molecules show high affinity to each other. With the maximal possible number of hydrogen bonds, a highly ordered structure is created. In ice crystals, the oxygen and hydrogen atoms of the water molecule are arranged in a regular hexagonal lattice. In water the structure is built up by continuous forming and breaking of hydrogen bonds. Through hydrogen bonds, water molecules can bind to biomolecules and stabilize their structure.

2.3 Formation of Ice Crystals

During freeze-drying, water is removed from the frozen product by sublimation. Therefore, the first step is a phase transition into the solid state by freezing. If a salt solution is cooled to the equilibrium freezing point, in many cases pure water does not crystallize, but the solution is undercooled resulting in a compositionally homogeneous but thermodynamically unstable phase. Undercooling finally is terminated by the very complex process of ice nucleation (HOBBS, 1974). Clusters of water molecules are formed, a process which can be described by statistical methods. If there exists a high probability of forming such clusters, additional water molecules may add to the clusters which finally grow into ice crystals and initiate freezing. This process is known as homogeneous nucleation. The nucleation rate $J(t)$ depends on the degree of undercooling given by FRANKS (1990)

$$J(t) = A \exp(B\,\Phi) \quad \text{where } \Phi = [(\Delta T)^2\, T^3]^{-1}$$

A and B are constants and ΔT is the degree of undercooling.

In practice, ice formation starts around -10 °C to -15 °C due to the presence of particulate impurities which catalyze the formation of ice crystals, a process known as heterogeneous nucleation. Ultrapure water can be undercooled to -40 °C, at which temperature the dependence of the nucleation rate on temperature changes dramatically (MICHELMORE and FRANKS, 1982).

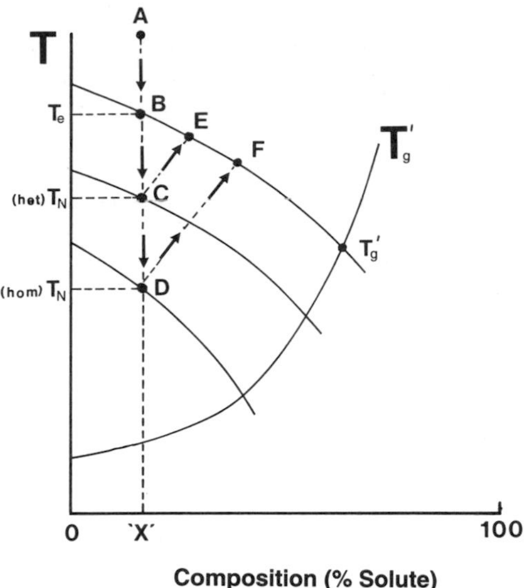

Fig. 1. Nucleation and freezing behavior (MACKENZIE, 1977). T_e, equilibrium freezing temperature; $T_{N(het)}$, temperature of heterogeneous nucleation; $T_{N(hom)}$, temperature of homogeneous nucleation; T_g', glass transition temperature.

Freezing of an aqueous solution is schematically described in Fig. 1: Starting at point A, temperature drops as the solution is chilled. At the equilibrium freezing temperature, point B, ice crystals are formed if the solution is "seeded" with ice crystals from an outside source. Otherwise the solution is undercooled and the temperature is lowered further. In this case crystallization starts at point C, where the heterogeneous nucleation temperature is reached. During freezing, latent heat is released and the temperature again rises to the equilibrium curve which is now reached at point E because of the higher concentration of the remaining solution due to the effect of freeze concentration. In the case of ultrapure solutions the system might even be cooled down to the homogeneous nucleation temperature at point D. Finally, the temperature of either the eutectic point or the "glass transition" point is reached. In the former case, both ice and solute crystallize to form the eutectic mixture. Most formulations in practical freeze-drying, however, are multicomponent systems which do not show a eutectic point. Instead, the solution becomes supersaturated and the liquidus curve shows no definite endpoint. The viscosity of the solution increases due to the increasing concentration of solutes. Finally, at a viscosity of about 10^{11}–10^{14} Pa s, ice crystallization is terminated because of the greatly restricted diffusion of the water molecules. This point is characterized by the glass transition temperature T_g' and the amount of unfrozen water w_g'. T_g' and w_g' depend on the nature and concentration of the different constituents of the mixture. The T_g' curve in Fig. 1 defined by these points does not represent a phase boundary but refers to a homogeneous solid state with high viscosity (isoviscosity curve). Below this curve, the amorphous phase exists as a stable glass (FRANKS, 1989, 1990).

The size and shape of ice crystals depend largely on the degree of undercooling, which again is a function of the cooling rate. The quality of the ice crystals, on the other hand, influences the properties of the freeze-dried product, e.g., texture and ease of rehydration.

2.4 Behavior of Solutions during Freezing

By removing water from the solution through formation of ice crystals all other components present in the solution become more concentrated. As an example, the concentration of a physiological sodium chloride solution increases from 0.9% to 23.3% by cooling the solution to the eutectic point at −21.2 °C (FRANKS, 1986). It was first shown by LOVELOCK, that the damage of erythrocytes and spermatozoa during freezing could be directly related to the concentration of NaCl in the solution (LOVELOCK, 1953a, b). The composition of the solution used in freeze-drying may also differ from the starting conditions due to preferable crystallization of single components. This may not only cause changes in salt concentrations but also shifts in pH values during freezing. In the case of buffer components, the molar composition at the eutectic point can vary considerably from the buffer composition, which is necessary for a certain pH value. For sodium and potassium phosphates, the observed pH shifts depend on the preferable precipitation of ice or solute components during cooling of the solution (VAN DEN BERG and ROSE, 1959). Differences in pH values are also observed for systems in the dry stage and in the hydrated stage (BELL and LABUZA, 1991). Using buffer substances in the formulation protocol, dramatic changes of the pH value can occur during freeze concentration. This is especially marked using phosphate and carbonate buffers, as well as buffers containing amino acids (FRANKS, 1985b). For instance, a sodium phosphate buffer at pH 7.0 has a mole ratio of

$$[NaH_2PO_4]/[Na_2HPO_4]=0.72$$

whereas the mole ratio at the ternary eutectic point is 57 which results in a large decrease of the pH value (FRANKS, 1990). These processes might exert dramatic effects on the behavior of the different compounds in the solution, finally leading to the deterioration of products. The pH value and also the ionic strength of a solution are very important for the conservation of the native structure of proteins. As a consequence, reversible denaturation or, in the worst case, irreversible denaturation by aggregation may occur. Inactivation by freezing is especially associated with proteins composed from subunits. Small molecules may undergo changes by chemical reactions like oxidation reactions or polymerization.

Typically, reaction rates slow down with decreasing temperature. However, the effects due to the very elevated concentrations of reactants in the frozen concentrated solution may far outweigh the retardation caused by low temperatures (KIOVSKI and PINCOCK, 1966; FENNEMA, 1975). As a consequence, unwanted chemical reactions may take place. The reaction rate of enzymatically catalyzed reactions, e.g., can increase by several orders of magnitude. Even the kinetic order of the reaction might change during freezing. For this reason, slow freezing of protein solutions is extremely dangerous and should be avoided whenever possible. Since the freeze concentration factor and the amount of ice formed are inversely proportional to the starting concentration of the solution, one should start with concentrations as high as possible. In some cases, proteins can be stabilized by storage at subzero temperatures in an undercooled liquid but are denatured by freezing. This denaturation is a complex function of the cooling rate and the final temperature (HATLEY et al., 1987). A common phenomenon seems to be the greater stability of proteins at higher concentrations.

2.5 Practical Freeze-Drying

The freeze-drying process can be divided into different steps:

- Freezing: The process is important with regard to the properties of the dried product.
- Primary drying: Water is removed from the ice by sublimation.
- Secondary drying: After sublimation of ice the residual moisture is removed by heating.

2.5.1 Freezing

The solution to be freeze-dried has first to be solidified by freezing. If the solutes crystallize, the solution is cooled to below the lowest equilibrium eutectic temperature of the mixture. The temperature may be 10–15 °C lower than the eutectic temperature because of undercooling of the liquid. The degree of undercooling influences the primary and secondary drying stages. A high degree of undercooling gives rise to small ice crystals with a large surface area which speeds up secondary drying. Large crystals, on the other hand, show large pores which facilitate mass transport during primary drying. For freeze-drying, the ice crystals formed should be as homogeneous as possible. The normal freezing process, however, leads to a broad spectrum of differently shaped ice crystals of varying size. Improvement can be achieved by rapid cooling ("flash freezing") or by an "annealing process", i.e., rewarming the ice crystals to temperatures near the eutectic temperature ("recrystallization").

The freezing pattern may also influence important properties of the dried product, e.g., the aroma of fruit juices and coffee. MALTINI (1975) showed that for fruit juices slow freezing preserves flavor components better than quick freezing, but the consistency and color of the product might be more optimal applying a quick freezing process. A higher degree of retention of flavor compounds performing slow freezing was also reported by FLINK (1975).

Freezing usually is performed in the freeze-drier on plates cooled to between −40 °C and −60 °C in trays or vials. This procedure allows only moderate temperature gradients because of the energy losses between the different boundaries (shelf–tray–liquid). As a rule of thumb, the typical time for freezing a 1 cm layer of a solution in glass bottles is 40–90 min at a temperature of −40 °C. Freezing outside the freeze-drying plant can be achieved by:

- Freezing on cooled plates, e.g., in trays or in the form of small particles on a drum cooler.
- Dropping in liquid nitrogen or some other cooling liquid.
- Co-spraying with liquid CO_2 or liquid nitrogen.
- Freezing with circulating cold air.

Separate freezing is necessary for the performance of continuous freeze-drying. Equipment producing small pellets by dropping the solution into liquid nitrogen is commercially available as Cryopel® process (BUCHMÜLLER and WEYERMANNS, 1990). Direct freezing inside the freeze-drying plant is preferable if the product has to be handled aseptically. It is easier to load the solution in a sterile way after filtration rather than to transfer material separately frozen in an aseptic way. Equipment is available for aseptic filling and drying by fully automated loading and unloading of the freeze dryer (e.g., KHAN, 1989).

2.5.2 Primary Drying

Energy has to be supplied to the frozen solution, in order to substitute for the heat of sublimation, which is of the order of 2850 kJ/kg of ice. The heat is supplied by contact, conduction or radiation to the sublimation front whereas at the same time the water vapor must be removed. That means freeze-drying is a process with coupled heat and mass transport. The driving force for the sublimation of ice is the difference of water vapor pressure at the surface of the sample (at a certain temperature) and the condensor (at a lower temperature). The drying process can proceed in different ways:

With heat transfer through the ice layer, drying proceeds from the surface of the sample, i.e., heat transfer and mass transfer have the same direction (Fig. 2A). Drying is complete when $x = d$. This type applies to a nonporous sample which is in close contact with a heated surface. The temperature of the heating plates must be low enough in order not to give rise to partial melting of the ice layer. Alternatively, in the case of a porous layer, for instance a granular material, the ice becomes insulated by an already dried layer (Fig. 2B). The heat must be provided through a barrier with low heat conductivity. Heat transfer and mass transfer occur in opposite directions.

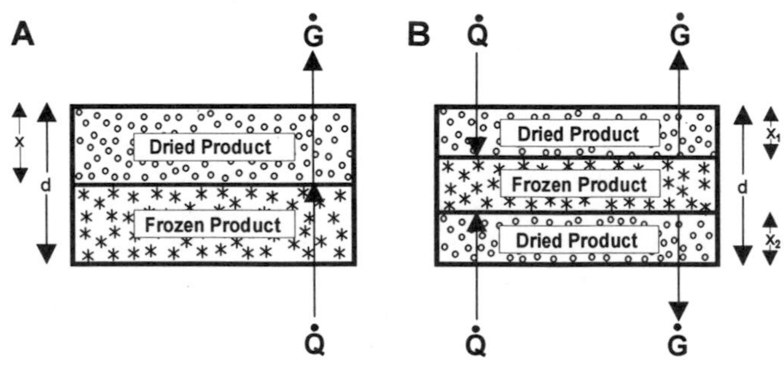

Fig. 2. Heat and mass transport during freeze-drying, (A) for a non-porous layer, (B) for a porous layer.

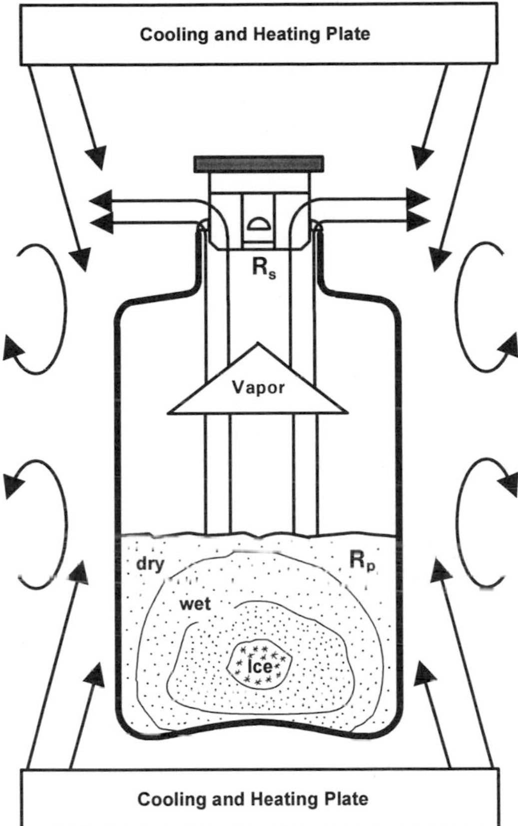

Fig. 3. Freeze-drying in vials. R_s, resistivity of the stopper; R_p, resistivity of the dried cake.

The drying time depends on various factors. Some of them are fairly constant and can be approximately given as: heat of sublimation L_s (about 2856 kJ per kg of ice), thermal conductivity of frozen substances λ_e (about 4.2 kJ/m/h/°C), and mass transfer coefficient b/μ (about 0.02 kg/m/h/mbar). Other factors, like temperature or pressure in the chamber might vary considerably. The mathematical solution for calculation of drying rates and drying time is given, e.g., by STEINBACH (1972).

Many pharmaceutical preparations are dried in ampoules or vials which are placed on heated shelves for drying (Fig. 3). The heat flow from the shelf to the product can be expressed as

$$\dot{Q} = A_v K_v (T_s - T_b)$$

where $\dot{Q}$ is the heat transfer rate, A_v the cross-sectional area of the vial, K_v the heat transfer coefficient to the vial, T_s the temperature of the shelf, and T_b the temperature of the product at the vial bottom.

The heat transfer coefficient results from different heat transfer mechanisms: heat transfer by direct contact between shelf and vial (K_c), heat transfer by conduction through the gas (K_g), and heat transfer by radiation (K_r), whereas K_c and K_r are pressure independent, K_g depends on the pressure inside the drying chamber. Heat transfer by conduction plays a substantial role only at higher pressures, where a substantial number of gas molecules are available. It follows that in order to increase the driving force for the sublimation process, the temperature and pressure should be as high as possible during the drying process, given the restraints that $P_c < P_0$ and $T_{sample} < T'_g$.

For freeze-drying in vials, the most severe restriction for heat transfer is the gap between the heated plate and the vial where no heat transfer by direct contact can take place.

The sublimation rate depends on the water vapor pressure of the ice, the resistance of the dried cake, and in the case of drying in vials also on the resistance of the stopper (Fig. 3):

$$\dot{G} = (P_0 - P_c)/(R_p + R_s)$$

where $\dot{G}$ is the sublimation rate, P_0 the equilibrium water vapor pressure of the ice, P_c the pressure in the drying chamber, R_p the resistivity of the dried cake, and R_s the resistivity of the stopper.

At higher product temperatures, P_0 increases more relative to P_c, which means a greater difference between P_0 and P_c, resulting in a higher driving force for sublimation.

In addition, the rate of sublimation depends on the temperature difference between the ice layer and the condenser. The temperature of the condenser is usually set at −60 °C. The pressure inside the chamber should not be lower than about half of the saturation vapor pressure at this temperature. Otherwise the vapor volume will increase too much and the diffusion rate will become unfavorable.

In most cases, one is concerned with multicomponent mixtures which do not have well defined eutectic points. Seldom, their freezing behavior can be predicted from equilibrium phase diagrams because the individual components usually crystallize incompletely or not at all. The limit temperature for a safe primary drying cycle is therefore the glass transition temperature T'_g. As already mentioned, above this temperature the frozen solid changes into a highly viscous fluid and the molecules regain their mobility. At higher temperatures partial melting can occur which may lead to "collapse" of the freeze-dried cake. Collapse phenomena are observed at temperatures very close to T'_g, and in many cases the collapse temperature (T_c) and the glass transition temperature (T'_g) are practically identical in a range of about 2–3 °C (e.g., LEVINE and SLADE, 1988; PIKAL and SHAH, 1990). Because a stable amorphous glass phase is reached below the glass transition temperature T'_g, the solution to be frozen should be cooled below this temperature as quick as possible.

Eutectic, collapse and glass transition temperatures, respectively, are best estimated by thermal analysis methods, measurement of the electrical resistance or by direct observation using the freeze microscope.

The critical parameter for a freeze-drying protocol therefore is the maximally tolerable temperature of the ice. But temperature measurement during freeze-drying is by no means a trivial task. Even if several temperature probes are placed at different locations inside the drying chamber, the results obtained might be ambiguous. Control of the freeze-drying process, for these reasons, is best achieved by following the partial water vapor pressure after equilibration. For this, the valve between the chamber and the ice condenser is closed for a fixed time and the temperature can be read from the equilibrium pressure.

The drying time depends on the quality of the ice crystals. Drying of an ice layer 1 cm in thickness takes about 10–20 h. Droplet-like particles with a diameter of 3 mm can be dried in a continuously working plant within 1–3 h (GÖLKER, 1989). One problem often encountered is the uneven distribution of residual water content for all the vials in the drying chamber. Rate and extent of drying can vary according to the position on the plates because of uneven distribution of heat and mass transport. Vials placed near the connection to the ice condenser will absorb moisture from the exhaust stream, especially in the case of hygroscopic materials, which can result in a higher residual water content compared to the remaining vials. The situation can be improved if drying is performed under conditions which reflect the sorption equilibrium at the desired water content.

The primary drying is finished when all ice is removed. Residual water which was not frozen has to be removed by evaporation during the secondary drying process.

2.5.3 Secondary Drying

After removal of the frozen water by sublimation, the "dry" substance still contains a high percentage of water. This water can be present as adsorbed water on the surface of the dried substance or may exist either as hydration water or dissolved in the dry amorphous phase. This residual water is difficult to remove because of the low vapor pressure inside the capillaries of freeze-dried material. The current practice for secondary drying is to increase the shelf temperature to 25–40 °C and to lower the chamber pressure as far as possible (e.g., KING, 1968; MELLOR, 1978). That a very low pressure inside the chamber does not speed up the rate of drying, was shown by PIKAL et al. (1990) by freeze-drying crystalline (mannitol) as well as amorphous substances (moxalactam disodium and povidon). At least

Tab. 2. Protein Inactivation and Denaturation

Irreversible Changes by Chemical Events	
like	Hydrolysis
	Oxidation (e.g., methionine)
	Deamidation (glutamine, asparagine)
	Disulfide bridges
	Crosslinking
Reversible (or Irreversible) Changes	
by	High temperature
	Extreme pH-values
	Detergents
	Shear forces
	Metal binding

up to a pressure of 0.26 mbar, the drying rate was found to be independent of the pressure. Mass transfer might also be influenced by the surface area and the structure of the pores of the dried product. At a first stage, the water content decreases very rapidly but only to values much higher than would be expected from the isotherm curves and the partial water vapor pressure in the chamber. Further lowering of the water content is possible only at increased temperatures.

2.5.4 Stability Improvement by Excipients

Due to stress situations during freezing, drying, or long-term storage, a number of unfavorable reactions can take place which may lead to deterioration and inactivation of biomolecules (Tab. 2). Structural changes during freezing and/or freeze-drying are especially marked for many proteins. For example, lactate dehydrogenase dissociates from the tetrameric native state into subunits during freezing and thawing (MARKERT, 1963; MASSARO, 1967); the structure of L-asparaginase from *Escherichia coli* is altered by freeze-drying resulting in a decrease in α-helical structure accompanied by an increase in the amount of β-structure (ROSENKRANZ and SCHOLTAN, 1971); freezing and thawing of myosin causes unfolding of the helical structure of the molecule; on the other hand, catalase, a globular protein, shows no conformational change during freeze-thawing but dissociates into subunits during freeze-drying (HANAFUSA, 1969; TANFORD and LOVRIEN, 1962).

The native state of biomolecules may be adversely affected by the different processes connected with freeze concentration. The activity of enzymes, for instance, may be negatively influenced by high salt concentrations. For lactate dehydrogenase, e.g., it was shown that the inactivation rate increased markedly with increasing salt concentration. A stabilizing effect at low temperatures, on the other hand, was observed with increasing protein concentration (SOLIMAN and VAN DEN BERG, 1971). This observation is also used for stabilization of labile proteins against freeze-denaturation by addition of albumin (TAMIYA et al., 1985).

Dehydration by freezing and dehydration by drying are qualitatively different processes (CROWE et al., 1990). Biomolecules like proteins, nucleic acids, membranes, or lipids contain a relatively constant amount of non-freezable water (about 0.25 g/g of dry substance). Freezing only removes the bulk water whereas drying also removes the non-freezable water. In the latter case significant changes in the physical properties, of, e.g., proteins are observed.

For prevention, or at least mitigation of unfavorable incidents, lyoprotectants or cryoprotectants are added before freezing or freeze-drying of labile biological molecules. Lyoprotectants are defined as substances which prevent damage by freezing, drying, and storage whereas cryoprotectants exhibit protection only during freezing. Substances like glycerol, dimethyl sulfoxide (DMSO), sugars, sugar derivatives, different polymeric compounds like dextran, polyvinyl pyrrolidone, Ficoll (hydrophilic copolymerisate of saccharose and epichlorohydrin), hydroxyethyl starch, polyethylene glycol, and various amino acids are reported as protective agents (e.g., MERYMAN, 1971; ASHWOOD-SMITH and WARBY, 1972; AKAHANE et al., 1981; GEKKO and TIMASHEFF, 1981; CARPENTER and CROWE, 1988a; TAMIYA et al., 1985).

The effect of a number of distinctly different chemical compounds as protective agents during freezing and freeze-drying of L-asparaginase from *Erwinia carotovora* was investigated by HELLMANN et al. (1983). The enzyme undergoes dissociation of the native tetrameric molecule, most likely in the freeze-drying stage. The conservation of activity during the freeze-drying process depends on the nature of the excipients included in the formulation as can be seen from Fig. 4. The effect of urea is understandable because this reagent normally denatures proteins. The relatively good effect of ammonium sulfate occurs with many proteins but there seems to be no special ionic effect since sodium chloride almost completely inactivates the protein. Sugars and sugar derivatives prevent dissociation, probably by affecting hydrophobic interactions important in the stabilization of the native tetramer.

Hydroxypropyl-β-cyclodextrin was shown to be an excellent excipient for the freeze-drying

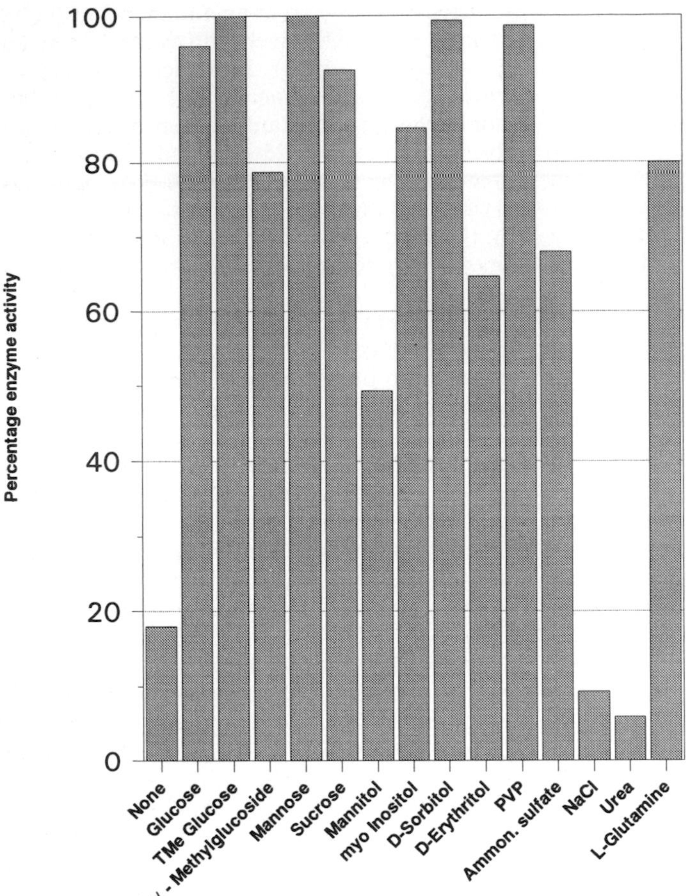

Fig. 4. Protection of L-asparaginase during freeze-drying by different excipients (data from HELLMANN et al., 1983). TME glucose, tetramethyl glucose; PVP, polyvinyl pyrrolidone.

of a mouse IgG_{2a} monoclonal antibody (RESSING et al., 1992).

In other cases, proteins can be freeze-dried without loss of activity but are not stable during storage. In this case, stabilizing compounds are necessary in order to improve the shelf life of the protein. Two examples are ribonuclease A and human growth hormone. Ribonuclease (RNase) loses activity during storage at different pH values as a function of time due to the formation of covalently bonded aggregates. The amino acid lysine plays the main role but asparagine/aspartic acid and glutamine/glutamic acid are also involved. Sucrose, Ficoll 70, and polyvinyl pyrrolidone (PVP) were used as protective substances against degradation during storage. Sucrose showed the best effect at neutral and alkaline and PVP at acidic pH values, whereas Ficoll 70 was an excellent excipient at all pH values studied (TOWNSEND and DE LUCA, 1988, 1991). Inactivation of human growth hormone (hGH) during storage is caused by oxidation and deamidation reactions followed by aggregation of the protein. The major breakdown products are Met-14 sulfoxide due to oxidation of methionine and Asp-149 caused by deamidation of asparagine. The influence of sodium chloride is of interest because this is included in many formulations for the sake of tonicity of the reconstituted solution. Not only with a physiological saline solution (0.9% NaCl) but also with a dilute saline (0.1% NaCl) the freeze-dried samples could not be reconstituted to form a clear solution. This effect was not shown with freeze-thawed samples, a result which suggests that denaturation occurs during the drying step. Without sodium chloride the losses during the freeze-drying process are minimal compared to the losses during storage. These losses can be minimized by addition of different excipients. The best system found was a combination of glycine and mannitol (PIKAL et al., 1991). No correlation was found between stability and water content in the studies with RNase and hGH.

Special conditions must be considered in cryo-conservation and freeze-drying of living cells. These methods are used, e.g., for the long-term storage of bacterial and yeast strains in the fermentation industry and for the conservation of microorganisms like starter cultures used in the dairy industry or in silage for agricultural purposes. Cell injury may occur, e.g., through the different effects connected with freeze concentration or by ice formation inside the cells. The survival rate of living cells during freezing and freeze-drying depends very much on the optimal cooling rate during freezing (see, e.g., MAZUR, 1970, 1977, 1984; LEIBO and MAZUR, 1971). Hydrophilic substances like polyethylene glycol (PEG) can influence the generation of ice crystals by perturbation of the diffusional motion of water molecules (MICHELMORE and FRANKS, 1982). Other substances like glycerol or dimethyl sulfoxide can penetrate and consequently change the permeability of the cell membrane. This facilitates dehydration of the cell by osmolysis. These compounds may also prevent the formation of large ice crystals during the freezing process (MAEDA et al., 1989).

The protective mechanism during freezing or freeze-drying is not well understood yet, but different theories can be found in the literature:

The deleterious effects due to freeze concentration may be minimized by additives, simply by dilution of the freeze-concentrated matrix surrounding the ice crystals. In this case the cryoprotectant should not crystallize during cooling, but form a glass-like phase.

There is considerable evidence that the instability of amorphous materials increases sharply at temperatures above the glass transition temperature (e.g., LEVINE and STADE, 1988; FRANKS et al., 1991). At temperatures below the glass transition temperature, the mobility of molecules and hence the reactivity is restricted, which in turn means improvement of product stability. The glass transition temperature depends on the nature and concentration of excipients. Low-molecular weight substances and salts stay in the amorphous phase and lower the temperature while polymers tend to raise the glass transition temperature. Ideal additives show high glass transition temperatures and low amounts of non-freezable water and are compounds for which water is a poor "plasticizer". The most stable formulation is thought to be one which is transformed to a glassy amorphous state and shows a glass transition temperature higher than the storage temperature.

The stabilizing effect of sugars and polyols and of amino acids on proteins in solution is well known. It has been ascribed to strengthening of hydrophobic interactions in the protein molecule (BACK et al., 1979) or to the preferential hydration of the protein due to the exclusion of the stabilizing solute from the surface of the macromolecule (ARAKAWA and TIMASHEFF, 1982, 1983). Addition of sucrose as well as other sugars increases the free energy of the system. It is supposed that this increase will be higher with an increase of the surface area of the macromolecule. As a result, the native conformation of the protein is stabilized, since thermodynamically unfolding becomes even less favorable in the presence of sucrose because of the increasing chemical potential with sugars added to the expanded denatured molecule. An important factor for governing the interaction of sugars with other macromolecules and therefore for their stabilizing effect, is the increase in surface tension of the water (LEE and TIMASHEFF, 1981; ARAKAWA and TIMASHEFF, 1982). Another reason for the exclusion of solutes could be steric effects as was described by ARAKAWA and TIMASHEFF (1985) for the interaction of polyethylene glycol with β-lactoglobulin. The preferential exclusion increases with increasing molecular weight of the polyethylene glycol which led to the suggestion that steric effects may be responsible for the interactions of polyethylene glycol and proteins. At higher concentrations, however, polyethylene glycol has a destabilizing effect, probably because of direct binding to the protein which becomes stronger in the case of the unfolded molecule. This mechanism of preferential hydration can also explain the stabilizing effects during freezing, as was discussed, e.g., by CARPENTER and CROWE (1988a, b).

The mechanism of preferential hydration might not be sufficient as an explanation of the stabilizing effect during dehydration by drying. Because, in contrast to other substances which are stabilizers during freezing, only sugars are able to stabilize phosphofructokinase (PFK) against inactivation during drying (CARPENTER et al., 1987), special interactions between sugar and protein may be necessary. By air-drying of PFK to different water levels, CARPENTER and CROWE (1988b) could show that the initial loss of enzymatic activity can be attributed to removal of bulk water. The complete loss of enzymatic activity, however, is connected to the dehydration of the protein itself. The inactivation observed during air-drying was irreversible and could be prevented completely in all stages of the drying process by addition of 100 mM trehalose. Similar effects, but to a lesser extent, were also observed by addition of other sugars. The authors suggest that the ability to protect PFK in the final drying stage may depend on direct binding of the sugar to the protein. The prevention of denaturation of catalase during freeze-drying by addition of sugars is also thought to be exerted by direct hydrogen bonding of sugar molecules to the protein molecule resulting in a monomolecular layer of the protective substance instead of the hydration monolayer (TANAKA et al., 1991).

Trehalose could play a special role as a protective agent for proteins. It could be excluded preferentially and form hydrogen bonds, thereby stabilizing the protein structure. Trehalose has recently found attention as a cryoprotectant because it is widely distributed in organisms that can withstand total dehydration (ROSER, 1991; COLAÇO et al., 1992). Excellent stabilizing effects are reported for sensitive enzymes like restriction enzymes, diverse monoclonal antibodies, and therapeutic proteins like factor VIII. But the outstanding properties of trehalose are controversial, and in fact thought to be characteristic for a number of carbohydrates (LEVINE and SLADE, 1992).

Although it is difficult to generalize rules for formulations, at least some conclusions can be drawn from data found in the literature:

1. To be effective as a lyoprotectant or cryoprotectant, the substance must remain at least partially amorphous. By dilution of the proteins to be freeze-dried with the amorphous excipient, the protein–protein interactions are minimized and hence denaturation by interaction of amino acid side chains or breaking of hydrogen bonds are minimized.
2. Some protective substances may interact directly with the protein structure by influencing the hydration of the protein.
3. Buffer substances and other salts have to be chosen carefully. Buffer substances may cause pH shifts during freezing by changed pK_a values or by differential crystallization. High salt concentrations may adversely affect the glass transition temperature and the collapse temperature.
4. From the importance of the glass transition temperature for the freeze-drying process and shelf life of the product, it follows that the formulation should have a glass transition temperature as high as possible.
5. The excipients must be safe and accepted in terms of regulations in the food and pharmaceutical industry.

2.5.5 Economical Considerations

For sensitive proteins, in many cases freeze-drying is the method of choice for converting them into stable formulations. This process is usually performed as a batch process in a chamber freeze-dryer. The main components of such a plant are outlined in Fig. 5. The fact that the sublimation of ice has to be performed at low temperatures and the condensation of the water vapor in the ice condensor at even lower temperatures, results in high expenditures for this process. The performance as a batch process causes additional energy losses because trays, silicone oil which is the heat exchange medium, and tons of steel which is the structural material of the equipment, have first to be cooled to low temperatures and subsequently rewarmed. The sublimation process needs a high energy input of 2900 kJ/kg of ice. Taking into consideration additional losses

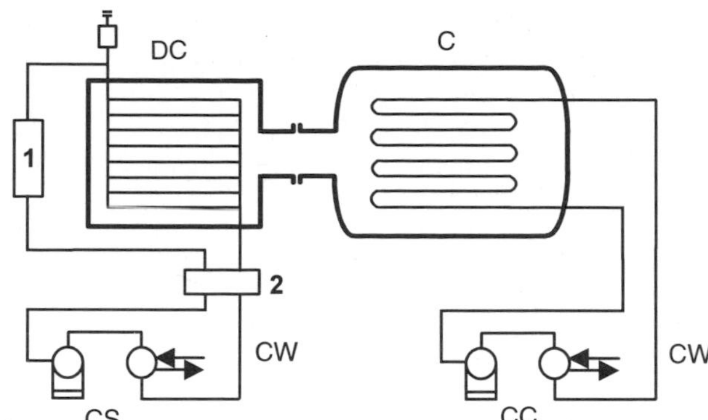

Fig. 5. Components of a freeze-drying plant. DC, drying chamber; C, condenser; CW, cold water; CC, compressor for the condenser; CS, compressor for the shelf; 1, 2, electrical heating systems.

due to temperature differences inside and outside the plant, the total energy consumption adds up to about 9000 kJ/kg of product depending on the size of the plant. In contrast, the specific heat consumption of spray-driers is of the order of 4000–8000 kJ/kg of product. The specific energy consumption for the process is shown in Tab. 3. It can be seen that up to 40% of the energy is lost for the drying process itself and will be consumed by the parts of the freeze-drying plant (WILK, 1989). One large cost factor in freeze-drying is the use of cooling water necessary for the compressors. The specific energy consumption of course depends on the size of the freeze-drying plant, and decreases with increasing plant capacity. Because of the high operational cost, every possibility to save energy should be used. One is the exact control of the process. If components not used during the whole process are switched off and the drying cycle is optimized with respect to temperature and vacuum, savings up to 30% may be achieved (WILK, 1989). For this reason, modern equipment is controlled by computer programs. Additional savings can result from the reuse of the heat from the ice condenser which normally is removed by cooling water.

2.6 Controlled Evaporative Drying

As discussed above, the degree of denaturation of biological molecules during freezing

Tab. 3. Energy Consumption during Freeze-Drying

Shelf Area (m^2)	5	10	20	40
Freezing (kJ)	76600	127800	211590	404760
(kJ/kg)	1900	1590	1320	1260
Sublimation (kJ)	162800	311000	593500	1165500
(kJ/kg)	4000	3800	3700	3600
Condensation (kJ)	129300	258300	501600	995600
(kJ/kg)	3200	3200	3100	3100
Total (kJ)	368900	697200	1306700	2565900
(kJ/kg)	9200	8700	8200	8000
Product (kJ)	269800	531900	1051000	2104400
Material[a] (kJ)	99000	165300	254600	461500
(%)	37	31	24	22

Capacity: 10 kg of product per m^2 of shelf area, product contains 20% of solid material.

[a] steel, glass, heat transfer liquid, etc.

and drying depends on the glass transition temperature and the time to reach this temperature during the process. If a formulation with a high T_g' is chosen, drying can be performed by evaporation at higher temperatures instead of sublimation from ice at very low temperatures. The most important evaporative technique for biological products is spray-drying. In this case, T_g' can be reached in a very short time. However, the product might be exposed to very high temperatures which might be dangerous for the product. On the other hand, in a freeze-drying process, the time to reach T_g' might expose the product to unfavorable conditions over extended periods. Controlled evaporative drying of small volumes removes water at a rate between these extremes (FRANKS et al., 1991). For this drying technique, formulation conditions should be found which show high intrinsic T_g' and for which water is a poor plasticizer. The effectiveness of the method could be shown by drying enzymes, where a high retention of residual enzyme activity was obtained.

3 Spray-Drying

Spray-drying is the second important process for the final recovery of biological products. The material is applied as a solution, suspension or a free-flowing wet substance. The amount of energy necessary for drying a substance depends on the type of water binding.

3.1 Binding of Water Molecules

Water can be associated with a substance in different ways (Tab. 4).

- Adhesive water: This is freely movable at the surface of substances so that no binding of any kind is observed.
- Capillary water: The behavior depends on the size of the capillary. In large capillaries ($r > 0.1$ µm) water is "bound" due to the capillary binding, whereas in small capillaries ($r < 0.1$ µm) water condenses due to the water vapor pressure depression from interfacial forces.
- Swelling water: Many biological products are hydrophilic polymers which can insert water molecules into the polymer layers. This leads to swelling and finally can result in a colloidal solution.
- Adsorptive water: The water molecules are dipoles which can interact with charged groups of, e.g., protein molecules. They also can form hydrogen bonds with suitable groups like hydroxyl, amino or carboxyl groups.
- Hydrate water: This can exist as coordination water (where water molecules are directly coordinated to ions, e.g., $Mg^{2+} \cdot 6H_2O$) or structural water (where water occupies discrete positions in the crystal lattice).
- Constitutional water: Water is covalently bound as a consequence of a chemical reaction.

Tab. 4. "Binding" of Water Molecules

Type	State of Binding	Mechanism	Mobility	Binding Energy (kJ/mol)
Adhesion	No bindings	Adhesion	Free	0
Capillary $r > 0.1$ µm	Mechanical	Capillary attraction	Free	0
Capillary $r < 0.1$ µm	Physical	Capillary condensation	Restricted	0–5
Hydration/swelling	Physicomechanical	Hydration osmosis	Restricted	2–20
Adsorption	Physicomechanical	H-bonds dipole–dipole interaction	Restricted	2–60
Hydrate	Chemical	Dipole–ion interaction (coordination)	Immovable	5–40
Constitutional	Chemical	Chemical reaction (covalent)	Immovable	20–100

3.2 Sorption Isotherms

The binding energy of water molecules ranges between 0 and 100 kJ/mol (STAHL, 1980). These values are reflected in the different shapes of sorption isotherms which are very useful for selecting drying conditions. For biological substances the situation is more complex because removal of water not only has a drying effect but changes the structure of the molecule, e.g., by breaking hydrogen bonds.

3.3 Heat and Mass Transfer

Spray-drying is an adiabatic process, and heat and mass transfer are coupled. At the beginning of the drying process, water which is not restricted in any form is removed from the surface. The drying time for this part of the process is constant and most of the water is removed in a relative short time during this period of the drying process. The drying time can be calculated according to BELTER et al. (1988):

$$t_c = \left(\frac{\rho_s \cdot d}{6k \cdot \rho}\right) \left(\frac{\lambda}{\hat{c}_p (T - T_i)}\right) (Y_0 - Y_c)$$

where ρ is the density, λ the heat of vaporization, c_p the heat capacity, Y the moisture content, Y_c the critical moisture content, k the mass coefficient, and T the temperature, with the conditions $t=0$, $Y=Y_0$, and $t=t_c$, $Y=Y_c$.

At the end of this drying phase the critical moisture content Y_c is reached. With increasing thickness of the dried layer the flow from the interior becomes restricted. Diffusion becomes more difficult and the drying rate is slowed down. This drying period may remove only a small amount of water but can take a large fraction of the total drying time. During this phase, the heat of evaporation λ exceeds that of free moving water and depends on the residual water content. With the assumption of $\mathrm{d}Y/\mathrm{d}t = -k\,Y$, where k is a rate constant and the initial conditions $t-t_c$, $Y=Y_c$, the integration of the equation results in

$$t - t_c = \left(\frac{1}{k}\right) \ln \left(\frac{Y_c}{Y}\right)$$

The energy needed is provided by a hot gas, in most cases hot air. The temperatures are in the range of 120 °C to 400 °C. Although this is far outside the range of the glass transition temperature, substances are stable because of a very short drying time in the spray-drying apparatus.

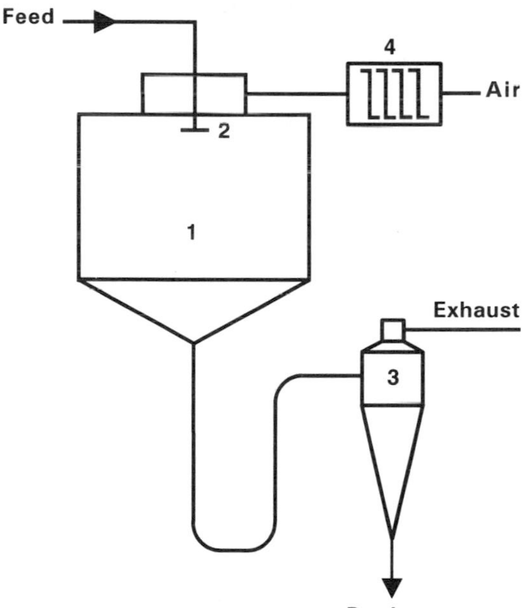

Fig. 6. Components of a spray-drying plant. 1, drying chamber; 2, dispersion device; 3, cyclon; 4, heat exchanger.

3.4 Equipment

The different parts of a spray-drying apparatus are shown in Fig. 6. The dispersion of solutions into small droplets can be achieved by means of rotating disks, different types of nozzles or ultrasound (SCHMIDT and WALZEL, 1980). Examples are shown in Fig. 7. The dispersion is influenced by

- interfacial tension of the liquid,
- density of the liquid,
- dynamic viscosity of the liquid,

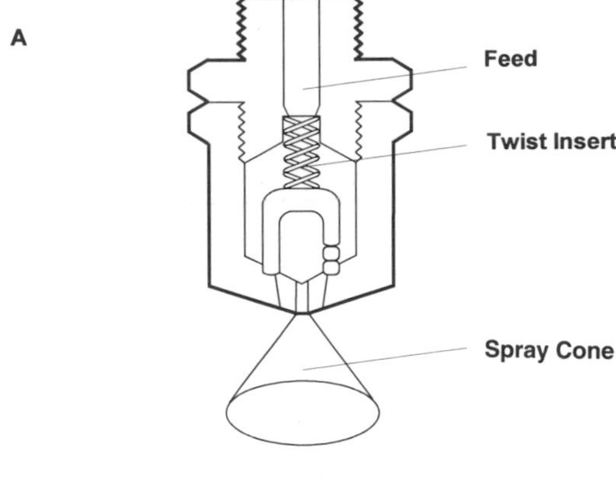

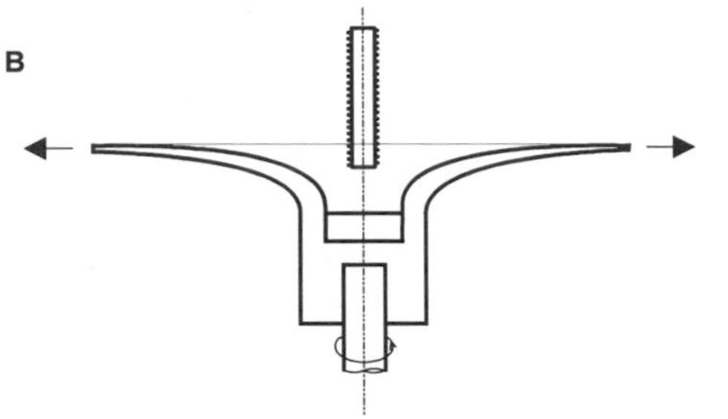

Fig. 7. Atomizers for spray-drying. (A) nozzle atomizer, (B) rotating disk atomizer.

- thickness of the liquid layer.
- relative velocities of liquid and air in the mixing zone.
- temperature.

Rotating Disk Atomizer

In this case dispersion is brought about by centrifugal forces. The liquid is applied at the center of the disk which rotates at a high velocity. On the edge of the disk, droplets tear off which pull threads of liquid. According to the operating conditions, the following events can take place: (1) If the disk is charged with too low an amount of liquid, primary droplets peel off from the disk in irregular intervals. (2) In the case of an optimally charged disk, division of the liquid threads into secondary droplets occurs which are smaller than the primary droplets. (3) Overloading the disk with liquid leads to the breakdown of the now thicker liquid thread and to periodic discharge of primary droplets in addition to secondary droplets, consequently the size distribution of the droplets is very uneven. (4) After overflowing the disk, the liquid drops off as a film and no regular droplets can be formed anymore.

The optimal volume of liquid for the creation of satisfactory droplets is defined (in $m^3 \cdot s^{-1}$) as

$$V = 4\frac{\sigma}{\rho_L}\left(\frac{R}{\omega^2 \cdot g}\right)^{1/4}$$

where σ is the surface tension (N/m), ρ_L the density of liquid (kg/m^3), R the dimension of disk (m), and ω the angle velocity of the disk (FRASER and EISENKLAM, 1956).

In most cases the disk is not completely flat but conically shaped which leads to better results with respect to size and shape of the droplets. The diameter varies between 5 cm and 35 cm with large industrial equipment; the rotational speed is between 4000 rpm and 50000 rpm. Centrifugal atomizers are especially suitable for dispersing suspensions which otherwise would tend to clog the nozzle. Droplet sizes achieved with these types of atomizers are between 25 and 950 µm.

Nozzle Atomizers

With pressure nozzles most of the applied pressure (of the order of 7 bar to maximal 500 bar) is converted into kinetic energy of the droplets which will be accelerated to 10–150 m/s. With two-component nozzles dispersion is achieved by compressed air which is ejected from the nozzle with a velocity of between 100 m/s and 200 m/s. Because the liquid cannot follow these high accelerations it is disintegrated into very tiny droplets creating large surfaces. For example, 1 kg of water, after dispersion in droplets of $d = 1$ µm, results in $n = 1.9 \cdot 10^{15}$ droplets exhibiting a surface of 6000 m^2. One-component nozzles are widely used for materials with low viscosity. The liquid is accelerated to high velocities and pressed through an orifice in the form of a conus and thereby dispersed in small droplets. With the more complicated two-component nozzles, the

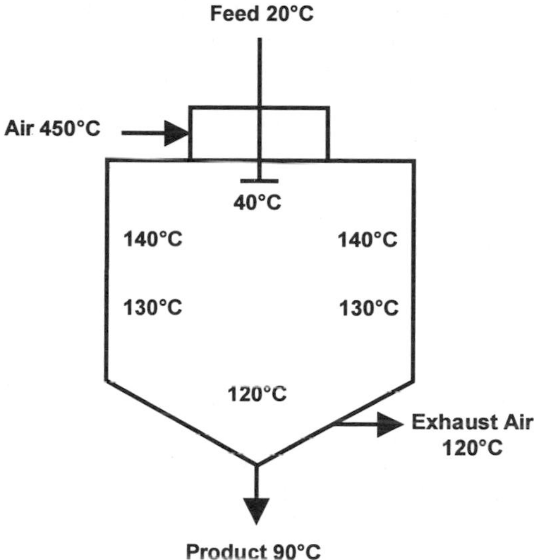

Fig. 8. Temperature distribution in a spray-dryer.

liquid is dispersed by means of a gas stream, usually air, which flows with much higher velocity from an outer ring channel and meets the liquid at the edge of an inner pipe. In this system, droplets usually have diameters between 15 and 150 µm.

The temperature distribution in a spray-dryer is shown in Fig. 8. The resulting drying time depends on the droplet size and is very short (of the order of seconds). For particles with diameters <100 µm the drying time is even less than a second. With a temperature difference of 150 °C between drying air and particle surface, the following values are valid (NÜRNBERG, 1980):

Droplet diameter (µm)	10	50	100	300
Drying time (s)	0.0032	0.08	0.32	2.88

Detailed information about spray-drying can be found in the literature (e.g., MASTERS, 1976; NÜRNBERG, 1980).

3.5 Application

Spray-drying is used for the production of technical enzymes for industrial use and as additives for washing detergents. It is also used as a last step for the production of single-cell protein. For protection purposes, proteins may be spray-dried in the presence of additives like galactomannan, polyvinyl pyrrolidone, methyl cellulose, or cellulose.

Embedding and coating of pharmaceuticals is done in order to achieve special effects:

- Chemical stabilization of compounds by preventing deterioration by oxygen, moisture, changes in pH value.
- Physical stabilization by preventing, for instance, crystallization of drug substances or changes of the crystalline state.
- Liquid substances like oils can be converted into solids.
- Changing of the solubility behavior of drugs. For theophylline, e.g., a slower drug dissolution rate was obtained after preparation of coated particles by spray-drying together with hydroxypropylmethyl cellulose as the coating polymer (WAN et al., 1991). On the other hand, the solubility of drugs with low solubility can be improved because of the extremely fine distribution of the drug.

Pharmaceutical preparations and biochemical products can also be spray-dried under aseptic conditions. For this task, the spray-drier must be connected to a filling line which allows the aseptic handling of the product. This technique is used for the production of, e.g., blood plasma and blood serum, culture media, different enzymes, antibiotics like bacitracin, streptomycin and tetracycline, vitamins and hormones.

The spray-drying technique is of great importance for the preservation of microorganisms, e.g., for the use as starter cultures in the dairy industry for the production of cheese or other milk products, and as silage cultures for agricultural purposes. The most important factor for these applications is the survival factor of the microorganisms and the retention of their enzymatic activity.

4 References

AKAHANE, T., TSUCHIYA, T., MATSUMOTO, J. J. (1981), Freeze denaturation of carp myosin and its prevention by sodium glutamate, *Cryobiology* **18**, 426–435.

ARAKAWA, T., TIMASHEFF, S. N. (1982), Stabilization of protein structure by sugars, *Biochemistry* **21**, 6536–6544.

ARAKAWA, T., TIMASHEFF, S. N. (1983), Preferential interaction of proteins with solvent components in aqueous amino acid solutions, *Arch. Biochem. Biophys.* **224**, 169–177.

ARAKAWA, T., TIMASHEFF, S. N. (1985), Mechanism of poly(ethylene glycol) interaction with proteins, *Biochemistry* **24**, 6756–6762.

ASHWOOD-SMITH, M. J., WARBY, C. (1972), Protective effect of low and high molecular weight compounds on the stability of catalase subjected to freezing and thawing, *Cryobiology* **9**, 137–140.

BACK, J. F., OAKENFULL, D., SMITH, M. (1979), Increased thermal stability of proteins in the presence of sugars and polyols, *Biochemistry* **18**, 5191–5196.

BELL, L. N., LABUZA, T. P. (1991), Potential pH implications in the freeze-dried state, *Cryo-Lett.* **12**, 235–244.

BELTER, A. P., CUSSLER, E. L., HU, W.-S. (1988), in: *Bioseparations – Downstream Processing for Biotechnology*, pp. 324–328, New York: John Wiley & Sons.

BUCHMÜLLER, J., WEYERMANNS, G. (1989), Cryopel®: Ein neues Verfahren zum Pelletieren und Frosten biologischer Substrate, *gas aktuell* **35**, 10–13.

CARPENTER, J. F., CROWE, L. M., CROWE, J. H. (1987), Stabilization of phosphofructokinase with sugars during freeze-drying: Characterization of enhanced protection in the presence of divalent cations, *Biochim. Biophys. Acta* **923**, 109–115.

CARPENTER, J. F., CROWE, J. H. (1988a), The mechanism of cryoprotection of proteins by solutes, *Cryobiology* **25**, 244–255.

CARPENTER, J. F., CROWE, J. H. (1988b), Modes of stabilization of a protein by organic solutes during desiccation, *Cryobiology* **25**, 459–470.

COLAÇO, C., SEN, S., THANGAVELU, M., PINDER, S., ROSER, B. (1992), Extraordinary stability of enzymes dried in trehalose: Simplified molecular biology, *Bio/Technology* **10**, 1007–1011.

CROWE, J. H., CARPENTER, J. F., CROWE, M. L., ANCHORDOGUY, T. J. (1990), Are freezing and dehydration similar stress vectors? A comparison of modes of interaction of stabilizing solutes with biomolecules, *Cryobiology* **27**, 219–231.

FENNEMA, O. (1975), Reaction kinetics in partially frozen aqueous systems, in: *Water Relations in Foods* (DUCKWORTH, R. B., Ed.), pp. 539–558, London: Academic Press.

FLINK, J. (1975), The retention of volatile components during freeze-drying: a structurally based mechanism, in: *Freeze-Drying and Advanced Food Technology* (GOLDBLITH, S. A., REY, L., ROTHMAYR, W. W., Eds.), pp. 351–372, London: Academic Press.

FRANKS, F. (1985a), in: *Biophysics and Biochemistry at Low Temperatures*, pp. 21–36, Cambridge: Cambridge University Press.

FRANKS, F. (1985b), in: *Biophysics and Biochemistry at Low Temperatures*, pp. 37–61, Cambridge: Cambridge University Press.

FRANKS, F. (1986), Molekulare Grundlagen der Kälteresistenz von Lebewesen. *Chem. Uns. Zeit* **20**, 146–155.

FRANKS, F. (1988), in: *Characterization of proteins* (FRANKS, F., Ed.), pp. 95–126, Clifton, NJ: Humana Press.

FRANKS, F. (1989), Improved freeze-drying. An analysis of the scientific principles, *Process Biochem.* **24**, 3–6.

FRANKS, F. (1990), Freeze-drying: From empiricism to predictability, *Cryo-Lett.* **11**, 93–110.

FRANKS, F., HATLEY, R. H. M. (1985), Low temperature unfolding of chymotrypsinogen, *Cryo-Lett.* **6**, 171–180.

FRANKS, F., HATLEY, R. H. M., MATHIAS, S. F. (1991), Materials science and the production of shelf-stable biologicals, *Pharm. Technol. Int.* **3**, 24–34.

FRASER, R. P., EISENKLAM, P. (1956), Liquid atomization and the drop size of sprays, *Trans. Inst. Chem. Eng.* **4**, 249–319 (cited in KNEULE, F., *Das Trocknen*, Verlag Sauerländer, Aarau, 1975).

GEKKO, K., TIMASHEFF, S. N. (1981), Thermodynamic and kinetic examination of protein stabilization by glycerol, *Biochemistry* **20**, 4467–4486.

GÖLKER, C. (1989), Gefriertrocknung biotechnologischer Produkte, in: *Gefriertrocknung in Entwicklung und Produktion* (Concept Heidelberg, Ed.), *J. Pharmatechnol.* **3**, 30–40.

HANAFUSA, N. (1969), Denaturation of enzyme protein by freeze-thawing and freeze-drying, in: *Freezing and Drying of Microorganisms* (NEI, T., Ed.), pp. 117–129.

HATLEY, R. H. M., FRANKS, F. (1986), Denaturation of lactate dehydrogenase at subzero temperatures, *Cryo-Lett.* **7**, 226–233.

HATLEY, R. H. M., FRANKS, F., MATHIAS, S. F. (1987), The stabilization of labile biochemicals by undercooling, *Process Biochem.* **22**, 169–172.

HELLMANN, K., MILLER, D. S., CAMMACK, K. A. (1983), The effect of freeze-drying on the quaternary structure of L-asparaginase from *Erwinia carotovora, Biochim. Biophys. Acta* **749**, 133–142.

HOBBS, P. V. (1974), *Ice Physics*, Oxford: Oxford University Press.

KHAN, S. (1989), Automatic flexible aseptic filling and freeze-drying of parenteral drugs, *Pharm. Technol.* **13**, 24–34.

KING, C. J. (1968), Rates of moisture sorption and desorption in porous, dried foodstuffs, *Food Technol.* **22**, 165–171.

KIOVSKI, T. E., PINCOCK, R. E. (1966), The mutarotation of glucose in frozen aqueous solutions, *J. Am. Chem. Soc.* **88**, 4704–4710.

LEE, J. C., TIMASHEFF, S. N. (1981), The stabilization of proteins by sucrose, *J. Biol. Chem.* **256**, 7193–7201.

LEIBO, S. P., MAZUR, P. (1971), The role of cooling rates in low-temperature preservation, *Cryobiology* **8**, 447–452.

LEVINE, H., SLADE, L. (1988), Principles of "cryostabilization" technology from structure/property relationships of carbohydrate/water systems – A review, *Cryo-Lett.* **9**, 21–63.

LEVINE, H., SLADE, L. (1992), Another view of trehalose for drying and stabilizing biological materials, *BioPharm* **5**, 36–40.

LOVELOCK, J. E. (1953a), The haemolysis of human red blood cells by freezing and thawing, *Biochim. Biophys. Acta* **10**, 414–426.

LOVELOCK, J. E. (1953b), The mechanism of the protective action of glycerol against haemolysis by freezing and thawing, *Biochim. Biophys. Acta* **11**, 28–36.

MACKENZIE, A. P. (1977), The physico-chemical basis for the freeze-drying process, *Dev. Biol. Stand.* **36**, 51–67.

MAEDA, T., TERADA, T., TSUTSUMI, Y. (1989), The role of glycerol and dimethyl sulfoxide on the freezing of fowl spermatozoa, *Cryo-Lett.* **10**, 393–400.

MALTINI, E. (1975), Thermal phenomena and structural behaviour of fruit juices in the processing stage of the freeze-drying process, in: *Freeze-drying and Advanced Food Technology* (GOLDBLITH, S. A., REY, L., ROTHMAYR, W. W., Eds.), pp. 121–139, London: Academic Press.

MARKERT, C. L. (1963), Lactic dehydrogenase isozymes: Dissociation and recombination of subunits, *Science* **140**, 1329–1330.

MASSARO, E. J. (1967), Urea-mediated freeze-thaw hybridization of lactate dehydrogenase, *Biochim. Biophys. Acta* **147**, 45–51.

MASTERS, K. (1976), *Spray Drying*, 2nd Ed., New York: John Wiley & Sons.

MAZUR, P. (1970), Cryobiology: the freezing of biological systems, *Science* **168**, 939–949.

MAZUR, P. (1977), The role of intracellular freezing in the death of cells cooled at supraoptimal rates, *Cryobiology* **14**, 251–272.

MAZUR, P. (1984), Freezing of living cells: Mechanisms and implications, *Am. J. Physiol.* **247**, C125–C142.

MELLOR, J. D. (1978), *Fundamentals of Freeze-drying*, London: Academic Press.

MERYMAN, H. T. (1971), Cryoprotective agents, *Cryobiology* **8**, 173–183.

MICHELMORE, R. W., FRANKS, F. (1982), Nucleation rates of ice in undercooled water and aqueous solutions of polyethylene glycol, *Cryobiology* **19**, 163–171.

NÜRNBERG, E. (1980), Darstellung und Eigenschaften pharmazeutisch relevanter Sprühtrocknungsprodukte, eine Übersicht, *Acta Pharm. Technol.* **26**, 39–67.

PIKAL, M. J. (1990a), Freeze-drying of proteins. Part I: Process design, *BioPharm* **3**, 18–27.

PIKAL, M. J. (1990b), Freeze-drying of proteins. Part II: Formulation selection, *BioPharm* **3**, 26–30.

PIKAL, M. J., SHAH, S. (1990), The collapse temperature in freeze-drying: dependence on measurement methodology and rate of water removal from the glassy phase, *Int. J. Pharm.* **62**, 165–186.

PIKAL, M. J., SHAH, S., ROY, M. L., PUTMAN, R. (1990), The secondary drying stage of freeze-drying: drying kinetics as a function of temperature and chamber pressure, *Int. J. Pharm.* **60**, 203–217.

PIKAL, M. J., DELLERMAN, K. M. ROY, K. L., RIGGIN, R. M. (1991), The effect of formulation variables on the stability of freeze-dried human growth hormone, *Pharm. Res.* **8**, 427–436.

PRIVALOV, P. L., GRIKO, YU. V., VENYAMINOV, S. YU., KUTYSHENKO, V. P. (1986), Cold denaturation of myoglobin, *J. Mol. Biol.* **190**, 487–498.

RESSING, M. E., JISKOOT, W., TALSMA, H., VAN INGEN, C. W., BEUVERY, E. C., CROMMELIN, D. J. A. (1992), The influence of sucrose, dextran, and hydroxypropyl-β-cyclodextrin as lyoprotectants for a freeze-dried mouse IgG_{2a} monoclonal antibody (MN12), *Pharm. Res.* **9**, 266–270.

ROSENKRANZ, H., SCHOLTAN, W. (1971), Circulardichroismus und Konformation der L-Asparaginase. *Hoppe-Seyler's Z. Physiol. Chem.* **352**, 1081–1090.

ROSER, B. (1991), Trehalose drying: A novel replacement for freeze-drying, *BioPharm* **4**, 47–53.

SCHMIDT, P., WALZEL, P. (1980), Zerstäuben von Flüssigkeiten, *Chem. Ing. Tech.* **52,** 304–311.

SOLIMAN, F. S., VAN DEN BERG, L. (1971), Factors affecting freezing damage of lactic dehydrogenase, *Cryobiology* **8,** 73–78.

STAHL, P. H. (1980), *Feuchtigkeit und Trocknen in der Pharmazeutischen Technologie.* Darmstadt: UTB, Verlag Steinkopff.

STEINBACH, G. (1972), Wärmeübergang und Stofftransport bei der Gefriertrocknung, Berechnung von Gefriertrocknungsprozessen, *VDI-Bildungswerk BW 1610,* Düsseldorf: VDI Verlag.

TAMIYA, T., OKAHASHI, N., SAKUMA, R., AOYAMA, T., AKAHANE, T., MATSUMOTO, J. J. (1985), Freeze denaturation of enzymes and its prevention with additives, *Cryobiology* **22,** 446–456.

TANAKA, K., TAKEDA, T., MIYAJIMA, K. (1991), Cryoprotective effect of saccharides on denaturation of catalase by freeze-drying, *Chem. Pharm. Bull.* **39,** 1091–1094.

TANFORD, C. (1980), *The Hydrophobic Effect.* New York: Wiley.

TANFORD, C., LOVRIEN, R. (1962), Dissociation of catalase into subunits. *J. Am. Chem. Soc.* **84,** 1892–1896.

TOWNSEND, M. W., DELUCA, P. P. (1988), Use of lyoprotectants in the freeze-drying of a model protein, ribonuclease A, *J. Parenter. Sci. Technol.* **42,** 190–199.

TOWNSEND, M. W., DELUCA, P. P. (1991), Nature of aggregates formed during storage of freeze-dried ribonuclease A. *J. Pharm. Sci.* **80,** 63–66.

VAN DEN BERG, L., ROSE, D. (1959), Effect of freezing on the pH and composition of sodium and potassium phosphate solutions: the reciprocal system KH_2PO_4-$Na_2HPO_4 \cdot H_2O$, *Arch. Biochem. Biophys.* **81,** 319–329.

WAN, L. S. C., HENG, P. W. S., CHIA, C. G. H. (1991), Preparation of coated particles using a spray drying process with an aqueous system, *Int. J. Pharm.* **77,** 183–191.

WILK, G. (1989), Die Kosten und einige Besonderheiten der Gefriertrocknung, in: Gefriertrocknung in Entwicklung und Produktion (Concept Heidelberg, Ed.), *J. Pharmatechnol.* **3,** 44–50.

WILLIAMS, N. A., POLLI, G. O. (1984), The lyophilization of pharmaceuticals: a literature review. *J. Parenter. Sci. Technol.* **38,** 48–59.

WOOG, H. (1989), Galenische Entwicklung von Lyophilisaten, in: Gefriertrocknung in Entwicklung und Produktion (Concept Heidelberg, Ed.), *J. Pharmatechnol.* **3,** 14–24.

IV. Process Validation, Regulatory Issues

27 Analytical Protein Chemistry

SRIKANTH SUNDARAM
DAVID M. YARMUSH
MARTIN L. YARMUSH

Piscataway, NJ 08854-0909, USA

1 Introduction

The final step in any protein purification scheme is the use of analytical methods to assess the purity and biological activity of the product protein. Determination of purity of a protein preparation generally involves an assessment of the quantity of particular types of impurities rather than a direct quantification of purity. Thus, in order to assess the purity of a given sample, one or more characteristic properties of the desired product must be identified that can be used to distinguish it from the putative impurities and contaminants. Since most isolation procedures are quite efficient at removing non-protein species, this chapter will focus on analytical methods that are useful in detecting contamination of the desired products by other proteins.

The criteria for choosing methods for purity analysis of proteins depend upon a number of factors such as: (1) the nature of the protein and its intended use, (2) the nature and spectrum of potential impurities and contaminants, (3) the accuracy and sensitivity needed, (4) the quantity of protein available for analysis, and (5) the ease of operation and cost/expense involved (RHODES and LAUE, 1990). Through the selection of appropriate analytical methods, a QC/QA (quality control/quality assurance) strategy can be devised to ensure product consistency and purity. The resulting purity determination is, however, only as accurate and complete as the analytical methods used.

Tab. 1 lists several sensitive analytical techniques used to detect the presence of impurities in a protein sample. Most of these methods rely on the separation of the potentially complex mixture using different properties such as molecular size or charge. Purity is demonstrated by showing that only one component is detectable and quantitative estimates are often arrived at by comparing the amount of impurities detected to that of the desired product. Some of these methods are also used routinely for the quantitative determination of various molecular properties or parameters. For example, polyacrylamide gel electrophoresis under reducing conditions can be used to determine size, molecular weight, and the presence of subunits in a protein. In this chapter, however, the discussion of these methods will be based mainly on their utility as anlytical tools to determine purity of complex protein samples.

Tab. 1. Methods of Detecting Impurities

Method	Property Used	Sensitivity
Electrophoretic Methods		
SDS-PAGE	Size	ng–μg
Native-PAGE	Charge, size	ng–μg
Isoelectric focusing	Isoelectric point	ng–μg
Chromatographic Methods		
Size exclusion	Size	ng–μg
Ion exchange	Charge, charge anisotropy	ng–μg
Affinity	Specific binding	ng–μg
Composition-Based Methods		
Amino acid composition	Chemical composition	mg
Peptide mapping	Primary sequence	μg
Protein sequencing	Primary sequence	μg
Activity-Based Methods		
Enzymatic activity	Catalytic activity	ng–μg
Immunoassay	Specific binding	ng–μg

2 General Strategies for Analytical Methods

The determination of purity and activity is a complex problem that depends very much upon the nature of the protein being assayed and the spectrum of potential impurities and contaminants. Impurities are usually defined as process-related substances present in raw materials which are not active material. Contaminants, on the other hand, are adventitious biological or chemical agents which have been accidentally introduced into the product (AMERICAN SOCIETY FOR TESTING AND MATERIALS, 1989; GARNICK et al., 1988).

Tab. 2 lists several typical impurities found in protein isolates and the detection methods most commonly used for their detection (adapted from ANICETTI et al., 1989). Impurities can be grouped into two main categories as either defined or undefined. Defined impurities can be assayed specifically using electrophoretic, chromatographic, or monoclonal-based immunochemical methods. Defined impurities that are intrinsic to the product are usually product variants such as products that have undergone deamidation and oxidation, aggregated forms, proteolytically modified products, and unintentionally amino-acid-substituted forms. Defined impurities can also include those that are extrinsic to the product such as components of culture media, or proteins which are used in the purification process. Undefined impurities are unknown or exceedingly complex materials for which detection is far more difficult. These include host proteins and host DNA. Host DNA, pyrogens, and microbial contaminants are of concern mainly in protein pharmaceuticals that are produced for *in vivo* use.

3 Analytical Methods

3.1 Electrophoretic Methods

Electrophoretic methods often provide a sensitive and simple approach for assessing protein purity, if not the most inexpensive. These methods can also be used to determine molecular weight, size, and subunit structure of a purified protein. Proteins carry a net charge at any pH other than their isoelectric point. Consequently, when placed in an electric field, proteins migrate at a rate propor-

Tab. 2. Typical Impurities in Protein Samples

Impurity	Detection Method
Defined Impurities	
Product variants	
Deamidation products	Isoelectric focusing
Oxidation products	Isoelectric focusing, HPLC
Amino acid substitutions	Peptide mapping, sequencing
Aggregated forms	HPLC size exclusion, SDS-PAGE
Proteolytic products	SDS-PAGE, HPLC-size exclusion
Homologous host proteins	Immunoassays
Production and processing proteins (e.g., monoclonals used for purification, etc.)	SDS-PAGE, immunoassays
Undefined Impurities	
Host cell and media proteins	SDS-PAGE, immunoassay 2D-gel electrophoresis
Endotoxins	Rabbit pyrogen, *Limulus* amebocyte lysate
DNA	Dot blot hybridization
Contaminants	
Microbes, viruses, mycoplasma	Electron microscopy, reverse transcriptase assay

tional to their charge density. Electrophoresis of proteins is rarely carried out in free solution; instead, it is usually carried out in a solution stabilized within a supporting medium (gel electrophoresis). This reduces the deleterious effects of convection and diffusion while at the same time enabling one to fix the separated proteins at their final positions soon after electrophoresis without loss of resolution.

Support media used for electrophoresis of proteins include paper, cellulose acetate, materials such as silica gel, cellulose or alumina, and gels of agarose, starch or polyacrylamide. Paper, cellulose acetate and thin-layer materials such as alumina, silica gel, serve mainly for support and to minimize convection. Thus, separation of proteins on these supports is dependent primarily on charge. By altering the porosity of agarose and polyacrylamide gels, one can introduce a molecular sieving effect into the process where separation is dependent on size and shape in addition to charge. Generally, pore size of gels are such that minimal sieving occurs. For polyacrylamide gels, however, the pore size can be of the same order as that of protein molecules, resulting in significant sieving effects.

Electrophoretic methods used for purity determination include native and SDS-polyacrylamide gel electrophoresis and isoelectric focusing. If the expected contaminant/impurity differs from the protein of interest in molecular weight, SDS gel electrophoresis is most commonly used. Native gel electrophoresis separates proteins based on both charge and size/shape. Isoelectric focusing techniques are used to separate proteins based strictly on their respective isoelectric points. Two-dimensional gel electrophoresis (in which separation is first carried out in one dimension using native gel electrophoresis or electrofocusing followed by separation in the second dimension using SDS gel electrophoresis) results in enhanced sensitivity in the determination of contaminants/impurities.

3.1.1 Polyacrylamide Gel Electrophoresis (PAGE)

Gel electrophoresis on polyacrylamide gels is the method of choice for electrophoretic analysis of protein samples (Hames, 1981; Garfin, 1990a; Laas, 1989a; Chrambach and Rodbard, 1981). Polyacrylamide gels are chemically inert, with good stability over wide ranges of pH, temperature, and ionic strength. Gels can be prepared easily from highly purified monomer reagents at reasonable cost.

Polyacrylamide (PAA) results from the polymerization of acrylamide into chains, with cross-linking of these chains by bifunctional compounds such as N,N′-methylene (bis)acrylamide. The polymerization is usually initiated by the addition of ammonium persulfate with N,N,N′,N′-tetramethylethylene diamine (TEMED) added as a catalyst. In the ammonium persulfate-TEMED system, TEMED catalyzes free radical formation from persulfate, which in turn initiates polymerization. Photopolymerization of polyacrylamide is achieved using the riboflavin-TEMED system. The effective pore size of polyacrylamide gels is dependent upon the acrylamide concentration. Increasing the monomer concentration results in decreasing the effective pore size. Also, the proportion of cross-linker (N,N′-methylene bisacrylamide) used will affect pore size and rigidity and swelling properties of the gel.

PAA gels are conveniently described by the "T" and "C" nomenclature. T is the weight percentage of total monomer and C is the proportion of cross-linker in the gel and they are given by:

$$\% T = 100(A+B)/V$$

$$\% C = 100B/(A+B)$$

where A and B are the amounts of acrylamide and cross-linker, respectively, and V is the volume of the gelling solution. Typically, values of T in the range of 10–20% and C in the range of 3–10% are utilized in electrophoretic applications where an added sieving effect of the gel is desired. When sieving is undesirable, a T value of 5% and a C value of 3% is a typical gel composition.

There are several issues that have to be considered while setting up a PAGE experiment: (1) the physical form of the gel, (2) whether to use a dissociating or non-dissociating buffer system, (3) whether to use a continuous or discontinuous buffer system, (4) the pH and ionic strength to use for the separation, and (5) the gel concentration that will be appropriate for the sample under investigation.

PAGE experiments were initially carried out in cylindrical rod gels in glass tubes. However, the vast majority of PAGE applications today involve the use of flat slab gels approximately 0.5–3 mm in thickness. Slab gels are characterized by better heat dissipation, ease of operation, and their ability to accomodate large numbers of samples. Rod gels are still used in two-dimensional electrophoresis, wherein the protein sample is separated in the first dimension in a rod gel. The rod is, in turn, attached to one edge of a slab gel for electrophoretic separation in the second dimension.

Continuous buffer systems are those in which the same buffer ions are present throughout the entire electrophoretic system at constant pH, i.e., in the sample, gel, and electrode reservoirs. In such systems, the protein sample is usually loaded directly onto the "resolving" gel where separation occurs (Fig. 1). Discontinuous (or multiphasic) buffer systems, on the other hand, employ different buffer ions in the gel compared to those in the electrode reservoirs. Differences occur in both buffer composition and pH. Also, loading of the sample is on a large-pore "stacking" gel polymerized on top of the small-pore "resolving" gel. The buffer discontinuity results in the concentration of large volume protein samples within the stacking gel. Subsequent separation in the resolving gel results in much higher resolution than that obtained with continuous buffer systems.

In its simplest form, the resolving gel and the stacking gel both contain a fast moving ion F (e.g., chloride) while the buffer in the upper electrode reservoir contains a slow moving ion S (e.g., glycine) as shown in Fig. 2. When the voltage is applied to the system, the fast moving ions leave the upper part of the stacking gel and are replaced by the slower moving ions entering the gel from the upper electrode reservoir. Since the fast moving ions migrate out at a rate faster than the one at which the slow moving ions enter the gel, a zone of "ionic vacuum" is created. The low ionic strength of this zone results in a higher field strength; thus, the slow moving ions entering this zone migrate faster. In other words, the dilution of slower ions in the zone between the slow and fast moving ions results in an increased field strength such that both the slow and fast ions reach the same migration speed. A self-regulating moving boundary is thus set up with the fast ions in the front and the slow ions in the rear. The proteins in the applied sample usually have electrophoretic mobilities between that of the slow and fast moving ions, and, therefore, are trapped between the two zones, sorted in the order of their mobilities, and enter the resolving gel as a well-defined thin zone of high protein density. Once in the resolving gel, the proteins separate due to the molecular sieving effect.

Two of the parameters that must be adjusted for optimizing the PAGE process are the buffer pH and gel porosity. If the aim of the PAGE experiment is simply to display the protein components in a given sample, standard procedures with a buffer pH of 8–9 and a

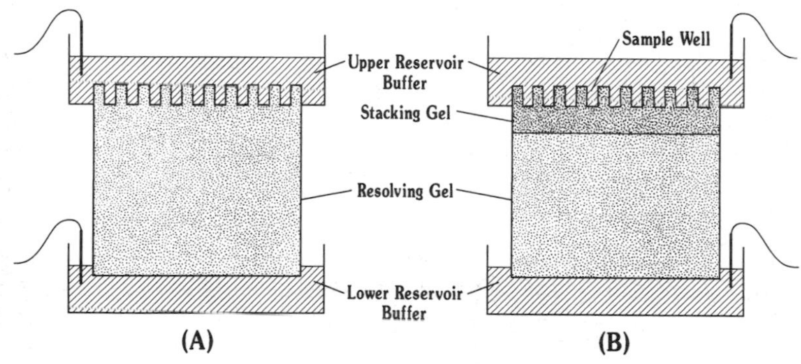

Fig. 1. Buffer systems used in polyacrylamide gel electrophoresis. (A) Continuous, (B) discontinuous.

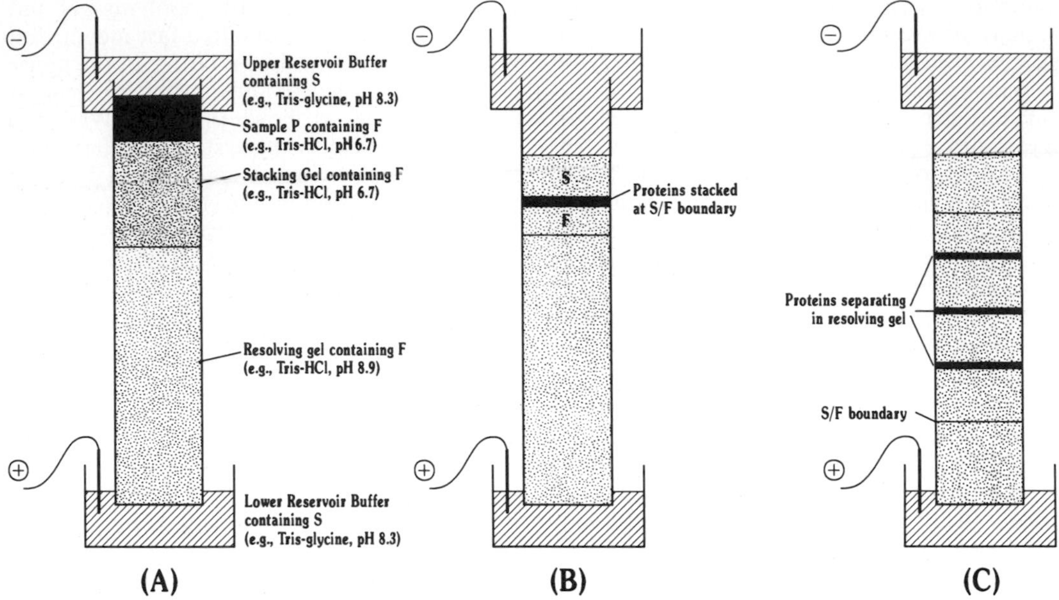

Fig. 2. Principle of the discontinuous buffer system. (A) Beginning of electrophoresis, (B) stacking, (C) separation.

gel concentration of 7–10% in the resolving gel should work fairly well for neutral and acidic proteins. For basic proteins, running the PAGE experiment with reversed polarity and a buffer pH of 4–5 should give acceptable results. On the other hand, if a separation problem is difficult and needs to be optimized, a technique called the Ferguson plot analysis can be used to establish optimal resolution conditions.

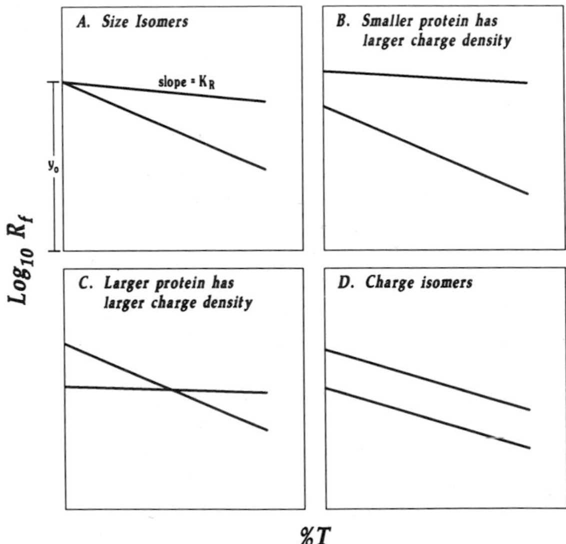

Fig. 3. Ferguson plot analysis. For details see text.

The Ferguson plot is a plot of the $\log_{10}$ relative mobility, R_f vs. gel concentration, % T. The relative mobility, R_f, refers to the mobility of the protein of interest measured with respect to a marker protein or a tracking dye. The slope of the Ferguson plot, K_R, is a measure of the retardation of the protein by the gel, and can be related to molecular size, while the ordinate intercept, Y_0, is a measure of the protein mobility in free solution. Four typical cases that are analyzed by the use of such plots have been described. Fig. 3A shows the case where the two components have identical charge densities and mobilities in free solution (size isomers). In a PAGE experiment, these proteins separate strictly according to size; thus, a high gel concentration will give the best separation. In Fig. 3B, the smaller component has the higher charge density and once again, a high gel concentration will give superior resolution. However, when the larger component has the higher charge density (Fig. 3C), it is best to run at a low or high gel concentration; typically the low gel concentrations is used because of faster separation times. In the case of charge isomers (Fig. 3D, i.e., components have the same size but different charge densities), the separation is independent of gel concentration. An IEF experiment would give much better results in such cases.

Gradient gels (gels with a continually varying acrylamide concentration) are generally better than homogeneous gels for several reasons. First, gradient gels are able to fractionate proteins over a wider range of molecular weights. Second, the protein bands obtained on gradient gels are sharper because the migration rate of any given component decreases and asymptotically reaches zero as it approaches the gel concentration whose porosity corresponds to its Stokes radius (or its "pore limit"). Casting of gradient gels is generally done by filling the gel cassette from a gradient mixer. Precast gradient gels suitable for most applications are available from several commercial sources, both for conventional vertical and automated electrophoresis equipment.

PAGE can be carried out in either dissociating or non-dissociating buffer conditions. Native PAGE is designed to fractionate protein samples under conditions that preserve subunit interaction, native protein conformation, and biological activity. The vast majority of PAGE applications, on the other hand, use a buffer system (i.e., with sodium dodecyl sulfate, SDS) designed to dissociate the proteins into their individual subunits. In the latter, the protein sample is denatured by heating at 100 °C in the presence of excess SDS and a thiol reagent such as mercaptoethanol. Under these conditions, most polypeptides bind SDS in a constant weight ratio of 1.4 g SDS per gram of polypeptide. The SDS-polypeptide complexes all have essentially identical charge densities; thus migration in SDS-PAGE applications is strictly according to size.

Native PAGE is carried out under buffer conditions that will not interfere with the native structure and biological activity of the components in the sample. The separation occurs on the basis of both molecular size and charge. There is no universal buffer system that is ideal for native PAGE. The selection is guided by the need to provide conditions that will maintain the activity of the individual components while at the same time enable good resolution. A problem with using discontinuous zone electrophoresis with native proteins is that, in some cases, the high concentration of the protein in the stacking gel may result in aggregation and precipitation of some or all of the components of the protein sample. In such cases, using a continuous buffer system under appropriate conditions (i.e., using small sample volumes at concentrations large enough to detect the protein and using buffers of ionic strengths that are lower than that of the gel or electrode reservoirs for sample preparation) can still provide adequate resolution.

SDS-PAGE is the most widely used method for the determination of the complexity and molecular weights of constituent polypeptides in a protein sample. As mentioned earlier, the separation of proteins takes place strictly according to molecular size and thus, when used with a calibrated set of standards, SDS-PAGE can be used to estimate molecular weights. The simplicity and speed of this method, along with its high resolution and low protein requirements (on the order of micro- to nanograms) are chiefly responsible for its popularity. It is also applicable to most separation or analytical problems since SDS solubilizes most proteins. The major disadvantage of SDS-PAGE is that the method destroys the biological activity of the individual proteins, which means that specific detection methods that use labeled antibodies or chromogenic substrates cannot be applied to visualization of proteins once they have been separated on the gel.

The set of conditions most commonly used for SDS-PAGE is that described by LAEMMLI (1970). The necessary components of the Laemmli SDS-PAGE system are the Tris-Cl gel buffer, Tris-glycine-SDS electrode buffer and the SDS-reducing sample buffer. SDS-PAGE can also be carried out under non-reducing conditions by omitting the thiol reagent from the sample buffer. This means that the intra- and interchain disulfide bridges remain intact. The latter technique is particularly useful in estimating the purities (and molecular weights) of oligomeric proteins. For single-chain proteins containing disulfide bonds, the presence of intrachain disulfide bridges would cause the complex to migrate faster than it would under reducing conditions, presumably because of the more extended structure brought about by reduction. When both SDS and the thiol reagent are left out of the Laemmli procedure, the protocol is the same as the classical native-PAGE system. Detailed procedures for running native- and SDS-PAGE gels are found in a number of literature sources (HAMES, 1981; GARFIN, 1990a; CHRAMBACH and RODBARD, 1981) as well as in literature supplied by those who sell gel electrophoresis equipment.

3.1.2 Isoelectric Focusing

Proteins are amphoteric molecules carrying a net charge that is dependent on the pH of their environment. The isoelectric pH value, termed p*I*, is a characteristic physicochemical property of a protein and is defined as the pH at which the molecule carries no net charge. Proteins are positively charged in solutions whose pH is below their p*I* values and negatively charged above their isoelectric points.

Isoelectric focusing (IEF) is a technique that separates proteins on the basis of differences in their p*I* values. It is generally carried out under non-denaturing conditions and is a very high resolution technique. Differences in p*I* values as small as 0.02 pH units can be quite easily resolved. Protein samples that appear homogeneous when tested by other analytical methods can often be separated into several components by IEF. This so-called microheterogeneity may be indicative of differences in primary structure, conformational isomers, differences in the type, and numbers of prosthetic groups (such as carbohydrates) or denaturation (GARFIN, 1990b; LAAS, 1989b).

IEF is performed by carrying out electrophoresis in a pH gradient set up between two electrodes, with the cathode being at a higher pH than the anode. As the protein moves through the pH gradient towards the electrode with opposite charge, its net charge and mobility will decrease as it approaches the point in the pH gradient equaling its p*I*. At this point, the protein has no net charge and will stop migrating. Thus, each protein in a complex sample will migrate to a unique point in the pH gradient depending upon its isoelectric point. Diffusion of the protein away from this point does not occur, since this would impart a charge to the molecule and the imposed electric field would drive it back to the point where its charge is zero. Thus, the protein condenses or focuses into a sharp band in the pH gradient at its individual characteristic isoelectric point (hence the term "focusing").

The key to the development of IEF into a widely used analytical technique was the establishment of pH gradients which remained stable in the presence of electric fields. This is usually accomplished with commercially available synthetic carrier ampholytes which are essentially mixtures of relatively small, multicharged amphoteric molecules with closely spaced p*I* values and high conductivity. During electrophoresis, the ampholytes separate to form smooth pH gradients. The pH interval spanned by the carrier ampholyte and the distance between the electrodes determines the slope of the pH gradient. Several varieties of carrier ampholytes are commercially available in many different pH ranges, including Ampholine (LKB), Pharmalyte (Pharmacia), Biolyte (Bio-Rad) and Servalyt (Serva). In general, these are mixed polymers, about 300–1000 Da in size, of aliphatic amino and carboxylic acids. Carrier ampholyte concentrations of about 2% are generally used.

In contrast to these "natural" gradients that develop automatically during electrophoresis, "artificial" gradients can be created in advance by immobilizing certain charged species to the gel matrix. Special derivatives of acrylamide containing amine or carboxylic acid functionalities, available under the trade name of Immobiline, are used to form these immobilized gradients. Several types of Immobiline molecules with different p*K* values are currently available. The advantages of the Immobiline gels over ampholyte gels are the lack of any gradient drift and better control of pH gradients. Precast Immobiline gels are commercially available for convenience. The disadvantages of these artificial gradients are long separation times due to hydrophobic interaction of the proteins with the Immobiline molecules and the limited stability of the Immobiline molecules.

Most IEF applications are carried out in polyacrylamide gels with large pores to eliminate any molecular sieving effects and are suitable for electrofocusing proteins up to about 500000 Da in size. Larger proteins usually require agarose gels; only non-charged agarose (i.e., which has low electroendoosmotic solvent flows) should be used. The resolution obtained with IEF gels depends upon the field strength and the slope of the pH gradient. The difference in p*I* between two adjacent protein bands is directly proportional to the square root of the ratio of the pH gradient to the voltage gradient. Thus, narrow pH ranges and high applied voltages give high resolution in IEF runs. The application of higher voltages shortens separation times; however, heat dissipation may be a problem.

3.1.3 Two-Dimensional Gel Electrophoresis

Two-dimensional gel electrophoresis is particularly useful in the analysis of complex protein mixtures and is especially important when the proteins of interest are present in low concentrations. 2-D gel electrophoresis is one of the most powerful methods available for protein mapping of cellular or sub-cellular homogenates (SINCLAIR and RICKWOOD, 1981; DUNBAR et al., 1990).

In high resolution 2-D gel electrophoresis, the proteins (after reduction) are separated in the first dimension according to their isoelectric points using IEF with carrier ampholytes. Following this separation, they are then subjected to a fractionation in the second dimension according to their molecular size using SDS-PAGE. The development and standardization of simplified equipment for reproducible and large-scale analysis and the availability of high quality reagents has made this powerful and versatile technique increasingly popular in the past decade. The use of 2-D gel electrophoresis with silver staining provides one of the best methods to estimate protein purity.

The success and reproducibility of 2-D gel electrophoresis is highly dependent on the conditions used for sample preparation. The conditions chosen must allow for the solubilization and disaggregation of the proteins in the sample without interfering with the formation of pH gradients needed for IEF. Typically, non-ionic detergents such as Triton X-100 are used in combination with a high concentration of urea and a suitable thiol reagent. The protein concentration must be high enough to allow detection of even minor components. Interfering substances such as particulates or insolubilized material, high salt concentrations, nucleic acids (removed by DNase and RNase treatment), and lipids (removed by extraction with ethanol) must be avoided.

It is common to run the IEF gel first before the SDS-PAGE, since SDS will interfere with the formation of pH gradients, and SDS-protein complexes cannot be differentiated by charge. Electrophoresis in the first dimension is carried out in a rod gel; after removal of the gel from the tube, it is introduced into the SDS-PAGE apparatus which is run in a slab gel format.

3.1.4 Analysis of Gels after Electrophoresis

The final step in any gel electrophoretic method is the visualization and quantitation of the separated protein bands (MERRIL, 1990). Except for some naturally colored proteins such as myoglobin or hemoglobin which can be visualized directly as long as their chromophores are not destroyed during the process, most proteins require organic or metal-based stains for visualization. Many organic stains, such as Coomassie Blue, Bromophenol Blue, etc. have been used for detection of protein bands. Of these, Coomassie Blue has proved to be the most sensitive. Metal-based stains such as the widely used silver stain offer high sensitivity and are over a 100-fold more sensitive than Coomassie Blue staining.

Coomassie Blue staining is used primarily in the detection of fairly abundant proteins; thus, it is not well suited for the detection of proteins present in trace quantities such as in purity determination. Coomassie Blue binds to amino groups on proteins, primarily through electrostatic and van der Waals interactions. Usually, 0.2–0.5 μg of protein (per band) can be detected and staining is quantitative up to 20 μg with an accuracy of ±10%.

Several protocols for staining gels with Coomassie Blue are available. Typically, the dye is dissolved in a water/methanol/acetic acid solution. The gel is placed in this staining solution for a certain period of time depending upon the gel thickness and concentration (usually overnight). Increasing the temperature to 40 to 50 °C increases the rate of staining. Excess stain is removed by transferring the gel to an isopropanol/water/acetic acid destaining solution and allowing the stain to diffuse out of the gel. Incorporating an anion-exchange resin in the destaining solution avoids having to constantly change the destaining solution. Electrophoretic destaining is also possible since Coomassie Blue is an anionic dye; transverse electrophoretic destaining equipment is available from several manufacturers. Other protocols which call for the use of TCA or perchloric acid as the fixing agent normally do not require destaining, but are less sensitive.

Silver staining is a highly sensitive method for visualization of trace proteins in gels; thus it is the preferred staining procedure for studies dealing with purity determinations (OAKLEY et al., 1980). Most silver staining protocols provide a linear relationship between stain density and protein concentration over a 40-fold range in concentration starting at 0.2 ng/band. As with organic stains, this relationship is protein-dependent. At protein concentrations over 2 ng/band, saturation occurs resulting in nonlinearity.

In diamine or ammoniacal silver stains, the silver ions are stabilized by the formation of diamine silver complexes with ammonium hydroxide. The gels are first washed with water/ethanol/acetic acid, followed by treatment with 10% glutaraldehyde for 30 min. After staining with the ammoniacal silver nitrate solution and washing, the image is developed using a citric acid/formaldehyde solution. The addition of citric acid lowers the concentration of ammonium ions, thereby liberating silver ions which are in turn reduced to metallic silver by formaldehyde. The staining reaction is stopped by washing with 5% acetic acid.

Non-diamine silver stains are simple and rapid. They rely on the reaction of silver nitrate with proteins under acidic conditions. Following the reaction, alkaline conditions are used to selectively reduce the silver ions to metallic silver by formaldehyde.

3.1.5 Capillary Electrophoresis

Although traditional modes of electrophoretic separations are widely practiced today, they are slow and labor-intensive, have limited quantitative capability, suffer from poor reproducibility, and are not easily automated. Capillary electrophoresis (CE) is a generic term applied to electrophoretic separations in narrow bore plastic tubes, glass capillaries, and thin films between parallel plates. In zone electrophoresis, the velocity of a migrating ion is directly proportional to the electric field. Thus, higher fields mean faster separations with higher efficiencies. One of the limits to increasing field strength is the Joule heating generated by the applied power. The result of Joule heating would be a temperature gradient from the center of the gel to the walls, leading to sample diffusion and band broadening. In CE, the use of thin-walled narrow bore glass or fused-silica capillary tube results in rapid heat dissipation, thereby permitting the separation to be run at much higher fields than traditional electrophoresis. Typical tube diameters used for CE are 50–100 μm, for which only small temperature effects have been estimated. The advantages of CE are numerous: (1) high resolution, (2) high degree of automation, (3) fast separation times, (4) on-column detection of sample via a wide variety of detectors, and (5) easy adaptation to micropreparative work. Furthermore, CE techniques can exploit numerous separation principles, giving enormous flexibility and wide applicability to a number of separation problems. Over the past two decades, CE has been used to separate a broad spectrum of molecules ranging from small organic and inorganic ions to large proteins and cells (GUZMAN et al., 1990; NIELSEN and RICKARD, 1990; KARGER et al., 1989). The major drawback of CE is the high cost of the specialized equipment needed.

3.2 Chromatographic Methods

Separation techniques which are based on the partitioning of the molecules to be separated between two phases, one stationary and the other mobile, are called chromatographic methods. Gel filtration (or size exclusion) chromatography separates proteins based on molecular size and shape and is one of the simplest methods for the detection of impurities in a protein sample. Ion exchange chromatography, which separates proteins based on net charge, can also be used to determine purity of proteins. Affinity chromatography, based on the specific interaction between a binding molecule (ligand) and the molecule of interest (ligate) is usually used when probing for a specific known impurity. Hydrophobic interaction chromatography (HIC), which involves the interaction of hydrophobic ligands with proteins, is being increasingly used as an analytical tool to determine purity. Affinity chromatography and HIC have been discussed in Chapter 24 and will not be discussed further.

The utility of chromatographic techniques in determining purity of a given protein sample has recently increased immensely as a result of high performance liquid chromatographic (HPLC) technology (CALAM, 1988). HPLC systems are characterized by high resolution and high flow rates, resulting in fast separation times. The availability of specialized HPLC equipment such as high speed pumps, sophisticated injection units and autosamplers, detectors and data-handling capabilities result in sensitivity of detection and ease of operation (i.e., a high degree of automation is available).

3.2.1 Size Exclusion Chromatography

In size exclusion chromatography (SEC), molecules are discriminated on the basis of molecular size (i.e., molecular weight and shape) by their differential permeation into matrices of controlled porosity (STELLWAGEN, 1990). The principle of SEC can be explained simply by defining two characteristic parameters of porous bed supports: the internal volume (V_i) consisting of the liquid within the pores of the beads and the external volume (given as the difference between the total volume of the bead and its internal volume) consisting of the volume between the beads (V_0). Very large molecules are, by virtue of their size, excluded from the pores of the support, and can thus equilibrate only with the external volume; such molecules elute first at the so-called void volume of the column (V_0). Very small molecules, on the other hand, totally permeate the pores of the beads and thus will elute last at the so-called total permeation volume ($V_t = V_i + V_0$). Molecules of intermediate size will elute between these two extremes with the smaller molecules eluting later than the larger ones. The separation of protein molecules in SEC occurs on the basis of size, not molecular weight. Thus, elongated proteins that have a larger hydrodynamic volume than globular proteins of the same molecular weight elute from SEC columns in a smaller elution volume.

The ideal matrix for SEC must be neutral and hydrophilic to minimize interaction between the support and the proteins. Supports used are conventional agarose, polyacrylamide and cross-linked dextran as well as combinations of these media. These traditional matrices are characterized by large bead sizes (100–250 μm), relative economy and slow flow rates. Matrices for high performance SEC are

Tab. 3. Size Exclusion Chromatography Matrices

Support	Name	Supplier	Bead Diameter (μm)	pH	Temperature (°C)
Conventional Gels					
Agarose	Sepharose	Pharmacia	40–165	4–9	40
	Biogel A	Bio-Rad	40–300	4–13	30
	Ultrogel A	IBF	60–140	3–10	36
Agarose/acrylamide	Ultrogel AcA	IBF	60–140	3–10	36
Cross-linked agarose	Sepharose-CL	Pharmacia	40–165	3–14	70
Dextran	Sephadex	Pharmacia	20–300	2–10	100
Dextran/polyacrylamide	Sephacryl	Pharmacia	25–75	2–11	100
Silica/CPG	Glycophase CPG	Pierce	30–80	<8	90
High-performance gels					
Agarose	Superose	Pharmacia	10–13	1–14	40
Vinyl polymer	Fractogel TSK	Merck	25–40	1–14	
Silica	TSK SW	Toyo Soda	10–13	3–8	45
	Shodex	Showa Denko	9	3–8	45
	Zorbax	Dupont	4–6	3–8	100
	Lichrosorb	Merck	10	3–8	100
	Synchropak	Synchrom	5–6.5	3–8	
	Protein Pak	Waters	10	3–8	90

usually based on silica, hydrophilized vinyl polymers, or highly cross-linked agarose with bead sizes between 5–50 μm. The smaller particle size enables faster and more efficient separations. Tab. 3 lists some conventional and high performance SEC matrices.

SEC is not as powerful as gel electrophoresis in the detection of impurities and size heterogeneity due to its diluting effect on the sample; consequently, it requires more material than a typical gel electrophoresis experiment. Typically, the sensitivity is of the order of micrograms when UV spectrophotometry is used to detect protein. Among chromatographic techniques, high performance SEC is the most widely used method for purity determination of protein samples.

3.2.2 Ion Exchange Chromatography

Ion exchange chromatography (IEC) is based on the interaction of proteins with a charged chromatographic support (KARLSSON et al., 1989). The interaction depends on: (1) the net charge and surface charge distribution of the protein, and (2) the ionic strength, chemical composition, pH, and the nature of ions in the solvent. The strength and the number of interactions, thus, determines the degree of retention of a protein in IEC. IEC has much better resolving power than gel electrophoresis techniques based on charge due to the fact that separation of proteins occurs on the basis of both charge density and charge anisotropy (surface charge distribution).

Ion exchangers consist of a matrix substituted with either basic or acidic groups (anion and cation exchangers, respectively). The supports used for IEC include hydrophilic supports such as agarose, dextran and cellulose, silica, and partially hydrophobic polystyrene-based and polymethacrylate-based polymers. These supports are available with a number of functional groups. Most commonly, amines such as diethylaminoethyl (DEAE) and quaternary aminoethyl (QAE) are used to functionalize anion exchangers, while carboxylic acids and sulfonates such as carboxymethyl (CM) and sulfoethyl (SE) are used with cation exchangers. Some functional groups used in ion exchangers as well as examples of conventional and high performance IEC supports are listed in Tab. 4.

Tab. 4. Ion-Exchange Chromatography Matrices

Name		Designation	p*K*
A. Functional Groups			
Anion Exchangers			
Diethyl aminoethyl		DEAE	9.0–9.5
Triethyl aminomethyl		TEAE	9.5
Triethyl aminopropyl		TEAP	
Polyethyleneimine		PEI	
Trimethyl hydroxypropyl		QA	
Quaternary aminoethyl diethyl-2-hydroxypropyl aminoethyl		QAE	
Quaternary aminomethyl		Q	
Cation Exchangers			
Methacrylate			6.5
Carboxymethyl		CM	3.5–4
Orthophosphate		P	3.0–6
Sulfonate		S	2
Sulfoethyl		SE	2
Sulfopropyl		SP	2.0–2.5
Matrix	**Name**	**Functional Groups**	**Supplier**
B. Ion Exchange Supports			
Conventional			
Agarose	Bio-gel A	DEAE, CM	Bio-Rad
Cross-linked agarose	Q-Sepharose	Q	Pharmacia
	S-Sepharose	S	Pharmacia
	Sepharose-CL 6B	DEAE, CM	Pharmacia
Dextran	Sephadex	DEAE, QAE, CM, SP	Pharmacia
Cellulose			
beaded	Sephacel	DEAE	Pharmacia
microgranular	Whatman 52, 32, 53	DEAE, CM	Whatman
fibrous	Whatman 23	DEAE, CM	Whatman
Trisacrylate polymer	Trisacryl	DEAE, SP, CM	IBF
High Performance			
Silica	Si 500, Si 300, Si 100	TEAP, CM, SP	Serva
Synthetic organic polymer	Mono Q, S	Q, S	Pharmacia
Synthetic organic polymer	PW	PET, SP	Bio-Rad
Non-porous synthetic polymer	HRLC MATP	PEI	Bio-Rad
	HRLC MATC	CM	Bio-Rad

Depending upon the functional groups employed, ion exchangers are generally classified into two categories: weak and strong ion exchangers. This terminology refers only to the state of ionization of the functional groups and has no bearing on the strength of the interaction. Weak ion-exchanging groups have variable charge density depending upon the pH of the solvent whereas strong ion-exchanging groups are permanently ionized and their charge densities are pH-independent. Weak anion exchangers are based on primary, secondary, and tertiary amines, while strong anion exchangers are based on aliphatic quaternary amines. Weak cation exchangers are based on carboxyl groups, while strong cation exchangers are based on sulfonic acid.

The choice of the ion exchanger depends on both the p*I* and charge distribution of the protein. Generally, cation exchangers are used to separate proteins at pH conditions below their p*I*, and anion exchangers are used for separations at pH conditions above the isoelectric point of the protein of interest. Elution in IEC is accomplished at constant pH by increasing mobile-phase ionic strength resulting in the displacement of the bound protein from the matrix in the order of increasing affinities for the charged support. NaCl is the most widely used displacing salt. Selectivity in IEC can be modified by changing either the pH of the mobile phase (which affects the charge density of the protein) or the type of displacing salt.

3.3 Activity-Based Methods

Activity-based methods use a specific biological property such as enzymatic activity or binding to antibodies to detect impurities in protein samples. In the case of enzymes, if the specific activity of the pure material is known, determination of unit activity can be used as a measure of relative purity. For other proteins, *in vitro* biochemical assays must be developed and performed to ensure that the product has not lost its biological activity as a result of purification. Immunoassays, based on the specific recognition by antibodies, is another activity-based technique useful in purity determination. In these assays, one probes either for the desired protein or more likely, a specific impurity. Typically, activity-based methods require two measurements: the total mass of protein in the sample and its biological activity. Relative purity is determined by comparing the measured activity to the expected amount of activity.

3.3.1 Enzyme Assays

The purity of enzymes produced for use in biotechnological applications can be monitored by the measurement of enzymatic activity. The accepted unit of enzyme activity, the International Unit (IU), is defined as the amount of enzyme required to convert 1 μmol of substrate to product in 1 min under specified conditions of pH, temperature, and buffer composition. The specific activity of an enzyme preparation is expressed as the catalytic activity per unit mass of the protein and is a convenient way of comparing the purity of enzyme preparations.

The progress of an enzymatic reaction is followed by measuring either the depletion of substrate or accumulation of product. In most cases, either the substrate or product show some measurable characteristic which is proportional to the concentration. The enzymatic activity of chymotrypsin, e.g., is determined by its action on N-*t*-Boc-L-phenylalanine *p*-nitrophenyl ester producing *p*-nitrophenol, which can be measured spectrophotometrically at 405 nm. Alkaline phosphatase which catalyzes the conversion of *p*-nitrophenylphosphate to *p*-nitrophenol is measured similarly. Fluorescence and luminescence have also been used in several enzyme assays such as *β*-galactosidase with its fluorogenic substrate fluorescein-di-*β*-D-galactopyranoside and firefly luciferase which produces light by the ATP-dependent oxidation of luciferin.

3.3.2 Immunoassays

Immunoassays are based on the specific interaction between an antigen and its antibody. The techniques of radioimmunoassay (RIA) and enzyme-linked immunosorbent assay (ELISA) are typically used in the detection of specific impurities and contaminants. They differ principally in the label used to monitor the binding of antigen to antibody – RIA uses radiolabeled proteins whereas ELISA uses antibody–enzyme conjugates.

ELISA uses chemical conjugates of enzymes and antibodies to detect the formation of antigen–antibody complexes on a solid-phase. Since most proteins bind to plastic surfaces such as the wells of a microtiter plate, this is the most widely used format for ELISA. The advantages of ELISA are that it is a versatile, robust, and easy technique to perform which utilizes stable reagents economically.

The strategies commonly used in ELISA are illustrated in Fig. 4. In direct antigen (Ag) ELISA, plates are prepared by coating with the protein (antigen) to be assayed. Antibodies (Ab) (poly- or monoclonal) that are specific to this antigen are labeled with a suitable enzyme in such a way as to preserve the biological activity of both the enzyme and the antibody. The Ag-coated plates are exposed to the labeled antibody, resulting in the formation of specific Ag-Ab complexes. After washing to

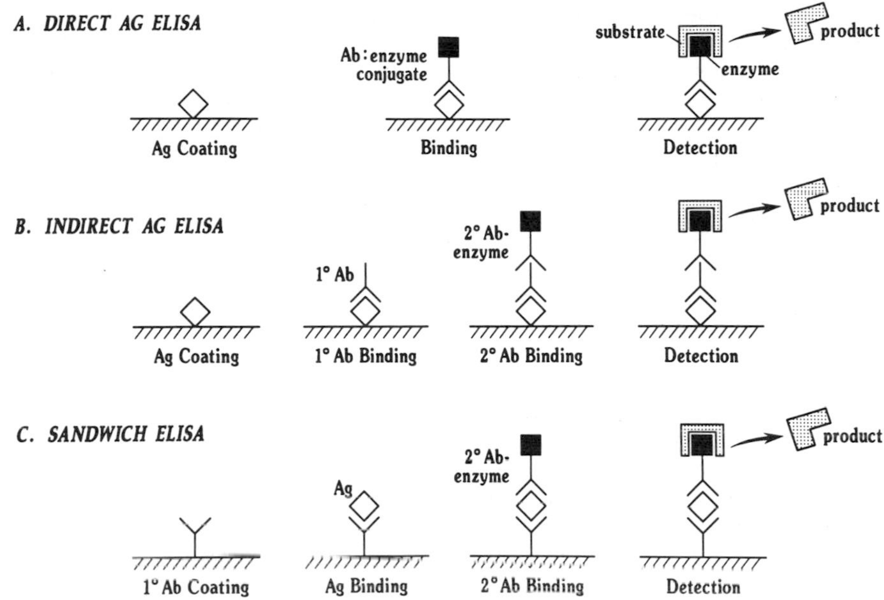

Fig. 4. Schematic diagram of enzyme immunoassay strategies. (A) Direct antigen ELISA, (B) indirect antigen ELISA, (C) sandwich ELISA.

remove excess reagents, the captured complex is treated with a suitable substrate which is acted upon by the enzyme to yield a product that can be determined spectrophotometrically. Direct Ag ELISA is generally not used except for characterization of antibody–enzyme conjugates since it entails the preparation of a specific antibody–enzyme conjugate for each application.

The more commonly used format for ELISA is the indirect Ag ELISA. Here, after the plates are coated with Ag, they are incubated with unlabeled specific antibodies (primary antibodies). The bound antibodies are then detected by using an anti-immunoglobulin–enzyme conjugate. The use of a primary antibody followed by a labeled secondary antibody enhances the sensitivity of this assay. However, the major advantage of this assay is that it utilizes "universal" anti-species immunoglobulin–enzyme conjugates, several of which are commercially available from a number of sources in a standardized and well characterized form.

The sandwich ELISA assay is used when specificity and low background characteristics are desired. The technique usually utilizes two monoclonal or polyclonal antibodies that bind specifically to the antigen but at different epitopes. The first step is the coating of the plate with the first antibody followed by incubation with the antigen. A second antibody conjugated to a suitable enzyme is then used to detect the bound antigen. This technique has enhanced specificity (due to the use of two different antibodies to bind and detect the Ag) and sensitivity (using an excess of the first antibody ensures that all of the Ag is captured).

Several enzymes have been used for the labeling of antibodies such as horseradish peroxidase (HRP), alkaline phosphatase (AP), β-galactosidase, urease, penicillinase. HRP and AP are the most commonly used detection systems. HRP catalyzes the reduction of H_2O_2 in conjunction with the oxidation of substrate, which results in a product that can be determined using optical density measurements. Some substrates that are used with HRP detection systems include *o*-phenylenediamine dihydrochloride (OPD, 492 nm), tetramethyl benzidine (TMB, 450 nm), 2,2′-azino-di(3-ethyl)-benzathiozalone sulfonic acid (ABTS, 650 nm), etc. Fluorescent and chemiluminescent assays are also available with this enzyme. AP is usually detected by the formation of *p*-nitrophenol (402–412 nm) from *p*-nitrophenyl phosphate (PNPP).

Conjugation methods generally vary with the enzyme used. Good quality conjugates are characterized by high antibody titer and avidity and specificity, high enzyme activity, and high proportion of single enzyme–single antibody conjugates. High quality conjugates are prepared by the periodate method or using heterobifunctional cross-linkers such as N-succinimidyl-3-(2-pyridyldithio)propionate (SPDP) or succinimidyl-N-4-carboxycyclohexylmethyl maleinimide. Conjugation of enzymes with antibodies can also be performed using glutaraldehyde as the cross-linking agent, but these reagents might be somewhat inferior to the former.

Immunoassays are particularly useful for the detection and/or quantitation of low levels of protein impurities introduced by the host cells (or production organism). Host cell proteins represent a large number of potential impurities. It is estimated, e.g., based on DNA content, that an *E. coli* cell can produce up to 3000–4000 different proteins. Furthermore, in highly purified products, such host cell impurities will be present only in fairly low levels. Assays for *E. coli* proteins and Chinese hamster ovary (CHO) proteins have been developed that can measure such impurities at the ppm level. The development of such multi-antigen immunoassays for host cell impurities requires a standard preparation of such impurities to serve as immunogen for the preparation of the polyclonal antibodies to be used in the assay. Typically, this is accomplished by the "blank run" approach in which a host cell line lacking the gene for the product of interest is used for a full-scale production run, to generate the reference impurity mixture (Jones, 1988). The assumptions involved in this approach are that the absence of product does not significantly affect the population of in-process impurities at either the expression stage or the purification step. Based on 2-D gel electrophoretic analysis, it is estimated that such a reference impurity mixture can contain up to 100–200 proteins. The usefulness of the assay is limited by the response elicited in an immunized animal by the least immunogenic component of this mixture. Thus, the antisera used in these assays must be rigorously characterized to ensure that antibodies have been produced to all the minor components of the reference mixture.

3.3.3 Total Protein Assays

Determination of the total protein content of a sample is an integral part of product characterization. Techniques such as quantitative amino acid analysis, Kjeldahl nitrogen assay, Folin–Lowry assay, dye binding assays using Coomassie Blue (Bradford assay), etc. have been developed for protein determination (Darbre, 1986). Routinely, however, the UV spectrophotometric technique, using a well defined extinction coefficient, is used with high purity products.

Quantitative amino acid analysis has often been used to determine protein content, especially in conjunction with extinction coefficient determination. Following hydrolysis, the amounts of well recovered amino acids such as alanine, leucine, and glycine are determined. Based on a knowledge of the amount of such amino acids expected per unit of pure protein, the amount of protein in the sample can then be calculated.

The Kjeldahl method is the one of the best methods for determining protein content and provides the most consistent results. It involves the quantitative decomposition of the protein to ammonium sulfate, carbon dioxide, and water with sulfuric acid and added catalysts, followed by the direct determination of ammonia. Typically, ammonia (and thus the nitrogen content) is determined by trapping in standard acid and back titration. The protein content can then be calculated using a conversion factor based on the nitrogen content of the pure protein.

The Folin–Lowry assay and its numerous modifications are probably the most commonly used methods for protein determination. They are based on the biuret reaction of proteins with copper(II) and the reduction of the Folin–Ciocalteu reagent (phosphomolybdic-phosphotungstic acid) to heteropolymolybdenum blue, which is monitored spectrophotometrically at 600 nm.

The determination of protein content by UV spectrophotometry is based on the absorption of proteins at 280 nm, which can be attributed to the tyrosine, tryptophan, and phenylalanine residues. The use of an accurately determined extinction coefficient is absolutely essential for the routine determination of protein content by this method.

3.4 Composition-Based Methods

Proteins are composed of amino acids, and the nature and composition of the various constituent amino acids determine the characteristic properties of a protein. The amino acid sequence of a protein (i.e., the order in which the constituent amino acids are arranged) determines its structure and function. Composition-based methods provide molar qualification of amino acids and specific prosthetic groups to assess the purity of the protein sample. Amino acid analysis is the most important of these methods and provides an important quantitative parameter in the characterization of purified proteins. Complete protein sequencing as well as limited N- and C-terminal sequence analysis are important in the initial characterization of a protein and can be utilized routinely in downstream processing, e.g., to assess fidelity of expression systems, to check for unintentional mutations, etc. An easier approach to routinely check for mutations such as single-amino acid substitutions and/or protein "clips" (i.e., shortened versions of the product protein) is through the use of peptide mapping.

3.4.1 Amino Acid Analysis

The first step in the determination of amino acid composition of proteins is releasing the amino acids quantitatively without concomitant degradation. The most common hydrolysis scheme utilizes 6 N HCl for 20–24 h at 110 °C under vacuum (Ozols, 1990). Several amino acids undergo modifications under these conditions. Asparagine and glutamine will be hydrolyzed quantitatively to aspartic and glutamic acid. Destruction of serine and threonine increases linearly with time with typical losses being around 5–10%. Cysteine cannot be determined directly from acid hydrolyzed samples and is usually either carboxymethylated or derivatized with 4-vinyl pyridine prior to hydrolysis. Tryptophan is not usually recovered in acid hydrolysates. Halogenation of tyrosine and oxidation of methionine are other side reactions that can occur during this step and are prevented by the inclusion of 0.1% (w/v) phenol and 2-mercaptoethanol, respectively (Ozols, 1990). The highly polar nature of the amino acids and the absence of strong UV absorbing chromophores in the majority of amino acids presents many problems in their separation and detection. Several methods can be used to separate and quantify amino acids in protein hydrolysates. Chromatographic techniques such as paper chromatography, thin-layer chromatography (TLC), gas chromatography and liquid chromatography (LC) using ion-exchange and reverse-phase columns have all been used for this purpose. The bulk of amino acid analysis these days is performed using high performance liquid chromatography (HPLC) systems which provide excellent speed and resolution in the separation of the constituent amino acids. Two alternative approaches have been used in HPLC analysis of protein hydrolysates: ion exchange separation of the free amino acids, followed by postcolumn derivatization with fluorescamine or precolumn derivatization with phenylisothiocyanate (PITC), dansyl and dabsyl chloride, etc. followed by separation on reverse-phase systems.

3.4.2 Protein Sequencing

Automated Edman degradation analysis (Edman sequencing) can provide valuable information about the primary structure and homogeneity of high purity proteins. The method is based on the coupling of the N-terminal amino group of the protein with phenylisothiocyanate (PITC), followed by the cleavage of the resulting phenylthiocarbonyl (PTC)-amino acid from the protein in the presence of trifluoroacetic acid (TFA). The cleaved PTC-amino acid is then converted, under acidic conditions, to a phenylthiohydantoin (PTH) derivative. The cleavage reaction also exposes the succeeding amino acid for subsequent cycles of coupling and cleavage. The PTH derivatives are then routinely assayed via thin layer chromatography (TLC) or more commonly via HPLC. The coupling of automated amino acid sequencers to reliable HPLC systems has greatly facilitated routine sequencing analysis of proteins.

The extent of sequence information that can be obtained from a single Edman degradation analysis of a large protein is limited due to the carry-over of amino acid residues between successive cycles as a result of incomplete coupling and/or cleavage, and the cleavage of internal peptide linkages during the TFA treatment that can provide new starting points for degradation. While it is now possible to obtain sequence information up to 60 or 70 residues in a single run, it is more common to first fragment the protein into smaller peptides and then perform sequencing analysis on these peptides. Also, through the use of sensitive, radioactive techniques wherein a radioisotope label is introduced either biosynthetically (intrinsic labeling) or extrinsically through a radiolabeled sequencing reagent, sequencing can be performed with as little as 1 pmol of protein.

C-terminal sequencing involves a sequential degradation procedure based on reaction with ammonium thiocyanate or the use of carboxypeptidases. More commonly, however, the C-terminal peptide is first isolated following protease or cyanogen bromide (CNBr) digestion, and then subjected to Edman degradation analysis.

3.4.3 Peptide Mapping

The principle of peptide mapping is to produce several small peptides from large proteins via peptide bond cleavage and then to compare the properties of these peptide mixtures. The fragmenting of the protein is usually achieved by using various specific enzymes to cleave the protein at specific sites. Commonly used proteases include the V8 protease, which cleaves after glutamic acid residues, and trypsin, which cleaves after lysine and arginine residues. The resultant peptides can then be analyzed by a number of techniques, including two-dimensional thin layer chromatography, gel electrophoresis, and more commonly, reverse-phase HPLC to produce a peptide map or "fingerprint" (HANCOCK et al., 1988).

Peptide mapping is one of the most powerful techniques for checking the primary structure of a protein. Peptide mapping, e.g., has been used to identify single amino acid changes in recombinant human growth hormone as well as recombinant human tissue-type plasminogen activator (KOHR et al., 1982; ANICETTI et al., 1989). However, problems regarding purity of the enzymes used, stability of the fragmented peptides and adequate resolution of these peptides as well as their structure and composition of all of the significant peptides must be worked out before this method can be used for evaluating consistency of primary structure on a routine basis.

3.5 Glycoproteins

Many proteins contain covalently linked oligosaccharides and are termed glycoproteins. The sugars commonly found include neutral sugars (D-galactose, D-mannose, etc.), the amino sugars (N-acetyl glucosamine, N-acetyl galactosamine) and the acidic sugars (sialic acid). Unlike the polypeptide chain of a glycoprotein which is synthesized exactly based on the information contained in the gene, the carbohydrates are not primary gene products, but are added on posttranslationally by enzymes termed glycosyltransferases. This leads to a mixture of products showing variability in carbohydrate content and primary structure, called microheterogeneity. This variability creates problems in establishing criteria for the assessment of purity and homogeneity of such molecules. In addition, enzymatic and/or chemical degradation of the carbohydrate portions of the molecule can occur during isolation, which further complicates the issue.

Glycoproteins are best evaluated on the basis of their carbohydrate attachment points and the consistency of their composition. The most commonly used procedure for determining carbohydrate composition involves the complete hydrolysis of the sugars from the protein. The carbohydrate component of the hydrolysate is analyzed by the following techniques: (1) capillary gas chromatography of alditol peracetate derivatives for the determination of the neutral sugar content, (2) amino acid analysis for the amino sugars, and (3) HPLC for the acidic sugars. The determination of the sites of attachment of the carbohydrate moiety is made as follows: (1) digestion of the protein by trypsin and/or other proteases, (2) isolation of the carbohydrate containing peptide fragments via affinity chromatography on a lectin affinity column, and (3) Edman degradation analysis of the peptides. Complete characterization of the carbohydrate units is an extremely complicated procedure and is not required for routine analysis of glycoproteins.

4 Summary

The determination of purity and activity of a biological product is a complex problem that depends very much upon the nature of the protein being assayed and the spectrum of potential impurities and contaminants in the product. The methods discussed in this chapter have been classified into the following categories: (1) electrophoretic methods such as native and SDS-PAGE and isoelectric focusing, (2) chromatographic methods such as size exclusion and ion exchange chromatography including HPLC modes, (3) activity-based methods such as enzyme assays and immunoassays, and (4) composition-based methods such as amino acid analysis, protein sequencing and peptide mapping. It should be clear from the variety of methods and spectrum of impurities discussed here that there are no universally applicable exact protocols for purity analysis. Determination of purity of a protein preparation generally involves an assessment of the quantity of particular types of impurities rather than a direct quantification of purity. Thus, the level of purity is clearly dependent on the particular assay used and its sensitivity. Electrophoretic methods are the most inexpensive, sensitive, and simple techniques for assessing protein purity. Separations can be carried out based on molecular size alone as in SDS-PAGE, or based on charge alone as in IEF, or based on both size and charge as in native PAGE. Two-dimensional gel electrophoresis is particularly useful in the analysis of complex protein mixtures. Sensitivity is fairly high, especially when used in conjunction with highly sensitive silver stains, and impurities can be detected at concentrations as low as 50 ppm. There are very few pitfalls associated with gel electrophoresis and thus, it is the method of first choice for assessing purity. The most commonly encountered one is the comigration of an impurity with the product band resulting in a false-negative. Sometimes, artifacts arise in IEF as a result of interaction of the protein molecules with the ampholytes used to establish the pH gradient.

Chromatographic methods are becoming increasingly popular as rapid and non-destructive methods of protein analysis. Separations can be carried out based on molecular size (size exclusion chromatography), charge (ion exchange), hydrophobicity (hydrophobic interaction or reverse-phase) and biological activity (affinity). Size exclusion chromatography cannot provide the same resolution as the corresponding electrophoretic method, but can still provide valuable information not obtainable from electrophoretic analysis. For example, if the sample contains oligomeric forms of the protein of interest, SDS-PAGE will not be able to provide this information. On the other hand, size exclusion chromatography, since it is carried out under native conditions, will show the sample to be heterogeneous and is probably the method of choice for detecting oligomeric impurities. Ion exchange chromatography has much better resolving power than gel electrophoresis techniques based on charge due to the fact that separation of proteins occurs on the basis of both charge density and charge anisotropy (surface charge distribution). Reverse-phase chromatography has become indispensable for amino acid analysis as well as peptide mapping, while analytical affinity chromatography provides the most direct method of assaying specific known impurities. The sensitivity, resolution, and separation times of traditional chromatographic techniques have been greatly improved by the increasing popularity of HPLC systems. The use of fluorescence and radioactivity detectors has improved the sensitivity of HPLC techniques from the microgram scale (for UV absorption) to even the picogram scale.

Activity-based assays suffer from the drawback that they provide very little information regarding the nature of impurities present in a given sample. Enzymatic assays arc only useful for products that have a well characterized, easily quantifiable catalytic activity. Immunoassays, on the other hand, are of wide applicability and are the most sensitive means to look for host cell protein impurities in a product (down to a few ppm). Immunoassays can also be designed to probe for specific impurities and/or contaminants in cases where the presence of such impurities is expected. In addition, *in vitro* biochemical assays must be developed and performed to ensure that the product has not lost its biological activity as a result of purification.

Composition-based methods include amino acid analysis, peptide mapping and protein sequencing. The main objective of such methods is to verify the primary structure of the protein of interest. Amino acid analysis is not a particularly sensitive method for this purpose; it is incapable of even detecting a 10% contamination of recombinant human tissue-type plasminogen activator with bovine serum albumin. It provides little information other than overall composition (within broad limits) and an amino acid profile. Peptide mapping is one of the most powerful techniques for ensuring correctness of primary structure; it is capable of detecting even single amino acid changes in the sequence of a protein but needs considerable workup and development before it can be used as a reliable tool for assessing primary structure homogeneity. Protein sequencing is the ultimate technique for assessing primary structure but is fairly involved and may require a considerable investment in automated equipment for routine analysis.

It is important to remember that no single assay can provide all the required information regarding the purity of a given protein product. The assays used are most effective if their respective advantages and weaknesses are complementary. For example, a 2-D silver stain gel electrophoresis analysis to detect host cell impurities must be used in conjunction with a multi-antigen immunoassay. In the 2-D gel electrophoresis experiment, some impurities may be concealed due to comigration of the bands, whereas in the immunoassay, impurities that failed to invoke an antibody response will remain undetected. Thus, using both methods along with a rigorous process validation program, will certainly enhance the quality of the purity determination.

5 References

AMERICAN SOCIETY FOR TESTING AND MATERIALS (1989), Standard guide for determination of purity, impurities, and contaminants in biological drug products (GARNICK, R. L., Chairman), *Annual Book of ASTM Standards,* Vol. 11.04, Designation E 1298-89, pp. 898-900, ASTM.

ANICETTI, V. R., KEYT, B. A., HANCOCK, W. S. (1989), Analysis of protein pharmaceuticals produced by recombinant DNA technology, *Trends Biotechnol.* 7, 342-349.

CALAM, D. H. (1988), HPLC and biotechnological products, in: *Biotechnologically Derived Medical Agents: The Scientific Basis of their Regulations* (GUERIGUIAN, J. L., FATTARUSSO, V., POGGIOLINI, D., Eds.), pp. 23-28, New York: Raven Press.

CHRAMBACH, A., RODBARD, D. (1981), Quantitative and preparative PAGE, in: *Gel Electrophoresis of Proteins: A Practical Approach* (HAMES, B. D., RICKWOOD, D., Eds.), pp. 93-143, Oxford: IRL Press.

DARBRE, A. (1986), Analytical methods, in: *Practical Protein Chemistry - A Handbook* (DARBRE, A., Ed.), pp. 227-337, New York: John Wiley.

DUNBAR, B. S., KIMURA, H., TIMMONS, T. M. (1990), Protein analysis using high-resolution two-dimensional polyacrylamide gel electrophoresis, *Methods Enzymol.* **182**, 441-458.

GARFIN, D. E. (1990a), One-dimensional gel electrophoresis, *Methods Enzymol.* **182,** 425-441.

GARFIN, D. E. (1990b), Isoelectric focusing, *Methods Enzymol.* **182,** 459-477.

GARNICK, R. L., ROSS, M. J., DUMEE, C. P. (1988), Analysis of recombinant biologicals, in: *Encyclopedia of Pharmaceutical Technology* (SWARBRICK, J., BOYLAN, J. C., Eds.) Vol. 1, pp. 253-313, New York: Marcel Dekker.

GUZMANN, N. A., HERNANDEZ, L., TERABE, S. (1990), High-resolution nanotechnique for separation, characterization, and quantitation of micro- and macromolecules, in: *Analytical Biotechnology: Capillary Electrophoresis and Chromato-*

graphy (HORVATH, C., NIKELLY, J. F., Eds.), pp. 1–36, Washington, DC: ACS.

HAMES, B. D. (1981), An introduction to polyacrylamide gel electrophoresis, in: *Gel Electrophoresis of Proteins: A Practical Approach* (HAMES, B. D., RICKWOOD, D., Eds.), pp. 1–91, Oxford: IRL Press.

HANCOCK, W. S., CONOVA-DAVIS, E., BATTERSBY, J., CHLOUPEK, R. (1988), The use and limitations of reverse phase HPLC for the analysis of recombinant proteins and their tryptic digests, in: *Biotechnologically Derived Medical Agents: The Scientific Basis of their Regulations* (GUERIGUIAN, J. L., FATTARUSSO, V., POGGIOLINI, D., Eds.), pp. 29–51. New York: Raven Press.

JONES, A. J. S. (1988), Process validation in the biotechnology industry, in: *The Impact of Chemistry on Biotechnology* (PHILLIPS, M., SHOEMAKER, S. P., MIDDLEKAUF, R. D., OTTENBRITE, R. M., Eds.), pp. 193–203, New York; ACS.

KARGER, B. L., COHEN, A. S., GUTTMAN, A. (1989), High-performance capillary electrophoresis in the biological sciences, *J. Chromatogr.* **492**, 585–614.

KARLSSON, E., RYDEN, L., BREWER, J. (1989), Ion-exchange chromatography, in: *Protein Purification: Principles, High Resolution Methods and Applications* (JANSON, J.-C., RYDEN, L., Eds.), pp. 107–148, New York: VCH Publishers.

KOHR, W. J., KECK, R., HARKINS, R. N. (1982), Characterization of intact and trypsin-digested biosynthetic human growth hormone by HPLC, *Anal. Biochem.* **122**, 348–359.

LAAS, T. (1989a) Electrophoresis in gels, in: *Protein Purification: Principles, High Resolution Methods and Applications* (JANSON, J.-C., RYDEN, L., Eds.), pp. 349–375, New York: VCH Publishers.

LAAS, T. (1989b) Isoelectric focusing in gels, in: *Protein Purification: Principles, High Resolution Methods and Applications* (JANSON, J.-C., RYDEN, L., Eds.), pp. 375–403, New York: VCH Publishers.

LAEMMLI, U. K. (1970), Cleavage of structural proteins during the assembly of the head of bacteriophage T4, *Nature* **227**, 680–685.

MERRIL, C. R. (1990), Gel staining techniques, *Methods Enzymol.* **182**, 477–487.

NIELSEN, R. G., RICKARD, E. C. (1990), Applications of capillary zone electrophoresis to quality control, in: *Analytical Biotechnology: Capillary Electrophoresis and Chromatography* (HORVATH, C., NIKELLY, J. F., Eds.), pp. 37–49. Washington, DC: ACS.

OAKLEY, B. R., KIRSCH, D. R., MORRIS, N. R. (1980), A simplified ultrasensitive method for detecting proteins in polyacrylamide gels, *Anal. Biochem.* **105**, 361–363.

OZOLS, J. (1990), Amino acid analysis, *Methods Enzymol.* **182**, 587–601.

RHODES, D. G., LAUE, T. M. (1990), Determination of purity, *Methods Enzymol.* **182**, 555–565.

SINCLAIR, J., RICKWOOD, D. (1981), Two-dimensional gel electrophoresis, in: *Gel Electrophoresis of Proteins: A Practical Approach* (HAMES, B. D., RICKWOOD, D., Eds.), pp. 189–218. Oxford: IRL Press.

STELLWAGEN, E. (1990), Gel filtration, *Methods Enzymol.* **182**, 317–328.

28 Biotechnology Facility Design and Process Validation

MICHAEL G. BEATRICE

Bethesda, MD 20892, USA

1 The Master Validation Plan

The qualification and validation of systems and equipment in support of biotechnology manufacturing should be designed into a facility prior to completion of the building plan and before any equipment order is placed or system is configured. Process validation cannot be initiated until the facility has been built and qualified and the necessary utilities are validated (CATTANEO, 1988).

A validated facility and process is one which meets specifications which have been empirically established based upon the design features of the facility or equipment and the specific product under investigation. When the proper sequence of validation studies is performed, productivity can actually increase. The concept of sequential validation studies has been well established. This chapter will define the proper sequence for validation of the facility and the process.

In terms of the biotech facility itself, validation protocols should be based upon the proposed use of all manufacturing, testing, and storage areas. This is of critical importance when the facility is intended for production of multiple biotechnology products (i. e., multi-use facility) or where several different manufacturing steps are intended to be conducted at the same time in the same area. Other scenarios include the manufacture of pilot lots and clinical materials in areas utilized or intended for commercial production, facilities in which both drugs and biologicals are produced simultaneously, and facilities where multiple products are made in the same areas on a campaigned basis. Product campaigning involves manufacturing of different products within the same equipment following a validated cleaning procedure. The benefits of a well designed validation program will be presented in general and related to their utility in a multi-use facility. This will be accomplished by taking a facility from initial design concept through construction, equipment qualification and validation and process validation. As our hypothetical facility develops, the proper framework for a Master Validation Plan (MVP) will be described. This plan can be utilized for project planning, economic and budgeting concerns, and to respond to questions raised by the regulatory authorities during facility review and inspection.

2 Benefits of Validation

When a facility and process has been adequately validated, benefits other than increased productivity become apparent. In addition, maintenance costs will be reduced for both systems and equipment as a result of working within well defined performance limits. These performance limits will have been initially established based upon the expected performance which may be estimated during the installation qualification (IQ) phase, further refined during the operational qualification (OQ) phase and "fine-tuned" during the process qualification (PQ) phase of the validation study. This scenario constitutes the basis for the most common validation sequence which has been popularized with the acronym IQ-OQ-PQ.

Reduced maintenance costs translate into reduced downtime for systems and equipment, since all operations will be maintained within established limits which are achievable and reproducible. In addition, if the operations have been defined and are maintained within reasonable limits, utility costs will be reduced and the process will be optimized.

3 Biologicals versus Pure Chemical Drugs

Prior to embarking on a discussion of the proper validation sequence, it is important to concentrate on the unique characteristics of the products that will be manufactured within these defined biotechnology systems. Most biotechnology products are derived from biological systems which are complex and inconsistent. The production sequence may include one or more of the following steps: Isolation and identification (if applicable), propagation of an organism or virus, growth on an expanded scale, cell disruption and/or product recovery, purification, concentration, formulation, filtration, and aseptic packaging.

Major problems can be encountered following the cell growth phase with regard to separation of the major biological component(s) from nutrients or contaminating proteins. If continuous cell culture techniques are utilized, the additional problem of the continuous elimination of contaminated matter requires initial validation, constant monitoring, and ongoing cyclic validation studies. When the complex growth, purification, and formulation process is completed, the resultant final mixture is often complex, and the clinically active biological component is frequently a very small proportion of the final product.

The above translates into a product which, in its final formulation, may be difficult or impossible to assay and quantitate on a consistent basis. In addition, direct correlation of a biological assay with the product's biological and clinical activity may be non-achievable. Therefore, if the manufacturing process has not been adequately validated, both quantifiable and non-quantifiable product characteristics may change with each lot or batch produced. The approval pathway for a product with this level of variability may be difficult to impossible.

Products of biotechnology may also be subject to inactivation or loss of activity as a result of chemical or physical factors encountered during a long and complicated manufacturing process. Factors such as exposure to heat and shear can lead to an irreversible decrease in activity due to loss of configuration or other reasons.

In addition, the residence time for the product in nutrient-rich media and the numerous transfers of additives during manufacturing increase the probability of multi-source contamination. The effects of contamination may not be detectable until several downstream steps have been completed and may challenge the capability of the purification steps beyond their validated capacities.

One should contrast the above concerns with production of a pure chemical compound from a defined active drug substance derived by chemical synthesis. Once the impurity profile of the drug substance has been established and sensitive and specific assay procedures developed, lot to lot consistency can be relatively easily quantified and maintained.

3.1 Scale-Up

The above concerns related to biotechnology products define the rationale for the differences in regulation between drugs and biological products. A biological product will not travel an easy path to approval, unless each critical manufacturing step has been defined and is conducted within an acceptable range. In addition, the manufacturer must demonstrate that the same manufacturing steps are maintained within the same limits for all subsequent production runs regardless of scale.

Scale-up is often mistakenly considered to consist only of a change of equipment volume. However, scale-up frequently requires equipment meeting differing design criteria, since reaction vessels and associated equipment tend to be limited. Therefore, it may be less costly and more productive to utilize a larger number of vessels of the same design rather than increasing to a single larger-scale reaction vessel of totally new design which will require re-validation of the manufacturing process. Increasing the volume of a vessel does not mean that elements such as composition of gasses, flow of nutrients, and speed of mixing can be increased by a corresponding quantity.

4 Initiating the Validation Program

The validation program begins with the development of acceptance criteria for each validation protocol. This section will describe important considerations which must be addressed prior to initiating any validation studies.

The validation program should begin at the time of initial facility design. Functional room placement should be based upon the proposed flow of personnel, components and product. This flow should be rational and proceed in a unidirectional manner from "dirty" to "clean" operations. The distinction between the two will be explained later in this chapter. In order to adequately evaluate the flow, it is necessary to utilize accurate and specific facility drawings with extensive detail for items, such as room finishes, air flow, pressure differentials, slopes for water systems, major equipment placement, product and personnel flow. These drawings should include specifications and tolerances where applicable. Possession of this type of detailed drawings will provide answers to many questions which arise during FDA (Food and Drug Administration) inspections and will also assist the manufacturer if and when facility renovation or equipment replacement is contemplated.

Drawings should proceed in a stepwise fashion after the purpose of the facility has been clearly defined. Considerations should also include potential problems such as power failures, alarm placement, emergency lighting and power systems which should be included in the initial design to eliminate the problems associated with retrofitting. For example, equipment not meeting exact initial design criteria may protrude into a controlled environment area and disrupt airflow by generation of turbulence and provide a "shelf" for collection of particulate contaminants. In order to ascertain that the design is appropriate, a theoretical simulation should be planned before product is introduced into the facility.

5 Production Variables and Good Manufacturing Practice

The thought process involved in designing a complete validation program for a biotech facility and process, requires understanding of the broad and variable nature of manufacturing processes for biological products.

An examination of the various product classes which constitute the category of biological products would document a range from the relatively simple techniques of crude extraction to the rigid controls required for cell culture or monoclonal antibody production. Some manufacturing pathways utilize a combination of simple and complex techniques to yield the desired biological drug substance. When this occurs, relationships between the processes utilized and their expected outcomes are not always clear.

One frequent misjudgment that is observed during review of a product and establishment license application or initial pre-license inspection is the failure to take production variables and their interactive effects into account when designing and validating the facility and process. This translates into the need to involve all departments in the validation planning from design to implementation.

An example of poor planning may assist in illustrating this point. One manufacturer proposed performing both fermentation and cell culture operations in separate areas of the same facility. The cell culture operation utilized complex media, long batch cycles and relatively low yields as compared to the fermentation process. The bioreactor vessels were of similar basic design for both processes.

An examination of the master validation plan which had been prepared by an outside contractor indicated that the two systems were to be validated by the same procedure. This was not an acceptable approach for the regulatory authorities. Concerns related to contamination control over the long batch cycles, method of harvesting, temperature, pH, shear and cell attachment need to be addressed in the cell culture operation.

The pre-license inspection of the facility will also focus on the degree of compliance with Good Manufacturing Practice Regulations (GMPRs) which are in Title 21 of the Code of Federal Regulations (21 CFR, 1992). The facility design should include adequate space to perform each production or testing operation and proper flow to alleviate mix-ups and cross-contamination. Other than economic concerns, facility design is dependent upon two major factors; the logistics related to manufacture of a specific product and whether the facility will be utilized for multiple products.

There is a distinction between Good Manufacturing Practice Regulations (GMPRs) and good manufacturing practice. The former encompass a set of regulations which apply to "finished pharmaceuticals" but are also pertinent to manufacturing from early steps in the process. Regulatory citations based upon these regulations are usually specific. The latter concept of good manufacturing practice includes the current procedures which have been accepted by the pharmaceutical industry and originate from various sources including national associations and societies. A manufacturing procedure of process can be held to industry standards which are not specified in the limited and specific framework of the current good manufacturing practice (cGMP) regulations. However, the GMPRs are broad enough to permit a general observation to encompass the details which are a part of good manufacturing practice.

6 Multiple-Use Facilities

A multiple-use facility can be defined as any facility in which more than one product is manufactured at the same or different time periods. This includes, but is not limited to the following examples;

- Production of multiple products at similar stages of manufacture in a common area
- Manufacture of a single product (multiple lots) at various stages of production within a common area
- The use of common areas or equipment for potentially infectious vs. inactivated products
- Use of a facility for manufacture of different products on a compaign basis

The most desirable facility in terms of ease of approval is obviously dedicated to the manufacture of a single product. This is not practical in the evolving biotechnology arena, since such a limitation would be economically unfeasible. The further one moves from this ideal situation, the more questions will be asked. It should be noted that in all examples stated below multiple use includes biological as well as drug products which are produced in the same facility.

The next best scenario involves the use of a facility for manufacture of different products on a campaign basis at different times – spatially separated using dedicated equipment each time. This involves appropriate labeling, accountability and storage of equipment as well as the development and documentation of appropriate cleaning validation studies. Cleaning validation will be discussed later in this chapter.

One step below this facility type is a campaigned facility in which different products are made in separate equipment at the same time in different production trains. In addition to the above concerns, one now must include documentation of well trained personnel to prevent mix-ups between production trains. It is also important to label all equipment, transfer lines, auxiliary tanks and glassware, etc., to insure dedication as well as to prevent mix-ups between production trains.

The process can become more complex and generate many questions for regulators when the same equipment is utilized for different products either on a campaign basis or in the same production rooms in parallel trains. This will require description of the specific procedures utilized to prevent product mix-ups, a clear definition of process controls and specific and sensitive cleaning validation which is designed to detect residual product and cleaning agents in addition to a general test such as residual protein.

6.1 Concerns for Multiple-Use Facilities

When planning a multiple-use facility, it should be kept in mind that a major concern is the introduction of contaminants that are difficult to detect. This is especially true if virus production will be performed in the facility. In addition, cross-contamination between products should be considered. Although rare in manufacturing operations with dedicated equipment, this problem becomes magnified when the same equipment is utilized for multiple products.

As the manufacturing processes for multiple-products interface, the question of workers alternating between production trains becomes more difficult to answer. It may be possible to provide an SOP for employee activities in the submission to regulatory authorities, however, this written document will not insure that workers are adequately trained and maintain operations within limited boundaries. In addition to training, the use of color-coded uniforms for employees working on specific production operations should be considered. This provides a visual alert, if an employee with the wrong uniform is seen handling equipment on another production line.

It has also been observed that multi-use facilities tend to take on varied production operations without full knowledge by senior management or departments such as production. This is due to the fact that parts of the organization may be involved in the production of research materials or clinical batches for themselves or on a contractual basis which are not a part of normal production scheduling. This may permit materials which are not the subject of the cleaning validation to be manufactured in production equipment.

6.2 Validation of Multiple-Use Facilities

Therefore, the validation of multi-use facilities starts with the manufacturer having full knowledge of what products will be produced in the facility, including research and clinical materials. All possible sources of contamination should be anticipated by thorough examination of the bioburden potential of product, excipients, and environment. The full range of possible adventitious agents should also be studied and validation efforts should focus on "worse case" conditions.

Equipment utilized to produce any materials at any scale must be validated to assure removal of residual product, microbial or viral contamination. The methods of analysis utilized in support of this validation effort should be specific and sensitive rather than general procedures designed to detect gross contaminants. Since the detection of potential viral contaminants is difficult to impossible, challenge validation studies may need to be performed.

7 Open versus Closed Systems

In order to address the concerns of microbial, adventitious or product to product contamination, firms usually maintain that all manufacturing operations will be conducted within "closed" systems.

In many situations, the manufacturers claim to be working within closed systems regardless of the number of actual breaks that are utilized during production. For example, one manufacturer claimed to utilize a closed system during all manufacturing steps. This alleged closed system process extended from inoculation of the seed reaction vessel up to and including final harvesting and inactivation. The company did not utilize any environmental controls because they maintained that the system was "closed" from start to finish.

An examination of the batch records combined with actually observing the manufacturing demonstrated that there were no less than 85 separate points of input and output in this "closed system". These included additive ports, sampling, and gas inlets. Upon examination of the procedures for sampling and addition of added substances, it was obvious that reliance was placed upon the microbiological technique of addition to open ports following flaming. This was not regarded as acceptable in light of the fact that the firm had had numerous failures due to contamination events. It should be noted that this information was not provided in the license application but had to be specifically requested upon inspection.

7.1 Concerns Related to Closed Systems

The following concerns emerge during the review and evaluation of a "closed system":

- The integrity of the system at all points where product addition, removal, or sampling may occur
- The manner and location of input and output from the system
- The manner of performing aseptic transfer operations
- Use of common vs. dedicated equipment in the process
- Cleaning validation for removal of residuals and contaminants

The above information should be presented in summary form on the application submitted to regulatory authorities for approval. In addition, detailed specifications and qualification data should be available to the investigator during facility inspections. More detail will be given relative to cleaning validation later in this chapter.

8 Presenting Plans to the Regulators

After consideration of all the previously discussed factors, the final facility design should be starting to come into focus. This is the time to present detailed drawings of the facility to the regulatory authorities, including engineering details, equipment placement, and flows of components personnel and product. Selection of the proper regulatory authority depends upon whether the product will be drug or biological produced by biotechnology.

If the product is considered to be a drug subject to approval under section 505 of the Federal Food, Drug and Cosmetic Act, the FDA field investigators will be responsible for conducting the inspection. A large part of their inspectional program centers on validation. Therefore, any discussion of facility design with the FDA field offices should include a well organized master validation plan (MVP) which includes facility, equipment, systems, and product. The formulation of an acceptable MVP will be emphasized in subsequent sections ot this chapter.

For manufacturing a biological product, an establishment license application will need to be submitted concurrently with the product license application. The reviewers and investigators for such products are from the FDA's Center for Biologics Evaluation and Research.

9 The Biologics Regulatory Process

Biological product regulations in 21 CFR Part 600 provide for the issuance of an establishment license to an applicant that has demonstrated the capability to manufacture a safe, pure, and potent biological product within the facility under consideration. No establishment license may be issued unless a product license is issued simultaneously with the establishment license, and the product is available for examination at the facility intended for licensure [21 CFR Part 601.10 (b)].

Demonstration of adequate control over the manufacturing of a biological product from start to finish has, in the past, been documented by the manufacturer (i. e., legal entity engaged in actual manufacture of the product) performing *all* steps in the production of a product within facilities under their direct supervision and control.

Even though the mechanism for regulation of biological products differs from that of other drug and device products, the basic philosophy of providing safe and effective products manufactured in establishments, which meet both Current Good Manufacturing Practice Regulations (21 CFR Part 211) and Biological Establishment Standards (21 CFR Part 600, Subpart B) Regulations, is all encompassing. Early in this century biological product development was regulated mainly by the requirement that manufacturing could only be performed in licensed and inspected laboratories under the strict control of the legal entity (licensee). This type of control seemed to contribute to a dramatic decrease of the death rates for pertussis, measles, and diphtheria due to better regulation of the manufacturing process.

From their historical inception, biological regulations have defined "manufacture" to mean all steps in the preparation of the product including, but not limited to, propagation, filling, testing, labeling, packaging, and storage by the manufacturer [21 CFR, 600.3(u)]. Further, "manufacturer" has been defined as the legal entity or person engaged in the manufacture of a licensable biological [21 CFR 600.3(t)]. These regulatory definitions formed the basis of a policy decision that required the legal entity (the legal preparer of the product which is intended to be licensed) to perform all steps in the "manufacture" of the biological in a licensed facility(s) under their direct supervision and control.

The above licensing policy interpretation is one example of a major difference between regulation of drugs and biologicals under the NDA and licensing systems, respectively. In order to receive a product and establishment license under the previous interpretation of licensing requirements, the sponsor of a biological product application must also be the owner or establish supervision and control over all facilities where manufacturing and testing operations for the biological product are performed. Some exceptions have been permitted for use of contract facilities to perform certain *limited* operations usually occurring after the preparation of a final purified concentrate (e. g., aseptic filling, lyophilization, labeling *or* limited product testing) provided that the licensee demonstrates adequate supervision and control over the contract operation.

In addition, regulatory policy interpretation has required new filings for both product and establishment license applications for a legal entity that moves manufacturing operations to another country or location where the local law requires that a separate legal entity be established. For example, if a large multinational company based in the U.S. holding a currently approved license for a biological product and establishment desires to move manufacturing operations to Ireland, a complete new filing for an establishment and product license would be requested since the Irish government would require that a distinct legal entity be established in Ireland. As a part of the new filing, the company may be asked to perform limited clinical testing in addition to complete product characterization.

The rationale behind this restrictive interpretation of the regulations can best be illustrated by reemphasizing the major difference in the analytical character of drugs and biologics.

Drugs, other than biological products, can for the most part be purified, characterized and identified by accurate and *reproducible* methodology. Traditional biologicals, on the other hand, are not uniform mixtures of easily characterized materials. The regulations, as originally interpreted, required strict control by the legal entity over all facilities utilized for manufacturing and testing the biological product at all stages in the process. From this point of view, it is easy to reason why the random "farming out" of manufacturing steps in the preparation of biological products has not been acceptable under licensure.

The above legal policy interpretation was predicated by the fact that traditional biological products are not easily characterized. However, recent advances in biotechnology have permitted more flexibility in interpretation of the licensing regulations, as products of biotechnology become more like traditional drugs in their indications for use as well as their ability to be easily characterized using more sensitive and specific methodology. The FDA Policy Statement Concerning Cooperative Manufacturing Arrangements for Licensed Biologics (*Federal Register*, 1992) decribes innovative arrangements among establishments who wish to cooperate in the manufacture of a licensed biological product. It should be noted that this flexibility may not be appropriate for products which best fit the traditional definition of a biological, (i. e., vaccines, allergenic extracts), or which exhibit difficulty or inconsistency in propagation, manufacturing or analysis.

Recognizing the need to develop a coordinated system for the regulation of biotechnology that would include all Federal regulatory agencies, the Office of Science and Technology Policy published the "Proposal for a Coordinated Framework for Regulation of Biotechnology" in a 1984 Federal Register publication (*Federal Register*, 1984). This document concluded that there is no need to extensively modify the existing regulations to accommodate the new technology products. However, innovative new approaches to deal more specifically with this technology explosion should be considered by regulatory agencies. This is the basis for the limited flexibility now afforded to the biotechnology industry through policy statements, guidance documents and points to consider.

10 Planning for Facility Expansion

Once all of the above design criteria and regulatory requirements have been addressed, a master validation plan should be structured to examine the critical elements of the design to determine that the systems and equipment meet and support anticipated production goals. This is also the time to build future expansion into the plan. A facility that is designed for a limited production capacity will not be assured of survival through the regulatory pathway and associated challenges from the marketplace. A "flexible facility" will be designed to include more production trains in separate suites with separate air and utility systems so that if contamination occurs in one system, other systems can be run without adverse effect. This requires careful planning and consideration of the economics involved with redundant design.

If the pre-designed system meets present and future production needs, one has to consider whether the system can be initially validated. This is best done by estimating production capacities and making a determination of what systems, equipment, and facility validation studies will be needed. This initial design should be incorporated into the pre-validation section of the MVP. The preceding considerations should be followed by a listing of the routine monitoring that will be needed on the multitude of systems and equipment. Possible system failures should also be considered, and appropriate back-up design should be provided in the design criteria.

11 Construction Validation Phase

Now that we have the basic design completed, and the pre-validation studies and anticipated monitoring program is described in the MVP, the contractor is selected, and the oversight phase, monitoring of construction operations, should begin. It should be kept in mind that traditional construction practices were not designed with the regulatory requirements of the pharmaceutical industry in mind.

The next section of the master validation plan should, therefore, describe a construction monitoring and validation program which ensures that all of the pre-determined design specifications will be included in the "as-built" facility and all equipment, including hard-piped transfer and water lines, distillation apparatus, autoclaves, dryheat tunnels, etc., is received in proper working order and meets all of the established pre-design specifications.

11.1 Equipment

For major equipment (i. e., distillation apparatus, autoclaves, fermentation vessels, etc.) there should be evidence of a pre-shipment inspection, and the purchaser should be aware of the method of conveyance to the plant site. This will provide assurance that the equipment meets specifications, will be packed and handled properly during shipment and will arrive at the correct time according to construction schedules. The major equipment is to be scheduled to arrive so that the packaging, post-shipment condition and specifications can be immediately verified. Damages or errors should be corrected immediately to ensure timely repair or replacement, if necessary.

Other, non-major components should also not be overlooked. For example, the stainless steel piping for the water systems and hard-piped transfer lines should be checked against acceptance criteria and condition upon arrival. To assure proper control it is recommended that components which will come into contact with the product at any stage in the manufacturing be quarantined upon receipt by written procedure and not released until accepted. Upon release for installation, there should be monitoring of the method of handling and installation.

11.2 Inspecting and Documentation of Ongoing Construction

As the construction is ongoing, constant inspections are warranted to verify that appropriate technique is utilized to avoid hidden damages to concealed systems and equipment and insure that engineering specifications are followed. The continuous inspection procedures should monitor the installation of all systems and insure that documentation is available for all critical specifications which will be difficult to physically check when all the walls are in place. Examples of this type of inspection include examination of the running slopes for water systems, condition and location of ductwork, placement of HVAC (heating, ventilation and air conditioning) fire cutoff dampers and type of welding of stainless steel piping.

Critical procedures such as welding of stainless steel piping in water for injection loop systems should be verified in several ways. The expertise of the welders should be pre-determined by requiring them to perform sample welds for inspection utilizing a "cold welding" (i. e., argon) procedure. Documentation in this instance may be comprised of retention samples of these "practice welds" cataloged by the welder along with photographs of actual welds and a summary of the background experience of the welders. Similar verification should be made of the installation expertise for other critical systems such as HVAC and clean steam systems.

At the completion of construction, the MVP should describe a program for assurance that compliance has been achieved with all prevalidation design specifications. This task will be facilitated if the measures described for verification during construction are well documented. For example, HVAC system interconnections can be verified by inspection along with slopes for water system loops, assurance of freedom from dead legs in water system use points, the quality and integrity of medicinal gasses as delivered to their points of use, and appropriate "fits" for equipment and room finishes.

11.3 Room Finishes

Room finishes should be appropriate to the processes and conditions expected. When examining finishes as a part of the post-construction inspection, the inspector qualifies the room as opposed to validating the room. The distinction is important, and the specifications examined during this procedure should be outlined in the master validation plan.

The inspection of room finishes should take into account the craftsmanship to determine if the "fits" are satisfactory. Any variances to expected specifications should be noted and corrected both physically and on the engineering diagrams, if appropriate. For example, if airwalls are used the finish around the intake vents as well as the sealing of the wallboard should be tight to prevent leakage. All grills should be flush and tight with spacers of sufficient size and design to permit cleaning. Room sealings in aseptic areas should utilize a caulk that does not promote growth, and nothing should protrude into aseptic areas to act as a particulate shelf such as HEPA (high efficiency particulate air) filters, frames, light fixtures, autoclaves, dry heat tunnels, etc.

The ceiling in an environmentally controlled area should also not permit escape of controlled air. Inspections have revealed leakage occurring at recessed light fixtures and ceiling grids in clean-room suspended ceilings. This is only evident when observed from above since, at full pressurization, a small degree of leakage is not sufficient to alter continuous monitoring data.

In the poorly designed facility, emergency lighting may be an afterthought. If this is the case, lighting and wiring that must be retrofitted could alter the integrity of room finishes. At the very least, exposed lighting and wiring could constitute a platform for the accumulation of particulates and disrupt what might otherwise be laminar flow within a controlled environment.

The construction section of the MVP should also consider the integrity and operating condition of items such as the following:

- Airlocks and interlocking mechanisms
- Fire safety systems including the positioning of firestats and location of dampers
- Alarm systems for fire, temperature control and HVAC failure

After all of the above specifications have been verified as a part of the MVP, the utility of all previously designed SOPs (standard operating procedures) should be reexamined. If additions, deletions, or other changes must be made, they should proceed through an approval mechanism which has been previously established and include the date that such changes were initiated and finalized.

12 Utilities

Since, as previously mentioned, a facility and its manufacturing processes cannot be validated until all of the utilities that are needed to support the operations are validated, the next section will describe some of the major utilities that are commonplace in a biotech facility.

In the following description of several major utility systems, the general acceptance criteria have been derived from several sources including the Good Manufacturing Practice Regulations For Large Volume Parenterals (LVP), hereafter described as the LVP document (*Federal Register*, 1976). Although these regulations were never finalized and are not currently in effect, they provide guidance that was adopted as good manufacturing practice for the pharmaceutical industry.

The following listing was excerpted from CATTANEO (1988) and includes an outline of the utility system and the general acceptance criteria which should be part of routine installation and operational qualification and validation procedures:

12.1 Plant Steam

- Reliability – the ability to deliver steam at the appropriate pressure to each point of use
- Chemical analysis does not show the presence of hydrazines or amines
- Process plant steam pressure should achieve the pressure specified for the corresponding point of use
- Plant steam should have limited, defined uses separate from clean steam

12.2 USP Purified Water Made by De-ionized (DI) Systems

- Reliability under all operating conditions
- Meet USP requirements for purified water (US PHARMACOPEIA, 1990)
- Meet conductivity and resistivity specifications
- Low level of pyrogens
- Low microorganism count (<50 colonies/100 mL)
- No *Pseudomonas cepacia* present
- Specified water pressure maintained at each drop

12.3 Water for Injection (WFI) Systems

- Reliability (can the system routinely deliver WFI meeting USP specifications?)
- Non-pyrogenic
- Meet conductivity and resistivity specifications
- Meet microbiology specifications (NMT 10 CFU/100 mL)
- Loop and storage system maintains a temperature of 80 °C ± 5 °C
- Water pressure, as specified, must be maintained at each drop and the end of the loop
- Rate of flow N.T.E. 5 fps

12.4 Clean Steam System

- Reliability (can the system routinely deliver clean steam at the appropriate pressure?)
- Non-pyrogenic
- Meets conductivity and resistivity specs
- Low micro-level less than 10 CFU/100 mL
- Insulated piping system to assure delivery of clean steam to point of use
- Steam pressure at each point of use should achieve specifications as previously determined

12.5 Nitrogen Systems

- Reliability of supply to operation
- Terminal filtration through 0.2 micron filter renders it sterile, as confirmed by sterility testing
- Purity - documentation of no foreign contamination

12.6 Compressed Air and Other Medical Gases

- Reliability of the system (delivers to point of use)
- Reliability of supply
- Oil-free (no detectable hydrocarbons in excess of 1 ppm are permitted)
- Dry
- Terminally sterilized through a 0.2 micron filter renders it sterile confirmed by sterility testing
- Purity - documentation of no foreign contamination

12.7 Heating, Ventilation and Air Conditioning (HVAC) System

This system has an impact on many other plant operations and will be discussed in greater detail later in the chapter.

Principal areas of system control are as follows:

- Temperature
- Relative humidity
- Number of air changes
- Environmental integrity (viable and non-viable particulates)
- Differential pressure
- Air flow pattern (is it adequate for room design?)
- Controls and alarms including computer control validation

12.8 Frequency of Monitoring for Utilities

It is recommended that the above critical elements of acceptance criteria should be measured at least daily for a minimum of 90 days consecutive use during operational validation procedures. All systems having multiple use points (e. g., air, water, and steam) should include testing of different use points each day provided that the entire system is monitored on at least a weekly basis. For water and steam systems, the daily monitoring should include sampling from the beginning and end of the system as well as at selected use points to determine if adequate control can be maintained. This initial validation period should be a minimum of 90 days unless arrangements for a shorter interval have been made in advance with regulatory authorities. During this period, there should be no unexplained failures in the system or at individual points of use. If failures are observed, information should be provided to describe corrective actions, follow-up test results and impact on the total system. In addition, a detailed discussion on the possibility of such a failure occurring during production should be prepared and available for review by the investigators.

12.9 Other Utilities

The above list is not exhaustive. All systems in a facility should be considered in relation to their interaction with each other. Examples of some other commonly employed systems include

- Clean-in-place
- Steam-in-place
- Vacuum and dust collection systems

Two systems, the heating, ventilation and air conditioning (HVAC) and water systems, are of particular importance and will be examined in greater detail. It is not possible to describe one design that exactly fits all facilities; the biotech industry is so diverse that each proposed system must be considered in terms of its intended function and use of the facility.

After having set the ground work for the facility design, the first of the two systems that will be discussed in detail, the HVAC system, shall be examined.

13 The HVAC System

Because HVAC systems are expensive at both the initial construction and daily operation stages, the system should be designed with flexibility to fit present and future needs. The HVAC system can account for 20–50% of initial construction costs according to recent literature on this topic. In addition, the space required to accommodate such systems ranges from 20 to 60% of the total floor area within a facility (KELTER, 1988).

The basic goal of an HVAC system is to bring fresh air into a facility, mix it with recirculated air (where practical), and then condition the mixture through filtration, heating, cooling, humidification, or dehumidification. The system also supplies air to work areas and then recovers it for partial or total exhaustion into the atmosphere (KELTER, 1988). Special considerations for biotech facilities include adequate airflow, contamination control, and appropriate validation.

Early critical design criteria to "engineer" into the HVAC plans for a new or retrofit facility includes the following listing (KELTER, 1988).

13.1 Adherance to Codes

- HVAC systems must meet lifetime safety code requirements established by local and state authorities
- FDA GMP requirements regarding documentation and scheduling of all maintenance must be designed into the MVP
- The environmental impact of exhausted and re-circulated air in research and production areas must be described and documented

13.2 Manufacturing Environment

- The design must incorporate all relevant data associated with the personnel, equipment, and processes to be served and should be flexible enough to accommodate addition of future products and technologies
- Written criteria should address contamination control, airflow volume and direction, particulate control, and the dilution and exhaust of toxic and explosive gases
- The initial bioburden should be defined as a baseline to be utilized as a part of the required investigation of future contamination incidents

13.3 System Flexibility

- Should be capable of accommodating research, contract and animal facilities
- In the future, biological regulations may be eased to permit virus and spore formers to be manufactured in "separate" designated areas or closed systems within the same facility if validation documentation is adequate

13.4 Design Reliability

- In order to determine the appropriate level of reliability needed in HVAC design, the cost of an interruption in the process should be determined
- For short-term drug discovery testing, the situation may be less critical than that needed in long-term safety assessment testing in later stages of research and in commercial production scheduling

- The "cost" of reliability is determined by the price of addition of redundant systems and equipment

13.5 Economics

- An economic model should be established and should include unit energy and operating costs as well as owner-specific financial considerations
- Using these data, a system designer can design a system that is both technically and economically appropriate

13.6 Airflow and Contamination in HVAC Systems

Airborne contaminant circulation must be minimized by relative pressurization between "clean" and "dirty" areas. In this regard, "clean" refers to areas where the product or components that must have limited bioburden are exposed or have the potential to be exposed to the environment. "Dirty" refers to areas in which equipment and residual components that have been used in the manufacturing process and are designated to be cleaned or discarded are located. An area or passageway can become "dirty" at the time of transit of used equipment and components.

The first consideration in the determination of which HVAC system will be suitable for a facility should be an examination of the physical location of the facility. An example: the inspection of a large manufacturer of parenteral products which was located in a rural setting surrounded by two farms. A review of environmental monitoring data indicated that contamination problems were cyclical. Upon further examination, it became evident that these contaminations (higher than normal environmental monitoring results combined with isolation of microorganisms which were not part of the baseline bioburden for the facility) coincided with the planting and harvesting seasons.

Location is also important in the relationship of fresh air inlets to external exhaust air ducts. In addition, the average seasonal temperature of the area where the facility is located is an important factor. The preceding governs the restrictions that one will place on the volume of air supplied from the outside, high filtration standards, and economics of the design. These design considerations directly oppose the criteria for most comfort-conditioning systems, which can establish airflow quantities solely on the basis of expected cooling loads (KELTER, 1988).

Filters in an HVAC system offer varying degrees of resistance to airflow, often as a function of their cleanliness. A significant correlation was observed between the condition of filters and the increase in established bioburden levels. When filters become obstructed, the fans within the system must be able to account for this variance and maintain relatively constant system airflow and pressure.

Upon inspection of a manufacturer that had numerous contamination results it became apparent that each incident was directly related to failure to conduct regular maintenance on filters and unbalanced airflow. This knowledge should have provided sufficient justification to open a dialogue with the maintenance personnel and to schedule regular meetings to discuss system design and status as well as to define each person's role in the upkeep of the facility. When questions regarding maintenance schedules or procedures arise, FDA investigators frequently interview the responsible maintenance personnel.

13.7 Environmental Control

FDA requires that a manufacturer establish and be able to demonstrate environmental control in critical areas of a parenteral manufacturing facility. If a manufacturer elects to design a facility with little or no environmental controls in areas where biotechnology products intended for parenteral administration *at any stage of manufacturing* either come into contact or *have the potential to come into contact with the environment*, a significant amount of validation data will be requested. The traditional response encountered is that all processing is performed within "closed" systems. Validation concerns for such systems have already been addressed.

Documentation that should be described in the MVP includes recording of all qualification and validation data and demonstration that a defined control level can be achieved and maintained. In addition, revalidation criteria should be established and followed. All manufacturing personnel should be aware of the specifications and limitations of the facility.

13.8 Distinct HVAC Systems

HVAC system designers have developed two distinct systems, each having their own advantages and disadvantages. They are the single-pass and recirculating systems.

The Single-Pass System

The single-pass or once-through system (100% outside air) is used most frequently in research and manufacturing applications having the greatest risks of cross-contamination (work with pathogens, spores, viruses, etc.). In these systems, air is taken from the outside environment, conditioned, distributed through the facility and exhausted. A constant level of relative space pressurization must be maintained (KELTER, 1988).

These systems usually require the use of reheating coils in individual zones or spaces to maintain space temperatures under all load conditions. Energy costs are typically high due to simultaneous heating and cooling requirements.

Because of these high costs, firms attempt to recover heat from system exhausts. If this is the situation, there may be some back-flow and mixing of air which is not intended for recirculation. This can be easily verified upon inspection of the engineering diagrams combined with a visual examination of all exposed ducts, filter banks, mixing chambers, and plenums.

The Recirculating System

In recirculating systems, exhaust air (some amount) is directed into a ducted return system where it usually is combined with air streams from other locations, drawn through various dust collectors and filtering systems and recirculated.

In this instance, only air that is loaded with moisture, solvents, or other contaminants is discharged directly into the atmosphere (or treated first, then discharged). This contaminated air may be replaced with fresh air to maintain proper ventilation and space pressurization. There is typically a reheat coil to maintain a constant air temperature.

Recirculating systems are generally preferred to once-through systems when no hazards or employee discomforts (smells) are likely from low levels of materials that may be present in recirculated air. Installation and operating costs can be significantly lower. However, from a containment standpoint, they are the most difficult to control. The costs of adequate control can outweigh the equipment and operating costs.

Further, as the outside air mixes with recirculated air, the recirculating system operates over a narrower temperature range and is thus more stable than the once-through system. It should be noted that many permutations and combinations of the above two systems may be observed within a single facility.

13.9 HVAC – System Control Considerations

Technical advances in airflow measurement stations and constant volume controllers allow today's HVAC systems to achieve new levels in space pressurization control. Utilization of microprocessor-based, direct digital controls rather than pneumatic control systems is becoming commonplace.

In the interest of economy, consideration is given to shutting down some HVAC systems when not in use. For example, exhaust hoods may only be needed in limited circumstances when weighing or other situations where dust generation or aerosolization may be problematic. Under previously designed systems, exhaust hoods usually remained running constantly to maintain the air balance. Today's more sophisticated control systems permit shut-down of individual components without adverse effect on system balance.

Maintenance Considerations

The costs of today's complex systems extend far beyond initial construction. Costs associated with maintenance should be considered a part of the systems control function. Automatic control systems require regular and vigilant maintenance. For example, periodic recalibration of sensors and readjustment of controllers.

A preventative maintenance program is, therefore, necessary to prevent system failure. GMPR inspectional provisions include examination of a maintenance log and the need to document regularly scheduled calibrations to insure that all systems and equipment operate within established specifications. Control system redundancy should be provided by having spare microprocessor or pneumatic components ready for action at a moment's notice. This will enhance the reliability of the control system.

Maintenance Training

Untrained maintenance staffs typically expect automatic control systems to operate without oversight. If this is allowed to happen, small problems could easily develop into large ones that could put a halt to research or production. This is another reason to include maintenance personnel in the design phase for the HVAC system (as well as total facility and system design).

14 Production Water Systems

The choice of a water system which is appropriate to meet current and anticipated future needs is a major facility design issue which should be considered only after careful review of the intended use of the facility and the regulatory requirements imposed on products of biotechnology. If professional literature is consulted, the issue becomes more confusing, since every water system vendor claims to have a system which is inexpensive and capable of delivering the best quality water for any pharmaceutical use.

Water is an essential component of parenteral manufacturing and probably accounts for the largest component volume utilized in biological manufacturing operations. Products of biotechnology require, from both a good manufacturing and regulatory viewpoint, strict control of all manufacturing steps including propagation of the source materials, manipulations of the crude bulk, purification and all steps through final filling and packaging. The introduction of contaminants at any stage, including bulk steps, such as fermentation and separation of the desired product from cellular debris, could compromise the final product in any of several manners.

Another major difference between traditional pharmaceutical products and biologicals is that there is no point in biological manufacturing where strict in-process controls are not required to be in place. A licensable biological product is defined to include the "virus, serum, toxin, blood, or blood derivative, allergenic product or *analogous product.*" Since even manufacturers who prepare the bulk for sale to another legal entity "For Further Manufacturing Use Only" are also required to be licensed, the term "active drug substance" as applied to traditional pharmaceuticals, is meaningless when applied to biologicals. Therefore, the traditional industry practice of utilizing high-quality water (such as Water for Injection) only for the final manufacturing steps utilizing the active drug substance is not directly applicable to biological product manufacturing.

In this section water systems from the basic definitions to their practical implementation in biotech facilities will be discussed. Since the majority of biotech products currently approved and anticipated to be approved in the future are parenteral products, the descriptions to follow will lead to a system designed to produce Water for Injection (WFI) which is recommended to be utilized from early production steps to minimize added bioburden to the final product. The quality of water utilized at each manufacturing step should be clearly described in the product license application.

14.1 Sources of Water

The source of incoming water to a manufacturing location is often limited depending upon availability in different areas. Many facilities have access to municipal water systems, while others must utilize sources such as well water. Municipal water systems often chemically treat the water (e. g., with chlorine, fluorine, settling agents), while well water carries with it the potential for contamination with heavy metals or other toxic substances.

The major distinction between well and municipal water is the fact that municipal water is often treated to meet potable water standards while well water may need extra treatment steps to meet the same initial standard. This assumption, however, should not be automatic. Inspection of numerous facilities worldwide has demonstrated that problems exist even in municipal water systems. Therefore, the manufacturer should have access to an incoming water sampling use point to test the water quality prior to pretreatment. One test that can serve as a useful point of reference is the level of chlorine at the initial plant sampling point which could be compared with the values from the municipal water source. This should be done regardless of whether or not the supplier of the water performs periodic testing and reports the results to the users. Water sampled prior to pretreatment should conform to the potable water specifications of the U.S. Public Health Service.

It should be noted that several water classification systems exist which add to the confusion regarding the quality of available water sources. For example, well water may also be referred to as "Level I" water which includes water from such sources as rivers and lakes and has been traditionally associated with utility use.

14.2 Pretreatment of Water

It is important to insure that the general usage system is protected by an appropriate backflow prevention device located within the production facility and tested routinely on a regular basis by a licensed testing company (Bjurnstrom and Coleman, 1987). The location of such a device and the details of the incoming water system should be diagramed on a schematic drawing which shows the exact location of all sampling ports as well as the slope of the pipe at each critical section.

The municipal or private water system should be constructed such that the potable water system utilized for general use and utilities is separate from the production water system and no cross-connections exist. This should be clearly evident on schematic diagrams which are available to FDA inspectors.

The next step for production water is to pass through a series of filters designed to reduce chlorine, chemicals, and bioburden prior to entering the various processing stages of the system. Activated carbon filters are microporous filters that function in reduction of chlorine and other chemicals. Prior to selection of an appropriate microporous filter, the specifications should be examined if they can withstand steam sterilization and allow adequate residence time for efficient performance.

Softening may be necessary as the next pretreatment step. The use of a zeolite softener is desirable in instances where on-site regeneration is permitted. The local codes should be consulted, since the on-site regeneration using zeolite softeners utilizes a heavy brine solution which has to be rejected to sewage (Bjurnstrom and Coleman, 1987).

14.3 Water Treatment Steps

14.3.1 Reverse Osmosis (RO)

The next logical component of a well designed system is the RO unit. The specifications for the unit selected should be checked with regard to the compatibility of the membranes with chemicals such as antiscalants which are utilized in some crude pretreatment steps.

It is important to realize that, since RO systems are cold and because RO filters are not absolute, microbiological contamination is not unusual (Avallone, 1989). In addition, it is usually recommended that, since RO filters are not absolute, two be operated in series. Even though the US Pharmacopeia indicates that WFI may be produced both by distillation and reverse osmosis (US Pharmacopeia, 1990), RO systems have been problematic when used as the final means for preparing WFI, since they provide excellent membrane sites for microbiological contamination which is difficult to monitor and control. The separation of clean, pyrogen-free water from contaminated feed water is via a thin membrane which may rupture or, at the very least, develop small leaks which could compromise the clean side.

When used as a method of pre-treatment, the user should review the currently available types of membranes and determine their compatibility with the entire purification system. Available membrane materials include hollow-fiber polyamides, cellulose acetate, and various composites. The use of RO as feed water for a distillation unit is common, however, the semipermeable RO membrane rejects highly charged salt ions to a greater extent than weakly ionized monovalent ions. Therefore, a deionization step may be employed prior to entering the distillation unit.

14.3.2 Deionization (DI)

Deionization should be viewed as one component of a total water system. When DI units follow the RO system, they can effectively remove residual ionic components. They do not remove organic or bacteriological contaminants. In fact, they can serve as excellent breeding areas for microorganisms. It was due to this concern that the FDA issued a "Letter to the Pharmaceutical Industry-RE: Validation and Control of Deionized Water Systems" (FDA, 1981). Therein, it was pointed out that the microbial population tends to increase with the length of time between DI service periods. The letter also pointed out the following:

> "Other factors which influence microbial growth include flow rates, temperature, surface area of the resin beds and, of course, the microbial quality of the feed water. Therefore, these factors should be considered in assessing the suitability of deionizing systems where microbial integrity of the product incorporating the purified water is significant. From this assessment a firm should be able to design a suitable routine water monitoring program and a program of other controls as necessary. It would be inappropriate for a firm to assess and monitor the suitability of the deionizer by relying solely upon the representations of the equipment manufacturer. Specifically, product quality could be compromised if a firm had a deionizer serviced at intervals based not on validation studies, but rather on the "recharge" indicator built into the unit. Unfortunately, such indicators are not triggered by microbial population. Typically they are triggered by measures of electrical conductivity or resistance. If a unit is infrequently used, sufficient time could elapse between recharging/sanitizing to allow the microbial population to increase significantly."

14.4 Water System Monitoring

This emphasizes the importance of a validation program for installation which also monitors continuous use. Considerations that should be given the system prior to installation include such factors as the microbial quality of the feed water (and residual chlorine levels where applicable), surface area of the ion exchange resin beds, temperature range of water during processing, operational range of flow rates, recirculation systems to minimize intermittent use and low flow, frequency of use, quality of regenerant chemicals, and frequency and method of sanitization.

Routine monitoring should include monitoring intervals for water quality and conductivity measurements which are based on results of validation studies, measurement of conditions and quality at significant stages through the deionizer (influent, post cation, post anion, post mixed-bed, etc.), microbial conditions of the bed, and specific methods for microbial testing (FDA, 1981).

Analytical methods should be sufficiently sensitive and reliable to detect those microorganisms that are objectionable in view of the product's intended use. Whatever analytical technique is selected, should be validated by established test validation methodology to be sufficiently sensitive and reliable (FDA, 1981).

14.5 Use of DI Water

Deionized water can be used to generate plant steam for boilers, since it minimizes the accumulation of scale on the boiler surfaces. It can, however, be corrosive to certain metallic surfaces such as cast iron. It is also a possible source of water for clean steam generators, if the sequence described in this section is followed, i. e., gross filtration of incoming water followed by a double-pass RO unit and then through a suitable DI storage and distribution system. Finally, using the previously described sequence as a model, the resultant DI water can serve as a source for the WFI distillation unit.

14.6 Composition of System Piping

If the system described up to this point were followed, the water would be of adequate quality to feed both the distillation system and clean steam generator, provided that the composition of the piping system does not add extractables to the water which can compromise its quality. The problem of extractables is especially acute during the "break-in" period for a newly constructed system that utilizes plastic piping. This should be taken into account during system validation studies.

Plastic pipe is less expensive and, therefore, has been desirable to hold down the cost of the system. There are several types of plastic available which utilize several different methods of attachment from heat welding to plastic glues. Available plastics include polyvinyl chloride (PVC), chlorinated PVC, polysulfone, and polypropylene. The latter which can be joined by sanitary heat molding is preferred over types which require priming and cementing.

Plastic pipes may be employed up to the final distillation unit, some most efficient distillation units will remove a large variety of contaminants. However, the efficiency of distillation units is not absolute, and some volatile contaminants can be carried over into the distillate. For this reason, the welded polypropylene system is suggested to keep extractables at a minimum.

From the distillation unit throughout the remainder of the WFI storage and delivery system, the recommended piping material has continued to be stainless steel of type 316 L or better. The designation "L" refers to low carbon grades which function to resist corrosion better than other grades. The low carbon grades also are more amenable to welding. Since high resistivity WFI is an aggressive solvent, corrosion resistance is important. This is especially true at the high temperatures required in a properly designed WFI storage and delivery system.

The LVP document referenced earlier indicated that piping for production water handling systems should be made of welded stainless steel (non-rusting grade) sterilizable with steam except that sanitary fittings capable of disassembly and cleaning may be immediately adjacent to the equipment or valves that must be removed from the lines for servicing.

14.6.1 Fittings for Piping Systems

Fittings for stainless piping systems may be categorized into two basic grades, industrial and sanitary. The fittings of choice for WFI systems are of sanitary design. Numerous valves and fittings are available for uses such as sampling and intake ports, flow restrictors and diverters. The choice of an appropriate fitting should be based upon its "cleanability" in both the open and closed positions. Sampling ports should also be capable of steaming prior to sample withdrawal to minimize operator error.

14.7 Production of WFI

Up to this point we have taken water through several pretreatment steps leading to the production of water for injection (WFI), the water quality is required for product contact as well as final rinsing of glassware and stoppers that will interface with aseptic product. The system has so far included gross filtration of the incoming source water, double-pass reverse osmosis and deionization. As previously described, distillation rather than reverse osmosis is regarded as the method of choice for WFI production.

The design of an efficient WFI system

should encompass the WFI purification unit, distribution, and storage systems. Several types of distillation apparatus are available including the single-effect, multi-effect, and vapor recompression units. Selection of the unit that is optimum for a particular facility should be based upon economy and the anticipated volumes of water needed for both final rinsing and product contact. One should keep in mind that buffers utilized in the purification procedures should be prepared with WFI to minimize bioburden and pyrogen levels.

The most efficient distillation apparatus at present is the multi-effect still in which the heated distillate from each successive stage heats the water in the next stage. This system remains hot and saves energy and time associated with cold start-up.

The composition and sloping of a distribution system has been previously described. The proposed LVPs have recommended a recirculating loop system maintained at a temperature of 80 °C ± 5 °C. If cooling is necessary for any use point, heat exchangers should be located as close to the point(s) of use as possible. The entire distribution system should be capable of withstanding periodic cleaning and sanitization usually employing clean steam.

Although a recirculating system is preferred, a batch system may also be utilized. If water is directed into a storage tank rather than a loop, storage should also be maintained at 80 °C, and the water should be tested and utilized by batching. Continuous circulation affords the advantage of constantly moving water which has an inhibitory effect on attachment and growth of microorganisms.

Storage tanks should be jacketed to maintain temperature and be constructed of the same polished stainless steel as the piping system. A system not capable of being maintained at 80 °C should have provisions for the water to be discarded within 24 hours unless sufficient validation data have been submitted and accepted by the regulatory authorities in support of a longer interval. This validation data should be supplemented by a cleaning and sanitization validation study at the end of a "worse-case" storage interval. The problem of condensation in a cold tank, if held for long periods of time without adequate filtration, could compromise the integrity of the water in the tank and/or lead, for instance, to collapse of the holding tank.

14.8 Validating the Water System

Once the design of the water system has been selected, the validation project design to be documented in the MVP should mimic the previously described steps which coincide with facility construction. The requirements for pre-validation of design criteria followed by consultation with regulatory authorities and monitoring during inspection are of vital importance to the production water system.

When the above are completed and the system is in place, the start-up validation may be broken down to the familiar IQ, OQ, and PQ steps. One should keep in mind that acceptance criteria for the water system should be formulated at the initial start-up validation cycle and refined during the operational qualification. The acceptance criteria should establish limits as defined to be low for chemical and microbial quality:

- Acceptance limit: The expected limit based upon the quality level that one needs to achieve. For WFI, the specifications have been established in the USP.
- Alert limit: A deviance from acceptable specifications which cannot be attributed to sampling error or other identifiable causes. A written plan of correction should be part of the SOPs.
- Action limit: The level of specifications which require immediate action to correct the situation. A written action procedure should also be available when this limit is reached.

15 Installation and Operational Qualification

Thus far, we have proceeded from preliminary facility design to selection of appropriate equipment and systems, pre and post construction documentation and installation of major utility systems. All of the above should have followed the sequence outlined in the MVP.

Once the above have been completed, the facility is ready for the more traditional IQ, OQ, and PQ for all equipment and systems. The fact that the adequacy of room finishes is a type of facility qualification should not be unrecognized. Documentation of room finishes, equipment "fits", and operational flows should be verified prior to facility and equipment qualification and process validation.

The next section of the MVP should now concentrate on installation and operational qualification of the equipment. Initial installation qualification should verify that all equipment meets pre-determined design specifications. The manufacturers' literature may be utilized for this purpose, provided that there is a written record of acceptance of the specifications that has proceeded through the standard SOP approval pathway.

Once the installation qualification has verified that the equipment is operating properly, the calibration and preventative maintenance schedules should be specified, in writing. There should be a maintenance log for each major piece of equipment in accordance with cGMP requirements. In addition, the individuals responsible for periodic calibration and maintenance should be specified in writing along with written procedures and schedules.

By the time that operational qualification has been completed, the operational limits should be defined. This coincides with the acceptance criteria which should be finalized following the process validation.

If any of the above qualification procedures are performed by an outside contractor, the raw data supporting the study conclusions should be available. In addition, the procedures utilized to perform these studies should be accepted through detailed SOP approval procedure. If, upon facility inspection, qualification and validation summary documents are available without supporting data or procedures, questions regarding the methods or adequacy of the studies may be raised. Another deficiency observed during pre-approval inspections of new facilities is the lack of coordination between persons responsible for calibration, validation, and maintenance. This frequently results in problems such as irrational acceptance criteria and failure to follow written schedules for calibration and maintenance.

16 Process Validation

Process validation should begin after the completion of installation and operational qualification and validation of the facility, utility systems and equipment. Written specifications should be proposed based upon manufacturing experience developed during the production of pilot lots. The experienced product development scientist will be aware of specific process steps where specifications should be given wide enough limits to accommodate scale-up. A sufficient number of validation runs should be conducted to assure that the final product can be consistenly produced within all established in-process specifications.

A successful process validation study will answer two questions. First, does the process actually do what it is supposed to do? In this regard, the validation protocol for each process should describe the intended outcomes based upon specifications that were established during pilot or other small-scale manufacturing procedures. One frequent deficiency noted during inspection of process validation data is the fact that specifications, limits, and in-process yields change as a result of larger-scale production without documentation of the exact reason for each change. It is not sufficient to change pre-determined in-process specifications based upon a limited number of process validation studies, unless a rationale for such change is documented in the validation study report.

The second question that will be answered is what are the acceptance limits for each critical in-process step? In this context, acceptance limits do not translate only into the normal operating ranges. The action and alert limits should also be established along with written SOPs detailing corrective actions, as previously described.

16.1 Process Validation Sequence

With the above goals in mind, process validation may be simplistically explained as a four-step process. Step one involves preparation of a *written protocol* which outlines the objectives of the study, methodology, pre-determined in-process specifications, yields and limits, and acceptance criteria. This protocol should be forwarded through the normal review and approval channels.

Performance of the validation study constitutes the primary activities for step two of the validation. This involves utilization of the actual product intended to be produced in the equipment and systems in question at as large a scale as can be practically accommodated. The most important concept is that process validation should mimic actual production conditions unless justification for other conditions can be documented. This includes use of the same number and types of personnel that will perform actual production operations. If this is done, the studies will serve as both a validation effort and employee training mechanism which can be documented on individual training records.

The primary justification for utilization of a smaller-scale process is the excessive cost of the product which may be expired prior to the end of the review process. In this situation, an acceptable approach may be that the MVP can allow the preliminary studies to validate final specifications, yields and limits with a commitment to perform concurrent process validation studies on the first two or three full-scale production lots. This option should be discussed with regulatory authorities prior to small-scale "representative" validation studies.

In some instances, manufacturers have attempted to rely on studies which utilize similar products or growth media to represent the preliminary process validation studies. This is not an acceptable approach for initial process validation studies since one of the primary purposes is to validate expected specifications, limits and acceptance criteria for the product which is the subject of an application for regulatory approval.

There are several critical points to keep in mind as studies are ongoing. First, all raw data should be retained for use in subsequent analyses. If an outside contractor is used to perform the studies, the raw data generated should become a part of the records. Validation summary reports or certificates of acceptance without supporting data are not sufficient. When deviations to expected specifications occur, a person or team should be designated the responsibility for immediate investigation and justification. No attempt should be made to delete outlying data points. Analysis of data and determination of whether a value is a true outlier should only be made in the data analysis portion of the study.

An important consideration during the performance of the validation study is the need to accurately record by diagram the actual placement of probes, load configurations, room locations for environmental sampling, and other physical parameters which are critical for assessment of study validity. In many situations the person(s) that actually performed the study are not available or cannot remember all of details of each study. Accurate location diagrams will help to confirm the accuracy and precision of the study.

The third step in the process validation study is *analysis of the results*. One frequent problem observed on inspection of study data is that the analysis portion was delegated to individuals without appropriate statistical background or data handling experience. This usually results in analyses that are difficult to interpret and conclusions that are not supported by actual data.

The method of data analysis should be documented along with justifications for any deviations from the original protocol. One section in the report should be devoted to a comparison of the actual and expected results with scientifically based rationale for changes. Final definition of all limits and action levels should follow this section.

The fourth and final step in the process validation package is a *description of the study conclusions*. This section should accurately and concisely summarize the data analysis and indicate if the results are within expected limits. Deviations from expected values should be explained and accompained by a short justification and reference to supporting data. Since replicate studies provide assurance that the product can consistently meet specifications, the "repeatability" of the process should be emphasized.

16.2 The Process Validation Report

This process validation study report should be reviewed and approved by the appropriate production departments, validation individuals, statisticians, quality control, and quality assurance. As required by cGMPs, all of the above functions should operate independently and be afforded sufficient authority to review and comment on the final report. If repeat studies are needed due to lack of adherence to pre-determined specifications, the departments noted above should provide for repeat testing utilizing the same or a revised protocol. The need to repeat all or part of a study should be reported in the conclusions section of the report. Once again, all of the above applies regardless of whether the study was performed by in-house personnel or an outside contractor.

The information presented up to this point constitutes the major portion of the MVP. However, several additional separate validation studies should also be a part of this plan. These separate validation studies are summarized in the following sections.

17 Cleaning Validation Studies

The importance of cleaning validation increases in proportion to the multi-use character of the facility. The starting point for adequate facility cleaning validation is the establishment of a baseline bioburden level. This should be performed prior to the introduction of product. The data will serve as a base from which to trend environmental bioburden data in an attempt to determine the cause of higher than expected contamination.

This baseline data may be utilized to determine possible sources of bioburden prior to facility start-up. One instance was observed where the environmental bioburden levels in a new facility were highest following regular cleaning cycles. An investigation revealed that the water utilized for cleaning with sterilized mopheads was de-chlorinated due to adsorption of the chlorine onto the walls of copper piping which supplied water to the new facility. This problem was corrected by thorough flushing of the lines prior to introduction of product into the facility.

If the data include identification of the microorganisms as well as analysis of the expected bioburden to be encountered from personnel and product, an effective selection of cleaning agents can be made. Following the selection of suitable agents, a schedule for rotation of the cleaning agents can be devised and complied in SOPs. The cleaning validation procedures should also include alert and action limits along with a description of the actions that will be taken if values fall outside of these limits.

17.1 Equipment Cleaning Validation

The need for equipment cleaning validation and its importance in multi-use facilities has already been discussed. The most significant consideration is the utilization of specific and sensitive assay procedures to detect both contamination as well as residual product. It is not enough to take a sample of final rinse water and look for total protein. The current configurations for reaction vessels and transfer lines are torturous and complex. Several mechanisms exist to document that all lines, valves, impellers, etc., have been adequately cleaned with the cleaning procedure.

One mechanism utilized effectively, for example, is the use of a removable dye in a trial run with product that travels through all equipment. This is followed by the normal cleaning procedure. Using this mechanism, one can visually examine surfaces of impellers, valves, joints, etc. If residual dye is observed, a swab sample can be taken and analyzed for product, protein, and other possible contaminants. The cleaning procedure can be adjusted until an acceptable level of cleaning has been reached.

18 Validation of Decontamination and Discharge Streams

Waste treatment and discharge information is needed in support of an environmental assessment statement which should be submitted for all new biotech facilities. In order to adequately characterize the efficiency of the processes in terms of environmental control, this assessment should list the quantities utilized for all hazardous chemicals and their chemical classification (i. e., potential or suspect carcinogen, etc.) as well as all active and inactive components.

The qualification portion of this study should also address the capacity of the reservoirs. Is the capacity of the reservoirs adequate for full-scale production? The discussion should include a description of the means of conveyance of discharge into waste treatment reservoirs as well as a description of the containment measures to protect against accidental discharge. A diagram of the location and capacity of containment wells in relation to the size of the reaction vessels should be provided. Written procedures should be in place to describe the actions that will be taken in the event of an accidental discharge.

All decontamination and inactivation procedures should be validated using appropriate "worse case" challenge conditions. One criticism of waste treatment systems is the failure to include appropriate challenge conditions in the decontamination validation study. All possible sources of waste disposal should be considered, including spent media, cell debris, purification residue, etc. Sampling should be performed from locations representing the end of the treatment process, and the waste streams should be tested by utilizing valid test parameters based upon the suspected composition and/or bioburden of the effluent.

19 Environmental Monitoring

Environmental monitoring is utilized in the validation of both systems and equipment. However, several important principles should be kept in mind in the design and conduct of environmental monitoring studies.

Environmental monitoring studies should be performed in all locations where product is or *may be* exposed to the environment. In many instances, a manufacturer will conclude that a controlled environment is not needed due to the use of closed system technology. In order for the regulatory authorities to accept this argument, the integrity of all closed systems must be validated. Refer to the previous discussion regarding validation of closed systems. In considering where product *may be* exposed to the environment, consider the integrity of operations performed in relationship to the level of environmental monitoring.

For example, one biotech facility was observed to perform purification steps within an environmentally controlled cold room. The environmental monitoring had been accomplished by utilization of active sampling devices at the level of introduction of product into the columns. However, the column effluent was observed to be collected in an open flask located on the floor under an open valve which was located less than 18 inches from floor level. There were no data to document the air flow pattern at this level including the degree of turbulence and disposition and quality of air after interaction with the floor. This was cited as an open, unvalidated collection procedure.

19.1 FDA Guideline on Aseptic Processing

Other general principles of environmental monitoring can be found in the FDA Guideline on Sterile Drug Products Produced by Aseptic Processing (FDA, 1987a). A listing of some of the key provisions follows:

- Active sampling devices should be utilized
- Both viable and non-viable particulates should be monitored
- Monitoring should be performed during work activity periods
- Monitoring should be performed at the level of product exposure

If the above general principles are followed, valid room air classifications can be selected based upon data generated and degree of product exposure.

20 Personnel Monitoring

Part of the environmental validation program should be the monitoring of personnel working in areas of actual or potential product exposure. Personnel gloves and facemasks are particular locations where contaminants can be retained and introduced into the environment.

When all of the environmental monitoring data are collected, written alert and action limits should be established along with procedures to implement when the environmental values are out of the target ranges. These data should be capable of collection and analysis by day, week, and year so that a trend analysis as required by cGMP regulation [21 CFR 211.180 (e)] can be performed.

20.1 Personnel Validation

Personnel generally have been recognized as the biggest single source of contamination in a manufacturing process. Therefore, documentation of the qualification and validation of personnel procedures should be provided in the MVP.

The following elements should be considered in a well designed personnel monitoring program:

- Is gowning, including level and procedures, appropriate for the function utilized? Consider the use of distinctive garments for maintenance and other non-production personnel. Color-coded garments may also be utilized to distinguish the various access levels for personnel (i. e., animal testing personnel, production, filling, packaging, etc.).
- Training should be appropriate for function and should be traceable to the individual's personnel file.
- Are personnel dedicated to one process or function at a time to minimize cross-contamination? This is especially important in multi-use facilities.
- What is the level of interaction of personnel working in "clean" and "dirty" production steps? What types of gowning change procedures are in effect for such interactions to take place?

21 Computer System Validation

This chapter would not be complete without a brief description of computer system validation. Both hardware and software should be validated. Refer to the FDA publications on validation of computer systems and software for an elaboration of the types of documentation required (FDA, 1983, 1987b).

Validation of hardware need not be elaborate, but should be sufficient to show confidence that it can consistently perform as expected. When systems or equipment are both controlled and monitored by programmable logic, sufficient documentation and back-up controls should be available to support all critical functions.

Validation of software should include simulated worse-case testing under both trial and actual production conditions. For systems which utilize biosensors to monitor critical process parameters and trigger the addition of nutrients or the opening of valves, the challenge should include a mechanical intervention. The system should be capable of alarm when challenged with a condition which is out of established specifications. It is recommended that a hard copy containing periodic monitoring of in-process variables be retained as part of production records during actual production conditions. Therefore, software validation studies should be designed to select the appropriate parameters and time intervals for record retention.

Once a computer control system has been validated and appears to operate within acceptable limits, the need for revalidation should be based upon either a process change or the results of a trend analysis which demonstrates difficulty in maintaining control of a computerized function.

22 Revalidation and Process Changes

Once all of the preceding validation studies and facility design criteria have been finalized, the Master Validation Plan should focus on the need for revalidation. Revalidation should be conducted whenever a significant change in equipment, systems, or process is made.

22.1 The Regulatory Impact of Process Changes

One of the most significant regulations for biological products in terms of impact on time and resources is 21 CFR 601.12 "Changes to be reported". This regulation has been in place for decades and requires reporting of "important" changes to manufacturing methods and labeling in advance of implementation. The key concept is the meaning of *important* which will be discussed in some detail.

The initial consideration should be given to definition of what constitutes a change to a facility or product. A change, from a regulatory viewpoint, can be broadly defined as anything that differs from the information in an approved product or establishment license application filing that is, or is intended to be, routinely done to a product and/or a facility. In the context of this regulation, it also includes important changes to personnel which has been interpreted to mean "key personnel", such as department or section heads with major decision-making responsibility in product manufacturing, quality assurance, and testing.

This is different from the situation where in the course of manufacturing a particular lot of a product, a deviation from approved SOP occurs due to personnel, equipment failure, or for some other explainable reason. This one-time occurrence does not always have to be reported to regulatory authorities as long as,

- the deviation is clearly documented and explained in the batch records for that particular lot,
- the deviation does not affect the safety, purity, potency, or effectiveness of the product, and
- it is a one-time occurrence.

This leads to a discussion of what constitutes an important change. From a regulatory viewpoint, "important" encompasses a large territory. An example of minor and major changes for both products and establishments may better serve to illustrate the meaning of "important".

22.2 Product Changes

When it comes to changes in a product, it is difficult to define an example of a minor change. Possibly the consolidation of steps in a complicated batch-production record, in the interest of clarity without changing any critical process operations which have defined operating parameters, may constitute a 30-day notice which may be submitted to the license file rather than an amendment which requires review and written approval.

It is easier to define a major product change, such as a change to the product indications or formulation, which always requires the filing of a full license application amendment for review and written approval prior to implementation. As a rule of thumb, any change to a product which has the potential of changing its characteristics, or which will have an undetermined effect on the product's safety, purity, potency, or effectiveness should be reported as full license application amendment including product comparability data demonstrating biochemical and clinical equivalence of the modified product to the previously approved product. The extent of data required (including clinical studies) depends upon the nature of the change proposed.

22.3 Facility Changes

Changes of a facility may be broadly broken down into equipment changes and location changes. Equipment changes may be minor, such as the replacement of an existing pH-meter with a newer, automated one, or major changes, such as the addition of a new reaction vessel which directly contacts the product.

Minor changes are at least reportable in accordance with the 30-day notice procedures. However, notification is not necessary every time a piece of equipment utilized for product testing is upgraded, only in cases where the upgrading may change the product or test specifications.

Major changes, on the other hand, usually require submission of a detailed establishment license application amendment for review and written approval prior to final implementation. This is especially true for equipment which, in the course of its use, makes direct contact with the product at any stage of manufacture, including during initial manipulations of organisms or viruses.

22.4 Revalidation Scheduling

Absent a significant change as described above, periodic revalidation should be scheduled for both systems and equipment to assure their continued operation within established specifications. For this purpose, concurrent validation studies may be performed provided that the process is not changed to accommodate the validation study. This precludes a change in scale, production time limits, or the elimination of steps to decrease validation study time.

23 Final Approval

A written procedure should be in place for the review and approval of all validation protocols, study results, and data summaries. These data should be reviewed by the validation, production, quality control, and quality assurance departments as a minimum. In accordance with cGMPs, all of the above departments should function independently and have the opportunity to review and comment on all aspects of the MVP. The above review function should be performed regardless of whether the validation studies were designed and performed by in-house personnel of an outside contractor.

24 Summary

This chapter was designed to take the reader through the thought process involved in the design, construction, and validation of a typical biotech facility. General guidance was provided on the unique regulatory requirements for products which require licensing under the U.S. Public Health Service Act. These requirements were contrasted with the requirements for approval of drug products under the Federal Food, Drug and Cosmetic Act. In addition, the scope of biological manufacturing processes was discussed and correlated with the concerns of regulatory authorities, when facilities designed for the manufacture of multiple products are utilized.

During the construction phase, the need for constant and vigilant monitoring combined with documentation of events was emphasized. Validation of systems and equipment was introduced by an in-depth discussion of two critical systems, the heating, ventilation, and air conditioning (HVAC) and the production water systems.

The entire validation package should be documented in a Master Validation Plan (MVP) which includes the following elements:

- Pre-validation engineering design
- Construction validation

- Facility qualification
- Utility systems qualification and validation
- Systems and equipment qualification
- Start-up validation
- Process validation
- Final approval

The selection of facility design, choice of systems, and validation package should be based upon the intended function of the facility. The problems associated with multi-use facilities have been emphasized. However, sufficient flexibility should be designed into the plans to provide an avenue for change, if the marketing potential of the proposed product is not realized.

25 References

AVALLONE, H. (1989), FDA, Talk at the *Johnson & Johnson Sterilization Sciences Seminar on Water Quality*, June 5, at Piscataway, NJ.

BJURSTROM, E. E., COLEMAN, D. (1987), *Biopharm,* 50–55.

CATTEANEO, D. J. (1988), Plant validation acceptance criteria, *Pharm. Eng.* **8**(4), 9–11.

21 CFR, Title 21, *Code of Federal Regulations*, Subchapter F: *Biologics*, Parts 600–680.

21 CFR (1992), Title 21, *Code of Federal Regulations* (United States), Part 211: *Current Good Manufacturing Practice for Finished Pharmaceuticals*, April 1.

FDA (1981), *Letter to the Pharmaceutical Industry RE: Validation and Control of Deionized Water Systems,* August.

FDA (1983), *Guide to Inspection of Computerized Systems in Drug Processing, Food and Drug Administration*, February.

FDA (1987a), *Guideline on Sterile Drug Products Produced by Aseptic Processing*, June.

FDA (1987b), *Software Development Activities, Food and Drug Administration*, July.

Federal Register (1976), Vol. **41**, No. 106, 22202–22219.

Federal Register (1984), Vol. **49**, No. 50, 856.

Federal Register (1992), Vol. **57**, No. 228, 55544–55546.

KELTER, S. L. (1988), *BioPharm. Manuf.* **1**(5), 30–37.

US PHARMACOPEIA (1990), Vol. XXII, *Official* from January 1.

29 Treatment of Biological Waste

DANIEL F. LIBERMAN

Cambridge, MA 02139, USA

1 Introduction

When the history of the 1970s and 1980s is written, historians will refer to this period as a true revolution in biological science. The ability to isolate, purify, modify, and express genes in microorganisms has revolutionized our understanding of the natural world and our ability to turn that understanding to practical application (CHURCH et al., 1989). These same historians will also identify this period as the time when public concerns over "genetic engineering" resulted in a number of governmental agencies establishing oversight authority on the development of the technology.

This volume contains "state of the art" information on the technical aspects or procedures which constitute "bioprocessing". While the various processes that comprise biotechnology vary substantially in their details, they all are based on the use of microorganisms or cells to serve as catalysts in the conversion of substrate to product.

While each technical process is associated with its own unique starting material (bacteria, mold, yeast, plant or animal cell, or enzyme), technical manipulation (cell or enzyme immobilization, microbial or cell genetic engineering, cell fusion), process (continuous, batch or fed-batch fermentation) and scale (small to very large), they all share the same common feature. They all produce wastes which either contain or have come in contact with living organisms.

It is ironic that it is not the research or production activities that are under the most scrutiny at this time, but rather, it is the wastes generated by these activities that continue to raise public and regulatory interest.

Why Process Bio-Waste?

Unless an organism or system is designed for environmental use (bio-pesticides, bio-mining, bio-remediation), it is unlikely that the organism will survive, or if it did survive, be reproductively fit. Since uncertainty exists for organisms whose environmental fate is not defined, regulatory authorities require the careful evaluation of bio-waste streams before disposal. The discharge of live organisms or even the discharge of unconsumed nutrients could have an environmental impact on human and animal populations as well as other organisms (microbes, algae, plants) that exist in an environment by changing the growth-limiting nutrient balance (BROWN et al., 1977; EVANS et al., 1981).

In this chapter, a series of discussions will introduce the reader to the various components of a bio-waste management/treatment program. First, the definition of bio-waste adopted by agencies in the U.S., National Institutes of Health (NIH) and the Environmental Protection Agency (EPA), will be presented. The discussion will then characterize the various classes of waste. The key elements of a waste management program will be identified, and the discussion will conclude with a review of current and proposed bio-waste treatment/disposal options.

2 Bio-Waste Generation

2.1 Definition of Bio-Waste

Waste can be defined as material that is worthless, defective or of no use; discarded as used, superfluous, or not fit for use; or debris resulting from a process (as of manufacture) that is of no further use in the system producing it. The term biological waste (bio-waste) is used when the material has come in contact with or contains some viable life-form. This type of waste is generated by a number of institutions as well as the general public. Since only institutional bio-wastes are regulated by National, Federal, state or municipal authorities, household wastes will be excluded from further discussion.

2.1.1 Biological Wastes and Hazardous Wastes

There is no universally accepted definition for biological waste. This has resulted in inconsistency in terminology used to define the waste stream generated by R&D and production facilities. While the terms infectious, pathological, biomedical, biohazardous, viable, bio-contaminated, rDNA, have been used to describe these wastes, the term that appears most frequently is "bio-waste". The failure to distinguish between genetically engineered organisms and pathogens has resulted in more confusion than anything else in bio-waste management. The fact that all bio-waste is not infectious is very difficult for regulators, legislators, and the general public to accept.

To avoid confusion, the following distinction will apply. Infectious bio-waste will contain organisms which can cause real or potential harm by infection or through disruption of the environment, while non-infectious bio-waste will not contain human, other animal, or environmental pathogens and therefore, will not cause illness in humans, or animal species, or pose a risk to the environment.

2.2 Waste Generators

The most common generators of bio-waste are listed in Tab. 1. While the data are based on domestic surveys (i.e., performed in the U.S.), it is highly unlikely that the results would be different if the surveys were from other countries. The bio-medical community is the largest generator. Their contribution to the bio-waste stream consists of material generated during the course of diagnosis, treatment of illness or disease, research and production activities and the production and testing of biologicals.

2.3 Bio-Waste Regulations

In the United States there are a number of Federal agencies which have developed or are considering regulatory initiatives which involve the generation, management and disposal of bio-waste (Tab. 2). The United States Environmental Protection Agency (USEPA) and the National Institutes of Health (NIH) have developed guidance documents to assist in the management and disposal of bio-waste. These documents have formed the basis for a number of Federal, state, and municipal regulations (HENDRICK, 1989).

2.4 Bio-Waste Characterization

2.4.1 RCRA

Within the United States, the Resource Conservation and Recovery Act (RCRA) of 1976 granted to the EPA the authority to define and regulate hazardous waste under RCRA, and the five characteristics of hazardous waste (toxicity, corrosive, activity, ignitability and infectious) were described (RCRA, 1976).

2.4.2 USEPA Definition of Infectious Waste

At present, infectious waste is defined as "waste capable of producing an infectious disease". The USEPA recognized that the following epidemiologic factors were necessary for

Tab. 1. Institutions which Generate Bio-Waste[a]

Type	Number of Facilities	SIC[b]
Medical laboratories	4563	8071
Commercial research	2341	8731
Non-commercial research	941	8733
Production of medicinal chemicals	228	2833
Production of pharmaceutical preparations	683	2834
Production of biological products	370	2836

[a] Source: USEPA (1988), *Proceedings of the Meeting on Medical Waste*, Annapolis, Maryland; Publication No. F-89-MTPF-FFFRF-88

[b] Systematic Industrial Code

Tab. 2. Federal Agencies Addressing Bio-Waste Issues

Agency	Authority	Activity
US Environmental Protection Agency (USEPA)	Guidance and Regulatory[a]	Issued guidelines; regulations to establish the medical waste tracking program, authority under RCRA to regulate the handling, storage and transport of medical wastes; establishment of standards for medical waste incinerators
Occupational Safety and Health Administration; US Department of Labor (OSHA)	Guidance and Regulatory[b]	Issues advisory notices and workplace standards focusing on occupational exposure to infectious materials and wastes
Centers for Disease Control; US Department of Health & Human Services (CDC)	Guidance and Recommendations[c]	Issues notices and advisories, sometimes jointly with other agencies focusing on infection and control issues
Agency for Toxic Substances and Disease Registry; US Department of Health and Human Services (ATSDR)	Study and Review[c]	Completed study required by Medical Waste Tracking Act focusing on evaluating health effects associated with medical wastes
National Institutes of Health; US Department of Health and Human Services (NIH)	Guidance and Recommendations[d]	Issues notices and advisories, sometimes jointly with other agencies focusing on genetic engineering/microorganisms, plants, and animals

[a] EPA's comprehensive authority to regulate medical waste management is granted under the Resource Conservation and Recovery Act. The Agency also has special regulatory authority to administer the Medical Waste Tracking Act of 1988 (42 USC 6901 et seq.).

[b] OSHA's primary authority is granted under the Occupational Safety and Health Act (29 USC 651 et seq.). Guidelines or regulations only apply to private facilities, unless a state extends coverage to employees of public facilities as well.

[c] Does not have the authority to issue regulations. Recommendations, however, are used by other agencies as the basis for regulations.

[d] Does not have the authority to issue regulations, however, states and local communities have incorporated NIH Guidelines directly into regulations.

disease induction: the presence of a pathogen, a dose sufficient to cause illness, a portal of entry, and a susceptible host (USEPA, 1986). For a waste to be infectious, it had to contain an agent with sufficient virulence and quantity so that when a susceptible host was exposed through ingestion, inhalation, contact with mucosal surfaces or after self-inoculation or skin penetration, illness would result. After considerable review, the USEPA (1986) defined six types of waste which were to be managed as infectious and four additional types of waste which could be designated as infectious depending on what or how much was known about its composition or history (Tab. 3).

2.4.3 Classes of Waste Subject to EPA Guidelines

Unless the organisms were known to be non-pathogenic, all waste which resulted from activity with organisms were to be managed as infectious bio-waste. The following types of waste were classified as infectious:

- Cultures and stocks of infectious agents and associated biologicals, including cultures from medical and pathological laboratories, cultures and stocks of infectious agents from research and industrial laboratories, wastes from the production of biologicals (fermentation), discarded live and attenuated vaccines, and culture dishes and devices used to transfer, inoculate, and mix cultures.
- Pathological wastes, including tissues, organs, and body parts that are removed during surgery or autopsy.
- Waste human blood and blood products, including serum, plasma, and other blood components.
- Sharps that have been used in patient care or in medical, research or industrial laboratories, including hypodermic needles, syringes, Pasteur pipettes, broken glass, and scalpel blades.
- Contaminated animal carcasses, body parts, and bedding of animals that were exposed to infectious agents during research, production of biologicals, or testing of pharmaceuticals.
- Wastes from surgery or autopsy that were in contact with infectious agents, including soiled dressings, sponges, drapes, lavage tubes, drainage sets, underpads, and surgical gloves.
- Laboratory wastes from medical, pathological, pharmaceutical or other research in commercial or industrial laboratories that were in contact with microorganisms including slides and cover slips, disposable gloves, laboratory coats and aprons, plastic tubing, paper, plastic.
- Dialysis wastes that were in contact with the blood of patients undergoing hemodialysis, including contaminated disposable equipment and supplies such as tubing, filters, disposable sheets, towels, gloves, aprons, and laboratory coats.
- Discarded medical equipment and parts that were in contact with microorganisms.
- Biological waste and discarded materials contaminated with blood, excretion, exudates, or secretion from human beings or animals which are isolated to protect others from communicable diseases.

Tab. 3. Types of Infectious Waste

Recommended
Isolation wastes
Cultures and stocks of infectious agents and associated biologicals
Human blood and blood products
Pathological wastes
Contaminated sharps
Contaminated animal carcasses, body parts, and bedding
Optional
Surgical/autopsy waste
Contaminated laboratory waste
Dialysis wastes
Contaminated equipment

Source: USEPA (1986), *Guide for Infectious Waste Management*

3 R & D Bio-Waste

3.1 Introduction

In the beginning, "few scientists believe that genetic engineering is free from risks". These prophetic words came from the summary statement issued at the end of the Asilomar Conference on recombinant DNA molecules (BERG et al., 1975). The uncertainty among the "experts" together with the emerging concern on the part of the public resulted in the development of the NIH Guidelines for Research Involving Recombinant DNA Molecules (NIH, 1976).

NIH responded to the concern that genetically engineered organisms would escape from the laboratory and be transferred to the community. NIH developed very specific language on the management of all surfaces and materials (including wastes) that come in contact with engineered organisms. Work surfaces were to be decontaminated after any spill of viable material; equipment was to be decontaminated prior to reuse, washing, or discard; and contaminated solid or liquid wastes were to be decontaminated before disposal. *Even if the organism was not pathogenic for humans, other animals or plants, or the environment, if it was genetically engineered, it was to be handled as if it were a pathogen.*

3.2 Large-Scale Processes

With the advent of commercial applications, the focus of concern expanded to include risks associated with industrial use of engineered strains and with the products obtained from these strains. NIH responded with the development of a set of large-scale (volumes which exceeded 10 liters) practices that established both the personnel practices and the containment practices that were appropriate to prevent release of recombinant DNA (rDNA) (NIH, 1983). Cultures of viable organisms containing recombinant DNA were to be handled in closed systems. Culture fluids could not be removed from these systems, unless the viable organisms containing rDNA molecules were inactivated by methods which are known to be effective for the organism containing the rDNA molecules. All equipment or materials which come in contact with viable organisms would have to be decontaminated prior to reuse, cleaning, or disposal.

3.3 Perception of Risk

The uncertainty associated with the behavior of genetically engineered organisms in the environment, if a breach in containment occurred, contributed to the public perception of risk. The public focused on whether genetic engineering could give an organism a selective advantage over unaltered members of the species. Some organisms could be demonstrated to pose little or no risk and, therefore, only minimal controls and containment procedures were appropriate. Experimentation produced organisms which were indistinguishable from those produced by standard genetic techniques. To avoid unnecessary overcontainment, the NIH developed a mechanism whereby the Guidelines were subject to periodic review and modification. This mechanism enabled scientists to petition for declassification of experiments. Over the past several years, the NIH has reduced containment requirements for the majority of experiments, and, in a number of instances, eliminated experiments from the Guidelines by exempting them from coverage.

3.4 Good Industrial Large-Scale Practices (GILSP)

Between 1983 and 1987, it became apparent that there were applications which used organisms considered to be of low or no appreciable risk. Since this was the case for the vast majority of genetically engineered organisms used in industrial production, a series of procedures and practices were developed for research and development activity with non-hazardous organisms.

The OECD developed a set of procedures and practices which they referred to as good industrial large-scale practices (GILSP) for organisms which by all available criteria were generally regarded as safe (OECD, 1987). NIH endorsed the concept of GILSP and advised all of its grantee institutions that the containment recommendations were to be based on the hazard or risk associated with the organism and not whether the organism was engineered. NIH finally admitted that it was inappropriate to classify an organism as being a potential hazard because it was engineered.

3.4.1 Basic GILSP Criteria

In order to determine if an experiment (or the use of a specific organism for that matter) meets GILSP criteria, it is necessary to review what is known about the host organism, if it is an engineered strain, the parental organism, the vector carries and facilitates the expression of the information (gene product) of interest.

3.4.2 Host

According to GILSP criteria, the host organism should be non-pathogenic; should not contain adventitious agents; and should have an extended history of safe industrial use, or a built-in environmental limitation that permits optimum growth in the industrial setting but limited survival without adverse consequences in the environment.

3.4.3 The Engineered Organism

The engineered organism should be non-pathogenic and as safe in the industrial setting as the host organism, and without adverse consequences in the environment.

3.4.4 The Vector/Insert

The vector/insert should be well-characterized and free from known harmful sequences; should be limited in size as much as possible to the DNA required to perform the intended function; should not increase the stability of the construct in the environment, unless that is a requirement of the intended function; should be poorly mobilized; and should not transfer any resistance markers to microorganisms not known to acquire them naturally, if such acquisition could compromise the use of a drug to control disease agents in human or veterinary medicine or agriculture.

3.4.5 Additional Criteria

It is clear that there are two other classes of non-pathogenic organisms that warrant the GILSP designation. The first consists of those constructed entirely from a single prokaryotic host (including its indigenous plasmids and viruses) or from a single eukaryotic host (including its chloroplasts, mitochondria or plasmids – *but excluding viruses*). The second consists entirely of DNA segments from different species that exchange DNA by known physiological processes. NIH, after considerable review, exempted both of these classes of organisms for experiments at small scale and has recently accepted this exemption for large scale as well.

4 Management of Wastes

4.1 Introduction

It is essential that each institution has a comprehensive waste management plan that addresses infectious, radioactive, chemical, and mixed wastes (e. g., biological/radioactive or biological/toxic, etc.) as well as solid waste. As part of this institutional plan, each laboratory or department unit should have detailed, written instructions for the management of the kind of wastes that it generates. Ideally, one person should be identified who will have overall responsibility for bio-waste management at the facility. The plan should specify the selected management options and procedures for the facility as a whole as well as for individual units within the facility. This is necessary because treatment techniques which are appropriate for research and development scale may be impractical or impossible at production levels. Therefore, the plan must detail the practices and procedures that are to be used with different types of bio-waste. Every facility that generates bio-waste must train its employees in the proper procedures for bio-waste management.

When we discuss waste management activities for bioprocessing associated wastes, we must consider two components, liquid wastes and the solids suspended in this liquid.

4.2 Selection of Management Options

The selection of available options depends upon a number of factors such as the nature of the waste, the quantity or volume of waste, the availability of on-site treatment, physical constraints, regulatory constraints, etc. Waste must be evaluated and categorized with regard to its composition and its potential to cause disease or damage. Treatment methods will vary and depend on waste type and characteristics such as chemical content, density, water content, and dissolved solids. Generally, it is desirable and efficient to handle all bio-waste in the same manner. However, if a selected option is not suitable for treatment of all wastes, then more than one treatment method must be included in the waste plan. Facilities may use a combination of treatment techniques for different components of the bio-waste stream, e. g., steam sterilization or pasteurization of fermentation fluids, incineration for contaminated solid wastes, or filtration/centrifugation techniques in conjunction with these or other treatments.

Regulation at the Federal, state, and local level may impact on the treatment of bio-waste. For example, Federal air pollution statutes regulate emission from incinerators, including those used to burn bio-waste. Water quality regulations have been applied to production facilities. Regulations and standards pertaining to chemical pollutants, thermal discharges, organic loads (biological oxygen demand), and particulates (total suspended solids) have been applied to bio-waste treatment systems that utilize chemical or thermal inactivation or grinding as part of the treatment process (BOYLAND et al., 1989). Hazardous waste regulations may also apply to waste treatment residue. For example, incinerator ash may be regulated as a hazardous waste, whereas chemical treatment could result in an altered pH or a toxic residue in the waste.

Another important factor in the selection of options for bio-waste management is the availability of on-site and off-site treatment. On-site treatment of bio-waste provides the advantage of a single facility or generator maintaining control over the waste. Off-site treatment may be the more cost-effective option when more than one facility can share a treatment option. Whenever off-site options are considered, generators should comply with all state and local regulations pertaining to transport, packaging, etc., to ensure that the waste is being handled and treated properly at the treatment site.

A final factor in the selection of management options is the prevailing community attitudes. These are expressed officially as local laws, ordinances, zoning restrictions, or health codes and unofficially as public opinion that can influence the official legal position. Community attitudes are especially relevant in citing issues of both on-site and off-site treatment facilities.

4.3 Plan Elements

4.3.1 Designation of Bio-Waste

Management begins with identifying: the types of bio-waste that are to be processed; the location in the facility where these wastes are generated; and the generators within the facility.

4.3.2 Segregation of the Waste Stream

In order to ensure proper treatment of the waste and to prevent unnecessary expenditure for special waste handling (as is the case for chemical or radioactive waste), bioprocessing waste should be segregated from other waste streams at the point of waste generation. If more than one type of treatment is necessary for a given category waste, then this waste must be segregated from other waste types at the facility. This is the case with mixed waste. Waste that is both radioactive and infectious must be processed in a way that eliminates one of the hazards before the waste can be disposed of properly. In this instance, the biological characteristics could be eliminated by steam sterilization and the waste could be subsequently disposed of as radioactive (STINSON et al., 1990). Waste that is sent off-site for treatment should be segregated and kept separate from waste that will be treated within the facility.

4.3.3 Packaging and Labeling

The purpose of properly packaging bio-waste is to provide protection to any personnel who may come in contact with it. If the waste is infectious, it should be discarded directly into appropriate containers. The type of packaging should be appropriate for the intended treatment method and not create any barriers to effective treatment.

Solid Waste

Bags should be used as the primary container for non-sharp, non-liquid waste. All such bags should be tear-resistant and leakproof. They should be able to contain waste during handling, movement, and treatment without breaking or spilling. For example, when waste is to be steam-sterilized, the steam must come in contact with the waste and the bag should be autoclavable. An autoclavable plastic bag is one that can withstand the temperature and pressure generated in an autoclave without losing its integrity. The bags should be unsealed when autoclaving as steam penetrates this type of bag poorly. If the bag is not autoclavable, then an outer container which will withstand steam sterilization must be used.

Liquid Waste

The volume of liquid to be processed determines whether or not a leakproof container is feasible. If one is dealing with small volumes, it is possible to use any leakproof container for transport purposes. For larger volumes of liquid waste, carboys or transport vessels which provide an appropriate leakproof environment can be used. Batch treatment in a closed system (e. g., a tank) is usually an appropriate procedure for larger volumes.

4.3.4 Handling and Movement of Bio-Waste

The purpose for developing handling procedures is to minimize exposure of waste handlers. The most important factor to consider is the integrity of the packaging. If the waste must be moved within the facility, devices, such as carts for solids or transport vessels for liquids, which are known to be appropriate can be used.

These transport aids should be washable (or otherwise cleanable), and should be disinfected periodically. It is important that each container be closed, tied or sealed by some appropriate means, before it is moved from the point of waste generation. The time and route of transfer within the facility should be selected so that contact with other personnel is minimized.

Waste must be packaged properly for transport off-site. All such waste must be labeled and handled in accordance with all applicable regulations.

4.3.5 Storage of Bio-Waste

The best way to deal with bio-waste storage is to avoid storage. It is best to treat it as soon as possible after it is generated. In the event that storage is necessary, storage intervals should be as short as possible. Storage conditions should minimize possible personnel exposure to the waste and prevent amplification of organisms associated with the waste. Refrigeration may be necessary to prevent organism replication. The storage location should be properly identified, secured, and entry-limited to waste management personnel. Storage locations should be cleaned and disinfected on a regular basis.

Waste placed in storage should be properly packaged so that the waste will not spill or leak. Containers should be labeled with information about type of waste, generation date (this is especially true for mixed biological/radioactive waste), and designated treatment. A log of the movement of containers into and out of storage is necessary to track the contents and to establish priorities for processing waste when treatment capacity becomes available. The old adage of "first-in-first-out" should be replaced by "most-hazardous-is-always-first-out".

5 Treatment of Bio-Waste

5.1 Introduction

The purpose of treating this waste is to change its biological character so as to reduce or eliminate its potential for causing either disease or a negative impact on the environment. The methods employed to decontaminate bio-waste are divided into three general categories: thermal treatment using dry heat (oven, incineration) and wet heat (steam sterilization, more widely known as autoclaving and pasteurization), chemical treatment with potassium or sodium hypochlorite or chlorine dioxide, etc., and, much less popular, radiation treatment. The efficacy of these methods depends on factors such as contact time, bio-load (number of microorganisms in the material to be treated), organic content, volume and physical state of the waste (liquid, solid). The presence of other waste products (radioisotopes or toxic chemicals) must also be taken into account when determining the proper method of waste treatment.

One important consideration in evaluating any decontamination procedure is the verification or validation of the process itself. Once the conditions are specified, it is necessary to verify that these conditions are met each time the decontamination procedure is used. This can be achieved through the use of biological, chemical, or mechanical indicators.

5.2 Thermal Inactivation

Thermal inactivation includes treatment methods that utilize heat transfer to provide conditions that reduce the presence of viable organisms in waste. Generally this method is used for treating larger volumes of infectious wastes (such as in industrial applications). Different thermal inactivation techniques are used for treatment of liquid and solid wastes (see Barton et al., 1989, for an extensive review).

5.2.1 Wet Heat (Steam Sterilization)

The most common decontamination method for bio-waste uses saturated steam in a pressure vessel to obtain elevated temperatures. For small volumes gravity displacement units or units using a pre-vacuum cycle are most common. Such units usually operate at a minimum pressure of 1.034 bar which yields a temperature of 121 °C. For larger volumes "in-place" sterilization procedures can be used. The fermenter vessel is used as the pressure chamber and steam is either injected into the fermenation fluid or the temperature of the fluid is raised to a temperature which is known to be lethal for the organism and maintained at that temperature for a predetermined time interval.

5.2.2 Decontamination Conditions

Successful decontamination requires an understanding that conditions to achieve decontamination vary with load characteristics, equipment design, and operational practices.

Common factors which can influence the effectiveness of the process include

- density of waste,
- configuration and size of waste load,
- organic content in the load,
- contact time for decontamination,
- temperature.

For steam sterilization to be effective the steam must come in contact with the waste stream. Any condition which interferes with this contact can and will affect treatment effectiveness. Alternatives to specific decontamination strategies should be considered for wastes with relatively high water content, such as fermentation fluids with or without high concentrations of suspended solids. Pasteurization or direct sewer discharge, if allowed, could be appropriate.

Incineration would be appropriate for animal carcasses, and pasteurization or direct sewer discharge could be appropriate for large fluid volumes.

Steam sterilization or pasteurization can be used for most bio-processing wastes. The technology can handle large volumes of waste; both have sufficiently reliable indicators (both biological and chemical indicators are available) to measure effectiveness. The processes can be easily verified and validated, and almost any load can be decontaminated if exposed to saturated steam or heat for the proper length of time.

5.2.3 Batch Treatment

Batch-type liquid waste treatment units consist of a vessel of sufficient size to contain the liquid waste generated during a specific operating period (9 hours, 24 hours, etc.). The system may include additional vessels to ensure continuous collection of waste without interruption of R&D or production activities that generate the waste. The waste may be preheated by heat exchangers, or heat may be applied by a steam jacket that envelopes the vessel. Heating is continued until a predetermined temperature is achieved and maintained for a designated period of time (similar to the steam sterilization cycle). Mixing may be required to assure that the temperature of the waste is uniform (i. e., no cold spots).

The decontamination temperature and the time required to inactivate the organism (holding time) depends on the nature of the organism(s) present in the waste. Since this treatment method is used most often in industrial applications, the identity of the organism is usually known. Time and temperature requirements can be selected on the basis of the organism present in the waste (*Escherichia coli*) or by using an organism that is more resistant to heat (such as *Bacillus stearothermophilus*).

After the treatment cycle is complete, the contents of the vessel/tank can be discharged to the environment. All discharges must comply with the local, state, or federal requirements. Since these requirements usually include temperature restriction, a second heat exchange may be necessary to remove excess heat from the effluent.

5.2.4 Continuous Treatment

The continuous treatment process for treating liquid waste is actually a semicontinuous process. The system can provide, on demand, thermal inactivation without the need for a large vessel or tank. A typical system consists of a small feed tank, an elaborate steam-based heat exchanger, associated piping, and a control and monitoring system. Liquid waste is introduced into the small feed tank, pumped across the heat exchanger at a constant fixed rate of flow, and recirculated until the required decontamination temperature has been achieved. The time of heat treatment is directly related to the diameter of piping and the length of the pipe run. The treated waste can be cooled by a second heat exchanger before discharge to the environment (i. e., sanitary sewer).

5.3 Dry Heat

5.3.1 Ovens

Dry heat is a less efficient treatment agent than steam and, therefore, higher temperatures and/or longer treatment cycles are necessary. A typical cycle for dry heat sterilization is treatment at 160 to 175 °C for two to four hours. The extensive time and energy requirements usually preclude the use of this technique for treatment of most forms of solid bio-waste. The one major exception is the use of heat to reduce the water content of waste materials.

5.3.2 Incineration

Incineration is a primary method for solid bio-waste disposal (Barbeito and Gremillian, 1968; Barbeito and Shapiro, 1977). This process can convert combustible bio-waste into non combustible residue or ash effectively. Volume reduction of the order of 90 percent can be achieved.

The three basic types of incinerators that are currently available for bio-waste disposal include multiple-chamber, rotary kiln, and controlled-air. Of these, the controlled-air unit is the most widely used.

Any combustible material, provided the material is kept at the proper temperature for an adequate period of time and mixed with the appropriate amount of oxygen, will burn (Boyland et al., 1989). Combustion is dependent on temperature, retention time, and proper mixing rate. These factors plus proper operation are required to ensure that the bio-waste is sterilized. The incinerator secondary chamber must be started and brought up to operating temperature before any waste material is fed in or allowed to ignite (Peterson and Stutzenberger, 1969). The correct waste material feed rate must be maintained to sustain the appropriate temperature, or the mixing rate will not be sustained. The feed rate should be adjusted to correct for the differing heat value of the waste. Overfeeding can overwhelm the combustion chamber with volatile products and lead to excess air emissions usually in the form of black smoke.

If either temperature, retention time, or mixing rate is not maintained, or if the incinerator is improperly operated, viable organisms could remain in the ash or could be released into the atmosphere (BARBEITO and SHAPIRO, 1977; ATSDR, 1990). In a correctly operated incinerator, the temperatures achieved by the incineration process are well above those that organisms can tolerate.

5.4 Chemical Disinfection

5.4.1 Introduction

Chemical disinfection should be viewed as an adjunct to steam sterilization within a bio-waste management program. Chemical decontamination should be considered the method of choice if the waste cannot be autoclaved because it contains antineoplastic agents, radioisotopes, or hazardous chemicals. The usefulness of chemical disinfection is limited by its chemical properties. The corrosivity, pH, incompatibility with other materials, etc., must be evaluated prior to use.

5.4.2 Chemical Disinfectants

Chemical disinfection processes, according to EPA, are appropriate for most liquid and some solid wastes (USEPA, 1986). As with all treatment options, the efficacy of the method must be demonstrated through the development of a testing and monitoring program. Such testing and monitoring should be performed periodically.

Test results indicate that chlorine is an effective disinfecting agent for wastes contaminated with vegetative bacteria, fungi, and viruses, but not for waste contaminated with bacterial spores (HANEL, 1989). Therefore, it is important to demonstrate that the chemical treatment is effective for each application of chemical disinfection.

5.4.3 Gaseous Decontamination

Ethylene oxide gas is occasionally used as a decontaminant, but its effectiveness varies considerably for different types of waste. It is sometimes used as a substitute for steam to sterilize products for surgical use or materials which will degrade if steam-sterilized.

Recently, a chemical decontamination system using chlorine dioxide with potential for more widespread application has been developed. The system processes bio-waste using an electrocatalytic oxidation system. The system purportedly will destroy any known living organism by the oxidizing solution's temperature, acidity, and chemical activity. The system requires no pressure vessels and only normal amounts of electric power (DHOUGE and ROGERS, 1989).

5.5 Radiation

Although gamma radiation will inactivate microorganisms, it is not widely used for routine bio-waste decontamination. Decontamination by this method requires exposure to a minimum dose of rads. It is a very expensive technology requiring highly trained operators. The radioactive source, usually cobalt-60, decays just as any other isotope; therefore, the source must be replaced on a regular basis or the time required for decontamination will become unacceptable. Replacing the source may nearly equal the system purchase price (FORSHAY, 1990).

Advantages to the irradiation process include nominal electricity use, no chemical exposure, no steam requirements, and no residual heat in the treated waste. But its disadvantages – high capital cost, the need for highly trained operators and support personnel, large space requirements, and the problem of timely disposal of the decayed radiation source – virtually preclude its use as a decontamination system for large-scale bio-waste.

5.6 Other Treatment and Disposal Options

Several other bio-waste treatment and disposal methods are under development. These methods include hydropulping, shredding, grinding, microwave technology, and recycling and/or reclamation (see SPURGIN, 1990, for review).

5.6.1 Hydropulping

Hydropulping is a process in which bio-waste is pulverized by a hammermill and submerged in a disinfection solution. Disinfected solids are dewatered and sent to a landfill for final disposal. Disinfected liquids are discharged into the sanitary sewer system. This process must be conducted under negative pressure with high efficiency particulate air (HEPA) filtered exhaust to avoid producing aerosols. This treatment method disinfects any material that can be pulverized.

5.6.2 Wet Grinding

Wet grinding of waste has been used in the United States on a very limited scale. This process grinds the waste material and discharges it into the sanitary sewer, where it is then treated as a sanitary waste. The disinfection methods used by sewage treatment plants will effectively treat bio-waste discharged to the sanitary sewer. The major drawback of this treatment method is that it increases the amount of solids that sewage treatment plants receive. Some sanitary sewer systems are unable to handle this increase in solids. Not all communities have sewer systems.

5.6.3 Microwaves

Microwave technology is currently being marketed in the United States as a bio-waste decontamination system. It is designed to treat all classes of solid bio-waste, even if it contains blood, secretions, bandages, and hypodermic needles with syringes (CUSAK, 1990). The waste is crushed before entering the microwave chamber, and then wetted down with a steam mist. It is then exposed to microwaves until decontamination temperatures are achieved. The decontamination process is controlled by an automatic temperature control system. This treatment must be conducted under negative pressure to avoid producing aerosols.

5.6.4 Recycling and Reclamation

Recycling and/or reclamation have been suggested as alternative methods of bio-waste management. The American health care industry has been using disposable material (e. g., needles and syringes) for many years, primarily to prevent cross infection between patients and because disposable items are believed to be more economical than reusable ones. A wide variety of plastics are used in the industry, and a single process is not available to reclaim them all. However, as recycling technologies improve, and as research and development on plastics progress, this situation may change.

5.7 Landfill Disposal

5.7.1 General Discussion

Landfills traditionally have been used in the United States for solid waste disposal. They generally fall into two types: dumps and sanitary landfills. Dumps are open pits into which solid waste is discarded. Very little monitoring, vector control (insect or rodent), or maintenance is performed at dumps. Sanitary landfills, on the other hand, are designed and constructed specifically for long-term solid waste storage. At a properly managed sanitary landfill, all waste material is covered daily, percolation of large amounts of rainwater through the waste material is prevented, access to the waste material is controlled, and migration of waste material or leachate (liquid material that has percolated out of the waste) from the landfill is prevented or controlled. Groundwater is generally monitored and the leachate collected.

5.7.2 Residential Waste

Bio-waste deposited in a landfill may be mixed with or placed next to other solid waste material. Many commercially operated landfills treat bio-waste as "special waste" with handling practices similar to requirements for asbestos. To determine whether disposing of untreated medical waste in a landfill poses any public health concerns, besides those normally associated with residential solid waste, a comparison of the microbiological activity of residential solid waste and medical waste is needed (DONELLY and SCARPINO, 1984).

Research conducted in West Germany indicates residential waste contains as much or more identical bacteria and fungi as bio-waste (KALNOWSKI et al., 1983; MOSE and REINTHALER, 1985; TROST and FILIP, 1985b). These results have been confirmed by a recent study at the University of Massachusetts Medical Center which showed that "... trash from patients on isolation precautions may be no more contaminated than trash from other patients ...". That study compared bacteria and enterovirus concentrations in bio-waste originating from isolation patients to the concentrations in medical waste of patients not treated in isolation (WALDERON, 1988). Enteroviruses are typically found in the gastrointestinal tract and are involved in respiratory ailments, gastroenteritis, meningitis, and neurological disorders.

Research on the presence and survival of enterovirus and polio virus in residential waste showed that those viruses tend to absorb to organic matter in solid waste and become deactivated. Analysis of leached field samples from properly operated sanitary landfills did not isolate any enterovirus or polio virus (SOBSEY et al., 1975).

Contaminants and microorganisms contained in solid waste deposited in dumps may enter the groundwater. This can occur when solid waste is not properly covered or when rainwater percolates through it. If this leachate is not collected or controlled, bacteria or viruses could, under certain conditions, enter the groundwater system and be transported away from the contamination source to enter local potable water wells. Research on virus survival in groundwater indicated that poliovirus, enterovirus, and hepatitis A virus can survive for up to 12 weeks in a groundwater system. This would be a sufficient amount of time for the microorganisms to move and enter local drinking water wells. However, bio-waste is only a small fraction of solid waste deposited in dumps or sanitary landfills and does not contain any significantly higher concentrations of microorganisms than solid waste. Thus it is unlikely to contribute any additional microorganism loading to aquifer systems (TROST and FILIP, 1985a).

5.8 Sanitary Sewer Disposal

A sanitary sewer system collects and treats waste material generated by humans. This waste material, sewage, contains microorganisms. In studies conducted on the microbial content of residential sewage, many infectious agents (including fungi, bacteria, and viruses) have been isolated. These agents are the result of human excretions, and if residential sewage is not properly treated, disease transmission of water-borne illness can occur (KABLER, 1959).

Most sewage from hospitals, clinics, laboratories, and blood banks originates from patients who do not have communicable diseases, from staff and from process waters (e. g., heating and cooling). Bio-wastes typically discharged to the sanitary sewer system by hospitals include blood and blood products, pathological and animal wastes. These waste materials constitute a small portion of the sanitary sewer discharges from those sources. Any blood and blood products discharged to the sanitary sewer are diluted by the large amounts of residential sewage to well below the concentration needed for blood-borne disease transmission.

Secondary treatment methods (trickling filters, activated sludge, anaerobic digestion, and stabilization ponds) are very effective in reducing the microbiological content of sewage. More than 90 percent of sewage microbiological content, including infectious agents, can be removed by secondary treatment followed by disinfection. Effective treatment of bio-waste can also be accomplished by septic tank systems because the environmental conditions of septic tanks are hostile to pathogens.

In general, while the EPA allows treated liquid bio-waste or ground-up solids to be discharged directly to the sanitary sewer, local ordinances may prohibit it. When allowed, personnel handling these wastes should use discretion to avoid possible clogging of drains. In some states, the landfilling of bio-waste is allowed while in others it is prohibited. The

EPA has historically recommended that only treated bio-waste be landfilled (USEPA, 1986). It must be noted that if treated waste is no longer infectious, it may be handled as ordinary waste provided it poses no hazard otherwise subject to regulation. If landfilling of bio-waste is allowed by the state, the EPA recommended that only well controlled sanitary landfills be used (WAGNER, 1989). Persons desiring to landfill untreated infectious waste should consult with state officials and the landfill operator prior to shipping these wastes to the facility.

6 Public Health Impacts

6.1 Populations Potentially at Risk

While bio-waste treatment and disposal may involve many individuals, e. g., health care workers, laboratory and production workers, animal care workers, janitorial and laundry workers usually are responsible for waste treatment or disposal. Refuse workers are involved with waste disposal in landfills; wastewater workers may come in contact with waste discharged to the sanitary sewer. Facility personnel, engineers, maintenance workers, and plumbers, may come into contact with untreated waste during the repair or maintenance of waste processing equipment.

6.2 Public Health Implications of Treatment and Disposal Methods

6.2.1 Steam Sterilization and Chemical Disinfection

Personnel must be protected from exposure during all decontamination procedures when an autoclave is to be used. Personnel must be protected from both the waste to be processed and thermal injury due to direct exposure or touching hot surfaces. When using chemical decontamination, workers must be protected from chemical burns (skin and eyes) which could occur if the chemical decontaminant was to splash onto unprotected areas of the body.

6.2.2 Incineration

Possible public health implications of treating waste by incineration include physical injuries, infectious agent emissions, and toxic chemical emissions. Waste incinerator operators can sustain burn or explosion injuries if standard incinertor safety precautions are not followed. In addition, operators may be exposed to infectous agent emissions from the incinerator. However, as previously discussed, it is unlikely that the emissions would contain an infective dose of agents.

At present, the public health implications related to contact with potentially hazardous chemical emissions and ash from bio-waste incinerators cannot be adequately evaluated, because available stack emission testing and ash sampling data are insufficient. However, the use of older medical pathological incinerators to burn plastic waste has considerable potential to produce chlorinated combustion products as well as other hazardous by-products, such as cadmium, lead, tetrachlorodibenzyl-*p*-dioxin, and chlorinated dibenzofurans (DOUCET, 1987; YASUHARA and MORITA, 1988).

6.2.3 Landfill Disposal

Disposing of untreated bio-waste in landfills is not likely to pose any additional public health implications other than those normally associated with landfilling residential solid waste. Scientific studies indicated that landfilled medical waste does not pose any additional microbiological dangers than those already associated with residential waste. As discussed previously in this section, infectious agents capable of blood-borne disease behave similarly to those normally found in residential solid waste.

6.2.4 Sanitary Sewer Disposal

The amount of bio-waste discharged into sanitary sewer systems represents only a small portion of the total waste processed in these systems. In addition, infectious agents are part of the normal flora of residential sewage. The treatment processes used by sanitary sewer systems in the United States have been shown to effectively treat infectious agents and to prevent infectious disease transmission. An epidemiological study of wastewater workers showed that these workers have no increased potential of becoming infected by blood-borne infectious agents (ATSDR, 1990). Therefore, bio-waste discarded into the sanitary sewer is highly unlikely to present any additional public health effects to wastewater workers or to the general public.

Wastewater workers could be injured by contaminated waste sharps discarded into the sanitary sewer. The frequency of waste sharp injuries to wastewater workers is anticipated to be less than the rate for refuse workers, because wastewater workers do not physically contact waste material as frequently as do refuse workers (ATSDR, 1990).

6.2.5 Other Treatment and Disposal Methods

Hydropulping, wet grinding, and microwave technologies shred waste material before further processing. This shredding process has the potential for associated physical injuries or infectious disease transmission. Physical injuries ranging from cuts to crush injuries could occur, if workers contact the hammermill or grinder during operation. These injuries are preventable, as machines are constructed to prevent human contact with operating hammermills or grinders.

These three processes also have the potential to produce aerosols during treatment. The aerosols may contain infectious agents that can be transmitted to a susceptible host. There is one documented case of aerosol transmission of infectious agents from a hospital hydropulping system, but this particular system is no longer in use in the United States (SPURGIN, 1990). The hydropulping system, microwave technology, and wet grinding systems currently being marketed are designed to prevent aerosol production by keeping the process under negative pressure and/or filtering the air before it is discharged.

7 Summary

Bio-waste can be effectively treated by chemical, physical, or biological means such as chemical decontamination, autoclaving, incineration, irradiation, and sanitary sewage treatment. To be effective, each of these treatment methods must be performed in a manner consistent with equipment manufacturers' instructions. Operators must follow accepted operating practices, and an operator training program must be part of a structure waste management system.

Research indicates that bio-waste does not contain any greater quantity or different types of microbiological agents than residential waste. Additionally, properly operated sanitary landfills provide microbiological environments hostile to most pathogenic agents. Therefore, untreated bio-waste could be safely disposed of in sanitary landfills, provided procedures are employed to prevent human contact with the waste during handling and disposal operations.

8 References

ATSDR (Agency for Toxic Substances and Disease Registry) (1990), *The Public Health Implications of Medical Waste : Report to Congress*, Atlanta, Georgia: U. S. Department of Health and Human Services.

BARBEITO, M. S., GREMILLIAN, C. G. (1968), Microbiological safety evaluation of an industrial refuse incinerator, *Appl. Microbiol.* **16**, 291–5.

BARBEITO, M. D., SHAPIRO, M. (1977), Microbiological safety evaluation of a solid and liquid pathological incinerator, *J. Med. Primatol.* **6**, 264–73.

BARTON, R., HASSEL, G., LANIER, W., SEEKER, W. (1989), *State of the Art Assessment of Medical Waste Thermal Treatment, EPA Contract 68-03-3365*, Energy and Environmental Research Corp.

BERG, P., BALTIMORE, D., BRENNER, S., ROBLIN, R. O., SINGER, M. F. (1975), Summary statement of the Asilomar Conference on rDNA molecules, *Science*, **188**, 991–4.

BOYLAND, J. L., GORDON, J., SPURGIN, R., LIBERMAN, D. F. (1989), Infectious waste management, in: *Biohazards Management Handbook*, (LIBERMAN, D. F., GORDON, J., Eds.) pp. 259–88. New York–Basel: Marcel Dekker Inc.

BROWN, C. M., ELLWOOD, D. C., HUNTER, J. R. (1977), Growth of bacteria at surfaces: Influence of nutrient limitation, *FEMS Microbiol. Lett.* **1**, 163–6.

CHURCH, T., COOPER, P., NAKAMURA, R. (1989), *The Political and Regulatory Environment of Medical Waste: Formation and Implementation of The Medical Waste Tracking Act, Report to the Medical Waste Policy Committee*, Albany, New York: State University of New York.

CUSAK, J. (1990), *ABB Sanitek Device, Report to Medical Waste Review Panel* (May 29).

DHOUGE, P., ROGERS, T. (1989), *Infectious Waste Processing with an Electro Catalytic Oxidation System, Abstract Hazwaste Expo 89*, Albuquerque, New Mexico.

DONNELLY, J. A., SCARPINO, P. V. (1984), *Isolation, Characterization and Identification of Microorganisms from Laboratory and Full Scale Landfills*, EPA Municipal Environmental Research Laboratory, *Publication No. 600/2-84-119*.

DOUCET, L. (1987), Institutional Waste Incineration, presentation at the *Environmental Management Workshop: An Integrated Approach to Hazardous Waste Management*, sponsored by the American Society for Hospital Engineers of the American Hospital Association, Peekskill, New York, (DOUCET, L., MAINKA, P. C., Eds.).

EVANS, C. G. T., PREECE, T. F., SARGEANT, K. (1981), *Microbial Plant Pathogens: Natural Spread and Possible Risks in Their Industrial Use. Publication of the Commission of the European Communities*, ECI. 724, United Kingdom.

FORSHAY, R. (1990), *Review Panel: Finding the Rx for Managing Medical Wastes*, OTA-0-459, Washington, DC: U.S. Government Painting Office.

GRADY, J. R., C. P. L., LIM, H. C. (1980), Ecology of biochemical reactors, in: *Biological Wastewater: Theory and Applications*, pp. 197–227, New York–Basel: Marcel Dekker, Inc.

HANEL, E. (1989), Liquid chemical germicides in: *Biohazards Management Handbook* (LIBERMAN, D. F., GORDON, J., Eds.), pp. 227–258, New York Basel: Marcel Dekker, Inc.

HENDRICK, E. (1989), What is infectious medical waste? *Plant Technol. Safe. Ser.* 4; 24–28, The Joint Commission on Accreditation of Healthcare Organizations.

KABLER, P. (1959), Removal of pathogenic microorganisms by sewage treatment processes, *Sewage Ind. Waste* **31**, 1373–82.

KALNOWSKI, G., WIEGAND, H., RUDEN, D. (1983), On the microbial contamination of hospital waste, *Zentralbl. Bakteriol. Mikrobiol. Hyg.* **178**, 364–79.

MOSE J. R., REINTHALER, F. (1985), Microbiology studies of the contamination of hospital waste and household refuse, *Zentralbl. Bakteriol. Mikrobiol. Hyg.* **181**, L98–110.

NIH (National Institutes of Health) (1976), *Guidelines for Research Involving Recombinant DNA Activity, Fed. Reg.* **41**, 27902.

NIH (National Institutes of Health) (1983), *Guidelines for Research Involving Recombinant DNA Activity, Fed. Reg.* **48**, 24555–24581.

OECD (Organization for Economic Co-operation and Development) (1987), *Recombinant DNA Safety Consideration*, Washington, DC: OECD, Publication and Information Center.

PETERSON, M. L., STUTZENBERGER, F. J. (1969), Microbiological evaluation of incinerator operations, *Appl. Microbiol.* **18**, 8–13.

RCRA (*Resource Conservation and Recovery Act*) (1976), Enacted by the 94th Congress, October 21.

RTI (Research Triangle Institute) (1988), *Review and Evaluation of Existing Literature on Generation, Management, and Potential Health Effects of Medical Waste*, prepared for the Health Assessment Section, Research Triangle Park, NC: Technical Assessment Branch, Office of Solid Wastes, USEPA.

SOBSEY, M. D., WALLIS, C., MELNICK, J. L. (1975), Studies of survival and fate of enteroviruses in an experimental model of a municipal solid waste landfill and leachate, *Appl. Microbiol.* **30**, 565–574.

SPURGIN, R. (1990), *Medical Waste Treatment Technologies*, Contract Report for the Office of Technology Assessment, March 16.

STINSON, M. C., GREEN, B. L., MARQUARDT, C. J., DUCATMAN, A. M. (1990), Autoclave inactivation of infectious radioactive laboratory waste contained within a charcoal filtration system, *Health Phys.* **61**, 137–142.

TROST, M., FILIP, Z. (1985a), Behavior of microorganisms in medical consulting room refuse and household refuse deposited in a model landfill, *Zentralbl. Bakteriol. Mikrobiol. Hyg.* **181**, 173–183.

TROST, M., FILIP, Z. (1985b), Microbiological investigation of medical consulting room refuse and municipal refuse, *Zentralbl. Bakteriol. Mikrobiol. Hyg.* **181**, 159–172.

USEPA (U. S. Environmental Protection Agency) (1986), *EPA Guide for Infectious Waste Management, EPA Document No. 530-SW-86014*, Washington, D. C.: USEPA, Office of Solid Waste and Emergency Response.

WAGNER, K. (1989), The Federal role in medical waste management, *Plant, Technol. Safe. Ser.* **4**, 12–16, The Joint Commission on Accreditation of Healthcare Organizations.

WALDERON, T. (1988), Study indicates isolation trash comparable to standard refuse, *Hosp. Infect. Control.* **15**, 118–119.

YASUHARA, A., MORITA, M. (1988), Formation of chlorinated aromatic hydrocarbons by thermal decomposition of vinylidene chloride polymer, *Environ. Sci. Technol.* **22**, 646–650.

Index

B

C

F

H

N

O